Selected Papers in

Multidimensional Digital Signal Processing

OTHER IEEE PRESS BOOKS

Adaptive Methods for Control Systems Design, *Edited by M. M. Gupta*
Residue Number System Arithmetic, *Edited by M. A. Soderstrand, W. K. Jenkins, G. A. Jullien, and F. J. Taylor*
Low-Temperature Electronics, *Edited by R. K. Kirschman*
Undersea Lightwave Communications, *Edited by P. K. Runge and P. R. Trischitta*
Singular Perturbations in Systems and Control, *Edited by P. V. Kokotovic and H. K. Khalil*
Getting the Picture, *By S. B. Weinstein*
Space Science and Applications, *Edited by J. H. McElroy*
Medical Applications of Microwave Imaging, *Edited by L. Larsen and J. H. Jacobi*
Modern Spectrum Analysis, *Edited by S. B. Kesler*
The Calculus Tutoring Book, *By C. Ash and R. Ash*
Imaging Technology, *Edited by H. Lee and G. Wade*
Phase-Locked Loops, *Edited by W. C. Lindsey and Chak M. Chie*
VLSI Circuit Layout: Theory and Design, *Edited by T. C. Hu and E. S. Kuh*
Monolithic Microwave Integrated Circuits, *Edited by R. A. Pucel*
Next Generation Computers, *Edited by E. A. Torrero*
Kalman Filtering: Theory and Application, *Edited by H. W. Sorenson*
Spectrum Management and Engineering, *Edited by F. Matos*
Digital VLSI Systems, *Edited by M. T. Elmasry*
Introduction to Magnetic Recording, *Edited by R. M. White*
Insights Into Personal Computers, *Edited by A. Gupta and H. D. Toong*
Television Technology Today, *Edited by T. S. Rzeszewski*
The Space Station: An Idea Whose Time Has Come, *By T. R. Simpson*
Marketing Technical Ideas and Products Successfully! *Edited by L. K. Moore and D. L. Plung*
The Making of a Profession: A Century of Electrical Engineering in America, *By A. M. McMahon*
Power Transistors: Device Design and Applications, *Edited by B. J. Baliga and D. Y. Chen*
VLSI: Technology Design, *Edited by O. G. Folberth and W. D. Grobman*
General and Industrial Management, *By H. Fayol; revised by I. Gray*
A Century of Honors, *An IEEE Centennial Directory*
MOS Switched-Capacitor Filters: Analysis and Design, *Edited by G. S. Moschytz*
Distributed Computing: Concepts and Implementations, *Edited by P. L. McEntire, J. G. O'Reilly, and R. E. Larson*
Engineers and Electrons, *By J. D. Ryder and D. G. Fink*
Land-Mobile Communications Engineering, *Edited by D. Bodson, G. F. McClure, and S. R. McConoughey*
Frequency Stability: Fundamentals and Measurement, *Edited by V. F. Kroupa*
Electronic Displays, *Edited by H. I. Refioglu*
Spread-Spectrum Communications, *Edited by C. E. Cook, F. W. Ellersick, L. B. Milstein, and D. L. Schilling*
Color Television, *Edited by T. Rzeszewski*
Advanced Microprocessors, *Edited by A. Gupta and H. D. Toong*
Biological Effects of Electromagnetic Radiation, *Edited by J. M. Osepchuk*
Engineering Contributions to Biophysical Electrocardiography, *Edited by T. C. Pilkington and R. Plonsey*
The World of Large Scale Systems, *Edited by J. D. Palmer and R. Saeks*
Electronic Switching: Digital Central Systems of the World, *Edited by A. E. Joel, Jr.*
A Guide for Writing Better Technical Papers, *Edited by C. Harkins and D. L. Plung*
Low-Noise Microwave Transistors and Amplifiers, *Edited by H. Fukui*
Digital MOS Integrated Circuits, *Edited by M. I. Elmasry*
Geometric Theory of Diffraction, *Edited by R. C. Hansen*
Modern Active Filter Design, *Edited by R. Schaumann, M. A. Soderstrand, and K. B. Laker*
Adjustable Speed AC Drive Systems, *Edited by B. K. Bose*
Optical Fiber Technology, II, *Edited by C. K. Kao*
Protective Relaying for Power Systems, *Edited by S. H. Horowitz*
Analog MOS Integrated Circuits, *Edited by P. R. Gray, D. A. Hodges, and R. W. Broderson*
Interference Analysis of Communication Systems, *Edited by P. Stavroulakis*
Integrated Injection Logic, *Edited by J. E. Smith*
Sensory Aids for the Hearing Impaired, *Edited by H. Levitt, J. M. Pickett, and R. A. Houde*
Data Conversion Integrated Circuits, *Edited by D. J. Dooley*
Semiconductor Injection Lasers, *Edited by J. K. Butler*
Satellite Communications, *Edited by H. L. Van Trees*
Frequency-Response Methods in Control Systems, *Edited by A. G. J. MacFarlane*

Selected Papers in

Multidimensional Digital Signal Processing

Edited by the

Multidimensional Signal Processing Committee

IEEE Acoustics, Speech, and Signal Processing Society

Committee Members:
Arthur Baggeroer, Massachusetts Institute of Technology
Rama Chellappa, University of Southern California
Bradley W. Dickinson, Princeton University
Dan E. Dudgeon, M.I.T. Lincoln Laboratory
Richard Gran, Grumman Aerospace
Lloyd J. Griffiths, University of Southern California
Thomas S. Huang, University of Illinois
Anil K. Jain, University of California at Davis
James Justice, University of Calgary
Avinash C. Kak, Purdue University
Roman Kuc, Yale University
Jae S. Lim, Massachusetts Institute of Technology
Thomas L. Marzetta, Schlumberger-Doll Research
James McClellan, Schlumberger Well Services
Russell M. Mersereau, Georgia Institute of Technology
David C. Munson, Jr., University of Illinois
Azriel Rosenfeld, University of Maryland
Paul L. Stoffa, University of Texas at Austin
John W. Woods (Chairman), Rensselaer Polytechnic Institute

A volume in the IEEE PRESS Selected Reprint Series, prepared under the sponsorship of the IEEE Acoustics, Speech, and Signal Processing Society

IEEE PRESS

The Institute of Electrical and Electronics Engineers, Inc., New York

PRINTED IN THE UNITED STATES OF AMERICA

IEEE Order Number: PC01974

Library of Congress Cataloging-in-Publication Data

Selected papers in multidimensional digital signal processing.

(IEEE Press selected reprint series)
Includes index.
1. Signal processing—Digital techniques.
I. IEEE Acoustics, Speech, and Signal Processing Society. Multidimensional Signal Processing Committee. II. Series.
TK5102.5.S4255 1986 621.38'043 86-10669

ISBN 0-89742-202-5

Contents

Preface

THE field of multidimensional digital signal processing (MDSP) has become increasingly important in recent years due to the confluence of a number of trends. Digital computational capability is growing at a tremendous rate, with a concommitant increase in parallel processing capability. In more and more applications, the data to be processed are inherently multidimensional in character. Finally, multidimensional sensor arrays are becoming widely available, e.g. CCD image sensors, X-ray and NMR computer tomography sensors, and seismic and acoustic arrays. As sensor and processing capabilities grow, multidimensional processing finds natural applications in such areas as image processing, geophysical signal processing, and sensor-array processing. Thus, MDSP is a natural topic for study from the point of view of both current and anticipated future applications to a wide variety of problems.

The processing of multidimensional (M-D) digital signals shares much similarity with that of 1-D digital signals. At the same time, there are many significant differences, e.g. Z-transforms almost never factor; poles and zeros are replaced by curves in four-dimensional space; maximum entropy spectral estimation requires the solution of a nonlinear optimization problem; the Fourier transform magnitude alone theoretically suffices to determine the image of a finite size 2-D object. The purpose of this volume is to bring together some recent papers in the MDSP area that will provide the best introduction to these unexpected differences, some of which, when seen from another viewpoint, can be regarded as new opportunities afforded by the advent of multidimensional data processing capabilities. Indeed, a number of interesting M-D problems discussed in this volume have no 1-D counterpart.

The MDSP Committee is one of six technical committees organized within the IEEE Acoustics, Speech and Signal Processing (ASSP) Society. The MDSP Committee was organized in 1980 and had its first meeting at the 2nd ASSP Workshop on Multidimensional Signal Processing, New Paltz, N.Y., in October 1981. The Committee has a membership of technical specialists in MDSP applications to problems arising in image processing, geophysical signal processing, and sensor-array processing. The first ASSP Workshop was held in Berkeley, CA in 1979, and since its organization, the MDSP Committee has taken over the planning and organization of the biennial series of workshops based on this theme: 3rd Workshop, Lake Tahoe, CA, 1983 and 4th Workshop, Leesburg, VA, 1985. Other technical activities support the ASSP Society in the multidimensional area, such as manuscript reviews for the Society's international conference, ICASSP, and recommendations of Associate Editors for the Society's Transactions. This reprint book is a natural outgrowth. It is patterned after the highly successful reprint book edited by the DSP Technical Committee [1].

In a rapidly growing area such as multidimensional signal processing, there are numerous candidate papers for inclusion in a reprint book such as this. Our first selection criterion was relevance to digital signal processing with (hopefully) some unique M-D feature of the problem. Our second criterion was excellence in either originality, i.e. a landmark paper, or in tutorial value. Finally, we did not include papers which have already been reprinted in [1] and [2], but have instead concentrated on more recent papers.

This book is organized into nine parts. The first four parts are mainly theoretical, while the latter five parts are more applications oriented. Each of the first eight parts has an introduction and supplementary bibliography. A number of excellent papers, which could not be reprinted because of space limitations, are listed in these bibliographies.

Part I treats the area of two-dimensional digital filter design concentrating on the less developed topic of recursive filters. Part II is concerned with multidimensional spectral estimation, emphasizing high resolution methods in the M-D setting. Part III deals with system theory issues arising in the case of a multidimensional (time) index parameter. Part IV is on the topic of multidimensional signal modeling, especially autoregressive type models.

Part V concerns the topic of signal reconstruction, both in a computer tomography setting and the more recent multidimensional phase retrieval problem. Part VI is a collection of papers on image processing applications. Part VII covers some geophysical processing applications. Part VIII treats the application of MDSP to the processing of signals from arrays of sensors. Finally, Part IX consists of a survey paper intended to introduce the reader with a signal-processing background to emerging applications in the area of image analysis and computer vision. This paper points the way towards relevant publications in this related area.

Acknowledgments

We would like to acknowledge the suggestions and comments received from our colleagues; J. Aggarwal, University of Texas at Austin, R. T. Lacoss, MIT Lincoln Laboratory, Lexington, MA, and the members of the IEEE PRESS Editorial Board.

Bibliography

[1] Digital Signal Processing Committee, *Selected Papers in Digital Signal Processing, II*. New York, NY: IEEE PRESS, 1976.

[2] S. K. Mitra and M. P. Ekstrom, Eds. *Two-Dimensional Digital Signal Processing*. Stroudsburg, PA: Dowden, Hutchinson, and Ross, 1978.

Part I
Digital Filter Design

THE state of the art in the design of multidimensional digital filters is fairly mature. This is not meant to suggest that all problems are solved or that significant improvements in existing algorithms cannot be found. It does mean, however, that a number of design procedures of proven usefulness have been published and used for a number of years, and much of this work has been digested into textbook form [1]. We have attempted to acknowledge this state of affairs in our compilation. We have not included papers which appeared in two excellent earlier reprint collections [2], [3]. Our intent is to complement these other collections as well as the text by Dudgeon and Mersereau [1].

This section contains five reprinted papers. One of these is concerned exclusively with FIR (Finite Impulse Response) digital filters, three are concerned with IIR (Infinite Impulse Response) or recursive filters, and the final one addresses both. This is an acknowledgment of the fact that the FIR problem is better understood and that effective design algorithms for FIR filters have been available for over ten years. The IIR problem, on the other hand, is mathematically richer but is considerably more difficult due to the need to guarantee the stability and control the phase response of the resultant designs.

The first paper, by Lodge and Fahmy, addresses the problem of designing linear phase FIR filters which minimize the integrated frequency domain error function raised to the pth power. These filters thus contain both least squares designs ($p = 2$) and equiripple designs (large p) as special cases. The algorithm proposed is iterative and has a better than linear rate of convergence. An important aspect of the Lodge and Fahmy algorithm is that the design time does not become impossibly large as the order of the filter increases, a difficulty encountered by other methods which seek to find an equiripple solution. Another approach to the same problem which designs equiripple FIR filters using projections onto convex sets has been published by Abo-Taleb and Fahmy [4]. Their method shows promise and may be of interest to the reader of this collection.

The second paper, by Aly and Fahmy, is concerned with the design of IIR filters. Unlike the others in the collection, it specifically addresses the problem of controlling the group delay (phase) response of the filter as well as the magnitude response. As with the previous paper a weighted least pth error is used for both the magnitude and group delay. The filters are realized as cascades of first and second-order sections, which makes it possible to test their stability at each iteration of the design procedure, but which restricts the class of possible solutions.

The third paper in the collection, by Ramamoorthy and Bruton, is representative of a large body of techniques for designing two-dimensional recursive filters by beginning with analog prototype filters. This indirect approach has two important advantages. First, it is relatively straightforward to assure that the analog filters will be stable if they are designed with only passive elements. This stability can be transfered to the digital design. Second, there are structures available for the analog filters which are extremely robust with respect to their performance under constraints of finite precision arithmetic. These properties too can often be transfered through the transformation process. A related, excellent body of work concerns the design and implementation of two-dimensional wave digital filters by Fettweis and his colleagues. A good summary of this work is found in [5].

The fourth paper in this section, by Ekstrom, Twogood, and Woods, presents a very powerful, but computationally intensive algorithm for the design of stable IIR filters. Their procedure is more general than most in that it produces a filter with non-symmetric half-plane support of arbitrary order without any restrictions on the factorability of the transfer function. Stability is guaranteed by inserting a stability constraint as a penalty function into a least squares minimization problem. The technique presented in this paper is a magnitude-only design in that there is no control over the phase response of the filter. An extension of this technique which allows for some control over the phase response is given in [6].

Often the simplest way to design a filter for a certain application is to transform or modify one which has already been designed. The final paper, by Chakrabarti and Mitra, is a tutorial concerned with a large variety of transformation techniques which take this point of view. It is concerned with transformations between continuous-time and discrete-time filters as well as transformations between one-dimensional and two-dimensional filters and transformations from one digital filter to another. Careful attention is given to properties of a transformation which preserve stability.

Limitations of space have forced us to be selective. We have chosen papers which are somewhat different from one another and from papers which have already been reprinted. We have also chosen papers which have tutorial value. Other papers could certainly been chosen to produce an equally interesting collection. The literature on this topic is extensive, reflecting the maturity of the area.

Bibliography

[1] D. E. Dudgeon and R. M. Mersereau, *Multidimensional Digital Signal Processing*. Englewood Cliffs, NJ: Prentice-Hall, 1984.

[2] S. K. Mitra and M. P. Ekstrom, Eds. *Two-Dimensional Digital Signal Processing.* Stroudsburg, PA: Dowden, Hutchinson, and Ross, 1978.

[3] Digital Signal Processing Committee, *Selected Papers in Digital Signal Processing, II.* New York, NY: IEEE PRESS, 1976.

[4] A. Abo-Taleb and M. M. Fahmy, "Design of FIR two-dimensional digital filters by successive projections," *IEEE Trans. Circ. Syst.,* vol. CAS-31, pp. 801–805, Sept. 1984.

[5] A. Fettweis, "Multidimensional wave digital filters," in *Proc. 1976 Europ. Conf. Circ. Th. Des., vol. II,* pp. 409–416.

[6] J. W. Woods, J-H Lee, and I. Paul, "Two-dimensional IIR filter design with magnitude and phase error criteria," *IEEE Trans. Acoust., Speech, Signal Processing,* vol. ASSP-31, pp. 886–894, Aug. 1983.

An Efficient l_p Optimization Technique for the Design of Two-Dimensional Linear-Phase FIR Digital Filters

JOHN H. LODGE, MEMBER, IEEE, AND MOUSTAFA M. FAHMY, SENIOR MEMBER, IEEE

Abstract—An iterative optimization technique, based on the method of parallel tangents coupled with an efficient line searching technique, is proposed for the design of two-dimensional linear-phase FIR digital filters. The performance index for the optimization procedure is shown to be convex and hence this technique will always converge towards the global minimum. An important feature is that the design time increases slowly as the number of filter coefficients increases. To illustrate the technique several circular low-pass filters were designed. These filters compare favorably with filters designed using the minimax technique that was developed by Harris. The technique can also be used for the design of one-dimensional and multidimensional linear-phase FIR digital filters.

I. Introduction

The design of two-dimensional digital filters has been of great interest lately. This is because these filters have many applications [1] in fields such as image enhancement, radar, sonar, and geophysical data processing. In many applications it is important to preserve the phase information [2], [3]. It is for this reason that filters with linear or zero phase are important. Such phase characteristics are easily achieved with finite impulse response (FIR) digital filters. In addition, fast transform techniques are available that can provide very efficient implementations for these filters.

Several design techniques for two-dimensional linear-phase FIR digital filters are well established. Perhaps the simplest technique and the technique with the shortest design time is the windowing technique [4]. However, this technique has a few disadvantages. In most cases the resulting filter is not optimal in any sense. Also, two-dimensional windows are not as well-behaved or well-understood as one-dimensional windows [5].

Equiripple designs can be achieved using McClellan transformations [6]. This technique enjoys a short design time and efficient implementations exist [7]. Although the resulting filter is equiripple, it is only suboptimal in the Chebyshev sense. Also, this technique cannot be used to closely approximate all magnitude functions.

Manuscript received August 30, 1979; revised January 14, 1980.

The authors are with the Department of Electrical Engineering, Queen's University, Kingston, Ont., Canada.

Reprinted from *IEEE Trans. Acoust., Speech, Signal Processing,* vol. ASSP-28, pp. 308–313, June 1980.

Several minimax design techniques are available. The first technique [8] was an algorithm based on linear programming. This method required an extremely long design time. Because of this, exchange ascent algorithms were developed [9], [10], [11]. A good comparison of these techniques can be found in [12]. Although they are more efficient than linear programming, their computational complexity limit them to the design of filters with input masks not greater than 15 X 15 [12]. Also, it is not difficult to see that the computational effort required for the error search and for the solution of the linear equations increases rapidly as the size of the filter increases. Moreover, a perturbation technique [9] is necessary, to handle the numerical difficulties that can be caused by degeneracy in the set of reference points, for the exchange algorithms.

This paper proposes an iterative technique for the design of two-dimensional linear-phase FIR digital filters that approximates a prescribed magnitude response using the l_p norm. This algorithm is globally convergent and has a short design time. An important feature is that the design time increases slowly as the number of filter coefficients increase. Also, designs extremely close to minimax designs can be achieved by using large values for p. The design technique can be used to closely approximate any magnitude function that exhibits half-plane symmetry.

II. Statement of the Problem

It can be easily shown that the frequency response of a two-dimensional linear-phase FIR digital filter can be given by

$$H(\omega_1, \omega_2) = \sum_{j} \sum_{k} b_{jk} \cdot \cos (j\omega_1 + k\omega_2) \quad (j,k) \in L \tag{1}$$

where L is the set of integer pairs defined by the input mask of the filter. A more general form that encompasses the special cases of octal and quadrantal symmetry is

$$H(\omega_1, \omega_2) = \sum_{i=1}^{K} a_i \cdot f_i(\omega_1, \omega_2) \tag{2}$$

where

$$a_i = b_{jk} \cdot l, \qquad l, j, k \in I,$$

$$f_i(\omega_1, \omega_2) = \cos j\omega_1 \cos k\omega_2 + \cos k\omega_1 \cos j\omega_2, \text{ for octal symmetry,}$$
$$= \cos j\omega_1 \cos k\omega_2, \text{ for quadrantal symmetry,}$$
$$= \cos (j\omega_1 + k\omega_2), \text{ for half-plane symmetry.}$$

I is the set of integers and the b_{jk} are simply related to the impulse response samples of the filter.

Let the desired amplitude characteristic $G(\omega_1, \omega_2)$ be defined on the discrete set of frequency pairs

$$(\omega_{1m}, \omega_{2n}), m = 1, \cdots, M: n = 1, \cdots, N.$$

Let

$$a = [a_1, \cdots, a_i, \cdots, a_K]^T$$

be the vector of parameters and

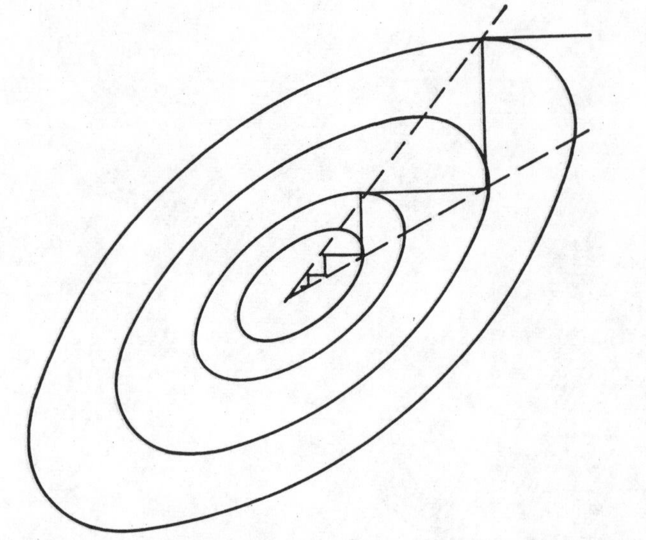

Fig. 1. The envelope of the steepest descent path points to the minimum.

$$f(\omega_1, \omega_2) = [f_1(\omega_1, \omega_2), \cdots, f_K(\omega_1, \omega_2)]^T$$

be the vector of frequency-dependent functions in the transfer function (2).

Let the performance index $J(a)$ be defined as

$$J(a) = \sum_{m=1}^{M} \sum_{n=1}^{N} W_{mn} \epsilon_{mn}^p \tag{3}$$

where

$$\epsilon_{mn} = [H(\omega_{1m}, \omega_{2n}) - G(\omega_{1m}, \omega_{2n})],$$

p is an even-positive integer, and W_{mn} is a weighting function to be specified by the designer. The problem is to select the elements of the vector a that minimize the performance index $J(a)$.

III. The Optimization Procedure

The iterative procedure used here is based upon the method of parallel tangents (PARTAN) [13]. This technique developed from the observation that in a two-parameter optimization problem the minimum point often lies along the lines that join the zigzags of the steepest descent path. This is illustrated in Fig. 1. PARTAN attempts to improve the convergence by adding an extra line search for each iteration, as is shown in Fig. 2. The g_i denotes the step resulting from the gradient line search, the t_i denotes the step resulting from the additional line search, and the a_i denotes the parameter vector at the ith iteration. PARTAN has the desirable property that it decreases the performance index every iteration unless the minimum is achieved. Also, it can be shown that PARTAN exhibits quadratic convergence. In fact, for a quadratic functional PARTAN is equivalent to the method of conjugate gradients [14].

The design algorithm begins by finding an initial approximation using one of the available windowing techniques. It will be shown later that the performance index is convex. Hence, a good initial approximation is not critical. A technique, such as the one of Manry and Aggarwal [15], can be used to determine the input mask of the filter to be designed.

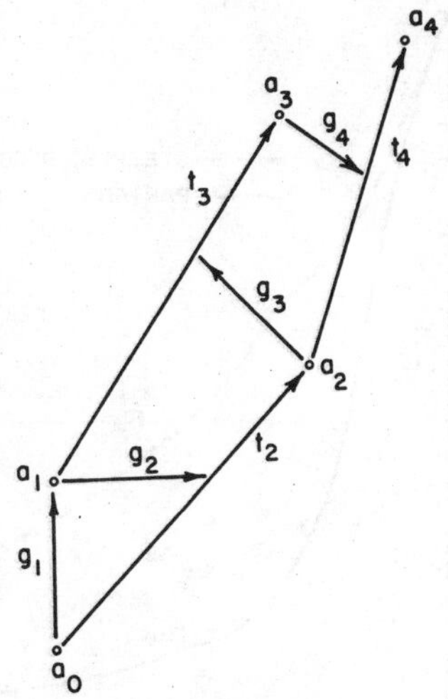

Fig. 2. The path of PARTAN.

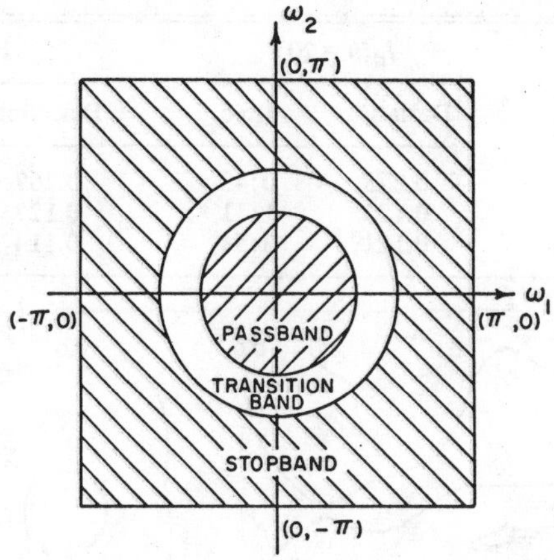

Fig. 3. Example 1: Desired magnitude response of the low-pass filter.

Next, the parameters are optimized using PARTAN. This requires the calculation of the gradient vector every iteration. The gradient vector is given by

$$\nabla J = p \sum_{m=1}^{M} \sum_{n=1}^{N} W_{mn} \epsilon_{mn}^{p-1} \cdot f(\omega_{1m}, \omega_{2n}). \tag{4}$$

An efficient line searching technique is desirable because two line searches are required every iteration except the first. Consider the performance index along the line $a + \Delta \cdot d$, where a is the parameter vector, d is the direction vector, and Δ is a real variable. The jth derivative of $J(a + \Delta \cdot d)$, with respect to Δ at the point given by $\Delta = 0$, is

$$\left.\frac{d^j J(a + \Delta \cdot d)}{d\Delta^j}\right|_{\Delta=0} = \frac{p!}{(p-j)!} \sum_{m=1}^{M} \sum_{n=1}^{N} W_{mn} \epsilon_{mn}^{p-j} V_{mn}^j, \quad 1 \leqslant j \leqslant p$$
$$= 0, \quad j > p \tag{5}$$

where

$$V_{mn} = d^T \cdot f(\omega_{1m}, \omega_{2n}).$$

Two important observations are made here. It can easily be seen that along any line $(a + \Delta \cdot d)$, at an arbitrary point (a), the second derivative of the performance index is greater than or equal to zero. This implies that the performance index is convex along all lines through the parameter space, which in turn implies that the performance index, is convex. Secondly, since only the first p derivatives are nonzero, the performance index along the line can be represented exactly by its Taylor series expansion, which is a pth order polynomial. Also, the first derivative along the line is given by

$$\frac{dJ(a + \Delta \cdot d)}{d\Delta} = \sum_{j=1}^{p} j \binom{p}{j} \left[\sum_{m=1}^{M} \sum_{n=1}^{N} W_{mn} \epsilon_{mn}^{p-j} V_{mn}^j \right] \Delta^{j-1}. \tag{6}$$

Since J is convex, (6) is a monotonic polynomial in Δ. Thus, the line search problem reduces to finding the real zero of (6). An efficient technique for locating the real zeros of a polynomial, and which has been used here, is the Newton-Raphson method.

A couple of steps can be taken to avoid numerical problems that can be encountered with large values of p. At each iteration the performance index (3) can be replaced by

$$\hat{J}(a) = \sum_{m=1}^{M} \sum_{n=1}^{N} W_{mn} [\epsilon_{mn}/\epsilon_{\max}]^p \tag{7}$$

where

$$\epsilon_{\max} = \max_{m,n} [\epsilon_{mn}].$$

This means that a new $\epsilon_{\max}$ must be found every iteration. Also, the gradient vector can be normalized before each line search in the gradient direction. This step is necessary because the gradient vector gets very small as the algorithm approaches the minimum point. By employing these steps, no numerical problems have been encountered in filter designs with values of p up to and including 20.

IV. Examples

Example 1: In this example we will design a low-pass filter. The desired filter response can be seen in Fig. 3. The prescribed magnitude response $G(\omega_{1m}, \omega_{2n})$ is 1 in the passband and 0 in the stopband. The sampling density here is 20. The weighting function W_{mn} was chosen to be 1 in the passband and stopband. If the performance index decreased by less than one percent from one iteration to the next, the optimization procedure was stopped. Minimax filters were designed to these specifications in [12] using various algorithms, the most efficient of which is due to Harris. Design times and maximum deviations achieved for three different sized filters, by the Harris algorithm and by the algorithm proposed here, are given in Table I. The l_p algorithm was run on a Burroughs B6700 computer. In all three filter designs the deviation achieved by the l_p algorithm was very close to that achieved by the minimax technique. This should be expected with a p as large as 20. Comparison of the absolute design times required by the two algorithms is meaningless because they

TABLE I
COMPARISON OF lp AND HARRIS ALGORITHMS

	$l_p(p = 20)$		Harris	
Mask Size	Deviation	Time	Deviation	Time
5 × 5	0.275	0:43	0.267	1:18
7 × 7	0.123	1:11	0.127	3:10
9 × 9	0.121	3:54	0.114	14:14

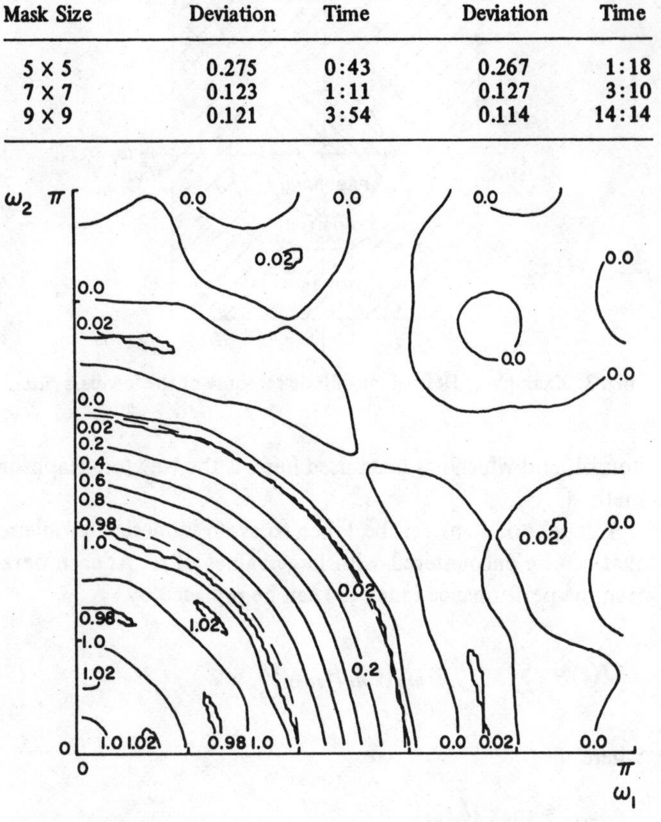

Fig. 4. Example 1: Magnitude response of the nonrecursive low-pass filter.

TABLE II
COMPARISON OF IMPLEMENTATIONS FOR A 1024 × 1024 IMAGE

	Maximum Deviation	Design Time (Min: Sec)	Implementation	#Multiplies per pixel	#Adds per pixel
49-coefficient recursive filter ($p = 8$)	0.022	11:00	zero phase (2 recursions)	96	96
17-coefficient diameter circular nonrecursive filter ($p = 20$)	0.021	3:29	direct convolution using half-plane symmetry	127	252
			direct convolution using octal symmetry	40	252
			FFT	41	61

were run on two entirely different computer systems. However, it is important to notice that, unlike the Harris algorithm, the design time of the l_p technique increases very slowly as the size of the filter increases. The design times given for the l_p technique include the 14 s required to obtain an initial approximation by windowing a 64 × 64 inverse FFT.

Twogood and Ekstrom [16] published an interesting paper comparing the merits of recursive and nonrecursive filters.

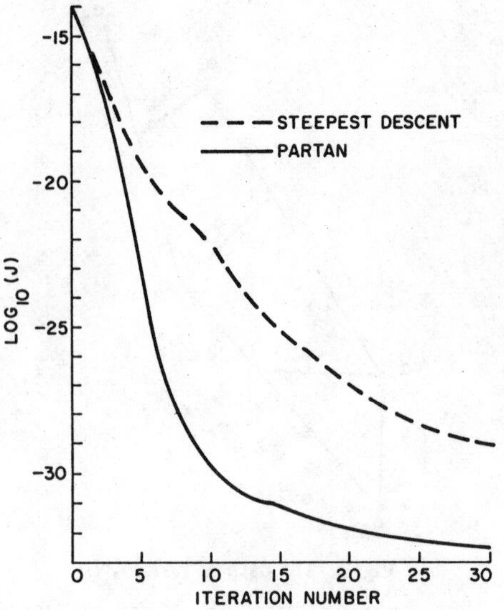

Fig. 5. Example 1: Convergence of the method of steepest descent and PARTAN.

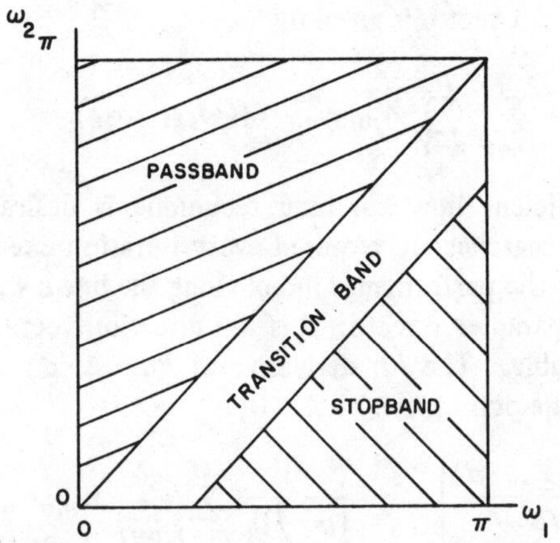

Fig. 6. Example 2: Desired magnitude response of the fan filter.

In that paper it is difficult to determine when the performance of the two filters is equivalent because the nonrecursive filter was designed using a McClellan transformation and the recursive filter was designed to minimize the sum of the squared errors. Here, an l_p technique to design linear phase nonrecursive filters is proposed. In a previous paper [17] an l_p technique to design recursive filters with zero phase implementation was developed. Example 2 of [17] has the same specifications as this example. A nonrecursive filter was designed that yielded a maximum deviation that is almost the same as the deviation achieved in [17]. This filter has a circular mask with a diameter of 17 coefficients. The frequency response of this filter is shown in Fig. 4. Table II summarizes some of the important features of these two filters.

Fig. 5 illustrates the convergence of the method of steepest descent, employing the same line searching technique, as compared with the algorithm including the extra PARTAN line search. It can be seen that the PARTAN algorithm converges much more quickly than does the method of steepest descent. The additional line search required by PARTAN resulted in an increase in design time of only 15 percent.

Example 2: In this example a fan filter is designed. The prescribed magnitude response $G(\omega_{1m}, \omega_{2n})$ is 1 in the passband,

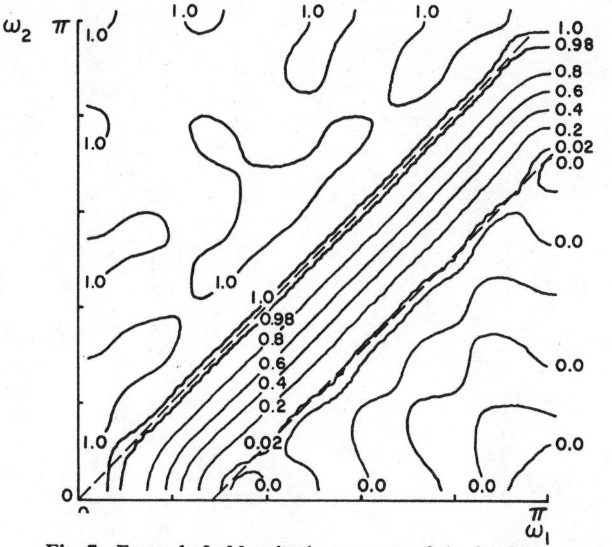

Fig. 7. Example 2: Magnitude response of the fan filter.

and 0 in the stopband. The weighting function W_{mn} was chosen to be 1 in the passband and stopband, and the sampling interval is 0.05π in both dimensions. This filter has a 17×17 input mask and required 4 min and 59 s of design time on a Burroughs B6700 computer. The frequency response can be seen in Fig. 7.

V. Conclusions

In this paper an iterative technique is proposed for the design of two-dimensional linear-phase FIR digital filters. This technique converges to the global minimum regardless of the initial approximation because the performance index is convex and the PARTAN algorithm decreases the performance index every iteration until the minimum is reached. Designs extremely close to minimax designs can be achieved. However, unlike present minimax techniques, design times and computational complexity increase very slowly as the filter size gets large. This technique is a general one in the following sense.

1) Any magnitude function with half-plane symmetry can be closely approximated.

2) The least pth norm is used. This includes least squares ($p = 2$) as a special case. Larger values of p can be used to put more emphasis on reducing the maximum error.

3) The proposed method is applicable to one-dimensional and multidimensional filter design.

4) Any input mask (not only rectangular) having half-plane symmetry can be used.

It is worth noting that it is possible to optimize both spatial and frequency domain specifications using this technique. This is because the output $y(m, n)$ of a two-dimensional linear-phase FIR digital filter can be given by

$$y(m,n) = \sum_{j} \sum_{k} b_{jk} \cdot [(x(m-j, n-k) + x(m+j, n+k))/2] \quad (j,k) \in L \tag{8}$$

where L is the set of integer pairs defined by the input mask of the filter and $x(m, n)$ is the input, which has the same form as (1).

References

[1] A. V. Oppenheim, Ed., *Applications of Digital Signal Processing.* Englewood Cliffs, NJ: Prentice-Hall, 1978.

[2] T. S. Huang, J. W. Burnett, and A. G. Deczky, "The importance of phase in image processing filters," *IEEE Trans. Acoust., Speech, Signal Processing*, vol. ASSP-23, pp. 529-542, Dec. 1975.

[3] A. V. Oppenheim, J. S. Lim, G. Kopec, and S. C. Pohlig, "Phase in speech and pictures," in *Proc. IEEE Int. Conf. Acoust., Speech, Signal Processing*, Washington, DC, 1979, pp. 632-637.

[4] T. S. Huang, "Two-dimensional windows," *IEEE Trans. Audio Electroacoust.*, vol. AU-20, pp. 80-90, Mar. 1972.

[5] T. C. Speake and R. M. Mersereau, "A comparison of different window formulations for two-dimensional FIR filter design," in *Proc. IEEE Int. Conf. Acoust., Speech, Signal Processing*, Washington, DC, 1979, pp. 5-8.

[6] J. H. McClellan, "The design of two-dimensional digital filters by transformations," in *Proc. 7th Annu. Princeton Conf. Inform. Sci. Syst.*, 1973, pp. 247-251.

[7] W. F. G. Mecklenbrauker and R. M. Mersereau, "McClellan transformations for two-dimensional digital filtering: Part II. Implementation," *IEEE Trans. Circuits Syst.*, vol. CAS-23, pp. 414-422, July 1976.

[8] J. V. Hu and L. R. Rabiner, "Design techniques for two-dimensional digital filters," *IEEE Trans. Audio Electroacoust.*, vol. AU-20, pp. 249-257, Oct. 1972.

[9] Y. Kamp and J. P. Thiran, "Chebyshev approximation for two-dimensional nonrecursive digital filters," *IEEE Trans. Circuits Syst.*, vol. CAS-22, pp. 208-218, Mar. 1975.

[10] H. S. Hersey and R. M. Mersereau, "An algorithm to perform minimax approximation in the absence of the Haar condition," M.I.T. Res. Lab. Electron., Cambridge, MA, Quarterly Progress Rep. No. 114, July 1974.

[11] D. B. Harris, "Iterative procedures for optimal Chebyshev design of FIR digital filters," S.M. thesis, Dep. Elec. Eng. Comput. Sci., Massachusetts Inst. Technol., Cambridge, Feb. 1976.

[12] D. B. Harris and R. M. Mersereau, "A comparison of algorithms for minimax design of two-dimensional linear-phase FIR digital filters," *IEEE Trans. Acoust., Speech, Signal Processing*, vol. ASSP-25, pp. 492-500, Dec. 1977.

[13] B. Shah, R. Buehler, and O. Kempthorne, "Some algorithms for minimizing a function of several variables," *J. Soc. Indust. Appl. Math.*, vol. 12, pp. 74-92, 1964.

[14] D. G. Luenberger, *Introduction to Linear and Nonlinear Programming.* Reading, MA: Addison-Wesley, 1965.

[15] M. T. Manry and J. K. Aggarwal, "Design and implementation of two-dimensional FIR digital filters with nonrectangular arrays," *IEEE Trans. Acoust., Speech, Signal Processing*, vol. ASSP-26, pp. 314-318, Aug. 1978.

[16] R. E. Twogood and M. P. Ekstrom, "Why filter recursively in two dimensions?," in *Proc. IEEE Int. Conf. Acoust., Speech, Signal Processing*, Washington, DC, 1979, pp. 20-23.

[17] J. H. Lodge and M. M. Fahmy, "An optimization technique for the design of half-plane 2-D recursive digital filters," to be published in *IEEE Trans. Circuits Syst.*

Design of Two-Dimensional Recursive Digital Filters with Specified Magnitude and Group Delay Characteristics

SAMY A. H. ALY AND MOUSTAFA M. FAHMY, SENIOR MEMBER, IEEE

***Abstract*—A technique is proposed for the design of two-dimensional (2-D) recursive filters with a response that best approximates, in the *lp* sense, prescribed magnitude and group delay specifications. The filter stability is guaranteed through the use of a frequency transformation. The optimization technique used is that of Davidon–Fletcher and Powell. Several examples are given to illustrate the proposed algorithm.**

Manuscript received February 16, 1978. This work was partially supported by the National Research Council of Canada under Grant A4149.

The authors are with the Department of Electrical Engineering, Queen's University, Kingston, Ont. K7L 3N6, Canada.

I. Introduction

In the last few years, several attempts have been made to design recursive digital filters that simultaneously meet magnitude and group delay specifications [1]–[3]. For the two-dimensional (2-D) case, in particular, Maria and Fahmy [4], [5] suggested starting with a 2-D filter that approximates the magnitude requirements and then cascading it by an all-pass 2-D filter to equalize the resulting group delay. The two parts of the filter are optimal in their own right, but the overall filter is

Reprinted from *IEEE Trans. Circuit Syst.*, vol. CAS-25, pp. 908–915, Nov. 1978.

not, in general, optimal in meeting simultaneously the magnitude the group delay specifications. Chottera and Jullien [3] have formulated the design problem of meeting simultaneously the two specifications, as a linear programming problem, to control the real and imaginary parts of the filter transfer function. This method restricts the filter realization to the direct form, which has relativeiy large roundoff error for high-order filters. Also, the implementation of the method requires relatively long computation time.

In this paper, the filter is assumed to be in the cascade form of (1) below. The filter coefficients are calculated to approximate both the magnitude and group delay characteristics simultaneously. The filter stability is ensured through the use of a frequency transformation and thus there is no need to check the filter stability after each iteration as is done in [4], [5].

II. Problem Statement

Consider a 2-D recursive filter with its transfer function, $H(z_1^{-1},z_2^{-1})$, being in the form

$$H(z_1^{-1},z_2^{-1})=A\prod_{k=1}^{K}\frac{\sum_{j=0}^{J_2^{(k)}}\sum_{i=0}^{J_1^{(k)}} b_{ij}^{(k)}z_1^{-i}z_2^{-j}}{\sum_{j=0}^{I_2^{(k)}}\sum_{i=0}^{I_1^{(k)}} d_{ij}^{(k)}z_1^{-i}z_2^{-j}}. \tag{1}$$

Here, A is the gain constant, $b_{ij}^{(k)}$ and $d_{ij}^{(k)}$ are the filter coefficients with $b_{00}^{(k)}=d_{00}^{(k)}=1$, K is the total number of sections cascaded, and $I_1^{(k)}$, $I_2^{(k)}$, $J_1^{(k)}$, and $J_2^{(k)}$ are integers each of which can be zero, one, or two.

The frequency response of the above filter may be expressed as

$$H(e^{-j\omega_1},e^{-j\omega_2})=|H(e^{-j\omega_1},e^{-j\omega_2})|e^{j\phi(\omega_1,\omega_2)} \tag{2a}$$

with the group delay functions defined as

$$\tau_i(\omega_1,\omega_2)=\frac{-\partial\phi(\omega_1,\omega_2)}{\partial\omega_i}, \qquad i=1,2. \tag{2b}$$

Let $(\omega_{1m},\omega_{2n})$, $m=1,\cdots,M$, $n=1,\cdots,N$ be $M\times N$ discrete points in the frequency plane $\omega_1-\omega_2$, and let $H(m,n)$ and $\tau_i(m,n)$ be defined as

$$H(m,n)=H(e^{-j\omega_{1m}},e^{-j\omega_{2n}})$$
$$\tau_i(m,n)=\tau_i(\omega_{1m},\omega_{2n}).$$

Let $\bar{X}$ be the parameter vector defined by

$$\bar{X}=\Big[(((d_{ij}^{(k)}, i=0,\cdots,I_1^{(k)}), j=0,\cdots,I_2^{(k)}), k=1,\cdots,K),$$
$$(((b_{ij}^{(k)}, i=0,\cdots,J_1^{(k)}), j=0,\cdots,J_2^{(k)}), k=1,\cdots,K), A\Big]^T$$

with $(i,j)\neq(0,0)$.

Let the performance index, $F(\bar{X})$, be defined as

$$F(\bar{X})=R_mF_m(\bar{X})+R_{\tau_1}F_{\tau_1}(\bar{X})+R_{\tau_2}F_{\tau_2}(\bar{X}) \tag{3}$$

where

$$F_m(\bar{X})=\sum_{m=1}^{M}\sum_{n=1}^{N}\lambda(m,n)(|H(m,n)|-Y(m,n))^P \tag{4a}$$

and

$$F_{\tau_i}(\bar{X})=\sum_{m\in M_\tau}\sum_{n\in N_\tau}\mu(m,n)(\tau_i(m,n)-\tau_{d_i}(m,n))^P. \tag{4b}$$

Here, F_m and F_{τ_i} are the error functions for the magnitude and group delays, respectively, $Y(m,n)$ and $\tau_{d_i}(m,n)$ are the desired magnitude and group delay filter characteristics, $\lambda(m,n)$ and $\mu(m,n)$ are frequency weighting functions, and R_m and R_{τ_i} represent the relative weights of the various components, $F_m(\bar{X})$ and $F_{\tau_i}(\bar{X})$, of the performance index. $M_\tau\times N_\tau$ is the discrete set of frequencies at which the group delay $\tau_i(m,n)$ is to approach $\tau_{d_i}(m,n)$.

The problem to be solved is to find the parameter vector $\bar{X}$ such that the performance index, $F(\bar{X})$, is minimum and the filter is stable.

III. Stability Requirements

Let $Q(s_1,s_2)$ be a two-variable polynomial given by

$$Q(s_1,s_2)=\sum_{j=0}^{L_2}\sum_{i=0}^{L_1}e_{ij}s_1^is_2^j$$

where s_1 and s_2 are two complex variables, and each of L_1 and L_2 is equal to either one or two. It has been shown in [6] that

$$Q(s_1,s_2)\neq 0, \qquad \text{for Re}[s_1]\geq 0 \text{ and Re}[s_2]\geq 0 \tag{5}$$

if the e_{ij}'s can be expressed in terms of another set of real (nonzero) parameters $[q_l]$ as follows:

Case I: $L_1=L_2=1$

$$\left.\begin{aligned}e_{00}&=q_1^2\\ e_{10}&=q_2^2\\ e_{01}&=q_3^2\\ e_{11}&=1\end{aligned}\right\}. \tag{6a}$$

Case II: $L_1=2$ and $L_2=1$

$$\left.\begin{aligned}e_{00}&=(q_1q_6-q_2q_5+q_3q_4)^2\\ e_{10}&=q_5^2+q_6^2\\ e_{20}&=q_3^2\\ e_{01}&=q_4^2\\ e_{11}&=q_1^2+q_2^2\\ e_{21}&=1\end{aligned}\right\} \tag{6b}$$

and similarly is the case of $L_1=1$ and $L_2=2$.

Case III: $L_1 = L_2 = 2$

$$\left.\begin{aligned} e_{00} &= (q_5q_{10} - q_6q_9 + q_7q_8)^2 \\ e_{10} &= (q_1q_{10} - q_3q_7 + q_4q_6)^2 \\ &\quad + (q_2q_{10} - q_3q_9 + q_4q_8)^2 \\ e_{20} &= q_{10}^2 \\ e_{01} &= (q_1q_8 - q_2q_6 + q_3q_5)^2 \\ &\quad + (q_1q_9 - q_2q_7 + q_4q_5)^2 \\ e_{11} &= q_6^2 + q_7^2 + q_8^2 + q_9^2 \\ e_{21} &= q_3^2 + q_4^2 \\ e_{02} &= q_5^2 \\ e_{12} &= q_1^2 + q_2^2 \\ e_{22} &= 1 \end{aligned}\right\} . \tag{6c}$$

From (5), it is clear that if the coefficients $[e_{ij}]$ satisfy (6) then

$$\hat{Q}(z_1^{-1}, z_2^{-1}) \triangleq Q(s_1, s_2)|_{s_1 = (1 - z_1^{-1})/(1 + z_1^{-1}),\ s_2 = (1 - z_2^{-1}/1 + z_2^{-1})} \neq 0, \quad \text{for } |z_1| \geq 1 \text{ and } |z_2| \geq 1. \tag{7}$$

Let (1) be rewritten in the form

$$H(z_1^{-1}, z_2^{-1}) = B \prod_{k=1}^{K} \frac{(1+z_1^{-1})^{J_1^{(k)}}(1+z_2^{-1})^{J_2^{(k)}} \sum_{j=0}^{J_2^{(k)}} \sum_{i=0}^{J_1^{(k)}} c_{ij}^{(k)} \left[\frac{1-z_1^{-1}}{1+z_1^{-1}}\right]^i \left[\frac{1-z_2^{-1}}{1+z_2^{-1}}\right]^j}{(1+z_1^{-1})^{I_1^{(k)}}(1+z_2^{-1})^{I_2^{(k)}} \sum_{j=0}^{I_2^{(k)}} \sum_{i=0}^{I_1^{(k)}} e_{ij}^{(k)} \left[\frac{1-z_1^{-1}}{1+z_1^{-1}}\right]^i \left[\frac{1-z_2^{-1}}{1+z_2^{-1}}\right]^j} \tag{8}$$

with $c_{00}^{(k)} = 1$.

It follows from the discussion above that if the $e_{ij}^{(k)}$'s are expressible in terms of the $q_l^{(k)}$'s as in (6), then the filter is always stable [7].

By noting that

$$\left.\frac{1-z^{-1}}{1+z^{-1}}\right|_{z=e^{j\omega}} = j \tan(\omega/2)$$

and

$$(1+z^{-1})|_{z=e^{j\omega}} = 2e^{-j\omega/2} \cos(\omega/2)$$

the above equation can be simplified as follows:

$$H(m,n) = B \cdot U(m,n) \cdot \prod_{k=1}^{K} H^{(k)}(m,n)$$

where

$$U(m,n) = [2\cos(\omega_{1m}/2)]^{v_1}[2\cos(\omega_{2n}/2)]^{v_2} e^{-j(v_1\omega_{1m} + v_2\omega_{2n})/2}$$

$$v_1 = \sum_{k=1}^{K} (J_1^{(k)} - I_1^{(k)})$$

$$v_2 = \sum_{k=1}^{K} (J_2^{(k)} - I_2^{(k)})$$

and

$$H^{(k)}(m,n) = \frac{N^{(k)}(m,n)}{D^{(k)}(m,n)} = \frac{\sum_{j=0}^{J_2^{(k)}} \sum_{i=0}^{J_1^{(k)}} c_{ij}^{(k)} [j\tan(\omega_{1m}/2)]^i [j\tan(\omega_{2n}/2)]^j}{\sum_{j=0}^{I_2^{(k)}} \sum_{i=0}^{I_1^{(k)}} e_{ij}^{(k)} [j\tan(\omega_{1m}/2)]^i [j\tan(\omega_{2n}/2)]^j}. \tag{9}$$

The relationships between the filter parameters $d_{ij}^{(k)}$, $b_{ij}^{(k)}$, and A in (1), and the modified parameters $e_{ij}^{(k)}$ (or $q_l^{(k)}$), $c_{ij}^{(k)}$, and B in (8) are straightforward. In the following, the components of the parameter vector $\bar{X}$ will be taken as follows:

$$\bar{X} \triangleq \Big[\big((q_l^{(k)}, l = 1, 2, \cdots), k = 1, 2, \cdots, K\big), \big(((c_{ij}^{(k)}, i = 0, 1, \cdots, J_1^{(k)}), j = 0, 1, \cdots, J_2^{(k)}), k = 1, 2, \cdots, K\big), B\Big]^T$$

with $(i,j) \neq (0,0)$. By choosing to work with the modified, rather than the original parameters, the stability of the filter is ensured without having to perform a stability test after each iteration as was done in [4], [5].

IV. Problem Analysis

Solving the resulting minimization problem, using the Fletcher and Powell optimization technique [8], [9], requires calculating the gradient vector, $\bar{G}(\bar{X})$, which is defined as

$$\bar{G}(\bar{X}) = \frac{\partial F(\bar{X})}{\partial \bar{X}} = \left[\frac{\partial F}{\partial X_1} \quad \cdots \quad \frac{\partial F}{\partial X_r} \quad \cdots \quad \frac{\partial F}{\partial X_{N_x}}\right]^T$$

where N_x is the dimension of the vector $\bar{X}$. From (3), $\bar{G}(\bar{X})$ can be written as

$$\bar{G}(\bar{X}) = R_m \bar{G}_m(\bar{X}) + R_{\tau_1} \bar{G}_{\tau_1}(\bar{X}) + R_{\tau_2} \bar{G}_{\tau_2}(\bar{X}). \tag{10}$$

The rth component of $\bar{G}_m(\bar{X})$, $g_{mr}(\bar{X})$, is defined by

$$g_{mr}(\bar{X}) = \frac{\partial F_m(\bar{X})}{\partial X_r} = p \sum_{m=1}^{M} \sum_{n=1}^{N} \lambda(m,n) \cdot (|H(m,n)| - Y(m,n))^{p-1} \frac{\partial |H(m,n)|}{\partial X_r} \tag{11}$$

with

$$\frac{\partial |H(m,n)|}{\partial X_r} = (|H(m,n)|)^{-1} \operatorname{Re}\left[H^*(m,n) \cdot \frac{\partial H(m,n)}{\partial X_r}\right]. \tag{12}$$

The rth component of $\bar{G}_{\tau_i}(\bar{X})$, $g_{\tau_i r}(\bar{X})$, is given by

$$g_{\tau_i r}(\bar{X})=\frac{\partial F_{\tau_i}(\bar{X})}{\partial X_r}=p\sum_{m\in M_\tau}\sum_{n\in N_\tau}\mu(m,n)\cdot\left(\tau_i(m,n)-\tau_{d_i}(m,n)\right)^{p-1}\frac{\partial \tau_i(m,n)}{\partial X_r} \quad (13)$$

with τ_i expressed as

$$\tau_i=\frac{v_i}{2}+\operatorname{cosec}(\omega_i)\sum_{k=1}^{K}\operatorname{Im}\left[\left(D_i^{(k)}/D^{(k)}\right)-\left(N_i^{(k)}/N^{(k)}\right)\right] \quad (14)$$

where

$$N_1^{(k)}=\sum_{l=0}^{J_2^{(k)}}\sum_{i=0}^{J_1^{(k)}} i c_{il}^{(k)}\cdot\left(j\tan(\omega_1/2)\right)^i\left(j\tan(\omega_2/2)\right)^l$$

$$N_2^{(k)}=\sum_{l=0}^{J_2^{(k)}}\sum_{i=0}^{J_1^{(k)}} l c_{il}^{(k)}\cdot\left(j\tan(\omega_1/2)\right)^i\left(j\tan(\omega_2/2)\right)^l \quad (15)$$

and $D_i^{(k)}$ has similar expressions. Equation (14) has been derived through noting the following equalities:

$$\log\left(H^{(k)}\right)=\log\left(|H^{(k)}|\right)+j\phi^{(k)}$$

$$\log(z_i)=j\omega_i$$

$$\operatorname{Re}\left[\frac{\partial\log\left(H^{(k)}\right)}{\partial\log(z_i)}\right]=\frac{\partial\phi^{(k)}}{\partial\omega_i}=-\tau_i^{(k)}.$$

V. Remarks

1) As defined in (4a, b) the error functions, $F_m(\bar{X})$ and $F_{\tau_i}(\bar{X})$, have different orders of magnitude. Hence, combining these functions in $F(\bar{X})$ may result in unintentional emphasis of one function at the expense of the others. This difficulty is circumvented by normalizing each function with respect to some selected reference. Thus in the proposed technique, the relative weights, R_m and R_{τ_i}, are chosen to have the following forms:

$$R_m=\alpha/E_m \quad (16a)$$

$$R_{\tau_i}=\left(\frac{1-\alpha}{2}\right)/E_{\tau_i} \quad (16b)$$

where E_m and E_{τ_i} are the reference functions defined by

$$E_m=\sum_{m=1}^{M}\sum_{n=1}^{N}\lambda(m,n)Y^p(m,n) \quad (17a)$$

$$E_{\tau_i}=\sum_{m\in M_\tau}\sum_{n\in N_\tau}\mu(m,n)\tau_{d_i}^p(m,n) \quad (17b)$$

and α is the magnitude to the group delay relative weighting factor.

Experience has shown that, to accelerate the convergence to the optimum, especially when the initial design is selected at random, α may be varied as the search for the optimum progresses [10]. At the start, α is set to unity, i.e., zero weight is assigned to the group delay error functions, and then is reduced gradually. When α attains the designer selected value, it is kept constant till the end of the computation.

2) If the level of $\tau_{d_i}(m,n)$, $i=1, 2$, is not of interest it is possible to replace $\tau_{d_i}(m,n)$ by the function $[\tau_{d_i}(m,n)+\tau_{0i}]$, where τ_{01} and τ_{02} are two additional parameters.

3) In the above analysis, the filter sections were assumed to have polynomials of orders zero, one, or two in both z_1^{-1} and z_2^{-1}. However, the same technique can be extended in a straightforward fashion to sections of any order.

4) Goodman [11] has shown that even if condition (5) is satisfied, $\hat{Q}(z_1^{-1},z_2^{-1})$ may still have zeros at $(z_1^{-1},-1)$ with $|z_1|\geq 1$ and $(-1,z_2^{-1})$ with $|z_2|\geq 1$. To avoid these situations, the following conditions must be satisfied:

a)

$$\sum_{i=0}^{L_1} e_{iL_2}(1-z_1^{-1})^i(1+z_1^{-1})^{L_2-i}\neq 0, \qquad \forall|z_1|\geq 1$$

b)

$$\sum_{j=0}^{L_2} e_{L_1j}(1-z_2^{-1})^j(1+z_2^{-1})^{L_1-j}\neq 0, \qquad \forall|z_2|\geq 1.$$

It is straightforward to show that the proposed choice for e_{ij} (as in (6)) will always satisfy the above conditions.

VI. Examples

Example 1 (LPF)

This example considers a 2-D LPF with the following circular symmetrical specifications:

$$Y(\omega_1,\omega_2)=1.0,\ 1.0,\ 0.8,\ 0.44,\ 0.14,\ 0.03,\ 0.002,\ 0.001,\ 0.001,\ 0.001,\ 0.001$$

for $\sqrt{\omega_1^2+\omega_2^2}/\pi=0.0, 0.1, 0.2,\cdots,1.0$, respectively. The group delay in the passband, $\sqrt{\omega_1^2+\omega_2^2}<0.4\pi$, is required to be constant (τ_{d_1} and τ_{d_2}). These specifications are approximated using one section (i.e., $K=1$) with $I_1^{(1)}=I_2^{(1)}=J_1^{(1)}=J_2^{(1)}=2$. In the performance index, the following values for the different parameters are used:

$$p=2$$

$$\alpha=0.5$$

$$\tau_{d_1}=\tau_{d_2}=2$$

$$\lambda(m,n)=\mu(m,n)=1.$$

The discrete frequency samples are taken equally spaced at $|\omega_i/\pi|=0.0, 0.1, 0.2,\cdots,1.0$, in the range $-\pi<\omega_1\leq\pi$ and $0\leq\omega_2\leq\pi$. The initial design is taken as $e_{ij}^{(1)}=c_{ij}^{(1)}=B=1$ for all i and j. The relative weighting factor α has the value 1.0 for the first five iterations and is decreased by a step of 0.1 every successive five iterations till it reaches the value of 0.5. After these 25 iterations the resulting filter is as follows (FILTER #1):

$$H(z_1^{-1},z_2^{-1})=0.035935\frac{(1\,z_1^{-1}z_1^{-2})\begin{bmatrix}1.0 & 0.428998 & 0.470506\\ 0.441454 & 0.316414 & 0.659983\\ 0.372886 & 0.412246 & 0.168825\end{bmatrix}\begin{bmatrix}1\\ z_2^{-1}\\ z_2^{-2}\end{bmatrix}}{(1\,z_1^{-1}z_1^{-2})\begin{bmatrix}1.0 & -0.535537 & 0.008103\\ -0.658556 & -0.119018 & 0.305373\\ 0.081547 & 0.279834 & -0.202243\end{bmatrix}\begin{bmatrix}1\\ z_2^{-1}\\ z_2^{-2}\end{bmatrix}}.$$

The resulting rms relative error values are 0.04808, 0.17054, and 0.12858 in the magnitude and group delays approximations, respectively.

The same specifications are approximated using two sections (i.e., $K=2$), with $I_1^{(k)}=I_2^{(k)}=J_1^{(k)}=J_2^{(k)}=2$ for $k=1, 2$. The performance index parameters are given the following values:

$$p=2, \quad \alpha=0.5$$
$$\tau_{d_1}=\tau_{d_2}=4$$
$$\lambda(m,n)=\mu(m,n)=1.$$

The above filter is used as an initial design for each section. After 40 iterations the search converges to the following filter (FILTER #2):

$$H(z_1^{-1},z_2^{-1})=0.134044\times10^{-2}\frac{(1\,z_1^{-1}z_1^{-2})\begin{bmatrix}1.0 & 0.201054 & 0.600823\\ 0.446691 & 1.162514 & 1.473812\\ 0.570455 & 1.120600 & 0.463334\end{bmatrix}\begin{bmatrix}1\\ z_2^{-1}\\ z_2^{-2}\end{bmatrix}}{(1\,z_1^{-1}z_1^{-2})\begin{bmatrix}1.0 & -0.490714 & 0.027371\\ -0.536990 & -0.127213 & 0.236525\\ 0.068811 & 0.211200 & -0.143242\end{bmatrix}\begin{bmatrix}1\\ z_2^{-1}\\ z_2^{-2}\end{bmatrix}}$$

$$\cdot\frac{(1\,z_1^{-1}z_1^{-2})\begin{bmatrix}1.0 & 0.201054 & 0.600822\\ 0.446690 & 1.162513 & 1.473811\\ 0.570454 & 1.120600 & 0.463335\end{bmatrix}\begin{bmatrix}1\\ z_2^{-1}\\ z_2^{-2}\end{bmatrix}}{(1\,z_1^{-1}z_1^{-2})\begin{bmatrix}1.0 & -0.491231 & 0.028179\\ -0.538819 & -0.128680 & 0.236831\\ 0.066576 & 0.209757 & -0.143378\end{bmatrix}\begin{bmatrix}1\\ z_2^{-1}\\ z_2^{-2}\end{bmatrix}}$$

The filter response, shown in Fig. 1, approximates the desired specifications with relative rms errors of 0.02341, 0.09322, and 0.08180 for the magnitude and group delays, respectively.

Example 2 (Narrow-Band LPF)

Here, we are going to consider a 2-D LPF with cutoff frequencies equal to one tenth the Nyquist frequencies. The specifications of this filter are as follows:

$$Y(\omega_1,\omega_2)=\begin{cases}1.0, & \text{for } \omega\leqslant 0.08\pi\\ 0.5, & \text{for } 0.08\pi<\omega<0.12\pi.\\ 0.0, & \text{for } \omega\geqslant 0.12\pi\end{cases}$$

The group delay is required to be constant (τ_{d_1} and τ_{d_2}) in the passband where $\omega=\sqrt{\omega_1^2+\omega_1^2}<0.1\pi$. These characteristics are approximated using one section (i.e., $K=1$) with $I_1^{(1)}=I_2^{(1)}=J_1^{(1)}=J_2^{(1)}=2$. The performance index parameters used are as follows:

$$\alpha=0.5, \quad P=2$$
$$\tau_{d_1}=\tau_{d_2}=5$$
$$\lambda(m,n)=\mu(m,n)=1.$$

The frequency samples are taken at $|\omega_i/\pi|=0.0, 0.02, 0.04,\cdots,0.2, 0.4,\cdots,1.0$ in the range $-\pi<\omega_1\leqslant\pi$ and $0\leqslant\omega_2\leqslant\pi$. The optimization starts from FILTER #1 and after 50 iterations it converges to the following (FILTER #3):

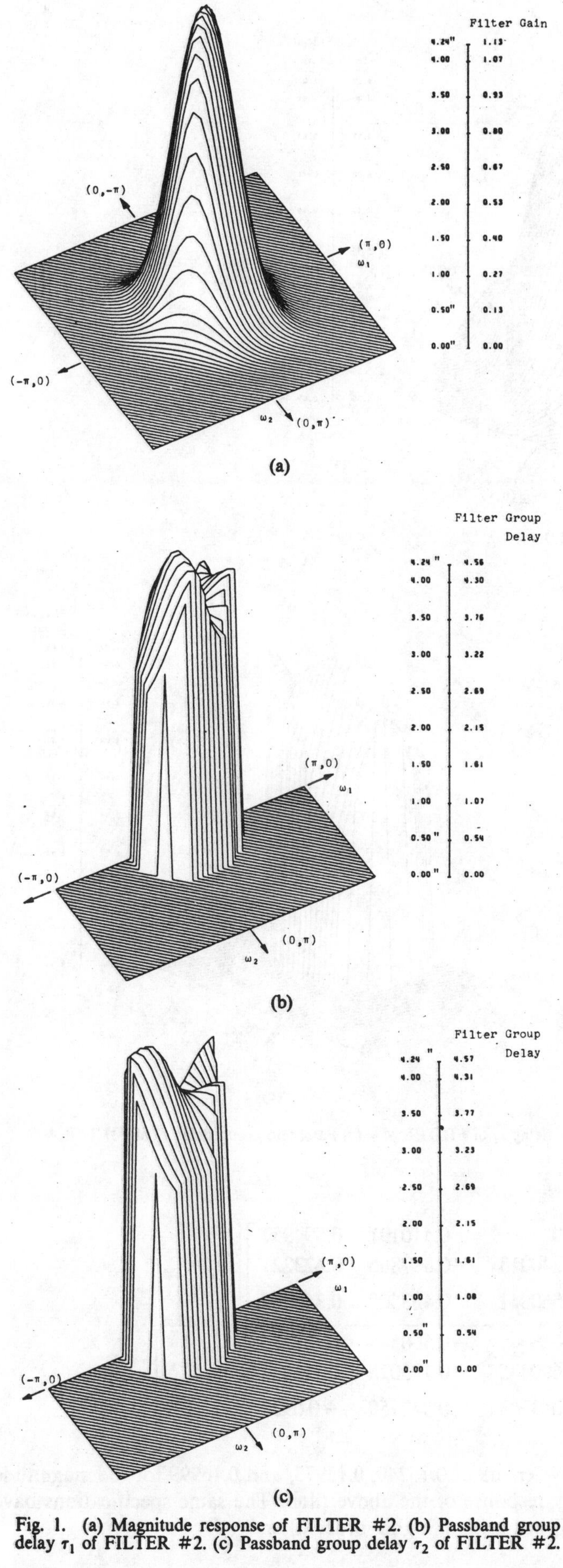

Fig. 1. (a) Magnitude response of FILTER #2. (b) Passband group delay τ_1 of FILTER #2. (c) Passband group delay τ_2 of FILTER #2.

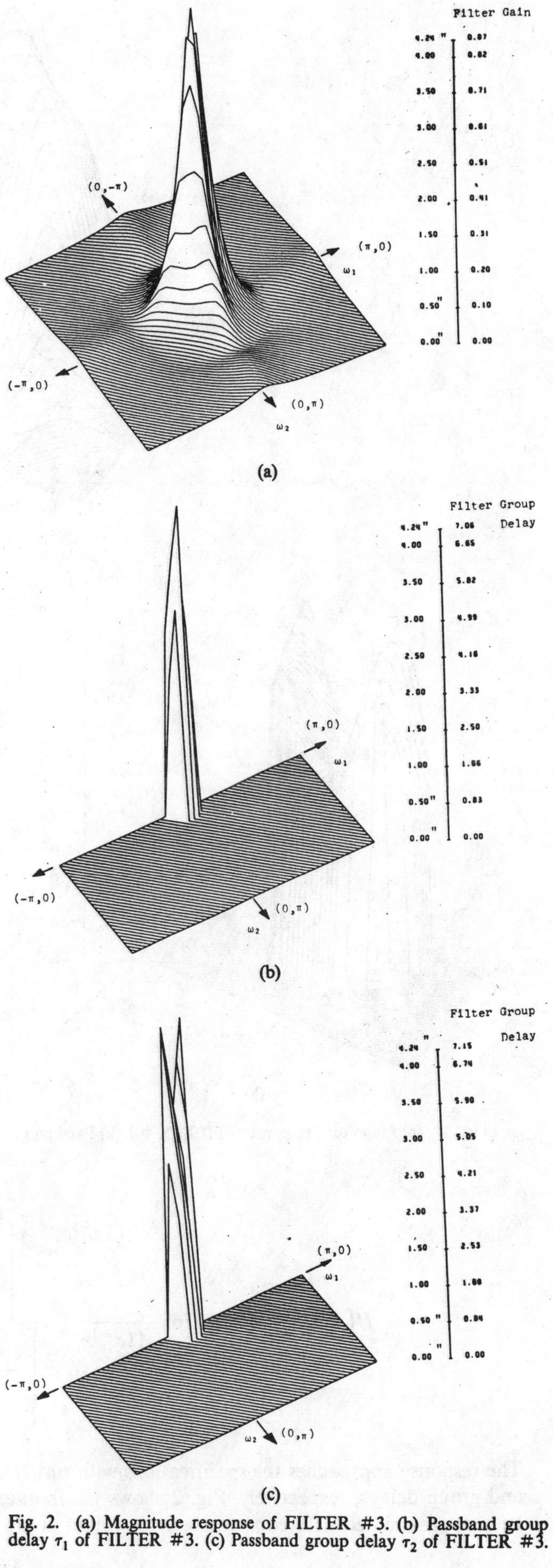

Fig. 2. (a) Magnitude response of FILTER #3. (b) Passband group delay τ_1 of FILTER #3. (c) Passband group delay τ_2 of FILTER #3.

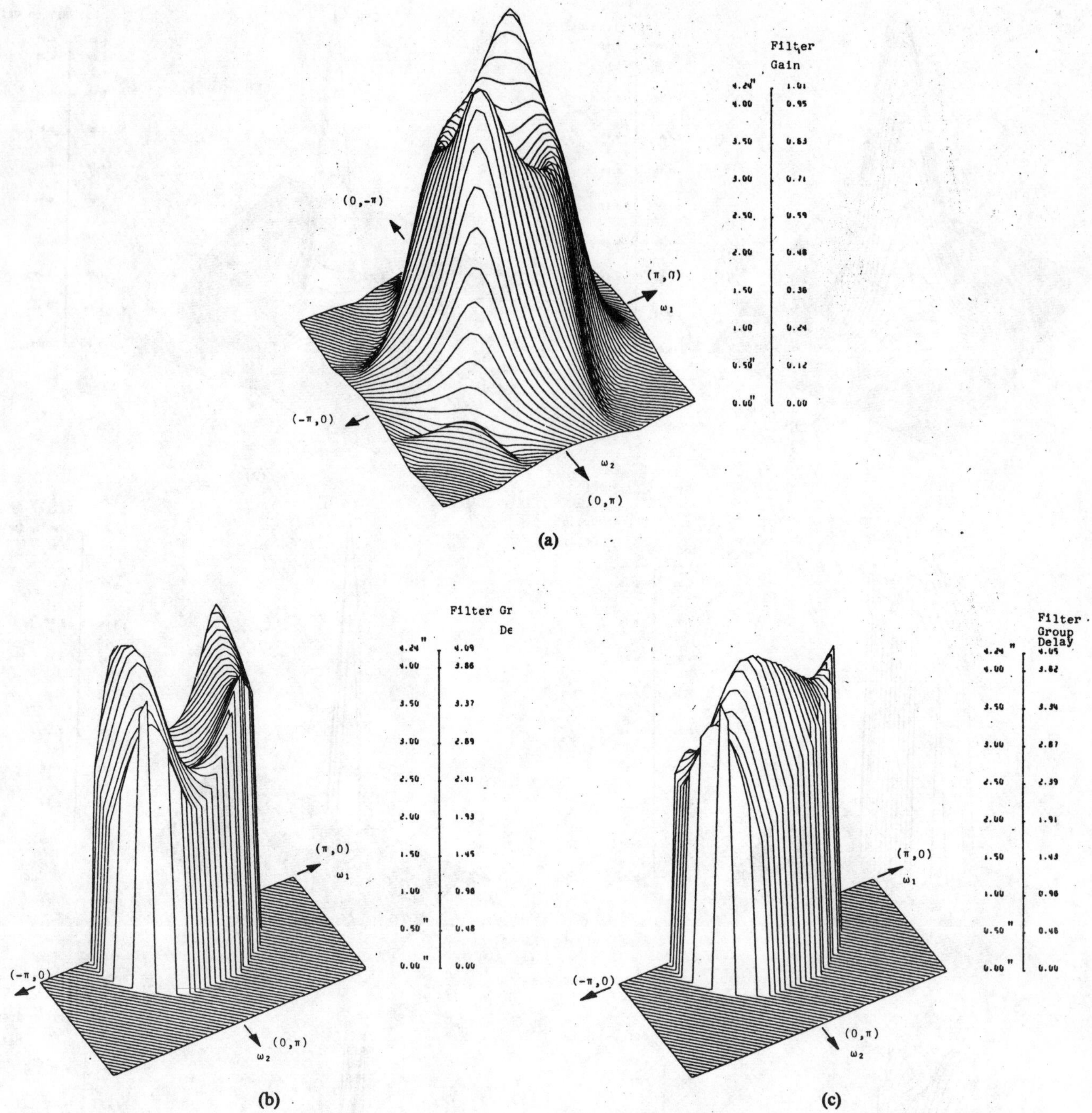

Fig. 3. (a) Magnitude response of FILTER #4. (b) Passband group delay τ_1 of FILTER #4. (c) Passband group delay τ_2 of FILTER #4.

$$H(z_1^{-1},z_2^{-1})=0.007765\frac{\left(1 z_1^{-1} z_1^{-2}\right)\begin{bmatrix}1.0 & 0.410191 & 0.594957\\ 0.240013 & -0.887865 & 0.423221\\ 0.560841 & 0.453500 & 0.360962\end{bmatrix}\begin{bmatrix}1\\ z_2^{-1}\\ z_2^{-2}\end{bmatrix}}{\left(1 z_1^{-1} z_1^{-2}\right)\begin{bmatrix}1.0 & -0.500549 & -0.138282\\ -0.690435 & -0.195020 & 0.346731\\ -0.043308 & 0.342758 & -0.093572\end{bmatrix}\begin{bmatrix}1\\ z_2^{-1}\\ z_2^{-2}\end{bmatrix}}.$$

The response approaches the specifications with rms relative errors of 0.15240, 0.15973, and 0.16999 for the magnitude and group delays, respectively. Fig. 2 shows the frequency response of the above filter. The same specifications have been approximated with comparable error values in [4], [5] but using a higher order filter.

Example 3 (Atmospheric Turbulence Deblurring Filter)

This is a circular symmetrical BPF with the following specifications:

$$Y(\omega_1,\omega_2)=\sqrt{E/(E^2+1/S)}$$

where $E=\sqrt{\pi/k}\cdot e^{-\omega^2/k}$, $S=\text{SNR}$ (assumed to be 100), k is the turbulence blur factor (assumed to be 0.5), and $\omega=\sqrt{\omega_1^2+\omega_2^2}$. The group delay is to be constant in the passband where $\omega<0.6\pi$. The filter is approximated using two sections (i.e., $K=2$), with $I_1^{(k)}=I_2^{(k)}=J_1^{(k)}=J_2^{(k)}=2$, $k=1,2$.

The initial design is taken as $x_r=1$ for $r=1,2,\cdots,N_x$. The parameters are varied in a strategy similar to that in example 1). After 60 iterations, the algorithm converges to the following (FILTER #4):

$$H(z_1^{-1},z_2^{-1})=4.88685\times10^{-3}\frac{(1\,z_1^{-1}z_1^{-2})\begin{bmatrix}1.0 & -3.045680 & 1.076602\\ -3.015179 & -2.259376 & -3.420389\\ 1.087298 & -3.421535 & -1.578152\end{bmatrix}\begin{bmatrix}1\\ z_2^{-1}\\ z_2^{-2}\end{bmatrix}}{(1\,z_1^{-1}z_1^{-2})\begin{bmatrix}1.0 & -0.440394 & 0.216301\\ -0.423759 & 0.318220 & -0.087699\\ 0.219000 & -0.092634 & 0.137613\end{bmatrix}\begin{bmatrix}1\\ z_2^{-1}\\ z_2^{-2}\end{bmatrix}}$$

$$\cdot\frac{(1\,z_1^{-1}z_1^{-2})\begin{bmatrix}1.0 & -0.052676 & 0.593751\\ -0.050184 & -2.254673 & -1.647444\\ 0.594803 & -1.645737 & -0.700552\end{bmatrix}\begin{bmatrix}1\\ z_2^{-1}\\ z_2^{-2}\end{bmatrix}}{(1\,z_1^{-1}z_1^{-2})\begin{bmatrix}1.0 & -0.376473 & 0.304989\\ -0.392962 & 0.292799 & -0.259586\\ 0.311390 & -0.257597 & 0.149088\end{bmatrix}\begin{bmatrix}1\\ z_2^{-1}\\ z_2^{-2}\end{bmatrix}}.$$

Fig. 3 shows a plot of the magnitude and group delay responses of the above filter. The response approaches the specifications with rms relative errors of 0.08477, 0.15233, and 0.14727 for the magnitude and group delays, respectively.

VII. Summary and Conclusion

In this paper, a technique has been proposed to design 2-D recursive filters with prescribed magnitude and group delay characteristics. The filter stability is guaranteed through the use of the bilinear frequency transformation. This technique can also be used in the one-dimensional (1-D) case (as a special case). For the sake of comparison, the example solved in [4], [5] has also been solved here. The proposed method gives a filter with comparable performance to that of the higher order filter obtained in [5].

References

[1] A. G. Deczkey, "Synthesis of recursive digital filters using the minimum p-error criterion," *IEEE Trans. Audio Electroacoust.*, vol. AU-20, no. 4, pp. 257–263 Oct. 1972.

[2] G. A. Maria and M. M. Fahmy, "A new design technique for recursive digital filters," IEEE Trans. *Circuits Syst.*, vol. CAS-23, no. 5, pp. 323–325, May 1976.

[3] A. Chottera and G. A. Jullien, "Designing near linear phase recursive filters using linear programming," in *Proc. IEEE Int. Conf. on ASSP* (May 1977), pp. 88–92.

[4] G. A. Maria and M. M. Fahmy, "An *lp* design technique to the two-dimensional filters," *IEEE Trans. Acoust., Speech, Signal Processing*, vol. ASSP-22, no. 1, pp. 15–21, Feb. 1974.

[5] ——, "*lp* approximation of the group delay response of one- and two-dimensional filters," *IEEE Trans. Circuits Syst.*, vol. CAS-21, no. 3, pp. 431–436, May 1974.

[6] P. A. Ramamoorthy and L. T. Bruton, "Frequency domain approximation of stable multi-dimensional discrete recursive filters," in *Proc. IEEE Int. Symp. Circuits and Systems* (Apr. 1977), pp. 654–657.

[7] J. L. Shanks, S. Treitel, and J. H. Justice, "Stability and synthesis of two-dimensional recursive filters," *IEEE Trans. Audio Electroacoust.*, vol. AU-20, pp. 115–128, June 1972.

[8] R. Fletcher and M. J. D. Powell, "A rapidly convergent descent method for minimization," *Comput. J.*, vol. 6, iss. 2, pp. 163–168, 1963.

[9] *System/360 Scientific Subroutine Package* (Version III, Programmer's Manual, 5th ed.), 1970, pp. 221–225.

[10] S. A. H. Aly, "Construction of an objective function for optimal circuit design," M.Sc. thesis, Department of Electrical Engineering, Cairo University, Giza, Egypt, 1977.

[11] D. Goodman, "Some difficulties with the double bilinear transformation in 2-D recursive filter design," to be published in *Proc. IEEE.*

CIRCUIT THEORY AND APPLICATIONS, VOL. 7, 229-245 (1979)

DESIGN OF STABLE TWO-DIMENSIONAL ANALOGUE AND DIGITAL FILTERS WITH APPLICATIONS IN IMAGE PROCESSING

P. A. RAMAMOORTHY* AND L. T. BRUTON†

Electrical Engineering Department, University of Calgary, Calgary, Alberta, Canada

SUMMARY

A method is proposed for the frequency domain design of linear two-dimensional analogue and digital filters with guaranteed stability. The technique used is based on the result that the numerator and the denominator of the input immittance of a two-variable network (which is passive and lossy) are strictly Hurwitz polynomials. One of these strictly Hurwitz polynomials is assigned to the denominator of a two-variable analogue transfer function and the network elements are then used as the variables of optimization thereby guaranteeing the stability of the analogue transfer function. The transfer function of the corresponding two-dimensional discrete (digital) filter is obtained from the analogue transfer function by the bilinear transformation. Examples illustrating the versatility of the technique in designing 2D digital filters of arbitrary order approximating a given magnitude and group delay response are presented. These filters are used to process a simple binary image. The results obtained demonstrate the importance of linear phase in image processing applications.

The method presented here can be extended to the design of stable *m*-dimensional analogue and digital filters.

1. INTRODUCTION

The subjects of two-variable (2V) network theory[1-4] and two-dimensional (2D) digital filter design[5] are of current interest because of the large number of applications which require 2D discrete data processing technology in fields such as geophysics, medicine and communications.[6-8] In particular, the design of 2D linear space-invariant or shift-invariant (LSI) filters has received considerable attention. In this paper we present a method of designing a class of 2D analogue and digital LSI filters. This class consists of those filters whose impulse response functions $h(t_1, t_2)$ (in the analogue or continuous case) or $h(n_1, n_2)$ (in the discrete case) are nonzero in the first quadrant only (or causal functions) and whose system functions $T(s_1, s_2)$ [$H(z_1, z_2)$ in the discrete case] are representable as ratios of two polynomials. That is

$$T(s_1, s_2) = \frac{\sum_{i=0}^{m_1}\sum_{j=0}^{m_2} p_{ij} s_1^i s_2^j}{\sum_{i=0}^{n_1}\sum_{j=0}^{n_2} q_{ij} s_1^i s_2^j}; \qquad m_1 \leq n_1, \qquad m_2 \leq n_2$$

$$\equiv \frac{P_1(s_1, s_2, p_{ij})}{Q(s_1, s_2, q_{ij})}; \qquad s_i = \sigma_i + j\omega_i, \qquad i = 1, 2, \text{ complex frequency variable} \tag{1}$$

in the 2D analogue filter case and

$$H(z_1, z_2) = \frac{\sum_{i=0}^{m_1}\sum_{j=0}^{m_2} n_{ij} z_1^i z_2^j}{\sum_{i=0}^{n_1}\sum_{j=0}^{n_2} d_{ij} z_1^i z_2^j} \tag{2}$$

---

* Student member, IEEE.
† Member, IEEE.

in the discrete filter case. Here $z_i[=\exp(s_iT_i)]$ represents the unit advance operator in the ith direction and T_i the sampling rate.

2. DESIGN OF 2D ANALOGUE FILTERS

In this section, the design of 2D causal analogue filters with a system function $T(s_1, s_2)$ [as given in equation (1)] is considered. In any frequency domain filter design, the coefficients of $T(s_1, s_2)$ have to be selected using optimization techniques such that $T(s_1, s_2)$ approximates a given frequency response. These coefficients also must be constrained such that the resulting filter is BIBO stable. For causal filters, stability is ensured if $Q(s_1, s_2, q_{ij})$ is strictly Hurwitz; that is if†

$$Q(s_1, s_2, q_{ij}) \neq 0 \quad \text{for} \quad \mathrm{Re}[s_1] \quad \text{and} \quad \mathrm{Re}[s_2] \geq 0 \tag{3}$$

The stability constraint in (3) is analogous to the requirement in 1D systems that the system function has no poles in the right-half plane. The 2D singularities defined by $Q(s_1, s_2, q_{ij}) = 0$, however, are not isolated points as poles are, but are rather multi-dimensional surfaces. Also, unlike the case of polynomials in a single variable, a general 2V polynomial cannot be expressed as a product of low-order factors. Therefore, there exist no obvious means to constrain the coefficients $\{q_{ij}\}$ in a manner that ensures equation (3) is satisfied.

The general 2D stability problem is solved here by transforming the coefficients $\{q_{ij}\}$ into functions of some other set of real coefficients $\{y_{kl}\}$, $l > k$, where k and l are integers in some finite range $2 \leq k, l \leq n$. Sufficiency conditions are established such that for all real values of $\{y_{kl}\}$, the corresponding transfer functions are stable. That is

$$\{q_{ij}\} = \mathrm{Fn}[y_{ki}]; \qquad \mathrm{Fn} \equiv \text{Function of} \tag{4}$$

such that

$$Q(s_1, s_2, q_{ij}) = Q(s_1, s_2, \mathrm{Fn}(y_{kl})) \neq 0 \quad \text{for} \quad \mathrm{Re}[s_1] \text{ and } \mathrm{Re}[s_2] \geq 0 \quad \text{and} \quad -\infty < y_{kl} < \infty \tag{5}$$

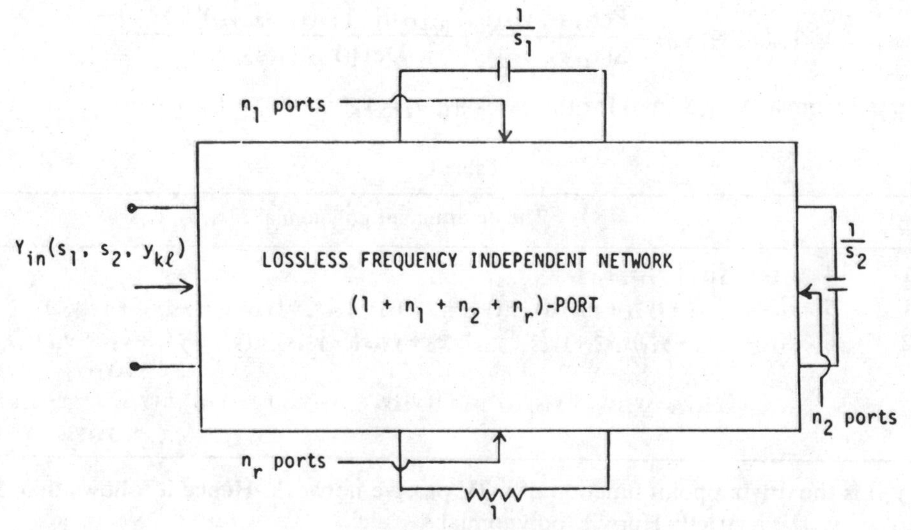

Figure 1. A two-variable passive network

Furthermore, since the values of the coefficients $\{y_{kl}\}$ can lie in the range $-\infty$ to ∞ they can be used as the variables of an unconstrained optimization technique.

The method proposed is based on some well-known results in 2V network theory. Although little work has been done in the area of 2D analogue filter design, passive 2V networks and their properties are well

† For quite some time this condition was also considered to be necessary for BIBO stability of 2D filters. It has been recently shown by Goodman[9] that $Q(s_1, s_2, q_{ij})$ can have nonessential singularities of the second kind on $\mathrm{Re}[s_1] = \mathrm{Re}[s_2] = 0$ and still the filter can be BIBO stable. In this paper we will consider only the design of filters satisfying (3).

known. In our design technique we will be applying the fact that both the numerator and the denominator of the driving point function of a 2V passive network are strictly Hurtwitz‡ polynomials.[1–4,10]

We propose the passive network structure in Figure 1 as a means of obtaining the polynomial $Q(s_1, s_2, y_{kl})$. The passive structure consists of a *lossless frequency independent multiport network* that is terminated at n_1 of its ports in unit capacitors s_1, n_2 of its ports in unit capacitors s_2, and n_r of its ports in unit resistors. Denoting the number of ports in the lossless network as n, it can easily be seen that $n = 1+n_1+n_2+n_r$. We can describe this lossless network in terms of its admittance matrix Y.§ Because the network is lossless and frequency independent, it immediately follows that *all the elements of Y are real and Y is skew symmetric.* We can therefore write Y as

$$Y = \left[\begin{array}{c|cccc} 0 & y_{12} & y_{13} & \cdots & y_{1n} \\ \hline -y_{12} & 0 & y_{23} & \cdots & y_{2n} \\ -y_{13} & -y_{23} & 0 & \cdots & y_{3n} \\ \cdot & \cdot & \cdot & \cdots & \cdot \\ \cdot & \cdot & \cdot & \cdots & \cdot \\ \cdot & \cdot & \cdot & \cdots & \cdot \\ -y_{1n} & -y_{2n} & -y_{3n} & \cdots & 0 \end{array}\right] = \left[\begin{array}{c|c} Y_{11} & Y_{12} \\ \hline -Y'_{12} & Y_{22} \end{array}\right] \qquad n = 1+n_1+n_2+n_r \tag{6}$$

Defining $\hat{Y}_{22}(s_1, s_2, y_{kl})$ as

$$\hat{Y}_{22}(s_1, s_2, y_{kl}) = Y_{22} + \mathrm{dia}[\overbrace{1 \ldots 1}^{n_r} \quad \overbrace{1 \quad s_1 \ldots s_1}^{n_1} \quad \overbrace{s_2 \ldots s_2}^{n_2}] \tag{7}$$

and denoting the determinant of $\hat{Y}_{22}(s_1, s_2, y_{kl})$ by $\Delta(s_1, s_2, y_{kl})$, we can show that the input admittance at port 1 is given by

$$Y_{\mathrm{in}}(s_1, s_2, y_{kl}) = \frac{P(s_1, s_2, y_{kl})}{\Delta(s_1, s_2, y_{kl})} \equiv \frac{Y_{12}[\mathrm{Adj}\,\{\hat{Y}_{22}(s_1, s_2, y_{kl})\}][Y'_{12}]}{\mathrm{Det}\,[\hat{Y}_{22}(s_1, s_2, y_{kl})]} \tag{8}$$

In Table I the polynomial $\Delta(s_1, s_2, y_{kl})$ for the cases $n_1, n_2 = 1, 2$ and $n_r = 1$ is given.

Table I

n_1	n_2	n	The determinant polynomial $\Delta(s_1, s_2, y_{kl})$
1	1	4	$s_1s_2 + y_{24}^2s_1 + y_{23}^2s_2 + y_{34}^2$
2	1	5	$s_1^2s_2 + y_{25}^2s_1^2 + (y_{23}^2 + y_{24}^2)s_1s_2 + (y_{35}^2 + y_{45}^2)s_1 + y_{34}^2s_2 + (y_{23}y_{45} - y_{24}y_{35} + y_{25}y_{34})^2$
2	2	6	$s_1^2s_2^2 + (y_{23}^2 + y_{24}^2)s_1s_2^2 + (y_{25}^2 + y_{26}^2)s_1^2s_2 + y_{56}^2s_1^2 + y_{34}^2s_2^2 + (y_{35}^2 + y_{36}^2 + y_{45}^2 + y_{46}^2)s_1s_2$ $+[(y_{23}y_{56} - y_{25}y_{36} + y_{26}y_{35})^2$ $+(y_{24}y_{56} - y_{25}y_{46} + y_{26}y_{45})^2]s_1 + [(y_{23}y_{45} - y_{24}y_{35} + y_{25}y_{34})^2 + (y_{23}y_{46} - y_{24}y_{36} + y_{26}y_{34})^2]s_2$ $+(y_{34}y_{56} - y_{35}y_{46} + y_{36}y_{45})^2$

$Y_{\mathrm{in}}(s_1, s_2, y_{kl})$ is the driving point function of a 2V passive network. Hence it follows that $\Delta(s_1, s_2, y_{kl})$ [and also $P(s_1, s_2, y_{kl})$] is a strictly Hurwitz polynomial.¶

We now have a polynomial $\Delta(s_1, s_2, y_{kl})$ which is strictly Hurwitz. The essential feature of the method is then to allocate the polynomial $\Delta(s_1, s_2, y_{kl})$ to the denominator $Q(s_1, s_2, q_{ij})$ of the analogue transfer function $T(s_1, s_2)$. The real variables y_{kl} may then be employed as variables of optimization with the

‡ This is true only if all irreducible factors common to the numerator and the denominator are first cancelled out, leaving mutually prime polynomials.

§ We assume here that an admittance matrix representation exists. However, this in no way affects the validity of the method, as we can always proceed using a scattering matrix representation.

¶ This statement excludes the degenerate case when some coefficients of $s_1^i s_2^j [0 \leq i(j) \leq n_1(n_2)]$ in $\Delta(s_1, s_2, y_{kl})$ are absent, in which case Δ becomes a modified- or non-Hurwitz polynomial.

assurance that the denominator $Q(s_1, S_2, q_{ij})$ is strictly Hurwitz over all real y_{kl}, guaranteeing the stability of the continuous transfer function over all real y_{kl}. That is, we choose

$$\begin{aligned}\Delta(s_1, s_2, y_{kl}) &= Q(s_1, s_2, q_{ij}) \\ &= Q[s_1, s_2, \mathrm{Fn}(y_{kl})]\end{aligned} \tag{9}$$

so that any 2V analogue transfer function of the form

$$T(s_1, s_2) = \frac{P_1(s_1, s_2)}{Q[s_1, s_2, \mathrm{Fn}\,(y_{kl})]} \tag{10}$$

will be stable for all real y_{kl}. Here P_1 is any 2V polynomial in s_1 and s_2 with degree in each variable less than or equal to that of Q.

In deriving the polynomial $\Delta(s_1, s_2, y_{kl})$ given in Table I, we have restricted the value of n_r to 1. In Appendix I it is shown that this value of $n_r = 1$ is sufficient to generate all stable 2D transfer functions $T(s_1, s_2)$ of the form given in equation (1).

Modifications in the $\hat{Y}_{22}(s_1, s_2, y_{kl})$ matrix

It is observed that the coefficients $\{y_{kl}\}$ of the matrix $\hat{Y}_{22}(s_1, s_2, y_{kl})$ in (10) can be used as the variables of an optimization technique in the design of stable 2D filters. Denoting the total number of independent coefficients of $\{y_{kl}\}$ as m, it is readily shown that

$$m = \frac{(n_1+n_2)(n_1+n_2+1)}{2} \tag{11}$$

Furthermore, the number of coefficients N of the denominator polynomial $Q(s_1, s_2, q_{ij})$ is given by $N = (n_1+1)(n_2+1)$, from which it follows by inspection of equation (11) that $m \geqslant N$. That is, the number of variables of optimization $\{y_{kl}\}$ is not less than the number of coefficients of $Q(s_1, s_2, q_{ij})$. For $n_{1,2} \gg 1$, the number of variables $\{y_{kl}\}$ becomes inordinately large, and from a computational viewpoint, leads to an excessively time-consuming algorithm. We therefore propose to alleviate this problem by suitably constraining some of the $\{y_{kl}\}$ to zero; a suitable reduced $\hat{Y}_{22}(s_1, s_2, y_{kl})$ is given by

$$\hat{Y}_{22}(s_1, s_2, y_{kl}) = \left[\begin{array}{cccccc|ccccc}
1 & y_{23} & & 0 & & & y_{2k} & & 0 & & \\
-y_{23} & s_1 & & & & & y_{3k} & \cdot & \cdot & \cdot & y_{3n} \\
 & & \cdot & \cdot & & & \cdot & \cdot & \cdot & \cdot & \cdot \\
 & & & \cdot & & & \cdot & \cdot & \cdot & \cdot & \cdot \\
0 & \cdot & \cdot & \cdot & & y_{ij} & \cdot & \cdot & \cdot & \cdot & \cdot \\
 & & & & -y_{ij} & s_1 & y_{jk} & \cdot & \cdot & \cdot & y_{jn} \\
\hline
-y_{2k} & -y_{3k} & \cdot & \cdot & \cdot & -y_{jk} & s_2 & y_{kl} & 0 & & \\
 & \cdot & \cdot & \cdot & \cdot & \cdot & -y_{kl} & \cdot & & & \\
0 & \cdot & \cdot & \cdot & \cdot & \cdot & & \cdot & \cdot & & \\
 & \cdot & \cdot & \cdot & \cdot & \cdot & 0 & & \cdot & \cdot & y_{mn} \\
 & -y_{3n} & \cdot & \cdot & \cdot & -y_{jn} & & & & -y_{mn} & s_2
\end{array}\right]; \tag{12}$$

$$i = n_1+1, \quad j = n_1+2, \quad k = n_1+3$$
$$l = n_1+4, \quad m = 1+n_1+n_2, \quad n = 2+n_1+n_2$$

for which it is observed that $m = N$. The determinant $\Delta(s_1, s_2, y_{kl})$ of this reduced 2V positive real matrix

$\hat{\hat{Y}}_{22}$ is used throughout the remainder of this paper. In Appendix II, we briefly discuss a method of obtaining the determinant polynomial for a given set of $\{y_{kl}\}$ using numerical techniques.

3. DESIGN OF 2D DIGITAL RECURSIVE FILTERS

In this section, we briefly discuss the use of the above technique in the design of 2D recursive filters and proceed to present numerical results. Similar to the discussion in Section 2, we can show that a digital filter with the system function $(H(z_1, z_2)$ will be BIBO stable if

$$D(z_1, z_2) \neq 0 \quad \text{for} \quad |z_1| \quad \text{and} \quad |z_2| \geqslant 1 \tag{13}$$

Our approach is to apply a double bilinear transformation to $T(s_1, s_2)$ to obtain $H(z_1, z_2)$. That is,

$$H(z_1, z_2) = T(s_1, s_2) = \left.\frac{P_1(s_1, s_2)}{\Delta(s_1, s_2, y_{kl})}\right|_{s_r = \frac{z_r - 1}{z_r + 1};\quad r = 1,2} \tag{14}$$

The coefficients $\{y_{kl}\}$ are used as the variables of optimization, thus guaranteeing the stability of the transfer function $H(z_1, z_2)$.

Example 1. Circularly symmetric lowpass filter design

In this example we design a digital filter approximating a circularly symmetric lowpass magnitude characteristic. The transfer function of the filter is assumed to be of allpole type and the order in each variable is set to five, i.e.,

$$T(\lambda_1, \lambda_2) \equiv \frac{A_N}{\sum_{i=0}^{5} \sum_{j=0}^{5} q_{ij} \lambda_1^i \lambda_2^j} \equiv \left.\frac{A_N}{\Delta(\lambda_1, \lambda_2, y_{kl})}\right|_{n_1 = n_2 = 5} \tag{15}$$

where λ_i, the warped frequency variable, is defined as

$$\lambda_i = \sigma_i + j\Omega_i \equiv \tanh\left(\frac{s_i T}{2}\right) \tag{16}$$

(T is assumed to be the same in both directions.)
The desired response is given by

$$\begin{aligned} M(j\omega_{1N}, j\omega_{2N}) &= 1 \quad \text{for} \quad \sqrt{(\omega_{1N}^2 + \omega_{2N}^2)} \leqslant 0{\cdot}2\pi \\ &= 0 \quad \text{otherwise} \end{aligned} \tag{17}$$

The frequency sample points are selected as

$$\begin{aligned} \omega_{1N}(I) &= 0, 0{\cdot}2\pi(0{\cdot}01\pi)\| \qquad & I &= 0, 1, \ldots, 20 \\ \omega_{1N}(I) &= 0{\cdot}3\pi, 0{\cdot}6\pi(0{\cdot}1\pi) \qquad & I &= 21, \ldots, 24 \\ \omega_{2N}(24 \pm I) &= \pm\omega_{1N}(I) \qquad & I &= 0, \ldots, 24 \end{aligned} \tag{18}$$

where the frequencies ω_{1N} and ω_{2N} are normalized with respect to the sampling frequency.

The coefficients $[q_{ij}]$ and A_N of the transfer function $T(\lambda_1, \lambda_2)$ in equation (15) are obtained by minimizing a suitable objective function; we select here the following objective function:

$$\Phi = \sum_I \sum_J \{M(j\omega_{1N}(I), j\omega_{2N}(J)) - |T(j\omega_{1N}(I), j\omega_{2N}(J))|\}^2 \tag{19}$$

The algorithm that has been employed uses a conventional quasi-Newton technique.[11] The variables of optimization in equation (19) are the parameters $\{y_{kl}\}$ of equation (15); the number of variables $\{y_{kl}\}$ is given

‖ The notation $x = x_1, x_2, (y)$ is used to denote that x takes the values of $x_1, x_1 + y, \ldots, x_2$.

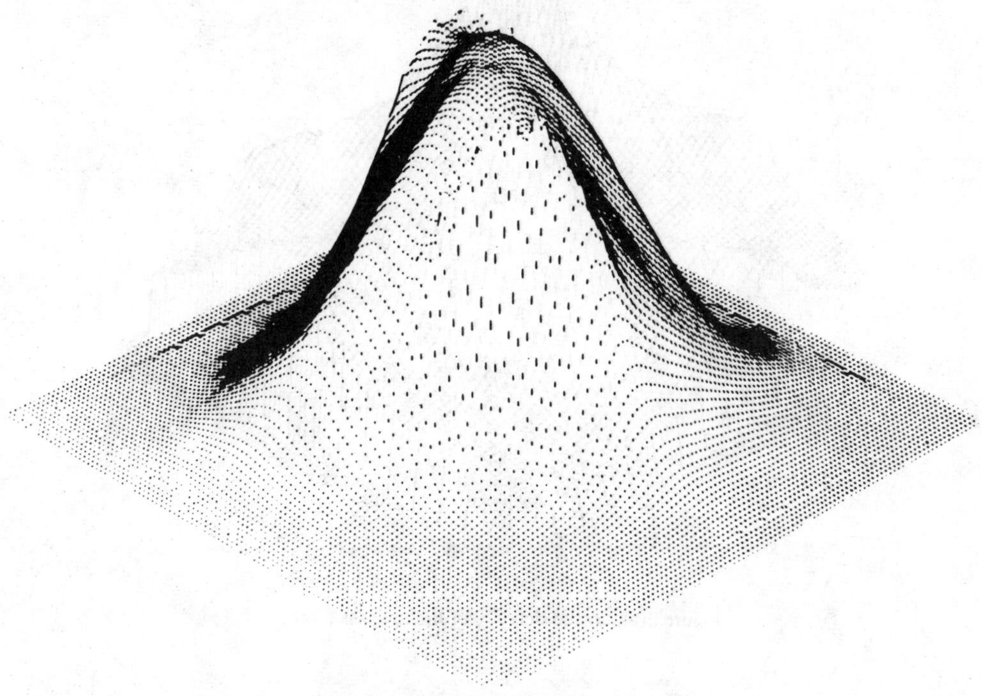

Figure 2. Magnitude response of the lowpass filter in Example 1

Table II. Coefficients of a fifth-order lowpass filter

$A_N = 0{\cdot}28627$			$B_N = 0{\cdot}28627$		
Matrix Q					
0·2549*E*−3	0·29161*E*−2	0·16603*E*−1	0·54040*E*−1	0·11484*E* 0	0·14855*E* 0
0·2791*E*−2	0·36484*E*−1	0·19949*E* 0	0·65486*E* 0	0·13220*E* 1	0·10905*E* 1
0·15414*E*−1	0·19228*E* 0	0·10676*E* 1	0·33900*E* 1	0·54523*E* 1	0·29362*E* 1
0·49205*E*−1	0·60178*E* 0	0·32211*E* 1	0·84794*E* 1	0·99470*E* 0	0·3722*E* 1
0·10035*E* 0	0·11554*E* 1	0·48891*E* 1	0·93392*E* 1	0·78224*E* 1	0·24344*E* 1
0·12327*E* 0	0·91318*E* 0	0·25125*E* 1	0·32694*E* 1	0·21511*E* 1	0·10000*E* 1
Matrix D					
0·65181*E*−1	−0·64500*E* 0	0·33632*E* 1	−0·48317*E* 1	0·32176*E* 0	−0·16445*E* 0
−0·79298*E* 0	0·78851*E* 1	−0·25871*E* 2	0·23838*E* 2	0·34048*E* 1	0·34667*E* 1
0·42941*E* 1	−0·28734*E* 2	0·61551*E* 2	−0·29302*E* 2	−0·13249*E* 2	−0·25519*E* 2
−0·63054*E* 1	0·28707*E* 2	−0·33487*E* 2	−0·72275*E* 1	−0·22705*E* 2	0·83011*E* 2
0·79070*E* 0	0·14820*E* 1	−0·74214*E* 1	−0·33313*E* 2	0·13676*E* 3	−0·12843*E* 3
−0·41341*E* 0	0·60739*E* 1	−0·36029*E* 2	0·10147*E* 3	0·14020*E* 3	0·78428*E* 2

by equation (11) which, in this case, is given by

$$m = N = (n_1 + 1)(n_2 + 1) = (5 + 1)(5 + 1) = 36$$

according to equation (12). The resultant optimized function $T(j\omega_{1N}(I), j\omega_{2N}(J))$ is described in terms of the corresponding matrix $[q_{ij}]$ [see equation (15)] in Table II.† This optimal stable analogue function

† The numerical results presented in this paper represent a local minimum only and not necessarily a global minimum.

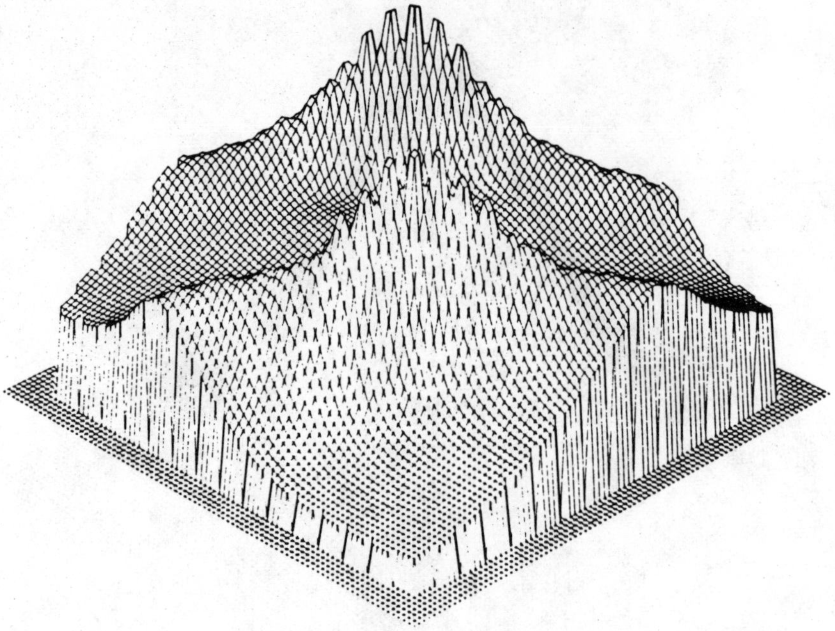

Figure 3(a). Group delay τ_1 of the lowpass filter

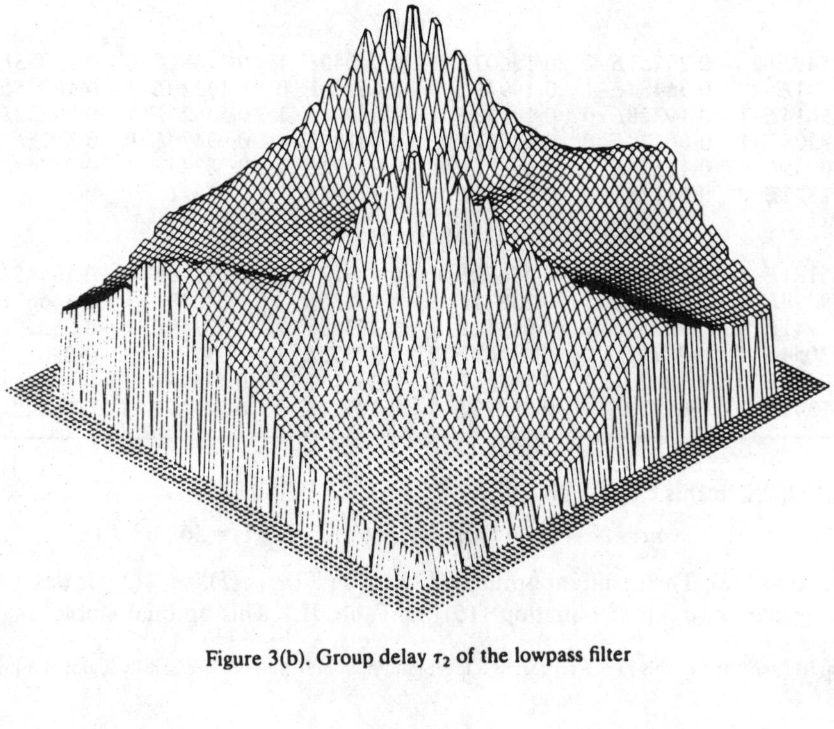

Figure 3(b). Group delay τ_2 of the lowpass filter

$T(\lambda_1, \lambda_2)$ is easily transformed to a corresponding stable discrete function by means of the double bilinear transformation [equation (14)] giving

$$H(z_1, z_2) = \frac{B_N(z_1+1)^5(z_2+1)^5}{\sum_{i=0}^{5}\sum_{j=0}^{5} d_{ij} z_1^i z_2^j} \tag{20}$$

where the values of B_N and $[d_{ij}]$ are given in Table II.

A 3D plot of the magnitude response $|H(e^{j\omega_{1N}}, e^{j\omega_{2N}})|$ is shown in Figure 2. The group delay responses $\tau_1(j\omega_{1N}, j\omega_{2N})$ and $\tau_2(j\omega_{1N}, j\omega_{2N})$ where

$$\tau_i(j\omega_{1N}, j\omega_{2N}) = \frac{\partial}{\partial\omega_{iN}}[\mathrm{Arg}\{H(e^{j\omega_{1N}}, e^{j\omega_{2N}})\}]; \qquad i = 1, 2 \tag{21}$$

are shown in Figures 3(a) and 3(b), respectively. From Figures 3(a) and 3(b) it can be observed that the group delays exhibit a large peak-to-peak variation near the band edges. The equalization of these group delays is considered in the next section.

4. GROUP DELAY APPROXIMATION

In one-dimensional digital filters, it is often sufficient to design a filter approximating the given magnitude response without regard to the phase characteristics. In multi-dimensional signal processing it is known that the phase characteristics are very important. Recently, Huang *et al.*[12] have demonstrated that in the processing of images it is necessary to design filters that have linear phase characteristics.

The design technique presented here can also be used for group delay approximation and therefore for approximating linear phase characteristics. We can achieve this either by including an error criterion for the delay in the function Φ that is being minimized or by cascading the filter that approximates the magnitude response with a second filter that has an allpass transfer function such that the overall delay response of the two cascaded filters approximates the required delay specifications. The latter approach is used in the Example 2 that follows: Denoting $T(\lambda_1, \lambda_2)$ as the analogue transfer function of the allpass filter, we have,

$$\begin{aligned} T(\lambda_1, \lambda_2) &= \frac{\sum_{i=0}^{n_1}\sum_{j=0}^{n_2} q_{ij}(-\lambda_1^i)(-\lambda_2^j)}{\sum_{i=0}^{n_1}\sum_{j=0}^{n_2} q_{ij}\lambda_1^i\lambda_2^j} \\ &\equiv \frac{Q(-\lambda_1, -\lambda_2)}{Q(\lambda_1, \lambda_2)} \end{aligned} \tag{22}$$

where

$$Q(\lambda_1, \lambda_2) \equiv \Delta(\lambda_1, \lambda_2, y_{kl}) \tag{23}$$

Defining $\tau_{1AP}(j\omega_{1N}, j\omega_{2N})$ and $\tau_{2AP}(j\omega_{1N}, j\omega_{2N})$ to be the group delay of the allpass filter in the directions ω_{1N} and ω_{2N}, respectively, we can show that

$$\begin{aligned} \tau_{1AP}(j\omega_{1N}, j\omega_{2N}) &\equiv \frac{\partial}{\partial\omega_{1N}}[\mathrm{Arg}\{T(j\Omega_1, j\Omega_2)\}] \\ &= \frac{\partial}{\partial\omega_{1N}}\left[-2\tan^{-1}\left\{\frac{\mathrm{Im}\,(T(j\Omega_1, j\Omega_2))}{\mathrm{Re}\,(T(j\Omega_1, j\Omega_2))}\right\}\right] \\ &= -\sec^2\left(\frac{\omega_{1N}}{2}\right)\mathrm{Re}\left[\frac{\partial Q(\lambda_1, \lambda_2)/\partial\lambda_1}{Q(\lambda_1, \lambda_2)}\right]\Bigg|_{\lambda_i = j\tan\left(\frac{\omega_{iN}}{2}\right)} \end{aligned} \tag{24}$$

and similarly

$$\tau_{2AP}(j\omega_{1N}, j\omega_{2N}) = -\sec^2\left(\frac{\omega_{2N}}{2}\right)\mathrm{Re}\left[\frac{\partial Q(\lambda_1, \lambda_2)/\partial\lambda_2}{Q(\lambda_1, \lambda_2)}\right]\Bigg|_{\lambda_i = j\tan\left(\frac{\omega_{iN}}{2}\right)} \tag{25}$$

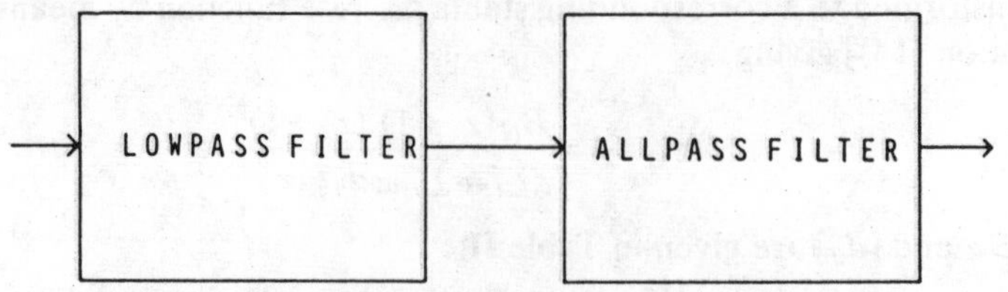

Figure 4. A cascaded system for group delay approximation

Example 2. Constant group delay approximation using a fifth-order allpass filter

In this example we approximate a constant group delay response by cascading an allpass filter of order 5 in each variable with the lowpass filter in Example 1. The overall system appears as shown in Figure 4. The frequency sample points are restricted to those inside of the lowpass filter. Thus we have

$$\omega_{1N}(I) = 0,\ 0{\cdot}2\pi(0{\cdot}01\pi) \qquad I = 0, 1, \ldots, 20$$

$$\omega_{2N}(20 \pm I) = \pm\omega_{1N}(I) \qquad I = 0, 1, \ldots, 20 \tag{26}$$

and

$$\sqrt{\omega_{1N}^2(I) + \omega_{2N}^2(20 \pm I)} \leq 0{\cdot}2\pi$$

We choose to obtain the coefficients q_{ij} of the transfer function $T(\lambda_1, \lambda_2)$ by minimizing an objective function given by

$$\Phi_{AP} = \sum_I \sum_J [\{\tau_{1spec}(j\omega_{1N}(I), j\omega_{2N}(J)) - \tau_{1LP}(j\omega_{1N}(I), j\omega_{2N}(J)) - \tau_{10}\}^2$$
$$+ \{\tau_{2spec}(j\omega_{1N}(I), j\omega_{2N}(J)) - \tau_{2LP}(j\omega_{1N}(I), j\omega_{2N}(J)) - \tau_{20}\}^2] \tag{27}$$

where $\tau_{1spec}(j\omega_{1N}, j\omega_{2N})$, $\tau_{2spec}(j\omega_{1N}, j\omega_{2N})$ are the specified constant group delays in direction of ω_{1N} and ω_{2N}, respectively, $\tau_{1LP}(j\omega_{1N}, j\omega_{2N})$, $\tau_{2LP}(j\omega_{1N}, j\omega_{2N})$ are the group delays of the lowpass filter and τ_{10}, τ_{20} are variables of optimization.

The same quasi-Newton technique is used to minimize the objective function given in equation (27) using $\{y_{kl}\}$ as the variables of optimization. The resulting matrix $[q_{ij}]$ is given in Table III. The corresponding discrete function obtained from $T(\lambda_1, \lambda_2)$ using the bilinear transformation is given by

$$H(z_1, z_2) = \frac{z_1^5 z_2^5 \sum_{i=0}^5 \sum_{j=0}^5 d_{ij} z_1^{-i} z_2^{-j}}{\sum_{i=0}^5 \sum_{j=0}^5 d_{ij} z_1^i z_2^j} \tag{28}$$

where the matrix $[d_{ij}]$ is given in Table III. The group delay of the allpass filter and the overall systems are

Table III. Coefficients of a fifth-order allpass filter

Matrix Q

0·14381*E*–3	0·21295*E*–2	0·12051*E*–1	0·43463*E*–1	0·73607*E*–1	0·91483*E*–1
0·21455*E*–2	0·26307*E*–1	0·15051*E* 0	0·41990*E* 0	0·7007 *E* 0	0·60229*E* 0
0·12484*E*–1	0·15241*E* 0	0·69423*E* 0	0·17485*E* 1	0·24387*E* 1	0·14851*E* 1
0·44607*E*–1	0·42561*E* 0	0·17645*E* 1	0·37006*E* 1	0·3903 *E* 1	0·18789*E* 1
0·76953*E*–1	0·72115*E* 0	0·23859*E* 1	0·37916*E* 1	0·32107*E* 1	0·15147*E* 1
0·90394*E*–1	0·54888*E* 0	0·12922*E* 1	0·16630*E* 1	0·12895*E* 1	0·1 *E* 1

Matrix D

0·38218*E* 0	–0·16162*E* 1	0·31629*E* 1	–0·43356*E* 1	0·19329*E* -	–0·10483*E* 1
–0·18426*E* 0	1·88920*E* 1	–0·21361*E* 2	0·27010*E* 2	–0·11965*E* 2	0·797601*E*1
0·44651*E* 1	–0·24323*E* 2	0·57765*E* 2	–0·63617*E* 3	0·30318*E* 2	–0·26644*E* 2
–0·59439*E* 1	0·30783*E* 2	–0·66084*E* 2	0·67362*E* 2	–0·47517*E* 2	0·51780*E* 2
0·29720*E* 1	–0·15753*E* 2	0·37714*E* 2	–0·55457*E* 2	0·70765*E* 2	–0·62890*E* 2
–0·16736*E* 1	0·11183*E* 2	–0·33841*E* 2	0·59643*E* 2	–0·66016*E* 2	0·37958*E* 2

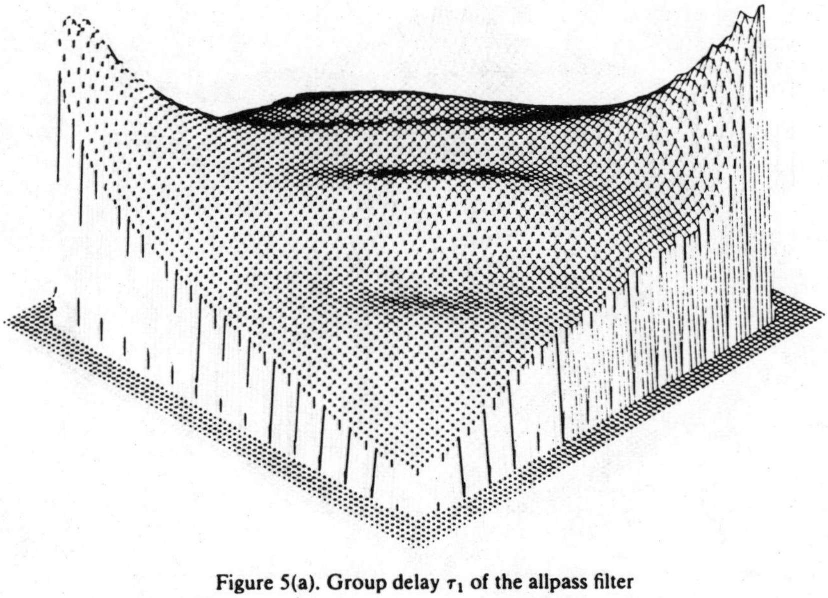

Figure 5(a). Group delay τ_1 of the allpass filter

shown in Figures 5(a), 5(b) and 6(a), 6(b), respectively. A comparison of Figures 6(a) and 6(b) with 3(a) and 3(b) reveals a significant improvement in the group delay characteristics of the overall system.

The group delay characteristics of the original fifth-order lowpass filter (Figures 3(a) and 3(b) exhibit a peak-to-peak variation of 8 units over the passband, whereas the cascade connection in Figure 4 and the corresponding group delay in Figures 6(a) and 6(b) exhibit a significantly reduced peak-to-peak variation of group delay equal to 3 units.

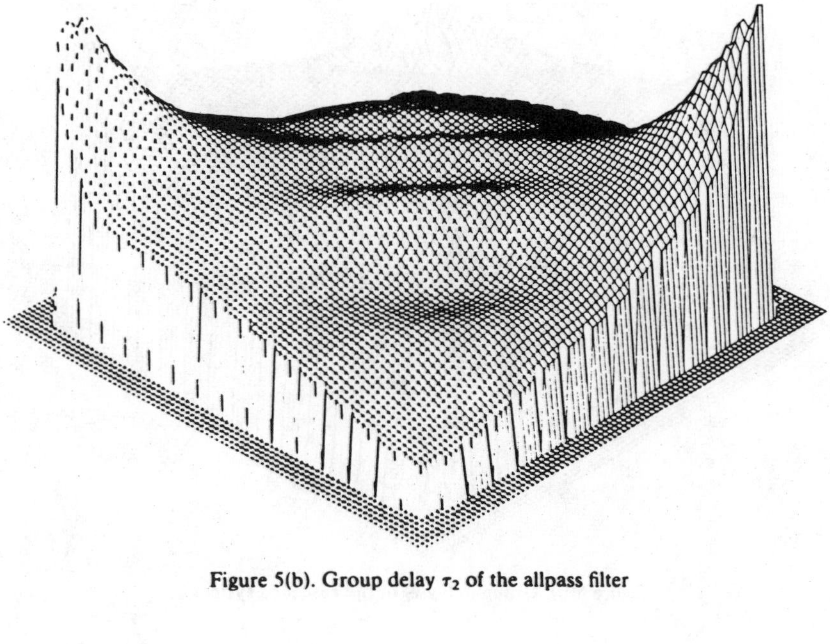

Figure 5(b). Group delay τ_2 of the allpass filter

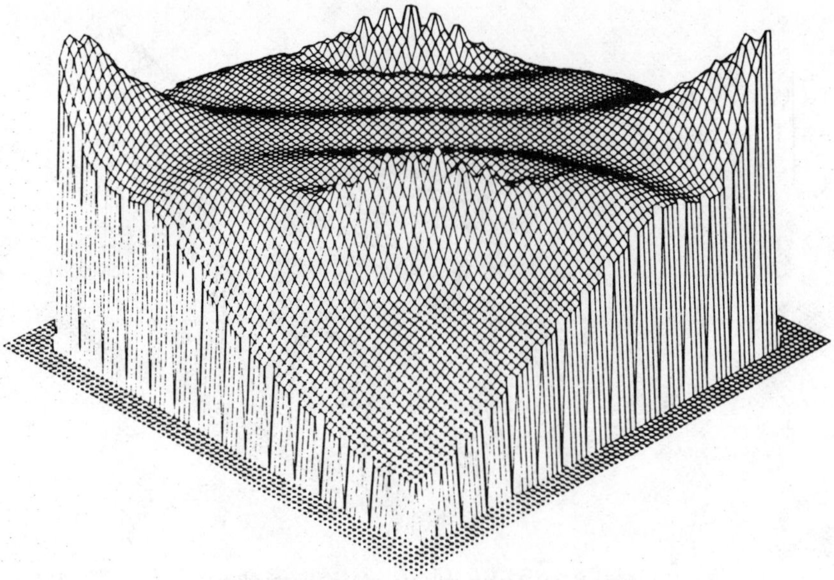

Figure 6(a). Group delay τ_1 of the cascaded system

5. APPLICATION TO IMAGE PROCESSING

Recently it has been shown[12,13] that the phase of the Fourier transform of an image is usually more important than the magnitude and while designing a filter for image processing it is necessary to take the phase specifications into consideration. In this section, we confirm the importance of linear phase in image processing by employing the two-dimensional filters that are described as Examples 1 and 2 to process a

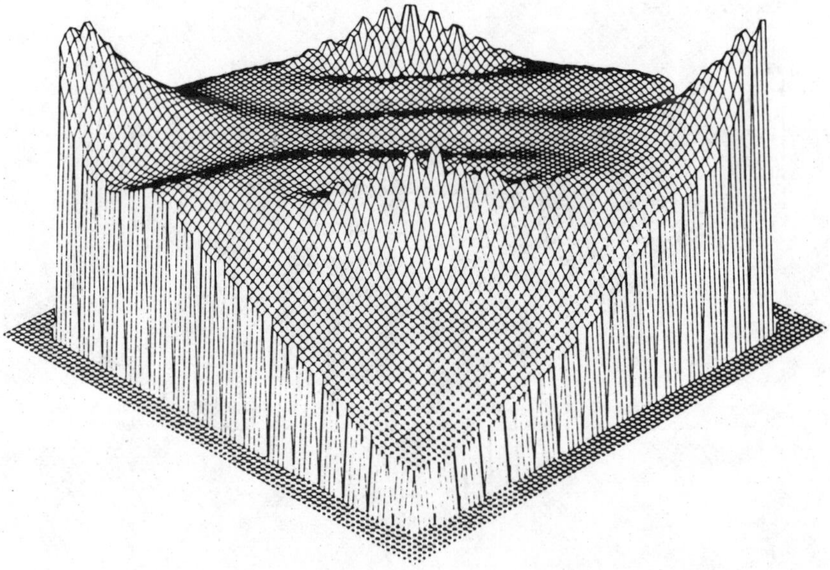

Figure 6(b). Group delay τ_2 of the cascaded system

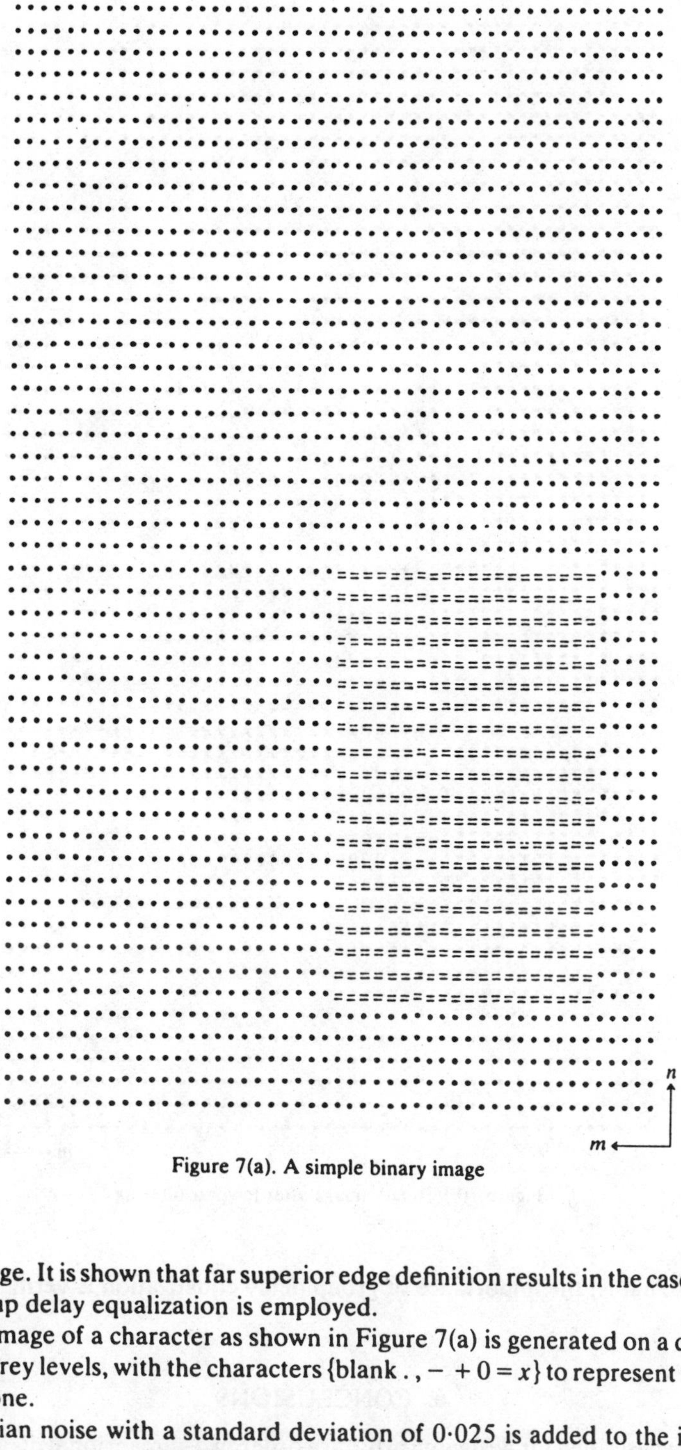

Figure 7(a). A simple binary image

simple rectangular image. It is shown that far superior edge definition results in the case of Examples 1 and 2 in cascade, where group delay equalization is employed.

A 50×50 sampled image of a character as shown in Figure 7(a) is generated on a digital computer. The line printer used nine grey levels, with the characters {blank . , − + 0 = x} to represent the values from −1 to 7 using a step size of one.

A zero mean Gaussian noise with a standard deviation of 0·025 is added to the image. The degraded image is passed through the lowpass filter in Example 1 and the output is shown in Figure 7(b). The transient overshoot in the image is clearly seen at the edges in this figure.

Now this lowpass filtered image is passed through the allpass filter in Example 2 which has equalized the group delay deviations in the lowpass filter for Example 1. The resulting output is shown in Figure 7(c). A comparison of Figures 7(b) and 7(c) reveals a significant improvement in the transient overshoot of the

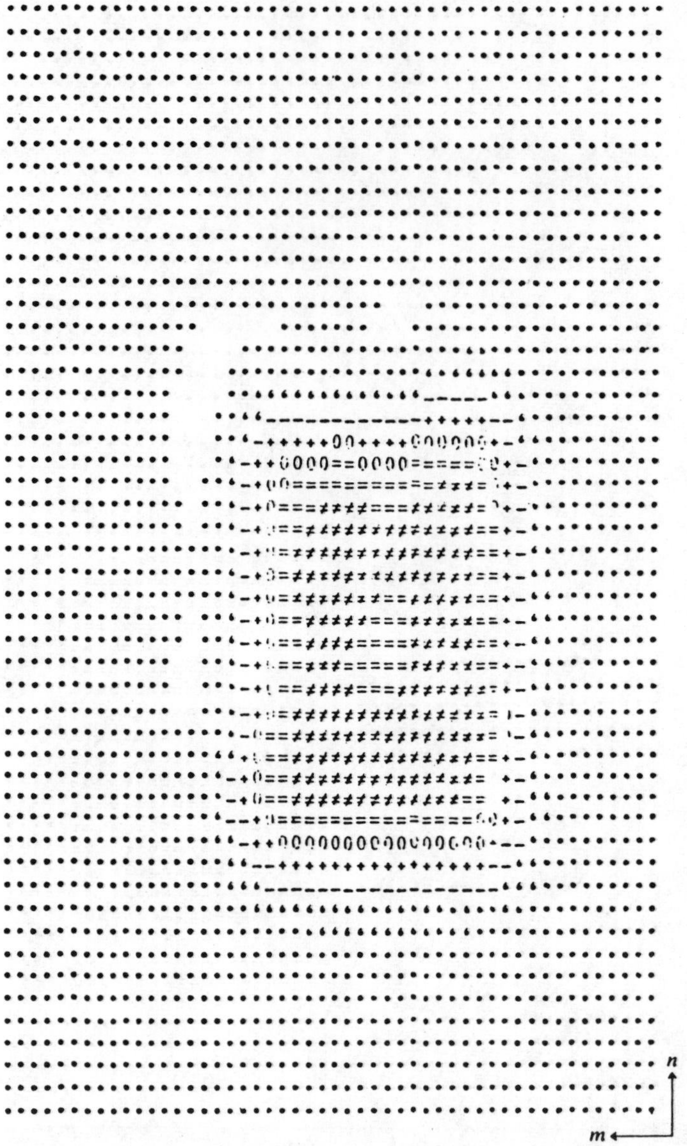

Figure 7(b). Binary image after lowpass filtering

image in Figure 7(c). That is, the importance of group delay equalization is verified for this image.

6. CONCLUSIONS

Techniques have been described for designing arbitrary order two-dimensional digital recursive filters with guaranteed stability. The method is based on the decomposition of a 2V polynomial $Q(s_1, s_2, q_{ij})$ as functions of independent real coefficients $\{y_{kl}\}$ $(l > k, 2 \leq k, l \geq n)$ such that Q remains strictly Hurwitz over all real $\{y_{kl}\}$. This strictly Hurwitz polynomial is assigned to the denominator of a 2V analogue transfer function $T(s_1, s_2)$, thereby guaranteeing the stability of $T(s_1, s_2)$ over all real $\{y_{kl}\}$. It is further shown that the generalized bilinear transformation can be used to obtain stable 2D digital filter transfer functions from

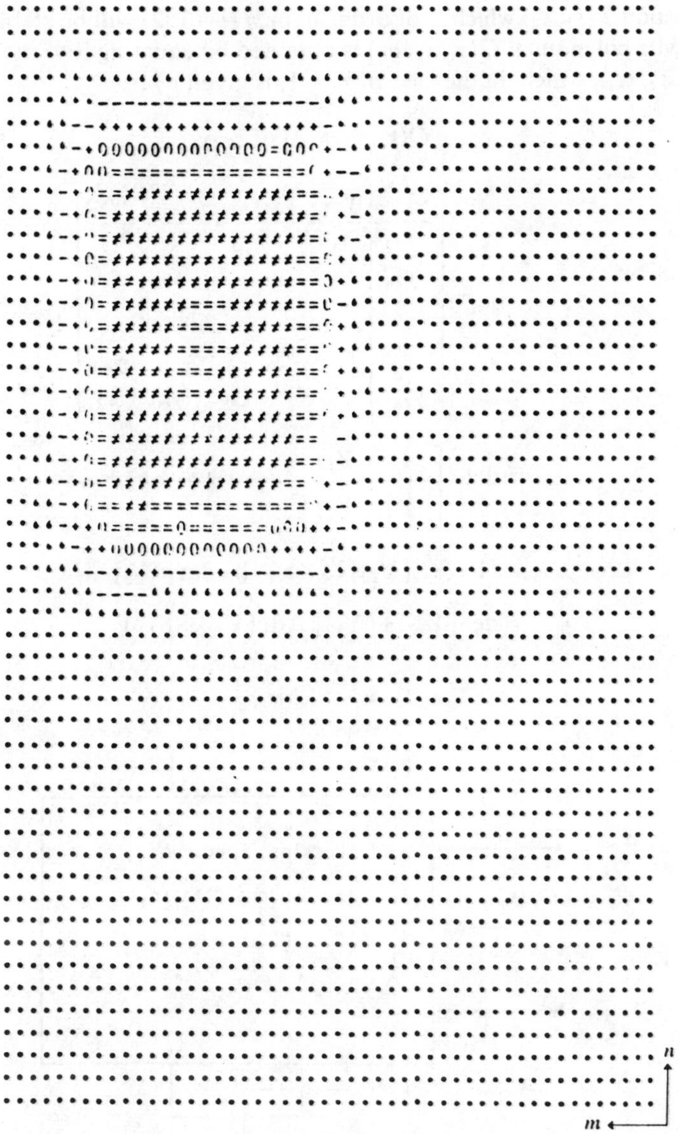

Figure 7(c). Binary image after filtering with an allpass filter for group delay equalization

$T(s_1, s_2)$. Numerical examples illustrating the use of this technique for the design of 2D digital filters approximating a given magnitude and group delay response are given. Finally the importance of linear phase in image processing is emphasized by way of an example.

APPENDIX I‡

In this section, it will be shown that the method presented in this paper can be used to generate *all* stable 2D transfer functions $T(s_1, s_2)$ as given in equation (1). Furthermore, it will be shown that the number of resistive terminations can be restricted to 1, corresponding to $n_r = 1$.

‡ In giving this proof, we assume that an admittance matrix representation for a frequency independent lossless multiport network exists. The results obtained, however, hold good for any general case as a similar proof can be presented using a scattering matrix description and the results of Youla[10] for the minimal synthesis of 2V reactive networks.

The transfer function $T(s_1, s_2)$ which is of order n_i in s_i $(i = 1, 2)$ will be stable if $Q(s_1, s_2, q_{ij})$ is strictly Hurwitz. The Hurwitz polynomial $Q(s_1, s_2, q_{ij})$ is obtained by equating it to the determinant of a positive real matrix $\hat{Y}_{22}(s_1, s_2, y_{kl})$, which, in the case of $n_r = 1$, is given by

$$Q(s_1, s_2) \equiv \Delta(s_1, s_2, y_{kl})$$

$$= \det \left[\begin{array}{c|cccccc} 1 & y_{23} & y_{24} & \cdot & \cdot & & y_{2n} \\ \hline -y_{23} & s_1 & \cdot & \cdot & \cdot & & \cdot \\ -y_{24} & \cdot & \cdot & \cdot & \cdot & & \cdot \\ \cdot & \cdot & \cdot & \cdot & \cdot & & \cdot \\ \cdot & \cdot & \cdot & \cdot & \cdot & & y_{mn} \\ y_{2n} & \cdot & \cdot & \cdot & -y_{mn} & & s_2 \end{array}\right] \tag{A1}$$

$$= \det \left[\begin{array}{c|c} 1 & Y_{12} \\ \hline -Y'_{12} & Y_{22} \end{array}\right] \quad n = 1 + n_1 + n_2 + n_r; \qquad n_r = 1, \qquad m = n - 1$$

$$= (1 + Y_{12}[Y_{22}]^{-1}[-Y'_{12}]) \cdot \det[Y_{22}]$$

$$= \det[Y_{22}] + [Y_{12}][\mathrm{Adj}\{Y_{22}\}][Y'_{12}]$$

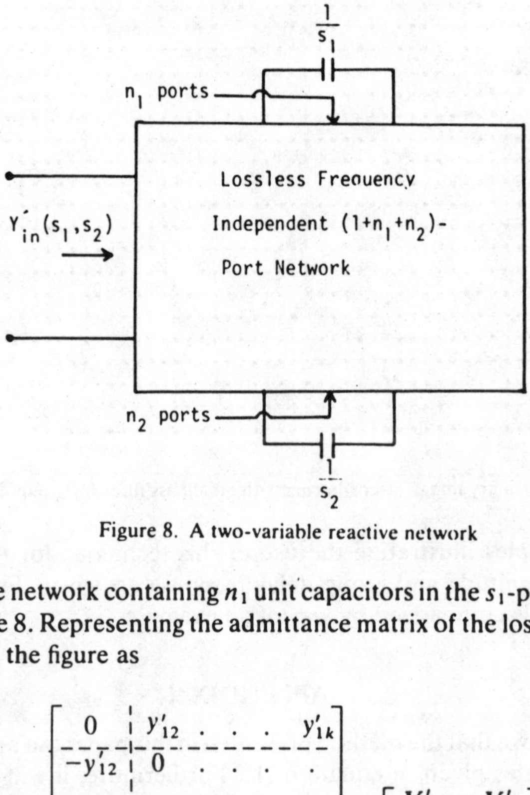

Figure 8. A two-variable reactive network

Now consider a 2V reactive network containing n_1 unit capacitors in the s_1-plane and n_2 unit capacitors in the s_2-plane as given in Figure 8. Representing the admittance matrix of the lossless frequency independent $(1 + n_1 + n_2)$-port network in the figure as

$$Y' = \left[\begin{array}{c|ccccc} 0 & y'_{12} & \cdot & \cdot & \cdot & y'_{1k} \\ \hline -y'_{12} & 0 & \cdot & \cdot & \cdot & \cdot \\ \cdot & \cdot & \cdot & \cdot & \cdot & \cdot \\ \cdot & \cdot & \cdot & \cdot & \cdot & \cdot \\ \cdot & \cdot & \cdot & \cdot & \cdot & \cdot \\ -y'_{1k} & \cdot & \cdot & \cdot & \cdot & 0 \end{array}\right] = \begin{bmatrix} Y'_{11} & Y'_{12} \\ -Y'^{t}_{12} & Y'_{22} \end{bmatrix} \quad k = 1 + n_1 + n_2 \tag{A2}$$

we can show that the input admittance at port 1 is given by

$$Y'_{\text{in}}(s_1, s_2) = Y'_{12}[Y'_{22} + \text{dia}\{s_1 \ldots s_1\ s_2 \ldots s_2\}]^{-1}[Y'_{12}]$$

$$= \frac{Y'_{12}[\text{Adj}\{Y'_{22} + \text{dia}(s_1 \ldots s_1\ s_2 \ldots s_2)\}][Y'_{12}]}{\det[Y'_{22} + \text{dia}(s_1 \ldots s_1\ s_2 \ldots s_2)]}$$

$$= M(s_1, s_2)/N(s_1, s_2) \quad \text{if} \quad (n_1 + n_2) \quad \text{is odd}$$

$$= N(s_1, s_2)/M(s_1, s_2) \quad \text{if} \quad (n_1 + n_2) \quad \text{is even}$$

where M and N are even and odd polynomials in s_1, s_2.
Now if we let

$$y'_{ij} = y_{i+1, j+1}; \qquad i, j = 1, 2, \ldots, (1 + n_1 + n_2) \tag{A4}$$

then, from (A1) and (A3), it can be observed that

$$Q(s_1, s_2, q_{ij}) = M(s_1, s_2) + N(s_1, s_2)$$

That is, we could have arrived at the same Hurwitz polynomial $Q(s_1, s_2, q_{ij})$ which is of degree n_i *in* s_i ($i = 1, 2$) by constructing the 2V reactive network in Figure 8.

By combining the preceding arguments and the results of Ansell[2] (that the ratio of odd to even part of any Hurwitz polynomial is a reactance function) and Rao[4] (that every 2V reactance function is realizable as the immittance of a 2V reactive network, where the network contains the minimum number of reactive elements in s_1 and s_2), we can conclude that the method proposed in this paper can be used to obtain *all* strictly Hurwitz polynomials and thereby all stable 2V analogue filter transfer functions $T(s_1, s_2)$ of the form given in equation (1). Such a method will require only one resistive termination corresponding to $n_r = 1$.

APPENDIX II

Before we proceed to the optimization process, it is necessary to evaluate the determinant polynomial $\Delta(s_1, s_2, y_{kl})$ of $\hat{Y}_{22}(s_1, s_2, y_{kl})$. Since it is very difficult to obtain $\Delta(s_1, s_2, y_{kl})$ in a closed form, that is as functions of the elements $\{y_{kl}\}$ of $\hat{Y}_{22}$, the following numerical technique is employed:

The determinant $\Delta(s_1, s_2, y_{kl})$ of $Y_{22}(s_1, s_2, y_{kl})$ is evaluated numerically over a grid of points in s_1 and s_2; we denote the ith sample value of $s_{1,2}$ by a superscript so that it is written $s_{1,2}^{(i)}$. Then $\Delta(s_1, s_2, y_{kl})$ is evaluated over a sample grid containing $(n_1 + 1)$ samples of s_1 and $(n_2 + 1)$ of s_2 and we write this set of samples to define the grid in the compact form

$$s_1 = \{s_1^{(0)}, s_1^{(1)}, \ldots, s_1^{(n_1)}\} \times s_2 = \{s_2^{(0)}, s_2^{(1)}, \ldots, s_2^{(n_2)}\} \tag{A5}$$

where $s_1^{(i)}(s_2^{(j)})$ are distinct for $i(j) = 0, 1, \ldots, n_1(n_2)$. We denote the resulting matrix of values of $\Delta(s_1^{(i-1)}, s_1^{(j-1)}, y_{kl})$ as B_{ij} so that

$$B_{ij} \equiv \Delta(s_1^{(i-1)}, s_2^{(j-1)}, y_{kl}) = \sum_{k=0}^{n_1} \sum_{l=0}^{n_2} q_{kl} s_1^k s_2^l \Big|_{s_1 = s_1^{(i-1)} \; s_2 = s_2^{(j-1)}} \tag{A6}$$

Then we can write B as

$$[B] = \begin{bmatrix} 1 & s_1^{(0)} & . & . & (s_1^{(0)})^{n_1} \\ 1 & . & . & . & . \\ . & . & . & . & . \\ . & . & . & . & . \\ 1 & s_1^{(n_1)} & . & . & (s_1^{(n_1)})^{n_1} \end{bmatrix} [q_{ij}] \begin{bmatrix} 1 & 1 & . & . & 1 \\ s_2^{(0)} & . & . & . & . \\ . & . & . & . & . \\ . & . & . & . & . \\ (s_2^{(0)})^{n_2} & . & . & . & (s_2^{(n_2)})^{n_2} \end{bmatrix} \tag{A7}$$

$$\equiv S_1[q_{ij}]S_2$$

so that

$$[q_{ij}] = S_1^{-1}{}' B S_2^{-1}{}' \qquad \text{(A8)}$$

The numerical technique is then to find the matrix $[q_{ij}]$ from equation (A8); this is always possible since the determinants of S_1 and S_2 are of the Vantermonde determinant form and hence are never zero.

REFERENCES

1. H. Ozaki and T. Kasami, 'Positive real functions of several variables and their applications to variable networks', *IRE Trans. Cir. Theor.* **CT-7**, 251–260 (1960).
2. H. G. Ansel. 'On certain two-variable generalizations of circuit theory, with applications to networks of transmission lines and lumped reactances;, *IEEE Trans. Cir. Theory.* **CT-11**, 214–223 (1964).
3. T. Koga, 'Synthesis of finite passive networks with prescribed two-variable reactance matrices', *IEEE Trans. Cir. Theor.* **CT-13**, 31–52 (1966).
4. T. N. Rao, 'Minimal synthesis of two-variable reactance matrices', *Bell Syst. Tech. J.* **48**, 163–199 (1969).
5. R. M. Mersereau and D. E. Dugeon, 'Two-dimensional digital filtering', *Proc. IEEE*, **63**, 610–623 (1975).
6. T. S. Huang, W. F. Schreiber and O. Tretiak, 'Image processing', *Proc. IEEE*, **59**, 1586–1609 (1969).
7. S. Treital, J. L. Shanks and C. W. Frasier, 'Some aspects of fanfiltering', *Geophysics*, **32**, 789–800 (1967).
8. 'Special issue on digital image processing', *Computer*, **7**, (1974).
9. D. Goodman, 'Some stability properties of two-dimensional linear shift-invariant digital filters', *IEEE Trans. Cir. Theor.* **CAS-24**, 201–208 (1977).
10. D. C. Youla, 'The synthesis of networks containing lumped and distributed elements', *Proc. PIB Symposium on Generalized Networks*, New York, 289–343 (1966).
11. R. Fletcher, 'Fortran subroutine for minimization by quasi-Newton methods', *Report R7125 AERE*, Harwell, England, 1972.
12. T. S. Huang, J. W. Burnett and A. G. Deczky, 'The importance of phase in image processing filters', *IEEE Trans. Acous. Speech Signal Process.* **ASSP-23** 529–542 (1975).
13. T. S. Huang, *Picture Processing and Digital Filtering*, Ed. T. S. Huang, Springer-Verlag, New York, 1975.

Two-Dimensional Recursive Filter Design—A Spectral Factorization Approach

MICHAEL P. EKSTROM, SENIOR MEMBER, IEEE, RICHARD E. TWOGOOD, MEMBER, IEEE, AND JOHN W. WOODS, MEMBER, IEEE

Abstract—This paper concerns development of an efficient method for the design of two-dimensional (2-D) recursive digital filters. The specific design problem addressed is that of obtaining half-plane recursive filters which satisfy prescribed frequency response characteristics. A novel design procedure is presented which incorporates a spectral factorization algorithm into a constrained, nonlinear optimization approach. A computational implementation of the design algorithm is described and its design capabilities demonstrated with several examples.

I. Introduction

RECENT technological advances admit the potential of using dedicated computer systems (or special-purpose hardware) to perform two-dimensional (2-D) signal processing tasks previously requiring large-scale scientific computers. In achieving this potential, such processing would be accessible to an enormously broad spectrum of applications. The principal characteristic of these applications is their need for "timely" data manipulation with affordable (small-scale) hardware.

In many aspects, recursive processors seem ideally suited to fulfill this need. Because of their reduced computational complexity over that of nonrecursive filter forms (fewer arithmetic operations, smaller memory requirements), recursive structures appear more compatible with small-scale hardware implementations, such as those involving mini- and microcomputers [1], [2]. That 2-D recursive structures have not enjoyed the success of their 1-D antecedents is attributable in large measure to difficulties in their design and analysis. Many of the basic techniques and tools used in analyzing 1-D filters rely rather fundamentally on the factorability of polynomials, and therefore cannot directly be applied to the 2-D problem. As a consequence, the issues of prime importance in recursive filtering, that of stability testing and filter design, are substantively less-tractable problems in the 2-D case [3].

While much progress has been made recently in developing stability theorems and practical tests based on these theorems [4]-[6], advances in recursive filter design have been less satisfying. The available literature features two prominent design approaches, those involving spectral transformations [7]-[9] and parameter optimization [10]-[12]. For the most part, these contributions have employed constrained filter forms (e.g., second-order, quarter plane) and addressed quite specific design problems (e.g., circularly-symmetric, lowpass). This appears to have been motivated by attendant simplifications derived from the special cases considered. These are simplifications involving stability testing and/or the availability of traditional 1-D filter design results. Despite this specificity (or perhaps because of it), their demonstrated design performance seems less than desirable, particularly for generic applications.

Manuscript received January 22, 1979; revised August 30, 1979. This work was performed under the auspices of the U.S. Department of Energy by the Lawrence Livermore Laboratory under Contract W-7405-ENG-48 and partially supported by the National Science Foundation under Grant ENG-78-04240.

M. P. Ekstrom and R. E. Twogood are with the Lawrence Livermore Laboratory, University of California, Livermore, CA 94550.

J. W. Woods is with the Department of Electrical and Systems Engineering, Rensselaer Polytechnic Institute, Troy, NY 12181.

Reprinted from *IEEE Trans. Acoust., Speech, Signal Processing,* vol. ASSP-28, pp. 16–26, Feb. 1980.

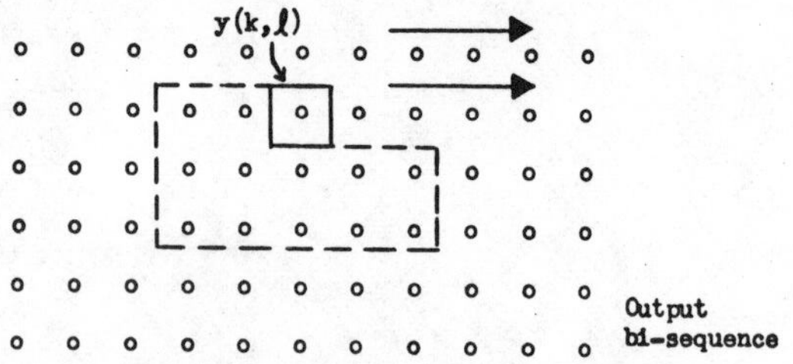

Fig. 1. Half-plane recursive filter.

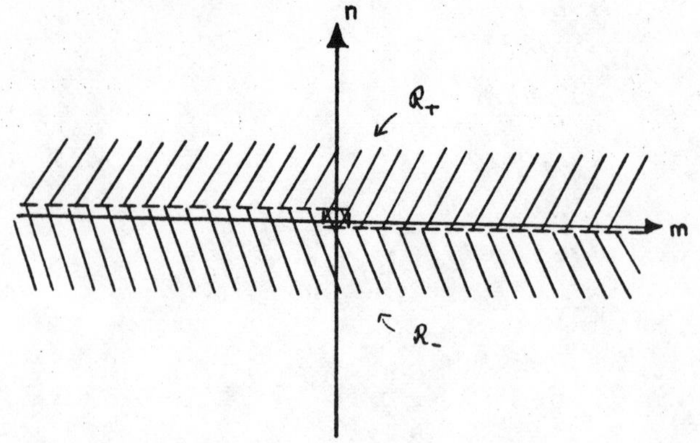

Fig. 2. Half-plane designations. ($\mathcal{R}_+$ and $\mathcal{R}_-$ overlap only at origin.)

The purpose of this paper is to present an efficient design algorithm for 2-D recursive digital filters, one which is suitable for satisfactorily approximating a *general* amplitude response specification. In order to achieve full design flexibility, half-plane recursive forms are adopted. Our design algorithm is an iterative, nonlinear optimization procedure with an unusual joint or dual constraint. One constraint in the frequency domain leads to satisfying the specified frequency responses, while the second constraint in the spatial domain leads to a stable finite-order filter. This last constraint is realized using a 2-D spectral factorization. The numerical realization of the design algorithm is described in detail, and a number of examples from the field of geophysical signal processing are presented to illustrate what we believe are its outstanding design capabilities.

II. Half-Plane Recursive Filters

To a certain extent, the structures traditionally used in 2-D recursive filter design have contributed to the limitations in design performance alluded to above. The 2-D canonical form adopted for our design method is both general and nonstandard, and avoids this limitation; it is the so-called half-plane recursive filter. Interestingly, it is defined over a half-plane (or rotations of the same), yet is recursively computable. The important property of this filter is its exclusive ability to realize an arbitrary magnitude response [4].

The specific half-plane filter to be considered here is of the form

$$y(k,l) = \sum_m \sum_n a(m,n)\, x(k-m, l-n) - \sum_{\substack{m=0\\(m,n)\neq(0,0)}}^{M} \sum_{n=0}^{N} b(m,n)\, y(k-m,l-n) - \sum_{m=-1}^{-M} \sum_{n=1}^{N} b(m,n)\, y(k-m,l-n) \tag{1}$$

and is illustrated in Fig. 1. The points within the dashed lines (the mask of the recursion) characterize the set of "previously" obtained outputs to be used in calculating the "present" output $y(k,l)$. The arrows indicate the sequence of computations, in this case a row-sequential ordering. Support $\mathcal{S}$ of the input mask is left unspecified, with the understanding that (1) may accommodate filtering with lag (i.e., the bisequence $a(k,l)$ may be double-sided in both variables).

To simplify the notation, we define the two half-planes shown in Fig. 2: $\mathcal{R}_+$ is the region $\{m \geq 0, n \geq 0\} \cup \{m < 0, n > 0\}$, and $\mathcal{R}_-$ is the region $\{m \leq 0, n \leq 0\} \cup \{m > 0, n < 0\}$. Finite-dimensional projections of these half-planes are indicated with a superscript "f." (The superscript "f" will be dropped when the distinction between the finite and infinite regions is unambiguous.) For example, $\mathcal{R}_+^f$ is the region $\{M \geq m \geq 0, N \geq n \geq 0\} \cup \{-M \leq m < 0, N \geq n > 0\}$. Using this notation, we can write (1) as

$$y(k,l) = \sum_{\mathcal{S}}\sum a(m,n)\, x(k-m,l-n) - \sum_{\mathcal{R}_+^f-(0,0)}\sum b(m,n)\, y(k-m,l-n) \tag{2}$$

and the half-plane filter transfer function as

$$H(z_1,z_2) = \frac{A(z_1,z_2)}{B(z_1,z_2)} = \frac{\sum_{\mathcal{S}}\sum a(m,n)\, z_1^{-m} z_2^{-n}}{\sum_{\mathcal{R}_+}\sum b(m,n)\, z_1^{-m} z_2^{-n}}. \tag{3}$$

We refer to this filter as being of $M \times N$-order.

As might be expected, the stability of the recursion in (2) is determined by the location of the singularities of $H(z_1,z_2)$. Necessary conditions for BIBO stability are

$$\begin{aligned} B(z_1,z_2) &\neq 0 \qquad \{|z_1| = 1, |z_2| \geq 1\} \\ B(z_1,\infty) &\neq 0 \qquad \{|z_1| \geq 1\}. \end{aligned} \tag{4}$$

Apparently, to directly ensure the stability of $H(z_1,z_2)$ requires the locations of the zeros of $B(z_1,z_2)$. That this cannot be accomplished is a determining factor in restricting the extension of standard 1-D design methodologies for the 2-D problem, as we will now see.

III. Spectral Factorization Design Approach

Traditional approaches to 1-D recursive filter design involve two distinct steps: approximation and stabilization [13]. Typically, a specified magnitude function is *approximated* by a ratio of cosine polynomials, either analytically or in an iterative manner, using mathematical optimization. This is followed by a factorization of the approximate into its causal and anticausal components for the purpose of *stabilization* (this sometimes involves a "pole-flipping" operation). The second step is essential to obtaining a realizable 1-D recursive filter.

In extending this approach to 2-D, one immediately encounters the problem of factoring 2-D polynomials. Because such polynomials are generally irreducible, the stabilization step in the analogous 2-D procedure is problematical, at least in its direct form.

The design procedure presented below is configured to avoid this dilemma while maintaining the basic spirit of this approach. It involves an integration of the approximation/stabilization steps and incorporates a recently developed 2-D spectral factorization algorithm. For our purposes, the crucial aspect of this factorization derives from the holomorphic properties of the factors. Because factorization is central to our development, we briefly review this algorithm prior to presenting our design procedure.

A. Review of 2-D Spectral Factorization [4]

The spectral factorization algorithm utilizes a homomorphic transform to compute the so-called 2-D cepstrum $\hat{c}(m,n)$ of a bisequence $c(m,n)$. Inasmuch as the support of $\hat{c}(m,n)$ is determined by the region of holomorphy of $\hat{C}(z_1,z_2)$, various factorizations can be achieved by simply sectioning the cepstrum into nonoverlapping (except at boundaries) regions, and performing the inverse homomorphic transform.

Specifically, a half-plane sectioning or projection corresponds to a factorization involving the region of homomorphy in (4). That is,

$$C(z_1,z_2)=B_+(z_1,z_2)\,B_-(z_1,z_2) \tag{5}$$

where

$$B_+(z_1,z_2)\neq 0 \quad \{|z_1|=1,\ |z_2|\geqslant 1\},$$
$$B_+(z_1,\infty)\neq 0 \quad \{|z_1|\geqslant 1\},$$

and

$$B_-(z_1,z_2)\neq 0 \quad \{|z_1|=1,\ |z_2|\leqslant 1\},$$
$$B_-(z_1,\infty)\neq 0 \quad \{|z_1|\geqslant 1\}.$$

Consequently, the factor $b_+(m,n)$ leads to a stable half-plane recursion of the form (2), and hence is said to be recursively stable. The two-factor, or half-plane, factorization is shown in Fig. 3. The difficulty here, which is unique to multidimensional factorizations and consistent with our remarks above, is that the factors are generally infinite dimensional. Thus, although $C(z_1,z_2)$ in (5) is a finite order polynomial, $B_+(z_1,z_2)$ and $B_-(z_1,z_2)$ are generally of infinite order. Of course, when $C(z_1,z_2)$ is reducible, finite dimensional terms result from the factorization algorithm. It is in utilizing this case that the factorization becomes a useful tool in recursive filter design.

B. Design Approach

Our design approach is fundamentally a mathematical optimization procedure, in which the filter numerator $a(m,n)$ and denominator $b_+(m,n)$ bisequences are iteratively derived by successive approximation. The ith iterates in this process are denoted by ${}^{(i)}a(m,n)$ and ${}^{(i)}b_+(m,n)$. It departs from classical approaches in the novel manner in which the guaranteed filter stability is woven into the optimization process.

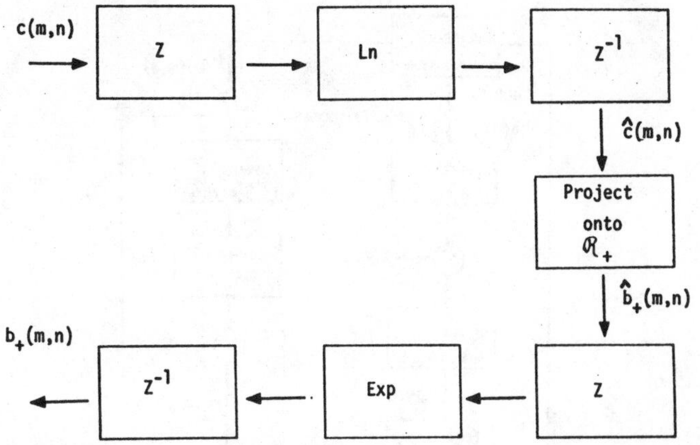

Fig. 3. 2-D spectral factorization with the half-plane projection operator.

The *approximation* aspect of the design problem is simply stated: choose the filter function whose magnitude response "best" matches the prescribed specification in some optimal sense. Letting the Fourier transform of the filter in (3) be denoted by

$$H(u,v)=\frac{A(u,v)}{B_+(u,v)}=H(z_1,z_2)\Big|_{z_1=e^{ju},\,z_2=e^{jv}} \tag{6}$$

and the specification (a magnitude spectrum) be $S(u,v)$, the weighted frequency-domain error in approximation is

$$\epsilon_a(u,v)=W(u,v)[S(u,v)-|H(u,v)|] \tag{7}$$

where $W(u,v)$ is an arbitrary weighting function.

Traditionally, an object function is defined to be a norm of $\epsilon_a(u,v)$ evaluated over a finite set of discrete frequencies,

$$\|\epsilon_a(u_i,v_j)\|=\|W(u_i,v_j)[S(u_i,v_j)-|H(u_i,v_j)|]\|. \tag{8}$$

It is in minimizing this norm that the filter optimally approximates or "best" matches the specified spectrum. Of course, it is essential that the minimization be carried out subject to stability constraints.

The *stabilization* constraint on the filter design is accomplished by augmenting the object function prior to minimization. The motivation for our procedure is as follows. Consider the ith approximate ${}^{(i)}b_+(m,n)$ of the filter denominator. In [4] a stability test for half-plane filters was presented which involved computing the magnitude function $|{}^{(i)}B(u,v)|$ spectrally factoring $|{}^{(i)}B(u,v)|$ to obtain $b_+(m,n)$ and comparing ${}^{(i)}b_+(m,n)$ to $b_+(m,n)$. If they are equal, ${}^{(i)}b_+(m,n)$ fulfills the conditions in (4) and is recursively stable. Alternately, if they are unequal, then ${}^{(i)}b_+(m,n)$ is unstable. As a consequence, the quantity

$$\epsilon_s(m,n)={}^{(i)}b_+(m,n)-b_+(m,n) \tag{9}$$

may be viewed as a "stability" error.

This error is incorporated into our design by constructing a new object function Q. We merge the approximation and

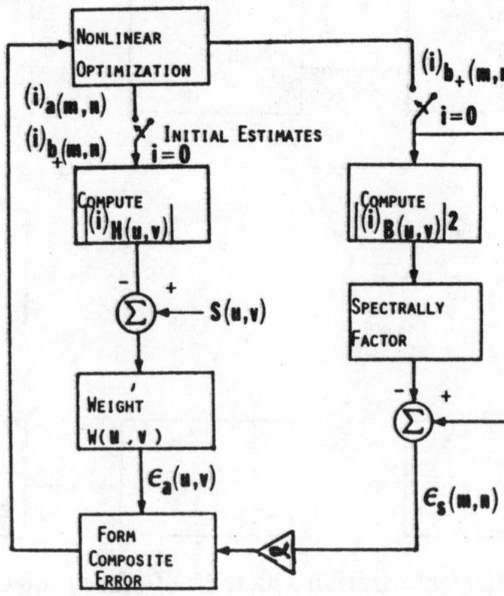

Fig. 4. Flow diagram for the design algorithm.

stability errors,

$$Q = \| \epsilon_a(u_i, v_j) + \alpha \epsilon_s(m_i, n_j) \| \tag{10}$$

where the stability errors are taken over a selected finite set of points in the m, n plane, and α is a scaling constant. Importantly, minimization of the object function Q involves a simultaneous minimization of both errors. The constant α is chosen to provide a suitable scaling between the two types of errors and to ensure an approximate "zeroing-out" of the stability error.

Because the numerator and denominator parameters enter into the errors nonlinearly, numerical optimization procedures must be employed to perform the required minimization. The gradients used in these procedures may be analytically and/or numerically computed and a variety of norms chosen in specifying Q.

The design algorithm which implements this approach follows and is illustrated in Fig. 4.

C. Design Algorithm

1) Specify the desired magnitude $S(u, v)$, DFT size, frequency weighting $W(u, v)$, scalar α, and filter order M and N.

2) Spectrally factor the inverse spectrum $1/S(u, v)$, and window this factor to $\mathcal{R}_+^f$, obtaining ${}^{(0)}b_+(m, n)$ (the zeroth iterate of the denominator bisequence); ${}^{(0)}a(m, n) \triangleq \delta(m, n)$.

3) Compute the magnitude responses $|{}^{(i)}A(u, v)|$ and $|{}^{(i)}B(u, v)|$ corresponding to ${}^{(i)}a(m, n)$ and ${}^{(i)}b_+(m, n)$, respectively, and form $|{}^{(i)}H(u, v)| = |{}^{(i)}A(u, v)|/|{}^{(i)}B(u, v)|$.

4) Compare $|{}^{(i)}H(u, v)|$ with the specification $S(u, v)$ over a set of discrete frequencies $\{u_i, v_j\}$, obtaining the weighted approximation errors $\epsilon_a(u_i, v_j)$.

5) Spectrally factor $|{}^{(i)}B(u, v)|^2$ and compare the resulting factor with ${}^{(i)}b_+(m, n)$ over a subset of points $\{m_i, n_j\}$ in $\mathcal{R}_+$, obtaining the stability errors $\epsilon_s(m_i, n_j)$.

6) Form the composite error vector ϵ by stacking the approximation errors, $\epsilon_a(u_i, v_j)$ and scaled stability errors $\alpha\epsilon_s(m_i, v_j)$.

7) Using a nonlinear optimization algorithm, compute the adjustments in ${}^{(i)}a(m, n)$ and ${}^{(i)}b_+(m, n)$ to obtain the $i + 1$ iterates, ${}^{(i+1)}a(m, n)$ and ${}^{(i+1)}b_+(m, n)$, respectively.

8) Repeat steps 3-7 (incrementing the iterate index by one per cycle) until a satisfactory convergence has been achieved.

D. Numerical Implementation

Despite its apparent complexity, the algorithm is reasonably simple in structure and quite straightforward to implement. We elaborate below on some of the practical aspects of its numerical implementation.

Step 1 consists of providing the information needed to define the design problem and initiate the algorithm. Because all Fourier transform computations are performed using 2-D discrete Fourier transforms, this information includes dimensions of the FFT's. In our designs we generally use 32×32 or 64×64 FFT's.

The initial approximate of the filter (the zeroth iterate) is taken to be an all-pole structure and is derived in step 2. Its denominator is obtained by spectrally factoring $1/S(u, v)$. The filter associated with this factor satisfies the specification (exactly) and is recursively stable, but is generally of infinite order. While the factor takes values on the entire half-plane, it typically has its energy concentrated in the finite region of the m, n plane, including the origin (a consequence of the "minimum-phase" properties of the factorization [4]). Thus, we truncate the factor to $\mathcal{R}_+^f$, obtaining the initial estimate of the denominator. As the truncated values are "small," it seems likely they contribute, at most, modest variations in the spectrum; so $S(u, v)$ may be adequately approximated by the magnitude of the resulting all-pole filter. Unfortunately, this approximate is not guaranteed to be recursively stable.

The magnitude response of the current filter iterate computed in step 3 is differenced with the specification $S(u, v)$ over a finite set of frequencies in step 4. These differences or errors are then weighted by a prescribed weighting function $W(u, v)$. Choice of the set of frequencies and weighting functions are highly dependent on the form and complexity of $S(u, v)$. To a great extent, both are used in quite standard ways to emphasize the relative importance of various frequency characteristics in the design. As a first cut, our practice has been to employ the DFT sampling frequencies in the error set, excluding those which fall into transition bands of $S(u, v)$. Refinement of this sampling may be appropriate, depending on the subsequent design performance and scale of the computations.

The so-called stability error is generated in step 5. It involves applying the spectral factorization stability test to the current denominator iterate. As in the previous step, the errors are selected differences, although here they are sampled in the spatial (m, n) domain, and typically over a small interval (for example, $\mathcal{R}_+^f$).

Unfortunately, replacement of the Fourier transforms with DFT's introduces two sources of error in $\epsilon_a(m, n)$ not directly related to the stability. The first is the numerical error associated with computing 2-D FFT's. It is a function of machine wordlength and DFT size, and generally quite small. The second source is the aliasing error associated with computing the 2-D cepstrum in the factorization. This error is a function of the size of DFT used and the spatial spread of the cepstral bisequence ${}^{(i)}\hat{b}(m,n)$. It is much larger than the first error, and to a certain extent controllable by varying the DFT size. We will return to this error in the following discussion.

Step 6 involves forming the composite error for the optimization. In this composition, the stability errors are scaled by α to establish a normalization relative to the approximation errors. While it is desired in the procedure to minimize the composite (total) error, it is essential that the stability error component be reduced to appropriately small values. (Of what value is a perfect approximate if it is unstable?) Through the α-scaling, error-significance weighting is introduced and partial control of the error components established.

The nonlinear optimization part of the design algorithm in step 7 can be formalized and implemented in a variety of ways. In our approach we choose to minimize a least squares error norm

$$Q = \| \epsilon_a(u_i, v_j) + \alpha \epsilon_s m_i, n_j) \|_2 \tag{11}$$

using a variate of the Levenberg-Marquardt (LM) algorithm [14] with a modified step size calculation. The $i+1$ iterates of the filter parameters are given by

$$\begin{aligned} {}^{(i+1)}a(m,n) &= {}^{(i)}a(m,n) + \gamma \Delta^{(i)} a(m,n) \\ {}^{(i+1)}b_+(m,n) &= {}^{(i)}b_+(m,n) + \gamma \Delta^{(i)} b_+(m,n) \end{aligned} \tag{12}$$

where the increments $(\Delta^{(i)}a(m,n), \Delta^{(i)}b_+(m,n))$ obtained from the basic Levenberg-Marquardt approach represent interpolations between Gauss-Newton and steepest descent results. The LM variate we use is a finite-difference analog in which the Jacobian is computed numerically rather than using explicit analytic forms for the derivatives [15]. The advantages of this algorithm are its extended convergence region and speed of convergence. It seems particularly well suited to problems involving difficult minimizations and/or potentially poor initial estimates.

The LM step size is modified by γ in (12) to permit a guaranteed minimization of the object function in the direction of the increments $\{{}^{(i)}\Delta a(m,n), {}^{(i)}\Delta b_+(m,n)\}$. This insures improvement in minimizing the object function per successive iteration. The parameter γ is determined by a one-dimensional "golden search" routine [16], or alternatively, by an interval-halving search. In practice, use of this modification appears to allow avoidance of many of the instabilities commonly encountered in large-scale parameter optimization problems.

Step 8 involves cycling through the design steps using successive iterates obtained in the previous step. This process terminates when the object function (composite error norm) has been reduced below a satisfactory level and the stability error is suitably small, as discussed above.

Some practical experience is helpful in making both judgments. Setting a prescribed level for the composite error is highly problem dependent and involves the traditional tradeoffs between design performance and filter complexity. Establishing acceptable limits for the stability error depends both on the design-at-hand and the size of the DFT used in the algorithm. Even in the ideal case, the optimization cannot drive the stability error identically to zero because of cepstral aliasing effects. While the magnitude of this aliasing is difficult to bound, our experience is that with errors of $0(10^{-8})$; the designed filters are stable. When uncertainty exists, subsequent stability tests can be performed *a posteriori* design with increasingly larger DFT sizes to verify stability.

When the iteration does not converge to an acceptable level, it is useful to examine both error components $\|\epsilon_a(u,v)\|_2$ and $\|\epsilon_s(m,n)\|_2$ of the object function. Three cases may occur. Each case is described in the following.

Case 1: Low approximation error, high stability error—This suggests the scaling is such to emphasize the approximation error over the stability error. This can be corrected by increasing the scaling factor α in (10), and recomputing the design. When α is appropriately adjusted, the object function (minimized) should be dominated by the approximation error.

Case 2: High approximation error, low stability error—Convergence of the optimization to a "local" rather than "global" minimum could result in this situation. The classical solution to this problem is to perturb the parameters, generating new initial estimates, and restart the iteration (assuming proper choice of α). Alternately, this problem could simply be due to insufficient degrees of freedom in the approximate. This calls for increasing the order of the numerator and/or denominator in (6) and restarting the design.

Case 3: High approximation error, high stability error—The high magnitude of both errors could be caused by a combination of Cases 1 and 2 above, and handled commensurately. However, this difficulty could also be due to a subtle effect of the cepstral aliasing. If the aliasing is excessive, $\epsilon_s(m,n)$ will be artificially large for *stable* designs whose cepstra extend far from the origin. The optimizer, in minimizing this error, adjusts the parameters to preclude this from occurring. Thus, in effect, it constrains the design to filters having denominators with short duration cepstra, hence, potentially larger approximation errors. It is our observation that the optimizer is unable to satisfactorily reduce either error under these circumstances. Clearly, this situation is correctable by simply increasing the size of the DFT used in the design algorithm, although this involves a subsequent increase in computations for the design algorithm.

E. Remarks

Remark 1: The distinguishing characteristics of our approach are its excellent approximation capabilities, extensive design versatility, and guaranteed stability of the designed filter. This is, of course, exactly what one would desire in a general-purpose design approach. The approximation performance derives from use of the half-plane structure and the fundamental superiority of designing by mathematical optimization when dealing with general design specifications. Although various difficulties may be encountered when using successive approximation procedures (e.g., slow convergence, local minima, sensitivity to initial estimates), they allow quite a direct control over the form and magnitude of the approximation error. This is accomplished through choice of error norm, frequency weighting function, and specific discrete frequencies used in constructing the object function, and, of course, the order of the approximate.

The design versatility available is indeed extensive. Structural symmetries, for example, can be incorporated into the design in a very routine manner. One may wish to pick filter forms commensurate with existing symmetries in the design specification. For instance, if a half-plane filter magnitude is to have quadrantal symmetry,

$$|H(u,v)| = |H(-u,v)| = |H(u,-v)| = |H(-u,-v)|, \tag{13}$$

it is known [17] that its numerator bisequence must be quadrantally symmetric, and the denominator bisequence of the form

$$b_+(m,n) = b_+^1(m) \circledast b_+^2(m,n) \tag{14}$$

where $b_+^1(m)$ is a causal 1-D sequence, $\circledast$ denotes the convolution operation, and

$$b_+^2(m,n) = \delta(m,n) \qquad m = 0$$
$$= b_+^2(-m,n) \quad \text{otherwise.} \tag{15}$$

Imposing this symmetry on the filter simply requires use of the relationships (14) and (15) in defining the filter function for the design algorithm. (In addition to satisfying the symmetry conditions, this form has the added advantages of reducing both the order of the optimization and the complexity of the filter realization.)

The algorithm can also be easily modified to accommodate "zero-phase" filter designs. Zero-phase filtering is realized recursively by processing the data with the $\mathfrak{R}_+$ half-plane filter *and* its corresponding $\mathfrak{R}_-$ form. This is the 2-D analog of forward- and reverse-time filtering. In this case, the magnitude spectrum realized in the processing is the magnitude squared of the individual half-plane filter. For this design, we simply replace $|H(u,v)|$ in the approximation error with $|H(u,v)|^2$ and proceed as before. Examples of this case will be given in the applications section following.

Remark 2: The substantive difficulty in designing with mathematical optimization is the inordinate complexity associated with solving for large numbers of parameters. What typically happens is that the basic extremal problem becomes quite ill-conditioned. Increases in the dimensionality of the parameter space lead to larger search volumes, increased numbers of local minima, higher sensitivity to initial estimates, unusual object function behavior, etc., all of which makes a convergence of the iteration to a satisfactory minimum more difficult.

The exact point where this complexity becomes excessive is, of course, dependent on both the problem and the computer system on which the design algorithm is run. We have successfully performed designs on a CDC 7600 involving up to 100 parameters, and find this provides ample design freedom for most problems we have encountered. (It is not uncommon for general optimization problems involving 80-100 parameters to be routinely computed at most major scientific computer facilities.)

For more complex problems, a reduction in the dimension of the parameter space would be required. This is accomplished quite directly in our case by partitioning the numerator and denominator designs. Thus, rather than optimizing over the numerator and denominator simultaneously, successive optimizations are performed. First, an all-pole filter is designed, for example, and this fixed denominator is then used in designing the filter numerator (alternately, design the numerator first, then denominator). In the numerator-only design, our basic algorithm is used with steps 5 and 6 deleted. When large numbers of parameters are involved, it is quite possible that this separation of numerator and denominator designs will, in fact, result in more satisfactory minima because of improved conditioning.

Several algorithmic approaches provide alternate partitionings which may be of interest [18]. They involve more extensive modification of the optimization calculation, but offer the potential of broad control over its required computations.

Remark 3: The computational burden of our design is heavily concentrated in the optimization (step 7), approximately 60 percent of the total computation. Of that 60 percent, slightly more than half is due to the numerical computation of the Jacobian via finite difference techniques; the remainder is principally the updating and solving of the matrix equations in the optimization. Most of the remaining 40 percent of the execution time is spent in steps 4-6 of the algorithm, i.e., the generation of the stability and approximation errors.

A typical execution time for a filter design problem is about two minutes on a CDC 7600[1] (this is for the design of a filter with 50 coefficients and using a 32 X 32 DFT). The execution times vary, of course, depending on the size of the filter, the number of iterations required to convergence (discussed below), and the size of the FFT used. Some potential savings could be obtained via assembly language programming (rather than Fortran) of the crucial stages of the optimization and/or the use of analytic derivatives in place of the numerical differencing techniques—neither of these approaches has yet been incorporated into our design program, however.

IV. Examples of Application

In this section we present the application of our algorithm to design problems in geophysical signal processing. These filter designs are, to some extent, uniquely 2-D in that they bear little direct analogy with any 1-D filtering operation. As such, they demonstrate the versatility of our approach, while giving some indication of the algorithm's achievable design performance.

Each of the three example designs to be discussed in the following are variations of what are called "fan filters." Such filters are useful in geophysical signal processing due to their ability to discriminate certain directional information in 2-D signals [19].

Example 1: 45° Fan—Our first example is the 45° fan specified by

$$S(u,v) = \begin{cases} 1 & \frac{\pi}{8} \leq \theta \leq \frac{3\pi}{8} \text{ or } -\frac{7\pi}{8} \leq \theta \leq -\frac{5\pi}{8} \\ 0.02 & \text{otherwise} \end{cases}$$

[1] Reference to a company or product names does not imply approval or recommendation of the product by the University of California or the U.S. Department of Energy to the exclusion of others that may be suitable.

where $\theta = \arctan v/u$. This specification is illustrated in Fig. 5. It is important to note that this specification possesses neither the circular, octant, or quadrantal symmetry commonly required of 2-D filter design procedures; this is an indication of the versatility of our technique.

Several designs were performed using the procedure described above, with various numerator and denominator supports. In each of the designs, a 64×64 DFT grid was used and a zero-phase implementation was assumed (i.e., two recursions of the filter in opposite directions would be required, necessitating that the magnitude squared of the iterate be calculated in the design procedure). (See Remark 1 above.) In addition, a weighting function $W(u, v)$ was chosen to minimize the large passband ripples caused by the narrowness of this fan: we let $W(u, v) = 5$ in the passband, $W(u, v) = 1$ in the stopband, and $W(u, v) = 0$ in the transition bands (defined to be any grid points within three grid points of the passband).

The results of these designs are summarized in Fig. 6, in which we have plotted the weighted RMS error as a function of iteration number. As shown in the figure, our approach was to first design the all-pole filters for 3×3, 4×4, 5×5, and 6×6 order half-planes. Note that the design algorithm converged rapidly in each case to a satisfactory solution within eight iterations. We then used the 3×3, 4×4, and 5×5 order half-planes as initial estimates in the general numerator-denominator design problem; in each of these designs we included a numerator of support 3×3. Note that although there was little improvement over the all-pole designs for the 3×3 and 4×4 half-plane denominators, there was a substantial improvement over the 5×5 half-plane denominator. The final filter coefficients for this 5×5 half-plane denominator, 3×3 numerator are

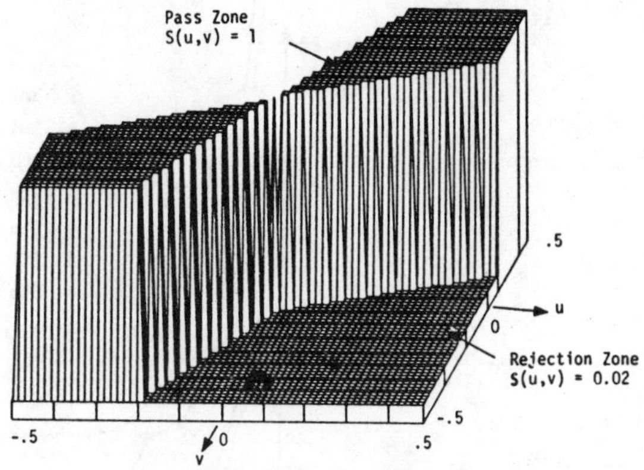

Fig. 5. Ideal 45° fan filter specification.

An alternate characterization of these designs is shown in Fig. 8, where the final weighted rms error (at convergence) is plotted as a function of the number of coefficients in the filter. The denominator-only (all-pole) designs have been connected in the figure with lines to distinguish them from the numerator-denominator designs. It is evident that the number of coefficients is not the sole determining factor in the quality of the approximation; the relative importance of the numerator and denominator is also significant. For example, the 4×4 half-plane denominator with a 3×3 numerator is essentially identical to the 4×4 all-pole design, but the 5×5 half-plane with a 3×3 numerator is clearly superior to the 5×5 all-pole design. We note also that, for this example, the 6×6 all-pole design is inferior to the 5×5 half-plane with a 3×3 numerator despite its larger number of filter coefficients.

Example 2: 90° Fan—Our second design example is the familiar 90° fan specified by (for $-\pi < \theta < \pi$)

n ↑	-0.6840	0.3739	0.0797
	-0.1706	0.1367	-0.0391
	0.6937	0.0227	-0.3478
→ m			

-0.1106	0.0149	-0.3270	-0.1226	0.0836	-0.1142	0.0725	0.0706	0.0081	-0.0016	0.0014
-0.0265	0.1237	-0.2752	1.0617	0.5161	0.0771	0.5067	-0.1950	-0.1671	-0.0307	-0.0082
-0.3389	-0.0604	-0.3222	0.8019	-0.8428	-0.9528	-0.5486	-0.8645	0.1273	0.1447	0.0392
0.1064	1.1057	0.5207	2.3028	-0.6831	-0.4464	1.4433	0.5773	0.8326	0.0520	-0.0788
0.0261	-0.1802	-1.0553	-0.5107	-4.2691	-0.0926	0.1365	-1.3895	-0.1824	-0.5885	-0.0221
				n ↑	3.3630	-0.0929	0.0985	0.5935	-0.0239	0.1842
				→ m						

or equivalently,

$$H(z_1, z_2) = \frac{0.6937 - 0.0227\, z_1^{-1} + \cdots + 0.0797\, z_1^{-2} z_2^{-2}}{3.360 - 0.0929\, z_1^{-1} + \cdots - 0.1106\, z_1^{5} z_2^{-5} + \cdots + 0.0014\, z_1^{-5} z_2^{-5}}.$$

The magnitude-squared response of this filter (i.e., two recursions) is shown in Fig. 7. Although the designed filter possesses only 70 coefficients, the approximation of its response to the desired $S(u, v)$ is quite good.

$$S(u, v) = \begin{cases} 1 & |\theta| \leqslant \pi/4 \text{ or } |\theta| \geqslant 3\pi/4 \\ 0.02 & \text{otherwise} \end{cases}$$

where θ is defined as above. This specification is illustrated in

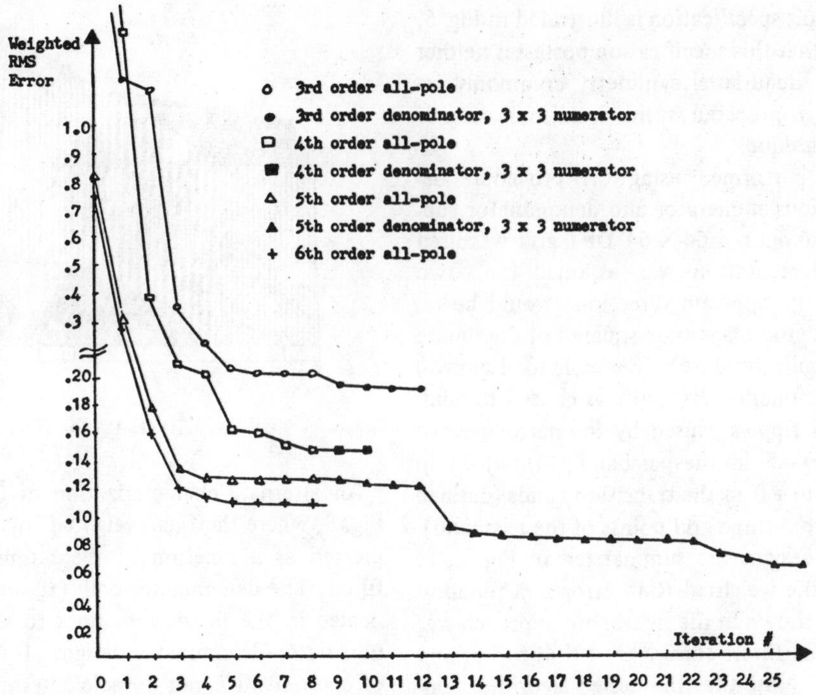

Fig. 6. Plot of weighted rms errors versus the iteration number for several 45° fan filter designs.

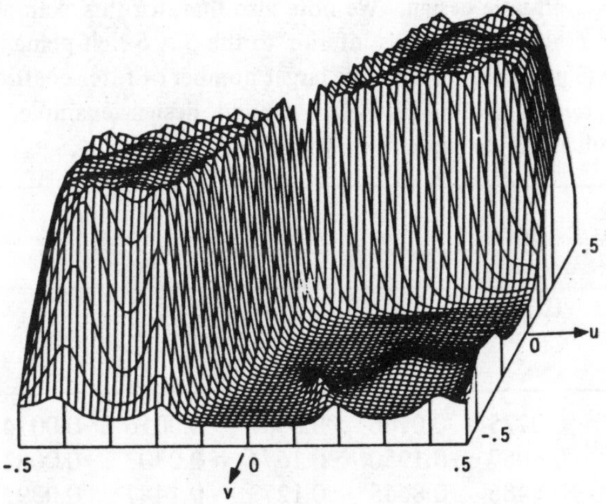

Fig. 7. Magnitude-squared response of the 4 X 4 half plane with a 3 X 3 numerator.

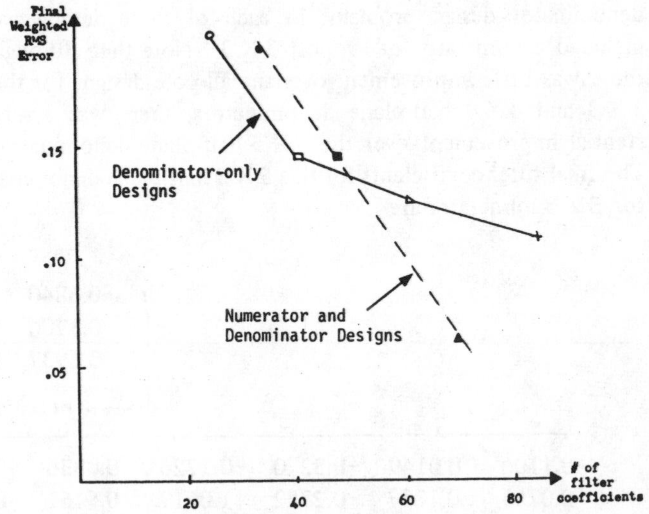

Fig. 8. Plot of the final weighted rms error (at convergence) versus the number of filter coefficients for the various 45° fan filter designs (same symbols as Fig. 7).

Fig. 9. Note that this spectrum possesses quadrantal symmetry, but it does not have octant or circular symmetry. This particular spectrum is commonly used as a benchmark test for 2-D digital filter (both recursive and nonrecursive) design procedures.

We again utilized our procedure to design a range of 2-D recursive half-plane filters (with and without numerators) for this specification. For each design, we had the following parameters: a 32 X 32 DFT grid size was used; a zero-phase implementation was assumed (so we again calculated the magnitude-squared response of the iterates, as per Remark 1 above); and a weighting function was incorporated with $W(u, v) = 4$ in the passband, $W(u, v) = 1$ in the stopband, and $W(u, v) = 0$ in the transition band (again defined to be any grid points within 3 grid points of the passband). The design results are summarized in Fig. 10, which again plots the weighted rms error versus the iteration number. The all-pole designs again converged rapidly in all cases (within 6 iterations), but the "pole-zero" designs generally required several more iterations. In this example, we observed that several of the designs gave outstanding approximations to the desired specification.

In particular, note that the 3 X 3 half-plane denominator with a 4 X 4 numerator, the 5 X 5 half-plane denominator

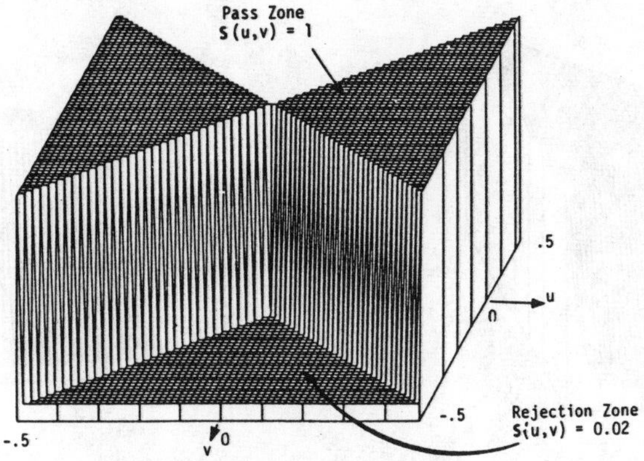

Fig. 9. Ideal 90° fan filter specification.

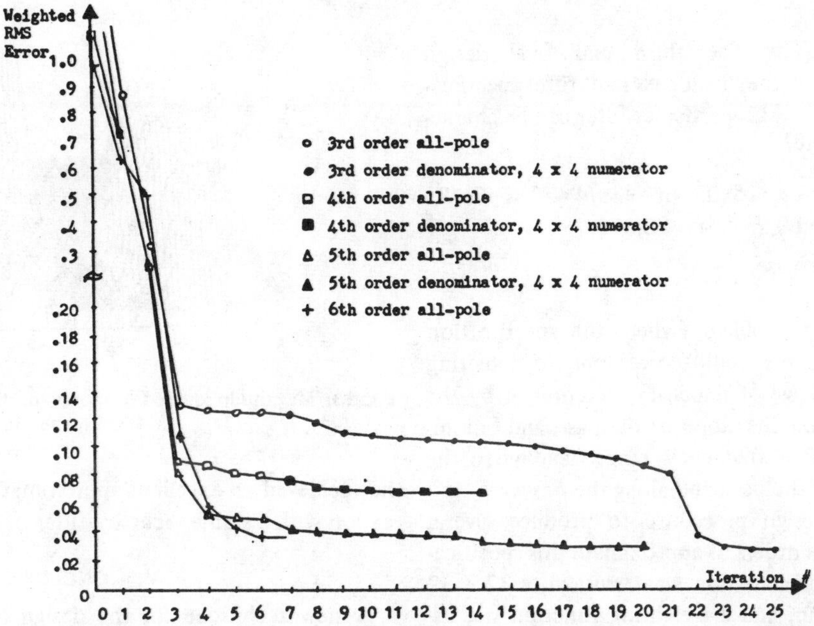

Fig. 10. Plot of weighted rms error versus the iteration number for several 90° fan filter designs.

with a 4 X 4 numerator, and the 6 X 6 half-plane all-pole designs yielded extremely low final rms errors. The final 3 X 3 half-plane denominator with a 4 X 4 numerator result (a total of only 41 coefficients) at convergence was

A plot of the magnitude (squared) response of this filter is given in Fig. 11. The remarkable accuracy of approximation demonstrated by this example is far superior to any comparable-order filter designs the authors have seen in the literature and is indicative of the power of our design technique.

n ↑	0.0364	0.0365	0.0369	-0.0077
	0.2045	-0.4309	0.0663	-0.0483
	0.6390	-0.7232	0.3232	0.0546
	0.4130	-0.4953	0.0627	0.0103
└→ m				

0.0012	0.0003	-0.0494	-0.0812	-0.0741	-0.0044	-0.0052
0.0024	0.1044	0.0665	0.6778	0.1526	0.2022	0.0144
-0.0083	-0.0721	-0.8787	0.0103	-1.0150	-0.2063	0.0024
		n	1.1825	0.1992	-0.0335	-0.0115
		└→ m				

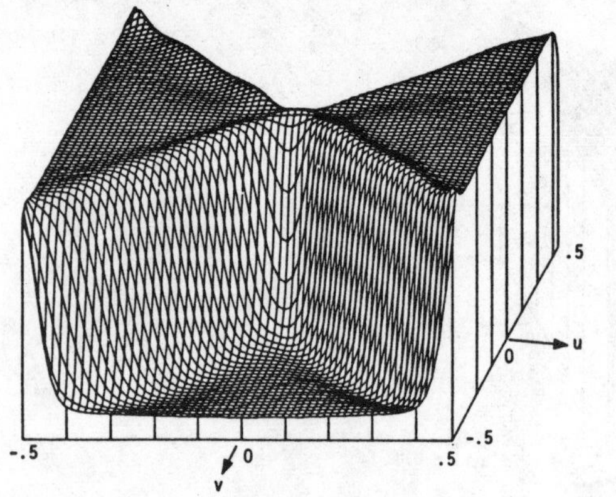

Fig. 11. Magnitude-squared response of the 3 × 3 half plane with a 4 × 4 numerator.

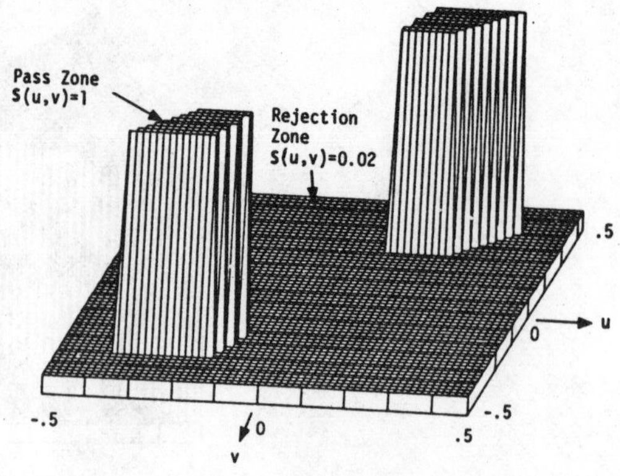

Fig. 12. Ideal bandpass fan filter specification.

Example 3: Bandpass Fan—The third and final design example to be presented is the "bandpass fan" filter specification illustrated in Fig. 12. This particular filter can be characterized (for $-\pi \leqslant \theta \leqslant \pi$) by

$$S(u, v) = \begin{cases} 1 & \pi/4 \leqslant \theta \leqslant 5\pi/8 \text{ or } -3\pi/4 \leqslant \theta \leqslant -7\pi/8, \\ & \text{and } 0.25 \leqslant |v| \leqslant 0.375 \\ 0.02 & \text{otherwise.} \end{cases}$$

In the seismic detection problem (where the specification would be in *k-f* space) this would correspond to detecting waves within a certain range of velocities (determined by the upper and lower limits on the slope of the passband boundaries) and within a certain frequency range (leading to the upper and lower limits of the passband along the *f* axis).

We again used our design procedure to produce several recursive filters of various orders to approximate this specification. A zero-phase implementation was assumed, a 32 × 32 DFT grid size was chosen, and a weighting function was assigned to be $W(u, v) = 10$ in the passband, $W(u, v) = 1$ in the stopband, and $W(u, v) = 0$ in the transition bands (chosen here to be any grid point within two points of a passband region). The best results obtained was the following 4 × 4 half-plane filter with a 4 × 4 numerator:

n ↑	0.1196	0.0696	-0.0337	-0.0178
	-0.1356	-0.0108	0.0590	0.0327
	0.0028	0.1480	-0.0232	0.0882
	0.1952	0.0098	-0.1704	-0.0571
→ m				

0.0763	0.0603	-0.0458	0.0167	0.0446	0.0126	-0.0064	0.0304	0.0021
-0.1037	-0.0687	-0.2761	0.0240	0.2607	-0.1487	0.0192	0.0746	-0.0203
0.0138	0.4862	0.1009	0.3431	-0.1561	-0.1183	0.2222	0.1495	-0.0039
-0.0006	-0.0888	-0.3235	-1.1471	0.2689	0.6300	-0.0264	0.3890	-0.1333
			n ↑ (→ m)	1.6791	-0.1460	-0.0500	0.0468	-0.1745

The magnitude-squared response of this filter is shown in Fig. 13, which again demonstrates that the design technique has achieved an excellent approximation to the desired specification with a stable recursive filter of low order.

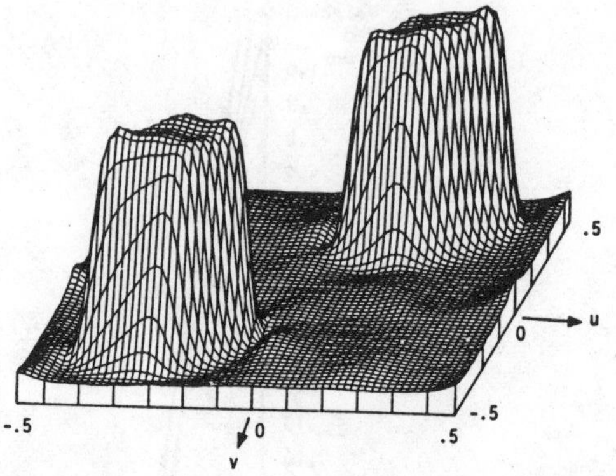

Fig. 13. Magnitude-squared response of the 4 × 4 half-plane with a 4 × 4 numerator.

V. Conclusions

A new technique for the design of two-dimensional (2-D) recursive digital filters has been presented. The design procedure incorporates a 2-D spectral factorization into a constrained, nonlinear optimization to yield stable half-plane recursive filters. The computational aspects of the implementation were discussed, and the versatility and excellent approximation capabilities of this approach were demonstrated via several design examples.

REFERENCES

[1] E. L. Hall, "A comparison of computations for spatial filtering," *Proc. IEEE*, vol. 60, pp. 887-891, July 1972.

[2] M. P. Ekstrom, "Computational complexity of two-dimensional digital signal processing," to be published.

[3] S. K. Mitra and M. P. Ekstrom, Eds., *Two-Dimensional Digital Signal Processing*. Stroudsberg, PA: Dowden, Hutchinson & Ross, 1978.

[4] M. P. Ekstrom and J. W. Woods, "Two-dimensional spectral factorization with applications in recursive filtering," *IEEE Trans. Acoust., Speech, Signal Processing*, vol. ASSP-24, pp. 115-128, Apr. 1976.

[5] R. A. Decarlo, J. Murray, and R. Saeks, "Multivariable Nyquist theory," *Int. J. Contr.*, vol. 25, pp. 657-675, May 1977.

[6] G. A. Shaw, "An algorithm for testing stability of two-dimensional digital recursive filters," in *Proc. IEEE Int. Conf. Acoust., Speech, Signal Processing*, Apr. 1978, pp. 769-772.

[7] J. M. Costa and A. N. Venetsanapoulos, "Design of circularly symmetric two-dimensional recursive filters," *IEEE Trans. Acoust., Speech, Signal Processing*, vol. ASSP-22, pp. 432-443, Dec. 1974.

[8] S. Chakrabarti and S. K. Mitra, "Design of two-dimensional digital filters via spectral transformations," *Proc. IEEE*, vol. 65, pp. 905-914, June 1977.

[9] A. M. Ali, "Design of inherently stable two-dimensional recursive filters imitating the behavior of one-dimensional analog filters," in *Proc. IEEE Int. Conf. Acoust., Speech, Signal Processing*, Apr. 1978, pp. 765-768.

[10] G. A. Maria and M. M. Fahmy, "An l_p design technique for two-dimensional recursive filters," *IEEE Trans. Acoust., Speech, Signal Processing*, vol. ASSP-22, pp. 15-22, Feb. 1974.

[11] J. B. Bednar, "Spatial recursive filter design via rational Chebyshev approximation," *IEEE Trans. Circuits Syst.*, vol. CAS-22, pp. 572-574, 1975.

[12] R. E. Twogood and S. K. Mitra, "Computer-aided design of separable two-dimensional digital filters," *IEEE Trans. Acoust., Speech, Signal Processing*, vol. ASSP-25, pp. 165-169, Apr. 1977.

[13] L. R. Rabiner and B. Gold, *Theory and Application of Digital Signal Processing*. Englewood Cliffs, NJ: Prentice-Hall, 1975.

[14] D. W. Marquardt, "An algorithm for least squares estimation of nonlinear parameters," *SIAM J. Appl. Math.*, vol. 11, pp. 431-441, June 1963.

[15] K. M. Brown and J. E. Dennis, Jr., "Derivative free analogues of the Levenberg-Marquardt and Gauss algorithm for nonlinear least-squares approximation," *Numer. Math.*, vol. 18, pp. 289-297, 1972.

[16] P. Whittle, *Optimization With Constraints*. London, England: Wiley-Interscience, 1971.

[17] D. M. Goodman, "Quadrantal symmetry conditions for non-symmetric half-plane filters," Lawrence Livermore Laboratory, Livermore, CA, Tech. Rep. UCRL-82443, Mar. 1979.

[18] L. S. Lasdon, *Optimization Theory for Large Systems*. London, England: Macmillian, 1970.

[19] P. Embree, J. P. Burg, and M. M. Backus, "Wide-band velocity filtering–The pie-slice process," *Geophysics*, vol. 28, pp. 948-974, 1963.

Design of Two-Dimensional Digital Filters via Spectral Transformations

SATYABRATA CHAKRABARTI, MEMBER, IEEE, AND SANJIT K. MITRA, FELLOW, IEEE

Abstract—Complex maps, with domains and codomains consisting of rational transfer functions, have often been used in designing two-dimensional (2-D) digital filters. Such maps, commonly known as spectral transformations, have the important property of carrying a stable rational transfer function to another stable rational function. This paper presents a unified framework for designing 2-D stable digital filters from prescribed magnitude-response specifications using spectral transformations such that the magnitude response of the resultant approximation is sort of a "best" approximation of the given specification. It is shown that there are two basic strategies for such designs, each with its own advantages and disadvantages. The design procedures are illustrated with practical examples. Further, it is also shown that all earlier results on 2-D filter design using spectral transformations follow as special cases of the general theory presented here.

I. Introduction

A TWO-DIMENSIONAL (2-D) digital filter is the realization of a computational procedure which converts a 2-D array (input) $\{x_{m,n}\}$ to another 2-D array (output) $\{y_{m,n}\}$. This paper deals with linear-shift-invariant (LSI) 2-D digital filters having both infinite and finite impulse responses. The 2-D infinite-impulse-response (IIR) LSI filters admit a transfer function representation of the following form:[1]

$$H(z_1, z_2) = \mathcal{Z}\{y_{m,n}\}/\mathcal{Z}\{x_{m,n}\} = \sum_{i=0}^{M_1}\sum_{j=0}^{N_1} a_{ij} z_1^i z_2^j \Big/ \sum_{i=0}^{M_2}\sum_{j=0}^{N_2} b_{ij} z_1^i z_2^j \quad (1)$$

where a_{ij} and b_{ij} are real constants. Similarly, a 2-D finite-impulse-response (FIR) LSI filter can be characterized by a transfer function of the following form:

$$H(z_1, z_2) = \sum_{k=0}^{M}\sum_{l=0}^{N} h_{kl} z_1^k z_2^l \quad (2)$$

where we have assumed that the transfer functions are first-quadrant [17], [36] functions, sometimes also termed "causal" in the literature. The functions $H(e^{-j\omega_1}, e^{-j\omega_2})$ and $|H(e^{-j\omega_1}, e^{-j\omega_2})|$ are known as the *frequency response* and the *magnitude response* of the 2-D digital filters, respectively.

Designing 2-D digital filters from prescribed magnitude-response specifications requires the determination of a FIR or an IIR transfer function $H(z_1, z_2)$ such that $|H(e^{-j\omega_1}, e^{-j\omega_2})|$ is the "best" approximation in a certain preassigned sense to the given specification. Note that the FIR filters are always stable. For IIR filters, however, the design procedure must include steps for guaranteeing stability.

The problem of approximating a 2-D magnitude-response specification using stable 2-D digital filters has been receiving considerable attention in the literature [1], [3]–[5], [12]–[14], [20], [24]–[29], [31], [32], [35], [38], [39]; even though, in most cases, the studies have been restricted to low-pass filters with circular symmetry. The purpose of this work is to develop a unified framework for designing 2-D digital filters from prescribed magnitude-response specifications using spectral transformations. By a spectral transformation we mean a complex map that carries a stable rational transfer function into another stable transfer function exhibiting a different frequency response, at the same time maintaining some desirable characteristics.

As a tool for designing filters, spectral transformations are known and used for 1-D continuous [41], [43] and digital filters [4], [9]–[11], [15], [33]. Various specific cases of spectral transformations for 2-D digital filters have also been investigated in the literature, which include both 1-D to 2-D transformations [1], [5], [6], [19], [26]–[29], [32] and 2-D to 2-D transformations [12], [13], [35], [39]. The motivations for studying spectral transformations are further strengthened by the fact that the use of such transformations reduces or bypasses the greater complexity and difficulty involved with stability checking [23, ch. 6].

II. Spectral Transformations and Design Methodology

In general, a spectral transformation (ST) must have the following characteristics: 1) it must produce an LSI stable transfer function from an LSI stable transfer function; 2) it must transform a real rational function into a real rational function; and 3) it must preserve some basic characteristics of the magnitude response (e.g., the ripple magnitude in the transmission and attenuation domains). The first two restrictions follow from the realizability requirements and the third restriction corresponds to an *a priori* guarantee of arriving at a "good" approximation.

The use of ST's leads to two fundamentally distinct design strategies for 2-D digital filters. In the first case, the ST's of the form $g(u_1, u_2)$ carry a 1-D transfer function $T(u)$ to a 2-D transfer function $T[g(u_1, u_2)]$ via the substitution of variables, $g(u_1, u_2) = u$, where u_1, u_2 and u are complex variables. If we use the "analog transformation" $g(s_1, s_2) = s$ on an analog transfer function $H(s)$, the resultant 2-D function is also an analog transfer function which can be converted to a 2-D digital transfer function via bilinear transformation; this leads to what we shall call the type I-A design approach. Otherwise, we can also start directly with a 1-D digital transfer function $H(z)$ and obtain a 2-D digital function $H[g(z_1, z_2)]$ via the

Manuscript received August 24, 1976; revised December 23, 1976. This work was supported by the National Science Foundation under Grant ENG 75-15027.

The authors are with the Department of Electrical Engineering, University of California, Davis, CA 95616.

[1] Note that the 2-D z-transform is assumed to be defined with positive exponents.

Reprinted from *Proc. IEEE,* vol. 65, pp. 905–914, June 1977.

substitution $g(z_1, z_2) = z$, which will be called the type I-B design approach.

It must also be noted that both type I-A and I-B design procedures yield digital filters with only a first-quadrant impulse response (recursing only in the positive directions). However, as Shanks *et al.* [39] have shown, given any 2-D first-quadrant transfer function $H(z_1, z_2)$, the transfer functions $H(z_1, z_2) \cdot H(1/z_1, 1/z_2)$ and $H(z_1, z_2) + H(1/z_1, 1/z_2)$ correspond to zero-phase filters with magnitude responses $|H(e^{-j\omega_1}, e^{-j\omega_2})|^2$ and $2 \operatorname{Re}[H(e^{-j\omega_1}, e^{-j\omega_2})]$, respectively. In both cases, the frequency response is symmetric about one of the frequency axes. Further, the transfer functions $H(z_1, z_2) \cdot H(1/z_1, z_2) \cdot H(z_1, 1/z_2) \cdot H(1/z_1, 1/z_2)$ and $H(z_1, z_2) + H(z_1, 1/z_2) + H(1/z_1, z_2) + H(1/z_1, 1/z_2)$ have zero-phase responses which are symmetrical about both frequency axes.

Hence, once we have a $H(z_1, z_2)$, we can "symmetrize" its frequency response and thus obtain 2-D digital filters exhibiting prescribed symmetry properties in the four quadrants of the (ω_1, ω_2)-plane.

The second strategy employs ST's which carry a 2-D transfer function $H(u_1, u_2)$ to another 2-D transfer function $H[G_1(v_1, v_2), G_2(v_1, v_2)]$, via the substitutions $G_1(v_1, v_2) = u_1$ and $G_2(v_1, v_2) = u_2$, where u_1, u_2, v_1, and v_2 are complex variables. The counterparts of type I-A and I-B designs, in this case, will be called type II-A (requires the use of double bilinear transformations) and II-B [direct (z_1, z_2)-domain approximation] design approaches for 2-D digital filters.

III. Type I-A Designs

The type I-A approach is one of the very first design methods proposed for 2-D digital filter design, and is a "natural" approach for the design of 2-D IIR filters. This design is essentially performed in the analog domain and then converted to the digital domain via bilinear transformation. All the usual advantages of design via bilinear transformation in the 1-D case [34], [37] are inherited by this approach for the 2-D cases as well.

In type I-A design, the basic problem is to design a stable linear lumped finite time-invariant 1-D transfer function $T(s)$ of an analog filter and to identify an ST $g(s_1, s_2)$ such that the magnitude response of $T[g(s_1, s_2)]$, under bilinear transformation in each variable, is some sort of a "best" approximation of a prescribed magnitude-response specification. Since a wealth of information already exists for 1-D analog filter design [41], our discussion will be concerned only with the identification of $g(s_1, s_2)$.

From the desired properties of the ST's mentioned in the beginning of Section II, it follows that $g(s_1, s_2)$ must be a two-variable real rational function. Further, since the frequency response of an analog transfer function $H(s_1, s_2)$ is

$$H(s_1, s_2)\Big|_{\substack{s_1 = j\omega_1 \\ s_2 = j\omega_2}}$$

the function g must map the imaginary axes of the (s_1, s_2)-plane in their entirety into the entire imaginary axis of the s-plane such that for every real $\omega_1, \omega_2 \geq 0$, there exists a real $\omega \geq 0$ satisfying [2], [21]

$$g(j\omega_1, j\omega_2) = j\omega \tag{3}$$

and vice-versa. Regarding the stability requirement, it can be shown that if $T(s)$ is stable, then $T[g(s_1, s_2)]$ is stable whenever $g(s_1, s_2)$ is stable [6].

Let $\Gamma(\omega_1, \omega_2)$ denote the function $(1/j)\, g(j\omega_1, j\omega_2)$. Let us also temporarily restrict our attention to the approximation of the prescribed magnitude specification by a purely analog 2-D filter. In that case, for fixed ω's, the equation $\Gamma(\omega_1, \omega_2) = \omega$ describes the constant-ω contours in the (ω_1, ω_2)-plane. In general it is possible to characterize stable ST's so that these will preserve the filter types[2] (i.e., carry a 1-LP to a 2-LP, a 1-HP to a 2-HP, a 1-BP to a 2-BP, etc.) or carry a 1-LP to a 2-HP, etc. We shall first study the general ST's and impose the stability and the realizability conditions afterward. Since we are operating in the analog domain, the frequency region of interest is the nonnegative orthant of the (ω_1, ω_2)-plane (R^2), which we shall denote as Ω. In special cases, however, the frequency region of interest, denoted by Ω_I, can merely be a "nice" (compact, and without holes) proper subset of Ω including the origin. Then the problem will be called a *local* problem, distinguishing it from the *global* case where $\Omega_I = \Omega$. Let $\Gamma(\Omega_I) \triangleq \{\Gamma(\omega_1, \omega_2) : \forall \langle \omega_1, \omega_2 \rangle \in \Omega_I\}$.

We can then characterize the ST's which preserve the filter type either locally or globally. The results are stated in the form of theorems.[3]

Theorem 1 (Locally Type Preserving)

Let $T(s)$ exhibit a particular filter type over $[0, \omega_0]$, $\omega_0 > 0$. Then $T[g(s_1 s_2)]$ is of the same filter type over Ω_I iff i) Γ (as defined before) is continuous in Ω_I, ii) $\Gamma(0, 0) = 0$ and $[0, \omega_0] \subseteq \Gamma(\Omega_I)$, and iii) Γ is a positive monotone increasing function in each variable in Ω_I.

Loosely speaking, condition i) guarantees that after transformation the resultant 1-D frequency "axis" does not have any gap; condition ii) insures that the origin of Ω is mapped into the origin of this "axis" and includes the domain of $|T(j\omega)|$; condition iii) guarantees that there is no "twist" in this axis. $g(\Omega_I)$ is really nothing but a closed interval of the form $[0, \omega]$, $\omega > 0$. It is easy to see that violation of any of these conditions results in the alteration of the filter type of the 1-D prototype. The global counterpart of this theorem is as follows.

Theorem 2 (Globally Type Preserving)

Given a $T(s)$ of a particular type, $T[g(s_1, s_2)]$ is of the same type iff: i) Γ is continuous on Ω; ii) $\Gamma(0, 0) = 0$; and iii) Γ is a positive monotone increasing function in each variable in Ω such that $\lim_{\omega_1 \to \infty} \Gamma(\omega_1, \omega_2) = \infty$ for any $\omega_2 < \infty$ and $\lim_{\omega_2 \to \infty} \Gamma(\omega_1, \omega_2) = \infty$, for any $\omega_1 < \infty$.

Evidently, any Γ which satisfies Theorem 2 satisfies Theorem 1 as well; however, the converse is not true, which is best illustrated by the following example. The function $g(s_1 s_2) = s_1 + s_2$ satisfies all conditions of Theorem 2 and hence all conditions of Theorem 1 for any $[0, \omega_1] \times [0, \omega_2] \subseteq \Omega$. On the other hand, $g(s_1, s_2) = (s_1 + s_2)/(s_1 s_2 + 1)$ with $\Gamma(\omega_1, \omega_2) = (\omega_1 + \omega_2)/(1 - \omega_1 \omega_2)$ satisfies condition iii) of Theorem 2 only inside the region $\{\langle \omega_1, \omega_2 \rangle : \omega_1 \omega_2 < 1; \omega_1, \omega_2 \geq 0\}$.

Any g, and equivalently any Γ, which satisfies Theorem 2 will be called *globally type preserving* (GTP). Those which satisfy Theorem 1 for a nonempty Ω_I but do not satisfy Theorem 2 will be called *locally type preserving* (LTP).

[2] For denoting the filter types, we shall use the following abbreviations: LP, HP, and BP will denote the low-pass, high-pass, and bandpass types, respectively; and an integer prefix, k, $k = 1, 2$, will denote the dimension.

[3] Proofs are omitted everywhere for the sake of brevity and to avoid mathematical technicalities. Some of the formal proofs are available in [6]; for others see [7].

We shall now present a global result for those ST's which carry a 1-LP and a 1-HP transfer function to a 2-HP and a 2-LP transfer function, respectively.

Theorem 3

Given a 1-LP (resp. 1-HP) $T(s)$, $T[g(s_1, s_2)]$ is 2-HP (resp. 2-LP) iff i) Γ is continuous on Ω, ii) $\Gamma(0, 0) = \infty$ and iii) Γ is a positive monotone decreasing function in each variable in Ω such that $\lim_{\omega_1 \to \infty} \Gamma(\omega_1, \omega_2) = 0$ for any $\omega_2 < \infty$ and $\lim_{\omega_2 \to \infty} \Gamma(\omega_1, \omega_2) = 0$ for any $\omega_1 < \infty$.

The local theorem for such ST's can be stated similarly.

As an example note that the function $g(s_1, s_2) = 1/(s_1 + s_2)$ satisfies the global theorem and, hence, the local theorem as well. (Caution! In this case $\Gamma(\omega_1, \omega_2) = -1/(\omega_1 + \omega_2)$. Such a transformation may not lead to stable first-quadrant filters.) However, the function $g(s_1, s_2) = (1 + s_1 s_2)/(s_1 + s_2)$ satisfies only the local theorem inside the region $\{\langle \omega_1, \omega_2 \rangle: \omega_1 \omega_2 < 1\} \subset \Omega$.

As far as the 1-BP to 2-BP transformations are concerned, the requirement $\Gamma(0, 0) = 0$ is unnecessarily severe in many cases. For example, consider the ST $g(s_1 s_2) = s_1 s_2/(s_1 + s_2)$. In that case $\Gamma(\omega_1, \omega_2) = \omega_1 \omega_2/(\omega_1 + \omega_2)$. This Γ is undefined at $\langle 0, 0 \rangle$; further, excluding $\langle 0, 0 \rangle$, it maps the entire ω_1-axis and the entire ω_2-axis to the origin. The local behavior of Γ includes both monotone increasing and monotone decreasing characteristics in each variable in appropriate regions. Thus, depending on a suitable 2-D region of transmission, g is a perfectly viable candidate for 1-BP to 2-BP transformation.

Further, for bandpass characteristics, only the local behavior of g is of interest, which is emphasized in the following theorem.

Theorem 4

Let $T(s)$ exhibit a bandpass characteristic over $[0, \omega_0]$ with a passband $[\omega_1, \omega_2] \subset [0, \omega_0]$. Then $T[g(s_1, s_2)]$ is 2-BP over Ω_I with a 2-D region of transmission Ω_T (which excludes both $\langle 0, 0 \rangle$ and $\langle \infty, \infty \rangle$) iff: i) Γ is continuous in Ω_I; ii) $[0, \omega_0] \subseteq \Gamma(\Omega_I)$ and $[\omega_1, \omega_2] = \Gamma(\Omega_T)$; and iii) either Γ is a positive monotone increasing function in both variables or Γ is a positive monotone decreasing function in both variables in Ω_I.

This local theorem is obviously false for type-preserving transformations in general and is also false for 1-LP to 2-HP and 1-HP to 2-LP type transformations. It must also be noted that, for nonideal 2-BP characteristics, Theorem 1 (i.e., the monotone increasing Γ) will always yield better approximations compared to those obtained under monotone decreasing Γ of Theorem 4, unless the magnitude response displays a special type of symmetry [6] in the transition band.

Clearly, the locally type-preserving maps of Theorem 1 are also locally bandpass type-preserving (LTP(BP)) maps of Theorem 4, but the converse is not true, i.e., the LTP(BP) maps are not necessarily LTP.

We shall now illustrate some very special types of LTP and LTP(BP) functions. Note that if g is a two-variable reactance function (2-RF) g satisfies (3). Denote the collection of such functions by the symbol $\mathfrak{F}_2$. Denote the collection of all single-variable reactance functions which do not have a pole at the origin by the symbol $\mathfrak{F}_1$. Let $\alpha_1, \alpha_2, > 0$.

Theorem 5

For any $F_1, F_2 \in \mathfrak{F}_1, g(s_1, s_2) \triangleq \alpha_1 F_1(s_1) + \alpha_2 F_2(s_2)$ is LTP whenever $g \in \mathfrak{F}_2$. Further, if g is LTP, not necessarily in $\mathfrak{F}_2$, then $g[F_1(s_1), F_2(s_2)]$ is also LTP. For any $g \in \mathfrak{F}_2$, if the intersection of $\Gamma(\Omega)$ and the set of positive reals is nonempty, g is LTP(BP) for a suitable choice of $[\omega_1, \omega_2]$ and Ω_T as in Theorem 4. For any $F_1, F_2 \in \mathfrak{F}_1$, $g(s_1, s_2) \triangleq \alpha_1 F_1(s_1) + \alpha_2 F_2(s_2)$ is LTP(BP). Further, if g is LTP(BP), not necessarily in $\mathfrak{F}_2$, so is $g[F_1(s_1), F_2(s_2)]$.

As an example, consider

$$g(s_1, s_2) = \frac{1}{s_1} + \frac{1}{s_2} + s_1 + s_2 + \frac{s_1}{4 + s_1^2} + \frac{s_2}{9 + s_2^2}.$$

In this case,

$$\Gamma(\omega_1, \omega_2) = \omega_1 + \omega_2 - \frac{1}{\omega_1} - \frac{1}{\omega_2} + \frac{\omega_1}{4 - \omega_1^2} + \frac{\omega_2}{9 - \omega_2^2}.$$

It is easy to verify that Γ is LTP(BP) on $[1, 1.5] \times [1, 1.5]$. Note, for any 2-RF g, with a realization as a driving-point function and a realizable transfer function $T(s)$, $T[g(s_1, s_2)]$ has an analog realization.

Going back to the general ST's, not necessarily 2-RF, we have the following.

Theorem 6

For *any* g satisfying (3), if ψ denotes a 1-LP to 1-HP or 1-LP to 1-BP transformation, the composite map ψg is a local (global) 1-LP to 2-HP or 1-LP to 2-BP transformation whenever g is LTP (GTP). (In fact, if ψ is a 1-LP to 1-BP transformation, g only needs to be LTP(BP).)

Note that the transformation $s \to ks$ (real and positive, corresponding to a frequency scaling $\omega \to k\omega$) is a GTP ST in one variable. Obviously, the LTP and LTP(BP) properties are invariant under frequency scaling.

From Theorem 6 it follows that any 2-HP design problem can be effectively reduced to a 2-LP design problem. Further, any 2-BP characteristic can be realized by cascading an appropriate number of suitable 2-LP and 2-HP filters (weak approximation). Hence, without much loss of generality, it is sufficient to restrict our attention to the design of 2-LP filters only.

We are now in a position to investigate the approximation problem involved with type I-A designs. The approximation problem formulated here can be used for type I-B and II-B designs with appropriate changes of domains. Since g must be a stable two-variable rational function, for the sake of mathematical expediency, at this stage we shall assume that g is 2-RF. [However, the approximation techniques presented here can be directly extended to any ST g which is not a 2-RF, but which satisfies (3).]

In practice, a 2-LP magnitude-response specification prescribes a 2-D region of transmission Ω_T, which is a proper subset of Ω and which includes $\langle 0, 0 \rangle$. Let the *band edge of transmission* $\mathcal{B}$ be defined as the part of the boundary of Ω_T which does not coincide either with the ω_1- or ω_2-axis. In practice $\mathcal{B}$ will be specified as a finite set of discrete points in Ω (if not, $\mathcal{B}$ can always be discretized).

As a first step of the approximation procedure, the elements of $\mathcal{B}$ must be prewarped. Let

$$\mathcal{B} = \{\langle \omega_{1_1}, \omega_{2_1} \rangle, \langle \omega_{1_2}, \omega_{2_2} \rangle, \cdots, \langle \omega_{1_n}, \omega_{2_n} \rangle\}.$$

From $\mathcal{B}$ a set of transformed frequency pairs $\mathcal{B}_D$ is computed by using the prewarping transformation as follows:

$$\omega_{i_k}^D = \tan\left(\frac{\omega_{i_k}}{2\nu_i}\right), \quad i = 1, 2, \; k = 1, 2, \cdots, n \tag{4}$$

where

ν_1, ν_2 sampling rates which are large enough to preclude any aliasing effect,

$\langle \omega_{1k}^D, \omega_{2k}^D \rangle$ elements of $\mathcal{B}_D$.

For the sake of concreteness, let us now assume that the 2-RF $g(s_1, s_2)$ can be represented by $\Gamma(\omega_1, \omega_2)$ as follows [using (3)]:

$$\Gamma(\omega_1, \omega_2) = \alpha\omega_1 + \beta\omega_2 + \sum_{i=1}^{n_1} \frac{2b_i\omega_1}{d_i^2 - \omega_1^2} + \sum_{k=1}^{n_2} \frac{2q_k\omega_2}{r_k^2 - \omega_2^2} \tag{5}$$

where α, β, the d_i's, and the r_k's are positive reals and the b_i's and q_k's are nonnegative reals. Without loss of generality we can assume that d_1 is the least element of the set $\{d_1, d_2, \cdots, d_{n_1}, r_1, r_2, \cdots, r_{n_2}\}$.

Let Ω_D denote the 2-D frequency region obtained from Ω_T after prewarping. *Select* d_1 to be such that $\Omega_D \subseteq [0, d_1] \times [0, d_1]$. It is easy to see that g is LTP on $[0, d_1] \times [0, d_1]$.

Let the $(n_1 + n_2 + 1)$-vector $\langle \alpha, \beta, d_2, d_3, \cdots, r_{n_2} \rangle$ be denoted by ξ and let the $(n_1 + n_2)$-vector $\langle b_1, b_2, \cdots, b_{n_1}, q_1, q_2, \cdots, q_{n_2} \rangle$ be denoted by ζ. Our objective is to determine ξ, ζ, and a $\omega \geqslant 0$ such that the function $\sigma: \omega_1 \to \omega_2$ determined by the equation $\Gamma(\xi, \zeta; \omega_1, \omega_2) = \omega$ for an LTP g on $[0, d_1] \times [0, d_1]$, of the form in (5), is the closest approximation to the function describing $\mathcal{B}_D$ in a suitable sense. The vectors ξ and ζ are included in g to show its dependence on these parameters.

Let m be the number of elements in $\mathcal{B}_D$ and $v \triangleq \langle v_1, v_2, \cdots, v_m \rangle \in R^m$, where, for fixed ξ, ζ, and ω, v_k is defined as

$$v_k = \sigma(\xi, \zeta, \omega; \omega_{1k}^D) - \omega_{2k}^D.$$

ω_{1k}^D (ω_{2k}^D) is the first (second) element of the kth ordered pair in $\mathcal{B}_D$, and $\sigma(\xi, \zeta, \omega; \omega_{1k}^D)$ denotes the positive real solution ω_2 of the equation $\Gamma(\xi, \zeta; \omega_{1k}^D, \omega_2) = \omega$, which is closest to ω_{2k}^D. Let $W \triangleq \{\omega_{1k}^D : k = 1, 2, \cdots, m\}$.

One of the possible formulations of our approximation problem may then be presented as the following constrained nonlinear minimization problem:

$$\text{minimize } \|v(\xi, \zeta, \omega)\|, \quad \omega_1 \in W \tag{6}$$

with respect to ξ, ζ, and ω, subject to the conditions that elements of ξ are reals greater than zero; elements of ξ are reals greater than or equal to zero; ω is a real greater than zero; and

$$d_1 < r_1, d_1 < d_2 < \cdots < d_{n_1}, r_1 < r_2 < \cdots < r_{n_2}$$

where $\|\cdot\|$ denotes a suitable vector norm on R^m.

Noting that W is a finite set, it can be shown that a solution to the above minimization problem indeed exists. Therefore, one of the standard constrained nonlinear minimization algorithms may be employed to obtain the desired function $\Gamma(\omega_1, \omega_2)$.

The solution to (13) automatically yields an optimal value for ω, say ω_c, which is the cutoff frequency of our 1-LP prototype filter. We can then obtain a stable 1-LP transfer function $\hat{F}_D(s)$, with cutoff frequency ω_c, which approximates the 1-D prototype magnitude response to an arbitrary level of accuracy. Then the overall approximation of the 2-LP specification is the stable transfer function $\hat{F}_D[g(s_1, s_2)_{\text{opt}}]$ which in turn yields the desired 2-D digital transfer function, via bilinear transformation in each variable. *Obviously, the error in the overall approximation is of the same order as the error in obtaining* $g(s_1, s_2)_{\text{opt}}$ [7].

It can also be shown that instead of considering the minimization problem of (6), the approximation problem may be formulated as a *suboptimal* problem, leading to a *linear* programming problem. However, the solution of such suboptimal linear programming formulation usually requires enormous amounts of computations and is not treated here.

At this stage we should also mention the limitations of the type I-A design using rational g's. Note that the rational functions are very bad approximants for closed curves in Ω. Consequently, for closed constant-ω contours in Ω (i.e., 2-BP specifications) we must use more than one rational function approximant, yielding constant-ω contours for 2-LP and 2-HP filters, which, in turn, when in cascade, will yield an approximation to the desired 2-BP characteristics. Depending on the nature of the closed contour, we may end up with a fairly pathological approximation problem.

Secondly, as mentioned before, guaranteeing the stability of a 2-D ST is a difficult problem. However, since 2-RF's are known to be stable, in view of Theorem 5, we can restrict our search for ST's entirely to the class $\mathcal{F}_2$. This is the approach suggested in [1], [6]. This implies that in most cases we are going to end up with even worse approximants than those we could have gotten otherwise from nonreactance function g's. However, as the problem of ensuring the stability of a 2-D rational function is quite time consuming, it is much more preferable to restrict the search merely to $\mathcal{F}_2$ or over sets of two-variable rational functions satisfying (3) which are not reactance functions, but are known to be stable. It must be born in mind however that all practical filters and all practical filter specifications will have transition bands bounded by, say, $\Gamma_1(\omega_1, \omega_2) = \omega_a$ and $\Gamma_2(\omega_1, \omega_2) = \omega_b$. The solution of the filter design problem will then require the determination of a 2-RF (or at least stable) LTP $\Gamma(\omega_1, \omega_2)$ such that the functions determined by $\Gamma(\omega_1, \omega_2) = \omega_a$ and $\Gamma(\omega_1, \omega_2) = \omega_b$ are the "best" approximations to $\Gamma_1(\omega_1, \omega_2) = \omega_a$ and $\Gamma_2(\omega_1, \omega_2) = \omega_b$, respectively. A solution to such a simultaneous approximation problem yielding LTP g's may not exist. In fact, even if $\Gamma_1 = \Gamma_2$, a solution to the simultaneous approximation problem will not necessarily exist.

In case of such "tight" specifications, the type I-A approach may fail completely. However, for most of the practical cases, the frequency domain specifications are likely to be amenable to type I-A design.

Fig. 1 illustrates the magnitude response of a 2-BP filter designed by this method from a 1-LP standard fourth-order Butterworth prototype using the ST $g(s_1, s_2) = (1 + s_1s_2)/(s_1 + s_2)$. The constant-$\omega$ contours are hyperbolic. The transfer function of this 2-D digital filter is given by

$$H(z_1, z_2) = \frac{1 - 4z_1z_2 + 6z_1^2z_2^2 - 4z_1^3z_2^3 + z_1^4z_2^4}{10.64 + 5.172z_1^2z_2^2 + 0.188z_1^4z_2^4}.$$

It must be pointed out that the design of 2-BP filters using 2-LP and 2-HP filters (or other such filters) in cascade is not as simple as it looks. Even with known 2-LP and 2-HP characteristics it is difficult to predict *a priori* the shape of the resulting transmission band edge and, more important, the effect on the transition regions, except in a very qualitative fashion. A lot of work still needs to be done to make this design approach truly viable.

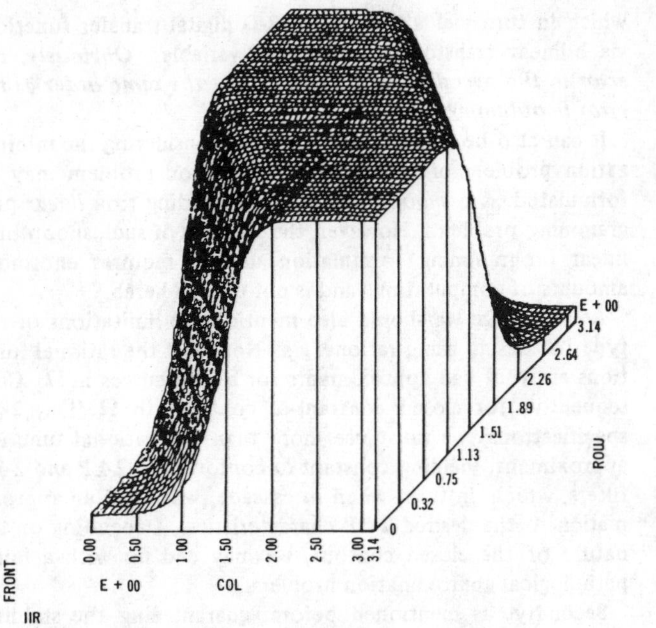

Fig. 1. The magnitude response of a 2-BP filter derived from the fourth-order 1-LP Butterworth prototype.

Special cases of type I-A design for recursive 2-D filters have been treated in the literature [1], [12], [13], [18], [39]. We shall now present a brief survey of these works in the framework of the theory of type I-A design.

The pioneering work on type I-A design was done by Farmer and Gooden [18] and Shanks *et al.* [39] in designing what has, since then, been called a rotated filter. They considered a *continuous 1-LP or 1-BP filter* $\hat{F}(s)$ as a 2-D filter $F(s, u)$ that varies in one dimension only; viz., s; u is a second complex variable such that $F(s, u) = \hat{F}(s)$. The ST's are defined as the rotation of the axes of the (s, u)-plane by an angle β, $0 < \beta < 2\pi$.

In the terminology of this paper, the approach of [39] consisted of using one of the following two transformations:

$$g_1(s_1, s_2) = s_1 \cos\beta + s_2 \sin\beta$$

$$g_2(s_1, s_2) = s_2 \cos\beta - s_1 \sin\beta.$$

Note that $\Gamma_i(0, 0) = 0$ and Γ_i is continuous in Ω in its entirety, $i = 1, 2$. Thus, by Theorem 2, the function Γ_1 (Γ_2) is a GTP spectral transformation whenever $0 < \beta < \pi/2$ ($3\pi/2 < \beta < 2\pi$). (In such cases, g_1 and g_2 are 2-RF.) Similarly, by Theorem 3, the function Γ_1 (Γ_2) is a 1-LP to 2-HP, or equivalently, a 1-HP to 2-LP transformation whenever $\pi < \beta < 3\pi/2$ ($\pi/2 < \beta < \pi$). For any β, $0 < \beta < 2\pi$, $\beta \neq \pi/2$, $\beta \neq \pi$, $\beta \neq 3\pi/2$, the transformations g_i, $i = 1, 2$, are LTP(BP). For $\beta = k\pi/2$, $k = 0, 1, 2, 3, 4$; of course, we only have single-variable frequency scalings.

Caution! For nondegenerate Γ_1 (Γ_2), whenever $\beta \in [0, \pi/2]$ ($[3\pi/2, 2\pi]$) the first-quadrant digital filters obtained via bilinear transformations on $F[g_1(s_1, s_2)]$ ($F[g_2(s_1, s_2)]$) may be unstable. Costa and Venetsanopoulos [13], in their work on circularly symmetric filters obtained by using rotational transformations, presented the stability regions for rotated filters for various values of β. In [13], only g_2 is studied; however, analogous results can be derived for g_1 in a straightforward manner.

Costa and Venetsanopoulos showed that for g_2, if $\hat{F}(s)$ is stable, the 2-D digital filter obtained by using bilinear transformation on $\hat{F}[g_2(s_1, s_2)]$ is stable whenever $(3\pi/2) < \beta < 2\pi$. They also considered the possibility of obtaining stable filters for other β's by changing the direction of recursion [13, Table I]. However, they recognize that all such possibilities result in a filter with the same magnitude response, i.e. the effective angle of rotation always satisfies $(3\pi/2) < \beta < 2\pi (0 < \beta < \pi/2$ for g_1).

All the examples presented in [12] and [39] vindicate the theory presented here. (For another recent work by Costa and Venetsanopoulos dealing with the rotations of 2-D filters, see [14].)

While Farmer and Gooden [18] and Shanks *et al.* [39] initiated the theory of recursive 2-D rotated digital filters, Costa and Venetsanopoulos [13] used it as a viable tool in the design of circularly symmetric 2-D recursive filters. This was achieved by cascading a number of rotated filters which approximate the circular constant-ω contours by a polygonal constant-ω contour. However, they did not mention any limitations of such cascaded designs. Some examples of circularly symmetric filters are given in [13].

Finally, Ahmadi *et al.* [1], in a very brief paper, employed a straightforward 2-RF ST for type I-A design of zero-phase 2-D digital filters. Only one example has been considered for the 2-RF:

$$g(s_1, s_2) = \frac{s_1 + s_2}{1 + b s_1 s_2}.$$

Then $\Gamma(0, 0) = 0$ and Γ is continuous in Ω except in the set $\{\langle\omega_1, \omega_2\rangle: b\omega_1\omega_2 = 1\}$. For any subset $\overline{\Omega}$ of Ω such that $\overline{\Omega} \cap \{\langle\omega_1, \omega_2\rangle: b\omega_1\omega_2 = 1\} \neq \emptyset$ it follows from Theorem 2 that g is LTP, which is corroborated by the 1-LP to 2-LP transformation illustrated in [1]. The corresponding digital filters are, obviously, stable.

IV. Type I-B Design: 2-D FIR Digital Filters

The type I-A approach, while quite suitable for the design of 2-D IIR filters, is not so useful for the design of 2-D FIR digital filters. On the other hand, the type I-B approach is applicable to the design of a restricted class of both types of 2-D filters. This design is performed entirely in the digital domain where an ST $g(z_1, z_2)$ carries a stable 1-D transfer function $T(z)$ into a 2-D transfer function $T[g(z_1, z_2)]$.

Let D_1 and D_2 denote the regions $D_1 = \{z: z \in C, |z| \leq 1\}$ and $D_2 = \{\langle z_1, z_2\rangle: \langle z_1, z_2\rangle \in C \times C, |z_1| \leq 1, |z_2| \leq 1\}$. Let ∂D_1 and ∂D_2 denote the boundaries of D_1 and D_2, respectively. The frequency response of a 1-D (2-D) digital filter is obtained by evaluating the transfer function over ∂D_1 (∂D_2) [35]. Since g must preserve the magnitude-response characteristic, we must have $g(\partial D_2) = \partial D_1$; further, for stable first-quadrant design, g must be such that for every $\langle\omega_1, \omega_2\rangle \in \Omega$ there exists an $\omega \geq 0$ such that

$$g(e^{-j\omega_1}, e^{-j\omega_2}) = e^{-j\omega}. \tag{7}$$

The corresponding $\Gamma(\omega_1, \omega_2)$ is now defined by

$$\Gamma(\omega_1, \omega_2) \triangleq j \ln g(e^{-j\omega_1}, e^{-j\omega_2}) \tag{8}$$

where we are only considering the principal values of the complex function $\ln(\cdot)$.

From (7) it follows that

$$|g(e^{-j\omega_1}, e^{-j\omega_2})| = 1. \tag{9}$$

It is possible to find rational g's (including polynomial g's) which satisfy this requirement. One good example of such g's is any 2-D all-pass function in z_1 and z_2. The general form of

2-D all-pass functions is well known [35]. It is not difficult to see that whenever $T(z)$ is stable $T[g(z_1, z_2)]$ is also stable iff $g(z_1, z_2)$ is stable [7].

It can be shown [7] that a straightforward application of this approach to the design of 2-D IIR filters usually results in extemely ill-behaved nonlinear approximation problems. It can also be shown [7] that the *only* class of admissible $g(z_1, z_2)$ for IIR design must be of the form

$$z_1^r \cdot z_2^t = z \tag{10}$$

where r and t are positive real rational numbers. Such g's are always stable and trivially realizable.

The type I-B design technique is considered by Chang and Aggarwal [8], who used an ST $g(z_1, z_2)$ of the form

$$g(z_1, z_2) = z_1 z_2^{m/n} \tag{11}$$

where m and n are integers. For causal transformation, of course, $(m/n) \geqslant 0$ must hold. Comparing (11) with (10), it follows that $g(z_1, z_2)$, as given above, can be obtained from (10) with $r = 1$ and $t = m/n$. Chang and Aggarwal studied only the transfer functions of separable filters which are realized by input/output signal array interpolations. However, their method is also applicable for the case of nonseparable 2-D transfer functions and can be directly extended to include g's of the form in (10) as well. Since the effect of such a transformation is again a rotation (as in the case of Shanks *et al.*) of the 2-D magnitude response in the frequency domain, details of the above mentioned generalizations are excluded here.

The situation, of course, is better for designing 2-D FIR filters. However, since the transfer functions of 1-D and 2-D FIR filters are polynomials, $g(z_1, z_2)$ *must* also be a polynomial in z_1 and z_2.

It is well known that for any transfer functions $T(z)$ and $H(z_1, z_2)$, the magnitude functions are periodic and doubly periodic in R and in R^2, respectively. In the 1-D case, the standard interval of frequency response is $[0, \pi]$, and, in the 2-D case, the standard region is $\Delta \triangleq [0, \pi] \times [0, \pi] \subset \Omega$. Using these notations, we can now state the conditions for which $g(z_1, z_2)$ will be GTP, LTP, LTP(BP), etc. However, we shall only mention the global results (the type I-A counterparts of these theorems are given inside the square brackets).

Theorem 7 [Theorem 2]

Given a $T(z)$ of a particular filter type, $T[g(z_1, z_2)]$ is of the same type iff i) Γ (as defined in (8)) is continuous on Δ, ii) $\Gamma(0, 0) = 0$, $\Gamma(\pi, \pi) = \pi$, and iii) Γ is a positive monotone increasing function in each variable in Δ.

Theorem 8 [Theorem 3]

Given a 1-LP (1-HP) $T(z)$, $T[g(z_1, z_2)]$ is 2-HP (2-LP) iff i) Γ is continuous in Δ, ii) $\Gamma(0, 0) = \pi$, $\Gamma(\pi, \pi) = 0$, and iii) Γ is a positive monotone decreasing function in each variable in Δ.

Similar results can be easily derived for other cases as well. For the class of so-called *linear-phase* 2-D FIR filters, the approximation of constant-ω contours becomes simple enough to be claimed to be a "natural" procedure. In designing such filters, the 1-D prototypes are approximated by means of *1-D linear-phase FIR transfer functions.*

A 1-D FIR filter with an impulse response of odd length, say $2n + 1$, and characterized by the transfer function

$$H(z) = \sum_{i=0}^{2n} h(k) z^k \tag{12}$$

has a zero-phase frequency response if the following symmetry condition is satisfied [34], [37]:

$$h(k) = h(2n - k), \qquad k = 0, 1, \cdots, n.$$

In that case, using standard trigonometrical relationships, it follows that [26]

$$|H(e^{-j\omega})| = \left| \sum_{i=0}^{n} a(k) (\cos \omega)^k \right| \tag{13}$$

for suitable real $a(k)$'s.

Similarly, for a 2-D FIR transfer function,

$$G(z_1, z_2) = \sum_{m=0}^{N_1-1} \sum_{p=0}^{N_2-1} h(m, p) z_1^m z_2^p$$

with $N_1 = 2n_1 + 1$, $N_2 = 2n_2 + 1$. The frequency response is zero phase whenever the elements of the impulse-response matrix satisfy the following symmetry conditions:

$$h(2n_1 - i, p) = h(i, p), \qquad i = 0, 1, \cdots, n_1$$
$$h(m, 2n_2 - i) = h(m, i), \qquad i = 0, 1, \cdots, n_2.$$

In that case it can be shown that [26]

$$|G(e^{-j\omega_1}, e^{-j\omega_2})| = \left| \sum_{m=0}^{n_1} \sum_{p=0}^{n_2} b(m, p) (\cos \omega_1)^m (\cos \omega_2)^p \right|$$

for a suitable choice of the real coefficients $b(m, p)$.

It is then clear that the problem of type I-B design of a zero phase 2-D FIR filter from a given magnitude-response specification requires the design of a 1-D prototype approximated by a function of the form given in (12). Such an approach implies that we are using the correspondence

$$\hat{\Gamma}(\omega_1, \omega_2) = \cos \omega \tag{14a}$$

where

$$\hat{\Gamma}(\omega_1, \omega_2) \triangleq \sum_{m=0}^{n_1} \sum_{p=0}^{n_2} b(m, p) (\cos \omega_1)^m (\cos \omega_2)^p. \tag{14b}$$

The transformation represented by (14) is called a ***generalized McClellan transformation*** [28], [29], [32]. The function $\Gamma(\omega_1, \omega_2)$ determined by (14) is obtained as arccos $[\hat{\Gamma}(\omega_1, \omega_2)]$, which is rather unwieldy for practical use. However letting $\xi_1 \triangleq \cos \omega_1$, $\xi_2 \triangleq \cos \omega_2$, and $\xi \triangleq \cos \omega$, we have

$$\hat{\Gamma}(\xi_1, \xi_2) \triangleq \sum_{m=0}^{n_1} \sum_{p=0}^{n_2} b(m, p) \xi_1^m \xi_2^p$$

and

$$\hat{\Gamma}(\xi_1, \xi_2) = \xi.$$

Using (12) and exploiting the recurrence formula for Chebychev polynomials [26], [27], we have

$$|H[\hat{\Gamma}(\xi_1, \xi_2)]| = \sum_{k=0}^{nn_1} \sum_{m=0}^{nn_2} d_{km} \xi_1^k \xi_2^m \tag{15}$$

where d_{km} are real numbers determined by the a's and b's of (13) and (14), respectively. Note that the coefficients $b(m, p)$ in (14) must be such that $-1 \leqslant \hat{\Gamma}(\xi_1, \xi_2) \leqslant 1$ for all $\langle \xi_1, \xi_2 \rangle \in [-1, 1] \times [-1, 1]$.

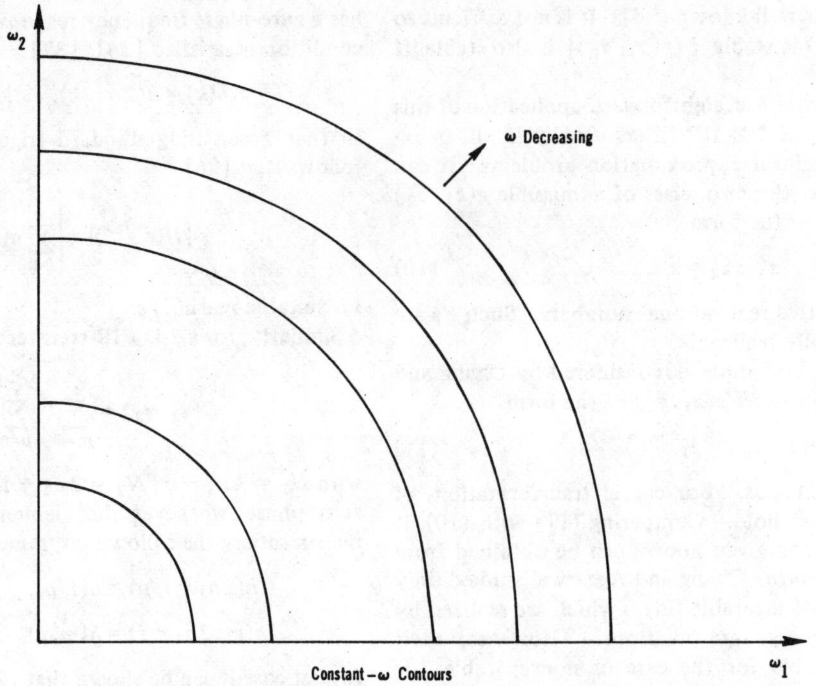

Fig. 2. The contour plot of an FIR 2-HP filter derived from a 1-LP FIR prototype of length 13 via the generalized McClellan transformation of (17).

The actual design procedure may be briefly stated as follows. Given a magnitude response function $F(\omega_1, \omega_2)$ in Ω, convert it to a function $\hat{F}(\xi_1, \xi_2)$ in $[-1, 1] \times [-1, 1]$ (which also involves prewarping). Then use type I-B design theory to obtain the desired approximation $H[\hat{\Gamma}(\xi_1, \xi_2)]$ to $\hat{F}(\xi_1, \xi_2)$. Using the technique of [27] or [32], the corresponding desired zero phase 2-D FIR digital transfer function can then be recovered from $H[\hat{\Gamma}(\xi_1, \xi_2)]$.

McClellan and Parks [26] and McClellan [25], in their pioneering studies, employed ST's of the form

$$\hat{\Gamma}(\xi_1, \xi_2) = A\xi_1 + B\xi_2 + C\xi_1\xi_2 + D \tag{16}$$

where $A, B, C, D \in R$.

Utilizing appropriate modifications of Theorems 7 and 8, it is then easy to verify that $\hat{\Gamma}(\xi_1, \xi_2)$ of (16) is as follows.

1) GTP iff

$$A + B + C + D = 1$$
$$A + B - C - D = 1$$
$$A - C\xi_2 \geqslant 0, \quad \forall \xi_2 \in [-1, 1]; \quad B - C\xi_1 \geqslant 0, \quad \forall \xi_1 \in [-1, 1]$$

or, equivalently,

$$A + B = 1 \qquad C + D = 0$$
$$A - C\xi_2 \geqslant 0, \quad \forall \xi_2 \in [-1, 1]; \quad B - C\xi_1 \geqslant 0, \quad \forall \xi_1 \in [-1, 1].$$

2) 1-LP (1-HP) to 2-HP (2-LP) globally iff

$$A + B + C + D = -1$$
$$A + B - C - D = -1$$
$$A - C\xi_2 < 0, \quad \forall \xi_2 \in [-1, 1]; \quad B - C\xi_1 < 0, \quad \forall \xi_1 \in [-1, 1]$$

or, equivalently,

$$A + B = -1 \qquad C + D = 0$$
$$A - C\xi_2 < 0, \quad \forall \xi_2 \in [-1, 1]; \quad B - C\xi_1 < 0, \quad \forall \xi_1 \in [-1, 1].$$

3) 1-LP (1-HP) to 2-BP, whenever, $A + B + C + D \neq 1$ $(A + B + C + D \neq -1)$.

In [26], McClellan considered the following cases: i) $A = B = C = -D = 0.5$, which carries a 1-LP filter to a "circularly symmetric" 2-LP filter, and, ii) $A = -B = 0.5$, $C = D = 0$, which carries a 1-LP filter to a "fan" filter which is a 2-BP filter.

According to the theoretical prediction, the choice $A = B = -0.5$, $C = -D = -0.4$, should yield a 1-LP (1-HP) to 2-HP (2-LP) transformation. This is illustrated in the following example. Consider the 1-LP FIR filter of length 13 designed by using the algorithm presented in [37, pp. 187–204], where the band edges are 0.0, 0.1, 0.15, and 0.5 (multiples of 2π) and the weighting factors are 1.7 and 1.0. In this case the ST corresponds to

$$\xi = -0.5\xi_1 - 0.5\xi_2 - 0.4\xi_1 \cdot \xi_2 + 0.4$$

or, rather,

$$\cos \omega = -0.5 \cos \omega_1 - 0.5 \cos \omega_2 - 0.4 \cos \omega_1 \cos \omega_2 + 0.4. \tag{17}$$

The contour plot corresponding to this transformation is shown in Fig. 2.

The general type I-B design using transformations of the form given in (17) is studied in great detail in [28], [29], and [32], which also includes a special case of zero phase filters with impulse responses of even length. In [32], the contour approximation problem is studied in such depth that with minor modifications the same techniques can be employed for *any* contour approximation problem with any well-behaved $\hat{\Gamma}(\omega_1, \omega_2)$, which does not have to correspond to (14). The algorithm presented in [32] can also include the ones which, at least theoretically, can design more than one contour of any shape simultaneously. The implementation of resultant digital filters has been studied very elegantly in [28], which also investigates the effects of coefficient quantization and arithmetic roundoff, and presents a detailed comparison between those implementations and other implementations of 2-D FIR filters.

However, the simultaneous contour approximation problem, as presented in [32], is incomplete. In order to formulate the contour approximation problems for two contours, the authors used parametrized representations of these two contours, which is applicable for problems involving more than two contours as well. While any smooth curve can be parametrized, determination of actual parametrization is difficult. Further, if the contour data are not supplied in the form of a smooth function, then they must be approximated by a smooth function before any parametrization attempt can be made, complicating the approximation problem enormously. Finally, given a finite set of smooth functions $f_1, f_2, \cdots, f_n$ (even in a single variable), simultaneous approximation of these functions by a smooth function f_*, say, involves simultaneous minimization of n functions of the form $\|f_* - f_i\|$, $i = 1, 2, \cdots, n$, which does not always admit a solution for any arbitrary choice of f_i's and a class of functions containing f_*. Any viable method for simultaneous contour approximation must resolve these problems satisfactorily.

There are, of course, other limitations inherent to the type I-B design approach as well, particularly when we are using polynomial g functions. Such functions are bad approximants for closed curves, and hence such an approach is not very useful in designing *multiband* bandpass filters where the band edges of the passbands (or stopbands) are closed curves, unless we use GTP or at least LTP transformations. Further, if the band edges fail to be unicursal curves, or if the band edges are closed curves which touch the ω_1- and ω_2-axis only at the origin, we can never hope for a nice approximation using *either* type I-A or I-B design. These design approaches also fail utterly in several other circumstances, but those are too pathological to arise in practice.

V. Type II Designs

In this section we shall start directly with the type II-B design. Since the type II-A design theory has never been used so far and has only limited scope for application, and, further, since the approximation technique for type II-A design is essentially the same as that for type II-B cases, this approach will not be discussed here.[4]

On the other hand, the type II-B design is perhaps the most convenient of the 2-D digital filter design methods we are going to discuss in this paper. The type II-B design technique has already been considered in depth in [35]. The present study will be concerned with the exposition of the type II-B theory in the context of the material presented so far. Since type II-B design yields a 2-D digital transfer function, the resultant filters are always realizable.

The basic idea is to convert a 2-D *digital* filter of *known* type to another 2-D digital filter meeting desired specifications. In other words, given a known stable transfer function $T(z_1, z_2)$, we would like to find a transformation $G: C^2 \to C^2$, with $G = \langle G_1, G_2 \rangle$, $v_1 = G_1(z_1, z_2)$, and $v_2 = G_2(z_1, z_2)$, $\langle v_1, v_2 \rangle \in C^2$, such that 1) if the digital transfer function $T(v_1, v_2)$ is stable and rational, $T[G_1(z_1, z_2), G_2(z_1, z_2)]$ is also stable and rational; 2) $G: \partial D_2 \to \partial D_2$ (also refer to Section IV and [35]). We also require that for every $\sigma_1, \sigma_2 \geq 0$ there must exist $\omega_1, \omega_2 \geq 0$ such that

$$G_1(e^{-j\sigma_1}, e^{-j\sigma_2}) = e^{-j\omega_1} \tag{18a}$$

$$G_2(e^{-j\sigma_1}, e^{-j\sigma_2}) = e^{-j\omega_2}. \tag{18b}$$

[4] This approach is studied in detail in [7].

Throughout this section, an ST G will be assumed to satisfy the conditions given above. It is easy to see that

$$|G_i(e^{-j\sigma_1}, e^{-j\sigma_2})| = 1, \qquad i = 1, 2. \tag{19}$$

As in Section IV, here also it may be pointed out that if both components of G are digital all-pass functions in two variables, then (19) is satisfied. Further, in this case, unlike Section IV, the all-pass functions (of special forms) are extremely useful as ST's. However, G_1 and G_2 can also be of the form in (10), or these may correspond to various McClellan transformations, etc.

The stability result in this case may be stated as follows. Let $T_G(z_1, z_2)$ denote $T[G_1(z_1, z_2), G_2(z_1, z_2)]$. Then whenever $T(v_1, v_2)$ is stable, $T_G(z_1, z_2)$ is stable iff both G_1 and G_2 are stable [35].

The following results are the appropriate counterparts of various theorems presented before (mentioned inside square brackets).

Theorem 9 [Theorem 7, Theorem 2]

Given a transfer function $T(v_1, v_2)$ of a particular filter type, the transfer function $T_G(z_1, z_2)$ is of the same type iff both G_1 and G_2 are GTP in the sense of Section IV.

Theorem 10 [Theorem 8, Theorem 3]

G is a local or global (vector) 2-LP (2-HP) to 2-HP (2-LP) ST iff both components G_1 and G_2 are local or global 1-LP (1-HP) to 2-HP (2-LP) ST's in the sense of Section IV.

As mentioned before, the component function G_i can be one of the following types:

i) generalized McClellan transformations ("natural" for zero phase designs);
ii) $G_i(z_1, z_2) = z_1^r z_2^t$, r and t are nonnegative rationals; (20)
iii) $G_i(z_1, z_2)$ is a 2-D digital all-pass function whose general form is given by [35]

$$G_i(z_1, z_2) = (\pm 1) \prod_{l=1}^{L_i} \frac{z_1^{N_l^{(i)}} z_2^{M_l^{(i)}} \left(\sum_{n=0}^{N_l^{(i)}} \sum_{m=0}^{M_l^{(i)}} a_{mn}^{(i)} z_1^{-n} z_2^{-m} \right)}{\left(\sum_{n=0}^{N_l^{(i)}} \sum_{m=0}^{M_l^{(i)}} a_{mn}^{(i)} z_1^{n} z_n^{m} \right)},$$

$$a_{00}^{(i)} = 1. \tag{21}$$

The effects of such transformations, particularly those characterized by (20) and (21), on the doubly periodic spatial frequency response of a 2-D digital filter is discussed in [35]. If the phase of G_i corresponds to the angle θ_i, then G maps $\langle 0, 0 \rangle$ into $\langle \theta_1, \theta_2 \rangle$. Further, if N_i (M_i) denotes the degree of z_1 (z_2) in the numerator of G_i, then the contour $(e^{-j\sigma_1}, 1)$ corresponding to the region $-\pi \leq \sigma_1, \sigma_2 \leq \pi$ is mapped N_i times around the unit circle in the $G_i(z_1, 1)$ plane. Similarly, the contour $(1, e^{-j\sigma_2})$ is mapped M_i times around the unit circle in the $G_i(1, z_2)$ plane. Thus the values of $N_i + M_i$, $i = 1, 2$, determine the number of transmission and attenuation regions included in Δ (defined in Section IV) after transforming, say, a 2-LP filter. The effects of interesting special cases of all-pass ST's on a given 2-LP characteristic are tabulated in [35]. Similar tables can be constructed for other cases as well.

The use of such tables simplifies the approximation problems considerably. Given any arbitrary 2-D frequency specification,

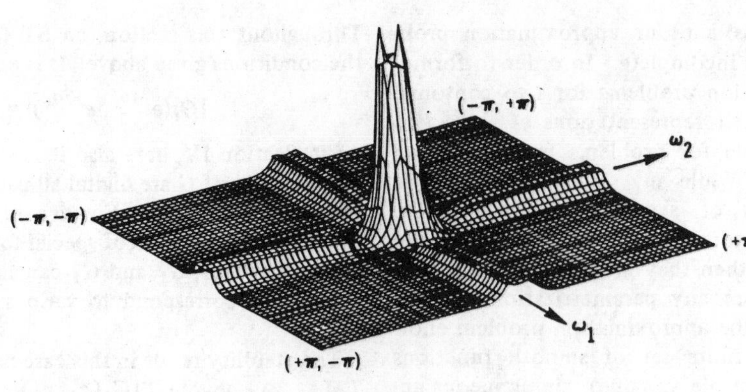

(a)

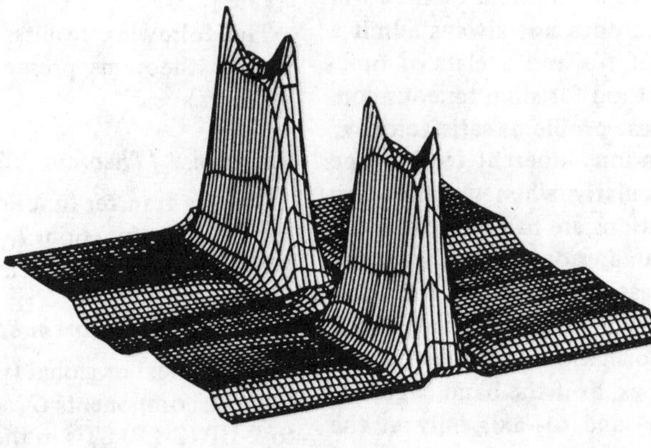

(b)

Fig. 3. (a) The magnitude of response of the 2-LP transfer functions of Maria and Fahmy. (b) The magnitude response of the 2-BP filter obtained by transforming the filter of (a).

we can select a $G = \langle G_1(z_1), G_2(z_2)\rangle$ such that a known 2-LP specification will be transformed into a filter with the same number of transmission and attenuation regions as the given characteristics. The resultant approximation using such G's, in most cases, will not necessarily yield the best possible approximation. However, this is offset by the simplicity of stability checking for single-variable functions. If we now define the error functions $e_i(\sigma_i) = \| \omega_i - j \ln G_i(e^{-j\sigma_i}) \| = 1, 2$ for a given norm $\|\cdot\|$, where σ_i's are taken from a finite set of points, then a possible formulation of the approximation problem may be presented as follows: minimize max $\{e_1(\sigma_1), e_2(\sigma_2)\}$ subject to $|G_i(e^{-j\sigma_i})| = 1$; G_i is stable, $i = 1, 2$.

In many cases, depending on the specification of the problem, the approximation can be simplified even more. An example taken from [35] is presented below.

Consider the 2-LP transfer function obtained by Maria and Fahmy [24]. The corresponding magnitude response is shown in Fig. 3(a). It has a value of 0.6 at $\pm 0.08\pi$ on each axis with peaks of approximate magnitude 1.27. Suppose that a bandpass filter is desired with two transmission regions on the ω_1-axis, each centered at $\pm 0.6\pi$ with a bandwidth of 0.6π between points on the ω_1-axis with magnitude 0.6. The width of the transmission region in the ω_2-direction between points of amplitude 0.6 should be 0.08π. A type VIII transformation as given in [35, Table I] will meet these requirements and is given by

$$u_1 = -\frac{z_1^2 + 0.052z_1 - 0.832}{-0.832z_1^2 + 0.052z_1 + 1}$$

$$u_2 = u_2.$$

This transformation produces the new transfer function whose magnitude response is shown in Fig. 3(b).

A simple application of the type II-B design theory can be made in obtaining the 2-D generalization of the 1-D variable-cutoff linear-phase digital filters of Oppenheim *et al.* [33]. Consider a 2-D zero phase filter $H(u_1, u_2)$ with a magnitude response of the following form:

$$|H(e^{-j\omega_1}, e^{-j\omega_2})| = \sum_{m=0}^{n_1} \sum_{n=0}^{n_2} b(m, p) (\cos \omega_1)^m (\cos \omega_2)^n$$

(using the earlier notations as much as possible). As in [33], we now introduce the following transformations:

$$\cos \omega_1 = A_1 + (1 - A_1) \cos \sigma_1 \tag{22a}$$

$$\cos \omega_2 = A_2 + (1 - A_2) \cos \sigma_2 \tag{22b}$$

with either $0 \leqslant A_1 < 1$, $0 \leqslant A_2 < 1$ or $-1 < A_1 \leqslant 0$, $-1 < A_2 \leqslant 0$.

From the earlier discussion in Section IV dealing with the analogous problems of McClellan transformations, it follows that the transformation, whose components are given by (22), is GTP in the sense of this section.

Let $\Gamma_c = \{\langle \sigma_{1c}, \sigma_{2c}\rangle\}$ denote the set of points lying on the cutoff curve of a 2-LP filter. Then the set of points lying on the cutoff curve of the transformed filter will be $\widetilde{\Gamma}_c = \{\langle \omega_{1c}, \omega_{2c}\rangle\}$ where

$$\omega_{i_c} = \cos^{-1}\left(\frac{\cos \sigma_{i_c} - A_i}{1 - A_i}\right), \quad i = 1, 2.$$

For $0 \leqslant A_i < 1$ $(-1 < A_i \leqslant 0)$, $i = 1, 2$, note that $\omega_{i_c} \geqslant \sigma_{i_c}$ $(\omega_{i_c} \leqslant \sigma_{i_c})$, $i = 1, 2$. Hence we can implement a 2-D zero phase digital filter for which the cutoff characteristic can be easily varied by changing only two parameters, A_1 and A_2.

Similar generalizations can be made for bandpass filters as well. A very flexible design technique for 2-D digital filters with tunable characteristics employing the type IV technique is presented in [35] which is applicable for more general classes of such filters.

VI. Conclusion

This paper is concerned with the study of spectral transformations in the design of 2-D digital filters. It has been shown that there are two basic strategies for such designs, each with its own advantages and limitations. Particular emphasis has been placed on the determination of the type-preserving characteristics of the general transformations. The existing works on spectral transformations, which are obtained as special cases, are also reviewed. It is also pointed out that a considerable amount of work is still required for the development of efficient algorithms for various contour approximation problems arising from attempts to apply spectral transformation to 2-D filter design.

Acknowledgment

The authors would like to thank E. Fields and K. Mondal for their assistance in computer programming. The authors are grateful to Prof. N. K. Bose, Dr. M. P. Ekstrom, Prof. E. I. Jury, and the reviewers for their critical comments and suggestions which helped to improve the clarity of the presentation substantially. The authors would also like to thank J. Jasiulek for discussions.

References

[1] M. Ahmadi, A. G. Constantinides, and R. A. King, "Design techniques for a class of stable two-dimensional recursive digital filters," presented at IEEE Int. Conf. Acoust., Speech, Signal Processing, Philadelphia, PA, Apr. 1976.

[2] H. G. Ansell, "On certain two-variable generalizations of circuit theory, with applications to networks of transmission lines and lumped reactances," *IEEE Trans. Circuit Theory*, vol. CT-11, pp. 214-223, June 1964.

[3] M. Bernabo *et al.*, "A method for designing 2-D recursive digital filters having a circular symmetry," presented at the Conf. Digital Signal Processing, Florence, Italy, Sept. 1975.

[4] P. Broome, "A frequency transformation for numerical filters," *Proc. IEEE*, vol. 54, pp. 326-327, 1966.

[5] F. Bernabo, P. L. Emiliani, and V. Cappelini, "Design of 2-dimensional recursive digital filters," *Electron. Lett.*, vol. 12, no. 11, pp. 288-289, May 1976.

[6] S. Chakrabarti, B. B. Bhattacharyya, and M. N. S. Swamy, "Approximation of two-variable filter specifications in analog domain," in *Proc. Int. Symp. Circuits and Systems*, Apr. 1977.

[7] S. Chakrabarti and S. K. Mitra, "Theory and applications of spectral transformations in designing 2-dimensional analog and digital filters," Dept. of Electrical Engineering, Univ. of California, David, CA, Tech. Rep. 139.

[8] H. Chang and J. K. Aggarwal, "Design of 2-dimensional recursive filters by interpolation," in *Proc. IEEE Int. Symp. Circuits and Systems*, pp. 369-372, Apr. 1976.

[9] A. G. Constaninides, "Frequency transformations for digital filters," *Electron. Lett.*, vol. 3, pp. 487, 1967.

[10] —, "Frequency transformations for digital filters," *Electron. Lett.*, vol. 4, pp. 115-116, 1968.

[11] —, "Spectral transformations for digital filters," *Proc. Inst. Elec. Eng.*, vol. 117, pp. 1585-1590, Aug. 1970.

[12] J. M. Costa, "Design and realization of stable two-dimensional recursive filters," M.A. Sc. thesis, Univ. Toronto, Toronto, Canada, Sept. 1973.

[13] J. M. Costa and A. N. Venetsanopolous, "Design of circularly symmetric two-dimensional recursive filters," *IEEE Trans. Acoust., Speech, Signal Processing*, vol. ASSP-22, pp. 432-443, Dec. 1974.

[14] —, "A group of linear spectral transformations for two-dimensional digital filters," *IEEE Trans. Acoust., Speech, Signal Processing*, vol. ASSP-24, no. 5, pp. 424-425, Oct. 1976.

[15] R. E. Crochiere and L. R. Rabiner, "On the properties of frequency transformations for variable cut-off linear phase digital filters," *IEEE Trans. Circuits and Systems*, vol. CAS-23, pp. 684-686, Nov. 1976.

[16] E. Dubois and M. L. Blostein, "A circuit analysis method for the design of recursive two-dimensional digital filters," presented at the IEEE Int. Symp. Circuits and Systems, Boston, MA, 1975.

[17] M. P. Ekstrom and J. W. Woods, "Two-dimensional spectral factorization with applications in recursive digital filtering," *IEEE Trans. Acoust., Speech, Signal Processing*, vol. ASSP-24, pp. 115-128, Apr. 1976.

[18] C. H. Farmer and D. S. Gooden, "Rotation and stability in a recursive digital filter," in *Proc. Two-Dimensional Digital Signal Processing Conf.*, Columbia, MO, pp. 1-2-1-1-2-12, Oct. 1971.

[19] O. Herrman, L. R. Rabiner, and D. S. K. Chan, "Practical design rules for optimum finite impulse response lowpass digital filters," *Bell Syst. Tech. J.*, vol. 52, no. 6, pp. 769-799, July-Aug. 1973.

[20] J. V. Hu and L. R. Rabiner, "Design techniques for two-dimensional digital filters," *IEEE Trans. Audio Electroacoust.* (*Special Issue on Digital Filtering*), vol. AU-20, pp. 249-251, Oct. 1972.

[21] T. S. Huang, "Stability of two-dimensional recursive filters," *IEEE Trans. Audio Electroacoust.*, vol. AU-20, pp. 158-163, June 1972.

[22] B. R. Hunt, "Digital image processing," *Proc. IEEE*, vol. 63, pp. 595-610, Apr. 1975.

[23] E. I. Jury, *Inners and Stability of Dynamic Systems*. New York: Wiley, 1974.

[24] G. A. Maria and M. M. Fahmy, "An *lp*-design technique for two-dimensional digital recursive filters," *IEEE Trans. Acoust., Speech, Signal Processing*, vol. ASSP-22, pp. 15-21, Feb. 1974.

[25] —, "*lp*-approximation of the group delay response of one- and two-dimensional filters," *IEEE Trans. Circuits and Systems*, vol. CAS-21, pp. 431-436, May 1974.

[26] J. H. McClellan, "The design of two-dimensional digital filters by transformation," in *Proc. Seventh Annu. Princeton Conf. Information Sciences and Systems*, pp. 247-251, Mar. 1973.

[27] J. H. McClellan and T. W. Parks, "Equiripple approximation of fan filters," *Geophysics*, vol. 37, pp. 573-583, 1972.

[28] W. F. G. Mecklenbräuker and R. M. Mersereau, "McClellan transformations for two-dimensional digital filtering: II–Implementation," *IEEE Trans. Circuits and Systems*, vol. CAS-23, pp. 414-422, July 1976.

[29] —, "Efficient design and implementation for two-dimensional FIR digital filters designed using generalized McClellan transformations," in *Proc. EASCON '75*, pp. 185-A-185-F, 1975.

[30] R. M. Mersereau and D. E. Dudgeon, "Two-dimensional digital filtering," *Proc. IEEE*, vol. 63, pp. 610-623, Apr. 1975.

[31] R. M. Mersereau, D. B. Harris, and H. S. Hersey, "An efficient algorithm for the design of equirriple two-dimensional FIR digital filters," in *Proc. IEEE Int. Symp. Circuits and Systems*, pp. 443-446, Apr. 1975.

[32] R. M. Mersereau, W. F. G. Mecklenbräuker, and T. F. Quatieri, Jr., "McClellan transformations for two-dimensional digital filtering: I–Design," *IEEE Trans. Circuits and Systems*, vol. CAS-23, pp. 405-414, July 1976.

[33] A. V. Oppenheim, W. F. G. Mecklenbräuker, and R. M. Mersereau, "Variable cutoff linear phase digital filters," *IEEE Trans. Circuits and Systems*, vol. CAS-23, pp. 199-203, Apr. 1976.

[34] A. V. Oppenheim and R. W. Schafer, *Digital Signal Processing*. Englewood Cliffs, NJ: Prentice-Hall, 1975.

[35] N. A. Pendergrass, S. K. Mitra, and E. I. Jury, "Spectral transformations for two-dimensional digital filters," *IEEE Trans. Circuits and Systems*, vol. CAS-23, pp. 26-35, Jan. 1976.

[36] P. Pistor, "Stability criterion for recursive filters," *IBM J. Res. Develop.*, vol. 18, pp. 59-71, Jan. 1974.

[37] L. R. Rabiner and B. Gold, *Theory and Applications of Digital Signal Processing*. Englewood Cliffs, NJ: Prentice Hall, Inc., 1975.

[38] L. R. Rabiner, J. H. McClellan, and T. W. Parks, "FIR digital filter design techniques using weighted Chebychev approximation," *Proc. IEEE*, vol. 63, pp. 595-610, Apr. 1975.

[39] J. L. Shanks, S. Treitel, and J. H. Justice, "Stability and synthesis of two-dimensional recursive filters," *IEEE Trans. Audio Electroacoust.*, vol. AU-20, pp. 158-163, June 1972.

[40] K. Steiglitz, "The equivalence of digital and analog signal processing," *Information and Control*, vol. 8, pp. 455-467, 1965.

[41] L. Weinberg, *Network Analysis and Synthesis*. New York: McGraw-Hill, 1962.

[42] L. C. Wood and S. Treitel, "Seismic signal processing," *Proc. IEEE*, vol. 63, pp. 649-661, Apr. 1975.

[43] W. C. Yengst, *Procedures of Modern Network Synthesis*. New York: Macmillan, 1964.

Part II
Multidimensional Spectral Estimation

THE spectral estimation problem is to estimate the power spectral density of a homogeneous random field, given samples from one or more independent realizations of the field. Much of the research on multidimensional (M-D) spectral estimation has been motivated by attempts to extend algorithms and theoretical results from one-dimensional (time-series) spectral estimation to the multidimensional case. The area has now reached a level of maturity such that the theoretical issues that distinguish the M-D spectral estimation problem from the 1-D spectral estimation problem are well understood.

The reprinted paper by McClellan surveys theoretical developments and computational algorithms in M-D spectral estimation up to 1982. Particular attention is paid to *correlation-based* spectral estimators, where the random field samples are first reduced to an estimate for samples of the autocorrelation function. In certain spectral estimation schemes, such as the Maximum Entropy Method or Pisarenko's Method, the spectral estimate is required to be consistent with the estimate of the correlation samples (the *correlation-matching* property). Although algebraic expressions for these estimators exist for the 1-D case, the M-D estimates must be obtained as solutions to optimization problems.

The reprinted paper by Lang and McClellan presents an M-D version of Pisarenko's Method and gives a test for M-D *extendibility,* namely whether there exists a non-negative power spectral density, of the required support, that is consistent with a given estimate of a set of correlation samples. That this property does not always hold in the M-D case was observed by Dickinson [1].

A detailed theoretical discussion of the M-D Maximum Entropy problem, and a solution based on a convex optimization problem is contained in the paper by Lang and McClellan [2]. An alternative Maximum Entropy solution based on an iterative algorithm (whose convergence properties, however, are not yet understood) is presented by Lim and Malik [3]. A number of numerical examples illustrating the high-resolution properties of the 2-D Maximum Entropy Method is contained in the reprinted paper by Malik and Lim.

The theme of the reprinted paper by Marzetta and Lang is that the spectral estimation problem is inherently ill-posed, in the sense that given an extendible set of correlation samples, there are an infinite number of different power spectral densities consistent with the correlation samples. It is shown, however, that even without any prior knowledge, it *is* possible to place nontrivial upper and lower bounds on certain linear functionals of the spectral density (for example, the average spectral density over some frequency interval). A new interpretation for Capon's High Resolution spectral estimator (sometimes called the Maximum Likelihood Method) as an upper bound of this type is obtained.

The problem of estimating a correlation matrix from independent realizations of a set of M-D random field samples is addressed by Burg, Luenberger, and Wenger [4]. The problem of incorporating the difficult extendibility constraint in the estimation procedure remains unsolved.

Given an imperfect estimate of a set of samples of the autocorrelation function (which may not be extendible), the correlation-matching constraint in the Maximum Entropy Method can be relaxed to permit only approximate consistency. A solution to this modified Maximum Entropy problem is obtained by Schott and McClellan [5].

Data-direct methods of spectral estimation such as autoregressive (AR) and autoregressive/moving-average (ARMA) modelling avoid the intermediate step of correlation estimation. Progress in developing M-D AR and ARMA methods has been linked to progress in the area of M-D recursive filter design. Tjostheim [6] discusses quarter-plane AR modelling for purposes of spectral estimation. A more general class of AR models have nonsymmetric half-plane support; alternatively, quarter-plane ARMA models may be used as in the paper by Cadzow and Ogino [7].

Another type of data-direct modelling is Markov modelling, which in the 1-D case is equivalent to AR modelling. This equivalence does not hold in the M-D case because M-D polynomials do not generally have finite factorizations. The reprinted paper by Chellappa, Hu, and Kung discusses the 2-D Markov modelling problem as a stochastic estimation problem in which the model parameters are estimated by maximizing a likelihood function.

While theoretical papers still dominate the M-D spectral estimation literature, the paper by Baggeroer and Falconer [8] applies M-D spectral estimation to marine refraction data. A number of real world issues and problems are apparent in this paper including a sparse sensor array with unevenly spaced hydrophones, and a pressure field that is evidently not a homogeneous random field.

Finally, with so much emphasis in the spectral estimation literature on high-resolution methods, it is important to employ objective criteria in evaluating the performance of these methods, as pointed out by Johnson [9] and by Kay and Demeure [10].

Bibliography

[1] B. W. Dickinson, "Two-dimensional Markov spectrum estimates need not exist," *IEEE Trans. Inform. Theory,* vol. IT-26, pp. 120–21, Jan. 1980.

[2] S. W. Lang and J. H. McClellan, "Multidimensional MEM spectral estimation," *IEEE Trans. Acoust., Speech, Signal Processing,* vol. ASSP-30, pp. 880–87, Dec. 1982.

[3] J. S. Lim and N. A. Malik, "A new algorithm for two-dimensional maximum entropy power spectrum estimation," *IEEE Trans. Acoust., Speech, Signal Processing,* vol. ASSP-29, pp. 401–13, June 1981.

[4] J. P. Burg, D. G. Luenberger, and D. L. Wenger, "Estimation of structured covariance matrices," *Proc. IEEE,* vol. 70, pp. 963–74, Sept. 1982.

[5] J.-P. Schott and J. H. McClellan, "Maximum entropy power spectrum estimation with uncertainty in correlation measurements," *IEEE Trans. Acoust., Speech, Signal Processing,* vol. ASSP-32, pp. 410–18, Apr. 1984.

[6] D. Tjostheim, "Autoregressive modeling and spectral analysis of array data in the plane," *IEEE Trans. Geosci. Remote Sensing,* vol. GE-19, pp. 15–24, Jan. 1981.

[7] J. A. Cadzow and K. Ogino, "Two-dimensional spectral estimation," *IEEE Trans. Acoust., Speech, Signal Processing,* vol. ASSP-29, pp. 396–401, June 1981.

[8] A. B. Baggeroer and R. Falconer, "Array refraction profiles and crustal models of the Canada Basin," *J. of Geophys. Res.,* vol. 87, no. B7, pp. 5461–76, July 1982.

[9] D. H. Johnson, "The application of spectral estimation methods to bearing estimation problems," *Proc. IEEE,* vol. 70, pp. 1018–1028, Sept. 1982.

[10] S. Kay and C. Demeure, "The high-resolution spectrum estimator—a subjective entity," *Proc. IEEE,* vol. 72, pp. 1815–16, Dec. 1984.

Multidimensional Spectral Estimation

JAMES H. McCLELLAN, SENIOR MEMBER, IEEE

Invited Paper

***Abstract*—Methods of multidimensional power spectral estimation are reviewed. Seven types of estimators are discussed: Fourier, separable, data extension, MLM, MEM, AR, and Pisarenko estimators. Particular emphasis is given to MEM where current research is quite active. Theoretical developments are reviewed and computational algorithms are discussed.**

I. Introduction

THE OPERATION OF spectral analysis arises in many fields of application. Situations in which the signals are inherently multidimensional can be found in geophysics, radio astronomy, sonar, and radar, to mention a few. These multidimensional problems present a challenging set of theoretical and computational difficulties that must be tackled. Conventional methods based on the discrete Fourier transform (DFT) generalize from one-dimensional (1-D) time series analysis in a straightforward manner when the sampling is uniform, but the recently popularized high-resolution methods such as MEM take a new and more complicated form. It is the purpose of this paper to review the present state of affairs for multidimensional (m-D) spectral estimation and to comment on current directions of research.

The available spectral estimators will be considered in seven separate classes. Five of these classes parallel the one-dimensional situation. These are the classical method based on the DFT, the maximum-likelihood method (MLM) [10], [14], the maximum-entropy method (MEM) due to Burg [6], the autoregressive method (AR), and Pisarenko's method [51]. Note that MEM and the AR method are distinct methods in the m-D case. It is only through special circumstances, unique to one dimension, that these two methods turn out to be identical in the 1-D case. The other two classes have no exact counterparts in 1-D spectral estimation. One is the class of separable estimators, where a 1-D estimator is employed along each of the individual dimensions. A mixture of 1-D estimators may be used (e.g., the DFT followed by MEM). The other is a set of methods in which the given data are extended, so that Fourier analysis can be used. The extension can be done by linear prediction, band-limited extrapolation, etc.

In the following, each type of estimator is discussed in detail. Particular attention is paid to the MEM, because that is where current research activity is centered.

Manuscript received January 7, 1982; revised April 21, 1982. This work was supported in part by the National Science Foundation under Grant ECS79-15226 and in part by the Army Research Office under Contract DAAG29-81-K-0073.

The author was with the Research Laboratory of Electronics, Massachusetts Institute of Technology, Cambridge, MA 02139. He is now with Schlumberger Well Services, Austin, TX.

II. Problem Statement

The spectral estimation problem may be stated briefly in the following way: Given samples of a stationary and homogeneous zero-mean random field $u(\boldsymbol{x})$, specified by its second-order statistics, estimate its power spectrum.[1] In this case, the autocorrelation function $r(\boldsymbol{\delta}) = E[u(\boldsymbol{x})u^*(\boldsymbol{x}+\boldsymbol{\delta})]$, is sufficient to describe the process. The power spectrum $S(\boldsymbol{k})$ of $u(\boldsymbol{x})$ is the multidimensional Fourier transform of $r(\boldsymbol{\delta})$. The spectral estimation process involves estimating $S(\boldsymbol{k})$ from a finite observation of one sample signal $u_1(\boldsymbol{x})$ of the random process. Usually, this finite observation takes the form of a finite number of samples of the signal $u_1(\boldsymbol{x})$ at $\boldsymbol{x} = \boldsymbol{x}_j, j = 1, 2, \cdots, N$. For obvious reasons, it is desirable to keep the number of samples as small as possible while maintaining a high-quality estimate. This tradeoff often forces one to operate near the resolution limit of a particular technique.

But there are more fundamental questions about such an estimation process within the theoretical definition of the random process. First, there must be an ergodicity property. Such a condition is necessary to justify the use of time averages over a single sample function of the process in place of the true statistical quantities. Second, it is the case that only a finite number of samples are available for time averaging, so even the assumption of ergodicity is not a sufficient justification. Third, the role of any underlying physical model must be taken into account. In spite of these theoretical difficulties and the fact that actual signals are rarely stationary in the strict sense of the word, spectral estimation has become a routinely used tool in many applications requiring signal analysis.

III. The Fourier Methods

Historically, methods based on the Fourier transform have been widely used for spectral estimation. The advent of the fast Fourier transform (FFT) algorithm helped to make these techniques very attractive. Although many variations exist, there are fundamentally only two distinct estimators of the Fourier type. One is based on the fact that the power spectrum and the autocorrelation function form a transform pair; the other is the periodogram estimate including its various smoothed versions. The use of these estimators in the multidimensional case is virtually identical to the well-known 1-D case. The following discussion is quite standard, but considerations unique to the m-D case will be emphasized.

[1] **Although the independent variable is labeled $\boldsymbol{x}$, it may not be wholly a spatial vector; one component is often time. In the spatial array case $\boldsymbol{x}$ might be four-dimensional with components x, y, z, and t.**

Reprinted from *Proc. IEEE,* vol. 70, pp. 1029–1039, Sept. 1982.

A. Windowing the Autocorrelation Function

The first method of Fourier estimation is that of windowing the autocorrelation function estimate. Since the autocorrelation function (acf) and the power spectrum form a Fourier transform pair, an obvious approach to spectral estimation is to first estimate the acf $r(\delta)$ at a finite number of separations δ_i, and then take the Fourier transform. If the acf estimate is statistically reliable, one would hope that the resulting spectral estimate will be acceptable.

One popular way of estimating the acf is to compute the spatial autocorrelation of the known signal values. This deterministic operation is simple to implement, especially if the data are given on a regularly spaced rectangular grid. Unfortunately, this acf estimate is not entirely reliable. At small separations (called lags in the time series case), many signal points enter into the estimate, making it quite robust. However, at large separations very few points enter into the estimate and the variability of the estimate can be quite high. The solution is obvious. The acf estimate must only be taken over small values of lag. This restriction can be implemented by windowing the acf estimate. The window choice is not entirely unconstrained. Since the expected value of the spectral estimate is equal to the convolution of the true power spectrum with the transform of the window function, it is sufficient, but not necessary, to choose the window to be nonnegative over all frequencies in order to keep this expected value positive. In the 1-D case, the triangular window satisfies this condition although windows of the raised cosine type are often used. In the two-dimensional (2-D) case (or higher dimensions), fewer windows are available. Huang [20] discusses the synthesis of nearly circularly symmetric 2-D windows from existing 1-D windows by rotating the 1-D window in the space domain. The frequency response of the window changes under such a transformation; it is not known whether positivity would be maintained. Finally, it is not clear that circular symmetry or any other fixed symmetry is what is called for in the spectral estimation problem.

B. Welch's Method

In the second Fourier method, one averages the (windowed) periodogram—the magnitude squared of the Fourier transform of the finite signal. The periodogram is known to be a poor estimate of the power spectrum of the process. For example, when applied to a time series consisting of samples from a white Gaussian noise process, the periodogram estimate has a variance that can be obtained analytically and is equal to the square of the variance of the white noise process. Bartlett [5] proposed a modification of the periodogram to obtain a statistically reliable spectral estimate. His method requires that the signal be segmented into blocks, the periodogram of each block be computed, and then the final spectral estimate obtained by averaging the individual periodograms. The averaging introduced serves to reduce the variance of the resulting estimate by a factor equal to the number of blocks, if the individual blocks are statistically independent; otherwise, the variance reduction is less. The price paid for this improvement in variance is a reduction in the frequency resolution.

Since the signal is effectively being windowed by a rectangular window, Bartlett's estimate has noticeably high sidelobes. Welch [58] proposed that, in addition to averaged periodograms, one could also implement different windowing schemes directly on the data to improve the sidelobe structure when estimating closely spaced narrow-band signals. For example, the Hamming window has a Fourier transform whose sidelobes are less than −42 dB relative to the peak, compared with −13 dB for the rectangular window. There is, however, a corresponding loss in resolution due to an increase in the width of the spectral peak, about 40 percent in the case of the Hamming versus the rectangular window.

The averaging scheme can also be changed when a tapered window such as the Hamming window is used. Since the tails of the Hamming window are quite small compared to the central portion of the window, the signal values near the middle of the window are more important in the statistical averaging. It seems intuitive, therefore, that the signal values near the edges of the window could be reused to increase their importance to the final estimate. This can be accomplished by taking the blocks of the signal to be overlapped. In the 1-D case, Welch analyzed the variance reduction that results from such overlapping. His results led to the common practice of using 50-percent overlap with windows such as the Hamming.

The foregoing discussion is primarily a review of the 1-D situation [47]. In the m-D case, there are three major areas of concern. First, there are few windows to choose from in the m-D case. Methods of m-D finite impulse response filter design would seem to offer the simplest way of generating new windows with prescribed frequency-domain characteristics [13]. Second, the data must be available on a regularly spaced grid (e.g., Cartesian or hexagonal) in order to implement efficiently the blocking and windowing of Welch's method. This complication applies also to the first method of windowing the acf estimate. In nonuniformly sampled situations, the use of the FFT is severely limited. Further, it may happen that even the correlation estimates at small lags are noisy, because only a few signal values are used in the average. The third problem is ubiquitous in Fourier analysis—the resolution limit.

C. The Uncertainty Principle

It is well known that the Fourier transform imposes certain limits on the resolution achievable from a given length of a time signal. Specifically, it can be shown that the frequency resolution is inversely proportional to the time duration of the signal (in 1-D). Likewise, the frequency resolution in m-D is inversely proportional to the extent of the signal.

For both Fourier methods, this resolution limit is serious. When windowing the acf estimate, the window size must be made large if the resolution is to be high, but this conflicts with the requirement that only the statistically stable estimates be used to obtain the spectrum. Likewise, the need for averaging in Welch's method is in direct conflict with the desire for high resolution. For the sake of resolution, each block must be as large as possible. It is often the case that there is a limited amount of data available, especially when sampling in space. For this reason the Fourier methods have met with limited success and much recent research has been directed at deriving high-resolution estimators. The remainder of this paper will be devoted to these techniques.

IV. Separable Spectral Estimation

A separable spectral estimator relies on the use of existing 1-D estimators to perform higher dimensional spectral estimation. Since the m-D Fourier transform operator has a separable kernel (the complex exponential), it is possible to view the transform as an operator that works successively on each dimension. This observation leads to the following strategy for spectral estimation. Along each dimension perform a 1-D Fourier analysis, but postpone the magnitude squared operation until all dimensions have been processed. It is crucial that the phase not be discarded at the intermediate steps.

The most common approach is to use the DFT along each dimension until the last, where a high-resolution estimator such as MEM would be employed and the magnitude squared would be taken [22]. This approach is popular in spatial array processing when the problem is 2-D, one dimension being time and the other linear distance along the array. Typically, there is a considerable amount of data available along the time dimension, so the Fourier analysis is not restricted by the resolution limit of the Uncertainty Principle. The time dimension is processed with the DFT whose output is complex-valued. Succeeding estimation algorithms must be able to handle complex data even when the original signal is purely real.

There are some other methods that might also fall under the heading of separable estimators, but they also involve data extension. These techniques are described in the next section.

V. Data Extension for Spectral Estimation

Another method of preserving the phase in a separable estimator is to artificially extend the signal in the spatial domain. This signal extension can be accomplished either by using a model to synthesize additional data points or by band-limited extrapolation.

A. Data Extension by Linear Prediction

Consider the 2-D case where the signal is sampled on a rectangular grid. Along one dimension (say the horizontal), a parametric model for an individual row can be computed. Such models are routinely computed as part of the MEM spectral estimate and are usually referred to as linear predictors. Since this model presumably represents the signal faithfully, it can be used to extend the data. It is only necessary to give the Pth-order all-pole model P initial conditions in order to begin its recursive operation. These initial conditions are taken to be the P signal points at the end of a row. The model is run as a zero-input system because the signal is assumed to be the impulse response of a linear time-invariant system. This should be a reasonable assumption in the case of estimating multiple sinusoids in noise. For more general spectra, especially those produced by filtering white noise, it is unlikely that this method would have merit.

Several researchers have proposed algorithms of this type. Frost and Sullivan [15] employed the Burg linear predictor to extend the data in both dimensions and then derived the 2-D spectral estimate by conventional Fourier techniques. The Burg predictor was chosen because it is important to guarantee the stability of the model which will generate the extended data. In a later paper, Frost [16] reported better results when the data prediction filter was derived using the modified covariance method. Here the sum of the forward and backward error energies is minimized, but the resulting filter is possibly unstable. A further simplification was made by assuming that the signals were plane waves and that each row of the data matrix would, therefore, be identical up to a phase factor, as would each column. As a result, only one predictor need be designed for all the rows and another for all the columns. Examples were presented to show the superior resolution property of the method over the DFT. These examples were of 30-dB sinusoids in noise, in which case the data extension works reliably and the periodogram estimate is not noisy. In low signal-to-noise ratio (SNR) environments, a need for averaging would probably arise.

In the same paper, Frost [16] proposed another data extension technique that is not a separable algorithm since it relies on a 2-D prediction filter to synthesize the additional data. The technique uses the data in the observation window to design four $M \times M$ 2-D least squares prediction error filters. The derivation of the filter coefficients requires the solution of a set of simultaneous linear equations—the normal equations of the appropriate least squares problem (see Section VII below where the AR method is discussed in more detail). The opposing filters (1st and 3rd quadrants, 2nd and 4th) have the same coefficients, conjugated and reversed in order. The extrapolation of the observed data must be done in a certain order. Each prediction filter is used to extend the signal in its corresponding quadrant. The order of computation must be such that the original data are expanded one layer at a time. In Frost's work, 3×3 and 6×6 prediction error filters were used to extrapolate a 16×16 data set to 64×64. The increase in the number of filter coefficients versus the 1-D extrapolators improved performance in low SNR situations.

Another technique due to Joyce [27] uses the data extension in a slightly different manner. A Burg prediction filter is used to extend the data in one of the dimensions, but then a 1-D DFT is taken along this dimension. Along the other dimension, the 1-D MEM algorithm (actually Burg's recursion for complex signals) is used to obtain the spectrum. Essentially, this technique uses the prediction filter to obtain a large amount of data in one of the dimensions and then proceeds as in the separable method above. As in the preceding examples, Joyce assumes that the nature of the signals is such that the same prediction filter could be used for all the rows.

B. Band-Limited Extrapolation

Data extension can also be carried out by band-limited extrapolation [17], [26], [48]. The band-limited extrapolator attempts to extend the signal consistent with an assumption that the given samples were from a band-limited (oversampled) signal. The algorithm iterates between the time and frequency domains, imposing the constraint of band-limitedness in the frequency domain and substituting the known signal values in the time domain until a solution has been reached. Effectively, the Fourier transform of the signal is being constructed as the algorithm proceeds. This algorithm works equally well in one or higher dimensions. Two approaches are possible: first, the signal can be extrapolated with a multidimensional version of this iteration; second, a 1-D version of the algorithm can be applied successively along each dimension in the spirit of a separable estimator. Examples of this type of processing have been presented by Jain [26]. Also presented are modifications of the basic iteration to improve its convergence on noisy data.

VI. MLM (Maximum-Likelihood Method) Spectral Estimation

Unlike many other spectral estimation techniques, the maximum-likelihood method (MLM) was originally proposed in an m-D setting as an array-processing technique [10], [14]. It was the first of the so-called high-resolution methods. The MLM philosophy is as follows: Given a spatial sampling of a wavefield $u(\boldsymbol{x})$ at an arbitrary set of locations, process these sensor outputs $u(\boldsymbol{x}_i)$ with a linear processor to obtain an estimate of the complex wave amplitude arriving from the direction specified by the wave vector $\boldsymbol{k}_0$. The array output $y(\boldsymbol{k}_0)$ is given by

$$y(\boldsymbol{k}_0) = \sum_i A_i(\boldsymbol{k}_0) u(\boldsymbol{x}_i). \tag{1}$$

The maximum likelihood estimate of the complex amplitude, given that there is a plane wave at $\boldsymbol{k}_0$, is obtained by minimizing the output power subject to the array passing undistorted a monochromatic plane wave with wave vector $\boldsymbol{k}_0$. Thus the following optimization problem is obtained:

$$\underset{A_i(\boldsymbol{k}_0)}{\text{minimize}}\ \boldsymbol{E}\left\{\left|\sum_i A_i(\boldsymbol{k}_0)u(\boldsymbol{x}_i)\right|^2\right\} \tag{2}$$

subject to

$$\sum_i A_i(\boldsymbol{k}_0)e^{-j\boldsymbol{k}_0\cdot\boldsymbol{x}_i} = 1. \tag{3}$$

More generally, (2) and (3) present a simple constrained optimization problem—quadratic performance function subject to linear constraints. Rewritten in vector form, the problem is

$$\underset{\boldsymbol{a}}{\text{minimize}}\ \boldsymbol{a}^H\boldsymbol{R}\boldsymbol{a} \tag{4}$$[2]

subject to

$$\boldsymbol{E}^H\boldsymbol{a} = 1 \tag{5}$$

where $\boldsymbol{a}$ is the vector of (complex) sensor weights $\{A_i(\boldsymbol{k}_0)\}$, $\boldsymbol{R}$ is a correlation matrix with entries $r(\boldsymbol{x}_i - \boldsymbol{x}_j)$, and $\boldsymbol{E} = \boldsymbol{E}(\boldsymbol{k}_0) = \{\exp[j\boldsymbol{k}_0 \cdot \boldsymbol{x}_i]\}$ is a steering vector corresponding to the hypothesized plane wave with wave vector $\boldsymbol{k}_0$. In a practical problem, the entries in the correlation matrix are estimates of the true correlation function sampled at vector separations given by the "co-array" [19] $\Delta = \{\boldsymbol{x}_i - \boldsymbol{x}_j$: for all i and $j\}$. Furthermore, the estimate is usually computed at a single temporal frequency so that the correlation matrix is actually frequency dependent. The exact nature of the correlation estimates will have an important impact on the nature of the MLM solution [2]. For example, lack of averaging in estimation of the correlation function has been observed to lead to degraded performance in MLM array processing. Often the estimate for $r(\boldsymbol{x}_i - \boldsymbol{x}_j)$ is formed by just computing $u(\boldsymbol{x}_i)u^*(\boldsymbol{x}_j)$ for a single pair (i, j). If the array geometry is nonuniform or the wavefield is nonhomogeneous, it is not possible to average over space. Some averaging to combat noise in the covariance estimates can be done by averaging independent "looks" in time. Finally, in the nonhomogeneous case, the covariance matrix $\boldsymbol{R}$ will contain entries $r(\boldsymbol{x}_i, \boldsymbol{x}_j)$, and the matrix will not have a Toeplitz structure even for a uniform linear array.

The general solution to (4) and (5) yields the optimal set of sensor weights

$$\boldsymbol{a} = \frac{\boldsymbol{R}^{-1}\boldsymbol{E}}{\boldsymbol{E}^H\boldsymbol{R}^{-1}\boldsymbol{E}} \tag{6}$$

and the output power which is used as the spectral estimate at wavenumber $\boldsymbol{k}_0$

$$P(\boldsymbol{k}_0) = [\boldsymbol{E}^H(\boldsymbol{k}_0)\boldsymbol{R}^{-1}\boldsymbol{E}(\boldsymbol{k}_0)]^{-1}. \tag{7}$$

[2]The superscript H denotes the conjugate transpose of a vector.

The MLM formulation is quite general [1] and will apply to m-D situations, even when the sensors are located at arbitrary positions. The computational burden of MLM is twofold. First, the inverse correlation matrix $\boldsymbol{R}^{-1}$ must be obtained. The matrix $\boldsymbol{R}$ may not have full rank, which frequently occurs with short duration measurements. Second, the power estimate (7) must be calculated over all wavenumbers of interest. Typically, this second operation is more time consuming, but fortunately some simplifications are possible [12]. In the 1-D equally spaced case, each diagonal of $\boldsymbol{R}^{-1}$ can be pre-summed and then multiplied by the correct complex exponential to form the denominator of (7). In addition, $\boldsymbol{R}^{-1}$ is Hermitian symmetric so that the diagonal sums are symmetric and the complex exponentials can be replaced by cosines. Finally, the FFT algorithm can be employed if the desired sampling in wavenumber is uniform. The generalization to 2-D rectangular grids (with uniform spacing) is straightforward. The aforementioned simplifications now occur in sub-blocks of $\boldsymbol{R}^{-1}$. The savings versus the direct evaluation of (7) can be impressive when either the array size is large or the sampling in wavenumber space is dense.

The performance of the MLM spectral estimate lies somewhere between the Fourier techniques and the MEM estimate when applied to peaked spectra. Nevertheless, its considerable flexibility and simplicity makes MLM an attractive m-D spectral estimate. Applications have included the processing of seismic data from the Large Aperture Seismic Array [10] and from experiments in the Arctic basin [3].

VII. MEM (Maximum-Entropy Method) Spectral Estimation

The maximum-entropy method (MEM) of spectral estimation has generated an enormous amount of interest in the field of time series analysis. This is primarily due to the fact that the MEM estimate can provide excellent frequency resolution, and to the fact that the 1-D MEM estimate can be computed from the linear equations of autoregressive (AR) signal modeling. In cases where the signal length is very short, MEM may be the only way to obtain the desired resolution. Since this situation of limited signal size is quite common in m-D signal processing, the motivation for a high-resolution spectral estimator is even stronger. Unfortunately, in the m-D case, the true maximum entropy estimate is distinctly different from the spectrum derived by AR modeling. This may be attributed in part to the lack of a natural "time" ordering in the 2-D plane [40]. Furthermore, computation of the general MEM estimate appears to require the solution of a nonlinear optimization problem. Recent research has been directed at characterization and computation of this true MEM estimate. Before discussing individual algorithms, the theory of the maximum entropy problem is presented and a necessary and sufficient condition for the existence of the optimal solution is stated.

A. Optimality Conditions for MEM

The maximum-entropy spectrum, introduced by Burg [6], is obtained as the solution to a constrained maximization problem. The entropy of a Gaussian process with power spectrum $S(\boldsymbol{k})$ is

$$H(S) = \int_K \ln S(\boldsymbol{k}) \tag{8}$$

where the set K is a region of the frequency–wavenumber

domain over which the power spectrum is assumed to be nonzero. The MEM estimate is that spectrum which maximizes $H(S)$ subject to the constraints given by the correlation measurements.

$$r_o(\boldsymbol{\delta}) = \int_K S(\boldsymbol{k})\, e^{j\boldsymbol{k}\cdot\boldsymbol{\delta}}, \quad \text{for } \boldsymbol{\delta} \in \Delta = \{\boldsymbol{0}, \pm\boldsymbol{\delta}_1, \cdots, \pm\boldsymbol{\delta}_M\}. \tag{9}$$ [3]

Equation (9) is usually referred to as the "correlation matching" constraint. These constraints are on samples of the inverse transform of the power spectrum and hence are linear. Two features of the problem definition need to be emphasized because they provide a generalization of the previous thinking about MEM. First, the set K over which the entropy integral is taken can be limited arbitrarily. The exact nature of K is problem dependent, but could be specified through a condition of band-limitedness for a time series or a cutoff wavenumber in spatial wave theory. Second, the co-array [19], Δ, that specifies the sampling of the correlation function may be nonuniform. Such nonuniformity will, in fact, complicate even the 1-D MEM, so that AR modeling no longer provides the solution [44]. In the m-D case, even with uniform sampling, the MEM is not AR modeling.

Let the optimal spectrum $S(\boldsymbol{k})$ be denoted as $S_o(\boldsymbol{k})$. In most situations of interest, when $S_o(\boldsymbol{k})$ exists it has the form

$$S_o(\boldsymbol{k}) = \frac{1}{P_o(\boldsymbol{k})} \tag{10}$$

where $P_o(\boldsymbol{k})$ is a positive Δ-polynomial

$$P_o(\boldsymbol{k}) = \sum_{\boldsymbol{\delta}\in\Delta} p_o(\boldsymbol{\delta})\, e^{-j\boldsymbol{k}\cdot\boldsymbol{\delta}} > 0, \qquad \text{for } \boldsymbol{k} \in K. \tag{11}$$

Note that $p_o(-\boldsymbol{\delta}) = p_o^*(\boldsymbol{\delta})$, since $P_o(\boldsymbol{k})$ is real. This result is familiar in the 1-D case where the MEM spectrum reduces to the magnitude squared of an all-pole model [7]. Woods [62] has proven the existence and uniqueness of the MEM estimate when the region K is the usual 2-D Nyquist square and the set Δ is a "nearest neighbors' array"—a set of points on a 2-D Cartesian grid, consisting of all points inside the outermost perimeter (i.e., no holes or gaps inside). The following theorem [30], [42] gives a necessary and sufficient condition in terms of the extendibility of the correlation measurements (see Section IX), and the behavior of Δ-polynomials that have zeros on K. In order to make the statement of the theorem more concise, it is convenient to think of the Δ-polynomial coefficients $p(\boldsymbol{\delta})$ and the correlation samples $r(\boldsymbol{\delta})$ as vectors in the vector space $\boldsymbol{R}^{2M+1}$, denoted as $\boldsymbol{p}$ and $\boldsymbol{r}$, respectively. In particular, $\boldsymbol{r}$ is the $2M + 1$-vector

$$\boldsymbol{r}^T = [r(\boldsymbol{0})\ \sqrt{2}\,\mathrm{Re}\, r(\boldsymbol{\delta}_1)\ \sqrt{2}\,\mathrm{Im}\, r(\boldsymbol{\delta}_1) \cdots \sqrt{2}\,\mathrm{Re}\, r(\boldsymbol{\delta}_M) \cdot \sqrt{2}\,\mathrm{Im}\, r(\boldsymbol{\delta}_M)]^T$$

likewise, for $\boldsymbol{p}$.

Theorem: If

$$\int_K \frac{1}{P(\boldsymbol{k})} = \infty$$

for every nonnegative Δ-polynomial that is zero somewhere on the set K, then, for every correlation vector $\boldsymbol{r}$ in the interior of the set of extendible correlations, there exists a strictly positive Δ-polynomial $P_o(\boldsymbol{k})$ such that

$$S(\boldsymbol{k}) = \frac{1}{P_o(\boldsymbol{k})}$$

satisfies the correlation-matching constraint (9) and maximizes the entropy (8). The converse is also true.

The condition that $\boldsymbol{r}_o$ lie in the interior of the set of extendible correlations is equivalent to the existence of at least one strictly positive power spectrum $S(\boldsymbol{k})$ that matches the correlations $\boldsymbol{r}_o$ via (9). The foregoing theorem says that one such $S(\boldsymbol{k})$ has the special form as the inverse of a positive Δ-polynomial and is the unique MEM spectrum. A detailed proof of this theorem together with a derivation of the following dual problem is available in [42].

If the constrained problem (8), (9) is attacked using Lagrange multipliers, the coefficients of the positive Δ-polynomial (11) will turn out to be the Lagrange multipliers. Thus any algorithm that works by trying to find the Δ-polynomial coefficients is actually solving a dual optimization problem. Most algorithms fall in this category because the dual problem is finite-dimensional.

B. Algorithms for the Dual Problem

In his doctoral thesis [7], Burg proposed a very general variational principle for estimating any function given a set of measurements of the function. He also outlined an iterative algorithm to compute the solution. In the case of MEM spectral estimation the result is the following dual optimization problem [30], [42]:

$$\underset{\boldsymbol{p}\in\boldsymbol{P}^+}{\text{minimize}}\ \{H(\boldsymbol{p}) + \boldsymbol{r}_o^T\boldsymbol{p}\} \tag{12a}$$

where

$$H(\boldsymbol{p}) = \int_K \ln\left[\frac{1}{P(\boldsymbol{k})}\right]. \tag{12b}$$

$H(\boldsymbol{p})$ is the entropy functional of a positive Δ-polynomial with coefficients $p(\boldsymbol{\delta})$, $\boldsymbol{P}^+$ is the set of vectors $\boldsymbol{p}$ corresponding to strictly positive Δ-polynomials, and $r_o(\boldsymbol{\delta})$, $\boldsymbol{\delta} \in \Delta$ are the given correlation measurements. The advantages of (12a, b) are: the problem is finite dimensional, the objective functional is convex, and (12a) can be treated essentially as an unconstrained minimization since $H(\boldsymbol{p})$ will be $-\infty$ when $P(\boldsymbol{k}) = 0$.

The gradient of $H(\boldsymbol{p})$ is

$$\nabla H(\boldsymbol{p}) = -\int_K \frac{\boldsymbol{E}(\boldsymbol{k})}{P(\boldsymbol{k})} = -Q(\boldsymbol{p}) \tag{13}$$

where the vector $\boldsymbol{E}(\boldsymbol{k})$ corresponds to the exponentials in (11).

$$\boldsymbol{E}^T(\boldsymbol{k}) = [1\ \ \sqrt{2}\cos(\boldsymbol{k}\cdot\boldsymbol{\delta}_1)\ \ \sqrt{2}\sin(\boldsymbol{k}\cdot\boldsymbol{\delta}_1) \cdots \sqrt{2}\cos(\boldsymbol{k}\cdot\boldsymbol{\delta}_M)\ \ \sqrt{2}\sin(\boldsymbol{k}\cdot\boldsymbol{\delta}_M)]. \tag{14}$$

Note that $P(\boldsymbol{k}) = \boldsymbol{p}^T\boldsymbol{E}(\boldsymbol{k})$ and that $Q(\boldsymbol{p})$ computes the correlation samples for the spectrum $1/P(\boldsymbol{k})$. The gradient of the functional to be minimized in (12a) is thus

$$-Q(\boldsymbol{p}) + \boldsymbol{r}_o. \tag{15}$$

At a zero of the gradient (15), the correlation matching constraint is satisfied and $\boldsymbol{p} = \boldsymbol{p}_o$. The convexity of the problem

[3] It is deliberate that $d\boldsymbol{k}$ is omitted from the integrals. Since the set K is unspecified and the dimension of $\boldsymbol{k}$ is unknown, the integrals (and a corresponding measure) will have to be defined to fit a particular problem. Also, to avoid certain degenerate situations, it is necessary to assume that each open set in K has nonzero measure.

guarantees that a zero of the gradient will give the global minimum in (12a).

Burg suggested using Newton's method to find the zero of (15). The Newton iteration can be written as

$$\boldsymbol{p}_{n+1} = \boldsymbol{p}_n - [H''(\boldsymbol{p}_n)]^{-1} \cdot [\boldsymbol{r}_o - Q(\boldsymbol{p}_n)] \tag{16}$$

where the matrix $H''(\boldsymbol{p})$ is the Hessian of the entropy functional $H(\boldsymbol{p})$.

$$H''(\boldsymbol{p}_n) = \int_K \frac{\boldsymbol{E}\boldsymbol{E}^T}{P^2(\boldsymbol{k})}. \tag{17}$$

Woods [62] has shown that the Hessian will always be positive definite and hence invertible. The fact that the objective functional in (12a) is strictly convex would guarantee the convergence of a suitable version of Newton's method, but not the one given by (16). In particular, a line search along the descent direction must be performed [38].

In the remaining discussion of iterative algorithms for the dual MEM problem, a common form will be found. Each of the algorithms can be expressed as a recursion

$$\boldsymbol{p}_{n+1} = \boldsymbol{p}_n - \alpha_n \boldsymbol{d}_n \tag{18}$$

where the update direction $\boldsymbol{d}_n$ is (hopefully) a descent direction and the step size α_n may or may not be adjusted at each iteration. Note that Newton's method corresponds to $\alpha_n = 1$ and $\boldsymbol{d}_n = [H''(\boldsymbol{p}_n)]^{-1} \cdot [\boldsymbol{r}_o - Q(\boldsymbol{p}_n)]$. In retrospect, this viewpoint affords an easy way to understand how the various methods compare. In the remainder of this section, four methods, developed over the past ten years, are described; in the next section, two recent algorithms based on steepest descent are presented.

Ong [46] proposed an algorithm that approximates $H''(\boldsymbol{p}_n)$ by performing a numerical Fourier inversion. The numerical integration on the $\boldsymbol{k}$-plane was found to be inadequate when the size of Δ was larger than 3×3. Furthermore, in the case of highly peaked spectra the convergence of (16) was observed to be quite slow, a fact that has been noted for all existing MEM algorithms.

Woods [62] proposed that the transformation $Q(\boldsymbol{p})$ be approximated by a power series expansion in the polynomial coefficients $\boldsymbol{p}$. For a positive polynomial $P(\boldsymbol{k})$, redefine the spectrum as

$$\frac{1}{P(\boldsymbol{k})} = \frac{c}{1 - A(\boldsymbol{k})}. \tag{19}$$

From the infinite series for $(1 - A(\boldsymbol{k}))^{-1}$ when $|A(\boldsymbol{k})| < 1$ on K, there is a direct computation of the correlation vector corresponding to $P^{-1}(\boldsymbol{k})$. Truncation of this series after a finite number of terms gives an approximation to the operator $Q(\boldsymbol{p})$, call it $Q_T(\boldsymbol{p})$. Furthermore, Woods argued that, for correlations consisting of plane waves, H'' can be approximated by the diagonal matrix $\epsilon^{-1} I$. In fact, the resulting algorithm is quite close to steepest descent with a fixed step size ϵ.

$$\boldsymbol{p}_{n+1} = \boldsymbol{p}_n - \epsilon (\boldsymbol{r}_o - Q_T(\boldsymbol{p}_n)). \tag{20}$$

The bulk of the computation lies in $Q_T(\boldsymbol{p}_n)$ which can be made efficient with FFT implementations. Convergence of the algorithm has been observed experimentally, but, due to all the approximations, it cannot be shown to converge to the correct MEM solution. Finally, in those cases where $|A(\boldsymbol{k})| \geqslant 1$, the algorithm will fail unless (19) is modified.

A third algorithm of this general type was proposed by Jain and Raganath [23]. Their proposal was to solve (12) by optimizing one coordinate $p(\boldsymbol{\delta}_i)$ at a time. A Newton-Raphson technique was employed to perform the minimization along each individual coordinate. Convergence is guaranteed in this case by the convexity of $H(\boldsymbol{p})$, but the running time can be excessive. The simplicity of each step is attractive but steepest descent is usually much quicker to converge [38].

Lim and Malik [36] have proposed an algorithm of the alternating projection type to solve the dual problem. On the surface this algorithm appears to be radically different from those previously described and from the steepest descent methods presented in the next section. However, this is not the case as it is also possible to write their iteration in the form (18), except that the update direction is neither that of Newton's method, nor is it the projected gradient. First, the algorithm will be described. Equations (9) and (10) can be viewed as constraints on $S(\boldsymbol{k})$ and $1/S(\boldsymbol{k})$. In particular, (9) constrains the inverse transform of $S(\boldsymbol{k})$ to be equal to $r_o(\boldsymbol{\delta})$ for $\boldsymbol{\delta} \in \Delta$; demands that the inverse transform of $1/S(\boldsymbol{k})$ be finite (i.e., zero outside $\boldsymbol{\delta} \in \Delta$). The following iterative algorithm is thus proposed: At any step of the procedure an estimate $S_n(\boldsymbol{k}) = 1/P_n(\boldsymbol{k})$ is available.

Step 1: Take the inverse transform of $S_n(\boldsymbol{k})$ and replace $r_n(\boldsymbol{\delta})$ with $r_o(\boldsymbol{\delta})$ for $\boldsymbol{\delta} \in \Delta$.

Step 2: Return to the spectral domain; call the new spectrum $\tilde{S}_n(\boldsymbol{k})$. If the correlation replacement leads to a spectrum $\tilde{S}_n(\boldsymbol{k})$ that is negative for some value of $\boldsymbol{k} \in K$, then $r_n(\boldsymbol{\delta})$ is replaced with $(1 - \alpha) r_o(\boldsymbol{\delta}) + \alpha r_n(\boldsymbol{\delta})$. The relaxation parameter α can be chosen to keep $\tilde{S}_n(\boldsymbol{k})$ positive.

Step 3: Take the inverse transform of $\tilde{S}_n^{-1}(\boldsymbol{k})$ and zero all coefficients outside $\boldsymbol{\delta} \in \Delta$.

Step 4: Return to the spectrum $\tilde{S}_{n+1}(\boldsymbol{k})$. If $\tilde{S}_{n+1}(\boldsymbol{k})$ is negative somewhere on K, then a second relaxation parameter β is introduced and $S_{n+1}(\boldsymbol{k})$ is given by $\beta \;\; S_n(\boldsymbol{k}) + (1 - \beta) \tilde{S}_{n+1}(\boldsymbol{k})$; $\beta = 0$, if $\tilde{S}(\boldsymbol{k}) > 0$.

These steps are repeated iteratively until the algorithm (hopefully) converges. If the iteration converges, then the MEM solution will be reached because (9) and (10) will be satisfied. However, the convergence properties of this algorithm are not yet understood.

In order to express the algorithm of Lim and Malik in the recursive form (18), note that the spectral estimate produced in step 4 is of the form of the inverse of a positive Δ-polynomial. Thus steps 1–4 can be viewed as one iteration of an algorithm to update the vector of polynomial coefficients.

$$\boldsymbol{p}_{n+1} = \boldsymbol{p}_n - (1 - \alpha)(1 - \beta) \boldsymbol{d}_n. \tag{21a}$$

The direction $\boldsymbol{d}_n$ can be written in terms of the gradient vector $\boldsymbol{g}_n = \boldsymbol{r}_o - Q(\boldsymbol{p}_n)$

$$\boldsymbol{d}_n = \int_K \left\{ \frac{P_n^2(\boldsymbol{k}) G_n(\boldsymbol{k}) \boldsymbol{E}(\boldsymbol{k})}{1 + (1 - \alpha) P_n(\boldsymbol{k}) G_n(\boldsymbol{k})} \right\}. \tag{21b}$$

The quantity $G_n(\boldsymbol{k})$ is the Fourier transform of the gradient vector $\boldsymbol{g}_n$. In this form, there are two comments that can be made. First, the update direction is a perturbed version of the steepest descent direction (i.e., the gradient). It is not clear from (21b) whether this direction is even a feasible descent direction. Experience shows that the algorithm usually converges and thus suggests that $\boldsymbol{d}_n$ is probably a feasible descent direction. The second comment relates to (21a) in which the step size is effectively $(1 - \alpha)(1 - \beta)$, the product being between 0 and 1. If the step size could be optimized as in (25) below, the convergence should be faster.

C. Two Steepest Descent Algorithms

Recently Lang and McClellan [32], [35] have shown that the MEM spectral estimate can be obtained as the solution to two additional finite-dimensional dual-optimization problems. Furthermore, each problem involves the optimization of a convex functional over a convex set. In this situation, one can employ any number of standard algorithms that are guaranteed to converge. Steepest descent and quasi-Newton methods have been implemented, but other approaches could be taken [38].

The first optimization problem follows from (12) and the observation that if $\boldsymbol{p}_o$ corresponds to the MEM spectrum then

$$\boldsymbol{r}_o^T \boldsymbol{p}_o = \int_K \frac{\boldsymbol{E}^T(\boldsymbol{k})}{P_o(\boldsymbol{k})} \boldsymbol{p}_o = \int_K \frac{P_o(\boldsymbol{k})}{P_o(\boldsymbol{k})} = \int_K 1. \tag{22}$$

Therefore, (22) states that the hyperplane $\boldsymbol{r}_o^T \boldsymbol{p} = \text{constant} = \int_K 1$ contains the optimum vector $\boldsymbol{p}_o$. The dual problem (12) is, therefore, equivalent to

$$\underset{\boldsymbol{p} \in \boldsymbol{P}^+}{\text{minimize}}\ H(\boldsymbol{p}) \tag{23a}$$

subject to $\boldsymbol{p}$ satisfying

$$\boldsymbol{r}_o^T \boldsymbol{p} = \int_K 1. \tag{23b}$$

The gradient projection method [38] can be used to solve (23a, b). In this method, the steepest descent iteration is modified so that the solution always lies in the hyperplane defined by the constraint (23b).

$$\boldsymbol{p}_{n+1} = \boldsymbol{p}_n - \alpha_n \bar{\boldsymbol{g}}_n. \tag{24}$$

The direction $\bar{\boldsymbol{g}}_n$ is the gradient, $\boldsymbol{g}_n = -Q(\boldsymbol{p}_n)$ projected onto the subspace $\boldsymbol{r}_o^T \boldsymbol{p} = 0$. Thus

$$\bar{\boldsymbol{g}}_n = \boldsymbol{g}_n - \frac{\boldsymbol{r}_o^T \boldsymbol{g}_n}{\boldsymbol{r}_o^T \boldsymbol{r}_o} \boldsymbol{r}_o. \tag{25}$$

The step size α_n is chosen to be a nonnegative scalar minimizing $H(\boldsymbol{p}_n - \alpha_n \bar{\boldsymbol{g}}_n)$. The convexity of H guarantees convergence. Finally, note that to start the recursion (24) an initial vector $\boldsymbol{p}_1$ must be chosen. This can be done by using the constraint to write $p_1(\boldsymbol{0})$ in terms of $p(\boldsymbol{\delta})$ for $\boldsymbol{\delta} \neq \boldsymbol{0}$ (see (26) below).

A slightly different descent algorithm was implemented by Lang [30]. As before the gradient direction is modified but the result is not (25). The constraint $\boldsymbol{r}_o^T \boldsymbol{p} = \int 1$ can be used to eliminate the parameter $p(0)$ via the equation

$$p(0) = \frac{1}{r_o(0)} \cdot \left[\int_K 1 - \sum_{\boldsymbol{\delta} \neq \boldsymbol{0}} r_o^*(\boldsymbol{\delta}) p(\boldsymbol{\delta}) \right]. \tag{26}$$

The reduced problem is now an unconstrained minimization problem in the $2M$ real variables $\{\text{Re}\, p(\boldsymbol{\delta}), \text{Im}\, p(\boldsymbol{\delta}) : \boldsymbol{\delta} \in \Delta, \boldsymbol{\delta} \neq \boldsymbol{0}\}$. The gradient vector with respect to these parameters is $\hat{\boldsymbol{g}}_n$

$$\hat{\boldsymbol{g}}_n = -\left[Q(\boldsymbol{p}_n) - \frac{r(0) \boldsymbol{r}_o}{r_o(0)} \right] \boldsymbol{\delta} \neq \boldsymbol{0}. \tag{27a}$$

$$\hat{g}_n(0) = \frac{1}{r_o(0)} \cdot \sum_{\boldsymbol{\delta} \neq \boldsymbol{0}} r_o^*(\boldsymbol{\delta}) \hat{g}_n(\boldsymbol{\delta}). \tag{27b}$$

The component $\hat{g}_n(\boldsymbol{0})$ chosen by (27b) is such that $\boldsymbol{r}_o^T \hat{\boldsymbol{g}}_n = 0$ and the correction keeps the solution within the hyperplane (27b).

Once again, the step size α_n is chosen by performing a line search from the point $\boldsymbol{p}_n$, in the direction $-\hat{\boldsymbol{g}}_n$, to a minimum of H. The minimum point is taken as $\boldsymbol{p}_{n+1}$. The convexity of H guarantees convergence. This technique is known as the reduced gradient method [38].

The second optimization problem, also resulting from (12), is the following:

$$\underset{\boldsymbol{p} \in \boldsymbol{P}^+}{\text{minimize}}\ \boldsymbol{r}_o^T \boldsymbol{p} \tag{28a}$$

subject to

$$H(\boldsymbol{p}) \leqslant H_o \tag{28b}$$

where H_o is the maximum entropy attainable in (8). It follows from (22) that the minimum in (28a) will be equal to $\int_K 1$. Furthermore, it is not necessary to know H_o beforehand. The problem can be solved with the constraint $H(\boldsymbol{p}) \leqslant 0$ to obtain a solution $\bar{\boldsymbol{p}}$. This solution can then be scaled to obtain $\boldsymbol{p}_o$ because the objective functional (28a) is linear and $H(\boldsymbol{p})$ is convex. The result is

$$\boldsymbol{p}_o = \frac{\bar{\boldsymbol{p}}}{\boldsymbol{r}_o^T \bar{\boldsymbol{p}}} \int_K 1. \tag{29}$$

Furthermore, the optimum $\bar{\boldsymbol{p}}$ lies on the boundary of the constraint set (28b) so that it is sufficient to solve (28a, b) with an equality constraint.

The justification for (28a, b) comes from the following reasoning. The hyperplane $\boldsymbol{r}_o^T \boldsymbol{p} = \int_K 1$ is tangent to the surface $H(\boldsymbol{p}) = H_o = H(\boldsymbol{p}_o)$ at the point $\boldsymbol{p} = \boldsymbol{p}_o$, where the correlation matching constraints are satisfied. The entropy $H(\boldsymbol{p})$ is a strictly convex function of $\boldsymbol{p}$. It follows that surfaces of constant entropy are convex, and that the intersection point $\boldsymbol{p}_o$ may be obtained as that point in the convex set $H(\boldsymbol{p}) \leqslant H_o$ having the minimum projection onto the vector $\boldsymbol{r}_o$. In the time-series case, this second optimization problem leads to the usual interpretation of the MEM estimate as a least squares predictor [30].

D. A Primal MEM Algorithm

It is possible to solve the MEM problem (8), (9) directly in the primal space. Wernecke and D'Addario [59] considered such a problem in image reconstruction rather than power spectrum estimation, but the two are equivalent if the image must be positive. The primal algorithm attempts to optimize the power spectrum $S(\boldsymbol{k})$ directly, or, at least, samples $S(\boldsymbol{k}_i)$ on a dense grid. The dimensionality of the primal problem can be quite large. In other respects, the problem is merely a constrained optimization problem to which any number of algorithms can be applied. One convenient feature of the primal formulation is that the correlation matching equality constraints (9) can be relaxed and replaced by an inequality constraint

$$\sum_{\boldsymbol{\delta} \in \Delta} \sigma_{\boldsymbol{\delta}}^2 \left| r_o(\boldsymbol{\delta}) - \int_K S(\boldsymbol{k}) e^{j\boldsymbol{k} \cdot \boldsymbol{\delta}} \right|^2 < \epsilon \tag{30}$$

that tolerates an error in the correlation measurements. After all, in practice, $r_o(\boldsymbol{\delta})$ would only be an estimate of the true acf so (30) should be more reasonable than (9). The parameters $\sigma_{\boldsymbol{\delta}}^2$ are introduced to weight the relative variance in the estimates. Newman [44], [45] has also considered this inequality constraint for the 1-D problem when there are missing lag estimates.

Wernecke and D'Addario actually implemented a penalty function [38] algorithm for maximizing entropy (8) subject to (30). In this method, the constraint is attached to (8) via a (large) fixed multiplier that penalizes errors in the constraint. The result is an unconstrained maximization problem that can be solved using steepest ascent or the conjugate gradient method. One simplification that the authors found useful was to deflect the true gradient direction to ignore variables $S(\boldsymbol{k}_i)$ below a certain threshold, unless they would be increased in the original search direction. Examples were given for image reconstruction on a 21 × 21 grid, in which case the optimization is performed over 441 variables.

A drawback of (30) is that it cannot weight correlation between errors in the covariance estimates. However, a general quadratic form based on a positive-definite matrix $\boldsymbol{W}$ can be introduced in place of the left-hand side of (30)

$$(\boldsymbol{r}_o - \int_K S(\boldsymbol{k})\boldsymbol{E}(\boldsymbol{k}))^T \boldsymbol{W}(\boldsymbol{r}_o - \int_K S(\boldsymbol{k})\boldsymbol{E}(\boldsymbol{k})) \leqslant 1. \quad (31)$$

At the time that the final version of this paper was prepared, the author succeeded in deriving a dual entropy minimization problem for constraints of the type (31).

$$\underset{\boldsymbol{p}\in P^+}{\text{minimize}} \ \{H(\boldsymbol{p}) + \boldsymbol{r}_o^T\boldsymbol{p} + (\boldsymbol{p}^T\boldsymbol{W}^{-1}\boldsymbol{p})^{1/2}\}. \quad (32)$$

The gradient for (32) is merely (15) plus an additional term, $\boldsymbol{W}^{-1}\boldsymbol{p}/(\boldsymbol{p}^T\boldsymbol{W}^{-1}\boldsymbol{p})^{1/2}$. Thus iterative algorithms based on steepest descent will still apply to this problem. Details of this method are being prepared for publication [42].

VIII. AR (Autoregressive) Spectral Estimation

In the previous section, the maximum entropy spectral estimator was shown to require the solution of a nonlinear optimization problem. Conspicuous in its absence was an autoregressive (AR) model for the spectrum. The optimum MEM spectrum was shown to be the inverse of a positive polynomial, but in higher dimensions (than 1-D) this polynomial may not be factorable as the magnitude squared of a finite-order polynomial. Since a spectrum based on an AR model will always be factorable, the MEM is, in the m-D case, more general than the AR spectral estimate [11]. In this section, the AR method is reviewed and several versions which have appeared in the literature are discussed.

The AR spectral estimate is derived by forming an AR model for the available signal samples. This AR model can be developed in different ways, but the most common is to choose the model coefficients so that the prediction error energy over the signal is minimized. For example, assume that the 2-D signal $s[m, n]$ is known for $0 \leqslant m, n \leqslant N - 1$, and let the AR model be $L \times L$

$$A(z, w) = \sum_{k=0}^{L-1}\sum_{l=0}^{L-1} a[k, l] z^{-k} w^{-l} \quad (33)$$

where $a[0, 0] = 1$. The prediction error is given by

$$E = \sum_{m=0}^{N-1}\sum_{n=0}^{N-1} |e[m, n]|^2 \quad (34)$$

where $e[m, n] = s[m, n] * a[m, n]$. Minimization of E with respect to $a[m, n]$ leads to the normal equations—a set of L^2 simultaneous linear equations.

$$\boldsymbol{Ra} = [E \ 0 \ 0 \ 0 \ \cdots \ 0]^T. \quad (35)$$

Solution of these equations is not difficult. In this 2-D situation with regular spacing the correlation matrix $\boldsymbol{R}$ will have a block Toeplitz structure for which a number of efficient matrix inversion algorithms exist [28].

The AR spectral estimate is quite tempting to use. It is easy to compute, and it is of the same form as the MEM estimate. However, the AR estimator usually does not satisfy the correlation matching property (9) of MEM. On the other hand, experiments with the AR estimate show that it does have potential as a high-resolution estimator [25].

A primary issue with m-D AR modeling is the choice of the prediction filter mask and the order of computation for the recursive model [41]. In (33) the denominator of the model was chosen to occupy only a rectangle in the first quadrant. Jackson and Chien [21] have shown that spectral estimates of plane waves based on such a first-quadrant model will exhibit peaks with a definite skew. Changing the filter mask to lie in a different quadrant will alter the skew. Therefore, they suggest combining these one-quadrant AR estimates either as $|A_1 \cdot A_2|^{-1}$ or $(|A_1|^2 + |A_2|^2)^{-1}$ to obtain an estimate that is more nearly circularly symmetric. Another option would be the nonsymmetric half-plane filter mask.

A simultaneous frequency and wavenumber estimator that uses 2-D linear prediction of a space-time signal has been described by Kumaresan and Tufts [29]. The method uses two quarter-plane filters, A_1 and A_2, as above. However, the order of A_1 (and A_2) is the highest allowable, i.e., an $M \times N$ prediction-error filter for an $M \times N$ array of data. There are only two prediction-error equations for each filter, so the solution is taken to be the minimum norm solution. The spectral estimate is formed as $(|A_1|^2 + |A_2|^2)^{-1}$. Simulation results show that this technique will locate plane waves in noise successfully.

Jain [24], [25] suggests the use of a semicausal model. The prediction filter mask includes samples in the past of one direction but uses the past and future in the other direction. It is also possible to use a totally noncausal model. This estimate is not strictly all-pole; a numerator term is also generated. Newman [43], Woods [61], and Tjostheim [57] have also proposed using AR modeling to obtain a spectral estimate.

A final paper that might be considered an AR estimate is the work of Roucos and Childers [52]. Their algorithm extrapolates the correlation function from its original finite set of measurements. At each step of the procedure, the new portions of the correlation function are obtained by maximizing the prediction error of an appropriate 2-D linear prediction filter. The result involves maximizing the determinant of the matrix of correlation samples and, hence, is akin to the entropy maximization for 1-D. The order of computation in the plane along which to extrapolate is left to the user. Enough new correlation points must be generated so that the DFT can be used to obtain the spectral estimate.

Two important points must be kept in mind about AR modeling. First, the technique is essentially limited to cases where the signal of acf is known on a regularly spaced grid. Nonuniform sampling renders the prediction-error filter mask useless. Second, the magnitude squared of the AR model is generally not the same as the correlation-matching MEM spectral estimate.

IX. (Generalized) Pisarenko Estimates

In Pisarenko's method [51] for 1-D signals, the spectral estimate is formed as the sum of line components in a background of white noise. The technique requires an eigenanalysis of the autocorrelation matrix $\boldsymbol{R} = [r(\boldsymbol{x}_i - \boldsymbol{x}_j)]$. The smallest eigenvalue $\lambda_{\min}$ is the noise power. When this smallest eigenvalue has multiplicity one, the location of the spectral lines can be determined by finding the zeros of a polynomial whose coefficients are the elements of the eigenvector corresponding to λ_{min}. When the multiplicity of λ_{min} is greater than one, the number of spectral lines is reduced, but a similar solution technique can still be applied to a reduced-order problem. Needless to say, extension of this method to higher dimensions would be unlikely if the same reliance on polynomial factorization were required. Recently, Lang and McClellan [30], [33], [34] have obtained a generalization of Pisarenko's estimate that is applicable in the general m-D situation. This new result is based on consideration of the extendibility question for positive functions.

A. The Extendibility Problem

Whenever spectral estimation is done from a set of correlation measurements under the constraint of correlation matching (9), the question of extendibility must be addressed. Specifically, for a finite set of correlation samples $\{r_o(\boldsymbol{\delta}):\boldsymbol{\delta}\in\Delta\}$, does there exist a positive spectrum $S(\boldsymbol{k})$ such that for $\boldsymbol{\delta}\in\Delta$

$$r_o(\boldsymbol{\delta}) = \int_K S(\boldsymbol{k})e^{j\boldsymbol{k}\cdot\boldsymbol{\delta}}. \tag{36}$$

If one such positive spectrum exists, then $\boldsymbol{r}_o$ (the vector of correlations) is said to be *extendible*. In 1-D regularly spaced situations, the extendibility of $\boldsymbol{r}_o$ is equivalent to the positive definiteness of a Toeplitz autocorrelation matrix. In higher dimensions, positive definiteness is not sufficient; a more general characterization is needed. Such a result was originally derived in 1952 by Calderón and Pepinsky [9] for a problem in X-ray crystallography and generalized by Rudin [53]. The result gives a geometric characterization of the set of extendible correlation vectors. Let $\boldsymbol{P}$ denote the set of vectors $\boldsymbol{p}$ (of polynomial coefficients) corresponding to positive Δ-polynomials $P(\boldsymbol{k})\geqslant 0$.[4] It turns out that $\boldsymbol{P}$ is a closed convex cone with its vertex at the origin.[5] The set $\boldsymbol{E}$ of extendible correlation vectors is characterized by the following theorem.

Extension Theorem: The correlation vector $\boldsymbol{r}$ is extendible (i.e., $\boldsymbol{r}\in\boldsymbol{E}$) if and only if $\boldsymbol{r}^T\boldsymbol{p}\geqslant 0$ for all $\boldsymbol{p}\in\boldsymbol{P}$. Furthermore, $\boldsymbol{E}$ is a closed convex cone with its vertex at the origin.

A more complete discussion of this theorem including a proof can be found in [30]. In 1-D, this theorem reduces to the positive definite condition because all positive 1-D polynomials are expressible as the magnitude squared of a lower order polynomial. The corresponding fact about positive polynomials in higher dimensions is false, as was shown by Hilbert in the solution to his seventeenth problem.

The extendible set $\boldsymbol{E}$ is important for testing the validity of a correlation vector $\boldsymbol{r}$ to be used in a correlation matching estimator. For instance, it would be useful to have a simple test for whether the given correlation vector $\boldsymbol{r}_o$ for MEM lies within the interior of $\boldsymbol{E}$, which consists of those vectors $\boldsymbol{r}$ satisfying $\boldsymbol{r}^T\boldsymbol{p}>0$ for all $\boldsymbol{p}\in\boldsymbol{P}$. The test of the extendibility theorem, as it stands, requires an exhaustive search through all positive polynomials. In the next section, a generalization of Pisarenko's method is presented. It will be shown that one possible use of this estimator would be as an extendibility test.

[4] Note the terminology "positive polynomial" actually refers to a nonnegative Δ-polynomial $P(\boldsymbol{k})\geqslant 0$, so that $P(\boldsymbol{k})$ may be zero. If $P(\boldsymbol{k})>0$ for all $\boldsymbol{k}\in K$, then the qualification strictly positive will be used.

[5] A cone with vertex zero is a set that contains all positive multiples of any element in the set; the vector $\boldsymbol{0}$ is thus a member of the set. The set is convex if for any two vectors, $\boldsymbol{x}$ and $\boldsymbol{y}$, in the set, the vector $(1-\alpha)\boldsymbol{x}+\alpha\boldsymbol{y}$ is also in the set for $0\leqslant\alpha\leqslant 1$.

B. A Generalization of Pisarenko's Estimate [30]

Pisarenko's 1-D spectral estimate decomposes the spectrum into a sum of discrete lines (sinusoids) plus white noise of unknown variance. The m-D generalization is obtained by observing that the boundary of the extendible set $\boldsymbol{E}$ contains the correlation vectors for line spectra. This follows from the fact that the boundary of $\boldsymbol{E}$ is characterized by

$$\boldsymbol{r}^T\boldsymbol{p} = 0 = \int_K S(\boldsymbol{k})P(\boldsymbol{k}). \tag{37}$$

Since $P(\boldsymbol{k})$ is nonnegative, $S(\boldsymbol{k})$ can only be nonzero where $P(\boldsymbol{k})=0$. The roots of $P(\boldsymbol{k})$ give the positions of the spectral lines in $S(\boldsymbol{k})$.

Thus given a measured correlation vector $\boldsymbol{r}_o$, and the (finite) correlation vector $\boldsymbol{n}$ of the known noise (for white noise $\boldsymbol{n}$ is the unit impulse), $\boldsymbol{r}_o$ can be decomposed as

$$\boldsymbol{r}_o = \boldsymbol{r}' + \gamma\boldsymbol{n} \tag{38}$$

with $\boldsymbol{r}'$ on the boundary of $\boldsymbol{E}$. If $\boldsymbol{n}\in\boldsymbol{E}$, the extendibility of $\boldsymbol{r}_o$ is equivalent to $\gamma\geqslant 0$. Thus the determination of γ will serve as an extendibility test. The solution for γ is obtained from the following dual optimization problem:

$$\gamma = \underset{\boldsymbol{p}\in\boldsymbol{P}}{\text{minimize}}\ \boldsymbol{r}_o^T\boldsymbol{p} \tag{39}$$

subject to

$$\boldsymbol{n}^T\boldsymbol{p} = 1. \tag{40}$$

This algorithm requires the minimization of a finite-dimensional convex functional over a convex set. When discretized in $\boldsymbol{k}$, the problem appears similar to 2-D FIR filter design, and linear programming is applicable. After the vector $\boldsymbol{p}$ is determined, the zeros of $P(\boldsymbol{k})$ are used to obtain the locations of the spectral lines. It can be shown that the number of lines is at most $2M$.

Representation Theorem: If $\boldsymbol{r}'$ is on the boundary of $\boldsymbol{E}$ then

$$r'(\boldsymbol{\delta}) = \sum_{i=1}^{2M} \mu_i e^{j\boldsymbol{k}_i\cdot\boldsymbol{\delta}} \tag{41}$$

for at most $2M$ nonnegative μ_i and $\boldsymbol{k}_i\in K$.

Note that in the 1-D time series case, (41) reduces to an expression with only M terms. However, there is a uniqueness problem that arises when $P(\boldsymbol{k})$ has zeros along a hyperplane in the m-D set K. A similar nonuniqueness question arises in 2-D FIR filter design. The severity of this nonuniqueness problem is not yet known for practical spectral estimation problems.

C. MUSIC

Variations of Pisarenko's eigenvector method have been published [50], [54], [55]. In 1979, Schmidt [54] published one which he called MUSIC–MUltiple SIgnal Classification. This estimate is formed from the eigenvectors of the estimated intersensor correlation matrix $\boldsymbol{R}=[r(\boldsymbol{x}_i-\boldsymbol{x}_j)]$. Corresponding to each eigenvalue λ_i (ordered by increasing magnitude) is a polynomial $V_i(\boldsymbol{k})$, whose coefficients are the elements of the eigenvector for λ_i. The MUSIC estimate is obtained by forming

$$\left(\sum_{i=1}^{L} |V_i(\boldsymbol{k})|^2\right)^{-1} \tag{42}$$

where the number of terms L is determined by subtracting the number of signals thought to be present from the dimension of the correlation matrix. The locations of the lines are obtained by picking the largest peaks of (42). Note that this step eliminates any need for polynomial factorization. Once the location of the spectral lines is known, the amplitude of each signal component can be calculated using least squares. By avoiding polynomial factorization, this technique should be applicable to m-D arbitrarily spaced array problems. As with the m-D Pisarenko estimate there are still potential uniqueness problems if (42) does not have a small number of discrete peaks.

The relationship between an m-D form of MUSIC and the generalized Pisarenko estimator of Section IX-B is not understood. There are clearly cases where the MUSIC estimate will not be the same, for example, when $P(\boldsymbol{k})$ in (40) is not factorable. Whether there are m-D problems where the two methods give the same answer is unknown.

X. Concluding Remarks

This paper presents a unified discussion of existing m-D spectral estimation techniques. An effort has been made to include reference to all methods, although the presentation is naturally colored by the author's own research experiences. Considerable emphasis has been placed on MEM and Pisarenko's method where it is likely that future research will be active. One area that is yet to be investigated for most m-D spectral estimators is that of asymptotic convergence with respect to the number of correlation measurements (or array sensors), number of time observations, etc. Such analyses would deal with the practical problem of calculating the spectral estimate from observed data. Finally, there will be continuing interest in the usefulness of these algorithms when applied to actual field data.

Acknowledgment

The author is indebted to S. Lang whose insights have been invaluable in preparing this paper. Comments by A. Baggeroer and G. Duckworth also helped to improve the manuscript.

References

[1] A. B. Baggeroer, "Recent signal processing advances in spectral and frequency wavenumber function estimation and their application to offshore structures," in *2nd Int. Conf. on Behaviour of Off-Shore Structures* (Imperial College, London, England, Aug. 1979), pp. 457-476.

[2] —, "Estimation errors in the measurement of cross spectral matrices for array processing," in *Workshop Italy-USA on Digital Processing* (Portovenere-La Spezia, Italy, Aug. 1981), pp. 45-61.

[3] A. B. Baggeroer and R. Falconer, "Array refraction profiles and crustal models," *J. Geophys. Res.* (accepted for publication).

[4] T. Barnard and J. P. Burg, "Analytical studies of techniques for the computation of high-resolution wavenumber spectra," *Advanced Array Res. Rep. 9*, Texas Instruments, Inc., May 1969.

[5] M. S. Bartlett, *An Introduction to Stochastic Processes with Special Reference to Methods and Applications*. New York: Cambridge Univ. Press, 1953.

[6] J. P. Burg, "Maximum entropy spectral analysis," presented at the 37th Int. Meet., Soc. Explor. Geophys., Oklahoma City, OK, Oct. 1968.

[7] —, "Maximum entropy spectral analysis," Ph.D. dissertation, Stanford University, Stanford, CA, May 1975.

[8] J. A. Cadzow and K. Ogino, "Two-dimensional spectral estimation," *IEEE Trans. Acoust., Speech Signal Proc.*, vol. ASSP-29, pp. 396-401, June 1981.

[9] A. Calderón and R. Pepinsky, "On the phases of Fourier coefficients for positive real functions," in *Computing Methods and the Phase Problem in X-Ray Crystal Analysis* (The X-Ray Crystal Analysis Laboratory, Dep. Physics, Pennsylvania State College, 1952), pp. 339-348.

[10] J. Capon, "High resolution frequency-wavenumber spectrum analysis," *Proc. IEEE*, vol. 57, pp. 1408-1418, Aug. 1969.

[11] B. W. Dickinson, "Two-dimensional Markov spectrum estimates need not exist," *IEEE Trans. Inform. Theory*, vol. IT-26, pp. 120-121, Jan. 1980.

[12] G. L. Duckworth, "Adaptive array processing for high-resolution acoustic imaging," *S. M. thesis*, M.I.T., Cambridge, MA, Sept. 1979.

[13] D. E. Dudgeon, "Fundamentals of digital array processing," *Proc. IEEE*, vol. 65, pp. 898-904, June 1977.

[14] D. J. Edelblute, J. M. Fisk, and G. L. Kinnison, "Criteria for optimum signal-detection theory for arrays," *J. Acoust. Soc. Amer.*, vol. 41, pp. 199-205, 1965.

[15] O. L. Frost and T. M. Sullivan, "High resolution two-dimensional spectral analysis," in *Proc. ICASSP 79* (Washington, DC, Apr. 1979), pp. 673-676.

[16] O. L. Frost, "High-resolution 2-D spectral analysis at low SNR," in *Proc. ICASSP 80* (Denver, CO, Apr. 1980), pp. 580-583.

[17] R. W. Gerchberg, "Super resolution through error energy reduction," *Optica Acta*, vol. 21, pp. 709-720, 1974.

[18] O. S. Halpeny and D. G. Childers, "Composite wavefront decomposition via multidimensional digital filtering of array data," *IEEE Trans. Circuits Syst.*, vol. CAS-22, pp. 552-563, June 1975.

[19] R. A. Haubrich, "Array design," *Bull. Seismological Soc. America*, pp. 977-991, June 1968.

[20] T. S. Huang, "Two-dimensional windows," *IEEE Trans. Audio Electroacoust.*, vol. AU-20, pp. 88-90, Mar. 1972.

[21] L. B. Jackson and H. C. Chien, "Frequency and bearing estimation by two-dimensional linear prediction," in *Proc. ICASSP 79* (Washington, DC, Apr. 1979), pp. 665-668.

[22] P. L. Jackson, L. S. Joyce, and G. B. Feldkamp, "Application of maximum entropy frequency analysis to synthetic aperture radar," in *Proc. RADC Spectrum Estimation Workshop* (Rome, NY, May 1978), pp. 217-225.

[23] A. K. Jain and S. Raganath, "Two-dimensional spectral estimation," in *Proc. RADC Spectrum Estimation Workshop* (Rome, NY, May 1978), pp. 151-157.

[24] A. K. Jain, "Spectral estimation and signal extrapolation in one and two dimensions," in *Proc. RADC Spectrum Estimation Workshop* (Rome, NY, Oct. 1979), pp. 195-214.

[25] A. K. Jain and S. Raganath, "High resolution spectrum estimation in two dimensions," in *Proc. 1st ASSP Workshop on Spectral Estimation* (Hamilton, Ont., Canada, Aug. 1981), pp. 3.4.1-3.4.5.

[26] —, "Extrapolation algorithms for discrete signals with application in spectral estimation," *IEEE Trans. Acoust., Speech Signal Proc.*, vol. ASSP-29, pp. 830-845, Aug. 1981.

[27] L. S. Joyce, "A separable 2-D autoregressive spectral estimation algorithm," in *Proc. ICASSP 79* (Washington, DC, Apr. 1979), pp. 677-680.

[28] J. H. Justice, "A Levinson-type algorithm for two-dimensional Wiener filtering using bivariate Szegö polynomials," *Proc. IEEE*, vol. 65, pp. 882-886, June 1977.

[29] R. Kumaresan and D. W. Tufts, "A two-dimensional technique for frequency-wavenumber estimation," *Proc. IEEE*, vol. 69, pp. 1515-1517, Nov. 1981.

[30] S. W. Lang, "Spectral estimation for sensor arrays," Ph.D. dissertation, M.I.T., Cambridge, MA., Aug. 1981.

[31] S. W. Lang and J. H. McClellan, "Spectral estimation for sensor arrays," in *Proc. 1st ASSP Workshop on Spectral Estimation* (Hamilton, Ont., Canada, Aug. 1981), pp. 3.2.1-3.2.7.

[32] —, "MEM spectral estimation for sensor arrays," in *Proc. Int. Conf. on Digital Signal Processing* (Florence, Italy, Sept. 1981), pp. 383-390.

[33] —, "The extension of Pisarenko's method to multiple dimensions," in *Proc. ICASSP 82* (Paris, France, May 1982), pp. 125-128.

[34] —, "Spectral estimation for sensor arrays," *IEEE Trans. Acoust., Speech, Signal Proc.*, (submitted for publication).

[35] —, "Multi-dimensional MEM spectral estimation," *IEEE Trans. Acoust., Speech, Signal Proc.* (accepted for publication).

[36] J. S. Lim and N. A. Malik, "A new algorithm for two-dimensional maximum entropy power spectrum estimation," *IEEE Trans. Acoust., Speech, Signal Proc.*, vol. ASSP-29, pp. 401-413, June 1981.

[37] D. G. Luenberger, *Optimization by Vector Space Methods*. New York: Wiley, 1969.

[38] —, *Introduction to Linear and Nonlinear Programming*. Reading, MA: Addison-Wesley, 1973.

[39] N. A. Malik, "One and two-dimensional maximum entropy spectral estimation," Ph.D. dissertation, M.I.T., Cambridge, MA, Nov. 1981.

[40] T. L. Marzetta, "A linear prediction approach to two-dimensional spectral factorization and spectral estimation," Ph.D. dissertation, M.I.T., Cambridge, MA., Feb. 1978.

[41] —, "Two-dimensional linear prediction: Autocorrelation arrays, minimum-phase prediction error filters, and reflection coefficient arrays," *IEEE Trans. Acoust., Speech, Signal Proc.*, vol. ASSP-28, pp. 725-733, Dec. 1980.

[42] J. H. McClellan and S. W. Lang, "Multi-dimensional MEM spectral estimation," presented at the *Int. Conf. on Spectral Analysis and its Use in Underwater Acoustics*, London, England, Apr. 1982.

[43] W. I. Newman, "A new method of multi-dimensional power spectral analysis," *Astrm. Astrophys.*, vol. 54, pp. 369-380, 1977.

[44] —, "Extension to the maximum entropy method II," *IEEE Trans. Inform. Theory*, vol. IT-25, Nov. 1979.

[45] —, "Extension to the maximum entropy method III," in *Proc.*

1st ASSP Workshop on Spectral Estimation (Hamilton, Ont., Canada, Aug. 1981), pp. 1.7.1–1.7.6.

[46] C. Ong, "An investigation of two new high-resolution two-dimensional spectral estimate techniques," in *Spec. Rep. 1, Long Period Array Processing Development*, Texas Instruments, Inc., Apr. 1971.

[47] A. V. Oppenheim and R. W. Schafer, *Digital Signal Processing*. Englewood Cliffs, NJ: Prentice-Hall, 1975.

[48] A. Papoulis, "A new algorithm in spectral analysis and band-limited extrapolation," *IEEE Trans. Circuits Syst.*, vol. CAS-22, pp. 735–742, Sept. 1975.

[49] J. V. Pendrell, "The maximum entropy principle in two-dimensional spectral analysis," Ph.D. dissertation, York Univ., Toronto, Ont., Canada, Nov. 1979.

[50] V. F. Pisarenko, "On the estimation of spectra by means of non-linear functions of the covariance matrix," *Geophys. J. Roy. Astron. Soc.*, vol. 28, pp. 511–531, 1972.

[51] ——, "The retrieval of harmonics from a covariance function," *Geophys. J. Roy. Astron. Soc.*, vol. 33, pp. 347–366, 1973.

[52] S. Roucos and D. G. Childers, "A two-dimensional maximum entropy spectral estimator," in *Proc. ICASSP 79* (Washington, DC, Apr. 1979), pp. 669–672.

[53] W. Rudin, "The extension problem for positive-definite functions," *Ill. J. Math.*, vol. 7, pp. 532–539, 1963.

[54] R. Schmidt, "Multiple emitter location and signal parameter estimation," in *Proc. RADC Spectrum Estimation Workshop* (Oct. 1979), pp. 243–258.

[55] R. O. Schmidt, "A signal subspace approach to emitter location and spectral estimation," Ph.D. dissertation, Stanford University, Stanford, CA, Aug. 1981.

[56] C. W. Therrien, "Relations between 2-D and multichannel linear prediction," *IEEE Trans. Acoust., Speech, Signal Processing*, vol. ASSP-29, pp. 454–456, June 1981.

[57] D. Tjostheim, "Autoregressive modeling and spectral analysis of array data in the plane," *IEEE Trans. Geosci. Remote Sensing*, vol. GE-19, pp. 15–24, Jan. 1981.

[58] P. D. Welch, "The use of FFT for the estimation of power spectra: A method based on time average over short, modified periodograms," *IEEE Trans. Audio Electroacoust.*, vol. AU-15, pp. 70–73, June 1967.

[59] S. J. Wernecke and L. R. D'Addario, "Maximum entropy image reconstruction," *IEEE Trans. Comput.*, vol. C-26, pp. 351–364, Apr. 1977.

[60] P. Whittle, "On stationary processes in the plane," *Biometrika*, vol. 41, pp. 434–449, Dec. 1954.

[61] J. W. Woods, "Two-dimensional discrete Markovian fields," *IEEE Trans. Inform. Theory*, vol. IT-18, pp. 232–240, Mar. 1972.

[62] ——, "Two-dimensional Markov spectral estimation," *IEEE Trans. Inform. Theory*, vol. IT-22, pp. 552–559, Sept. 1976.

Spectral Estimation for Sensor Arrays

STEPHEN W. LANG, MEMBER, IEEE, AND JAMES H. McCLELLAN, SENIOR MEMBER, IEEE

Abstract—The array processing problem is briefly discussed and an abstract spectral estimation problem is formulated. This problem involves the estimation of a multidimensional frequency-wave vector power spectrum from measurements of the correlation function and knowledge of the spectral support.

The investigation of correlation-matching spectral estimates leads to the extendibility question: does there exist any positive spectrum on the spectral support that exactly matches a given set of correlation samples? In answering this question, a mathematical framework is developed in which to analyze and design spectral estimation algorithms.

Pisarenko's method of spectral estimation, which models the spectrum as a sum of impulses plus a noise component, is extended from the time series case to the more general array processing case. Pisarenko's estimate is obtained as the solution of a linear optimization problem, which can be solved using a linear programming algorithm such as the simplex method.

I. INTRODUCTION

JUST as the power spectrum of a stationary time series describes a distribution of power versus frequency, the frequency-wave vector power spectrum of a homogeneous and stationary wavefield describes a distribution of power versus wave vector and temporal frequency, or equivalently, versus propagation direction and temporal frequency. The frequency-wave vector spectrum, or information that can be derived from it, is important in many applications areas. In radio astronomy, the formation of an image is equivalent to the estimation of a power spectrum. The detection and bearing estimation of targets in radar and sonar can be based upon information contained in a power spectral estimate. Hence, power spectral estimation, from data provided by sensor arrays, is of great practical interest.

Section II contains a synopsis of wavefields and sensor arrays and an introduction to the spectral estimation problem. Alternative mathematical representations of power spectra, as measures and as spectral density functions, are discussed. Section II introduces the *coarray*, the set of vector separations and time lags for which correlation samples are available, and the *spectral support*, the region of frequency-wave vector space containing power to which the sensors are sensitive. No particular structure is assumed for either the coarray or the spectral support. Section II concludes with the formulation of an abstract problem: the estimation of a power spectrum given only that it is positive on the spectral support, zero outside, and has certain known correlations for separations in the coarray. Although simpler than many problems encountered in practice, the key features that distinguish the array problem from the problem of time series power spectral estimation are retained: the multidimensionality of the frequency variable and the nonuniformity of the coarray.

Given this problem formulation, it is natural to consider spectral estimates that match the known information: spectral estimates that are positive on the spectral support, zero outside, and exactly match the measured correlations. The investigation of such *correlation-matching* spectral estimates raises two important questions. The first and more fundamental question concerns the existence of any such estimate. This *extendibility* problem has deep historical roots [1], and was recently raised by Dickinson [2] with reference to the 2-D maximum entropy spectral estimation method, and is the subject of some recent work by Cybenko [3]-[4]. The extendibility problem is explored in Section III. Extendible sets of correlation measurements are characterized. Their dependence on the spectral support and the effect of discretizing the spectral support is also considered. In answering the extendibility question, the necessary mathematical framework is developed in which to analyze specific spectral estimation methods and to design algorithms for their computation.

The second question raised is that of uniqueness: is there a unique correlation-matching spectral estimate, and if not, how can a specific one be chosen? In fact, a unique estimate does not exist except in very special cases; the task of a spectral estimation method is the selection of one out of an ensemble of spectra satisfying the correlation matching, positivity, and spectral support constraints. Section IV concerns Pisarenko's method [5], which involves modeling the correlation measurements as a sum of two components. One, a noise component of known spectral shape but unknown amplitude, is made as large as possible without rendering the other component nonextendible. The spectral estimate resulting from Pisarenko's method is shown to solve a linear optimization problem. A solution to this optimization problem will always exist if the correlation measurements are extendible. In fact, Pisarenko's method is shown to be intimately related to the extendibility question and an algorithm for the computation of Pisarenko's estimate will also serve as an extendibility test. It is shown that Pisarenko's estimate is not always unique in the general case, although it is unique in the time series case, where the linear optimization problem reduces to an eigenvalue problem.

II. THE ARRAY PROCESSING PROBLEM

Imagine a multidimensional homogeneous medium supporting a complex-valued wavefield $u(\boldsymbol{x}, t)$ and containing an array of sensors. The wavefield will be assumed to be homogeneous

Manuscript received January 25, 1982; revised September 1, 1982.

This work was supported in part by a Fannie and John Hertz Foundation Fellowship, in part by the National Science Foundation under Grant ECS79-15226, and in part by the Army Research Office under Contract DAAG29-81-K-0073.

S. W. Lang is with Schlumberger-Doll Research, Ridgefield, CT 06877.

J. H. McClellan is with Schlumberger Well Services, Austin Engineering Center, Austin, TX 78759.

Reprinted from *IEEE Trans. Acoust., Speech, Signal Processing,* vol. ASSP-31, pp. 349–358, Apr. 1983.

and stationary so that its second order statistics are described by a correlation function r, or equivalently, by a power spectrum μ [6].

$$r(\boldsymbol{\delta}, \tau) = E[u^*(\boldsymbol{x}, t)\, u(\boldsymbol{x} + \boldsymbol{\delta}, t + \tau)] = \int e^{j(\boldsymbol{k} \cdot \boldsymbol{\delta} + \omega\tau)}\, d\mu. \tag{2.1}$$

The representation of the power spectrum by a positive measure μ[1] provides the flexibility needed to deal with a range of spectral supports in a unified manner and to handle spectra that contain impulses: finite power at a single wave vector.

It is more common, in the engineering literature, to represent a power spectrum by means of a positive spectral density function $S(\boldsymbol{k}, \omega)$. In this representation

$$r(\boldsymbol{\delta}, \tau) = \int e^{j(\boldsymbol{k} \cdot \boldsymbol{\delta} + \omega\tau)}\, S(\boldsymbol{k}, \omega)\, d\nu, \tag{2.2}$$

where ν is some fixed measure which allows (2.2) to be interpreted as a multidimensional surface or volume integral, possibly weighted, over frequency-wave vector space.

Given a power spectral density function $S(\boldsymbol{k}, \omega)$, it is possible to define a corresponding positive measure by requiring the measure of a subset B of frequency-wave vector space to equal the integral of the spectral density function over B:

$$\mu(B) = \int_B d\mu = \int_B S(\boldsymbol{k}, \omega)\, d\nu. \tag{2.3}$$

A simple spectral estimation problem will now be formulated. Particular attention will be paid to modeling features of the data collection process that are common to many array processing problems. These features include measurement of the correlation function at a finite number of nonuniformly distributed points, and constraints on the region of frequency-wave vector space in which power may be present.

The sensors each produce a time function that is the wavefield u sampled at a point in space. The collection of time functions produced by all the sensors, the array output or response, is to be processed so as to provide an estimate of the frequency-wave vector power spectrum. The stochastic character of the wavefield invariably leads to random variations of any spectral estimate based on the array output. To combat this effect, spectral estimates are often based on stable statistics derived from the array output. A common example of such a statistic is a correlation estimate calculated by multiplying the output of one sensor with the time-delayed output of a second sensor, and averaging over time. This process results in an estimate of the correlation function at a temporal lag corresponding to the delay time and a spatial separation that is the vector distance between the two sensors. The averaging process provides statistically stable correlation estimates and results in the statistical stability of a spectral estimate based on these correlation estimates. It is important to note that estimates of the correlations are only available for a finite set of intersensor separations and time delays, the *coarray* [8]. The topic of error in the correlation estimates will not be addressed. Rather, this paper is concerned with the properties of sets of actual correlation samples, and with spectral estimates based on correlation samples.

It is assumed that the spectrum is known to be confined to a bounded region of frequency-wave vector space, the *spectral support*. Outside of this support the spectrum is assumed to be zero. A bounded spectral support can arise naturally in several ways. For example, in a medium that supports scalar waves, known source, medium, and sensor characteristics can be used to construct an appropriate spectral support. The source may have a known temporal bandwidth, or a known finite angular extent. The dispersion relation and attenuation in the medium limits the region of frequency-wave vector space in which power may be present. The sensors may have finite temporal bandwidth and may be directional. All of these effects can be modeled by assuming that no power is present outside of a certain region of frequency-wave vector space. A known spectral support, based on the physics of a particular problem, constitutes important prior information that can be brought to bear on the spectral estimation problem.

In many applications much more data is available in the time dimension than in the space dimension. In these cases it is convenient to separate out the time variable by Fourier analyzing the time series output of each sensor and then doing a separate wave vector spectral estimate for each temporal frequency by using the Fourier coefficients as data for a wave vector spectral estimator. Thus the estimation problem is formulated for complex data even though physical wavefields are real valued. Fortunately, conventional Fourier analysis is often satisfactory when data are abundant, as well as being implicit in the narrow-band character of many sensors. Where limited data in the time dimension make the above approach impractical and wide-band sensor arrays are available, the full problem may be treated by including the temporal variables τ and ω in the vectors $\boldsymbol{\delta}$ and $\boldsymbol{k}$. Thus $\boldsymbol{\delta}$ would describe a separation in both space and time, and $\boldsymbol{k}$ a space-time wave vector. It shall be assumed that one of these two approaches has been taken; hence, the temporal variables τ and ω will be dropped.

A simple example of the spectral estimation model developed above is provided by a sensor array composed of uniformly oriented dish antennas.

Example 2.1: A three dish array. Imagine that an array of dish antennas, shown in Fig. 1, is used to receive a single temporal frequency ω_0, corresponding to a wavelength λ.

A dish antenna of diameter d has a passband that is roughly described by

$$\sqrt{k_2^2 + k_3^2} \leqslant \frac{0.61}{d}.$$

Assuming that the wavefield satisfies the dispersion relation for a homogeneous, nondispersive medium, the support for the

[1] A *positive measure* is a set function that assigns a nonnegative power to each measurable subset of frequency-wave vector space [7]. The power in a subset B of frequency-wave vector space is denoted $\int_B d\mu = \mu(B)$.

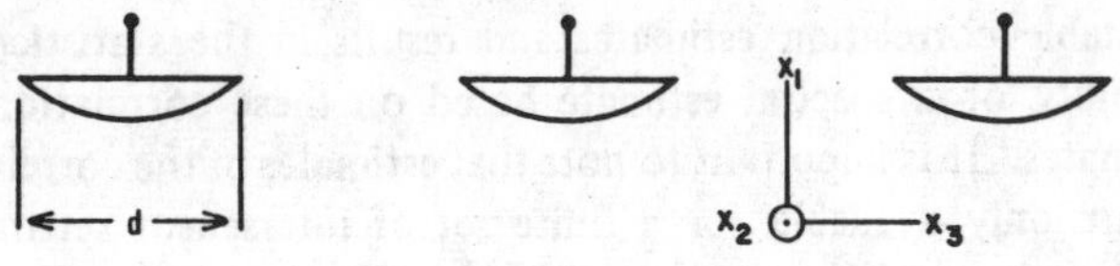

Fig. 1. A three dish array.

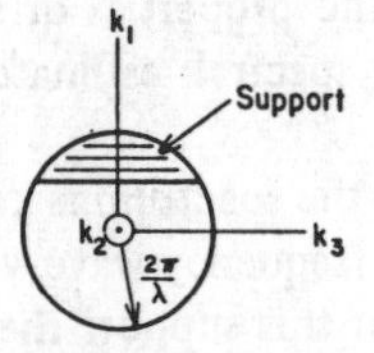

Fig. 2. Spectral support for an array of dish antennas.

spectral estimate should be the polar cap described by the two equations

$$k_1^2 + k_2^2 + k_3^2 = \left[\frac{2\pi}{\lambda}\right]^2$$

$$\sqrt{k_2^2 + k_3^2} \leqslant \frac{0.61}{d}$$

and shown in Fig. 2.

The coarray for this problem is just the set of all 3-dimensional spatial separations between antennas in the array.

III. Extendibility

A simple model of the array processing problem was constructed in the last section: given certain correlation measurements and a spectral support, produce a spectral estimate. It is natural to use the known information about the spectrum to constrain the spectral estimate by requiring that the spectral estimate match the measured correlations, be positive, and be confined to the spectral support. Such spectral estimates are called *correlation-matching* spectral estimates.

The investigation of correlation-matching spectral estimates raises a fundamental existence question. Given a finite collection of measured correlations and a spectral support, does there exist at least one correlation-matching spectral estimate? If such a spectral estimate exists, the measured correlations are said to be *extendible*.[2] After some necessary mathematical definitions, this existence question will be answered by characterizing the set of extendible correlation measurements.

A. Spectral Supports and Coarrays

It is first necessary to define more carefully the terms spectral support and coarray. The spectral support K is assumed to be a compact subset of R^D, i.e., K is closed and bounded. Assuming that K is compact leads to a certain technical advantage: a continuous function on a compact set attains its infimum and supremum. Furthermore, compactness should always hold in a physical problem. As discussed in the previous section, knowledge of source, medium, and sensor characteristics can be used to construct an appropriate spectral support.

The coarray Δ will be defined as a finite subset of R^D with the properties

i) $\mathbf{0} \in \Delta$;
ii) if $\boldsymbol{\delta} \in \Delta$, then $-\boldsymbol{\delta} \in \Delta$;
iii) $\{e^{j\mathbf{k}\cdot\boldsymbol{\delta}}: \boldsymbol{\delta} \in \Delta\}$ is a set of linearly independent functions on $K \subset R^D$.

Condition i) implies knowledge of $r(\mathbf{0})$, the total power in the spectrum. Condition ii) reflects the fact that the correlation function is always conjugate symmetric; thus if $r(\boldsymbol{\delta})$ is known, so to is $r(-\boldsymbol{\delta})$. Together, conditions i) and ii) imply that Δ is of the form

$$\Delta = \{\mathbf{0}, \pm\boldsymbol{\delta}_1, \cdots, \pm\boldsymbol{\delta}_M\}. \tag{3.1}$$

Condition iii) guarantees that the correlation measurements are independent; each measurement gives new information about the spectrum.

If $D > 1$ then the spectral estimation problem is *multidimensional*. If $D = 1$, $K = [-\pi, \pi]$, and $\Delta = \{0, \pm 1, \cdots, \pm M\}$ then the spectral estimation problem is that of the familiar *time series case* and the extendibility question reduces to the famous trigonometric moment problem [9].

B. Conjugate-Symmetric Functions and Their Vector Representation

The spectral support and the coarray naturally suggest a vector-space setting for the spectral estimation problem, one in which conjugate-symmetric complex-valued functions on Δ will play a central role. A *conjugate-symmetric* function f on Δ is one for which $f(-\boldsymbol{\delta}) = f^*(\boldsymbol{\delta})$ for all $\boldsymbol{\delta} \in \Delta$. Correlation samples, from which spectral estimates are to be made, are such functions. (Because of this symmetry, many of the expressions to follow are real valued even though, for the sake of simplicity, they have been written in a form which suggests that they might be complex valued.) The coarray Δ has $2M + 1$ elements, and so a conjugate-symmetric function on Δ is characterized by $2M + 1$ independent real numbers. Thus a conjugate-symmetric function on Δ may be thought of as a vector in R^{2M+1}.[3] Both the functional notation $f(\boldsymbol{\delta})$ and the vector notation $\boldsymbol{f}$ will be used.

Since $\{e^{j\mathbf{k}\cdot\boldsymbol{\delta}}: \boldsymbol{\delta} \in \Delta\}$ is a linearly independent set of functions on K, it follows that each vector $\boldsymbol{p}$ in R^{2M+1} can be uniquely associated with a real-valued *Δ-polynomial* $P(\mathbf{k})$ on K through the relation

$$P(\mathbf{k}) = \sum_{\boldsymbol{\delta} \in \Delta} p(\boldsymbol{\delta})\, e^{-j\mathbf{k}\cdot\boldsymbol{\delta}}. \tag{3.2}$$

The vector $\boldsymbol{p}$ shall be termed *positive* if $P(\mathbf{k}) \geqslant 0$ on K. P shall denote the set of those vectors associated with positive Δ-polynomials. From the compactness of K, it can be shown that P is a closed convex cone with vertex at the origin.[4]

[2] The correlation function obtained through the inverse Fourier transform of a correlation-matching spectral estimate is a suitable *extension* of the correlation measurements to all spatial separations.

[3] A vector space over the real numbers is chosen because it is only multiplication by a *real* number which sends a correlation function into another correlation function.

[4] A set C is a *cone with vertex at the origin* if $\boldsymbol{x} \in C$ implies $\alpha\boldsymbol{x} \in C$ for all $\alpha > 0$ [10]. Cones are important kinds of sets in the spectral estimation problem because it is only multiplication by *positive* real numbers that sends a correlation function into another correlation function, and a Δ-polynomial into another Δ-polynomial.

The inner product between a vector $\boldsymbol{r}$ of correlation samples and a vector $\boldsymbol{p}$ of polynomial coefficients shall be defined as

$$(\boldsymbol{r}, \boldsymbol{p}) = \sum_{\boldsymbol{\delta} \in \Delta} r^*(\boldsymbol{\delta})\, p(\boldsymbol{\delta}).$$

$$= r(\mathbf{0})\, p(\mathbf{0}) + 2 \sum_{i=1}^{M}$$

$$\cdot\, [\mathrm{Re}\, r(\boldsymbol{\delta}_i)\, \mathrm{Re}\, p(\boldsymbol{\delta}_i) + \mathrm{Im}\, r(\boldsymbol{\delta}_i)\, \mathrm{Im}\, p(\boldsymbol{\delta}_i)]. \qquad (3.3)$$

This inner product gives a new way of writing a Δ-polynomial: $P(\boldsymbol{k}) = (\boldsymbol{\phi_k}, \boldsymbol{p})$, where $\boldsymbol{\phi_k}$ denotes the vector with components $\phi_{\boldsymbol{k}}(\boldsymbol{\delta}) = e^{j\boldsymbol{k}\cdot\boldsymbol{\delta}}$. Also note that if $\boldsymbol{r} = \int_K \boldsymbol{\phi_k}\, d\mu$ then $(\boldsymbol{r}, \boldsymbol{p}) = \int_K P(\boldsymbol{k})\, d\mu$, an expression of Parseval's relation.

C. Characterizations of Extendibility

Let E denote the set of extendible correlation vectors. That is, $\boldsymbol{r} \in E$ if

$$\boldsymbol{r} = \int_K \boldsymbol{\phi_k}\, d\mu \qquad (3.4)$$

for some positive measure μ on K. From the properties of the integral, it follows that E is a closed convex cone with vertex at the origin. Furthermore, a section through E at $r(\mathbf{0}) = 1$;

$$E' = \{\boldsymbol{r} \in E\colon\ r(\mathbf{0}) = 1\} \qquad (3.5)$$

is the convex hull of the compact set

$$A = \{\boldsymbol{\phi_k}\colon\ \boldsymbol{k} \in K\}. \qquad (3.6)$$

Thus E is the closed convex cone, with vertex at the origin, generated by A. This characterization of extendible correlations is similar to that given originally by Carathéodory in 1907 for the trigonometric moment problem [1]. It is important in that the set of extendible correlation vectors is described in terms of the simple set A. It also gives a clear geometric picture of extendibility, and will be useful in proofs.

A second characterization of extendibility which is more useful in the development of spectral estimation methods results from expressing E as the intersection of all the closed half-spaces containing it [10]. This characterization involves duality, since half-spaces are defined by linear functionals, i.e., elements of the dual space. A closed half space is defined by a vector $\boldsymbol{q}$ and a real number c as the set

$$\{\boldsymbol{r}\colon\ (\boldsymbol{r}, \boldsymbol{q}) \geqslant c\}. \qquad (3.7)$$

To determine the particular half-spaces containing E, it is sufficient to consider those correlation vectors that generate E: positive multiples of vectors in the set A. A closed half-space contains E if and only if $(\alpha\boldsymbol{\phi_k}, \boldsymbol{q}) = \alpha Q(\boldsymbol{k}) \geqslant c$ for every $\boldsymbol{k} \in K$ and every $\alpha \geqslant 0$. Since α may be made arbitrarily large, it must be true that $Q(\boldsymbol{k}) \geqslant 0$, i.e., $\boldsymbol{q}$ is a member of the cone P. The smallest half-space containing E for such a $\boldsymbol{q}$ corresponds to choosing $c = 0$. Thus

$$E = \bigcap_{\boldsymbol{p} \in P} \{\boldsymbol{r}\colon\ (\boldsymbol{r}, \boldsymbol{p}) \geqslant 0\} \qquad (3.8)$$

or, in words, the following.

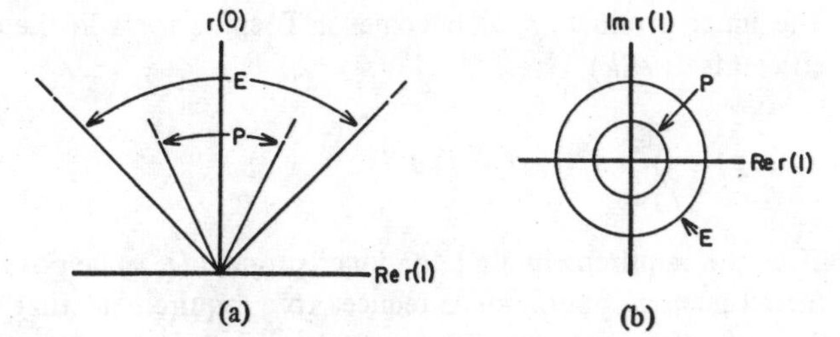

Fig. 3. E and P for $K = [-\pi, \pi]$ and $\Delta = \{0, \pm 1\}$. (a) A section through E and P at $\mathrm{Im}\, r(1) = 0$, and (b) a section through E and P at $r(0) = 1$.

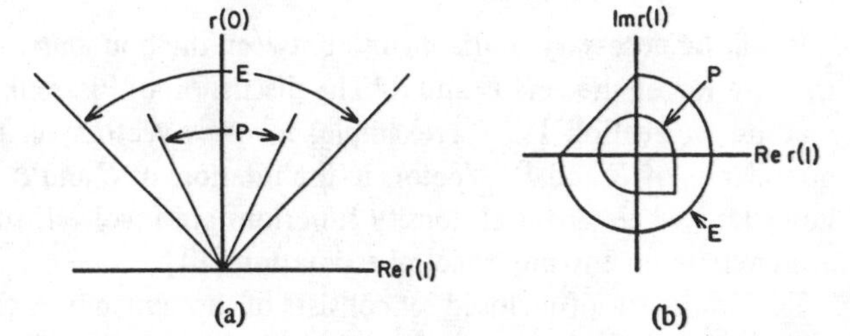

Fig. 4. E and P for $K = [-\pi, \pi/2]$ and $\Delta = \{0, \pm 1\}$. (a) A section through E and P at $\mathrm{Im}\, r(1) = 0$, and (b) a section through E and P at $r(0) = 1$.

The Extension Theorem: The vector $\boldsymbol{r}$ is extendible if and only if $(\boldsymbol{r}, \boldsymbol{p}) \geqslant 0$ for all positive $\boldsymbol{p}$.

Positive polynomials thus occur naturally in the extendibility problem, since they define the supporting hyperplanes of the set E of extendible correlation vectors. In the language of functional analysis, the extension theorem, which is a form of Farkas' Lemma [11], simply states that E and P are positive conjugate cones [10]. This theorem has the important effect of transferring the simple characterization of P, in terms of positivity, to a characterization of E. Although the incorporation of a spectral support into the problem is new, essentially the same characterization of extendibility was originally used by Calderón and Pepinsky [12], and Rudin [13].

Fig. 4 demonstrates the dependence of E on the spectral support. There are two ways of looking at this dependence. The direct way is to note that E is the convex cone generated by A; because K has been reduced, A has shrunk and E is smaller than in Fig. 3. The indirect way involves constraints; the set K constrains the set P via the positivity condition and the set P constrains the set E via the extendibility theorem. Thus when K shrinks, P grows, and E shrinks.

In the time series case, the extendibility theorem reduces to a test of the positive definiteness of a Toeplitz matrix formed from the correlation samples. Hence, extendibility may be thought of as a general analog of positive-definiteness.

Example 3.1: The Time Series Case; $D = 1$, $K = [-\pi, \pi]$, $\Delta = \{0, \pm 1, \cdots, \pm M\}$. In this case, the extendibility problem reduces to the trigonometric moment problem [9]. Although not generally true, it follows in the time series case, from the fundamental theorem of algebra, that a positive polynomial may be factored as the squared magnitude of an Mth degree trigonometric polynomial

$$P(k) = |A(k)|^2.$$

The inner product $(\boldsymbol{r}, \boldsymbol{p})$ becomes a Toeplitz form in the coefficients of $A(k)$

$$(\boldsymbol{r}, \boldsymbol{p}) = \sum_{i,j=0}^{M} a^*(i)\, r(i-j)\, a(j).$$

Thus the requirement that the inner product $(\boldsymbol{r}, \boldsymbol{p})$ be positive for all positive polynomials reduces to a requirement that the Toeplitz form corresponding to the correlation measurements be positive definite.

D. Boundary and Interior

It will be necessary to distinguish between the boundary and the interior of the sets E and P. The discussion of Pisarenko's method in Section IV, for example, involves vectors on the boundaries of E and P. Vectors in the interiors of E and P are important where spectral density functions are involved, such as in maximum entropy spectral estimation [14].

The *boundary* of a closed set consists of those members that are arbitrarily close to some vector outside the set. The *interior* of a closed set consists of those members that are not on the boundary. The boundary and the interior of a finite dimensional set do not depend upon a particular choice of a vector norm [15]. Moreover, since P and E are convex sets, they have interiors and boundaries which are particularly simple to characterize.

The boundary of P, denoted ∂P, consists of those positive polynomials which are zero for some $\boldsymbol{k} \in K$. The interior of P, denoted P°, consists of those polynomials which are strictly positive on K.

Positive polynomials may be used to define the boundary and the interior of E. The boundary of E, denoted ∂E, consists of those extendible correlation vectors which make a zero inner product with some nonzero positive polynomial. The interior of E, denoted E°, consists of those correlation vectors which make strictly positive inner products with every nonzero positive polynomial.

E. Power Spectral Density Functions

Many spectral estimation methods represent the power spectrum, not as a measure, but as a spectral density function. This leads to a modification of the extendibility problem: given a fixed finite positive measure ν, which defines the integral

$$\boldsymbol{r} = \int_K \boldsymbol{\phi_k}\, S(\boldsymbol{k})\, d\nu, \tag{3.9}$$

which correlation vectors $\boldsymbol{r}$ can be derived from some strictly positive function $S(\boldsymbol{k})$? Under one additional constraint on ν, easily satisfied in practice, it can be shown that vectors which can be represented in this fashion are exactly those vectors in the interior of E. Furthermore, it can be shown that any vector in the interior of E can be represented in the form (3.9) for some continuous, strictly positive $S(\boldsymbol{k})$.

An Extension Theorem for Spectral Density Functions: If every neighborhood of every point in K has strictly positive ν-measure, then

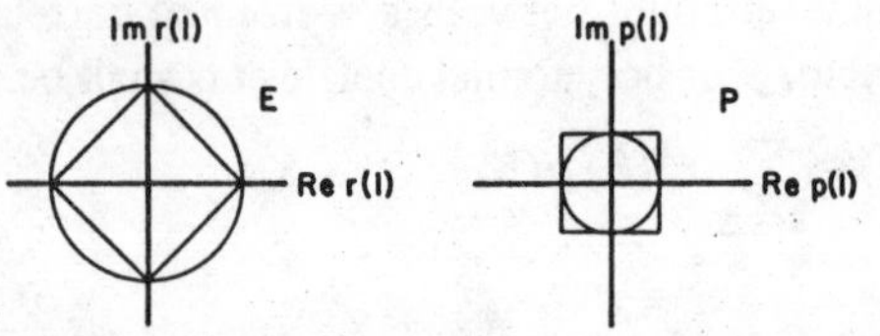

Fig. 5. Approximation of a spectral support by sampling; a section at $r(0) = 1$.

1) if $S(\boldsymbol{k})$ is uniformly bounded away from zero over K, then

$$\boldsymbol{r} = \int_K \boldsymbol{\phi_k}\, S(\boldsymbol{k})\, d\nu \in E^\circ;$$

2) if $\boldsymbol{r} \in E^\circ$ then

$$\boldsymbol{r} = \int_K \boldsymbol{\phi_k}\, S(\boldsymbol{k})\, d\nu$$

for some continuous, strictly positive function $S(\boldsymbol{k})$.

A proof of this theorem is contained in Appendix A.

F. Discretization of the Spectral Support

Many spectral supports of interest contain an infinite number of points. These spectral supports must often be approximated, in computational algorithms, by a finite number of points. It is important, therefore, to understand the effects of such approximation.

Consider the discrete spectral support

$$K = \{\boldsymbol{k}_i \in R^D \colon i = 0, \cdots, N-1\}. \tag{3.10}$$

A measure μ on a discrete support is completely characterized by its value $\mu(\boldsymbol{k}_i)$ at each point. Thus the inverse Fourier integral reduces to a finite sum

$$\int_K \boldsymbol{\phi_k}\, d\mu = \sum_{i=0}^{N-1} \boldsymbol{\phi}_{\boldsymbol{k}_i}\, \mu(\boldsymbol{k}_i). \tag{3.11}$$

Similarly, for spectral density functions

$$\int_K \boldsymbol{\phi_k}\, S(\boldsymbol{k})\, d\nu = \sum_{i=0}^{N-1} \boldsymbol{\phi}_{\boldsymbol{k}_i}\, S(\boldsymbol{k}_i)\, \nu(\boldsymbol{k}_i). \tag{3.12}$$

The measure ν can be considered to define a quadrature rule for integrals over the spectral support.

From the definitions of extendible correlation vectors and of positive polynomials, it can be seen that if a spectral support is formed by choosing a finite number of points out of some original spectral support, then the new set E is a convex polytope inscribed within the original set E and the new set P is a convex polytope circumscribed about the original set P. Hence, the new E is smaller than the original E and the new P is larger than the original P. A sufficiently dense sampling of the original spectral support will result in polytopes which approximate the original sets to arbitrary precision. For example, Fig. 5 shows the effect of approximating the spectral support $[-\pi, \pi]$ by the four samples $\{0, \pm(\pi/2), \pi\}$ for $\Delta = \{0, \pm 1\}$. The original E and P cones have a circular cross section at

$r(0) = 1$, as shown in Fig. 3. The cones corresponding to the sampled support both have a square cross section. The boundaries of the new and old cones intersect at vectors corresponding to the sample points.

IV. Pisarenko's Method

Pisarenko described a time series spectral estimation method in which the spectrum is modeled as a sum of impulses plus a white noise component [5]. If the white noise component is chosen as large as possible, he showed that the position and amplitudes of the impulses needed to match the measured correlations are uniquely determined. Pisarenko's method will be derived in the more general array setting and for a more general noise component. The relationship of Pisarenko's method to the extendibility question will be demonstrated.

The extended Pisarenko's estimate will be derived as the solution of an optimization problem involving the minimization of a linear functional over a convex region defined by linear constraints. A solution to this optimization problem always exists, but it may not be unique. A dual optimization problem is derived which, in the time series case, leads to the familiar interpretation of Pisarenko's method as the design of a constrained least squares smoothing filter. Again, a solution to this dual problem always exists, but may not be unique.

Algorithms for the computation of Pisarenko's method are discussed. A primal optimization problem is written, for a spectral support composed of a finite number of points, as a standard form linear program. The application of the simplex method to the solution of this primal linear program is discussed. A dual linear program is presented. The possibility of computational algorithms faster than the simplex method is also discussed.

A. Pisarenko's Method for Sensor Arrays

The basis of Pisarenko's method is the unique decomposition (Fig. 6) of a correlation vector $\boldsymbol{r}$ into the sum of a scaled noise correlation vector $\boldsymbol{n}$, in the interior of E, plus a remainder $\boldsymbol{r}'$, on the boundary of E

$$\boldsymbol{r} = \boldsymbol{r}' + \alpha \boldsymbol{n}. \tag{4.1}$$

The assumption that $\boldsymbol{n}$ is in E° implies that such a decomposition of an arbitrary vector $\boldsymbol{r}$ exists and is unique. Consider the one-parameter family of correlation vectors

$$\boldsymbol{r}_c = \boldsymbol{r} - c\boldsymbol{n}. \tag{4.2}$$

For c sufficiently positive, $\boldsymbol{r}_c$ must be nonextendible and, for c sufficiently negative, $\boldsymbol{r}_c$ must be extendible, since the assumption that $\boldsymbol{n} \in E^\circ$ implies that E contains a neighborhood of $\boldsymbol{n}$. The convexity of E implies that there is some greatest number α such that $\boldsymbol{r}' = \boldsymbol{r} - \alpha\boldsymbol{n}$ is extendible. Since there are nonextendible vectors arbitrarily close to $\boldsymbol{r}'$, $\boldsymbol{r}'$ must be on the boundary of E. Furthermore, since $\alpha \geq 0$ if and only if $\boldsymbol{r}$ is extendible, this decomposition of $\boldsymbol{r}$ can also be used as an extendibility test.

This unique decomposition of $\boldsymbol{r}$ can be formulated as a primal linear optimization problem over all positive power

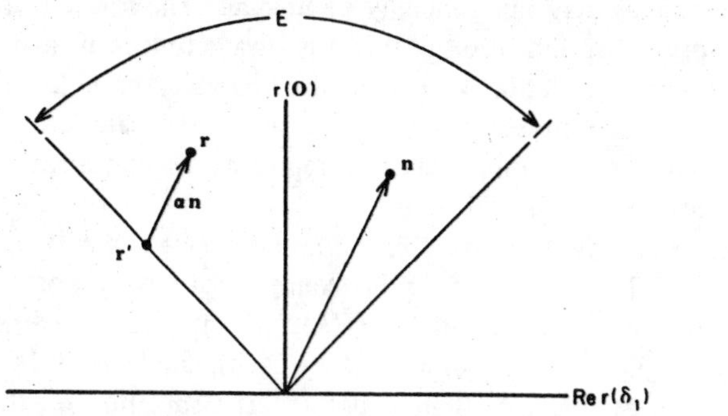

Fig. 6. Decomposition of a vector $\boldsymbol{r}$ into vector $\boldsymbol{r}'$ on the boundary of E plus a multiple of a given vector $\boldsymbol{n}$.

spectra. Note that $\boldsymbol{r}'$ has at least one positive spectral representation μ' and that, from (4.1) for $\boldsymbol{\delta} = \boldsymbol{0}$

$$\alpha = \left[\frac{r(0)}{n(0)} - \frac{1}{n(0)} \int_K d\mu'\right]. \tag{4.3}$$

The statement that α is the largest number such that the remainder $\boldsymbol{r}' = \boldsymbol{r} - \alpha\boldsymbol{n}$ is extendible leads to the linear optimization problem

$$\max_{\mu \geq 0} \left[\frac{r(0)}{n(0)} - \frac{1}{n(0)} \int_K d\mu\right] \tag{4.4a}$$

such that

$$\boldsymbol{r} = \int_K \boldsymbol{\phi}_{\boldsymbol{k}}\, d\mu + \left[\frac{r(0)}{n(0)} - \frac{1}{n(0)} \int_K d\mu\right] \boldsymbol{n}. \tag{4.4b}$$

The maximum is α and is attained at $\mu = \mu'$.

Since $\boldsymbol{n}$ is extendible, it corresponds to some positive measure μ_n. Hence, (4.1) becomes

$$\boldsymbol{r} = \int_K \boldsymbol{\phi}_{\boldsymbol{k}} [d\mu' + \alpha d\mu_n]. \tag{4.5}$$

If $\alpha \geq 0$ then $\mu' + \alpha\mu_n$ is a positive measure which matches the correlation measurements and which has the largest possible noise component.

Some further information about the remainder $\boldsymbol{r}'$ and its spectral representation can be derived. $\boldsymbol{r}'$ is on the boundary of E; hence, it makes a zero inner product with some nonzero positive polynomial

$$(\boldsymbol{r}', \boldsymbol{p}') = \int_K P'(\boldsymbol{k})\, d\mu' = 0. \tag{4.6}$$

It follows that the support of μ' must be on the zero set of $P'(\boldsymbol{k})$. More precisely, the support of any spectral representation of $\boldsymbol{r}'$ must be on the intersection of the zero sets of all positive polynomials that make a zero inner product with $\boldsymbol{r}'$. This suggests the final step in the derivation of Pisarenko's method; namely the association of the remainder $\boldsymbol{r}'$ with an impulsive spectrum.

The fact that the objective functional of the primal optimization problem is not strictly convex suggests that the solution μ' may not generally be unique. The solution μ' to the primal optimization problem is always unique if, and only if, every correlation vector on the boundary of E has a unique spectral representation. In the time series case, every such $\boldsymbol{r}'$ does have a unique spectral representation, as a sum of M or fewer impulses [5].

Example 4.1: The Time Series Case, $D = 1, K = [-\pi, \pi], \Delta = \{0, \pm 1, \cdots, \pm M\}$. As in example 3.1, every positive polynomial can be factored as $P'(k) = |A(k)|^2$ for some Mth degree trigonometric polynomial $A(k)$. $A(k)$, and hence $P'(k)$, can be zero at no more than M points. The spectrum μ', therefore, must be a sum of impulses at these points. Furthermore, since it is possible to construct a positive polynomial that is zero at $N \leqslant M$ arbitrarily selected points, and nowhere else, it follows that $\boldsymbol{r}'$ has a unique spectral representation as a sum of impulses at the common zeros of all positive polynomials $\boldsymbol{p}'$ such that $(\boldsymbol{r}, \boldsymbol{p}') = 0$.

More generally, the extension theorem combined with Carathéodory's theorem [16] shows that there is at least one spectral representation of $\boldsymbol{r}'$ as a sum of no more than $2M$ impulses.

The Representation Theorem: If $\boldsymbol{r}' \in \partial E$, then there exist $a(i) \geqslant 0$ and $\boldsymbol{k}_i \in K$ such that

$$\boldsymbol{r}' = \sum_{i=1}^{2M} a(i) \boldsymbol{\phi}_{\boldsymbol{k}_i}. \tag{4.7}$$

A proof of the representation theorem can be found in Appendix B. This representation, and thus the solution to the primal optimization problem may not be unique. A further discussion of this uniqueness problem can be found in Appendix C.

If α and the locations of the impulses in a unique solution μ' could be determined for a given $\boldsymbol{r}'$, then the impulse amplitudes could be calculated simply by solving a set of linear equations. A dual optimization problem will now be derived which gives α and $\boldsymbol{p}'$ such that $(\boldsymbol{r}', \boldsymbol{p}') = 0$. Then, if $\boldsymbol{r}'$ has a unique spectral representation, the impulse locations can be determined from the zeros of $P(\boldsymbol{k})$. From the extendibility theorem

$$\begin{aligned}(\boldsymbol{r}', \boldsymbol{p}) &= (\boldsymbol{r} - \alpha \boldsymbol{n}, \boldsymbol{p}) \\ &= (\boldsymbol{n}, \boldsymbol{p}) \, [(\boldsymbol{r}, \boldsymbol{p})/(\boldsymbol{n}, \boldsymbol{p}) - \alpha] \geqslant 0.\end{aligned} \tag{4.8}$$

Since $\boldsymbol{n} \in E^\circ$ and $\boldsymbol{r}' \in \partial E$, it follows that $(\boldsymbol{n}, \boldsymbol{p}) > 0$ and $(\boldsymbol{r}', \boldsymbol{p}) \geqslant 0$ for all $\boldsymbol{p} \in P$. Furthermore, since $(\boldsymbol{r}', \boldsymbol{p}) = 0$ for some $\boldsymbol{p}' \in P$, it follows that

$$\alpha = \min\, (\boldsymbol{r}, \boldsymbol{p}) \tag{4.9a}$$

over the set

$$\{\boldsymbol{p} \in P\colon\ (\boldsymbol{n}, \boldsymbol{p}) = 1\} \tag{4.9b}$$

and that the minimum is attained at $\boldsymbol{p}'$. The solution to this dual problem may not be unique even in the time series case, where it reduces to the eigenvector problem derived by Pisarenko and leads to the interpretation of Pisarenko's method as determining a constrained least squares smoothing filter.

Example 4.2: The Time Series Case, $D = 1, K = [-\pi, \pi]$, $\Delta = \{0, \pm 1, \cdots, \pm M\}$. As in example (3.1)

$$(\boldsymbol{r}, \boldsymbol{p}) = \sum_{i,j=0}^{M} a^*(i)\, r(i-j)\, a(j).$$

Furthermore, if $\boldsymbol{n}$ corresponds to white noise of unit power,

$$(\boldsymbol{n}, \boldsymbol{p}) = p(0) = \sum_{i=0}^{M} |a(i)|^2.$$

Thus the dual optimization problem reduces to finding the eigenvector of the Toeplitz matrix associated with $\boldsymbol{r}$ corresponding to the smallest eigenvalue. If there are several such eigenvectors, the impulses are located at the common zeros of the corresponding polynomials. Any normalized eigenvector corresponding to the minimal eigenvalue gives the coefficients of a smoothing filter, the sum of whose squared magnitudes is constrained to be one, which gives the least output power when fed an input process whose correlations are described by $\boldsymbol{r}$ [17].

B. The Computation of Pisarenko's Estimate

In the design of algorithms to compute Pisarenko's estimate, one may be concerned with a discrete spectral support

$$K = \{\boldsymbol{k}_i \in R^D\colon\ i = 0, \cdots, N-1\}. \tag{4.10}$$

On such a support, the primal problem (4.4) can be rewritten as the standard form *linear program*

$$\min_{\nu \geqslant 0} \sum_{i=0}^{N-1} \nu(\boldsymbol{k}_i) \tag{4.11a}$$

such that, for $\boldsymbol{\delta} \in \Delta$, $\boldsymbol{\delta} \neq 0$,

$$\sum_{i=0}^{N-1} \left[e^{j\boldsymbol{k}_i \cdot \boldsymbol{\delta}} - \frac{n(\boldsymbol{\delta})}{n(\boldsymbol{0})} \right] \nu(\boldsymbol{k}_i) = \frac{r(\boldsymbol{\delta})}{r(\boldsymbol{0})} - \frac{n(\boldsymbol{\delta})}{n(\boldsymbol{0})} \tag{4.11b}$$

with N variables and $2M$ constraints. The minimum is $1 - (n(\boldsymbol{0})/r(\boldsymbol{0}))\, \alpha$ and is attained for $\nu' = (1/r(\boldsymbol{0}))\, \mu'$. The fundamental theorem of linear programming [18] is equivalent to the representation theorem in this case. Given that a solution exists to this linear program, as shown in the previous section, the fundamental theorem guarantees a solution in which no more than $2M$ of the $\nu(\boldsymbol{k}_i)$'s are nonzero, a so-called *basic* solution.

The dual linear program [15]

$$\min_{q} \sum_{\boldsymbol{\delta} \neq 0} \left[\frac{r^*(\boldsymbol{\delta})}{r(\boldsymbol{0})} - \frac{n^*(\boldsymbol{\delta})}{n(\boldsymbol{0})} \right] q(\boldsymbol{\delta}) \tag{4.12a}$$

such that, for $i = 0, \cdots, N-1$

$$\sum_{\boldsymbol{\delta} \neq 0} \left[e^{-j\boldsymbol{k}_i \cdot \boldsymbol{\delta}} - \frac{n^*(\boldsymbol{\delta})}{n(\boldsymbol{0})} \right] \boldsymbol{q}(\boldsymbol{\delta}) \geqslant -1 \tag{4.12b}$$

is equivalent to the dual problem (4.9) for a discrete spectral support, where the constraint

$$(\boldsymbol{n}, \boldsymbol{p}) = 1 \tag{4.13}$$

has been used to eliminate $p(\mathbf{0})$ and where $\boldsymbol{q} = n(\mathbf{0})\,\boldsymbol{p}$. Its minimum is $(n(\mathbf{0})/r(\mathbf{0}))\,\alpha - 1$ and is attained at $\boldsymbol{q}' = n(\mathbf{0})\,\boldsymbol{p}'$.

The primal problem can be solved using the *simplex method* [18]. The application of the simplex method to the primal problem results in essentially the same computational algorithm as the application of the (single) *exchange method* to the dual problem [19]. By incorporating a technique to avoid cycling [20], an algorithm can be obtained which is guaranteed to converge to an optimal solution in a finite number of steps, although implementations have typically been slow.

The problem of Chebyshev approximation is related to the computation of Pisarenko's estimate; it also can be formulated as the minimization of a linear functional over a convex space defined by linear inequality constraints [16]. It also has been solved using the simplex (single exchange) method. However, for the particular problem of the Chebyshev approximation of continuous functions by polynomials in one variable, a computational method exists which is significantly faster than the simplex method, a *multiple-exchange* method due to Remes. Although attempts have been made to extend this method to more general problems [21], the resulting algorithms are not well understood; in particular, there is no proof of convergence.

Finally, the undiscretized optimization problems involved in the computation of Pisarenko's estimate, (4.4) and (4.9), are a form known as *semi-infinite programs*. Both theoretical and computational aspects of such programs are discussed in a collection of papers edited by Hettich [22].

V. Summary

This paper has been concerned with what is probably the simplest interesting problem in array processing; the estimation of a power spectrum with a known support, given certain samples of its correlation function. Although simple, this problem retains several features which are common to many array processing problems: multidimensional spectra, nonuniformly sampled correlation functions, and arbitrary spectral supports.

The investigation of correlation-matching spectral estimates led to the extendibility problem. Two characterizations of extendibility were given. This problem, in the time series case, is known as the trigonometric moment problem and its solution involves consideration of the positive definiteness of the correlation samples. Positive definiteness can therefore be considered as a special case of extendibility.

Building on the theoretical framework developed in solving the extendibility problem, Pisarenko's method was extended from the time series case to the array processing problem. Pisarenko's method was shown to be intimately related to the extendibility problem. The computation of Pisarenko's estimate was shown to involve the solution of a linear optimization problem. The solution of this problem was shown not to be unique in general, although it is unique in the time-series case, where the linear optimization problem reduces to an eigenvalue problem.

Although the spectral estimation problem considered in this paper was developed for array processing, the theoretical framework and the resulting algorithms should be useful in other multidimensional problems, such as image processing.

Appendix A
The Extension Theorem for Spectral Density Functions

This appendix concerns the extension theorem for spectral density functions discussed in Section III-E. It is assumed that every neighborhood of every point in K has strictly positive ν-measure. This condition guarantees that correlation vectors corresponding to impulses in K can be approximated by correlation vectors corresponding to continuous, strictly positive spectral density functions.

An Extension Theorem for Spectral Density Functions: If every neighborhood of every point in K has strictly-positive ν-measure, then

1) if $S(\boldsymbol{k})$ is uniformly bounded away from zero over K, then

$$\boldsymbol{r} = \int_K \boldsymbol{\phi}_{\boldsymbol{k}}\, S(\boldsymbol{k})\, d\nu \in E^\circ,$$

2) if $\boldsymbol{r} \in E^\circ$ then

$$\boldsymbol{r} = \int_K \boldsymbol{\phi}_{\boldsymbol{k}}\, S(\boldsymbol{k})\, d\nu,$$

for some continuous, strictly positive functions $S(\boldsymbol{k})$.

Proof: The first statement may be proved by consideration of the mapping from a bounded function $Q(\boldsymbol{k})$ to a vector $\boldsymbol{r}$ defined by

$$\boldsymbol{r} = \int_K \boldsymbol{\phi}_{\boldsymbol{k}}\, Q(\boldsymbol{k})\, d\nu. \tag{A1}$$

That $S(\boldsymbol{k})$ is uniformly bounded away from zero means that, for some $\epsilon > 0$, $S(\boldsymbol{k}) > \epsilon$ for all $\boldsymbol{k} \in K$. Because the functions $\{e^{j\boldsymbol{k}\cdot\boldsymbol{\delta}}\colon \boldsymbol{\delta} \in \Delta\}$ are linearly independent functions on K, and since every neighborhood of every point in K contains a set of strictly positive measure, it follows that the image of the set of bounded Δ-polynomials

$$\{\boldsymbol{q} \in R^{2M+1}\colon |Q(\boldsymbol{k})| < \epsilon\} \tag{A2}$$

under (A1), is a neighborhood of $\mathbf{0}$. Therefore the image of

$$\{S(\boldsymbol{k}) + Q(\boldsymbol{k})\colon |Q(\boldsymbol{k})| < \epsilon\} \tag{A3}$$

is a subset of E which is a neighborhood of $\boldsymbol{r}$. Hence, $\boldsymbol{r} \in E^\circ$.

The second statement may be proved by considering the set E_B of correlation vectors corresponding to spectral density functions which are integrable, continuous, and strictly positive (hence, bounded away from zero),

$$\{S(\boldsymbol{k}) \in C(K)\colon S(\boldsymbol{k}) > 0\}. \tag{A4}$$

E_B is convex and, from the argument above, it follows that E_B is open. It is easily shown that the vectors $\boldsymbol{\phi}_{\boldsymbol{k}}$, for $\boldsymbol{k} \in K$, are in the closure of E_B. From Carathéodory's theorem [16], it follows that every $\boldsymbol{r} \in E$ can be written as a positive sum of some $2M+1$ such $\boldsymbol{\phi}_{\boldsymbol{k}_i}$. Since each $\boldsymbol{\phi}_{\boldsymbol{k}_i}$ is in the closure of E_B, it follows that each $\boldsymbol{r} \in E$ is also. Therefore, the closure of

E_B is E. Two open convex sets with the same closure must be identical. Since E is the closure of both E_B and E°, it follows that $E_B = E^\circ$.

APPENDIX B
THE REPRESENTATION THEOREM

The representation theorem of Section IV-A is a simple extension of Carathéodory's theorem [16] for correlation vectors on the boundary of E, making use of the extension theorem. It is the generalization of the "A theorem of C." Carathéodory [9, ch. 4] to multiple dimensions. In view of the derivation of Pisarenko's method, in Section IV, as a linear program, the representation theorem may also be viewed as a form of the fundamental theorem of linear programming [18].

The Representation Theorem: If $\boldsymbol{r}'$ is on the boundary of E, then for some $2M$ nonnegative $a(i)$ and some $\boldsymbol{k}_i \in K$:

$$\boldsymbol{r}' = \sum_{i=1}^{2M} a(i)\boldsymbol{\phi}_{\boldsymbol{k}_i}. \tag{B1}$$

Proof: Consider the compact convex set $E' = \{\boldsymbol{r} \in E: \boldsymbol{r}(\boldsymbol{0}) = 1\} \subset R^{2M}$, which is the convex hull of $A = \{\boldsymbol{\phi}_{\boldsymbol{k}}: \boldsymbol{k} \in K\}$. From Carathéodory's theorem, any element in E' can be expressed as a convex combination of $2M + 1$ elements of A

$$\boldsymbol{r} = \sum_{i=1}^{2M+1} \theta_i \boldsymbol{\phi}_{\boldsymbol{k}_i} \tag{B2}$$

with $\theta_i \geqslant 0$, $\Sigma_{i=1}^{2M+1} \theta_i = 1$, and $\boldsymbol{k}_i \in K$. If one of the θ_i is zero, the proof is complete. Otherwise, since $\boldsymbol{r}$ is on the boundary of E', there is some nonzero $\boldsymbol{p} \in P$ such that

$$0 = (\boldsymbol{r}, \boldsymbol{p}) = \sum_{i=1}^{2M+1} \theta_i(\boldsymbol{\phi}_{\boldsymbol{k}_i}, \boldsymbol{p}). \tag{B3}$$

Thus, for each i, $(\boldsymbol{\phi}_{\boldsymbol{k}_i}, \boldsymbol{p}) = 0$. The $\boldsymbol{\phi}_{\boldsymbol{k}_i}$'s must be linearly dependent, so there are some $\beta_i \in R$, not all zero, such that $\Sigma_{i=1}^{2M+1} \beta_i \boldsymbol{\phi}_{\boldsymbol{k}_i} = 0$. Let λ be the number with the smallest magnitude such that $\theta_i + \lambda\beta_i = 0$ for some i. Then

$$\boldsymbol{r} = \sum_{i=1}^{2M+1} (\theta_i + \lambda\beta_i)\boldsymbol{\phi}_{\boldsymbol{k}_i}. \tag{B4}$$

One of the coefficients is zero, reducing this to a sum over only $2M$ terms. Recognizing that any element of E is a scaled version of an element of E' completes the proof.

Note that, in the times series case, $\boldsymbol{r}'$ could be expressed as a sum of no more than M complex exponentials while the above theorem only guarantees a representation in terms of $2M$ exponentials. This is not a deficiency in the proof, but a genuine feature of the problem, as the following one-dimensional example shows.

Example B.1: $D = 1$, $K = [-\pi, \pi/2]$, $\Delta = \{0, \pm 1\}$. Suppose that $\boldsymbol{r}$ is on the straight portion of the boundary of E, as indicated in Fig. 7. Clearly, $\boldsymbol{r}$ has a unique representation as a convex sum of members of A in terms of the *two* correlation vectors corresponding to $k = \pi/2$ and $k = -\pi$,

$$r(\delta) = \tfrac{1}{2} e^{j(\pi/2)\delta} + \tfrac{1}{2} e^{-j\pi\delta}, \quad \delta \in \Delta.$$

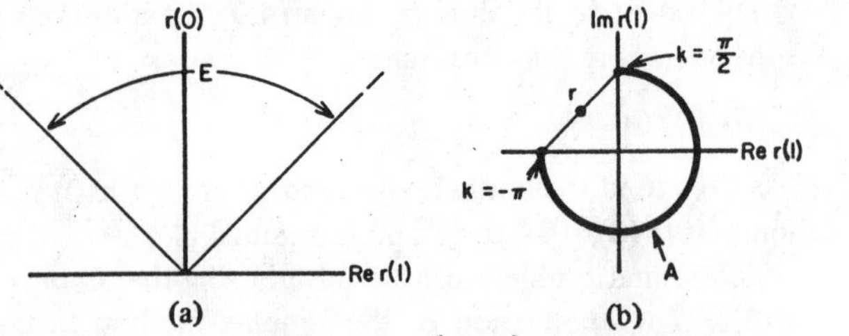

Fig. 7. E for $K = [-\pi, \pi/2]$ and $\Delta = \{0, \pm 1\}$. (a) A section through E at $\text{Im}\, r(1) = 0$, and (b) a section through E at $r(0) = 1$.

APPENDIX C
THE UNIQUENESS OF PISARENKO'S ESTIMATE

As discussed in Section IV-A, Pisarenko's estimate is unique if one and only one spectrum can be associated with each correlation vector on the boundary of E. Trivial uniqueness problems result if two distinct $\boldsymbol{k}$'s in K give rise to the same $\boldsymbol{\phi}_{\boldsymbol{k}}$'s. More generally, consider the set of correlation vectors corresponding to the zero set of some nonzero positive polynomial $\boldsymbol{p}$

$$Z = \{\boldsymbol{\phi}_{\boldsymbol{k}}: (\boldsymbol{\phi}_{\boldsymbol{k}}, \boldsymbol{p}) = P(\boldsymbol{k}) = 0, \quad \boldsymbol{k} \in K\}. \tag{C1}$$

Any vector $\boldsymbol{r}' \in E$ which makes a zero inner product with $\boldsymbol{p}$ can be expressed as a sum of positive multiples of vectors from the set Z. It follows that, if this set is linearly independent the representation is unique. Conversely, if this set is linearly dependent, then an $\boldsymbol{r}'$ on the boundary of E can be constructed which has more than one spectral representation. If the set is linearly dependent then there is a finite collection of nonzero real numbers $c(i)$ and $\boldsymbol{\phi}_{\boldsymbol{k}_i} \in Z$ such that

$$\sum_i c(i)\boldsymbol{\phi}_{\boldsymbol{k}_i} = 0. \tag{C2}$$

Because $\boldsymbol{\phi}_{\boldsymbol{k}_i}(\boldsymbol{0}) = 1$ for all i, there must be at least one $c(i)$ which is strictly positive and one which is strictly negative. Thus

$$\boldsymbol{r}' = \sum_{c(i)>0} c(i)\boldsymbol{\phi}_{\boldsymbol{k}_i} = \sum_{c(i)<0} [-c(i)]\, \boldsymbol{\phi}_{\boldsymbol{k}_i} \tag{C3}$$

is a nonzero correlation vector on the boundary of E with at least two spectral representations.

Therefore, Pisarenko's estimate is unique if and only if the set of correlation vectors corresponding to the zero of each nonzero positive polynomial is linearly independent. In particular, for Pisarenko's estimate to be unique, no nonzero positive polynomial can have more than $2M$ zeros. This condition is similar to, though not quite as strict as the Haar condition [23], which involves all polynomials, not just positive ones.

The factorability of polynomials, in the time series case, leads to a strong result. In the time series case, a nonzero positive polynomial can have no more than M zeros. Furthermore, a nonzero positive polynomial can be constructed that is zero at M or fewer arbitrary locations, and nowhere else. This implies (Example 4.1) that a correlation vector in ∂E has a unique spectral representation and that this spectrum is composed of M or fewer impulses. Furthermore, it implies that any spectrum composed of M or fewer impulses has a correlation vector in ∂E.

However, a simple example shows that Pisarenko's estimate

is not guaranteed to be unique in most multidimensional situations. Consider the nonzero positive polynomial

$$P(\boldsymbol{k}) = 1 - \cos(\boldsymbol{k} \cdot \boldsymbol{\delta}) \geqslant 0 \tag{C4}$$

for some nonzero $\delta \in \Delta$. The zero set of $P(k)$ includes the portion of the hyperplane

$$\{\boldsymbol{k}: \boldsymbol{k} \cdot \boldsymbol{\delta} = 0\} \tag{C5}$$

that is in K. Many spectral supports, of practical interest, intersect this hyperplane at an infinite number of points, implying the existence of some correlation vector on the boundary of E with a nonunique spectral representation. This nonuniqueness problem is similar to the nonuniqueness problem in multi-dimensional Chebyshev approximation [24].

ACKNOWLEDGMENT

The work of B. Dickinson was instrumental in starting the authors on this course of research, and they are grateful to G. Cybenko for references to the semi-infinite programming literature.

REFERENCES

[1] J. Stewart, "Positive definite functions and generalizations, an historical survey," *Rocky Mountain J. Math.*, vol. 6, pp. 409-434, Summer 1976.
[2] B. W. Dickinson, "Two-dimensional Markov spectrum estimates need not exist," *IEEE Trans. Inform. Theory*, vol. IT-26, pp. 120-121, Jan. 1980.
[3] G. Cybenko, "Moment problems and low rank Toeplitz approximations," presented at the Int. Symp. on Rational Approximation Syst., Katholieke Universiteit Leuven, Leuven, Holland, Aug. 31-Sept. 1, 1981 (to appear in *Circuits, Syst., Signal Processing*).
[4] —, "Affine minimax problems and semi-infinite programming," *Math. Programming*, to be published.
[5] V. F. Pisarenko, "The retrieval of harmonics from a covariance function," *Geophys. J. R. Astr. Soc.*, vol. 33, pp. 347-366, 1973.
[6] A. B. Baggeroer, "Space/time random processes and optimum array processing," Naval Undersea Center, San Diego, CA., Rep. NUC TP 506, Apr. 1976.
[7] H. L. Royden, *Real Analysis*. New York: Macmillan, 1968.
[8] R. A. Haubrich, "Array design," *Bull. Seismological Soc. Amer.*, vol. 58, pp. 977-991, June 1968.
[9] U. Grenander and G. Szegö, *Toeplitz Forms and Their Applications*. Berkeley and Los Angeles: Univ. of California Press, 1958.
[10] D. G. Luenberger, *Optimization by Vector Space Methods*. New York: Wiley, 1969.
[11] M. Avriel, *Nonlinear Programming*. Englewood Cliffs, NJ: Prentice-Hall, 1976.
[12] A. Calderón and R. Pepinsky, "On the phases of Fourier coefficients for positive real functions," *Computing Methods and the Phase Problem in X-Ray Crystal Analysis*, The X-Ray Crystal Analysis Laboratory, Dep. Physics, Pennsylvania State College, pp. 339-348, 1952.
[13] W. Rudin, "The extension problem for positive-definite functions," *Ill. J. Math.*, vol. 7, pp. 532-539, 1963.
[14] S. W. Lang and J. H. McClellan, "Multidimensional MEM spectral estimation," *IEEE Trans. Acoust., Speech, Signal Processing*, vol. ASSP-30, pp. 880-887, Dec. 1982.
[15] K. Hoffman, *Analysis in Euclidean Space*. Englewood Cliffs, NJ: Prentice-Hall, 1975.
[16] E. W. Cheney, *Introduction to Approximation Theory*. New York: McGraw-Hill. 1966.
[17] T. M. Sullivan, O. L. Frost, and J. R. Treichler, "High resolution signal estimation," Argosystems, Inc., Sunnyvale, CA, June 1978.
[18] D. G. Luenberger, *Introduction to Linear and Nonlinear Programming*. Reading, MA: Addison-Wesley, 1973.
[19] E. Stiefel, "Note on Jordan elimination, linear programming, and Tchebycheff approximation," *Numerische Mathematik*, vol. 2, pp. 1-17, 1960.
[20] C. H. Papadimitriou and K. Steiglitz, *Combinatorial Optimization*. Englewood, NJ: Prentice-Hall, 1982.
[21] D. B. Harris and R. M. Mersereau, "A comparison of algorithms for minimax design of two-dimensional linear phase FIR digital filters," *IEEE Trans. Acoust., Speech, Signal Processing*, vol. ASSP-25, pp. 492-500, Dec. 1977.
[22] R. Hettich, Ed., *Semi-Infinite Programming*. Berlin: Springer-Verlag, 1979.
[23] J. R. Rice, *The Approximation of Functions, Vol. 1–Linear Theory*. Reading, MA: Addison-Wesley, 1964.
[24] —, *The Approximation of Functions, vol. 2–Nonlinear and Multivariate Theory*. Reading, MA: Addison-Wesley, 1969.
[25] S. W. Lang and J. H. McClellan, "Spectral estimation for sensor arrays," in *Proc. 1st Acoust., Speech, Signal Processing Workshop Spectral Estimation*, Hamilton, Ont., Canada, Aug. 17-18, 1981, pp. 3.2.1-3.2.7.
[26] S. W. Lang, "Spectral estimation for sensor arrays," Ph.D. dissertation, Massachusetts Institute of Technology, Cambridge, MA, Aug. 1981.

Properties of Two-Dimensional Maximum Entropy Power Spectrum Estimates

NAVEED A. MALIK AND JAE S. LIM

Abstract—In this paper we present the results of a number of experiments that have been performed to study the properties of two-dimensional maximum entropy spectral estimates. The results presented include studies on the resolution differences for real and complex data, resolution properties of the spectral estimates, and the determination of the relative power of the sinusoids from the spectral estimates. The results also include the effects of signal-to-noise ratio, size and shape of the known autocorrelation region, and data length and initial phase of a sinusoid, on the spectral estimates. In most cases studied, the properties of two-dimensional maximum entropy spectral estimates can be viewed as simple extensions of their one-dimensional counterparts.

I. Introduction

The maximum entropy (ME) method of power spectrum estimation (PSE), originally suggested by Burg [1], has proven to be a powerful technique of spectral analysis, primarily due to its high resolution properties. In the one-dimensional (1-D) case, when consecutive correlation values are available, the ME problem is identical to autoregressive (AR) signal modeling. This problem is linear, and its properties have been widely investigated in the literature [2], [3]. The two-dimensional (2-D) ME problem, however, is highly nonlinear, and no closed-form solution has yet been found. Recently, a new iterative algorithm [4]-[6] for solving the 2-D ME problem has been developed, and this paper presents the results of several experiments which were conducted using this algorithm to investigate the properties of 2-D ME spectral estimates.

II. The Maximum Entropy Problem

The ME method of spectral estimation is aimed at obtaining correlation-matching spectral estimates given a finite segment of the autocorrelation function (ACF) of the signal. Using standard notation, given the ACF

$$R_x(\boldsymbol{n}) \quad \text{for } \boldsymbol{n} \in A \tag{1}$$

where A is a region symmetric about and including the origin, the ME spectral estimate is obtained by maximizing the entropy

$$H = \int_{\omega=-\pi}^{\pi} \ln P_x(\boldsymbol{\omega})\, d\boldsymbol{\omega} \tag{2}$$

under the constraint that the spectral estimate $\hat{P}_x(\boldsymbol{\omega})$ be consistent with the known information, that is,

$$F^{-1}[\hat{P}_x(\boldsymbol{\omega})] = R_x(\boldsymbol{n}) \quad \text{for } \boldsymbol{n} \in A \tag{3}$$

where F^{-1} denotes the inverse Fourier transform operation. Maximizing the entropy H in (2) is equivalent to requiring the spectral estimate to be in the following form [4], [5]:

$$\hat{P}_x(\boldsymbol{\omega}) = \frac{1}{\sum_{\boldsymbol{n}\in A} \lambda(\boldsymbol{n})\, e^{-j\boldsymbol{\omega}\cdot\boldsymbol{n}}} \tag{4}$$

where $\lambda(\boldsymbol{n})$ represents a finite extent sequence which is zero outside $\boldsymbol{n} \in A$.

From (3) and (4), the ME problem can be stated concisely as follows:

Given $R_x[\boldsymbol{n}]$ for $\boldsymbol{n} \in A$, determine $\hat{P}_x(\boldsymbol{\omega})$ such that

i) $F^{-1}[\hat{P}_x(\boldsymbol{\omega})] = R_x[\boldsymbol{n}] \quad$ for $\boldsymbol{n} \in A$

and

ii) $\hat{P}_x(\boldsymbol{\omega}) = \dfrac{1}{\sum_{\boldsymbol{n}\in A} \lambda[\boldsymbol{n}]\, e^{-j\boldsymbol{\omega}\cdot\boldsymbol{n}}}$.

This problem has no known closed-form solution for the 2-D case. The algorithm developed by Lim and Malik [4] to solve this problem is an iterative "alternating projections" type of algorithm, and the results presented in this paper have been obtained using this algorithm.

III. Experimental Results

The data used in studying the properties of 2-D ME spectral estimates are 2-D sinusoids buried in white Gaussian noise. For one set of experiments, it is assumed that the exact correlation values are available over the region A. The region A, unless otherwise noted, is a square, symmetric about the origin. For the case of M real sinusoids, the exact ACF values are given by

$$R_x(n_1, n_2) = \sum_{i=1}^{M} a_i^2 \cos(\omega_{i1} n_1 + \omega_{i2} n_2) + \sigma^2 \delta(n_1, n_2) \tag{5}$$

where a_i^2 is the power of the ith sinusoid, ω_{i1} and ω_{i2} give its frequency location, and σ^2 represents the noise power.

Manuscript received January 4, 1982; revised April 20, 1982. This work was supported in part by the National Science Foundation under Grant ECS80-07102 and in part by the Advanced Research Projects Agency monitored by the ONR under Contract N00014-81-K-0742 NR-049-506.

The authors are with the Department of Electrical Engineering and Computer Science, Massachusetts Institute of Technology, Cambridge, MA 02139.

Reprinted from *IEEE Trans. Acoust., Speech, Signal Processing,* vol. ASSP-30, pp. 788–797, Oct. 1982.

For the case of M complex exponentials, the exact ACF is given by

$$R_x(n_1, n_2) = \sum_{i=1}^{M} a_i^2 \, e^{j(\omega_{i1} n_1 + \omega_{i2} n_2)} + \sigma^2 \delta(n_1, n_2). \tag{6}$$

For both (5) and (6), $R_x(n_1, n_2)$ is assumed known for $(n_1, n_2) \in A$.

A parallel set of experiments uses biased ACF values estimated from synthetic data sets. If the data are available in a square array of size $P \times P$, the ACF is estimated as

$$R_x(n_1, n_2) = \frac{1}{P2} \sum_{k_1=1}^{P} \sum_{k_2=1}^{P} x(k_1, k_2) \cdot x^*(k_1 + n_1, k_2 + n_2) \tag{7}$$

where $x(n_1, n_2)$ represents the synthetic data set given by

$$x(n_1, n_2) = \sum_{i=1}^{M} \sqrt{2}\, a_i \cos(\omega_{i1} \cdot n_1 + \omega_{i2} \cdot n_2 + \phi_i) + w(n_1, n_2) \tag{8}$$

where the number of sinusoids is M, $w(n_1, n_2)$ represents white noise of power σ^2, ϕ_i is the phase term associated with the ith sinusoid, and the sums in (7) run over known data values only. For the case of complex data, the cosine in (8) is replaced by a complex exponential and $\sqrt{2}\, a_i$ is replaced by a_i.

The signal-to-noise (S/N) ratio is defined as the sum of the powers of each sinusoid divided by the noise power. Specifically, for the case of M sinusoids with a_i^2 representing the power of the ith sinusoid, the S/N ratio is given by

$$S/N \text{ ratio} \triangleq \frac{\sum_{i=1}^{M} a_i^2}{\sigma^2}$$

where σ^2 is the noise power.

The 2-D spectral estimates are displayed in the form of contour plots with the maximum value normalized to zero dB. The contours are labeled, when possible, with the nearest integer value of the contour level, in dB below the maximum (0 dB). The 0 dB contour is a point and the center of the smallest contour (which sometimes looks like the symbol ◊) represents this point. The increment between contours (CINC), in dB, is always noted, and is chosen so that there are five equally spaced contours between the maximum and minimum value. The notation X marks the true peak location. For real data, the spectral estimates are symmetric about the origin, and thus only half the 2-D frequency plane is displayed. The full 2-D plane is displayed for spectra of complex signals. The frequency axes, and all frequency values, are in terms of the normalized frequency units of $\omega/2\pi$. Thus, for example, the interval $(-\pi, \pi)$ is represented by $(-0.5, 0.5)$ and the peak location of $(\omega_1, \omega_2) = (0.2\pi, 0.3\pi)$ is represented by the ordered pair $(0.1, 0.15)$.

The 2-D ME spectral estimates are compared in some cases with the maximum likelihood (ML) and the Bartlett estimates. The ML estimate for 2-D signals is obtained by inverting the matrix of 2-D autocorrelations Φ_{NM} [7]. The estimate is given by

$$P_{ML}(\omega_1, \omega_2) \triangleq \frac{NM}{E^\dagger \Phi_{NM}^{-1} E} \tag{9}$$

where Φ_{NM}^{-1} represents the inverse of the block Toeplitz matrix of autocorrelations and $E^\dagger$ is the complex conjugate transpose of the vector

$$E \triangleq \text{COL}\,[1, e^{-j\omega_1}, e^{-j2\omega_1}, \cdots, e^{-j(N-1)\omega_1}, e^{-j\omega_2}, \cdots, e^{-j\{(N-1)\omega_1 + \omega_2\}}, \cdots, e^{-j(M-1)\omega_2}, \cdots, e^{-j\{(N-1)\omega_1 + (M-1)\omega_2\}}]. \tag{10}$$

The Bartlett estimate is obtained by taking the Fourier transform of the ACF values which are known over the region A. The ACF is first windowed by a 2-D separable triangular window to prevent the spectrum from becoming negative regions.

A. Special Regions in the 2-D Frequency Plane for Real Data

For 1-D real signals, it is well known [2], [8] that the symmetry and periodicity of the power spectrum causes errors in the peak location near $\omega/2\pi = 0$ and $\omega/2\pi = 0.5$. This is due to the interference with the correlated mirror peaks that occur at negative frequencies for real data. For example, if the data are given by

$$x(n) = \sin(\omega_0 n) + w(n) \tag{11}$$

the power spectrum consists of two peaks located at $\omega = \omega_0$ and $\omega = -\omega_0$. If $\omega_0/2\pi$ is close in value to 0 or 0.5, the interference between the peaks causes them to move closer in the spectral estimate, initially causing errors in the location of the spectral peaks and eventually a complete merging of the two.

The case for a 2-D real sinusoid was found to be similar except that the two-dimensional periodicity of the spectrum combined with its symmetry results in errors in the peak location at several points in the 2-D frequency plane. Fig. 1(a) illustrates the symmetries and the location of the mirror peaks for the case of 2-D real sinusoids. The upper half-plane, which completely specifies the power spectrum, is indicated by bold lines, and the small geometric shapes show the locations of mirror peaks introduced by the symmetry and the periodicity. The shaded regions in Fig. 1(b) indicate the special regions in the upper half-plane where the estimate of the peak location for real data can be expected to suffer.

A number of examples using both real and complex data have been studied. While the spectral estimates from the real data suffered resolution loss in terms of both sharpness and location of spectral peaks in the special frequency regions, shown in Fig. 1(b), relative to other frequency regions, spectral estimates from the complex data showed no such resolution loss. These results have been observed for all three methods, ME, ML, and Bartlett.

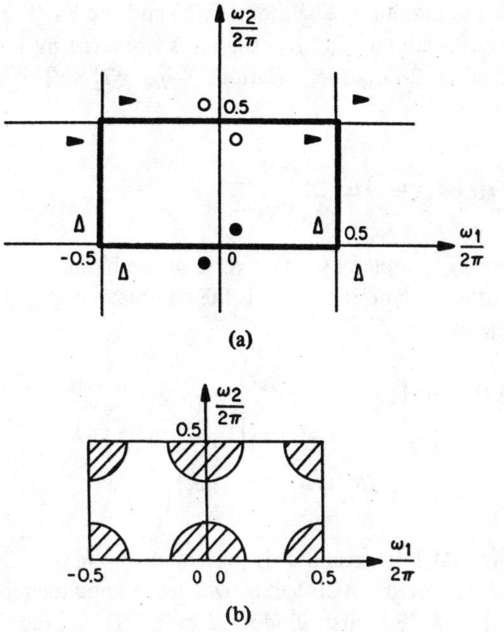

Fig. 1. The 2-D frequency plane for real data. (a) Symmetry and periodicity. (b) Special regions for real data.

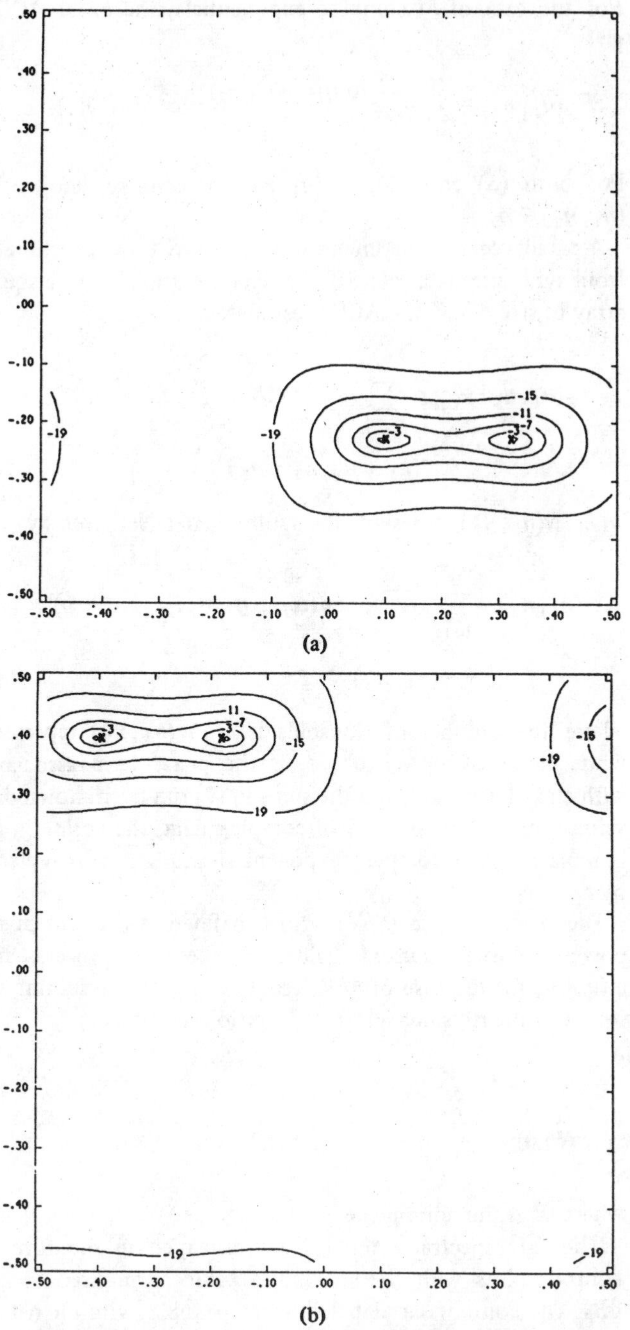

Fig. 2. The ME PS estimates do not depend on absolute peak location for complex signals. S/N ratio = 0 dB, 5 × 5 ACF, $a_1^2 = 1.0$ and $a_2^2 = 1.0$, $\sigma^2 = 2.0$. (a) $\omega/2\pi = (0.1, -0.23)$ and $\omega/2\pi = (0.325, -0.23)$, CINC = 3.83. (b) $\omega/2\pi = (-0.4, 0.4)$ and $\omega/2\pi = (-0.175, 0.4)$, CINC = 3.83.

B. Resolution of ME Spectral Estimates

To study the resolution properties of the ME spectral estimates, two sets of experiments were performed. In both sets of experiments we have used the exact ACF values and complex data to separate the issues of resolution from the effects of the data length and the special regions for real data as discussed in Section III-A.

The first set of experiments was directed towards studying the resolvability of two sinusoids buried in noise. To develop a quantitative measure of the resolvability, we have made the observation that for a given size and shape of A, and a given S/N ratio, the spectral estimates for the ME, ML, and Bartlett methods do not depend on the absolute location of the peaks in the 2-D frequency plane. That is, the shape and size of the estimated spectral peaks remain the same regardless of where the complex sinusoids are located, if the same relative distance and orientation of the peaks is maintained. Fig. 2(a) and (b) illustrates this phenomenon for ME spectral estimates. In these cases, the frequency separation between the peaks is held constant and the orientation of the peaks is kept horizontal. The results clearly show the invariance of the ME spectral estimates under these conditions. Several other examples support this conclusion, and similar results have been observed for the ML and Bartlett estimates. In addition to this observation, we have found for the three methods that a larger separation in the frequencies of two sinusoids always produces "more resolved" spectral estimates at a given S/N ratio, size and shape of the region A, and orientation of the peaks. Based on the above, a reasonable quantitative measure of the resolvability would be the minimum frequency separation, denoted by d, above which the two sinusoids are resolved and below which they are not resolved. With this measure d, smaller values imply higher resolution, while larger values imply lower resolution.

A number of examples have been studied to determine the minimum distance d between two peaks such that they are resolved in the sense that the estimated power spectra display two distinct peaks. One peak location is held constant, while the location of the second peak is varied over a range such that initially the peaks are not resolved, and as the distance between the peaks is increased, the two peaks are resolved in the spectral estimate. Fig. 3(a)-(c) is representative of the results obtained by the ME method as the separation between the peaks is increased. Initially the two peaks are not resolved, and the spectral estimate consists of a single spectral peak,

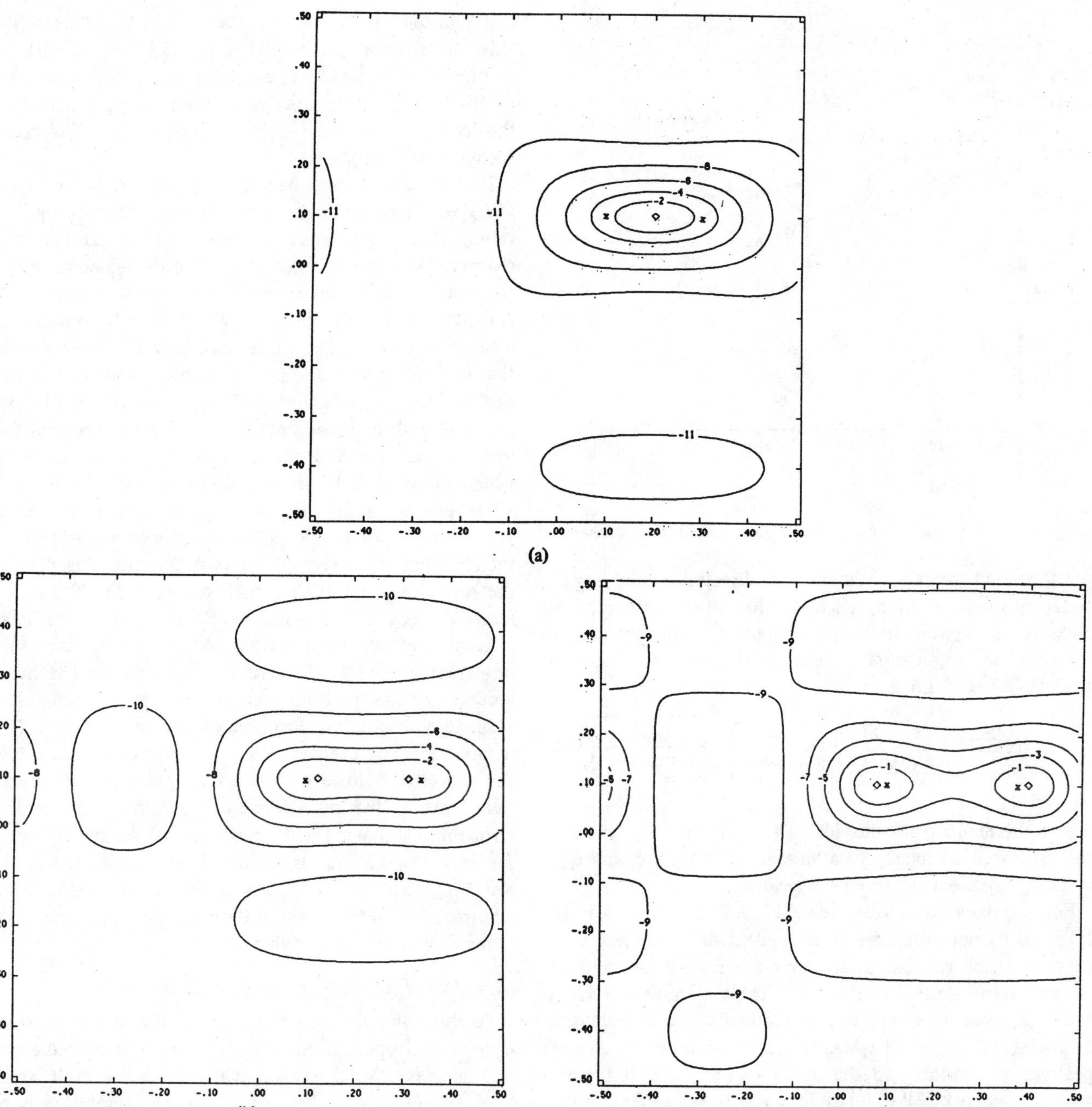

Fig. 3. Change in the ME spectral estimates as the separation between two peaks is increased. S/N ratio = −5 dB, 3 × 3 ACF, $a_1^2 = 1.0$ and $a_2^2 = 1.0$, $\sigma^2 = 6.32$. (a) $\omega/2\pi$ = 0.1, 0.1) and $\omega/2\pi$ = (0.3, 0.1), CINC = 2.24 dB. (b) $\omega/2\pi$ = (0.1, 0.1) and $\omega/2\pi$ = (0.34, 0.1), CINC = 2.00 dB. (b) $\omega/2\pi$ = (0.1, 0.1) and $\omega/2\pi$ = (0.38, 0.1), CINC = 1.99 dB.

located approximately at the midpoint of the line joining the true peak locations. As the distance between the peaks increases, the spectral estimate shows a distortion or stretching in the direction of the peaks, and eventually, the two peaks are resolved. Fig. 4 summarizes the resolution performance of the three techniques. It is clear that as in the 1-D case, the ME method affords higher resolution than the other two methods. It may be noted here that the resolution performance of the Bartlett estimates is determined only by the size of the ACF array available for analysis, and is independent of the S/N ratio, as far as the resolution measure d is concerned. This is due to the fact that changing the noise level only affects the dc level in the estimated power spectra when exact ACF values are used [9]. The minimum distance d for the peaks to be resolved in the ME and ML estimates decreases with increasing S/N ratio, with the ME method consistently outperforming the ML method.

At this point, it is necessary to point out that the measure adopted for the resolution performance evaluation is fairly arbitrary, and is used only to study the relative performance of the different techniques under the same set of conditions. The minimum resolution distance between two peaks also depends on their orientation in the 2-D frequency plane and on the shape of A. Thus, the resolution measure d is only an indicator of the relative performance, and should not be considered as an absolute measure.

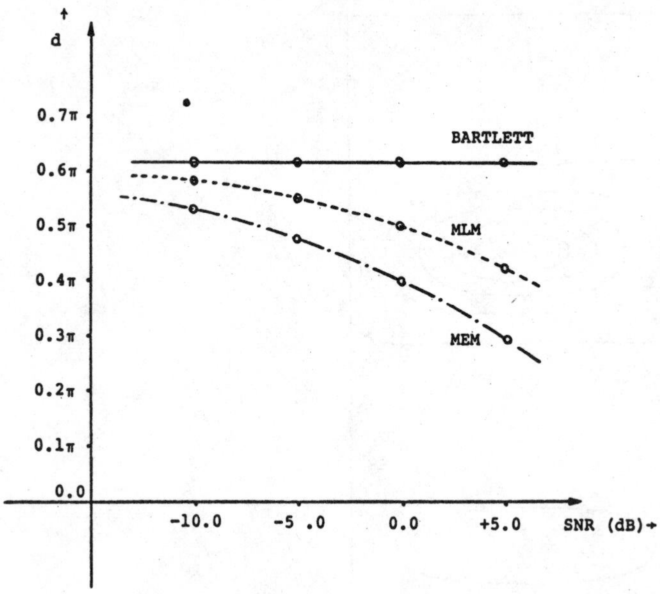

Fig. 4. Resolution measure d of the ME, ML, and Bartlett estimates.

The second set of experiments was directed to studying the accuracy of the resulting peak location when the number of sinusoids present is accurately estimated. The quantitative measure used to measure the error in the location of spectral peak (LOSP) is defined as

$$\text{error(LOSP)} = \sum_{i=1}^{M} \sqrt{\left(\frac{\omega_{i1e}}{2\pi} - \frac{\omega_{i1t}}{2\pi}\right)^2 + \left(\frac{\omega_{i2e}}{2\pi} - \frac{\omega_{i2t}}{2\pi}\right)^2} \tag{12}$$

where the number of sinusoids is M, ω_{i1e} and ω_{i2e} represent the estimated frequency locations of the ith peak, and ω_{i1t} and ω_{i2t} represent the true peak location.

For the one-sinusoid case using 3 X 3 ACF values, the ME, ML, and Bartlett estimates all showed LOSP errors very close to zero. For two sinusoids and a 5 X 5 ACF, all methods showed some finite LOSP error. Table I shows the LOSP error of some representative one-sinusoid and two-sinusoid examples. Although the ME estimates exhibit much sharper peaks than the other two, the table shows that the ME method gives the worst LOSP estimate for the two-sinusoid examples. The ML and Bartlett estimates give LOSP estimates of approximately the same magnitude.

C. *Effect of S/N Ratio and ACF Size on Spectral Estimates*

To study the effect of S/N ratio and ACF support size on the ME spectral estimates, a number of examples have been investigated. In all cases, we have used the exact ACF.

As the S/N ratio increases, the peak in the spectral estimate becomes sharper. This leads to two distinct effects. First, two peaks located close together in the 2-D frequency plane, which are not resolved at low S/N ratios, become resolvable at high S/N ratios. This is clear from Fig. 4 which shows that the minimum distance d for the ME spectral estimates decreases as the S/N ratio increases. Second, for real data, the interference between mirror peaks decreases and this leads to an enlarged "region of resolution." To illustrate this, various examples have been studied. Fig. 5(a) and (b) is representative of the effect of increasing the S/N ratio for the case of a single real sinusoid. It is clear that the peak in the estimate is considerably sharper for the higher S/N ratio case (5 dB), shown in Fig. 5(b). In fact, the peak location for the lower S/N ratio (0 dB), shown in Fig. 5(a), is quite erroneous. This is because the location lies in the special region for this S/N ratio and the given ACF support.

The effect of increasing the ACF support region A on the ME spectral estimates is similar to that of increasing the S/N ratio. That is, the peaks in the spectral estimates grow much sharper. As a consequence, the size of the region of resolution for real data increases as the size of the ACF support region is increased. Results similar to Fig. 5 have been observed for a single real sinusoid in white noise as we increase the size of the ACF support. As another consequence of the sharper peaks in the estimate, the resolving power of the ME estimate increases with increasing size of A. This has been verified by various examples, and one such example is shown in Fig. 6, which shows the ME spectral estimate for three sinusoids in white noise. When the ACF support region has size 5 X 5 [Fig. 6(a)], two of the peaks merge into a single peak and the resulting estimate only shows two peaks. Fig. 6(b) shows the result of using a 7 X 7 ACF support size. The peak estimates are seen to be sharper, and all three peaks are resolved.

Another effect which is common to increasing the S/N ratio or increasing the size of the ACF support region A is the higher accuracy of the resulting peak location in the estimates. With a single sinusoid, the location of the spectral peak (LOSP) is fairly accurate even for low S/N ratios or for small sizes of the region A. However, when multiple peaks are present in the spectrum, the interference between the peaks can lead to erroneous estimates for the spectral peak locations, especially for very low S/N ratios, or for small sizes of the region A. Referring back to Fig. 5, it is seen that the LOSP is totally incorrect for the S/N ratio of 0 dB and becomes more accurate as the S/N ratio increases to 5 dB.

D. *Effect of ACF Shape on Spectral Estimates*

To study the effect of the shape of the ACF support region A on the ME spectral estimates, we have investigated a number of examples using the exact ACF. The main result which we have expected and observed is that the resolution is better along the direction of more correlation points. This effect is shown in Fig. 7. The shape of the region A in these examples is a rectangle of size 3 X 5. Fig. 7(a) shows the ME spectral estimate when the orientation of the peaks is along the longer dimension of the region A, and the peaks are seen to be resolved. In Fig. 7(b), the orientation of the peaks is in the direction of the shorter dimension, and the resulting spectral estimate shows only a single peak, illustrating a poorer resolution in the direction of the shorter dimension.

E. *Estimation of Sinusoidal Power from ME Spectral Estimates*

For 1-D signals, it is well known that the power of a sinusoid is linearly proportional to the area under the peak corresponding to the sinusoid in the ME spectral estimate [10]. For 2-D signals, we have observed that the power of a 2-D sinusoid is linearly proportional to the volume under the peak in the ME spectral estimate. To illustrate this, Table II shows six examples in which the ratio of the true power of two sinusoids is com-

TABLE I

COMPARISON OF ME, ML, AND BARTLETT ESTIMATES FOR PEAK LOCATION ACCURACY USING EXACT AUTOCORRELATION VALUES. THE PEAK LOCATIONS ARE LISTED AS THE PAIRS ($\omega_1/2\pi$, $\omega_2/2\pi$), AND THE PEAK LOCATION ERROR (LOSP ERROR) IS ALSO IN UNITS OF 2π. S/N RATIO = 5 dB, ONE SINUSOID CASES: 3 × 3 ACF, TWO SINUSOID CASES: 5 × 5 ACF.

		MAXIMUM ENTROPY		MAXIMUM LIKELIHOOD		BARTLETT ESTIMATE	
	TRUE LOCATION	ESTIMATED LOCATION	LOSP ERROR	ESTIMATED LOCATION	LOSP ERROR	ESTIMATED LOCATION	LOSP ERROR
ONE SINUSOID	-.4000,0.4000	-.4000,0.4000	0.0000	-.4000,0.4000	0.0000	-.4000,0.4000	0.0000
	0.0745,-.4456	0.0745,-.4456	0.0000	0.0745,-.4456	0.0000	0.0745,-.4456	0.0000
	-.3000,-.3000	-.3000,-.3000	0.0000	-.3000,-.3000	0.0000	-.3000,-.3000	0.0000
	-.0500,-.0500	-.0500,-.0500	0.0000	-.0500,-.0500	0.0000	-.0500,-.0500	0.0000
	-.3125,0.3000	-.3125,0.3000	0.0000	-.3125,0.3000	0.0000	-.3125,0.3000	0.0000
TWO SINUSOIDS	-.4000,0.0000 0.0745,-.4456	-.4010,0.0040 0.0755,-.4496	0.0082	-.4010,0.0000 0.0755,-.4456	0.0020	-.4010,0.0020 0.0755,-.4476	0.0045
	0.3000,-.3000 -.3000,-.3000	0.2760,-.3000 -.2760,-.3000	0.0480	0.2970,-.3000 -.2970,-.3000	0.0059	0.2810,-.3000 -.2810,-.3000	0.0379
	0.3000,0.4120 -.0500,-.0500	0.3050,0.4060 -.0550,-.0440	0.0156	0.2990,0.4110 -.0490,-.0490	0.0028	0.3010,0.4120 -.0510,-.0500	0.0019
	0.1234,0.3456 -.3125,0.3000	0.1374,0.3396 -.3265,0.3060	0.0304	0.1304,0.3476 -.3195,0.2980	0.0146	0.1374,0.3416 -.3265,0.3040	0.0291
	0.2000,0.3125 -.1125,0.0330	0.1950,0.3135 -.1075,0.0320	0.0102	0.1990,0.3115 -.1115,0.0340	0.0028	0.1990,0.3125 -.1115,0.0330	0.0019
	0.3300,0.0000 0.0000,0.3333	0.3230,0.0070 0.0070,0.3263	0.0197	0.3300,0.0000 0.0000,0.3333	0.0000	0.3300,0.0000 0.0000,0.3333	0.0000
	-.3000,-.2000 0.1000,0.4430	-.2900,-.2040 0.0900,0.4470	0.0215	-.3000,-.2000 0.1000,0.4430	0.0000	-.3000,-.2000 0.1000,0.4430	0.0000
	-.1000,-.1000 0.3900,0.4000	-.1010,-.1000 0.3910,0.4000	0.0019	-.1000,-.1000 0.3900,0.4000	0.0000	-.1000,-.1000 0.3900,0.4000	0.0000

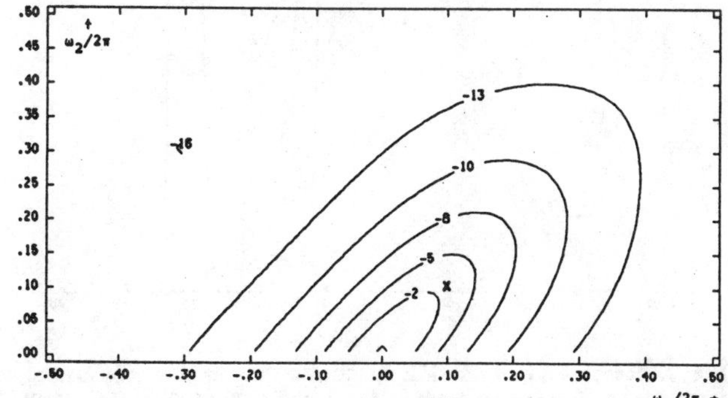

(a)

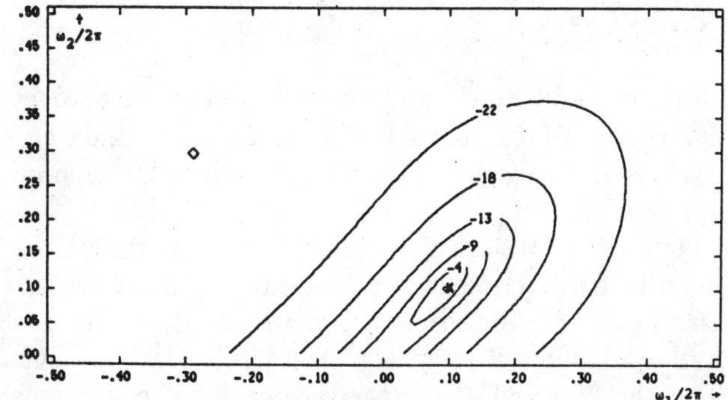

(b)

Fig. 5. Improvement in the ME spectral estimates with increasing S/N ratio, 3 × 3 ACF. (a) $\omega/2\pi = (0.1, 0.1)$, S/N ratio = 0 dB, $a_1^2 = 1.0$, $\sigma^2 = 1.0$, CINC = 2.7 dB. (b) $\omega/2\pi = (0.1, 0.1)$, S/N ratio = 5 dB, $a_1^2 = 1.0$, $\sigma^2 = 0.316$, CINC = 4.57 dB.

pared to the volume ratio and the spectral amplitude ratio. The volume is determined by dividing the frequency plane into two half-planes along the perpendicular bisector of the line joining the two peaks and then integrating the spectral estimates above a certain threshold in each half plane. In these examples, complex data with exact ACF values on a support region of size 5 × 5 have been used. The frequencies of the two sinusoids used are well separated to have a well-defined volume for each peak estimate. The results clearly show that the power of a 2-D sinusoid is linearly proportional to the volume under the peak in the ME spectral estimate.

F. Data Length Versus Spectral Estimates

In most applications of power spectrum estimation, it is the actual data rather than its ACF that is available for analysis. In such cases, the ACF has to be estimated from the data and then used to obtain the ME spectral estimates. The biased estimator for the ACF is used in all cases here, since the unbiased estimator can lead to nonpositive definite autocorrelation estimates.

One important issue that arises is the effect of the size of the data segment on the spectral estimates. It is clear that if one has a large amount of data, the ACF estimates will be very good, and hence the ME spectral estimates will be better also. Similarly, the smaller the amount of data, the poorer the ACF estimate, and hence the spectral estimates can be expected to suffer. To study these effects, several examples have been investigated where complex synthetic data are generated and the autocorrelation values are estimated from the data. The

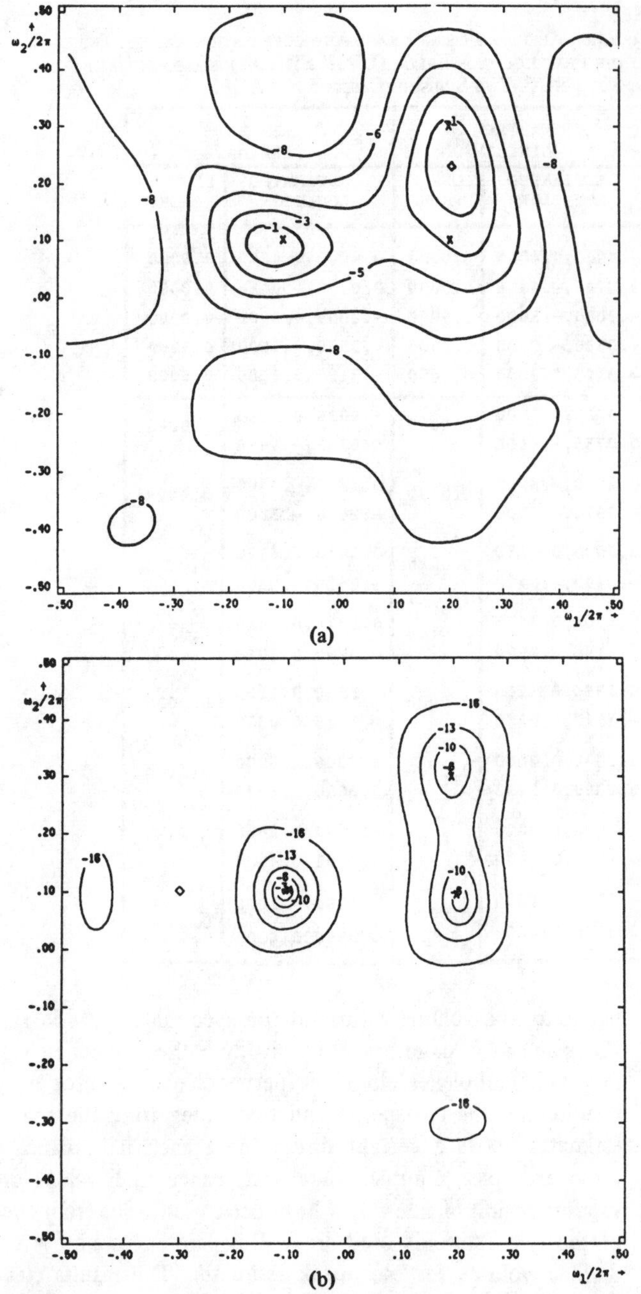

Fig. 6. Improved resolution for multiple sinusoids with increasing ACF support size A. $\omega/2\pi = (-0.1, 0.1)$, $(0.2, 0.1)$, and $(0.2, 0.3)$, S/N ratio = −5 dB, $a_1^2 = a_2^2 = a_3^2 = 1.0$, $\sigma^2 = 9.48$. (a) 5 × 5 ACF, CINC = 1.74 dB. (b) 7 × 7 ACF, CINC = 3.35 dB.

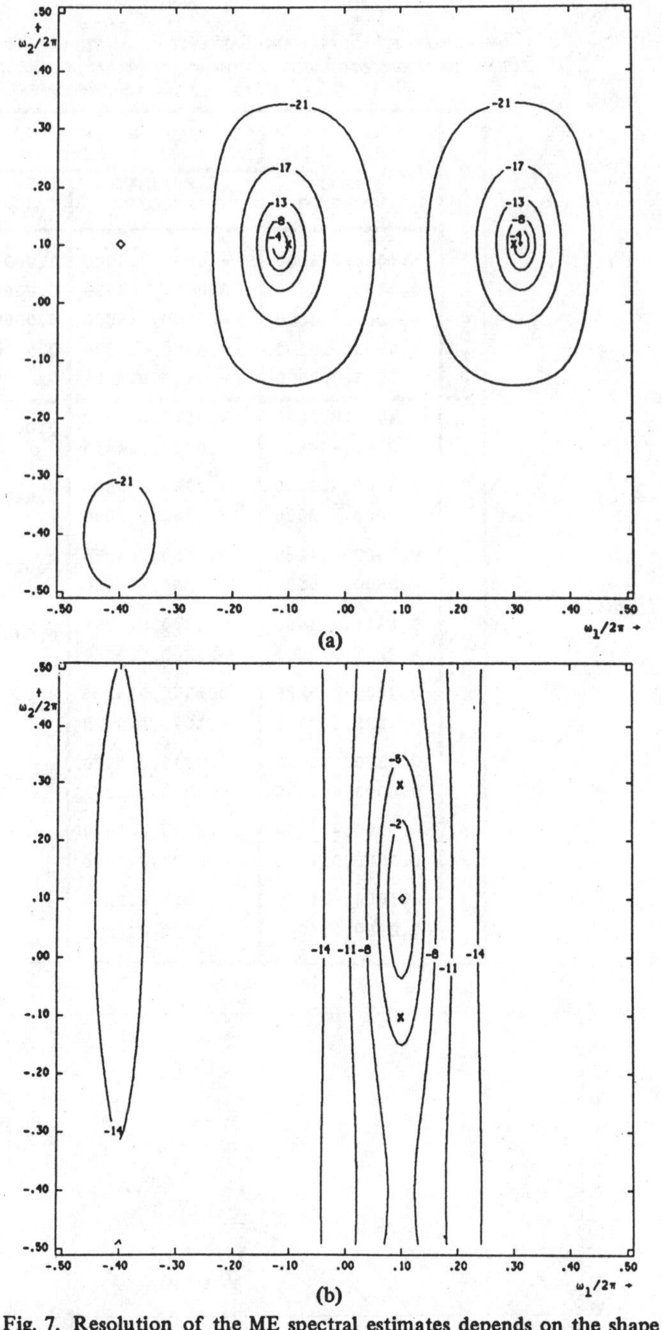

Fig. 7. Resolution of the ME spectral estimates depends on the shape of the ACF support region A. 3 × 5 ACF, S/N ratio = 0 dB, $a_1^2 = 1.0$ and $a_2^2 = 1.0$, $\sigma^2 = 2.0$. (a) $\omega/2\pi = (-0.1, 0.1)$ and $\omega/2\pi = (0.3, 0.1)$, CINC = 4.38 dB. (b) $\omega/2\pi = (0.1, -0.1)$ and $\omega/2\pi = (0.1, 0.3)$, CINC = 2.83 dB.

shape of the 2-D data segment and the ACF support region used are squares in all cases. Fig. 8 shows the effect of changing the size of the known data set on the ME spectral estimates, for the case of two sinusoids and a 5 × 5 region of support for the ACF. Data lengths ranging from 4 × 4 up to 60 × 60 have been considered, and the two examples in the figure correspond to the data lengths of 4 × 4 and 60 × 60, respectively. As is clear from the figures, the shorter data length gives a spectral estimate that seems distorted (stretched) and the location of the spectral peak (LOSP) is not very accurate. As the data length is increased, the shape of the spectral peak becomes more symmetric and the accuracy of the LOSP improves. The improvement of the spectral estimates has been observed to be very rapid, and the difference between the estimates obtained from a 12 × 12 data segment and a 60 × 60 segment is negligible. Several other examples support this observation.

Figs. 9 and 10 show the results obtained by the ML method and the Bartlett method respectively for the same data set used in Fig. 8. The spectral estimates clearly are not as sharp as the ME spectral estimates. In addition, we have observed that the ML and Bartlett estimates show continuous improvement up to the data length of 44 × 44, and the spectral contours do not achieve the same symmetry as the ME estimates until a data length of about 60 × 60 is reached.

TABLE II
RELATIVE POWER ESTIMATION FROM ME SPECTRA OF 2-D SINUSOIDS

POWER	$\omega_1/2\pi$	$\omega_2/2\pi$	S/N RATIO	TRUE POWER RATIO	ESTIMATED VOLUME RATIO	ESTIMATED AMPLITUDE RATIO
2.00	0.20	0.20	0 dB	2.00	2.097	4.022
1.00	-0.3	-0.4				
2.00	0.20	0.20	-3 dB	2.00	2.072	4.168
1.00	-0.3	-0.4				
2.00	0.20	0.20	-5 dB	2.00	2.038	2.663
1.00	-0.3	-0.4				
1.50	0.20	0.20	0 dB	1.50	1.502	5.815
1.00	-0.3	-0.4				
1.50	0.20	0.20	-3 dB	1.50	1.513	2.297
1.00	-0.3	-0.4				
1.50	0.20	0.20	-5 dB	1.50	1.511	1.765
1.00	-0.3	-0.4				

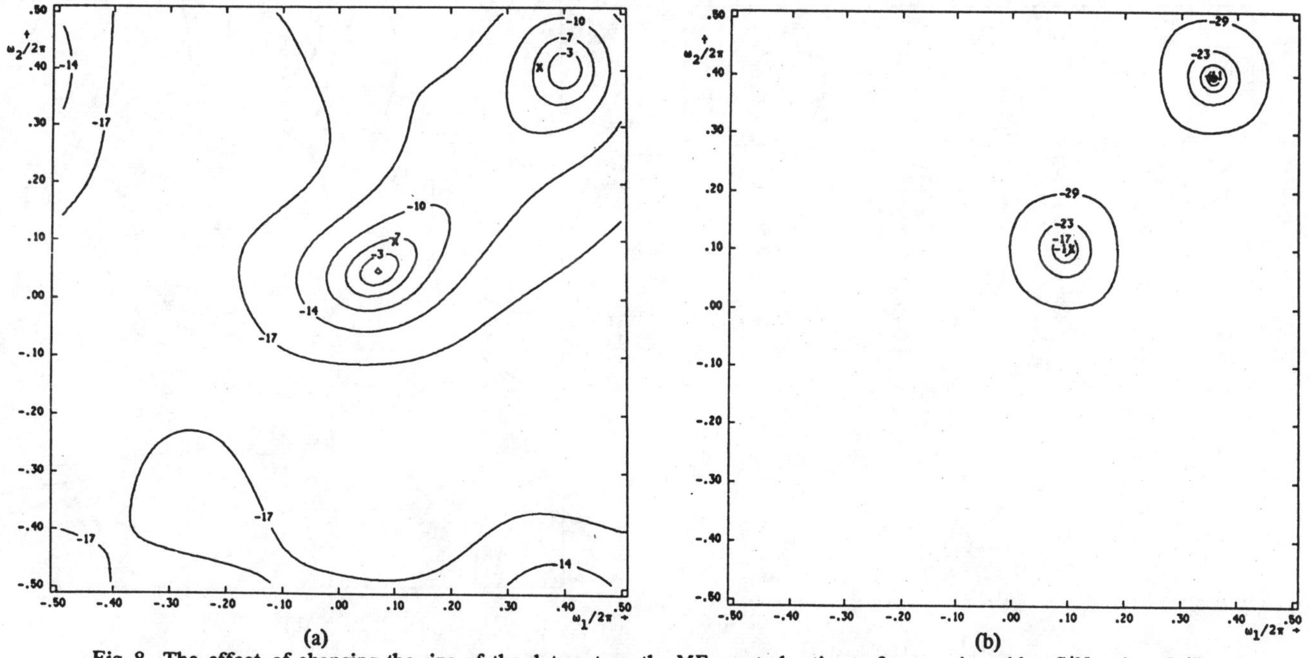

(a) (b)

Fig. 8. The effect of changing the size of the data set on the ME spectral estimate for two sinusoids. S/N ratio = 5 dB, $a_1^2 = 1.0$ and $a_2^2 = 1.0$, $\sigma^2 = 0.632$, 3 × 3 ACF, relative phase = 0°, true peak locations are (0.1, 0.1) and (0.35, 0.4), (a) Data length = 4 × 4, CINC = 3.50 dB. (b) Data length = 60 × 60, CINC = 5.99 dB.

This is in contrast with the ME spectral estimate, which has been observed to show a rapid improvement, with very little improvement visible after a data length of 12 × 12. This result indicates that the ME method may have a significant advantage relative to the ML and Bartlett methods in applications such as array processing, where the number of sensors is small and thus only a small amount of data is available.

G. Effect of Initial Phase on Spectral Estimates

When the data consist of a sinusoid in noise, and the ACF has to be estimated from the data themselves, it becomes important to know the effect of the initial phase of the sinusoid on the spectral estimate. In the 1-D case, it has been noted that the phase causes a shift in the LOSP in the periodogram and the direct data Burg methods of PSE [8], [11]. Although the algorithm used here is not a direct data method, the ACF must be estimated from the data, and therefore the starting phase will have an effect on the ACF values, and hence on the spectral estimates. To determine the effect of the initial phase on the ME spectral estimates, we have studied several examples with the data consisting of a single complex sinusoid in white noise, using a 3 × 3 region of support for the ACF. The results are similar to the 1-D case in that the LOSP shows an oscillation about the true peak location, the amplitude of the oscillation decreasing with increasing data length and increasing S/N ratio, as expected. Fig. 11(a) shows the

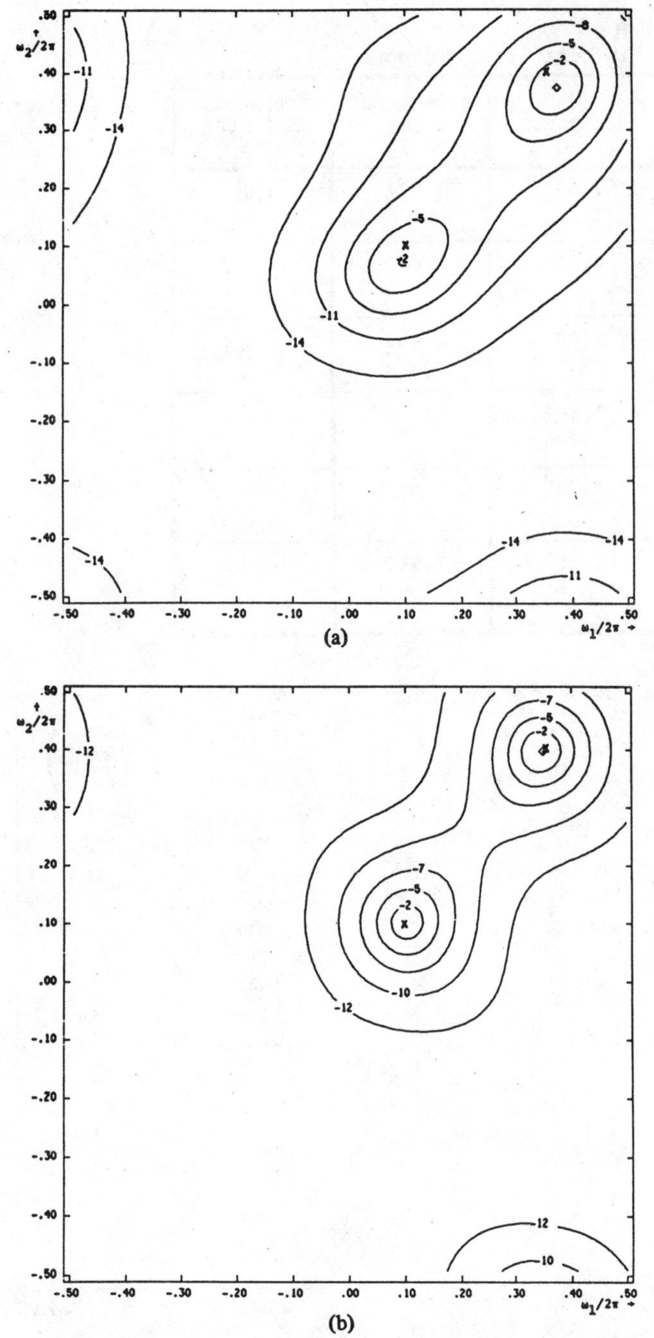

Fig. 9. The effect of data length on ML estimates for two sinusoids. Same data parameters as Fig. 8. (a) ML estimate; data length = 4 X 4, CINC = 2.90 dB. (b) ML estimate; data length = 60 X 60, CINC = 2.53 dB.

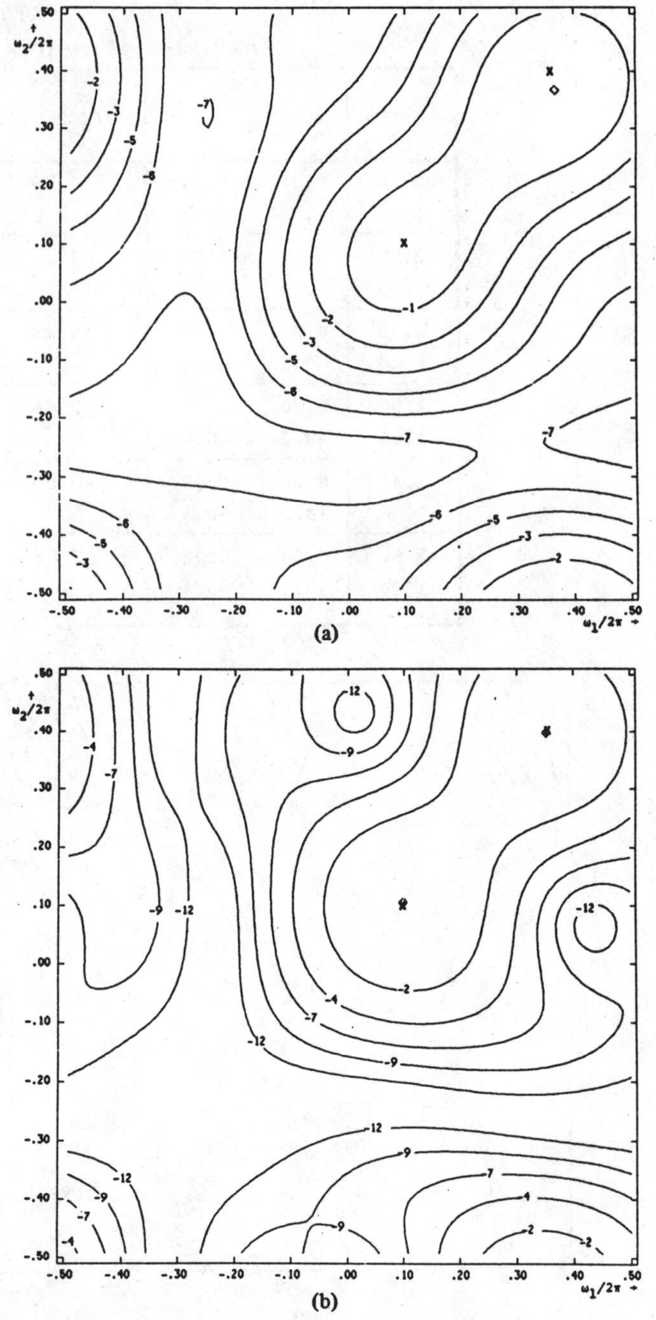

Fig. 10. The effect of data length on Bartlett estimates for two sinusoids. Same data parameters as Fig. 8. (a) Bartlett estimate; data length = 4 X 4, CINC = 1.272 dB. (b) Bartlett estimate; data length = 60 X 60, CINC = 2.40 dB.

oscillation in the LOSP for one example where the size of the data segment used is 12 X 12, and the initial phase of the sinusoid is varied from zero to 2π (corresponding to 1.0 in the figure). The size of the region A is 3 X 3 and the S/N ratio is 5 dB. Fig. 11(b) shows the corresponding result for the ML method. The result for the Bartlett method is very similar to that shown in Fig. 11(b). All the points in the figure are based on a single realization of the noise. The oscillations in the LOSP for the three techniques are seen to be very similar in amplitude. The behavior of the ME method is different, however, from the ML and Bartlett techniques.

IV. Conclusion

This paper has been concerned with obtaining a characterization of the ME method of PSE for two-dimensional signals. The experiments performed were for data sets consisting of sinusoids in white Gaussian noise, using estimated ACF values as well as exact ACF values. It was found that, like the one-dimensional case, the estimation of power spectra for real data gives erroneous peak location estimates when the peaks are located in certain regions of the 2-D frequency plane. The special regions are caused by the periodicity and symmetry of the spectrum.

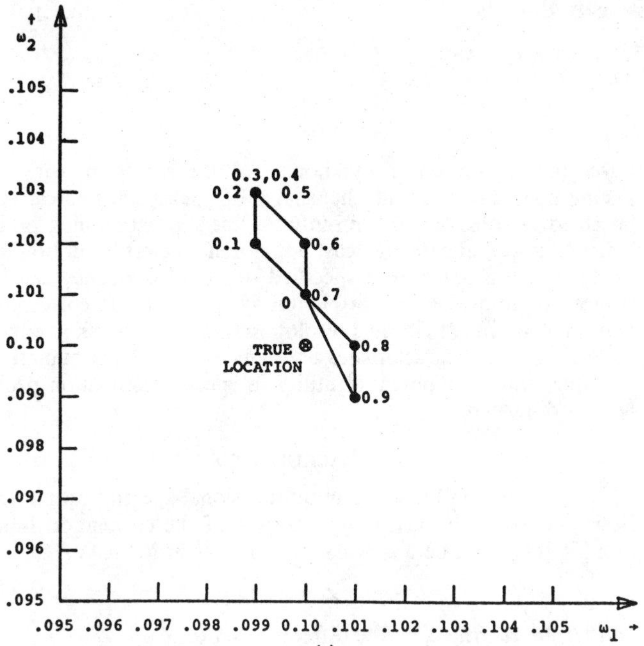

(a)

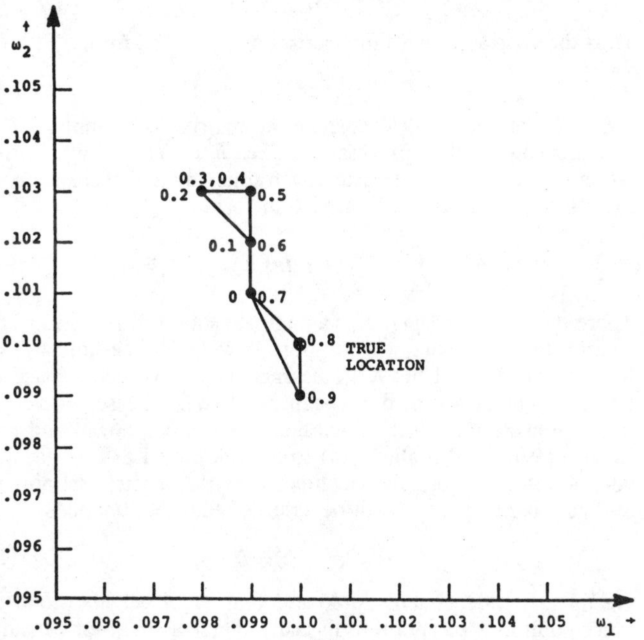

(b)

Fig. 11. The variation of estimated peak location as a function of initial phase. 3 X 3 ACF, true peak location = (0.1, 0.1), S/N ratio = 5 dB, $a_1^2 = 1.0$, $\sigma^2 = 0.316$. (a) ME method. (b) ML method.

To study the resolution performance, we have studied two measures, one representing the resolvability of two sinusoids in noise and the other representing the accuracy of the estimated LOSP. It was found that like the 1-D case, the ME method affords the highest resolution among the ME, ML, and Bartlett methods. When the two sinusoids are well resolved by all three techniques, the accuracy of the estimated LOSP is higher for the ML and Bartlett methods than for the ME method.

Increasing the S/N ratio increases the resolution of the ME spectral estimates. For the case of real data, the special regions discussed above were found to decrease in size with increasing S/N ratio. For both real and complex data, the spectral peaks grow sharper in the spectral estimates and the peak location accuracy improves with increasing S/N ratio. It was found that two peaks located close together that cannot be resolved at low S/N ratios can be resolved at high S/N ratios.

The effect of increasing the size of the known ACF gave results similar to those obtained by increasing the S/N ratio. The main result associated with changing the shape of the ACF support region was the higher resolution along the direction of more correlation points.

Determining the power of a sinusoid from the ME spectral estimate requires evaluating the volume under the estimated spectrum. Specifically, we have observed that the power of a sinusoid is linearly proportional to the volume under the peak in the estimated spectrum. This is similar to the 1-D case, where the power is proportional to the area under the peak in the ME spectral estimate.

For the case of estimated ACF values, it was found that the ME spectral estimate improves very rapidly with increasing data length. The effect of the starting phase of sinusoidal data was investigated and it was found that the location of the spectral peak oscillates about the true position for different values of the initial phase, similar to the 1-D case.

In summary, even though the algorithm that solves the 1-D ME PSE problem by autoregressive signal modeling does not extend to the 2-D case, most properties of 2-D ME spectral estimates can be considered as straightforward extensions of the corresponding 1-D results.

Acknowledgment

The authors gratefully acknowledge the help of M. Glaser, who provided the data for Table II.

References

[1] J. P. Burg, "Maximum entropy spectral analysis," presented at the 37th Meet. Soc. Exploration Geophysicists, 1967.

[2] W. Y. Chen and G. R. Stegen, "Experiments with maximum entropy power spectra of sinusoids," *J. Geophys. Res.*, vol. 79, pp. 3019-3022, July 10, 1974.

[3] T. J. Ulrych and T. N. Bishop, "Maximum entropy spectral analysis and autoregressive decomposition," *Rev. Geophys. Space Phys.*, vol. 13, pp. 183-200, Feb. 1975.

[4] J. S. Lim and N. A. Malik, "A new algorithm for two-dimensional maximum entropy power spectrum estimation," *IEEE Trans. Acoust., Speech, Signal Processing*, vol. ASSP-29, pp. 401-413, June 1981.

[5] N. A. Malik, "One and two dimensional maximum entropy spectral estimation," Sc.D. thesis, Massachusetts Inst. Technol., Cambridge, Nov. 1981.

[6] J. S. Lim and N. A. Malik, "Maximum entropy power spectrum estimation of signals with missing correlation points," *IEEE Trans. Acoust., Speech, Signal Processing*, vol. ASSP-29, pp. 1215-1217, Dec. 1981.

[7] J. V. Pendrell, "The maximum entropy principle in two-dimensional spectral analysis," Ph.D. dissertation, York Univ., Toronto, Ont., Canada, Nov. 1979.

[8] D. N. Swingler, "Burg's maximum entropy algorithm versus the discrete Fourier transform as a frequency estimator for truncated real sinusoids," *J. Geophys. Res.*, vol. 85, pp. 1435-1438, Mar. 10, 1980.

[9] A. V. Oppenheim and R. W. Schafer, *Digital Signal Processing.* Englewood Cliffs, NJ: Prentice-Hall, 1975, pp. 532-571.

[10] R. T. Lacoss, "Data adaptive spectral analysis methods," *Geophys.*, vol. 36, pp. 661-675, Aug. 1971.

[11] P. L. Jackson, "Truncation and phase relationships of sinusoids," *J. Geophys. Res.*, vol. 72, pp. 1400-1403, 1967.

Power Spectral Density Bounds

THOMAS L. MARZETTA, MEMBER, IEEE, AND
STEPHEN W. LANG, MEMBER, IEEE

Abstract—**The determination of a power density spectrum from a finite set of correlation samples is an ill-posed problem. Furthermore, it is not possible even to bound the values that consistent power density spectra can take on at a particular point. A more reasonable problem is to try to determine the total spectral power in some frequency interval. Although this power cannot be determined exactly, upper and lower bounds on its possible values can be determined. This observation leads to a unified treatment of certain classical and modern spectral estimation techniques and to new interpretations for two data adaptive spectral estimators, maximum likelihood method (MLM) and data adaptive spectral estimator (DASE). According to these new interpretations, MLM and DASE provide upper bounds on spectral power in a specified frequency region subject to the assumption that the spectral density is constant in that region. These methods make no use of an extendibility constraint that can be used to obtain tight upper bounds, as well as nontrivial lower bounds on power. Cybenko has studied a related problem of bounding windowed power, for an arbitrary window, with no assumptions about the form of the spectral density. A new type of classical resolution limit for these bounds is derived and a numerical example is presented.**

I. Introduction

A large class of spectral estimation techniques is based on a two-step procedure of first reducing observed random process samples to a finite set of estimates of correlation function samples and then computing a spectral estimate in terms of the estimated correlation samples. The performance of these methods is degraded by two sources of error: the statistically induced errors in the estimated correlation samples and the nonuniqueness of the mapping from a finite set of correlation samples to the power density spectrum. It follows that even if the true correlation samples were available, the problem of estimating the spectrum would still be ill-posed, in that it is possible to find an infinite number of power density spectra that are consistent with the correlation estimates [1]. Furthermore, consistent power density spectra exist that take on arbitrary values at a particular point. Thus values of the spectral density function at a point cannot be determined without further prior knowledge.

Instead of attempting to estimate power density at a single point, window-type spectral estimation methods estimate the total spectral power in some frequency region. Although the determination of total spectral power in a region is also an ill-posed problem, upper and lower bounds can be determined from correlation samples.

This observation leads to a unified treatment of window-type spectral estimators. In particular, two modern methods, Davis and Regier's data adaptive spectral estimator (DASE) [2] and the earlier maximum likelihood method (MLM) of Capon [3] (a special case of DASE), are shown to provide upper bounds on spectral power in a specified region of frequency, subject to the assumption that the spectral density is constant in that region. It is shown that the assumptions and constraints that determine the DASE upper bound yield a trivial lower bound of zero. Furthermore, it is shown that the DASE method ignores an available constraint that can be used to obtain a least upper bound on spectral power, as well as a nontrivial greatest lower bound.

In a recent paper [4], Cybenko suggests that tight upper and lower bounds on linear functionals of the spectrum, with no assumptions made about the shape of the spectrum, can be obtained as solutions of semi-infinite linear programming problems. As a special case, the least upper and greatest lower bounds on the spectral power in a specified region of frequency can be computed from a set of correlation samples. In this correspondence a new type of classical resolution limit for Cybenko bounds is derived, their characteristics are demonstrated by a numerical example, and their potential utility in spectral estimation problems is discussed.

II. Background

Consider the problem of multidimensional spectral estimation from nonuniformly distributed samples of the correlation function [5], [6]. Let X be a set of sample points in R^D, an *array*

$$X = \{x_1, \cdots, x_N\}, \tag{1}$$

and let the *coarray* Δ be the difference set of X

$$\Delta = \{x_i - x_j\}. \tag{2}$$

Thus the coarray Δ is a finite subset of R^D of the form

$$\Delta = \{\mathbf{0}, \pm\delta_1, \cdots, \pm\delta_M\}. \tag{3}$$

Let R be the $N \times N$ correlation matrix for samples of a homogeneous random process on X, i.e., $R_{ij} = r(x_i - x_j)$. Correlation measurements are linear constraints on the power spectrum $S(k)$ through its inverse Fourier transform:

$$r(\delta) = \int_K e^{jk\cdot\delta} S(k)\, d\nu(k), \qquad \delta \in \Delta, \tag{4}$$

where the *spectral support* K is a compact subset of R^D, and ν is a positive finite measure, with support K, which defines integrals of functions over K. Thus K is the region of wavevector space in which power is assumed to be present, Δ is the set of vector separations at which the correlation function is known, and ν is some measure which allows (4) to be interpreted as a surface or volume integral (possibly weighted) over the spectral support in D-dimensional space. It will be assumed that the functions

$$e^{jk\cdot\delta}, \qquad \delta \in \Delta, \tag{5}$$

are linearly independent on K and that every neighborhood of every point in K has nonzero measure. These assumptions ensure, respectively, that the correlation measurements are independent and that $S(k)$ contributes to the correlation measurements at each point in the spectral support, through (4).

The collection of correlation measurements on the coarray can be considered a vector in R^{2M+1}.[1] Defining ϕ_k as the correlation vector with components $e^{jk\cdot\delta}$, $\delta \in \Delta$, corresponding to an impulse at k, (1) becomes, in vector notation,

$$r = \int_K \phi_k S(k)\, d\nu(k). \tag{6}$$

Let E be the set of *extendible*[2] correlation vectors, those which can be written as (4) for some nonnegative $S(k)$. The interior of

[1] In general, $r(0)$ is real, $r(\delta_i)$, $i = 1, \cdots, M$, are complex, and $r(-\delta_i) = r^*(\delta_i)$, giving $2M + 1$ real numbers.

[2] The correlation function obtained through the inverse Fourier transform of $S(k)$ is a suitable *extension* of the correlation measurements to all spatial separations.

Manuscript received September 23, 1982; revised April 6, 1983.

The authors are with Schlumberger-Doll Research, P.O. Box 307, Ridgefield, CT 06877.

Reprinted from *IEEE Trans. Inform. Theory,* vol. IT-30, pp. 117–122, Jan. 1984.

this set consists of exactly those correlation vectors which can be written as (4) for some spectral density function $S(k)$ which is uniformly bounded away from zero:

$$S(k) > \epsilon > 0, \qquad k \in K. \tag{7}$$

Vectors on the boundary of E can be realized by impulsive spectra [5], [6].

Since the observed correlations are correlations of a random process, the correlation matrix R must be positive-definite. Moreover, since they are correlations of a homogeneous random process, the correlation vector r must be extendible. In the *time series case*,[3] extendibility is equivalent to positive-definiteness of the correlation samples. More generally, extendibility is a stricter condition than positive-definiteness [9].

III. Classical Techniques

Classical spectral estimation techniques are associated with spectral windows that have the form of a Δ-polynomial:[4]

$$W(k) = \sum_{\delta \in \Delta} w(\delta) e^{-jk\cdot\delta}, \qquad w(\delta) = w^*(-\delta). \tag{8}$$

For Δ-polynomial windows, the windowed power can be exactly determined from the correlation samples:

$$\int_K W(k) S(k)\, d\nu(k) = \sum_{\delta \in \Delta} r^*(\delta) w(\delta). \tag{9}$$

A classical window[5] implicitly divides the spectral support into two regions, a mainlobe region B, where the window is large, and a sidelobe region $K - B$, where the window is small (Fig. 1). The mainlobe width cannot be made arbitrarily small while retaining this simple division of the spectral support. Generally, the mainlobe width must be larger than some limit: the classical or Rayleigh resolution limit. The sidelobe level can only be reduced by increasing the mainlobe width beyond this lower bound.

Under the assumption that the power density spectrum is constant over the mainlobe region, and assuming that the sidelobe leakage is small, the windowed power, properly normalized, can be interpretated as the power in the mainlobe region. For a window with its mainlobe centered at k_0:

$$\int_K W(k) S(K)\, d\nu(k) \approx S(k_0) \int_B W(k)\, d\nu(k). \tag{10}$$

Alternatively, for positive windows under the assumption that the power density spectrum is constant over the mainlobe region, the normalized windowed power can be interpreted as an upper bound on the power density in the mainlobe region.

$$\left[\int_B W(k)\, d\nu(k)\right]^{-1} \int_K W(k) S(k)\, d\nu(k) \geqslant S(k_0). \tag{11}$$

Because the windows used in classical spectral estimation are typically translated versions of one another, the characteristics of a single window determine the characteristics of the entire spectral estimate.

IV. MLM and DASE

MLM and DASE are conventionally interpreted as procedures for designing windows that vary from wavevector to wavevector [3], [2]. Hence, in using these procedures, one loses the simple characterization of the spectral estimate by a single window shape. Furthermore, the windows are often strangely shaped, not

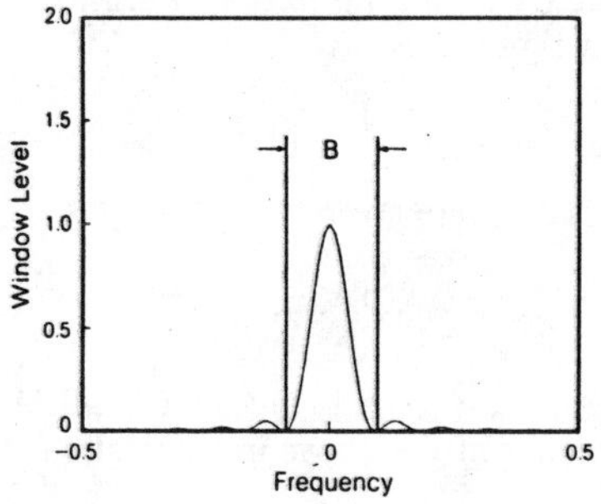

Fig. 1. Classical window.

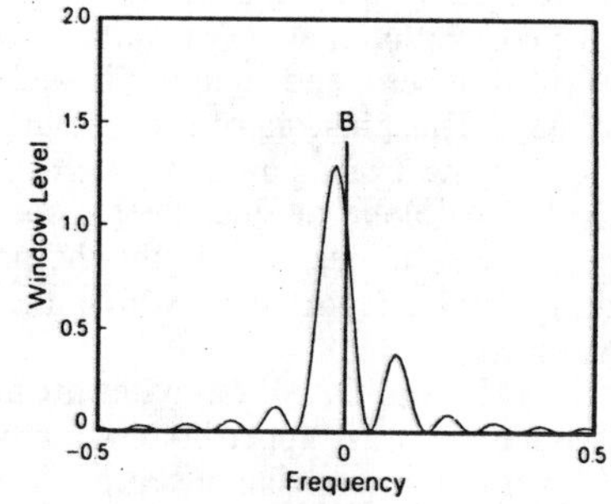

Fig. 2. Data adaptive window.

lending themselves to a simple division of the spectral support into a mainlobe and a sidelobe region (Fig. 2). However the estimates can be simply characterized if MLM and DASE are interpreted as upper bounds on the power in certain regions of the spectral support.

Under the assumption that the power density spectrum is constant in some subset B of the spectral support, the correlation matrix R can be decomposed as

$$R = p\Gamma + Q, \tag{12}$$

where p is the total power in the region B and Q is the correlation matrix corresponding to the part of the spectrum outside of B. Γ is defined as the correlation matrix corresponding to a constant power density in B and a total power of unity:

$$\Gamma = \frac{1}{\nu(B)} \int_B \gamma_k \gamma_k^\dagger\, d\nu(k) \tag{13}$$

where the steering vector γ_k is defined by

$$\gamma_k^\dagger = \left[e^{-jk\cdot x_1}, \cdots, e^{-jk\cdot x_N}\right], \tag{14}$$

and $\nu(B)$ is the measure (D-dimensional volume) of the region B.

A key observation is that this decomposition is nonunique; there is a continuous range of values for p and Q such that (12) is satisfied. Since power is a nonnegative quantity, it follows that p has a lower bound of zero. An upper bound on p is imposed by the requirement that the correlation matrix Q be nonnegative-definite.

Specifically, it is shown in the Appendix that

$$0 \leqslant p \leqslant \frac{1}{\lambda_{\max}}, \tag{15}$$

where $\lambda_{\max}$ is the maximum eigenvalue of $R^{-1}\Gamma$. Any value of p in this range is consistent with the nonnegative-definiteness constraint on Q. This upper bound corresponds to Davis and Regier's DASE estimate, proving the equivalence of the above interpretation to their own adaptive window interpretation [2].

The MLM estimate can be derived as a special case of the

[3] $D = 1$, $K = [-\pi, \pi]$, $X = \{0, 1, \cdots, M\}$, and $\Delta = \{0, \pm 1, \cdots, \pm M\}$.
[4] In the time series case these are ordinary trigonometric polynomials.
[5] Standard designs exist for the time series case [7].

DASE estimate by letting the region B shrink to a single point $\boldsymbol{k}_0$. The matrix Γ becomes

$$\Gamma = \gamma_{k_0}\gamma_{k_0}^{\dagger}, \tag{16}$$

and the bounds on p are given by

$$0 \leqslant p \leqslant \frac{1}{\gamma_{k_0}^{\dagger} R^{-1} \gamma_{k_0}}. \tag{17}$$

The upper bound corresponds to Capon's MLM estimate, proving the equivalence of the above interpretation to his own interpretation [3].[6]

To summarize, given a positive-definite correlation matrix R, and a subset B of the spectral support over which the spectral density is assumed to be constant, the nonnegative-definiteness condition on the correlation matrix Q results in a lower bound on the total power in B of zero, and an upper bound that is given by the DASE estimate. Thus, instead of interpretating the MLM or DASE estimates as the exact power through variable windows with no particular mainlobe or sidelobe regions, they are more simply interpretated as upper bounds on the power in a fixed region of the spectral support over which the power density spectrum is assumed constant.

Although the MLM and DASE estimates are upper bounds on power, they are not the least upper bounds on power. Since the extendibility constraint on correlation samples is more strict than the positive-definiteness constraint on the corresponding correlation matrix [9], tighter bounds could be obtained by requiring that the matrix Q be extendible instead of merely nonnegative-definite. Furthermore, the extendibility condition would lead to nontrivial lower bounds. A comparison of lower to upper bounds would provide an indication of the tightness of the bounds. It can be seen that these new bounds would be based on only two assumptions: the given correlation samples are correct, and the power density spectrum is constant in the region B. It is clear that the latter assumption is unwarranted, in general, since there is no *a priori* reason to assume that the spectrum is constant anywhere. By abandoning this assumption, and imposing the extendibility constraint, more realistic upper and lower bounds can be obtained. Such bounds were recently proposed by Cybenko [4].

V. Cybenko Bounds

Let $W(\boldsymbol{k})$ denote an arbitrary window function, not necessarily a Δ-polynomial. The Cybenko bounds can be obtained by solving the following constrained maximization and minimization problems [4].

Choose $S(\boldsymbol{k})$ to maximize (minimize) the windowed power:

$$\int_K W(\boldsymbol{k}) S(\boldsymbol{k})\, d\nu(\boldsymbol{k}), \tag{18a}$$

subject to the following two constraints:

$$\boldsymbol{r} = \int_K \phi_k S(\boldsymbol{k})\, d\nu(\boldsymbol{k}), \tag{18b}$$

$$S(\boldsymbol{k}) \geqslant 0. \tag{18c}$$

The upper bound is obtained from the solution of the maximization problem and the lower bound is obtained from solution of the minimization problem. These bounds do not require a smoothness assumption, as did the classical, MLM, and DASE estimates. The semi-infinite programming problems (18) can be solved approximately by discretizing the spectral support and using a linear programming algorithm, although more efficient methods are available [10].

Of particular interest are windows that are constant on some subset B of the spectral support and zero elsewhere. For this case, the Cybenko bounds have the following properties:

1) The bounds increase monotonically as the region increases.
2) When the region B is identical to the spectral support K, then both the upper and lower Cybenko bounds are equal to the variance of the random process $r(\mathbf{0})$.
3) For the time series case, when the region B consists of a single point, the Cybenko upper bound is equal to the MLM power estimate, and the Cybenko lower bound is zero, a result of the equivalence of the extendibility and nonnegative-definiteness conditions for this special case.

Qualitatively, when the size of the region B is sufficiently large, a nontrivial lower bound is obtained, and the upper and lower bounds are close to one another, implying that the spectral power in B can be inferred reliably. On the other hand, as the size of B is decreased, the lower bound eventually goes to zero, and the spectral power in B can take on any value between zero and the upper bound. In this regime, the upper bound as a function of the center frequency of the window is potentially a "high resolution" spectral estimate, in the sense that adjacent spectral peaks not resolvable by classical techniques may be resolved. This property has been demonstrated for the MLM technique [3], [11].

As stated above, the Cybenko lower bound approaches zero as the size of the region B decreases. This statement is supported quantitatively by the following resolution limit theorem, which provides a *sufficient* condition for the Cybenko lower bound to be zero for at least one wavevector.

Theorem: Suppose that, for each $\boldsymbol{k}' \in K$, the Cybenko lower bound is computed with a nonnegative window $W_{k'}(\boldsymbol{k})$, where $\boldsymbol{k}'$ denotes the window position. (The size and shape of the window may change with $\boldsymbol{k}'$.) In addition it is required that, for all $\boldsymbol{k}' \in K$, $W_{k'}(\boldsymbol{k}) > 0$ if and only if $W_k(\boldsymbol{k}') > 0$.[7] If, for all $\boldsymbol{k}' \in K$, the ν-measure of the support of $W_{k'}(\boldsymbol{k})$ is strictly less than $\nu(K)/(2M+1)$, then a trivial lower bound of zero is obtained at some $\boldsymbol{k}' \in K$.

The proof, given in the Appendix, relies on the fact that there exists an impulsive spectrum of no more than $2M+1$ impulses that matches any extendible correlation vector [5], [6], [12]. For the time series case, in which $\nu(K)$ is equal to 2π, this condition is equivalent to the classical resolution limit. For window widths less than this limit, the Cybenko lower bound is zero for at least one frequency.

VI. Example

An example is presented in this section of Cybenko bounds in the time series case with ν the Lebesque measure on $[-\pi, \pi]$. Consider

$$\boldsymbol{r} = \int_{-\pi}^{\pi} \phi_k S(k) \frac{dk}{2\pi}, \tag{19}$$

where the correlation vector $\boldsymbol{r}$ corresponds to a coarray

$$\Delta = \{0, \pm 1, \cdots, \pm 10\}, \tag{20}$$

and the spectral density function is as given in Fig. 3. The Cybenko bounds were computed for a family of rectangular windows, parameterized by their center locations and widths. The semi-infinite programming problems (18) were solved by discretizing the interval $[-\pi, \pi]$, using about 100 grid points, and solving the resulting linear programs using the simplex algorithm.

In Fig. 4 are shown the resulting upper and lower bounds on power in an interval of variable width, centered at zero frequency. There are several things to note about this result. First note that

[6]This new interpretation was first derived, for the special case of MLM, by Marzetta [8].

[7]A special case where this condition is satisfied is that of a symmetric window of fixed shape with $W_{k'}(\boldsymbol{k}) = W(\boldsymbol{k} - \boldsymbol{k}') = W(\boldsymbol{k}' - \boldsymbol{k})$, for all $\boldsymbol{k}' \in K$.

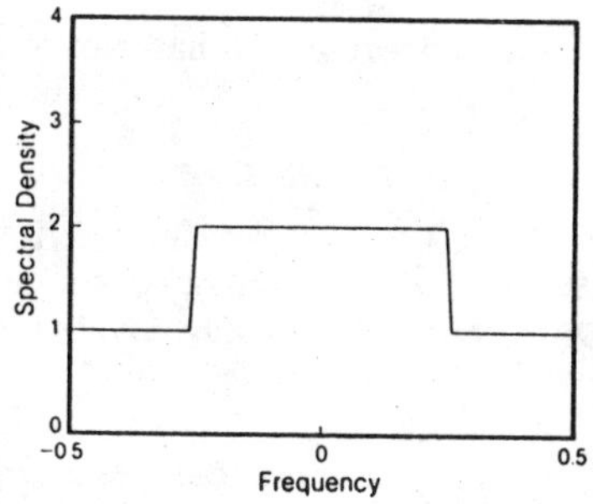

Fig. 3. Power density spectrum.

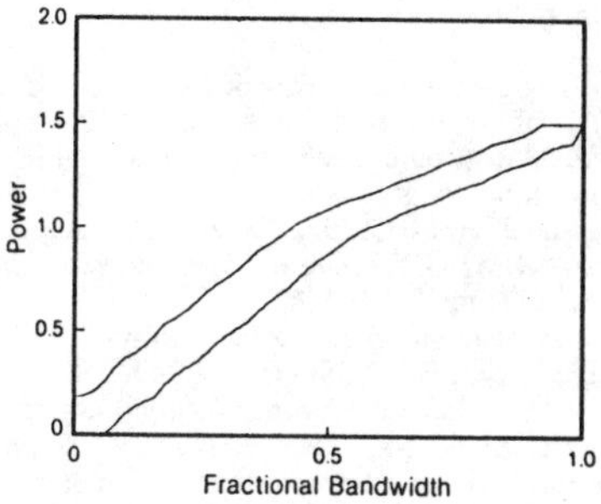

Fig. 4. Bounds on power in an interval about zero versus interval width.

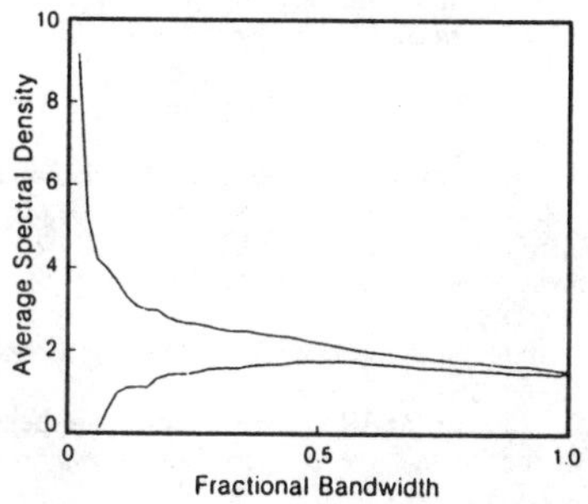

Fig. 5. Bounds on average spectral density versus bandwidth.

the zero bandwidth limit of the upper bound is nonzero. This is the MLM case, as mentioned in the previous section. Also note that the upper bound saturates at a bandwidth smaller than the full spectral support. This occurs when r becomes extendible on the window support. Thus extendibility, as distinct from positive-definiteness, can be an issue even in the time series case. Note that the lower bound is zero for sufficiently small bandwidths, verifying the resolution limit discussed in the previous section. It is zero until r is no longer extendible on the portion of the spectral support outside of the window support. Finally, note that when the window support is the full spectral support, the upper bound equals the lower bound, which equals $r(0)$, the total spectral power.

It is instructive to normalize the bounds by the window width, resulting in a bound on the *average spectral density* over the window support. Fig. 5 shows the effect of this normalization on the bounds shown in Fig. 4. It is clear that the uncertainty about the average spectral density in an interval, defined as the distance between the upper and lower bounds in Fig. 5, is infinite for zero bandwidth (the MLM case) and decreases as the size of the interval increases.

Fig. 6 shows upper and lower bounds on average spectral density, computed for intervals which are 1/10 the size of the spectral support and centered at various frequencies. The actual average power spectral density, computed by smoothing the spectrum of Fig. 3 with the same rectangular window, is also shown. It lies, of necessity, between the upper and lower bounds.

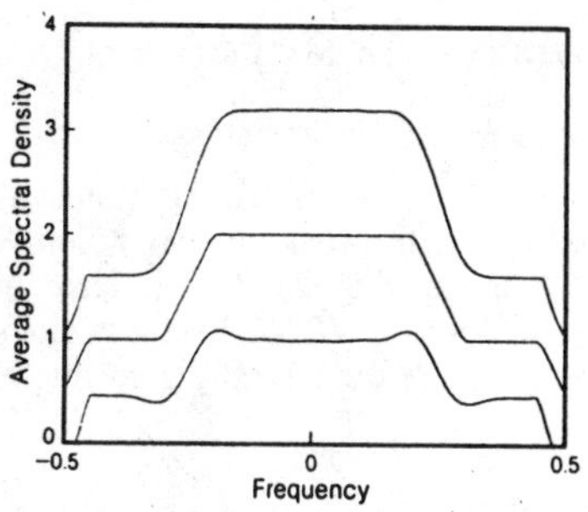

Fig. 6. Bounds on average spectral density versus center frequency.

VII. Conclusion

Although it is not possible to bound the value that the spectral density function can take on at a particular point, it is possible to place nontrival upper and lower bounds on linear functionals of the power spectrum. Viewing spectral estimates as bounds on the power in certain regions of the spectral support leads to a simple and unified interpretation of certain classical and modern spectral estimation techniques.

In any practical spectral estimation problem, the exact values of the correlation samples are not available and correlation estimates must be used to compute the various spectral bounds. The bounds obtained from estimates of the correlation samples are not, strictly speaking, deterministic bounds; they are bounds subject to the assumption that the correlation estimates are correct. Either confidence intervals or a method of allowing for error in the correlation estimates would be desirable for Cybenko bounds. Furthermore, a practical difficulty which can occur when Cybenko bounds are derived from correlation estimates is that, in order to obtain the bounds, the correlation estimate vector must be extendible.[8] For the case of multidimensional spectral estimation, this is a potentially serious problem, since none of the existing methods of estimating the correlation function are known to always produce an extendible correlation vector. Both of these problems can be ameliorated by generalizing the Cybenko bounds to allow uncertainties in the correlation estimates. Replacing the correlation matching constraint of (18b) with the constraint that the inverse Fourier transform lie within a convex set around the estimated correlation vector might allow information about uncertainties in the correlation estimates to be incorporated into the computation of the bounds without seriously affecting the computability of the bounds.

Although interesting theoretically, and representing perhaps the ultimate in nonparametric spectral estimation techniques, the practical application of Cybenko bounds will depend upon the development of reliable and reasonably fast algorithms for their computation.

It is interesting to note that the general problem of inferring an unknown function, $F(k)$, given a finite set of linear functionals of F, has been extensively studied by the geophysical community [13]. It has long been recognized that this is an ill-posed problem, but that a restricted class of linear functionals of F can be obtained. This technique is known as the Backus–Gilbert method, and it contains the classical spectral estimation techniques as a special case. More recently, the inverse gravity problem was investigated using approach closely analogous to Cybenko's approach to the spectral estimation problem [14]. Specifically, it was shown that upper and lower bounds on linear functionals of the density variation of the earth can be computed from a finite set of measurements of the gravitational field, by solving semi-infinite linear programming problems.

[8] In contrast, the Capon bound and the Davis–Regier bound require only that the correlation matrix be Hermitian positive-definite; no other structure is assumed.

Appendix: Some Mathematical Details

A. The DASE and MLM Upper Bounds

From (12), $Q = R - p\Gamma$, which is nonnegative-definite for $p \leqslant p_{max}$, where p_{max} is the value of p at which $R - p\Gamma$ becomes singular. Since R is nonsingular, this is also the value of p at which $R^{-1}(R - p\Gamma) = I - pR^{-1}\Gamma$ becomes singular. Hence $p_{max} = 1/\lambda_{max}$, where λ_{max} is the largest eigenvalue of $R^{-1}\Gamma$.

B. The Resolution Limit Theorem

The resolution limit theorem of Section V is proved below.

The Cybenko lower bound as a function of window position is devoted by $C_L(k')$, $k' \in K$. The proof consists of showing that the ν-measure of the support of C_L is less than $\nu(K)$, the measure of K. If this is true then there must be some wavevector $k'' \in K$ that is not included in the support of C_L, i.e., $C_L(k'') = 0$, a trivial lower bound.

The extendibility of the correlation samples implies that there exists an impulsive spectrum on K, of no more than $2M + 1$ impulses, that matches the correlation vector [5], [6], [12]. Let the locations of these impulses be denoted by $\{k_i \in K: i = 1,\cdots,2M+1\}$ and their respective powers by $\{a_i \geqslant 0: i = 1,\cdots,2M+1\}$. The Cybenko lower bound on windowed power is less than or equal to the windowed power of this feasible spectrum. Thus

$$0 \leqslant C_L(k') \leqslant \sum_{i=1}^{2M+1} a_i W_{k'}(k_i), \qquad k' \in K.$$

By assumption, whenever $W_{k'}(k_i) > 0$, $W_{k_i}(k') > 0$. Thus at any point k' such that $C_L(k') > 0$ we have $\sum_{i=1}^{2M+1} a_i W_{k_i}(k') > 0$, implying that any point contained in the support of C_L is also contained in the support of $\sum_{i=1}^{2M+1} a_i W_{k_i}(k')$, so the ν-measure of the support of C_L is less than or equal to the ν-measure of support of $\sum_{i=1}^{2M+1} a_i W_{k_i}(k')$. By the finite subadditivity property of measures, this in turn implies that the ν-measure of the support of C_L must be less than or equal to the sum of the ν-measures of the supports of the windows W_{k_i}. The measure of each of these windows is, by assumption, strictly less than $\nu(K)/(2M + 1)$. Hence the measure support of C_L has ν-measure strictly than $\nu(K)$.

Acknowledgment

The authors thank G. Cybenko for referring them to the literature on semi-infinite programming and for pointing out some deficiencies in an earlier statement and proof of the resolution limit theorem.

References

[1] V. F. Pisarenko, "The retrieval of harmonics from a covariance function," *Geophys. J. R. Astr. Soc.*, vol. 33, pp. 347–366, 1973.
[2] R. E. Davis and L. A. Regier, "Methods for estimating directional wave spectra from multi-element arrays," *J. Marine Res.*, vol. 35, pp. 453–477, 1977.
[3] J. Capon, "High-resolution frequency-wavenumber spectrum analysis," *Proc. IEEE*, vol. 57, pp. 1408–1418, Aug. 1969.
[4] G. Cybenko, "Affine minimax problems and semi-infinite programming," to appear in *Math. Programming*.
[5] S. W. Lang and J. H. McClellan, "Spectral estimation for sensor arrays," in *Proc. First ASSP Workshop on Spectral Estimation*, Hamilton, ON, Aug. 17–18, 1981, pp. 3.2.1–3.2.7.
[6] ——, "Spectral estimation for sensor arrays," *IEEE Trans. Acoustics, Speech, Signal Proc.*, vol. ASSP-31, pp. 349–358, Apr. 1983.
[7] F. J. Harris, "On the use of windows for harmonic analysis with the discrete Fourier transform," *Proc. IEEE*, vol. 66, pp. 51–83, Jan. 1978.
[8] T. L. Marzetta, "A new interpretation for Capon's maximum likelihood method of frequency-wavenumber spectral estimation," *IEEE Trans. Acoustics, Speech, and Signal Proc.*, vol. ASSP-31, pp. 445–449, Apr. 1983.
[9] B. W. Dickinson, "Two-dimensional Markov spectrum estimates need not exist," *IEEE Trans. Inform. Theory*, vol. IT-26, pp. 120–121, Jan. 1980.
[10] R. Hettich, Ed., *Semi-Infinite Programming.* New York: Springer-Verlag, 1979.
[11] H. Cox, "Resolving power and sensitivity to mismatch of optimum array processors," *J. Acoust. Soc. Amer.*, vol. 54, no. 3, pp. 771–785, 1973.
[12] G. Cybenko, "Moment problems and low rank Toeplitz approximations," *Circuits, Systems, and Signal Proc.*, vol. 1, no. 3–4, pp. 345–366, Dec. 1982.
[13] R. Parker, "Understanding inverse theory," *Ann. Rev. Earth Planet Sci.*, vol. 5, pp. 35–64, 1977.
[14] C. Safon, G. Vasseur, and M. Cuer, "Some applications of linear programming to the inverse gravity problem," *Geophys.*, vol. 41, no. 6, pp. 1215–1229, Oct. 1977.

On Two-Dimensional Markov Spectral Estimation

R. CHELLAPPA, SENIOR MEMBER, IEEE, YU-HEN HU, AND SUN-YUAN KUNG, MEMBER, IEEE

Abstract—We give systematic methods for two-dimensional (2-D) spectral estimation from raw data using Gaussian Markov random field (MRF) models. The MRF models considered in this paper characterize the statistical dependence of the observation $y(s)$ on its neighbors in *all* directions. Due to the modeling assumption, the 2-D spectrum is an explicit function of the parameters of the MRF model. Thus, the spectral estimation problem reduces to that of estimating the appropriate structure and the parameters of the MRF model. A tractable algorithm for maximum likelihood (ML) estimation of parameters is obtained by using doubly periodic boundary conditions for the given data. Asymptotically consistent decision rules are used to choose the appropriate neighbor set of the MRF model. The MRF spectral estimate obtained using the doubly periodic boundary conditions has the property that its inverse discrete Fourier transforms are in perfect agreement with the periodic sample correlation values of the given observations, over a symmetric window identical to the structure of the MRF model. Using this result, and the fact that the 2-D maximum entropy spectral (MES) estimate has a structure similar to the MRF power spectrum, we show that the MRF spectral estimate developed in this paper is a good approximation to the MES estimate. This result provides a reasonable basis (i.e., maximum likelihood) on which to justify the use of the MES estimate with correlation estimates obtained from a finite piece of data samples.

I. INTRODUCTION

IN THIS paper, we shall present a *model based* approach for two-dimensional (2-D) spectral estimation of given random sample data. Two-dimensional power spectral estimation is of interest in frequency wavenumber analysis [1], image restoration [2], filtering of radar images [3], texture classification [4], and other important applications. Given an array of 2-D random data $\{y(s), s \text{ in } \Omega\}$ where

$$\Omega = \{(i,j): 0 \leqslant i,j \leqslant M-1\}$$

is a finite 2-D lattice region, the problem is to estimate the 2-D power spectrum corresponding to the given data. One of the earliest methods proposed by Cote [5] is to extend the 1-D Blackman-Tukey type approach to 2-D case. A more direct and popular approach in image processing literature [2, ch. 7], [4] is to calculate the periodogram estimate. Although computationally attractive, the raw periodogram estimate is not a consistent estimate of the power spectrum.

Recently, a class of methods known as the parametric methods using 2-D causal autoregressive (AR) [6] and causal autoregressive and moving average (ARMA) [7] models has become very popular. These methods can be extended to include AR or ARMA models defined with nonsymmetric half plane (NSHP) [8] neighbor sets. It is our contention that the causal models or their generalization to NSHP neighbor sets are only a subclass of more general spatial models. Unlike in the 1-D case, the notion of causality or one-sided Markovianity in 2-D is rather restrictive. It is possible that an observation at a location may depend upon observations at neighboring locations in all directions, not necessarily only on those belonging to quarter plane or NSHP neighbor sets. Therefore, in this paper, we shall represent the given array of random observations by a noncausal Markov random field (MRF) model.

In the MRF model the observation $y(s)$ obeys a Markovian property

$$\Pr\{y(s)| \quad \text{all } y(r), r \neq s\} = \Pr\{y(s)| \quad \text{all } y(s+r), r \text{ in } N\}$$

where N is a (noncausal) neighbor set excluding the origin (0, 0). The set, N, is symmetric such that if the point "r" is in N, then "$-r$" is also in N. For example, the simplest bilateral MRF model is obtained when $N = \{(0,-1), (0,1), (-1,0), (1,0)\}$. For the Gaussian MRF random field, $y(s)$ can be represented by [9], [10]

$$y(s) = \sum_{r \in N} \theta_r y(s+r) + e(s) \tag{1}$$

where $\{\theta_r; r \text{ in } N\}$ are MRF parameters and $\{e(s)\}$ is a correlated 2-D noise sequence. In [9], [10], it is shown that the power spectrum corresponding to such a model has a particular structure

$$S(\lambda) = \nu \Big/ \left[1 - \sum_{r \in N} \theta_r e^{j\lambda^t r}\right] \tag{2}$$

where $\lambda = (\lambda_1, \lambda_2)$ in $\Lambda = \{(\lambda_1, \lambda_2); 0 \leqslant \lambda_1, \lambda_2 \leqslant 2\pi\}$ and ν = variance of $\{e(s)\}$.

In the view of the parametric model used in (1), the desired spectrum is a function of the parameters $\{\nu, \theta_r; r \text{ in } N\}$ and the neighbor set N characterizing the model. This assumption reduces the spectral estimation problem to two subtasks: 1) to estimate the parameters of the MRF model, and 2) to choose an appropriate neighbor set N for the MRF model.

For the estimation of parameters in MRF models, there are three known methods, viz. the coding method [11], the least square method [12], [13], and the maximum likelihood (ML) [14] method.[1] Among them, the ML method yields an asymp-

Manuscript received November 2, 1981; revised November 1, 1982. This work was supported in part by the Office of Naval Research under Contract N00014-81-K-0191 and by the National Science Foundation under Grant ECS-82-04181.

R. Chellappa and S.-Y. Kung are with the Department of Electrical Engineering–Systems, University of Southern California, Los Angeles, CA 90007.

Y.-H. Hu is with the Department of Electrical Engineering, Southern Methodist University, Dallas, TX 75275.

[1] Note that this is different from the ML spectrum estimates [1].

Reprinted from *IEEE Trans. Acoust., Speech, Signal Processing,* vol. ASSP-31, pp. 836–841, Aug. 1983.

totically consistent and efficient estimate. In general, due to the fact that the Jacobian of the transformation matrix from noise variate to $\{y(s)\}$ is not unity in noncausal MRF models, the log-likelihood function required in ML estimation is a complicated function of parameters. But by using a doubly periodic extension of the finite lattice region Ω ("toroidal lattice," [14], [15]), explicit expressions can be written down for the log-likelihood function to obtain the ML estimate of the MRF parameters. On the other hand, as a result of the toroidal assumption, the MRF spectral estimate obtained will be a discrete power spectrum instead of a continuous spectrum. Nevertheless, the discrete spectrum can be related to the continuous spectrum via a simple relation [16] and, hence, can be regarded as a good approximation of the continuous MRF spectral estimate.

In addition to the parameter estimation issue, a second problem to be tackled in a model based approach is the choice of appropriate model. In this paper, an asymptotically consistent transitive decision rule developed elsewhere [12], [13], [17] is briefly discussed for choosing an appropriate neighbor set of the MRF model.

The MRF spectral estimate derived by substituting the Gaussian ML estimates of parameters with toroidal lattice assumption has several interesting properties. First, we show that the (periodic) sample correlations obtained with toroidal lattice assumption are in perfect agreement with the inverse discrete Fourier transform (IDFT) of the discrete MRF spectrum estimate. This sample correlation matching property, together with the fact that the structure of the MRF estimate is similar to the 2-D maximum entropy spectral (MES) estimate [10], leads to the conclusion that the discrete MRF spectrum estimate obtained in this paper is also a good approximation of the 2-D MES estimate corresponding to the same set of random data. Second, we show that the MRF spectrum derived in this paper is similar to the conventional periodogram spectrum [2].

The noncausal MRF model used in this paper has also been used in [18] for 2-D spectral estimation. In this work, the MRF model parameters are estimated by solving a set of 2-D Yule-Walker equations. As pointed out in [19], such an approach may lead to nonunique solutions and the resulting estimates are not statistically efficient as the ML estimates. Furthermore, the choice of the particular MRF model used in [18] is arbitrary.

The contributions of this paper can be summarized as follows. 1) A systematic procedure for estimating 2-D spectrum from raw data is given by using a general class of 2-D noncausal MRF models. 2) A reasonable basis using the principle of maximum likelihood is given to justify the use of MES estimates with correlation estimates obtained from a finite piece of data samples.

A. Organization

The organization of the paper is as follows. In Section II, the MRF spectral estimation on toroidal lattices is discussed. The MRF parameter estimation using ML estimator is derived and a criterion for neighbor selection is mentioned. In Section III, it is shown that the MRF spectrum estimate obtained satisfies the sample correlation matching property and, hence, is good approximation to the 2-D MES estimate.

II. The Markov Random Field Spectral Estimation

A. The MRF Spectrum on Toroidal Lattices [16]

We are interested in fitting a Gaussian MRF model characterized by the neighbor set N as represented in (1). Knowing that the lattice region Ω in which the data samples are specified contains only finite lattice points, it is possible that for some point "s" near the boundary of Ω, there exists some point "r" in N such that $(s+r)$ is not in Ω, and thus, $y(s+r)$ in (1) is not defined. This difficulty can be resolved by assuming a doubly periodic extension of the given *random data* samples such that

$$y(s) = \sum_{r \in N} \theta_r y(s \oplus r) + e(s) \tag{3}$$

where $(s \oplus r) = ([(s_1 + r_1) \bmod M, (s_2 + r_2) \bmod M])$. Such periodic extension shall be termed "toroidal lattice" in the rest of this paper. In (3), $\{e(s)\}$ is a correlated noise sequence with the correlation function

$$E(e(s)e(r)) = \begin{cases} -\theta_{r-s}\nu & \text{if } (s-r) \bmod M \text{ is in } N \\ \nu & \text{if } s = r \bmod M \\ 0 & \text{otherwise} \end{cases} \tag{4}$$

where $\nu > 0$ is the variance of $e(s)$. We also note that in (3), different correlation regions N characterize different MRF models. Suppose that (3) is represented in vector matrix notation; we have

$$A(\boldsymbol{\theta})\boldsymbol{y} = \boldsymbol{e} \tag{5}$$

where $\boldsymbol{y} = \text{column}[y(s);\ s \text{ in } \Omega]$, $\boldsymbol{e} = \text{column}\ [e(s);\ s \text{ in } \Omega]$, and $A(\boldsymbol{\theta})$ is a symmetric block-circulant matrix

$$A(\boldsymbol{\theta}) = \begin{bmatrix} A_{11} & A_{12} & \cdots & A_{12} \\ A_{12} & A_{11} & \cdots & A_{13} \\ \cdots & \cdots & \cdots & \cdots \\ A_{12} & A_{13} & \cdots & A_{11} \end{bmatrix} \tag{6}$$

with each $A_{i,j}$ itself being an M-by-M symmetric circulant matrix whose (m, n)th element $a_{i,j}(m, n)$ is [16]

$$a_{i,j}(m,n) = \begin{cases} -\theta_r & \text{if } r = (|m-n| \bmod M, \\ & \quad \cdot |i-j| \bmod M) \text{ in } N \\ 1 & \text{if } i = j \bmod M \text{ and } m = n \bmod M; \\ 0 & \text{otherwise.} \end{cases} \tag{7}$$

Clearly, from (7), $A(\boldsymbol{\theta})$ matrix is a function of $\boldsymbol{\theta} = \text{column}\ [\theta_r;\ r \text{ in } N]$. Equation (4) can be rewritten as

$$E\{\boldsymbol{e}\boldsymbol{e}^t\} = \nu A(\boldsymbol{\theta}). \tag{8}$$

Using (8) and the orthogonality condition [9], [10], we have

$$E\{\boldsymbol{y}\boldsymbol{e}^t\} = \nu I$$

and

$$E\{\boldsymbol{y}\boldsymbol{y}^t\} = \nu[A(\boldsymbol{\theta})]^{-1}. \tag{9}$$

Since a covariance matrix must be positive definite, from (8) and (9), the eigenvalues of $A(\theta)$, denoted by, say μ_k, for $k = (k_1, k_2)$ in $K = \{k; 0 \leq k_1, k_2 \leq M - 1\}$, must be greater than zero. Since $A(\theta)$ is a block circulant matrix, its eigenvalues equal the discrete Fourier transform of its first row. Using (6), (7), we have a sufficient condition for bounded input, bounded output stability of $\{y(s)\}$.

$$\mu_k = 1 - \theta^t \phi_k > 0, \quad k \text{ in K} \tag{10}$$

where ϕ_k = column $[\exp(j2\pi k^t r/M); r \text{ in } N]$.

As a result of the finite toroidal lattice assumption, the correlation function $R_d(r) = E\{y(s \oplus r) y(s)\}$ is also periodic, where the subscript "d" denotes the periodic extension. The periodicity of the correlation function leads to a discrete power spectrum $S_d(\lambda_k)$ which is defined only at discrete frequencies $\lambda_k = (2\pi/M)k$, for k in K

$$S_d(\lambda_k) = \nu/(1 - \theta^t \phi_k) \quad k \text{ in K}. \tag{11}$$

Clearly, $S_d(\lambda_k)$ can be characterized by the set of MRF model parameters $\{\theta_r; r \text{ in } N\}$. Note also that $S_d(\lambda_k)$ corresponds to the kth eigenvalue of the covariance matrix of y, $\nu[A(\theta)]^{-1}$. Hence, the positive definiteness of the covariance implies the positivity of the discrete spectral estimate.

B. ML Estimates of MRF Model Parameters

In the previous section, an MRF model on a toroidal lattice is studied. Since the MRF spectrum is parameterized by the parameters θ, in order to obtain an MRF spectral estimate $\hat{S}_d(\lambda_k)$, it is sufficient to estimate the model parameters. We shall assume, for a moment, that the correlation region N is specified.

As mentioned earlier, there are three methods available for estimation of MRF parameters: the coding method, the least square (LS) method, and the ML method. The first two methods are easy to compute. However, the coding method does not have minimum variance in view of its partial utilization of data [14]. Also, it is not unique and often yields several correlated estimates which can differ from each other considerably. The LS method is known to be inefficient and not have minimum variance property [12]. The ML estimate, however, does have minimum variance property but, in general, is very difficult to compute. The major obstacle lies in that the Jacobian of the transformation between the noise variate vector e and the data vector y is a complicate function of θ for noncausal MRF model. Therefore, the likelihood function of the parameters cannot be computed easily.

Suppose that the MRF model is represented on a toroidal lattice. The use of such a model enables us to write an explicit expression for the log-likelihood function. If e is an (M^2-variate) zero mean Gaussian random vector, then the log-likelihood function of y can be written as

$$\log P(y \mid \theta, \nu) = (\tfrac{1}{2}) \sum_{k \in K} \log(1 - {}^t\phi_k) - (M^2/2) \log 2\pi\nu - \left(\frac{1}{2\nu}\right) y^T A(\theta) y. \tag{13}$$

Now define a periodic sample correlation function on the toroidal lattice

$$C_d(r) = M^{-2} \sum_{s \in \Omega} y(s) y(s \oplus r). \tag{14}$$

The quadratic form $y^T A(\theta) y$ in (13) then can be simplified as [24]

$$y^T A(\theta) y = M^2 [C_d(0) - \sum_{r \in N} \theta_r C_d(r)] = M^2 (C_d(0) - \theta^t C_d) \tag{15}$$

where C_d = column $[C_d(r); r \text{ in } N]$. Thus, the log-likelihood function to be maximized in (13) becomes [14]

$$\log p(Y \mid \theta, \nu) = \sum_{k \in K} \log(1 - \theta^t \phi_k) - M^2 \log 2\pi\nu - (M^2/\nu)(C_d(0) - \theta^t C_d). \tag{16}$$

By differentiating (16) with respect to ν and equating to zero, we have

$$\nu^* = C_d(0) - \theta^{*t} C_d \tag{17}$$

where θ^* are the θ at the maximum of (16). Equation (17) may be substituted into (16) (without "*") as an additional constraint to be satisfied during the course of the maximization of (16). In doing so, the task of maximization of (16) becomes a task of *minimizing*

$$\log[(C_d(0) - \theta^t C_d)] - M^{-2} \sum_{k \in K} \log(1 - \theta^t \phi_k). \tag{18}$$

This task can be accomplished by using conventional iterative optimization procedures such as the gradient method or the Newton-Raphson method.

In implementing the minimization, we note that the minimization of (18) must be subject to two implicit requirements: 1) $(C_d(0) - \theta^t C_d) > 0$ and 2) $1 - \theta^t \phi_k > 0$. This is consistent with (4), (17), and (10), which state that the noise variance should be positive and the noise covariance matrix must be positive definite.

C. Decision Rule for Choosing an Appropriate Gaussian MRF Model

In the derivation of the 2-D MRF power spectrum estimate, we assumed that the particular neighbor set N of the Gaussian MRF model is known. In practice, however, this should also be estimated from the given data. Detailed discussions regarding the neighbor set selection problem in Gaussian MRF models may be found in [12], [13]. We only give a brief account of the relevant results in this paper. The neighbor set selection problem comes under the category of multiple decision problem involving compound hypotheses. A method that is well suited for this problem is to compute a test statistic for different models and choose the one corresponding to the minimum. The Akaike information criterion (AIC) [21] and the Bayes method [17] are two such procedures. The AIC statistics can be written down from the expression for the likelihood function in (13): the best model is the one which minimizes the AIC statistics. The AIC method, in general, gives transitive decision rules, but is not consistent even for one-dimensional autoregressive models [22]. We give test statistics which can be derived using Bayes decision theoretic arguments.

Suppose we have three sets N_1, N_2, N_3 of neighbors containing m_1, m_2, m_3 lattice points, respectively. Corresponding to each N_i, we write the MRF model as

$$y(s) = \sum_{r \in N_i} \theta_r y(s \oplus r) + e(s).$$

Then the decision rule [12], [13] for the choice of appropriate neighbors can be summarized below.

Choose the neighbor set N_{i_0} if $i_0 = \text{argument} \min_i \{g_i\}$, where

$$g_i = -\sum_{k \in K} \log(1 - \theta_i^{*T} \phi_{ik}) + M^2 \log \nu_i^* + m_i \log M^2 \quad (19)$$

$$\theta_i^* = \text{column } [\theta_{ir}^*, r \text{ in } N_i]$$

$$\phi_{ik} = \text{column } [\exp(j2\pi k^t r/M), r \text{ in } N_i].$$

The model selection procedure consists of computing g_k for different models and choosing the one corresponding to the minimum. The test statistics in (19) lead to asymptotically consistent transitive decision rules. The first term in (19) is due to the noncausal nature of the MRF model. Equation (19) can be used to determine the appropriate Gaussian MRF model for the data whose MRF spectral estimate is desired.

III. Relation to Other 2-D Spectral Estimates

A. Relations to MRF Spectrum on Infinite Lattice Model

In Section II-B, an MRF spectral estimate using an ML estimate of the MRF parameters is derived. The use of the toroidal lattice model leads to a discrete spectral estimate $S_d(\lambda_k)$, $k \in K$. It is natural to question how this discrete spectral estimate relates to the continuous MRF spectrum in (2) without using the toroidal lattice model (i.e., with the infinite lattice model).

Provided that the true MRF parameters are known, comparing (2) to (11), it is clear that [16], [23]

$$S_d(\lambda_k) = S(\lambda)\big|_{\lambda = \lambda_k}.$$

Furthermore, for M being sufficiently large compared to the size of neighbor set[2] N, the periodic correlation $R_d(r)$ and the true correlation $R(r)$ are close to each other. To illustrate the numerical proximity, suppose that an isotropic MRF model with $\theta_r = 0.24$ for every r in $N = \{(0,1), (1,0), (0,-1), (-1,0)\}$ is given. Then correlation functions $R(r)$ and $R_d(r)$ are computed by using the IFT (inverse Fourier transform) of $S(\lambda)$ and the IDFT of $S_d(\lambda_k)$, respectively. We have for the correlation function $R(r)$ computed with infinite lattice model $R(1,0) = 0.434$, $R(1,1) = 0.295$, $R(0,2) = 0.220$, $R(2,1) = 0.180$. The corresponding quantities computed using the toroidal lattice ($M = 64$) assumption are $R_d(1,0) = 0.4329$, $R_d(1,1) = 0.2935$, $R_d(0,2) = 0.2181$, $R_d(2,1) = 0.1785$, which are numerically close to those of the infinite lattice model. Hence, the toroidal lattice model is an excellent approximation of the infinite lattice model.

[2] In practice, the size of the neighbor set N is much smaller than that of the data sample region Ω.

We note that in [24], a 2-D MRF spectral estimate is given where the sample correlation function is estimated using an infinite lattice array. In view of the above example, we expect that the discrete MRF spectrum obtained by using toroidal lattice will be a good approximation of that obtained in [24] at $\lambda = \lambda_k$.

B. Approximation of 2-D Maximum Entropy Spectrum

In this section, we shall show that the (discrete) MRF spectral estimate obtained in Section II-B is a good approximation of 2-D maximum entropy spectral (MES) estimate. MES method has been a popular spectrum estimation method with a variety of applications. Due to its high resolution property, there are many efforts to generalize the 1-D ME method to the 2-D case [10], [19], [25]-[27]. To make this paper self-contained, the 2-D MES estimation problem is briefly reviewed.

2-D ME Problem Formulation: Denote $R(r)$ to be the autocorrelation function of $\{y(s)\}$, $S(\lambda)$ to be the power spectrum of $\{y(s)\}$, and A to be a set of lattice points on which $R(r)$ is known. The 2-D ME problem is to determine $\hat{S}(\lambda)$, an estimate of $S(\lambda)$, such that the entropy H defined by

$$H = \int \log \hat{S}(\lambda)\, d\lambda \quad (20)$$

is maximized and

$$R(r) = IFT\{\hat{S}(\lambda)\} \qquad r \text{ in } A \quad (21)$$

where IFT{·} denotes the inverse 2-D Fourier transform.

It is shown [28] that the MES estimate, $\hat{S}(\lambda)$, if it exists, has the following structure:

$$S_{\text{MES}}(\lambda) = \nu' \Big/ \left[1 - \sum_{r \in A} \theta'_r \exp(j\lambda^t r)\right]. \quad (22)$$

Consequently, to compute the MES estimate, it would suffice to compute the parameters $\{\nu', \theta'_r; r \text{ in } A\}$ such that (21) is satisfied. In doing so, iterative optimization procedures are often required [10], [20], [25], [26].

As observed in [10], the 2-D continuous MES estimate in (22) has the same structure as the Gaussian MRF spectrum (2) with the given correlation region A replaced by N. Thus, one possible method for 2-D MES estimation from the random data is to estimate the MRF parameters in (2) from the given data such that when the estimates are used in (2), the resulting spectral estimate satisfies (21). Although it is not clear that such a method exists, we show below that the discrete MRF spectral estimate obtained by using Gaussian ML estimates of parameters on the toroidal lattice is a good approximation to 2-D MES estimate. The first step in this direction is to show that $\hat{S}_d(\lambda_k)$ obtained in Section II satisfies a sample correlation matching property.

Theorem 1 (Sample Correlation Matching Property):

$$R_d(r) = \text{IDFT}\{\hat{S}_d(\lambda_k)\} \qquad \text{for } r \text{ in } N. \quad (23)$$

Proof: Taking the gradient of (18) with respect to $\boldsymbol{\theta}$ and setting the gradient to zero at $\boldsymbol{\theta} = \boldsymbol{\theta}^*$, we have

$$0 = \boldsymbol{C}_d / [C_d(0) - \boldsymbol{\theta}^{*t} \boldsymbol{C}_d] - (1/M^2) \sum_{k \in K} [\boldsymbol{\phi}_k / (1 - \boldsymbol{\theta}^{*t} \boldsymbol{\phi}_k)].$$

Using (17) in the above expression and rearranging terms, it yields

$$\mathbf{C}_d = M^{-2} \sum_{k \in K} \phi_k [\nu^*/(1 - \theta^{*t}\phi_k)] = \text{IDFT}[\hat{S}_d(\lambda_k)].$$

Thus, the theorem is proved. Q.E.D.

In view of the sample correlation matching property and the fact that MES has the same structure as the MRF spectrum, it is hoped that $\hat{S}_d(\lambda_k)$ is a valid 2-D MES spectral estimate. Indeed, as the sample size M^2 gets sufficiently large compared to the correlation region N, we have $C_d(r) \to C(r)$ where $C(r)$ is the sample correlation function estimated using infinite lattice model. Hence, we have

$$\hat{S}_d(\lambda_k) \to S_{\text{MES}}(\lambda = \lambda_k) \quad \text{as} \quad M \to \infty,$$

The above argument indicating the asymptotic equivalence of the MRF estimate developed in this paper and the classical 2-D MES estimate can be strengthened by showing the asymptotic equivalence of the criterion functions minimized in the two approaches. It can be shown [20] that the criterion function $J(\cdot)$ minimized in the classical 2-D MES analysis is

$$J(\theta, \nu) = (R_0 - \theta^t R) + \int \log [\nu/(1 - 2\theta^t \psi_\lambda)] \, d\lambda \tag{24}$$

where

$$\psi_\lambda = \text{column} [\exp(j\lambda^t r), r \in A]$$

and $\mathbf{R} = \text{column}[R_r, r \in A]$ are the given theoretical correlations. By comparison, the criterion function minimized in our approach is the minus log-likelihood function

$$L(\theta, \nu) = \frac{1}{M^2} \sum_{k \in K} \log \frac{\nu}{(1 - 2\theta^t \phi_k)} + (1/\nu)(C_d(0) - \theta^t \mathbf{C}_d). \tag{25}$$

The first term in (25) can be considered as the discrete approximation to the entropy integral in (24); and the second term in (25) can be considered as the estimate of the first term in (24) from the samples. As the lattice size M tends to ∞, to the extent the integral and summation and the true and sample correlations in (24) and (25) can be interchanged, the criterion functions (24) and (25) are asymptotically equivalent.

Theorem 1 also provides a sound theoretical basis justifying the use of the sample correlation function to find the MES estimate when data samples are available. With a doubly periodic extension of the finite data array, Theorem 1 shows that the ML estimates of the model parameters obey a sample correlation matching property and, hence, can be taken as parameters of the corresponding approximate MES estimate.

C. On the Connection to Conventional Spectral Estimates

Due to the toroidal lattice assumption, the discrete MRF spectrum $S_d(\lambda_k)$ can be related to other conventional spectral estimates. Let us define a 2-D Fourier transform vector η_k by

$$\eta_k = \text{column}\{\exp(js^t\lambda_k; s \text{ in } \Omega)\}.$$

Then, by (10), we have

$$\eta_k^t A(\theta) \eta_k = \mu_k = 1 - \theta^t \phi_k. \tag{26}$$

Now, using (9), (11), and (26), it can be shown that

$$S_d(\lambda_k) = \eta_k^t \nu^* [A(\theta^*)]^{-1} \eta_k \sim \eta_k^t E\{\mathbf{y}\mathbf{y}^t\} \eta_k = E\{|\eta_k^t \mathbf{y}|^2\}. \tag{27}$$

Therefore, $S_d(\lambda_k)$ can be related to the 2-D periodogram estimate via (27).

IV. Conclusion

In this paper, we have presented a model based 2-D spectral estimation procedure by using an MRF model of the given 2-D random data. We have assumed a doubly periodical extension of the finite data array and made use of Gaussian ML estimates of MRF parameters to obtain the MRF spectrum estimate. We have shown that the MRF spectrum obtained in this paper is a good approximate of the MRF spectrum on an infinite lattice model. More importantly, we have shown that the MRF spectrum obtained obeys a sample correlation matching property, and hence can be taken as a good approximation of the 2-D MES estimate as $M \to \infty$. This sample correlation matching property also serves a theoretical basis (maximum likelihood) for justifying the use of the MES estimate with sample correlation function obtained from a finite piece of data samples.

Acknowledgment

The authors wish to thank Prof. R. L. Kashyap and the anonymous reviewers for many helpful comments.

References

[1] J. Capon, "High-resolution frequency-wavenumber spectrum analysis," *Proc. IEEE*, vol. 57, pp. 1408–1418, Aug. 1969.

[2] H. C. Andrews and B. R. Hunt, *Digital Image Restoration*. Englewood Cliffs, NJ: Prentice-Hall, 1977.

[3] V. S. Frost *et al.*, "A system model for imaging radars and its application to minimum mean square filtering," in *Proc. 5th Int. Conf. Pattern Recog.*, Miami, FL, Dec. 1980, pp. 121–133.

[4] J. W. Weszka *et al.*, "A comparative study of texture measures for terrain classification," *IEEE Trans. Syst., Man, Cybern.*, vol. SMC-6, pp. 269–285, Apr. 1976.

[5] L. Cote, "Two-dimensional spectral analysis," Dep. Stat., Purdue Univ. W. Lafayette, IN, Mimeograph Series 83.

[6] D. Tjostheim, "Autoregressive modeling and spectral analysis of array data in the plane," *IEEE Trans. Geosci. Remote Sensing*, vol. GE-19, pp. 15–23, Jan. 1981.

[7] J. A. Cadzow and K. Ogino, "Two-dimensional spectrum estimation," *IEEE Trans. Acoust., Speech, Signal Processing*, vol. ASSP-29, pp. 396–401, June 1981.

[8] D. M. Goodman and M. P. Ekstrom, "Multidimensional spectral factorization and univariate AR models," *IEEE Trans. Automat. Contr.*, vol. AC-25, pp. 158–262, Apr. 1980.

[9] J. W. Woods, "Two-dimensional discrete Markov random fields," *IEEE Trans. Inform. Theory*, vol. IT-18, pp. 232–249, Mar. 1972.

[10] ——, "Two-dimensional Markov spectral estimation," *IEEE Trans. Inform. Theory*, vol. IT-22, pp. 552–559, Sept. 1976.

[11] J. E. Besag, "Spatial interaction and statistical analysis of lattice systems," *J. Royal Stat. Soc. Series B*, vol. B-26, pp. 192–236, 1974.

[12] R. Chellappa and R. L. Kashyap, "Statistical inference in Gaussian Markov random field models," in *Proc. IEEE Conf. Pattern Recog. Image Processing*, Las Vegas, NV, June 1982, pp. 77–80.

[13] R. L. Kashyap and R. Chellappa, "Estimation and choice of neighbors in spatial interaction models of images," *IEEE Trans. Inform. Theory*, vol. IT-29, pp. 60–72, 1983.

[14] P.A.P. Moran and J. E. Besag, "On the estimation and testing of spatial interaction in Gaussian lattices," *Biometrika*, vol. 62, pp. 555-562.
[15] P.A.P. Moran, "A Gaussian Markovian process on a square lattice," *J. Appl. Probability*, vol. 10, pp. 605-612, 1973.
[16] R. L. Kashyap, "Random field models on finite lattices for finite images," presented at the Conf. on Inform. Sci., Johns Hopkins Univ., Baltimore, MD, May 1981.
[17] R. L. Kashyap and R. Chellappa, "Decision rules for the choice of neighbors in random field models of images," *Comput. Graphics Image Processing*, vol. 15, pp. 301-318, Apr. 1981.
[18] A. K. Jain and S. Ranganath, "High resolution spectrum estimation in two dimensions," in *Proc. 1st ASSP Workshop on Spectrum Estimation*, McMaster Univ., Hamilton, Ont. Canada, Aug. 1981.
[19] A. K. Jain, "Advances in mathematical models for image processing," *Proc. IEEE*, vol. 69, May 1981.
[20] S. W. Lang, "Spectral estimation for sensor array," Ph.D. dissertation, Massachusetts Inst. Technology, Cambridge, Aug. 1981.
[21] H. Akaike, "A new look at the statistical model identification," *IEEE Trans. Automat. Contr.*, vol. AC-19, pp. 716-722, Dec. 1974.
[22] R. L. Kashyap, "Inconsistency of the AIC rule for estimating the order of autoregressive models," *IEEE Trans. Automat. Contr.*, vol. AC-25, pp. 996-998, Oct. 1980.
[23] —, "Analysis and synthesis of image patterns by spatial interaction models," *Progress in Pattern Recognition*, vol. 3, L. N. Kand and A. Rosenfeld, Eds. Amsterdam, The Netherlands: North-Holland, 1981, pp. 149-186.
[24] T. Ulrych and C. J. Walker, "High resolution two dimensional power spectral estimation," in *Applied Time Series Analysis II*, Findley, Ed. New York: Academic, 1981, pp. 71-99.
[25] J. S. Lim and N. A. Malik, "A new algorithm for two-dimensional maximum entropy power spectrum estimation," *IEEE Trans. Acoust., Speech, Signal Processing*, vol. ASSP-29, pp. 401-413, June 1981.
[26] J. H. McClellan, "Multidimensional spectral estimation," *Proc. IEEE*, vol. 70, pp. 1029-1039, Sept. 1982.
[27] J. P. Burg, "Maximum entropy spectral analysis," Ph.D. dissertation, Stanford Univ., Stanford, CA, 1975.

Part III
Systems

THE papers in this part of the book concentrate on system theoretic aspects mainly for the class of two-dimensional (2-D) signals and systems. These aspects include stability of 2-D recursive difference equations, the computation of the complex cepstrum, and the choice of sampling lattice for M-D signals. Also considered in Part III are some stochastic aspects of 2-D systems including a generalization of linear prediction to two dimensions, and a treatment of realizable Wiener filtering for homogeneous (= stationary) random fields (= processes). In this introduction we will attempt to set these papers in some perspective and at the same time refer to other relevant papers listed in the bibliography which follows.

The first paper, by O'Connor and Huang, concerns the stability of general 2-D recursive digital filters. This paper generalizes and unifies a whole body of work on the stability of 2-D recursive filters dating from 1978 back to Huang's initial paper [1] on stability in 1972. O'Connor and Huang relate stability of a 2-D recursive filter to the filters' 2-D phase function. They also develop a number of practical stability tests. Two stability topics not covered in this paper are the theoretical importance of nonessential singularities of the second kind as shown by Goodman [2] and the connection with asymptotic stability of the unforced 2-D system as shown by Kamen in 1980 [3].

The second paper, by Dudgeon, concerns the complex cepstrum in two dimensions. The cepstrum has been used in the calculation of 2-D spectral factorizations [4], [5] and should prove equally useful for more general convolution operator factorizations. Also, the concept of the complex cepstrum provides valuable insight into the inherent structure of 2-D signals. Recursive equations are developed for the cepstrum of 2-D minimum phase signals.

The third paper, by Mersereau and Speake, concerns the generalization of multidimensional signal processing to noncartesian grids or lattices. The most common noncartesian lattice is the hexagonal lattice which arises in temporal-vertical sampling [6] of interlaced TV. In general, hexagonal sampling results in more efficient signal processing for isotropically band-limited signals. This paper also treats the generalized DFT and decimation/interpolation for these regular periodic lattices.

The fourth paper, by Allebach, concerns a still more general type of multidimensional sampling which he calls time-sequential sampling. Unlike the regular periodic spatial lattices, the time-sequential spatial lattice cannot be generated by integer multiples of a set of basis vectors. A pairwise exchange algorithm is used to minimize the aliasing error given a structural constraint on the sampling sequences. This theory has application to the planning of satellite track geometries as well as airborne and ship surveys.

Papers five and six are concerned with stochastic 2-D signal processing. The fifth paper, by Marzetta, concerns 2-D linear prediction, including minimum energy-delay [7] prediction error filters and reflection coefficient arrays. The 2-D reflection coefficient arrays corresponding to a finite-support autocorrelation array are almost always of infinite support in the direction parallel to the scanning. In the orthogonal direction the size of the reflection coefficient array equals half the size of the autocorrelation array as expected from the 1-D theory.

The sixth paper, by Ekstrom, is on 2-D Wiener filtering with a causality constraint imposed by scanning, the so-called causal Wiener filter. A generalization to the M-D case is suggested. The development rests largely on an earlier paper on 2-D spectral factorization [5]. Fixed-delay estimators are also presented. These ideal Wiener filters can be the basis for approximation by more practical finite-order filters [8]. Also, these linear space-invariant filters can be incorporated in more subjectively relevant space-invariant and nonlinear filters.

Bibliography

[1] T. S. Huang, "Stability of two-dimensional recursive filters," *IEEE Trans. Audio Electroacoust.*, vol. AU-20, pp. 158–163, June 1972.

[2] D. Goodman, "Some stability properties of two-dimensional linear shift-invariant digital filters," *IEEE Trans. Circuit Syst.*, vol. CAS-24, p. 201–208, Apr. 1977.

[3] E. W. Kamen, "Asymptotic stability of linear shift-invariant two-dimensional digital filters," *IEEE Trans. Circuit Syst.*, vol. CAS-27, p. 1234–1240, Dec. 1980.

[4] D. E. Dudgeon, *Two-Dimensional Recursive Filtering*, Sc.D. Thesis, Dept. Elec. Eng., MIT, May 1974.

[5] M. P. Ekstrom and J. W. Woods, "Two-dimensional spectral factorization with applications in recursive digital filtering," *IEEE Trans. Acoust., Speech, Signal Processing*, vol. ASSP-24, p. 115–128, Apr. 1976.

[6] J.-Y. Ouellet and E. Dubois, "Sampling and reconstruction of NTSC video signals at twice the color subcarrier frequency," *IEEE Trans. Commun.*, vol. COM-29, p. 1823–1832, Dec. 1981.

[7] T. L. Marzetta, "The minimum energy-delay property of 2-D minimum phase filters," *IEEE Trans. Acoust., Speech, Signal Processing*, vol. ASSP-30, p. 658–659, Aug. 1982.

[8] J.-H. Lee and J. W. Woods, "Design and implementation of two-dimensional fully recursive digital filters," *IEEE Trans. Acoust., Speech, Signal Process.*, vol. ASSP-34, pp. 178–191, Feb. 1986.

Stability of General Two-Dimensional Recursive Digital Filters

BRIAN T. O'CONNOR AND THOMAS S. HUANG, SENIOR MEMBER, IEEE

Abstract–Two-dimensional recursive filters are defined from a different point of view. A general stability preserving mapping theorem is presented which allows most recursive filters of a particular type to be mapped into any other type of recursive filter. In particular, any type of filter can be mapped into a first-quadrant filter. This mapping is used to prove a number of general stability theorems. Among these is a theorem which relates the stability of any digital filter to its two-dimensional phase function. Furthermore, other stability theorems which are valid for any type of recursive filter are presented. Finally, a number of practical stability tests are developed including one which requires the testing of only several one-dimensional polynomial root distributions with respect to the unit circle.

I. Introduction

Two-dimensional recursive digital filters have generated much interest lately. They have the potential of saving computer time and memory. In virtually all developments of recursive filtering, one starts with a two-dimensional system function from which a two-dimensional difference equation is obtained. Then a particular solution of this equation which is acquired by an iterative procedure is considered. This iterative procedure represents the two-dimensional filter and also acts as a computational realization of the system. However, this method of developing recursive filters is ambiguous. In general, a difference equation can have many solutions which can be acquired by iterative procedures [1], [2]. Researchers have tried to avoid this ambiguity by considering only the causal recursive formula. However, this approach is very limiting because in two dimensions causality has little importance. The important point is whether the iterative procedure can be utilized to consistently calculate any output from a given set of initial conditions.

For these reasons, in this paper we shall define recursive filters from a different point of view. Then, we will present a number of stability theorems for general recursive filters. Among these is a theorem which relates the stability of any digital filter to its two-dimensional phase function. Futhermore, it will be shown that any general recursive filter can be mapped into a first-quadrant filter. Next, a number of efficient and practical stability tests will be developed including one which reduces to testing the zero distribution with respect to the unit circle of several one-dimensional polynomials. Finally, several examples of determining the stability of general two-dimensional recursive filters will be presented.

II. General Recursive Filters

Linear constant coefficient recursive equations can be utilized to implement a class of linear shift-invariant (LSI) systems on many input arrays. The form that will be considered is

$$o(m,n) = \sum_{(r,s) \in \alpha} a(r,s)\, i(m-r, n-s)$$

$$\sum_{(k,l) \in \beta - (0,0)} b(k,l)\, o(m-k, n-l) \qquad (1)$$

Manuscript received October 28, 1977; revised March 28, 1978 and July 18, 1978.

B. T. O'Connor was with the School of Electrical Engineering, Purdue University, West Lafayette, IN 47907. He is now with the TRW Defense and Space System Group, Redondo Beach, CA 90278.

T. S. Huang is with the School of Electrical Engineering, Purdue University, West Lafayette, IN 47907.

Reprinted from *IEEE Trans. Acoust., Speech, Signal Processing,* vol. ASSP-26, pp. 550–560, Dec. 1978.

where $a(r, s)$ and $b(k, l)$ are real finite extent arrays with respective lattice supports of α and β and $i(m, n)$ and $o(m, n)$ are the respective input and output arrays. Furthermore, we will assume that $b(0, 0) = 1$ and that β is contained in a lattice sector with vertex (0, 0) of angle less than π. These conditions guarantee that for all finite extent input arrays and a class of infinite extent inputs, (1) can be solved by incrementing the values of the indices (m, n) in such a fashion that all values of the output can be computed in turn from a given set of initial conditions [1], [3]. In other words, the stated conditions imply that (1) is recursively computable for many inputs. For the moment we will assume that the input array has finite support and hence, (1) will always be recursively computable. If this is the case, then it is possible to compute $o(m, n)$ iteratively by using the right-hand side of (1) to obtain new output values. We plan to interpret (1) in this manner, that is, as a recursive formula rather than a rewritten difference equation derived from the system function. However, it is important to note that the solution obtained by iterating (1) is a particular solution of the two-dimensional difference equation that (1) also represents. The reader is referred to [1] for a more detailed account on the relationship of difference equations to recursive equations. It may be helpful for the reader to view the equal sign in (1) as an assignment of the calculated value from the right side to the appropriate output value on the left-hand side.

If zero initial conditions are assumed in implementing (1), it then becomes a computational realization of an LSI system with the following impulse response:

$$h(m,n) = a(m,n) - \sum_{(r,s) \in \beta - (0,0)} b(r,s)\, h(m-r, n-s). \tag{2}$$

Again, this will be viewed as a recursive formula rather than a rewritten difference equation.

If $a(m, n)$ and $b(m, n)$ are assumed to be first-quadrant arrays, then (1) becomes the standard quarter-plane filter which has been extensively studied in the literature [2]-[6]. Futhermore, $a(m, n)$ and $b(m, n)$ can be redefined to represent any type of nonsymmetric half-plane filter considered by Ekstrom and Woods [7] and Dudgeon [1], [3]. Therefore, the development of recursive filters in this paper includes all the previously considered filters and more.

At first glance it may seem that an even more general formulation can be given by removing the restriction on β. However, if β is not a subset of a lattice sector of angle less than π, then both (1) and (2) become inconsistent. No matter how the filtering is performed, numerical values for a number of output points must be assumed *before* they are calculated. In other words, (1) and (2) will not be recursively computable.

For the rest of this work, it will be assumed that (1) implements an LSI system on a given input array. If the input is allowed to have an infinite support, iterative problems may occur. However, it can readily be shown that a necessary and sufficient condition for (1) to be recursively computable is that the minimum angle lattice sector, say β^*, containing β never intersects the support of $i(m - r, n - s)$ (function of r and s) at an infinite number of points for any (m, n). The following theorem summarizes the above discussion. It also follows from results given in [1] and [8].

Theorem 1: Let $b(m, n)$ be a real finite extent array satisfying: 1) $b(0, 0) = 1$, and 2) there exists an $(m, n) \neq (0, 0)$ such that $b(m, n) \neq 0$; then (1) is recursively computable if and only if:

a) $\beta = \{(m, n): b(m, n) \neq 0\}$ is a subset of a lattice sector with vertex (0, 0) of angle less than π;

b) the support of $i(m - r, n - s)$ when considered as a function of (r, s) does not intersect β^* (which is the minimum angle lattice sector containing β) at an infinite number of points for any (m, n).

If $b(m, n)$ satisfies the conditions in the above theorem, that is 1), 2), and a), then it will be called a recursive filter array. Associated with a recursive filter array is a lattice sector called β^* which consists of all lattice points in the minimum angle sector containing β. β^* can be uniquely defined by two vectors $\bar{\beta}_1 = (M_1, N_1) \in Z^2$ and $\bar{\beta}_2 = (M_2, N_2) \in Z^2$ as $\beta^* = S[(M_1, N_1), (M_2, N_2)] = \{(m, n) \in Z^2: (m, n) = r_1\bar{\beta}_1 + r_2\bar{\beta}_2$, with r_1 and r_2 nonnegative real numbers$\}$ where Z is the set of integers and M_1 and N_1 along with M_2 and N_2 are mutually prime integers. If $M_1 \cdot N_2 - N_1 \cdot M_2 \neq 0$ then $\bar{\beta}_1$ and $\bar{\beta}_2$ are not colinear vectors, and hence, β is composed of noncolinear points. However, if $M_1 N_2 - N_1 M_2 = 0$ then $\bar{\beta}_1$ and $\bar{\beta}_2$ are colinear (in fact, equal) and β is composed entirely of colinear points. Fig. 1 illustrates the relationship between β and β^*.

Since the $a(m, n)$ array does not affect recursibility, it will be assigned the value $\delta(m, n)$ (the unit impulse) for the remainder of the discussion. This simplification will allow us to avoid some rather unpleasant problems due to nonessential singularities of the second kind which may arise when discussing stability [9]. Equations (1) and (2) now become

$$o(m,n) = i(m,n) - \sum_{(r,s) \in \beta - (0,0)} b(r,s)\, o(m-r, n-s) \tag{3}$$

$$h(m,n) = \delta(m,n) - \sum_{(r,s) \in \beta - (0,0)} b(r,s)\, h(m-r, n-s). \tag{4}$$

So far it has been stated that under appropriate conditions, (1)-(4) are recursively computable; however, no mention has been made about how they should be iterated. For simplicity, we shall consider only (4). First, a close study of this equation will reveal that $h(m, n) = 0$ for $(m, n) \in Z^2 - \beta^*$ is a consistent set of initial conditions. Next, a sufficient condition for calculating $h(m, n)$ with $(m, n) \in \beta^*$ is to know all $h(m, n)$ in the parallelogram formed by connecting this point to the rays formed from $\bar{\beta}_1$ and $\bar{\beta}_2$ [see Fig. 1(b)]. Here, points and lines are considered to be degenerate parallelograms. Therefore, starting with $h(0, 0) = 1$ we can build up any parallelogram by a number of smaller parallelograms, and hence, we can find any $h(m, n)$ in this manner. The "parallelogram rule" presents one approach of computing $h(m, n)$ for any $(m, n) \in \beta^*$. This rule can be extended to apply to (1)-(3). It readily follows from the definitions of β and β^* and (4) that there exist positive integers r_1 and r_2 such that $h(r_1 M_1, r_1 N_1) \neq 0$ and $h(r_2 M_2, r_2 N_2) \neq 0$, and thus, β^* is the minimum angle sector

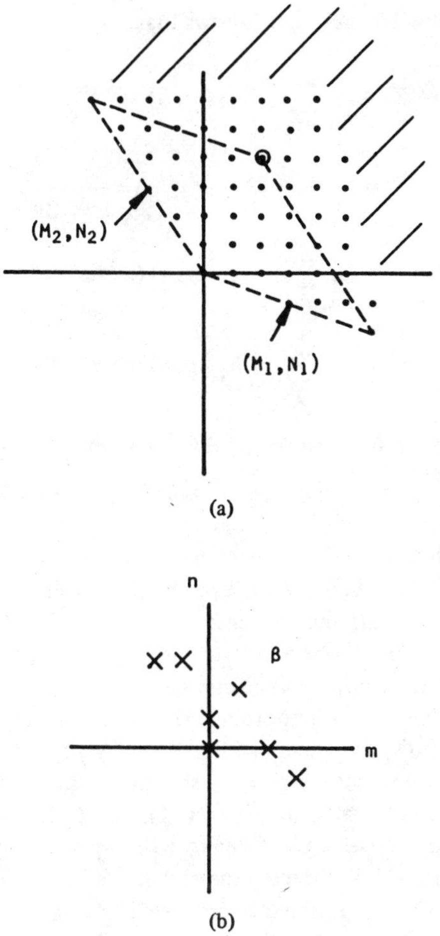

Fig. 1. Example depicting the relationship of β, the support of $b(m,n)$ to β^*, the minimum angle sector containing β, where $\beta^* = S[(M_1, N_1), (M_2, N_2)] = S[(3, -1), (-2, 3)]$. (b) also exemplifies the parallelogram rule.

containing the infinite extent impulse response $h(m,n)$. In general, (1) and (3) usually implement LSI systems with infinite extent impulse responses which have supports which are subsets of shifted sectors of angle less than π [1], [3].

There exists another method of finding the recursive solution of (4) which will prove useful in the forthcoming stability discussion. Define $B(w,z)$ for a recursive filter array $b(m,n)$ as

$$B(w,z) = \sum_{(r,s)\in\beta} b(r,s)\, w^r z^s \tag{5}$$

where r and s can be negative as well as positive integers. Since $b(0,0) = 1$, (5) can be written as $B(w,z) = 1 - C(w,z)$, where $C(w,z)$ is a polynomial (possibily in both negative and positive powers of w and z) with no constant term. A formal expansion for $1/B(w,z)$ can be obtained by

$$\frac{1}{B(w,z)} = \frac{1}{1 - C(w,z)} = \sum_{n=0}^{\infty} [C(w,z)]^n$$

$$= \sum_{(m,n)\in\beta^*} h(m,n)\, w^m z^n. \tag{6}$$

This is strictly a formal power series and no claim of any type of convergence is given. Since $C(w,z)$ has no constant term and its coefficient array is confined to a sector of angle less than π, this implies that for any $n > 0$, $[C(w,z)]^n$ can only contribute to terms $h(m,n)$ supported in β^*. Furthermore, given any $(m',n') \in \beta^*$ there exists an $n_1 > 0$ such that $\sum_{n=n_1}^{\infty} [C(w,z)]^n$ does not contribute to $h(m',n')w^{m'}z^{n'}$ term of $h(m,n)w^m z^n$. In other words, any $h(m,n)$ can be found in a finite number of calculations. We claim that the sequence $h(m,n)$ obtained via (6) is equal to that obtained from (4). The following theorem establishes this.

Theorem 2: Assume that $b(m,n)$ is a recursive filter array, $h_1(m,n)$ is the solution of (4) [remember, this is the solution acquired by appropriately iterating (4)], and that $h_2(m,n)$ is the sequence acquired from (6), then $h_1(m,n) = h_2(m,n)$.

Proof: First, both sequences were constructed so that they are solutions to the following difference equation:

$$\sum_{(r,s)\in\beta} b(r,s)\, h(m-r, n-s) = \delta(m,n). \tag{7}$$

Therefore, $f(m,n) = h_1(m,n) - h_2(m,n)$ must be a particular solution of the homogeneous equation

$$\sum_{(r,s)\in\beta} b(r,s)\, f(m-r, n-s) = 0. \tag{8}$$

Secondly, since h_1 and h_2 have support in β^*, $f(m,n)$ must be supported there also. Therefore, (8) means that convolution of the two arrays $b(r,s)$ and $f(r,s)$ both with supports in β^*, is always zero. Recalling that $b(0,0) = 1$ and that β^* is a lattice sector of angle less than π, it follows that $f(0,0) = 0$ because β intersects the support of $f(-r,-s)$ only at $(0,0)$. Next, (8) can be used to show that $f(M_1,N_1)$ and $f(M_2,N_2)$ must also be zero. Proceeding successively in this fashion, it can be shown that $f(m,n) = 0$ for all $(m,n) \in \beta^*$. Therefore, $h_1(m,n) = h_2(m,n)$.

III. Stability Theorems

In most applications only bounded-input bounded-output (BIBO) stable filters are of interest. For an LSI system a necessary and sufficient condition for BIBO stability is that $\sum_{(m,n)} |h(m,n)| < \infty$. Therefore, the recursive filter represented by the recursive filter array $b(m,n)$ is stable if and only if the impulse response obtained from (4) is absolutely summable, that is, $\sum_{(m,n)\in\beta^*} |h(m,n)| < \infty$. This impulse response can be rotated by a multiple of 90°, transposed or flipped, and the resulting sequences are still absolutely summable. Equation (6) can be employed to identify the resulting recursive filter array. These facts can be utilized to readily prove the following theorem which is a generalization of a result in [10].

Theorem 3: If $b(m,n)$ is a stable recursive filter array, then all the following recursive filter arrays are also stable.

a) $b(m,n)$ identity.
b) $b(n,-m)$ clockwise 90° rotation.
c) $b(-m,-n)$ clockwise 180° rotation.
d) $b(-n,m)$ clockwise 270° rotation.
e) $b(-m,n)$ n-axis mirror image.
f) $b(-n,-m)$ transpose with respect to $m = -n$ diagonal.
g) $b(m,-n)$ m-axis mirror image.
h) $b(n,m)$ transpose with respect to $m = n$ diagonal.

This theorem is a special case of a more general mapping theorem which will be developed below.

Consider two sectors $S_1[(M_1, N_1), (M_2, N_2)]$ and $S_2[(P_1, Q_1), (P_2, Q_2)]$ with $D = M_1N_2 - M_2N_1 \neq 0$ and $E = P_1Q_2 - P_2Q_1 \neq 0$. There exists many linear mappings of the form

$$\begin{aligned} m &= k_1m' + k_2n' \\ & \qquad (k_i \in Z) \\ n &= k_3m' + k_4n' \end{aligned} \tag{9}$$

that map S_1 one-to-one (not necessarily onto) into S_2. For example, one mapping is as follows:

$$\begin{aligned} k_1 &= (N_2P_1 - N_1P_2)\,\mathrm{sgn}\,(D) \\ k_2 &= (M_1P_2 - M_2P_1)\,\mathrm{sgn}\,(D) \\ k_3 &= (N_2Q_1 - N_1Q_2)\,\mathrm{sgn}\,(D) \\ k_4 &= (M_1Q_2 - M_2Q_1)\,\mathrm{sgn}\,(D). \end{aligned} \tag{10}$$

With this mapping S_2 is the smallest sector containing the mapped points. If S_2 is the first quadrant in the lattice plane, then (10) can be used to derive the following mapping of S_1 into S_2 by setting $P_1 = 1, Q_1 = 0, P_2 = 0$, and $Q_2 = 1$:

$$\begin{aligned} k_1 &= \mathrm{sgn}\,(D)N_2 \\ k_2 &= -\mathrm{sgn}\,(D)M_2 \\ k_3 &= -\mathrm{sgn}\,(D)N_1 \\ k_4 &= \mathrm{sgn}\,(D)M_1. \end{aligned} \tag{11}$$

These sector transformations are similar to mappings in [7], [8]. In [8], Dudgeon employs a transformation similar to (9) to compute the cepstra of nonfirst-quadrant arrays. The above ideas will be utilized to prove the following important theorem.

Theorem 4: Let $b(m, n)$ be a recursive filter array with support β. Then $b(m, n)$ is stable if and only if for any $k_i \in Z$ with $K = k_1 \cdot k_4 - k_2 \cdot k_3 \neq 0$, the recursive filter array

$$g(m, n) = g(k_1m' + k_2n', k_3m' + k_4n') = b(m', n') \tag{12}$$

with $(m', n') \in \beta$ is stable.

Proof: Since $K \neq 0$, then $g(m, n)$ is a recursive filter array with support $\gamma = \{(m, n) \in Z^2: m = k_1m' + k_2n', n = k_3m' + k_4n', (m', n') \in \beta\}$. The requirement that $K \neq 0$ also implies that the mapping (12) of $b(m, n)$ to $g(m, n)$ is one-to-one and onto. Let $h_1(m, n)$ and $h_2(m, n)$ be the impulse responses of filters $b(m, n)$ and $g(m, n)$, respectively, with respective supports of β' and γ'. (Note that β' and γ' are subsets of β^* and γ^*, respectively.) It will first be shown that $h_2(k_1m + k_2n, k_3m + k_4n) = h_1(m, n)$ for $(m, n) \in \beta'$. Define $G(w, z) = \sum_{(m,n) \in \gamma} g(m, n)w^mz^n$. Therefore, if the primed variables are suppressed, we have

$$\begin{aligned} G(w, z) &= \sum_{(m,n) \in \gamma} g(m, n)\,w^mz^n = \sum_{(m,n) \in \beta} g(k_1m \\ &\quad + k_2n, k_3m + k_4m)\,w^{k_1m + k_2n}z^{k_3m + k_4n} \\ &= \sum_{(m,n) \in \beta} b(m, n)\,(w^{k_1}z^{k_3})^m\,(w^{k_2}z^{k_4})^n \\ &= B(w^{k_1}z^{k_3}, w^{k_2}z^{k_4}). \end{aligned} \tag{13}$$

By (6) and Theorem 2 it follows that

$$\begin{aligned} 1/G(W, Z) &= \sum_{(m,n) \in \gamma'} h_2(m, n)\,w^mz^n \\ &= 1/B(w^{k_1}z^{k_3}, w^{k_2}z^{k_4}) = \frac{1}{1 - C(w^{k_1}z^{k_3}, w^{k_2}z^{k_4})} \\ &= \sum_{(m,n) \in \beta'} h_1(m, n)\,(w^{k_1}z^{k_3})^m\,(w^{k_2}z^{k_4})^n \\ &= \sum_{(m,n) \in \beta'} h_1(m, n)\,w^{k_1m + k_2n}z^{k_3m + k_4n}. \end{aligned}$$

Since $K \neq 0$, for each $(m, n) \in \beta'$, the term

$$h_1(m, n)\,w^{k_1m + k_2n}z^{k_3m + k_4n}$$

of the last series only occurs once. Hence, it follows that $h_1(m, n) = h_2(k_1m + k_2n, k_3m + k_4n)$ for all $(m, n) \in \beta'$. This implies that impulse responses $h_1(m, n)$ and $h_2(m, n)$ are related by a one-to-one and onto mapping, so if one is absolutely summable the other will be and visa versa.

This theorem is important because it allows virtually any type of filter to be mapped into virtually any other type while at the same time preserving stability. In particular, if $D = M_1N_2 - M_2N_1 \neq 0$, then (9), (11), and (12) can be used to map a general recursive filter to a first-quadrant quarter-plane filter with $K = D$. Hence, the stability of a nonfirst-quadrant array can be determined by checking the stability of the first-quadrant filter obtained by mapping $b(m, n)$ into the first quadrant. Equation (11) offers one such mapping if β has at least three noncolinear points, that is, $D \neq 0$. However, if β is composed of entirely colinear points on the line defined by (M_1, N_1) (mutually prime integers), $b(m, n)$ can be mapped into a one-dimensional filter sequence $c(m)$ by $c(m) = b(mM_1, mN_1)$. It is not difficult to show that the resulting one-dimensional filter is stable if and only if the original filter is stable.

As an example, consider the recursive filter array $b(m, n)$ defined as

$$B(w, z) = 0.5w^{-1}z + 1 + 0.85w + 0.1wz + 0.5w^2z^{-1}.$$

Note that $\beta^* = S[(2, -1), (-1, 1)]$. $b(m, n)$ can be mapped into a first-quadrant filter, say $g(m, n)$, using (11) and (12). This yields $g(m, n) = g(m' + n', m' + 2n') = b(m', n')$ where $(m', n') \in \beta$. Therefore,

$$G(w, z) = 1 + 0.5w + 0.5z + 0.85wz + 0.1w^2z^3.$$

By Theorem 4, the stability of $b(m, n)$ can be determined by checking the stability of $g(m, n)$, a first-quadrant filter.

There exists a number of tests for determining the stability of first-quadrant quarter-plane filters. All of them relate stability to the zero set of the equation $B(w, z) = 0$. The following theorem summarizes these results.

Theorem 5: Let $b(m, n)$ be a first-quadrant quarter-plane recursive filter array and define $B(w, z) = \sum_{(m,n) \in \beta} b(m, n)w^mz^n$. The two-dimensional recursive filter that $b(m, n)$ represents is stable if and only if

1) $B(w,z) \neq 0 \quad |w| \leq 1, \quad |z| \leq 1$

if and only (iff)

2a) $B(w,z) \neq 0 \quad |w| \leq 1, \quad |z| = 1$

2b) $B(a,z) \neq 0 \quad |z| \leq 1$ for some a, $|a| \leq 1$

iff

3a) $B(w,z) \neq 0 \quad |w| = 1, \quad |z| \leq 1$

3b) $B(w,b) \neq 0 \quad |w| \leq 1$ for some b, $|b| \leq 1$

iff

4a) $B(w,z) \neq 0$ on $T^2 = \{(w,z): |w| = 1 = |z|\}$

4b) $B(a,z) \neq 0 \quad |z| \leq 1$ for some a, $|a| \leq 1$

4c) $B(w,b) \neq 0 \quad |w| \leq 1$ for some b, $|b| = 1$

iff

5a) $B(w,z) \neq 0$ on T^2

5b) $B(z,z) \neq 0 \quad |z| \leq 1$

iff

6a) $B(w,z) \neq 0$ on T^2

6b) $B(z^{l_1}, z^{l_2}) \neq 0 \quad |z| \leq 1$

for some $l_1, l_2 \in Z^+$ where Z^+ denotes the positive integers.

Proof:

1) See [2], [4].

2)-3) See [2], [12], [13].

4) See [13]-[15].

5) See [14].

6) This is a generalization of part 5 and follows from an application of a theorem in [16, p. 87].

The literature is full of procedures for determining stability of first-quadrant filters by checking the conditions of 1), 2), or 3) of the above theorem [2]-[7]. There are basically two types of methods, mapping and algebraic tests. Mapping tests are generally inefficient and stability cannot be guaranteed [2], [4]. Later, we will present some efficient tests which are somewhat similar to mapping procedures and which are valid for any type of filters. In theory, algebraic methods can exactly determine stability in a finite number of steps. However, problems may arise in practice due to finite precision arithmetic. In the literature [2], [5], [6], [17], the given algebraic methods are used to determine if $B(w,z) = 0$ on $|w| \leq 1, |z| = 1$ (or $|z| \leq 1, |w| = 1$). However, a test can be modified to check for zeros on T^2. It will be shown later that this modified test is applicable to any type of filter.

In the remainder of this section several theorems which directly relate the stability of a general recursive filter, represented by a recursive filter array $b(m,n)$, to the zero distribution of $B(w,z)$ will be derived. In the first theorem, stability of any type of recursive filter can be determined by checking two conditions. First, $B(w,z)$ must be free of zeros on T^2. Next, one one-dimensional polynomial derived from $B(w,z)$ must be free of zeros inside and on the unit circle.

Theorem 6: Let $b(m,n)$ be a recursive filter array with support β and associated minimum angle sector $\beta^* = S[(M_1,N_1), (M_2,N_2)]$. Then $b(m,n)$ is stable if and only if

Case 1: $D = M_1N_2 - M_2N_1 \neq 0$

a) $B(w,z) \neq 0$ on T^2

b) $B(\lambda^{P_1}, \lambda^{P_2}) \neq 0$ for $|\lambda| \leq 1$ for some P_1 and P_2 where $P_1 = \operatorname{sgn}(D)N_2l_1 - \operatorname{sgn}(D)N_1l_2$, $P_2 = -\operatorname{sgn}(D)M_2l_1 + \operatorname{sgn}(D)M_1l_2$, and $l_i \in Z^+$.

Case 2: $D = 0$

a) for some $l_1 \in Z^+$, $C(\lambda) = \Sigma_{m=0}\, c(m)\, \lambda^{l_1 m} \neq 0$ for $|\lambda| \leq 1$ where $c(m) = b(mM_1, mN_1)$.

Proof:

Case 1: $D \neq 0$. By Theorem 4, the first-quadrant filter $g(m,n) = g(k_1m' + k_2n', k_3m' + k_4n') = b(m',n')$ where k_i, $i = 1, \cdots, 4$ are given in (11) is stable iff $b(m,n)$ is stable. By part 6) of Theorem 5 $g(m,n)$ is stable iff: i) $G(w,z) \neq 0$ on T^2, and ii) $G(\lambda^{l_1}, \lambda^{l_2}) \neq 0$ for $|\lambda| \leq 1$ for some $l_1, l_2 \in Z^+$. Equation (13) implies that $G(w,z) = B(w^{k_1}z^{k_3}, w^{k_2}z^{k_4})$, and therefore, since $K = k_1 \cdot k_4 - k_2 \cdot k_3 = D \neq 0$ with $k_i \in Z$ then $G(w,z) \neq 0$ on T^2 iff $B(w,z) \neq 0$ on T^2. Moreover, part b) follows directly from ii) after the appropriate change of variables.

Case 2: $D = 0$. This follows directly from the discussion after Theorem 4 and results on the stability of one-dimensional recursive filters [20].

Before stating the next theorem a review of some facts concerning one-dimensional sequences will help clarify our terminology. Assume that $d(n)$ is a finite sequence of real numbers $d(M), \cdots, d(N)$ with $M \leq N$, $d(M) \neq 0$, and $d(N) \neq 0$. A polynomial (possibly in both z and z^{-1}) $D(z) = \Sigma_{n=M}^{N} d(n) z^n$ can be formed which has R roots counting multiple roots and roots at infinity where

$$R = |\min(0,M)| + \max(0,N). \tag{14}$$

Let $M^1 = |\min(0,M)|$ and $N^1 = \max(0,N)$ so $R = M^1 + N^1$. If $D(z)$ has no roots on the unit circle, then we can define on $[-\pi, \pi]$ a continuous, odd phase function associated with the Fourier transform of the sequence. This phase function contains no linear phase component if and only if $D(z)$ has exactly N^1 zeros outside the unit circle and M^1 zeros inside. Thus, anytime we say that a sequence or a polynomial has no linear phase component we mean it satisfies the above conditions. The fact that the phase of the Fourier transform has no linear component can be reinterpreted with the aid of Nyquist theory. If the unit circle is defined to be the Nyquist contour, the above phase condition means that the Nyquist plot does not encircle nor pass through the origin. In other words, if $\operatorname{Ind}(D(z))$ is defined to be the number of encirclements of zero of the Nyquist plot of D along the Nyquist contour $|z| = 1$, then $d(n)$ [or $D(z)$] has no linear phase component if and only if $\operatorname{Ind}(D(z)) = 0$. ($D(z)$ must not have any roots on the unit circle.) Both of these interpretations will be used in the statement and proof of the following theorem.

Theorem 7: $b(m,n)$ is a stable recursive filter array if and only if

a) $B(w,z) \neq 0$ on T^2

b) $B(\lambda, 1)$ and $B(1, \lambda)$ have no linear phase terms [i.e., $\operatorname{Ind}(B(\lambda,1)) = 0 = \operatorname{Ind}(B(1,\lambda))$].

Proof:

Sufficiency: If part a) holds, then by following the argument in [16, pp. 87-89] we have

$$\text{Ind}(B(\lambda^{l_1}, \lambda^{l_2})) = l_1 \cdot \text{Ind}(B(\lambda, 1)) + l_2 \cdot \text{Ind}(B(1, \lambda)) \quad (15)$$

for any $l_1, l_2 \in Z$. Condition b) implies that $\text{Ind}(B(\lambda^{l_1}, \lambda^{l_2})) = 0$ for any $l_1, l_2 \in Z$, and, in particular, for $l_1 = P_1$ and $l_2 = P_2$ of Theorem 6. Therefore, since $\beta(\lambda^{P_1}, \lambda^{P_2})$ is a polynomial in positive powers of λ, this implies that it has no zeros in $|\lambda| \leq 1$. Hence, conditions a) and b) imply the conditions of Theorem 6, so they are sufficient for stability.

Necessity: From Theorem 6, a) is necessary. Furthermore, Theorem 6 implies that there exists integers P_1, P_2, P_1', and P_2' with $P_1 P_2' \neq P_1' P_2$ such that $B(\lambda^{P_1}, \lambda^{P_2}) \neq 0$ and $B(\lambda^{P_1'}, \lambda^{P_2'}) \neq 0$ on $|\lambda| \leq 1$. Since these are polynomials in positive powers of λ, it follows that they both have no linear phase terms. Employing (15) gives

$$\text{Ind}(B(\lambda^{P_1}, \lambda^{P_2})) = P_1 \cdot \text{Ind}(B(\lambda, 1)) + P_2 \cdot \text{Ind}(B(1, \lambda)) = 0$$

$$\text{Ind}(B(\lambda^{P_1'}, \lambda^{P_2'})) = P_1' \cdot \text{Ind}(B(\lambda, 1)) + P_2' \cdot \text{Ind}(B(1, \lambda)) = 0.$$

Since $P_1 P_2' \neq P_1' P_2$, this implies $\text{Ind}(B(1, \lambda)) = 0 = \text{Ind}(B(\lambda, 1))$. Of course, the "1" in condition b) can be replaced by any complex constant b satisfying $|b| = 1$.

The conditions of this theorem are equivalent to those given in part 4) of Theorem 5. However, this theorem is much more general because it is true for any type of filter. Hence, no longer is it necessary to check different zero region conditions of $B(w, z)$ for each different type of filter as done in the past [2], [7].

Conditions a) and b) of Theorem 7 can be restated respectively as: 1) the Fourier transform of $b(m, n)$ is never equal to zero, and 2) there exist no linear phase components in the phase function. These conditions are very similar to restrictions Dudgeon [21] placed on the Fourier transforms of arrays having rational z transforms to ensure that these sequences have well-defined 2-D complex cepstra. Dudgeon also proved that these conditions are equivalent to the phase of the Fourier transform being continuous, odd, and periodic. Here, the phase function [1], [21] is defined as the contour integral

$$\phi(u, v) = \text{Im}\left[\int C_u'(z)/C_u(z)\, dz\right] + \phi(u, 0) \quad (16)$$

where $C_u(z) = \Sigma_n [\Sigma_m c(m, n) \exp(-jum)] z^n$, the prime denotes differentiation with respect to z, and the contour integral starts at $z = 1$ and proceeds around unit circle to $z = \exp(jv)$. This is called the unwrapped phase of the sequence $c(m, n)$. Other equivalent definitions for the two-dimensional unwrapped phase exist. Theorem 7 along with the above observations suggests the following theorem.

Theorem 8: A recursive filter represented by a recursive filter array is stable iff the unwrapped phase is continuous, odd, and periodic.

Proof: If the recursive filter is stable then by Theorem 7 and a result due to Dudgeon [1], [21], the unwrapped phase is continuous, odd, and periodic. If the phase is defined as continuous, odd, and periodic, then $B(w, z) \neq 0$ on T^2 because otherwise the phase would not be defined at the point where $B(w, z) = 0$. Moreover, the phase of any projection, i.e., $B(\lambda^{l_1}, \lambda^{l_2})$, $l_i \in Z$, must be continuous, odd, and periodic so, in particular, $B(\lambda, 1)$ and $B(1, \lambda)$ must have continuous, odd, and periodic phases. This, in turn, implies that they have no linear phase component.

This theorem is important because it relates the phase of a two-dimensional recursive filter to its stability. This will aid in our development of efficient, practical stability testing algorithms.

IV. Practical Stability Tests

Theorem 7 requires one extra zero distribution test than Theorem 6. However, it is usually more efficient to determine the zero distribution of both $B(\lambda, 1)$ and $B(1, \lambda)$ than just one $B(\lambda^{P_1}, \lambda^{P_2})$. Therefore, we shall only consider methods of determining the validity of the stability conditions of Theorem 7. As mentioned earlier, one approach for determining the stability of a general recursive filter is to map it into a first-quadrant filter and to apply any of the numerous algebraic tests which check if $G(w, z)$ has any zeros in $|z| \leq 1, |w| = 1$ (or $|w| \leq 1$, $|z| = 1$). However, in this section we will take a different approach. Here, we will consider methods which directly check the conditions in Theorem 7, that is, a) $B(w, z) \neq 0$ on T^2, and b) $B(\lambda, 1)$ and $B(1, \lambda)$ have no linear phase components.

The first step in testing stability should be validating condition b). If either of these terms contain a linear phase component the filter is unstable. There exists a number of efficient methods for checking this condition. For instance, the Nyquist plot of the polynomial can be calculated and inspected for zero encirclements. This information can be obtained by unwrapping the phase of the sequences which form these one-dimensional polynomials. A very efficient and accurate phase unwrapping method has been recently developed [22]. However, it has been our observation that for polynomials of order less than a hundred or so, analytic methods can be applied to obtain this information more efficiently. In fact, the Marden-Jury table [20], [23] test probably offers the least amount of computation for a polynomial of order less than a hundred or so.

If condition b) is satisfied then condition a) must be checked next. Here, two different approaches can be taken. First, an algebraic approach can be formulated which is a modification of first-quadrant algebraic methods. Secondly, an efficient algorithm somewhat similar to the mapping tests in [2], [4], and [18] can be developed.

One possible algebraic procedure which uses the Schur-Cohn matrix [5] is listed below. Given a filter $b(m, n)$:

1) Define $E(w, z) = w^M z^N B(w, z)$ so that $E(w, z)$ has only nonnegative powers in w and z.

2) Form Schur-Cohn matrix $C(w)$ with the polynomial coefficients of $E(w, z)$ viewed as a polynomial in w. This is a Hermitian matrix.

3) Calculate the determinant $D(w) = \det[C(w)]$ assuming $|w| = 1$ to obtain a self-inversive polynomial in w.

4) Find all zeros w_i of $D(w)$ along $|w| = 1$.

5) For each w_i determine if $E(w_i, z)$ has any roots on $|z| = 1$.
   i) direct calculation
   ii) apply Schur's theorems [5], [20].

Steps 4) and 5) are considerably different than the original Schur-Cohn stability procedure. No longer is sign definiteness of $D(w)$ a necessary condition for stability. Because if there exists a w_i with $|w_i| = 1$ such that $D(w_i) = 0$, then $E(w_i, z)$ may either have a root on the unit circle *or* just reciprocal roots with respect to the unit circle. This implies that steps 4) and 5) must be included. For these reasons this test is somewhat more complicated than those which determine the stability of quarter-plane filters. However, this does not imply that it is more efficient to map a filter to the first quadrant and apply the Schur-Cohn test given in [5] because the mapped filter may have a higher degree.

In conclusion, algebraic stability tests can be developed which are valid for any type of filter. These tests are somewhat similar to quarter-plane schemes and hence, they suffer from the same defects, such as excessive computer storage and time along with finite arithmetic problems. Furthermore, it is very difficult to program these methods efficiently. Therefore, it is our belief that nonalgebraic methods are the only feasible techniques for determining the stability of moderate and high-order filters. (For instance, if $b(m, n)$ is a first-quadrant filter, then nonalgebraic methods should definitely be used if $b(m, n)$ has a support larger than 4×4.) The rest of this paper is devoted to the development of efficient nonalgebraic methods.

Again, it is assumed that condition b) has been verified and it is desired to determine if $B(w, z) = 0$ on $T^2 = \{(w, z)\colon |w| = 1, |z| = 1\}$. Our approach is based on Theorem 8 which states that a filter is stable if and only if its phase is continuous, odd, and periodic. If there exists a point where the Fourier transform is zero, this implies that the phase is discontinuous at that point. Now, in most cases the existence of such a discontinuity can be determined without actually finding this point. This important observation can be demonstrated by studying the zero region of $B(w, z)$, that is, $\{(w, z)\colon B(w, z) = 0, (w, z) \in C^2\}$. This region consists of a finite number of two-dimensional "sheets" in C^2 [16].

In most cases if a zero sheet intersects T^2 at a point (w_0, z_0), it then intersects $D_{01} = \{(w, z)\colon |w| = 1, |z| < 1\}$ and $D_{0-1} = \{(w, z)\colon |w| = 1, |z| > 1\}$ (also $D_{10} = \{(w, z)\colon |w| < 1, |z| = 1\}$ and $D_{-10} = \{(w, z)\colon |w| > 1, |z| = 1\}$) in some neighborhood of (w_0, z_0). Roughly speaking, this is a consequence of viewing the solutions of $B(w, z) = 0$ as algebraic functions [24] and applying the implicit function theorem [25]. For instance, Fig. 2 depicts the solution of this equation in the neighborhood of (w_0, z_0). Case A represents the above observation, while case B depicts the rarely occurring event where the zero set does not intersect both D_{0-1} and D_{01} in the neighborhood of (w_0, z_0).

If case A occurs, the solution of $B(w, z) = 0$ in the neighborhood of (w_0, z_0) and in regions D_{01} and D_{0-1} is a one-dimensional curve. If $|w| = 1$ is parameterized as $w = e^{-ju}$ $(0 \leq u < 2\pi)$ with $w_0 = e^{-ju_0}$, then this curve is one of the root maps of

$$B_u(z) = \sum_n \left(\sum_m b(m, n)\, e^{-jum} \right) z^n \tag{17}$$

which passes through $(w_0, z_0) = (e^{-ju_0}, e^{-jv_0})$. Therefore, there exist two open intervals $U_{10} = (u_1, u_0)$ and $U_{02} = (u_0, u_2)$ such that if $u_a \in U_{10}$ and $u_b \in U_{02}$, then $B_{u_a}(z)$ and $B_{u_b}(z)$ have a different number of zeros inside (outside) the unit circle. This same argument can be applied to $B_v(w)$ where

$$B_v(w) = \sum_m \left(\sum_n b(m, n)\, e^{-jvn} \right) w^m. \tag{18}$$

Conversely, a continuity argument can be employed to show that if the number of roots of $B_u(z)$ or $B_v(w)$ inside the unit circle varies as a function of u or v, respectively, then this implies there exists a $(w_0, z_0) \in T^2$ such that $B(w_0, z_0) = 0$. Thus, if $0 \leq u < 2\pi$ is sampled fine enough, so there exists a sample in U_{10} and in U_{02} (or V_{10} and V_{02}), then the existence of the point $(w_0, z_0) = (e^{-ju_0}, e^{-jv_0})$ is substantiated, although the values for u_0 and v_0 are not explicitly known. Since $b(m, n)$ is assumed real here, it is only necessary to check the interval $0 \leq u \leq \pi$ $(0 \leq v \leq \pi)$.

The above discussion suggests the following practical test for condition a) of Theorem 7. First, sample either $0 \leq u \leq \pi$ or $0 \leq v \leq \pi$ or both on a fine grid. The intersample distance will determine the accuracy of the test. Next, evaluate the coefficients of $B_u(z)$ or $B_v(w)$ for all grid samples. This can be done quite efficiently by employing the FFT algorithm on the rows of $b(m, n)$ to obtain the coefficients of $B_u(z)$ for $u_i = 2\pi i/L$ and on the columns of $b(m, n)$ to obtain the coefficients of $B_v(w)$ for $v_i = 2\pi i/L$. (Here, L is usually a power of two.) Next, some method is used to determine the root distribution of each $B_{u_i}(z)$ or $B_{v_i}(w)$ with respect to the unit circle for $i = 0, \cdots, L/2$. For virtually all digital filter applications these polynomials will have orders less than 100 and in most cases less than ten. Thus, this implies that the Marden-Jury table method should be one of the most efficient methods for determining this. If, however, the Nyquist plot of each polynomial via Tribolet's phase unwrapping algorithm is used to obtain this root distribution information, then this procedure becomes a very efficient implementation and generalization of DeCarlo's [18], [19] Nyquist-like stability test. Furthermore, if the roots of these polynomials are found then this procedure becomes a generalization and an efficient implementation of Huang's and Shanks *et al.*'s root tests [2], [4]. At any rate, the table method should be the most efficient. The last step of this procedure is to check whether this zero distribution ever changes as a function of u_i or v_i. If it does change, then the filter is definitely unstable. On the other hand, if it does not change, then the filter may or may not be stable. However, if a very dense sampling grid is used, then we can be "almost sure" that the filter is stable.

A close study of (17) will reveal that it can be used to define the two-dimensional unwrapped phase when substituted into (16). Equation (17) defines the phase on a series of vertical lines in the u-v plane while (18) allows the phase to be computed on a set of horizontal lines [See Fig. 3(a) and (b)]. If along any of these lines the phase contains a linear component, then by the above discussion and Theorem 8 the filter is unstable (and, in fact, the two-dimensional phase is not well defined). As mentioned earlier, it is more efficient to obtain this information via an analytic method such as Marden-Jury table than actually unwrapping the phase along each line.

The above tests require that a number of one-dimensional FFT's be taken; more precisely, one for each column or row. Furthermore, a series of root distribution tests must then be executed. However, we can derive an interesting procedure which requires only one root distribution test with at most one FFT computation. This method was motivated by Mersereau's One-Projection Theorem and the Projection-Slice Theorem [26], [27]. In essence, it relies on the fact that if a two-dimensional sequence has a phase which is continuous,

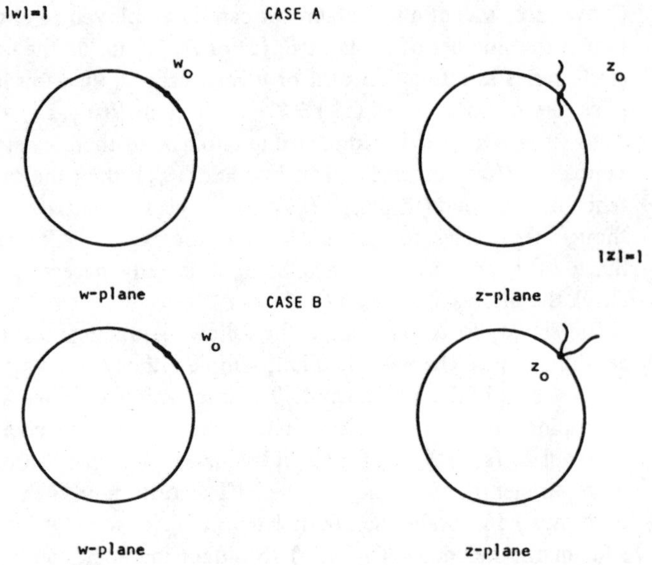

Fig. 2. Several possible solutions of $B(w, z) = 0$ in the neighborhood of a point $(w_0, z_0) \in T^2$.

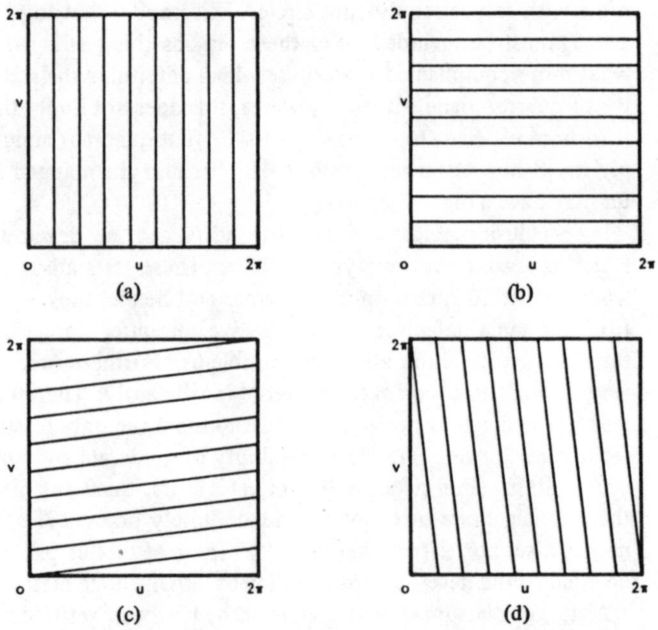

Fig. 3. Parameterization of torus T^2 using different phase unwrapping methods.

odd, and periodic, then any projection of this array has a phase which is also continuous, odd, and periodic and it is equal to some slice of the two-dimensional phase. By decreasing the angle of the projection, the resulting phase can be calculated on as fine a grid as desired. Fig. 3(c) and (d) shows over which values of the u-v plane the phase is calculated for two different projections. In this case, the toroid T^2 is approximately parameterized as a tightly wound helix as opposed to circles which result if (17) or (18) are used. There exists a number of different projections which will give a desired accuracy. Here, for simplicity, we only consider two. First, the phase of the sequence formed by concatenating the zero padded rows will equal the two-dimensional phase calculated on a slice of angle $\tan^{-1}(1/M)$ depicted in Fig. 3(c) where M is the length of the zero padded rows [26], [27]. Moreover, by concatenating the zero padded columns of length N, the above procedure will yield a slice of angle $\tan^{-1}(1/N)$ from the vertical as shown in Fig. 3(c). The grid on which the slice is computed can be made finer and finer, thus increasing the accuracy of the stability test by adding more zeros to the original array's columns or rows.

Since only the existence of a linear phase term, or equivalently, the appropriate zero distribution with respect to the unit circle of this one-dimensional sequence need be determined, then either an analytic method or a phase unwrapping algorithm can be employed to extract this information. Generally, this one-dimensional sequence will have several hundred or more elements. It has been our experience that for such large sequences, a modified version of Tribolet's phase unwrapping algorithm [22], [28] works faster and more accurately than any other procedure.

So far no mention has been made about the relation of the zero distribution with respect to the unit circle of the projection to that required by continuity of the two-dimensional phase. The one-dimensional sequence associated with column concatenation has a representation of $B(\lambda^N, \lambda)$, while row concatenation has one of $B(\lambda, \lambda^M)$. Note that these sequences can have both positive and negative indices. By (15) and Theorem 7, if $b(m, n)$ is stable, then for any integers M and N

$$\mathrm{Ind}(B(\lambda^N, \lambda)) = \mathrm{Ind}(B(\lambda, \lambda^M)) = 0. \tag{19}$$

However, if $b(m, n)$ is not stable and if $\mathrm{Ind}(B(\lambda, 1)) = \mathrm{Ind}(B(1, \lambda)) = 0$, then in most cases if N or M is large enough in magnitude, we will have

$$\mathrm{Ind}(B(\lambda^N, \lambda)) \neq 0$$

and/or

$$\mathrm{Ind}(B(\lambda, \lambda^M)) \neq 0. \tag{20}$$

Therefore, if for any integers M or N, (19) does not hold, then $b(m, n)$ is unstable. Moreover, if (19) holds for very large values of M or N, then we can be "almost sure" that $b(m, n)$ is stable.

Notice that these row and column concatenation stability tests like the row and column algorithms presented earlier cannot guarantee the stability of a filter. However, the accuracy of these tests can be increased to almost any desired level. On the other hand, there still will exist examples which are unstable but will be indicated as stable by these tests. For example, a zero sheet may just touch T^2 at one point, say (e^{-ju_0}, e^{-jv_0}), where u_0 and v_0 are irrational numbers, and not pass between D_{01} and D_{0-1}. However, in critical cases as this, the more accurate (at least theoretically) algebraic tests when implemented on a computer with its finite precision arithmetic may also fail. Moreover, for filters of order greater than four or so in either dimension, the algebraic methods are very difficult to program and computational time becomes excessive. Hence, we believe that for moderate and high-order filters, these approximate or practical tests are the only feasible choice. Of all the practical tests, the above algorithms, in particular, those based on (17) and (18) appear to be the most efficient. Figs. 4 and 5 summarize these algorithms.

In the next section several examples are given to illustrate our algorithms.

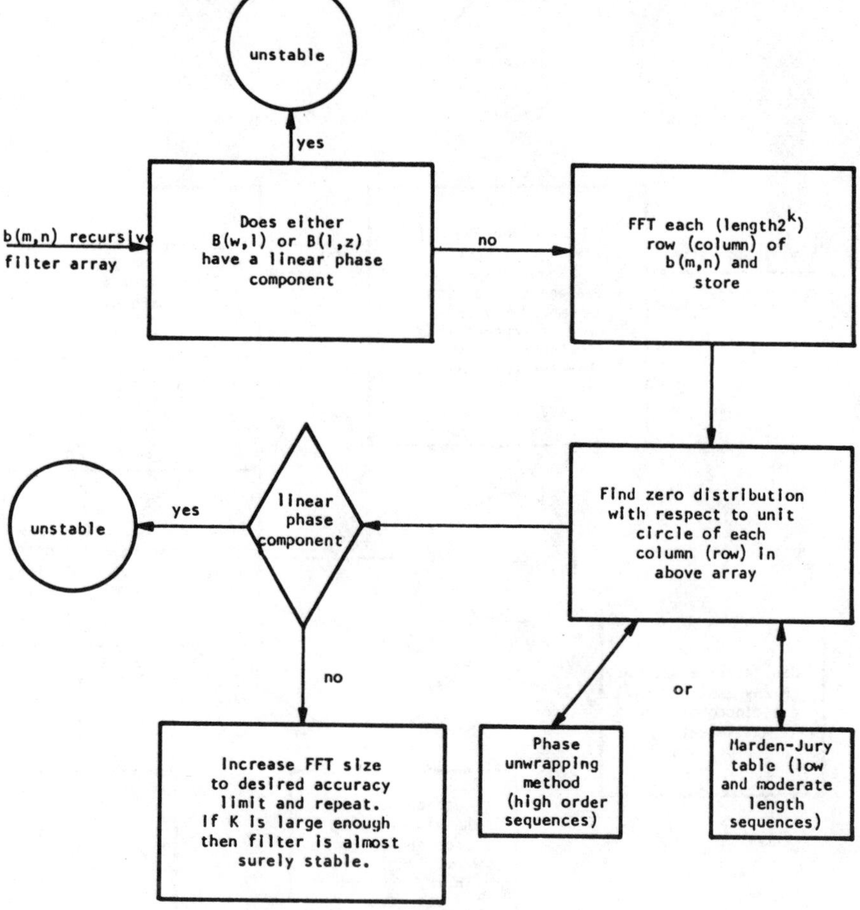

Fig. 4. Efficient stability algorithms for two-dimensional recursive filters (row and column algorithms).

V. Examples

Example: $B_1(w,z) = 0.5w^{-1}z + 1 + 0.85w + 0.1wz + 0.5w^2z^{-1}$.

Condition 7b):

$$B_1(w,1) = 0.5w^{-1} + 1 + 0.95w + 0.5w^2$$

$$B_1(1,z) = 0.5z^{-1} + 1.85 + 0.6z.$$

Both have no linear phase components.

Condition 7a) (*Row Test*):

Each row was transformed via a 1024 length FFT and the root distribution of the resulting columns revealed that, at least for this accuracy, $B_1(w,z) \neq 0$ on T^2. (Since $b(m,n)$ is real, only 513 root distribution tests are needed.) Therefore, $b(m,n)$ is "most likely" a stable filter array. A similar result is obtained if the column algorithm is used.

Condition 7a) (*Concatenation Tests*):

i) $B_1(\lambda,\lambda^M) = 0.5\lambda^{2-M} + 1 + 0.85\lambda + 0.5\lambda^{M-1} + 0.1\lambda^{M+1}$

ii) $B_1(\lambda^N,\lambda) = 0.5\lambda^{1-N} + 1 + 0.85\lambda^N + 0.1\lambda^{N+1} + 0.5\lambda^{2N-1}$.

For these to have no linear phase terms, the number of roots inside the unit circle for i) should equal $M - 2$ and $N - 1$ for ii) where M and N are the lengths of the zero padded rows and columns, respectively. Table I lists results of using the concatenation algorithm for different values of M and N. The root distributions were determined via a modified adaptive phase unwrapping algorithm [22], [28]. This algorithm efficiently implements the Nyquist test. Our experience has been that this algorithm gives accurate root distribution information (also phase information) for sequences as long as 600 on a PDP-11-45 with a floating-point mantissa of 24 bits. However, for longer sequences a larger floating-point word size should be used to minimize FFT noise problems. Notice that the zero distributions always correspond to that of a stable filter.

Example: $B_2(w,z) = 0.5w^{-1}z + 1 + 0.89w + 0.1wz + 0.5w^2z^{-1}$. $B_2(w,1)$ and $B_2(1,z)$ have no linear phase term, so condition 7b) is satisfied. When the row algorithm is employed, a phase discontinuity occurs for FFT sizes of 32 or larger. Therefore, $b_2(m,n)$ is definitely unstable. Alternately, this fact can be determined by using the column algorithm.

Condition 7a) (*Concatenation Tests*):

Concatenations of zero padded rows and columns yield

i) $B_2(\lambda,\lambda^M) = 0.5\lambda^{2-M} + 1 + 0.89\lambda + 0.5\lambda^{M-1} + 0.1\lambda^{M+1}$

ii) $B_2(\lambda^N,\lambda) = 0.5\lambda^{1-N} + 1 + 0.89\lambda^N + 0.1\lambda^{N+1} + 0.5\lambda^{2N-1}$.

If $b(m,n)$ is stable, then the number of zeros inside the unit circle of the above polynomials should equal $M - 2$ and $N - 1$, respectively, for all values of M and N. Table II indicates results for different values of M and N. A phase discontinuity appears when $M \geqslant 12$ and $N \geqslant 18$, and hence, this test also in-

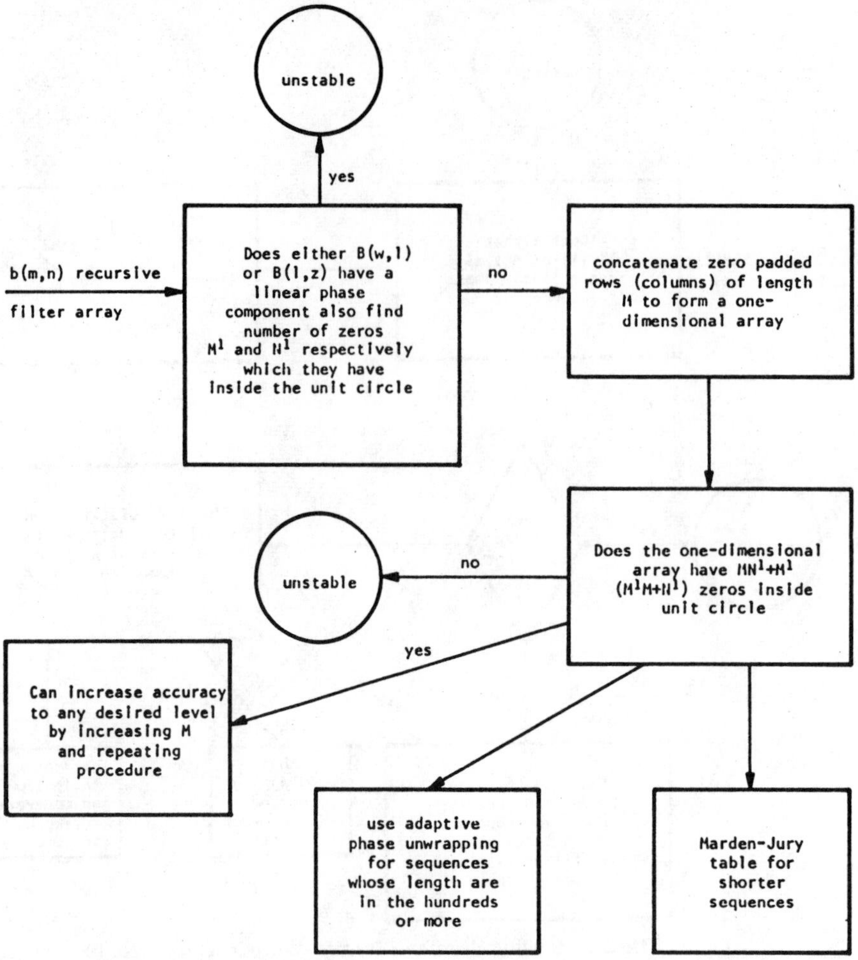

Fig. 5. Stability algorithms for two-dimensional filters which require only one one-dimensional root distribution test (concatenation algorithms).

TABLE I

M or N value	Length of row concatenated sequence	Length of column concatenated sequence	Ideal zero distribution inside unit circle row-column	M and N Sequence zero distribution row-column
12	24	35	10-11	10-11
18	36	39	16-17	16-17
36	72	107	34-35	34-35
72	144	215	70-71	70-71
144	288	431	142-143	142-143
288	576	863	286-287	286- -

TABLE II

M or N value	Length of row concatenated sequence	Length of column concatenated sequence	Ideal zero distribution inside unit circle row-column	M and N Sequence zero distribution row-column
6	12	17	4-5	4-7
12	24	35	10-11	8-11
18	36	53	16-17	14-19
36	72	107	34-35	32-41
72	144	215	70-71	68-81
144	288	431	142-143	138-165

dicates that there exists a $(w_0, z_0) \in T^2$ such that $B(w_0, z_0) = 0$, so $b(m, n)$ is unstable.

Example: $B_3(w, z) = 1 - 1.2w + 0.5w^2 - 1.5z + 1.8wz - 0.75w^2 z + 0.6z^2 - 0.72wz^2 + 0.2718w^2 z^2$. The conditions of Theorem 7b) are satisfied. The row algorithm detects a phase discontinuity for FFT sizes of 128 or larger while the column algorithm detects a discontinuity for FFT sizes of 32 or larger. The concatenation algorithms indicate a phase discontinuity for $B_3(\lambda, \lambda^M)$ for $M = 18$ and for $B_3(\lambda^N, \lambda)$ for $N = 36$. Therefore, all these algorithms indicate that $b(m, n)$ is unstable. It is interesting to note that $b_3(m, n)$ can be made stable by changing $b_3(2, 2)$ by 0.0001 to 0.2719.

It should be stressed that, in practice, it is only necessary to apply the row, column, row concatenation, or column concatenation algorithms once for some suitable large FFT size L or values of M or N. This choice is dependent on the accuracy desired by the user. Indeed, there is a tradeoff between the accuracy of the test and computation time.

VI. Concluding Remarks

We have presented recursive filtering from a somewhat different point of view. A general mapping theorem has been formulated which allows any type of filter to be mapped into a first-quadrant filter. This first-quadrant filter is stable if and only if the original filter is stable. Several general stability theorems which relate stability to the zero set of $B(w, z)$ have been presented. These theorems led to the conclusion that a filter is stable if and only if its phase function is continuous, odd, and periodic. This observation suggested several practical stability testing algorithms. Among these are several methods which appear to be extremely efficient for higher order filters. All the results in this paper can be generalized to n-dimensional filters. Moreover, the practical stability tests can be applied to any finite two-dimensional array to determine if its cepstrum exists by determining if the Fourier transform of the array ever equals zero.

References

[1] D. Dudgeon, "Two-dimensional recursive filtering," Sc.D. thesis, Dep. Elec. Eng., M.I.T., Cambridge, May 1974.

[2] T. Huang, "Stability of two-dimensional recursive filters," *IEEE Trans. Audio Electroacoust.*, vol. AU-20, pp. 158-163, June 1972.

[3] R. Mersereau and D. Dudgeon, "Two-dimensional digital filtering," *Proc. IEEE*, vol. 63, pp. 610-623, Apr. 1975.

[4] J. Shanks, S. Treitel, and J. H. Justice, "Stability and synthesis of two-dimensional recursive filters," *IEEE Trans. Audio Electroacoust.*, vol. AU-20, pp. 115-128, June 1972.

[5] B. Anderson and E. Jury, "Stability test for two-dimensional recursive filters," *IEEE Trans. Audio Electroacoust.*, vol. AU-21, pp. 366-372, Aug. 1973.

[6] G. Maria and M. Fahmy, "On the stability of two-dimensional digital filters," *IEEE Trans. Acoust., Speech, Signal Processing*, vol. ASSP-21, pp. 470-472, Oct. 1973.

[7] M. Ekstrom and J. Woods, "Two-dimensional spectral factorization with applications to recursive filtering," *IEEE Trans. Acoust., Speech, Signal Processing*, vol. ASSP-24, pp. 115-128, Apr. 1976.

[8] D. Dudgeon, "The computation of two-dimensional cepstra," *IEEE Trans. Acoust., Speech, Signal Processing*, vol. ASSP-25, pp. 476-483, Dec. 1977.

[9] D. Goodman, "Some stability properties of two-dimensional linear shift-invariant digital filters," *IEEE Trans. Circuits Syst.*, vol. CAS-24, pp. 201-208, Apr. 1977.

[10] J. Costa and A. Venetsanopoulous, "A group of linear spectral transformations for two-dimensional digital filters," *IEEE Trans. Acoust., Speech, Signal Processing*, vol. ASSP-24, pp. 424-425, Oct. 1976.

[11] J. Justice and J. Shanks, "Stability criterion for n-dimensional digital filters," *IEEE Trans. Automat. Contr.*, vol. AC-18, pp. 284-286, June 1973.

[12] D. Goodman, "An alternate proof of Huang's stability theorem," *IEEE Trans. Acoust., Speech, Signal Processing*, vol. ASSP-24, pp. 426-427, Oct. 1976.

[13] M. Strintzis, "Tests of stability of multidimensional filters," *IEEE Trans. Circuits Syst.*, vol. CAS-24, pp. 432-437, Aug. 1977.

[14] R. DeCarlo, J. Murray, and R. Saeks, "Multivariable Nyquist theory," *Int. J. Contr.*, vol. 25, pp. 657-675, May 1977.

[15] N. Bose, "Problems and progress in multidimensional systems theory," *Proc. IEEE*, vol. 65, pp. 824-840, June 1977.

[16] W. Rudin, *Function Theory in Polydiscs.* New York: Benjamin, 1969.

[17] N. Bose, "Implementation of a new stability test for two-dimensional filters," *IEEE Trans. Acoust., Speech, Signal Processing*, vol. ASSP-25, pp. 117-120, Apr. 1977.

[18] R. DeCarlo, "An algebraic topological approach to stability theory," Ph.D. dissertation, Dep. Elec. Eng., Texas Tech., Lubbock, 1976.

[19] R. DeCarlo, R. Saeks, and J. Murray, "A Nyquist-like test for the stability of two-dimensional digital filters," *Proc. IEEE*, vol. 65, pp. 978-979, June 1977.

[20] E. Jury, *Theory and Applications of the z-Transform Method.* New York: Wiley, 1964.

[21] D. Dudgeon, "The existence of cepstra of two-dimensional rational polynomials," *IEEE Trans. Acoust., Speech, Signal Processing*, vol. ASSP-23, pp. 242-243, Apr. 1975.

[22] J. Tribolet, "A new phase unwrapping algorithm," *IEEE Trans. Acoust., Speech, Signal Processing*, vol. ASSP-25, pp. 170-177, Apr. 1977.

[23] M. Marden, *The Geometry of the Zeros of a Polynomial in a Complex Variable.* New York: Amer. Math. Soc., 1949.

[24] G. Bliss, *Algebraic Functions.* New York: Amer. Math. Soc., 1933.

[25] R. Gunning and H. Rossi, *Analytic Functions of Several Complex Variables.* Englewood Cliffs, NJ: Prentice-Hall, 1965.

[26] R. Mersereau and A. Oppenheim, "Digital reconstruction of multidimensional singnals from their projections," *Proc. IEEE*, vol. 62, pp. 1319-1338, Oct. 1974.

[27] R. Mersereau and D. Dudgeon, "The representation of two-dimensional sequences as one-dimensional sequences," *IEEE Trans. Acoust., Speech, Signal Processing*, vol. ASSP-22, pp. 320-325, Oct. 1974.

[28] B. O'Connor and T. S. Huang, "Phase unwrapping," Purdue Univ., W. Lafayette, IN, TR-EE 77-35, pp. 87-110 (for work done in 1976), 1977.

The Computation of Two-Dimensional Cepstra

DAN E. DUDGEON, MEMBER, IEEE

Abstract—**In this paper we shall explore two methods of computing the complex cepstrum of a two-dimensional (2-D) signal. The two principal methods for computing 1-D cepstra, using discrete Fourier transforms (DFT's) and the complex logarithm function or using a recursion relation for minimum-phase signals, may be extended to two dimensions. These two algorithms are developed and simple examples of their use are given. As a matter of course, we shall also be drawn into considering the definitions of 2-D causality and 2-D minimum-phase signals. In addition, we shall explore the relationship among the nonzero regions of a signal, its inverse, and its cepstrum.**

I. Introduction

THE CEPSTRUM is an entity derivable from a digital signal (or system impulse response) in the same sense that a spectrum is derivable from a digital signal. The term "cepstrum" was coined by Bogert *et al.* [1] in describing the inverse Fourier transform of the real logarithm of the power spectrum of a signal. Later, Oppenheim *et al.* [2] used the term "complex cepstrum" to describe the inverse z transform of the complex logarithm of the z transform of a signal. Though this terminology was needed to distinguish between the cepstrum and the complex cepstrum of a signal, it was somewhat confusing since the complex cepstrum of a real signal is real. Consequently, in this paper we shall use the term "cepstrum" interchangeably with "complex cepstrum" to mean the inverse transform of the complex logarithm of the transform of a signal. The "cepstrum" of Bogert *et al.* can be easily expressed in terms of the "complex cepstrum" of Oppenheim *et al.* We shall write

$$\hat{s}(n) = Z^{-1}\{\ln[Z[s(n)]]\} \tag{1}$$

where $\hat{s}(n)$ is the cepstrum, $s(n)$ is the signal, $Z[\cdot]$ is the z-transform operator, and $Z^{-1}[\cdot]$ is its inverse. As discussed in Oppenheim and Schafer [3, ch. 10], to define the cepstrum properly the function

$$\hat{S}(z) = \ln[Z[s(n)]] = \ln\left[\sum_n s(n)z^{-n}\right]$$

must be a valid z transform. We shall restrict our attention to signals which are stable (i.e., the region of convergence of their z transforms includes the contour $|z| = 1$) and whose cepstra are also stable. This implies

$$0 < |S(e^{j\omega})| < \infty \quad \text{for} \quad -\pi \leqslant \omega < \pi.$$

In addition, care must be taken in properly defining the complex logarithm function so that $\hat{S}(z)$ is analytic. In general, using the principal value of the complex logarithm will not suffice because of the discontinuities in the imaginary part which it can introduce. Furthermore, the requirement that $\hat{S}(z)$ be analytic means that the phase of $s(n)$ (that is, the imaginary part of $\hat{S}(e^{j\omega})$) must be continuous and periodic. Continuity is ensured by the proper definition of the complex logarithm, but periodicity must be ensured by removing any linear phase components by redefining the origin of the signal's independent variable.

The cepstrum has gained some notoriety, especially in speech research (see [2], [3], for example) because of the following property. If two functions are combined by convolution (for example, the input to a filter and the filter's impulse response), then their cepstra (if they exist) are combined by addition. Consequently, the concept of a cepstrum is useful in dealing with problems involving convolution and deconvolution. In addition, the concept of a cepstrum is useful when thinking about log spectra, since the real part of the Fourier transform of the cepstrum is the log magnitude of the signal's spectrum and the imaginary part is the phase.

The concept of a two-dimensional (2-D) cepstrum has already proved important in attacking the problem of 2-D spectral factorization [4], [5]; a difficult problem because of the absence of the ability to factor a general 2-D polynomial. Recently, the use of the 2-D cepstrum as a relatively fast and easily implementable test for the stability of 2-D recursive filters has been demonstrated [11]. Other possible applications for the 2-D cepstrum in the context of signal processing include characterization of reverberance in pictures or array signals and separating seismic signal components combined by convolution.

In this paper, we shall consider two algorithms for the computation of a 2-D cepstrum from a 2-D signal. The first method is based on the discrete Fourier transform (DFT) and the complex logarithm function, and it requires that the linear-

Manuscript received March 1, 1976; revised March 3, 1977, and July 20, 1977.

The author is with the Computer Systems Division, Bolt Beranek and Newman Inc., Cambridge, MA 02138.

Reprinted from *IEEE Trans. Acoust., Speech, Signal Processing,* vol. ASSP-25, pp. 476–484, Dec. 1977.

phase component of the 2-D signal be removed. The second method is based on the derivation of recursion relations which hold when the signal and its cepstrum are zero outside of a region. To study when this condition is true, we shall examine the relationship among the nonzero regions of the signal, its inverse, and its cepstrum. A brief discussion of the two methods of computing a 2-D cepstrum concludes the paper.

II. Two-Dimensional Cepstra and the Removal of the Linear-Phase Component

Because of these interesting properties of the 1-D cepstrum, it is desirable to extend the concept of the cepstrum to two-dimensional signals; that is, signals indexed on two independent variables.

The author has shown previously that a suitable 2-D cepstrum can be defined for a certain class of 2-D signals [6]. Specifically, it was shown that a stable 2-D signal $s(m, n)$ whose z transform is a ratio of 2-D polynomials, that is

$$S(w,z) = \sum_m \sum_n s(m,n) w^{-m} z^{-n} \triangleq \frac{\sum_p \sum_q a(p,q) w^{-p} z^{-q}}{\sum_p \sum_q b(p,q) w^{-p} z^{-q}}$$

$$= \frac{A(w,z)}{B(w,z)}$$

(where the sums on p, q have a finite number of terms) will have a stable cepstrum provided

$$|S(e^{j\mu}, e^{j\nu})| \neq 0 \quad \text{for} \quad -\pi \leq \mu, \nu < \pi$$

and provided that the origin of the signal has been adjusted to ensure that the phase is continuous and periodic in both frequency variables μ and ν. (That this can be done in the first place is at the heart of the existence proof in [6].) In order to be able to compute the cepstra of signals from this important class of functions, we must be able to determine the amount of linear phase associated with a particular signal so that it can be eliminated.

Let us start with a signal $s(m, n)$ whose z transform is tl ratio of two polynomials

$$S(w,z) = \frac{A(w,z)}{B(w,z)}$$

and which is not zero for $|w| = |z| = 1$. The 2-D Fourier transform of $s(m, n)$ is given by

$$S(e^{j\mu}, e^{j\nu}) = |S(e^{j\mu}, e^{j\nu})| e^{j\phi(\mu,\nu)}$$

$$= \sum_m \sum_n s(m,n) e^{-j\mu m - j\nu n}$$

where we have denoted the phase by $\phi(\mu, \nu)$. We can define $\phi(\mu, \nu)$ in such a way so that it is a continuous function of μ and ν [6]. Having done this, we will find that

$$\phi(\mu, \nu + 2\pi) = \phi(\mu, \nu) + 2\pi J$$

$$\phi(\mu + 2\pi, \nu) = \phi(\mu, \nu) + 2\pi I$$

in general. Thus we can write the phase function as

$$\phi(\mu, \nu) = \phi_p(\mu, \nu) + I\mu + J\nu.$$

To eliminate the linear-phase terms leaving only the periodic part $\phi_p(\mu, \nu)$, we can define a new signal by shifting the origin of the original signal

$$s_p(m,n) = s(m - I, n - J).$$

The signal $s_p(m, n)$ will have a continuous and periodic phase function ϕ_p when it is computed by integrating the phase derivative as in [6]. (It is important to note that if the phase is computed using the complex logarithm or arctangent function, only the principal value of ϕ_p will be obtained. The principal value can still exhibit discontinuities of 2π in some cases despite the elimination of the linear-phase component.)

Note that the problem of determining the coefficients of the linear-phase component (I and J) is really two separate one-variable problems. For example, in determining the value of I, it suffices to examine the original phase curve for $\nu = 0$, since

$$\phi(\mu, 0) = \phi(\mu + 2\pi, 0) + 2\pi I.$$

It can be shown that this problem is identical to determining the coefficient of the linear-phase component of the one-dimensional signal

$$x(m) = \sum_n s(m,n).$$

Similarly, the parameter J can be determined by examining the degree of phase linearity of the one-dimensional signal

$$y(n) = \sum_m s(m,n).$$

Consequently, any techniques which are developed to facilitate the determination of the coefficient of the linear-phase component of a 1-D signal can be used for 2-D signals as well. (This assumes, of course, that the 2-D signal belongs to the set of 2-D rational polynomials.)

The need to make $\hat{S}(w, z)$ analytic in a region containing the curve $|w| = |z| = 1$ means that we can define the 2-D cepstrum only for signals whose phase is continuous and periodic. An arbitrary signal from the set of signals which have rational 2-D z transforms and which are nonzero for $|w| = |z| = 1$ will not have periodic phase, in general. But, by shifting the signal origin without perturbing its shape or amplitude, we can create a signal for which the 2-D cepstrum can be defined. If we take the attitude that the cepstrum of an arbitrary member of the set is defined as the cepstrum of the signal shifted so as to eliminate the linear-phase components, then many signals will have the same cepstrum. These signals will be identical to one another, except for the location of their origin. Thus, in some sense, the cepstrum contains signal shape and amplitude information, but does not contain any "origin information." (In practical problems, this is usually no handicap since the signal origin can be specified by other constraints.)

III. Computing the Two-Dimensional Cepstrum Using the DFT

An approximation to the 2-D cepstrum can be computed by using a 2-D DFT (usually implemented by an FFT, or fast

Fourier transform, algorithm). We shall assume that our input signal $s(m, n)$ is zero outside of a $M \times M$ square. (The results are easily generalized to a rectangle.)

$$s(m,n) = 0 \quad \text{for} \quad m,n < 0 \quad \text{and for} \quad m,n \geq M$$

then we may calculate the N-point DFT of $s(m, n)$ where $N \geq M$

$$S(k,l) \triangleq \sum_{m=0}^{M-1} \sum_{n=0}^{M-1} s(m,n) e^{-j(2\pi km/N) - j(2\pi ln/N)},$$

$$0 \leq k,l < N.$$

(Note that the DFT is the z transform of the signal evaluated at $w = e^{j(2\pi k/N)}$, $z = e^{j(2\pi l/N)}$ for k, l from 0 to $N - 1$.) The complex logarithm can then be computed as

$$\hat{S}(k,l) = \ln |S(k,l)| + j\phi_p(k,l).$$

(In this section, we shall use the more concise notation $\phi_p(k, l)$ to represent the samples of the phase function $\phi_p[(2\pi k/N), (2\pi l/N)]$.) Again we face the problem of correctly defining the phase ϕ_p so that it is continuous and it has no linear components, except that here we have only samples of the phase, which are inherently discontinuous. The standard computer complex-logarithm function uses the principal value for the phase, and consequently "jumps" of 2π in the value of the phase may be seen. But if N is large enough so that the change in phase over the interval of $2\pi/N$ is small compared to π, then it is fairly easy to generate an algorithm to detect the jumps of 2π in the phase. More sophisticated phase-unwrapping algorithms are also available [7].

The phase continuity must be checked in both dimensions, but this can be done with a 1-D phase-unwrapping algorithm applied to either the rows or columns of $S(k, l)$. It is possible to do this because the phase for a particular row, for example, can be formulated as the integral of the partial derivative of the phase for that row as in [6].

At this point, the inverse 2-D DFT can be performed on $\hat{S}(k, l)$ to yield $\hat{s}_d(m, n)$, an aliased version of the true cepstrum $\hat{s}(m, n)$. Since $\hat{s}(m, n)$ is not constrained to be zero outside the $N \times N$ square, $\hat{s}_d(m, n)$ is equal to the sum of shifted versions of $\hat{s}(m, n)$. This relationship may be written as

$$\hat{s}_d(m,n) = \sum_{i=-\infty}^{\infty} \sum_{j=-\infty}^{\infty} s(m + iN, n + jN).$$

Clearly, the approximation

$$\hat{s}_d(m,n) \simeq \hat{s}(m,n)$$

is a good one only if $\hat{s}(m,n)$ goes to zero fast enough as either m or n approach N. This also argues for using as large a value for N as is practical.

To verify this method of computing a 2-D cepstrum, a subroutine called UNWRAP, based on Tribolet's method [7], was written to compute the N-point DFT $X(k)$, the spectral log magnitude $\ln |X(k)|$, and the unwrapped phase $\phi_p(k)$ with the linear component removed given the input sequence $x(m)$ and an initial phase $\phi(0)$. It is assumed that $x(m) = 0$ for $m < 0$ and for $m \geq M$, and that $N \geq M$ and N is a power of 2. The subroutine UNWRAP was written so that $x(m)$ may be real or complex, the degree of linear phase removed is indicated, and the initial phase may be specified externally (rather than relying on internal arctangent computation which has an ambiguity). The algorithm used for computing a 2-D cepstrum is as follows.

1) Compute the initial phase at the origin $\phi(0, 0)$ by

$$\phi(0,0) = \tan^{-1}(S_I/S_R)$$

where S_R and S_I are the real and imaginary parts, respectively, of

$$S_R + jS_I = \sum_{m=0}^{M-1} \sum_{n=0}^{M-1} s(m,n).$$

There is an ambiguity in the value of $\phi(0, 0)$ which cannot be resolved, but it is unimportant. If $s(m, n)$ is a real signal and $S_R > 0$, then $\phi(0, 0)$ can be taken as zero. If $s(m, n)$ is real and $S_R < 0$, then $\phi(0, 0)$ can be taken to be π.

2) Form the sequence $x(m) = \Sigma_{n=0}^{M-1} s(m, n)$. Using UNWRAP, compute $X(k)$, $\ln |X(k)|$, and $\phi_p(k)$ given the input sequence $x(m)$ and the initial phase $\phi(0, 0)$ computed in Step 1. Save I, the degree of linear phase removed by UNWRAP for use in Step 3.

3) For each n from 0 to $M - 1$, circularly shift the nth row of $s(m, n)$ by I to form

$$s_p(m,n) = s((m - I) \bmod N, n)$$

and then compute the intermediate transform

$$S_n(k) = \sum_{m=0}^{M-1} s_p(m,n) e^{-j(2\pi km/N)}.$$

4) For each k from $k = 0$ to $N - 1$, use UNWRAP with $S_n(k)$ (as a function of n) as the input and $\phi_p(k)$ (computed in Step 2) as the initial phase. The output of UNWRAP will be the spectral log-magnitude function $\ln |S(k, l)|$ and the unwrapped phase function $\phi_p(k, l)$ which has had its linear component removed. (According to the theory [6], the amount of linear phase removed from each row by UNWRAP should be identical. This was verified in several examples.) Form the 2-D complex log spectrum

$$\hat{S}(k,l) = \ln |S(k,l)| + j\phi_p(k,l).$$

5) Take the inverse 2-D DFT ($N \times N$ points) to yield $\hat{s}_d(m,n)$, the approximation to $s(m,n)$.

$$\hat{s}_d(m,n) = \frac{1}{N^2} \sum_{k=0}^{N-1} \sum_{l=0}^{N-1} S(k,l) e^{j \frac{2\pi}{N}(km + ln)}.$$

The subroutine FOUR2, written by Brenner [8], was used to compute both the 1-D and 2-D DFT's needed in the cepstral computation. The previous algorithm could have also been written to perform the column operations first and the row operations second. This would correspond to interchanging m and n on input, performing the computation, and interchanging m and n again on output.

We shall now discuss two simple examples which illustrate

TABLE I

$\hat{s}(m, n)$ FOR $a = -0.4$, $b = -0.5$

(m,n)	True	Computed by DFT, N=8	Computed by DFT, N=64
(-1,-1)	0.	-3.8681000×10^{-3}	-5.2608988×10^{-10}
(-1, 0)	0.	3.5850657×10^{-3}	-5.0145865×10^{-9}
(0,-1)	0.	4.0086840×10^{-3}	-5.7959178×10^{-10}
(0, 0)	0.	-3.1978840×10^{-3}	5.5733835×10^{-9}
(0, 1)	5.0×10^{-1}	5.0284887×10^{-1}	5.0000001×10^{-1}
(0, 2)	-1.25×10^{-1}	-1.2779548×10^{-1}	-1.2500000×10^{-1}
(0, 3)	$4.166\bar{6}\times10^{-2}$	4.4567834×10^{-2}	4.1666667×10^{-2}
(1, 0)	4.0×10^{-1}	4.0359472×10^{-1}	4.0000002×10^{-1}
(1, 1)	-2.0×10^{-1}	-2.0273839×10^{-1}	-2.0000001×10^{-1}
(1, 2)	1.0×10^{-1}	1.0233378×10^{-1}	1.0000001×10^{-1}
(2, 2)	-6.0×10^{-2}	-6.2296076×10^{-2}	-6.0000003×10^{-2}
(2, 3)	4.0×10^{-2}	4.1907853×10^{-2}	4.0000002×10^{-2}
(3, 0)	$2.133\bar{3}\times10^{-2}$	2.6236880×10^{-2}	2.1333330×10^{-2}
(3, 2)	3.2×10^{-2}	3.4455027×10^{-2}	3.2000002×10^{-2}

Note: The overbar denotes an infinitely repeated digit.

this method of computing a 2-D cepstrum. Consider the following signal:

$$s(0,0) = 1$$

$$s(1,0) = -a$$

$$s(0,1) = -b$$

$$|a| + |b| < 1. \tag{2}$$

For this signal it is possible to explicitly write the 2-D cepstrum as

$$\hat{s}(m,n) = -\frac{(m+n-1)!}{m!n!} a^m b^n$$

for m and $n \geqslant 0$, but $(m,n) \neq (0,0)$

$$\hat{s}(m,n) = 0 \tag{3}$$

for m or $n < 0$ and $(m,n) = (0,0)$. In this case, we have a known cepstrum with which to compare the computed cepstrum.

Example 1: The cepstrum of a test signal with $a = -0.4$ and $b = -0.5$ was computed. The results for typical values of $\hat{s}_d(m, n)$ computed by the DFT method for $N = 8$ and for $N = 64$ are shown in Table I along with the true cepstral value. All the computations were done on a PDP-10 computer (36-bit words) in single precision floating-point arithmetic (27-bit fraction). The effects of aliasing are obvious in Table I. When $N = 8$, the size of the errors was about 3×10^{-3}, but when $N = 64$, the size of the errors was the order of 10^{-9}. (These errors were due to the precision of the arithmetic since the error caused by aliasing is the order of 10^{-21} for $N = 64$ in this example.)

Example 2: To test the ability of the UNWRAP subroutine to correctly remove linear phase components, the cepstrum of the following signal was computed:

$$s(3,4) = -0.3$$

$$s(4,4) = 0.06$$

$$s(2,5) = -0.4$$

$$s(3,5) = 1.11$$

$$s(4,5) = -0.2$$

$$s(2,6) = 0.04$$

$$s(3,6) = -0.1.$$

TABLE II

(m,n)	True $\hat{s}$ (m,n)	Computed $\hat{s}_d$(m,n) for N=64
(-2,-2)	-2.16×10^{-2}	$-2.1600002 \times 10^{-2}$
(-2,-1)	-4.8×10^{-2}	$-4.8000005 \times 10^{-2}$
(-2, 0)	-8.0×10^{-2}	$-8.0000008 \times 10^{-2}$
(-2, 1)	0.	$-1.6991844 \times 10^{-9}$
(-2, 2)	0.	$-5.1903761 \times 10^{-10}$
(-1.-1)	-1.2×10^{-1}	$-1.2000005 \times 10^{-1}$
(-1, 0)	-4.0×10^{-1}	$-4.0000032 \times 10^{-1}$
(-1, 1)	0.	$-4.2762646 \times 10^{-8}$
(0,-1)	-3.0×10^{-1}	$-2.9999972 \times 10^{-1}$
(0, 0)	0.	8.1781764×10^{-9}
(0, 1)	-1.0×10^{-1}	$-1.0000030 \times 10^{-1}$
(0, 2)	-5.0×10^{-3}	$-5.0001037 \times 10^{-3}$
(0, 5)	-2.0×10^{-6}	$-2.0444075 \times 10^{-6}$
(1, 0)	-2.0×10^{-1}	$-1.9999972 \times 10^{-1}$
(1, 1)	-2.0×10^{-2}	$-1.9999961 \times 10^{-2}$
(5, 5)	-8.064×10^{-8}	$-8.4116778 \times 10^{-8}$

This signal is composed of a signal of the form of (2) with $a = 0.2$, $b = 0.1$ convolved with another signal of the form of (2) with $a = 0.4$, $b = 0.3$ but rotated 180°. Finally, the output of the convolution was centered at $(m, n) = (3, 5)$. The cepstral computation correctly calculated and removed the linear phase component corresponding to this shift. The true cepstrum is the sum of a cepstrum of the form of (3) and another cepstrum of the form of (3) rotated 180°. The true and computed cepstra for $N = 64$ are given in Table II for some representative values. Values for $m < 0$, for example, are obtained from the computed cepstrum at $N + m$ because of aliasing.

IV. Region of Support of the Two-Dimensional Cepstrum

In this section, we shall examine under what conditions the 2-D cepstrum is nonzero only in a region of the (m, n) plane, often called a "region of support." In succeeding sections a recursive method of computing the 2-D cepstrum will be detailed based on these conditions. To begin, we shall demonstrate, as did Pistor [9], that the 2-D cepstrum may be written as a convolution. Recall the z-transform pair [3]

$$z \frac{\partial}{\partial z} S(w,z) = -ns(m,n). \tag{4}$$

Since

$$\hat{S}(w,z) = \ln [S(w,z)]$$

we can write

$$z \frac{\partial}{\partial z} S(w,z) = z \frac{\partial}{\partial z} \ln [S(w,z)]$$

$$= zV(w,z) \frac{\partial}{\partial z} S(w,z) \tag{5}$$

where $V(w, z) = 1/S(w, z)$. Taking the inverse 2-D z transform of both sides, using $|w| = |z| = 1$ as the contour of integration, we get the convolution

$$n\hat{s}(m, n) = \sum_k \sum_l l s(k, l) v(m - k, n - l). \tag{6}$$

The function $v(m, n)$ is the stable inverse of the signal $s(m, n)$. It is obtained by taking the inverse 2-D z transform of one over the 2-D z transform of $s(m, n)$, using $|w| = |z| = 1$ as the contour of integration. Since $S(w, z)$ is analytic and nonzero on this contour, it follows that $V(w, z)$ will also be analytic and nonzero there, and $v(m, n)$ will be stable necessarily.

In a similar manner to the derivation of (6), it can be shown that the following equation also holds:

$$m\hat{s}(m, n) = \sum_k \sum_l k s(k, l) v(m - k, n - l). \tag{7}$$

It becomes obvious now that the region of support of the cepstrum is determined by the regions of support of the signal and its stable inverse. Intuitively, the shape of the cepstrum's region of support is the convolution of the shape of the signal's region of support with the shape of its inverse's region of support.

We can now define a "closed" region of support R. If the convolution of any two signals with support on R is another signal with support on R, then R is a closed region of support. An example of a closed region of support would be a convex sector in the (m, n) plane with its vertex at the origin [5]. Using this definition, we see that a cepstrum will have support on R if both the signal and its stable inverse have support on R. The converse statement is also true and is easily proved by series expansion of the exponential function as in [4], [5], [9].

The following working definitions suggest themselves at this point. A 2-D signal is said to be causal if it has support on a closed region. A 2-D signal is said to be a minimum-phase signal if it and its inverse are stable and have support on the same closed region R. Using these definitions, we can say the cepstrum of a minimum-phase signal is causal, just as in 1-D theory. There is an obvious problem with these definitions which runs counter to our intuitive perception of a minimum-phase signal. The problem is that the entire (m, n)-plane is a valid closed region of support, and so any stable signal may lay claim to being causal and minimum phase since it and its stable inverse have support on the entire plane. Consequently, it seems that the notions of causality and minimum phase are inappropriate, and that closed regions of support of signals and their inverse are more central to the problems in multidimensional theory.

V. A Recursive Computation Procedure for the Cepstrum of a Two-Dimensional Minimum-Phase Signal

In this section we shall consider a method of computing the cepstrum of a 2-D signal $s(m, n)$, which has minimum phase and has support only on the first quadrant Q_1 in the (m, n) plane. (The positive m and n axes are included in Q_1 as well as the origin.) Our working definition of minimum phase implies that both the signal and its stable inverse have support only on Q_1. From (6) and (7) we know the cepstrum $\hat{s}(m, n)$ has support only on Q_1 also. In this case, a recursion formula analogous to the recursion formula for 1-D cepstra [3] can be easily derived. If we rewrite (5) as

$$z \frac{\partial}{\partial z} \hat{S}(w, z) = \frac{z \frac{\partial}{\partial z} S(w, z)}{S(w, z)} \tag{8}$$

then

$$[S(w, z)] \left[z \frac{\partial}{\partial z} \hat{S}(w, z) \right] = z \frac{\partial}{\partial z} S(w, z). \tag{9}$$

Taking the inverse 2-D z transform of (9), using $|w| = |z| = 1$ as the contour of integration, gives

$$-\sum_k \sum_l s(m - k, n - l) l \hat{s}(k, l) = -ns(m, n).$$

In general, the indices (k, l) take on all integer values, but since $s(m, n)$ and $\hat{s}(m, n)$ we support on Q_1 only, the recursion relationship can be written

$$\sum_{k=0}^{m} \sum_{l=0}^{n} s(m - k, n - l) l \hat{s}(k, l) = ns(m, n). \tag{10}$$

Finally (assuming that $s(0, 0) \neq 0$), the $\hat{s}(m, n)$ term can be pulled out of the sum on the left side to give

$$\hat{s}(m, n) = \frac{s(m, n)}{s(0, 0)} - \frac{1}{ns(0, 0)} \sum_{\substack{k=0 \\ (k,l) \neq (m,n)}}^{m} \sum_{l=0}^{n} s(m - k, n - l) \cdot l \hat{s}(k, l) \tag{11}$$

for $n \neq 0$. Similarly, the formula

$$\hat{s}(m, n) = \frac{s(m, n)}{s(0, 0)} - \frac{1}{ms(0, 0)} \sum_{\substack{k=0 \\ (k,l) \neq (m,n)}}^{m} \sum_{l=0}^{n} s(m - k, n - l) \cdot k \hat{s}(k, l) \tag{12}$$

can be shown to hold for $m \neq 0$. By applying the 2-D analog of the initial value theorem, it can be shown that

$$\hat{s}(0, 0) = \ln [s(0, 0)].$$

By using formula (11), the values of $\hat{s}(0, n)$ may be calculated. Then by using formula (12), the remaining values of $\hat{s}(m, n)$ may be calculated. (The reader will note that formula (12) could have been used to calculate $\hat{s}(m, 0)$ first, and that either (11) or (12) could be used to calculate $\hat{s}(m, n)$ for $m \cdot n \neq 0$ once the values on the axes have been calculated.)

If $s(0, 0)$ is a complex or a real negative number, then the complex logarithm must be used to compute $\hat{s}(0, 0)$. Consequently, it is possible for a real signal with $s(0, 0) = -1$, for example, to have a real cepstrum except for the one complex point $\hat{s}(0, 0)$. In general, if we let

$$h(m, n) = cu_0(m, n)$$

where $u_0(m, n)$ is unity if $m = n = 0$ and zero otherwise and c is a complex constant, then

$$g(m,n) = cs(m,n) = \sum_k \sum_l s(m-k, n-l)h(k,l).$$

Since the cepstrum of a convolution is the sum of the cepstra,

$$\hat{g}(m,n) = \hat{s}(m,n) + \hat{h}(m,n)$$
$$= \hat{s}(m,n) + [\ln(c)]\, u_0(m,n).$$

In the case where $c = -1$, $\ln(c) = j\pi$ and that value is added to the cepstral point at $\hat{s}(0, 0)$, making it complex.

Because of the use of the complex logarithm, there is still an ambiguity in the imaginary part of $\hat{s}(0, 0)$. This ambiguity is the same as that in computing the initial value of the phase $\phi(0, 0)$ in Section III.

To verify the recursive method of computing a cepstrum for a minimum-phase signal with support on Q_1, a subroutine was written to compute $\hat{s}(m, n)$. Rather than store the values of $\hat{s}(m, n)$ for use in the recursion, it is computationally more efficient to save the values $m\hat{s}(m, n)$ for $m \neq 0$ and $n\hat{s}(m, n)$ for $m = 0$. If we let

$$f(m,n) = m\hat{s}(m,n) \quad \text{for} \quad m \neq 0$$
$$= n\hat{s}(m,n) \quad \text{for} \quad m = 0$$

and then rearrange (11) and (12), we can compute

$$f(m,n) = \frac{1}{s(0,0)}\left[ms(m,n) - \sum_{\substack{k=0 \\ (k,l)\neq(m,n)}}^{m} \sum_{l=0}^{n} s(m-k, n-l)\cdot f(k,l)\right]$$

for $m \neq 0$, and

$$f(0,n) = \frac{1}{s(0,0)}\left[ns(0,n) - \sum_{l=1}^{n-1} s(0, n-l) f(0,l)\right]$$

when $m = 0$. Finally, the cepstral point at (m, n) is given by

$$\hat{s}(m,n) = \frac{1}{m} f(m,n) \quad \text{for} \quad m \neq 0$$

$$\hat{s}(m,n) = \frac{1}{n} f(m,n) \quad \text{for} \quad m = 0, n \neq 0.$$

If the values of $\hat{s}(m, n)$ are to be stored locally (as opposed to being transmitted to some other processing or storage unit), then the method just presented will require more storage than a direct implementation of (11) and (12).

Example 3: The following minimum-phase test signal was used as an input to the recursive algorithm previously described:

$s(0,0) = 1$

$s(1,0) = 1.3$

$s(2,0) = 0.44$

$s(3,0) = 0.032$

$s(0,1) = 1.3$

$s(1,1) = 1.47$

$s(2,1) = 0.368$

$s(3,1) = 0.0192$

$s(0,2) = 0.52$

$s(1,2) = 0.494$

$s(2,2) = 0.0624$

$s(0,3) = 0.06$

$s(1,3) = 0.048.$

This test signal was generated by convolving four signals of the form of (2). Consequently, the cepstrum will be the sum of four cepstra of the form of (3). The parameters of the four signals were

$s_1 - a = -0.4, b = -0.5$

$s_2 - a = -0.1, b = -0.2$

$s_3 - a = -0.8, b = 0$

$s_4 - a = 0, \quad b = -0.6.$

In Table III, the values of the true cepstrum, the cepstrum computed by the recursive method, and the cepstrum computed by the DFT method for both $N = 16$ and $N = 64$ are given for some typical points.

One difficulty with the recursive method of computation is that it is generally not known beforehand if a particular 2-D signal possesses the minimum-phase property. If the signal $s(m, n)$ with support on Q_1 is not a minimum-phase signal, then the cepstrum defined by (1) will not be supported on Q_1 alone. Consequently, the recursive computation cannot be applied to compute $\hat{s}(m, n)$. However, if we proceed blindly and use the recursive computation anyway, we will compute the unstable cepstrum $\hat{s}_u(m, n)$ defined below.

If the stable inverse defined in Section IV is used in (6), the cepstrum computed will be stable but nonzero outside of Q_1. It will be approximated by the $\hat{s}_d(m, n)$ derived in Section III. On the other hand, if the unstable inverse $v_u(m, n)$ defined by the recursion

$$v_u(m,n) = \frac{1}{s(0,0)}\left[u_0(m,n) - \sum_{\substack{k=0 \\ (k,l)\neq(m,n)}}^{m} \sum_{l=0}^{n} s(k,l)v(m-k, n-l)\right] \tag{13}$$

is used for $v(m, n)$ in (6), the cepstrum computed will be the unstable cepstrum $\hat{s}_u(m, n)$ and it will be equal to the cepstrum computed by the recursion relations (11), (12). (Note that $v_u(m, n)$ has support on Q_1 because of its derivation and the fact that $s(m, n)$ has support on Q_1.)

The difference between the stable and unstable cepstra computed in the case where $s(m, n)$ is not a minimum-phase signal is not unlike the situation where two 1-D signals satisfy the same difference equation, one being causal but unstable, the other being stable but not causal.

TABLE III

(m,n)	True $\hat{s}$ (m,n)	Recursive $\hat{s}$ (m,n)	DFT, N=16 $\hat{s}_d$ (m,n)	DFT, N=64 $\hat{s}_d$ (m,n)
(-1, 0)	0.		2.5431043×10^{-3}	4.1327439×10^{-9}
(0,-1)	0.		1.6846394×10^{-4}	$-7.7319993 \times 10^{-10}$
(0, 0)	0.	0.	$-1.9324345 \times 10^{-3}$	1.0652002×10^{-8}
(0, 1)	1.3	1.3	1.3001315	1.3000000
(0, 2)	-3.25×10^{-1}	-3.25×10^{-1}	$-3.2511696 \times 10^{-1}$	$-3.2500001 \times 10^{-1}$
(0, 5)	2.1866×10^{-2}	2.1866×10^{-2}	2.1953726×10^{-2}	2.1865988×10^{-2}
(1, 0)	1.3	1.3	1.3014496	1.3000001
(1, 1)	-2.2×10^{-1}	-2.2×10^{-1}	$-2.2009827 \times 10^{-1}$	$-2.2000001 \times 10^{-1}$
(1, 5)	-1.2532×10^{-2}	-1.2532×10^{-2}	$-1.2602926 \times 10^{-2}$	$-1.2531985 \times 10^{-2}$
(2, 0)	-4.05×10^{-1}	-4.05×10^{-1}	$-4.0611151 \times 10^{-1}$	$-4.0500002 \times 10^{-1}$
(2, 1)	8.2×10^{-2}	8.1999999×10^{-2}	8.2084349×10^{-2}	8.2000003×10^{-2}
(2, 5)	1.50096×10^{-2}	1.500959×10^{-2}	1.5068206×10^{-2}	1.5009591×10^{-2}
(3, 0)	$1.9233\overline{3} \times 10^{-1}$	1.9233333×10^{-1}	1.9320900×10^{-1}	1.9233334×10^{-1}
(3, 3)	$-2.6693\overline{3} \times 10^{-2}$	-2.669332×10^{-2}	$-2.6754379 \times 10^{-2}$	$-2.6693299 \times 10^{-2}$
(4, 4)	-1.40014×10^{-2}	-1.400138×10^{-2}	$-1.4051512 \times 10^{-2}$	$-1.4001368 \times 10^{-2}$
(5, 5)	$-8.06408064 \times 10^{-3}$	-8.064065×10^{-3}	$-8.1059900 \times 10^{-3}$	$-8.0640477 \times 10^{-3}$

An alternate way of thinking about the unstable cepstrum is to recall that in the derivation of the recursion relations, it was necessary to perform a 2-D inverse z transform. The contour of integration for the inverse z transform must be contained in the region of convergence. Different answers will result if different regions of convergence are assumed. If both the signal and its cepstrum are assumed stable, the region of convergence will include the contour $|w| = |z| = 1$. If the signal's inverse does not have support on Q_1 alone, then the resulting cepstrum will not have support on Q_1 alone. However, if the cepstrum is assumed to have support only on Q_1, then the contour $|w| = |z| = 1$ may not lie in the region of convergence and it will be necessary to choose some other contour of integration in order to perform the 2-D inverse z transform. In this case, the cepstrum will be unstable.

The blind use of the recursive algorithm may be regarded as a test for the minimum-phase property. In [11], the 2-D cepstrum computed by the DFT method is checked to see if it has support on Q_1 alone. Because of aliasing, it is sometimes difficult to say with certainty that the computed cepstrum is identically zero outside of the first quadrant. Similarly, the cepstrum could be computed by the recursive algorithm and checked for exponential growth as m and n increase. The problem of aliasing is circumvented, but the more difficult test for exponential growth is substituted for the test for zero.

Example 4: The nonminimum phase signal

$$s(0, 0) = 0.1$$

$$s(1, 0) = 1.0$$

$$s(1, 1) = 0.5$$

was used as input to the recursive cepstral computation. It is straightforward to show that the stable cepstrum corresponding to this input signal (after removal of the linear-phase component) is given analytically by (3) with $a = -0.1$ and $b = -0.5$, except that m is replaced by $-m$. These values were verified by using the DFT cepstral computation method.

The unstable inverse given by (13) can be written analytically as

$$v_u(m, n) = \frac{m!}{n!(m-n)!}(-1)^m (10)^{m+1} (0.5)^n, \quad m \geq 0, \ 0 \leq n \leq m$$
$$= 0, \quad \text{otherwise.}$$

Since the input signal and its unstable inverse are zero outside the 0°-45° sector, the unstable cepstrum given by the convolution (6) will also be zero outside of this sector. Analytically, the unstable cepstrum can be written as

$$\hat{s}_u(m, n) = \frac{(m-1)!}{n!(m-n)!} 10^m (-1)^{m-1} (0.5)^n, \quad m \geq 1, \quad 0 \leq n \leq m$$
$$= 0, \quad \text{otherwise.}$$

These values verified the values computed by the recursive method. Note the rapid growth of $\hat{s}_u(m, n)$ due to the factor 10^m.

VI. Sector Minimum-Phase Signals

The method of computation illustrated in Section V can be extended to minimum-phase signals whose regions of support are certain sectors in the (m, n) plane. To be specific, let us consider a 2-D signal of the following form: $s(m, n) = 0$ for $m < 0$ and for $n + mN < 0$ (n, m, and N are integers). We shall say such signals have sector support since they are zero outside of the sector bounded by the positive n axis and the ray extending from the origin with a slope of $-N$. Such signals can be reversibly mapped into the first quadrant by skewing the m axis. We can visualize pushing up each row of $s(m, n)$ until its nonzero part lies entirely in the first quadrant (as in Fig. 1) to achieve the transformation

$$s'(m', n') \triangleq s'(m, n + mN) = s(m, n)$$

and the relationships

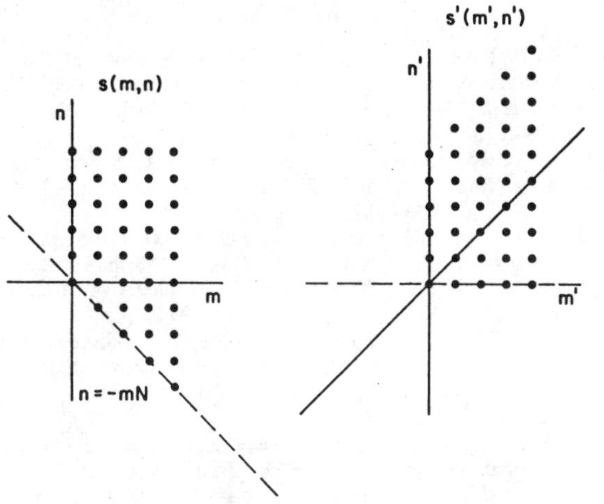

Fig. 1. Transformation to primed coordinate system.

$$m' = m \qquad m = m'$$
$$n' = n + mN \quad n = n' - m'N.$$

Consider the case where $s(m, n)$ is minimum phase and has sector support. Then its inverse $v(m, n)$ is also stable and has sector support. Using this transformation, we create a signal $s'(m', n')$ and its inverse $v'(m', n')$ which are stable and have support only in the first quadrant Q_1 of the (m', n') plane. (It is straightforward to verify that the v' created by transforming v is indeed the inverse of the s' created by transforming s.) The cepstrum $\hat{s}'(m', n')$ may now be computed by the recursive formulation of Section V, since $s'(m', n')$ is a minimum-phase signal with support only in the first quadrant of the (m', n') plane. We can apply the inverse transformation to $\hat{s}'(m', n')$ to obtain the cepstrum $\hat{s}(m, n)$. To verify that $\hat{s}(m, n)$ obtained this way is indeed the 2-D cepstrum of $s(m, n)$, we could rederive the recursion relations (11) and (12) for signals and cepstra which have sector support. Then the inverse transformation could be applied to the recursive relations (11) and (12) and the results shown to be identical to the rederived recursive relations. Alternatively, we can examine the 2-D z transforms of $\hat{s}'(m', n')$ and $\hat{s}(m, n)$. If we let $S(w, z)$ be the 2-D z transform of $s(m, n)$ and $S'(w, z)$ be the 2-D z transform of $s'(m', n')$, then

$$S'(w, z) = \sum_{m'} \sum_{n'} s'(m', n') w^{-m'} z^{-n'}$$
$$= \sum_{m} \sum_{n} s(m, n) w^{-m} z^{-(n+mN)}$$
$$= S(wz^N, z).$$

Taking complex logarithms we get

$$\hat{S}'(w, z) = \hat{S}(wz^N, z).$$

Taking the inverse z transforms yields

$$\hat{s}'(m', n') = \frac{1}{(2\pi j)^2} \oint\oint \hat{S}(wz^N, z) w^{m'-1} z^{n'-1} \, dw \, dz.$$

Now let $y = wz^N$

$$\hat{s}'(m', n') = \frac{1}{(2\pi j)^2} \oint\oint \hat{S}(y, z) y^{m'-1} z^{n'-m'N-1} \, dy \, dz$$
$$= \hat{s}(m', n' - m'N)$$
$$= \hat{s}(m, n)$$

as expected.

The reader will note that similar generalizations can be made by considering reflections in and rotations about the m axis and the n axis, but they offer nothing conceptually different than what has been discussed here. Also, by allowing N to take on negative values, we can create sectors which are smaller than a quadrant (that is, contained by an acute rather than an obtuse angle) without altering any of our derivations.

VII. Comparison of the DFT and Recursive Methods for Computing Two-Dimensional Cepstra

It is difficult to reach a conclusion as to which method of computing a 2-D cepstrum is to be preferred, the DFT method or the recursive method, since they differ in a variety of ways. For example, the DFT method requires many evaluations of the complex-logarithm function which may be performed by computation, by looking up values in a precomputed table, or by some combination of the two. Consequently, one cannot state that a complex-logarithm evaluation is, say 10 times costlier than a multiply without being more specific about how the complex-logarithm evaluation is to be done. The DFT method also requires, obviously, the evaluations of many DFT's. The cost in computation (and sometimes storage) can vary somewhat depending on what radix FFT's are used. Furthermore, new methods of computing DFT's may significantly reduce computation [10]. The logic to control the DFT method is more complicated than that needed to control the recursive method because of the need to do phase unwrapping. It is difficult, within the scope of this paper, to examine how much cost this control logic adds to each method.

The most significant differences between the two methods are their limitations. The recursive method works exactly (ignoring the limits imposed by finite wordlength) if the input signal is minimum phase with quadrant (or sector) support. If the input signal is not minimum phase, but has quadrant (or sector) support, the recursive method will yield the unstable cepstrum $\hat{s}_u(m, n)$. On the other hand, the input signal need not be minimum phase for the DFT method to work. The DFT method can compute and remove the linear-phase component and then proceed to compute its result, $\hat{s}_d(m, n)$. But $\hat{s}_d(m, n)$ is not equal to the desired $\hat{s}(m, n)$ even in theory; it is an aliased approximation. In some cases, as illustrated in the examples, the aliasing can introduce a significant error. To reduce the aliasing problem requires using larger DFT's and hence increasing the amount of computation done.

As a rough guide to the amount of computation, consider the following situation. We desire the N^2 cepstral points inside the square

$$0 \leq m \leq N - 1$$
$$0 \leq n \leq N - 1.$$

We assume that the complex signal $s(m, n)$ is zero outside of the square

$$0 \leqslant m \leqslant M - 1$$

$$0 \leqslant n \leqslant M - 1$$

and that M is significantly less than N, say M is 10 percent of N. Using the algorithm described in Section III and using Tribolet's algorithm [7] to do the phase unwrapping, we find that we need to do $4N + M + 2$ N-point FFT's plus $N^2 + N$ complex-logarithm evaluations. The number of real multiplies in an N-point FFT is roughly proportional to $2N \log_2 N$, so that the number of real multiplies in the DFT method of computing the cepstrum is proportional to $8N^2 \log_2 N$ (roughly) with complex-logarithm evaluations going up as $N^2 + N$.

Using the recursive method and taking advantage of the knowledge that $s(m, n)$ is zero outside of the $M \times M$ square, the recursive method needs $M^2 - 1$ complex and 2 real multiplies to compute a cepstral point. We desire N^2 such points, so the number of real multiplies for the recursive method goes up as $N^2(4M^2 - 2)$. (Because of edge effects, this estimate is somewhat pessimistic, but it indicates the order of the dependence of the number of multiplies on M and N.)

References

[1] B. P. Bogert, M. J. R. Healy, and J. W. Tukey, "The quefrency alanysis of time series for echoes," in *Proc. Symp. Time Series Analysis*, M. Rosenblatt, Ed. New York: Wiley, 1963, pp. 209-243.

[2] A. V. Oppenheim, R. W. Schafer, and T. G. Stockham, Jr., "Nonlinear filtering of multiplied and convolved signals," *Proc. IEEE*, vol. 56, pp. 1264-1291, Aug. 1968.

[3] A. V. Oppenheim and R. W. Schafer, *Digital Signal Processing*. Englewood Cliffs, NJ: Prentice-Hall, 1975, pp. 480-531.

[4] D. E. Dudgeon, "Two-dimensional recursive filtering," Sc.D. thesis, Dep. Elec. Eng., M.I.T., May 1974.

[5] M. P. Ekstrom and J. W. Woods, "Two-dimensional spectral factorization with applications in recursive digital filtering," *IEEE Trans. Acoust., Speech, Signal Processing*, vol. ASSP-24, pp. 115-128, Apr. 1976.

[6] D. E. Dudgeon, "The existence of cepstra for two-dimensional rational polynomials," *IEEE Trans. Acoust., Speech, Signal Processing*, vol. ASSP-23, pp. 242-243, Apr. 1975.

[7] J. M. Tribolet, "A new phase unwrapping algorithm," *IEEE Trans. Acoust., Speech, Signal Processing*, vol. ASSP-25, pp. 170-177, Apr. 1977.

[8] N. Brenner, "Three Fortran programs that perform the Cooley-Tukey Fourier transform," Lincoln Lab. Tech. Rep. 1967-2, July 28, 1967.

[9] P. Pistor, "Stability criterion for recursive filters," *IBM. Res. Develop.*, vol. 18, pp. 59-71, Jan. 1974.

[10] S. Winograd, "On computing the discrete Fourier transform," *Proc. Nat. Acad. Sci. U.S.*, vol. 73, pp. 1005-1006, Apr. 1976.

[11] M. P. Ekstrom and R. E. Twogood, "A stability test for 2-D recursive digital filters using the complex cepstrum," in *Proc. 1977 IEEE Int. Conf. Acoustics, Speech, and Signal Processing*, pp. 535-538.

The Processing of Periodically Sampled Multidimensional Signals

RUSSELL M. MERSEREAU, FELLOW, IEEE, AND THERESA C. SPEAKE

Abstract—This paper discusses algorithms for processing multidimensional signals which are sampled on regular, but nonrectangular sampling lattices. Such sampling lattices are dictated by some applications and may be chosen for others because of their resulting symmetric responses or computational efficiencies. We show that any operation which can be performed on a rectangular lattice can be performed on any regular periodic lattice, including FIR and IIR filtering, DFT calculation, and decimation and interpolation. This paper also discusses how generalized decimators and interpolators can be used to convert from one sampling lattice to another.

I. Introduction

THERE are many generalizations of one-dimensional (1-D) periodic sampling which permit the exact representation of a band-limited continuous function of several independent variables. By far, the most common of these is rectangular sampling, which corresponds to evaluating the function at sample locations that form a regular hypercubic lattice. Although there are a number of good reasons for using rectangular sampling to represent band-limited signals, the freedom to choose an alternative sampling lattice is not one that should be abandoned casually. It was shown in [1] that one type of nonrectangular sampling, hexagonal sampling and its higher dimensional generalizations, resulted in a lower sampling density and more efficient signal processing algorithms than rectangular sampling for isotropically band-limited signals. Nonrectangular systems can be designed to exhibit more and different symmetries than their rectangular counterparts.

There are applications for which nonrectangular sampling may be more suited to a problem than rectangular sampling. Phased array antennas, for example, are typically designed with a hexagonal arrangement of elements because of the desire for a highly symmetric response and because such an arrangement requires fewer elements. One procedure for testing these antennas requires the measurement of samples of the electric field on a plane in the near field of the antenna. Because of the geometry of the phased array, the resulting processing is more accurate if hexagonal sampling is used for these measurements. Similar results might be expected in measuring fields in or near crystals and processing the results. In this case the sampling lattice should be related to the crystal lattice.

The purpose of this paper is to show that most common signal processing operations, such as convolution, recursive and nonrecursive filtering, decimation and interpolation, Fourier analysis, and discrete Fourier analysis can be performed on any regular (i.e., periodic) sampling lattice and to derive these algorithms. By retaining sampling information in the formulation of the signal processing algorithms, we have a means by which to weigh sampling alternatives. On occasion, this even leads to different algorithms for rectangularly sampled signals. An example of this is shown in the development of very

Manuscript received February 24, 1982; revised July 13, 1982. This work was supported in part by the National Science Foundation under Grant ECS-7817201 and the Joint Services Electronics Program under Contract DAAG29-81-K-0024.

R. M. Mersereau is with the School of Electrical Engineering, Georgia Institute of Technology, Atlanta, GA 30332.

T. C. Speake is with the Department of Electrical Engineering Technology, Southern Technical Institute, Marietta, GA, 30060.

Reprinted from *IEEE Trans. Acoust., Speech, Signal Processing,* vol. ASSP-31, pp. 188–194, Feb. 1983.

general Cooley-Tukey algorithms in [2]. This paper should logically have preceded [2], since the DFT algorithm which is derived in this paper is evaluated using the generalized Cooley-Tukey algorithm in [2]. This current paper and [2] taken together are a generalization of the algorithms for hexagonal sampling which are presented in [1].

This paper will begin by discussing multidimensional sampling itself. While most of this material appears in the key paper by Petersen and Middleton [3], a review of the major results is unavoidable since the sampling operation plays such a central role in what is to follow. This review will also allow us to establish a matrix notation that will not only simplify our mathematical expressions, but will also make the analogy with the 1-*D* case clearer. The following section will concern linear shift-invariant systems and will discuss convolution, difference equations, and Fourier transforms. Section IV will discuss generalized discrete Fourier transforms and discrete spectral analysis. Finally, in the last section we will discuss the problems of decimation and interpolation on multidimensional sampling lattices. By using general decimators and interpolators we can solve the problem of interpolating from one lattice (e.g., a 3-*D* body-centered cubic lattice) to another (e.g., a 3-*D* cubic lattice). The approach to this problem discussed in this paper represents a better solution to this problem than the algorithm discussed in [4].

II. Multidimensional Periodic Sampling

One-dimensional periodic sampling of a band-limited signal $x_a(t)$ corresponds to forming a sequence of numbers by evaluating $x_a(t)$ at equispaced values of its argument. Thus, if $x(n)$ is used to denote the sequence of values, we form

$$x(n) = x_a(nT) \tag{1}$$

for all integer values of n. (The subscript "a" in (1) denotes the fact that $x_a(t)$ is an analog waveform.) The parameter T is called the *sampling period* and $1/T$ is called the *sampling rate*. In extending this concept to permit the sampling of *D*-dimensional signals, the set of equispaced sample locations becomes a *D*-dimensional lattice.

A *D*-dimensional lattice is formed by taking all integer linear combinations of a set of D linearly independent (column) vectors $\{\boldsymbol{v}_1, \boldsymbol{v}_2, \cdots, \boldsymbol{v}_D\}$. An example of a 2-*D* lattice is shown in Fig. 1. Collectively, these vectors are known as the *basis* of the lattice. The *sampling matrix*, $\boldsymbol{V}$, is a $D \times D$ matrix whose columns form the basis. Thus,

$$\boldsymbol{V} = [\boldsymbol{v}_1 \,\vdots\, \boldsymbol{v}_2 \,\vdots\, \cdots \,\vdots\, \boldsymbol{v}_D]. \tag{2}$$

Each location in the sampling lattice can then be expressed as

$$\boldsymbol{t} = \boldsymbol{V}\boldsymbol{n} \tag{3}$$

where $\boldsymbol{n}$ is a vector with integer entries. The most common sampling lattice is the rectangular one, for which $\boldsymbol{V}$ is diagonal.

For a given lattice neither the basis nor the sampling matrix is unique. If $\boldsymbol{E}$ is a $D \times D$ matrix of integers such that $|\det \boldsymbol{E}| = 1$ (such a matrix is called *unimodular*), then $\hat{\boldsymbol{V}}$, given by

$$\hat{\boldsymbol{V}} = \boldsymbol{E}\boldsymbol{V} \tag{4}$$

and $\boldsymbol{V}$ define the same lattice. The quantity $|\det \boldsymbol{V}|$, however, is unique for a given lattice and it physically corresponds to the reciprocal of the sampling density (sampling rate) [4].

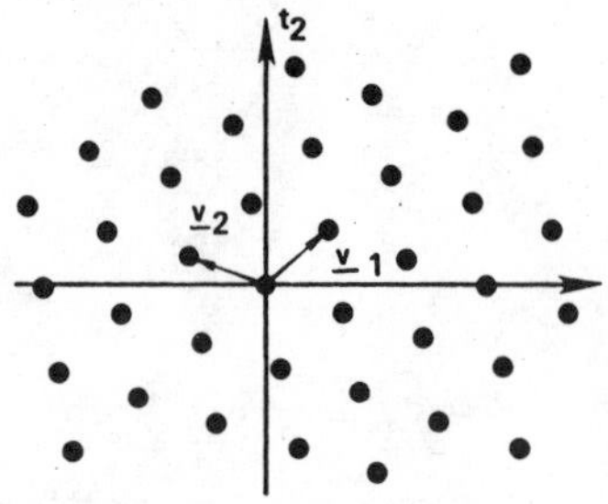

Fig. 1. A two-dimensional sampling lattice.

If $x_a(\boldsymbol{t})$ is sampled on the lattice which is defined by (3), the sequence of samples $x(\boldsymbol{n})$ is given by

$$x(\boldsymbol{n}) = x_a(\boldsymbol{V}\boldsymbol{n}) \tag{5}$$

which bears an obvious resemblance to (1).

Recall that in the 1-*D* case, if $x(n) = x_a(nT)$, then

$$X(\omega) = \frac{1}{T} \sum_{r=-\infty}^{\infty} X_a\left(\omega - \frac{2\pi r}{T}\right) \tag{6}$$

which says that the spectrum of the sequence is an aliased version of the spectrum of $x_a(t)$. If $X_a(\omega) \equiv 0$ for $|\omega| \geqslant \pi/T$, $X_a(\omega)$ can be recovered exactly from $X(\omega)$. An analogous situation occurs in the *D*-dimensional case. Here, however, the degree of aliasing depends not only on the sampling density, but also upon the geometry of the sampling lattice.

Let the *D*-dimensional Fourier transform of $x_a(\boldsymbol{t})$ be defined as

$$X_a(\boldsymbol{\omega}) = \int_{-\infty}^{\infty} x_a(\boldsymbol{t}) \exp\,[-j\boldsymbol{\omega}^T \boldsymbol{t}]\; d\boldsymbol{t} \tag{7}$$

where $\boldsymbol{\omega}^T$ denotes the transpose of $\boldsymbol{\omega}$, and where the integral is evaluated over all of $\boldsymbol{t}$-space. Similarly, let the Fourier transform of the sequence $x(\boldsymbol{n})$ be defined as

$$X(\boldsymbol{\omega}) = \sum_{\boldsymbol{n}=-\infty}^{\infty} x(\boldsymbol{n}) \exp\,[-j\boldsymbol{\omega}^T \boldsymbol{V}\boldsymbol{n}]. \tag{8}$$

Then, if $x(\boldsymbol{n}) = x_a(\boldsymbol{V}\boldsymbol{n})$, it can be shown using techniques enumerated in [3] that

$$X(\boldsymbol{\omega}) = \frac{1}{|\det \boldsymbol{V}|} \sum_{\boldsymbol{r}=-\infty}^{\infty} X_a(\boldsymbol{\omega} - \boldsymbol{U}\boldsymbol{r}) \tag{9}$$

where

$$\boldsymbol{U}^T \boldsymbol{V} = 2\pi \boldsymbol{I} \tag{10}$$

and $\boldsymbol{I}$ is the $D \times D$ identity matrix. An aliased spectrum is depicted in Fig. 2. In the area of mathematics known as the geometry of numbers, the lattice formed by $\boldsymbol{U}$ is known as the *polar lattice* [4]. It is also known as the *reciprocal lattice*, a term we prefer. The matrix $\boldsymbol{U}$ will be called the *aliasing matrix*.

$X_a(\boldsymbol{\omega})$ is *band-limited* with a spectrum limited to a region W of Fourier space if

$$X_a(\boldsymbol{\omega}) = 0 \qquad \boldsymbol{\omega} \notin W. \tag{11}$$

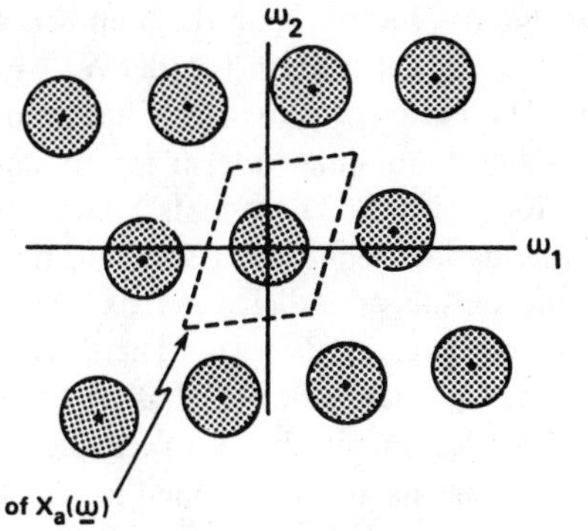

Fig. 2. An aliased spectrum resulting from the sampling of a band-limited signal on a periodic lattice.

Band-limited signals can sometimes be recovered exactly from their sample values. Such an exact recovery is possible if the functions $X_a(\boldsymbol{\omega} - \boldsymbol{Ur})$, for all integer vectors $\boldsymbol{r}$, do not possess overlapping regions of support. Whether or not this condition is met depends both upon the shape of W and on the geometry of the reciprocal lattice. If, however, this condition is met, then

$$X_a(\boldsymbol{\omega}) = \begin{cases} |\det \boldsymbol{V}|\, X(\boldsymbol{\omega}), & \boldsymbol{\omega} \in W \\ 0, & \text{otherwise} \end{cases} \tag{12}$$

and

$$x_a(\boldsymbol{t}) = \sum_{\boldsymbol{n}} x_a(\boldsymbol{Vn})\, \phi(\boldsymbol{t} - \boldsymbol{Vn}) \tag{13}$$

$$\phi(\boldsymbol{t}) = \frac{1}{(2\pi)^D} \int_W \exp\,[-j\boldsymbol{\omega}^T \boldsymbol{t}]\, d\boldsymbol{\omega}. \tag{14}$$

Two special 2-D cases are worthy of mention at this point. The first is rectangular sampling, for which

$$\boldsymbol{V}_R = \begin{bmatrix} T_1 & 0 \\ 0 & T_2 \end{bmatrix} \tag{15}$$

and

$$\boldsymbol{U}_R = \begin{bmatrix} \frac{2\pi}{T_1} & 0 \\ 0 & \frac{2\pi}{T_2} \end{bmatrix}. \tag{16}$$

In this case both the sampling and reciprocal lattices are rectangular. The second special case corresponds to hexagonal sampling, for which

$$\boldsymbol{V}_H = \begin{bmatrix} \frac{T_1}{2} & \frac{T_1}{2} \\ T_2 & -T_2 \end{bmatrix} \tag{17}$$

$$\boldsymbol{U}_H = \begin{bmatrix} \frac{2\pi}{T_1} & \frac{2\pi}{T_1} \\ \frac{\pi}{T_2} & -\frac{\pi}{T_2} \end{bmatrix}. \tag{18}$$

III. Linear Shift-Invariant Systems

Consider a D-dimensional linear shift-invariant system with input array $x(\boldsymbol{n})$ and output array $y(\boldsymbol{n})$, where both arrays are defined over identical lattices. The input and output arrays are then related by the convolution sum

$$y(\boldsymbol{n}) = \sum_{\boldsymbol{k}} x(\boldsymbol{k})\, h(\boldsymbol{n} - \boldsymbol{k}) \tag{19}$$

where $h(\boldsymbol{n})$, the *impulse response*, is the response of the system to the signal

$$\delta(\boldsymbol{n}) = \begin{cases} 1, & \boldsymbol{n} = \boldsymbol{0} \\ 0, & \boldsymbol{n} \neq \boldsymbol{0} \end{cases}. \tag{20}$$

It should be observed that the form of the convolution sum is independent of the sampling matrix $\boldsymbol{V}$.

A. Frequency Response

If we let the input to a linear shift-invariant system be a sampled complex sinusoid of the form

$$x(\boldsymbol{n}) = \exp\,[j\boldsymbol{\omega}^T \boldsymbol{Vn}] \tag{21}$$

then, using (19), the system output is seen to be

$$y(\boldsymbol{n}) = \exp\,[j\boldsymbol{\omega}^T \boldsymbol{Vn}] \sum_{\boldsymbol{k}} h(\boldsymbol{k}) \exp\,[-j\boldsymbol{\omega}^T \boldsymbol{Vn}]. \tag{22}$$

Sampled complex sinusoids of the form of (20) are thus eigenfunctions of periodically sampled linear shift-invariant systems. This leads us to define the *frequency response* of an LSI system as the corresponding eigenvalue

$$H(\boldsymbol{\omega}) = \sum_{\boldsymbol{n}} h(\boldsymbol{n}) \exp\,[-j\boldsymbol{\omega}^T \boldsymbol{Vn}]. \tag{23}$$

The frequency response $H(\boldsymbol{\omega})$ is a periodic function of $\boldsymbol{\omega}$ with periodicity matrix $\boldsymbol{U}$, where $\boldsymbol{U}^T \boldsymbol{V} = 2\pi \boldsymbol{I}$. By this we mean that

$$H(\boldsymbol{\omega}) = H(\boldsymbol{\omega} + \boldsymbol{Ur}) \tag{24}$$

for any integer vector $\boldsymbol{r}$.

Let I_H denote any period of $H(\boldsymbol{\omega})$. This period has a volume of $|\det \boldsymbol{U}|$. The frequency response can then be inverted by performing the integral

$$h(\boldsymbol{n}) = \frac{1}{|\det \boldsymbol{U}|} \int_{I_H} H(\boldsymbol{\omega}) \exp\,[j\boldsymbol{\omega}^T \boldsymbol{Vn}]\, d\boldsymbol{\omega}. \tag{25}$$

To verify the validity of this relation we can substitute (23) into (25) and exploit the periodicity of the exponentials.

B. The Fourier Transform

We have already seen that if the Fourier transform of a sequence $x(\boldsymbol{n})$ is defined as

$$X(\boldsymbol{\omega}) = \sum_{\boldsymbol{n}=-\infty}^{\infty} x(\boldsymbol{n}) \exp\,[-j\boldsymbol{\omega}^T \boldsymbol{Vn}] \tag{26}$$

then the Fourier transform of a sequence and the Fourier transform of the band-limited signal from which it is derived are simply related by aliasing. This definition is consistent with the one-dimensional one. Since the frequency response is the Fourier transform of the impulse response, like the frequency response, the Fourier transform $X(\boldsymbol{\omega})$ is periodic with periodicity matrix $\boldsymbol{U} = 2\pi(\boldsymbol{V}^T)^{-1}$ and it can be inverted by using (25).

The general Fourier transform has properties which are identical to the corresponding properties of 1-*D* Fourier transforms, except for the fine details of the resulting expression. In particular, the transform is linear and the convolution theorem holds. Thus, if $x(\boldsymbol{n})$ is the input to a linear shift-invariant system and $y(\boldsymbol{n})$ is the corresponding output, we can write

$$Y(\boldsymbol{\omega}) = X(\boldsymbol{\omega})\, H(\boldsymbol{\omega}). \tag{27}$$

Two Fourier transform properties which look slightly different are the shift-property and Parseval's relation. The former states that

$$F[x(\boldsymbol{n} - \boldsymbol{n}_0)] = \exp\,[-j\boldsymbol{\omega}^T \boldsymbol{V}\boldsymbol{n}_0]\; X(\boldsymbol{\omega}) \tag{28}$$

where $F[\cdot]$ denotes the Fourier transform operator. Parseval's relation can be written as

$$\sum_{\boldsymbol{n}} x(\boldsymbol{n})\, y^*(\boldsymbol{n}) = \frac{1}{|\det \boldsymbol{U}|} \int_{I_V} X(\boldsymbol{\omega})\, Y^*(\boldsymbol{\omega})\, d\omega. \tag{29}$$

where I_V is one period of $X(\boldsymbol{\omega})\, Y^*(\boldsymbol{\omega})$.

C. Difference Equations

A periodically sampled LSI system can be defined implicitly by means of a linear constant coefficient difference equation of the form

$$\sum_{\boldsymbol{k}} b(\boldsymbol{k})\, y(\boldsymbol{n} - \boldsymbol{k}) = \sum_{\boldsymbol{k}} a(\boldsymbol{k})\, x(\boldsymbol{n} - \boldsymbol{k}). \tag{30}$$

where the sums each involve only a finite number of terms. Since the form of this difference equation is completely independent of the sampling matrix $\boldsymbol{V}$, the same hardware or software realization of the filter can be used whether the input and output arrays are rectangularly sampled, hexagonally sampled, or whether a more general sampling matrix is used. The stability of the difference equation is also independent of $\boldsymbol{V}$.

IV. Discrete Fourier Transforms (DFT)

The DFT is an exact Fourier representation for periodically sampled arrays with a finite number of nonzero samples. It assumes the form of a periodically sampled Fourier transform. As in the 1-*D* case, the DFT can be interpreted as a Fourier series representation for one period of a periodic sequence. It is thus easiest if we begin with a discussion of periodic sequences.

A sequence $\tilde{x}(\boldsymbol{n})$ is periodic if it satisfies the relation

$$\tilde{x}(\boldsymbol{n} + \boldsymbol{N}\boldsymbol{r}) = \tilde{x}(\boldsymbol{n})$$
$$\det \boldsymbol{N} \neq 0 \tag{31}$$

for all integer vectors $\boldsymbol{n}$ and $\boldsymbol{r}$. The integer matrix $\boldsymbol{N}$ is called the *periodicity matrix* and the number of samples in one period of the signal is given by $|\det \boldsymbol{N}|$. For a given periodic sequence the periodicity matrix is not unique; it can be multiplied by any unimodular integer matrix and still describe the same periodic signal. These facts follow by analogy from the corresponding facts concerning sampling matrices.

The role of the periodicity matrix can be explained somewhat by reference to Fig. 3. There we see three different periodic signals each of which has as its fundamental period an $N_1 \times N_2$ rectangular block of samples. What distinguishes these three signals is the manner in which these blocks are abutted together. The columns of the periodicity matrix are vectors which denote the displacement from a sample on one period to the corresponding sample on another period. Thus, for the rectangular arrangement in Fig. 3(a) an acceptable periodicity matrix is

$$\boldsymbol{N} = \begin{bmatrix} N_1 & 0 \\ 0 & N_2 \end{bmatrix} \tag{32}$$

For the hexagonal arrangement in Fig. 3(b) one possible periodicity matrix is

$$\boldsymbol{N} = \begin{bmatrix} N_1 & N_2/2 \\ N_1 & -N_2/2 \end{bmatrix} \tag{33}$$

(where N_2 is assumed to be divisible by 2) and for the final arrangement, a periodicity matrix is given by

$$\boldsymbol{N} = \begin{bmatrix} N_1 & -1 \\ 0 & N_2 \end{bmatrix}. \tag{34}$$

Any periodic sequence $\tilde{x}(n)$ can be represented as a sum of harmonically related complex exponentials. To see this we can first take the Fourier transform of both sides of (31) using the shift property (28). This yields

$$\tilde{X}(\boldsymbol{\omega}) = \exp\,[-j\boldsymbol{\omega}^T \boldsymbol{V}\boldsymbol{N}\boldsymbol{r}]\; \tilde{X}(\boldsymbol{\omega}). \tag{35}$$

Thus, for all frequencies at which $\tilde{X}(\boldsymbol{\omega}) \neq 0$ we have

$$\exp\,[-j\boldsymbol{\omega}^T \boldsymbol{V}\boldsymbol{N}\boldsymbol{r}] = 1 \tag{36}$$

or

$$\boldsymbol{\omega}^T = 2\pi \boldsymbol{k}^T (\boldsymbol{V}\boldsymbol{N})^{-1} \tag{37}$$

where $\boldsymbol{k}$ is a vector of integers. This says that the only frequency components that a periodic array can have are harmonically related. Thus, we can write

$$\tilde{x}(\boldsymbol{n}) = \sum_{\boldsymbol{k} \in J_N} \tilde{X}(\boldsymbol{k}) \exp\,[j(2\pi \boldsymbol{k}^T (\boldsymbol{V}\boldsymbol{N})^{-1}\, \boldsymbol{V}\boldsymbol{n})] \tag{38}$$

$$= \sum_{\boldsymbol{k} \in J_N} \tilde{X}(\boldsymbol{k}) \exp\,[j(2\pi \boldsymbol{k}^T \boldsymbol{N}^{-1} \boldsymbol{n})]. \tag{39}$$

The exponential $\exp\,[j(2\pi \boldsymbol{k}^T \boldsymbol{N}^{-1} \boldsymbol{n})]$ is periodic in $\boldsymbol{n}$ with periodicity matrix $\boldsymbol{N}$ and is periodic in $\boldsymbol{k}$ with periodicity matrix $\boldsymbol{N}^T$. As a consequence, these exponentials are distinct for only $|\det \boldsymbol{N}|$ values of $\boldsymbol{k}$. It is sufficient, therefore, to limit the sum to a region J_N which contains one period of exp $[j(2\pi \boldsymbol{k}^T \boldsymbol{N}^{-1} \boldsymbol{n})]$ considered as a function of $\boldsymbol{k}$.

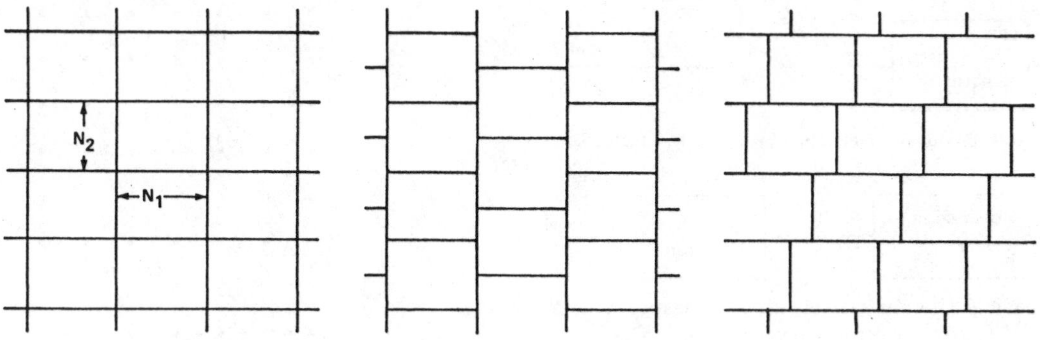

Fig. 3. Three different periodic signals which have the same fundamental period.

The Fourier series coefficients $\tilde{X}(\boldsymbol{k})$ can be evaluated using the relation

$$\tilde{X}(\boldsymbol{k}) = \frac{1}{|\det \boldsymbol{N}|} \sum_{\boldsymbol{n} \in I_{\boldsymbol{N}}} \tilde{x}(\boldsymbol{n}) \exp\left[-j\boldsymbol{k}^T(2\pi \boldsymbol{N}^{-1})\,\boldsymbol{n}\right] \tag{40}$$

where $I_{\boldsymbol{N}}$ denotes one period of $\tilde{x}(n)$.

If we now consider $x(\boldsymbol{n})$ to be a sequence with finite support limited to $I_{\boldsymbol{N}}$, $\tilde{x}(\boldsymbol{n})$ can be considered to be its periodic extension. By varying $\boldsymbol{N}$ we vary the manner by which the periodic extension is formed. This results in the discrete Fourier transform for periodically sampled signals.

$$X(\boldsymbol{k}) = \sum_{\boldsymbol{n} \in I_{\boldsymbol{N}}} x(\boldsymbol{n}) \exp\left[-j\boldsymbol{k}^T(2\pi \boldsymbol{N}^{-1})\,\boldsymbol{n}\right] \tag{41}$$

$$x(\boldsymbol{n}) = \frac{1}{|\det \boldsymbol{N}|} \sum_{\boldsymbol{k} \in J_{\boldsymbol{N}}} X(\boldsymbol{k}) \exp\left[j\boldsymbol{k}^T(2\pi \boldsymbol{N}^{-1})\,\boldsymbol{n}\right]. \tag{42}$$

In keeping with standard usage we have moved the normalization constant from the forward to the inverse transform. It should be noted that this reduces to the normal DFT in the 1-D case and to the familiar rectangular multidimensional DFT when $\boldsymbol{N}$ is diagonal.

The numbers $X(\boldsymbol{k})$ can be interpreted as samples of the Fourier transform of $x(\boldsymbol{n})$. We can see this by comparing (26) with (41). The DFT corresponds to evaluating the Fourier transform at these values of $\boldsymbol{\omega}$ for which

$$\boldsymbol{k}^T(2\pi \boldsymbol{N}^{-1}) = \boldsymbol{\omega}^T \boldsymbol{V}$$

or

$$\boldsymbol{\omega} = (\boldsymbol{V}^{-1})^T(2\pi \boldsymbol{N}^{-1})^T \boldsymbol{k}$$
$$= \boldsymbol{U}(\boldsymbol{N}^{-1})^T \boldsymbol{k}. \tag{43}$$

Thus, the matrix $\boldsymbol{R} = \boldsymbol{U}(\boldsymbol{N}^{-1})^T$ serves as a Fourier domain sampling matrix. This provides one method for choosing the periodicity matrix $\boldsymbol{N}$. It must contain integer entries and it must be consistent with the region of support of $x(\boldsymbol{n})$. Apart from these conditions, $\boldsymbol{N}$ can be chosen to place the DFT samples where desired.

As an illustration of this fact, we might consider the three periodicity matrices

$$\boldsymbol{N}_1 = \begin{bmatrix} N_1 & 0 \\ 0 & N_2 \end{bmatrix} \tag{44}$$

$$\boldsymbol{N}_2 = \begin{bmatrix} N_1 & N_1 \\ N_2 & -N_2 \end{bmatrix} \tag{45}$$

$$\boldsymbol{N}_3 = \begin{bmatrix} N_1 & N_2 \\ N_2 & N_1 \end{bmatrix}. \tag{46}$$

$\boldsymbol{N}_1$ transforms a rectangularly sampled $N_1 \times N_2$ 2-D array into rectangular samples of its Fourier transform. It thus represents the traditional DFT. $\boldsymbol{N}_3$ transforms a hexagonally sampled sequence into hexagonal samples of its Fourier transform. This transform was discussed extensively in [1]. The middle periodicity matrix $\boldsymbol{N}_2$ is a hybrid. It transforms a rectangularly sampled sequence into hexagonal samples of its Fourier transform.

Almost any algorithm for evaluating the 1-D DFT can be generalized to permit the evaluation of the DFT in (41). In [1] the Cooley-Tukey algorithm is generalized. The Cooley-Tukey algorithm exploits structure which is present in the 1-D DFT when N, the length of the 1-D DFT is a highly composite integer. A similar exploitable structure exists in the multidimensional case when $\boldsymbol{N}$, the periodicity matrix can be factored into nontrivial integer matrix factors.

V. Decimation and Interpolation

The problems of decimation and interpolation for periodically sampled multidimensional sequences are similar in some respects to their one-dimensional counterparts [6], but their description is hampered by the fact that multidimensional bandwidths have shapes as well as size and by our desire to be completely general. In addition, in the multidimensional case, decimators and interpolators affect not only the sampling rate, but also the geometry of the sampling lattice. We will demonstrate this at the end of this section with a specific example which converts from rectangular to hexagonal samples.

We can begin by considering the problem of decimation. Let $x_a(\boldsymbol{t})$ be a band-limited analog signal which has been sampled using the sampling matrix $\boldsymbol{V}$ to produce the sequence $x(\boldsymbol{n}) = x_a(\boldsymbol{V}\boldsymbol{n})$. Let I_X denote the region of support of the band-limited signal and let I_V denote one period of the Fourier transform of the sequence. If the conditions of the sampling theorem have been satisfied, then $I_X \subseteq I_V$ and

$$X(\boldsymbol{\omega}) = \frac{1}{|\det \boldsymbol{V}|} \sum_{\boldsymbol{r}=-\infty}^{\infty} X_a(\boldsymbol{\omega} - \boldsymbol{U}\boldsymbol{r}). \tag{47}$$

Now let $y(\boldsymbol{m}) = x(\boldsymbol{D}\boldsymbol{m})$ be defined, which is a downsampled version of $x(\boldsymbol{n})$. The *downsampling matrix* $\boldsymbol{D}$ is an integer matrix with $\det \boldsymbol{D} \neq 0$. Since we can also write $y(\boldsymbol{m}) = x_a(\boldsymbol{V}\boldsymbol{D}\boldsymbol{m})$, relating $y(\boldsymbol{m})$ to the original analog signal $x_a(\boldsymbol{t})$, it follows that

$$Y(\boldsymbol{\omega}) = \frac{1}{|\det \boldsymbol{V}| \cdot |\det \boldsymbol{D}|} \sum_{\boldsymbol{r}=-\infty}^{\infty} X_a(\boldsymbol{\omega} - (\boldsymbol{D}^T)^{-1} \boldsymbol{U}\boldsymbol{r}). \tag{48}$$

If $\boldsymbol{D}$ is unimodular, then $\boldsymbol{V}$ and $\boldsymbol{V}\boldsymbol{D}$ define the same sampling

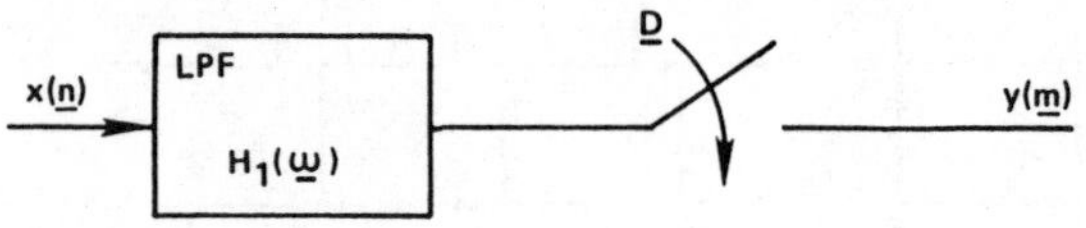

Fig. 4. A decimator for decimating by the matrix factor $\boldsymbol{D}$.

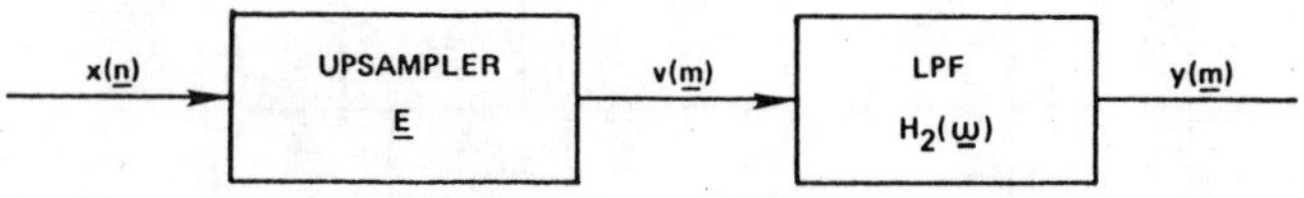

Fig. 5. An interpolator for interpolating by the matrix factor $\boldsymbol{E}$.

lattice. This represents a degenerate case in which decimation becomes simply a reordering of the input samples. In the more interesting case $|\det \boldsymbol{D}| \geqslant 2$. Let I_{VD} denote the region containing one period of $Y(\boldsymbol{\omega})$. It is clear that $I_{VD} \subseteq I_V$. In fact, the volume of I_{VD} will be smaller than the volume of I_V by a factor of $|\det \boldsymbol{D}|$. If $x_a(\boldsymbol{t})$ is recoverable from $y(\boldsymbol{m})$ then we must also have $I_X \subseteq I_{VD}$. This leads to the structure for a decimator which is shown in Fig. 4. The signal $x(\boldsymbol{n})$ is passed through a low-pass filter whose passband is confined to I_{VD} and is then downsampled by the decimation matrix $\boldsymbol{D}$. The frequency response of the low-pass filter is given by

$$H(\boldsymbol{\omega}) = \begin{cases} \dfrac{1}{|\det \boldsymbol{D}|}, & \boldsymbol{\omega} \in I_{VD} \\ 0, & \text{otherwise.} \end{cases} \tag{49}$$

Interpolation is the reverse problem by which we try to increase the number of samples taken from an underlying band-limited analog signal by filtering. Let $\boldsymbol{E}$ be a matrix of integers, known as the *interpolation matrix*, such that

$$x(\boldsymbol{n}) = y(\boldsymbol{E}\boldsymbol{n}). \tag{50}$$

Here $x(\boldsymbol{n})$ is the original low-rate signal and $y(\boldsymbol{n})$ is the higher rate interpolated signal. Let V be the sampling matrix for $y(\boldsymbol{m})$; then $\boldsymbol{E}\boldsymbol{V}$ is the sampling matrix for $x(\boldsymbol{n})$. From our earlier results it then follows that $X(\boldsymbol{\omega})$ is periodic with periodicity matrix $2\pi(\boldsymbol{E}\boldsymbol{V})^{T^{-1}}$ and that $Y(\boldsymbol{\omega})$ is periodic with periodicity matrix $2\pi(\boldsymbol{V}^T)^{-1}$. If I_{EV} and I_V denote one period of the respective Fourier transforms, then

$$I_X \subseteq I_{EV} \subseteq I_V.$$

The regions of support for $X(\boldsymbol{\omega})$ and $Y(\boldsymbol{\omega})$ are both I_X which is a subset of their respective periods.

Interpolation can be performed using the two-step procedure which is illustrated in Fig. 5. The sequence $x(\boldsymbol{n})$ is first passed through an up-sampler whose operation is described by the rule

$$v(\boldsymbol{m}) = \begin{cases} x(\boldsymbol{n}), & \boldsymbol{m} = \boldsymbol{E}\boldsymbol{n} \\ 0, & \text{otherwise.} \end{cases} \tag{51}$$

The signal $v(\boldsymbol{m})$ has the proper lattice geometry, but since

$$V(\boldsymbol{\omega}) = \sum_{\boldsymbol{m}} v(\boldsymbol{m})\, e^{-j\boldsymbol{\omega}^T \boldsymbol{V}\boldsymbol{m}} \tag{52}$$

$$= \sum_{\boldsymbol{n}} v(\boldsymbol{E}\boldsymbol{n})\, e^{-j\boldsymbol{\omega}^T \boldsymbol{V}\boldsymbol{E}\boldsymbol{n}}$$

$$= \sum_{\boldsymbol{n}} x(\boldsymbol{n})\, e^{-j\boldsymbol{\omega}^T \boldsymbol{V}\boldsymbol{E}\boldsymbol{n}} = X(\boldsymbol{\omega}) \tag{53}$$

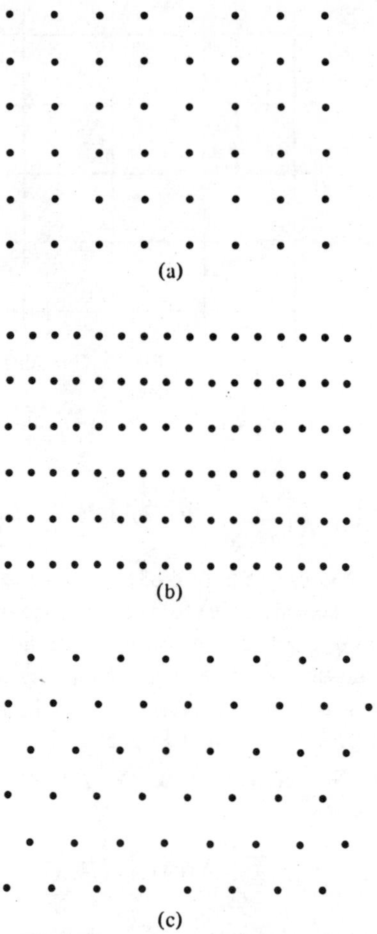

Fig. 6. An example of resampling. (a) Original sampling lattice. (b) High density intermediate lattice. (c) Final sampling lattice.

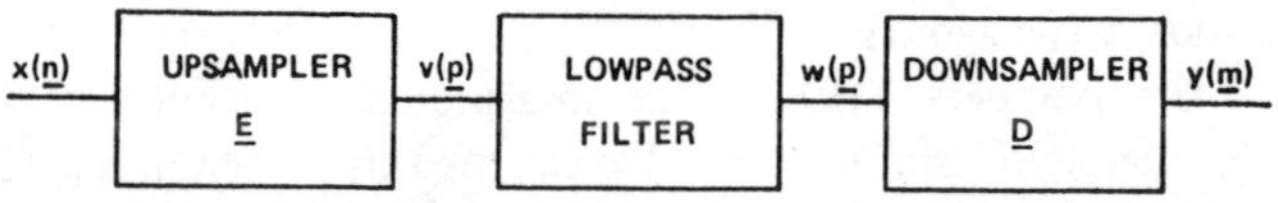

Fig. 7. A system for performing a general change of sampling lattice using an interpolator and a decimator.

it is periodic with periodicity matrix $2\pi(\boldsymbol{V}\boldsymbol{E})^{T^{-1}}$. This signal must then be passed through a low-pass filter with the frequency response

$$H(\boldsymbol{\omega}) = \begin{cases} |\det \boldsymbol{E}|, & \boldsymbol{\omega} \in I_{EV} \\ 0, & \text{otherwise} \end{cases} \tag{54}$$

to remove all periods except for the central one.

As an example, let us consider the interpolation from the rectangular lattice shown in Fig. 6(a) to the hexagonal lattice shown in Fig. 6(c). This can be accomplished by first interpolating the rectangularly sampled signal to the lattice shown in Fig. 6(b) and then decimating the result. The overall system is shown in Fig. 7. There we have cascaded an interpolator and a decimator; the two low-pass filters have been combined into a single low-pass filter. One appropriate sampling matrix for the lattice in Fig. 6(a) is

$$\boldsymbol{V}_R = \begin{bmatrix} T_1 & 0 \\ 0 & T_2 \end{bmatrix}. \tag{55}$$

Similarly, for the intermediate lattice and the hexagonal lattice we can use

$$\boldsymbol{V}_I = \begin{bmatrix} \frac{T_1}{2} & 0 \\ 0 & T_2 \end{bmatrix} \tag{56}$$

$$\boldsymbol{V}_H = \begin{bmatrix} T_1 & \frac{T_1}{2} \\ 0 & T_2 \end{bmatrix}. \tag{57}$$

A sufficient pair of interpolation and decimation matrices to accomplish the desired transformations are given by

$$\boldsymbol{E} = \begin{bmatrix} 2 & 0 \\ 0 & 1 \end{bmatrix} \tag{58}$$

$$\boldsymbol{D} = \begin{bmatrix} 2 & -1 \\ 0 & 1 \end{bmatrix}. \tag{59}$$

Summary

In this paper we have presented a very general framework for processing multidimensional signals which have been sampled on any given periodic sampling lattice. The general conclusion seems to be that almost anything that can be done in the one-dimensional case or the rectangular multidimensional case can be generalized. This notation has been greatly aided by the introduction of a number of simple matrix operators which act on the indexes of the signal. Not only do these matrices help to make the analogy with the 1-D case clearer, but they provide a compact means of describing the geometry of a sampling lattice. It is of particular interest to note that the problem of interpolating from one lattice to another can be handled simply with generalized decimators and interpolators.

References

[1] R. M. Mersereau, "The processing of hexagonally sampled two-dimensional signals," *Proc. IEEE*, vol. 67, pp. 930–949, May 1979.

[2] R. M. Mersereau and T. C. Speake, "A unified treatment of Cooley-Tukey algorithms for the evaluation of the multi-dimensional DFT," *IEEE Trans. Acoust., Speech, Signal Processing*, vol. ASSP-29, pp. 1011–1018, Oct. 1981.

[3] D. P. Petersen and D. Middleton, "Sampling and reconstruction of wavenumber-limited functions in N-dimensional Euclidean spaces," *Inform. Contr.*, vol. 5, pp. 279–323, 1962.

[4] T. C. Speake and R. M. Mersereau, "An interpolation technique for periodically sampled two-dimensional signals," in *1981 IEEE Int. Conf. Acoust., Speech, Signal Processing*, Apr. 1981, pp. 1010–1013.

[5] C. G. Lekkerkerker, *Geometry of Numbers*. Groningen: The Netherlands: Wolters-Noordhoff, 1969.

[6] R. W. Schafer and L. R. Rabiner, "A digital signal processing approach to interpolation," *Proc. IEEE*, vol. 61, pp. 692–702, June 1973.

Design of Antialiasing Patterns for Time-Sequential Sampling of Spatiotemporal Signals

JAN P. ALLEBACH

Abstract—The aliasing that results from time-sequential sampling of spatiotemporal signals is strongly dependent on the order in which the spatial points are sampled. To design sampling patterns that reduce aliasing, the sequence of sampling points is mapped into several shorter subsequences via the chinese remainder theorem. A pairwise exchange algorithm then finds the best ordering of each subsequence. The patterns obtained with this procedure perform substantially better than those known previously, and perform as well as the optimal patterns that can be expressed in closed form when the signal is termporally undersampled by less than a factor of 2.

I. Introduction

MANY signal processing and communications problems involve time-varying images. To be processed digitally, these signals must be sampled in space and time. It is common practice to do the sampling in a time-sequential fashion, collecting a frame of samples one-by-one from the spatial region and then repeating this process. The scanning action may be generated electromechanically or by multiplexing the outputs from an array of sensors.

Manuscript received February 14, 1983; revised June 22, 1983. This work was supported by the National Science Foundation under Grants ENG 78-25840 and ECS-8120759.

The author was with the Department of Electrical Engineering, University of Delaware, Newark, DE 19711. He is now with the School of Electrical Engineering, Purdue University, West Lafayette, IN 47907.

With some systems such as sensor arrays, the spatial points may be sampled in any order. With other systems, the ordering may be partially constrained by the scanning mechanism. In either case, the points are most frequently taken in lexicographic order which in 2 spatial dimensions results in line-by-line scanning. A number of researchers have experimented with other orderings of the spatial points [1]-[5]. In particular, Deutsch [1], [3] proposed an ordering which tends to distribute the samples taken during any time interval of duration less than the frame period uniformly over the spatial region. Since the ordering may be generated by a mapping from the bit reversed output of a binary counter, we refer to it as the bit reversed sampling pattern.

Fig. 1 shows the lexicographic and bit reversed sampling patterns for one spatial dimension. During each frame period of duration B, M samples are taken uniformly over the spatial region at interval X. With either pattern, we would expect to resolve signal components with temporal frequency $f_0 < 1/(2B)$ and spatial frequency $u_0 < 1/(2X)$. With the bit reversed pattern samples taken during a time interval B/λ, $1 \leqslant \lambda \leqslant M$ are distributed at approximately a spatial interval of λX. With this pattern, we might expect to also resolve signal components with $f_0 < \lambda/(2B)$ and $u_0 < 1/(2\lambda X)$. As λ increases, we trade spatial resolution for temporal resolution.

The experimental results that have been reported in the

Reprinted from *IEEE Trans. Acoust., Speech, Signal Processing,* vol. ASSP-32, pp. 137–144, Feb. 1984.

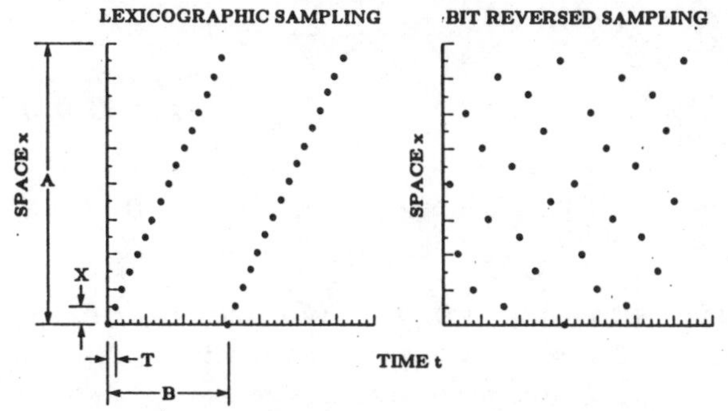

Fig. 1. Time-sequential sampling in one spatial dimension using the lexicographic and bit reversed patterns. Every T seconds, one sample is taken somewhere in the spatial region which has extent A. After the frame period B seconds, the spatial region has been sampled uniformly at interval X; and the pattern is repeated.

literature and our discussion of Fig. 1 suggest that it may be useful to consider sampling patterns other than the lexicographic pattern; but they do not provide the basis for a complete understanding of time-sequential sampling and the effect of the ordering of the spatial sampling points. Fourier analysis has contributed greatly to our understanding of problems in sampling and processing one- and multidimensional signals with various nontime-sequential sampling point geometries [6]-[9]. Although spatiotemporal sampling is only a special case of the general multidimensional sampling problem analyzed by Petersen and Middleton [7], their analysis does not apply to time-sequential sampling because the lattice of sample points cannot be generated by sums of integer multiples of vectors from a basis set.

Recently, we investigated time-sequential sampling [10] for both deterministic and stochastic signals. For stochastic signals, we assumed spatial and temporal wide-sense stationarity and zero mean. As in the general multidimensional case, time-sequential sampling causes replication of the spectrum or spectral density, depending on the signal type, at the points on a lattice that is reciprocal to the sampling lattice. However, in contrast to the general case, the replications are individually weighted by coefficients that are the 3D discrete Fourier transform (DFT) of the sampling pattern. These coefficients are the key to sampling pattern performance. For any sampling pattern, signals bandlimited to $1/(2B)$ temporally and $1/(2X)$ spatially can be recovered from their time-sequentially obtained samples with no error in the deterministic case and with zero mean-squared error in the stochastic case.

We also considered the case where the signal is undersampled temporally. Under these conditions, the reconstructed signal is degraded by signal-dependent aliasing. For stochastic signals, we took the signal-to-aliasing noise power ratio (SNR) as a measure of sampling pattern performance averaged over the signal ensemble. We found significant differences between the SNR's for the lexicographic and bit reversed sampling patterns, particularly when the signal was oversampled spatially, but concluded that other patterns probably exist which would perform even better than the bit reversed pattern. These results have implications for bandwidth compression and robustness of the sampling patterns. In the spatially oversampled case, the results illustrate the tradeoff between spatial and temporal resolution. They also suggest that significant bandwidth compression may be possible if the signal is oversampled spatially to relax the spatial specifications that processing filters must satisfy, or to obtain better image quality with displays that provide poor spatial interpolation or have only binary capability.

In the present paper, we attempt to identify patterns that improve on the performance of the bit reversed pattern. Specifically, we consider the design of sampling patterns that maximize the SNR developed in Section II. Since it is not feasible to search all $(MN)!$ possible orderings of the $M \times N$ spatial sampling points, the optimization task is, at the outset, a difficult one. In addition, a completely arbitrary pattern may be undesirable in terms of the hardware required to implement it, or the subsequent processing that the signal is to undergo. In Section III, we impose structure on the sampling pattern by mapping single indexes to multiple indexes. These mappings are based on the Chinese remainder theorem from number theory which has been used in the past to simplify certain signal processing computations [11, pp. 106-125], [12], [13]. In our case, the index mappings reduce the size of the space that must be searched for the optimal pattern, and simplify the computation of the SNR, thereby speeding up the search.

Even with these improvements, however, an exhaustive search is not feasible. To perform the actual search, we use a pairwise exchange algorithm that has been applied to the problem of assigning poles and zeros to 2nd order stages to minimize roundoff error in digital filters, and to the problem of designing dither patterns for binary display of images [14]-[16]. The algorithm searches locally for an optimum from many different starting points. It is described in Section IV. Section V contains two examples that illustrate the overall design procedure.

In Section VI, we show that provided the signal is not undersampled temporally by more than a factor of 2, it is possible to obtain the optimal sampling pattern in closed form. Finally, in Section VII, we compare SNR's of patterns found by the search procedure with those of patterns obtained in closed form.

II. Preliminaries: Description of Time-Sequential Sampling and the Signal-to-Aliasing-Noise Power Ratio

Let $g(x,y,t)$ be the continuous-parameter spatiotemporal signal, and assume that the spatial region of interest is $0 \leqslant x < A_1$, $0 \leqslant y < A_2$. Throughout, we shall consider only the case of two spatial dimensions. The analysis may easily be generalized to any number of spatial dimensions. We define the time-sequentially sampled version of g by

$$h(x,y,t) = X^2 B \sum_l g(\alpha_l X, \beta_l X, lT) \cdot \delta(x - \alpha_l X, y - \beta_l X, t - lT). \quad (1)$$

As before, X is the spatial sampling interval; B is the frame period; T is the temporal sampling interval; and $\delta(x,y,t)$ is the 3D Dirac delta function. The volume of each sample cell is X^2B. The spatial coordinates of the sample taken at time lT are $(\alpha_l X, \beta_l X)$. To assure that each spatial point is sampled once in every time interval of duration B, we require that $M = A_1/X$ and $N = A_2/X$ be integers; that (α_l, β_l), $l = 0, \cdots, MN - 1$ be a permutation of all pairs of integers (a, b), $0 \leqslant a \leqslant M - 1$ and $0 \leqslant b \leqslant N - 1$; and that $(\alpha_{l+MN}, \beta_{l+MN}) = (\alpha_l, \beta_l)$ for all l. Then $B = MNT$. The (α_l, β_l) completely define the sampling pattern. For example, with lexicographic sampling,

$$\alpha_l = [l/N], \beta_l = l \bmod N, l = 0, \cdots, MN - 1,$$

where $[x]$ is the greatest integer $\leqslant x$.

In addition, let us assume that $g(x, y, t)$ is a spatially and temporally wide-sense stationary random process with zero mean and power spectral density $S_{gg}(u, v, f)$. While the process does not have finite spatial support, we limit our interest to an $A_1 \times A_2$ region. Following the usual practice of randomizing the origin of the sampling pattern, we obtain [10] for the power spectrum of h

$$S_{hh}(u, v, f) = \sum_m \sum_n \sum_p |Q_{mnp}|^2 \cdot S_{gg}(u - m/A_1, v - n/A_2, f - p/B), \tag{2}$$

$$Q_{mnp} = \frac{1}{MN} \sum_{l=0}^{MN-1} \exp\left[-i2\pi(m\alpha_l/M + n\beta_l/N + pl/(MN))\right]. \tag{3}$$

Thus, the power spectrum of the time-sequentially sampled signal consists of replications of the power spectrum of the original signal at the points in a lattice that is reciprocal to the period of the sampling pattern. The replications are weighted by the squared magnitude of coefficients Q_{mnp} that are the 3D discrete Fourier transform (DFT) of the sampling pattern. For any pattern

$$\begin{aligned} &Q_{000} = 1, \\ &Q_{mn0} = \delta_{m \bmod M}\,\delta_{n \bmod N}, \\ &Q_{00p} = \delta_{p \bmod MN}, \\ &Q_{m+aM,\, n+bN,\, p+cMN} = Q_{mnp} \end{aligned} \tag{4}$$

for all integers a, b, and c.

Let g be bandlimited to some region Ω, the support of S_{gg}; and assume that the smallest rectangular region that contains Ω is $\{(u, v, f)\colon |u|, |v| < U;\ |f| < W\}$. The second line of (4) indicates that the spectral replications with nonzero weight in the $p = 0$ plane of (2) are separated by M/A_1 and N/A_2 in the spatial frequency directions u and v, respectively. Since $M/A_1 = N/A_2 = 1/X$, none of the replications will overlap the spectrum of the original signal if the conditions

$$U \leqslant 1/(2X), \quad W \leqslant 1/(2B) \tag{5}$$

are satisfied. In this case, we would expect to be able to recover the original signal from its samples without aliasing.

Letting $e(x, y, t)$ be the error between $h(x, y, t)$ and $g(x, y, t)$ after all frequency components outside Ω have been filtered out, we find [10] that

$$\sigma_e^2 = E\{|e(x, y, t)|^2\} = \sum_m \sum_n \sum_p |Q_{mnp}|^2\, V_{mnp}; \quad (m, n, p) \neq (0, 0, 0) \tag{6}$$

$$V_{mnp} = \iiint_\Omega S_{gg}(u - m/A_1, v - n/A_2, f - p/B)\, du\, dv\, df. \tag{7}$$

If (5) is satisfied, $\sigma_e^2 = 0$. Otherwise, the recovered signal will contain additive noise that is uncorrelated with g and has power σ_e^2. The signal-to-aliasing-noise power ratio is σ_g^2/σ_e^2.

When we need to assume a specific model for g in subsequent sections, we shall take it to be white and bandlimited to an ellipsoidal region in the frequency domain:

$$S_{gg}(u, v, f) = \begin{cases} (3\sigma_g^2)/(4\pi U^2 W), & (u, v, f) \in \Omega \\ 0, & \text{else}; \end{cases}$$

$$\Omega = \{(u, v, f)\colon (u/U)^2 + (v/U)^2 + (f/W)^2 < 1\}. \tag{8}$$

III. Application of Index Mappings to Sampling Pattern Design

For our problem, a number of possible approaches exist for mapping the indexes to multiple indexes. We choose to require that

$$M = M_1 \cdots M_r$$
$$N = N_1 \cdots N_s$$

where the factors $M_1, \cdots, M_r, N_1, \cdots, N_s$ are all relatively prime. We define $q = r + s$ and $L = L_1 \cdots L_q$, where

$$L_j = \begin{cases} M_j, & j = 1, \cdots, r \\ N_{j-r}, & j = r + 1, \cdots, q. \end{cases}$$

We then establish the following mappings for the indexes in (3):

$$\begin{aligned} &\alpha_l \leftrightarrow \alpha_1(l), \cdots, \alpha_r(l) \qquad && m \leftrightarrow m_1, \cdots, m_r \\ &\beta_l \leftrightarrow \beta_1(l), \cdots, \beta_s(l) && n \leftrightarrow n_1, \cdots, n_s \\ &l \leftrightarrow l_1, \cdots, l_q && p \leftrightarrow p_1, \cdots, p_q \end{aligned}$$

For α_l, the mapping is given by

$$\alpha_j(l) = (\bar{M}_j^{-1}\alpha_l) \bmod M_j \tag{9}$$

and the inverse mapping is given by

$$\alpha_l = \left[\sum_{j=1}^{r} \bar{M}_j \alpha_j(l)\right] \bmod M \tag{10}$$

where $\bar{M}_j = M/M_j$ and $\bar{M}_j^{-1}$ is the inverse of $\bar{M}_j$ mod M_j; so there exists an integer b such that

$$\bar{M}_j \bar{M}_j^{-1} + bM_j = 1.$$

The mappings for β_l and l are similar to (9) and (10) except that they are in terms of N and the N_j's for β_l, and L and the L_j's for l. For m, n, and p, we use a different form for the mappings. For m,

$$m_j = m \bmod M_j; \tag{11}$$

and

$$m = \left[\sum_{j=1}^{r} \bar{M}_j \bar{M}_j^{-1} m_j\right] \bmod M. \tag{12}$$

Again, the mappings for n and p are similar to (11) and (12) except that they are in terms of N and the N_j's for n, and L and the L_j's for p. Both types of mappings used here are versions of the Chinese remainder theorem [11, pp. 117, 118].

When these mappings are substituted into (3), it reduces rather straightforwardly to

$$Q_{mnp} = \frac{1}{L} \sum_{l_1=0}^{L_1-1} \cdots \sum_{l_q=0}^{L_q-1} \exp\left\{-i2\pi\left[\sum_{j=1}^{r} m_j \alpha_j(l)/M_j + \sum_{j=1}^{s} n_j \beta_j(l)/N_j + \sum_{j=1}^{q} p_j l_j/L_j\right]\right\} \tag{13}$$

So far, we have not really gained anything. We have not restricted the form that the sequences may take; hence we have not reduced the size of the space that must be searched for an optimal sampling pattern. Since $\alpha_j(l) = \alpha_j(l_1, \cdots, l_q)$ and $\beta_j(l) = \beta_j(l_1, \cdots, l_q)$, the multiple sum over the l_j's may not be factored into a product of sums, so we have not simplified the computation of the Q_{mnp}'s either.

We can, however, accomplish both these objectives by constraining $\alpha_j(l)$ to depend only on l_j, and $\beta_j(l)$ to depend only on l_{j+r}. Then (13) reduces to

$$Q_{mnp} = \prod_{j=1}^{q} Q^j_{m_j n_j p_j}; \tag{14}$$

$$Q^j_{m_j n_j p_j} = \begin{cases} \dfrac{1}{L_j} \displaystyle\sum_{l_j=0}^{L_j-1} \exp\{-i2\pi[m_j \alpha_j(l_j) + p_j l_j]/L_j\} & j = 1, \cdots, r, \\ \dfrac{1}{L_j} \displaystyle\sum_{l_j=0}^{L_j-1} \exp\{-i2\pi[n_{j-r} \beta_{j-r}(l_j) + p_j l_j]/L_j\} & j = r+1, \cdots, q. \end{cases} \tag{15}$$

We have thus factored the sequence $\{(\alpha_l \beta_l)\}, l = 0, \cdots, L-1$ into q short 1D subsequences:

$$\{\alpha_j(l_j)\}, \quad l_j = 0, \cdots, L_{j-1}, \quad j = 1, \cdots, r;$$

$$\{\beta_{j-r}(l_j)\}, \quad l_j = 0, \cdots, L_{j-1}, \quad j = r+1, \cdots, q.$$

The coefficient Q_{mnp} for the full sequence is a product of the coefficients Q^j_{mnp} for each of the subsequences.

We are limiting our search to a relatively small and rather arbitrarily chosen subset of $(L_1!) \cdots (L_q!)$ sampling patterns from the set of all $(L_1 \cdots L_q)!$ possible patterns. At this point, we have no guarantee that the subset contains sampling patterns with high SNR's. However, the subsequences are unconstrained; and the coefficients for these sequences have a structure that is similar to those of the full sequence. To the extent that reordering the full sequence effectively shapes the spectral coefficients, this suggests that reordering the subsequences should accomplish the same thing. In Section VII, we will present stronger evidence supporting this hypothesis when we compare SNR's of patterns found by searching through the subset defined above, with SNR's of optimal patterns that can be obtained in closed form under certain conditions.

IV. Pairwise Exchange Search Algorithm

The pairwise exchange algorithm searches in the space of permutations of the subsequences for a minimum of σ_e^2 in some locality of the starting point. By repeating the search from many different random initial points, a number of local minima are found; and the best of these may be taken as a global minimum. Of course, there is no assurance that this is the true global minimum. However, if a sufficient number of random starting points are tried so that several lead to approximately the same minimum value for σ_e^2, one may feel with some confidence that this is near the global minimum.

The algorithm may be described as follows.

1) Randomly order the points within each subsequence.

2) Enumerate all possible pairwise exchanges of the points within each subsequence.

  a) If an exchange does not decrease σ_e^2, restore the pair to their original positions and proceed to the next pair.

  b) If an exchange does decrease σ_e^2, keep it.

3) Stop when no pairwise exchange will further decrease σ_e^2.

4) Repeat steps 1)–3) a fixed number of times and retain the best of these orderings.

Since the algorithm continues to search as long as improvements are obtained, it does not converge after a fixed number of steps.

V. Performance of the Design Procedure

We used the pairwise exchange search algorithm in conjunction with the index mappings to design several sampling patterns. These results are discussed in Section VII. Here we only consider two examples—one to demonstrate the effect of reducing the size of the search space on the pairwise exchange algorithm, and another to illustrate how the subsequences go together to produce the full sequence.

For the first example, we designed a 15 × 13 sampling pattern without using index mappings by searching directly through sequences with length 195. We then repeated the design using index mappings and searching through subsequences with lengths 5, 3, and 13. In both cases, the signal was sampled at the Nyquist rate spatially and at $\frac{2}{3}$ the Nyquist rate temporally. Also, 10 random starts were tried in both cases.

From each start, an average of 349 274 exchanges were tried and 6302 kept in 19 complete enumerations when searching through the full sequence. At the start, the average noise power was − 8.0 dB. The minimum noise power obtained overall by the search was −17.6 dB. When searching through the subsequences, an average of 419 exchanges were tried and 27 kept in 5 complete enumerations. At the start, the average noise power was − 8.2 dB. The minimum noise power obtained was −17.8 dB.

These results show that the noise power is reduced significantly by the pairwise exchange search. They also show that the same improvement in noise power can be obtained, but with a very substantial reduction in computational effort, by using the index mappings.

For the second example, we designed a 14 × 15 sampling pattern using index mappings. Tables I and II and Fig. 2 give the parameters for the mappings as defined in (9)–(12), the best orderings for the subsequences produced by the search, and the resulting full sequence. To show how the subsequences define the full sequence, we determine the spatial coordinate indices (α_l, β_l) at the lth sampling instant for $l = 19$. From (9) and Table I, we have $l_1 = (4 \cdot 19) \bmod 7 = 6$, $l_2 = (1 \cdot 19) \bmod 2 = 1$, $l_3 = (3 \cdot 19) \bmod 5 = 2$, and $l_4 = (1 \cdot 19) \bmod 3 = 1$. From Table II, $\alpha_1(6) = 1$, $\alpha_2(1) = 0$, $\beta_1(2) = 4$, and $\beta_2(1) = 2$. From (10) and Table I, $\alpha_{19} = (1 \cdot 2 + 0 \cdot 7) \bmod 14 = 2$ and $\beta_{19} = (4 \cdot 3 + 2 \cdot 5) \bmod 15 = 7$. Thus, the entry in Fig. 2 for the coordinates (2, 7) is 19.

VI. Optimal Sampling Patterns

In this section, we obtain a closed-form expression for the sampling pattern that minimizes the mean-squared error σ_e^2 given by (6) and (7), under the condition that the signal not be temporally undersampled by more than a factor of 2. We also assume Nyquist spatial sampling; so $U = 1/(2X)$, and $W \leqslant 1/B$. Consequently, $V_{mnp} = 0$, if $|m| \geqslant M$, $|n| \geqslant N$, or $|p| > 1$.

In addition, we have the following symmetries for any S_{gg} and any sampling pattern:

$$|Q_{mnp}| = |Q_{-m,-n,-p}|$$

$$V_{mnp} = V_{|m|,|n|,|p|}. \tag{16}$$

Combining these with the properties of Q_{mnp} indicated by the second and fourth lines of (4), we find that (6) reduces to

$$\sigma_e^2 = 2 \sum_{m=0}^{M-1} \sum_{n=0}^{N-1} |Q_{mn1}|^2 \mathcal{V}_{mn}; \tag{17}$$

$$\mathcal{V}_{mn} = V_{m,n,1} + V_{m-M,n,1} + V_{m,n-N,1} + V_{m-M,n-N,1}. \tag{18}$$

We approach the problem of finding the sampling pattern that minimizes σ_e^2 by first considering what we would want the coefficients Q_{mnp} for that pattern to be. These are given by (3).

Rather than define the sampling pattern by the spatial coordinates (α_l, β_l) of the sample taken at the lth time instant, we may also define it by the time instant γ_{ab} at which we sample at the spatial coordinates (a, b). In terms of γ_{ab}, we have

$$Q_{mnp} = \frac{1}{MN} \sum_{a=0}^{M-1} \sum_{b=0}^{N-1} \exp \{-i2\pi [ma/M + nb/N + p\gamma_{ab}/(MN)]\}. \tag{19}$$

TABLE I
Parameters for Mappings When $M = 14$ and $N = 15$

$M_1 = 7$	$\bar{M}_1 = 2$	$\bar{M}_1^{-1} = 4$
$M_2 = 2$	$\bar{M}_2 = 7$	$\bar{M}_2^{-1} = 1$
$N_1 = 5$	$\bar{N}_1 = 3$	$\bar{N}_1^{-1} = 2$
$N_2 = 3$	$\bar{N}_2 = 5$	$\bar{N}_2^{-1} = 2$
$L_1 = 7$	$\bar{L}_1 = 30$	$\bar{L}_1^{-1} = 4$
$L_2 = 2$	$\bar{L}_2 = 105$	$\bar{L}_2^{-1} = 1$
$L_3 = 5$	$\bar{L}_3 = 42$	$\bar{L}_3^{-1} = 3$
$L_4 = 3$	$\bar{L}_4 = 70$	$\bar{L}_4^{-1} = 1$

TABLE II
Best Ordering of the Subsequences Found by the Pariwise Exchange Algorithm

l_1	0	1	2	3	4	5	6
$\alpha_1(l_1)$	4	0	3	6	2	5	1
l_2	0	1					
$\alpha_2(l_2)$	1	0					
l_3	0	1	2	3	4		
$\beta_1(l_3)$	0	2	4	1	3		
l_4	0	1	2				
$\beta_2(l_4)$	0	2	1				

α_l = \ β_l =	0	1	2	3	4	5	6	7	8	9	10	11	12	12	14
0	135	37	149	51	163	65	177	79	191	93	205	107	9	121	23
1	(0)	112	14	126	28	140	42	154	56	168	70	182	84	196	98
2	75	187	89	201	103	5	117	19	131	33	145	47	159	61	173
3	150	42	164	66	178	80	192	94	206	108	10	122	24	136	38
4	15	127	29	141	43	155	57	169	71	183	85	197	99	1	113
5	90	202	104	6	118	20	132	34	146	48	160	62	174	76	188
6	165	67	179	81	193	95	207	109	11	123	25	137	39	151	53
7	30	143	44	156	58	170	72	184	86	198	100	2	114	16	128
8	105	7	119	21	133	35	147	49	161	63	175	77	189	91	203
9	180	82	194	96	208	110	12	124	26	138	40	152	54	166	68
10	45	157	59	171	73	185	87	199	101	3	115	17	129	31	143
11	120	22	134	36	148	50	162	64	176	78	190	92	204	106	(8)
12	195	97	209	111	13	125	27	139	41	153	55	167	69	181	83
13	60	172	74	186	88	200	102	4	116	18	130	32	144	46	158

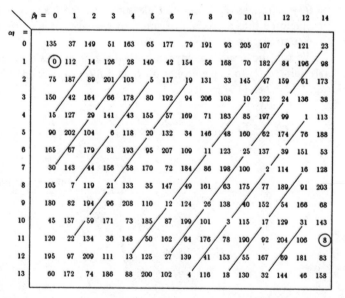

Fig. 2. Full 14 × 15 sampling pattern defined by the subsequences in Table II. Each entry is the index of the sampling time l. To show the structure in the pattern, the first 32 samples have been marked.

This is the 2D DFT of the sequence $\exp[-i2\pi p\gamma_{ab}/(MN)]$. Consequently, we find from Parseval's relation that

$$\sum_{m=0}^{M-1} \sum_{n=0}^{N-1} |Q_{mnp}|^2 = 1 \ \forall p. \tag{20}$$

Considering (17) and (20) together we see that our problem is to distribute one unit of energy over the weights $\mathcal{V}_{mn}$ in such a way as to minimize the weighted sum. An optimal way to do this would be to have

$$Q_{mn1} = \delta_{(m-m^*) \bmod M} \, \delta_{(n-n^*) \bmod N} \tag{21}$$

where (m^*, n^*) satisfies

$\mho_{m^*,n^*} \leqslant \mho_{m,n}\ \forall (m,n).$

The inverse of (19) is

$$\exp\left[-i2\pi p\gamma_{ab}/(MN)\right] = \sum_{m=0}^{M-1}\sum_{n=0}^{N-1} Q_{mnp} \cdot \exp\left[i2\pi(ma/M + nb/N)\right]. \quad (22)$$

Not every possible choice for Q_{mn1} will yield a sequence of the form given by the left side of the above equation with $p = 1$, where γ_{ab} is a permutation of the integers $0, \cdots, MN - 1$.

However, substituting (21) into (22) with $p = 1$ leads to

$$\exp\left[-i2\pi\gamma_{ab}/(MN)\right] = \exp\left[i2\pi(m^*a/M + n^*b/N)\right]. \quad (23)$$

Letting γ^*_{ab} denote the solution to this equation, we find that

$$\gamma^*_{ab} = -[m^*aN + n^*bM] \bmod (MN). \quad (24)$$

For any pair (a, b), γ^*_{ab} is well defined. Suppose there exist two pairs (a_1, b_1) and (a_2, b_2) such that $\gamma^*_{a_1b_1} = \gamma^*_{a_2b_2}$. If the conditions

$$\gcd(m^*N, M) = 1,$$
$$\gcd(n^*M, N) = 1, \quad (25)$$

where gcd denotes greatest common divisor, are satisfied, it is easily shown that $a_2 = a_1 \bmod M$ and $b_2 = b_1 \bmod N$. Therefore, (24) defines an isomorphism between the integers $0 \leqslant \gamma \leqslant MN - 1$ and the pairs (a, b), $0 \leqslant a \leqslant M - 1$, $0 \leqslant b \leqslant N - 1$. Note that (25) implies that $\gcd(M, N) = 1$.

For implementation of the sampling pattern, it can be shown that the inverse of (24) is given by

$$\alpha^*_l = -(m^*N)^{-1} l \bmod M,$$
$$\beta^*_l = -(n^*M)^{-1} l \bmod N \quad (26)$$

where $(m^*N)^{-1}$ is the inverse of $m^*N \bmod M$ and $(n^*M)^{-1}$ is the inverse of $n^*M \bmod N$ as defined in Section III.

Having determined the general form of the optimal sampling pattern, it remains for us to find the coordinates (m^*, n^*) that minimize $\mho$ subject to the conditions given by (25). Assuming that $\gcd(M, N) = 1$, it is shown in the Appendix that

$$m^* = \begin{cases} M/2 + 1/2, & M \text{ odd};\\ M/2 + 1, & M \text{ even}, M/2 \text{ even};\\ M/2 + 2, & M \text{ even}, M/2 \text{ odd}; \end{cases} \quad (27)$$

for white, ellipsoidally bandlimited signals with power spectrum given by (8). The expression for n^* is identical to (27) with M replaced by N. Since (m^*, n^*) does not depend on B or W, the same sampling pattern is optimal for all $W \leqslant 1/B$.

With the optimal sampling pattern, the mean-squared error given by (17) reduces to $\sigma_e^2 = 2\mho_{m^*,n^*}$. For white, ellipsoidally bandlimited signals, we find in the Appendix that this is approximately

$$\sigma_e^2 \cong 32\pi/3(1 - 3s/4 + s^3/16), \quad 0 \leqslant s \lesssim 2,$$
$$\sigma_e^2 = 0, \quad 2 \lesssim s,$$
$$s = [2 + 1/(BW)^2]^{1/2}. \quad (28)$$

The approximation improves as M and N increase. The noise power depends only on BW—not on M and N, and is zero for $W < 1/(\sqrt{2}B)$. Therefore, with an optimal sampling pattern, the temporal Nyquist rate is a factor of $1/\sqrt{2}$ less than it is with an arbitrary pattern such as the lexicographic or bit reversed patterns.

For other signal classes, the optimal sampling pattern is still given by (24) or (26); however (27) may not be valid for (m^*, n^*). In general, (m^*, n^*) may be found by a simple search over the values that satisfy (25).

VII. Experimental Results

We used the pairwise exchange search algorithm with index mappings to design patterns with 56 X 55, 78 X 77, and 104 X 99 spatial points. We assumed a white, ellipsoidally bandlimited signal sampled spatially at the Nyquist rate and temporally at $\frac{2}{3}$ the Nyquist rate. For each size, we tried 100 random starts. We also generated sampling patterns of these sizes using the closed form expression from Section VI. The SNR's for all these patterns are plotted in Fig. 3 as a function of the time-bandwidth product BW. For comparison, the SNR's for bit reversed and lexicographic sampling are also included.

The curves for the patterns obtained via the search algorithm and those obtained from the closed form solution are all tightly clustered. For the closed form patterns, this verifies that the SNR is approximately independent of M and N as discussed in Section VI. For the patterns obtained via the search, this shows that the design procedure is capable of producing near optimal patterns as was speculated in Section III. Finally, the optimal patterns are seen to perform substantially better than both the lexicographic and bit reversed patterns with a higher temporal Nyquist rate indicated by the fact that the value of BW below which the SNR becomes infinite has increased from 0.5 to approximately 0.707.

For the second part of the experimental work, we used the pairwise exchange algorithm with index mappings to design sampling patterns for white, ellipsoidally bandlimited signals that were oversampled spatially, and undersampled temporally by more than a factor of 2. The closed form expression for the optimal pattern is not valid under these conditions. We designed 36 X 35 patterns for signals that were spatially oversampled by factors of 1, 2, and 4. The values of BW used for the search were 0.67, 2.5, and 10.0, respectively. The SNR's are plotted in Fig. 4 along with those for the lexicographic and bit reversed patterns. The patterns obtained via the search improve upon both the lexicographic and bit reversed patterns. Interestingly, the pattern designed for signals spatially oversampled by a factor of 4 shows evidence of tailoring to be optimal at $BW = 10.0$. When the signal is oversampled spatially, it is possible with these patterns to maintain a constant SNR while increasing BW significantly.

VIII. Conclusions

In many systems for processing spatiotemporal signals, time-sequential sampling is an unavoidable consequence of hardware limitations. Whether or not it is avoidable, time-sequential sampling has the potential to improve signal processor performance. To realize this potential, however, we must sample the spatial points in the proper order. Previously, sampling patterns were only chosen on an ad hoc basis.

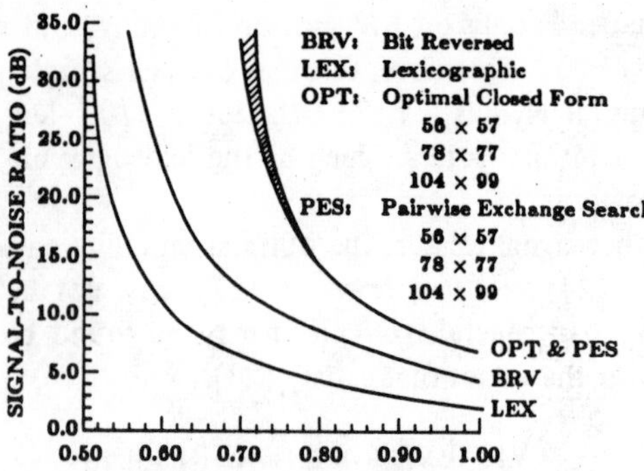

Fig. 3. Signal-to-noise ratio for time-sequential sampling of a white, ellipsoidally bandlimited signal with several different sampling patterns. In the two spatial dimensions, sampling was at the Nyquist rate. All the curves for the optimal closed form and pairwise exchange search patterns lie in the shaded region.

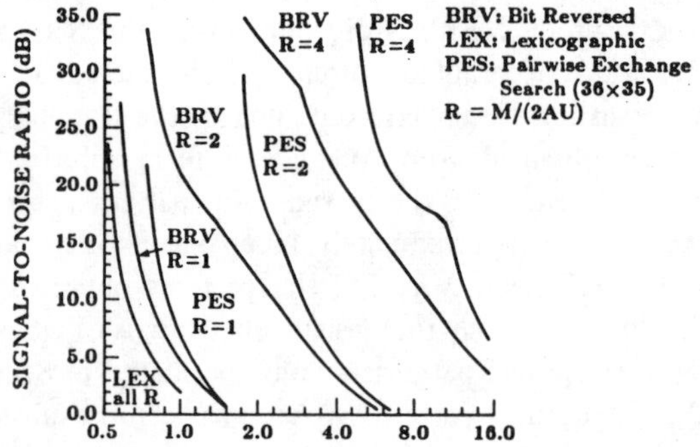

Fig. 4. Signal-to-noise ratio for time-sequential sampling of a white, ellipsoidally bandlimited signal with several different sampling patterns when the spatial Nyquist rate is exceeded by the factor R.

We have described a procedure for design of sampling patterns that partially constrains the structure of the pattern by mapping single indexes to multiple indexes. The mapping is based on the Chinese remainder theorem. A pairwise exchange algorithm is then used to find the best pattern from among those in the reduced set. We applied the procedure specifically to the design of sampling patterns that maximize the signal-to-aliasing-noise power ratio (SNR) of temporally undersampled signals that are white and ellipsoidally bandlimited. The procedure can be applied directly to the design of sampling patterns for other classes of signals provided the integrals involved in calculating the SNR can be evaluated either in closed form or numerically. The procedure may also be generalized for use with performance measures other than SNR.

For the special case where the signal is temporally undersampled by less than a factor of 2, we obtained a simple closed form expression for the optimal sampling pattern. Comparing the performance of these patterns with those obtained via the more generally applicable design procedure, we conclude that the latter is indeed capable of generating near optimal patterns.

Appendix
Derivation of (m^*, n^*) and σ_e^2 for White, Ellipsoidally Bandlimited Signals

We seek the minimum of $\mho_{mn}$, given by (7) and (18), for $0 \leq m \leq M-1$ and $0 \leq n \leq N-1$. Let us begin by defining continuous-parameter functions

$$\tilde{\mho}(\mu, \nu) = \tilde{V}(\mu, \nu) + \tilde{V}(\mu - 2U, \nu) + \tilde{V}(\mu, \nu - 2U) + \tilde{V}(\mu - 2U, \nu - 2U);$$

$$\tilde{V}(\mu, \nu) = \iiint_{\Omega} S_{gg}(u - \mu, v - \nu, f - 1/B)\, du\, dv\, df. \qquad \text{(A1)}$$

Under the conditions of Nyquist spatial sampling, $M = 2A_1 U$ and $N = 2A_2 U$; so $\tilde{\mho}(m/A_1, n/A_2) = \mho_{mn}$. For white, ellipsoidally bandlimited signals with power spectrum given by (8), we have [10]

$$\tilde{V}(\mu, \nu) = \begin{cases} (4\pi/3)(1 - 3s/4 + s^3/16), & 0 \leq s \leq 2, \\ 0, & 2 \leq s; \end{cases}$$

$$s = [(\mu/U)^2 + (\nu/U)^2 + 1/(BW)^2]^{1/2}. \qquad \text{(A2)}$$

We now show that $(\mu, \nu) = (U, U)$ is a local minimum of $\tilde{\mho}$ for white, ellipsoidally bandlimited signals. For this to be true, it is sufficient that $\tilde{\mho}_\mu = 0$, $\tilde{\mho}_\nu = 0$, $\tilde{\mho}_{\mu\mu} > 0$, $\tilde{\mho}_{\nu\nu} > 0$, and $\tilde{\mho}_{\mu\mu}\tilde{\mho}_{\nu\nu} > \tilde{\mho}_{\mu\nu}$ at $(\mu, \nu) = (U, U)$. Here, the subscript μ or ν denotes the partial derivative with respect to that variable. $\tilde{V}$ is an even function of μ and ν; so $\tilde{V}_\mu$ is odd in μ and even in ν, $\tilde{V}_\nu$ is even in μ and odd in ν; $\tilde{V}_{\mu\mu}$ and $\tilde{V}_{\nu\nu}$ are even in both μ and ν; and $\tilde{V}_{\mu\nu}$ satisfies $\tilde{V}_{\mu\nu}(\mu, \nu) = -\tilde{V}_{\mu\nu}(-\mu, \nu) = -\tilde{V}_{\mu\nu}(\mu, -\nu) = \tilde{V}_{\mu\nu}(-\mu, -\nu)$. Taking partial derivatives of the expression for $\tilde{\mho}$ on the first line of (A1) and using these symmetry properties leads to $\tilde{\mho}_\mu = 0$, $\tilde{\mho}_\nu = 0$, $\tilde{\mho}_{\mu\mu} = 4\ \tilde{\mho}_{\mu\mu}$, $\tilde{\mho}_{\nu\nu} = 4\ \tilde{V}_{\nu\nu}$, and $\tilde{\mho}_{\mu\nu} = 0$ at $(\mu, \nu) = (U, U)$. Thus, we see that the first two conditions for a minimum of $\tilde{\mho}$ are satisfied directly, and that the fifth condition is satisfied if the third and fourth conditions are.

To verify that the third and fourth conditions are indeed satisfied, we differentiate (A2) twice with respect to μ obtaining for $BW > 1/\sqrt{2}$

$$\tilde{\mho}_{\mu\mu}(U, U) = \pi(s^4 - 3s^2 + 4)/(U^2 s^3);$$

$$s = [2 + 1/(BW)^2]^{1/2}. \qquad \text{(A3)}$$

Since the quadratic factor in s^2 has no real roots, $\tilde{\mho}_{\mu\mu}(U, U) > 0$ for all values of BW in this range. It follows similarly that $\tilde{\mho}_{\nu\nu}(U, U) > 0$. For $BW < 1/\sqrt{2}$, $\tilde{\mho}(U, U) = 0$, which is the minimum possible value for $\tilde{\mho}$. This concludes the proof that $(\mu, \nu) = (U, U)$ is a local minimum of $\tilde{\mho}$. At present, we have no formal proof that it is global. However, we have verified by computer search over (m, n) the optimality of the (m^*, n^*) that we derive next from this minimum.

The indices (m, n) of $\mho$ that correspond to the point (U, U) of $\tilde{\mho}$ are $(M/2, N/2)$. However, m^* and n^* must be integer-valued and must satisfy (25). Let us consider how to pick m^* as close as possible to $M/2$ subject to these conditions. If M is odd, $(M-1)/2$ and $(M+1)/2$ are the closest integers to $M/2$. Since $\tilde{\mho}$ is even in μ about the point U, we can choose either value. Let $m^* = (M+1)/2$; and $d = \gcd(m^* N, M)$. Assuming $\gcd(M, N) = 1$, $d | m^*$ and $d | M$ where $|$ denotes "divides". Therefore, $d | (2m^* - M)$ which equals 1; so $d = 1$; and (25) is satisfied.

If M is even, $M/2$ is an integer; but $\gcd((M/2)N, M) = M/2 > 1$.

The integers closest to $M/2$ are $(M-2)/2$ and $(M+2)/2$. Let $m^* = (M+2)/2$ and d be as before. Following the same approach, we conclude that $d|2$. If $M/2$ is even, m^* is odd, and $d \neq 2$, so (25) is satisfied. If $M/2$ is odd, m^* is even; and $d = 2$. In this case, we move out to $m^* = (M+4)/2$. Reasoning as before, we find that $d|4$. Since m^* is odd, $d \neq 2$ or 4; and (25) is satisfied. Putting everything together, we obtain (27). The expression for n^* follows similarly.

To obtain (28), we note that for large M and N, $(m^*, n^*) \cong (M/2, N/2)$; so $\mathcal{O}_{m^*n^*} \cong \tilde{\mathcal{O}}(U, U)$.

References

[1] S. Deutsch, "Pseudorandom dot scan television systems," *IEEE Trans. Broadcast*, vol. BC-11, pp. 11-21, July 1965.

[2] R. C. Brainard, F. W. Mounts, and B. Prasada, "Low resolution TV: Subjective effects of frame repetition and picture replenishment," *Bell Syst. Tech. J.*, vol. 46, pp. 261-271, Jan. 1967.

[3] S. Deutsch, "Visual displays using pseudorandom dot scan," *IEEE Trans. Commun. Technol.*, vol. COM-21, pp. 65-75, Jan. 1973.

[4] R. F. Stone, "A practical narrow-band television system," *IEEE Trans. Broadcast.*, vol. BC-22, pp. 21-32, June 1976.

[5] K. Takikawa, "Simplified 6.3 Mbit/s codec for video conferencing," *IEEE Trans. Commun.*, vol. COM-29, pp. 1877-1882, Dec. 1981.

[6] A. J. Jerri, "The Shannon sampling theorem—Its various extensions and applications: A tutorial review," *Proc. IEEE*, vol. 65, pp. 1565-1596, Nov. 1977.

[7] D. P. Petersen and D. Middleton, "Sampling and reconstruction of wave number-limited functions in *N*-dimensional Euclidean spaces," *Inf. Control*, vol. 5, pp. 279-323, Dec. 1962.

[8] R. M. Mersereau, "The processing of hexagonally sampled two-dimensional signals," *Proc. IEEE*, vol. 67, pp. 930-949, June 1979.

[9] H. Stark, "Sampling theorems in polar coordinates," *J. Opt. Soc. Amer.*, vol. 69, pp. 1519-1525, Nov. 1979.

[10] J. P. Allebach, "Analysis of sampling-pattern dependence in time-sequential sampling of spatiotemporal signals," *J. Opt. Soc. Amer.*, vol. 71, pp. 99-105, Jan. 1981.

[11] T. M. Apostol, *Introduction to Analytic Number Theory.* New York: Springer-Verlag, 1976.

[12] C. S. Burrus, "Index mappings for multidimensional formulation of the DFT and convolution," *IEEE Trans. Acoust., Speech, Signal Processing*, vol. ASSP-25, pp. 239-242, June 1977.

[13] D. P. Kolba and T. W. Parks, "A prime factor FFT algorithm using high-speed convolution," *IEEE Trans. Acoust., Speech, Signal Processing*, vol. ASSP-25, pp. 281-294, Aug. 1977.

[14] B. Liu and A. Peled, "Heuristic optimization of the cascade realization of fixed-point digital filters," *IEEE Trans. Acoust., Speech, Signal Processing*, vol. ASSP-23, pp. 464-473, Oct. 1975.

[15] K. Steiglitz and B. Liu, "An improved algorithm for ordering poles and zeros of fixed point recursive digital filters," *IEEE Trans. Acoust., Speech, Signal Processing*, vol. ASSP-24, pp. 341-343, Aug. 1976.

[16] J. P. Allebach and R. N. Stradling, "Computer-aided design of dither signals for binary display of images," *Appl. Opt.*, vol. 18, pp. 2708-2713, Aug. 1979.

Two-Dimensional Linear Prediction: Autocorrelation Arrays, Minimum-Phase Prediction Error Filters, and Reflection Coefficient Arrays

THOMAS L. MARZETTA

Abstract—In this paper, a number of results in one-dimensional (1-D) linear prediction theory are extended to the two-dimensional (2-D) case. It is shown that the class of 2-D minimum mean-square linear prediction error filters with continuous support have the minimum-phase property and the correlation-matching property, and that they can be solved by means of a 2-D Levinson algorithm. A significant practical result to emerge from this theory is a reflection coefficient representation for 2-D minimum-phase filters. This representation provides a domain in which to construct 2-D filters, such that the minimum-phase condition is automatically satisfied.

I. Introduction

ONE of the important results in one-dimensional (1-D) linear prediction theory is the minimum-phase property [2], [3] of one-step minimum mean-square linear prediction error filters. The operation of computing a finite prediction error filter (PEF) figures in some techniques for designing 1-D recursive filters [4]-[6]; the fact that the PEF is minimum phase guarantees the stability of the resulting recursive filter.

A closely related property of these filters is the *correlation-matching property* [7]-[9], [15], which says that the operation of solving for a PEF and mean-square prediction error, given a positive definite autocorrelation sequence, is invertible. The one-to-one relation between the class of positive definite autocorrelation sequences, and the class of minimum-phase PEF's and positive mean-square prediction errors is the basis for the maximum entropy method of spectral estimation. In this method, the original finite autocorrelation sequence is extrapolated to infinity so as to maximize the entropy of the spectral estimate. This is done by computing a finite PEF to obtain an autoregressive (AR) model. The minimum-phase property ensures a stable AR model, and the correlation-matching property ensures that the spectral estimate is consistent with the original autocorrelation sequence.

The computation of a PEF involves the solution of a Toeplitz system of linear equations. These equations can be solved efficiently by means of the Levinson algorithm [10], [11], in which a sequence of PEF's of increasing order is computed successively. At each step of the recursion, the previously computed filter is used to compute a number, called a *reflection coefficient* (or *partial correlation coefficient*), whose magnitude is less than one. The updated filter is obtained directly from the reflection coefficient and the previous filter.

The relation between reflection coefficient sequences and minimum-phase filters is one-to-one [3], and it is the basis for the Burg algorithm for AR modeling [7], the lattice structures for minimum-phase filters [12], [13], and the Schur-Cohn stability test [3], [14]. At any rate, the properties of PEF's are closely related to a number of theoretical and practical results in time-series analysis [15].

Given the success of time-series linear prediction theory, there is considerable motivation for attempting to develop a similar theory for two-dimensional (2-D) random processes. The efforts of researchers to do this have been largely unsuccessful. Shanks conjectured that 2-D PEF's with finite, quarter-plane support were guaranteed to be minimum phase [16]; this conjecture was subsequently shown to be false when Genin and Kamp constructed a counterexample [17]. Moreover, it can be shown by inspection that the correlation-matching property cannot hold for this class of filters [1].

More general classes of 2-D PEF's whose support consist of finite portions of two quadrants, have been considered [18], but they also fail to have the desired properties.

Several so-called 2-D Levinson algorithms have appeared in the literature [19]-[21]; these are fast algorithms for solving for various 2-D PEF's. Although these algorithms are useful from a computational standpoint, they bear little or no resemblance to the 1-D Levinson algorithm. More importantly, they do not contain anything analogous to a reflection coefficient sequence, so unlike the 1-D Levinson algorithm, they do not provide a representation for minimum-phase filters.

In this paper we show how to extend virtually all of the 1-D linear prediction theories to the 2-D case. Our treatment of 2-D linear prediction is based on a definition of causality which totally orders the points in the plane. Adopting this notion of causality, we are led to consider a class of 2-D PEF's whose support includes the origin, the point (N, M), where N and M are integers, with N nonnegative, and *all* points simultaneously in the future of the origin and in the past of (N, M); we say that this type of filter has *continuous support.* Because of the definition of 2-D causality that we adopt, these filters always have *infinite* support in one of the two variables (except in the trivial 1-D case where $N = 0$). It can be shown that 2-D PEF's with continuous support have both the

Manuscript received December 17, 1979; revised May 27, 1980. This work was supported in part by a Vinton Hayes Fellowship in Communications and is based on the author's Ph.D. dissertation, Department of Electrical Engineering, Massachusetts Institute of Technology, Cambridge, MA, February 1978.

The author is with Schlumberger-Doll Research, Ridgefield, CT 06877.

Reprinted from *IEEE Trans. Acoust., Speech, Signal Processing,* vol. ASSP-28, pp. 725–733, Dec. 1980.

minimum-phase property and the correlation-matching property. Of course, these results cannot be applied directly to any practical problem because of the infinite support of the filters. Nevertheless, these results do lead indirectly to a new, computationally useful representation for 2-D minimum-phase filters.

It turns out that 2-D PEF's with continuous support can be solved for by means of a 2-D Levinson algorithm which has precisely the same structure as the 1-D Levinson algorithm. Specifically, the updated filter is obtained directly in terms of the previously computed filter and a "reflection coefficient." This implied relation between 2-D minimum-phase filters and 2-D reflection coefficient arrays (that is, 2-D one-sided arrays of numbers, whose magnitudes are less than one) can be shown to be one-to-one. The most important properties of this representation are as follows. 1) Given a 2-D minimum-phase filter with *finite* support, there is a unique reflection coefficient array which almost always has *infinite* support in one of the two variables. 2) Given a 2-D reflection coefficient array with *finite* support, there is a unique 2-D minimum-phase filter, also with *finite* support, which is obtainable by means of a *finite* recursion. Consequently, this representation cannot be used as a stability test because an infinite number of reflection coefficients must be checked; on the other hand, it does provide a new domain in which to design 2-D minimum-phase filters. Our point is, that by formulating 2-D AR modeling problems, and 2-D recursive filter design problems in the 2-D reflection coefficient domain, the difficult minimum-phase condition is automatically satisfied merely by constraining the reflection coefficient magnitudes to be less than one. In addition, 2-D minimum-phase filters with finite reflection coefficient support have some useful properties, including 2-D lattice implementations [22], which are not shared by more general types of minimum-phase filters.

This paper is chiefly concerned with the theoretical aspects of 2-D linear prediction. It is assumed that the reader is already familiar with 1-D linear prediction. In Section II, we briefly discuss 2-D causality and the 2-D minimum-phase condition. In Section III we discuss the properties of 2-D PEF's with continuous support: the minimum-phase property, the correlation-matching property, and the 2-D Levinson algorithm. Finally, in Section IV we discuss the 2-D reflection coefficient representation.

II. 2-D Causality and the 2-D Minimum-Phase Condition

A. 2-D Causality

The notion of causality is based on a definition of "past," "present," and "future." For any point (s, t) we define the *past* to be the set of points

$$\{(k, l) \mid k = s, l < t; k < s, -\infty \leq l \leq \infty\}$$

and the *future* to be the set

$$\{(k, l) \mid k = s, l > t; k > s, -\infty \leq l \leq \infty\}.$$

This is illustrated in Fig. 1. It is straightforward to verify two implications of this definition.

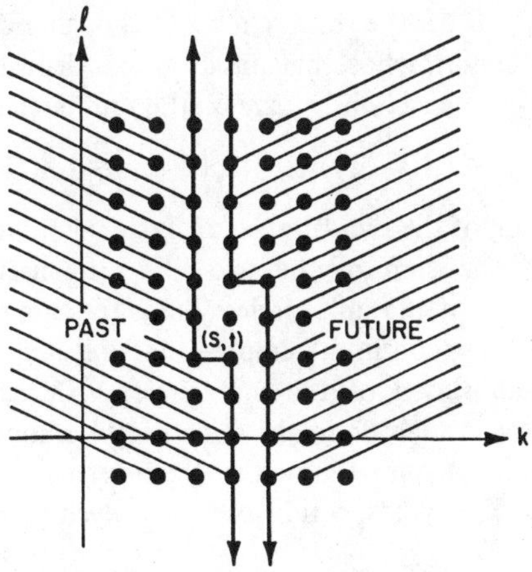

Fig. 1. Definition of past, present, and future.

1) If (k_1, l_1) is in the past of (k_2, l_2), then (k_2, l_2) is in the future of (k_1, l_1).

2) If (k_1, l_1) is in the past of (k_2, l_2) and (k_2, l_2) is in the past of (k_3, l_3), then (k_1, l_1) is in the past of (k_3, l_3).

As a matter of notation, if (k_1, l_1) is in the past of (k_2, l_2), we denote this by

$$(k_1, l_1) \prec (k_2, l_2)$$

or equivalently by

$$(k_2, l_2) \succ (k_1, l_1).$$

Accordingly, we define a 2-D *one-sided*, linear, shift-invariant filter to be one whose unit sample response vanishes at all points in the past of the origin. Equivalently, a 2-D filter, $a(k, l)$, is one-sided if its z-transform has the form

$$A(z_1, z_2) = \sum_{l=0}^{\infty} a(0, l) z_2^{-l} + \sum_{k=1}^{\infty} \sum_{l=-\infty}^{\infty} a(k, l) z_1^{-k} z_2^{-l} = \sum_{(k,l) \succeq (0,0)} \sum a(k, l) z_1^{-k} z_2^{-l}. \tag{1}$$

B. The 2-D Minimum-Phase Condition

We define a 2-D *stable*, linear, shift-invariant filter to be one whose unit sample response is absolutely summable.

We define a 2-D *minimum-phase* filter to be a 2-D, one-sided, stable, linear, shift-invariant filter which has a one-sided, stable inverse. In this paper, we will be concerned with a special class of 2-D minimum-phase filters that we call analytic minimum phase. We define a 2-D filter to be *analytic minimum phase* if 1) the filter is minimum phase and 2) the filter is analytic in some neighborhood of the unit circles [that is, for some $(1 - \epsilon) < |z_1|, |z_2| < (1 + \epsilon)$]. We note that any filter which is a ratio of two minimum-phase filters having finite support is analytic minimum phase.

C. A New Theorem for 2-D Minimum-Phase Filters

For future reference, we now state and prove a new theorem, which is a generalization of a well-known 1-D theorem [23].

Theorem 1 [1]: If $A(z_1, z_2)$ is an analytic minimum-phase filter, and $\delta(z_1, z_2)$ is a one-sided filter, analytic in some neighborhood of the unit circles (although not necessarily minimum phase), whose magnitude is less than the magnitude of $A(z_1, z_2)$ when z_1 and z_2 are on the unit circles,

$$|\delta(z_1, z_2| < |A(z_1, z_2)|, \quad |z_1| = |z_2| = 1, \tag{2}$$

then the sum of the two filters is analytic minimum phase.

Proof: The sum of the two filters is analytic in some neighborhood of the unit circles. Since the sum is nonzero on the unit circles, continuity implies that the sum is nonzero in some neighborhood of the unit circles. Consequently, the inverse filter is analytic in the same neighborhood, and therefore is stable. We now show that the inverse can be written in a form, which by inspection can be proved to be one-sided. We have

$$\begin{aligned}[A(z_1, z_2) + \delta(z_1, z_2)]^{-1} &= A^{-1}(z_1, z_2)[1 + A^{-1}(z_1, z_2)\,\delta(z_1, z_2)]^{-1} \\ &= A^{-1}(z_1, z_2) \sum_{n=0}^{\infty} (-1)^n [A^{-1}(z_1, z_2)\,\delta(z_1, z_2)]^n \end{aligned} \tag{3}$$

where, because of (2) and continuity, the geometric series converges uniformly in some neighborhood of the unit circles. Since a product of one-sided filters is one-sided, it follows that each term of the series is one-sided, so the uniform limit is one-sided.

III. 2-D Prediction Error Filters

A. PEF's with Discontinuous Support

A key observation in developing a theory of 2-D linear prediction is that the minimum-phase property does not hold for all 1-D PEF's; in particular, it is guaranteed only for a very special class of filters: those involving the prediction of the sample $x(t)$, given the M immediately preceding samples. These filters are said to have continuous support. As an example of a PEF with *discontinuous* support consider the following three-step predictor:

$$[\hat{x}(t) \,|\, x(t-3), x(t-4)] = -[h_3 x(t-3) + h_4 x(t-4)] \tag{4}$$

based on the discontinuous autocorrelation sequence

$$\{r(0), r(1), r(3), r(4)\} = \{1, 0, -0.2, -0.9\}. \tag{5}$$

Solving the "normal equations," we obtain a *nonminimum-phase* PEF

$$H(z) = (1 + 0.2z^{-3} + 0.9z^{-4}) \tag{6}$$

with mean-square prediction error $P = \frac{3}{20}$. Furthermore, in this example, even if the filter were minimum-phase, the correlation-matching property would not be satisfied. The autocorrelation sequence contains four parameters, while the PEF and mean-square prediction error together contain a total of three parameters, so an infinite number of different autocorrelation sequences can yield the same $H(z)$ and P.

Given the fact that the minimum-phase property and the correlation-matching property are guaranteed only for a special class of 1-D PEF's, it is reasonable to expect that these properties are shared only by a special class of 2-D PEF's. In particular, we should not expect them to be guaranteed for 2-D filters with discontinuous support. Now if we accept the definition of causality adopted in this paper, then it follows that all 2-D filters with *finite* support have *discontinuous* support. Consider, for example, a typical one-quadrant filter:

$$H(z_1, z_2) = (1 + h(0,1)\,z_2^{-1} + h(1,0)\,z_1^{-1} + h(1,1)\,z_1^{-1}z_2^{-1}). \tag{7}$$

The support for this filter includes the points (0, 0) and (1, 1), and two intermediate points (0, 1) and (1, 0). But there are an *infinite* number of points simultaneously in the future of (0, 0) and in the past of (1, 1): $\{(0, l), l \geq 1\}$ and $\{(1, l), l \leq 0\}$, and all but two of these points are excluded from the support of the filter.

Our claim is that Shanks' conjecture is incorrect, not because his filters are two-dimensional, but rather because his filters have discontinuous support. We show in the next section that if a 2-D PEF has continuous support, then it *is* guaranteed to be minimum phase. On the other hand, if a PEF, either 1-D or 2-D, has discontinuous support, then it is not guaranteed, in general, to be minimum phase.

This does not eliminate the possibility that special classes of PEF's with discontinuous support might have the minimum-phase property. For example, it is proved in [1] that the 2-D PEF for the sample $x(k, l)$ given $\{x(k-s, l-t);\ 1 \leq s \leq N, -\infty \leq t \leq \infty\}$ is always minimum phase despite its discontinuous support. (This PEF is denoted $F_N(z_1, z_2)$ in the Appendix of this paper.) Jury [26] has conjectured that a special class of quarter-plane PEF's with discontinuous support are always minimum phase, but no one has yet succeeded in proving or disproving his conjecture. (Jury's conjecture states that if $A(z_1, z_2)$ is a finite quarter-plane filter with rectangular-shaped support, and if $B(z_1, z_2)$ is a PEF based on the autocorrelation of $A(z_1, z_2)$, then $B(z_1, z_2)$ is minimum phase if it has the same support as $A(z_1, z_2)$. However, quarter-plane filters do not seem to have any special properties not shared by other 2-D filters with discontinuous support. Consequently, if Jury's conjecture were proved correct, then it would be reasonable to conjecture that if $B(z_1, z_2)$ is *any* finite 2-D PEF with discontinuous support, then it is minimum phase provided that $A(z_1, z_2)$ has the *same* support as $B(z_1, z_2)$. This more general conjecture would also apply to 1-D PEF's with discontinuous support.)

B. 2-D PEF's with Continuous Support

Consider the following 2-D prediction problem (corresponding to a PEF with continuous support):

$$[\hat{x}(k, l) \,|\, \{x(k-s, l-t); (0,0) \prec (s,t) \preccurlyeq (N, M)\}] = -\sum_{(0,0) \prec (s,t) \preccurlyeq (N,M)} \sum h(N, M; s, t)\, x(k-s, l-t); \tag{8}$$

the geometry of the predictor is illustrated in Fig. 2. The support of the corresponding PEF, denoted

$$H_{N,M}(z_1, z_2) = \left[1 + \sum_{(0,0) \prec (k,l) \preccurlyeq (N,M)} \sum h(N, M; k, l)\, z_1^{-k} z_2^{-l}\right], \tag{9}$$

is shown in Fig. 3.

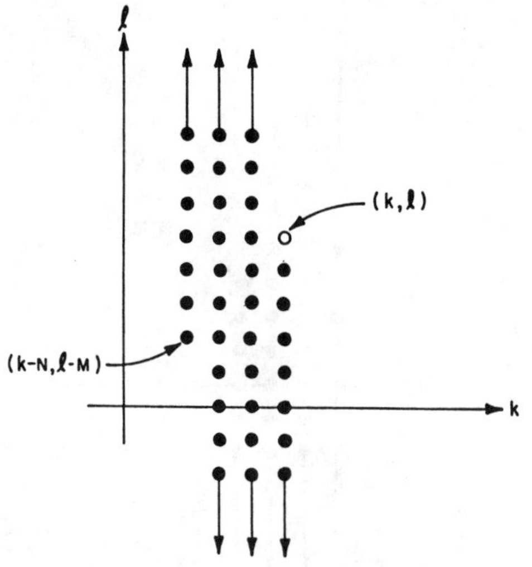

Fig. 2. Geometry of a one-step predictor with continuous support.

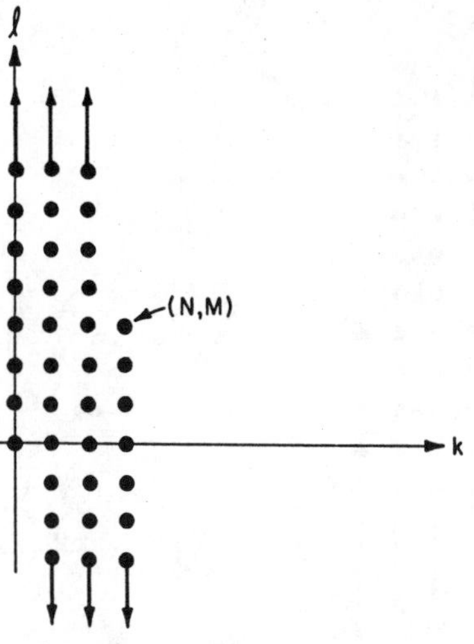

Fig. 3. Geometry of a prediction error filter with continuous support.

Applying the orthogonality principle, we obtain the "normal equations" satisfied by the filter coefficients, and by the mean-square prediction error $P_{N,M}$:

$$\left[r(s,t) + \sum_{(0,0)\prec(k,l)\preccurlyeq(N,M)}\sum h(N,M;k,l)\,r(s-k,t-l)\right] = P_{N,M}\delta_s\delta_t, \quad (0,0)\preccurlyeq(s,t)\preccurlyeq(N,M). \tag{10}$$

These are an infinite set of simultaneous linear equations, and unless we impose certain conditions on the autocorrelation array, there might not be a stable (that is, absolutely summable) solution for the filter coefficients. In our treatment of the problem, we require that the autocorrelation array be analytic positive definite. We define the autocorrelation array, $\{r(s,t);\ (0,0)\preccurlyeq(s,t)\preccurlyeq(N,m)\}$, to be *analytic positive definite* if

1) the autocorrelation function $r(s,t)$ decays exponentially fast to zero as t goes to plus or minus infinity;

2) the following Toeplitz matrix is positive definite for all z_2 on the unit circle:

$$\begin{bmatrix} R_0(z_2) & R_1(1/z_2) & \cdots & R_{N-1}(1/z_2) \\ R_1(z_2) & R_0(z_2) & \cdots & R_{N-2}(1/z_2) \\ \vdots & & & \\ R_{N-1}(z_2) & R_{N-2}(z_2) & \cdots & R_0(z_2) \end{bmatrix} \tag{11}$$

where

$$R_s(z_2) = \sum_{t=-\infty}^{\infty} r(s,t)\,z_2^{-t}, \quad 0 \leqslant s \leqslant (N-1); \tag{12}$$

3) the mean-square prediction errors associated with the class of *finite* PEF's with support (9) have a positive lower bound.

A sufficient condition for an autocorrelation array to be analytic positive definite is that it correspond to a 2-D power density spectrum which is analytic in some neighborhood of the unit circles, and strictly positive on the unit circles.

The following theorems can be proved [1].

Theorem 2a: Given an analytic positive definite autocorrelation array $\{r(s,t), (0,0)\preccurlyeq(s,t)\preccurlyeq(N,M)\}$, then

1) the normal equations (10) have a unique solution for the PEF, $H_{N,M}(z_1,z_2)$, and the mean-square prediction error $P_{N,M}$;

2) $P_{N,M}$ is positive;

3) $H_{N,M}(z_1,z_2)$ is analytic minimum phase.

Theorem 2b: Given any analytic minimum phase filter $H_{N,M}(z_1,z_2)$ of the form (9) for some (N,M) where N is finite, and any positive $P_{N,M}$, there is exactly one analytic positive definite autocorrelation array, $\{r(s,t);\ (0,0)\preccurlyeq(s,t)\preccurlyeq(N,M)\}$, such that the normal equations (10) are satisfied. The autocorrelation array is given by the formula

$$r(s,t) = \frac{1}{(2\pi j)^2}\oint_{|z_1|=1}\oint_{|z_2|=1} \cdot \frac{P_{N,M}\,z_1^{s-1}z_2^{t-1}\,dz_1\,dz_2}{H_{N,M}(z_1,z_2)\,H_{N,M}(1/z_1,1/z_2)}$$
$$(0,0)\preccurlyeq(s,t)\preccurlyeq(N,M). \tag{13}$$

The rigorous proofs for these theorems are rather complicated [1], and we will not duplicate them here. An outline of the proof of Theorem 2a is given in the Appendix in order to illustrate the nature of the proofs.

We note that (13), evaluated for $(s,t)\succ(N,M)$, gives the maximum-entropy extrapolation of the autocorrelation array $\{r(s,t);\ (0,0)\preccurlyeq(s,t)\preccurlyeq(N,M)\}$. Therefore, Theorems 2a and 2b solve the problem of *extrapolating* a 2-D autocorrelation array with continuous support. Unfortunately, in any practical problem we start with an autocorrelation array with finite and therefore *discontinuous* support, so we are faced with an interpolation problem, which even in the 1-D case has no algebraic solution. (In fact, it has been shown that a 2-D positive definite autocorrelation array with discontinuous support does not necessarily have a positive definite extension over the entire plane [27]. An obvious conjecture, therefore, is that a 1-D positive definite autocorrelation sequence with discontinuous support does not necessarily have a positive definite extension. In practice, a 1-D autocorrelation sequence of this type can be obtained from measurements taken with a linear

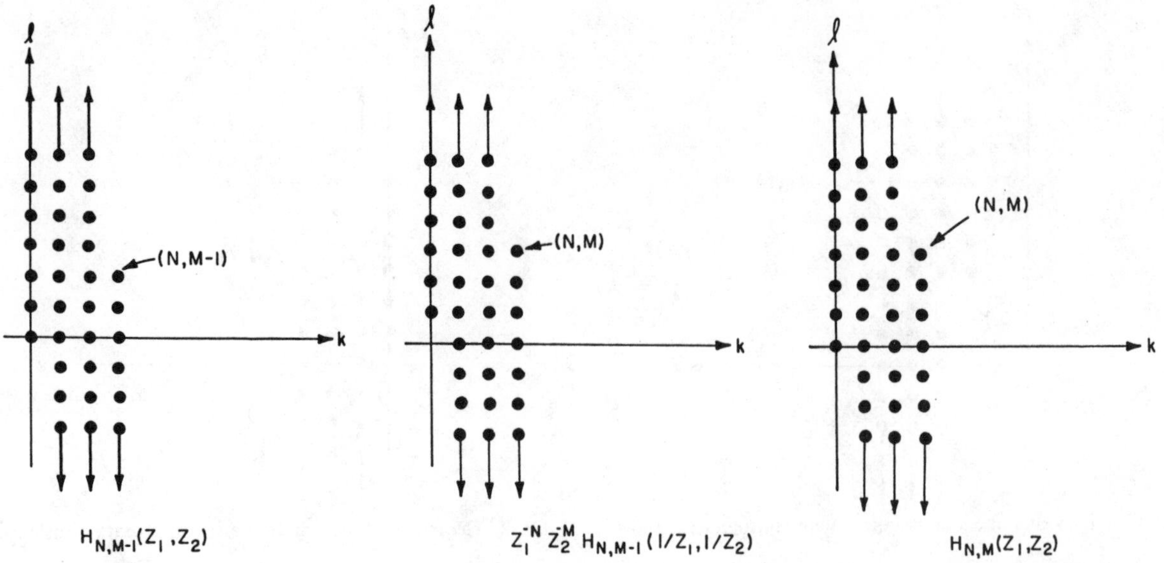

(a) (b) (c)

Fig. 4. Structure of 2-D Levinson recursion.

array of nonuniformly spaced sensors, resulting in a discontinuous sampling of the spatial autocorrelation function.)

C. The 2-D Levinson Algorithm

The geometry of the PEF, $H_{N,M}(z_1,z_2)$ allows it to be solved recursively by a Levinson-type algorithm. This algorithm is a key part of the proof of Theorems 2a and 2b.

Theorem 3 [1]: Suppose that we have an analytic positive definite autocorrelation array $\{r(s,t); (0,0) \preccurlyeq (s,t) \preccurlyeq (N,M)\}$. We further assume that we have a solution to the normal equations for $H_{N,M-1}(z_1,z_2)$ and $P_{N,M-1}$. Then the solution for $H_{N,M}(z_1,z_2)$ and $P_{N,M}$ is given by the formulas

$$H_{N,M}(z_1,z_2) = [H_{N,M-1}(z_1,z_2) + \rho(N,M)\, z_1^{-N} z_2^{-M}\, H_{N,M-1}(1/z_1, 1/z_2)] \tag{14}$$

and

$$P_{N,M} = P_{N,M-1}[1 - \rho^2(N,M)] \tag{15}$$

where

$$\rho(N,M) = -\frac{1}{P_{N,M-1}}\left[r(N,M) + \sum_{(0,0)\prec(s,t)\preccurlyeq(N,M-1)}\sum h(N,M-1;s,t) \cdot r(N-s,M-t)\right]. \tag{16}$$

Furthermore, if $H_{N,M-1}(z_1,z_2)$ is analytic minimum phase, then $H_{N,M}(z_1,z_2)$ is also analytic minimum phase.

Proof: The geometry of the recursion is illustrated in Fig. 4. The structure of the 2-D Levinson algorithm is identical to that of the 1-D Levinson algorithm. Given that $H_{N,M-1}(z_1,z_2)$ and $P_{N,M-1}$ satisfy the normal equations for $(N,M-1)$:

$$\left[r(s,t) + \sum_{(0,0)\prec(k,l)\preccurlyeq(N,M-1)}\sum h(N,M-1;k,l)\, r(s-k,t-l)\right] = P_{N,M-1}\delta_s\delta_t, \quad (0,0) \preccurlyeq (s,t) \preccurlyeq (N,M-1), \tag{17}$$

it can be shown by direct substitution that the solution for $H_{N,M}(z_1,z_2)$ and $P_{N,M}$ as given by (14)-(16) satisfies the normal equations for (N,M). (Alternatively, this can be derived directly by a simple innovations theory argument.) The number $\rho(N,M)$ given by (16) is called a "reflection coefficient;" using Schwarz's inequality it can be shown to have a magnitude less than one. Now if $H_{N,M-1}(z_1,z_2)$ happens to be analytic minimum phase, we can apply Theorem 1 to show that $H_{N,M}(z_1,z_2)$ is also analytic minimum phase.

Conceptually, the 2-D Levinson algorithm provides a recursive procedure for solving for $H_{N,M}(z_1,z_2)$. We begin, as in the 1-D case, with the filter $H_{0,0}(z_1,z_2)=1$. We then solve for the sequence of 1-D PEF's, $\{H_{0,t}(z_1,z_2); 1 \leqslant t \leqslant \infty\}$ using (14)-(16) for $N=0$ and $M=t$, which is equivalent to the 1-D Levinson algorithm. We then solve for the sequence of 2-D PEF's, $\{H_{1,t}(z_1, z_2); -\infty \leqslant t \leqslant \infty\}$, beginning with $H_{1,-\infty}(z_1,z_2) = H_{0,+\infty}(z_1,z_2)$ and applying (14)-(16) for $N=1$ and $M=t$. The procedure continues in a similar manner, thereby sequentially generating an array of PEF's, $\{H_{s,t}(z_1,z_2); (0,0) \preccurlyeq (s,t) \preccurlyeq (N,M)\}$, where the index s is updated by the condition that $H_{s,-\infty}(z_1,z_2) = H_{s-1,+\infty}(z_1,z_2)$, and the index t is updated by (14)-(16). Given this recursion, the minimum-phase property of $H_{N,M}(z_1,z_2)$ is intuitively reasonable. If $H_{s,t-1}(z_1,z_2)$ is minimum phase, then so is $H_{s,t}(z_1,z_2)$. But since the PEF's are computed sequentially, starting with the minimum-phase filter $H_{0,0}(z_1,z_2)=1$, this strongly suggests that all of the $H_{s,t}(z_1,z_2)$ are minimum phase. (Although this is an effective plausibility argument for Theorem 2a, the rigorous proof is considerably more difficult, since beginning with $H_{0,0}(z_1,z_2)$, an infinite number of intermediate PEF's must be computed before the 2-D Levinson algorithm yields $H_{N,M}(z_1, z_2)$.)

It is well known that the 1-D Levinson algorithm is closely related to the problem of normal incidence wave-propagation in a medium consisting of equal travel-time layers [24], [8]; this is the reason for the term "reflection coefficient." For convenience we also call the number $\rho(N,M)$ a reflection coefficient, even though there is no physical interpretation for the 2-D Levinson algorithm.

IV. The 2-D Reflection Coefficient Representation

Equation (14) of the 2-D Levinson algorithm suggests a relation between 2-D minimum-phase filters and 2-D reflection coefficient arrays. We define a *reflection coefficient array* to be a one-sided array of real numbers, $\{\rho(k,l); (0,1) \preccurlyeq (k,l) \preccurlyeq (N,M)\}$, whose magnitudes are less than one. In addition, in our treatment, we require that $\rho(k,l)$ decay exponentially fast to zero as l goes to plus or minus infinity. It can be shown that there is a one-to-one relation between 2-D reflection coefficient arrays of this type, and 2-D analytic minimum-phase filters of the form (9) [1].

Theorem 4a: Given a 2-D reflection coefficient array, $\{\rho(k,l); (0,1) \preccurlyeq (k,l) \preccurlyeq (N,M)\}$, for some finite N, such that

$$|\rho(k,l)| < (1+\epsilon)^{-|l|}, \tag{18}$$

for some positive constant ϵ, there is a unique set of 2-D analytic minimum-phase filters, $\{H_{s,t}(z_1,z_2); (0,0) \preccurlyeq (s,t) \preccurlyeq (N,M)\}$, of the form

$$H_{s,t}(z_1,z_2) = \left[1 + \sum_{(0,0) \prec (k,l) \preccurlyeq (s,t)} \sum h(s,t;k,l) z_1^{-k} z_2^{-l}\right] \tag{19}$$

such that

1) $H_{0,0}(z_1,z_2) = 1;$ (20)

2) $H_{s,t}(z_1,z_2) = [H_{s,t-1}(z_1,z_2) + \rho(s,t) z_1^{-s} z_2^{-t} H_{s,t-1}(1/z_1, 1/z_2)],$ $(0,1) \preccurlyeq (s,t) \preccurlyeq (N,M);$ (21)

3) $H_{s,-\infty}(z_1,z_2) = H_{s-1,+\infty}(z_1,z_2), \quad 1 \leqslant s \leqslant N;$ (22)

4) $\lim_{t \to +\infty} H_{s,t}(z_1,z_2) = H_{s,+\infty}(z_1,z_2), \quad 0 \leqslant s \leqslant (N-1);$ (23)

5) $\lim_{t \to -\infty} H_{s,t}(z_1,z_2) = H_{s,-\infty}(z_1,z_2), \quad 1 \leqslant s \leqslant N;$ (24)

(the convergence of (23) and (24) is uniform in some neighborhood of the unit circles.)

Theorem 4b: Given any analytic minimum-phase filter, $H_{N,M}(z_1,z_2)$, for some finite N, of the form (19), there is a unique reflection coefficient array, $\{\rho(k,l); (0,1) \preccurlyeq (k,l) \preccurlyeq (N,M)\}$, and a unique set of analytic minimum-phase filters, $\{H_{s,t}(z_1,z_2); (0,0) \preccurlyeq (s,t) \preccurlyeq (N,M-1)\}$ such that (18)–(24) are satisfied.

Given the reflection coefficient array, the implicit order in which the filters are computed begins with $(s,t) = (0,0)$, followed by $\{s = 0, 1 \leqslant t \leqslant \infty\}$, followed by $\{s = 1, -\infty \leqslant t \leqslant \infty\}$, and so on. Equation (21) updates the index t, (22) updates the index s.

Some important properties of this representation are as follows.

1) If the filter has finite nonzero support, then the reflection coefficient array almost always has infinite nonzero support in the z_2 direction. For example, the minimum-phase filter,

$$H_{1,1}(z_1,z_2) = (1 + 0.5z_2^{-1} + 0.5z_1^{-1} + 0.25z_1^{-1}z_2^{-1}), \tag{25}$$

can be shown to have the following reflection coefficient array:

$$\rho(k,l) = \begin{cases} \frac{1}{2}, & (k,l) = (0,1) \\ \frac{1}{4}, & (k,l) = (1,1) \\ \frac{6(-2)^{-l}}{(16 \cdot 4^{-l} - 1)} & k = 1,\ l \leqslant 0 \\ 0, & \text{otherwise.} \end{cases} \tag{26}$$

This can be verified by means of a "backwards" version of (21), whereby $H_{s,t-1}(z_1,z_2)$ and $\rho(s,t)$ are expressed in terms of $H_{s,t}(z_1,z_2)$.

2) If the reflection coefficient array has finite nonzero support, then the filter also has finite nonzero support, and it can be obtained by means of a finite recursion. For example, given the finite reflection coefficient array,

$$\rho(k,l) = \begin{cases} 0.8, & (k,l) = (0,1) \\ -0.6, & (k,l) = (1,-1) \\ 0.2, & (k,l) = (1,0) \\ 0.1, & (k,l) = (1,1) \\ 0, & \text{otherwise} \end{cases} \tag{27}$$

it can be shown that the resulting minimum-phase filter is

$$H_{1,1}(z_1,z_2) = [1 + 0.7z_2^{-1} - 0.14z_2^{-2} - 0.048z_2^{-3} - 0.48z_1^{-1}z_2^{2} - 0.4496z_1^{-1}z_2 + 0.268z_1^{-1} + 0.1z_1^{-1}z_2^{-1}]. \tag{28}$$

The recursion is performed by first applying (21) for $(s,t) = (0,1)$, then by using (21), (22), and the fact that $\{\rho(k,l) = 0; (0,2) \preccurlyeq (k,l) \preccurlyeq (1,-2)\}$ to give $H_{1,-2}(z_1,z_2)$. Finally, (21) is successively applied for $\{s = 1, -1 \leqslant t \leqslant 1\}$ to obtain $H_{1,1}(z_1,z_2)$. It will be noted that the nonzero support for the filter is greater than that of the reflection coefficient array. This is an unavoidable property of the 2-D reflection coefficient representation. The same phenomenon occurs in the 1-D case when the reflection coefficient sequence has discontinuous support. In either case, a lattice structure can be a more efficient way to implement the filter, than a direct structure [22].

3) If $H_{N,M}(z_1,z_2)$ has infinite nonzero reflection coefficient support, then it can be approximated by a finite filter, $H^{(L)}(z_1,z_2)$, with finite reflection coefficient support, $\rho^{(L)}(k,l)$, where

$$\rho^{(L)}(k,l) = \begin{cases} \rho(k,l), & |l| \leq L \\ 0, & \text{otherwise} \end{cases} \tag{29}$$

and

$$\lim_{L\to\infty} H^{(L)}(z_1,z_2) = H_{N,M}(z_1,z_2) \tag{30}$$

where the convergence is uniform in some neighborhood of the unit circles.

4) If $H_{N,M}(z_1,z_2)$ has finite reflection coefficient support, then it can be implemented by means of a 2-D lattice structure. Likewise the inverse (recursive) filter $H_{N,M}^{-1}(z_1,z_2)$ can be implemented by a lattice structure. The only constants to appear in the lattice structures are the reflection coefficients. (The discovery of these lattice structures is due to S. W. Lang [22].)

5) If $H_{N,M}(z_1,z_2)$ has finite reflection coefficient support, then the autocorrelation function associated with the autoregressive spectrum,

$$S(z_1,z_2) = \frac{P_{N,M}}{H_{N,M}(z_1,z_2)\,H_{N,M}(1/z_1,1/z_2)}, \tag{31}$$

can be obtained algebraically, by means of a backwards version of the 2-D Levinson algorithm. In contrast, if the filter has infinite reflection coefficient support, even if the filter has finite support, then the autocorrelation function can only be obtained by numerical evaluation of the inverse Fourier transform.

The practical implication of these properties is that 2-D minimum-phase filters can be constructed in the reflection coefficient domain, and that by using a sufficiently big reflection coefficient array, and by choosing the reflection coefficients appropriately, any desired frequency response can be approximated arbitrarily closely. The fundamental advantage of working in the reflection coefficient domain is that the minimum-phase condition is automatically guaranteed.

The 2-D reflection coefficient representation has been used for recursive filter design [1], [25]. Further development of design algorithms is needed before the method becomes competitive with the best of the other available filter design techniques.

The representation is potentially useful in 2-D autoregressive spectral estimation. Some possible computational algorithms are presented in [1], but no numerical experiments have been performed.

V. Conclusions

We have shown that by working with arrays having continuous support, all of the major theoretical results of 1-D linear prediction can be extended to the 2-D case. The significant practical result to emerge from this new theory is the 2-D reflection coefficient representation for 2-D minimum-phase filters.

We have given a simple explanation for the failure of Shanks' conjecture; likewise we have shown that the 2-D covariance extension problem is really an interpolation problem.

Two-dimensional filters with continuous support behave like 1-D filters with continuous support, and 2-D filters with discontinuous support behave like 1-D filters with discontinuous support. Most of theoretical difficulties with 2-D filtering occur because any practical problem involves filters, autocorrelation arrays, and arrays of data which have *finite* and therefore *discontinuous* support. This suggests that research efforts would be profitably directed towards 1-D problems involving sequences with discontinuous support. These 1-D problems would be useful to solve in their own right; furthermore, any breakthroughs on these problems may lead to similar breakthroughs on 2-D problems.

It is appropriate to end this paper by mentioning some results of other researchers in the area of 2-D linear prediction. Suppose that we have a 2-D random process whose power spectrum is analytic in some neighborhood of the unit circles and strictly positive on the unit circles. Then according to the 2-D spectral factorization theorem, the PEF for the sample $x(k,l)$ based on all past samples is analytic minimum phase, and it is a whitening filter for the random process. The 2-D spectral factorization theorem was first discovered by Whittle [28]. It was rediscovered by Helson and Lowdenslager [29], Dudgeon [30], and Ekstrom and Woods [31]. Helson and Lowdenslager proved the theorem under more general conditions than the positive analytic condition. Intuitively we expect that the analytic minimum-phase PEF $H_{N,M}(z_1,z_2)$, discussed in this paper, converges uniformly to the minimum-phase whitening filter as N goes to plus infinity. In fact, this was proved recently by Delsarte, Genin, and Kamp [32]. In the same paper, Delsarte *et al.* established the uniform convergence of the PEF's of Chang and Agarwal [18] (which have discontinuous support) to the minimum-phase whitening filter, thereby proving that these filters are guaranteed to be minimum phase if they are made sufficiently big.

Appendix

Outline of Proof of Theorem 2a

1) We first obtain a constructive solution to the normal equations (10) for the special case $M = -\infty$. Our solution is of the form

$$\begin{aligned} H_{N,-\infty}(z_1,z_2) &= H_{N-1,+\infty}(z_1,z_2) \\ &= G_{N-1}(z_2)\,F_{N-1}(z_1,z_2) \end{aligned} \tag{A1}$$

where $G_{N-1}(z_2)$ and $F_{N-1}(z_1,z_2)$ are of the form

$$G_{N-1}(z_2) = \left[1 + \sum_{l=1}^{\infty} g(N-1;l)\,z_2^{-l}\right] \tag{A2}$$

and

$$\begin{aligned} F_{N-1}(z_1,z_2) &= \left[1 + \sum_{k=1}^{N-1} F_{N-1;k}(z_2)\,z_1^{-k}\right] \\ &= \left[1 + \sum_{k=1}^{N-1}\sum_{l=-\infty}^{\infty} f(N-1;k,l)\,z_1^{-k} z_2^{-l}\right]. \end{aligned} \tag{A3}$$

$G_{N-1}(z_2)$ is the analytic minimum-phase solution to the following 1-D spectral factorization problem:

$$Q_{N-1}(z_2) = \frac{P_{N-1,+\infty}}{G_{N-1}(z_2)\, G_{N-1}(1/z_2)} \tag{A4}$$

where $Q_{N-1}(z_2)$ is a positive analytic spectrum. In turn, $Q_{N-1}(z_2)$, along with $F_{N-1}(z_1, z_2)$, is obtained as the solution to the following set of linear equations:

$$\left[R_s(z_2) + \sum_{k=1}^{N-1} F_{N-1;k}(z_2)\, R_{s-k}(z_2)\right] = Q_{N-1}(z_2)\, \delta_s,$$
$$0 \leqslant s \leqslant (N-1), \tag{A5}$$

where the $R_s(z_2)$ are defined by (12).

These equations have a recursive solution, analogous to the 1-D Levinson algorithm, in terms of reflection coefficient functions of z_2. Using Theorem 1, we can prove inductively that $F_{N-1}(z_1, z_2)$ is analytic minimum phase. Therefore, since $G_{N-1}(z_2)$ and $F_{N-1}(z_1, z_2)$ are analytic minimum phase, it follows that $H_{N-1,+\infty}(z_1, z_2)$ is also analytic minimum phase.

2) We next obtain a constructive solution to the normal equations, for $H_{N,t}(z_1, z_2)$, valid for all *sufficiently small t*, $(t \to -\infty)$ of the form

$$H_{N,t}(z_1, z_2) = [A_{N,t}(z_2)\, H_{N-1,+\infty}(z_1, z_2) + B_{N,t}(1/z_2)\, z_1^{-N} z_2^{-t}\, H_{N-1,+\infty}(1/z_1, 1/z_2)] \tag{A6}$$

where

$$A_{N,t}(z_2) = \left[1 + \sum_{l=1}^{\infty} a(N, t; l)\, z_2^{-l}\right], \tag{A7}$$

and

$$B_{N,t}(z_2) = \sum_{l=0}^{\infty} b(N, t; l)\, z_2^{-l}. \tag{A8}$$

(The form of this solution, as well as (A1), can be formally derived by simple innovations theory arguments.)

Solutions for $A_{N,t}(z_2)$ and $B_{N,t}(z_2)$ are obtainable by Neumann-type series. In addition, using the series solutions we can show that

$$\lim_{t \to -\infty} A_{N,t}(z_2) = 1 \tag{A9}$$

and

$$\lim_{t \to -\infty} B_{N,t}(z_2) = 0 \tag{A10}$$

where the convergence is uniform in some neighborhood of the unit circle.

Therefore, we have a solution to $H_{N,t}(z_1, z_2)$ for all sufficiently small t, and we can prove the uniform convergence of the sequence of filters to $H_{N-1,+\infty}(z_1, z_2)$ as t goes to minus infinity. Therefore, we can apply Theorem 1 to show that $H_{N,t}(z_1, z_2)$ is analytic minimum phase for all sufficiently small t.

3) Having proved the existence of an analytic minimum-phase solution for $H_{N,t}(z_1, z_2)$ for all sufficiently small t, we can successively apply the 2-D Levinson algorithm to obtain analytic minimum-phase solutions for $H_{N,t}(z_1, z_2)$ for all $t \leqslant M$.

Acknowledgment

The author would like to thank A. Baggeroer, J. McClellan, and A. Willsky for their valuable comments during the course of this research. Also, the author thanks the referees for their thorough reviews of this paper.

References

[1] T. Marzetta, "A linear prediction approach to two-dimensional spectral factorization and spectral estimation," Ph.D. dissertation, Dep. Elec. Eng. & Comput. Sci., Mass. Inst. Technol., Cambridge, MA, Feb. 1978.

[2] E. Robinson and H. Wold, "Minimum-delay structure of least-square/eo-ipso predicting systems for stationary stochastic processes," in *Proc. Symp. Time Series Analysis*, M. Rosenblatt, Ed. New York: Wiley, 1962, pp. 192–196.

[3] J. Markel and A. Gray "On autocorrelation equations as applied to speech analysis," *IEEE Trans. Audio Electroacoust.*, vol. AU-21, pp. 69–79, Apr. 1973.

[4] J. Shanks, "Recursion filters for digital processing," *Geophys.*, vol. 32, pp. 33–51, Feb. 1967.

[5] C. Burrus and T. Parks, "Time domain design of recursive digital filters," *IEEE Trans. Audio Electroacoust.*, vol. AU-18, pp. 137–141, June 1970.

[6] L. Scharf and J. Luby, "Statistical design of autoregressive-moving average digital filters," *IEEE Trans. Acoust., Speech, Signal Processing*, vol. ASSP-27, pp. 240–247, June 1979.

[7] J. Burg, "Maximum entropy spectral analysis," Ph.D. dissertation, Dep. Geophys., Stanford Univ., Stanford, CA, May 1975.

[8] L. Pusey, "An innovations approach to spectral estimation and wave propagation," Sc.D. thesis, Dep. Elec. Eng., Mass. Inst. Technol., Cambridge, MA, June 1975.

[9] R. Dubroff, "The effective autocorrelation function of maximum entropy spectra," *Proc. IEEE*, pp. 1622–1623, Nov. 1975.

[10] N. Levinson, "The Wiener RMS error in filter design and prediction," in *Extrapolation, Interpolation, and Smoothing of Stationary Time Series*, N. Wiener, Ed. New York: Wiley, 1949, Appendix B.

[11] J. Durbin, "The fitting of time-series models," *Rev. Int. Inst. Statist.*, vol. 28, pp. 233–244, 1960.

[12] A. Gray and J. Markel, "Digital lattice and ladder filter synthesis," *IEEE Trans. Audio Electroacoust.*, vol. AU-21, pp. 491–500, Dec. 1973.

[13] J. Makhoul, "A class of all-zero lattice digital filters: Properties and applications," *IEEE Trans. Acoust., Speech, Signal Processing*, vol. ASSP-26, pp. 304–314, Aug. 1978.

[14] A. Vieira and T. Kailath, "On another approach to the Schur-Cohn criterion," *IEEE Trans. Circuits Syst.*, vol. CAS-24, pp. 218–220, Apr. 1977.

[15] J. Makhoul, "Linear prediction: A tutorial review," *Proc. IEEE*, vol. 63, pp. 561–580, Apr. 1975.

[16] J. Shanks, S. Treitel, and J. Justice, "Stability and synthesis of two-dimensional recursive filters," *IEEE Trans. Audio Electroacoust.*, vol. AU-20, pp. 115–128, June 1972.

[17] Y. Genin and Y. Kamp, "Counterexample in the least square inverse stabilization of 2D-recursive filters," *Electron. Lett.*, vol. 11, pp. 330–331, 1975.

[18] H. Chang and J. Aggarwal, "Design of two-dimensional semi-causal recursive filters," *IEEE Trans. Circuits Syst.*, vol. CAS-25, pp. 1051–1059, Dec. 1978.

[19] J. Bednar and C. Farmer, "An algorithm for the inversion of finite block Toeplitz matrices with application to spatial digital filters," in *Proc. Comput. Image Processing and Recognition*, vol. 2, Columbia, MO, Aug. 1972, pp. 9-3-1–9-3-10.

[20] J. Justice, "A Levinson-type algorithm for two-dimensional Wiener filtering using bivariate Szego polynomials," *Proc. IEEE*, vol. 65, pp. 882–886, June 1977.

[21] B. Levy, M. Morf, and S. Kung, "New results in 2-D systems theory III: Recursive realization and estimation for 2-D systems," in *Proc. 20th Midwest Symp. Circuits Syst.* North Hollywood, CA: Western Periodicals, 1977.

[22] S. Lang, personal communication, July 14, 1978.

[23] J. Claerbout, *Fundamentals of Geophysical Data Processing With Applications to Petroleum Prospecting.* New York: McGraw-Hill, 1976, pp. 29–31.

[24] J. Claerbout, "Synthesis of a layered medium from its acoustic transmission response," *Geophys.*, vol. 33, pp. 264–269, Apr. 1968.
[25] T. Marzetta, "The design of 2-D recursive filters in the 2-D reflection coefficient domain," in *Proc. 1979 Int. Conf. Acoust., Speech, Signal Processing*, 79CH1379-7 ASSP, 1979, pp. 32–35.
[26] E. Jury, "Stability of multidimensional scalar and matrix polynomials," *Proc. IEEE*, vol. 66, pp. 1018–1047, Sept. 1978.
[27] B. Dickinson, "Two-dimensional Markov spectrum estimates need not exist," *IEEE Trans. Inform. Theory*, vol. IT-26, pp. 120–121, Jan. 1980.
[28] P. Whittle, "On stationary processes in the plane," *Biometrika*, vol. 41, pp. 434–449, Dec. 1954.
[29] H. Helson and D. Lowdenslager, "Prediction theory and Fourier series in several variables," *Acta Mathematica*, vol. 99, pp. 165–202, 1958.
[30] D. Dudgeon, "Two-dimensional recursive filtering," Sc.D. thesis, Dep. Elec. Eng., Mass. Inst. Technol., Cambridge, MA, May 1974.
[31] M. Ekstrom and J. Woods, "Two-dimensional spectral factorization with applications in recursive digital filtering," *IEEE Trans. Acoust., Speech, Signal Processing*, vol. ASSP-24, pp. 115–127, Apr. 1976.
[32] P. Delsarte, Y. Genin, and Y. Kamp, "Half-plane Toeplitz systems," MBLE Research Lab., Brussels, Belgium, Rep. R391, Feb. 1979.

Realizable Wiener Filtering in Two Dimensions

MICHAEL P. EKSTROM, SENIOR MEMBER, IEEE

Abstract—**The extension of Wiener's classical mean-square estimation theory to a two-dimensional setting is presented. In analogy with the one-dimensional problem, the optimal realizable filter is derived by solution of a two-dimensional, discrete Wiener-Hopf equation using a spectral factorization procedure. Filters are developed for the cases of prediction, filtering, and smoothing, and appropriate error expressions are derived to characterize their performance.**

I. Introduction

WHILE developed in antiquity, Wiener's minimum mean-square error (MMSE) estimation theory [1] continues to play a prominent role in modern time series analysis. Indeed, its extension to new applications is a topic of active research interest.

The representative problem addressed by Wiener was that of optimally estimating an unobserved time signal $s(t)$ given a noise corrupted observation, $s(t) + w(t)$, where $w(t)$ is a noise process. Both signal and noise are modeled as wide-sense stationary processes; the estimator is constrained to be linear, and derived optimal in the sense of achieving the MMSE associated with the estimate. Two classes of estimators were described by Wiener: the so-called noncausal (unrealizable, bilateral) filter which uses all past, present, and future observations in forming the estimate, and the so-called causal (realizable, unilateral) filter which uses only past and present observations.

With appropriate generalization, the fundamental estimation problem addressed by Wiener, that of estimating signals from

Manuscript received May 28, 1981. This work was supported in part by a Senior Scientist Award of the Alexander von Humboldt Foundation, while the author was Visiting Scientist at the Lehrstuhl für Nachrichtentechnik, Universität Erlangen-Nürnberg, Erlangen, West Germany.

The author is with Schlumberger-Doll Research, Ridgefield, CT 06877.

Reprinted from *IEEE Trans. Acoust., Speech, Signal Processing,* vol. ASSP-30, pp. 31–40, Feb. 1982.

noisy observations, occurs in many applications involving two-dimensional (2-D) data sets. An example problem which has received much attention in the literature is that of image restoration [2], [3]. Here a defocused or blurred image is recorded, and it is desired to remove the blur effects, thereby "restoring" the fidelity of the object being imaged. In this case, the observed image is modeled as a linear transformation (this transformation characterizes the blurring) plus additive noise. The object field is, of course, the signal to be estimated. Image restoration problems of this type occur in a wide variety of fields, including atmospheric physics [4], biomedical imaging [5], and laser fusion experimentation [6].

As a consequence of its general form, efforts have been made for some time to extend Wiener's formalism to 2-D problems. Gabor [7] apparently first derived the 2-D Wiener "noncausal" filter for continuous fields, with the discrete or digital version being derived by Helstrom [8]. Both filters have been extensively and successfully used in optical and digital signal processing applications, respectively.

In contrast, despite the forty year lapse since Wiener's original work, the 2-D (or multidimensional) generalization of his "causal" filter has not been previously published. This appears to be due in large measure to the mathematical complexity associated with the theory of 2-D random fields, and the absence of major applications requiring "real-time" 2-D digital signal processing. This situation has changed remarkably, however, with recent developments in the theory of 2-D systems [9] and the promise of practical implementations of 2-D digital processors offered by VLSI. In digital image processing, for example, the design of numerous "real-time" processors are now being attempted for tasks thought to be overly complex for digital implementation only a few years ago.

The purpose of this paper is to present the 2-D generalization of Wiener's "causal" MMSE filtering theory, thereby completing the extension of his basic approach to a multidimensional setting. Because of applications interests, the problem of optimally estimating 2-D *discrete* random fields is addressed. The presentation begins with a discussion of the issues of physical realizability and causality as they arise in 2-D. The filtering problem is then formulated, and the optimal filter derived by solving a 2-D discrete Wiener-Hopf equation. A 2-D spectral factorization procedure is used for this purpose [10]. Canonical versions of the filter are derived for the cases of prediction, filtering, and smoothing.

II. Physical Realizability in Two Dimensions

Typically, in 1-D the causality and realizability descriptors are used synonymously to distinguish system functions having the property of being physically realizable, that is, they can be realized by stable systems constructed with physically-real components. More precisely, a 1-D LSI system with unit sample response $h(n)$ is said to be causal (realizable) iff

$$h(n) = 0, \quad n < 0. \tag{1}$$

Such systems are also said to be unilateral because of their one-sided response.

How do physical realizability conditions manifest themselves in 2-D? Efforts to develop a description in analogy to (1) for 2-D systems have seemingly been hindered by the fact that causality is traditionally a time-domain concept, while 2-D filtering applications are typically defined in terms of spatial not temporal variables. For example, in imaging applications the 2-D signals are light-intensity fields defined over spatial regions (spatial independent variables). This has led to the adoption of needlessly restrictive definitions of causality (physical realizability) [11], [12], and the characterization of some 2-D filter forms as "semicausal" or "partially anticipative" [13]. Inasmuch as causality is a binary attribute (in this regard much like pregnancy), the incongruity of these last descriptors is apparent.

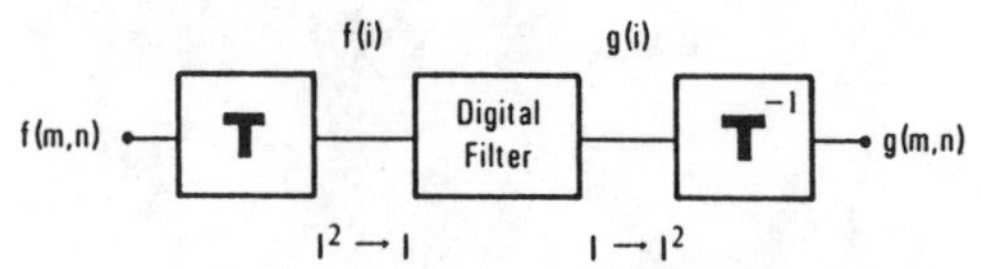

Fig. 1. Block diagram of 2-D digital filter model.

With some care, however, simple and intuitive characterizations of 2-D realizability (based on the support of the filter's point spread response) can be developed. When dealing with image fields, it is well known that analog signal processors having bilateral or double-sided response functions are realizable using optical components. In this case, the optimal, *realizable* 2-D Wiener filter is the direct 2-D extension of Wiener's 1-D *unrealizable* filter (see [7]).

Of more specific interest here are 2-D signal processors employing digital filtering elements. In this case, the input/output relations of a 2-D LSI system are described by

$$g(m,n) = \sum_{m',n'}\sum f(m-m', n-n')\, h(m',n') \tag{2}$$

where $f(m,n)$ and $g(m,n)$ are the input and output bisequences and $h(m,n)$ the system unit point spread response. The independent variables m, n are taken to be integers. All the bisequences in (2) are assumed to be absolutely summable.

Although this digital filtering operation is conceptualized in the 2-D spatial domain, there is a time sequency associated with its realization. When implementing the filtering, the 2-D signal is first mapped into a 1-D sequence by a mapping $T: I^2 \to I$, the filter output is computed according to the lagged products and sums in (2), and the result then mapped to 2-D via T^{-1} (see Fig. 1). As an illustration of this implementation structure, consider the "real-time" processing of imagery from a video sensor. The image illuminating the vidicon face is raster scanned and converted to a time signal. This signal (or sequence, following sampling and A/D conversion) is filtered according to (2). The filtered signal then drives a raster graphics device, displaying the processed image.

A. Spatial Conditions

It is the form of the mapping T which imposes realizability conditions on the spatial support of the filter response $h(m,n)$. For specific mappings, these conditions can be straightforwardly developed as 2-D analogs of (1). The most commonly used mappings in applications involve so-called row-, column-, and diagonal-sequential scanning. To illustrate the result, the case of row-sequential scanning will be treated in detail, with the results stated for the remaining two cases.

Definition: The bisequence $f(m,n)$ is said to be row-sequentially scanned by the transformation

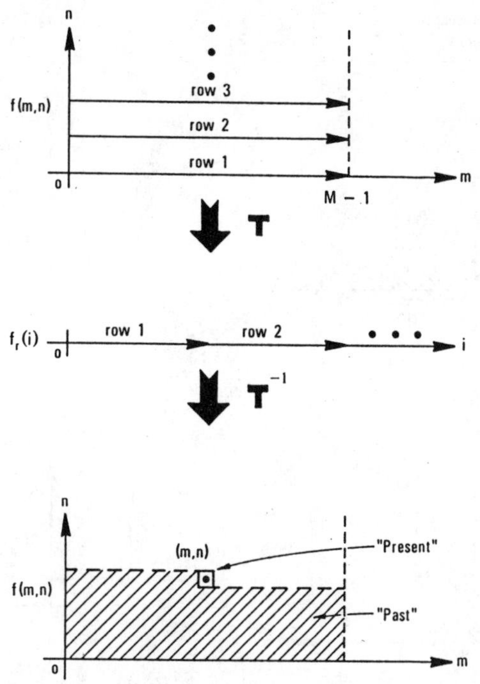

Fig. 2. Row sequential mapping of $f(m, n) \to f_r(i)$.

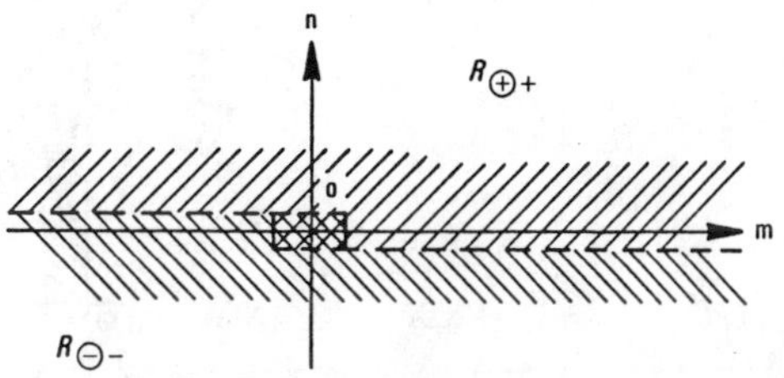

Fig. 3. Regions of support for the upper half-plane ($R_{\oplus+}$) and lower half-plane ($R_{\ominus-}$). (The half-planes intersect at the origin.)

$$T:\ i = m + nM \quad \forall\, 0 \leqslant m < M,\ \ 0 \leqslant n < N$$

resulting in the 1-D sequence

$$f_r(i) = f_r(m + nM) = f(m, n)$$

where $f(m, n)$ is taken to be defined on the region $0 \leqslant m < M$, $0 \leqslant n < N$.

This mapping involves a simple concatenation of rows, and totally orders the bisequence, with i indexing the time sequency of the processing. From the fundamental causality relations, any physically realizable processor may use only present and past values of the input sequence. Under T^{-1}, these values of $f(m, n)$ are shown in Fig. 2. Thus, the following theorem can be stated.

Theorem 1: Under row-sequential scanning, the system in (2) is physically realizable iff $h(m, n)$ is absolutely summable and its support is contained in the region $R_{\oplus+}$,[1] where

$$R_{\oplus+} \triangleq \{(m,n):\ 0 \leqslant m \leqslant \infty, 0 \leqslant n \leqslant \infty\} \cup \{(m,n):\ -\infty \leqslant m < 0, 0 < n \leqslant \infty\}. \tag{4}$$

The region $R_{\oplus+}$ is called an upper half-plane or nonsymmetric half-plane [10], and is illustrated in Fig. 3. The lower half-plane is correspondingly defined

$$R_{\ominus-} = \{(m,n):\ -\infty \leqslant m \leqslant 0, -\infty \leqslant n \leqslant 0\} \cup \{(m,n):\ 0 < m \leqslant \infty, -\infty \leqslant n < 0\}. \tag{5}$$

On reflection, these realizability results may appear intuitively evident. Although there exist more elegant ways of characterizing the ordering induced by T [14], the above approach does provide considerable practical insight into their origin and their subsequent modification for alternate applications. The transparency of the half-plane genesis is an example of this insight.

B. Transform Conditions

Recall that 1-D physical realizability as described in (1) also invokes certain requirements on the regions of analyticity of the system transfer function $H(z)$. As might be expected, this also carries over to the 2-D case, the conditions of interest here being those for half-plane bisequences.

The 2-D z transform of $h(m, n)$ is written as

$$H(z_1, z_2) = \sum_{m,n=-\infty}^{+\infty}\sum h(m,n) z_1^{-m} z_2^{-n} \tag{6}$$

and is defined over a region of convergence in C^2 such that

$$\sum_{m,n=-\infty}^{+\infty}\sum |h(m,n) z_1^{-m} z_2^{-n}| < \infty. \tag{7}$$

[1]For a description of the "⊕+" notation, see [10].

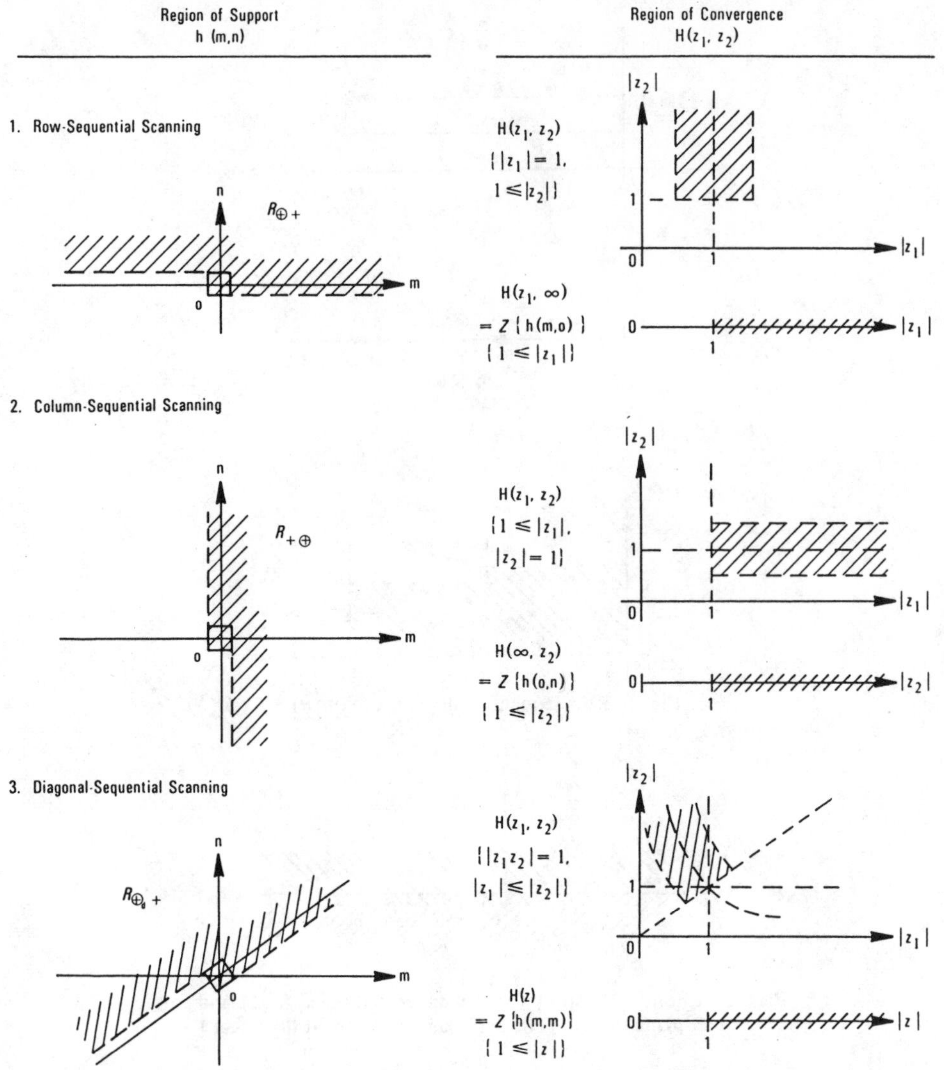

Fig. 4. Examples of stable bisequences with prescribed regions of support and their z transforms with corresponding regions of holomorphy.

A transform pair is indicated by the relation

$$h(m,n) \Longleftrightarrow H(z_1,z_2). \tag{8}$$

The relations of interest for upper (and lower) half-plane bisequences are as follows.

Theorem 2 [10]: An absolutely summable bisequence $h(m,n)$, taking support on $R_{\oplus+}(R_{\ominus-})$ has a z transform satisfying the set of holomorphic conditions $D_{\oplus+}(D_{\ominus-})$ where

$D_{\oplus+}$: *Condition 1*–$H(z_1,z_2)$ is homomorphic on an open region including $\{(z_1,z_2): |z_1| = 1, 1 \leqslant |z_2|\}$.

Condition 2–$H(z_1,\infty)$ is analytic on a region including $\{z_1: 1 \leqslant |z_1|\}$.

Likewise

$D_{\ominus-}$: *Condition 1*–$H(z_1,z_2)$ is holomorphic on open region including $\{(z_1,z_2): |z_1| = 1, |z_2| \leqslant 1\}$.

Condition 2–$H(z_1,\infty)$ is analytic on a region including $\{z_1: |z_1| \leqslant 1\}$.

These conditions follow naturally from the concept of "one-sidedness." For the $R_{\oplus+}$ case, $h(m,n)$ takes values only for $n \geqslant 0$ (the *full*, upper half-plane); consequently, it is one-sided with respect to n. This leads to Condition 1 of Theorem 2. The second condition arises because $h(m, 0)$ is a unilateral sequence, one-sided with respect to m. A corresponding realizability theorem can be stated.

Theorem 3: Under row-sequential mapping, the system in (2) is physically realizable iff $H(z_1,z_2)$ satisfies the set of holomorphic conditions $D_{\oplus+}$.

This result and those for column- and diagonal-sequential scanning are summarized in Fig. 4. The last two sets are derived by noting the coordinate transformations between the various half-planes and $R_{\oplus+}$.

III. Optimal Wiener Filtering

Theorems 1 and 2 will assume a fundamental role in developing the 2-D MMSE filtering formalism. The first theorem formally extends the concepts of causality/physical realizability to 2-D in a manner consistent with Wiener's approach, and in so doing, establishes the relevant canonical filter form. The second theorem serves as a basis for both deriving the optimal filter and performing the operations required in its calculation. Because of the general irreducibility of 2-D polynomials, this last step must involve procedures which do not explicitly depend on the root-finding techniques employed so extensively by Wiener. With these results in hand, the filtering problem can now be addressed.

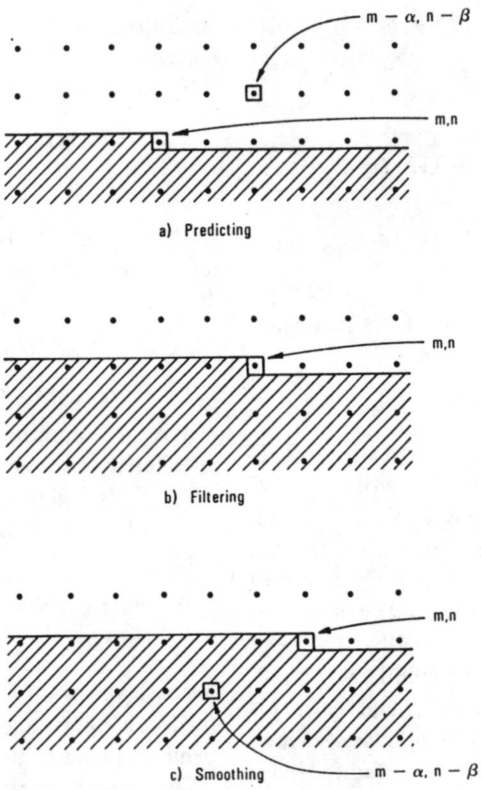

Fig. 5. Regions of support for observations $a(m, n)$ used in forming estimate $\hat{s}(m - \alpha, n - \beta)$. (Cross-hatched area is "past.")

A. Problem Formulation

The formulation of the 2-D discrete MMSE estimation problem is as follows. An arbitrary bisequence $a(m, n)$ is assumed to be observed or measured, and it is of interest to infer or estimate a signal $s(m, n)$ from this set of observations. Random field models are assumed for both bisequences: $a(m, n)$ and $s(m, n)$ are taken to be sample functions from homogeneous random fields with 2-D autocovariances $R_a(m, n)$ and $R_s(m, n)$, respectively. Their crosscovariance is $C_{sa}(m, n)$.

The estimator is chosen to be discrete, linear, and shift-invariant, and the bisequences scanned in a row-sequential format. The point estimates are given in the general form

$$\hat{s}(m - \alpha, n - \beta) = \sum_{m', n' \in R_{\oplus +}}\sum a(m - m', n - n')\, h(m', n') \quad (9)$$

where the weights $h(m, n)$ constitute the unit sample response of the 2-D physically realizable estimator or filter. Thus, an appropriate choice of α and β yields the three cases of practical interest (see Fig. 5).

*Case 1: Prediction—$\alpha + \beta M < 0$—*In the prediction case, only "past" values of $a(m, n)$ (i.e., with respect to the point estimated) are used in forming the estimate. For row-sequential ordering, this corresponds to the condition $\alpha + \beta M < 0$.

*Case 2: Filtering—$\alpha + \beta M = 0$—*The filtering case involves using all "past" and the "present" observations. Here, the condition $\alpha + \beta M = 0$ implies $\alpha = \beta = 0$.

*Case 3: Smoothing—$\alpha + \beta M > 0$—*This is the case of "filtering with delay." "Future" observations are included in constructing the estimate. This corresponds to the condition $\alpha + \beta M > 0$.

Commensurate with the statistical models adopted, the classical minimum mean-square error is used for design. This error is

$$\xi = E\{[s(m - \alpha, n - \beta) - \hat{s}(m - \alpha, n - \beta)]^2\} \quad (10)$$

where $E\{\cdot\}$ is an expectation operator taken over the ensemble of possible measurement and signal bisequences. The optimal filter is designed to minimize this error, subject to the constraint that $h(m, n)$ be physically realizable:

$$\text{minimize } \xi \quad \forall m, n \quad (11)$$

subject to $h(m, n)$ taking support in $R_{\oplus +}$.

As in the 1-D case, the most concise method of performing this minimization involves use of the so-called orthogonal projection theorem [15].

Theorem 3: With $\hat{s}(m - \alpha, n - \beta)$ given as in (9), the optimal filter $h_o(m, n)$ which minimizes ξ in (10) is such that

$$E\{[s(m - \alpha, n - \beta) - \hat{s}(m - \alpha, n - \beta)]\, a(m - k, n - l)\} = 0 \quad \forall k, l \in R_{\oplus +}. \quad (12)$$

This states that the filter weights are optimal when the estimate error is orthogonal to the observations used in forming the estimate.

As a first step in deriving $h_o(m, n)$, $\hat{s}(m - \alpha, n - \beta)$ in (9) is substituted into (12) yielding

$$E\{s(m - \alpha, n - \beta)\, a(m - k, n - l)\} = E\left\{\sum_{\tau, \lambda \in R_{\oplus +}}\sum a(m - \tau, n - \lambda)\, h_o(\tau, \lambda)\, a(m - k, n - l)\right\} \quad \forall k, l \in R_{\oplus +}. \quad (13)$$

Interchanging the order of expectation and summation and using the covariance relations

$$C_{sa}(k-\alpha, l-\beta) = E\{s(m-\alpha, n-\beta)\, a(m-k, n-l)\} \quad (14a)$$

$$R_a(k-\tau, l-\lambda) = E\{a(m-\tau, n-\lambda)\, a(m-k, n-l)\} \quad (14b)$$

(13) can be written as

$$C_{sa}(k-\alpha, l-\beta) = \sum_{\tau,\lambda \in R_{\oplus+}}\sum R_a(k-\tau, l-\lambda)\, h_o(\tau, \lambda) \qquad \forall k, l \in R_{\oplus+}. \quad (14)$$

In accordance with traditional nomenclature, this relation is called a 2-D discrete Wiener-Hopf equation of the first kind. It must, of course, be solved to obtain $h_o(m, n)$.

B. Solution for the Optimal Filter

Solution of this Wiener-Hopf equation is greatly complicated by two features. First, the system of equations in (14) is not defined for all k, l, rather only for $k, l \in R_{\oplus+}$. Second, $h_o(m, n)$ is constrained to be physically realizable. Despite this complexity, classical procedures can be carefully extended to solve (14) [16]. Here, however, a more direct and simpler approach will be presented.

In this approach, the two features above are initially ignored, that is, (14) is assumed to be defined for *all* k, l and $h_o(k, l)$ is taken to be unconstrained in form. (As will be seen below, both assumptions are to be subsequently corrected.) Under these conditions, the system of equations to solve becomes

$$C_{sa}(k-\alpha, l-\beta) = \sum_{\tau,\lambda=-\infty}^{+\infty}\sum R_a(k-\tau, l-\lambda)\, h_o(\tau, \lambda) \qquad \forall k, l. \quad (15)$$

Using the convolution and shift properties of the 2-D z transform, (15) is transformed to obtain

$$z_1^{-\alpha} z_2^{-\beta} \Phi_{sa}(z_1, z_2) = \Phi_a(z_1, z_2)\, H_o(z_1, z_2) \quad (16)$$

where

$$C_{sa}(k, l) \Longleftrightarrow \Phi_{sa}(z_1, z_2)$$

$$R_a(k, l) \Longleftrightarrow \Phi_a(z_1, z_2)$$

$$h_o(k, l) \Longleftrightarrow H_o(z_1, z_2).$$

The 2-D power spectral density $\Phi_a(z_1, z_2)$ can be spectrally factored [10] into the form

$$\Phi_a(z_1, z_2) = \Phi_a^{\oplus+}(z_1, z_2)\, \Phi_a^{\ominus-}(z_1, z_2) \quad (17)$$

where $\Phi_a^{\oplus+}(z_1, z_2)$ and its inverse satisfy the holomorphic conditions $D_{\oplus+}$, and $\Phi^{\ominus-}(z_1, z_2)$ and its inverse satisfy the holomorphic conditions $D_{\ominus-}$. Thus, for example, $\Phi_a^{\oplus+}(z_1, z_2)$ has no zeros or poles (nonessential singularities of the first kind) in a region including $\{(z_1, z_2): |z_1| = 1, 1 \leqslant |z_2|\}$ and its inverse 2-D z transform takes support on $R_{\oplus+}$. Consistent with the even symmetry of $\Phi_a(z_1, z_2)$, the constraint

$$\Phi_a^{\ominus-}(z_1, z_2) = \Phi_a^{\oplus+}(z_1^{-1}, z_2^{-1}) \quad (18)$$

is chosen to ensure the uniqueness of (17).

The decomposition in (17) is similar in spirit to Wiener's 1-D spectral factorization in that it equivalently factors the covariance into the convolutional product of two terms, one of which is "causal" or physically realizable and the other which is "anticausal" or antirealizable. The holomorphic conditions of the individual terms in (17) guarantee this behavior.

Incorporating this factorization into (16) yields

$$z_1^{-\alpha} z_2^{-\beta} \Phi_{sa}(z_1, z_2) = \Phi_a^{\oplus+}(z_1, z_2)\, \Phi_a^{\ominus-}(z_1, z_2) \cdot H_o(z_1, z_2). \quad (19)$$

At this point, corrections to the assumptions made above are introduced into the formalism. First, because (14) was only defined for $k, l \in R_{\oplus+}$, (19) is projected onto $R_{\oplus+}$ to obtain

$$[z_1^{-\alpha} z_2^{-\beta} \Phi_{sa}(z_1, z_2)]_{\oplus+} = [\Phi_a^{\oplus+}(z_1, z_2)\, \Phi_a^{\ominus-}(z_1, z_2)\, H_o(z_1, z_2]_{\oplus+} \quad (20)$$

where the notation $[\cdot]_{\oplus+}$ indicates the component of the bracketed term whose inverse 2-D z transform takes support on $R_{\oplus+}$. Recall that the inverse transforms of both $\Phi_a^{\ominus-}(z_1, z_2)$ and its inverse are antirealizable bisequences (i.e., take support on $R_{\ominus-}$). Hence, (20) can also be written as

$$\left[\frac{z_1^{-\alpha} z_2^{-\beta} \Phi_{sa}(z_1, z_2)}{\Phi_a^{\ominus-}(z_1, z_2)}\right]_{\oplus+} = [\Phi_a^{\oplus+}(z_1, z_2)\, H_o(z_1, z_2)]_{\oplus+} \quad (21)$$

because values on $R_{\oplus+}$ for the convolution product of an antirealizable bisequence and an arbitrary bisequence, only depend on values of the arbitrary bisequence on $R_{\oplus+}$.

The second correction involves constraining $h_o(k, l)$ to be realizable. As $\Phi_a^{\oplus+}(z_1, z_2)$ has an inverse transform taking support on $R_{\oplus+}$, it follows that

$$[\Phi_a^{\oplus+}(z_1, z_2)\, H_o(z_1, z_2)]_{\oplus+} = \Phi_a^{\oplus+}(z_1, z_2)\, H_o(z_1, z_2) \quad (22)$$

because the convolution product of two realizable bisequences is itself realizable. Consequently, (21) can be written as

$$\left[\frac{z_1^{-\alpha} z_2^{-\beta} \Phi_{sa}(z_1, z_2)}{\Phi_a^{\ominus-}(z_1, z_2)}\right]_{\oplus+} = \Phi_a^{\oplus+}(z_1, z_2)\, H_o(z_1, z_2). \quad (23)$$

Solving (23), the transfer function of the 2-D physically realizable Wiener filter is given by

$$H_o(z_1, z_2) = \frac{1}{\Phi_a^{\oplus+}(z_1, z_2)} \left[\frac{z_1^{-\alpha} z_2^{-\beta} \Phi_{sa}(z_1, z_2)}{\Phi_a^{\ominus-}(z_1, z_2)}\right]_{\oplus+}. \quad (24)$$

This is the main result of the paper. The model, criteria, and general viewpoint adopted here for the 2-D optimal filtering problem are essentially identical in context to those employed by Wiener in his classical work. With a careful generalization of physical realizability to 2-D, both the canonical, realizable filter form and spectral factorization follow as natural extensions of their 1-D antecedents. Thus, it is intuitively pleasing that the form of $H_o(z_1, z_2)$ in (24) appears similar to that of the 1-D result [17].

C. Expressions for the Optimal Filtering Error

The performance of the optimal realizable filter can be characterized by its MMSE. Recall from the orthogonal projection theorem (Theorem 3) that the point estimator error is orthogonal to all observations, hence to the optimal estimate

$$\begin{aligned}\xi &= E\{[s(m-\alpha, n-\beta) - \hat{s}(m-\alpha, n-\beta)]^2\} \\ &= E\{[s(m-\alpha, n-\beta) - \hat{s}(m-\alpha, n-\beta)]\, s(m-\alpha, n-\beta)\}.\end{aligned} \quad (25)$$

Using (9) and (14a), this can also be written as

$$\xi = R_s(0,0) - \sum\sum_{m',n' \in R_{\oplus+}} h_o(m',n')\, C_{sa}(m'-\alpha, n'-\beta) \quad (26)$$

$$= R_s(0,0) - \sum\sum_{m',n'=-\infty}^{+\infty} h_o(m',n')\, C_{sa}(m'-\alpha, n'-\beta) \quad (27)$$

where, because $h_o(m,n)$ is realizable, the summations may be taken over the entire plane.

The second term in (27) is expanded using the optimal filter form to give

$$\sum\sum_{m',n'=-\infty}^{+\infty} h_o(m',n')\, C_{sa}(m'-\alpha, n'-\beta)$$

$$= \sum\sum_{m',n'=-\infty}^{+\infty} C_{sa}(m'-\alpha, n'-\beta) \frac{1}{(2\pi j)^2} \oint\oint z_1^{m'-1} z_2^{n'-1} \cdot \frac{1}{\Phi_a^{\oplus+}(z_1,z_2)} \left[\frac{z_1^{-\alpha} z_2^{-\beta} \Phi_{sa}(z_1,z_2)}{\Phi_a^{\ominus-}(z_1,z_2)}\right]_{\oplus+} dz_1\, dz_2. \quad (28)$$

Defining the bisequence $p(m,n)$ via the transform relation

$$p(m,n) \Longleftrightarrow \frac{\Phi_{sa}(z_1,z_2)}{\Phi_a^{\ominus-}(z_1,z_2)} \quad (29)$$

it follows that

$$\left[\frac{z_1^{-\alpha} z_2^{-\beta} \Phi(z_1,z_2)}{\Phi_a^{\ominus-}(z_1,z_2)}\right]_{\oplus+} = \sum\sum_{k,l \in R_{\oplus+}} p(k-\alpha, l-\beta)\, z_1^{-k} z_2^{-l}. \quad (30)$$

Equation (28) can now be written as

$$\sum\sum_{m',n'=-\infty}^{+\infty} h_o(m',n')\, C_{sa}(m'-\alpha, n'-\beta)$$

$$= \sum\sum_{m',n'=-\infty}^{+\infty} C_{sa}(m'-\alpha, n'-\beta) \frac{1}{(2\pi j)^2} \oint\oint z_1^{m'-1} z_2^{n'-1} \cdot \frac{1}{\Phi_a^{\oplus+}(z_1;z_2)} \sum\sum_{k,l \in R_{\oplus+}} p(k-\alpha, l-\beta)\, z_1^{-k} z_2^{-l}\, dz_1\, dz_2. \quad (31)$$

Interchanging the orders of summation and integration gives

$$\sum\sum_{m',n'=-\infty}^{+\infty} h_o(m',n')\, C_{sa}(m'-\alpha, n'-\beta)$$

$$= \sum\sum_{k,l \in R_{\oplus+}} p(k-\alpha, l-\beta) \left\{\frac{1}{(2\pi j)^2} \oint\oint z_1^{-k} z_2^{-l} \frac{1}{\Phi_a^{\oplus+}(z_1,z_2)} \cdot \sum\sum_{m',n'=-\infty}^{+\infty} C_{sa}(m'-\alpha, n'-\beta)\, z_1^{m'-1} z_2^{n'-1}\, dz_1\, dz_2\right\} \quad (32)$$

$$= \sum\sum_{k,l \in R_{\oplus+}} p(k-\alpha, l-\beta) \frac{1}{(2\pi j)^2} \oint\oint z_1^{-k-1} z_2^{-l-1} \cdot \frac{z_1^{+\alpha} z_2^{+\beta} \Phi_{sa}(z_1^{-1}, z_2^{-1})}{\Phi_a^{\oplus+}(z_1,z_2)}\, dz_1\, dz_2 \quad (33)$$

$$= \sum\sum_{k,l \in R_{\oplus+}} [p(k-\alpha, l-\beta)]^2. \quad (34)$$

Consequently, the MMSE associated with (24) can be written as

$$\xi = R_s(0,0) - \sum\sum_{k,l \in R_{\oplus+}} [p(k-\alpha, l-\beta)]^2 \quad (35)$$

where $p(k,l)$ can be evaluated from (29). As the region of homomorphy of $\Phi_{sa}(z_1,z_2)/\Phi_a^{\ominus-}(z_1,z_2)$ is an annular region including the unit bicircle, this can be accomplished numerically using Fourier transforms.

It is also of interest to note the variation in MMSE with α, β. This is facilitated by writing the summations in (35) as

$$\xi = R_s(0,0) - \left\{\sum_{k=0}^{\infty} [p(k-\alpha, -\beta)]^2 + \sum_{k=-\infty}^{+\infty} \sum_{l=1}^{\infty} [p(k-\alpha, l-\beta)]^2\right\}$$

$$= R_s(0,0) - \left\{\sum_{m=-\alpha}^{\infty} [p(m,-\beta)]^2 + \sum_{m=-\infty}^{+\infty} \sum_{n=1-\beta}^{\infty} [p(m,n)]^2\right\}. \quad (36)$$

A careful examination of the terms in (36) indicates that ξ is monotone nonincreasing in α and β. Thus, generally speaking, increases in either the row or column "lag" lead to improvement of the MMSE. Furthermore, achieving the absolute minimum ξ (the so-called irreducible error) requires *both* row and column "lags" $\rightarrow \infty$. For this case

$$\xi_i = R_s(0,0) - \sum\sum_{m,n=-\infty}^{+\infty} [p(m,n)]^2 \quad (37)$$

which is indeed the direct analog to the classical 1-D result [18].

In a practical design problem, both (36) and (37) would be used in determining the appropriate values of α and β for the optimal smoother. A reasonable strategy would be to choose a minimum value of $\alpha + \beta M$ such that the realizable error ξ was acceptably close to the irreducible (unrealizable) error, ξ_i. Developing the tradeoffs between column and row "lags" (α and β, respectively) could be accomplished via a simple parameter search procedure.

D. Special Case: Filter Signals from Additive White Noise

As an example of the application of (24), consider the case of filtering an additive signal and white, zero-mean noise (uncorrelated). Here

$$a(m,n) = s(m,n) + w(m,n) \quad (38)$$

where the noise spectrum is given by W_o. Thus

$$\Phi_{sa}(z_1,z_2) = \Phi_s(z_1,z_2) \quad (39)$$

$$\Phi_a(z_1,z_2) = \Phi_s(z_1,z_2) + W_o. \quad (40)$$

Plugging these relations into the optimal filter form [(24) with $\alpha = \beta = 0$] gives

$$H_o(z_1, z_2) = \frac{1}{\Phi_a^{\oplus+}(z_1, z_2)} \left[\frac{\Phi_s(z_1, z_2)}{\Phi_a^{\ominus-}(z_1, z_2)}\right]_{\oplus+}. \tag{41}$$

Following some algebraic manipulation, this reduces to

$$H_o(z_1, z_2) = 1 - \frac{1}{\Phi_a^{\oplus+}(z_1, z_2)} \left[\frac{W_o}{\Phi_a^{\ominus-}(z_1, z_2)}\right]_{\oplus+} \tag{42}$$

and its final form

$$H_o(z_1, z_2) = 1 - \frac{W_o'}{\Phi_a^{\oplus+}(z_1, z_2)} \tag{43}$$

where

$$W_o' = \frac{W_o}{\phi_a(0, 0)} \tag{44}$$

and

$$\phi_a(m, n) \Longleftrightarrow \Phi_a^{\oplus+}(z_1, z_2).$$

The corresponding MMSE is given from (35) as

$$\xi = R_s(0, 0) - \sum_{m, n \in R_{\oplus+}}\sum [p(m, n)]^2 \tag{45}$$

which can also be written as

$$\xi = R_s(0, 0) - \sum_{m,n=-\infty}^{+\infty}\sum [p_{\oplus+}(m, n)]^2 \tag{46}$$

where $p_{\oplus+}(m, n)$ is the projection of $p(m, n)$ onto $R_{\oplus+}$. This quantity is obtained from the transform pair

$$p_{\oplus+}(m, n) \Longleftrightarrow \Phi_a^{\oplus+}(z_1, z_2) - W_o' \tag{47}$$

either directly or through Parseval's formula.

The filter in (43) has an interesting and simple implementation structure. Assume that the observation spectrum is factored to obtain

$$\Phi_a^{\oplus+}(z_1, z_2) = \frac{N(z_1, z_2)}{D(z_1, z_2)} \tag{48}$$

where $N(z_1, z_2)$ and $D(z_1, z_2)$ are polynomials. This corresponds to adopting an autoregressive, moving-average (ARMA) model for the 2-D bisequence $a(m, n)$. The optimal filter takes the form

$$H_o(z_1, z_2) = 1 - \frac{W_o' D(z_1, z_2)}{N(z_1, z_2)}. \tag{49}$$

Hence, it can be implemented in a *recursive* structure. The order of the recursion is determined by the order of the numerator polynomial $N(z_1, z_2)$; its stability is guaranteed by the properties of the spectral factorization. Alternatively, if $a(m, n)$ is modeled as an autoregressive (AR) process, $\Phi_a^{\oplus+}(z_1, z_2)$ is an "all-pole" transform and $H_o(z_1, z_2)$ can be implemented in a *nonrecursive* structure. Its order is determined, in turn, by the order of the denominator polynomial $D(z_1, z_2)$.

IV. Remarks

Remark 1: The traditional motivations for using the Wiener formalism essentially extend to the 2-D case. These include the compatible statistical description of many estimation/detection problems, suitability of the error measure, availability of covariance information, global optimality for the Gaussian case, and ease of filter derivation. Of these, only the universality of the error measure appears to be subject to any notable qualification (in the sense that it may be applications-dependent).

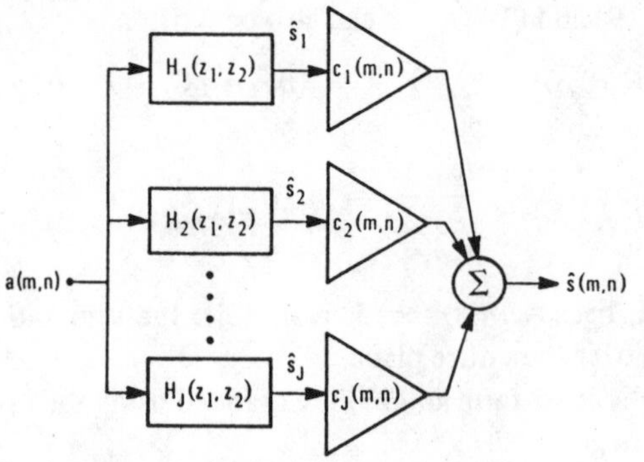

Fig. 6. Block diagram for multicategory Wiener filter.

This qualification frequently arises in image processing applications involving the human visual system. There, it is often-quoted that Wiener filtering (as implemented by optical processors, for example) does not always improve the *visual* quality of image data. A simplistic characterization of the filtering suggests the reasons for this situation. Because images are positive functions, their power spectra normally have sizable low-frequency spectral content. As noise is broad band, the corresponding Wiener filter has a low-pass type characteristic; hence, it tends to filter out the image high-frequency content, such as edges and other fine detail. This results in a visual blurring of the image.

Although the difficulties encountered with such filtering may be due in some measure to the MMSE, they appear more likely attributable to limitations in the underlying image models. While the second-order statistical properties of an image may be adequately modeled as being globally homogeneous, much of its *visual* information-bearing character is contained in the local, nonhomogeneous structure.

Based on this premise, Lebedev and Mirkin [19] proposed a composite image model which adapts the basic Wiener formalism to account for this local structure. Their approach is to model an image as composed of fragments of J classes of 2-D random fields. Each class is assumed homogeneous and distinguished by its correlation function; the probability distribution of the class occurrence is assumed known (or measurable). The estimator based on this model is a so-called multicategory Wiener filter (see Fig. 6) comprised of a bank of Wiener filters, one per class. The point estimate $\hat{s}(m, n)$ is formed as a weighted sum of individual filter outputs $\hat{s}_j(m, n)$, the weights being the computed *a posteriori* probabilities.

The important feature of this structure is its combination of truly adaptive filtering and LSI component estimators. Variations of the Lebedev/Mirkin processor based on 2-D Kalman filters have been shown, in fact, to give quite impressive results in image processing application [20]. With the individual filters $\{H_i(z_1, z_2)\}$ designed as realizable Wiener estimators, one should expect comparible performance, real-time operation, and more attractive levels of implementation complexity.

Remark 2: The most notable difference between the 1-D and 2-D cases arises from the irreducibility of 2-D polynomials.

The spectral factorization and realizability projection operations called for in (24) cannot be performed using polynomial algebra, as they are in the 1-D case. Consequently, the 2-D filter derivation must be approached from an approximation viewpoint, using procedures applicable to nonrational spectra.

This difference is more one of technique than concept. In most practical 1-D applications, the relevant covariance information is obtained from experimental measurements; hence, the underlying power spectra are in nonrational form. This difficulty is circumvented by applying approximation procedures to obtain rational spectra density estimates *a priori* their use in filter derivation. In the 2-D case, the derivation must proceed using nonrational models, and applying rational approximation techniques *a posteriori* in a filter design step.

Fortunately, it is possible to perform the factorization/projection operations in (24) with nonrational spectra, using algorithms based on sectioning cepstral bisequences [10]. These algorithms are computationally straightforward to implement, and generally can be adjusted to provide desired degrees of numerical precision simply by increasing the size of DFT's used in the calculations.

Specific approaches to the filter design depend to some extent on the particular power spectra involved (symmetry conditions), desired realization structure (recursive and/or nonrecursive), and canonical form of the estimator (predictor, filter, or smoother). The choice of recursive or nonrecursive structure involves evaluating tradeoffs associated with both the filter design performance and realization complexity, and is a complex question [21]. Recursive forms are considered generally to offer more potential advantages for "real-time" 2-D processing problems, but this is, by no means, a unanimous opinion among researchers in the field. However, both the prediction and filtering estimator forms are compatible with this structure, and effective algorithms exist for suitably performing the design [22], [23]. The smoothing estimator may involve both recursive and nonrecursive components as might be expected from the shift operation on $p(m, n)$ [see (24) and (30)].

Remark 3: The flexibility of choosing scanning patterns when dealing with 2-D signals is one of the interesting features not present in the 1-D case. Depending on the covariance information, there may be significant advantages in implementation complexity available with a particular ordering. Natural images tend to have strong horizontal and vertical correlation, and in processing such data, the row- and column-sequential scannings are apparent choices.

Adaptation of the estimator formalism for alternate scannings follows directly from the development of realizability conditions above. For example, the optimal 2-D estimator for the cases of column- and diagonal-scanning are obtained from (24) by simply replacing the "$\oplus +$" notation with "$+\oplus$" and "$\oplus_d +$", respectively. The MMSE expression (35) may be similarly modified.

Remark 4: The formalism can also be analogously extended to a full multidimensional (M-D) setting. The concepts of unilateral modeling and corresponding half-plane support regions generalize in M-D to "half-spaces" and associated M-D spectral factorizations [24]. Avoiding details, the M-D realizable filter is given by

$$H_o(\mathbf{z}) = \frac{1}{\Phi_a^+(\mathbf{z})}\left[\frac{z_1^{-\alpha} z_2^{-\beta} \cdots z_M^{-\Omega}\, \Phi_{sa}(\mathbf{z})}{\Phi_a^-(\mathbf{z})}\right]_+ \tag{50}$$

where

$$\mathbf{z} = (z_1 z_2 \cdots z_M)$$

and "+" denotes a "half-space" factorization and/or projection as defined in [24]. The numerical procedures for deriving the 2-D optimal filter form also extend to the M-D problem, although the computations may be realistically prohibitive for dimensions >3 or 4.

V. Concluding Comments

A reasonable conclusion, derived from the results of recent M-D signal processing literature, is that almost all the important time-series methodologies can be extended to multidimensions. The extension of Wiener's realizable filtering theory presented here suggests that when the underlying concepts are carefully generalized, the extension itself may be beautifully simple in structure. Algorithms for computationally deriving the realizable filter are more complex in the 2-D case, but perhaps not to the extent one might expect. Details of these numerical calculations and specific examples in applying the formalism will be reported in a subsequent paper now in preparation.

Acknowledgment

The warm hospitality and consideration of Professor W. Schüssler, Lehrstuhl für Nachrichtentechnik, Universität Erlangen-Nürnberg, and his staff during this period are most gratefully appreciated.

The author also wishes to acknowledge some helpful discussions with T. Marzetta.

References

[1] N. Wiener, *The Extrapolation, Interpolation, and Smoothing of Stationary Time Series.* New York: Wiley, 1949.

[2] H. C. Andrews and B. R. Hunt, *Digital Image Restoration.* Englewood Cliffs, NJ: Prentice-Hall, 1976.

[3] W. K. Pratt, *Digital Image Processing.* New York: Wiley, 1978.

[4] B. R. Frieden, "Statistical models for the image restoration problem," *Comput. Graphics, Image Processing*, vol. 12, pp. 40-59, Jan. 1980.

[5] W. L. Rogers, K. F. Koral, R. Mayans, P. F. Leonard, and J. W. Keyes, "Coded aperture imaging of the heart," *J. Nucl. Med.*, vol. 19, pp. 730-750, 1978.

[6] E. E. Fenimore, T. M. Cannon, D. B. Van Hulsteyn, and P. Lee, "Uniformly redundant array imaging of laser driven compressions: Preliminary results," *Appl. Opt.*, vol. 18, pp. 945-947, Apr. 1979.

[7] D. Gabor, "The smoothing and filtering of two-dimensional images," *Progress Biocybern.* (Amsterdam), vol. 2, 1965.

[8] C. W. Helstrom, "Image restoration by the method of least squares," *J. Opt. Soc. Amer.*, vol. 57, pp. 297-303, Mar. 1967.

[9] S. K. Mitra and M. P. Ekstrom, Ed., *Two-Dimensional Digital Signal Processing.* Stroudsburg, PA: Dowden, Hutchinson, and Ross, 1978.

[10] M. P. Ekstrom and J. W. Woods, "Two-dimensional spectral factorization with applications in recursive digital filtering," *IEEE Trans. Acoust., Speech, Signal Processing*, vol. ASSP-24, pp. 115-128, Apr. 1976.

[11] T. S. Huang, "Stability of two-dimensional recursive filters," *IEEE Trans. Audio Electroacoust.*, vol. AU-20, pp. 158-162, June 1972.

[12] D. M. Goodman, "Some stability properties of two-dimensional linear shift-invariant digital filters," *IEEE Trans. Circuits Syst.*, vol. CAS-24, pp. 201-209, April 1977.

[13] H. Chang and J. K. Aggarwal, "Design of two-dimensional semi-

causal recursive filters," *IEEE Trans. Circuits Syst.*, vol. CAS-26, pp. 1051–1059, Dec. 1978.

[14] D. S. K. Chan, "The structure of recursible multidimensional discrete systems," *IEEE Trans. Automat. Contr.*, vol. AC-25, pp. 663–673, Aug. 1980.

[15] A. Papoulis, *Probability, Random Variables, and Stochastic Processes.* New York: McGraw-Hill, 1965.

[16] M. P. Ekstrom, "Digital, realizable Wiener filtering in two-dimensions," in *Proc. 1979 Int. Symp. Math. Theory Networks, Syst.*, Delft, The Netherlands, July 1979.

[17] H. Freeman, *Discrete-Time Systems.* New York: Wiley, 1965.

[18] H. L. Van Trees, *Detection, Estimation, and Modulation Theory: Part I.* New York: Wiley, 1968.

[19] D. S. Lebedev and L. I. Mirkin, "Smoothing of two-dimensional images using the 'composite' model of a fragment," in *Iconics, Digital Holography, and Image Processing.* Institute for Problems in Information Transmission, Academy of Sciences, U.S.S.R., 1975, pp. 57–62; also in [9].

[20] V. K. Ingle and J. W. Woods, "Multiple model recursive estimation of images," in *ICASSP '79 Conf. Rec.*, Apr. 1979, pp. 642–645.

[21] R. E. Twogood and M. P. Ekstrom, "Why filter recursively in two-dimensions?," in *ICASSP '79 Conf. Rec.*, Apr. 1979, pp. 20–23.

[22] M. P. Ekstrom, R. E. Twogood, and J. W. Woods, "Two-dimensional recursive filter design—a spectral factorization approach," *IEEE Trans. Acoust., Speech, Signal Processing*, vol. ASSP-28, pp. 16–26, Feb. 1980.

[23] J. W. Woods, I. Paul, and N. Sangal, "2-D direct form, recursive filter design with magnitude and phase specifications," in *Proc. ICASSP 81*, Apr. 1981, pp. 700–703.

[24] D. M. Goodman and M. P. Ekstrom, "Multidimensional spectral factorization and unilateral autoregressive models," *IEEE Trans. Automat. Contr.*, vol. AC-25, pp. 258–262, Apr. 1980.

Part IV
Signal Modeling

THE papers covered in this section deal with signal models that are useful in two-dimensional signal processing. The primary emphasis is on stochastic modeling techniques. The basic underlying problem considered in these papers is to find a suitable two-dimensional linear system whose output provides a good match to certain given data. This problem is called *spectral factorization* when the input to the desired system is a white noise or a moving average random field and the output is required to realize a given power spectral density function. It is well known now that this problem is considerably more complex than the corresponding one-dimensional case due to the fact that it is not possible, in general, to reduce a two-dimensional polynomial as a product of lower order polynomials. This was noted as the main difficulty in fitting two-dimensional dynamic models to observed topographical data by Whittle. In our first reprint paper, Whittle introduces a *unilateral representation* which is nothing but the *nonsymmetric half plane* (NSHP) model of [1]. Whittle also suggests spectral factorization by causal and anticausal decomposition of the log of the power spectrum. This approach is the basis of the so-called homomorphic decomposition method used for two-dimensional filter design and spectral factorization using the DFTs [1], [2], [3].

The second paper, by Roesser, is not directly concerned with spectral factorization but introduces a two-dimensional state-space model whose structure readily admits generalizations of several key concepts from the state variable theory for one-dimensional systems. For this model, it is easy to characterize two-dimensional concepts of state transition matrix, observability, controlability, etc. A significant advantage of this model is due to the fact that every finite order causal model (NSHP) can be transformed to the particular model of Roesser. This allows one to analyze all finite order causal models via the rich state-variable theory. Some applications in image processing and extensions concerning the realizability of this model have been considered in [4], [5].

The paper by Jain contains a comprehensive survey of mathematical models from Karhunen-Loeve representations and state variable models to linear prediction models. This paper generalizes the concepts of linear prediction to causal, semicausal, and noncausal representations. These three classes of models are motivated by the three canonical classes of partial differential equations viz. hyperbolic, parabolic, and elliptic, which have been studied earlier for representing random fields [6], [7]. As it turns out, certain stable finite difference approximations of these three classes of partial differential equations result in the three different prediction geometries [8], [9]. Linear prediction-based spectral factorization techniques to realize these three forms are discussed in this paper. A similar approach for spectral factorization and modeling of two-dimensional causal systems has also been developed in [10]. By adopting a definition of causality that totally orders the points in the plane (similar to the NSHP definition), a reflection coefficient representation has been obtained for causal prediction error filters. This result is useful in designing stable filters which can realize a given power spectral density arbitrarily closely. The paper by Jain and a subsequent extension [11] contain algorithms for designing stable and finite order, causal, semicausal, and noncausal models such that a given power spectral density is realized arbitrarily closely. Applications of these models in image processing and spectral estimation are shown. Other related results are available in [3], [12].

While the first three papers are primarily concerned with either spectral factorization algorithms or model properties, the fourth paper, by Kashyap and Chellappa, is concerned with estimation of model parameters from observed data and the choice of neighbors (or order) in what are called the simultaneous autoregressive (SAR) and conditional Markov (CM) models. These models belong to the noncausal class of models driven by white noise fields. The paper shows that the classical least squares method for identifying model parameters leads to inconsistent estimates for nonunilateral neighbor sets (i.e., noncausal models). The maximum likelihood criterion, on the other hand, leads to nonlinear algorithms which yield asymptotically consistent and efficient estimates. For additional results see [12], [13], [14].

Bibliography

[1] M. P. Ekstrom and J. W. Woods, "Two-dimensional spectral factorization with application in recursive digital filtering," *IEEE Trans. Acoust., Speech, Signal Processing*, vol. ASSP-24, pp. 115–128, Apr. 1976.

[2] M. P. Ekstrom, R. E. Twogood, and J. W. Woods, "Design of 2-D all pole recursive filters using spectral factorization," *IEEE Trans. Acoust., Speech, Signal Processing*, vol. ASSP-28, pp. 16–26, Feb. 1980.

[3] H. Chang and J. K. Aggarwal, "Design of two-dimensional semicausal recursive filters," *IEEE Trans. Circuit Syst.*, vol. CAS-25, pp. 1051–1059, Dec. 1978.

[4] P. E. Barry, R. Gran, and C. R. Waters, "Two dimensional filtering—a state estimator approach," in *Proc. Conf. Decision and Control*, Clearwater Beach, FL, pp. 613–618, Dec. 1976.

[5] E. Fornasini and G. Marchesini, "State space realization theory of two-dimensional filters," *IEEE Trans. Automat. Contr.*, vol. AC-21, pp. 484–492, Aug. 1976.

[6] V. Heine, "Models for two dimensional stationary stochastic processes," *Biometrika*, vol. 42, pp. 170–178, 1955.

[7] E. Wong, "Two dimensional random fields and representation of images," *SIAM J. Appl. Math.*, vol. 16, pp. 756–770, July 1968.

[8] A. K. Jain, ''Partial differential equations and finite difference methods in image processing, Part I—image representation,'' *J. Optimiz. Theory Appl.*, vol. 23, pp. 65–91, Sept. 1977.
[9] A. K. Jain and J. R. Jain, ''Partial differential equations and finite difference methods in image processing—Part II-image restoration,'' *IEEE Trans. Automat. Contr.*, vol. AC-23, pp. 817–834, Oct. 1978.
[10] T. L. Marzetta, ''Two dimensional linear prediction: Autocorrelation, arrays, minimum phase prediction error filter, and reflection coefficient arrays,'' *IEEE Trans. Acoust., Speech, Signal Processing*, vol. ASSP-28, pp. 725–733, Dec. 1980.
[11] S. Ranganath and A. K. Jain, ''Two dimensional linear prediction models, Part I: Spectral factorization and realization,'' *IEEE Trans. Acoust., Speech, Signal Processing*, vol. ASSP-33, pp. 280–299, Feb. 1985.
[12] W. E. Larimore, ''Statistical inference on stationary random fields,'' *Proc. IEEE*, vol. 65, pp. 961–968, June 1977.
[13] J. E. Besag, ''Spatial interaction and statistical analysis of lattice systems,'' *J. Roy. Stat. Soc.*, Ser. B., vol. B-36, pp. 192–236, 1974.
[14] R. Chellappa, ''Two-dimensional discrete Gaussian Markov random field models for image processing,'' in *Progress in Pattern Recognition*, L. Kanal and A. Rosenfeld, Eds. New York, NY: North-Holland, 1985, vol. 2, pp. 79–112.

ON STATIONARY PROCESSES IN THE PLANE

By P. WHITTLE

Applied Mathematics Laboratory, New Zealand Department of Scientific and Industrial Research

The sampling theory of stationary processes in space is not completely analogous to that of stationary time series, due to the fact that the variate of a time series is influenced only by past values, while for a spatial process dependence extends in all directions. This point is elaborated in §§2–4. The estimation and test theory developed in §7 is applied in §8 to uniformity data for wheat and oranges. The final section is devoted to an examination of some particular two-dimensional processes.

1. Introduction

The disturbing effect of topographic correlation on the results of field experiments, forest and crop surveys, sampling surveys of populated areas, etc., is well known, and it is recognized that we have here examples of two-dimensional stochastic processes. The physicists also encounter higher dimensional processes (for instance, in the studies of turbulence and of systems of particles) and have indeed been the principal investigators of the subject.

The processes we mentioned can only as a first approximation be regarded as stationary, if they can be so regarded at all. However, the approximation is satisfactory sufficiently often to make the study of the stationary type of process worth while.

Much of the two-dimensional theory is no more than a formal extension of that used in the study of time series, so we shall treat these aspects as briefly as possible. There is one interesting new feature, however, which is mentioned in the summary above, and we shall consider it at length.

The difficulty of examining particular models is in general greater in two dimensions than in one, simply because there are more mathematical obstacles of a technical nature. One seems to have left the domain of the elementary functions entirely. We shall indicate these obstacles, but take advantage of the fact that for many purposes they may be avoided.

For many applications it is sufficient to consider only purely non-deterministic processes, and we shall restrict our attention to processes of this type, more particularly to linear autoregressions.

2. The line transect

It is useful to begin with a discussion of the simple *line transect* (i.e. a straight line laid over an area, along which observations are taken equidistantly). The observations of the transect may be regarded as being generated by a one-dimensional process, just as are the terms of a time series. However, there is an important difference between the two cases. At any instant in a time series we have the natural distinction of past and future, and the value of the observation at that instant depends only upon *past* values. That is, the dependence extends only in one direction: backwards. In the case of the transect, however, there is no such day and night distinction between the two directions, and dependence will extend both ways. We can consider an example for the more general two-dimensional case of a field: a dab of fertilizer applied at any point in the field will ultimately affect soil fertility in *all* directions. (Exceptions are of course possible; the field may have so strong a slope that only the area on the downward side of the spot will be affected.)

Reprinted with permission from *Biometrika,* vol. 41, pp. 434–449, 1954.

Denoting now observation and 'error' variates by ξ_t, ϵ_t respectively ($t = \ldots, -2, -1, 0, 1, 2, \ldots$), the simplest realistic time series model is perhaps the first-order autoregression

$$\xi_t = a\xi_{t-1} + \epsilon_t. \tag{1}$$

For the transect, however, model (1) would have to be regarded as constituting a degenerate case, in which dependence extended only in one direction. The simplest non-degenerate transect model would be

$$\xi_t = a\xi_{t-1} + b\xi_{t+1} + \epsilon_t, \tag{2}$$

where it is intuitively clear that a and b must not be too large. We shall term (2) a bilateral autoregression, in distinction to (1) which is a unilateral autoregression.

Consequences of the admission of the bilateral type of scheme become apparent when one considers the estimation of parameters. Whereas the parameter a in (1) may be estimated consistently by minimizing the residual sum of squares $\sum_t (\xi_t - a\xi_{t-1})^2$, an attempt to estimate a and b in (2) simply by minimizing

$$U = \sum_t (\xi_t - a\xi_{t-1} - b\xi_{t+1})^2, \tag{3}$$

leads to nonsensical results. One could explain the breakdown by saying that it is not legitimate to include ξ_{t+1} in the conditional mean of ξ_t when the value of ξ_{t+1} itself depends upon ξ_t. Formally, we may say that the Jacobian of the transformation from the ϵ_t to the ξ_t is not unity for relation (2) as it would be for a unilateral relation such as (1). It may be shown, however (and in § 7 will be proved for the more general two dimensional case), that the correct equations for the least-square estimates are obtained by minimizing kU, where k is a certain function of the parameters given by

$$\log k = -\frac{1}{2\pi}\int_0^{2\pi} \log(a\,e^{i\omega} - 1 + b\,e^{-i\omega})\,(a\,e^{-i\omega} - 1 + b\,e^{i\omega})\,d\omega. \tag{4}$$

3. Formal properties of the bilateral scheme

We shall briefly examine the properties of the transect model as a preparation for the discussion of the two-dimensional case. Let us consider the general bilateral linear autoregression

$$L(T)\,\xi_t = \epsilon_t, \tag{5}$$

where $L(T) = \Sigma a_j T^j$ and T is the translation operator

$$T\xi_t = \xi_{t+1}. \tag{6}$$

Equation (5) has the solution

$$\xi_t = \frac{\epsilon_t}{L(T)} = \Sigma b_j \epsilon_{t+j}, \tag{7}$$

where b_j is the coefficient of $e^{ij\omega}$ in the Fourier expansion of $[L(e^{i\omega})]^{-1}$ (see Bartlett, 1946, p. 60). Using Fourier transforms it may similarly be shown that the spectral function of scheme (5) is

$$F(\omega) = \frac{\sigma^2(\epsilon)}{L(e^{i\omega})\,L(e^{-i\omega})} \tag{8}$$

(cf. Doob, 1944; Daniell, 1946). The autocovariances

$$\phi(j) = \operatorname{cov}(\xi_t, \xi_{t+j}) \quad (j = 0, \pm 1, \pm 2, \ldots) \tag{9}$$

are generated as the coefficients in the Fourier expansion of $F(\omega)$. Thus, writing for convenience $e^{i\omega} = z$, $\sigma^2(\epsilon) = v$, we have

$$F(\omega) = \frac{v}{L(z)\,L(z^{-1})}. \tag{10}$$

It is as well to consider the conditions under which expressions (7) and (8) are valid, since a number of essential points are involved. If (5) is to represent a unilateral scheme, expressing 'dependence only on the past', then the Laurent expansion of $[L(z)]^{-1}$ on $|z| = 1$ must not involve positive powers of z, and we are led to the usual condition for the stability of a time-series autoregression: that all roots of $L(z) = 0$ fall *inside* the unit circle (Wold, 1938). If, however, one is prepared to admit bilateral schemes, then all that is required is that $[L(z)]^{-1}$ possess a convergent Laurent expansion—coefficients otherwise unrestricted. The stationarity condition is in this case greatly weakened; all that is required is that no root of $L(z) = 0$ fall *on* the unit circle.

While this is the only condition which need be imposed upon a particular $L(z)$ to secure stationarity, it is desirable that $L(z)$ be such that in relation (5) it is ξ_t which is the dependent variate (not, for example, ξ_{t+1}, as would be the case if relation (1) were modified to $a\xi_t = \xi_{t+1} - \epsilon_t$). This may be achieved by multiplying $L(z)$ by some integral power of z (or, what is the same thing, redefining the ϵ sequence by translation so that ϵ_t comes into correspondence with ξ_t). For, suppose that as z describes the unit circle once in a positive direction $L(z)$ encircles the origin in the complex plane r times (where r is necessarily integral). The normalization will then be achieved by translating the ϵ sequence r steps backwards, so that $L(T)$ is replaced by the normalized operator $T^{-r}L(T)$, or $L^*(T)$, say. Since the origin is not a branch point for $\log L^*(z)$ (as it was for $\log L(z)$ if r was not zero) then $\log L^*(z)$ may be expanded in a Laurent series in z, and the operator $L^*(T)$ may be represented

$$L^*(T) = e^{\Sigma c_j T^j}. \tag{11}$$

It will be assumed in the sequel that this normalization has been performed, so that the representation (11) is always possible.

4. Indeterminacies of estimation

If one is prepared to admit bilateral schemes, then the autoregression exhibits the same kind of ambiguity that Wold (1938) has shown to obtain for the moving average. Thus, suppose that one is given a set of autocovariances which could have been generated by an autoregression of order p, and that one is set the task of determining the autoregression, i.e. of determining the polynomial $L(T)$ of (5). It is thus necessary to find an $L(z)$ which will satisfy equation (10), where $F(\omega)$ is determined by the given autocovariances. Such an $L(z)$ may be chosen in anything up to 2^p ways, since if a particular $L(z)$ has roots $\alpha_1, \alpha_2, \ldots, \alpha_p$, then there are 2^p possible finite autoregressions, corresponding to the 2^p possible representations of $F(\omega)$:

$$F(\omega) = \frac{\text{const.}}{|(z^{\pm 1} - \alpha_1)(z^{\pm 1} - \alpha_2)\ldots(z^{\pm 1} - \alpha_p)|^2}. \tag{12}$$

Two of these schemes are unilateral, and differ only in that the 'time' axis runs in opposite directions.

A similar argument applies for the least square fitting of a pth order autoregression to a set of data, since here again one works from the autocovariances, which could as easily have stemmed from one of the 2^p fitted autoregressions as from another.

For the sake of example, let us examine once more the bilateral scheme (2). If the scheme has been normalized according to the last paragraph of § 3 then the equation

$$a - z + bz^2 = 0 \tag{13}$$

will have one root inside the unit circle and one outside. Let us denote the roots α, β^{-1}, $[|\alpha| < 1, |\beta| < 1]$. Defining coefficients A and B by the relation

$$(z-\alpha)(z-\beta) = z^2 + Az + B, \tag{14}$$

we can construct an autoregression

$$\xi_t + A\xi_{t-1} + B\xi_{t-2} = \epsilon_t, \tag{15}$$

which is unilateral and yet generates the same autocorrelations as (2).

Evaluating integral (4) we find that a and b will be estimated by minimizing the quantity

$$\begin{aligned}[1+\sqrt{(1-4ab)}]^{-2}\,\Sigma(\xi_t - a\xi_{t-1} - b\xi_{t-2})^2 \\ \approx [1+\sqrt{(1-4ab)}]^{-2}[(1+a^2+b^2)C_0 - 2(a+b)C_1 + 2abC_2],\end{aligned} \tag{16}$$

where C_s is the observed autocovariance of lag s. Transforming to parameters A and B we find expression (16) proportional to

$$(1+A^2+B^2)C_0 + 2(A+AB)C_1 + 2BC_2 \approx \text{const.}\,\Sigma(\xi_t + A\xi_{t-1} + B\xi_{t-2})^2, \tag{17}$$

which is another indication of the equivalence of schemes (2) and (15). The practical significance of this equivalence is that the minimization of the rather awkward expression (16) may be replaced by that of the much simpler expression (17). The estimates of a and b may then be calculated from those of A and B by means of relation (14).

It may seem unnecessary to introduce the bilateral type of scheme when any such scheme may effectively be reduced to a unilateral one. Indeed, the step appears positively undesirable when we consider how much more complicated and indeterminate the parameter estimates are for a bilateral model. We shall see, however, that in the two-dimensional case the reduction to a unilateral scheme complicates matters very much, in contrast to the one-dimensional case we have just considered. There is thus no escaping the explicit introduction of a dependence in all directions. A further point is relevant: it is the multilateral scheme which in general corresponds to reality, even in those cases for which the formal work of estimation, etc., is simplest performed using an equivalent unilateral model.

5. Generalities on the two-dimensional process

One can consider continuous processes in which the variate has a value at every point of the plane, or discrete processes, which are usually such that the variate is observed only at the points of a rectangular plane lattice. We shall for the most part consider the discrete case, as being simpler mathematically and of greater practical interest. The observed and error variates will be denoted ξ_{st} and ϵ_{st} respectively ($s, t = \ldots, -2, -1, 0, 1, 2, \ldots$). When discussing continuous processes this notation will be modified to $\xi(x, y)$ and $\epsilon(x, y)$ (x and y assuming all real values).

The particular model which we shall consider almost exclusively is the two-dimensional linear autoregression, which we shall write

$$L(T_s, T_t)\,\xi_{st} = \epsilon_{st}, \tag{18}$$

where T_s and T_t are translation operators defined by

$$T_s \xi_{st} = \xi_{s+1,t}, \quad T_t \xi_{st} = \xi_{s,t+1} \tag{19}$$

and

$$L(T_s, T_t) = \sum_j \sum_k a_{jk} T_s^j T_t^k. \tag{20}$$

Corresponding to equations (7)–(10) we have

$$\xi_{st} = \frac{\epsilon_{st}}{L(T_s, T_t)} = \sum_j \sum_k b_{jk} \epsilon_{s+j,t+k}, \tag{21}$$

$$F(\omega_1, \omega_2) = \frac{v}{L(z_1, z_2) L(z_1^{-1}, z_2^{-1})} \tag{22}$$

$$= \sum_j \sum_k \phi(j, k) z_1^j z_1^k, \tag{23}$$

where b_{jk} is the coefficient of $z_1^j z_2^k$ in the Fourier expansion of $[L(z_1, z_2)]^{-1}$, ($z_1 = e^{i\omega_1}, z_2 = e^{i\omega_2}$), v is the variance of the ϵ's, $\phi(j, k)$ the covariance of ξ_{st} and $\xi_{s+j,t+k}$, and $F(\omega_1, \omega_2)$ the corresponding spectral function, as defined by (23).

As before, it is necessary and sufficient for the validity of (21) and (22) that $L(z_1, z_2)$ be not zero for any z_1 and z_2 which simultaneously satisfy $|z_1| = 1$, $|z_2| = 1$ (at least when the autoregression (18) is finite). We shall assume that the operator L has been so normalized that $L(z_1, z_2)$ will not circle the origin in the complex plane as either z_1 or z_2 moves around the unit circle. (If L does not circle the origin as z_1 traces out the unit circle and z_2 holds a particular value, neither will it do so for any other value of z_2. For if it should do so there would be some intermediate value of z_2 for which L would move *through* the origin, against hypothesis.)

6. Unilateral representation of the two-dimensional process

We shall prove, by a type of argument which is now well known (due to Wiener; see, for example, Wiener, 1949, p. 78), that a unique process may be found which will generate a given set of autocorrelations and in which ξ_{st} is expressed as an autoregression upon ξ_{su} ($u > t$) and ξ_{vw} ($v > s$, w unrestricted). That is, in the lattice of Fig. 1 the value at the white dot may be expressed in terms of the values at the black dots. The given autocorrelogram may obviously not be completely arbitrary, but the necessary conditions are very mild.

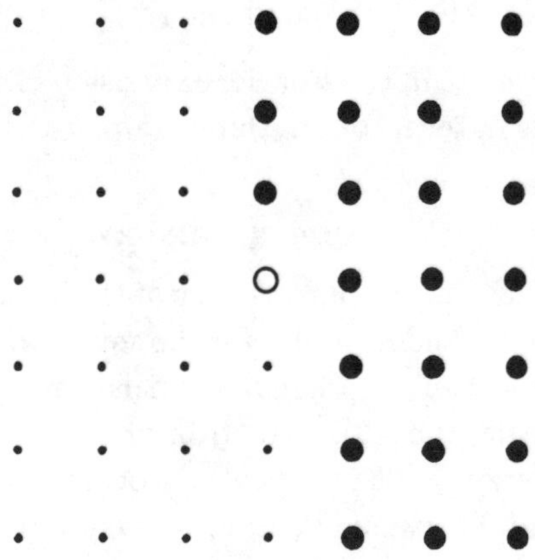

Fig. 1

The first condition is that the autocorrelations be such that they could have been generated by a purely non-deterministic process, which implies among other things that log $F(\omega_1, \omega_2)$ has a Fourier expansion, so that $F(\omega_1, \omega_2)$ may be represented

$$F(\omega_1, \omega_2) = \exp\left[\sum_j \sum_k \alpha_{jk} z_1^j z_2^k\right]. \tag{24}$$

Define now the function

$$P(z_1, z_2) = \exp -\left[\frac{\alpha_{00}}{2} + \sum_{k=1}^{\infty} \alpha_{0k} z_2^k + \sum_{j=1}^{\infty} \sum_{k=-\infty}^{\infty} \alpha_{jk} z_1^j z_2^k\right]. \tag{25}$$

Since
$$F(\omega_1, \omega_2) = \frac{1}{P(z_1, z_2)\, P(z_1^{-1}, z_2^{-1})}, \tag{26}$$

the autoregression
$$P(T_s, T_t)\, \xi_{st} = \epsilon'_{st} \quad (\sigma^2(\epsilon') = 1) \tag{27}$$

will have spectral function $F(\omega_1, \omega_2)$, as required. Further, the expansion of $P(T_s, T_t)$ will involve the same powers of T_s and T_t as occur of z_1 and z_2 in expression (25), so that the autoregression is of the required form. To ensure that (27) may validly be written out as an autoregression we include a second condition: that $P(e^{i\omega_1}, e^{i\omega_2})$ possess a Fourier expansion.

Representation (27) corresponds to the unilateral representation in the case of a transect. It is by no means as useful, however, and for such purposes as estimation one is in general better advised to work with the original model. For one thing, it is not true in two dimensions as it is in one, that the unilateral representation of a finite autoregression is also a finite autoregression. This is confirmed by the following simple example: the finite autoregression

$$(1+\beta^2)\, \xi_{st} = \beta(\xi_{s+1,t} + \xi_{s,t+1} + \xi_{s,t-1}) + \epsilon_{st} \tag{28}$$

has a unilateral representation which is infinite:

$$\xi_{st} = 2\beta\xi_{s,t+1} - \beta^2\xi_{s,t+2} - \beta^2\xi_{s+1,t+1} + \beta(1-\beta^2)\sum_0^{\infty} \beta^j \xi_{s+1,t-j} + \epsilon'_{st}. \tag{29}$$

Further, the real usefulness of the unilateral representation is that it suggests a simplifying change of parameters (although the same parameter transformation is usually suggested in the evaluation of integral (4) or its two-dimensional equivalent). For most two-dimensional models, however, the appropriate transformation, even if evident, is so complicated that nothing is gained by performing it. For example, to calculate the unilateral representation of the model

$$\xi_{st} = \alpha(\xi_{s+1,t} + \xi_{s-1,t} + \xi_{s,t+1} + \xi_{s,t-1}) + \epsilon_{st} \tag{30}$$

one must evaluate the coefficients in the Fourier expansion of $\log[1 - \alpha(z_1 + z_1^{-1} + z_2 + z_2^{-1})]$, and these are not expressible in terms of anything simpler than elliptic integrals.

7. SAMPLING THEORY

In this section we shall consider the sampling theory of the discrete process, deriving general estimation equations, tests of fit, and formulae for the asymptotic variances and covariances of parameter estimates. The pattern of development is similar to that of previous articles on time-series analysis (Whittle, 1951–3), apart from the fact that special attention must be paid to the effect of multilaterality. The particular processes considered are those which are purely non-deterministic and for which $[F(\omega_1, \omega_2)]^{-1}$ is not zero for any real ω_1, ω_2, so that the process is representable as an autoregression. We shall also in general assume the variates normally distributed. Estimation equations thus obtained on the maximum-likelihood criterion can be regarded as least-square estimation equations when the variate distribution is not normal, but the extent to which remaining results are valid in such a case is a matter for further inquiry.

Let us suppose that we have a series of mn observations ξ_{st} ($s = 1, 2, \ldots, m$; $t = 1, 2, \ldots, n$), so that the empirical covariance of lag j, k is

$$C_{jk} = \frac{1}{mn} \sum_{s=1}^{m-j} \sum_{t=1}^{n-k} \xi_{st}\xi_{s+j,t+k}. \tag{31}$$

We shall uniformly neglect end-effects, and so have divided in (31) by mn instead of $(m-j)(n-k)$. Let us further define the quantity

$$\begin{aligned} f(\omega_1, \omega_2) &= \frac{1}{mn}\{[\Sigma\Sigma\xi_{st}\cos(s\omega_1 + t\omega_2)]^2 + [\Sigma\Sigma\xi_{st}\sin(s\omega_1 + t\omega_2)]^2\} \\ &= \sum_{-m}^{m} \sum_{-n}^{n} C_{jk}\cos(j\omega_1 + k\omega_2) \\ &= \sum_{-m}^{m} \sum_{-n}^{n} C_{jk}e^{i(j\omega_1 + k\omega_2)}, \end{aligned} \tag{32}$$

which we see on comparison with (23) to be the *empirical spectral function*, and in fact the appropriate extension of the Schuster periodogram to two dimensions.

At the end of the section we shall prove the following basic result: that if the variates ξ_{st} are normally distributed with zero mean, and are generated by a stationary process of the type considered with spectral function $F(\omega_1, \omega_2)$, then their joint likelihood is given, apart from end-effects, by the expression

$$p(\xi) = \frac{1}{(2\pi V)^{\frac{1}{2}(mn)}} \exp\left[-\frac{mn}{8\pi^2}\int_0^{2\pi}\int_0^{2\pi}\frac{f}{F}\,d\omega_1\,d\omega_2\right], \tag{33}$$

where

$$V = \exp\left[\frac{1}{4\pi^2}\int_0^{2\pi}\int_0^{2\pi}\log F\,d\omega_1\,d\omega_2\right]. \tag{34}$$

Taking the logarithm of (33) we see that the maximum-likelihood estimates of the parameters of F are obtained by minimizing

$$\mathfrak{L} = \frac{1}{4\pi^2}\iint\left[\log F + \frac{f}{F}\right]d\omega_1\,d\omega_2. \tag{35}$$

Even if the variates are not considered to be normally distributed, expression (35) will upon minimization yield the least-square estimates (see Whittle, 1953, p. 132).

Expression (35) actually does not lead to the most convenient form of the estimation equations, this is obtained by expressing $\mathfrak{L}$ in terms of the autocovariances rather than the periodogram

$$\mathfrak{L} = \frac{1}{4\pi^2}\iint\log F\,d\omega_1\,d\omega_2 + \Sigma\Sigma c_{jk}C_{jk}, \tag{36}$$

where the coefficients c are given by

$$\Sigma\Sigma c_{jk}z_1^j z_2^k = \frac{1}{F(\omega_1, \omega_2)}. \tag{37}$$

Consider now the special case of a process generated by a stochastic difference equation of type (18) so that

$$F = \frac{v}{L(z_1, z_2)\,L(z_1^{-1}, z_2^{-1})}. \tag{38}$$

Inserting this relation in (36) we obtain

$$\mathfrak{L} = \log v + \log k + U/mnv, \tag{39}$$

where

$$U = \sum_1^m \sum_1^n \epsilon_{st}^2 = \sum_1^m \sum_1^n [L(T_s, T_t)\,\xi_{st}]^2 = \sum_j \sum_k c_{jk} C_{jk} \tag{40}$$

and

$$\log k = -\frac{1}{4\pi^2}\iint \log [L(z_1, z_2)\,L(z_1^{-1}, z_2^{-1})]\,d\omega_1\,d\omega_2. \tag{41}$$

The minimized value of (39) with respect to the irrelevant parameter v is $\log (kU/mn)$, so that the quantity to be minimized with respect to the parameters of L is kU. This is the result which is of most use in investigations of the present type: that the least-square estimates are yielded by the minimization of the usual 'residual sum of squares' times a function of the parameters, k. We see from (41) that $-\frac{1}{2}\log k$ may be interpreted as the absolute term in the double Fourier expansion of $L(e^{i\omega_1}, e^{i\omega_2})$, a fact that may sometimes ease its evaluation.

Suppose that F contains the unknown parameters $\theta_1, \theta_2, \ldots, \theta_p$ (v being included among these, if, as is usual, its value is unknown). Then, following a familar line of reasoning (see Whittle, 1953, p. 135, for an application similar to the present one) that the asymptotic covariance matrix of the least square estimates of $\theta_1, \theta_2, \ldots, \theta_p$ is

$$\frac{2}{mn}\left[\frac{1}{4\pi^2}\iint \frac{\partial \log F}{\partial \theta_j}\,\frac{\partial \log F}{\partial \theta_k}\,d\omega_1\,d\omega_2\right]^{-1}. \tag{42}$$

Suppose again that when the p parameters have been fitted the minimized value of kU is $(kU)_p$, and that when an additional q parameters have been fitted the quantity falls to $(kU)_{p+q}$. Then we can test the improvement attained by the introduction of the extra q parameters by using the fact that, if the initial p-parameter hypothesis were correct, then

$$\psi^2 = (mn-p-q)\log \frac{(kU)_p}{(kU)_{p+q}} \tag{43}$$

would be asymptotically distributed as χ^2 with q degrees of freedom. The argument is again an old one (cf. Whittle, 1952, 1953).

Proof of relation (33). Let the process under consideration have the unilateral representation (27). The joint frequency function of the mn residuals ϵ'_{st} ($s = 1, 2, \ldots, m$; $t = 1, 2, \ldots, n$) is

$$p(\epsilon') = \frac{1}{(2\pi)^{\frac{1}{2}(mn)}}\exp\left[-\frac{1}{2}\sum_1^m \sum_1^n (\epsilon'_{st})^2\right]. \tag{44}$$

Performing the linear transformation to which (27) is equivalent, we obtain (neglecting end effects) the following expression for the frequency function of the ξ_{st}:

$$\begin{aligned} p(\xi) &\approx \frac{A^{mn}}{(2\pi)^{\frac{1}{2}(mn)}}\exp\left[-\frac{1}{2}\sum_1^m \sum_1^n [P(T_s, T_t)\,\xi_{st}]^2\right] \\ &\approx \frac{A^{mn}}{(2\pi)^{\frac{1}{2}(mn)}}\exp\left[-\frac{mn}{2}\sum\sum c_{jk} C_{jk}\right] \\ &\approx \frac{A^{mn}}{(2\pi)^{\frac{1}{2}(mn)}}\exp\left[-\frac{mn}{8\pi^2}\iint \frac{f}{F}\,d\omega_1\,d\omega_2\right], \end{aligned} \tag{45}$$

where A is the coefficient of ξ_{st} in (27). By (24) and (25)

$$A = e^{-\frac{1}{2}\alpha_{00}} = \exp\left[-\frac{1}{8\pi^2}\iint \log F\, d\omega_1 d\omega_2\right], \tag{46}$$

which on insertion in (45) leads to (33).

The nature of the approximate equality in (45) may be specified more exactly. If the right-hand member in (45) is denoted p', then the relation obtaining is that $\log p$ and $\log p'$ are asymptotically equal (neglect of end-effects leading to neglect of terms in the exponent). This kind of approximation is quite satisfactory, since the addition of a term of relative order m^{-1} or n^{-1} to the logarithm of the likelihood will for large m and n have little effect on either the values of the maximum-likelihood estimates, or the significance points of likelihood ratios.

8. Numerical examples

Neither of the two sets of numerical data to be examined provides a picture book example of the particular models we have been studying, but they are perhaps the more valuable for just that reason.

The first set of data concern a uniformity trial on wheat (yield of grain) conducted by Mercer & Hall (1911). It involves 500 plots, each 11 ft by 10·82 ft., arranged in a 20×25 rectangle, plot totals constituting the observations. The first and fourth quadrants of the correlation field are given in Table 1; the second and third quadrants can be filled in by means of the relation

$$\phi(-j, -k) = \phi(j, k).$$

We note that correlations along the north-south (s) axis are considerably stronger than those along the east-west (t) axis, at least part of the explanation presumably being that the plots are not square. Correlations are generally higher in the north-east direction than in the north-west, indicating a definite directional effect. Another fact worthy of note is that correlations do not decrease monotonically as one moves out from the origin, instead they dip to a minimum and then rise again. One could imagine competition between neighbouring plants producing such an effect if the distances were smaller, but the more likely explanation in the present case is that there are 'waves of fertility' of the kind often remarked in a ploughed field (see, for example, Neyman, 1952, p. 75).

In Table 2 we summarize the various schemes considered, giving the fitted coefficients and the corresponding values of k, U and kU.

Since all kU are near to 0·7 we can say roughly that if the introduction of a new parameter depresses kU by Δ, then by equation (43) Δ is significant at the 5 % level if

$$3{\cdot}841 < 500 \log\left(\frac{0{\cdot}7+\Delta}{0{\cdot}7}\right)$$

or $\Delta > 0{\cdot}0054$. The corresponding limits if two or three parameters have been introduced are 0·0084 and 0·0110. These limits provide useful rules of thumb for judging significances of differences in kU in the table. Thus, hypotheses 1 and 3 do not differ significantly in degree of fit, while 1 and 4 very definitely do. Hypotheses such as 1 and 5 cannot strictly be compared in this manner, since neither hypothesis is a special instance of the other. However, one would be very much inclined to say that the fit of 1 is superior to that of 5, and this conclusion is borne out if one compares each in turn with the more general hypothesis 7.

The surprising feature of the results is that the simple unilateral scheme 1 fits the data so much better than the symmetric second order schemes 5 and 6. Thus, although the theme of this article has largely been that spatial processes will in general be multilateral, our first example appears to be dominantly unilateral. The reason for this becomes apparent upon a re-examination of the correlations of Table 1. These correlations fall steeply as soon as a lag is introduced (to 0·52 on the s axis and 0·29 on the t axis), but decrease only slowly as the lag is increased (the corresponding figures for a lag of two units are thus 0·41 and 0·15). However, it will be seen in § 9 that the correlogram of a scheme such as 5 or 6 decays smoothly right from the origin, having in fact zero derivative there (cf. Fig. 2), so that neither 5 nor 6 can possibly fit the observations well. There are at least two possible explanations of this behaviour of the observed correlogram. The underlying scheme may in fact be a unilateral

Table 1. *Autocorrelations for the wheat data, $s = 0$ to 4, $t = -3$ to $+3$*

t	$s=0$	$s=1$	$s=2$	$s=3$	$s=4$
−3	0·1880	0·1602	0·1509	0·1276	0·1352
−2	0·1510	0·0234	0·0020	−0·0137	−0·1039
−1	0·2923	0·1853	0·1349	0·0788	0·0878
0	1·0000	0·5252	0·4055	0·3639	0·3561
1	0·2923	0·2354	0·1799	0·1205	0·1399
2	0·1510	0·1285	0·0999	0·0749	0·0859
3	0·1880	0·1935	0·2483	0·2415	0·2284

Table 2. *Details of the models fitted*

Model no.	$L(T_s, T_t)$	k	U	kU
1	$1-0{\cdot}488T_s-0{\cdot}202T_t$	1	0·6848	0·6848
2	$1-0{\cdot}483T_s-0{\cdot}179T_t^{-1}$	1	0·6940	0·6940
3	$1-0{\cdot}492T_s-0{\cdot}211T_t+0{\cdot}019T_sT_t$	1	0·6845	0·6845
4	$1-0{\cdot}402T_s-0{\cdot}168T_t-0{\cdot}172T_s^2-0{\cdot}092T_t^2$	1	0·6564	0·6564
5	$1-0{\cdot}159(T_s+T_s^{-1}+T_t+T_t^{-1})$	1·1240	0·6508	0·7314
6	$1-0{\cdot}213(T_s+T_s^{-1})-0{\cdot}102(T_t+T_t^{-1})$	1·1332	0·6217	0·7045
7	$1-0{\cdot}488T_s+0{\cdot}030T_s^{-1}-0{\cdot}202T_t-0{\cdot}034T_t^{-1}$	0·9843	0·6816	0·6709

one, due to e.g. the presence of a slope in the ground or a prevailing wind. Alternatively, we must recall that the observations are not *point* observations of growth, but *integrated* observations of the growth over an area. An integration such as this will enhance the auto-covariance of zero lag relative to the others, and the correlogram of a scheme such as 6, when thus distorted, would not be dissimilar to the observed correlogram. However, the consideration of such an effect would lead us too far for present purposes, and we shall take the observations at their face value.

Scheme 1 is not completely satisfactory, in that it does not explain the dip and rise in the correlation for increasing distance. The inclusion of second-order terms in the same direction as the first-order terms improves the fit considerably, see scheme 4.

Unilateral schemes such as 1–4 can be fitted directly by least square regression methods, since for them $k = 1$. The fitting of multilateral schemes such as 5–7 is more difficult, however. The evaluation of U according to (40) is direct, it is the evaluation of k from (41) which presents new problems. For the present calculations the pedestrian methods of series expansion and numerical integration have been employed, although it is likely that better methods exist. Thus, for a scheme

$$\xi_{st} = \alpha\xi_{s+1,t} + \beta\xi_{s-1,t} + \gamma\xi_{s,t+1} + \delta\xi_{s,t-1} + \epsilon_{st}, \tag{47}$$

we have $\log(k) =$ minus twice absolute term in $\log[1 - \alpha z_1 - \beta z_1^{-1} - \gamma z_2 - \delta z_2^{-1}]$

$$= \sum_{j=1}^{\infty} \sum_{k=0}^{j} \frac{(2j)!}{j[k!\,(j-k)!]^2} (\alpha\beta)^k (\gamma\delta)^{j-k}. \tag{48}$$

For $\alpha\beta = \gamma\delta$ this reduces to

$$\log(k) = \sum_{j=1}^{\infty} \frac{1}{j} \binom{2j}{j}^2 (\alpha\beta)^j. \tag{49}$$

Expansions (48) and (49) are useful as long as $\alpha\beta$ and $\gamma\delta$ are very small, but convergence becomes slow as these quantities approach their maximum value of $\frac{1}{16}$. In Table 3 the value of $\log k$ in (49) is given for some specimen values of $\theta = \sqrt{(\alpha\beta)}$; these values have been obtained by numerical integration, and vary slowly enough to permit quite a fair graphical interpolation.

Table 3. *Specimen values of the correcting factor* $\log(k)$

θ	0·00	0·05	0·10	0·15	0·20	0·22	0·25
$\log(k)$	0·0000	0·0076	0·0420	0·1010	0·2028	0·2656	0·4406

Table 4. *Autocorrelations for the orange data,* $s = 0$ to 9, $t = -4$ to 4.

t	Values of s									
	0	1	2	3	4	5	6	7	8	9
−4	0·3912	0·3902	0·3771	0·3581	0·3710	0·2923	0·2549	0·2546	0·2440	0·2131
−3	·4609	·3956	·3930	·3696	·3484	·3353	·3137	·2834	·2894	·2870
−2	·4667	·4159	·3982	·3470	·3243	·3225	·3250	·2801	·2721	·2752
−1	·5462	·4669	·4336	·3854	·3880	·3761	·3495	·2914	·2567	·2606
0	1·0000	·5403	·5052	·4840	·4585	·4233	·3960	·3246	·3210	·3150
1	0·5462	·4458	·4280	·3751	·3914	·3622	·3716	·2948	·2958	·2925
2	·4667	·3858	·3799	·3459	·3603	·3475	·3401	·2905	·2766	·2230
3	·4609	·4190	·3716	·3764	·3509	·3336	·3369	·2765	·2433	·2741
4	·3912	·3543	·3882	·3392	·3537	·3446	·3305	·2893	·2685	·2321

The models of Table 2 were fitted simply by inserting trial values of the parameters and calculating the corresponding kU. Improved values were then found by approximating kU in the neighbourhood of its minimum by a paraboloid, and then locating this approximate minimum, whereupon the process was repeated. The work is tedious, but not unreasonably so if one considers the effort which the data themselves represent.

Our second example is derived from a uniformity trial of 1000 orange trees by Batchelor & Reed (1924), the trees being arranged in a 20×50 rectangular lattice. The correlation field $\rho(s, t)$ is given in Table 4, and the correlations along the s axis plotted in Fig. 2. The correlations in the two quadrants follow a very similar pattern, indicating a high degree of symmetry in the original experimental field. A feature of interest is, that while the correlations $\rho(s, t)$ for the most part change smoothly in value as one moves through adjacent values of s and t, there is a sharp discontinuity at the origin. The correlation surface thus consists of a smoothly rounded broad dome with a spike at the apex. This suggests that the yield of an individual tree, Y_{st}, may be represented

$$Y_{st} = \xi_{st} + \eta_{st}, \tag{50}$$

where ξ_{st} obeys a process with smooth correlogram, while distinct η_{st} are uncorrelated with one another (or with the ξ_{st}) so that the η correlogram consists simply of a 'spike' at the origin. A natural interpretation of (50) springs to mind: that ξ represents the fertility of the

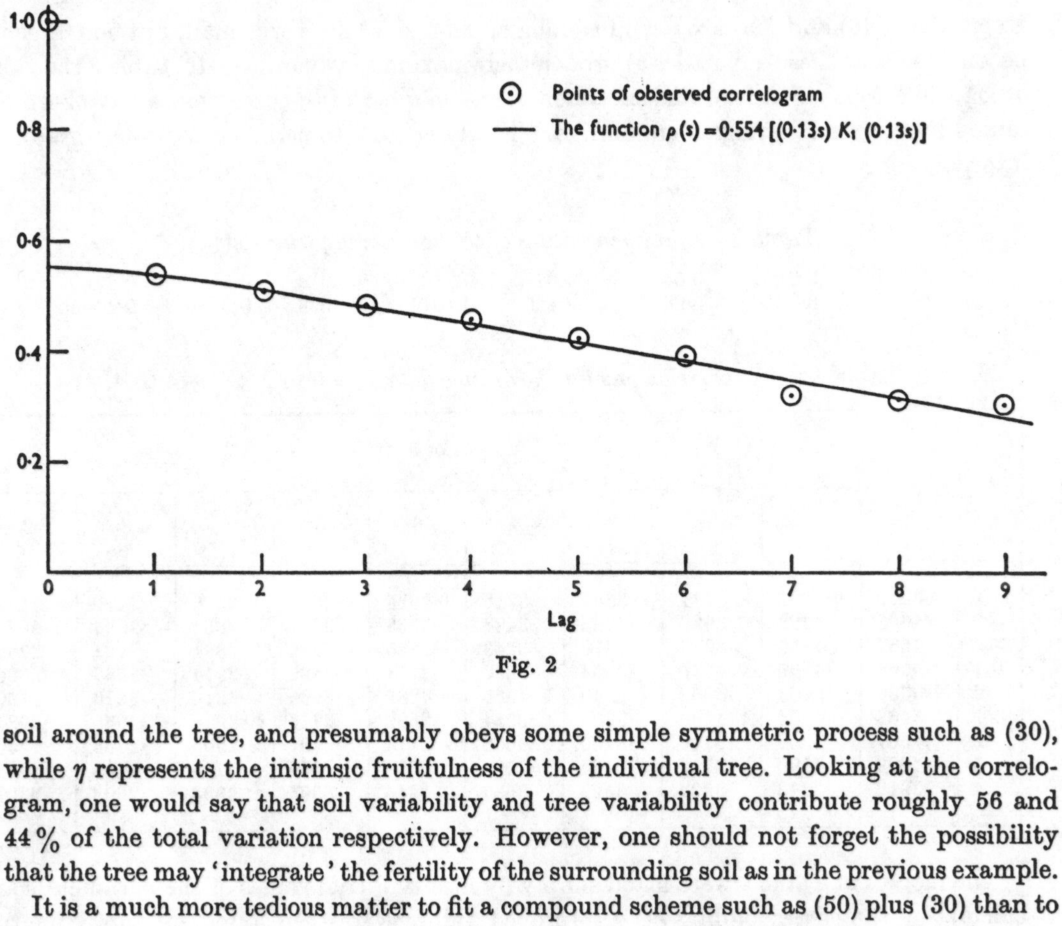

Fig. 2

soil around the tree, and presumably obeys some simple symmetric process such as (30), while η represents the intrinsic fruitfulness of the individual tree. Looking at the correlogram, one would say that soil variability and tree variability contribute roughly 56 and 44 % of the total variation respectively. However, one should not forget the possibility that the tree may 'integrate' the fertility of the surrounding soil as in the previous example.

It is a much more tedious matter to fit a compound scheme such as (50) plus (30) than to fit a simple autoregression such as (30). Supposing var $(\eta) = A$ and var $(\epsilon) = B$ we have

$$F(\omega_1, \omega_2) = A + \frac{B}{[1 - \alpha(z_1 + z_1^{-1} + z_2 + z_2^{-1})]^2}. \tag{51}$$

The quantities required for estimation purposes are $\log V$ and the coefficients in the expansion of F^{-1} (see (36)). We find

$$\log V = +\frac{1}{4\pi^2}\iint \log F\, d\omega_1 d\omega_2 = \log(A+B) + \sum_{j=1}^{\infty} \frac{1}{j}\binom{2j}{j}\alpha^{2j}t$$
$$+4(\alpha\beta)^2[1-2\beta]$$
$$+36(\alpha\beta)^4[-\tfrac{1}{2}+4\beta-4\beta^2]$$
$$+400(\alpha\beta)^6[\tfrac{1}{3}-6\beta+16\beta^2-\tfrac{32}{3}\beta^3]$$
$$+4900(\alpha\beta)^8[-\tfrac{1}{4}+8\beta-40\beta^2+60\beta^3-32\beta^4]$$
$$+63504(\alpha\beta)^{10}[\tfrac{1}{5}-10\beta+80\beta^2-224\beta^3+256\beta^4-\tfrac{1024}{10}\beta^5]$$
$$+\ldots; \tag{52}$$

and

$$F^{-1} = \Sigma\Sigma c_{jk} z_1^j z_2^k$$
$$= \frac{1}{A+B}[1+(2\beta-2)S+(1-5\beta+4\beta^2)S^2+(4\beta-12\beta^2+8\beta^3)S^3$$
$$+(-\beta+13\beta^2-28\beta^3+16\beta^4)S^4+\ldots], \tag{53}$$

where $$\beta = \frac{A}{A+B}, \quad S = \alpha(z_1+z_1^{-1}+z_2+z_2^{-1}).$$

Since the scheme is not a pure autoregression, F^{-1} is not a polynomial and the coefficients c_{jk} do not terminate, so that in the present case we find that we should have to include correlations of up to about lag 10 in order to obtain a sufficient approximation to the sum $\Sigma\Sigma c_{jk} C_{jk}$ of (36). However, correlations are only available up to $t = 4$, to continue would be a very arduous matter.

In general, if the scheme is not purely an autoregressive one it will be possible to apply least square methods only if the observed correlogram decays quite quickly. In a case such as the present one it becomes necessary to use more primitive methods of fitting: e.g. equation of observed and theoretical autocorrelation coefficients.

On the other hand, a slow rate of decay has the advantage that the discrete model may be approximated by a continuous one, which is in most cases a simplification. Thus, if in scheme (30) α has a value approaching 0·25 (corresponding to a very slow decay of dependence) then it is shown in § 9 that the scheme may be approximated by a continuous one with autocovariance function

$$\rho(r) = \text{const.}\, rK_1(\kappa r), \tag{54}$$

where r is the distance between the two points considered, K_1 a modified Bessel function, and

$$\kappa = \sqrt{\left(\frac{1}{\alpha}-4\right)}. \tag{55}$$

The thin line of the correlogram (Fig. 2) indicates the values of function (55) when the constants have been adjusted for coincidence at r_1 and r_8. It is found that $\kappa = 0{\cdot}13$, whence $\alpha = 0{\cdot}2489$.

The agreement looks impressive, but must be discounted considerably, since almost any monotone decreasing function would fit the observed curve reasonably well if only the end-points were arrange to coincide. However, if one fits, for example, an exponential curve in the same fashion, agreement is not at all as good, the exponential curve sagging too much in the middle.

9. SPECIAL PROCESSES

Strictly, one does not need to be acquainted with the properties of a model in order to fit it to a set of data (cf. the examples of the previous section). However, a knowledge of these properties does help one to arrive at a good model quickly, to say the least. In the present context 'properties' is usually synonymous with 'correlogram', since it is most often the correlogram one uses to characterize a given set of data. We shall now examine a few special models and their correlograms, but in a rather unsystematic fashion, since few results of an explicit nature are obtainable.

The simplest model of interest is the degenerate autoregression

$$\xi_{st} = \alpha\xi_{s+1,t} + \beta\xi_{s,t+1} + \epsilon_{st}, \tag{56}$$

with 'solution'

$$\xi_{st} = \sum_{j=0}^{\infty}\sum_{k=0}^{\infty}\binom{j+k}{j}\alpha^j\beta^k\epsilon_{s+j,t+k}. \tag{57}$$

For the process to be stationary, ξ_{st} being taken as the dependent variate, one must have $|\alpha|+|\beta|<1$ (see § 5). The autocovariances are generated by

$$(1-\alpha z_1-\beta z_2)^{-1}(1-\alpha z_1^{-1}-\beta z_2^{-1})^{-1},$$

and yield correlations

$$\rho(s,0)=A^s, \quad \rho(0,t)=B^t,$$

$$\rho(s,t) = \left[\sum_{j=0}^{s}\binom{s+t-j}{t}\alpha^{s-j}\beta^t A^j + \sum_{k=0}^{t}\binom{s+t-k}{s}\alpha^s\beta^{t-k}B^k\right], \tag{58}$$

where

$$\left.\begin{aligned} A &= \frac{1+\alpha^2-\beta^2-\Delta}{2\alpha}, \quad B = \frac{1+\beta^2-\alpha^2-\Delta}{2\beta}, \\ \Delta &= \sqrt{\{(1+\alpha+\beta)(1+\alpha-\beta)(1-\alpha+\beta)(1-\alpha-\beta)\}}.\end{aligned}\right\} \tag{59}$$

Formula (58) holds for $s,t>0$.

The continuous analogue of scheme (56) would be the first-order stochastic differential equation, formally written

$$\left(\alpha\frac{\partial}{\partial x}+\beta\frac{\partial}{\partial y}+\gamma\right)\xi(x,y) = \epsilon(x,y). \tag{60}$$

Rotating the plane to new axes $X=0$, $Y=0$, inclined at an angle $\tan^{-1}(\beta/\alpha)$ to the old ones, we find the relation may be written

$$\left(\sqrt{(\alpha^2+\beta^2)}\frac{\partial}{\partial X}+\gamma\right)\xi'(X,Y) = \epsilon'(X,Y), \tag{61}$$

where $\xi'(X,Y)=\xi(x,y)$, $\epsilon'(X,Y)=\epsilon(x,y)$. That is, scheme (60) may be regarded as a series of Markoff processes running side by side in the direction $\tan^{-1}(\beta/\alpha)$ and independent one of the other. This indicates the degenerate nature of the first-order scheme.

The simplest second-order scheme (and thus the simplest non-degenerate scheme) is the symmetric autoregression

$$\xi_{st} = \alpha(\xi_{s+1,t}+\xi_{s-1,t}+\xi_{s,t+1}+\xi_{s,t-1})+\epsilon_{st}. \tag{62}$$

Written in the form

$$\left[\Delta_s^2+\Delta_t^2+\left(4-\frac{1}{\alpha}\right)\right]\xi_{st} = \epsilon_{st} \tag{63}$$

(where Δ is the central difference operator) the analogy with the continuous relation, a stochastic Laplace equation, becomes apparent:

$$\left[\left(\frac{\partial}{\partial x}\right)^2 + \left(\frac{\partial}{\partial y}\right)^2 - \kappa^2\right] \xi(x,y) = \epsilon(x,y). \tag{64}$$

The continuous relation is the easier to handle (as was also the case for relations (56), (60)) due to its essentially less artificial nature. Explicit results have been obtained for scheme (62) (Van der Pol & Bremmer, 1950; Stöhr, 1950), but these are not simple. From the formula (Titchmarsh, 1948, p. 201)

$$\frac{1}{4\pi^2}\iint \frac{e^{i(X\omega_1+Y\omega_2)}\,d\omega_1\,d\omega_2}{(\omega_1^2+\omega_2^2+\kappa^2)^{\mu+1}} = \left(\frac{r}{2\kappa}\right)^{\mu} \frac{K_\mu(\kappa r)}{\Gamma(\mu+1)} \quad (r = \sqrt{(X^2+Y^2)}), \tag{65}$$

we deduce that for scheme (64)

$$\xi(x,y) = \iint_{-\infty}^{+\infty} \epsilon(x+X, y+Y)\, K_0(\kappa r)\, dX\, dY \tag{66}$$

and

$$\phi(X,Y) = \frac{r}{2\kappa} K_1(\kappa r), \tag{67}$$

where K_ν is the modified Bessel function of the second kind, order ν. Since $\lim rK_1(r) = 1$ the correlation coefficient corresponding to (67) is

$$\rho(r) = \kappa r K_1(\kappa r). \tag{68}$$

Scheme (64) is a special case of the general second-order stochastic difference equation. According to the usual classification into elliptic, parabolic, and hyperbolic forms, (64) would be regarded as a circular form with centre at the origin. An investigation of the remaining second-order schemes would be of great interest, both theoretically and practically.

The correlation function (68) is of interest in that it may be regarded as the 'elementary' correlation in two dimensions, similar to the exponential $e^{-\alpha|x|}$ in one dimension. Both correlation curves are monotone decreasing, but (68) differs in that it is flat at the origin, and that its rate of decay is slower than exponential.

Attempts have been made to represent two-dimensional correlograms as sums of exponentials, largely because the exponential has proved itself the natural choice in one dimension, and the observed curves display the same monotonic decay. However, it is apparent from the previous paragraphs that the exponential function has no divine right in two dimensions, while the example of the last section indicated that a K_1 function fitted the observations better than did an exponential.

Two-dimensional processes can be constructed which have exponential correlation functions, but these are very artificial. For example, Matérn (1947) has shown that the correlation function $\exp - \alpha\sqrt{(X^2+Y^2)}$ corresponds to a spectral function $(\omega_1^2+\omega_2^2-\alpha^2)^{-\frac{3}{2}}$. The simplest process with such a spectral function may be formally written

$$\left[\left(\frac{\partial}{\partial x}\right)^2 + \left(\frac{\partial}{\partial y}\right)^2 - \alpha^2\right]^{\frac{3}{4}} \xi(x,y) = \epsilon(x,y) \tag{69}$$

and it is difficult to visualize a physical mechanism which would lead to such a relation.

The examination of the two sets of uniformity data was begun by Mr I. D. Dick of the New Zealand Department of Scientific and Industrial Research, who also in large degree stimulated the author to the present investigation. Thanks are also due to the supervisor of the punched card calculations, Mr E. W. Jones.

REFERENCES

BARTLETT, M. S. (1946). Stochastic processes. Mimeographed N. Carolina lecture notes.

BATCHELOR, L. D. & REED, H. S. (1924). Relation of the variability of yields of fruit trees to the accuracy of field trials. *J. Agric. Res.* **12**, 245–83.

DANIELL, P. J. (1946). Contribution to discussion on stochastic processes. *J. R. Statist. Soc.* B, **8**, 88–90.

DOOB, J. L. (1944). The elementary Gaussian processes. *Ann. Math. Statist.* **15**, 229–82.

MATÉRN, B. (1947). Metoder att uppskatta noggrannheten vid linje- och provytetaxering. *Medd. SkogsforsknInst.* **36**, no. 1.

MERCER, W. B. & HALL, A. D. (1911). The experimental error of field trials. *J. Agric. Sci.* **4**, 107–32.

MORAN, P. A. P. (1950). Notes on continuous stochastic phenomena. *Biometrika*, **37**, 17–23.

NEYMAN, J. (1952). *Lectures and Conferences on Mathematical Statistics and Probability.* Graduate School, U.S. Dep. of Agriculture.

QUENOUILLE, M. H. (1949). Problems in plane sampling. *Ann. Math. Statist.* **20**, 355–75.

STÖHR, A. (1950). Über einige lineare partielle Differenzgleichungen mit konstanten Koeffizienten. *Math. Nachr.* **3**, 295–315.

TITCHMARSH, E. C. (1948). *Introduction to the Theory of Fourier Integrals.* Oxford University Press.

VAN DER POL, B. & BREMMER, H. (1950). *Operational Calculus.* Cambridge University Press.

WHITTLE, P. (1952). Tests of fit in time series. *Biometrika*, **39**, 309–18.

WHITTLE, P. (1953). The analysis of multiple stationary time series. *J. R. Statist. Soc*, B, **15**, 125–39.

WIENER, N. (1949). *The Extrapolation, Interpolation and Smoothing of Stationary Time-Series.* New York, Wiley.

WILLIAMS, R. M. (1954). The choice of sampling intervals for systematic samples from populations with stationary correlation. (Unpublished paper.)

WOLD, H. (1938). *A Study in the Analysis of Stationary Time Series.* Upsala.

[EDITORIAL NOTE. The preceding paper by Whittle (on p. 443) and the following paper by Patankar (on pp. 459–60) both use as an illustration of theory Mercer & Hall's (1911) wheat-plot data. On a first inspection it might be thought that the serial correlations in (i) the column $s = 0$ and (ii) in the row $t = 0$ of Whittle's Table 1, p. 443, should correspond with Patankar's correlations described as (i) 'along the rows, results over the whole data, original' (last column of Table 5·2, p. 460) and (ii) 'along the columns, results over the whole data' (last column of Table 5·1). Reference to the authors has, however, confirmed that the quantities computed are differently defined. Thus if

$i = 1, 2, \ldots, 25$ denotes columns running from west to east,

$j = 1, 2, \ldots, 20$ denotes rows running from north to south,

the co-variance in Whittle's expression for r_{s0} is

$$\sum_{i=1}^{25}\sum_{j=1}^{20-s} x_{i,j}x_{i,j+s} - \sum_{i=1}^{25}\sum_{j=1}^{20-s} x_{i,j} \times \sum_{i=1}^{25}\sum_{j=1}^{20-s} x_{i,j+s}/\{25(20-s)\},$$

with similar expressions for the variances. The corresponding covariance used by Patankar is, however,

$$\sum_{i=1}^{25}\left\{\sum_{j=1}^{20-s} x_{i,j}x_{i,j+s} - \sum_{j=1}^{20-s} x_{i,j} \times \sum_{j=1}^{20-s} x_{i,j+s}/(20-s)\right\},$$

with similar expressions for the variances. Thus Whittle bases his serial correlations on the total variation and covariation while Patankar's correlations are based on within-column (or within-row) variation and covariation. As might be expected, Whittle's correlations are larger than Patankar's because of the variation between rows and columns.]

29-2

A Discrete State-Space Model for Linear Image Processing

ROBERT P. ROESSER, MEMBER, IEEE

Abstract—**The linear time-discrete state-space model is generalized from single-dimensional time to two-dimensional space. The generalization includes extending certain basic known concepts from one to two dimensions. These concepts include the general response formula, state-transition matrix, Cayley–Hamilton theorem, observability, and controllability.**

I. Introduction

IMAGE processing by nonoptical means has been receiving extensive attention in the last few years. Several books, e.g., [1]–[3] and many papers, e.g., [4]–[6], have been published that have established nonoptical image processing as a viable area of research. A large portion of this research emphasizes the linear processing of images for two main reasons: 1) Many image processing tasks are linear in nature. These tasks include image enhancement, image restoration, picture coding, linear pattern recognition, and TV bandwidth reduction. 2) There are many known linear techniques that may be brought to bear in the treatment of linear image processing, and therefore simplify such treatment. These techniques include transform theory, matrix theory, superposition, etc.

Several ways are commonly used to represent the operations involved in image processing. These include transfer functions, partial difference (recursive) equations, and convolution summations. For example, VanderLugt [7], [11] has presented an extensive development of linear optics based on transfer functions. The transfer functions relate the two-dimensional Fourier transform of an output image to that of the input image. Complex optical systems are easily described by combinations of transfer functions that correspond to individual components of the optical system.

Partial difference equations are used by Habibi [6] to describe a model for estimating images corrupted by noise. The model corresponds to a two-dimensional extension of Kalman filters.

Convolution summations are discussed by Fryer and Richmond [5] in work that involves simplifying a two-dimensional filter to a single-dimensional filter.

The time-discrete state-space model offers great utility in the formulation and analysis of linear systems. Linear systems that are described by transfer functions, difference equations, or convolution summations are easily formulated into a state-space representation. Once so formulated, many known techniques may be applied to systematically analyze the model. Consequently, the state-space model is a general and powerful tool that is used to unify the research and study of time-discrete linear systems.

This paper develops a discrete model for linear image processing that closely parallels the well-known state-space model for time-discrete systems. Because of this parallel many of the concepts that are known for the temporal model may be carried over to the spatial model. This is done by generalizing from a single coordinate in time to two coordinates in space. The spatial model will hopefully have some of the same utility in unifying the study of two-dimensional linear systems as does the temporal model for one-dimensional linear systems.

Temporal systems are inherently nonanticipatory and are often treated as such for the sake of physical realizability in real time; whereas spatial systems do not have causality as an inherent limitation. That is, an image processor may have right to left dependency as well as left to right dependency. Causality is built into the temporal state-space model if an initial state is assumed to be fully specified. In order to establish a close parallel for the spatial model the same built-in causality will be intentionally assumed despite the fact that causality is not necessary for physical realizability in real space. Such an image processor is said to be unilateral. If the constraint of causality is removed, then an image processor is said to be bilateral.

Concepts that are developed in this paper include 1) formulation of the state-space model, 2) the definition of a state-transition matrix, 3) the derivation of a general response formula, 4) a two-dimensional parallel to the Cayley–Hamilton theorem, 5) observability and controlability, and 6) computation of the state-transition matrix. Some of these concepts are based upon an extension of published material on linear iterative circuits coauthored by this writer [8], [9]. A finite field is assumed in the case of iterative circuits, whereas a real field is assumed for image processing. One particular concept, the two-dimensional Cayley–Hamilton theorem, is treated in [9]. An interesting alternative proof to the theorem is the topic of a paper published by Vilfan [10].

Manuscript received February 5, 1974. Paper recommended by A. S. Morse, Chairman of the IEEE S-CS Linear Systems Committee.
The author is with the Department of Electrical Engineering, Wayne State University, Detroit, Mich. 48202.

Reprinted from *IEEE Trans. Automat. Contr.*, vol. AC-20, pp. 1–10, Feb. 1975.

II. The Model

An image is a generalization of a temporal signal in that it is defined over two spatial dimensions instead of a single temporal dimension. Consequently, two space coordinates i and j take the place of time t. Also two-state sets are introduced to replace the single-state set. The following definitions are made for the model.

i	An integer-valued vertical coordinate.
j	An integer-valued horizontal coordinate.
$\{R\}$	A set of real n_1-vectors which convey information vertically.
$\{S\}$	A set of real n_2-vectors which convey information horizontally.
$\{u\}$	A set of real p-vectors that act as inputs.
$\{y\}$	A set of real m-vectors that act as outputs.

A specific image processor is then defined as a 6-tuple $<\{R\}, \{S\}, \{u\}, \{y\}, f, g>$ where f is the next state function;

$$f: \{R\} \times \{S\} \times \{u\} \rightarrow \{R\} \times \{S\}$$

and g is the output function;

$$g: \{R\} \times \{S\} \times \{u\} \rightarrow \{y\}.$$

Now, since f and g are to be linear functions they may be represented by the following matrix equations:

$$R(i+1,j) = A_1 R(i,j) + A_2 S(i,j) + B_1 u(i,j)$$

$$S(i,j+1) = A_3 R(i,j) + A_4 S(i,j) + B_2 u(i,j)$$

$$y(i,j) = C_1 R(i,j) + C_2 S(i,j) + D u(i,j), \qquad i,j \geqslant 0.$$

A_1, A_2, A_3, A_4, B_1, B_2, C_1, C_2, D are matrices of appropriate dimensions. Boundary conditions $R(0,j)$ and $S(i,0)$ and also the input $u(i,j)$ are externally specified. In the next section a computational rule is obtained that uniquely determines the states $R(i,j)$ and $S(i,j)$ and also the output $y(i,j)$ (for $i,j \geqslant 0$) from the boundary conditions and inputs. Thus given values for the boundary conditions (such as all zero) the equations produce an output vector image from an input vector image. This formulation is general so that any discrete linear image process may be so represented. Notation is condensed somewhat by introducing the following matrices and vectors:

$$A = \begin{bmatrix} A_1 & A_2 \\ A_3 & A_4 \end{bmatrix} \quad B = \begin{bmatrix} B_1 \\ B_2 \end{bmatrix} \quad C = [C_1 \; C_2]$$

$$T(i,j) = \begin{bmatrix} R(i,j) \\ S(i,j) \end{bmatrix} \quad T'(i,j) = \begin{bmatrix} R(i+1,j) \\ S(i,j+1) \end{bmatrix}.$$

Then

$$T'(i,j) = AT(i,j) + Bu(i,j)$$

$$y(i,j) = CT(i,j) + Du(i,j).$$

III. General Response Formula

A state-transition matrix $A^{i,j}$ is defined as follows.

Definition: For

$$A = \begin{bmatrix} A_1 & A_2 \\ A_3 & A_4 \end{bmatrix}$$

$$A^{i,j} = A^{1,0}A^{i-1,j} + A^{0,1}A^{i,j-1}, \qquad (i,j) > (0,0)$$

$$A^{0,0} = I$$

$$A^{-i,j} = A^{i,-j} = 0, \qquad \text{for } j \geqslant 1, i \geqslant 1.$$

Examination of this definition bears out that it is an effective recursive definition for integer values of i and j, such that either $i>0$ or $j>0$ or $(i,j)=(0,0)$. It parallels the definition of the time-discrete state-transition matrix $A^t = A^1A^{t-1}$.

It is now to be shown that this state-transition matrix, $A^{i,j}$, may be used in an expression for the response of the model in terms of the inputs and boundary conditions. The term boundary conditions is used here to refer to the states along the edges of the model. Specifically, the set of boundary conditions consist of $R(0,j)$ for $j \geqslant 0$ and $S(i,0)$ for $i \geqslant 0$.

Definition: The following partial ordering is used for integer pairs:

$$(h,k) \leqslant (i,j), \qquad \text{iff } h \leqslant i \text{ and } k \leqslant j$$

$$(h,k) = (1,j), \qquad \text{iff } h = i \text{ and } k = j$$

$$(h,k) < (i,j), \qquad \text{iff } (h,k) \leqslant (i,j) \text{ and } (h,k) \neq (i,j).$$

Lemma: Let the input, $u(i,j)$, for all (i,j) and the boundary conditions, $R(0,j)$ and $S(i,0)$, for $(i,j) \neq (0,0)$ be equal to zero. Then $T(i,j) = A^{i,j}T(0,0)$.

Proof: The proof is accomplished by induction.

First, $T(0,0) = IT(0,0) = A^{0,0}T(0,0)$. This implies the hypothesis is true for $(i,j)=(0,0)$. Now assume the hypothesis is true for all (h,k) such that $(0,0) \leqslant (h,k) < (i,j)$, and show that it is true for (i,j).

$$T(i,j) = \begin{bmatrix} R(i,j) \\ S(i,j) \end{bmatrix} = \begin{bmatrix} A_1R(i-1,j) + A_2S(i-1,j) + B_1 \cdot 0 \\ A_3R(i,j-1) + A_4S(i,j-1) + B_2 \cdot 0 \end{bmatrix}$$

$$= \begin{bmatrix} A_1 & A_2 \\ 0 & 0 \end{bmatrix} T(i-1,j) + \begin{bmatrix} 0 & 0 \\ A_3 & A_4 \end{bmatrix} T(i,j-1)$$

$$= A^{1,0}A^{i-1,j}T(0,0) + A^{0,1}A^{i,j-1}T(0,0)$$

$$= A^{i,j}T(0,0).$$

This is an effective inductive proof because an enumeration can be found such that all (i,j) are reached but not before all $(h,k)<(i,j)$. Such an enumeration is the diagonal enumeration (0,0), (0,1), (1,0), (0,2), (1,1), (2,0), $\cdots$ Q.E.D.

It is to be noted that since the matrices are not functions of (i,j) the model is spatially invariant. The effect then of $T(h,k)$ on $T(i,j)$ is $A^{i-h,j-k}T(h,k)$. The superposition property of linear systems may be used to obtain a more general expression for $T(i,j)$ in which $u(i,j)$ and the other boundary conditions are not assumed to be zero.

Effect of u(h,k): Assume $u(h,k)$ for some $(h,k)<(i,j)$ is the only nonzero input and all boundary conditions are zero. Then

$$T(h+1,k)=\begin{bmatrix} A_1\cdot 0+A_2\cdot 0+B_1\cdot u(h,k) \\ 0 \end{bmatrix}=\begin{bmatrix} B_1 \\ 0 \end{bmatrix}u(h,k)$$

and

$$T(h,k+1)=\begin{bmatrix} 0 \\ A_3\cdot 0+A_4\cdot 0+B_2\cdot u(h,k) \end{bmatrix}=\begin{bmatrix} 0 \\ B_2 \end{bmatrix}u(h,k).$$

Then

$$T(i,j)=A^{i-(h+1),j-k}T(h+1,k)+A^{i-h,j-(k+1)}T(h,k+1)$$

$$=\left(A^{i-h-1,j-k}\begin{bmatrix} B_1 \\ 0 \end{bmatrix}+A^{i-h,j-k-1}\begin{bmatrix} 0 \\ B_2 \end{bmatrix}\right)u(h,k).$$

Effect of R(0,k): Assume that $R(0,k)$ is the only nonzero boundary condition and that all inputs are zero.

$$T(0,k)=\begin{bmatrix} R(0,k) \\ 0 \end{bmatrix}.$$

Then

$$T(i,j)=A^{i,j-k}T(0,k)$$

$$=A^{i,j-k}\begin{bmatrix} R(0,k) \\ 0 \end{bmatrix}.$$

Effect of S(h,0): Similarly to $R(0,k)$ the effect of $S(h,0)$ is

$$T(i,j)=A^{i-h,j}\begin{bmatrix} 0 \\ S(h,0) \end{bmatrix}.$$

We thus have the following theorem.

Theorem: For all $(i,j)\geqslant 0$.

$$T(i,j)=\sum_{k=0}^{j}A^{i,j-k}\begin{bmatrix} R(0,k) \\ 0 \end{bmatrix}+\sum_{h=0}^{i}A^{i-h,j}\begin{bmatrix} 0 \\ S(h,0) \end{bmatrix}$$

$$+\sum_{(0,0)<(h,k)<(i,j)}\sum$$

$$\cdot\left(A^{i-h-1,j-k}\begin{bmatrix} B_1 \\ 0 \end{bmatrix}+A^{i-h,j-k-1}\begin{bmatrix} 0 \\ B_2 \end{bmatrix}\right)u(h,k).$$

Proof: By superposition of the effects of all inputs and boundary conditions. Q.E.D.

An expression for $y(i,j)$, called the *general response formula*, may now be written

$$y(i,j)=[C_1\ C_2]\left(\sum_{k=0}^{j}A^{i,j-k}\begin{bmatrix} R(0,k) \\ 0 \end{bmatrix}\right.$$

$$+\sum_{h=0}^{i}A^{i-h,j}\begin{bmatrix} 0 \\ S(h,0) \end{bmatrix}$$

$$+\sum_{(0,0)<(h,k)<(i,j)}\sum\left(A^{i-h-1,j-k}\begin{bmatrix} B_1 \\ 0 \end{bmatrix}\right.$$

$$\left.\left.+A^{i-h,j-k-1}\begin{bmatrix} 0 \\ B_2 \end{bmatrix}\right)u(h,k)\right)+Du(i,j).$$

IV. Properties of $A^{i,j}$

Some properties of $A^{i,j}$ are

1)

$$A=\begin{bmatrix} A_1 & A_2 \\ A_3 & A_4 \end{bmatrix}=\begin{bmatrix} A_1 & A_2 \\ 0 & 0 \end{bmatrix}+\begin{bmatrix} 0 & 0 \\ A_3 & A_4 \end{bmatrix}.$$

Thus $A=A^{1,0}+A^{0,1}$.

2)

$$A^{i,0}=A^{1,0}A^{i-1,0}+A^{0,1}A^{i,-1}=A^{1,0}A^{i-1,0}.$$

Thus $A^{i,0}=(A^{1,0})^i$. Similarly $A^{0,j}=(A^{0,1})^j$.

3)

$$I=\begin{bmatrix} I & 0 \\ 0 & I \end{bmatrix}$$

where I is the identity matrix with appropriate dimensions. Thus

$$I^{1,0} = \begin{bmatrix} I & 0 \\ 0 & 0 \end{bmatrix} \text{ and } I^{0,1} = \begin{bmatrix} 0 & 0 \\ 0 & I \end{bmatrix}.$$

4)

$$I^{1,0}A = \begin{bmatrix} I & 0 \\ 0 & 0 \end{bmatrix}\begin{bmatrix} A_1 & A_2 \\ A_3 & A_4 \end{bmatrix} = \begin{bmatrix} A_1 & A_2 \\ 0 & 0 \end{bmatrix} = A^{1,0}.$$

Briefly $I^{1,0}A = I^{1,0}A^{1,0} = A^{1,0}$. Similarly $I^{0,1}A = I^{0,1}A^{0,1} = A^{0,1}$.

5)

$$I^{0,1}A^{1,0} = \begin{bmatrix} 0 & 0 \\ 0 & I \end{bmatrix}\begin{bmatrix} A_1 & A_2 \\ 0 & 0 \end{bmatrix} = 0.$$

Briefly $I^{0,1}A^{1,0} = 0$. Similarly $I^{1,0}A^{0,1} = 0$.

V. Characteristic Function of a Matrix

If the primary inputs and outputs are neglected in the model equations, a representation arises for the state behavior of the circuit, having the form

$$R(i+1,j) = A_1R(i,j) + A_2S(i,j)$$

$$S(i,j+1) = A_3R(i,j) + A_4S(i,j).$$

These equations are useful in the development of a form for a two-dimensional characteristic matrix of A. Operators are first introduced that advance a particular coordinate of their operand.

Definition: Let E be an operator that has the effect of advancing the vertical coordinate or the first subscript of the function upon which it is operating. Likewise, let F be an operator that has the effect of advancing the horizontal coordinate or second subscript of the function upon which it is operating.

The effect of these operators on the state vectors is

$$R(i+1,j) = ER(i,j)$$

$$S(i,j+1) = FS(i,j).$$

The state equations can be rewritten using these advance operators.

$$(EI - A_1)R(i,j) - A_2S(i,j) = 0$$

$$-A_3R(i,j) + (FI - A_4)S(i,j) = 0.$$

These equations are equivalently represented in the overall matrix form.

$$\begin{bmatrix} (EI - A_1) & -A_2 \\ -A_3 & (FI - A_4) \end{bmatrix} T(i,j) = 0.$$

The above equation represents a system of homogeneous equations in the elements of $T(i,j)$. If the system is to have a nontrivial solution for $T(i,j)$, then the transformation represented by the matrix must be singular.

The above matrix is said to be the two-dimensional characteristic matrix of the partitioned matrix A where

$$A = \begin{bmatrix} A_1 & A_2 \\ A_3 & A_4 \end{bmatrix}.$$

The characteristic matrix of A is denoted by $cm(A)$ and may be represented as

$$cm(A) = EI^{1,0} + FI^{0,1} - A.$$

Now since $cm(A)$ must be singular, its determinate must be equal to zero.

$$|cm(A)| = 0.$$

If E and F are placed in the above by general indeterminates x and y, respectively, the result is an expression called the two-dimensional characteristic equation for A. The determinate of $cm(A)$ with x and y replacing E and F is called the two-dimensional characteristic function of the matrix X, and is denoted by $f(x,y)$.

$$|cm(A)| = f(x,y) = 0.$$

$f(x,y)$ will be a monic multinomial in x and y with degree n_1 in x and degree n_2 in y; where n_1 is the dimension of R and n_2 is the dimension of S. $f(x,y)$ has the form

$$f(x,y) = \sum_{(0,0) < (i,j) < (n_1,n_2)} \sum a_{i,j}x^iy^j, \qquad \text{where } a_{n_1,n_2} = 1.$$

Comparing these concepts to the one-dimensional case, it is observed that they are correspondingly analogous to the one-dimensional characteristic matrix, equation and function of a matrix. $xI - A$ is the one-dimensional characteristic matrix of A and $f(x) = |xI - A| = 0$ is the one-dimensional characteristic equation.

The Cayley–Hamilton theorem in the one-dimensional case states that a matrix A satisfies its own characteristic equation, i.e., $f(A) = 0$. The following theorem extends this to the two-dimensional case.

Definition: $E^iF^jA = F^jE^iA = A^{i,j}$ for any 2×2 partition of A.

Two-Dimensional Cayley–Hamilton Theorem: Every partitioned matrix

$$A = \begin{bmatrix} A_1 & A_2 \\ A_3 & A_4 \end{bmatrix}$$

satisfies its own characteristic equation. That is $f(E,F)A=0$.

Proof: Let $B=(xI^{1,0}+yI^{0,1}-A)$, so that $f(x,y)=\det B$. Cramer's rule for computing the inverse of a matrix, in this case matrix B, states

$$\operatorname{adj} B\cdot(xI^{1,0}+yI^{0,1}-A)=\det B\cdot I=f(x,y)I \tag{1}$$

where adj B is the transpose of the cofactor matrix of B. The elements of adj B transpose are computed by taking the determinate of the matrix formed by deleting the row and column containing the corresponding element in B. The elements of adj B will consequently be multinomials in x and y having degrees in x and y not greater than n_1 and n_2, respectively, where n_1 is the rank of $I^{1,0}$ and n_2 is the rank of $I^{0,1}$. Therefore, adj B may be written in the form of a matrix multinomial.

$$\operatorname{adj} B=\sum_{i=0}^{n_1}\sum_{j=0}^{n_2} B_{i,j}x^iy^j. \tag{2}$$

Represent the characteristic multinomial $f(x,y)$ as

$$f(x,y)=\sum_{i=0}^{n_1}\sum_{j=0}^{n_2} b_{i,j}x^iy^j, \qquad \text{where } b_{n_1,n_2}=1. \tag{3}$$

Substituting (2) and (3) into (1), we have,

$$\sum_{i=0}^{n_1}\sum_{j=0}^{n_2} B_{i,j}x^iy^j(xI^{1,0}+yI^{0,1}-A)=\sum_{i=0}^{n_1}\sum_{j=0}^{n_2} b_{i,j}x^iy^jI.$$

Expand the left side and adjust i and j.

$$\sum_{i=1}^{n_1+1}\sum_{j=0}^{n_2} B_{i-1,j}I^{1,0}x^iy^j+\sum_{i=0}^{n_1}\sum_{j=1}^{n_2+1} B_{i,j-1}I^{0,1}x^iy^j$$
$$-\sum_{i=0}^{n_1}\sum_{j=0}^{n_2} B_{i,j}Ax^iy^j=\sum_{i=0}^{n_1}\sum_{j=0}^{n_2} b_{i,j}x^iy^jI.$$

The coefficients of each term x^iy^j on both sides of the equation must be equal. Equating these coefficients yields the following:

$$B_{n_1,j}I^{1,0}=0, \qquad \text{for } 0\leqslant j\leqslant n_2 \tag{4}$$

$$B_{i,n_2}I^{0,1}=0, \qquad \text{for } 0\leqslant i\leqslant n_1 \tag{5}$$

$$B_{i-1,j}I^{1,0}+B_{i,j-1}I^{0,1}-B_{i,j}A=b_{i,j}I, \qquad \text{for } i\neq 0\, j\neq 0 \tag{6}$$

$$B_{0,j-1}I^{0,1}-B_{0,j}A=b_{0,j}I, \qquad \text{for } j\neq 0 \tag{7}$$

$$B_{i-1,j}I^{1,0}-B_{i,0}A=b_{i,0}I, \qquad \text{for } i\neq 0 \tag{8}$$

$$-B_{0,0}A=b_{0,0}I. \tag{9}$$

Multiply each of the equations (6)–(8) on the right by $A^{i,j}$ and sum equations (6)–(9) over all i,j.

$$\sum_{i=1}^{n_1}\sum_{j=1}^{n_2}(B_{i-1,j}I^{1,0}+B_{i,j-1}I^{0,1}-B_{i,j}A)A^{i,j}$$
$$+\sum_{j=1}^{n_2}(B_{0,j-1}I^{0,1}-B_{0,j}A)A^{0,j}$$
$$+\sum_{i=1}^{n_1}(B_{i-1,0}I^{1,0}-B_{i,0}A)A^{i,0}-B_{0,0}A$$
$$=\sum_{i=0}^{n_1}\sum_{j=0}^{n_2} b_{i,j}IA^{i,j}.$$

Collect coefficients of $B_{i,j}$ on the left side of the equation.

$$B_{i,j}(I^{1,0}A^{i+1,j}+I^{0,1}A^{i,j+1}-AA^{i,j}), \qquad \text{for } 1\leqslant i\leqslant n_1\; 1\leqslant j\leqslant n_2 \tag{10}$$

$$B_{n_1,j}(I^{0,1}A^{n_1,j+1}-AA^{n_1,j}), \qquad \text{for } j\neq 0\, j\neq n_2 \tag{11}$$

$$B_{i,n_2}(I^{1,0}A^{i+1,n_2}-AA^{i,n_2}), \qquad \text{for } i\neq 0\, i\neq n_1 \tag{12}$$

$$B_{0,j}(I^{1,0}A^{1,j}+I^{0,1}A^{0,j+1}-AA^{0,j}), \qquad \text{for } j\neq 0\, j\neq n_2 \tag{13}$$

$$B_{i,0}(I^{0,1}A^{i,1}+I^{1,0}A^{i+1,0}-AA^{i,0}), \qquad \text{for } i\neq 0\, i\neq n_1 \tag{14}$$

$$B_{n_1,n_2}(AA^{n_1,n_2}) \tag{15}$$

$$B_{0,0}(I^{0,1}A^{0,1}+I^{1,0}A^{1,0}-A). \tag{16}$$

Expressions (10)–(16) exhaust all i,j combinations. We will now evaluate each in turn.

$$I^{1,0}A^{i+1,j}=I^{1,0}A^{1,0}A^{i,j}+I^{1,0}A^{0,1}A^{i+1,j-1}$$
$$I^{0,1}A^{i,j+1}=I^{0,1}A^{1,0}A^{i-1,j+1}+I^{0,1}A^{0,1}A^{i,j} \tag{10}$$

but $I^{1,0}A^{0,1}=0$ and $I^{0,1}A^{1,0}=0$.

Thus,

$$I^{1,0}A^{i+1,j}+I^{0,1}A^{i,j+1}-AA^{i,j}=A^{1,0}A^{i,j}+A^{0,1}A^{i,j}-AA^{i,j}$$
$$=AA^{i,j}-AA^{i,j}=\mathbf{0}.$$

$$I^{0,1}A^{n_1,j+1}=I^{0,1}A^{1,0}A^{n_1-1,j+1}+I^{0,1}A^{0,1}A^{n_1,j}$$
$$=A^{0,1}A^{n_1,j}. \tag{11}$$

Thus,

$$B_{n_1,j}(I^{0,1}A^{n_1,j+1}-AA^{n_1,j})$$
$$=B_{n_1,j}(A^{0,1}A^{n_1,j}-A^{0,1}A^{n_1,j}-A^{1,0}A^{n_1,j})$$
$$=-B_{n_1,j}A^{1,0}A^{n_1,j}=-B_{n_1,j}I^{1,0}A^{1,0}A^{n_1,j}=\mathbf{0}$$

since $B_{n_1,j}I^{1,0}=0$ from (4).

$$\text{Similarly}=\mathbf{0} \text{ as in (11).} \tag{12}$$

$$AA^{0,j}=A^{1,0}A^{0,j}+A^{0,1}A^{0,j}=A^{1,0}A^{0,j}+A^{0,j+1}$$
$$I^{1,0}A^{1,j}=I^{1,0}A^{1,0}A^{0,j}+I^{1,0}A^{0,1}A^{1,j-1}=A^{1,0}A^{0,j}. \tag{13}$$

Thus,

$$I^{1,0}A^{1,j}+I^{0,1}A^{0,j+1}-AA^{0,j}$$
$$=A^{1,0}A^{0,j}+A^{0,j+1-A1,0}A^{0,j}-A^{0,j+1}=\mathbf{0}.$$

$$\text{Similarly}=\mathbf{0} \text{ as in (13).} \tag{14}$$

$$B_{n1,n2}=\mathbf{0}, \text{ since adj } B \text{ can have no term in } x^{n1}y^{n2}. \tag{15}$$

$$A=A^{1,0}+A^{0,1}, \text{ thus, } I^{0,1}A^{0,1}+I^{1,0}A^{1,0}-A=\mathbf{0}. \tag{16}$$

Briefly, the left side of the equation being considered is zero, giving

$$0=\sum_{i=0}^{n_1}\sum_{j=0}^{n_2} b_{i,j}A^{i,j}=f(E,F)A$$

which completes the proof.

VI. Observability and Controllability

The notions of observability and controllability for time-discrete systems carry over to parallel notions for discrete-space image processors.

Definition: A state T_0 is observable iff whenever it appears as the initial state and all other boundary conditions are zero, there exists a pattern of inputs and a pair $(i,j) \geqslant (0,0)$ such that $y(i,j)$ is not the same as when the initial state is zero, and the same pattern of inputs is applied.

Definition: A state T_0 is controllable iff when all boundary conditions are zero there exists some pair $(i,j) \geqslant 0$ and some input pattern such that $T(i,j) = T_0$.

An image processing model is said to be observable (controllable) iff all states are observable (controllable).

It is often desirable to reduce a model to an equivalent model that is observable and controllable. Here equivalence between models will be taken to mean no pattern of inputs exist so that the output from one model is different at some pair $(i,j) \geqslant (0,0)$ than the output from the other model at (i,j) when the boundary conditions of both models are zero.

To test for observability the output $y(i,j)$ for each possible initial state is compared with the output $y_0(i,j)$ with zero initial state, for all $(i,j) \geqslant (0,0)$ and all input patterns. If $y(i,j) = y_0(i,j)$ for all $(i,j) \geqslant 0$ and all input conditions, then the model is not observable. Using the general response formula this condition reduces to

$$CA^{i,j}T(0,0) = 0, \qquad \text{for all } (i,j) \geqslant 0.$$

The two-dimensional Cayley–Hamilton theorem implies that any two-tuple power of A is linearly dependent upon those $A^{i,j}$ for which $(0,0) \leqslant (i,j) < (n_1, n_2)$ so that the condition for nonobservability can be limited to those (i,j) such that $(0,0) \leqslant (i,j) < (n_1, n_2)$. The condition may then be put into matrix form $KT = 0$, where K is the diagnostic matrix defined as follows:

$$K = \begin{bmatrix} C \\ CA^{0,1} \\ CA^{0,2} \\ \vdots \\ CA^{0,n_2} \\ CA^{1,0} \\ CA^{1,1} \\ \vdots \\ CA^{n_1,0} \\ CA^{n_1,1} \\ \vdots \\ CA^{n_1,n_2-1} \end{bmatrix}.$$

If $KT = 0$ then the model is not observable, but may be reduced to an observable model. Let K_1 be a matrix consisting of the first n_1 columns of K and let K_2 be a matrix consisting of the last n_2 columns of K. The conditions $KT = 0$ may then be split into the two conditions $K_1 R = 0$ and $K_2 S = 0$. A reduced model may then be formed by using the equivalence classes of $\{R\}$ modulo the null space of K_1 as the new vertical state set, and the equivalence classes of $\{S\}$ modulo the null space of K_2 as the new horizontal state set. As a result only the new zero vector will satisfy $K_1 R = 0$ and $K_2 S = 0$, implying that the new model is observable. It remains to find the characterizing matrices for the reduced model. First vector representation for the equivalence classes of vertical and horizontal states is found, then the original characterizing matrices are modified so that the behavior of the model remains the same. K_1 and K_2 may be reduced so that their new dimension agrees with their rank. This is done by forming matrices G_1 and G_2 from the first complete set of linearly independent rows of K_1 and K_2, respectively. $G_{(\,)}$ and $K_{(\,)}$ have the same null space. Furthermore each equivalence class of $\{R\}$ and $\{S\}$ will correspond to a single vector $G_1 R$ and $G_2 S$, respectively. The equivalence classes of $\{T\}$ may then be represented as

$$\bar{T} = GT, \qquad \text{where } G = \begin{bmatrix} G_1 & 0 \\ 0 & G_2 \end{bmatrix}.$$

Now let

$$H = \begin{bmatrix} H_1 & 0 \\ 0 & H_2 \end{bmatrix}$$

be a right inverse to G, i.e., $GH = I$. The reduced model will then have the following characterizing matrices:

$$\bar{A} = GAH \qquad \bar{B} = GB \qquad \bar{C} = CH \qquad \bar{D} = D.$$

To test for controllability the state $T(i,j)$ is examined with all boundary conditions equal to zero for all input patterns. If $T(i,j)$ doesn't equal T_0 for some T_0, for any $(i,j) \geqslant (0,0)$ and any input pattern, then the model is uncontrollable. Using the general formula for $T(i,j)$ this condition becomes

$$\sum_{(0,0) \leqslant (h,k) < (i,j)} \sum \left(A^{i-h-1,j-k} \begin{bmatrix} B_1 \\ 0 \end{bmatrix} + A^{i-h,j-k-1} \begin{bmatrix} 0 \\ B_2 \end{bmatrix} \right) u(k,k) \neq T_0$$

for all (i,j) and all $u(h,k)$. As for observability, the two-dimension Cayley–Hamilton theorem allows us to limit the condition for noncontrollability to those (i,j) such that $(0,0) \leqslant (i,j) < (n_1,n_2)$. The condition may then be put into matrix form.

$QU \neq T_0$ for all U and some T_0 where U and Q are defined as follows:

$$U = \begin{bmatrix} u(0,0) \\ u(0,1) \\ \vdots \\ u(0,n_2) \\ u(1,0) \\ \vdots \\ u(n_1,n_2-1) \end{bmatrix}$$

$$Q = [M(0,0), M(0,1), \ldots, M(0,n_2), \\ M(1,0), \ldots, M(n_1,n_2-1)]$$

$$M(i,j) = A^{i-1,j}\begin{bmatrix} B_1 \\ 0 \end{bmatrix} + A^{i,j-1}\begin{bmatrix} 0 \\ B_2 \end{bmatrix}.$$

The maximum rank of Q is equal to n_1+n_2 (the number of rows), which is equal to the dimension of $\{T\}$. If Q has n_1+n_2 as its rank, then its range space must equal $\{T\}$ so that there would be no T_0 such that $QU \neq T_0$ for all U. The model would then be controllable. If however the rank of Q is less than n_1+n_2, then the model may be reduced to a controllable model. Let Q_1 be formed from the first n_1 rows of Q and Q_2 be formed from the last n_2 rows of Q.

Q_1 and Q_2 may be reduced so that their column dimension agrees with their rank. This is done by forming matrices G_1 and G_2 from the first complete set of linearly independent columns of Q_1 and Q_2, respectively.

The controllable states T will be those formed from the direct sum of the vectors in the range spaces of G_1 and G_2. A state set for the reduced model may now be specified as a set of vectors

$$\left\{ \bar{T} = \begin{bmatrix} \bar{R} \\ \bar{S} \end{bmatrix} \right\}$$

which are mapped into the controllable states under the direct sum of G_1 and G_2. That is $\{\bar{T}\}$ is the domain of the linear mapping

$$T = \begin{bmatrix} G_1 & 0 \\ 0 & G_2 \end{bmatrix}\begin{bmatrix} \bar{R} \\ \bar{S} \end{bmatrix} = G\bar{T}.$$

Let

$$H = \begin{bmatrix} H_1 & 0 \\ 0 & H_2 \end{bmatrix}$$

be a left inverse of G, i.e., $HG = I$. Then,

$$\bar{T} = HG\bar{T} = HT.$$

Note that the states in the range space of G are controllable and that the dimension of $\{\bar{T}\}$ is equal to the rank of G. This implies that each $\bar{T}$ is mapped by G uniquely into a controllable T. Therefore, each $\bar{T}$ of the reduced model will be controllable. The characterizing matrices of the reduced model can now be found by noting the effect of the mappings G and H on the original characterizing matrices

$$\bar{A} = HAG \qquad \bar{B} = HB$$

$$\bar{C} = CG \qquad \bar{D} = D.$$

VIII. Computation of the Transition Matrix

Computing the transition matrix $A^{i,j}$ using the recursive definition becomes quite tedious as i and j become large. It is therefore desirable to extend the techniques that are known for computing a single power of a matrix, such as the Cayley-Hamilton technique or Sylvester's theorem, to parallel techniques for $A^{i,j}$. The Cayley–Hamilton technique will be treated in what follows. Other methods may be extended in a similar fashion.

From the two-dimensional Cayley–Hamilton theorem we have

$$f(E,F)A = 0.$$

Letting $a_{h,k}$ be the coefficient of E^hF^k in $f(E,F)$ the last equation becomes

$$\sum_{(0,0) \leqslant (h,k) \leqslant (n_1,n_2)} \sum a_{h,k} A^{h,k} = 0.$$

If both sides of this equation are operated upon by E^iF^j it becomes

$$\sum_{(0,0) \leqslant (h,k) \leqslant (n_1,n_2)} \sum a_{h,k} A^{h+i,k+j} = 0.$$

Consider this equation for the set of pairs $\{(i,j) | (-n_1 \leqslant i \leqslant 0 \text{ and } j=1) \text{ or } (-n_2 \leqslant j \leqslant 0 \text{ and } i=1)\}$. These pairs have the property that whenever one exponent of $A^{h+i,k+j}$ is negative the other will be positive for $(0,0) \leqslant (h,k) \leqslant (n_1,n_2)$. From the definition of $A^{i,j}$ each such $A^{h+i,k+j}$ will be equal to zero. The following set of equations is produced, each corresponding to a pair (i,j) in the set above:

$$\sum_{(i,0)<(h,k)<(n_1,n_2)} \sum a_{h,k} A^{h-i,k+1} = 0, \qquad 0 \leq i \leq n_1$$

and

$$\sum_{(0,j)<(h,k)<(n_1,n_2)} \sum a_{h,k} A^{h+1,k-j}, \qquad 0 \leq j \leq n_2.$$

Each of these equations can be written in terms of a function of E and F obtained by modifying the two-dimensional characteristic function $f(E,F)$. The modification in the first case consists of multiplying $f(E,F)$ by $E^{-i}F$ and then deleting all terms involving negative exponents of E. The modification for the second case consists of multiplying $f(E,F)$ by EF^{-j} and then deleting all terms involving negative exponents of F. Let

$$f_{i,j}(E,F) = E^{-i}F^{-j}f(E,F)$$

(with deletion of terms having negative exponents). Matrix A then must satisfy the following set of equations:

$$f_{i,-1}(E,F)A = 0, \qquad 0 \leq i \leq n_1 - 1$$

$$f_{-1,j}(E,F)A = 0, \qquad 0 \leq j \leq n_2 - 1.$$

That is, not only must A satisfy the two-dimension characteristic equation, but it must also satisfy the above set of $n_1 + n_2$ equations as well. It will be shown that these equations may be used to reduce $A^{i,j}$ to a linear combination of those $A^{h,k}$ where $(0,0) \leq (h,k) < (n_1,n_2)$. This leads to the definition of two-dimensional eigenvalues.

Definition: The two-dimensional eigenvalues of a partitioned matrix

$$A = \begin{bmatrix} A_1 & A_2 \\ A_3 & A_4 \end{bmatrix}$$

are the pairs (x,y) that simultaneously solve the following set of equations:

$$f_{i,-1}(x,y) = 0, \qquad 0 \leq i \leq n_1 - 1$$

$$f_{-1,j}(x,y) = 0, \qquad 0 \leq j \leq n_2 - 1$$

$$f(x,y) = 0.$$

The equations in the above definition may be used to reduce a multinomial in x,y to one of degree $(h,k) < (n_1,n_2)$. The next theorem establishes this result.

Theorem: Two-dimensional division algorithm. Any multinomial $g(x,y)$ of degree $\geq (0,0)$ may be expressed as follows:

$$g(x,y) = \sum_{j=0}^{n_2} p_j(x) f_{-1,j}(x,y) + \sum_{i=0}^{n_1} q_i(y) f_{i,-1}(x,y) + m(x,y)f(x,y) + r(x,y)$$

where the degree (h,k) of $r(x,y)$ is less than (n_1,n_2).

Proof: Each term in $g(x,y)$ of degree $(i,j) \geq (n_1,n_2)$, having the coefficient $b_{i,j}$, may be reduced to a sum of terms of degree less than (k,j) by subtracting $b_{i,j}x^{i-n_1}y^{j-n_2}f(x,y)$ from it. This is repeated until there are no terms of degree greater than (n_1,n_2). The remainder will have the form

$$g(x,y) - m(x,y)f(x,y).$$

Each term of this remainder of degree $(i,n_2-1) \geq (n_1,n_2-1)$ having the coefficient $c_{i,j}$, may then be reduced to a sum of terms of degree less than (i,n_2-1) by subtracting $c_{i,j}f_{-1,n_2-1}(x,y)$ from it. This is repeated until there are no terms $(h,n_2-1) \geq (n_1,n_2-1)$. A similar process is used for the other functions $f_{-1,j}$ and $f_{i,-1}$. The result will be a remainder of the form

$$r(x,y) = g(x,y) - \sum_{j=0}^{n_2} p_j(x) f_{-1,j}(x,y) + \sum_{i=0}^{n_1} q_i(y) f_{i,-1}(x,y) + m(x,y)f(x,y)$$

where the degree of $r(x,y)$ is less than (n_1,n_2). Q.E.D.

If the equation $f_{i,1}(E,F)\,A = 0$ is operated upon by F^j no additional terms are generated so that the result will be a valid equation. Likewise, $E^i f_{1,j}(E,F)A = 0$ is a valid equation. Consequently, the two-dimensional division algorithm may be applied to matrix multinomials and in particular to $A^{i,j}$. Thus,

$$A^{i,j} = \sum_{j=0}^{n_2} p_j(E) f_{-1,j}(E,F)A + \sum_{i=0}^{n_1} q_i(F) f_{i,-1}(E,F)A + M(E,F)f(E,F)A + r(E,F)A$$

where the degree of $r(E,F)A$ is less than (n_1,n_2). The first three terms are zero because A satisfies the previously-mentioned equations. Then $A^{i,j} = r(E,F)A$. Likewise, if (x,y) is an eigenvalue of A, then from the division algorithm $g(x,y) = r(x,y)$.

Since the operations are the same for obtaining $r(x,y)$ and $r(E,F)A$ they will have the same coefficients. $A^{i,j}$ may then be computed by determining $r(x,y)$ from $x^i y^j$. $r(x,y)$ has $(n_1+1)(n_2+1)-1$ coefficients. These are determined by solving the set of simultaneous equations resulting from $x^i y^j = r(x,y)$, using different eigenvalues (x,y). The same number of eigenvalues as coefficients are necessary.

The determination of two-dimensional eigenvalues of a matrix A is not, in general, obvious. However, if the characteristic function $f(x,y)$ is factorable into linear factors then the two-dimensional eigenvalues are easily identified.

Theorem: Suppose the two-dimensional characteristic function of a matrix A is factorable into linear factors as follows:

$$f(x,y)=(x-a_1)(x-a_2)\cdots(x-a_{n_1})(y-b_1)(y-b_2)\cdots(y-b_{n_2}).$$

Let $a_0=b_0=0$. The set of two-dimensional eigenvalues is

$$\{(a_i,b_j)|(0,0)<(i,j)\leqslant(n_1,n_2)\}.$$

Proof: From the hypothesis $f(x,y)$ may be expressed as the product $p(x)q(y)$. Consequently, $f_{i,-1}(x,y)$ may be expressed as $p_i(x)yq(y)$, where $p_i(x)=x^{-i}p(x)$ (with negative powers of x deleted). Thus, $f_{i,-1}(x,y)$ contains each $(y-b_j)$ as a factor, which implies that $f_{i,-1}(x,b_j)=0$. Likewise $f_{-1,j}(a_i,y)=0$. Thus, for $(x,y)=(a_i,b_j)$ all functions, including f, will be zero. Q.E.D.

If the factors of $f(x,y)$ in the previous theorem are distinct then there will be a total of $(n_1+1)(n_2+1)-1$ eigenvalues, which is the same as the number of coefficients in $r(x,y)$. There is, therefore, just enough equations $a_h^i b_k^j=r(a_h,b_k)$ to solve for the coefficients of $r(x,y)$.

Example: Let

$$A=\begin{bmatrix} A_1 & A_2 \\ A_3 & A_4\end{bmatrix}=\begin{bmatrix} a & b \\ 0 & d\end{bmatrix};$$

then

$$cm(A)=\begin{bmatrix} x-a & -b \\ 0 & y-d\end{bmatrix}$$

$$f(x,y)=|cm(A)|=(x-a)(y-d).$$

The two-dimensional eigenvalues are $(0,d)$, $(a,0)$, and (a,d).

$$A^{i,j}=r(E,F)A \quad \text{where deg } r(E,F)<(1,1)$$
$$=a_{0,0}I+a_{1,0}A^{1,0}+a_{0,1}A^{0,1}.$$

To find the coefficients substitute the eigenvalues into

$$x^iy^j=r(x,y)=a_{1,1}xy+a_{1,0}x+a_{0,1}y$$

$$0^id^j=a_{0,0}+a_{1,0}0+a_{0,1}d$$

$$a^i0^j=a_{0,0}+a_{1,0}a+a_{0,1}0$$

$$a^id^j=a_{0,0}+a_{1,0}a+a_{0,1}d.$$

The solution to these equations is

$$a_{0,0}=-a^id^j \quad a_{1,0}=a^{i-1}d^j \quad a_{0,1}=a^id^{j-1}$$

Note that

$$a^{1,0}=\begin{bmatrix} a & b \\ 0 & 0\end{bmatrix}$$

$$a^{0,1}=\begin{bmatrix} 0 & 0 \\ 0 & d\end{bmatrix}.$$

Then

$$a^{i,j}=-a^id^j\begin{bmatrix} 1 & 0 \\ 0 & 1\end{bmatrix}+a^{i-1}d^j\begin{bmatrix} a & b \\ 0 & 0\end{bmatrix}$$

$$+a^id^{j-1}\begin{bmatrix} 0 & 0 \\ 0 & d\end{bmatrix}$$

$$=\begin{bmatrix} 0 & a^{i-1}d^jb \\ 0 & 0\end{bmatrix}.$$

To check:

$$A^{1,1}=A^{1,0}A^{0,1}+A^{0,1}A^{1,0}$$

$$=\begin{bmatrix} a & b \\ 0 & 0\end{bmatrix}\begin{bmatrix} 0 & 0 \\ 0 & d\end{bmatrix}+\begin{bmatrix} 0 & 0 \\ 0 & d\end{bmatrix}\begin{bmatrix} a & b \\ 0 & 0\end{bmatrix}$$

$$=\begin{bmatrix} 0 & bd \\ 0 & 0\end{bmatrix}+\begin{bmatrix} 0 & 0 \\ 0 & 0\end{bmatrix}=\begin{bmatrix} 0 & bd \\ 0 & 0\end{bmatrix}.$$

This checks with the previous result.

VIII. Conclusion

This paper is an attempt to establish a parallel of the linear discrete-time state-space model for linear discrete-space image processing. However, it can only be assumed to be an initial attempt. Only the more basic and well-known concepts have been extended. Thus, there is much room for future research to be done along this line. Specifically this research should include:

1) Generalization to bilateral models.

2) Methods for programming a spatial transfer function into a state-space model.

3) Discovery of a general method for factoring multinomials and a method for finding two-dimensional eigenvalues.

4) Finding methods for obtaining canonical forms.

5) Establishing criteria for stability.
6) Application of estimation theory.

Finally, of a more general nature, the techniques and concepts of optimal control could be extended to the spatial model.

References

[1] A. Rosenfeld, *Picture Processing by Computer*. New York: Academic, 1969.
[2] H. C. Andrews, *Computer Techniques in Image Processing*. New York: Academic, 1970.
[3] B. S. Lipkin and A. Rosenfeld, *Picture Processing and Psychopictorics*. New York: Academic, 1970.
[4] T. S. Huang, W. F. Schreiber, and O. J. Tretiak, "Image processing," *Proc. IEEE*, vol. 59, pp. 1586–1609, Nov. 1971.
[5] W.D. Fryer and G.E. Richmond, "Two-dimensional spatial filtering and computers," in *Proc. Nat. Electron. Conf.*, Oct. 19, 1962.
[6] A. Habibi, "Two-dimensional Bayesian estimate of images," *Proc. IEEE (Special Issue on Digital Picture Processing)*, vol. 60, pp. 878–883, July 1972.
[7] A. Vander Lugt, "Operational notation for the analysis and synthesis of optical data-processing systems," *Proc. IEEE*, vol. 54, pp. 1055–1063, Aug. 1966.
[8] D. D. Givone and R. P. Roesser, "Multidimensional linear iterative circuits—general properties," *IEEE Trans. Comput.*, vol. C-21, pp. 1067–1073, Oct. 1972.
[9] ——, "Minimization of multidimensional linear iterative circuits," *IEEE Trans. Comput.*, vol. C-22, pp. 673–678, July 1973.
[10] B. Vilfan, "Another proof of the two-dimensional Cayley–Hamilton theorem," *IEEE Trans. Comput.* (Corresp.), vol. C-22, p. 1140, Dec. 1973.
[11] F. T. S. Yu, *Introduction to Diffraction, Information Processing, and Holography*. Cambridge, Mass. : M.I.T. Press, 1973.

Advances in Mathematical Models for Image Processing

ANIL K. JAIN, MEMBER, IEEE

Invited Paper

Abstract—Several state-of-the-art mathematical models useful in image processing are considered. These models include the traditional fast unitary transforms, autoregressive and state variable models as well as two-dimensional linear prediction models. These models introduced earlier [51], [52] as low-order finite difference approximations of partial differential equations are generalized and extended to higher order in the framework of linear prediction theory. Applications in several image processing problems, including image restoration, smoothing, enhancement, data compression, spectral estimation, and filter design, are discussed and examples given.

I. Introduction

MATHEMATICAL models are becoming increasingly important because of their role in the development of useful algorithms for image processing. Virtually all applications of image processing utilize some sort of mathematical models. The continuing advances in high-speed digital processors, digital memories, and very-large-scale integration (VLSI) have led to successful algorithms for many difficult problems. Table I gives a description of some of the typical problems in image processing and their associated modeling requirements. A typical algorithm requires quantification of the processing criterion and a model such as bandwidth, power spectrum, etc., of the data (input to the algorithm). While most of the problems listed in Table I also occur in one-dimensional signal processing, special care is needed in the development of two- (and higher) dimensional algorithms. The major difference besides the higher dimensionality is that of causality. A large number of one-dimensional signal processing methods are based on the fact that the observed data is the output of a causal system. For two-dimensional images the data coordinates are spatial and any causality associated with an image is purely due to its scanning or acquisition technique. Therefore, it is not surprising that a large number of image processing algorithms for edge extraction, enhancement, restoration, data compression, etc., are noncausal.

The computational efficiency of algorithms is often measured by their memory and operation count requirements. The most efficient algorithms would be such that the required number of operations per pel would be independent of the size of the image. Unfortunately, a large number of algorithms require an operation count which is proportional to log N, N, or higher for $N \times N$ images. Table II lists some of the desirable properties of two-dimensional models and algorithms which tend to minimize their computational complexity. In this paper, we will consider mathematical models which are useful for solving image processing problems such as listed in Table I and their tradeoffs with respect to the desirable characteristics of Table II.

We start by series expansion models in Section II, followed by one-dimensional stochastic models in Section III. In Sections IV, V, and VI two-dimensional causal, semicausal, and noncausal models, respectively, are discussed. In Section VII, we consider applications of these models in several image processing problems.

Notation and Definitions

1) We will denote two-dimensional sequences defined on a rectangular grid by $u_{i,j}$, $x_{i,j}$, etc. Upper case letters such as U, X, etc. will denote matrices. For example

$$U \triangleq \{u_{i,j};\, 1 \leqslant i \leqslant N,\, 1 \leqslant j \leqslant M\}$$

denotes an $N \times M$ matrix whose elements are $u_{i,j}$. Whenever necessary we will also use $[U]_{i,j}$, to denote the (i,j)th element of a matrix U. The jth column of U will be denoted as u_j.

2) If A is a matrix then A^T is its transpose and A^* is its complex conjugate.

3) Let $\{u_{i,j}\}$ be an $N \times M$ sequence. We define an $NM \times 1$ lexicographic ordered vector u_r obtained from this sequence as

$$u_r = \mathcal{O}_r(u_{i,j}) \triangleq [u_{1,1} u_{1,2} \cdots u_{1,M}, u_{2,1} \cdots u_{2,M} \cdots u_{N,1} \cdots u_{N,M}]^T$$

where $\mathcal{O}_r$ signifies the row by row ordering operation. Similarly we define $\mathcal{O}_c$ as a column by column ordering operator and write

$$u_c = \mathcal{O}_c(u_{i,j}) \triangleq [u_{1,1} u_{2,1} \cdots u_{N,1}, u_{1,2} \cdots u_{N,2}, \cdots u_{1,M} \cdots u_{N,M}]^T.$$

4) When each $u_{i,j}$ is a random variable, we will call $\{u_{i,j}\}$, a discrete *random field*. A given image could be considered as a sample function of this random field.

5) The mean and covariances of a random field $\{u_{i,j}\}$ are defined as

$$Eu_{i,j} = \mu_{i,j}$$

$$\text{Cov}\,[u_{i,j}, u_{m,n}] \triangleq E(u_{i,j} - \mu_{i,j})(u_{m,n} - \mu_{m,n}) = r_u(i,j;m,n) \quad (1)$$

Manuscript received May 28, 1980; revised October 30, 1980. This work was supported by the U.S. Army Research Office, Durham, NC, under Grant DAAG29-78-G-0206.

The author is with the Signal and Image Processing Laboratory, Department of Electrical and Computer Engineering, University of California, Davis, CA 95616.

Reprinted from *Proc. IEEE,* vol. 69, pp. 502–528, May 1981.

TABLE I
TYPICAL PROBLEMS IN IMAGE PROCESSING

	Problem	Description	Models
1.	SMOOTHING:	Given noisy image data filter it to smooth out the noise variations.	Noise & image, Power Spectra
2.	ENHANCEMENT	Bring out or enhance certain features of the image e.g., edge enhancement, contrast stretching, etc.	Features
3.	RESTORATION & FILTERING	Restore an image with known (or unknown) degradation as close to its original form as possible, e.g., image deblurring, image reconstruction, image registration, geometric correction etc.	Degradations, Criterion of "closeness".
4.	DATA COMPRESSION	Minimize the Number of Bits required to store/transmit an image for a given level of Distortion.	Distortion Criterion, Image as an Information source
5.	FEATURE EXTRACTION	Extract certain features from an image, e.g., edges.	Features, detection criterion
6.	DETECTION AND IDENTIFICATION	Detect and identify the presence of an object from a scene e.g., matched filter, pattern recognition and image segmentation, texture analysis etc.	Detection criterion, object and scene
7.	INTERPOLATION/ AND EXTRAPOLATION	Given image data at certain points in a region, estimate the image values of all other points inside this region (interpolation) and also at points outside this region (extrapolation).	Estimation Criterion, and Degree of smoothness of the data
8.	SPECTRAL ESTIMATION	Given image data in a region, estimate its power spectrum.	Criterion of Estimation,A-priori model for data
9.	SPECTRAL FACTORIZATION	Given the magnitude of the frequency response of a filter-design a realizable filter e.g., a stable 'causal' filter.	Criterion of realizability;
10.	SYNTHESIS	Given a description or some features of an image, design a system which reproduces a replica of that image; e.g., texture synthesis.	Features, Criterion of reproduction.

TABLE II
DESIRABLE PROPERTIES OF IMAGE PROCESSING ALGORITHMS

Property	Description
Linearity	Linear operations on data
Separability	Independent Row and Column operations
Shift Invariance	Operations leading to Toeplitz and Circulant Matrix Manipulations
Markovian or Finite Memory	Only local and/or sparse operation required in each pixel e.g., FIR Filters.

Often we will consider the special case when

$$\mu_{i,j} = \mu = \text{constant}$$
$$r_u(i, j; m, n) = r_u(i - m; j - n). \tag{2}$$

For notational simplicity, whenever there is no confusion, we will drop the subscript u. A random field satisfying (2) is also called *translational* (*or spatial*) *invariant*, *homogeneous*, *or wide-sense stationary*. For random fields with Gaussian statistics this also implies strict sense stationarity. Unless otherwise mentioned, the term "*stationary*" refers to wide-sense stationarity.

6) A random field $\{x_{i,j}\}$ will be called a *white noise field* whenever the random variables $\{x_{i,j}\}$ are mutually uncorrelated, i.e., its covariance function is of the form

$$r_x(i, j; m, n) = \sigma_x^2(i, j)\, \delta_{i,m}\delta_{j,n}$$

where $\delta_{i,m}$ is the Kronecker delta function and $\sigma_x^2(i, j)$ is the variance of $x_{i,j}$.

II. SERIES EXPANSION MODELS AND IMAGE TRANSFORMS

A. Unitary Transforms

A classical way of analyzing a function is by its series expansion in terms of a set of complete orthonormal functions. In the context of image processing a general orthogonal series expansion for an $N \times N$ image[1] $\{u_{i,j}\}$ is a pair of *unitary transformations* of the form

$$u_{i,j} = \sum_{k=1}^{N} \sum_{l=1}^{N} v_{k,l}\, a^*(i, j; k, l) \tag{3}$$

$$v_{k,l} = \sum_{i=1}^{N} \sum_{j=1}^{N} u_{i,j}\, a(i, j; k, l) \tag{4}$$

where $\{a\,(i, j, k, l)\}$, usually called the *image transform*, is a

[1] Much of the subsequent analysis is easily generalized to rectangular, $N \times M$ images. But for simplicity we shall only consider square images here.

TABLE III
TYPICAL FAST UNITARY TRANSFORM USED IN IMAGE PROCESSING

Transform	Formula
Discrete Fourier (DFT)	$a_{m,n} = \frac{1}{\sqrt{N}} \exp\left\{-j\frac{2\pi(m-1)(n-1)}{N}\right\}, 1 \le m,n \le N$
Discrete Cosine (DCT)	$a_{m,n} = \begin{cases} \frac{1}{\sqrt{N}}, & m=1,\ 1 \le n \le N \\ \sqrt{\frac{2}{N}} \cos\frac{(m-1)(2n-1)\pi}{2N}, & 2 \le m \le N,\ 1 \le n \le N \end{cases}$
Discrete Sine (DST)	$a_{m,n} = \sqrt{\frac{2}{N+1}} \sin \frac{mn\pi}{N+1}, \quad 1 \le m,\ n \le N$
Walsh-Hadamard (WHT)	$a_{m,n} = (-1)^{\sum_{i=0}^{p-1} m_i n_i}, \quad 1 \le m,n \le N=2^p$; m_i, n_i = ith binary digit (0 or 1) in the binary expansion of $(m-1)$ and $(n-1)$ respectively.

set of complete orthonormal basis functions satisfying the properties

$$\sum_{k=1}^{N}\sum_{l=1}^{N} a(i,j;k,l)\, a^*(i',j';k,l) = \delta_{i,i'}\delta_{j,j'} \quad \text{(Orthonormality)}$$

$$\sum_{i=1}^{N}\sum_{j=1}^{N} a(i,j;k,l)\, a^*(i,j;k',l') = \delta_{k,k'}\delta_{l,l'} \quad \text{(Completeness).}$$

(5)

The elements $v_{k,l}$ are called the *transform coefficients* and $\{v_{k,l}\}$ the *transformed image.* Equation (3) is a deterministic representation of an image considered as an $N^2 \times 1$ vector. Alternatively, the image $u_{i,j}$ is the output of a linear *spatially variant* (SV) *finite impulse response* (FIR) system whose *impulse response* (IR) or *point spread function* (PSF) is

$$h(i,j;k,l) = a^*(i,j;k,l), \qquad 1 \leqslant i,j,k,l \leqslant N \tag{6}$$

to a unit impulse $\delta_{i,k}\delta_{j,l}$. For each (i,j) the array $\{a^*(i,j;k,l)\}$ is also called a *basis image.* It is readily seen from (3) and (4) that the general unitary transformation would require $O(N^4)$ operations, one operation being a multiplication and a summation. For typical size ($N = 256$) images this means over a billion operations would be needed to compute the transform coefficients. To reduce dimensionality, the unitary transformations in (3) and (4) are restricted to the *product-separable* class, satisfying the condition

$$a(i,j;k,l) = a_{i,k}b_{j,l} \tag{7}$$

where $A = \{a_{i,j}\}$ and $B = \{b_{i,j}\}$ are unitary matrices (i.e., $A^{-1} = A^{*T}$). Often in image processing one chooses $B = A$ so that (3) and (4) yield

$$V = AUA^T \tag{8}$$

$$U = A^{*T}VA^*. \tag{9}$$

Now the transformations require column operations followed by row operations on the result, reducing the computations to $O(N^3)$ operations. Even this reduction is insufficient and the choice of image transforms is further restricted to *fast transforms.* Typically, these transform matrices have structural properties which lead to fast Fourier transform (FFT) type algorithms. Hence a transformation of the type $y = Ax$, for an $N \times 1$ vector x could be performed in $O(N \log N)$ operations so that for images the operation count is $O(N^2 \log N)$ or $\log N$ per pel. Examples of common fast unitary transforms are the discrete Fourier (DFT), cosine (DCT), sine (DST), Walsh-Hadamard (WHT) transforms [1]-[8] (see Table III), etc. Fig. 1 shows the basis images of the DCT. Other fast transforms include the Haar, Slant [1]-[3], and a family of sinusoidal transforms [9], [10]. A useful property of all unitary transforms is their energy conservation property

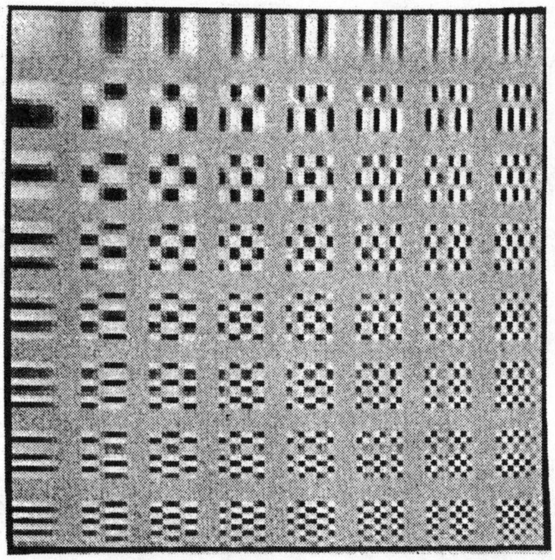

Fig. 1. 8 X 8 basis images of the DCT. Each 8 X 8 basis image if $B_{i,j}$ orthogonal to the rest and represents a spatially varying FIR of a system whose input is an impulse at (i,j).

$$\sum_{i=1}^{N}\sum_{j=1}^{N} |u_{i,j}|^2 = \sum_{k=1}^{N}\sum_{l=1}^{N} |v_{k,l}|^2 \tag{10}$$

known as Parseval's relation. This follows from the fact that a unitary transformation is simply a rotation of the image viewed as a vector in an N^2 dimensional vector space so that the length of the vector remains unchanged.

B. The Karhunen-Loeve Transform

Of particular significance among unitary transforms is the so-called Karhunen-Loeve transform (KLT) for random fields. Without loss of generality we are assuming zero-mean random fields. More generally, one could consider the autocorrelation function instead. It is the complete orthonormal set of basis images $\phi(i,j;k,l)$ determined from the eigenvalue equation

$$\sum_{m=1}^{N}\sum_{n=1}^{N} r(k,l;m,n)\,\phi(i,j;m,n) = \lambda_{i,j}\phi(i,j;k,l) \tag{11}$$

where $r(\cdot)$ is the image covariance function. For separable covariance functions, the KLT is also separable. Two significant properties which make the KLT very desirable are as follows [1], [11]-[13].

1) It completely decorrelates the transform coefficients, i.e.,

$$\text{Cov}\,[v_{k,l}(\mathcal{A}), v_{m,n}(\mathcal{A})] = \sigma^2_{k,l}(\mathcal{A})\,\delta_{k,m}\delta_{l,n}, \qquad \text{for } \mathcal{A} = \Phi \tag{12}$$

where $\mathcal{A}$ denotes an arbitrary $N^2 \times N^2$ unitary transform Φ

TABLE IV
TYPICAL COVARIANCE FUNCTION MODELS USED IN IMAGE PROCESSING

Model Description	Covariance Function $r(k,\ell) \triangleq \mathrm{Cov}[u_{i,j}, u_{i+k,j+\ell}]$	Comments
Separable	$\sigma^2 \rho_1^{\lvert k\rvert} \rho_2^{\lvert \ell\rvert}$	Typically $\rho_1 \simeq \rho_2 = 0.95$ ρ_1, ρ_2 are one step correlation parameters
Nonseparable Exponential or Isotropic	$\sigma^2 \exp\{-\sqrt{\alpha_1 k^2 + \alpha_2 \ell^2}\}$	For $\alpha_1 = \alpha_2$, this is called the Isotropic model. $\rho_1 = \exp(-\alpha_1)$, $\rho_2 = \exp(-\alpha_2)$.

is the KLT and $\sigma^2_{k,l}(\mathcal{A})$ are the variances of the $\mathcal{A}$-transform coefficients $v_{k,l}(\mathcal{A})$.

2) Compared to all other unitary transforms, the KLT packs the maximum expected energy in a given number of samples M, i.e.,

$$\sum_{k,l \in \mathcal{M}(\Phi)}\sum \sigma^2_{k,l}(\Phi) > \sum_{k,l \in \mathcal{M}(\mathcal{A})}\sum \sigma^2_{k,l}(\mathcal{A}), \qquad \forall 1 \leqslant M \leqslant N^2 \tag{13}$$

where $\mathcal{M}(\mathcal{A})$ is the set containing M index pairs (k, l) corresponding to the largest M variances in the $\mathcal{A}$-transform domain. This property serves as a basis for transform data compression techniques.

Table IV shows the two commonly used stationary covariance models used in image processing. For the separable model, the KLT is given by

$$\phi(i, j; k, l) = \psi_1(i, k)\,\psi_2(j, l) \tag{14}$$

where $\psi_n(i, j)$ is the KLT corresponding to the one dimensional covariance function $\rho_n^{|i-j|}$, $n = 1, 2$. Unfortunately, $\boldsymbol{\Psi}_n$ is not a fast transform. Depending on the value of ρ_n, it has been shown that a suitable fast sinusoidal transform [9], [10] could be found as a good approximation to the KLT. For example for $-0.5 \leqslant \rho_n \leqslant 0.5$, the sine transform (see Table III) and for $0.5 \leqslant \rho_n \leqslant 1$, the cosine transform [10], [14] are good substitutes for the KLT. For common monochrome images, the correlation parameters ρ_1 and ρ_2 are close to unity so that the DCT is the preferred fast transform. It has also been shown [10], [15] that the sinusoidal transforms have equivalent performance as $N \to \infty$. In image processing N can be quite large, and one often processes smaller blocks (typically 16×16) of an image at a time. The performance differences between the various transforms are significant enough to warrant the use of the KLT or a reasonable substitute of it. Recently it has been shown [16] that the separable DCT is a good substitute for the nonseparable KLT of other stationary random fields also, including those modeled by the nonseparable exponential covariance function shown in Table IV. Fig. 2 shows the data compression efficiency of the various unitary transforms for random fields modeled by the nonseparable exponential covariance function of Table IV. Here we are plotting the residual expected energy (also called the basis restriction error [10])

$$\sigma_e^2(\mathcal{A}, M) = 1 - \left(\sum_{k,l \in \mathcal{M}}\sum \sigma^2_{k,l}(\mathcal{A}) \Big/ \sum_{k=1}^{N}\sum_{l=1}^{N} \sigma^2_{k,l}(\mathcal{A})\right) \tag{15}$$

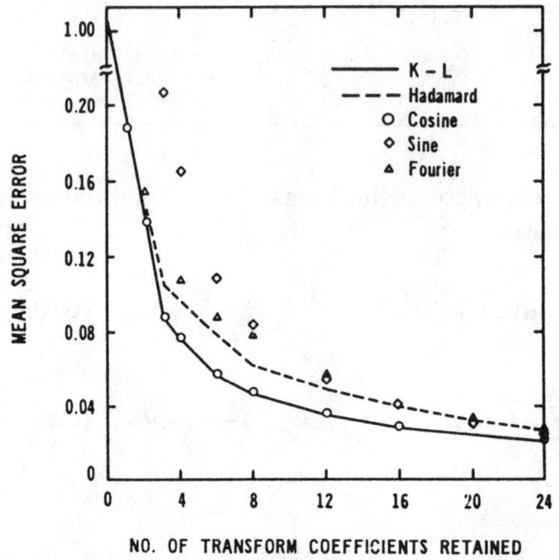

Fig. 2. Data compression efficiency of various transforms for 8×8 random fields with covariance function $r(k, l) = (0.95)^{\sqrt{k^2 + l^2}}$.

remaining in the $N^2 - M$ samples of the $\mathcal{A}$-transform. Clearly we see that the DCT has the best performance for the chosen values of ρ_1 and ρ_2. In the later sections we will see that the DCT and DST are useful in other statistical representations of images also. Image transforms have been applied extensively in data compression, noise smoothing, and restoration of images [1]-[3], [7], [17]-[19].

C. The Singular-Value Decomposition (SVD) Representation

Considering an $N \times N$ image as a matrix U of real numbers, it is possible to express it as [20]-[22]

$$U = \sum_{m=1}^{K} \lambda_m^{1/2}\, \boldsymbol{\psi}_m \boldsymbol{\phi}_m^T \tag{16}$$

where K is the rank of U and $\boldsymbol{\psi}_m$ and $\boldsymbol{\phi}_m$ are the orthonormal eigenvector solutions of

$$U^T U \boldsymbol{\phi}_m = \lambda_m \boldsymbol{\phi}_m$$
$$U U^T \boldsymbol{\psi}_m = \lambda_m \boldsymbol{\psi}_m. \tag{17}$$

The quantities λ_m, $m = 1, \cdots, K$ are positive and (16) is called the singular value or the outer product expansion of U and can also be written as

$$U = \boldsymbol{\Psi} \boldsymbol{\Lambda}^{1/2} \boldsymbol{\Phi}^T \tag{18}$$

where $\boldsymbol{\Psi}$ and $\boldsymbol{\Phi}$ are $N \times K$ matrices whose column vectors are $\boldsymbol{\psi}_m$ and $\boldsymbol{\phi}_m$, respectively, and $\boldsymbol{\Lambda}^{1/2}$ is the diagonal matrix of elements $\{\lambda_m^{1/2}\}$. If the singular values λ_m are arranged in decreasing order then the partial sum

$$U_M^{\cdot} \triangleq \sum_{m=1}^{M} \lambda_m^{1/2} \psi_m \phi_m^T \tag{19}$$

is the best least squares approximation of U for any $M \geqslant 0$, i.e., the error

$$e_M = \sum_{i=1}^{N} \sum_{j=1}^{N} (u_{i,j} - u_{i,j}^{\cdot})^2 \tag{20}$$

is minimum for any fixed M. This means the energy packed in the M coefficients $\{\lambda_m, 1 \leqslant m \leqslant M\}$ is maximized by the SVD transform. We note that the KLT maximizes the *average energy* packed in M samples for an ensemble of images. Thus, for any given image, the SVD transform would be more efficient in terms of its data compression (and other least squares processing) ability. However, in view of (17) this transform would be different for each different image. Hence, unlike the KLT, a good fast transform substitute, independent of the image, cannot be found for the SVD. Although SVD transform has potential applications in image data compression and restoration problems [23]–[26], the computational burden introduced by it overwhelms its desirable least squares property. Fortunately, some of the iterative least squares algorithms [95] could be employed to achieve practically the same results. The SVD has other applications [27], e.g., in approximation of a two-dimensional power spectrum by a separable product of one-dimensional power spectra (see Section V), in the design of digital filters [22], [28] and also in texture analysis of images [29].

III. Image Representation by One-Dimensional Stochastic Processes

Often it is desired to design image processing algorithms for an ensemble of images. For practical reasons this ensemble is generally characterized by the mean and covariance functions. These functions could be specified by a mathematical formula (e.g., as in Table IV) or via the SDF, or simply as arrays of numerical values. An alternative is to consider the image ensemble as being generated by a linear system forced by white noise or a random sequence of known SDF. The impulse response of this linear system is often specified by a difference equation. Computational complexity as well as performance of various processing algorithms can be studied in terms of this difference equation.

A simple way to characterize an image is to consider it as a collection of one-dimensional signals, e.g., as an output of a raster scanner, or as a sequence of rows (or columns) ignoring the interrow (or column) dependencies. For such cases, one-dimensional representations of stochastic processes are useful. One dimensional stochastic models have been applied in line by line processing of images for DPCM coding, hybrid coding, recursive filtering and restoration, etc. [17], [38]–[42], [46], [52], [58].

A. Autoregressive Representations

If $\{u_k\}$ is a zero-mean stationary Gaussian random sequence, then a causal representation of the type

$$u_k, \forall k = \sum_{n=1}^{p} a_n u_{k-n} + \epsilon_k \qquad E\epsilon_k = 0 \qquad E\epsilon_k \epsilon_l = \beta^2 \delta_{k,l} \tag{21}$$

is called a (one-sided) autoregressive (AR) representation. The sequence $\{\epsilon_k\}$ is a zero-mean white-noise random process independent of the past outputs. AR models have the following important properties.

1) The quantity

$$\bar{u}_k = \sum_{n=1}^{p} a_n u_{k-n}$$
$$= E[u_k | u_n, \forall n \leqslant k-1] \tag{22}$$

is the best mean-square predictor of u_k based on all of its past and depends only on the past p samples. Thus (21) becomes

$$u_k = \bar{u}_k + \epsilon_k \tag{23}$$

which says the sample at k is the sum of its minimum variance causal prediction estimate plus the prediction error. This is also called the *innovations representation.* The sequence $\{u_k\}$ defined by (21) is called a pth-order Markov process.

2) The AR process is stationary and causally stable (in the usual bounded-input–bounded-output (BIBO) sense) if and only if the roots of the polynomial

$$A_p(z) = 1 - \sum_{n=1}^{p} a_n z^{-n} \tag{24}$$

lie inside the unit circle. If stationary, then its spectral density function (SDF) is given by

$$S_u(z) = \beta^2 / [A_p(z) A_p(z^{-1})], \qquad z = e^{j\omega}, \; -\pi < \omega < \pi. \tag{25}$$

From (21), it is seen that the transfer function of an AR representation is $1/A_p(z)$ which is an all pole model.

3) *Theorem 1:* Given an arbitrary set of positive definite real covariances $\{r_k\}$ on a window $W = \{-p \leqslant k \leqslant p\}$, there exists a unique AR model (21) whose parameters are identified by solving the linear Toeplitz system of equations

$$r_k - \sum_{n=1}^{p} a_n r_{k-n} = \beta^2 \delta_{k,0}, \qquad 0 \leqslant k \leqslant p. \tag{26}$$

Moreover, this model is causally stable and stationary.

The significance of this theorem is that the covariances generated by the model (obtained by the Fourier inverse of (25)) would match exactly the given covariances on the window W. The model covariances outside this window provide a positive definite *extrapolation* of the given covariance sequence.

An important consequence of this result is that any given SDF can be approximated arbitrarily closely by a finite-order AR spectrum. In other words, if $\{r_k\}$ is a covariance sequence corresponding to a given positive and analytic SDF $S(z)$, and if $\{a_n^p, n = 1, \cdots, p\}$ and β_p^2 denote the solution of (26), then

$$\lim_{p \to \infty} \beta_p^2 / [A_p(z) A_p(z^{-1})] = S(z), \qquad z = \exp(j\omega)$$

where $A_p(z)$ is given by (24) with a_n replaced by a_n^p. This relation states that the solution of (26), as $p \to \infty$, provides the spectral factorization of $S(z)$. If $S(z)$ happens to be a rational all pole spectrum, the a_n^p will be zero when $n > p$, for some $p < \infty$. In general (26) can be used to find a rational, all

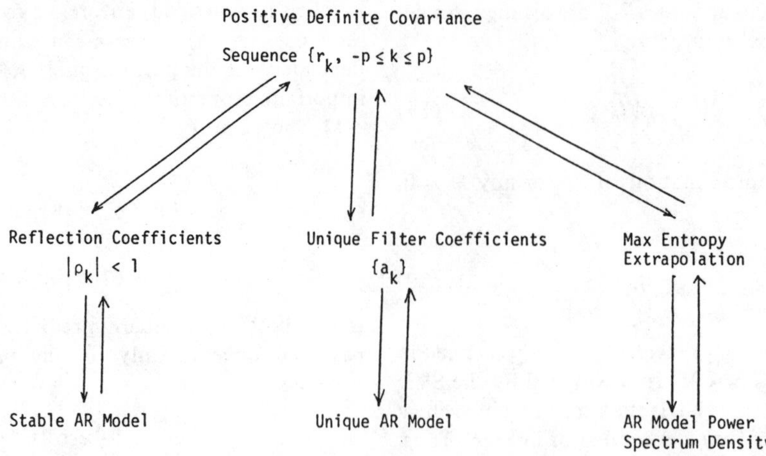

Fig. 3. Properties of the AR models.

pole spectral approximation of an arbitrary positive SDF. For these and other related ideas see Whittle [30] and Astrom [31].

4) *Levinson Algorithm [32], [33]:* The Toeplitz system of equations (26), also called the *normal equations*, can be solved in $O(p^2)$ operations by the recursions

$$a_{n+1,k} = \begin{cases} a_{n,k} - \rho_{n+1} a_{n,n+1-k}, & a_{n,0} = 1, \quad 1 \leq k \leq n \\ \rho_{n+1}, & k = n+1 \end{cases}$$

$$\beta_{n+1}^2 = \beta_n^2 (1 - \rho_{n+1}^2), \qquad \beta_0^2 = r_0$$

$$\rho_{n+1} = \frac{1}{\beta_n^2} \left[r_{n+1} - \sum_{k=1}^{n} a_{n,k} r_{n+1-k} \right], \quad \rho_1 = r_1/r_0 \tag{27}$$

where $\beta^2 = \beta_p^2$, $a_n = a_{p,n}$ are the coefficients required in (26). The elements $\{\rho_n, 1 \leq n \leq p\}$ are called the *reflection coefficients*. If the covariance matrix $\{r_{k-n}, 0 \leq k, n \leq p\}$ is positive definite then it could be shown that $|\rho_n| < 1$. Moreover the sequences $\{r_k\}$, $\{a_k\}$ and $\{\rho_k\}$ are unique.

5) A necessary and sufficient condition for the stability of the AR model is that $|\rho_n| < 1$, $\forall n$, or equivalently that the matrix $\{r_{k-n}, 0 \leq k, n \leq p\}$ be positive definite.

6) *Theorem 2 (Maximum Entropy Extrapolation):* The given positive definite covariance sequence $\{r_k, -p \leq k \leq p\}$ has a unique maximum entropy extrapolation and is given by the covariances generated by the AR model of (21).

Although the proof of this result is available at various places in the literature [34], [36], the following simple proof is offered. We define as entropy of a covariance sequence $\{r_k, -N \leq k \leq N\}$ the average entropy of $(N+1)$ Gaussian random variables which is given by (within an additive constant)

$$H_N = \frac{1}{N+1} \log |R_N| \tag{28}$$

where $|R_N|$ is the determinant of the Toeplitz covariance matrix associated with the covariance sequence. From the theory of Toeplitz matrices it is easy to show that the determinant of R_N follows the simple recursion [35]

$$|R_N| = |R_{N-1}| \beta_N^2, \qquad \beta_N^2 \leq \beta_{N-1}^2$$

From this, H_N is monotonically nonincreasing function of N. For $N \geq p$, to maximize H_N, we therefore simply need to have $\beta_k^2 = \beta_p^2$ for $N \geq k \geq p$. This means the optimum ρ_k are zero for $k > p$ and one obtains

$$a_{n+p,k} = \begin{cases} a_{p,k}, & \text{for } n \geq 0, \quad 1 \leq k \leq p \\ 0, & k \geq p+1. \end{cases} \tag{29}$$

Thus the pth-order AR model maximizes the entropy. By setting $\rho_{n+p} = 0$ in (27) the extrapolated covariances are obtained as

$$r_{n+p} = \sum_{k=1}^{p} a_k r_{n+p-k}, \qquad n \geq 1. \tag{30}$$

The SDF given by (25), determined via the coefficients $\{a_k\}$ is called the *maximum entropy spectrum* of the given covariance sequence. It is interesting to note that (29) provides the necessary conditions for extrapolation of $\{r_k\}$ from $k = p+1$ to $k = N$, for any $N > p$.

7) The reflection coefficients are such that at recursion step n in (27), the sum of the forward and backward predictive errors is minimized, i.e., if we define

$$\sigma_e^2 \triangleq E(\epsilon_{m,k}^+)^2 + E(\epsilon_{n,k}^-)^2 \tag{31}$$

$$\epsilon_{n,k}^+ \triangleq x_k - \sum_{m=1}^{n} a_{n,m} x_{k-m} = \epsilon_{n-1,k}^+ - \rho_n \epsilon_{n-1,k}^- \tag{32}$$

$$\epsilon_{n,k}^- \triangleq x_{k-n} - \sum_{m=1}^{n} a_{n,m} x_{k-n+m} = \epsilon_{n-1,k-1}^- - \rho_n \epsilon_{n-1,k}^+ \tag{33}$$

then ρ_n is such that σ_e^2 is minimized. This relation is useful in finding ρ_n directly when observed samples of the random process rather than their covariances are available [34].

Fig. 3 summarizes the properties of the AR models. In image processing, AR models have been found useful in modeling images line by line, as illustrated by the following examples.

Example 1: Consider zero-mean monochrome images whose covariance function models are listed in Table IV. If the image is processed one column at a time then the covariance function of pels on any column is of the form

$$r_k = E u_{i,j} u_{i+k,j} = \sigma^2 \rho_1^{|k|} \tag{34}$$

for both the models of Table IV. The corresponding AR representation is

$$x_i = \rho_1 x_{i-1} + \epsilon_i \qquad \beta^2 = E\epsilon_i^2 = \sigma^2 (1 - \rho_1^2). \tag{35}$$

Example 2: In certain image processing applications, each column (or row) of an image is first unitarily transformed, i.e., for the jth column

$$v_j = Au_j, \qquad j = 1, \cdots, N \tag{36}$$

and each element $v_j(i)$ is modeled as a first-order AR process, i.e.,

$$v_j(i) = a_1(i)\, v_{j-1}(i) + e_j(i), \qquad i = 1, \cdots, N. \tag{37}$$

If A is the KLT, i.e., the elements $v_j(i)$ and $v_j(k)$ are uncorrelated (independent under Gaussian assumptions) for $i \neq k$, then (37) represents a set of N decoupled AR models. Such algorithms have been called *hybrid* (i.e., a combination of nonrecursive and recursive procedures) and have been used in image restoration and coding [17], [52], [58]. For a tutorial review of AR models see [37]. Practical examples are discussed in section VI.

Example 3: Common images represent a nonnegative luminance function. In many applications from the physics of image formation it is possible to model the image as a power spectrum. For example, in high-resolution radar imaging a target can be considered as a set of distributed scatterers. The overall radar cross section (RCS) of the target measured as a function of aspect angle has the properties of a power spectrum. Let $S(x), -\frac{1}{2} < x < \frac{1}{2}$ represent a line of the image. Assume $S(x) > 0$ and that the Fourier series of $S(x)$ is uniformly convergent. Then its inverse Fourier transform is a valid covariance sequence. For a sampled image, if s_k, $\{0 \leqslant k \leqslant N - 1\}$ represents a set of positive and bounded real numbers, it can be shown [38] that its DCT coefficients given by

$$r_{-k} \triangleq r_k = \frac{1}{N} \sum_{m=0}^{N-1} s_m \cos \frac{\pi k}{N}\left(m + \frac{1}{2}\right), \qquad 0 \leqslant k \leqslant N - 1 \tag{38}$$

is a covariance sequence. Hence it is possible to find a pth-order AR model whose output covariances would match the $\{r_k\}$ over $[-p, p]$. In other words, one could synthesize each line of the image arbitrarily closely by a suitable AR model. Note that the image itself is being modeled by a deterministic function even though the AR model may characterize a random process whose covariances are related to the image. Fig. 4 shows an original and a synthesized image line by line by segments of $N = 32$ pels, generated by an eighth-order AR model. This figure also shows the image obtained by cosine transform coding when the same number of transform coefficients are retained. The higher resolution provided by the AR model approach is the result of the covariance extrapolation performed by that model. For other examples and details see [38].

B. State Variable Models

State variable models have been used to represent two-dimensional images considered as the output of a raster scanner. The scanner output is modeled as a one-dimensional random process and is characterized by a set of state variable equations of the form

$$\frac{dx(t)}{dt} = A(t)\, x(t) + B(t)\, \varepsilon(t)$$

$$y(t) = C(t)\, x(t) \tag{39}$$

where $y(t)$ is the scanner output at time t, and $x(t)$ is an $n \times 1$ vector, $\varepsilon(t)$ is a $p \times 1$ zero-mean white-noise vector such that

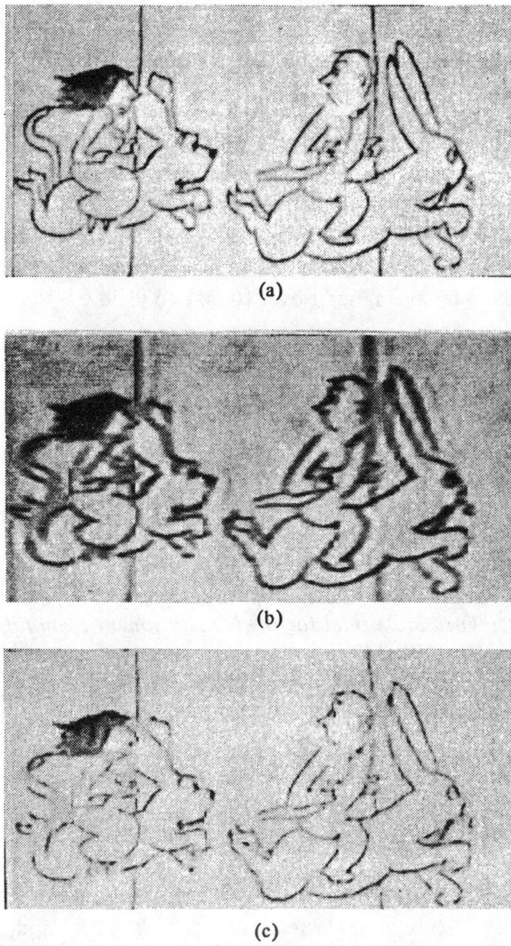

Fig. 4. Line by line synthesis of an image by eighth-order AR models. Each line has been synthesized by line segments of 32 pixels. (a) Original. (b) Synthesis using Cosine transform, 4 to 1 sample reduction. (c) Synthesis using AR model, 4 to 1 sample reduction.

$$E\varepsilon(t)\varepsilon^T(t') = K\delta(t - t'). \tag{40}$$

A, B, C, K are appropriate matrices which are determined such that $y(t)$ satisfies (approximately, if not exactly) the statistics of the scanner output.

The first attempt to model images by state variable techniques was made by Nahi and Asseffi [39], [40]. Although their final model has limitations because of several approximations, their modeling procedure does expose several difficulties in representing two-dimensional random field by one-dimensional models.

Consider an image being raster scanned from left to right and top to bottom with instantaneous repositioning so that the scanner output is continuous. Let $u(i, x)$ denote the brightness at a point x of the ith scanned line of the image, where $i = 0, \pm 1, \pm 2, \cdots, \pm\infty$, and x lies in the continuous interval $[0, M]$. The scanner output at any instant

$$t = jM + x, \qquad 0 \leqslant x \leqslant M \tag{41}$$

is $s(t) = u(j, x)$, where the scanner speed is assumed to be unity. Thus the scanner acts like a stacking operator which stacks one row after another. It is easy to show that the scanner output $s(t)$ is a nonstationary process even though $u(i, x)$ is a two-dimensional stationary random field. Defining

$$\tau = iM + y, \quad 0 \leq y \leq M, \quad i = 0, 1, 2, \cdots \tag{42}$$

the covariance of the scanner output is given by [39]

$$r_s(x, y) = Es(t)s(t+\tau) = \begin{cases} r(i, y), & x + y \leq M. \\ r(i+1, M-y), & x + y > M. \end{cases} \tag{43}$$

Since r_s depends on both x and y, it is a nonstationary covariance function. It could be *approximated* as a stationary covariance by averaging it over $[0, M]$ to yield

$$\tilde{r}(\tau) = \frac{M-y}{M} r(i, y) + \frac{y}{M} r(i+1, M-y). \tag{44}$$

Now $\tilde{r}$ depends only in i and y (the parameters of τ). Thus it is a function of τ only and is therefore stationary. Now a rational approximation of $\tilde{r}(\tau)$ is made from which a state variable model can be determined using conventional spectral factorization techniques in one dimension. For an example see [39].

C. State Variable Model for the Nonstationary Scanning Process s(t) [41], [42]

Instead of approximating the covariance function $r_s(x, y)$ by a stationary covariance model, a time varying state variable model can be determined for the scanning process. Suppose the image has N scan lines and its covariance function is separable, i.e.,

$$r(n, \tau) = r_1(n) r_2(\tau) \tag{45}$$

where r_1, r_2 are stationary covariance functions. Suppose $r_2(\tau)$ has a state variable realization (A, B, C, K) and the $N \times N$ covariance matrix $R_1 = \{r_1(i-j), 1 \leq i, j \leq N\}$ has a lower-upper triangular (LU) factorization $R_1 = HH^T$. Then $r(n, \tau)$ has the realization

$$\dot{x} = Ax + B\epsilon$$

$$x(kM) = x_k$$

$$y = Cx$$

$$E\epsilon(t)\epsilon^T(\tau) = K\delta(t-\tau)$$

$$s(t) = s_k(t) = \sum_{j=0}^{k} h_{k,j} y(t - (k-j)M),$$

$$kM < t \leq (k+1)M. \tag{46}$$

These equations give a nonstationary or the so-called *cyclostationary* representation of $s(t)$. This is because (46) is to be reinitialized after every M time units. Also, because of the delays involved, this representation is non-Markovian.

The derivation of (46) is quite simple, especially in comparison to the elaborate procedure of the earlier development. However, it is time varying, and being non-Markovian its usefulness as such is limited.

D. A Vector Scanning Model

Suppose the image considered above is scanned by a column of N raster scanners and define

$$\hat{s}(t) = [s_0(t) s_1(t) \cdots s_{N-1}(t)]^T$$

$$\hat{x}(t) = [x^T(t), \quad x^T(t-(N-2)M), \cdots, x^T(t-(N-1)M)]^T,$$

$$(N-1)M \leq t \leq NM.$$

Then the vector scanner $s(t)$ has a state variable *Markovian representation*

$$\frac{d\hat{x}}{dt} = \mathfrak{A}\hat{x} + \mathfrak{B}\hat{\epsilon}(t)$$

$$E\hat{x}_0\hat{x}_0^T \triangleq P_0 = R_1$$

$$E\hat{\epsilon}(t)\hat{\epsilon}^T(\tau) = (I \otimes K)\delta(t-\tau), \qquad (N-1)M \leq t, \ \tau \leq NM$$

$$\hat{s}(t) = (H \otimes C)\hat{x} \tag{47}$$

where $\mathfrak{A} = I \otimes A$, $\mathfrak{B} = I \otimes B$, and $\otimes$ denotes the Kronecker product. Now we have a Markov model, but its dimensional has increased.

State variable models have been found useful in restoration of images degraded by spatially varying PSF's where Fourier techniques are not applicable and particularly when the PSF can be modeled as a finite impulse response and/or the degradation is a causal process (e.g., motion blur). For details see [45].

E. Noncausal Models [18], [46]

Earlier we saw that a causal AR representation is of the type

$$u_k = \bar{u}_k + \epsilon_k, \qquad E\epsilon_k = 0$$

where $\bar{u}_k$ is the best linear mean square predictor of u_k based on the past values $\{u_l, l < k\}$ and $\{\epsilon_k\}$ is a white-noise sequence. Thus $\bar{u}_k$ is a minimum variance causal predictor of u_k. In an analogous fashion, we can define a minimum variance noncausal predictor $\bar{u}_k$ which depends on the past as well as the future values of u_k. Let $\{u_k\}$ be any zero-mean Gaussian random sequence and let $\bar{u}_k$ denote the best linear mean-square estimate of u_k based on all $\{u_l, l \neq k\}$. Writing

$$\bar{u}_k \triangleq \sum_{l \neq k} a_{k,l} u_l \tag{48}$$

we determine coefficients $a_{k,l}$ by minimizing the mean-square error $E[(u_k - \bar{u}_k)^2]$. This minimization gives the result as

$$r_{k,l} - \sum_{j \neq k} a_{k,j} r_{j,l} = \beta_k^2 \delta_{k,l}, \qquad \beta_k^2 = \min E(u_k - \bar{u}_k)^2. \tag{49}$$

In matrix form after defining $a_{k,k} = -1$, this becomes

$$-AR = B \qquad \text{or} \qquad -A = BR^{-1} \tag{50}$$

where B is a diagonal matrix of elements β_k^2 and R is assumed to be positive definite. The noncausal *minimum variance representation* (MVR) of the (scalar) nonstationary random process $\{u_k\}$ is now defined as

$$u_k - \sum_{l \neq k} a_{k,l} u_l = \nu_k \qquad \text{or} \qquad -Au = \nu \tag{51}$$

where u and ν are vectors of elements $\{u_k\}$ and $\{\nu_k\}$ respectively and ν is a random process that represents the noncausal prediction error. Using (50) and (51) we get

$$R^{-1}u = B^{-1}\nu. \tag{52}$$

It is interesting that the minimum variance noncausal representation of (52) does not require any spectral factorization, but only needs the inversion of the covariance matrix R. Note that the random process $\{\nu_k\}$ is not a white-noise sequence. Instead, it is a "colored"-noise sequence whose covariance is obtained via (52) to give

$$R_\nu \triangleq E\nu\nu^T = BR^{-1}B. \tag{53}$$

The elements of the diagonal matrix B are determined by noting that $B \equiv \text{diag}\,[R_\nu]$ and are given by

$$\beta_k^2 = \frac{1}{[R^{-1}]_{k,k}}. \tag{54}$$

Hence, the noncausal representation (51) is complete by specifying R^{-1}. For an infinite stationary process with analytic SDF $S(\omega)$, the noncausal MVR becomes

$$r_0^+ u_k + \sum_{\substack{l=-\infty \\ l\neq 0}}^{\infty} r_l^+ u_{k-l} = \beta^{-2}\nu_k \tag{55}$$

where

$$r_k^+ = \frac{1}{2\pi}\int_{-\pi}^{\pi} S^{-1}(\omega)\exp(jk\omega)\,d\omega$$

is the covariance sequence corresponding to $1/S(\omega)$ and

$$\beta^2 = 1/r_0^+. \tag{56}$$

Example 4: A special case of interest is to find the noncausal MVR for causal AR processes. From (55), it follows that for the pth-order AR model of (21), the noncausal MVR is given by

$$u_k - \sum_{n=1}^{p} h_n(u_{k-n} + u_{k+n}) = \nu_k, \qquad \forall k$$

$$h_n = \frac{-\sum_{k=0}^{p} a_k a_{n+k}}{\sum_{k=0}^{p} a_k^2}, \qquad a_0 \triangleq -1$$

$$r_\nu(k) \triangleq E\nu_j\nu_{j+k} = \beta_\nu^2\left[\delta_{k,0} - \sum_{n=1}^{p} h_n(\delta_{k,n} + \delta_{k,-n})\right]$$

$$\beta_\nu^2 = \frac{\beta^2}{\sum_{k=1}^{p} a_k^2}. \tag{57}$$

This is a $2p$th-order stochastic difference equation forced by a pth-order moving average process $\{\nu_k\}$.

If we define $N \times 1$ vectors $u = \{u_k\}$, $\nu = \{\nu_k\}$, etc., (57) can be written as

$$Hu = \nu + b \qquad E\nu b^T = 0$$

$$R_\nu \triangleq E\nu\nu^T = \beta_\nu^2 H \tag{58}$$

where H is the $N \times N$ symmetric banded Toeplitz matrix whose entries along the mth subdiagonal are $-h_m$, and the $N \times 1$ vector b contains $2p$ nonzero terms involving the boundary variables $\{u_{1-k}, u_{N+k}, 1 \leqslant k \leqslant p\}$. Defining

$$u^0 = H^{-1}\nu \qquad u^b = H^{-1}b \tag{59}$$

we obtain an orthogonal decomposition of u

$$u = u^0 + u^b \qquad E[u^0(u^b)^T] = 0 \tag{60}$$

which is such that the covariance matrix of u^0 is

$$R^0 \triangleq Eu^0u^{0T} = \beta_\nu^2 H^{-1} \tag{61}$$

and the process u^b is completely determined by the $2p$ boundary variables $\{u_{1-k}, u_{N+k}, 1 \leqslant k \leqslant p\}$. This representation provides the following result which is useful in developing the so-called fast KLT algorithms [18].

Theorem 3: Let $\{u_k\}$ be a stationary, pth-order AR sequence. For any $N > 1$ if the $2p$ boundary values $\{u_{1-k}, u_{N+k}, 1 \leqslant k \leqslant p\}$ are given then the KLT of the sequence $\{u_k, 1 \leqslant k \leqslant N\}$ conditioned on these boundary values is the orthonormal set of eigenvectors of the $N \times N$ banded Toeplitz matrix H.

This theorem could be proven following [18] where the special case $p = 1$ was considered. It has been shown [10] that for a given H, a fast sinusoidal transform could be found as a good approximation to its KLT. For the case $p = 1$, the KLT is the fast sine transform (see Table III). Noncausal representations are useful in data compression and restoration of images using image transforms [46]–[53]. The boundary variables are processed first (e.g., filtered in image estimation problems) and $\tilde{u}^b$, an estimate of u^b is determined. The residual process $\tilde{u}^0 = u - u^b$ is then processed by the KLT of u^0. For example it has been shown for $p = 1$ that the noncausal representation can be used for block by block coding of a sequence in such a way that interblock redundancy could be exploited (via the boundary values) leading to an algorithm more efficient than the conventional block by block KLT coding where one does not consider the interblock effects. For details see [9], [17], [19]. Algorithms in other applications [93] will require manipulation of the banded Toeplitz matrix and certain circulant [54] and other [10], [49] decompositions have been found useful. In the sequel we will find the above theory useful in developing semicausal and noncausal representations for two-dimensional images.

IV. Linear Prediction Models in Two Dimensions

One important property of many one-dimensional systems is that of causality. For two-dimensional images, causality is not inherent in the data. Moreover, the data could be such that a causal realization by a finite-order linear system is not possible, even if the SDF is a rational function. This is because it is generally not possible to factorize a two-dimensional polynomial as a product of lower order polynomials. In general, one can think of causal, semicausal, and noncausal representations for two-dimensional images. These representations are the discrete equivalent of the classical categories, viz., initial-value (or hyperbolic), initial-boundary value (or parabolic) and boundary value (or elliptic), of two-dimensional linear systems characterized by partial differential equations. In this section we define such models in the framework of linear prediction. In the next section we will consider their realization from a given SDF.

Linear prediction models in two dimensions are useful in image data transmission and storage via DPCM coding and hybrid coding, design of recursive, semirecursive and nonrecursive filters for image estimation, restoration and filtering and in image analysis. Examples are considered in Section VI.

Let $\{u_{i,j}\}$ be an arbitrary zero-mean Gaussian random field and let $\bar{u}_{i,j}$ denote a prediction estimate of the random variable $u_{i,j}$.

A. Causal Prediction

Suppose the samples of the random field $\{u_{i,j}\}$ are arranged in any desired, one-dimensional ordered sequence $\{u(k)\}$. Then $\bar{u}_{i,j}$ is defined as a *causal prediction* of $u_{i,j}$ if it depends only

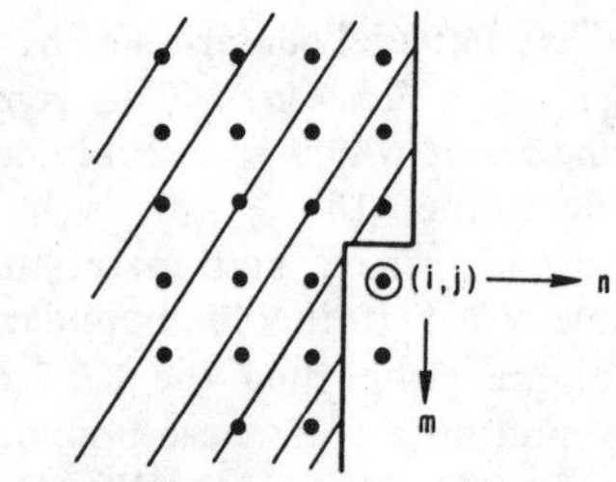

Fig. 5. Prediction region for causal models.

on the elements that occur before the element $u_{i,j}$. A common example occurs when an image is raster scanned, say, column by column, and $\bar{u}_{i,j}$ is a linear estimate based on all the elements scanned before arriving at (i, j), i.e.,

$$\bar{u}_{i,j} = \sum_{m,n \in \mathcal{S}} a(i, j; m, n)\, u_{m,n},$$

$$\mathcal{S} = \{m, n: n < j, \forall m\} \bigcup \{m, n: n = j, m < i\}. \quad (62)$$

Fig. 5 shows the set $\mathcal{S}$ for causal prediction at (i, j). This definition of causality includes, as a special case, single quadrant causal predictors of the type

$$\bar{u}_{i,j} = \sum_{\substack{m=0 \\ (m,n) \neq (0,0)}}^{\infty} \sum_{n=0}^{\infty} a(i, j; m, n)\, u_{i-m,j-n}\,. \quad (63)$$

If u_j denotes the jth column vector of the image, (62) gives

$$\bar{u}_j = A^0_{j,j} u_j + \sum_{n<j} A_{j,n} u_n \quad (64)$$

where $A^0_{j,j}$ is a lower triangular matrix whose diagonal entries are zero.

B. Semicausal Prediction[2]

If the estimate $\bar{u}_{i,j}$ is causal in one of the coordinates and noncausal in the other, it is called a *semicausal predictor*. For example, a linear semicausal predictor which is causal in "j" and noncausal in "i" would be of the form

$$\bar{u}_{i,j} = \sum_{m,n \in \mathcal{S}} a(i, j; m, n)\, u_{m,n},$$

$$\mathcal{S} = \{m, n: n < j, \forall m\} \bigcup \{m, n: n = j, \forall m \neq i\} \quad (65)$$

where $\mathcal{S}$ is shown in Fig. 6. In vector notation this becomes

$$\bar{u}_j = A^0_{j,j} u_j + \sum_{n<j} A_{j,n} u_n \quad (66)$$

where $\{A_{j,n}\}$ are matrices of elements $\{a(i, j; m, n)\ \forall i, m\}$ and $A^0_{j,j}$ has zeros at its diagonal elements.

C. Noncausal Prediction

The quantity $\bar{u}_{i,j}$ is defined as a noncausal prediction of $u_{i,j}$ if it can be written as a linear combination of possibly all the

[2]We note that in the two-dimensional literature the term 'causal' is often used for single quadrant models only and the model of (62) is sometimes called 'semicausal' or nonsymmetric half-plane (NSHP) model. Our definition of semicausality includes all of the half-plane as indicated in (65).

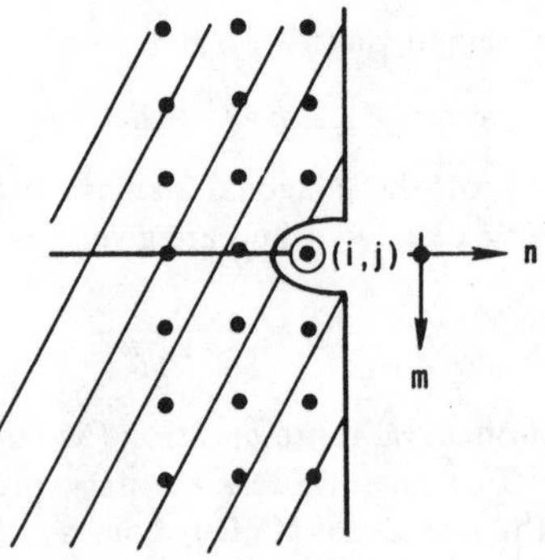

Fig. 6. Prediction region for semicausal models.

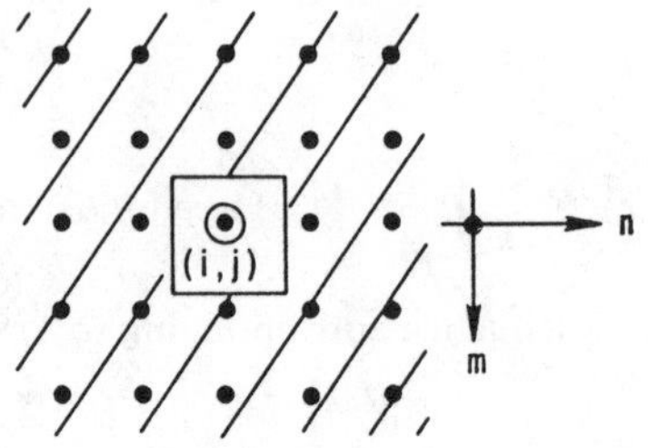

Fig. 7. Prediction region for noncausal models.

variables in the random field, except $u_{i,j}$ itself. For example, a *linear noncausal predictor* would be of the type

$$\bar{u}_{i,j} = \sum_{m,n \in \mathcal{S}} a(i, j; m, n)\, u_{m,n},$$

$$\mathcal{S} = \{m, n; (m, n) \neq (i, j)\} \quad (67)$$

and is shown in Fig. 7. Note that $\bar{u}_{i,j}$ contains terms from all the four quadrants about the point (i, j).

Example 4: The following are examples of causal, semicausal, and noncausal predictors.

Causal:

$$\bar{u}_{i,j} = a_1(i,j)\, u_{i-1,j} + a_2(i,j)\, u_{i,j-1} + a_3(i,j)\, u_{i-1,j-1}\,.$$

Semicausal:

$$\bar{u}_{i,j} = a_1(i,j)\, u_{i-1,j} + a_2(i,j)\, u_{i+1,j} + a_3(i,j)\, u_{i,j-1}\,.$$

Noncausal:

$$\bar{u}_{i,j} = a_1(i,j)\, u_{i-1,j} + a_2(i,j)\, u_{i+1,j} + a_3(i,j)\, u_{i,j-1} + a_4(i,j)\, u_{i,j+1}\,.$$

D. Minimum Variance Prediction

A minimum variance prediction estimate is one that minimizes the mean-square error

$$e_{i,j} = E(u_{i,j} - \bar{u}_{i,j})^2 \quad (68)$$

at each i, j. For a Gaussian random field, the minimum variance predictors would be linear, whose coefficients $a(\cdot\cdot;\cdot\cdot)$ could be determined from its covariance function. Minimization of (68) and use of (62), (65), or (67) yields the orthogonality relation

$$E\left[u_{i,j} - \sum_{m,n \in \mathcal{S}} a(i, j; m, n)\, u_{m,n}\right] u_{p,q} = \beta^2_{i,j} \delta_{i,p} \delta_{j,q},$$

$$p, q \in \mathcal{S}_0 \quad \mathcal{S}_0 \triangleq \mathcal{S} \bigcup [i, j] \quad (69)$$

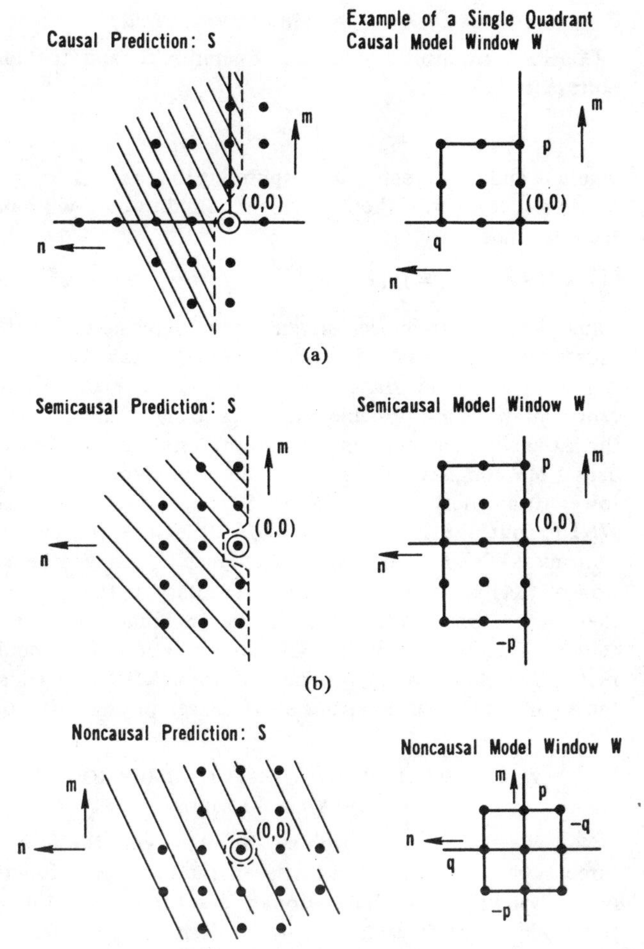

Fig. 8. Prediction regions and windows for stationary models.

Written in terms of covariances, we obtain

$$r(i, p; j, q) - \sum_{m,n \in \mathcal{S}} a(i, m; j, n)\, r(m, p; n, q) = \beta_{i,j}^2 \delta_{i,p} \delta_{j,q}, \qquad p, q \in \mathcal{S}_0 \quad (70)$$

where $\mathcal{S}$ depends on whether $\bar{u}_{i,j}$ is causal, semicausal, or noncausal, and $\beta_{i,j}^2$ is the minimized value of $e_{i,j}$. A solution of the simultaneous equations (70) gives the unknowns $a(\cdot\cdot;\cdot\cdot)$ and $\beta_{i,j}^2$.

E. Stochastic Representation of Gaussian Fields

Let $\bar{u}_{i,j}$ be an arbitrary prediction of $u_{i,j}$. Then we define a stochastic representation of the random field $\{u_{i,j}\}$ as

$$u_{i,j} = \bar{u}_{i,j} + \epsilon_{i,j} \quad (71)$$

where $\{\epsilon_{i,j}\}$ is another random field such that the given covariance properties of $\{u_{i,j}\}$ are satisfied. There are three types of representations that we would be interested in considering here. These are as follows:

(i) minimum variance representations (MVR)
(ii) white-noise-driven representations (WNDR)
(iii) autoregressive moving average (ARMA) representations.

For minimum variance representations, $\bar{u}_{i,j}$ is chosen to be a minimum variance predictor. These representations could be causal, semicausal, or noncausal in structure. One basic difference among these representations is their spatial structure which leads to different types of processing algorithms. For WNDR $\{\epsilon_{i,j}\}$ is chosen to be a white-noise field resulting in several types of models including the Karhunen–Loeve representations. In ARMA representations, $\{\epsilon_{i,j}\}$ is a colored noise field with a truncated covariance function, i.e.,

$$E\epsilon_{i,j}\epsilon_{m,n} = 0, \qquad \forall |i-m| > K, \qquad |j-n| > L \quad (72a)$$

for some fixed integers $K > 0, L > 0$. For example the function

$$r_\epsilon(i, j; m, n) = \begin{cases} 1, & i = m, \; j = n \\ \alpha, & |i-m| = 1, \; |j-n| = 1 \\ 0, & \text{otherwise} \end{cases} \quad (72b)$$

represents a stationary moving average field.

F. Stationary Models

For stationary random fields, the covariances become a function of two variables, i.e.,

$$r(i, j; m, n) = r(i-m, j-n)$$

and the various predictors become spatially invariant (or shift invariant) yielding the representation

$$u_{i,j} = \sum\sum_{m,n \in \hat{\mathcal{S}}} a_{m,n} u_{i-m,j-n} + \epsilon_{i,j} \quad (73)$$

where $\hat{\mathcal{S}}$ is a window of index pairs (m, n) which is now independent of i, j and is defined as

$$\hat{\mathcal{S}} = \begin{cases} \{n \geq 1, \forall m\} \cup \{n = 0, m \geq 1\}, & \text{for causal models} \\ \{n \geq 1, \forall m\} \cup \{n = 0, \forall m \neq 0\}, & \text{for semicausal models} \\ \{\forall (m, n) \neq (0, 0)\}, & \text{for noncausal models.} \end{cases} \quad (74)$$

Note that $\hat{\mathcal{S}}$ does not contain the origin $(0, 0)$. Fig. 8 shows $\hat{\mathcal{S}}$ for the various cases above. Often one is only interested in representations where $a_{m,n}$ are nonzero only over a finite window W, (see Fig. 8 for examples) called the *prediction window*, which is a subset of $\hat{\mathcal{S}}$, so that

$$u_{i,j} = \sum\sum_{m,n \in W} a_{m,n} u_{i-m,j-n} + \epsilon_{i,j}. \quad (75)$$

In that event (75) becomes a constant coefficient stochastic difference equation with a rational transfer function

$$H(z_1^{-1}, z_2^{-1}) = \frac{1}{A(z_1^{-1}, z_2^{-1})} \triangleq \left[1 - \sum\sum_{m,n \in W} a_{m,n} z_1^{-m} z_2^{-n}\right]^{-1}. \quad (76)$$

The SDF of random fields represented by these stationary

models becomes

$$S_u(z_1, z_2) = \frac{S_\epsilon(z_1, z_2)}{\left[1 - \sum\sum_{m,n \in W} a_{m,n} z_1^{-m} z_2^{-n}\right]\left[1 - \sum\sum_{m,n \in W} a_{m,n} z_1^{m} z_2^{n}\right]} \quad (77)$$

where $S_\epsilon(z_1, z_2)$ is the SDF of $\{\epsilon_{i,j}\}$. In general, $\{\epsilon_{i,j}\}$ would be a moving average field so that (77) is a rational function expressed as a ratio of two-dimensional polynomials in z_1 and z_2.

Minimum Variance Models: If (75) is a minimum variance representation of a stationary field, then the orthogonality condition (69) becomes

$$r(k, l) - \sum\sum_{m,n \in W} a_{m,n} r(k - m, l - n) = \beta^2 \delta_{k,0} \delta_{l,0},$$

$$k, l \in \hat{\mathcal{S}}_0 \qquad \hat{\mathcal{S}}_0 = \mathcal{S} \cup (0, 0). \quad (78)$$

Defining

$$a_{0,0} \triangleq -1 \qquad W_0 = W \cup (0, 0) \quad (79)$$

(78) reduces to

$$-\sum\sum_{m,n \in W_0} a_{m,n} r(k - m, l - n) = \beta^2 \delta_{k,0} \delta_{l,0}, \qquad (k, l) \in \hat{\mathcal{S}}_0. \quad (80)$$

Mapping the arrays $\{a_{m,n}\}$, $\{u_{m,n}\}$, $\{\delta_{k,0}\delta_{l,0}\}$ with support on W_0 into vectors $\boldsymbol{a}$, $\boldsymbol{u}$, $\boldsymbol{l}$, respectively, by column ordering, we obtain

$$\mathcal{R}\boldsymbol{a} = -\beta^2 \boldsymbol{l} \quad (81)$$

where $\mathcal{R}$ is the covariance matrix of the vector $\boldsymbol{u}$. For example, if W_0 is the noncausal rectangular window $[-p, p] \times [-q, q]$ then $\boldsymbol{a}$ is of size $(2p + 1)(2q + 1)$ and $\mathcal{R}$ is a $(2p + 1) \times (2q + 1)$ doubly block Toeplitz matrix with basic dimension $(2p + 1) \times (2p + 1)$. The unit vector $\boldsymbol{l}$ takes the value 1 at the location i_0, which corresponds to the (0, 0) location in the window W_0. Equations (79) and (81) can be solved to give

$$\begin{aligned} \boldsymbol{a} &= -\beta^2 \mathcal{R}^{-1} \boldsymbol{l} \\ &\triangleq -\beta^2 \boldsymbol{k}_{i_0} \\ \beta^2 &= 1/\boldsymbol{k}_{i_0}(i_0) \\ &= 1/[\mathcal{R}^{-1}]_{i_0, i_0}. \end{aligned} \quad (82)$$

These equations are much easier to solve than (70) because the dimensionality of $\mathcal{R}$ equals the square of the number of points in the window W_0 rather than the square of the number of points in the image. For a given positive definite $\mathcal{R}$, a unique solution of (81) is obtained. However, for a given admissible set of predictor coefficients $a_{m,n}$, there is not a unique $\mathcal{R}$ which will yield these coefficients as the solution of (81). Moreover, for an arbitrary positive definite $\mathcal{R}$, the solution of (81) does not assure prediction coefficients which would yield stable models. In spite of these shortcomings, equation (81) is useful because i) it is a linear (Toeplitz block) system of equations and ii) it could be used in finding approximate causal, semicausal and noncausal MVR's which are stable.

G. White-Noise-Driven Representations (WNDR)

Consider an arbitrary matrix operator $\mathcal{A}$ and the linear representation

$$\mathcal{A}\boldsymbol{u} = \boldsymbol{e} \quad (83)$$

where $\boldsymbol{u}$ and $\boldsymbol{e}$ are vectors corresponding to the fields $\{u_{i,j}\}$ and $\{\epsilon_{i,j}\}$, respectively. When $\{\epsilon_{i,j}\}$ is required to be a white-noise field, we must have

$$\mathcal{A}\mathcal{R}\mathcal{A}^T = \mathcal{B} = \text{Diagonal} \qquad \mathcal{R} = E\boldsymbol{u}\boldsymbol{u}^T, \quad \mathcal{B} = E\boldsymbol{e}\boldsymbol{e}^T. \quad (84)$$

Thus, any white noise driven linear representation must satisfy the factorization equation (84). However, it does not have a unique solution $\mathcal{A}$ and therefore many different representations are possible. If $\mathcal{A}$ is a lower triangular matrix, then (83) is a causal representation. As we shall see, the causal MV representations are of this type. When $\mathcal{A}$ is a block lower triangular matrix, then (83) becomes a semicausal WNDR (but not necessarily MVR). Otherwise, it would be a noncausal WNDR. Since $\mathcal{R}$ is a covariance matrix, another solution of (84) is obtained when $\mathcal{A}$ is a unitary matrix containing the eigenvectors of $\mathcal{R}$ and $\mathcal{B}$ is the diagonal matrix of its eigenvalues. This would yield the KL representation of the random field. The choice of the operator $\mathcal{A}$ for WNDR's depends on the algorithmic considerations or on a given physical situation.

V. Realization of the Two-Dimensional Prediction Models

Now we consider the problem of identifying the foregoing three types of representations given the covariance function, or equivalently, (for the stationary case) the spectral density function of the image. When the desired representation is required to be causal and stable, the above problem is also called the *spectral factorization problem*. Let $S(z_1, z_2)$ represent the two-dimensional z-transform of a covariance sequence $r(m, n)$. When $z_1 = \exp(j\omega_1)$, $z_2 = \exp(j\omega_2)$, this becomes the SDF and will also be written as $S(\omega_1, \omega_2)$. The main result of this section is that it is possible to find finite order, stable causal, semicausal, and noncausal MVR's which would match a given well behaved SDF arbitrarily closely. Details are given in the Appendix.

A. Separable Models

When the covariance function is separable, e.g., in the stationary case, if $r(k - m, l - n) = r_1(k - m)\, r_2(l - n)$, the solution of (80) reduces to two independent one-dimensional models. For example, if $r_1(m)$ and $r_2(n)$ have the one-dimensional realizations

$$\sum_{m \in W_1} a_1(m) x(k - m) = e_1(k)$$

$$\sum_{m \in W_2} a_2(n) y(l - n) = e_2(l).$$

Then $r_1(m)\, r_2(n)$ has the realization

$$\sum_{m \in W_1,} \sum_{n \in W_2} a_1(m)\, a_2(n)\, u_{k-m, l-n} = \epsilon_{k,l}$$

$$r_\epsilon(m, n) = r_{e_1}(m)\, r_{e_2}(n). \quad (85)$$

n \ m	-1	0	1
1	0	0	0
0	$-\alpha_1$	1	$-\alpha_1$
-1	0	0	0

n \ m	-1	0	1
1	$\alpha_1\alpha_2$	$-\alpha_2$	$\alpha_1\alpha_2$
0	$-\alpha_1$	1	$-\alpha_1$
-1	$\alpha_1\alpha_2$	$-\alpha_2$	$\alpha_1\alpha_2$

Fig. 9. Correlation arrays $\rho_\epsilon(m, n)$ for semicausal and noncausal model prediction errors.

Example 5: Consider the separable covariance model shown in Table IV. Using the above result the minimum variance causal, semicausal, and noncausal models are obtained as follows [51], [52].

Causal model (C1):

$$u_{i,j} = \rho_1 u_{i-1,j} + \rho_2 u_{i,j-1} - \rho_1 \rho_2 u_{i-1,j-1} + \epsilon_{i,j}$$
$$r_\epsilon(m, n) = \sigma^2 (1 - \rho_1^2)(1 - \rho_2^2)\, \delta_{m,0} \delta_{n,0}. \tag{86}$$

Semicausal MVR (SC2):

$$u_{i,j} = \alpha_1 (u_{i-1,j} + u_{i+1,j}) + \rho_2 u_{i,j-1} - \rho_2 \alpha_1 (u_{i-1,j-1} + u_{i+1,j+1}) + \epsilon_{i,j}$$
$$r_\epsilon(m, n) = \sigma^2 \frac{(1 - \rho_1^2)(1 - \rho_2^2)}{(1 + \rho_2^2)} \rho_\epsilon(m, n)$$
$$\alpha_1 \triangleq \frac{\rho_1}{(1 + \rho_1^2)}. \tag{87}$$

Noncausal MVR (NC3):

$$u_{i,j} = \alpha_1 (u_{i-1,j} + u_{i+1,j}) + \alpha_2 (u_{i,j-1} + u_{i,j+1}) - \alpha_1 \alpha_2 (u_{i-1,j-1} + u_{i+1,j-1} + u_{i-1,j+1} + u_{i+1,j+1}) + \epsilon_{i,j}$$
$$r_\epsilon(m, n) = \sigma^2 \frac{(1 - \rho_1^2)(1 - \rho_2^2)}{(1 + \rho_1^2)(1 + \rho_2^2)} \cdot \rho_\epsilon(m, n). \tag{88}$$

Fig. 9 shows the correlation arrays $\rho_\epsilon(m, n)$ for the above two models.

Remarks:

1) The prediction window is necessarily rectangular for separable models. Thus causal MVR's are necessarily quarter-plane models for separable covariance functions although the converse is not true.

2) All the three models in the above example represent the same stationary random field. The spatial structural differences yield different types of algorithms. For example, it has been shown [9], [17], that the causal, semicausal and the noncausal models yield naturally the predictive, hybrid and transform coding algorithms for data compression of images. For other applications see [1], [52], [53], [55].

3) While the causal MVR's are white noise driven, the semicausal and noncausal MVR's are not.

4) If the given SDF $S(\omega_1, \omega_2)$ is not separable, then its best least squares separable approximation can be found via its singular value decomposition as

$$S(\omega_1, \omega_2) \simeq \hat{S}(\omega_1, \omega_2) \triangleq \lambda S_1(\omega_1) S_2(\omega_2)$$

where $S_1(\omega_1)$ and $S_2(\omega_2)$ are the solutions corresponding to the largest eigenvalue λ of the following eigenvalue equations

$$\int_{-\pi}^{\pi} K_1(\omega_1, y) S_1(y)\, dy = \lambda S_1(\omega_1), \qquad \int_{-\pi}^{\pi} S_1^2(\omega)\, d\omega = 1$$

$$\int_{-\pi}^{\pi} K_2(x, \omega_2) S_2(x)\, dx = \lambda S_2(\omega_2), \qquad \int_{-\pi}^{\pi} S_2^2(\omega)\, d\omega = 1$$

where

$$K_1(x, y) = \int_{-\pi}^{\pi} S(x, y') S(y, y')\, dy'$$

$$K_2(x, y) = \int_{-\pi}^{\pi} S(x', x) S(x', y)\, dx'.$$

If $S(\omega_1, \omega_2) > 0$, then it is easy to show that $\lambda > 0$, $S_1(\omega_1) > 0$, $S_2(\omega_2) > 0$ and S_1, S_2 would be valid one-dimensional SDF's if S is a valid two-dimensional SDF. Now $\hat{S}$ is a separable SDF and therefore an appropriate causal, semicausal or noncausal realization of $\hat{S}$ could be found. A similar method of approximating a SDF (or equivalently, the magnitude of the frequency response of a filter) by a positive separable function has been used in [27] for the design of separable two-dimensional filters when the given SDF is discrete. In that case, the above equations reduce to the matrix SVD relations of Section II-C.

B. Causal Minimum Variance Representations

It can be shown that the causal MVR's are also white-noise driven representations. In general identification of a causal MVR requires a two-stage factorization. For example if the SDF is a rational function of the form

$$S(z_1, z_2) = \frac{C_0}{\left[\sum_{n=-q}^{q} \sum_{m=-p_1}^{p_1} \alpha_{m,n} z_1^{-m} z_2^{-n}\right]} \tag{89}$$

where C_0 is a constant, then, the first-stage factorization is achieved by solving a set of normal equations (see (A12) in Appendix) parametric in z_1 and the result is

$$S(z_1, z_2) = \frac{\sum_{n=0}^{q} \hat{a}_n^q(z_1) r_n(z_1)}{\left[1 + \sum_{n=1}^{q} \hat{a}_n^q(z_1) z_2^{-n}\right]\left[1 + \sum_{n=1}^{q} \hat{a}_n^q(z_1^{-1}) z_2^{n}\right]}. \tag{90}$$

Then a second stage factorization of S (equation (A13)) gives

$$S(z_1, z_2) = \frac{\beta^2}{\left[\sum_{n=0}^{q} a_n^q(z_1) z_2^{-n}\right]\left[\sum_{n=0}^{q} a_n^q(z_1^{-1}) z_2^{n}\right]}. \tag{91}$$

In general $a_n^q(z_1)$ will be irrational functions expressed by their infinite Laurent series. Under certain conditions of positivity and analyticity of the SDF the causal model so determined would be stable and causally invertible (i.e., "minimum-phase"). However, one is confronted with the problem of solving an infinity of sets of q simultaneous equations ((A12)

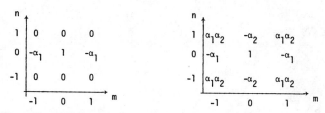

n \ m	1	0
1	-.0805	.532
0	.532	-1

	n = 2	n = 1	n = 0
m = 2	-.0247	-.0503	.0802
m = 1	-.0503	.001	.534
m = 0	.0802	.534	-1

Fig. 10. Single quadrant causal MVR coefficients.

and (A13)) for the rational SDF of (89). An *approximate* rational factorization of S could be obtained by finding a rational approximation for $a_n^q(z_1)$. For example, let

$$a_n^q(z_1) \simeq \sum_{m=-p}^{p} a_{m,n} z_1^{-m}, \qquad n \geqslant 1$$

$$a_0^q(z_1) \simeq \sum_{m=0}^{p} a_{m,0} z_1^{-m}.$$

Then with the analyticity constraint on S, it is possible to find a suitable integer $p < \infty$, such that $\tilde{S}$, the rational factorized form obtained by replacing $a_n^q(z_1)$ by their approximations above would be of the form given in (77) and would be arbitrarily close to S. The realization of $\tilde{S}$ would now be of the form of (75). Thus it is possible to find finite-order causal MVR's which would come arbitrarily close to the infinite order causal MVR's of a given SDF. It can also be shown that the solution of (82) for a causal window W_0 and for suitably large p and q should yield a causal MVR which would realize the given SDF arbitrarily closely. The proof of this and related considerations such as stability will be considered elsewhere. Another approach of rationalizing (91) and retaining stability is via a reflection coefficient design method proposed by Marzetta [56].

Example 6: Consider the nonseparable exponential covariance model in Table IV with $\alpha_1 = \alpha_2 = 0.05$. Single quadrant causal MVR coefficients $a_{m,n}$ over $[0, 1] \times [0, 1]$ and $[0, 2] \times [0, 2]$ regions are obtained as shown in Fig. 10.

Fig. 11 compares the given covariance array with the one generated by the model on the $[4 \times 4]$ grid for the $[2 \times 2]$ model. Although both the models above turn out to be stable, the given and the model covariances do not match exactly. However, as the model order is increased, e.g., from $[1 \times 1]$ to $[2 \times 2]$ the covariance match has been found to improve.

$$r(k,\ell) = \exp\left\{-.05\sqrt{k^2+\ell^2}\right\}$$

.868	.894	.905	.894	.868
.894	.932	.951	.932	.894
.905	.951	1.000	.951	.905
.894	.932	.951	.932	.894
.868	.894	.905	.894	.868

ACTUAL COVARIANCES ON W_c

.696	.730	.697	.579	.479
.730	.819	.840	.701	.579
.697	.840	1.000	.840	.697
.579	.701	.840	.819	.730
.479	.579	.697	.730	.696

ESTIMATED COVARIANCES BY p=q=2 CAUSAL MODEL

.821	.853	.866	.853	.821
.858	.907	.932	.907	.858
.874	.934	1.000	.934	.874
.858	.907	.932	.907	.858
.821	.853	.866	.853	.821

ESTIMATED COVARIANCES BY p=q=2 SEMICAUSAL MODEL

Fig. 11. Covariances estimated by the causal and semicausal models.

C. Semicausal Minimum Variance Representations

For these models only a single stage factorization of the SDF is required. Unlike causal MVR's, semicausal MVR's are not white-noise-driven models. However, for a given semicausal MVR, a causal MVR can always be found under some mild restrictions [see Appendix]. But the causal realization may have a higher (even infinite) order and/or may become spatially varying as shown by the following example.

Example 7: Consider a spatially invariant, semicausal MVR which is noncausal in the "i" variable and causal in the "j" variable

$$u_{i,j} = a(u_{i-1,j} + u_{i+1,j}) + bu_{i,j-1} + \epsilon_{i,j}.$$

Let $\{u_{i,j}\}$ be a random field with $1 \leqslant i, j \leqslant N$, $u_{0,j} = u_{N+1,j} = 0$, $0 < a < \frac{1}{2}$, $|2a + b| < 1$. In vector notation this becomes

$$Qu_j = bu_{j-1} + \boldsymbol{\epsilon}_j \qquad (92)$$

where Q is a symmetric tridiagonal Toeplitz matrix whose diagonal elements are unity, and the subdiagonal elements equal $-a$. For this to be a minimum variance representation, the orthogonality conditions require (see (A14))

$$E\boldsymbol{\epsilon}_j\boldsymbol{\epsilon}_k^T = \beta^2 Q \delta_{j,k}$$

where β^2, the variance of $\epsilon_{i,j}$ is assumed to be constant. The matrix Q has an lower-upper factorization $Q = L^T \boldsymbol{\Gamma}^{-1} L$, where

$$L = \begin{bmatrix} 1 & & & & \\ -r_2 & 1 & & & \\ & -r_3 & 1 & & \\ & & \ddots & \ddots & \\ & & & -r_N & 1 \end{bmatrix}$$

$$\boldsymbol{\Gamma}^{-1} = \text{Diag}\left\{\frac{a}{r_1}, \frac{a}{r_2}, \cdots, \frac{a}{r_N}\right\}$$

and $\{r_k\}$ are given by the backward recursion

$$r_k = \frac{a}{1 - ar_{k+1}}, \qquad r_{N+1} = 0, \qquad 1 \leqslant k \leqslant N.$$

Following Appendix A.2, we get a causal MVR as

$$u_{i,j} = r_i u_{i-1,j} + \frac{br_i}{a} \sum_{k=0}^{N-i+1} \alpha_{i,k} u_{i+k,j-1} + \nu_{i,j}$$

where $\alpha_{i,k}$ is the element in the kth upper diagonal and the ith row of $(L^T)^{-1}$. Now the sparse structure of the semicausal representation has been lost by this causal representation. Also, it is no longer a constant coefficient model. In the steady state, i.e., as i and $N \to \infty$, the asymptotic values of r and $\alpha_{i,k}$ are given by

$$r \triangleq r_\infty = \frac{1 - \sqrt{1 - 4a^2}}{2a}, \qquad \alpha_{i,k} = r^k, \qquad 0 < r < 1$$

and we obtain a constant coefficient, infinite order, causal MVR

$$u_{i,j} = ru_{i-1,j} + \frac{b}{a} \sum_{k=0}^{\infty} r^{k-1} u_{i+k,j-1} + \nu_{i,j}$$

$$E\nu_{i,j}^2 = \beta^2(1 + r^2).$$

In a practical situation, one would truncate the summation to a finite number of terms to obtain an approximate, finite-order causal model. However, the number of terms that may have to be retained to achieve stability could well be quite large resulting in a high-order model.

For a rational SDF as in (89), one needs to solve (A25) for the given value of q from which a_n^q are determined. This gives the first stage of factorization in the z_2 variable and we obtain

$$S = \frac{-\beta^2 a_0^q(z_1)}{\left[\sum_{n=0}^{q} a_n^q(z_1)\, z_2^{-n}\right]\left[\sum_{n=0}^{q} a_n^q(z_1^{-1})\, z_2^{n}\right]}. \tag{93}$$

No second stage factorization is required for semicausal MVR's. As in the case of causal models $\{a_n^q(z_1)\}$ will be generally irrational but analytic functions in the neighborhood of $|z_1| = 1$. When approximated by suitable rational functions, e.g., letting

$$a_n^q(z_1) \simeq \sum_{m=-p}^{p} a_{m,n} z_1^{-m}, \qquad a_{0,0} = -1, \qquad n > 0 \tag{94}$$

we obtain a finite-order semicausal MVR

$$u_{i,j} = \sum_{\substack{m=-p \\ m \neq 0}}^{p} a_{m,0} u_{i-m,j} + \sum_{n=1}^{q} \sum_{m=-p}^{p} a_{m,n} u_{i-m,j-n} + \epsilon_{i,j} \tag{95}$$

where $\{\epsilon_{i,j}\}$ is a moving average with SDF

$$S_\epsilon(z_1, z_2) = \beta^2 \left[1 - \sum_{m=-p}^{p} a_{m,0} z_1^{-m}\right]. \tag{96}$$

The SDF of this semicausal MVR is

$$S_u = \frac{S_\epsilon}{\left[\sum_{n=0}^{q} \sum_{m=-p}^{p} a_{m,n} z_1^{-m} z_2^{-n}\right]\left[\sum_{n=0}^{q} \sum_{m=-p}^{p} a_{m,n} z_1^{m} z_2^{n}\right]}. \tag{97}$$

m \ n	1	0
1	-.01844	.39316
0	.25368	-1.0
-1	-.01844	.39326

$\beta^2 = .0451$

m \ n	2	1	0
2	-.10294	-.02726	.0048
1	-.01931	.03832	.39006
0	-.03115	.27624	-1.0
-1	-.01931	.03832	.39006
-2	-.10294	-.02726	.0048

$\beta^2 = .0444$

Fig. 12. Semicausal MVR coefficients.

Thus it is possible to find finite-order semicausal MVR's which would realize a given positive and analytic SDF arbitrarily closely. It can also be shown that for a semicausal prediction window with suitably large but finite values of p and q, the finite-order semicausal MVR obtained via the solution of (82) would realize a given positive and analytic SDF arbitrarily closely. Thus, even though the solution of (82) corresponding to a fixed size window W need not ensure an admissible (i.e., stable) representation, solving it successively (or recursively) for increasing size windows should eventually lead to an admissible as well as a reasonably accurate representation. The advantage, of course, is that one would be solving only finite-order equations, whereas the approach via equations (A24), (A25), in general requires solving an infinite set of equations (i.e., for every $|z_1| = 1$). In many examples (see below and in the next section), the acceptable values of p and q have been found not to be very large.

Remarks

1) The finite-order semicausal MVR's realize SDF's which contain both numerator and denominator polynomials (see (97)). However, the numerator polynomial is one dimensional and is in the noncausal dimension.

2) The above remark implies semicausal MVR's are driven by colored noise. In fact $\{\epsilon_{i,j}\}$ is a moving average process in the "i" variable and is white in the "j" variable.

3) Semicausal MVR's were introduced in [9] and have been found useful in developing semirecursive or the so-called hybrid algorithms which are recursive in one direction and are transform based in the other directions. See examples in Section VI and [51], [52], [55], [57]–[59].

4) The semicausal models are recursive (or causal) only in the j variable. A re-indexing in the i-variable to attempt to represent it as a two-dimensional causal model would yield an unstable model. Thus the model in Example 7 written as

$$u_{i,j} = \frac{1}{a} u_{i-1,j} - u_{i-2,j-1} - \frac{b}{a} u_{i-1,j-1} - \frac{1}{a} \epsilon_{i-1,j}$$

would be unstable if solved recursively in i and j.

Example 8: We return to the nonseparable exponential covariance function considered in Example 5. Fig. 12 shows the semicausal MVR coefficients for $p = q = 1$ and $p = q = 2$.

Fig. 11 shows the covariance match achieved by the semicausal MVR model. Compared to single quadrant causal model for $p = q = 2$, we find a better covariance fit is provided by the semicausal model (see Fig. 13).

D. Noncausal Representations

The results for two-dimensional noncausal MVR's are analogous to the one dimensional results discussed in Section III-E. All the relevant equations for the two-dimensional case are given in the Appendix. The following conclusions are made.

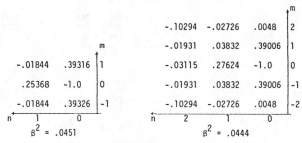

Actual covariances r(m,n), r(-m,n)=r(m,-n)=r(-m,-n) (rows: n ↑, columns: → m)

.905	.894	.868
.951	.932	.894
1.00	.951	.905

Causal Model Covariance Mismatch (rows: n ↑, columns: → m)

.209	.315	.389
.111	.231	.315
0.	.111	.208

Semicausal Model Covariance Mismatch (rows: n ↑, columns: → m)

.039	.041	.047
.019	.025	.036
0.	.017	.031

Fig. 13. Covariance mismatch of the causal and semicausal models.

1) An arbitrary positive SDF can be realized arbitrarily closely by a stationary finite-order noncausal MVR provided S^{-1} has a uniformly convergent Fourier series. Conversely, if the given SDF S is a rational function with a zeroth-order numerator polynomial, then there exists a unique finite-order noncausal MVR realization of S.

2) As in the case of causal and semicausal models, an admissible finite-order noncausal MVR with a specified spectral mismatch error could be identified via the finite-order block Toeplitz equation (82).

Example 9: Consider the SDF

$$S(z_1, z_2) = [1 - \alpha(z_1 + z_1^{-1} + z_2 + z_2^{-1})]^{-1}, \qquad 0 < \alpha < \tfrac{1}{4}. \tag{98}$$

Since S^{-1} is a two-dimensional polynomial, the *noncausal MVR* is

$$u_{i,j} = \alpha(u_{i+1,j} + u_{i-1,j} + u_{i,j+1} + u_{i,j-1}) + \epsilon_{i,j} \tag{99}$$

$$r_\epsilon(k, l) = \begin{cases} 1, & (k, l) = (0, 0) \\ -\alpha, & (k, l) = (\pm 1, 0), (0, \pm 1) \\ 0, & \text{otherwise.} \end{cases}$$

Now let us consider finding a semicausal MVR. Comparing S with (89) we see $q = 1$. The equation for $r_n(z)$ is obtained via (A11) and (98) as

$$[1 - \alpha(z_1 + z_1^{-1})]\, r_n(z_1) - \alpha[r_{n-1}(z_1) + r_{n+1}(z_1)] = \delta_{n,0}$$

which solves to give

$$r_n(z_1) = r_n(z_1^{-1}) = AC^{|n|}$$
$$A = 1/\sqrt{\alpha_0^2 - 4\alpha^2}$$
$$\alpha_0 = 1 - \alpha(z_1 + z_1^{-1})$$
$$C = [\alpha_0 - \sqrt{\alpha_0^2 - 4\alpha^2}\,]/2\alpha.$$

Using this in (A25) and (A26) for $q = 1$ and simplifying we obtain

$$\hat{a}_1 = \hat{a}_1^1 = -C(z_1^{-1})$$
$$h(z_1) = A^{-1}[1 - C(z_1)\, C(z_1^{-1})]^{-1} = \tfrac{1}{2}\{\alpha_0 + \sqrt{\alpha_0^2 - 4\alpha^2}\}$$
$$a_0 = a_0^1 = -\beta^2 h(z_1)$$
$$a_1 = \beta^2 h(z_1)\, C(z_1^{-1}) = \alpha\beta^2.$$

Note $h(z_1)$ is an irrational function of z_1. Assuming $0 < \alpha \ll 1$, we obtain the rational approximation

$$h(z_1) = 1 - \alpha(z_1 + z_1^{-1}) + 0(\alpha^2).$$

Ignoring $0(\alpha^2)$ terms we obtain

$$\beta^2 \simeq 1, \; a_0(z_1) = -1 + \alpha(z_1 + z_1^{-1}), \; a_1(z_1) = \alpha, \; S_\epsilon = -a_0(z_1)$$

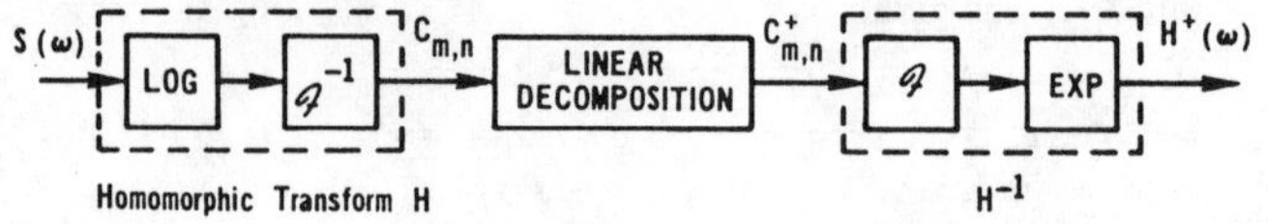

Fig. 14. Homomorphic transform method for two-dimensional spectral factorization.

and the finite-order semicausal MVR

$$u_{i,j} = \alpha(u_{i-1,j} + u_{i+1,j}) + \alpha u_{i,j-1} + \epsilon_{i,j} \tag{100}$$

$$r_\epsilon(k, l) = \delta_{l,0} \begin{cases} 1, & k = 0 \\ -\alpha, & |k| = 1 \\ 0, & \text{otherwise.} \end{cases}$$

For $0 < \alpha < \frac{1}{3}$, equation (100) can be shown to be a stable semicausal model. The SDF corresponding to this model is

$$S_u = \frac{[1 - \alpha(z_1 + z_1^{-1})]}{[1 - \alpha(z_1 + z_1^{-1}) - \alpha z_2^{-1}][1 - \alpha(z_1^{-1} + z_1) - \alpha z_2]}.$$

For $a = b = \alpha$ in Example 7, we see (100) and the model considered there are identical. A causal MVR approximation of (100) is obtained as discussed in that example.

E. Causal and Semicausal Model Realizations Via Homomorphic Transformation

The factorization required in the causal and semicausal models could be achieved via the so-called homomorphic transformation[3] shown in Fig. 14. One starts with the Fourier series of $\log S$ (the homomorphic transform of S) as

$$\tilde{S}(z_1, z_2) \triangleq \log S(z_1, z_2) = \sum_{m=-\infty}^{\infty} \sum_{n=-\infty}^{\infty} c_{m,n} z_1^{-m} z_2^{-n},$$
$$|z_1| = 1, \; |z_2| = 1 \tag{101}$$

where it is assumed the sequence $\{c_{m,n}\}$, also called the *Cepstrum*, is absolutely summable, i.e.,

$$\sum_m \sum_n |c_{m,n}| < \infty. \tag{102}$$

Now, if $\tilde{S}$ is decomposed as a sum of, say, three components as

$$\tilde{S} \triangleq \tilde{S}_\epsilon + \tilde{H}^+ + \tilde{H}^- \tag{103}$$

then

$$S = \frac{e^{\tilde{S}_\epsilon}}{[e^{-\tilde{H}^+} e^{-\tilde{H}^-}]} \triangleq S_\epsilon H^+ H^- \tag{104}$$

is a product of three factors. If the decomposition is such that

[3]Also called Wiener-Doob factorization [61], [62] or Hilbert transformation [56].

$H^+(z_1, z_2) = H^-(z_1^{-1}, z_2^{-1})$ and $S_\epsilon(z_1, z_2) = S_\epsilon(z_1^{-1}, z_2^{-1})$ then there exists a stable two-dimensional linear system with transfer function $H^+(z_1, z_2)$ such that the SDF of the output is S if the SDF of the input is S_ϵ.

This algorithm is due to Ekstrom and Woods [64] and is a direct extension in two dimensions of a method by Wiener [61]. The causal and the semicausal models can be realized by finding their respective decompositions in such a way that the impulse response $h^+_{m,n} = \mathcal{F}^{-1}\{H^+(\omega_1, \omega_2)\}$ has causal and semicausal regions of support $\hat{\mathcal{S}}_0 = \hat{\mathcal{S}} \cup (0, 0)$ where $\hat{\mathcal{S}}$ is defined in (74). We give the specific decompositions for causal MVR, semicausal WNDR, and semicausal MVR models.

Causal MVR:

$$\tilde{H}^+ = \sum_{m=1}^{\infty} c_{m,0} z_1^{-m} + \sum_{m=-\infty}^{\infty}\sum_{n=1}^{\infty} c_{m,n} z_1^{-m} z_2^{-n}$$

$$\tilde{H}^- = \sum_{m=-\infty}^{-1} c_{m,0} z_1^{-m} + \sum_{m=-\infty}^{\infty}\sum_{n=-\infty}^{-1} c_{m,n} z_1^{-m} z_2^{-n}$$

$$\tilde{S}_\epsilon = c_{0,0}. \qquad (105)$$

In view of (102), $\tilde{H}^+(z_1, z_2)$ is analytic in the region $\{|z_1| = 1, |z_2| \geqslant 1\} \cup \{|z_1| \geqslant 1, z_2 = \infty\}$. Hence $H^+(z_1, z_2)$ will be analytic in the same region. Therefore, the impulse response $h^+_{m,n}$ will be zero for $[-\infty \leqslant m \leqslant -1, n = 0] \cup [n < 0, \forall m]$, i.e., $\{h^+_{m,n}\}$ is causal.

Semicausal WNDR:

$$\tilde{H}^+ = \frac{1}{2}\sum_{m=-\infty}^{\infty} c_{m,0} z_1^{-m} + \sum_{m=-\infty}^{\infty}\sum_{n=1}^{\infty} c_{m,n} z_1^{-m} z_2^{-n}$$

$$\tilde{H}^- = \frac{1}{2}\sum_{m=-\infty}^{\infty} c_{m,0} z_1^{-m} + \sum_{m=-\infty}^{\infty}\sum_{n=-\infty}^{-1} c_{m,n} z_1^{-m} z_2^{-n}$$

$$\tilde{S}_\epsilon = 0. \qquad (106)$$

Now $\tilde{H}^+$ and hence H^+ will be analytic in the region $\{|z_1| = 1, |z_2| \geqslant 1\}$ and impulse response $h^+_{m,n}$ will be zero for $\{\forall m, n < 0\}$, i.e., $h^+_{m,n}$ is semicausal.

Semicausal MVR: A simple examination of (95)–(97) reveals that semicausal MVR's require that within a constant multiplier S_ϵ should equal $1/H^+(z_1, \infty)$. Using this condition, we obtain

$$\tilde{H}^+ = \sum_{m=-\infty}^{\infty} c_{m,0} z_1^{-m} + \sum_{m=-\infty}^{\infty}\sum_{n=1}^{\infty} c_{m,n} z_1^{-m} z_2^{-n}$$

$$\tilde{H}^- = \sum_{m=-\infty}^{\infty} c_{m,0} z_1^{-m} + \sum_{m=-\infty}^{\infty}\sum_{n=-\infty}^{-1} c_{m,n} z_1^{-m} z_2^{-n}$$

$$\tilde{S}_\epsilon = -\sum_{m=-\infty}^{\infty} c_{m,0} z_1^{-m}. \qquad (107)$$

The region of analyticity of $\tilde{H}^+$ and H^+ is $\{|z_1| = 1, |z_2| \geqslant 1\}$ and that of $\tilde{S}_\epsilon$ and S_ϵ is $\{|z_1| = 1, \forall z_2\}$. Hence $h^+_{m,n}$ is semicausal and S_ϵ is a SDF.

Example 10: Consider the SDF of Example 9. Assuming $0 < \alpha \ll 1$, we obtain

$$\tilde{S} \simeq \alpha(z_1 + z_1^{-1} + z_2 + z_2^{-1}) + 0(\alpha^2).$$

Ignoring $0(\alpha^2)$ terms we obtain

$$H^+ = \begin{cases} \alpha(z_1^{-1} + z_2^{-1}), & \text{causal MVR} \\ \frac{1}{2}\alpha(z_1 + z_1^{-1}) + \alpha z_2^{-1}, & \text{semicausal WNDR} \\ \alpha(z_1 + z_1^{-1}) + \alpha z_2^{-1}, & \text{semicausal MVR} \end{cases}$$

$$\tilde{S}_\epsilon = \begin{cases} 0, & \text{for causal MVR and semicausal WNDR} \\ -\alpha(z_1 + z^{-1}), & \text{for semicausal MVR.} \end{cases}$$

This gives

$$H^+ = 1/e^{-H^+} \simeq \begin{cases} 1/[1 - \alpha z_1^{-1} - \alpha z_2^{-1}], & \text{causal MVR} \\ 1/[1 - \frac{\alpha}{2}(z_1 + z_1^{-1}) - \alpha z_2^{-1}], & \text{semicausal WNDR} \\ 1/[1 - \alpha(z_1 + z_1^{-1}) - \alpha z_2^{-1}], & \text{semicausal MVR} \end{cases}$$

$$S_\epsilon \simeq \begin{cases} 1, & \text{causal MVR and semicausal WNDR} \\ 1 - (z_1 + z_1^{-1}), & \text{semicausal MVR.} \end{cases}$$

Remarks

1) We note that for noncausal models, one need not go through the above procedure. A suitable truncation of the Fourier series of S^{-1} yields the desired model.

2) In general, the Fourier series of $\tilde{S}$ and H^+ will contain an infinite number of terms even if S and $\tilde{H}^+$ (or its approximation) respectively were (finite order) polynomials. Therefore, a practical algorithm requires two stages of numerical approximations. First, in the evaluation of $\tilde{S}$ and then in the evaluation of H^+. An algorithm utilizing the DFT to obtain such an approximate spectral factorization and its error analysis has been considered by Rino [63] for one-dimensional problems. Extension of the algorithm in two dimensions is given in [64].

3) From the foregoing discussion the BIBO stability condition for finite-order causal and the semicausal models can be stated quite simply as follows.[4]

Theorem 4: A causal MVR whose transfer function H is given by

$$H(z_1, z_2) = 1/A(z_1, z_2)$$

$$A(z_1, z_2) = 1 - \sum_{m=1}^{p} a_{m,0} z_1^{-m} - \sum_{m=-p}^{p}\sum_{n=1}^{q} a_{m,n} z_1^{-m} z_2^{-n}$$

is stable if and only if i) $A(z_1, z_2) \neq 0$, $|z_1| \geqslant 1$, $z_2 = \infty$, ii) $A(z_1, z_2) \neq 0$, $|z_1| = 1$, $|z_2| \geqslant 1$.

Theorem 5: A semicausal WNDR or MVR whose transfer function H is given by

$$H = 1/A$$

$$A(z_1, z_2) = 1 - \sum_{\substack{m=-p \\ m\neq 0}}^{p} a_{m,0} z_1^{-m} - \dagger \sum_{m=-p}^{p}\sum_{n=1}^{q} a_{m,n} z_1^{-m} z_2^{-n}$$

[4]It is assumed that the causal, semicausal, and the noncausal model equations are solved, respectively, as initial, initial-boundary, and boundary value problems.

is stable if and only if $A(z_1, z_2) \neq 0$, $|z_1| = 1$, $|z_2| \geqslant 1$.

Theorem 6: A noncausal MVR or WNDR whose transfer function is given by

$$H = 1/A, \; A(z_1, z_2) = 1 - \sum_{m=-p}^{p} \sum_{n=-q}^{q} a_{m,n} z_1^{-m} z_2^{-n}$$

is stable if and only if $A(z_1, z_2) \neq 0$, $|z_1| = 1$, $|z_2| = 1$.

The conditions stated in these theorems assure H to be analytic in the appropriate regions in the z_1, z_2 hyperplanes so that its stability conditions are satisfied. Two-dimensional stability has been discussed almost invariably for causal models and occasionally for noncausal models. For details see [65]-[68].

F. Other Methods

Since exact factorization of a two-dimensional SDF $S(\omega_1, \omega_2)$ (rational or not) as the magnitude square of a rational function $H(\omega_1, \omega_2)$ is not possible in general, any number of other methods which could give reasonable rational approximations should be possible. The advantage of model identification via the prediction model orthogonality equations (Sections V-A to V-D) is that the equations to be solved are linear. The advantage of the homomorphic transform approach is that a suitable FFT (even if large in size) could be used to obtain the approximate models. In both procedures, however, situations exist where the model order may get very large in order to be stable.

In image processing applications it is generally desirable that the model order not be very large. This is because the image data is often processed block by block (often 16 X 16) and if the model order approaches the size of a column (or row), then nonrecursive or transform based methods become more efficient (even computationally). Therefore, identification techniques which give reasonable (low-) order models with assurance of stability would be useful in many applications such as restoration and data compression. High-order models, on the other hand, could be useful in image synthesis, object identification, spectral estimation, and digital filter design applications.

Among the various other methods for identifying image models, perhaps the most direct is the parameter identification method via the maximum-likelihood approach [69]. Sugimoto *et al.* [70], [71] have applied this method for identification of finite order semicausal models. Similar approaches are possible for causal and noncausal models although the algorithmic details would differ considerably.

Another method suggested in [51], [52] is based on the fact that the causal, semicausal, and noncausal models can also be considered as finite difference approximations of hyperbolic, parabolic, and elliptic partial differential equations, respectively. Hence, low-order stable models can be determined by identifying the parameters of the finite difference approximations subject to their stability constraints which are known from the theory of their parent partial differential equations.

Since the semicausal MVR and WNDR require only factorization in only one variable, one-dimensional system identification techniques [72] could potentially be applied for identifying these models. Once the semicausal MVR is known, the causal realization (or an approximation of it), if desired, could be found via the method of Section V-C.

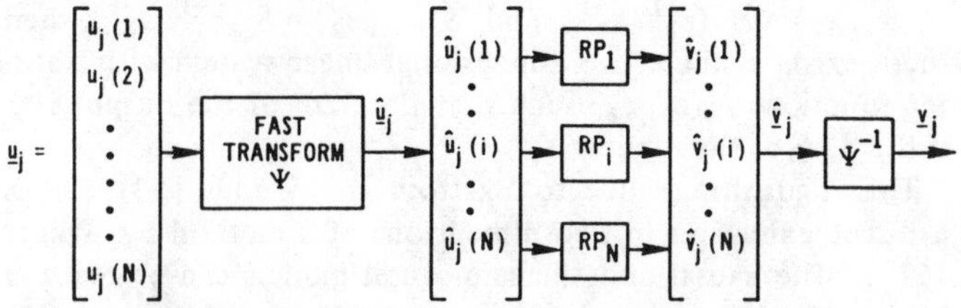

Fig. 15. Parallel structure of semicausal algorithms. RP_i is the ith channel recursive processor such as a recursive filter for image restoration or a DPCM encoder for data compression.

VI. Applications in Image Processing

The three types of models discussed in the foregoing have applications in most of the image processing problems listed in Table I. As mentioned before the choice of a particular model depends not only on the accuracy of the model but also on the associated algorithmic architecture. The three models yield three quite different processing techniques which will be described briefly.

A. Algorithmic Structures

The causal models can be easily viewed to represent images as the output of a line scanner and hence yield, quite naturally, algorithms which are recursive in nature. Causal MVR's have been found most useful in data compression of images [17] for real time transmission where the successive transmitted samples represent the prediction error between the new imagepel and its predicted value based on the past transmission. Causal MVR's have usefulness in the design of recursive filters for noise smoothing and restoration of blurred images (especially when the blurring process is also causal, e.g., in motion blur). It should be mentioned, however, that two-dimensional images generally do not have any causality (or dynamics) associated with them and the usual solutions of smoothing and most restoration problems are boundary valued (or noncausal). The recursive methods therefore have to sweep back and forth between the boundaries. This, together with the algorithmic complexities of two-dimensional recursions, have kept the users favor with more direct Fourier transform based algorithms [6], [7], [73], [75]. However, potential for these algorithms seems to be substantial for space variant restoration problems where Fourier techniques do not apply.

The semicausal models yield hybrid algorithms which are transform based (nonrecursive) in one dimension and recursive in the other. The finite-order semicausal MVR's have a structure where the model coefficients (e.g., matrices in (66)) be represented by banded Toeplitz matrices. It is possible to find a fast unitary sinusoidal transform [10] which would reduce these matrices to their diagonal forms (or nearly so). For example, the semicausal model of (92) reduces to a set of N decoupled first-order Markov processes

$$\lambda_k \hat{u}_j(k) = b\hat{u}_{j-1}(k) + \hat{\epsilon}_j(k), \qquad k = 1, \cdots, N \qquad (108)$$

under the transformation $\hat{u}_j = \Psi u_j$, $\Psi = \text{DST}$ (see Table III), where λ_k are the eigenvalues of Q. Now, several image processing problems, e.g., noise smoothing, data compression, restoration, system identification etc. can be formulated as N one-dimensional problems, which could be solved in parallel. Fig. 15 shows the parallel structure of semicausal algorithms. Each line u_j of the image is first unitarily transformed and each transformed element $\hat{u}_j(k)$ is processed recursively in the 'j' variable. The output vector $\hat{v}_j$ is inverse transformed to give the processed vector v_j.

Semicausal models have been used in real-time data compression of images for transmission from remotely piloted vehicles (RPV) [17], [58], [74], in restoration and noise smoothing problems [52], [55], [70], [71] spectral estimation [59], etc. The associated algorithms combine the advantages of low memory and hardware complexity of recursive methods and the relatively high performance of transform processing methods.

Noncausal models have been used most widely either as moving average models or via the transform methods. A large number of commonly used image processing operators, e.g., spatial smoothing, gradient and compass edge detectors, bilinear interpolation, unsharp masks, etc., are moving average operators. It has been shown that noncausal MVR's naturally lead to transform domain algorithms and have been found useful in most of the problems listed in Table I.

B. Smoothing, Enhancement, and Restoration of Images

The common smoothing problem is to find the best linear mean-square estimate of the image $u_{i,j}$ given the noisy observations

$$y_{i,j} = u_{i,j} + n_{i,j}$$

where $\{n_{i,j}\}$ is a white-noise field independent of $\{u_{i,j}\}$. Causal models have been used by several authors [52], [76]–[79] to develop recursive filter implementations. Semicausal and vector scanning models have been considered in [52], [55], [70], [71], [80] to develop semirecursive or line to line recursive algorithms. Noncausal models have been shown to yield fast transform based nonrecursive algorithms [52], [73] and have also been found useful in developing moving average FIR filters.

The following two examples show how some of the commonly used operators in image processing are related to the linear prediction models. Use of appropriate prediction models (depending on the images) can lead to similar but better operators.

Example 11: Consider an image represented by the noncausal MVR of (99). The optimum filter transfer function is given by another noncausal MVR as

$$G(z_1, z_2) = \beta^2 / \{\beta^2 + \sigma_n^2[1 - \alpha(z_1 + z_1^{-1} + z_2 + z_2^{-1})]\}$$

where σ_n^2 is the variance of the noise. At signal-to-noise ratios (SNR) of approximately 5, G can be approximated by the FIR filter [52]

$$G \simeq T(z_1, z_2) = \tfrac{1}{2}\,[1 + \tfrac{1}{4}\,(z_1 + z_1^{-1} + z_2 + z_2^{-1})]$$

which is simply a commonly used spatial averaging filter. At a different signal to noise ratios another suitable moving average approximation can be found whose region of support would increase with the reciprocal of SNR. In fact, moving average filters can be adapted (via the noncausal models) to local changes in SNR. Fig. 16 shows that a noisy (SNR = 5) image optimally filtered via an appropriate noncausal model or its approximate spatial averaging filter is better than using the optimal (Wiener) filter based on the usual separable covariance model [52].

Example 12 (Enhancement): The prediction operator (causal, semicausal, or noncausal) denoted by $A(z_1, z_2)$ generally performs some sort of differentiation. When applied to a real-world image, the prediction error generally contains a nonstationary component which generally represents high spatial frequencies, e.g., edges. Thus the operator

$$C(z_1, z_2) = 1 + \lambda A(z_1, z_2)$$

would add to the image a quantity proportional to high spatial frequencies (or gradient). For example, in the case of the noncausal MVR considered above, with $\alpha \simeq \frac{1}{4}$

$$C = (1 + \lambda) - \frac{\lambda}{4}(z_1 + z_1^{-1} + z_2 + z_2^{-1}).$$

Now the operator A is simply a discrete Laplacian and C is called the "*Unsharp Mask*" used often for edge enhancement of images.

Image models have also been used in restoration of images blurred due to motion, atmospheric turbulence and other shift invariant PSF. State variable, line recursive as well as frame-recursive models have been used to develop recursive algorithms for deblurring of images [60]. Noncausal models have been used to develop transform-domain deblurring algorithms. Other restoration techniques such as iterative gradient methods, singular value decomposition have mostly been used [7], [73], [75], [95] for solving deterministic (e.g., least squares interpolation) problems although image models could be used (especially to improve their performance in the presence of noise).

C. Image Data Compression

This is one application where image models have probably had the most significant impact. Causal prediction models have been used most widely in the design of intraframe and interframe predictive or the so-called DPCM coders. Semicausal models have been employed in transform/DPCM coding for RPV and teleconferencing applications. Noncausal models give rise to transform coding algorithms which have been found to give high performance. The prediction model parameters determine the coder design details such as quantizer design, bit allocation, optimum transform, etc. These details and other related considerations for image data compression may be found in the recent survey articles [17], [81] and in [16], [57].

Example 13: Consider the semicausal model of (92) when it is white noise driven. It reduces to (108) after sine transformation of the vectors u_j. Fig. 17 shows images encoded via this model at an average rate of 1 bit/pel. This model has also been used effectively to simultaneously filter and encode noisy images. For details see [17], [57], [58].

D. Edge Extraction

Example 14 (Edge Extraction): An image can be considered as being composed of two components, i.e.,

$$u_{i,j} = u_{i,j}^s + u_{i,j}^b$$

where $u_{i,j}^s$ and $u_{i,j}^b$ represent the stationary and nonstationary (boundaries and edges) components respectively. Such two source models for images have been considered by Schreiber, Yan and Sakrison, Jain and Wang, and others [82]–[84], [57]. Using a WNDR for the stationary component, the image can be expressed as

$$\epsilon_{i,j} = \epsilon_{i,j}^s + \epsilon_{i,j}^b$$

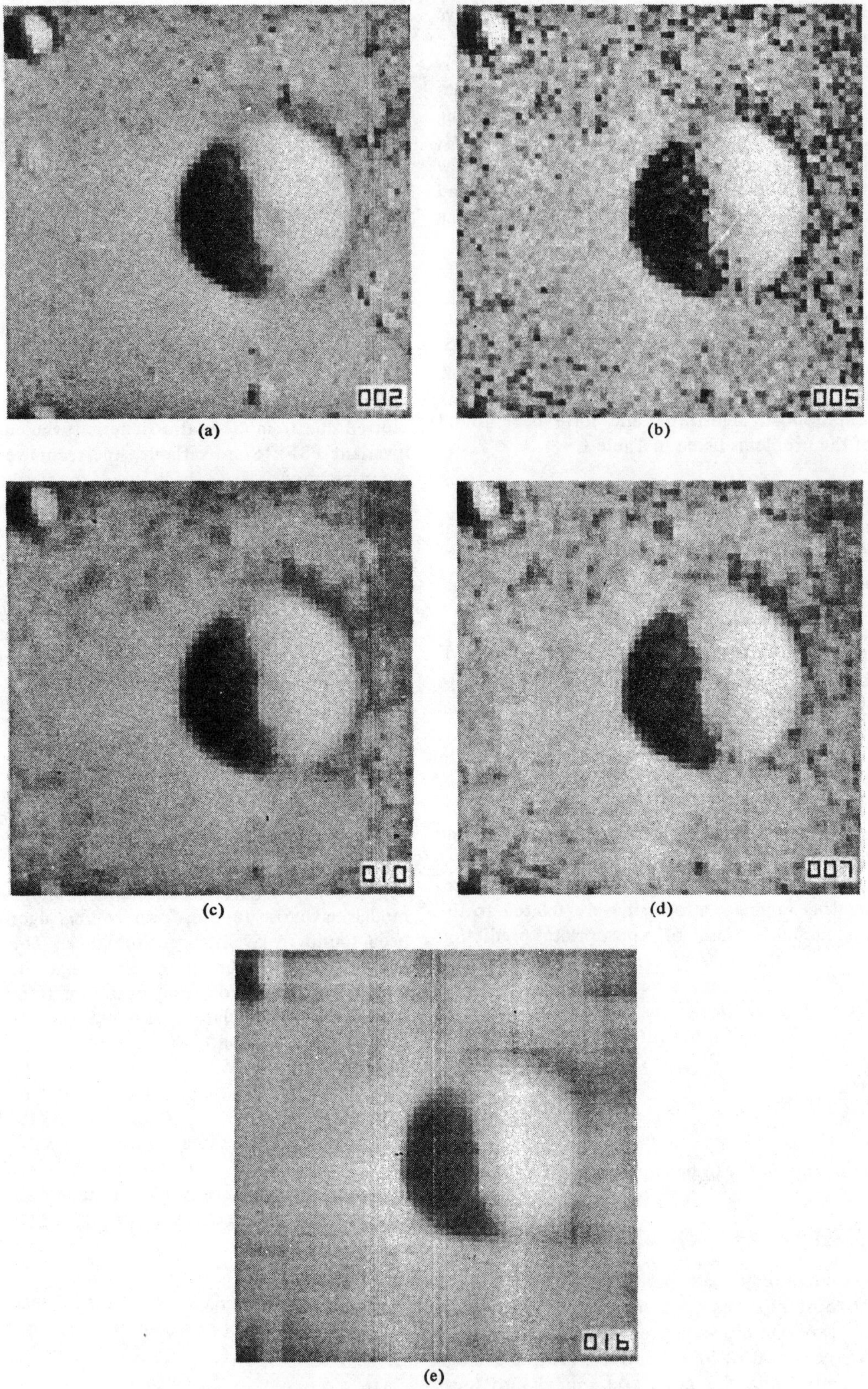

Fig. 16. Smoothing of a noisy image. (a) Original. (b) Noisy. (c) Optimum filtered using a noncausal model. (d) Spatial averaged. (e) Optimum filtered based on the separable covariance model.

where $\epsilon = \mathfrak{L}[u]$, and $\mathfrak{L}$ is the whitening prediction operator (causal, semicausal or noncausal). Now, $\epsilon_{i,j}$ can be viewed as a sum of signal $\epsilon_{i,j}^{b}$ (edges) and white noise $\epsilon_{i,j}^{s}$. The edge detection problem is therefore reduced to the detection of a signal in the presence of noise. Fig. 18 shows the edges detected using a noncausal model. Such two source models are useful in designing "edge preserving" image-processing algorithms.

(a)

(b)

(c)

Fig. 17. Data compression of images via a semicausal model. (a) Original. (b) Encoded. 1 bit/pel. (c) Adaptive codings, 1 bit/pel. In the adaptive method, the model parameters are updated after every 16 × 16 block of pels. Mean-square error improvement due to adaptation is in excess of 3 dB.

E. Two-Dimensional Spectral Estimation

Let $r(m, n)$ be a two-dimensional positive definite covariance sequence given on a rectangular window $W_c = [-p_c, p_c] \times [-q_c, q_c]$. The problem is to find an estimate of the SDF

(a)

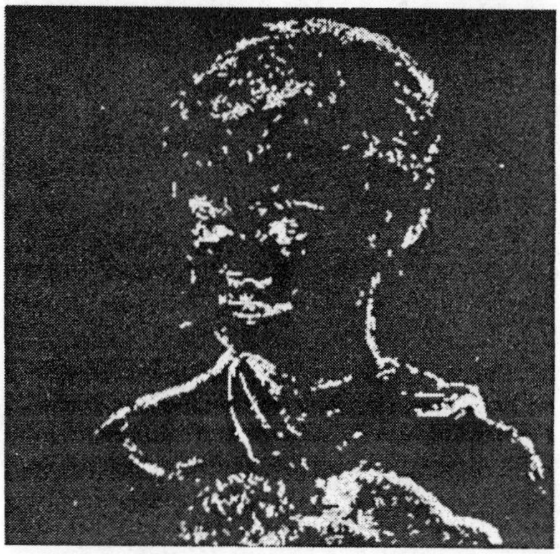

(b)

Fig. 18. Edges extracted using a noncausal model.

associated with this covariance sequence. In two dimensions, the maximum entropy spectral estimation problem is to *find a positive definite extension of* $r(m, n)$ *outside* W_c *such that the entropy*

$$H = \frac{1}{4\pi^2} \int_{-\pi}^{\pi} \int_{-\pi}^{\pi} \log S(\omega_1, \omega_2)\, d\omega_1\, d\omega_2$$

is maximized. Note that it is required that the Fourier coefficients of the estimated spectrum S match the given covariances exactly on W_c. The solution, if it exists, requires that the Fourier series of $1/S$ should be truncated outside the window W_c, i.e., S should be the SDF of a noncausal MVR, i.e., of the form

$$S(z_1, z_2) = \beta^2 \left[\sum_{m,n} \sum_{\in W_c} \alpha_{m,n} z_1^{-m} z_2^{-n} \right]^{-1}, \qquad |z_1| = 1, \quad |z_2| = 1 \quad (109)$$

where the coefficients $\alpha_{m,n}$, β^2 are determined by matching the given covariances to the Fourier coefficients of S on W_c. Unfortunately, this maximum entropy solution need not exist because, in two (and higher) dimensions, a positive definite sequence on a rectangular window need not have a positive definite extension [85], [86]. However, if $r(m, n)$ has at least one positive definite extension then the maximum entropy solution (109) exists and is unique [94]. When the solution exists, the problem of finding $\alpha_{m,n}$ is nonlinear and cannot be, in general, reduced to a linear problem as in the one-dimensional case. This is because of the spectral factorization difficulty mentioned earlier.

However, it is possible to obtain positive SDF estimates if we relax the condition of exact covariance match on W_c. This is motivated by the fact that often the available covariances are estimated from data and are noisy. The foregoing finite-order causal, semicausal, and noncausal MVR's could be used to estimate the SDF. The algorithm simply requires solving (80) for $(k, l) \in W_0$ or equivalently (81) and (82) where W_0 is a subset of $\hat{\mathcal{S}}_0$ and could be a causal, semicausal, or noncausal prediction filter window. Note that the W_0 and W_c are not of equal size. For example, for a causal MVR, if $W_0 = [0, p] \times [0, q]$ then the covariance sequence is needed over $W_c = [-p, p] \times [-q, q]$. After having solved (80), the estimated SDF (if admissible) is given as follows.

Causal MVR:

$$S(z_1, z_2) = \beta^2 \Big/ \left| 1 - \sum\sum_{m,n \in W} a_{m,n} z_1^{-m} z_2^{-n} \right|^2, \qquad |z_1| = 1, \quad |z_2| = 1. \tag{110}$$

Semicausal MVR:

$$W_0 = [-p, p] \times [0, q], \qquad W_c = [-2p, 2p] \times [-q, q], \qquad W = W_0 - (0, 0)$$

$$S(z_1, z_2) = S_\epsilon(z_1, z_2) \Big/ \left| 1 - \sum\sum_{m,n \in W} a_{m,n} z_1^{-m} z_2^{-n} \right|^2, \qquad |z_1| = 1, \quad |z_2| = 1$$

$$S_\epsilon(z_1, z_2) = \beta^2 \left[1 - \sum_{m=1}^{p} a_{m,0} (z_1^{-m} + z_1^{m}) \right]. \tag{111}$$

Noncausal MVR:

$$W_0 = [-p, p] \times [-q, q],$$

$$W_c = [-2p, 2p] \times [-2q \times 2q], \qquad W = W_0 - (0, 0)$$

$$S(z_1, z_2) = \beta^2 \left[1 - \sum\sum_{m,n \in W} a_{m,n} z_1^{-m} z_2^{-n} \right]^{-1}, \qquad |z_1| = 1, \quad |z_2| = 1. \tag{112}$$

It should be noted that the solution of (81) and (82), while it guarantees admissible predictor coefficients, the estimated spectrum may not always be positive (i.e., admissible) for any arbitrary positive definite $\mathcal{R}$. This is because the solution yields coefficients which satisfy (80) only partially, i.e., for $k, l \in W_0$ (and not necessarily for all $k, l \in \hat{\mathcal{S}}_0$). However, for a large enough W_0 and covariances that come from an analytic SDF, an admissible SDF can be found. In many practical examples, the window size W_0 has been found to be not large and in many cases it is actually quite small for admissible SDF.

Example 15: Consider the covariance sequence

$$r(m, n) = \frac{\cos 2\pi(m + n)}{8} + \frac{\cos 2\pi(m + n)}{12} + 0.05\delta_{m,0}\delta_{n,0}$$

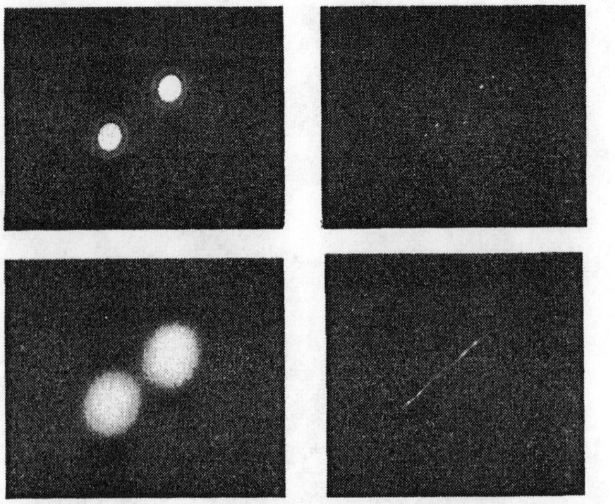

Fig. 19. Spectral estimation comparison of windowed DFT versus semicausal MVR methods. Top row; from left windowed DFT spectrum amplitude versus semicausal MVR spectrum amplitude. Bottom row; same quantities shown on dB scale.

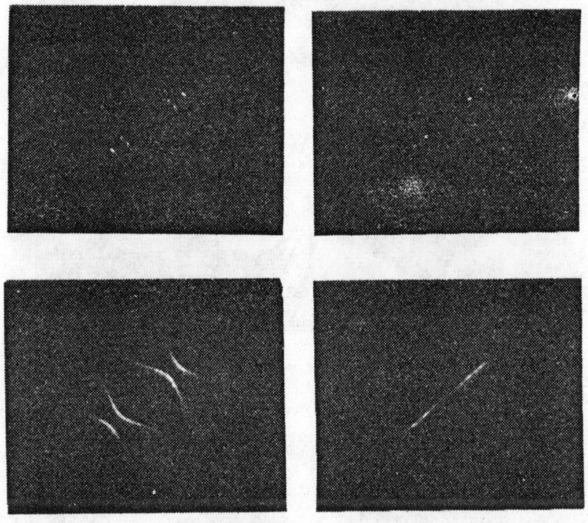

Fig. 20. Spectral estimation comparison of causal versus semicausal models. Top row from left; model spectrum amplitude versus semicausal model spectrum amplitude. Bottom row; same quantities shown on dB scale.

whose SDF has line spectra at $\pm(\frac{1}{8}, \frac{1}{8})$ and $\pm(\frac{1}{12}, \frac{1}{12})$ in the frequency plane. The covariances are available over $[-4, 4] \times [-2, 2]$. Figs. 19 and 20 shows the spectra estimated by a) Bartlett windowing and taking DFT, b) single quadrant causal MVR for $p = q = 5$, c) semicausal MVR, $p = 3, q = 4$. Note that objects not resolved by the DFT method are resolved by the causal and semicausal estimates. Spurious peaks appear in the causal model and the covariance match over the window W_c is not very good. The semicausal estimate is much more accurate both in terms of resolution, peak location and its covariance match. For other examples and more discussion see [59].

Remarks:

1) The causal and noncausal MVR spectra (110) and (112) are of the same form as the maximum entropy spectrum (109). However, they would not be equal in general. If the given covariance sequence is separable, then the maximum entropy and the causal, the semicausal and noncausal spectra would be identical.

2) The semicausal MVR spectra is ARMA type and is therefore not a maximum entropy spectra.

3) All the equations for calculating the prediction model coefficients are linear.

4) As expected from the theory of these two-dimensional prediction models, experimentally, it has been observed that the covariance match on the window W_c improves as p, q are increased.

F. Extension to Adaptive Models and Other Applications

Other applications of image models are in adaptive processing where the model parameters are updated with the changes in image properties—Fig. 17, for example, shows the result of adapting the semicausal model of (108) block by block. In general adaptations could be made continuously and recursively from pel to pel, as is done in many DPCM techniques for image coding. However, it has been observed that models which are adapted at every pel tend to be sensitive to noise and rarely show any significant improvement over more practical block by block adaptive models.

Applications also exist in identification, recognition and synthesis of images of texture. Image models also provide a priori information concerning performance bounds such as achievable signal to noise ratio, compression versus distortion, etc., [16], [52].

VII. Summary and Conclusions

We have presented several mathematical models which have been used and are of potential use in image processing. Included in our discussion were several traditional models (e.g., series expansion, one dimensional AR, state variable etc.) and some new models (viz. the causal, semicausal and noncausal prediction models). Not included in our discussion were considerations of the Markovian property, stability tests, controllability, observability and other such notions. Also, not considered were models for vector random fields which have potential applications in multispectral and color image processing. Theoretical results for these and related topics can be found in [66], [87]–[92]. It was shown however, that the models considered do have a large number of applications in image (and two-dimensional signal) processing. Among the prediction models considered, the semicausal and noncausal models seem to offer several algorithmic and performance tradeoffs against the purely causal (or recursive) methods.

Appendix

A.1 Causal Minimum Variance Representations

First we consider the general nonstationary case for arbitrary random fields. Let us define a block matrix $\mathcal{R}$ whose elements are the matrices $R_{j,k}$ which are the cross covariances $E u_j u_k^T$. Then $\mathcal{R}$ is the covariance matrix of the column ordered array vector $\boldsymbol{u}$. Let $\{\boldsymbol{u}(k)\}$, $\{\bar{\boldsymbol{u}}(k)\}$ and $\{\boldsymbol{\epsilon}(k)\}$ be the column ordered sequences corresponding to the arrays $\{u_{i,j}\}$, $\{\bar{u}_{i,j}\}$ and $\{\epsilon_{i,j}\}$, respectively. From our definition of causality, we can write

$$\boldsymbol{u}(k) = \bar{\boldsymbol{u}}(k) + \boldsymbol{\epsilon}(k).$$

The orthogonality condition (69) requires that $\boldsymbol{\epsilon}(k)$ be orthogonal to $\boldsymbol{u}(l)$ for $\forall l < k$. This means $\{\boldsymbol{\epsilon}(k)\}$ must be a white-noise sequence because

$$E\boldsymbol{\epsilon}(l)\,\boldsymbol{\epsilon}(k) = E[\boldsymbol{u}(l) - \bar{\boldsymbol{u}}(l)]\,\boldsymbol{\epsilon}(k) = 0, \qquad \forall l < k.$$

Since $E\boldsymbol{\epsilon}(l)\boldsymbol{\epsilon}(k) = E\boldsymbol{\epsilon}(k)\boldsymbol{\epsilon}(l)$, the above is true also for $k < l$. Thus *causal MVR's are also white-noise driven representations*. If $\beta_{i,j}^2$ denotes the variance of $\epsilon_{i,j}$, we have

$$E\epsilon_{i,j}\epsilon_{i+m,j+n} = \beta_{i,j,n}^2 \delta_{m,0}\delta_{n,0}. \tag{A1}$$

In matrix notation, equations (64) and (71) become

$$-A_{j,j}u_j = \sum_{n<j} A_{j,n}u_n + \epsilon_j, \qquad \forall j$$

where

$$-A_{j,j} \triangleq I - A_{j,j}^0 \tag{A2}$$

is a unit lower triangular matrix (i.e., the diagonal entries are 1). This equation can also be written in terms of a *block lower triangular* matrix $\mathcal{A}$ whose block elements are the matrices $A_{j,n}$, as

$$-\mathcal{A}\boldsymbol{u} = \boldsymbol{\epsilon} \qquad \mathcal{A} = \{A_{j,n}\}. \tag{A3}$$

Since the diagonal blocks of $\mathcal{A}$ are $\{A_{j,j}\}$, which are lower triangular matrices themselves, the matrix $\mathcal{A}$ is a strictly lower triangular matrix. Equation (A3) implies

$$\mathcal{A}[E\boldsymbol{u}\boldsymbol{u}^T]\mathcal{A}^T = E\boldsymbol{\epsilon}\boldsymbol{\epsilon}^T \quad \text{or} \quad \mathcal{R} = \mathcal{A}^{-1}\mathcal{R}_{\epsilon}(\mathcal{A}^{-1})^T$$

$$\text{or} \quad \mathcal{R}^{-1} = \mathcal{A}^T\mathcal{R}_{\epsilon}^{-1}\mathcal{A} \tag{A4}$$

where $\mathcal{R}_{\epsilon}$ is a diagonal matrix of the variances of $\boldsymbol{\epsilon}(k)$. This is a *lower-upper* factorization of $\mathcal{R}$. Thus, the causal MVR requires a factorization of the block covariance matrix $\mathcal{R}$ by a strictly lower triangular matrix. For a positive definite $\mathcal{R}$, such a factorization always exists and can be obtained in two stages of factorization. The first stage requires a block-lower triangular factorization of $\mathcal{R}$. The second stage requires lower triangular factorization of a sequence of matrix blocks. For example, for $N \times N$ images, define

$$\hat{A}_{j,n} \triangleq A_{j,j}^{-1}A_{j,n} \qquad R_{n,k} \triangleq Eu_n u_k^T. \tag{A5}$$

Then the desired solution of (A4) is obtained by first solving for $\hat{A}_{j,n}$ the so called normal equation

$$\sum_{n=1}^{j-1} \hat{A}_{j,n}R_{n,k} = -R_{j,k}, \qquad 1 \leq k \leq j-1, \qquad j = 2, \cdots, N \tag{A6}$$

which is a set of $(j-1) \times (j-1)$ block matrix equations.

Given $\{\hat{A}_{j,n}, n \leq j\}$, the unit lower triangular matrix $A_{j,j}$ is obtained by the lower–upper factorization

$$\left[\sum_{n=1}^{j} \hat{A}_{j,n}R_{n,j}\right]^{-1} = A_{j,j}^T(R_j^{\epsilon})^{-1}A_{j,j} \tag{A7}$$

where R_j^{ϵ} is the covariance matrix of the prediction error vector $\boldsymbol{\epsilon}_j$ and is diagonal. Once $A_{j,j}$ is known, $A_{j,n}$ is obtained via (A5). In practice the above algorithm will be quite tedious because the size of $\mathcal{R}$ is $N^2 \times N^2$ and the number of operations would be $0(N^6)$.

In the case of stationary models defined on the infinite plane if one starts with (73), a parallel result corresponding to (A1), (A5)–(A7), is obtained as

$$-A_0u_j = \sum_{n=1}^{\infty} A_n u_{j-n} + \boldsymbol{\epsilon}_j \qquad E\boldsymbol{\epsilon}_j\boldsymbol{\epsilon}_k^T = \beta^2 I\delta_{j,k} \tag{A8}$$

and

$$-\sum_{n=1}^{q} \hat{A}_n^q R_{n-j} = R_{-j}, \quad 1 \leq j \leq q, \quad q = 1, 2, \cdots$$

$$\left[\sum_{n=1}^{q} \hat{A}_n^q R_n\right]^{-1} = \frac{1}{\beta^2}[A_0^q]^T[A_0^q], \quad R_n \triangleq E u_j u_{j+n}^T \tag{A9}$$

$$A_n^q \triangleq A_0 \hat{A}_n^q, \quad A_n = \lim_{q\to\infty} A_n^q, \quad \forall n \geq 0 \tag{A10}$$

where A_0 is a unit lower triangular matrix and $\hat{A}_n^q$, A_n^q, R_n, etc., are all doubly infinite Toeplitz matrices. If the given SDF $S(\omega_1, \omega_2)$ is positive and analytic, then following [56], it could be shown that there exists a unique[5] causal realization such as (A8). Moreover, the coefficients A_n could be obtained by solving the sequence of normal equations of (A9) and taking the limit $q \to \infty$ as in (A10). In terms of Z_1-transforms

$$r_n(z_1) \triangleq \sum_{m=-\infty}^{\infty} r(m, n) z_1^{-m}$$

$$a_n(z_1) \triangleq \sum_{m=-\infty}^{\infty} a_{m,n} z_1^{-m}, \quad n \geq 1$$

$$a_0(z_1) \triangleq \sum_{m=0}^{\infty} a_{m,0} z_1^{-m}, \quad \text{etc.} \tag{A11}$$

Equations (A9), (A10) reduce to scalar qth-order AR model equations, parametric in z_1

$$-\sum_{n=1}^{q} \hat{a}_n^q(z_1) r_{n-j}(z_1) = r_j(z_1^{-1}), \quad j = 1, \cdots, q \tag{A12}$$

where

$$-\sum_{n=0}^{q} \hat{a}_n^q(z_1) r_n(z) = \beta^2/[a_0^q(z_1^{-1}) a_0^q(z_1)]$$

$$a_n^q(z_1) = a_0^q(z_1)\hat{a}_n^q(z_1), \quad \hat{a}_0^q(z_1) = 1, \quad n = 1, \cdots q \tag{A13}$$

where $a_0^q(z_1)$ is such that all its roots are inside the unit circle $|z_1| = 1$. If the SDF is a rational function of the form of (89), these equations can be used to obtain the causal MVR for the given value of q. In general this representation will be of infinite order. A finite-order approximation may be obtained as explained in Section V-B.

A.2 Semicausal Minimum Variance Representation

For semicausal models the orthogonality conditions require that for $\forall k < j$, the prediction error vector $\boldsymbol{\epsilon}_j$ be orthogonal to u_k, and at $k = j$, the elements $\epsilon_j(i)$ be orthogonal to $u_j(m)$, $\forall m \neq i$. These conditions yield the representation equation

$$-A_{j,j} u_j = \sum_{n<j} A_{j,n} u_n + \boldsymbol{\epsilon}_j, \quad A_{j,j} \triangleq -I + A_{j,j}^0, \quad \forall j$$

$$E u_k \boldsymbol{\epsilon}_j^T = B_j \delta_{j,k}, \quad k \leq j$$

$$E \boldsymbol{\epsilon}_k \boldsymbol{\epsilon}_j^T = E \boldsymbol{\epsilon}_j \boldsymbol{\epsilon}_k^T = -A_{j,j} B_j \delta_{j,k}, \quad \forall j, k \tag{A14}$$

where

$$B_j = \text{Diag}\,\{E\epsilon_{i,j}^2\} \triangleq \text{Diag}\,\{\beta_{i,j}^2\}.$$

From above $-A_{j,j}B_j$ is a covariance matrix and must therefore be symmetric. Since B_j is diagonal, $A_{j,j}$ could not be lower or upper triangular.[6] Also, $\{\boldsymbol{\epsilon}_j\}$ is a white-noise vector sequence. However, $\{\epsilon_{i,j}\}$ need not be a white-noise field because the elements of $\boldsymbol{\epsilon}_j$ could be mutually correlated. Recall that in the case of causal MVR's, $\{\epsilon_{i,j}\}$ was a white-noise field. Now (A14) can be written as a block lower triangular matrix equation

$$-\mathcal{A}\boldsymbol{u} = \boldsymbol{\epsilon} \qquad \mathcal{A} = \{A_{j,n}\} \tag{A15}$$

where $A_{j,j}$ is no longer lower triangular (compare with (A3)). Once again the $\mathcal{A}$ satisfies an equation similar to (A4), the difference being $\mathcal{R}_{\boldsymbol{\epsilon}}$ would now be block diagonal and $\mathcal{A}$ is block lower triangular (rather than also being lower triangular). Now, the matrices $A_{j,n}$ could be found by a single stage lower-upper factorization by solving

$$-\sum_{n=1}^{j} A_{j,n} R_{n,k} = B_j \delta_{j,k}, \quad k \leq j. \tag{A16}$$

The solution is given by

$$A_{j,n} = A_{j,j}\hat{A}_{j,n}$$

$$A_{j,j} = -B_j H_j \quad H_j \triangleq \left[\sum_{n=1}^{j} \hat{A}_{j,n} R_{n,j}\right]^{-1}$$

$$B_j(i,i) \triangleq \beta_{i,j}^2 = 1/H_j(i,i) \tag{A17}$$

where $\hat{A}_{j,n}$ is the solution of (A6).

It is now worth noting that a causal MVR could always be obtained from a semicausal MVR whenever the covariance matrix $R_j^\epsilon = E\boldsymbol{\epsilon}_j\boldsymbol{\epsilon}_j^T = -A_{j,j}B_j$ is positive definite so that it has a factorization

$$R_j^\epsilon = L_j^T \boldsymbol{\Gamma}_j^{-1} L_j \tag{A18}$$

where $\boldsymbol{\Gamma}_j = \{\gamma_j(k)\}$, is a diagonal matrix and L_j is a unit-lower triangular matrix. Then the first equation in (A14) can be written as

$$-\tilde{A}_{j,j} u_j = \sum_{n<j} \tilde{A}_{j,n} u_n + \boldsymbol{\nu}_j \tag{A19}$$

where

$$\tilde{A}_{j,j} \triangleq B_j L_j B_j^{-1} \qquad \tilde{A}_{j,n} \triangleq B_j \boldsymbol{\Gamma}_j (L_j^T)^{-1} A_{j,n}$$

$$\boldsymbol{\nu}_j \triangleq B_j \boldsymbol{\Gamma}_j (L_j^T)^{-1} \boldsymbol{\epsilon}_j. \tag{A20}$$

From these it follows that $\tilde{A}_{j,j}$ is also a unit-lower triangular matrix and $\{\nu_{i,j}\}$ is a white-noise random field since

$$R_j^\nu = E\boldsymbol{\nu}_j\boldsymbol{\nu}_j^T = B_j \boldsymbol{\Gamma}_j B_j \tag{A21}$$

is a diagonal matrix. Comparison of (A19) with (A2) shows that it is now a causal MVR. Given a semicausal MVR, a causal MVR is guaranteed provided the lower-upper factorization of (A18) exists, i.e., if R_j^ϵ is positive definite. However, it may not always be desirable to get to a causal representation because the spatial structure of this representation may be-

[5] This means the infinite predictor coefficients $a_{m,n}$ and the covariance sequence $r(m, n)$ are unique with respect to one another.

[6] Except when $A_{j,j}$ is diagonal. In that case however, equation (A14) will reduce to a causal model and would not be admissible here.

come more cumbersome than the semicausal case as shown by Example 7 in Section V-C.

In the case of stationary semicausal MVR's defined on the infinite plane the equations corresponding to (A16) and (A17) become

$$-A_0 u_j = \sum_{n=1}^{\infty} A_n u_{j-n} + \varepsilon_j \qquad E\varepsilon_j \varepsilon_k^T = -\beta^2 A_0 \delta_{j,k} \quad \text{(A22)}$$

$$-\sum_{n=1}^{q} \hat{A}_n^q R_{n-j} = R_{-j}, \qquad 1 \leqslant j \leqslant q, \qquad q = 1, 2, \cdots$$

$$-A_0^q = \beta^2 \left[R_0 + \sum_{n=1}^{q} \hat{A}_n^q R_n \right]^{-1}$$

$$A_n^q = A_0^q \hat{A}_n^q, \qquad \lim_{q \to \infty} A_n^q = A_n \quad \text{(A23)}$$

where $-A_0$ now is a symmetric, positive definite Toeplitz matrix. Written in terms of Z_1-transform variables defined in (A11) with

$$a_0(z_1) = \sum_{m=-\infty}^{\infty} a_{m,0} z_1^{-m} \quad \text{(A24)}$$

we obtain, again, the scalar qth-order AR model equations

$$-\sum_{n=1}^{q} \hat{a}_n^q(z_1) r_{n-j}(z_1) = r_j(z_1^{-1}), \qquad j = 1, \cdots, q \quad \text{(A25)}$$

and

$$-a_0^q(z_1) = \beta^2 h(z_1), \quad h(z_1) \triangleq \left[r_0(z_1) + \sum_{n=1}^{q} \hat{a}_n^q(z_1) r_n(z_1) \right]^{-1}$$

$$\beta^2 = 1 \Bigg/ \left(\frac{1}{2\pi} \int_{-\pi}^{\pi} h(z_1)\, d\omega \right), \qquad z_1 = e^{j\omega},$$

$$S_\epsilon(z_1, z_2) = -\beta^2 a_0^q(z_1), \qquad a_n^q(z_1) = a_0^q(z_1) \hat{a}_n^q(z_1),$$

$$a_n(z_1) = \lim_{q \to \infty} a_n^q(z_1). \quad \text{(A26)}$$

For a positive and analytic SDF, a unique semicausal MVR such as (A22) can be shown to exist and its coefficients could be identified via (A23) or equivalently via (A25) and (A26).

A.3 Noncausal Representations

For noncausal MVR's, the orthogonality condition gives

$$r(i,j;p,q) - \sum\sum_{(m,n) \neq (i,j)} a(i,j;m,n) r(m,n;p,q) = \beta_{i,j}^2 \delta_{i,p} \delta_{j,q}.$$

Define

$$a(i,j;i,j) = -1, \; A_{j,n} = \{a(i,j;m,n), \forall i,m\}, \; B_j = \text{Diag}\, \{\beta_{i,j}^2, \forall i\}.$$

Using these we obtain the matrix equations similar to (50)

$$-\sum_n A_{j,n} R_{n,q} = B_j \delta_{j,q}, \qquad \text{or} \qquad -\mathcal{A}\mathcal{R} = \mathcal{B},$$

$$\text{or} \qquad -\mathcal{A} = \mathcal{B}\mathcal{R}^{-1} \quad \text{(A27)}$$

where $\mathcal{A}$, $\mathcal{R}$, and $\mathcal{B}$ are block matrices defined as

$$\mathcal{A} = \{A_{j,n}, \forall j,n\}, \qquad \mathcal{R} = \{R_{j,n}, \forall j,n\}, \qquad \mathcal{B} = \text{Diag}\, \{B_j, \forall j\}.$$

From this, we can obtain, as in (54)

$$\beta_{i,j}^2 = 1/[(\mathcal{R}^{-1})_{i,i}]_{j,j}. \quad \text{(A28)}$$

Thus $\beta_{i,j}^2$ is obtained by finding the jth diagonal block of $\mathcal{R}^{-1}$. Then the ith diagonal term of this matrix block equals $\beta_{i,j}^{-2}$. $\mathcal{A}$ is now obtained directly from (A27).

The orthogonality condition also requires

$$Eu_{i,j}\epsilon_{p,q} = \beta_{i,j}^2 \delta_{i,p} \delta_{j,q}, \qquad \forall i,j$$

which gives

$$E\epsilon_{i,j}\epsilon_{p,q} = r_\epsilon(i,p;j,q) = \begin{cases} \beta_{i,j}^2, & (i,j) = (p,q) \\ -a(i,j;p,q)\beta_{i,j}^2, & (i,j) \neq (p,q). \end{cases} \quad \text{(A29)}$$

Hence, the noncausal MVR of an arbitrary Gaussian field $\{u_{i,j}\}$ is given by

$$u_{i,j} = \sum\sum_{(m,n) \neq (i,j)} a(i,j;m,n) u_{m,n} + \epsilon_{i,j} \quad \text{(A30)}$$

where $\{\epsilon_{i,j}\}$ is a zero-mean *nonwhite* Gaussian random field whose covariances are given by (A29). For stationary random fields defined on the infinite plane, the two-dimensional noncausal MVR is obtained by a direct extension of the one-dimensional result (see (55) and (56)) as

$$r_{0,0}^+ u_{i,j} + \sum_{\substack{k=-\infty \\ (k,l) \neq (0,0)}}^{\infty} \sum_{l=-\infty}^{\infty} r_{k,l}^+ u_{i-k,j-l} = r_{0,0}^+ \epsilon_{i,j} \quad \text{(A31)}$$

$$r_{k,l}^+ = \frac{1}{2\pi} \int_{-\pi}^{\pi} S^{-1}(\omega_1, \omega_2) \exp\{j(k\omega_1 + l\omega_2)\}\, d\omega_1\, d\omega_2, \qquad \forall k,l \quad \text{(A32)}$$

$$E\epsilon_{i,j}\epsilon_{i+k,j+l} = r_\epsilon(k,l) = \begin{cases} 1/r_{0,0}^+, & (k,l) = (0,0) \\ r_{k,l}^+/(r_{0,0}^+)^2, & (k,l) \neq (0,0). \end{cases} \quad \text{(A33)}$$

A rational approximation of $S^{-1}(z_1, z_2)$ will yield a finite-order (approximate) realization. Clearly factorization of S is not required in the case of noncausal MVR's. Thus if S^{-1} is a two-dimensional polynomial

$$S^{-1}(z_1, z_2) = \sum_{k=-p}^{p} \sum_{l=-q}^{q} r_{k,l}^+ z_1^{-p} z_2^{-q} \quad \text{(A34)}$$

then the noncausal MVR is

$$\sum_{k=-p}^{p} \sum_{l=-q}^{q} r_{k,l}^+ u_{i-k,j-l} = r_{0,0}^+ \epsilon_{i,j} \quad \text{(A35)}$$

where $\{\epsilon_{i,j}\}$ is a zero mean moving average field with SDF

$$S_\epsilon(z_1, z_2) = \left(\sum_{k=-p}^{p} \sum_{l=-q}^{q} r_{k,l}^+ z_1^{-p} z_2^{-q} \right) \Bigg/ (r_{0,0}^+)^2. \quad \text{(A36)}$$

REFERENCES

[1] A. K. Jain, *Multidimensional Techniques in Digital Image Processing*. To be published.

[2] W. K. Pratt, *Digital Image Processing*. New York: Wiley, 1978.

[3] N. Ahmed and K. R. Rao, *Orthogonal Transforms for Digital Signal Processing*. New York: Springer Verlag, 1975.

[4] E. O. Brigham, *The Fast Fourier Transform*. Englewood Cliffs, NJ: Prentice-Hall, 1974.

[5] H. F. Harmuth, *Transmission of Information by Orthogonal Signals*. New York: Springer Verlag, 1970.

[6] H. C. Andrews, *Computer Techniques in Image Processing*. New York: Academic Press, 1970.

[7] H. C. Andrews and B. R. Hunt, *Digital Image Restoration*. Englewood Cliffs, NJ: Prentice-Hall, 1977.

[8] A. Rosenfeld and A. Kak, *Digital Picture Processing*. New York: Academic Press, 1976.

[9] A. K. Jain, "Some new techniques in image processing," in *Proc. Symp. Current Math Problems in Image Science* (Naval Post Graduate School, Monterey, CA). North Hollywood, CA: Western Periodicals Co., Nov. 1976.

[10] A. K. Jain, "A sinusoidal family of unitary transforms," *IEEE Trans. Pattern Anal. Mach. Intelligence*, vol. PAMI-1, pp. 356–365, Oct. 1979.

[11] H. Hotelling, "Analysis of a complex of statistical variables into principal components," *J. Educ. Psychol.*, vol. 24, pp. 417–441, and 498–520, 1933.

[12] S. Watanabe, "Karhunen-Loeve expansion and factor analysis, theoretical remarks and applications," *Trans. 4th Prague Conf. Inform. Theory, Statist. Decision Functions, and Random Processes* (Prague, Czechoslovakia) pp. 635–660, 1965.

[13] H. P. Kramer and M. V. Mathews, "A linear coding for transmitting a set of correlated signals," *IRE Trans. Inform. Theory*, vol. IT-2, pp. 41–46, Sept. 1956.

[14] N. Ahmed, T. Natarajan, and K. R. Rao, "Discrete cosine transform," *IEEE Trans. Computers*, vol. C-23, pp. 90–93, Jan. 1974.

[15] M. Hamidi and J. Pearl, "Comparison of the cosine and Fourier transforms of Markov-1 signals," *IEEE Trans. Acoust. Speech, Signal Processing*, vol. ASSP-24, pp. 428–429, Oct. 1976.

[16] J. R. Jain and A. K. Jain, "Interframe adaptive data compression techniques for images," Tech. Rep. SIPL-79-2, Signal and Image Processing Lab., Dep. Elec., Comput. Eng., Univ. California, Davis, Aug. 1979.

[17] A. K. Jain, "Image data compression—A review," *Proc. IEEE*, vol. 69, pp. 349–389, Mar. 1981.

[18] —, "A fast Karhunen Loeve transform for a class of random processes," *IEEE Trans. Commun*, vol. COM–24, pp. 1023–1029, Sept. 1976.

[19] A. K. Jain, S. H. Wang, and Y. Z. Liao, "Fast Karhunen Loeve transform data compression studies," presented at Nat. Telecommun. Conf., Dallas, TX, Nov.–Dec. 1976.

[20] G. E. Forsythe and P. Henrici, "The cyclic Jacobi method for computing the principal values of a complex matrix," *Proc. Amer. Math. Soc.*, vol. 94, pp. 1–23, 1960.

[21] G. H. Golub and C. Reinsch, "Singular value decomposition and least squares solutions," *Numer. Math.*, vol. 14, pp. 403–420, 1970.

[22] S. Treitel and J. L. Shanks, "The design of multistage separable planar filters," *IEEE Trans. Geosci. Electron.*, vol. GE-9, pp. 10–27, Jan. 1971.

[23] T. S. Huang, W. F. Schreiber, and O. J. Tretiak, "Image processing," *Proc. IEEE*, vol. 59, pp. 1586–1609, Nov. 1971.

[24] M. M. Sondhi, "Image restoration: The removal of spatially invariant degradations," *Proc. IEEE*, vol. 60, pp. 842–853, July 1972.

[25] H. C. Andrews, "Two dimensional transforms," in *Picture Processing and Digital Filtering*, T. S. Huang, Ed. (Topics in Applied Physics Series, vol. 6), Berlin, Germany: Springer-Verlag, 1975.

[26] H. C. Andrews and C. L. Patterson, "Singular value decomposition (SVD) image coding," *IEEE Trans. Commun.*, vol. COM-24, pp. 425–432, Apr. 1976.

[27] R. E. Twoogood and S. K. Mitra, "Computer aided design of separable two-dimensional digital filters," *IEEE Trans. Acoust. Speech Signal Processing*, vol. ASSP-25, pp. 165–169, Apr. 1977.

[28] V. C. Klema and A. J. Laub, "The singular value decomposition: Its computation and some applications," *IEEE Trans. Automat. Contr.*, vol. AC-25, pp. 164–176, Apr. 1980.

[29] B. Ashjari and W. K. Pratt, "Supervised classification with singular value decomposition texture," USCIPI Rep. 860, Image Proc. Inst., USC, Los Angeles, CA. Mar. 1979.

[30] P. Whittle, *Prediction and Regulation by Linear Least-Squares Methods*. London, England: English Univ. Press, 1954.

[31] K. Astrom, *Introduction to Stochastic Control Theory*. New York: Academic Press, 1970.

[32] N. Levinson, "The Wiener RMS error criterion in filter design and prediction," *J. Math. Phys.*, vol. 25, pp. 261–278, Jan. 1947.

[33] J. Durbin, "The filtering of time series models," *Rev. Int. Inst. Stat.*, vol. 28, pp. 233–244, 1960.

[34] J. Burg, "Maximum entropy spectral analysis," Ph.D. dissertation, Dep. Geophysics, Stanford Univ., Stanford, CA, 1975.

[35] S. Zohar, "Toeplitz matrix inversion: The algorithm of W. F. Trench," *J. Assoc. Comput. Mach.*, vol. 16, pp. 592–601, Oct. 1969.

[36] T. J. Ulrych and T. N. Bishop, "Maximum entropy spectral analysis and autoregressive decomposition," *Rev. Geophys. Space Phys.*, vol. 13, pp. 183–200, Feb. 1975.

[37] J. Makhoul, "Linear prediction: A tutorial review," *Proc. IEEE*, vol. 63, pp. 561–580, Apr. 1975.

[38] A. K. Jain and S. Ranganath, "Image coding by autoregressive synthesis," *Proc. IEEE ICASSP '80* (Denver, CO), pp. 770–773, Apr. 1980.

[39] N. E. Nahi and T. Assefi, "Bayesian recursive image estimation," *IEEE Trans. Comput.* (Short Notes), vol. C-21, pp. 734–738, July 1972.

[40] T. Assefi, "Two dimensional signal processing with application to image restoration," Tech. Rep. 32-1596, JPL, Caltech, Pasadena, CA, Sept. 1, 1974.

[41] N. E. Nahi and C. A. Franco, "Application of Kalman filtering to image enhancement," in *Proc. IEEE Conf. Decision and Control* (New Orleans, LA), pp. 63–65, Dec. 1972.

[42] S. R. Powell and L. M. Silverman, "Modeling of two dimensional covariance functions with application to image restoration," *IEEE Trans. Automat. Cont.*, vol. AC-19, pp. 8–12, Feb. 1974.

[43] N. E. Nahi, *Estimation Theory and Applications*. New York: Wiley, 1969.

[44] A. H. Jazwinsky, *Stochastic Processes and Filtering Theory*. New York: Academic Press, pp. 70–92, 1970.

[45] A. O. Aboutalib and L. M. Silverman, "Restoration of motion degraded images," *IEEE Trans. Circuits Syst.*, vol. CAS-22, pp. 278–286, Mar. 1975.

[46] A. K. Jain, "Noncausal representations for finite discrete signals," in *Proc. IEEE Conf. Decision and Control* (Tucson, AZ), 1974.

[47] A. K. Jain and E. Angel, "Image restoration, modeling and reduction of dimensionality," *IEEE Trans. Comput.*, vol. C-23, pp. 470–476, May 1974.

[48] A. K. Jain, "A fast Karhunen-Loeve transform for recursive filtering of images corrupted by white and colored noise," *IEEE Trans. Comput.*, vol. C-26, pp. 560–571, June 1977.

[49] —, "An operator factorization method for restoration of blurred images," *IEEE Trans. Comput.*, vol. C-25, pp. 1061–1071, Nov. 1977.

[50] —, "Image coding via a nearest neighbors image model," *IEEE Trans. Commun.*, vol. COM-23, pp. 318–331, Mar. 1975.

[51] —, "Partial differential equations and finite difference methods in image processing, Part I—Image representation," *J. Optimization Theory Appl.*, vol. 23, no. 1, pp. 65–91, Sept. 1977.

[52] A. K. Jain and J. R. Jain, "Partial differential equations and finite difference methods in image processing, Part II: Image restoration," *IEEE Trans. Automat. Contr.*, vol. AC-23, pp. 817–834, Oct. 1978.

[53] A. Zvi Meiri, "The pinned Karhunen Loeve transform of a two dimensional Gauss-Markov field," in *Proc. 1976 SPIE Meeting* (San Diego, CA), 1976.

[54] A. K. Jain, "Fast inversion of banded Toeplitz matrices circular decompositions," *IEEE Trans. ASSP*, vol. ASSP-26, pp. 121–126, Oct. 1978.

[55] —, "A semicausal model for recursive filtering of two dimensional images," *IEEE Trans. Comput.*, vol. C-26, pp. 343–350, Apr. 1977.

[56] T. L. Marzetta, "A linear prediction approach to two dimensional spectral factorization and spectral estimation," Ph.D. dissertation, Dep. Elec. Eng. Comput. Sci., MIT, Cambridge, MA, Feb. 1978.

[57] S. H. Wang, "Applications of stochastic models for image data compression," Ph.D. dissertation, Dep. Elec. Eng., SUNY Buffalo, NY, Sept. 1979; Also see S. H. Wang and A. K. Jain, Tech. Rep. SIPL-79-6, Signal and Image Processing Lab., Dep. Elec. Comput. Eng., Univ. California Davis, Sept. 1979.

[58] A. K. Jain and S. H. Wang, "Stochastic image models and hybrid coding," Final Rep., NOSC Contract N00953-77-C-003MJE, Dep. Elec. Eng., SUNY Buffalo, NY, Oct. 1977.

[59] A. K. Jain, "Spectral estimation and signal extrapolation in one and two dimensions," in *Proc. RADC Spectrum Estimation Workshop* (Rome, NY), pp. 195–214, Oct. 1979.

[60] E. Angel and A. K. Jain, "Frame to frame restoration of diffusion images," *IEEE Trans. Automat. Contr.*, vol. AC-23, pp. 850–855, Oct. 1978.

[61] N. Wiener, *Extrapolation, Interpolation and Smoothing of Stationary Time Series.* New York: Wiley, 1949.
[62] J. L. Doob, *Stochastic Processes.* New York: Wiley, 1953.
[63] C. L. Rino, "Factorization of spectra by discrete Fourier transforms," *IEEE Trans. Inform. Theory*, vol. IT-16, pp. 484-485, July 1970.
[64] M. P. Ekstrom and J. W. Woods, "Two dimensional spectral factorization with application in recursive digital filtering," *IEEE Trans. Acoust. Speech Signal Processing*, vol. ASSP-24, pp. 115-128, Apr. 1976.
[65] T. S. Huang, "Stability of two dimensional recursive filters," *IEEE Trans. Audio Electroacoust.*, vol. AU-20, pp. 158-163, June 1972.
[66] *Proceedings IEEE*, vol. 65, Special Issue on Multidimensional Systems (Guest Editor N. K. Bose), June 1977.
[67] D. Goodman, "Some stability properties of two dimensional linear shift invariant filters," *IEEE Trans. Circuits Syst.*, vol. CAS-24, pp. 201-208, Apr. 1977.
[68] J. W. Woods, "Two dimensional discrete Markov fields," *IEEE Trans. Inform. Theory*, vol. IT-18, pp. 232-240, Mar. 1972.
[69] W. E. Larimore, "Statistical inference on stationary random fields," *Proc. IEEE*, vol. 65, pp. 961-970, June 1977.
[70] S. Sugimoto, H. Mizutani and T. Mizokawa, "Causality and recursive estimation of two dimensional random image fields," in *Proc. 8th SICE Symp. Control Theory*, pp. 145-150, Hachigi, Tokyo, 1979.
[71] H. Mizutani and S. Sugimoto, "Semicausal models and smoothing for 2-D random image fields," in *Proc. 11th JAACE Symp. Stoch. Syst.* (Tokyo, Japan), pp. 161-164, Nov. 1979.
[72] T. Kailath, D. O. Mayne, and R. K. Mehra, Eds., *IEEE Trans. Aut. Contr.*, Special Issue on System Identification and Time Series Analysis, vol. AC-19, Dec. 1974.
[73] B. R. Hunt, "The application of constrained least squares estimation to image restoration by digital computer," *IEEE Trans. Comput.*, vol. C-22, pp. 805-812, Sept. 1973.
[74] R. W. Means, E. H. Wrench, and H. J. Whitehouse, "Image transmission via spread spectrum techniques," ARPA Quart. Tech. Rep. ARPA-QR6, Nav. Ocean Syst. Cent., San Diego, CA, Jan.-Dec. 1975; Also see ARPA-QR8, Annu. Rep., Jan.-Dec. 1975.
[75] B. R. Hunt, "Digital image processing," *Proc. IEEE*, vol. 63, pp. 693-708, Apr. 1975.
[76] A. Habibi, "Two dimensional Bayesian estimate of images," *Proc. IEEE*, vol. 60, pp. 878-883, July 1972.
[77] M. Strintzis, "Comments on two dimensional Bayesian estimate of images," *Proc. IEEE*, vol. 64, pp. 1255-1257, Aug. 1976.
[78] N. E. Nahi and A. Habibi, "Decision directed recursive image enhancement," *IEEE Trans. Circuits Syst.*, vol. CAS-22, pp. 286-293, Mar. 1975.
[79] J. W. Woods and C. H. Radewan, "Kalman filtering in two dimensions," *IEEE Trans. Inform. Theory*, vol. IT-23, pp. 473-482, July 1977.
[80] M. S. Murphy and L. M. Silverman, "Image model representation and line by line recursive restoration," in *Proc. Conf. Decision Contr.* (Clearwater Beach, FL), pp. 601-606, Dec. 1976.
[81] A. Netravali and J. Limb, "Picture coding: A survey," *Proc. IEEE*, vol. 68, pp. 366-406, Mar. 1980.
[82] W. F. Schreiber, C. F. Knapp, and N. D. Kay, "Synthetic highs: An experimental TV bandwidth reduction system," *J. Soc. Motion Pic. Television Engrs.*, vol. 68, pp. 525-537, Aug. 1959.
[83] D. N. Graham, "Image transmission by two dimensional contour coding," *Proc. IEEE*, vol. 55, pp. 336-346, Mar. 1967.
[84] J. K. Yan and D. J. Sakrison, "Encoding of images based on a two component source model," *IEEE Trans. Commun.*, vol. COM-25, pp. 1315-1322, Nov. 1977.
[85] W. Rudin, "The extension problem for positive definite functions," *Ill. J. Math.*, vol. 7, pp. 532-539, 1963.
[86] B. W. Dickinson, "Two dimensional Markov spectrum estimates need not exist," *IEEE Trans. Inform. Theory*, vol. IT-26, pp. 120-121, Jan. 1980.
[87] R. P. Roesser, "A discrete state space model for linear image processing," *IEEE Trans. Automat. Contr.*, vol. AC-20, pp. 1-10, Feb. 1975.
[88] S. Attasi, "Modelling and recursive estimation for double indexed sequences," IRIA Rep. IA/129, Domaine de Volvceau, Racquencourt, 78150 Le Chesnay, B.P. 5, France, July 1975.
[89] E. Fornasini and G. Marchesini, "State space realization theory of two dimensional filters," *IEEE Trans. Automat. Contr.*, vol. AC-21, pp. 484-492, Aug. 1976.
[90] E. Wong, "Recursive causal linear filtering for two dimensional random fields," *IEEE Trans. Inform. Theory*, vol. IT-24, pp. 50-59, Jan. 1978.
[91] M. Morf, B. Levy, and S. Y. Kung, "New results in 2-D systems theory, Part I: 2-D polynomial matrices, factorization and coprimeness," *Proc. IEEE*, vol. 65, pp. 861-872, June 1977.
[92] S. Y. Kung, B. C. Levy, M. Morf, and T. Kailath, "New results in 2-D systems theory, Part II: Realization and the notions of controllability, observability and minimality," *Proc. IEEE*, vol. 65, pp. 945-960, June 1977.
[93] A. K. Jain and K. W. Au, "On linear estimation via fast Fourier transform," in *Proc. Conf. Decision and Control* (Ft. Lauderdale, FL), Dec. 1979.
[94] J. W. Woods, "Two dimensional Markov spectral estimation," *IEEE Trans. Inform. Theory*, vol. IT-22, pp. 552-559, Sept. 1976; also see vol. IT-26, pp. 129-130, Jan. 1980.
[95] E. S. Angel and A. K. Jain, "Restoration of images degraded by spatially varying point spread functions by a conjugate gradient method," *Appl. Opt.*, vol. 17, pp. 2186-2190, July 1980.

Estimation and Choice of Neighbors in Spatial-Interaction Models of Images

RANGASAMI L. KASHYAP, FELLOW, IEEE, AND RAMALINGAM CHELLAPPA, MEMBER, IEEE

Abstract—Some aspects of statistical inference for a class of spatial-interaction models for finite images are presented: primarily the simultaneous autoregressive (SAR) models and conditional Markov (CM) models. Each of these models is characterized by a set of neighbors, a set of coefficients, and a noise sequence of specified characteristics. We are concerned with two problems: the estimation of the unknown parameters in both SAR and CM models and the choice of an appropriate model from a class of such competing models. Assuming Gaussian-distributed variables, we discuss maximum likelihood (ML) estimation methods. In general, the ML scheme leads to nonlinear optimization problems. To avoid excessive computation, an iterative scheme is given for SAR models, which gives approximate ML estimates in the Gaussian case and reasonably good estimates in some non-Gaussian situations as well. Likewise, for CM models, an easily computable consistent estimate is given. The asymptotic mean-squared error (mse) of this estimate for a four-neighbor CM model is shown to be substantially less than the mse of the popular coding estimate. Asymptotically consistent decision rules are given for choosing an appropriate SAR or CM model. The usefulness of the estimation scheme and the decision rule for the choice of neighbors is illustrated by using synthetic patterns. Synthetic patterns obeying known SAR and CM models are generated, and the models corresponding to true and several competing neighbor sets are fitted. The estimation scheme yields estimates close to the parameters of the true models, and the decision rule for the choice of neighbors picks up the true model from the class of competing models.

Manuscript received June 4, 1981; revised March 15, 1982. This paper was supported in part by the National Science Foundation under Grant ECS-80-09041 and in part by the Air Force Office of Scientific Research under Grant AFOSR-77-3271. This paper was presented in part at the IEEE Computer Society Conference on Pattern Recognition and Image Processing, Dallas, TX, August 981.

R. L. Kashyap is with the School of Electrical Engineering, Purdue University, West Lafayette, IN 47907.

R. Chellappa was with the Computer Vision Laboratory, University of Maryland, College Park, MD and the School of Electrical Engineering, Purdue University, West Lafayette, IN. He is now with the Department of Electrical Engineering Systems and Image Processing Institute, University of Southern California, Los Angeles, CA 90007.

I. Introduction

SPATIAL-interaction models (often known as random field models) have many applications in image processing and analysis. For instance, they can be used for the design of image restoration algorithms [1]–[3], for image coding [4], and for texture analysis [5]–[8]. Typically, an image is represented by two-dimensional scalar data, the gray level variations defined over a rectangular or square lattice. One of the important characteristics of such data is the special nature of the statistical dependence of the gray level at a lattice point on those of its neighbors.

Reprinted from *IEEE Trans. Inform. Theory,* vol. IT-29, pp. 60–72, Jan. 1983.

The spatial-interaction models characterize this statistical dependency by representing $y(s)$, the gray level at location s, as a linear combination of the gray levels $\{y(s+s'), s' \in N\}$ and an additive noise, where N is called the neighbor set which does not include (0, 0).

Specific restrictions on the members of neighbor set N yield representations familiar in the image processing literature. For instance, the familiar *causal* models [4]–[6] are obtained when N is defined as $N = \{(i, j): i \leq 0, j \leq 0, (i, j) \neq (0, 0)\}$. Likewise, the unilateral neighbor set used in [3] result when N is a finite subset of a half plane defined in [9]. Neighbor sets more general than the causal or the unilateral have been of much interest in image processing [1], [2], [7], [8], [10] in the analysis of wheat data on fields [11], [12], in navigational problems [13], and in the analysis of geographic data [14].

There are two nonequivalent ways [12] of specifying the underlying interaction among the given observations leading to two classes of models, the simultaneous models [12], [15], [16] and the conditional Markov models [7], [8], [10], [12], [16]. Within the class of simultaneous models there are simultaneous autoregressive (SAR) models [11]–[16], simultaneous moving average (SMA) models [17], and simultaneous autoregressive and moving average (SARMA) [17] models, as in one-dimensional time series models. The class of SAR models is a subset of the class of CM models, i.e., for every SAR model there exists a unique conditional Markov (CM) model with equivalent spectral density function, but the converse is not always true. Still the class of SAR models deserves detailed study for the following reasons. SAR models are parsimonious, i.e., the CM model in general is characterized by more parameters than the equivalent SAR model (if one exists). Secondly, the study of SAR models can be extended to include SMA and SARMA models which are not subsets of CM models. Hence the study of elementary SAR models considered in this paper will be of considerable use.

Given a finite image we are interested in fitting an appropriate model from the class of SAR or CM models. Given a SAR or CM model the finite image can be analyzed either as a finite block of an infinite image defined over an infinite lattice, or as an image defined only on a finite lattice. In general, the estimation methods are complicated when infinite lattice representations are used for SAR and CM models. To overcome this problem we will consider some finite lattice representations. The finite lattice models can be justified in their own right or as approximations to infinite lattice models. After choosing a finite lattice representation, two problems have to be tackled in fitting an appropriate model, namely, a method of estimating the parameters of the model given the structure of the model, and a criterion to choose between different possible structures.

The classical least square (LS) estimates are asymptotically consistent and efficient for SAR models characterized by *unilateral* neighbor sets, but they are not even consistent [11], [14] for bilateral neighbor sets. Asymptotically consistent and efficient estimates can be obtained via the maximum likelihood (ML) estimation method with an appropriate distribution assumption for $\{y(s)\}$. In general due to the fact that the Jacobian of the transformation matrix for bilateral SAR or CM models is not unity, the log-likelihood function is a complicated function even for Gaussian variables. An asymptotic approximation developed in [11] for infinite lattice SAR models will be of use for very simple models only.

However, by using a finite lattice representation [14]–[17] for SAR models an explicit expression can be written for the log-likelihood function for any arbitrary distribution. Specifically, a toroidal lattice representation given in Section II-B yields a transformation matrix that is block-circulant whose eigenvalues, and hence the log-likelihood function can be written down explicitly. Unfortunately, the resulting function is nonquadratic in the parameters, requiring the use of nonlinear numerical optimization algorithms. To avoid excessive computations, we give an iterative scheme that yields approximate ML estimates for Gaussian distributions. These estimates are reasonably good in some non-Gaussian situations as well.

For CM models we use the consistent estimation scheme suggested in [10] which involves inverting an $m \times m$ symmetric matrix, where m is the number of independent parameters to be estimated. However, this estimate is not efficient. Actual comparison of efficiencies for a simple isotropic CM model with $N = \{(0,1),(0,-1),(-1,0),(1,0)\}$ indicates that the efficiency of this estimate is superior to that of the popular coding estimate [8], [12].

The second problem considered is the choice of appropriate neighbors in SAR and CM models. As illustrated in [18], [19], different neighbor sets in SAR or CM models account for different image patterns and the quality of regeneration is dependent upon the particular underlying model. From one-dimensional time series analysis, it is known that the use of an appropriate model leads to good results in forecasting and similar applications. We give asymptotically consistent decision rules for the appropriate choice of neighbors in conditional or simultaneous models. The derivation of this decision rule for SAR models may be found in [20]. The decision rule for choosing between different neighbor sets in CM models is similar in structure to that in simultaneous models. The usefulness of the estimation scheme and the decision rule for the choice of neighbors can be demonstrated by applying them to synthetic patterns for which the underlying true model of the synthetic pattern is known.

The organization of the paper is as follows. In Section II, we consider the estimation problem in SAR models and give an iterative estimation scheme. The results of applying this estimation scheme to synthetic data generated by known models are also given. A consistent estimation scheme for CM models is presented in Section III along with the results for synthetic data. An expression for the MSE of this estimate is also derived. The need for using an appropriate SAR or CM model is discussed, and decision rules for the choice of neighbors in SAR or CM models are given in Section IV.

II. Estimation Schemes in SAR Models

A. Model Representation and Estimation in Infinite Lattice SAR Models

Assume that the stationary image $\{y(s)\}$ obeys the infinite lattice SAR model in (2.1), with associated neighbor set N

$$y(s) = \sum_{r \in N} \theta_r y(s+r) + \sqrt{\rho}\,\omega(s). \tag{2.1}$$

In (2.1), $(\theta_r, r \in N)$ and ρ are unknown parameters, and $\omega(\cdot)$ is an independent and identically distributed (i.i.d.) noise sequence with zero mean and unit variance. By stationarity $E(y(s)) = 0$. A sufficient condition on $\boldsymbol{\theta}^T = \{\theta_r, r \in N\}$ to ensure stationarity of $y(\cdot)$ is

$$\left\{1 - \sum_{(i,j) \in N} \theta_{i,j} z_1^i z_2^j \neq 0\right\},$$
$$\text{for all } z_1, z_2 \text{ such that } |z_1| = 1, |z_2| = 1; \tag{2.2}$$

N need not be symmetric. If N is symmetric we have to assume that the coefficients of the symmetrically opposite neighbors are equal, i.e., $\theta_{k,1} = \theta_{-k,-1}$. Otherwise, the parameters may not be identifiable [12]. Note that $y(\cdot)$ is not Markov with respect to any arbitrary bilateral neighbor set N, i.e.,

$$p(y(s) \mid \text{all } y(r), s \neq r) \neq p(y(s) \mid \text{all } y(s+r), r \in N). \tag{2.3}$$

Given a finite image defined on a square $M \times M$ grid Ω, we are interested in estimating the parameters of the model characterizing the image. A popular method of estimation is that of least squares (LS), which yields the estimate in (2.4)

$$\hat{\theta} = \left[\sum_s z(s) z^T(s)\right]^{-1} \left(\sum_s z(s) y(s)\right), \tag{2.4}$$

$$\hat{\rho} = \frac{1}{M^2} \sum_s \left(y(s) - \hat{\theta}^T z(s)\right)^2, \tag{2.5}$$

$$z(s) = \operatorname{col}\left[y(s+r), r \in N\right]. \tag{2.6}$$

One of the drawbacks of the LS is that in general $\hat{\theta}$ is not consistent for nonunilateral neighbor sets [14]. A qualitative explanation for the inconsistency of $\hat{\theta}$ can be given as follows. Substitute (2.1) into (2.4) to obtain after simplification,

$$\boldsymbol{S}(\hat{\boldsymbol{\theta}} - \boldsymbol{\theta}) = \sum_s z(s) \omega(s), \tag{2.7}$$

where $\boldsymbol{S} = \sum_s z(s) z^T(s)$. Multiplying the left-hand side (LHS) of (2.7) by its transpose and taking expectations, we get

$$E\left(\boldsymbol{S}(\hat{\boldsymbol{\theta}} - \boldsymbol{\theta})(\hat{\boldsymbol{\theta}} - \boldsymbol{\theta})^T \boldsymbol{S}\right) = E\left[\sum_r \sum_s z(s)\omega(s)\omega(r) z^T(r)\right]. \tag{2.8}$$

Since

$$E(z(s)\omega(s)) \neq 0 \tag{2.9}$$

with nonunilateral neighbor sets, the right-hand side (RHS) of (2.8) does not go to zero even as M tends to infinity leading to the inconsistency of $\hat{\boldsymbol{\theta}}$. Another popular method is the ML method, which yields asymptotically consistent and efficient estimates. As mentioned earlier, the derivation of the log-likelihood function is extremely difficult, even in the Gaussian case, for SAR models with nonunilateral neighbor sets since the Jacobian of the transformation matrix is a complicated function of the model parameters. It has been shown [11] that for large values of M, an approximate expression can be obtained for the determinant of the transformation matrix. Typically, for a SAR model with spectrum inversely proportional to

$$\left\|1 - \sum_{r \in N} \theta_r \exp\left(\sqrt{-1}\, w^T r\right)\right\|^2$$

the log of the determinant is approximately equivalent to the absolute term in the expansion of

$$-2 \log\left(1 - \sum_{r \in N} \theta_r \exp\left(\sqrt{-1}\, w^T r\right)\right).$$

This approximation is useful only for very simple SAR models. These considerations motivate us to use another class of representations, known as finite lattice SAR models. The finite lattice models considered here can be analyzed either as approximations to infinite lattice models or as an independent class of models.

B. SAR Model Representation on Finite Lattices: [15], [16]

Suppose we partition the finite lattice Ω into mutually exclusive and totally inclusive subsets Ω_I, the interior set, and Ω_B, the boundary set, such that

$$\Omega_B = \{s = (i,j): s \in \Omega \text{ and } (s+r) \notin \Omega \text{ for at least one } r \in N\},$$

and

$$\Omega_I = \Omega - \Omega_B.$$

For every $s \in \Omega_B$, there exists a $r \in N$ so that $(s+r) \notin \Omega$ and consequently $y(s+r)$ is not defined by (2.1). Hence (2.1) needs modification.

The toroidal lattice SAR model for a *finite* image $\{y(s), s \in \Omega\}$ is defined by two equations for s in Ω_I and Ω_B as in (2.10) and (2.11).

$$y(s) = \sum_{r \in N} \theta_r y(s+r) + \sqrt{\rho}\,\omega(s), \qquad s \in \Omega_I, \tag{2.10}$$

$$y(s) = \sum_{r \in N} \theta_r y_1(s+r) + \sqrt{\rho}\,\omega(s), \qquad s \in \Omega_B, \tag{2.11}$$

$$\begin{aligned} y_1(s + (k,l)) &\text{ with } s = (i,j), \\ &= y(s + (k,l)), \quad \text{if } (s+(k,l)) \in \Omega \\ &= y[(k+i-1) \bmod M + 1, (l+j-1) \bmod M + 1], \\ &\qquad \text{if } (s + (k,l)) \notin \Omega. \end{aligned} \tag{2.12}$$

In the RHS of (2.11), y_1 takes the role of y in (2.10). $y_1(s)$ in (2.12) is a function of $y(r)$, $r \in \Omega$ even when $s \notin \Omega$. If the image $y(\cdot)$ were folded into a torus, $y_1(s) = y(s)$.

Equations (2.10) and (2.11) give M^2 equations relating the image variables $\{y(s)\}$ and i.i.d. random variables $\{\omega(s)\}$. Denoting $\boldsymbol{y}$ and $\boldsymbol{\omega}$ as $M^2 \times 1$ vectors of lexicographic ordered arrays $\{y(\cdot)\}$ and $\{\omega(\cdot)\}$, (2.10)–(2.11) can be rewritten as $B(\boldsymbol{\theta})\boldsymbol{y} = \sqrt{\rho}\,\boldsymbol{\omega}$, where $B(\boldsymbol{\theta})$ is a block-circulant matrix:

$$B(\boldsymbol{\theta}) = \begin{bmatrix} B_{1,1} & B_{1,2} & \cdots & B_{1,M} \\ B_{1,M} & B_{1,1} & \cdots & B_{1,M-1} \\ \cdots & \cdots & & \cdots \\ B_{1,2} & \cdots & \cdots & B_{1,1} \end{bmatrix}.$$

For instance when the neighbor set of dependence N is $\{(-1,0),(0,1),(1,0),(0,-1)\}$ we have

$$B_{1,1} = \text{circulant}\,(1, -\theta_{0,1}, 0, \cdots, -\theta_{0,-1}),$$

$$B_{1,2} = \text{circulant}\,(-\theta_{1,0}, 0, \cdots, 0),$$

$$B_{1,M} = \text{circulant}\,(-\theta_{-1,0}, 0, \cdots, 0),$$

and

$$B_{1,j} = \mathbf{0} \qquad j \neq 1, 2, M.$$

To ensure stationarity, the coefficients $\{\theta_r, r \in N\}$ must obey [17]:

$$\mu_s(\boldsymbol{\theta}) \triangleq (1 - \boldsymbol{\theta}^T \boldsymbol{\psi}_s) \neq 0, \qquad s \in \Omega, \qquad (2.13)$$

where

$$\boldsymbol{\psi}_s = \text{col}\left[\exp\left[\sqrt{-1}\,\frac{2\pi}{M}(s - \mathbf{1})^T r; r \in N\right]\right], \qquad \mathbf{1} = (1,1).$$

The finite lattice representation on a toroidal lattice is only one of several possible finite lattice representations [17].

C. Estimation in Toroidal Lattice SAR Models

1) *Least Squares and ML Estimates:* The LS estimate in (2.4) is not consistent for toroidal lattice SAR model. The ML estimation method yields asymptotically consistent and efficient estimates. To obtain an expression for the log-likelihood function, we impose a Gaussian structure on the noise sequence $\omega(\cdot)$. Then the likelihood of the observations can be written as

$$\ln p(\boldsymbol{y} \mid \boldsymbol{\theta}, \rho) = \ln \det B(\boldsymbol{\theta}) - (M^2/2) \ln 2\pi\rho - \frac{1}{2\rho} \sum_{\Omega} \left(y(s) - \boldsymbol{\theta}^T \boldsymbol{z}(s)\right)^2, \quad (2.14)$$

Since $\det B(\boldsymbol{\theta}) = \Pi_\Omega (1 - \boldsymbol{\theta}^T \boldsymbol{\psi}_s)$,

$$\ln p(\boldsymbol{y} \mid \boldsymbol{\theta}, \rho) = \sum_{\Omega} \ln(1 - \boldsymbol{\theta}^T \boldsymbol{\psi}_s) - (M^2/2) \ln 2\pi\rho - \frac{1}{2\rho} \sum_{\Omega} \left(y(s) - \boldsymbol{\theta}^T \boldsymbol{z}(s)\right)^2, \quad (2.15)$$

The ML estimates are obtained by maximizing (2.15) with respect to $\boldsymbol{\theta}$ and ρ. Since the log-likelihood function is nonquadratic in $\boldsymbol{\theta}$, the estimation involves the use of numerical optimization methods, such as Newton–Raphson approach, which are computationally expensive.

We give an iterative method which yields estimates close to the ML estimates with a faster convergence rate.

2) *An Iterative Estimation Scheme:* This method is derived by replacing the term $\ln(1 - \boldsymbol{\theta}^T \boldsymbol{\psi}_s)$ in (2.15) with the approximation $\ln(1 + a) = a - a^2/2$. The corresponding approximation of $\ln p(\boldsymbol{y} \mid \boldsymbol{\theta}, \rho)$ denoted as $J(\boldsymbol{\theta}, \rho)$ can be written as

$$J(\boldsymbol{\theta}, \rho) = -\boldsymbol{V}^T \boldsymbol{\theta} + 0.5 \boldsymbol{\theta}^T \boldsymbol{R} \boldsymbol{\theta} - (M^2/2) \ln 2\pi\rho - \frac{1}{2\rho} \sum_{s \in \Omega} \left(y(s) - \boldsymbol{\theta}^T \boldsymbol{z}(s)\right)^2, \quad (2.16)$$

where

$$\boldsymbol{V} = \sum_{\Omega} \boldsymbol{C}_s, \quad m \times 1 \text{ vector}, \qquad (2.17)$$

$$\boldsymbol{R} = \sum_{\Omega} (\boldsymbol{S}_s \boldsymbol{S}_s^T - \boldsymbol{C}_s \boldsymbol{C}_s^T), \quad m \times m \text{ matrix}, \qquad (2.18)$$

$$\boldsymbol{C}_s = \text{col}\left[\cos \frac{2\pi}{M}\left((s - \mathbf{1})^T r\right), r \in N\right], \qquad (2.19)$$

$$\boldsymbol{S}_s = \text{col}\left[\sin \frac{2\pi}{M}\left((s - \mathbf{1})^T r\right), r \in N\right]. \qquad (2.20)$$

Theorem 1: The estimates $\bar{\boldsymbol{\theta}}, \bar{\rho}$ maximizing $J(\boldsymbol{\theta}, \rho)$ are obtained as the limits of $\boldsymbol{\theta}_t, \rho_t$ defined by

$$\boldsymbol{\theta}_{t+1} = \left(\boldsymbol{R} - \frac{1}{\rho_t}\boldsymbol{S}\right)^{-1}\left(\boldsymbol{V} - \frac{1}{\rho_t}\boldsymbol{U}\right), \qquad t = 0, 1, 2, \cdots \qquad (2.21)$$

and

$$\rho_t = \frac{1}{M^2} \sum_{\Omega} \left(y(s) - \boldsymbol{\theta}_t^T \boldsymbol{z}(s)\right)^2, \qquad t = 0, 1, 2, 3, \cdots, \qquad (2.22)$$

where

$$\boldsymbol{S} = \sum_{\Omega} \boldsymbol{z}(s) \boldsymbol{z}^T(s), \quad m \times m \text{ matrix}, \qquad (2.23)$$

and

$$\boldsymbol{U} = \sum_{\Omega} \boldsymbol{z}(s) y(s), \quad m \times 1 \text{ vector}. \qquad (2.24)$$

The initial value $\boldsymbol{\theta}_0$ is chosen as $\boldsymbol{\theta}_0 = \boldsymbol{S}^{-1}\boldsymbol{U}$. All the summations in (2.22)–(2.24) are over $s \in \Omega$ and m is the dimension of $\boldsymbol{\theta}$.

$\boldsymbol{\theta}_{t+1}$ in (2.21) is the value of $\boldsymbol{\theta}$ maximizing $J(\boldsymbol{\theta}, \rho_t)$. ρ_t in (2.22) is the value of ρ maximizing $J(\boldsymbol{\theta}_t, \rho)$. These facts allow us to establish Theorem 1.

Comments

1) Usually, the numerical optimization methods require the evaluation of the first and possibly the second derivatives of the likelihood function with respect to the unknown parameters at each step of the iteration. In our scheme we approximate the second derivative by a function of matrices $\boldsymbol{R}$ and $\boldsymbol{S}$ which are indepen-

dent of iterations and can be evaluated once the neighbor set N, and the image data are known.

2) It is difficult to analyze the asymptotic distribution properties of the estimate $\bar{\theta}$ because of the approximations involved. Numerical experiments indicate that these estimates are close to the ML estimates.

3) The iterative estimation method has been obtained by using the toroidal lattice representation. It is a natural question to ask if the toroidal lattice assumption is reasonable. Simulation results reported elsewhere [18] indicate that the toroidal lattice models are good approximations to the infinite lattice models. Specifically, second-order properties, like the correlation function, of the infinite lattice model and the toroidal lattice model are numerically very close to each other.

D. Experimental Results

We give the results of applying the estimation scheme in Section II-B to some synthetic data generated from known SAR models. The estimates of the model should be very close to the true parameters. The scheme used to generate the synthetic data is in [16].

Experiment I: (Gaussian data and a SAR model with $N = \{(-1,0), (1,0), (-1,1), (1,-1), (1,1), (-1,-1)\}$.)

The parameters of that true model are $\alpha = 30.00$, $\rho = 1.1111$, and $\theta_{-1,0} = \theta_{1,0} = 0.12$, $\theta_{1,1} = \theta_{-1,-1} = -0.14$, and $\theta_{-1,1} = \theta_{1,-1} = 0.28$. The parameter α is the expected value of $\{y(s)\}$. For simplicity we assumed $\alpha = 0$ in Section II-A. To facilitate easy display of the data generated by a SAR model as an image with gray level variations in (0–63) range, we have used $\alpha = 30.0$. α can be estimated from the given image as the sample mean. Using this model the synthetic data is generated using a Gaussian pseudorandom number generator. For estimation of the parameters the sample mean of the window was subtracted and the iterative scheme in (2.21)–(2.22) is used. The numerical values of the estimates corresponding to the model with neighbor set N are $\bar{\theta}_{-1,0} = \bar{\theta}_{1,0} = 0.1262$, $\bar{\theta}_{1,1} = \bar{\theta}_{-1,-1} = -0.1613$, $\theta_{-1,1} = \bar{\theta}_{1,-1} = 0.3116$ and $\bar{\rho} = 1.0520$. On the other hand the LS estimates obtained from (2.4) and (2.5) are

$$\hat{\theta}_{-1,0} = \hat{\theta}_{1,0} = 0.1470, \qquad \hat{\theta}_{1,1} = \hat{\theta}_{-1,-1} = -0.1526,$$
$$\hat{\theta}_{-1,-1} = \hat{\theta}_{1,1} = 0.3680, \qquad \hat{\rho} = 1.0126.$$

Note the approximate ML estimates $(\bar{\theta}, \bar{\rho})$ are much closer to the true parameters than the LS estimates $(\hat{\theta}, \hat{\rho})$.

Experiment II: (Synthetic data generated by using uniform random numbers and SAR model with $N = \{(-1,0), (1,0), (-1,1), (1,-1), (1,1), (-1,-1)\}$.)

The approximate ML estimation scheme has been developed by using Gaussian assumption for the variables. It should be of interest to see how good the estimation scheme is for non-Gaussian variables, as it would be unreasonable to assume that all images are generated by Gaussian random fields. Synthetic data obeying the same SAR model as in Experiment I but with $\omega(\cdot)$ obeying a uniform distribution was generated and the estimation scheme in Theorem 1 was used to estimate the parameters. The numerical values of the estimates are

$$\bar{\theta}_{-1,0} = \bar{\theta}_{1,0} = 0.1343, \qquad \bar{\theta}_{-1,-1} = \bar{\theta}_{1,1} = -0.1562,$$
$$\bar{\theta}_{-1,1} = \bar{\theta}_{1,-1} = 0.3187, \qquad \bar{\rho} = 1.0113.$$

The LS estimates are

$$\hat{\theta}_{-1,0} = \hat{\theta}_{1,0} = 0.1541, \qquad \hat{\theta}_{-1,-1} = \hat{\theta}_{1,1} = -0.1530,$$
$$\hat{\theta}_{-1,1} = \hat{\theta}_{1,-1} = 0.3711, \qquad \hat{\rho} = 0.9732.$$

Compared to the case of Gaussian data, the estimates $(\bar{\theta}, \bar{\rho})$ obtained for uniform data are inferior, understandably so. Nevertheless, the estimates are not too far from the true parameter values. Further the iterative estimate $\bar{\theta}$ is superior to the LS estimate.

III. Estimation Schemes in CM Models

A. Model Representation

Infinite Lattice Models [10]: Assume that the observations $\{y(s),\ s \in \Omega\}$ have zero mean and obey the CM model on an infinite lattice,

$$y(s) = \sum_{r \in N} \theta_r y(s+r) + e(s). \tag{3.1}$$

The neighbor set N is symmetric:

$$\theta_r = \theta_{-r}, \qquad \text{for all } r \in N.$$

The stationary Gaussian noise sequence $\{e(s)\}$ is defined by

$$E(e(s) \mid \text{all } y(r), r \neq s) = 0,$$
$$E[e(s)] = 0,$$
$$E[e^2(s)] = \nu, \tag{3.2}$$

Using (3.1) and (3.2) one can prove that the noise sequence $\{e(s)\}$ has the correlation structure given below:

$$E(e(s)e(r)) = \begin{cases} \nu, & s = r, \\ -\theta_{s-r}\nu, & (s-r) \in N, \\ 0, & \text{otherwise.} \end{cases} \tag{3.3}$$

It can be shown [10] that the observation $y(s)$ obeying (3.1) satisfies the Markov condition,

$$p(y(s) \mid y(r), \text{all } r, r \neq s) = p(y(s) \mid y(s+r), \forall r \in N).$$

2) *Finite Lattice Models* [7], [16], [17], [21]: The finite lattice models used in this paper assume a toroidal lattice representation. The representation is characterized by (3.1) defined over $s \in \Omega_I$ together with

$$y(s) = \sum_{r \in N} \theta_r y_1(s+r) + e(s), \qquad s \in \Omega_B, \tag{3.4}$$

where $y_1(\cdot)$ is defined in Section II-B. Equations (3.1) and (3.4) yield M^2 equations relating $\{e(s)\}$ and $\{y(s)\}$ through

$$H(\theta) y = e, \tag{3.5}$$

where $H(\theta)$ is a block-circulant symmetric matrix. To

ensure stationarity, we need

$$\mu'_s \triangleq (1 - 2\theta^T\phi_s) > 0, \qquad \text{for all } s \in \Omega$$
$$\theta = \text{col}\,[\theta_r, r \in N_s],$$

and

$$\phi_s = \text{col}\left[\cos\frac{2\pi}{M}\left((s-1)^T r, r \in N_S\right.\right]$$

where N_S is the asymmetrical half of N, i.e.,

$$N = N_S \cup \bar{N}_S,$$
$$\bar{N}_S = \{r: -r \in N_S\},$$
$$N_S \cap \bar{N}_S = \varnothing$$

B. Estimation in CM Models

1) *Coding and ML Methods:* Given an image and the infinite lattice conditional Markov model (3.1), we are interested in estimating the model's parameters. This problem has received some attention in the image processing literature [7]–[8]. The coding method [12] developed in statistical literature has been used for modeling the textures in [7], for the case of binary and binomial variables. Consider the case of a Gaussian CM model with $N = \{(0,1), (1,0), (0,-1), (-1,0)\}$, characterized by $\theta = \text{col}(\theta_{0,1}, \theta_{1,0})$. The conditional probability density of $y(\cdot)$ is

$$p(y(s) \mid y(s+r), r \in N)$$
$$= \frac{1}{(2\pi\nu)^{1/2}} \exp\left\{-\frac{1}{2\nu}\left[y(s) - \sum_{r\in N} \theta_r y(s+r)\right]^2\right\}.$$

The coding estimate θ'_C of θ is given by

$$\theta'_C = \left[\sum_{\Omega_0} q(s) q^T(s)\right]^{-1} \sum_{\Omega_0} q(s) y(s), \tag{3.6}$$

where

$$q(s) = \text{col}[y(s+r) + y(s-r);\, r = (0,1), (1,0)],$$

and Ω_0 is a subset of Ω with every other site of Ω skipped. One of the main disadvantages of this method is that the estimates thus obtained are not efficient [21] due to a partial utilization (50 percent) of the data. The coding estimate is not unique. The coding scheme yields another estimate, say θ''_C, obtained from (3.6), where s is summed over $\Omega - \Omega_0$. The different coding estimates for the same parameter can differ considerably. For instance, in the Mercer–Hall wheat data and a CM model with $N = \{(0,1), (0,-1), (1,0), (-1,0), (1,1), (-1,-1), (-1,1), (1,-1)\}$ the estimates of $\theta_{0,1}$ obtained by four possible coding schemes are 0.043, 0.085, 0.243, 0.236 [12]. These estimates are dependent and hence a simple averaging of these highly dependent estimates is not satisfactory.

An estimate with good asymptotic properties like consistency and efficiency can be obtained by the ML procedure. This method involves assuming an appropriate distribution for $\{e(s)\}$. But it is difficult to derive an explicit expression for the log-likelihood function even in Gaussian case due to the problem of evaluating the Jacobian of the transformation matrix. The ML estimate for general CM models can be obtained by assuming the toroidal lattice representation for $\{y(s)\}$ and Gaussian structure for $\{e(s)\}$ and maximizing the log-likelihood function. For the toroidal representation corresponding to (3.5) the log-likelihood function $\ln p(y \mid \theta, \nu)$ can be written as [21]

$$\ln p(y \mid \theta, \nu) = \sum (1/2) \ln(1 - 2\theta^T\phi_s)$$
$$- (M^2/2) \ln 2\pi\nu - \frac{1}{2\nu} y^T H(\theta) y. \tag{3.7}$$

Note that the contribution of the exponential term of the probability density function is linear in θ unlike the SAR models. Numerical optimization procedures like the Newton–Raphson approach can be applied to obtain the ML estimates. The ML estimates are consistent and efficient, but computationally unattractive. We give below an estimate which is consistent and computationally efficient.

2) *A Consistent Estimation Scheme:* Consider the estimate

$$\theta^* = \left[\sum_{\Omega_I} q(s) q^T(s)\right]^{-1} \left(\sum_{\Omega_I} q(s) y(s)\right) \tag{3.8}$$

and

$$\nu^* = \frac{1}{M^2} \sum_{\Omega_I} (y(s) - \theta^{*T} q(s))^2, \tag{3.9}$$

where Ω_I is as in Section II-B. The estimate θ^* is an improvement over the coding estimates like θ'_C, θ''_C, etc. We will state a theorem regarding the consistency of the estimate θ^* and give an expression for the asymptotic variance of the estimate θ^*. Although this estimate was suggested in [10], no results are known regarding its statistical behavior. An expression for the asymptotic variance of θ^* for an isotropic conditional model with $N = \{(0,1), (1,0), (0,-1), (-1,0)\}$ is given.

Theorem 2: Let $y(s)$, $s \in \Omega$ be the set of observations obeying the CM model (3.1). Then

a) the estimate θ^* is asymptotically consistent;
b) the asymptotic covariance matrix of θ^* is

$$E(\theta - \theta^*)(\theta - \theta^*)^T$$
$$= \frac{1}{M^2}\left[\nu Q^{-1} + 2\nu^2 Q^{-2} - \frac{\nu}{M^2} Q^{-1} \sum_{s} \sum_{\substack{r \\ (s-r)\in N}} \theta_{(s-r)} T_{r,s} Q^{-1}\right], \tag{3.10}$$

where $Q = E[q(s)q^T(s)]$, and $T_{r,s} = E[q(r)q^T(s)]$;
c) for the isotropic conditional model with $N = \{(0,1), (1,0), (0,-1), (-1,0)\}$ and $\theta_r = \theta$, for all $r \in N$ the asymptotic expected mean square error is

$$E(\theta - \theta^*)^2 = \frac{2\theta^2(1 - 4\theta\alpha_{1,0})^2}{4M^2\alpha_{1,0}^2}, \tag{3.11}$$

TABLE I
COMPUTATION OF ASYMPTOTIC VARIANCES AND EFFICIENCIES OF DIFFERENT ESTIMATES IN ISOTROPIC CONDITIONAL MODEL WITH $N = \{(0,1),(0,-1),(-1,0),(1,0)\}$; COLUMN 5 IS FROM [21]

4θ	$M^2 \mathrm{var}(\hat{\theta}_{ML})$	$M^2 \mathrm{var}(\theta_C)$	$M^2 \mathrm{var}(\underset{\sim}{\theta}^*)$	$\mathrm{eff}(\theta_C)$	$\mathrm{eff}(\underset{\sim}{\theta}^*)$
.1	.4928	.497	.494	.991	.9975
.2	.472	.489	.478	.965	.987
.3	.437	.474	.450	.921	.971
.4	.390	.454	.412	.859	.946
.5	.333	.427	.365	.779	.912
.6	.267	.393	.309	.681	.864
.7	.197	.349	.244	.564	.807
.8	.1243	.296	.1753	.419	.709
.9	.0556	.224	.1004	.248	.553

where

$$\alpha_{k,l} = \frac{\mathrm{cov}\left(y(s), y(s+(k,l))\right)}{\mathrm{cov}\left(y^2(s)\right)}. \tag{3.12}$$

The elements of matrices Q and $T_{r,s}$ are functions of normalized autocorrelation coefficients $\alpha_{k,l}$. The proof is given in Appendix II.

Equation (3.8) can also be used for estimation in toroidal lattice representations by summing over Ω instead of Ω_I. The resulting difference in the numerical values is negligible for sufficiently large M.

C. Comparison of Estimates

We compare the asymptotic variance of the estimate (3.8), with the asymptotic variances of the coding estimate and the ML estimate for the isotropic conditional model in part c) of Theorem 2. From [21], the asymptotic variance of the coding estimate is

$$M^2 \mathrm{var}(\theta_C^*) = \frac{\theta(1-4\theta\alpha_{1,0})}{2\alpha_{1,0}}. \tag{3.13}$$

Also from [21], the variance of the ML estimate $\hat{\theta}_{\mathrm{ML}}$ is

$$\mathrm{var}\left(\hat{\theta}_{ML}\right) = \frac{0.5}{M^2\left(I(\theta) - 4V_{10}^2(\theta)\right)}, \tag{3.14}$$

where

$$I(\theta) = \frac{1}{4\pi^2} \int\int_0^{2\pi} \frac{(\cos x + \cos y)^2\, dx\, dy}{(1 - 2\theta(\cos x + \cos y))^2}$$

and

$$V_{st}(\theta) = \frac{1}{4\pi^2} \int\int_0^{2\pi} \frac{\cos(sx + ty)\, dx\, dy}{(1 - 2\theta(\cos x + \cos y))}.$$

Tabulated values of V_{10}, (θ), $\alpha_{1,0}$, and $I(\theta)$ are available in [21] for different values of θ. Using these values, (3.11), (3.15), and (3.16), the columns 2–4 of Table I are computed. The asymptotic efficiencies, in columns 5 and 6 are defined by

$$\mathit{eff}(\theta_C') = \frac{\mathrm{var}(\theta_{ML})}{\mathrm{var}(\theta_C')}$$

and

$$\mathit{eff}(\theta^*) = \frac{\mathrm{var}(\theta_{ML})}{\mathrm{var}(\theta^*)}.$$

It is evident that the estimate $\boldsymbol{\theta}^*$ computed using (3.8) is more efficient than the coding estimate for all $\boldsymbol{\theta}$. For small values of θ, the efficiencies of both θ_C' and θ^* are nearly one. But as θ approaches 0.25, the efficiency of θ_C falls compared to that of θ^*. For $\theta = 0.9/4$, the efficiency of θ^* is more than twice that of θ_C'. Note that column 5 is available in [21].

D. Experimental Results

We give the results of applying the estimation scheme (3.8)–(3.9) to some synthetic data generated from known CM models. The synthetic generation scheme is given in Appendix I and the results of the estimation scheme are given below.

Experiment III: (Gaussian Data and a CM model with $N = \{(-1,1),(1,1),(1,-1),(-1,-1)\}$.)

In the true model the values of α and ν are 30.0 and 1.1111 respectively, $\theta_{-1,1} = -0.14$, and $\theta_{1,1} = 0.28$. Using this model, synthetic data was generated. For estimation of the parameters, the sample mean was subtracted and (3.13)–(3.14) were used. The actual values of the estimates of the estimates are $\theta_{-1,1}^* = -0.1410$, $\theta_{1,1}^* = 0.2787$, and $\nu^* = 1.1033$ which are quite close to the true parameters.

Experiment IV: (Synthetic data generated by uniform distribution and the neighbor set in experiment III.)

The estimate $\boldsymbol{\theta}^*$ was not derived using any specific assumption regarding the underlying distribution of y and consequently it should be fairly robust. To test this hypothesis $y(\cdot)$ was synthesized for the model in experiment III, the sequence $\eta(\cdot)$ being obtained from a uniform random number generator with zero mean and unit variance. The values of estimates are $\theta^*_{-1,1} = -0.14282$, $\theta^*_{1,1} = 0.27572$, and $\nu^* = 1.146$ which are not much different from the true $\boldsymbol{\theta}$ compared to the estimates for Gaussian data.

IV. Choice of Neighbors in Simultaneous and Conditional Models

A. Motivation and Possible Approaches

We briefly discuss the need for choosing appropriate SAR or CM models, consider possible approaches and suggest decision rules. From one-dimensional time series analysis, it is known that a model of appropriate order should be fitted to obtain good results in applications like forecasting and control. A similar situation is true in the case of two-dimensional models. The problem becomes more difficult due to the rich variety of model structures. In the two-dimensional case, within the same class of (say) SAR models, different neighbor sets account for different patterns as shown in [18], [19]. Thus the quality of the reconstructed image varies considerably depending on how similar the underlying model is to the true model, and so the use of appropriate neighbor set is important.

Suppose we have an original image, (say) a 64×64 window from one of the Brodatz textures, and fit different SAR of CM models to the texture. It may be argued that by visual inspection of the reconstructed patterns corresponding to the different fitted models, a decision can be made regarding the appropriate model. There are several criticisms of this procedure. First the decision rule is subjective and no quantitative measure of possible error in the decision is given. More significantly, the reconstructed patterns corresponding to an original model and another model which includes the original model and some extra neighbors look very similar. Hence, a decision based on visual inspection is unreliable. Given an arbitrary image it should be possible to choose on a quantitative basis without visual inspection an appropriate model from a family of such models. In the context of this paper, different models should be interpreted as representing different neighbor sets N. We discuss the possible statistical procedures for both SAR and CM models and suggest decision rules separately.

The possible approaches are using pairwise hypothesis testing [8], [11], [12], Akaike's information criterion (AIC) [13], and the Bayes approach [20]. The pairwise hypothesis method has been used for SAR models in [11], [12] for wheat and orange data, for conditional models in field data [12] and texture modeling [8]. The main criticisms of this approach are that the resulting decision rules are not transitive, i.e., if a model C_1 is preferred to C_2 and C_2 is preferred to C_3, then it does not follow that C_1 is preferred to C_3 [22]. Also the decision rules are not consistent, i.e., the probability of choosing an incorrect model does not go to zero even as the number of observations goes to infinity. In the literature on the choice of an appropriate CM model [8], [12] the test statistics are computed using the estimate derived by coding methods. As discussed in Section III, the coding estimates are not efficient and in general more than one estimate results for a given CM model. The use of coding estimates in the choice of neighbor sets leads to several problems [8] and some *ad hoc* decision rules are used.

The model selection problem comes under the category of a multiple decision problem. A method that is well-suited to this problem is to compute a test statistic for different models and choose the one corresponding to the minimum. The AIC criterion and the Bayes method are two such procedures. The AIC method has been used for the choice of SAR models in the literature [13]. The AIC method, in general, gives transitive decision rules but is not consistent even for one-dimensional autoregressive models [23].

B. Bayes Decision Rules for Choice of the SAR Model

We formulate the problem and give the test statistic. The actual derivation of the test statistic can be done by using standard Bayes decision theory as in [20] for SAR models.

Suppose we have three sets N_1, N_2, and N_3 of neighbors containing m_1, m_2, and m_3 neighbors respectively. Corresponding to each N_i, we have a toroidal SAR model C_i

$$y(s) = \sum_{r \in N_k} \theta_{kr} y(s+r) + \sqrt{\rho_k}\,\omega(s), \qquad s \in \Omega_I, \tag{4.1}$$

$$y(s) = \sum_{r \in N_k} \theta_{kr} y_1(s+r) + \sqrt{\rho_k}\,\omega(s), \qquad s \in \Omega_B, \tag{4.2}$$

where $y_1(\cdot)$ is as in Section II-B, $\theta_{k,r} \neq 0$, $r \in N_k$, and $\rho_k > 0$, $k = 1, 2, 3$ and the noise sequence $\{\omega(s)\}$ is Gaussian.

The models C_i, $i = 1, 2, \cdots$ are mutually exclusive. According to Bayesian theory, the optimal decision rule for minimizing the average probability of decision error chooses the model C_i which maximizes the posterior probability $P(C_i/y)$, where y is the vector of all the observations. The quantity $P(C_i/y)$ is computed from the Bayes rule, $P(C_i/y) = p(y/C_i)\,P(C_i)/P(y)$. We will set $P(C_i)$ same for all i in the absence of any contrary information, so that

$$p(y/C_k) = \int p(y/\boldsymbol{\theta}, \rho)\, p(\boldsymbol{\theta}, \rho/C_k)\, |d\boldsymbol{\theta}|\, d\rho.$$

The representation (4.1)–(4.2) together with the Gaussian assumption yields an expression for $p(y/\boldsymbol{\theta}, \rho)$ similar to (2.14) in Section II-C. By assuming that $p(\boldsymbol{\theta}, \rho/C_i)$ is a regular prior density and using asymptotic integration results [24] for large values of M, it can be shown that [20]

$$\begin{aligned} p(y/C_k) \approx {} & -(M^2/2)\ln \bar{\rho}_k \\ & + (1/2) \sum_{s \in \Omega} \ln\left(1 - \bar{\boldsymbol{\theta}}_k^T C_{ks} + \bar{\boldsymbol{\theta}}_k^T Q_{ks} \bar{\boldsymbol{\theta}}_k\right) \\ & - (m_k/2)\ln(M^2) + \ln p(\bar{\boldsymbol{\theta}}_k, \bar{\rho}_k/C_k). \end{aligned}$$

If the prior densities $p(\boldsymbol{\theta}, \rho/C_k)$ are known they can be substituted in $p(y/C_k)$ given above. In general, for the model selection problem the prior densities are hard to

TABLE II
DETAILS OF SAR MODELS FITTED TO GAUSSIAN DATA GENERATED BY $N = \{(-1,0), (1,0), (-1,1), (1,-1), (-1,-1), (1,1)\}$, $\alpha = 30.0$, $\rho = 1.1111$, $\theta_{-1,0} = \theta_{1,0} = 0.12$, $\theta_{-1,1} = \theta_{1,-1} = 0.28$, $\theta_{-1,-1} = \theta_{1,1} = -0.14$[1]

Number	Neighbor Set N	$\hat{\rho}$	Estimate of Coefficients	Test Statistic C_k
1	(-1,0), (1,0), (-1,1), (1,-1), (-1,-1), (1,1)	1.0520	$\theta_{-1,0} = \theta_{1,0} = .1262$ $\theta_{-1,1} = \theta_{1,-1} = .3116$ $\theta_{-1,-1} = \theta_{1,1} = -.1613$	2088.3*
2	(0,-1), (0,1), (-1,1), (1,-1), (-1,-1), (1,1)	1.1540	$\theta_{0,-1} = \theta_{0,1} = .0396$ $\theta_{-1,1} = \theta_{1,-1} = .3254$ $\theta_{-1,-1} = \theta_{1,1} = -.1691$	2480.
3	(-1,0), (0,-1), (-1,-1)	2.5713	$\theta_{0,-1} = .0504$ $\theta_{-1,0} = .2826$ $\theta_{-1,-1} = -.5625$	3863.3
4	(0,-1) (0,1), (-1,0), (1,0)	3.4651	$\theta_{0,-1} = \theta_{0,1} = .0524$ $\theta_{-1,0} = \theta_{1,0} = .1705$	5381.7
5	(-1,1), (1,-1), (-1,-1), (1,1)	1.1595	$\theta_{-1,1} = \theta_{1,-1} = .3332$ $\theta_{-1,-1} = \theta_{1,1} = -.1638$	2523.5
6	(0,-1), (0,1), (-1,0), (1,0), (-1,1), (1,-1), (-1,-1), (1,1)	1.0535	$\theta_{0,-1} = \theta_{0,1} = .0169$ $\theta_{-1,0} = \theta_{1,0} = .1233$ $\theta_{-1,1} = \theta_{1,-1} = .3086$ $\theta_{-1,-1} = \theta_{1,1} = -.1638$	2103.1
7	(-1,0), (1,0), (-1,1), (1,-1), (-1,-1), (1,1) (-2,0), (2,0)	1.0558	$\theta_{-1,0} = \theta_{1,0} = .1276$ $\theta_{-1,1} = \theta_{1,-1} = .3095$ $\theta_{-1,-1} = \theta_{1,1} = -.1578$ $\theta_{-2,0} = \theta_{2,0} = .0079$	2114.7

[1]The estimate of $\alpha = 30.034$.

specify. Hence we derive an approximate test statistic by dropping the terms due to prior densities. Then the decision rule for the choice of appropriate neighbor set is the following.

Choose the neighbor set N_{k*} if

$$k^* = \arg\min_k \{C_k\}, \tag{4.3}$$

where

$$C_k = \left\{ -\sum_{s\in\Omega} \ln\left(1 - 2\bar{\theta}^T C_{ks} + \bar{\theta}_k^T Q_{ks} \bar{Q}\right) + M^2 \ln \bar{\rho}_k + m_k \ln(M^2) \right\},$$

$$Q_{ks} = S_{ks}S_{ks}^T + C_{ks}C_{ks}^T,$$

$$C_{ks} = \operatorname{col}\left[\cos\frac{2\pi}{M}\left((s-1)^T r\right), r \in N_k\right], \tag{4.4}$$

and

$$S_{ks} = \operatorname{col}\left[\sin\frac{2\pi}{M}\left((s-1)^T r\right), r \in N_k\right].$$

Note that if the kth CM model has a unilateral SAR representation, then the decision statistics reduce to

$$C_k = M^2 \ln \bar{\rho}_k + m_k \ln M^2. \tag{4.5}$$

This expression follows from the fact that the Jacobian of the transformation matrix $B(\theta)$ from the noise variates to the observations is approximately unity. The model selection procedure consists of computing C_k for different models, which may be causal or noncausal, and choosing the one corresponding to the lowest C_k. The difference between the Akaike rule and the rule in (4.5) is that the coefficient of m_k in (4.5) is $\ln M^2$ whereas it is 2 (independent of M) in the Akaike rule.

C. Experimental Results

To illustrate the usefulness of the decision rule in (4.2), we consider the synthetic Gaussian data generated by the SAR model $N = N_0 = \{(-1,0), (1,0), (-1,1), (1,-1), (-1,-1), (1,1)\}$. The test statistics C_k for each of the fitted models are given in Table II together with the model description and estimates of parameters. The decision rule correctly picks up the true model. Note that the numerical values corresponding to closely competing models are quite close.

It would be interesting to see if the conclusion reached by the decision rule (4.4) derived using Gaussian assumption for $\{\omega(\cdot)\}$ agrees with visual inspection. To answer

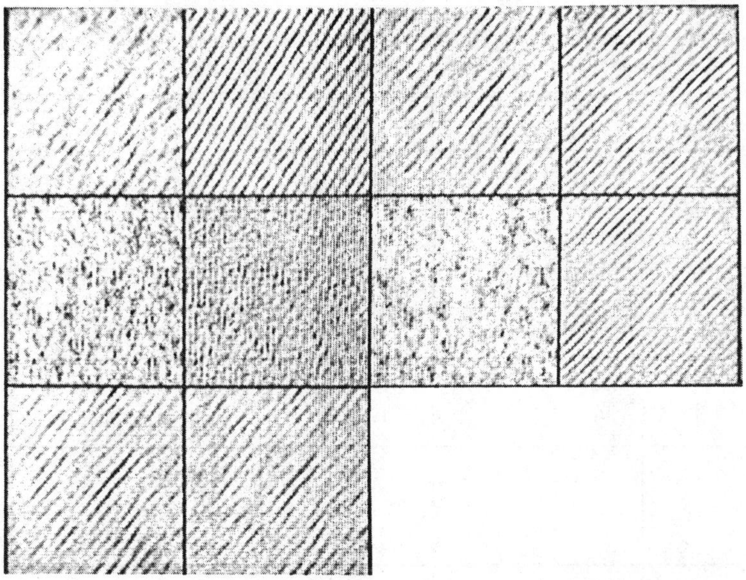

Fig. 1. Synthetic images generated by different SAR models fitted to data generated by the SAR model with true $N = \{(-1,0), (1,0), (-1,1), (1,-1), (-1,-1), (1,1)\}$ in Table II. Windows (1, 1), (1,2), and (1,3) are generated by true N, with true parameters, LS estimates and approximate ML estimates respectively. Windows (2,2) through (3,2) are generated by models 3 through 7 in Table II and Window (1,4) from model 2 in Table II. Window (2, 1) is generated by a SAR model with $N = \{(-1,0), (1,0), (0,-1)\}$.

this query, synthetic patterns corresponding to some of the fitted models in Table II are given in Fig. 1. We shall use matrix notation in referring to the windows in Fig. 1. The image (1, 1) corresponds to the true model in Table II and (1, 2) is generated by the true neighbor set N, with the LS estimates replacing the true parameters. Note that the quality of the image in (1,2) is poor, due to the inconsistency of the LS estimate. The window (1,3) corresponds to the true neighbor set with the parameters estimated by the approximate ML scheme and is similar to the original pattern. The windows (2,2) through (3,2) are generated by models 3 through 7 in Table II. Approximate ML estimation method is used for parameter estimation in these windows. The image (2, 1) is generated by the model with $N = \{(-1,0), (0,-1), (-1,-0)\}$ and is not at all good. The window (2,3) is generated by $N = \{(0,1), (0,-1), (-1,0), (1,0)\}$ and does not possess the diagonal patterns present in the original image in (1, 1). The window (2,4) is generated by a SAR model with $N = \{(-1,1), (1,-1), (-1,-1), (1,1)\}$, a subset of the neighbor set of the true model and is not as good as the image (1,3). Lastly, the image (3, 1) depends on the eight nearest neighbors and looks very similar to image (1,3).

In general, the patterns corresponding to two SAR models with neighbor sets N_1 and N_0, such that $N_0 \subseteq N_1$, the common coefficients being same, others being negligible, are visually very similar to one another but the decision rule picks up the true model. Similar results are noted for the case, when the underlying distribution of $\omega(\cdot)$ is uniform. The quality of synthesized pictures is inferior compared to the Gaussian case, due to the violation of the distribution assumption.

D. Bayes Decision Rule for the Choice of CM Models

Suppose we have three sets N_1, N_2, N_3 of neighbors containing $2m_1$, $2m_2$, $2m_3$ members respectively. Corresponding to each N_k, we write the CM model as

$$y(s) = \sum_{r \in N_k} \theta_{kr} y(s+r) + e(s), \qquad s \in \Omega_I, \quad (4.6)$$

$$y(s) = \sum_{r \in N_k} \theta_{kr} y_1(s+r) + e(s), \qquad s \in \Omega_B, \quad (4.7)$$

where $y_1(\cdot)$ is defined in Section II-B, $\theta_{kr} \neq 0$, $r \in N_k$, $\nu_k > 0$, $k = 1,2,3$, and $\{e(s)\}$ is Gaussian. Then the decision rule for the choice of appropriate neighbors is: choose the neighbor set N_{k*} if,

$$k^* = \arg\min_k \{C_k\}, \quad (4.8)$$

where

$$C_k = -\sum_{S \in \Omega} \ln(1 - 2\theta^{*T}_k \phi_{ks}) + M^2 \ln \nu_k + m_k \ln(M^2), \quad (4.9)$$

$$\theta_k^* = \text{col}[\theta_r^*, r \in N_{Sk}],$$

and

$$\phi_{k,s} = \text{col}\left[\cos\frac{2\pi}{M}((s-1)^T r, r \in N_{Sk}\right]$$

where N_{Sk} is the asymmetrical half of N_k, i.e.,

$$N_k = N_{Sk} \cup \bar{N}_{Sk},$$

$$\bar{N}_{Sk} = \{r: -r \in N_{Sk}\}, N_{Sk} \cap \bar{N}_{Sk} = \varnothing.$$

The model selection procedure consists of computing C_k

TABLE III
DETAILS OF CM MODELS FITTED TO THE GAUSSIAN DATA GENERATED BY $N = \{(-1,1),(1,1),(1,-1),(-1,-1)\}$, $\alpha = 30.0$, $\nu = 1.1111$, $\theta_{-1,1} = \theta_{1,-1} = -0.14$, AND $\theta_{1,1} = \theta_{-1,1} = 0.28$[2]

	Number	Neighbor Set N_l	$\hat{\nu}$	Estimate of Coefficients	Test Statistics C_k
	1	(1,1)	1.1638	$\theta_{1,1}$ = .3116	1575.3
	2	(-1,1)	1.3520	$\theta_{-1,1}$ = .2093	1628.9
True Model	3	(-1,1), (1,1)	1.1033	$\theta_{-1,1}$ = -.1410 $\theta_{1,1}$ = .27875	1464.30*
	4	(0,1), (1,0)	1.4934	$\theta_{0,1}$ = -.0052 $\theta_{1,0}$ = -.0020	1659.7
	5	(-1,1), (1,1) (1,0)	1.1033	$\theta_{-1,1}$ = -.14101, $\theta_{1,1}$ = .27877, $\theta_{1,0}$ = -.0051	1472.7
	6	(-1,1), (1,1), (0,1)	1.1033	$\theta_{-1,1}$ = -.1410 , $\theta_{1,1}$ = .27873, $\theta_{0,1}$ = -.001717	1472.2
	7	(-1,1), (1,1) (0,1), (1,0)	1.1033	$\theta_{-1,1}$ = -.14101, $\theta_{1,1}$ = .27876 $\theta_{1,0}$ = .0049, $\theta_{0,1}$ = -.0009	1481.0

[2]The estimate of $\hat{\alpha} = 30.00$.

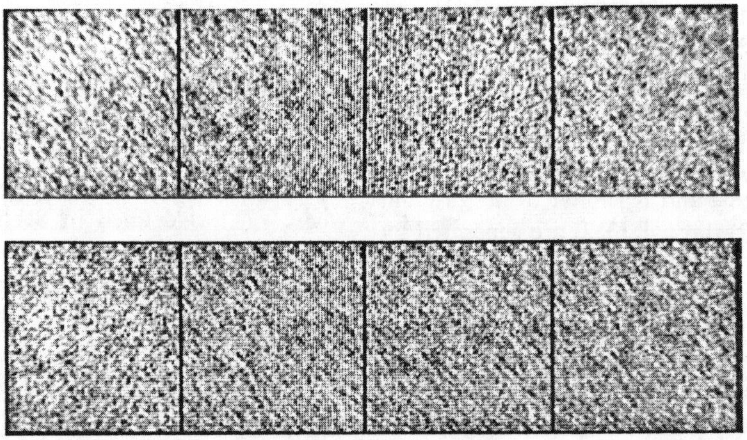

Fig. 2. Synthetic images generated by different CM models fitted to data generated by the CM model with true $N = \{(-1,1),(1,-1),(-1,-1),(1,1)\}$ in Table III. Window (1, 1) is generated by the true model in Table III. The images (1, 2) through (2, 4) are generated by models 1 through 7 in Table III. Windows (2, 2), (2, 3), and (2, 4) corresponding to models which include true N and some extra neighbors look visually similar to the window (1, 4) corresponding to true N but the decision rule (4.9) correctly eliminates the former models.

for different models and choosing the one corresponding to the lowest C_k.

E. Experimental Results

To illustrate the usefulness of the decision rule in (4.8), we consider the synthetic data generated by the CM model with $N = \{(-1,1),(1,1),(1,-1),(-1,-1)\}$ considered in experiment III. The test statistics C_k in (4.9) was computed for each of the fitted models in Table III. The decision rule correctly picks up the true model.

The synthetic patterns corresponding to some of the fitted models in Table III are given in Fig. 2. The window (1, 1) corresponds to the pattern generated by the neighbor set N mentioned above and parameters $\theta_{-1,1} = -0.14$ and $\theta_{1,1} = 0.28$. The images (1, 2) through (2, 4) are generated by models 1 through 7 in Table III. The parameters are estimated by least squares method. Image (1, 4) is generated with the true neighbor set and estimated parameters and is very similar to the image (1, 1). The window (2, 1) corresponds to the inappropriate model with $N = \{(0,1),(1,0),(0,-1),(-1,0)\}$ and is comparatively of poor quality. Image (2,4) is generated by a CM model with $N = \{(1,0), (0,1), (-1,1), (1,1), (-1,0), (0,-1), (1,-1), (-1,-1)\}$ and is very similar to the original. However, the decision rule correctly eliminates this model in preference to correct model.

V. Conclusion

We have considered some aspects of statistical inference in spatial interaction models. Specifically, we have considered some estimation schemes in both SAR and CM models. We have also given decision rules for the choice of neighbors in these models. We have not suggested any preference of simultaneous AR models over conditional models or vice versa. The appropriateness of one class of models over the other for the given data should be inferred from the data itself.

Acknowledgment

The authors would like to thank the reviewers for their comments which have improved the readability of the paper. The second author would like to sincerely thank Professor Azriel Rosenfeld for his continued interest and encouragement.

Appendix I

Synthetic Generation of Data Obeying Known CM Models

Equation (3.5) can be equivalently written as [17] $\sqrt{H(\boldsymbol{\theta})}\,y = \sqrt{\nu}\,\boldsymbol{\eta}$, where $\boldsymbol{\eta}$ is an i.i.d. noise sequence with zero mean and unit variance, and is of known distribution, not necessarily Gaussian. The synthetic generation is done by assigning some arbitrary values in the stationary region to $\boldsymbol{\theta}$ and the noise sequence is generated from a random number generator. Since $\sqrt{H(\boldsymbol{\theta})}$ is a block-circulant matrix Fourier computations can be used. The synthetic generation scheme is as follows [17]:

$$y = \sum_{\Omega}\left(f_s x_s / \sqrt{\mu_s'(\boldsymbol{\theta})}\right) + \alpha(\mathbf{1}), \tag{A1}$$

where $x_s = (\sqrt{\nu}/M^2) f_s^{*T}\boldsymbol{\eta}$, $\mathbf{1} = (1,1,\cdots,1)$, M^2-vector, and $\mu_s'(\boldsymbol{\theta})$, and $s \in \Omega$ are defined in Section III-A2).

We generate the vector $\boldsymbol{\eta}$ from pseudorandom numbers, generate its Fourier transform $\{x_s\}$, and finally use (A1).

Appendix II

Proof of Theorem 2: a) We have from (3.8)

$$\boldsymbol{\theta}^* = \left[\sum_{\Omega} \boldsymbol{q}(s)\boldsymbol{q}^T(s)\right]^{-1}\left(\sum_{\Omega}\boldsymbol{q}(s)y(s)\right). \tag{A2}$$

Substituting for $y(s)$ from (3.1) and simplifying, we have

$$\left[\sum_{\Omega}\boldsymbol{q}(s)\boldsymbol{q}^T(s)\right](\boldsymbol{\theta}^* - \boldsymbol{\theta}) = \sum_{\Omega}\boldsymbol{q}(s)e(s). \tag{A3}$$

Since $E(\boldsymbol{q}(s)e(s)) = 0$ by (3.2) and $[\Sigma_{\Omega}\boldsymbol{q}(s)\boldsymbol{q}^T(s)]$ is a positive definite matrix, the consistency of the estimate $\boldsymbol{\theta}$ follows.

b) To make our calculations easy we assume from now on that $e(s)$ is normally distributed. Multiplying (A3) by its transpose and taking expectations, we have

$$E\left[\sum_s \boldsymbol{q}(s)\boldsymbol{q}^T(s)(\boldsymbol{\theta}^* - \boldsymbol{\theta})(\boldsymbol{\theta}^* - \boldsymbol{\theta}^T)\left(\sum_s \boldsymbol{q}(s)\boldsymbol{q}^T(s)\right)\right] = E\left[\sum_s \boldsymbol{q}(s)e(s)\sum_r(\boldsymbol{q}(r)e(r))^T\right]. \tag{A4}$$

The RHS of (A4)

$$\begin{aligned} &= \sum_s\sum_r E(\boldsymbol{q}(s)e(s))E(\boldsymbol{q}^T(r)e(r)) \\ &\quad + \sum_s\sum_r E(\boldsymbol{q}(s)e(r))E(\boldsymbol{q}^T(r)e(s)) \\ &\quad + \sum_s\sum_r E(e(s)e(r))E(\boldsymbol{q}(r)\boldsymbol{q}^T(s)) \\ &= I + II + III, \end{aligned} \tag{A5}$$

where

$$I = 0, \quad \text{by using (3.2)}, \tag{A6}$$

$$II = 2M^2\nu^2 I_{m\times m}, \quad \text{by using (3.2) and } E(e(s)y(s)) = \nu, \tag{A7}$$

$$III = \nu\sum_s E(\boldsymbol{q}(s)\boldsymbol{q}^T(s)) - \nu \sum_{\substack{r\ s \\ (s-r)\in N}}\sum \theta(s-r)E(\boldsymbol{q}(r)\boldsymbol{q}^T(s)), \quad \text{using (3.3)}. \tag{A8}$$

Defining

$$E(\boldsymbol{q}(s)\boldsymbol{q}^T(s)) = Q, \quad m\times m \text{ matrix},$$
$$E(\boldsymbol{q}(r)\boldsymbol{q}^T(s)) = \boldsymbol{T}_{r,s}, \quad m\times m \text{ matrix},$$
$$III = M^2\nu Q - \nu\sum_{\substack{s\ r \\ (s-r)\in N}}\sum \theta(s-r)\boldsymbol{T}_{r,s}. \tag{A9}$$

Substituting (A6), (A7), and (A9) the RHS of (A4)

$$= M^2\left[\nu Q + 2\nu^2 I_{m\times m} - \nu\sum_{\substack{s\ r \\ (s-r)\in N}}\sum \theta_{(s-r)}\boldsymbol{T}_{r,s}\right]. \tag{A10}$$

For large values of M,

$$\frac{1}{M^2}\sum_s \boldsymbol{q}(s)\boldsymbol{q}^T(s) = Q + \xi(M), \tag{A11}$$

where $\xi(M)$ is such that

$$E(\xi^2(M)) = O(1/M^2). \tag{A12}$$

Using (A11)

$$\begin{aligned}&(\text{LHS of (A4)})/M^4 \\ &= E\left[(Q + \xi(M))(\boldsymbol{\theta}^* - \boldsymbol{\theta})(\boldsymbol{\theta}^* - \boldsymbol{\theta})^T(\boldsymbol{\theta} + \xi(M))^T\right]\end{aligned} \tag{A13}$$

$$\simeq QE\left[(\boldsymbol{\theta}^* - \boldsymbol{\theta})(\boldsymbol{\theta}^* - \boldsymbol{\theta})^T\right]Q + O(1/M^2), \quad \text{by (A12)}. \tag{A14}$$

Substitution of (A10) and (A13) into (A4) yields (3.10). Q.E.D.

c) Consider the isotropic conditional model with $N = \{(0,1), (1,0), (0,-1), (-1,0)\}$, we have

$$y(s) = \theta q(s) + e(s), \tag{A15}$$

where $q(s) = y(s + (0,1)) + y(s + (0,-1)) + y(s + (1,0)) + y(s + (-1,0))$.

The part b) of Theorem 2 yields

$$E(\theta^* - \theta)^2 = \frac{1}{M^2} \frac{1}{\left(E(q^2(s))\right)^2} \cdot \left[4\nu^2 + \nu E(q^2(s)) - \theta\nu \sum_{s} \sum_{\substack{r \\ (s-r)\in N}} E(q(s)q(r))\right]. \tag{A16}$$

Let

$$\gamma_{k,l} = E[y(s)y(s + (k,l))] \tag{A17}$$

and

$$\gamma_{k,l} = \gamma_{|k|,|l|} = \gamma_{l,k}. \tag{A18}$$

Express the higher order correlations $\gamma_{2,1}$, $\gamma_{2,0}$, and $\gamma_{3,0}$ in terms of $\gamma_{0,0}$, $\gamma_{1,0}$, and $\gamma_{1,1}$ by the following formulas

$$\gamma_{0,0} = \nu/(1 - 4\theta\alpha_{1,0}), \tag{A19}$$

$$\gamma_{2,1} = \frac{1}{2\theta}\gamma_{1,1} - \gamma_{1,0}, \tag{A20}$$

$$\gamma_{3,0} = \gamma_{1,0}\left(1 + \frac{1}{\theta^2}\right) - \frac{\gamma_{0,0}}{\theta} - \frac{3\gamma_{1,1}}{\theta}, \tag{A21}$$

$$\gamma_{2,0} = \frac{1}{\theta}\gamma_{1,0} - \gamma_{0,0} - 2\gamma_{1,1}. \tag{A22}$$

Equations (A19)–(A22) can be obtained by multiplying $y(s)$ by appropriately chosen $y(s + (k,l))$ and taking expectations on both sides of the equation.

Consider the various terms in (A16).

$$E(q^2(s)) = \frac{1}{\theta}4\gamma_{1,0}, \quad \text{using (A22)}, \tag{A23}$$

$$\sum_{s} \sum_{\substack{r \\ (s-r)\in N}} E(q(s)q(r)) = 4[9\gamma_{1,0} + \gamma_{3,0} + 6\gamma_{2,1}]$$

$$= 4\left[4\gamma_{1,0} + \frac{1}{\theta^2}\gamma_{1,0} - \frac{1}{\theta}\gamma_{0,0}\right], \quad \times \text{by (A20)–(A21)}, \tag{A24}$$

Substitution of (A23) and (A24) in (A16) yields

$$E(\theta^* - \theta)^2 = \frac{1}{M^2} \frac{\theta^2}{16\gamma_{1,0}^2}\left[4\nu^2 + 4\nu\gamma_{0,0} - 16\theta\gamma_{1,0}\nu\right],$$

which on using (A19) and $\alpha_{1,0} = \gamma_{1,0}\,\gamma_{0,0}$ yields

$$E(\theta^* - \theta)^2 = \frac{2\theta^2(1 - 4\theta\alpha_{1,0})^2}{4M^2\alpha_{1,0}^2}.$$

References

[1] A. K. Jain and J. R. Jain, "Partial difference equations and finite difference methods in image processing—Part 2: image restoration," *IEEE Trans. Automat. Contr.*, vol. AC-23, pp. 817–833, Oct. 1978.

[2] R. L. Kashyap and R. Chellappa, "Image restoration using random fields," in *Proc. Eighteenth Ann. Allerton Conf. Commn., Contr., Comput.*, University of Illinois, Urbana, IL, pp. 956–965, Oct. 1980.

[3] J. W. Woods, "Markov image modeling," *IEEE Trans. Automat. Contr.*, vol. AC-23, pp. 846–850, Oct. 1978.

[4] E. J. Delp, R. L. Kashyap, and O. R. Mitchell, "Image data compression using autoregressive time series models," *Pattern Recog.*, vol. 11, pp. 313–323, Dec. 1979.

[5] B. H. McCormick and S. N. Jayaramamurthy, "Time series models for texture synthesis," *Int. J. Comput. Inform. Sci.*, vol. 3, no. 4, pp. 329–343, 1974.

[6] J. T. Tou, "Pictorial feature extraction and recognition via image modeling," *Comput. Graphics Image Proc.*, vol. 12, pp. 376–406, Apr. 1980.

[7] M. Hassner and J. Sklansky, "The use of Markov random fields as models of textures," *Comput. Graphics Image Proc.*, vol. 12, pp. 357–370, Apr. 1980.

[8] G. R. Cross and A. K. Jain, "Markov random field texture models," in *Proc. IEEE Comput. Soc. Conf. Pattern Recog. Image Proc.*, Dallas, TX, pp. 597–601, Aug. 1981.

[9] D. M. Goodman and M. P. Ekstrom, "Multi-dimensional spectral factorization and unilateral AR models," *IEEE Trans. Automat. Contr.*, vol. AC-25, pp. 258–262, Apr. 1980.

[10] J. W. Woods, "Two-dimensional discrete Markov random fields," *IEEE Trans. Inform. Theory*, vol. 18, pp. 232–240, Mar. 1972.

[11] P. Whittle, "On stationary processes in the plane," *Biometrika*, vol. 41, pp. 434–449, 1954.

[12] J. E. Besag, "Spatial interaction and statistical analysis of lattice systems," *J. Royal Stat. Soc., Ser. B*, vol. B-36, pp. 192–236, 1974.

[13] W. E. Larimore, "Statistical inference on stationary random fields," in *Proc. IEEE*, vol. 65, pp. 961–960, June 1977.

[14] K. Ord, "Estimation methods for models of spatial interaction," *J. Amer. Stat. Ass.*, vol. 70, pp. 120–126, Mar. 1975.

[15] R. L. Kashyap, "Univariate and multivariate random field models for images," *Comput. Graphics Image Proc.*, vol. 12, pp. 257–270, Mar. 1980.

[16] ——, "Random field models on finite lattices for finite images," in *Proc. 5th Int. Conf. Pattern Recog.*, Miami, FL, Dec. 1980.

[17] ——, "Finite lattice random field models for finite images," presented at the Conf. Inform. Sci. Syst., Baltimore, MD, Mar. 1981.

[18] R. Chellappa, "Stochastic models for image analysis and processing," Ph.D. dissertation, Purdue University, W. Lafayette, IN, Aug. 1981.

[19] R. Chellappa and R. L. Kashyap, "Synthetic generation and estimation in random field models of images," in *Proc. IEEE Comput. Soc. Conf. Pattern Recog. Image Proc.*, Dallas, TX, pp. 577–582, Aug. 1981.

[20] R. L. Kashyap, R. Chellappa, and N. Ahuja, "Decision rules for the choice of neighbors in random field models of images," *Comput. Graphics Image Proc.*, vol. 15, pp. 301–318, Apr. 1981.

[21] P. A. P. Moran and J. E. Besag, "On the estimation and testing of spatial interaction in Gaussian Lattices," *Biometrika*, vol. 62, no. 3, pp. 555–562, 1975.

[22] R. L. Kashyap, "A Bayesian comparison of different classes of models using empirical data," *IEEE Trans. Automat. Contr.*, vol. AC-22, pp. 715–727, Oct. 1977.

[23] ——, "Inconsistency of the AIC rule for estimating the order of autoregressive models," *IEEE Trans. Automat. Contr.*, vol. AC-25, pp. 996–998, Oct. 1980.

[24] D. V. Lindley, "The use of prior probability distributions in statistical inference and decisions," in *Proc. Fourth Berkeley Symp. Math. Statist. Prob.*, vol. 1, pp. 453–468, 1961.

[25] R. Chellappa and R. L. Kashyap, "Digital image restoration using spatial interaction models," *IEEE Trans. Acous., Speech, Signal Processing*, pp. 461–472 June 1982.

[26] A. K. Jain, "Advances in mathematical models for image processing," in *Proc. IEEE*, vol. 69, no. 5, pp. 512–528, May 1981.

[27] N. K. Bose, *Applied Multidimensional Systems Theory*. New York: Van Nostrand Reinhold, 1982.

Part V
Reconstruction

ONE of the most important applications of multidimensional digital signal processing is the reconstruction of three-dimensional structures (signals) from two-dimensional observations. Typically, a three-dimensional structure is illuminated, and the intensity distribution of the illumination after transmission through the structure is observed by a two-dimensional sensor. The problem is to reconstruct the three-dimensional structure from such two-dimensional observed data. There are two cases which involve different mathematical formulations. In the first case, the wavelength of the illumination is comparable to the sizes of the details of interest in the three-dimensional structure. A diffraction model is needed, and it can be shown that properly recorded two-dimensional data generates the three-dimensional Fourier transform of the structure under study.

In the second case, the wavelength of the illumination is much smaller than the sizes of the details in the three-dimensional structure. Then, a geometrical optics model can be used, and the recorded two-dimensional data, after proper scaling, turn out to be a projection (i.e., line integral) of the three-dimensional structure. The problem of the reconstruction of a three-dimensional structure from its projections is usually referred to as computerized tomography (CT) because a major application of such reconstruction is in diagnostic radiology, where images of sections of a human body are reconstructed.

Phase Retrieval

The two important issues are uniqueness of solutions, and algorithms for obtaining solutions. It is obvious that without additional constraints on the signal, the phase angle can be arbitrary. Most work in the phase retrieval area has imposed the limited support constraint on the signal. More precisely, the support of the signal is assumed to be compact. In the early 1960's, Walther [16] and Hofstetter [8] showed that for one-dimensional signals, the solution to the phase retrieval problem is still not unique under the compact-support constraint. Then, in the late 1970's, Bruck and Sodin [1] pointed out the surprising fact that the solution is nevertheless almost always unique for signals of two or more dimensions. A rather complete uniqueness theory for discrete signals is presented in the paper by Hayes, which is the first paper in our collection here. The corresponding theory for continuous signals is presented in a paper by J. L. C. Sanz and T. S. Huang in [10].

Hayes' paper also describes an iterative algorithm for obtaining solutions, and in a paper in [14], Fienup discussed in more detail a number of iterative algorithms for phase retrieval and related problems. All of these algorithms can be considered as modifications of the basic algorithm of Gerchberg and Saxton [4]. Additional papers on algorithms (especially those of Bates and coworkers) can be found in [9], [10] and [14]. It is important to note that none of the algorithms reported in the literature is guaranteed to converge to the true solution, although they often do if the initial guess solution is chosen properly, and if one imposes additional constraints (such as non-negativity) on the signal.

Finally, we mention the interesting result of Van Hove *et al.* [15]: A signal can be reconstructed uniquely, if one is given (1) the magnitude of the Fourier transform of the signal, and (2) the phase of the transform quantized to 1 bit.

Computerized Tomography

Although the main impetus behind the development of techniques for three-dimensional reconstruction from projections has been, at least until recently, the area of X-ray computerized tomography, there are many other imaging modalities which give rise to the same mathematical problem. The papers in our collection treat CT imaging with X-ray, ultrasound, NMR, and radar. An important omission is the reconstruction of molecular structures from electron micrographs [2]. The reader is also referred to [6], [7] and [13] for further applications.

The second paper in our collection, by Munson, O'Brien, and Jenkins, reviews the basic mathematics of X-ray CT and shows that the same mathematical formulation underlies the imaging process of spotlight-mode synthetic aperture radar.

The third paper by Pan and Kak shows that for reconstruction with diffracting radiation, two-dimensional interpolation in the frequency domain may be superior to filtered-backpropagation. For a rather extensive discussion comparing reconstruction algorithms for straight-ray and diffraction tomography, the reader is referred to Kak's chapter in [5]. The reader is also referred to [11] where some signal processing issues in diffraction tomography are highlighted.

The fourth and last paper introduces the exciting new area of NMR (nuclear magnetic resonance) imaging to the novice. It is to be emphasized that the physics and the mathematics of NMR are quite different from other CT modalities. This paper presents the mathematics behind NMR imaging, but stops short of discussing the reconstruction algorithms. It is intended as a resource for those who wish to develop their own reconstruction algorithms.

An important issue not discussed in our collection of papers is: How do we deal with the problem of missing data? For example, in applying X-ray CT to nondestructive testing, it may not be possible to take projections for some viewing angle ranges. Then, we have to reconstruct with incomplete data. One approach is to extrapolate. Interested readers are referred to [3], [9], [10], [12], and [14]. In particular, the chapter by Sanz and Huang in [9] gives an overview of the problem of support-limited signal extrapolation.

Bibliography

[1] Y. H. Bruck and L. G. Sodin, "On the ambiguity of the image reconstruction problem" *Opt. Comm.*, pp. 304–308, Sept. 1979.

[2] D. J. deRosier and A. Klug, "Reconstruction of three-dimensional structures from electron micrographs," *Nature*, vol. 217, pp. 130–134, 1968.

[3] R. W. Gerchberg, "Super-resolution through error energy reduction," *Opt. Acta*, vol. 21, pp. 709–720, 1974.

[4] R. W. Gerchberg and W. O. Saxton, "A practical algorithm for the determination of phase from image and diffraction plane pictures," *Optik*, vol. 35, pp. 237 ff, 1972.

[5] S. Haykin, Ed., *Array Signal Processing*. Englewood Cliffs, NJ: Prentice-Hall, 1985.

[6] G. T. Herman, Ed., *Image Reconstruction from Projections*. Heidelberg, W. Germany: Springer Verlag, 1979.

[7] G. T. Herman and F. Natterer, Eds., *Mathematical Aspects of Computerized Tomography*. Heidelberg, W. Germany: Springer Verlag, 1981.

[8] E. M. Hofstetter, "Construction of time-limited functions with specified autocorrelation functions," *IEEE Trans. Inform. Theory*, vol. IT-10, pp. 119–126, Apr. 1964.

[9] T. S. Huang, Ed. *Advances in Computer Vision and Image Processing, Vol. 1: Image Reconstruction from Incomplete Observations*. Greenwich, CT: JAI Press, 1984.

[10] J. Opt. Soc. Amer., *Special Issue on Signal Recovery*, vol. 73, pp. 1409–1616, Nov. 1983.

[11] M. Kaveh, M. Soumekh, and J. F. Greenleaf, "Signal processing for diffraction tomography," *IEEE Trans. Sonics Ultrason.*, vol. SU-31, pp. 230–238, July 1984.

[12] A. Papoulis, "A new algorithm in spectral analysis and band-limited extrapolation," *IEEE Trans. Circuit Syst.*, vol. CAS-22, pp. 735–742, Sept. 1975.

[13] Proc. IEEE, *Special Issue on Computerized Tomography*, vol. 71, pp. 289–448, Mar. 1983.

[14] W. T. Rhodes, J. R. Fienup, and B. E. A. Saleh, Eds. *Transformations in Optical Signal Processing*, Proc. SPIE 373, 1981.

[15] P. L. Van Hove, M. H. Hayes, J. S. Lim, and A. V. Oppenheim, "Signal reconstruction from signed Fourier transform magnitude," *IEEE Trans. Acoust., Speech, Signal Processing*, vol. ASSP-31, pp. 1286–1293, Oct. 1983.

[16] A. Walther, "The question of phase retrieval in optics," *Opt. Acta*, vol. 10, pp. 41 ff, 1963.

The Reconstruction of a Multidimensional Sequence from the Phase or Magnitude of Its Fourier Transform

MONSON H. HAYES, MEMBER, IEEE

Abstract—This paper addresses two fundamental issues involved in the reconstruction of a multidimensional sequence from either the phase or magnitude of its Fourier transform. The first issue relates to the uniqueness of a multidimensional sequence in terms of its phase or magnitude. Although phase or magnitude information alone is not sufficient, in general, to uniquely specify a sequence, a large class of sequences are shown to be recoverable from their phase or magnitude. The second issue which is addressed in this paper concerns the actual reconstruction of a multidimensional sequence from its phase or magnitude. For those sequences which are uniquely specified by their phase, several practical algorithms are described which may be used to reconstruct a sequence from its phase. Several examples of phase-only reconstruction are also presented. Unfortunately, however, even for those sequences which are uniquely defined by their magnitude, it appears that a practical algorithm is yet to be developed for reconstructing a sequence from only its magnitude. Nevertheless, an iterative procedure which has been proposed is briefly discussed and evaluated.

I. Introduction

UNDER a variety of conditions, a multidimensional sequence may be reconstructed from partial information about its Fourier transform. For example, if an m-dimensional (m-D) sequence $x(n_1, \cdots, n_m)$ is zero whenever any of the indexes n_k for $k = 1, \cdots, m$ are negative, then $x(n_1, \cdots, n_m)$ may be recovered from the real part or, except for $x(0, \cdots, 0)$, from the imaginary part of its Fourier transform [1]. If, on the other hand, $x(n_1, \cdots, n_m)$ is minimum phase [2], then it can be recovered from the magnitude or, to within a scale factor, from the phase of its Fourier transform.

Although Fourier transform phase or magnitude information alone is not, in general, sufficient to uniquely specify a signal, the ability to reconstruct a signal from only phase or magnitude information would be useful in number of important applications. For example, in many problems which arise in X-ray crystallography, electron microscopy, coherence theory, and optics [3], only the magnitude of the Fourier transform of an electromagnetic wave may be recorded or is available for measurement. Therefore, a complete specification of the electromagnetic wave depends upon the retrieval of Fourier transform phase information from only spectral magnitude information. In other applications, either the spectral magnitude or phase of a signal may be severely distorted so that the restoration of the signal must rely only on the undistorted component. For example, in the class of problems referred to as blind deconvolution, a desired signal is to be recovered from an observation which is the convolution of the desired signal with some unknown signal [4]. Since little is usually known about either the desired signal or the distorting signal, deconvolution of the two signals is generally a very difficult problem. However, in the special case in which the distorting signal is known to have

Manuscript received December 19, 1980; revised September 28, 1981. This work was performed at the Lincoln Laboratory, a center for research operated by the Massachusetts Institute of Technology, Cambridge, MA, with the support of the Department of the Air Force.

The author is with the Department of Electrical Engineering, Georgia Institute of Technology, Atlanta, GA 30332.

Reprinted from *IEEE Trans. Acoust., Speech, Signal Processing,* vol. ASSP-30, pp. 140–154, Apr. 1982.

a phase which is identically zero, the phase of the signal is undistorted. Such a situation occurs, at least approximately, in long-term exposure to atmospheric turbulence or when images are blurred by severely defocused lenses with circular aperture stops [5]. In this case, except for phase reversals, the phase of the observed signal is approximately the same as the phase of the desired signal and, therefore, it is of interest to consider signal reconstruction from phase information alone.

In this paper, two fundamental issues are addressed which relate to the problem of reconstructing an m-D sequence from either the phase or magnitude of its Fourier transform. The first is concerned with the development of conditions under which an m-D sequence is uniquely defined by its Fourier transform phase or magnitude. In general, of course, phase or magnitude information alone is not sufficient to uniquely specify an m-D sequence. For example, any m-D sequence may be convolved with a zero phase sequence to produce another m-D sequence with the same phase. Similarly, any m-D sequence may be convolved with an all-pass sequence to produce another m-D sequence with the same magnitude. Therefore, without any additional information or constraints, the Fourier transform phase or magnitude may, at best, uniquely specify an m-D sequence to within a zero phase or all-pass convolutional factor, respectively. Nevertheless, with a few basic results from the theory of polynomials in several variables, some useful conditions may be derived under which an m-D sequence is uniquely defined by the phase or magnitude of its Fourier transform. The second question which is addressed in this paper concerns the development of algorithms for reconstructing an m-D sequence from either its phase or magnitude when the sequence satisfies the appropriate uniqueness conditions.

This paper is organized as follows. In Section II, the basic results from the algebra of polynomials in more than one variable which are required to develop the ideas in this paper are briefly reviewed. Then, in Section III, the notation and terminology related to multidimensional signals are presented. The uniqueness question is then addressed in Section IV where conditions are presented under which an m-D sequence is uniquely defined in terms of the phase or magnitude of its Fourier transform. Finally, Section V presents some algorithms for reconstructing an m-D sequence from either the phase or magnitude of its Fourier transform. Also included in this section are several examples which illustrate this reconstruction.

II. Polynomials in Two or More Variables

In order to consider the uniqueness of a multidimensional sequence in terms of its Fourier transform phase or magnitude, some theory from the algebra of polynomials in two or more variables is required. This section is therefore intended to provide the necessary background. Specifically, some notation and terminology related to the algebra of polynomials in two or more variables are briefly reviewed. In addition, two theorems are presented which are of considerable importance in many multidimensional signal processing applications and will be referred to frequently in this paper. Proofs of these theorems as well as a detailed treatment of many topics not presented in this section may be found in [6].

A. Definitions

A polynomial in the m variables $\mathbf{z} = (z_1, z_2, \cdots, z_m)$ is a function of the form

$$p(\mathbf{z}) = p(z_1, \cdots, z_m) = \sum_{k_1 + \cdots + k_m \leq N} \cdots \sum c(k_1, \cdots, k_m) z_1^{k_1} \cdots z_m^{k_m} \tag{1}$$

where $k_1, k_2, \cdots, k_m$ are nonnegative integers, and where $c(k_1, \cdots, k_m)$ are arbitrary numbers which are referred to as the coefficients of the polynomial. Each term in the sum (1) is called a monomial. Thus, monomials are functions of the form

$$f(\mathbf{z}) = f(z_1, \cdots, z_m) = c z_1^{k_1} z_2^{k_2} \cdots z_m^{k_m}. \tag{2}$$

The degree of the monomial in (2) is defined to be

$$d(f) = k_1 + k_2 + \cdots + k_m \tag{3}$$

and the degree of the polynomial in (1) is defined to be equal to the degree of the monomial which has the largest degree and a nonzero coefficient. Although not standard terminology, polynomials which consist of a sum of two or more monomials will be referred to as nontrivial polynomials. A trivial polynomial is therefore either a constant or a monomial of nonzero degree.

It is often useful to consider $p(\mathbf{z})$ in (1) as a polynomial in one variable, say z_k, with coefficients which are polynomials in the remaining $(m - 1)$ variables, e.g.,

$$p(\mathbf{z}) = \sum_{n=0}^{N} \xi_k(n) z_k^n \tag{4}$$

where $\xi_k(n)$ for $n = 0, 1, \cdots, N$ are polynomials in the $(m - 1)$ variables z_i for $i \neq k$. Written in this form, the largest value of n for which $\xi_k(n)$ is nonzero is referred to as the degree of $p(\mathbf{z})$ with respect to the variable z_k. The polynomial $p(\mathbf{z})$ will therefore be said to be of degree $\mathbf{N} = (N_1, \cdots, N_m)$ in $\mathbf{z} = (z_1, \cdots, z_m)$ if $p(\mathbf{z})$ is of degree N_k with respect to the variable z_k for $k = 1, \cdots, m$.

If all the coefficients of a polynomial $p(\mathbf{z})$ belong to a particular number field $\mathfrak{F}$ then $p(\mathbf{z})$ is called a polynomial over $\mathfrak{F}$. The set of all polynomials in m variables over $\mathfrak{F}$ will be denoted by $\mathfrak{F}(\mathbf{z})$. If two polynomials $p_1(\mathbf{z})$ and $p_2(\mathbf{z})$ in $\mathfrak{F}(\mathbf{z})$ are equal to within a factor of zero degree, i.e., $p_1(\mathbf{z}) = c p_2(\mathbf{z})$ where $c \in \mathfrak{F}$ and $c \neq 0$, then $p_1(\mathbf{z})$ and $p_2(\mathbf{z})$ are called associated polynomials. A polynomial $p \in \mathfrak{F}(\mathbf{z})$ with $d(p) > 0$ is called a reducible polynomial over $\mathfrak{F}$ if there are polynomials $p_1, p_2 \in \mathfrak{F}(\mathbf{z})$ with $d(p_1) > 0$ and $d(p_2) > 0$ such that $p(\mathbf{z}) = p_1(\mathbf{z}) p_2(\mathbf{z})$. If no such decomposition is possible, then $p(\mathbf{z})$ is called an irreducible polynomial. It may be noted that a polynomial which is irreducible over one field is not necessarily irreducible over another field. For example, although the polynomial $p(z_1, z_2) = z_1^2 + z_2^2$ is irreducible over the field of real numbers, it is not irreducible over the field of complex numbers since $p(z_1, z_2) = (z_1 + jz_2)(z_1 - jz_2)$.

B. Factorization and Interpolation

In this section, two important theorems from the algebra of polynomials in two or more variables are presented. The first theorem concerns the factorization of polynomials whereas the second concerns the uniqueness of a polynomial in terms of its values over a finite set of points. Both of these theorems play a key role in the development of conditions under which a multidimensional sequence is uniquely defined by the phase or magnitude of its Fourier transform.

It is well known that a polynomial in a single variable defined over the field of complex numbers may always be factored into a product of first degree polynomials [6]. This result is not true, however, for polynomials in two or more variables. Nevertheless, a polynomial of nonzero degree may always be uniquely decomposed, to within factors of zero degree, into a product of irreducible polynomials. More specifically,

Theorem 1: Any polynomial $p \in \mathcal{F}(z)$ of nonzero degree may always be factored into a product of polynomials which are irreducible in $\mathcal{F}$. Furthermore, if $p(\mathbf{z})$ has two different factorizations:

$$p(\mathbf{z}) = f_1(\mathbf{z}) f_2(\mathbf{z}) \cdots f_k(\mathbf{z}) = g_1(\mathbf{z}) g_2(\mathbf{z}) \cdots g_l(\mathbf{z}) \tag{5}$$

then $k = l$ and the factors $f_i(\mathbf{z})$ and $g_i(\mathbf{z})$ may be ordered in such a way that the factors are associated.

The second theorem of interest in this section concerns the uniqueness of a polynomial in terms of its values over a finite set of points. It is well known that a polynomial $p(z)$ in one variable of degree N is uniquely defined in terms of its values over a set $A = \{a_1, \cdots, a_{N+1}\}$ of $N + 1$ distinct points and may be reconstructed from these points by, for example, the Lagrange or Newton interpolation formulas [6]. This result may be extended to polynomials in m variables if the set of points A is replaced with an m-dimensional lattice of points, $L(A_1, \cdots, A_m)$. More specifically, let A_k be a set of N_k distinct points in the field $\mathcal{F}$ for $k = 1, \cdots, m$. The m-dimensional lattice $L(A_1, \cdots, A_m)$ is then defined as the m-fold Cartesian product of these m sets of points, i.e.,

$$L(\mathbf{A}) = L(A_1, \cdots, A_m) = \prod_{k=1}^{m} A_k = A_1 \times A_2 \times \cdots \times A_m. \tag{6}$$

The result of interest is the following

Theorem 2: Suppose $p_1, p_2 \in \mathcal{F}(z)$ are polynomials of degree at most N_k in the variable z_k for $k = 1, 2, \cdots, m$. If, for each k, A_k is a set of N_k distinct numbers in the field $\mathcal{F}$, and if

$$p_1(\mathbf{z}) = p_2(\mathbf{z}) \qquad \text{for all } \mathbf{z} \in L(\mathbf{A}) \tag{7}$$

then

$$p_1(\mathbf{z}) = p_2(\mathbf{z}) \qquad \text{for all } \mathbf{z}.$$

Although Theorem 2 establishes the uniqueness of a polynomial in terms of its values over a lattice, it does not provide a method for recovering a polynomial from its samples. The theory of interpolation for polynomials in two or more variables, however, is well established [7] and, since it will not be required in this paper, will not be discussed.

III. Notation and Framework

This paper is concerned with some of the issues involved in reconstructing a real multidimensional sequence from either the phase or magnitude of its Fourier transform. In this section, some notation and terminology related to multidimensional sequences is presented and the general framework of the reconstruction problem is established.

An m-dimensional sequence is a function of m integer-valued variables, n_1 through n_m, which will be denoted by $x(n_1, \cdots, n_m)$. The z-transform of the m-dimensional sequence $x(n_1, \cdots, n_m)$ is defined by

$$X(z_1, \cdots, z_m) = \sum_{n_1} \cdots \sum_{n_m} x(n_1, \cdots, n_m) z_1^{-n_1} \cdots z_m^{-n_m}. \tag{8}$$

In order to express (8) as well as some later results more succinctly, vector notation will be used whenever possible. For example, an m-dimensional sequence and its z-transform will be written as $x(\mathbf{n})$ and $X(\mathbf{z})$, respectively. In addition, with $\mathbf{n} = (n_1, \cdots, n_m)$ an integer-valued vector, $\mathbf{z}^{\mathbf{n}}$ will be defined by

$$\mathbf{z}^{-\mathbf{n}} = z_1^{-n_1} z_2^{-n_2} \cdots z_m^{-n_m}. \tag{9}$$

With this notation, (8) becomes

$$X(\mathbf{z}) = \sum_{\mathbf{n}} x(\mathbf{n}) \mathbf{z}^{-\mathbf{n}}. \tag{10}$$

Throughout this paper, all sequences will be assumed to have rational z-transforms with a region of convergence which includes the unit polydisk: $|z_k| = 1$ for $k = 1, 2, \cdots, m$. In this case, the Fourier transform of $x(\mathbf{n})$ exists and is given by

$$X(\boldsymbol{\omega}) = X(\mathbf{z})\big|_{\mathbf{z} = \exp[j\boldsymbol{\omega}]} = \sum_{\mathbf{n}} x(\mathbf{n}) e^{-j\mathbf{n}\cdot\boldsymbol{\omega}}. \tag{11}$$

Written in polar form, $X(\boldsymbol{\omega})$ is represented in terms of its magnitude and phase as

$$X(\boldsymbol{\omega}) = |X(\boldsymbol{\omega})| \exp [j\phi_x(\boldsymbol{\omega})]. \tag{12}$$

Most of the sequences which are considered in this paper have finite support, i.e., $x(\mathbf{n})$ is nonzero only for finitely many values of $\mathbf{n}$. For convenience, it will be assumed, without any loss in generality, that a sequence with finite support is equal to zero for all $\mathbf{n} < \mathbf{0}$.[1] In the general case, a sequence with finite support may always be shifted to satisfy this assumption. If a multidimensional sequence is zero outside the region $\mathbf{0} \leqslant \mathbf{n} < \mathbf{N}$, i.e., $x(n_1, \cdots, n_m) = 0$ whenever $n_k < 0$ or $n_k \geqslant N_k$ for $k = 1, \cdots, m$, then the region of support will be denoted by $R(\mathbf{N}) = R(N_1, \cdots, N_m)$. Furthermore, $F(\mathbf{n})$ will be used to denote the set of all real m-dimensional sequences which have, for some $\mathbf{N}$, a region of support equal to $R(\mathbf{N})$. Thus, $x \in F(\mathbf{n})$ will be taken to mean that $x(\mathbf{n})$ is a real m-D sequence with finite support which is nonzero only when $\mathbf{n} \geqslant \mathbf{0}$.

[1] If $\mathbf{m}$ and $\mathbf{n}$ are two vectors, then $\mathbf{m} < \mathbf{n}$ means that $m_k < n_k$ for each k. A similar interpretation is to be assumed for any other inequality.

Since the z-transform of a sequence $x \in F(\boldsymbol{n})$ is a polynomial in z^{-1}, $X(z)$ may always be uniquely written, to within factors of zero degree, as

$$X(z) = \alpha z^{-n_0} \prod_{k=1}^{p} X_k(z) \tag{13}$$

where α is a real number, $\boldsymbol{n}_0$ is a vector of nonnegative integers, and where $X_k(\boldsymbol{z})$ are nontrivial irreducible polynomials in $\boldsymbol{z}^{-1}$. The irreducible factors $X_k(\boldsymbol{z})$ may be viewed as the m-dimensional counterpart of the zeros of the z-transform of a one-dimensional finite-length sequence.

If $x \in F(\boldsymbol{n})$ has a z-transform $X(\boldsymbol{z})$ of degree N in z^{-1} with no trivial factors, i.e., $\boldsymbol{n}_0 = \boldsymbol{0}$ in (13), then $\tilde{X}(\boldsymbol{z})$ will be defined by

$$\tilde{X}(\boldsymbol{z}) \equiv \boldsymbol{z}^{-N} X(\boldsymbol{z}^{-1}). \tag{14}$$

Note that $\tilde{X}(\boldsymbol{z})$ is also a polynomial of degree N in $\boldsymbol{z}^{-1}$ with no trivial factors and is the z-transform of a sequence $\tilde{x} \in F(\boldsymbol{n})$. The sequence $\tilde{x}(\boldsymbol{n})$ is simply a "time-reversed" version of $x(\boldsymbol{n})$. The sequence $\tilde{x}(\boldsymbol{n})$ and its z-transform $\tilde{X}(\boldsymbol{z})$ will be referred to frequently in the following sections.

An important property of a sequence $x \in F(\boldsymbol{n})$ is that its z-transform need only be known over a finite set of points in order to uniquely specify the sequence. Although these points cannot be chosen arbitrarily, Theorem 2 in Section II provides one set of points which is sufficient for this unique specification. More specifically, the following is a direct consequence of Theorem 2.

Lemma: Suppose $x, y \in F(\boldsymbol{n})$ with support $R(\boldsymbol{N})$. Let A_k be a set of M_k distinct complex numbers with $M_k \geqslant N_k$ for $k = 1, \cdots, m$. If

$$X(\boldsymbol{z})|_{L(A)} = Y(\boldsymbol{z})|_{L(A)} \tag{15}$$

then

$$x(\boldsymbol{n}) = y(\boldsymbol{n}) \qquad \text{for all } \boldsymbol{n}.$$

Note that if the elements of the sets A_k are complex numbers with unit magnitude, then $X(\boldsymbol{z})$ evaluated over the lattice $L(\boldsymbol{A})$ is equal to $X(\boldsymbol{\omega})$ evaluated over a lattice in the $\boldsymbol{\omega}$-plane. Specifically, let

$$\Omega_k = \{\beta_{k,l}\}_{l=1}^{M_k} \qquad \text{with} \quad 0 \leqslant \beta_{k,l} < 2\pi \tag{16a}$$

and

$$A_k = \{\exp(j\beta_{k,l})\}_{l=1}^{M_k} \tag{16b}$$

where the elements in the set Ω_k are assumed to be distinct. Then

$$X(\boldsymbol{z})|_{L(A)} = X(\boldsymbol{\omega})|_{L(\Omega)}. \tag{17}$$

Finally, note that if the numbers $\beta_{k,l}$ are equally spaced between zero and 2π, i.e., $\beta_{k,l} = 2\pi l/M_k$, then $X(\boldsymbol{z})|_{L(A)}$ is equal to the $\boldsymbol{M}$-point discrete Fourier transform (DFT) of $x(\boldsymbol{n})$. In this case, the $\boldsymbol{M}$-point DFT of a sequence $x(\boldsymbol{n})$ will be denoted by $X(\boldsymbol{k})_M$ which, in polar form, will be written as

$$X(\boldsymbol{k})_M = |X(\boldsymbol{k})|_M \exp[j\phi_x(\boldsymbol{k})_M]. \tag{18}$$

IV. Multidimensional Uniqueness Constraints

This section is concerned with the uniqueness of a multidimensional sequence in terms of the phase or magnitude of its Fourier transform. Specifically, in Section IV-A, conditions are given under which a multidimensional sequence with finite support is uniquely specified by the phase of its Fourier transform. A similar set of conditions is given in Section IV-B for the uniqueness of a multidimensional sequence with finite support in terms of its Fourier transform magnitude. Finally, in Section IV-C the results in Section IV-A and IV-B are used to generate a set of conditions under which a multidimensional sequence with finite support is uniquely specified in terms of either its Fourier transform phase or magnitude. In addition, the extension of these results to multidimensional sequences whose convolutional inverses have finite support is described.

A. Uniqueness in Terms of Phase

It has recently been shown [8] that a 1-D (one-dimensional) finite length sequence is uniquely specified to within a scale factor by either the phase or the tangent of the phase of its Fourier transform if its z-transform has no zeros on the unit circle or in reciprocal pairs. By imposing a similar constraint on a multidimensional sequence, this result may be directly extended to the multidimensional case. This constraint involves the notion of a symmetric z-transform[2] which is defined as follows.

Definition: The z-transform of a sequence $x \in F(\boldsymbol{n})$ will be defined to be symmetric if, for some vector $\boldsymbol{k}$ of positive integers

$$X(\boldsymbol{z}) = \pm \boldsymbol{z}^{-k} X(\boldsymbol{z}^{-1}). \tag{19}$$

Note that a 1-D sequence which has a z-transform with all of its zeros on the unit circle or in reciprocal pairs is symmetric. Therefore, (19) represents an extension of this property to multidimensional sequences. Furthermore, note that if $x(\boldsymbol{n})$ has a symmetric z-transform, then its Fourier transform $X(\boldsymbol{\omega})$ has linear phase. Finally, note that if a sequence $x \in F(\boldsymbol{n})$ has a z-transform with no trivial factors, i.e., $\boldsymbol{n}_0 = 0$ in (13), then $X(z)$ is symmetric if

$$X(\boldsymbol{z}) = \pm \tilde{X}(\boldsymbol{z}). \tag{20}$$

It should be pointed out that a symmetric z-transform may be reducible, irreducible, trivial, or nontrivial. For example, $X(z_1, z_2) = a + bz_1^{-1} + bz_2^{-1} + a(z_1z_2)^{-1}$ is symmetric and, in addition, is irreducible if $a \neq b$. If, on the other hand, $A(\boldsymbol{z})$ is an arbitrary polynomial in $\boldsymbol{z}^{-1}$, then

$$X(\boldsymbol{z}) = A(\boldsymbol{z}) \cdot \tilde{A}(\boldsymbol{z}) \tag{21}$$

is symmetric and reducible. Finally, note that any trivial polynomial in $\boldsymbol{z}^{-1}$, i.e., $X(\boldsymbol{z}) = \beta \boldsymbol{z}^{-k}$, is symmetric.

The multidimensional version of the one-dimensional uniqueness theorem in [8] is as follows (see the Appendix for the proof).

[2] A symmetric z-transform should not be confused with the algebraic definition of a symmetric polynomial [6].

Theorem 3: Let $x, y \in F(\boldsymbol{n})$. If $X(\boldsymbol{z})$ and $Y(\boldsymbol{z})$ have no nontrivial symmetric factors and $\phi_x(\boldsymbol{\omega}) = \phi_y(\boldsymbol{\omega})$ for all $\boldsymbol{\omega}$, then $x(\boldsymbol{n}) = \beta y(\boldsymbol{n})$ for some positive real number β. If $\tan[\phi_x(\boldsymbol{\omega})] = \tan[\phi_y(\boldsymbol{\omega})]$ for all $\boldsymbol{\omega}$, then $x(\boldsymbol{n}) = \beta y(\boldsymbol{n})$ for some real number β.

Note that the theorem hypothesis excludes only nontrivial symmetric factors. Therefore, $X(\boldsymbol{z})$ and $Y(\boldsymbol{z})$ may contain trivial (linear-phase) factors. It should be emphasized, however, that the nontrivial symmetric factors which are excluded from $X(\boldsymbol{z})$ and $Y(\boldsymbol{z})$ need not be irreducible. For example, if $X(\boldsymbol{z}) = P(\boldsymbol{z})\,Q(\boldsymbol{z})$ with $P(\boldsymbol{z}) = A(\boldsymbol{z})\tilde{A}(\boldsymbol{z})$, then $x(\boldsymbol{n})$ does not satisfy the constraints of the theorem since $P(\boldsymbol{z})$ is a (reducible) symmetric factor of $X(\boldsymbol{z})$. In effect, the exclusion of symmetric factors from $X(\boldsymbol{z})$ is equivalent to the constraint that if $A(\boldsymbol{z})$ is an irreducible factor of $X(\boldsymbol{z})$ then $\tilde{A}(\boldsymbol{z})$ is not a factor of $X(\boldsymbol{z})$.

Theorem 3 provides a set of conditions under which a multidimensional sequence is uniquely defined to within a scale factor by the phase of its Fourier transform. Note, however, that it is assumed that the phase of the Fourier transform is known for all $\boldsymbol{\omega}$. Nevertheless, it is possible to show that a sequence with support $R(\boldsymbol{N})$ is uniquely defined to within a scale factor by the phase of its $\boldsymbol{M}$-point DFT provided $\boldsymbol{M} \geq 2\boldsymbol{N} - 1$. Specifically, the following theorem may be derived from Theorem 3 above and Theorem 2 in Section II (see the Appendix for the proof).

Theorem 4: Let $x, y \in F(\boldsymbol{n})$ with support $R(\boldsymbol{N})$ and let $\boldsymbol{M} \geq 2\boldsymbol{N} - 1$. If $X(\boldsymbol{z})$ and $Y(\boldsymbol{z})$ have no nontrivial symmetric factors and if $\phi_x(\boldsymbol{k})_{\boldsymbol{M}} = \phi_y(\boldsymbol{k})_{\boldsymbol{M}}$, then $y(\boldsymbol{n}) = \beta x(\boldsymbol{n})$ for some positive number β. If $\tan[\phi_x(\boldsymbol{k})_{\boldsymbol{M}}] = \tan[\phi_y(\boldsymbol{k})_{\boldsymbol{M}}]$, then $y(\boldsymbol{n}) = \beta x(\boldsymbol{n})$ for some real number β.

Theorem 4 asserts that, within the set of all multidimensional sequences with support $R(\boldsymbol{N})$ which have z-transforms with no nontrivial symmetric factors, a multidimensional sequence is uniquely defined to within a scale factor by the phase of its $\boldsymbol{M}$-point DFT when $\boldsymbol{M} \geq 2\boldsymbol{N} - 1$. The importance of this theorem lies in the fact that it allows for the development of practical algorithms for reconstructing a sequence from the phase of its DFT. However, in reconstructing a sequence which satisfies the constraints of Theorem 4 from $\phi_x(\boldsymbol{k})_{\boldsymbol{M}}$, it is not sufficient to simply find a sequence with support $R(\boldsymbol{N})$ and the correct phase since the reconstructed sequence may have nontrivial symmetric factors and, thus, will not represent the correct solution. Therefore, since the factorization of a multidimensional polynomial to check for the presence of nontrivial symmetric factors is, in general, a very difficult problem, it will be useful to include some additional information in order to guarantee that the reconstructed sequence has no nontrivial symmetric factors. The additional information which will be assumed is the value of the linear delay, $\boldsymbol{n}_0$, in (13). Specifically, suppose that $x(\boldsymbol{n})$ has support $R(\boldsymbol{N})$ and a z-transform with no symmetric factors, trivial or nontrivial. Note that the exclusion of trivial symmetric factors implies that $\boldsymbol{n}_0$ in (13) must be equal to zero. In this case, if $\boldsymbol{M} \geq 2\boldsymbol{N} - 1$ then scaled versions of $x(\boldsymbol{n})$ are the only sequences with support $R(\boldsymbol{N})$ and an $\boldsymbol{M}$-point DFT with phase $\phi_x(\boldsymbol{k})_{\boldsymbol{M}}$. This result is stated formally in the following theorem, a proof of which may be found in the Appendix.

Theorem 5: Let $x, y \in F(\boldsymbol{n})$ with support $R(\boldsymbol{N})$ and let $\boldsymbol{M} \geq 2\boldsymbol{N} - 1$. If $X(\boldsymbol{z})$ has no symmetric factors and $\phi_x(\boldsymbol{k})_{\boldsymbol{M}} = \phi_y(\boldsymbol{k})_{\boldsymbol{M}}$ then $y(\boldsymbol{n}) = \beta x(\boldsymbol{n})$ for some positive number β. If $\tan[\phi_x(\boldsymbol{k})_{\boldsymbol{M}}] = \tan[\phi_y(\boldsymbol{k})_{\boldsymbol{M}}]$, then $y(\boldsymbol{n}) = \beta x(\boldsymbol{n})$ for some real number β.

Note that, in contrast to Theorem 4, there are no constraints on the z-transform of $y(\boldsymbol{n})$. Therefore, $y(\boldsymbol{n})$ may be any multidimensional sequence with support $R(\boldsymbol{N})$. Implied by the exclusion of any symmetric factors in $X(\boldsymbol{z})$ is the additional constraint that $\boldsymbol{n}_0 = 0$ in (13). Finally, it should be pointed out that this theorem may easily be generalized to include the case in which $\boldsymbol{n}_0$ is nonzero but known.

B. Uniqueness in Terms of Magnitude

In Section IV-A, the uniqueness of a multidimensional sequence in terms of the phase of its Fourier transform was considered. This section addresses the dual problem related to the uniqueness of a multidimensional sequence in terms of its Fourier transform magnitude. It appears that the first treatment of this uniqueness question was provided by Bruck and Sodin [9] who postulated that the uniqueness of a 2-D sequence with finite support is related to the irreducibility of its z-transform. In this section, a slightly more general result is derived which includes sequences with irreducible z-transforms as a special case. Even more importantly, however, as in Section IV-A, the uniqueness of a multidimensional sequence in terms of a finite set (lattice) of samples of its Fourier transform magnitude is considered.

Let $x(\boldsymbol{n})$ be a sequence in $F(\boldsymbol{n})$ for which $|X(\boldsymbol{\omega})|$ is known for all $\boldsymbol{\omega}$. Since the inverse Fourier transform of $|X(\boldsymbol{\omega})|^2$ is the autocorrelation, $r_x(\boldsymbol{n})$, of $x(\boldsymbol{n})$:

$$r_x(\boldsymbol{n}) = x(\boldsymbol{n}) * x(-\boldsymbol{n}) \tag{22}$$

the specification of $|X(\boldsymbol{\omega})|$ is equivalent to the knowledge of $r_x(\boldsymbol{n})$ or its z-transform $R_x(\boldsymbol{z})$:

$$R_x(\boldsymbol{z}) = X(\boldsymbol{z})\,X(\boldsymbol{z}^{-1}). \tag{23}$$

Since the most general form for the z-transform of a sequence $x \in F(\boldsymbol{n})$ is given by (13), then

$$R_x(\boldsymbol{z}) = \alpha^2 \prod_{k=1}^{p} X_k(\boldsymbol{z}) \cdot X_k(\boldsymbol{z}^{-1}). \tag{24}$$

Now suppose that the polynomial

$$P(\boldsymbol{z}) = \prod_{k=1}^{p} X_k(\boldsymbol{z}) \tag{25}$$

is of degree $\boldsymbol{N}$ in $\boldsymbol{z}^{-1}$. Multiplying $R_x(\boldsymbol{z})$ by $\boldsymbol{z}^{-\boldsymbol{N}}$ yields a polynomial in $\boldsymbol{z}^{-1}$ which is of degree $2\boldsymbol{N}$ in $\boldsymbol{z}^{-1}$:

$$Q_x(\boldsymbol{z}) = \boldsymbol{z}^{-\boldsymbol{N}} R_x(\boldsymbol{z}) = \alpha^2 \prod_{k=1}^{p} X_k(\boldsymbol{z}) \cdot \tilde{X}_k(\boldsymbol{z}). \tag{26}$$

It is apparent that $Q_x(\boldsymbol{z})$ and $|X(\boldsymbol{\omega})|$ contain exactly the same information about $x(\boldsymbol{n})$ since one may be uniquely derived from the other. Therefore, the ability to uniquely recover $x(\boldsymbol{n})$ from $|X(\boldsymbol{\omega})|$ is equivalent to the ability to uniquely recover $X(\boldsymbol{z})$ from $Q_x(\boldsymbol{z})$. With this in mind, it follows that

$x(\boldsymbol{n})$ cannot be unambiguously recovered from only the magnitude of its Fourier transform. For example, the sign of α as well as the linear phase term $\boldsymbol{z}^{\boldsymbol{n}_0}$ in $X(\boldsymbol{z})$ are not recoverable from $Q_x(\boldsymbol{z})$. Even more important is the observation that, without additional information, it is not possible to determine whether $X_k(\boldsymbol{z})$ or $\tilde{X}_k(\boldsymbol{z})$ is a factor of $X(\boldsymbol{z})$. This ambiguity is not surprising, however, since it represents the multidimensional extension of a result which is familiar for 1-D sequences [10]. Specifically, for any finite duration sequence $x(n)$, another sequence $y(n)$ may be generated which has the same Fourier transform magnitude as $x(n)$ by simply reflecting a zero of $X(z)$ about the unit circle. For m-D sequences, $\tilde{X}_k(\boldsymbol{z})$ represents the reflection of the zero contour of $X_k(\boldsymbol{z})$ about the unit polydisk.

It will be useful in the following discussion to define an equivalence relation on the set $F(\boldsymbol{n})$ as follows:

$$y(\boldsymbol{n}) \sim x(\boldsymbol{n}) \quad \text{if} \quad y(\boldsymbol{n}) = \pm x(\boldsymbol{k} \pm \boldsymbol{n}) \tag{27}$$

for some integer-valued vector $\boldsymbol{k}$. In other words, the equivalence class generated by a sequence $x \in F(\boldsymbol{n})$ is defined to be the set of all sequences which may be derived from $x(\boldsymbol{n})$ by a linear shift, a "time-reversal," or a change in the sign of the sequence. Note that all of the sequences within a given equivalance class have the same Fourier transform magnitude. Thus, it will be convenient to refer to the Fourier transform magnitude of the sequences within an equivalence class as the Fourier transform magnitude of the class.

In general, there will be more than one equivalence class having the same Fourier transform magnitude. More specifically, given a sequence $x \in F(\boldsymbol{n})$, there may exist another sequence $y \in F(\boldsymbol{n})$ with the same Fourier transform magnitude as $x(\boldsymbol{n})$ but which is not in the same equivalence class as $x(\boldsymbol{n})$. Therefore, the goal of this section is to develop a set of conditions which guarantee the existence of only one equivalence class with a given Fourier transform magnitude. The first question to be addressed, however, concerns the number of equivalence classes which have a given Fourier transform magnitude. Once this has been established, conditions which guarantee the existence of only one equivalence class may then be easily determined. The answer to the first question is implied by the following theorem, a proof of which may be found in the Appendix.

Theorem 6: Let $x \in F(\boldsymbol{n})$ have a z-transform given by

$$X(\boldsymbol{z}) = \alpha \boldsymbol{z}^{-\boldsymbol{n}_0} \prod_{k=1}^{p} X_k(\boldsymbol{z}) \tag{28}$$

where $X_k(\boldsymbol{z})$ for $k = 1, \cdots, p$ are nontrivial irreducible polynomials. If $y \in F(\boldsymbol{n})$ and $|X(\boldsymbol{\omega})| = |Y(\boldsymbol{\omega})|$ for all $\boldsymbol{\omega}$, then $Y(\boldsymbol{z})$ must be of the form

$$Y(\boldsymbol{z}) = \pm \alpha \boldsymbol{z}^{-\boldsymbol{m}} \prod_{k \in I} X_k(\boldsymbol{z}) \cdot \prod_{k \notin I} \tilde{X}_k(\boldsymbol{z}) \tag{29}$$

where I is a subset of the integers in the interval $[1, p]$.

This theorem actually follows from the fact that the only way to generate a new sequence $y(\boldsymbol{n})$, which has the same Fourier transform magnitude as $x(\boldsymbol{n})$, is to convolve $x(\boldsymbol{n})$ with an all-pass sequence $g(\boldsymbol{n})$, i.e., a sequence for which $|G(\boldsymbol{\omega})| = 1$ for all $\boldsymbol{\omega}$. Specifically, if $|G(\boldsymbol{\omega})| = 1$ and $G(\boldsymbol{z})$ is rational, then $G(\boldsymbol{z})$ must be of the form

$$G(\boldsymbol{z}) = \pm \boldsymbol{z}^{-\boldsymbol{l}} \prod_{k=1}^{q} G_k^{-1}(\boldsymbol{z}) \cdot \tilde{G}_k(\boldsymbol{z}) \tag{30}$$

where $G_k(\boldsymbol{z})$ for $k = 1, \cdots, q$ are nontrivial irreducible polynomials in $\boldsymbol{z}^{-1}$. Therefore, given a sequence $x \in F(\boldsymbol{n})$ with a z-transform of the form (28), then $y(\boldsymbol{n}) = x(\boldsymbol{n}) * g(\boldsymbol{n})$ has finite support if and only if for each k, $G_k(\boldsymbol{z}) = X_i(\boldsymbol{z})$ for some $i \in [1, p]$. Consequently, it follows that $Y(\boldsymbol{z})$ must be of the form given by (29).

As a consequence of Theorem 6, if $x \in F(\boldsymbol{n})$ is a sequence with a z-transform given by (28), and if $y \in F(\boldsymbol{n})$ with $|Y(\boldsymbol{\omega})| = |X(\boldsymbol{\omega})|$, then $Y(\boldsymbol{z})$ must have the same number p of nontrivial irreducible factors. Furthermore, except for a scale factor of (-1) and linear shifts, the only way to generate another sequence $y \in F(\boldsymbol{n})$ for which $|Y(\boldsymbol{\omega})| = |X(\boldsymbol{\omega})|$ is to replace one or more nontrivial factors $X_k(\boldsymbol{z})$ of $X(\boldsymbol{z})$ with $\tilde{X}_k(\boldsymbol{z})$. However, if $X_k(\boldsymbol{z})$ is symmetric, then this replacement may only change $X(\boldsymbol{z})$ by a factor of (-1). Therefore, it follows that the number of equivalence classes with magnitude $|X(\boldsymbol{\omega})|$ is at most $2^{(p'-1)}$ where p' is the number of nonsymmetric irreducible factors in $X(\boldsymbol{z})$. Thus, the following is an immediate consequence of Theorem 6.

Theorem 7: Let $x \in F(\boldsymbol{n})$ have a z-transform with at most one irreducible nonsymmetric factor, i.e.,

$$X(\boldsymbol{z}) = P(\boldsymbol{z}) \prod_{k=1}^{p} X_k(\boldsymbol{z}) \tag{31}$$

where $P(\boldsymbol{z})$ is irreducible and where $X_k(\boldsymbol{z})$ for $k = 1, \cdots, p$ are irreducible and symmetric. If $y \in F(\boldsymbol{n})$ with $|X(\boldsymbol{\omega})| = |Y(\boldsymbol{\omega})|$ for all $\boldsymbol{\omega}$, then $y(\boldsymbol{n}) \sim x(\boldsymbol{n})$.

As in Section IV-A, it may be shown that the magnitude of the Fourier transform of a sequence $x \in F(\boldsymbol{n})$ with support $R(\boldsymbol{N})$ need only be known over a lattice of points for the results in Theorem 7 to remain valid. Specifically, the following theorem may be derived from Theorem 7 above and Theorem 2 in Section II (see the Appendix for the proof).

Theorem 8: Let $x, y \in F(\boldsymbol{n})$ with support $R(\boldsymbol{N})$ and let Ω_k be a set of M_k distinct real numbers in the interval $(0, 2\pi)$ with $M_k \geq 2N_k - 1$ for $k = 1, \cdots, m$. If $X(\boldsymbol{z})$ has at most one irreducible nonsymmetric factor and

$$|X(\boldsymbol{\omega})|_{L(\boldsymbol{\Omega})} = |Y(\boldsymbol{\omega})|_{L(\boldsymbol{\Omega})} \tag{32}$$

then $y(\boldsymbol{n}) \sim x(\boldsymbol{n})$.

A special case of this theorem results when the points in the sets Ω_k are equally spaced between 0 and 2π. In this instance, $X(\boldsymbol{\omega})|_{L(\boldsymbol{\Omega})}$ is equal to the $\boldsymbol{M}$-point DFT of $x(\boldsymbol{n})$, i.e.,

$$X(\boldsymbol{\omega})|_{L(\boldsymbol{\Omega})} = X(\boldsymbol{k})_{\boldsymbol{M}} \tag{33}$$

Therefore, (32) may be replaced with the constraint that

$$|X(\boldsymbol{k})|_{\boldsymbol{M}} = |Y(\boldsymbol{k})|_{\boldsymbol{M}} \tag{34}$$

with $\boldsymbol{M} \geq 2\boldsymbol{N} - 1$.

C. Extensions

In Section IV-A, conditions are presented under which an m-D sequence is uniquely defined to within a scale factor by the phase of its Fourier transform. A similar set of conditions are presented in Section IV-B which allow an m-D sequence to be uniquely specified by the magnitude of its Fourier transform to within a delay, a sign, and a "time-reversal." It is of interest to note that these uniqueness constraints are not mutually exclusive. Specifically, suppose that $x \in F(\boldsymbol{n})$ has a z-transform which, except for trivial factors, is nonsymmetric and irreducible, i.e.,

$$X(z) = z^k P(z) \tag{35}$$

where $P(z)$ is an irreducible nonsymmetric polynomial in z^{-1}. It then follows that $x(\boldsymbol{n})$ satisfies the constraints of both Theorems 3 and 7. Furthermore, if $x(\boldsymbol{n})$ is known to have support $R(\boldsymbol{N})$, then the constraints of Theorems 4 and 8 are also satisfied. Therefore, the following is a direct consequence of these theorems.

Theorem 9: If $x \in F(\boldsymbol{n})$ has a z-transform which, except for trivial factors, is irreducible and nonsymmetric, then $x(\boldsymbol{n})$ is uniquely specified (in the sense of Theorems 3 and 7) by either the phase or magnitude of its Fourier transform. If, in addition, $x(\boldsymbol{n})$ is known to have support $R(\boldsymbol{N})$, then the phase or magnitude of the $\boldsymbol{M}$-point DFT of $x(\boldsymbol{n})$ is sufficient for this unique specification provided $\boldsymbol{M} \geqslant 2\boldsymbol{N} - 1$.

It should be pointed out that the constraint that $X(z)$ is irreducible and nonsymmetric is not a particularly strict requirement for sequences in two or more variables. Specifically, it may be shown that within the set of all polynomials in $k > 1$ variables, the subset of reducible polynomials is a set of measure zero [11]. It may similarly be shown that the set of all symmetric polynomials corresponds to a set of measure zero. Therefore, since the set of reducible polynomials and the set of symmetric polynomials are both sets of measure zero, then so is the union of these sets. However, since the complement of this union corresponds to the set of irreducible and nonsymmetric polynomials, it follows that almost all polynomials in two or more variables are irreducible and nonsymmetric. Consequently, almost all sequences with finite support satisfy the constraints of Theorem 9 and, therefore, are uniquely defined to within a scale factor by the phase of their Fourier transform or to within a sign, a linear shift, and a time-reversal by the magnitude of their Fourier transform.

Although the results which have been presented thus far have been confined to sequences with finite support, an extension is easily made to those sequences whose convolutional inverses have finite support. Specifically, let $x_i(\boldsymbol{n})$ denote the convolutional inverse of an m-dimensional sequence $x(\boldsymbol{n})$, i.e.,

$$x(\boldsymbol{n}) * x_i(\boldsymbol{n}) = \delta(\boldsymbol{n}) \tag{36}$$

where $\delta(\boldsymbol{n})$ is the m-dimensional unit sample function. Now suppose that $x(\boldsymbol{n})$ is a stable sequence which as a z-transform of the form

$$X(z) = 1/P(z) \tag{37}$$

where $P(z)$ is a polynomial in z^{-1}. In this case, the convolutional inverse of $x(\boldsymbol{n})$ has a z-transform given by

$$X_i(z) = P(z) \tag{38}$$

so that $x_i \in F(\boldsymbol{n})$. In addition, the phase or magnitude of the Fourier transform of $x_i(\boldsymbol{n})$ is uniquely specified by the phase or magnitude, respectively, of the Fourier transform of $x(\boldsymbol{n})$, i.e.,

$$|X_i(\boldsymbol{\omega})| = |X(\boldsymbol{\omega})|^{-1} \tag{39}$$

and

$$\phi_{x_i}(\boldsymbol{\omega}) = -\phi_x(\boldsymbol{\omega}). \tag{40}$$

Therefore, if $x(\boldsymbol{n})$ is a stable sequence with a z-transform given by (37), then $x(\boldsymbol{n})$ is uniquely defined by the phase or magnitude of its Fourier transform (in the sense of Theorems 3 or 7) if the polynomial $P(z)$ satisfies the appropriate uniqueness constraints.

V. Algorithms

In Section IV, the uniqueness of a multidimensional sequence in terms of the phase or magnitude of its Fourier transform was addressed. In this section, the reconstruction problem is considered for the case in which the desired sequence satisfies the appropriate uniqueness constraints. In Section V-A, where the phase-only reconstruction problem is discussed, several 1-D reconstruction algorithms [8], [12] are extended to the multidimensional case and some examples are presented. The problem of reconstructing a multidimensional sequence from the magnitude of its Fourier transform is discussed in Section V-B. Unfortunately, however, unlike phase-only reconstruction, it appears that a practical algorithm for magnitude-only signal reconstruction is yet to be realized.

A. Phase-Only Reconstruction

In this section, the problem of reconstructing a multidimensional sequence from the phase of its Fourier transform is addressed. In order to ensure that there is a unique solution to the reconstruction problem, it will be assumed that $x(\boldsymbol{n})$ is an m-D sequence with support $R(\boldsymbol{N})$ which has a z-transform with no symmetric factors. For convenience, it will also be assumed that $x(\boldsymbol{0}) = \alpha_0$ is nonzero and known. In this case, α_0 along with the phase of the $\boldsymbol{M}$-point DFT of $x(\boldsymbol{n})$ uniquely specify $x(\boldsymbol{n})$ provided $\boldsymbol{M} > 2\boldsymbol{N} - 1$ (Theorem 5).

One algorithm for reconstructing $x(\boldsymbol{n})$ from the phase of its Fourier transform involves finding the solution to a set of linear equations. Specifically, from the definition of $\phi_x(\boldsymbol{\omega})$, it may be shown, as in [8], that $x(\boldsymbol{n})$ satisfies the equation

$$\sum_{\boldsymbol{n} \neq \boldsymbol{0}} x(\boldsymbol{n}) \sin[\phi_x(\boldsymbol{\omega}) + \boldsymbol{n} \cdot \boldsymbol{\omega}] = -\alpha_0 \sin \phi_x(\boldsymbol{\omega}) \tag{41a}$$

provided $\phi_x(\boldsymbol{\omega}) \neq \pm\pi/2$. For the case in which $\phi_x(\boldsymbol{\omega}) = \pm\pi/2$, then $x(\boldsymbol{n})$ satisfies the equation

$$\sum_{\boldsymbol{n} \neq \boldsymbol{0}} x(\boldsymbol{n}) \cos(\boldsymbol{n} \cdot \boldsymbol{\omega}) = -\alpha_0. \tag{41b}$$

Substituting the values of the phase of the $\boldsymbol{M}$-point DFT of $x(\boldsymbol{n})$ into (41) leads to a set of M_0 linear equations in $N_0 - 1$ unknowns where $M_0 = M_1 \times M_2 \times \cdots \times M_m$ and $N_0 = N_1 \times N_2 \times \cdots \times N_m$ [due to the symmetry of the phase, half of these equations are redundant and may be eliminated]. Arranging the elements of $x(\boldsymbol{n})$ into a vector, v_x, of length N_0, the linear equations in (41), when augmented with the equation $x(\boldsymbol{0}) = \alpha_0$, may be written as

$$A v_x = -\alpha_0 \boldsymbol{b} \tag{42}$$

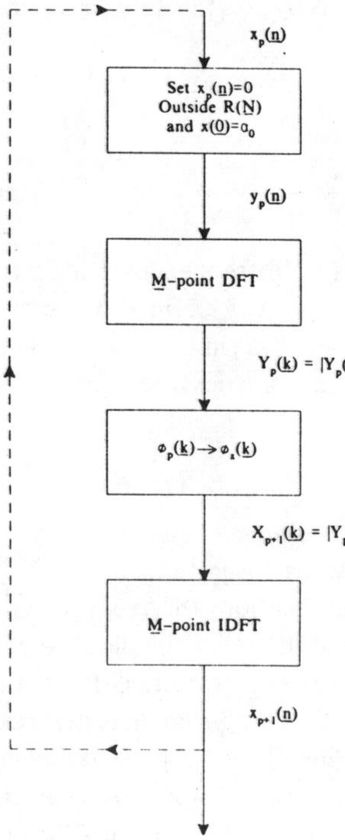

Fig. 1. Block diagram of the phase-only iteration.

where A is a matrix with N_0 columns and M_0 rows, and where $\boldsymbol{b}$ is a vector of length M_0. Since any solution to (42) corresponds to an m-dimensional sequence with support $R(\boldsymbol{N})$ which has an $\boldsymbol{M}$-point DFT with phase whose tangent equals that of $x(\boldsymbol{n})$, it follows from Theorem 5 that the columns of A are linearly independent (otherwise, multiple solutions would exist). Therefore, the matrix $S = A^T A$ is nonsingular and the desired sequence may be reconstructed from

$$v_x = -\alpha_0 (A^T A)^{-1} A^T \boldsymbol{b}. \tag{43}$$

Although this algorithm has been used successfully in reconstructing 2-D sequences of moderate size ($N_0 \leqslant 256$), its application in practice is limited by the computational difficulties encountered in solving linear equations when the number of unknowns becomes large. Therefore, it is of interest to consider alternative solutions to the reconstruction problem. One such solution is the multidimensional extension of the 1-D phase-only iteration [8], [12]. This algorithm is characterized by the repeated transformation between the time and frequency domains where, in each domain, the known information about the desired sequence is imposed on the current estimate. More specifically, let $\boldsymbol{M} \geqslant 2\boldsymbol{N} - 1$ and let $x_p(\boldsymbol{n})$ be the estimate of $x(\boldsymbol{n})$ after p iterations and suppose that the $\boldsymbol{M}$-point DFT's of $x(\boldsymbol{n})$ and $x_p(\boldsymbol{n})$ have the same phase. If $x_p(\boldsymbol{n})$ has support $R(\boldsymbol{N})$, then from Theorem 5, $x_p(\boldsymbol{n})$ is a scaled version of $x(\boldsymbol{n})$ and the iteration may be terminated. Otherwise, the sequence $y_p(\boldsymbol{n})$ is formed by setting $x_p(\boldsymbol{n})$ equal to zero outside $R(\boldsymbol{N})$ and equal to α_0 at $\boldsymbol{n} = 0$. Since the phase of $y_p(\boldsymbol{n})$ no longer equals the phase of $x(\boldsymbol{n})$, a new estimate, $x_{p+1}(\boldsymbol{n})$, is formed by taking the $\boldsymbol{M}$-point DFT of $y_p(\boldsymbol{n})$, replacing the phase with the correct phase, and taking the inverse DFT. Repeating this procedure defines the iteration which is illustrated, in terms of a block diagram, in Fig. 1.

A fundamental question of considerable practical and theoretical importance concerns the conditions under which $x_p(\boldsymbol{n})$ converges to $x(\boldsymbol{n})$. For 1-D sequences, it has been shown that this iteration will always converge to the desired sequence $x(n)$ for any initial estimate $x_0(n)$ provided that $x(n)$ is uniquely defined in terms of $x(0) = \alpha_0$ and the phase of its M-point DFT. By a straightforward extension of this result to multidimensional sequences, it may similarly be shown that $x_p(\boldsymbol{n})$ converges to $x(\boldsymbol{n})$ for any initial estimate $x_0(\boldsymbol{n})$ provided that $x(\boldsymbol{n})$ is uniquely specified by $x(\boldsymbol{0}) = \alpha_0$ and the phase of its $\boldsymbol{M}$-point DFT. Furthermore, even if $x(0)$ is not known, if $x_p(0)$ is set equal to an arbitrary nonzero constant in the iteration, and if $x(\boldsymbol{n})$ satisfies the constraints of Theorem 5, then

$$\lim_{p \to \infty} x_p(\boldsymbol{n}) = \beta x(\boldsymbol{n}) \tag{44}$$

where β is a scaling factor.

Consistent with this theoretical result, in all of the examples which have been considered, the iteration has always converged to the desired sequence. An example is presented in Figs. 2

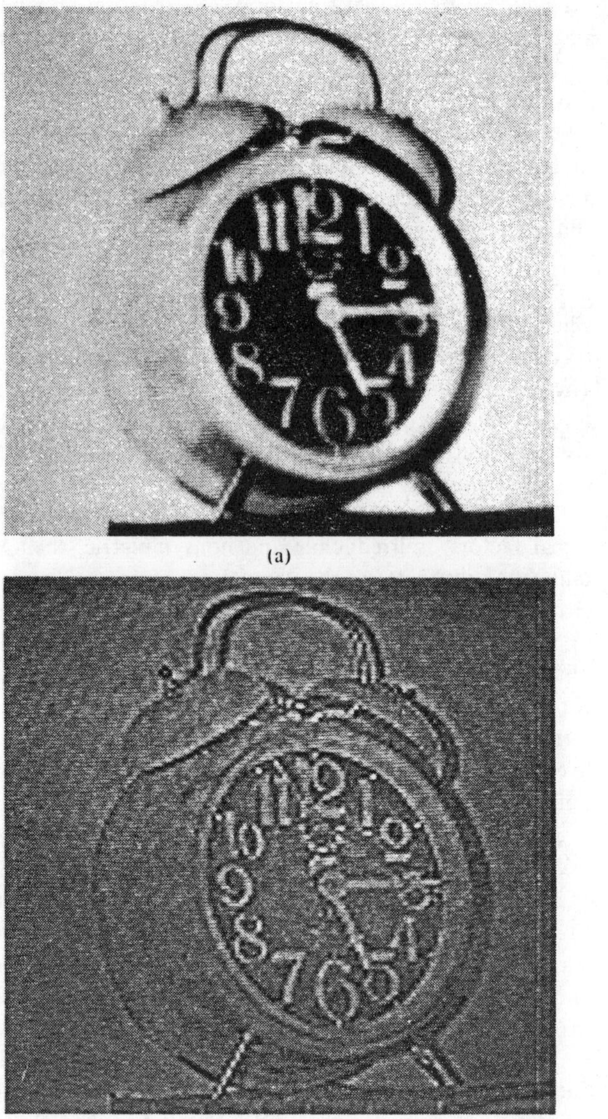

(a)

(b)

Fig. 2. Original image and its phase-only synthesis. (a) Original image. (b) Phase-only image formed by combining the phase of the Fourier transform of image (a) with a constant magnitude.

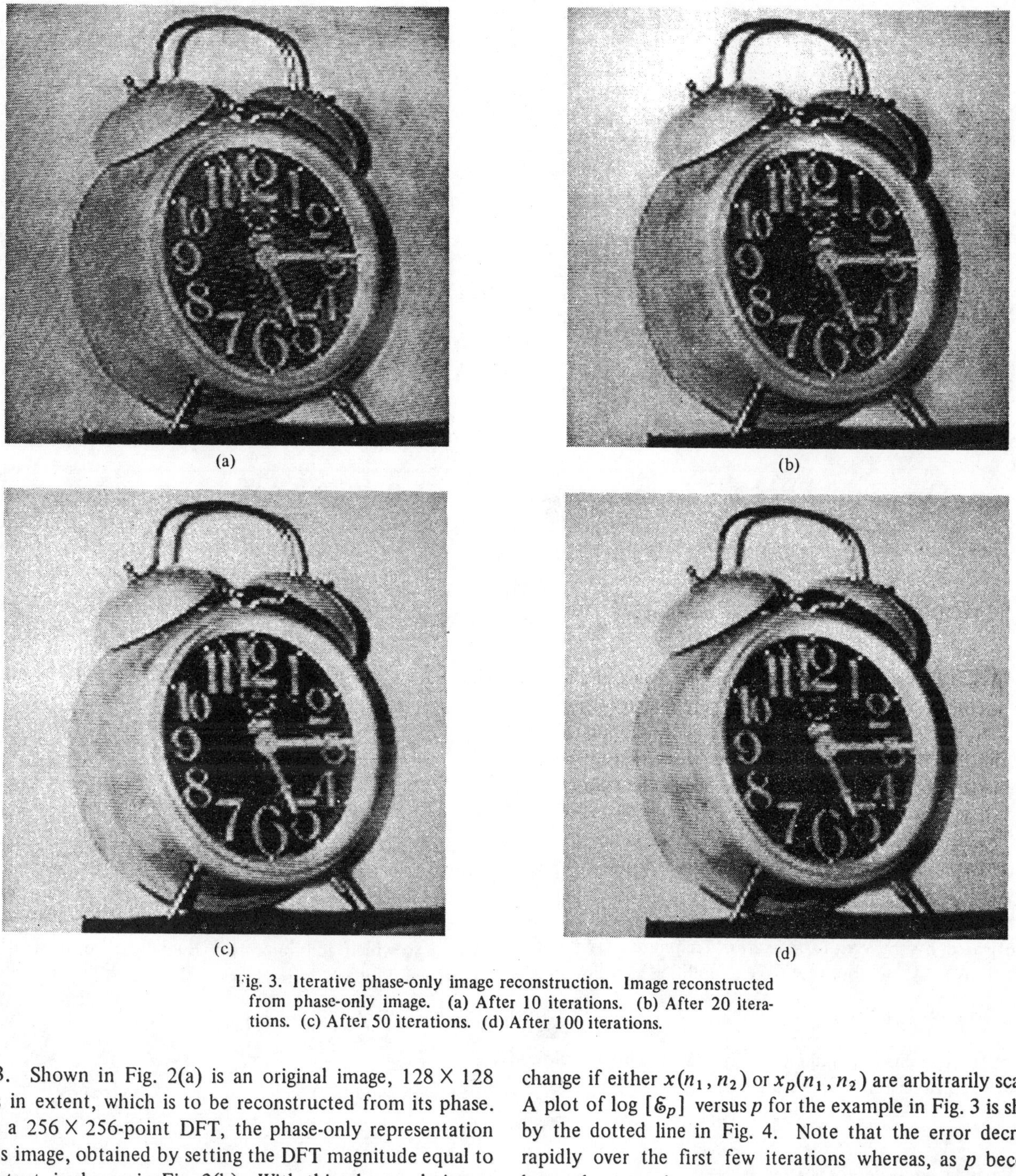

Fig. 3. Iterative phase-only image reconstruction. Image reconstructed from phase-only image. (a) After 10 iterations. (b) After 20 iterations. (c) After 50 iterations. (d) After 100 iterations.

and 3. Shown in Fig. 2(a) is an original image, 128 × 128 pixels in extent, which is to be reconstructed from its phase. Using a 256 × 256-point DFT, the phase-only representation of this image, obtained by setting the DFT magnitude equal to a constant, is shown in Fig. 2(b). With this phase-only image as the initial estimate in the iteration, the results which are obtained after 10, 20, 50, and 100 iterations are shown in Fig. 3 (each image has been appropriately scaled for display).

As a measure of the error between the estimate $x_p(n_1, n_2)$ after p iterations and the original image $x(n_1, n_2)$, consider the normalized mean-square error $\mathcal{E}_p$ defined by

$$\mathcal{E}_p = \frac{1}{N_1 N_2} \sum_{n_1=0}^{N_1-1} \sum_{n_2=0}^{N_2-1} \left[\frac{1}{\sigma_x} x(n_1, n_2) - \frac{1}{\sigma_p} x_p(n_1, n_2)\right]^2 \tag{45}$$

where σ_x^2 and σ_p^2 are the variances of $x(n_1, n_2)$ and $x_p(n_1, n_2)$, respectively (this error criterion was selected since $\mathcal{E}_p$ does not change if either $x(n_1, n_2)$ or $x_p(n_1, n_2)$ are arbitrarily scaled). A plot of log $[\mathcal{E}_p]$ versus p for the example in Fig. 3 is shown by the dotted line in Fig. 4. Note that the error decreases rapidly over the first few iterations whereas, as p becomes large, the error decreases more slowly. Since this behavior has been observed to be typical in all of the examples which have been considered, the computation time required to achieve a small mean-square error may be quite large, particularly when the support of the desired sequence is large. Therefore, it is of interest to consider methods for increasing the rate of convergence of the iteration.

One possibility for increasing the rate of convergence is as follows. With $\boldsymbol{M} \geqslant 2\boldsymbol{N} - 1$, suppose that the elements of $x(\boldsymbol{n})$ within the region $R(\boldsymbol{M})$ are arranged into a vector v_x in such a way that v_x may be partitioned as

$$v_x = \begin{bmatrix} v_x^{(1)} \\ \hline v_x^{(2)} \end{bmatrix} \tag{46}$$

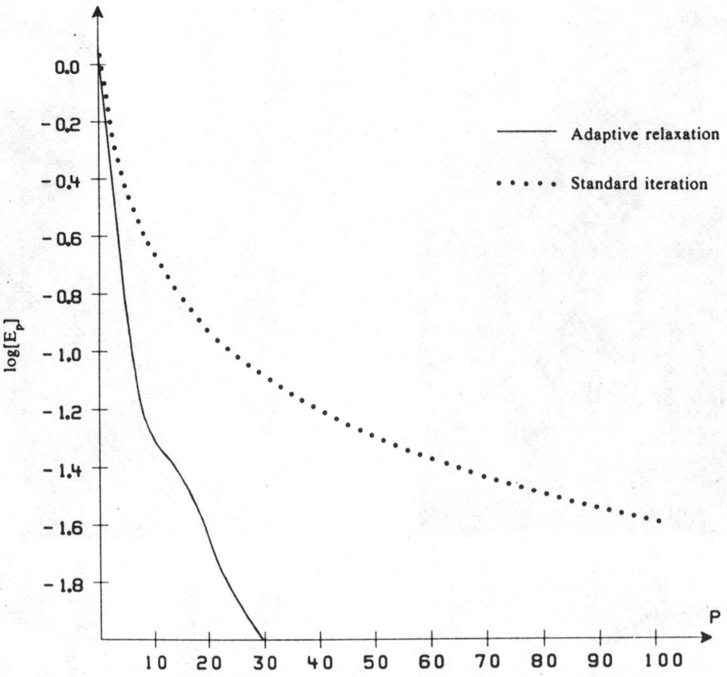

Fig. 4. Plot of normalized mean-square error for the phase-only iteration.

where $v_x^{(1)}$ is a vector of length $N_0 = N_1 \times N_2 \times \cdots \times N_m$ which contains all of the elements of $x(\boldsymbol{n})$ within the region $R(\boldsymbol{N})$ and where $v_x^{(2)}$ contains only elements of $x(\boldsymbol{n})$ which lie outside $R(\boldsymbol{N})$. With this representation, the phase-only iteration may be expressed mathematically as

$$v_{p+1} = F(v_p) \tag{47}$$

where v_{p+1} and v_p are the vectors which correspond to the sequences $x_{p+1}(\boldsymbol{n})$ and $x_p(\boldsymbol{n})$, respectively, and where F is the nonlinear mapping which maps $x_p(\boldsymbol{n})$ to $x_{p+1}(\boldsymbol{n})$.

Motivated by the various relaxation techniques for iterative algorithms [14], suppose that (47) is modified as follows:

$$v_{p+1} = (1 - \lambda_p)\, v_p + \lambda_p F(v_p) \tag{48}$$

where the relaxation parameter λ_p is a scaler which is allowed to vary as a function of p. With the vector $\boldsymbol{r}_p$ defined by

$$\boldsymbol{r}_p = F(v_p) - v_p \tag{49}$$

(48) may be equivalently written as

$$v_{p+1} = v_p + \lambda_p \boldsymbol{r}_p. \tag{50}$$

Several special cases of (50) are immediately apparent. If λ_p is a fixed constant $\lambda_p = \lambda$, then (50) corresponds to the original iteration (47) when $\lambda = 1$, whereas $\lambda = 0$ produces the trivial result $v_{p+1} = v_p$. Intermediate values of λ, i.e., $0 < \lambda < 1$, correspond to what is commonly referred to as the underrelaxed version of (47).

A common difficulty encountered with iterations of the form (50) is in the determination of the optimum value of λ_p which maximizes the rate of convergence of the iteration. However, for the phase-only iteration, it is relatively straightforward to derive an expression for λ_p which is optimum in a certain sense. Specifically, if (50) is partitioned as in (46), then

$$\begin{bmatrix} v_{p+1}^{(1)} \\ \hline v_{p+1}^{(2)} \end{bmatrix} = \begin{bmatrix} v_p^{(1)} \\ \hline v_p^{(2)} \end{bmatrix} + \lambda_p \begin{bmatrix} r_p^{(1)} \\ \hline r_p^{(2)} \end{bmatrix}. \tag{51}$$

Now note that for any λ_p, v_{p+1} corresponds to a sequence, $x_{p+1}(\boldsymbol{n})$ which has an $\boldsymbol{M}$-point DFT with a phase whose tangent equals that of $x(\boldsymbol{n})$. Therefore, it follows that a convergent solution is obtained if and only if $v_{p+1}^{(2)} = \boldsymbol{0}$. Consequently, a reasonable approach for selecting λ_p is to choose that value, $\lambda_p = \lambda_p^*$, which minimizes the Euclidean norm of $v_{p+1}^{(2)}$. In other words, λ_p^* is defined by

$$\left[\frac{d}{d\lambda_p} \|v_{p+1}^{(2)}\|^2\right]_{\lambda_p = \lambda_p^*} = 0. \tag{52}$$

From (51) and (52), it follows that λ_p^* is given by

$$\lambda_p^* = -\frac{\langle v_p^{(2)}, r_p^{(2)} \rangle}{\|r_p^{(2)}\|^2} \tag{53}$$

where $\langle -, - \rangle$ denotes the standard inner product between two vectors.

Although it has not yet been shown theoretically that this procedure of "adaptive relaxation" will always lead to a convergent solution, in all of the examples which have been considered, the iteration has always converged to the correct sequence. Furthermore, it has been observed that the number of iterations required to achieve a given mean-square error is, in general, substantially reduced when adaptive relaxation is used. Specifically, consider again the image in Fig. 2(a) which is to be reconstructed from the phase of its Fourier transform. With the phase-only image in Fig. 2(b) as the initial estimate of the original image, the results obtained after 5, 10, 15, and 20

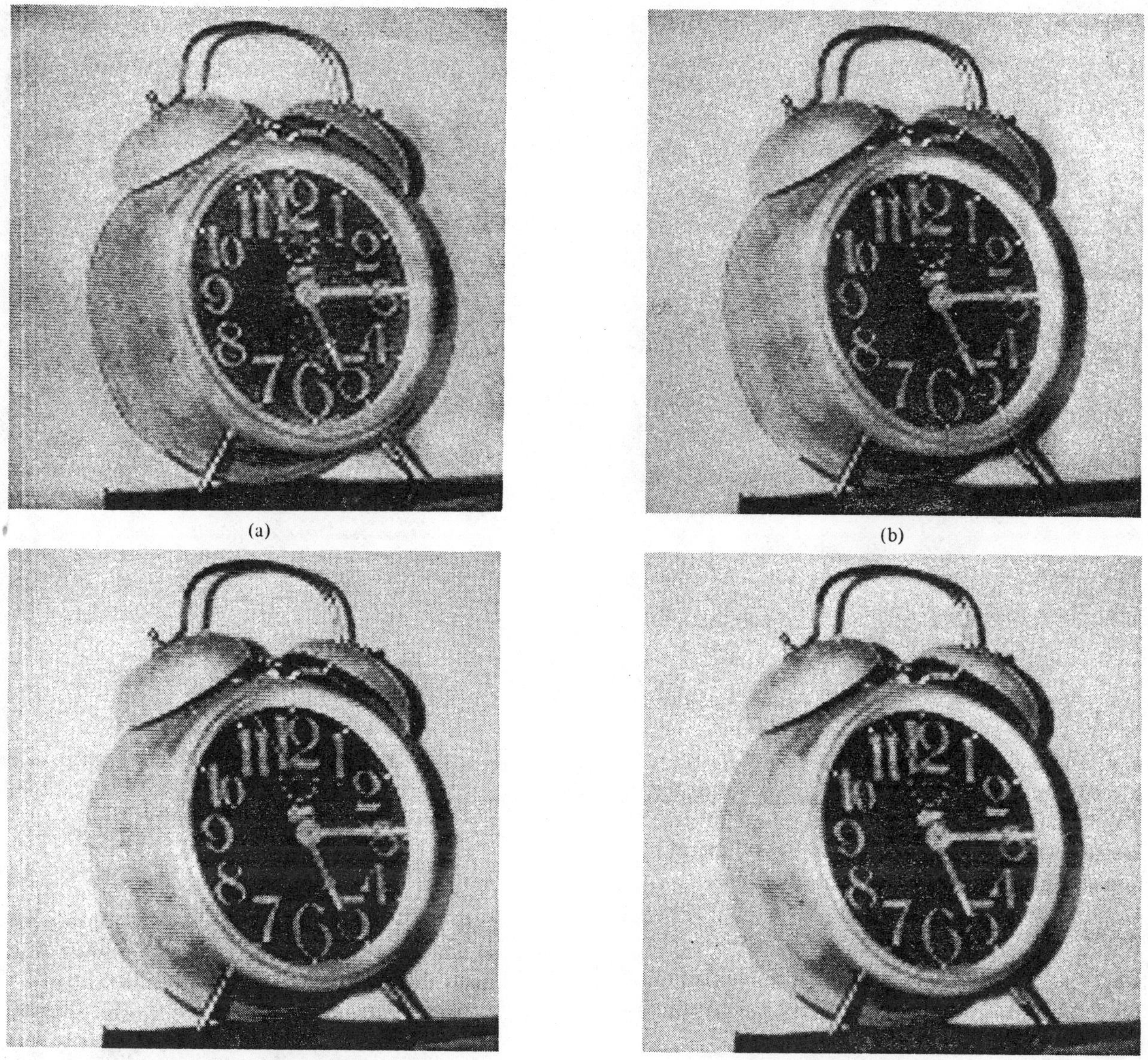

(a) (b) (c) (d)

Fig. 5. Iterative phase-only image reconstruction with adaptive relaxation. Image reconstructed from phase-only image. (a) After 5 iterations. (b) 10 iterations. (c) 15 iterations. (d) 20 iterations.

iterations with adaptive relaxation are shown in Fig. 5. For a quantitative comparison of these results with those shown in Fig. 3, a plot of log $[\mathcal{E}_p]$ versus p is shown in Fig. 4. As evidenced in this figure, adaptive relaxation tends to maintain a rapid decrease in the mean-square error, even for relatively large values of p.

B. Magnitude-Only Reconstruction

Due to its importance in the phase retrieval problem for wave amplitudes and coherence functions, an algorithm for reconstructing a multidimensional sequence from the magnitude of its Fourier transform has been the objective of many research efforts and the subject of a considerable number of published papers [3]. Nevertheless, in the absence of any additional knowledge of the desired sequence, there is, as yet, no practical algorithm which will always recover the correct phase from only magnitude information. It appears that there are at least two problems which have made the development of such an algorithm difficult. The first is the nonlinear relationship between the coefficients of a multidimensional sequence and the magnitude of its Fourier transform. Recall, for example, that the Fourier transform magnitude of a sequence $x(\mathbf{n})$ may be used to obtain the autocorrelation $r_x(\mathbf{n})$ of $x(\mathbf{n})$. Therefore, with magnitude information alone, it is always possible to define a set of second-order equations which relate the known values of the autocorrelation of $x(\mathbf{n})$ at various lags with the unknown coefficients of the desired multidimensional sequence, e.g.,

$$r_x(\mathbf{n}) = \sum_{\mathbf{k}} x(\mathbf{n} + \mathbf{k})\, x(\mathbf{k}). \tag{54}$$

One possible solution to the magnitude-only reconstruction problem may thus consist of solving these nonlinear equations for $x(\mathbf{n})$. Although there exist techniques for finding solutions to a set of simultaneous nonlinear equations in more than one unknown, when the number of equations and the number of

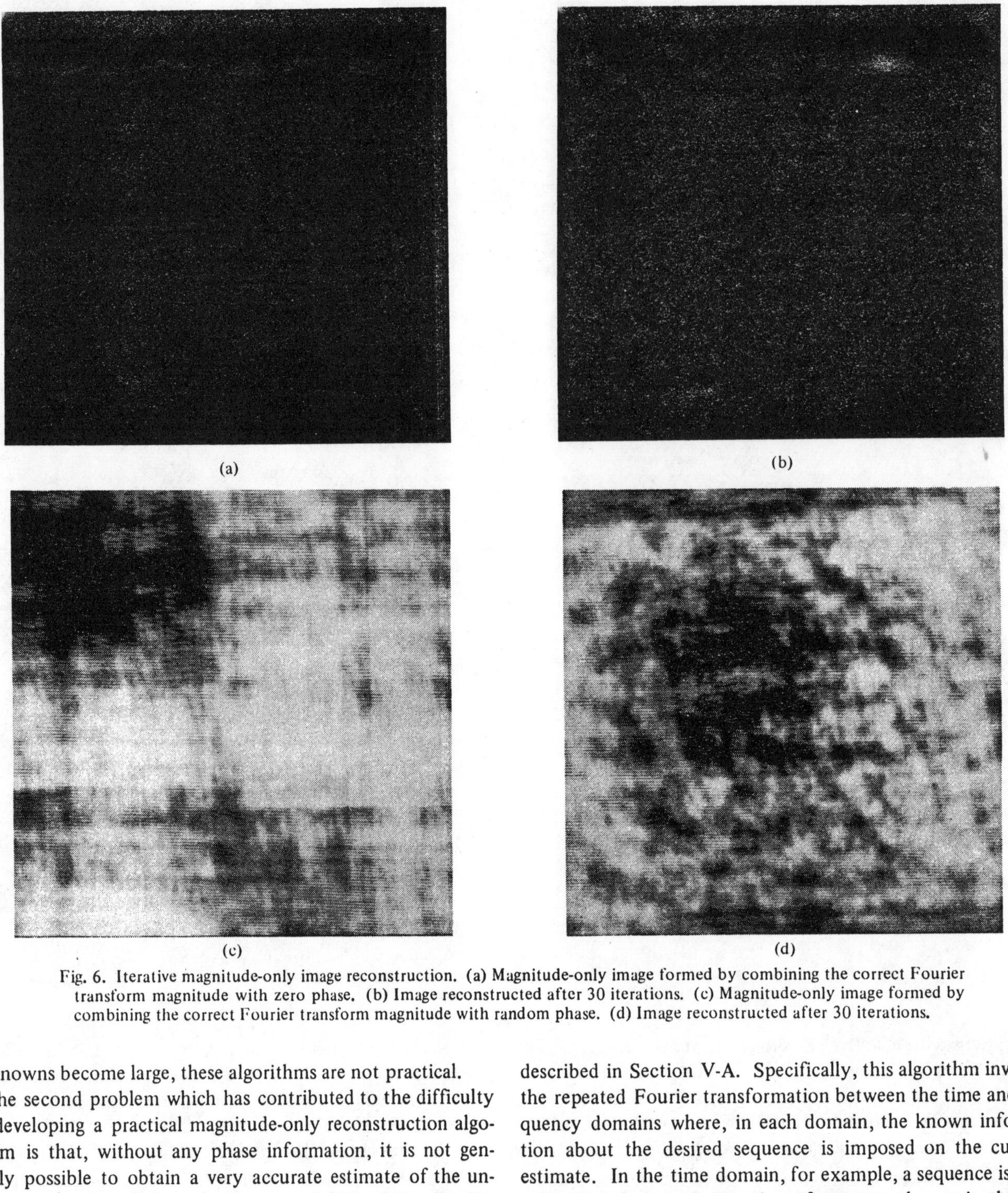

Fig. 6. Iterative magnitude-only image reconstruction. (a) Magnitude-only image formed by combining the correct Fourier transform magnitude with zero phase. (b) Image reconstructed after 30 iterations. (c) Magnitude-only image formed by combining the correct Fourier transform magnitude with random phase. (d) Image reconstructed after 30 iterations.

unknowns become large, these algorithms are not practical.

The second problem which has contributed to the difficulty in developing a practical magnitude-only reconstruction algorithm is that, without any phase information, it is not generally possible to obtain a very accurate estimate of the unknown sequence from only the magnitude of its Fourier transform. For example, if the Fourier transform magnitude of an image is combined with either zero or random phase, the result which is obtained upon an inverse Fourier transformation is an image which generally does not contain any recognizable features (see Fig. 6). In contrast, a phase-only image which has a Fourier transform with the correct phase and a constant magnitude contains many of the important characteristics of the original image (see Fig. 2).

In spite of these difficulties, an algorithm which has been considered for 2-D magnitude-only reconstruction [16] is an iterative procedure similar in style to the phase-only algorithm described in Section V-A. Specifically, this algorithm involves the repeated Fourier transformation between the time and frequency domains where, in each domain, the known information about the desired sequence is imposed on the current estimate. In the time domain, for example, a sequence is constrained to have a given region of support whereas in the frequency domain, the sequence is constrained to have a given Fourier transform magnitude. Unlike the phase-only iteration, however, it has been observed that the magnitude-only iteration will not generally converge to the correct solution even if the desired sequence satisfies the uniqueness constraints of Theorem 7. There appear to be two factors which determine whether or not the iteration converges to the correct solution. The first pertains to the ability to obtain an initial estimate to begin the iteration which is sufficiently close to the correct solution. It has been observed, for example, that for a 2-D sequence with support $R(N_1, N_2)$, if the initial estimate used in

(a)

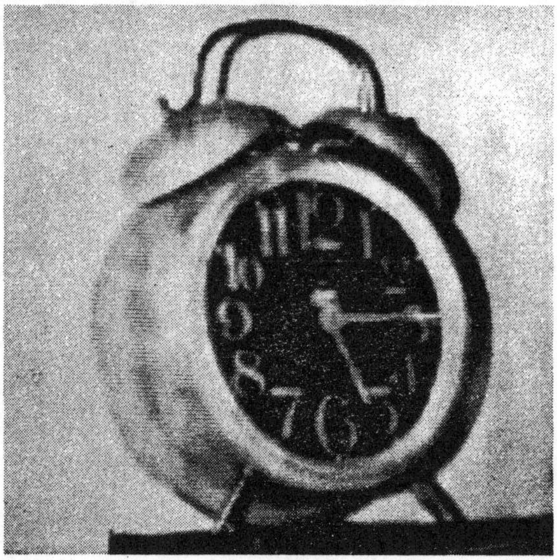

(b)

Fig. 7. Iterative amplitude-only image reconstruction. (a) Amplitude-only image formed by combining the correct Fourier transform magnitude with one bit of phase information. (b) Image reconstructed after 20 iterations.

the iteration has a Fourier transform with the correct magnitude and either zero phase or random phase, then the iteration will not generally converge to the correct sequence. This conclusion is based on many attempts to reconstruct a 2-D sequence from the magnitude of its Fourier transform. The sequences considered had regions of support which varied from $R(2, 2)$ to $R(128, 128)$ and, even in those cases for which the desired sequence was known to have an irreducible z-transform (and thus satisfied the constraints of Theorem 7), the correct solution was not generally obtained. Two examples are shown in Fig. 6 for an image which is 128×128 pixels in extent. In Fig. 6(a) is the magnitude-only synthesis of the image in Fig. 2(a) which was formed by taking a 256×256-point DFT of the original image, setting the phase equal to zero, and taking the inverse DFT. Using this image as the initial estimate in the magnitude-only iteration, the result which is obtained after 30 iterations is shown in Fig. 6(b). If, instead of zero phase, a random phase is used in the magnitude-only synthesis, the result is the image shown in Fig. 6(c). Using this image as the initial estimate in the magnitude-only iteration, the result which is obtained after 30 iterations is shown in Fig. 6(d). Although the results after only 30 iterations are shown, in both cases there is virtually no change in the reconstructed image from one iteration to the next after the first 10 or 20 iterations. Similar results have been observed for 2-D sequences with smaller support, e.g., $R(4, 4)$, in which the magnitude-only iteration was run for up to 1000 iterations.

Whereas Fig. 6 illustrates the effects which typically occur when the magnitude-only iteration is initialized with an estimate which is not a close approximation to the correct solution, Fig. 7 shows that, with the appropriate initial conditions, the iteration tends to converge to the correct solution. Specifically, shown in Fig. 7(a) is an "amplitude-only" image, $x_0(n_1, n_2)$, which was obtained from the Fourier transform of the image in Fig. 2(a) by quantizing the phase to one bit so that $X_0(\omega_1, \omega_2)$ has the correct magnitude and a phase which, for each frequency, is equal to either zero or π. In other words

$$X_0(\omega_1, \omega_2) = \begin{cases} |X(\omega_1, \omega_2)| & \text{if } -\pi/2 < \phi_x(\omega_1, \omega_2) \leq \pi/2 \\ -|X(\omega_1, \omega_2)| & \text{otherwise} \end{cases} \tag{55}$$

where $X(\omega_1, \omega_2) = |X(\omega_1, \omega_2)| \exp[j\phi_x(\omega_1, \omega_2)]$ is the Fourier transform of the image in Fig. 2(a). With $x_0(n_1, n_2)$ used as the initial estimate in the iteration, the result which is obtained after 20 iterations is shown in Fig. 7(b). Similar results have been observed in other examples in which the initial estimates used to begin the iteration were sufficiently close to the correct solution.

The second factor which appears to have an effect on the convergence of the iteration concerns the shape of the known region of support of the sequence. Specifically, recall from Theorem 8 that magnitude information alone may only uniquely specify a multidimensional sequence to within a sign, a linear shift, and a time-reversal. Therefore, if a 2-D sequence $x(n_1, n_2)$ is known to have support $R(N_1, N_2)$, even if information were available to resolve the sign and linear shift ambiguities, there still would be two possible solutions to the magnitude-only reconstruction problem, namely $x(n_1, n_2)$ and $\tilde{x}(n_1, n_2) = x(N_1 - n_1, N_2 - n_2)$. This ambiguity is a result of the fact that $R(N_1, N_2)$ is a "symmetric" region of support so that for any sequence $x(n_1, n_2)$ which has support $R(N_1, N_2)$, then $\tilde{x}(n_1, n_2)$ also has support $R(N_1, N_2)$. This property, however, is not true for "nonsymmetric" regions of support. For example, suppose that $x(n_1, n_2)$ is known to have a triangular region of support $T(N)$, i.e., $x(n_1, n_2) = 0$ whenever $n_1 < 0$, $n_2 < 0$, or $n_1 + n_2 \geq N$. Assuming that no smaller triangular region of support would contain the nonzero values of $x(n_1, n_2)$, the only sequences with support $T(N)$ and a Fourier transform magnitude equal to $|X(\omega_1, \omega_2)|$ are $\pm x(n_1, n_2)$. In this case, therefore, if the triangular support constraint is imposed in the iteration along with a possible scaling by a factor of (-1) to force an arbitrary but fixed nonzero value of $x_p(n_1, n_2)$ to be positive, then there is only one solution consistent with the imposed time and frequency domain constraints. As may be expected, it has been observed that the convergence of the magnitude-only iteration is more likely to occur with a nonsymmetric support constraint than with a symmetric support constraint. Furthermore, in each case, the convergence of the iteration may be attributed to the fact that a nonsymmetric region of support is imposed in the iteration. Specifically, note that if $x(n_1, n_2)$ has support $T(N)$, then it also has support $R(N, N)$. Therefore, while the iteration usually converges to $x(n_1, n_2)$ when the time domain constraint sets $x_p(n_1, n_2)$ equal to zero outside $T(N)$, when $R(N, N)$ is used as the support constraint, convergence of the iteration is not generally obtained.

VI. Summary

In this paper, conditions have been developed under which a multidimensional sequence is uniquely specified by the phase or magnitude of its Fourier transform. Although it was initially assumed that either the phase or magnitude was known for all frequencies, the uniqueness conditions were then extended to include the case in which the phase or magnitude was specified over a particular finite set of points. Under the specified uniqueness constraints, several algorithms were described for reconstructing a multidimensional sequence from the phase of its DFT.

Although the results reported in this paper answer some of the important questions on the general problem of multidimensional signal reconstruction from phase or magnitude, there are a variety of issues that remain to be investigated. One such issue relates to the development of a practical algorithm for reconstructing a multidimensional sequence from the magnitude of its Fourier transform. Another issue which requires further investigation is an understanding of the sensitivity of a reconstructed sequence to inaccurate information about the original (unknown) sequence. For example, in most practical problems of interest, either the phase or magnitude may not be known exactly due to errors such as measurement noise and it is important to understand the effects of these errors on the reconstructed sequence. These and other important issues are currently under investigation.

Appendix

Proof of Theorem 3: Let $x, y \in F(n)$ and let N be an integer-valued vector which is large enough so that both $x(\boldsymbol{n})$ and $y(\boldsymbol{n})$ are equal to zero outside $R(N)$. Now consider the sequence $g(\boldsymbol{n})$ defined by

$$g(\boldsymbol{n}) = x(\boldsymbol{n}) * y(-\boldsymbol{n}) \tag{A1}$$

which has a z-transform

$$G(z) = X(z)\, Y(z^{-1}). \tag{A2}$$

Since the phase of the Fourier transform of $g(\boldsymbol{n})$ is equal to

$$\phi_g(\boldsymbol{\omega}) = \phi_x(\boldsymbol{\omega}) - \phi_y(\boldsymbol{\omega}) \tag{A3}$$

it follows that if $\phi_x(\boldsymbol{\omega}) = \phi_y(\boldsymbol{\omega})$ then $\phi_g(\boldsymbol{\omega}) = 0$ or if

$$\tan[\phi_x(\boldsymbol{\omega})] = \tan[\phi_y(\boldsymbol{\omega})]$$

then $\tan[\phi_g(\boldsymbol{\omega})] = 0$. In either case, the Fourier transform of $g(\boldsymbol{n})$ is real which implies that

$$G(z) = G(z^{-1}). \tag{A4}$$

Therefore, from (A2) and (A4)

$$X(z)\, Y(z^{-1}) = X(z^{-1})\, Y(z). \tag{A5}$$

Multiplying both sides of (A5) by z^{-N} results in the following polynomial equation in z^{-1}:

$$X(z)\, \tilde{Y}(z)\, z^{-m} = \tilde{X}(z)\, Y(z)\, z^{-n} \tag{A6}$$

where $\boldsymbol{m}$ and $\boldsymbol{n}$ are integer-valued vectors with $\boldsymbol{m} \geqslant 0$ and $\boldsymbol{n} \geqslant 0$. Now consider an arbitrary nontrivial irreducible factor $X_k(z)$ of $X(z)$. From Theorem 1 in Section II, it follows that $X_k(z)$ must be associated either with a factor of $\tilde{X}(z)$ or with a factor of $Y(z)$. However, if $X_k(z)$ is associated with a factor of $\tilde{X}(z)$, then

$$X_k(z) = \alpha \tilde{X}_i(z) \tag{A7}$$

for some i. If $i = k$, then (A7) implies that

$$X_k(z) = \alpha^2 X_k(z). \tag{A8}$$

Therefore, $\alpha = \pm 1$ and $X_k(z)$ is symmetric. If, on the other hand, $i \neq k$, then

$$X_k(z)\, X_i(z) = \alpha \tilde{X}_i(z)\, X_i(z) \tag{A9}$$

and $A(z) = \tilde{X}_i(z)\, X_i(z)$ is a symmetric factor of $X(z)$. Both cases, however, are excluded by the theorem hypothesis. Consequently, each nontrivial irreducible factor of $X(z)$ must be associated with a factor of $Y(z)$.

Similarly, it follows from (A6) and Theorem 1 that any nontrivial irreducible factor $Y_k(z)$ of $Y(z)$ must be associated either with a factor of $X(z)$ or with a factor of $\tilde{Y}(z)$. Furthermore, since $Y(z)$ contains no nontrivial symmetric factors, it follows that each nontrivial irreducible factor of $Y(z)$ must be associated with a factor of $X(z)$. Therefore, $X(z)$ and $Y(z)$ may differ by at most a trivial factor, i.e.,

$$Y(z) = \beta z^{k} X(z). \tag{A10}$$

However, if the phase or the tangent of the phase of $x(\boldsymbol{n})$ and $y(\boldsymbol{n})$ are equal, then $\boldsymbol{k} = 0$. The theorem then follows by noting that β must be positive if $\phi_x(\boldsymbol{\omega}) = \phi_y(\boldsymbol{\omega})$.

Proof of Theorem 4: Let $x(\boldsymbol{n})$ and $y(\boldsymbol{n})$ satisfy the constraints of the theorem and consider the sequence $g(\boldsymbol{n}) = x(\boldsymbol{n}) * y(-\boldsymbol{n})$. Since $x(\boldsymbol{n})$ and $y(\boldsymbol{n})$ have support $R(N)$, then $g(\boldsymbol{n})$ is zero outside the region $-N < \boldsymbol{n} < N$. Therefore, if $\boldsymbol{M} \geqslant 2N - 1$ then the $\boldsymbol{M}$-point DFT of $g(\boldsymbol{n})$ is equal to the product of the $\boldsymbol{M}$-point DFT's of $x(\boldsymbol{n})$ and $y(\boldsymbol{n})$,

$$G(\boldsymbol{k})_M = X(\boldsymbol{k})_M \cdot Y(-\boldsymbol{k})_M. \tag{A11}$$

Thus, if $\phi_x(\boldsymbol{k})_M = \phi_y(\boldsymbol{k})_M$ or $\tan[\phi_x(\boldsymbol{k})_M] = \tan[\phi_y(\boldsymbol{k})_M]$ then $G(\boldsymbol{k})_M$ must be real. Therefore, since $g(\boldsymbol{n})$ is nonzero only for $-N < \boldsymbol{n} < N$, it follows that $g(\boldsymbol{n})$ must be even and, consequently, the Fourier transform of $g(\boldsymbol{n})$ must be real. Thus, repeating the steps in the proof of Theorem 3, the desired results follows.

Proof of Theorem 5: If $x(\boldsymbol{n})$ and $y(\boldsymbol{n})$ satisfy the constraints of the theorem it follows, as in the proof of Theorem 4, that $g(\boldsymbol{n}) = x(\boldsymbol{n}) * y(-\boldsymbol{n})$ is an even sequence. Therefore, as in the proof of Theorem 3, it follows from (A5) that since $X(z)$ contains no symmetric factors, then each nontrivial irreducible factor of $X(z)$ must be associated with a factor of $Y(z)$. Consequently, $X(z)$ and $Y(z)$ must be related by

$$Y(z) = z^{\mathbf{m}} P(z) X(z) \tag{A12}$$

where $P(z)$ is a polynomial in z^{-1} and $\mathbf{m}$ is an integer-valued vector. However, since $Y(z)$ and $X(z)$ are both polynomials in z^{-1}, and since $X(z)$ contains no trivial factors, then $Q(z) = z^{\mathbf{m}} P(z)$ must also be a polynomial in z^{-1}. Furthermore, in order for the phase or tangent of the phase of $x(\mathbf{n})$ and $y(\mathbf{n})$ to be equal, $q(\mathbf{n})$ must be an even sequence, i.e., $Q(z) = Q(z^{-1})$. Therefore, $Q(z) = \beta$ and the theorem follows by noting that β must be positive if $\phi_x(\mathbf{k})_M = \phi_y(\mathbf{k})_M$.

Proof of Theorem 6: Let $x, y \in F(\mathbf{n})$ with $X(z)$ given by (28) and let $\mathbf{N}$ be large enough so that both $x(\mathbf{n})$ and $y(\mathbf{n})$ are zero outside the region $R(\mathbf{N})$. If $|X(\boldsymbol{\omega})| = |Y(\boldsymbol{\omega})|$, then it follows that

$$X(z) X(z^{-1}) = Y(z) Y(z^{-1}). \tag{A13}$$

Therefore, let the z-transform of $y(\mathbf{n})$ be given by

$$Y(z) = \beta z^{\mathbf{m}} \prod_{k=1}^{q} Y_k(z) \tag{A14}$$

where $Y_k(z)$ are nontrivial irreducible factors for $k = 1, \cdots, q$. Substituting (28) and (A14) into (A13) and multiplying by $z^{-\mathbf{N}}$ yields the following equation in z^{-1}:

$$\alpha^2 z^{-\mathbf{m}_1} \prod_{k=1}^{p} X_k(z) \tilde{X}_k(z) = \beta^2 z^{-\mathbf{m}_2} \prod_{k=1}^{q} \tilde{Y}_k(z) Y_k(z) \tag{A15}$$

where $\mathbf{m}_1 \geqslant 0$ and $\mathbf{m}_2 \geqslant 0$. From Theorem 1 in Section II it follows that $\mathbf{m}_1 = \mathbf{m}_2$ and $p = q$:

$$\alpha^2 \prod_{k=1}^{p} X_k(z) \tilde{X}_k(z) = \beta^2 \prod_{k=1}^{p} Y_k(z) \tilde{Y}_k(z). \tag{A16}$$

Again from Theorem 1 it follows that the factors $Y_k(z)$ may be ordered in such a way that, for each k, $Y_k(z)$ is associated with either $X_k(z)$ or $\tilde{X}_k(z)$. Therefore, with I the set of integers k in the interval $[1, p]$ for which $Y_k(z)$ is associated with $X_k(z)$, it follows that

$$Y(z) = \eta z^{\mathbf{m}} \prod_{k \in I} X_k(z) \prod_{k \notin I} \tilde{X}_k(z). \tag{A17}$$

Finally, since $|X(\boldsymbol{\omega})| = |Y(\boldsymbol{\omega})|$ implies that $\eta = \pm\alpha$, the desired result (29) follows.

Proof of Theorem 8: Suppose $x, y \in F(\mathbf{n})$ and have support $R(\mathbf{N})$. Let Ω_k and A_k be sets of M_k distinct points as defined in (14) for $k = 1, \cdots, m$, and let $L(\boldsymbol{\Omega})$ and $L(A)$ be the lattices generated by these sets. Since

$$|X(\boldsymbol{\omega})|^2_{L(\boldsymbol{\Omega})} = R_x(z)|_{L(A)} \tag{A18}$$

it follows that if

$$|X(\boldsymbol{\omega})|^2_{L(\boldsymbol{\Omega})} = |Y(\boldsymbol{\omega})|^2_{L(\boldsymbol{\Omega})} \tag{A19}$$

then

$$R_x(z)|_{L(A)} = R_y(z)|_{L(A)}. \tag{A20}$$

Therefore, (A19) implies that

$$Q_x(z)|_{L(A)} = Q_y(z)|_{L(A)} \tag{A21}$$

where $Q_x(z)$ and $Q_y(z)$ are polynomials in z^{-1} as defined in (26) which have degree at most $2(\mathbf{N} - 1)$. Thus, if $\mathbf{M} \geqslant 2\mathbf{N} - 1$, it follows from Theorem 2 that $Q_x(z) = Q_y(z)$ for all z and, consequently, that $|X(\boldsymbol{\omega})| = |Y(\boldsymbol{\omega})|$ for all $\boldsymbol{\omega}$. Therefore, since $X(z)$ has at most one irreducible nonsymmetric factor, it follows from Theorem 7 that $y(\mathbf{n}) \sim x(\mathbf{n})$.

References

[1] R. R. Read and S. Treitel, "The stabilization of two-dimensional recursive filters via the discrete Hilbert transform," *IEEE Trans. Geosci. Electron.*, pp. 153-160, July 1973.

[2] M. P. Ekstrom and J. W. Woods, "Two-dimensional spectral factorization with applications in recursive digital filtering," *IEEE Trans. Acoust., Speech, Signal Processing*, vol. ASSP-24, pp. 115-128, Apr. 1976.

[3] H. A. Ferwerda, "The phase reconstruction problem for wave amplitudes and coherence functions," in *Inverse Source Problems in Optics*, H. P. Bates, Ed. Berlin, Germany: Springer-Verlag, 1978, ch. 2.

[4] T. G. Stockham, T. M. Cannon, and R. B. Ingebretson, "Blind deconvolution through digital signal processing," *Proc. IEEE*, pp. 678-692, Apr. 1975.

[5] H. C. Andrews and B. R. Hunt, *Digital Image Restoration*. Englewood Cliffs, NJ: Prentice Hall, 1977.

[6] A. Mostowski and M. Stark, *Introduction to Higher Algebra*. New York: Pergamon, 1964.

[7] S. Narumi, "Some formulas in the theory of interpolation of many independent variables," *Tohoku Math. J.*, vol. 18, pp. 309-321, 1920.

[8] M. H. Hayes, J. S. Lim, and A. V. Oppenheim, "Signal reconstruction from phase or magnitude," *IEEE Trans. Acoust., Speech, Signal Processing*, vol. ASSP-28, pp. 672-680, Dec. 1980.

[9] Yu. H. Bruck and L. G. Sodin, "On the ambiguity of the image reconstruction problem," *Opt. Commun.*, pp. 304-308, Sept. 1979.

[10] A. V. Oppenheim and R. W. Schafer, *Digital Signal Processing*. Englewood Cliffs, NJ: Prentice-Hall, 1975.

[11] M. H. Hayes and J. H. McClellan, "The number of irreducible polynomials of a given order in more than one variable," to be published.

[12] A. V. Oppenheim, M. H. Hayes, and J. S. Lim, "Iterative procedures for signal reconstruction from phase," in *Proc. 1980 Int. Opt. Comput. Conf.*, vol. SPIE-231, Apr. 1980, pp. 121-129.

[13] V. T. Tom, T. F. Quatieri, M. H. Hayes, and J. H. McClellan, "Convergence of iterative nonexpansive signal reconstruction algorithms," *IEEE Trans. Acoust., Speech, Signal Processing*, pp. 1052-1058, Oct. 1981.

[14] J. M. Ortega and W. C. Rheinboldt, *Iterative Solution of Nonlinear Equations in Several Variables*. New York: Academic, 1970.

[15] J. R. Fienup, "Space object imaging through the turbulent atmosphere," *Opt. Eng.*, pp. 529-534, Sept. 1979.

A Tomographic Formulation of Spotlight-Mode Synthetic Aperture Radar

DAVID C. MUNSON, JR., MEMBER, IEEE, JAMES DENNIS O'BRIEN, STUDENT MEMBER, IEEE, AND W. KENNETH JENKINS, SENIOR MEMBER, IEEE

Abstract—**Spotlight-mode synthetic aperture radar (spotlight-mode SAR) synthesizes high-resolution terrain maps using data gathered from multiple observation angles. This paper shows that spotlight-mode SAR can be interpreted as a tomographic reconstruction problem and analyzed using the projection-slice theorem from computer-aided tomography (CAT). The signal recorded at each SAR transmission point is modeled as a portion of the Fourier transform of a central projection of the imaged ground area. Reconstruction of a SAR image may then be accomplished using algorithms from CAT. This model permits a simple understanding of SAR imaging, not based on Doppler shifts. Resolution, sampling rates, waveform curvature, the Doppler effect, and other issues are also discussed within the context of this interpretation of SAR.**

I. Introduction

BOTH computer-aided tomography (CAT) and synthetic aperture radar (SAR) are well-known techniques for constructing high-resolution images by processing data obtained from many different perspective views of a target area. The CAT scan, as it is familiarly termed now, is an X-ray technique which enables the imaging of two-dimensional cross sections of solid objects [1], [2]. In particular, tomography is used extensively for noninvasive medical examination of internal organs and in nondestructive testing of manufactured items. Although SAR is well known to a more exclusive community, it too is a well-developed technique for producing high-resolution images. In a SAR system, the desired image is a terrain map. The data are collected by means of an airborne or spaceborne microwave radar which illuminates the target area from different perspectives.

An early form of SAR, known as unfocused strip mapping, was demonstrated experimentally at the University of Illinois as far back as the early 1950's [3]. In strip-mapping SAR (both focused and unfocused), the position of the antenna remains fixed relative to the aircraft, thereby illuminating a strip of terrain as the aircraft flies. Proper processing of the returned signals allows the effective synthesis of a very large antenna, providing high resolution [3]–[6]. Extensive developmental work on optical processing of data collected in strip-mapping SAR was subsequently carried out by Brown and coworkers at what is now the Environmental Research Institute of Michigan. Walker [7], working with Brown, made a breakthrough in characterizing the requirements for optically processing coherent radar data collected from targets placed on a rotating platform, an experimental setup designed to simulate an airborne radar flying around a stationary ground patch—this was, in essence, the first spotlight-mode SAR. In this paper, we will focus on this latter type of SAR, in which the physical antenna is steered so that the same terrain area remains illuminated during a long data-collection interval [5], [7]–[9]. Spotlight-mode SAR is able to provide higher resolution of a more limited area than strip-mapping SAR, because the same terrain area is observed from many different angles.

The fact that CAT and spotlight-mode SAR have developed independently—CAT having attracted the interest of biomedical researchers, and SAR that of communications and radar specialists—has obscured a remarkable similarity of principle which they share. CAT processing has generally been characterized by the projection-slice theorem [1], [2], [10], [11], while spotlight-mode SAR processing has been described in the radar language of Doppler filtering [7], [8], [12]. A survey of published literature on SAR may not lead one to suspect that SAR is a distant cousin of CAT. The major objective of this paper is to demonstrate that CAT and SAR are, in fact, very similar concepts. It will be shown that spotlight-mode SAR can be interpreted as a tomographic reconstruction problem, and that the signal processing theory can be characterized in terms of the projection-slice theorem. The advantages in developing this connection are several. First, research into mathematical methods and algorithms developed in one field may be transferred to the other. Second, the tomographic interpretation simplifies the understanding of SAR, especially for those not versed in radar terminology. A third advantage is that the role of coherence is underscored and the speckle phenomenon is easily explained when SAR is interpreted as a narrow-band version of CAT.

Some of these concepts have been examined in relation to acoustic imaging using sources emitting short pulses or polychromatic continuous-wave (CW) illumination over circular apertures [13], [14]. Several authors have also discussed SAR utilizing polychromatic CW illumination over circular apertures (see, for instance, [15]). In practice, though, airborne radars are not coherent over extended flight trajectories due to unavoidable phase errors (caused by atmospheric granularity and motion error) and due to the angular dependence of the reflectors being imaged. Chen and Andrews [16] discuss the spotlight-mode SAR problem for sources emitting sinusoidal waveforms. Herman [2], [17] discusses related work in other fields, including radio astronomy.[1]

To introduce our basic approach, Section II reviews the projection-slice theorem and summarizes the fundamental principles of tomographic reconstruction. The mathematical formulation of SAR in Section III contains the significant contribution of this paper. If linear FM waveforms are transmitted during spotlight-mode operation, it is demonstrated that the demodulated wave-

Manuscript received July 6, 1982; revised March 28, 1983. This work was supported by the Joint Services Electronics Program under Contract N00014-79-C-0424. Portions of this manuscript were presented at the Fifteenth Asilomar Conference on Circuits, Systems, and Computers, Pacific Grove, CA, November 9–11, 1981.

The authors are with the Coordinated Science Laboratory and the Department of Electrical Engineering, University of Illinois, Urbana, IL 61801.

[1]Mensa *et al.* [18] also present some closely related work on microwave tomographic imaging. Reference [18] was published while this paper was still in review.

Reprinted from *Proc. IEEE,* vol. 71, pp. 917–925, Aug. 1983.

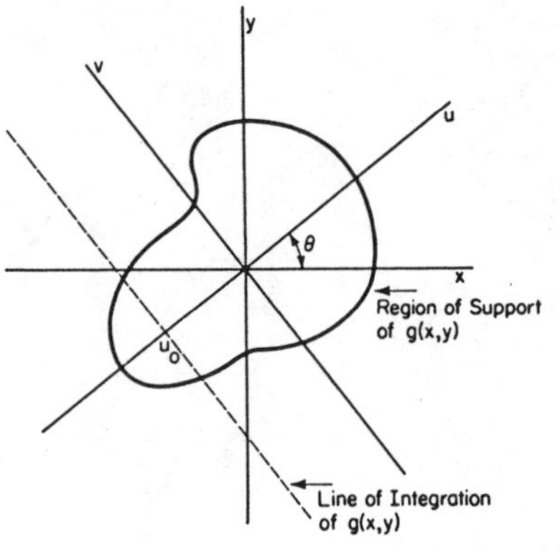

Fig. 1. Line of integration for determining the projection $p_\theta(u_0)$.

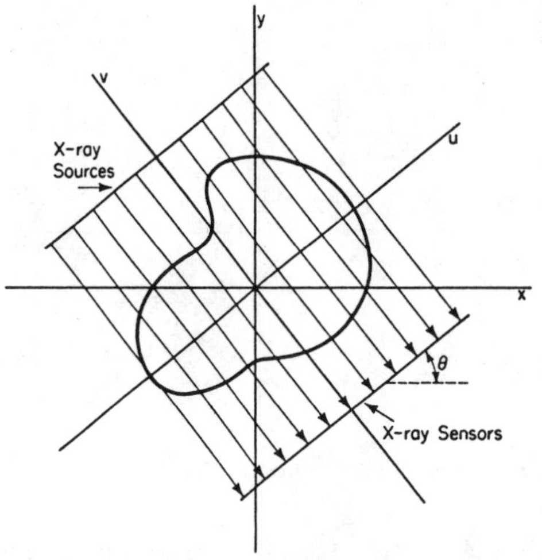

Fig. 2. A parallel-beam X-ray tomography system.

form obtained at each look angle approximates a piece of the one-dimensional (1-D) Fourier transform of a central projection of the ground patch at a corresponding projection angle. A similar statement has been made by Brown [19] in his interpretation and generalization of the work by Walker [7]. Together with the projection-slice theorem, this enables SAR and the processing of SAR data to be viewed in terms similar to CAT, although the physical constraints and data-gathering limitations are somewhat different. SAR is then discussed as a narrow-band version of CAT and the possibility of using CAT processing algorithms, such as convolution backprojection, for SAR is presented. Finally, for completeness, Section IV addresses the issues of resolution, sampling rates, wavefront curvature, quadratic phase errors, effect of Doppler and time-varying range, and the modification of results for slant-plane geometry.

II. Brief Review of Computer-Aided Tomography

Computer-aided tomography is a technique for providing a two-dimensional (2-D) cross-sectional view of a three-dimensional (3-D) object through digital processing of many 1-D projectional views taken from different look angles. These projectional views are obtained by passing sets of narrow X-ray beams through the object and detecting their intensities using an array of sensors. If desired, a 3-D reconstruction may be obtained by this technique as a collection of parallel cross sections.

A. Projection-Slice Theorem

The principle underlying the theory of CAT is the projection-slice theorem [1], [2], [10], [11]. Let $g(x, y)$ be an unknown signal that is to be reconstructed from its projections. The Fourier transform of g is defined as

$$G(X, Y) = \int_{-\infty}^{\infty}\int_{-\infty}^{\infty} g(x, y)\, e^{-j(xX+yY)}\, dx\, dy$$

so that

$$g(x, y) = \frac{1}{4\Pi^2}\int_{-\infty}^{\infty}\int_{-\infty}^{\infty} G(X, Y)\, e^{j(xX+yY)}\, dX\, dY. \quad (1)$$

The projection of g at angle θ is formally given by

$$p_\theta(u) = \int_{-\infty}^{\infty} g(u\cos\theta - v\sin\theta,\, u\sin\theta + v\cos\theta)\, dv \quad (2)$$

where $p_\theta(u)$ evaluated at $u = u_0$ is simply a line integral in the direction of the v axis as illustrated in Fig. 1. Note that the u axis forms an angle θ with the x axis. The function $p_\theta(u)$ represents a series of such line integrals for each value of θ.

The 1-D Fourier transform of $p_\theta(u)$ is given by

$$P_\theta(U) = \int_{-\infty}^{\infty} p_\theta(u)\, e^{-juU}\, du.$$

Using this notation, the statement of the projection-slice theorem is simply

$$P_\theta(U) = G(U\cos\theta, U\sin\theta) \quad (3)$$

that is, the Fourier transform of the projection at angle θ is a "slice" of the 2-D transform $G(X, Y)$ taken at an angle θ with respect to the X axis.

B. Tomographic Reconstruction

In X-ray tomography, $g(x, y)$ is an unknown cross-sectional attenuation coefficient which is to be measured. Samples of the projection $p_\theta(u)$ are obtained in a parallel-beam system with an array of X-ray sources and detectors oriented at an angle θ with respect to the x axis, as shown in Fig. 2. Since the intensity of a received X-ray beam exhibits exponential dependence on the line integral of g [2], projections of g are found in terms of the logarithms of the measured intensities, i.e.,

$$p_\theta(u) = -\log\frac{I_\theta(u)}{I_0}$$

where I_0 is the intensity of the X-ray source and $I_\theta(u)$ is the received intensity at the detector. Projections $p_{\theta_i}(u)$ are obtained at equally spaced angles $\theta = \theta_i$ by rotating either the object or the array of X-ray sources and detectors through a set of discrete angles spanning 360°.

In present tomographic systems, it is common to reconstruct the attenuation coefficient $g(x, y)$ from the projections $p_{\theta_i}(u)$ via the convolution backprojection method [1]. This method will be described later in conjunction with a modification for SAR. An alternate method is direct Fourier domain reconstruction. It is the direct Fourier technique which is used primarily in SAR and it has also been considered for use in tomography [11], [20]. Using the direct Fourier technique, $g(x, y)$ is reconstructed from samples $p_{\theta_i}(u_j)$ of its projections as follows. For each angle θ_i, uniformly spaced samples of the Fourier transform $P_{\theta_i}(U)$ are computed from $\{p_{\theta_i}(u_j)\}_{j=1}^{N}$ via the fast Fourier transform (FFT). From the projection-slice theorem, (3), the samples of $P_{\theta_i}(U)$ are samples of $G(X, Y)$ along a line at angle θ_i with the X axis. Thus the series of 1-D FFT's for the various θ_i provides samples of $G(X, Y)$ on the polar grid shown in Fig. 3. Interpolating the polar samples of G to a Cartesian grid allows the efficiencies of the 2-D inverse FFT to be utilized to approximate g on a

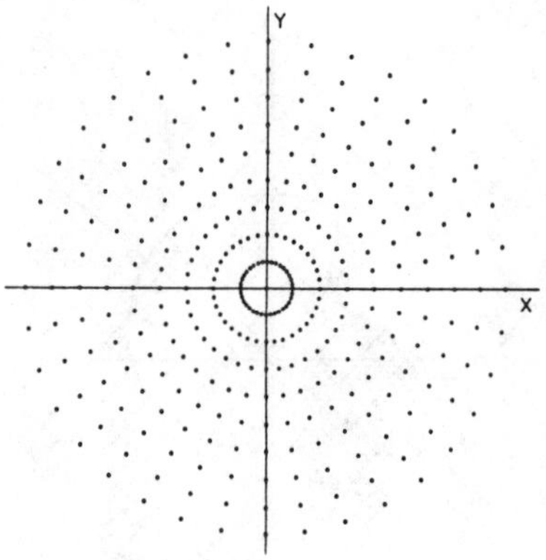

Fig. 3. Locus of known samples of $G(X, Y)$ (Fourier domain).

Cartesian grid [9], [11], [20]. It must be said, though, that no really "fast" algorithm has yet been devised to compute g from the polar samples of G (although this is an area of research [2], [11], [17], [20]–[23]); unless a time-consuming interpolation is carried out prior to employing the FFT, the image quality produced by the Fourier technique will be noticeably worse than that provided by the convolution backprojection method.

III. Spotlight-Mode Synthetic Aperture Radar

A. Introduction

The limiting factor in the azimuthal (cross-range) resolving capabilities of an ordinary ranging radar is the antenna beamwidth in the distant field. For airborne operation, a narrow beamwidth requires an impractically large antenna. Spotlight-mode SAR effectively avoids this requirement by collecting radar returns from many different angular views of a target. By properly processing the return signals, very high resolution in azimuth can be achieved. High resolution in range may be achieved by transmitting high-bandwidth pulses as in a conventional radar.

The geometry for data collection in a spotlight-mode SAR is shown in Fig. 4. The x–y coordinate system (denoting *range* and *azimuth* coordinates, respectively) is centered on a relatively small patch of ground illuminated by a narrow RF beam from the moving radar. As the radar traverses the flight path, the radar beam, as operating in the spotlight mode, is continuously pointed in the direction (angle θ) of the ground patch. At points corresponding to equal increments of θ, high-bandwidth pulses (such as linear FM) are transmitted to the ground patch and echoes are then received and processed. The radar return yields a projectional view of the target, provided the phase front of the radio waves exhibits no significant curvature. This view turns out, in fact, to be a band-pass filtered projection of the ground-patch reflectivity, since the radio waves are narrow-band.

As the aircraft moves with respect to the target patch, the received signal undergoes a slight Doppler shift (to be discussed later) which varies according to the observation angle. The SAR imaging equations can be derived as a function of either the Doppler shift or the underlying change in viewing angle. It is important to emphasize, however, that the imaging principle employed in spotlight-mode SAR is tomographic, rather than Doppler based. That is, although the radar antenna must be moved from point to point to obtain different viewing angles, successful imaging is not dependent on a difference in relative velocity between the antenna and ground patch during pulse transmission and reception. As far as the imaging mechanism is concerned, the aircraft could completely stop at each transmission point in space and the SAR would still work properly.

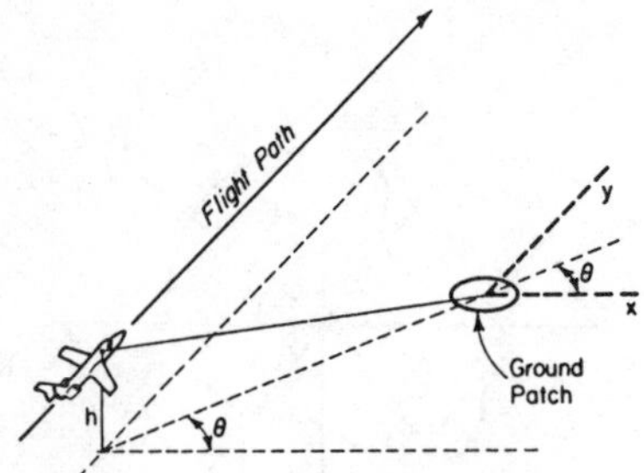

Fig. 4. Geometry for data collection in spotlight-mode SAR.

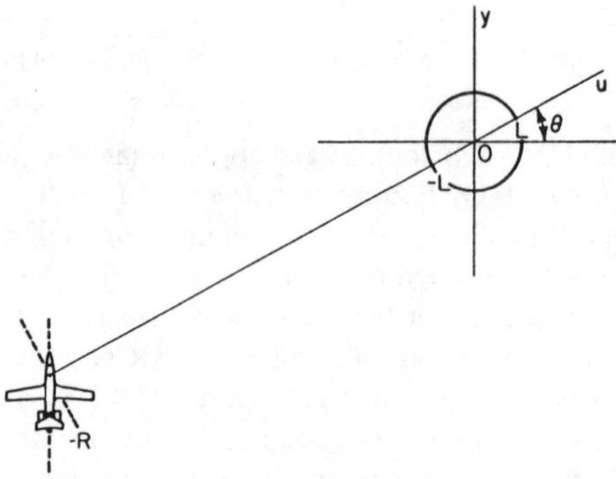

Fig. 5. Ground-plane geometry for data collection in spotlight-mode SAR.

B. Basic SAR Derivation

If it is assumed that the depression angle from the radar to the ground patch is zero, the geometry reduces to Fig. 5. Later, it will be indicated how the results can be modified for the case of a nonzero depression angle (i.e., the height $h \neq 0$). In Fig. 5, the reflectivity density of the ground patch is modeled by the complex function[2] $g(x, y)$ where a sinusoid reflected from a point (x_0, y_0) is scaled in amplitude by $|g(x_0, y_0)|$ and shifted in phase by $\angle g(x_0, y_0)$ where $g(x, y) = |g(x, y)|\exp(j\angle g(x, y))$. Furthermore, it is assumed that $g(x, y)$ is constant over the range of frequencies and range of viewing angles θ employed by the radar.

At angle θ, let the radar transmit a linear FM chirp pulse $\mathrm{Re}\{s(t)\}$, where

$$s(t) = \begin{cases} e^{j(\omega_0 t + \alpha t^2)}, & |t| \leq \dfrac{T}{2} \\ 0, & \text{otherwise} \end{cases} \tag{4}$$

ω_0 is the RF carrier frequency and 2α is the FM rate. The return signal from a differential area centered on the point (x_0, y_0) at a distance R_0 from the radar will be

$$r_0(t) = A|g(x_0, y_0)|\cos\left(\omega_0\left(t - \frac{2R_0}{c}\right) + \alpha\left(t - \frac{2R_0}{c}\right)^2 + \angle g(x_0, y_0)\right) dx\, dy$$

[2]The amplitude scaling occurs because only a fraction of the incident radiation is reflected back to the radar. The phase shift of the reflected wave may be caused by several factors; foremost is probably the shift at the air/target interface due to the difference between the dielectric constants of air and the target material. This effect is similar to that observed at the boundary of two waveguides having different characteristic impedances where at least one of the waveguides is dissipative [24]. The phase shift is also due to the tendency of the RF radiation to creep around target surfaces and its ability to penetrate soft objects and be reflected from within.

where A accounts for propagation attenuation, c is the speed of light, and $2R_0/c$ accounts for the two-way travel time from radar to target. The return $r_0(t)$ is written more simply as

$$r_0(t) = A \cdot \mathrm{Re}\left\{g(x_0, y_0)s\left(t - \frac{2R_0}{c}\right)\right\} dx\, dy. \quad (5)$$

In Fig. 5, points in the ground patch equidistant from the radar lie on an arc, but for a typical system $R \gg L$, so that this arc is nearly a straight line. Therefore, taking $p_\theta(u)$ to be the line integral given by (2), it follows from (5), by superposition,[3] that the return signal from a differential line of scatterers normal to the u axis at $u = u_0$ is given by

$$r_1(t) = A \cdot \mathrm{Re}\left\{p_\theta(u_0)s\left(t - \frac{2(R+u_0)}{c}\right)\right\} du.$$

If $R \gg L$, the attenuation A may be taken as a constant. Therefore, the return from the entire ground patch can be approximated by the integral of r_1 over u, given by

$$r_\theta(t) = A \cdot \mathrm{Re}\left\{\int_{-L}^{L} p_\theta(u)s\left(t - \frac{2(R+u)}{c}\right) du\right\}. \quad (6)$$

This expression has the form of a convolution, thus the transform of p_θ can be obtained by Fourier methods over a range of frequencies determined by the bandwidth of $s(\cdot)$. For our case though, $s(\cdot)$ is a chirp pulse, and a different form of processing is customarily used. Substituting (4) into (6) gives

$$r_\theta(t) = A \cdot \mathrm{Re}\left\{\int_{-L}^{L} p_\theta(u)\exp\left\{j\left[\omega_0\left(t - \frac{2(R+u)}{c}\right) + \alpha\left(t - \frac{2(R+u)}{c}\right)^2\right]\right\} du\right\} \quad (7)$$

on the interval[4]

$$-\frac{T}{2} + \frac{2(R+L)}{c} \leq t \leq \frac{T}{2} + \frac{2(R-L)}{c}. \quad (8)$$

Letting

$$\tau_0 = \frac{2R}{c}$$

be the round-trip delay to the center of the ground patch and mixing (multiplying) $r_\theta(t)$ with the reference chirp

$$\cos\left[\omega_0(t - \tau_0) + \alpha(t - \tau_0)^2\right] \quad (9)$$

yields

$$\bar{r}_\theta(t) = \frac{A}{2}\mathrm{Re}\left\{\int_{-L}^{L} p_\theta(u)\left[\exp\left\{j\left[\omega_0\left(2t - \frac{2u}{c} - 2\tau_0\right) + \alpha\left((t-\tau_0)^2 + \left(t - \tau_0 - \frac{2u}{c}\right)^2\right)\right]\right\} + \exp\left\{j\left[\frac{4\alpha u^2}{c^2} - \frac{2u}{c}(\omega_0 + 2\alpha(t - \tau_0))\right]\right\}\right] du\right\}.$$

The first term is centered on the RF carrier frequency ω_0, whereas the second term is not, so low-pass filtering $\bar{r}_\theta(t)$ gives

$$\hat{r}_\theta(t) = \frac{A}{2}\mathrm{Re}\left\{\int_{-L}^{L} p_\theta(u) \cdot \exp\left\{j\left[\frac{4\alpha u^2}{c^2} - \frac{2u}{c}(\omega_0 + 2\alpha(t - \tau_0))\right]\right\} du\right\}.$$

Similarly, it can be shown that mixing $r_\theta(t)$ with

$$\sin\left[\omega_0(t - \tau_0) + \alpha(t - \tau_0)^2\right]$$

and low-pass filtering gives the quadrature component

$$\tilde{r}_\theta(t) = \frac{A}{2}\mathrm{Im}\left\{\int_{-L}^{L} p_\theta(u) \cdot \exp\left\{j\left[\frac{4\alpha u^2}{c^2} - \frac{2u}{c}(\omega_0 + 2\alpha(t - \tau_0))\right]\right\} du\right\}.$$

The real signals $\hat{r}_\theta(t)$ and $\tilde{r}_\theta(t)$, therefore, determine the real and imaginary components of the complex signal

$$C_\theta(t) = \frac{A}{2}\int_{-L}^{L} p_\theta(u)\exp\left\{j\frac{4\alpha u^2}{c^2}\right\} \cdot \exp\left\{-j\frac{2}{c}(\omega_0 + 2\alpha(t - \tau_0))u\right\} du. \quad (10)$$

The effect of the quadratic phase term in (10) will be considered later; for now let us assume that this factor can be removed to obtain

$$\bar{C}_\theta(t) = \frac{A}{2}\int_{-L}^{L} p_\theta(u)\exp\left\{-j\frac{2}{c}[\omega_0 + 2\alpha(t - \tau_0)]u\right\} du.$$

This last expression can be identified as the Fourier transform of the projection $p_\theta(u)$, that is,

$$\bar{C}_\theta(t) = \frac{A}{2}P_\theta\left[\frac{2}{c}(\omega_0 + 2\alpha(t - \tau_0))\right]. \quad (11)$$

The net result is that, at least within the time interval considered, the processed return signal $\bar{C}_\theta(t)$ is the Fourier transform of a projection.

According to the projection-slice theorem, $\bar{C}_\theta(t)$ is a slice at angle θ of the 2-D transform G of the unknown reflectivity density. From (8), the processed return $\bar{C}_\theta(t)$ is available for

$$-\frac{T}{2} + \frac{2(R+L)}{c} \leq t \leq \frac{T}{2} + \frac{2(R-L)}{c}.$$

So, from (11), $P_\theta(X)$ is determined for $X_1 \leq X \leq X_2$ with

$$X_1 = \frac{2}{c}\left(\omega_0 - \alpha T + \frac{4\alpha L}{c}\right)$$
$$X_2 = \frac{2}{c}\left(\omega_0 + \alpha T - \frac{4\alpha L}{c}\right). \quad (12)$$

Thus only the segment of width $X_2 - X_1$ of each slice of G is actually determined. Processing returns from angles satisfying $|\theta| \leq \theta_M$ provides samples of $G(X, Y)$ on the polar grid in the annulus segment shown in Fig. 6. Transform points at angle θ are determined from the radar return collected at angle θ. The inner and outer radii, X_1 and X_2, are proportional to the lowest and highest frequencies in the transmitted chirp pulse [7]. This observation follows easily since for a typical SAR, $\omega_0 \pm \alpha T \gg 4\alpha L/c$, so that (12) reduces to

$$X_1 = \frac{2}{c}(\omega_0 - \alpha T)$$
$$X_2 = \frac{2}{c}(\omega_0 + \alpha T). \quad (13)$$

[3]The assumption of superposition is properly questioned. The extent of its validity is carefully discussed by Rihaczek [25].

[4]For t not satisfying (8), the argument of $s(\cdot)$ in (6) can lie outside $\pm T/2$ so that $r_\theta(t)$ will not be correctly given by (7). It is possible, however, to process $r_\theta(t)$ over the slightly larger interval $-(T/2) + (2R/c) \leq t \leq (T/2) + (2R/c)$ to obtain somewhat improved resolution.

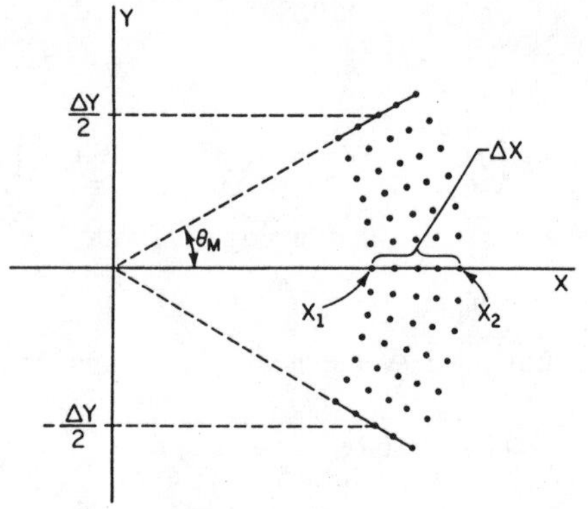

Fig. 6. Annulus segment containing known samples of $G(X, Y)$.

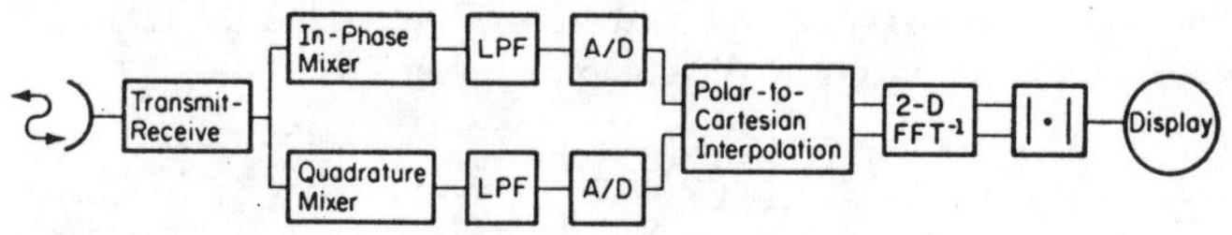

Fig. 7. Required processing for spotlight-mode SAR.

From (4), the lowest and highest frequencies in the transmitted chirp pulse are $\omega_0 - \alpha T$ and $\omega_0 + \alpha T$ which are proportional to X_1 and X_2 in (13).

As remarked earlier, there is no known fast FFT-type algorithm for computing approximate samples of g from polar samples of G. In the case of SAR, an approximation to g is obtained by interpolating the known samples of G to a Cartesian grid, assuming unknown samples (those outside the annulus segment in Fig. 6) to be zero. This may be accomplished in a number of ways [11], e.g., by first linearly interpolating the samples of G on each radial line in Fig. 6 to a set of uniformly spaced vertical lines (resulting in a "keystone" grid [26]). These data are then linearly interpolated in the vertical dimension giving approximate samples of G on a Cartesian grid. After interpolation, an inverse 2-D FFT is employed and the amplitude of g is displayed for viewing [9], [26]. Before FFT processing, the samples of G are windowed to reduce sidelobe levels and are translated to the origin to save computation. The translation has no effect on the amplitude of the FFT output. The required processing for SAR is similar to that for tomography and is summarized in Fig. 7.

C. SAR as a Narrow-Band Version of CAT

The mathematical models for SAR and CAT bear enough resemblance to each other to allow both systems to be explained in terms of the projection-slice theorem. There are important differences, however. In SAR, the line integral involved in the projection is taken perpendicular to the direction in which the radio waves travel. This is in contrast to tomography, where the line integral is taken along the path of the X-rays. Also, a SAR system, utilizing a chirp waveform, determines the transform of the projection rather than the projection itself. Although it is not difficult to convert back and forth between the projection and its transform, this factor may partially explain why spatial domain reconstruction algorithms have not been considered for SAR.

The most striking difference between SAR and tomographic imaging is that SAR data are necessarily narrow-band. As we have seen, the transform domain data in a SAR system are restricted to lie in a small annulus segment with inner and outer edges that are determined by the frequency content of the transmitted chirp. Indeed, if the transmitted waveform is a sinusoid (zero bandwidth) rather than a chirp, data are obtained at only a single ring in the transform domain [16]. However, if waveforms such as pseudo-random signals or even very short pulses are transmitted, the frequency content of the data in the transform domain is significant. (In fact, according to the model presented, if the SAR *could* transmit an impulse, $s(t) = \delta(t)$, with infinite bandwidth, the SAR imaging equation (6) would provide the projection $p_\theta(u)$ directly, as in tomography.) In general, the range information measured by a coherent SAR may be thought of as consisting of two parts: coarse range information providing the range resolution of the SAR, and fine range information (fractional wavelength range) that makes it possible to obtain high azimuth resolution through coherent processing.

Considering the narrow-band feature of SAR, it may seem surprising that acceptable imagery can be obtained. For example, an edge oriented at an angle $\theta_0 + 90°$ in the ground patch with real reflectivity will have significant frequency content in the transform domain along a line at an angle θ_0. If θ_0 does not fall within the look angle of the radar, such an edge will be obscured. Yet, unlike the attenuation coefficient in CAT, the reflectivity density in SAR is complex. Although the *magnitude* of the observed reflectivity density of resolution cells in the final processed image will not have significant frequency content within the annulus segment in Fig. 6, a random complex reflectivity can be attributed to an assumed distribution of point scatterers (each having random phase) in each resolution cell, giving rise to frequency components over much of the transform plane. Thus a segment of the transform plane displaced from the origin can contain significant information about the magnitude of the overall reflectivity density. This property is similar to holography, where a piece of a hologram contains the information essential for reconstructing a recognizable image. Indeed, a strong connection between SAR and holography has been established [27], [28].

Although SAR systems can produce high-quality imagery, various aberrations, such as coherent speckle, may be observed due to the narrow-band feature. For example, if the target is illuminated with a sinusoid of wavelength λ, then two point reflectors with approximately in-phase reflectivity and displaced in range by an odd multiple of $\lambda/4$ will result in destructive interference.[5] For the case of a transmitted chirp, interference can still occur. Here the response of the SAR system to a point reflector will be approximately a modulated 2-D sinc pulse having the width of a resolution cell (assuming an approximately rectangular region in Fig. 6); responses from adjacent point reflectors in the same resolution cell will thus overlap and interfere either constructively or destructively (see Fig. 8). At the expense of resolution, coherent speckle may be reduced by dividing the Fourier space into several sections and incoherently summing the magnitudes of the responses from each section.

Because SAR is essentially a narrow-band version of CAT, reconstruction algorithms used in one system may be used with only slight modification in the other. In the next section, a modification of the tomographic convolution-backprojection algorithm is introduced for use in SAR. This algorithm permits computation to proceed in step with data collection, since the final integration is performed over θ.

D. Reconstruction via the Backprojection Algorithm

The reconstruction of an image via convolution backprojection is accomplished on the basis of writing the Fourier integral (1) in polar form

[5]This factor is $\lambda/4$ rather than $\lambda/2$ due to two-way travel.

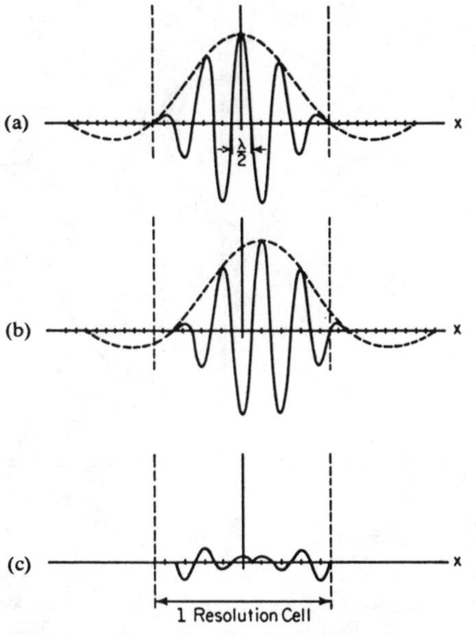

Fig. 8. One-dimensional example of destructive interference. (a) Real part of response from point target at $x = 0$. (b) Real part of response from point target at $x = \lambda/4$. (c) Sum of the responses. (Imaginary parts also destructively interfere.)

$$g(\rho\cos\phi, \rho\sin\phi) = \frac{1}{4\pi^2}\int_{-\pi/2}^{\pi/2}\int_{-\infty}^{\infty} G(r\cos\theta, r\sin\theta)|r| \cdot \exp[jr\rho\cos(\phi-\theta)]\,dr\,d\theta$$
$$= \frac{1}{4\pi^2}\int_{-\pi/2}^{\pi/2}\int_{-\infty}^{\infty} P_\theta(r)|r| \cdot \exp[jr\rho\cos(\phi-\theta)]\,dr\,d\theta \qquad (14)$$

where the second equality follows by the projection-slice theorem (3). The integral over r may be identified as an inverse Fourier transform with argument $\rho\cos(\phi-\theta)$, allowing (14) to be rewritten

$$g(\rho\cos\phi, \rho\sin\phi) = \frac{1}{2\pi}\int_{-\pi/2}^{\pi/2}[p_\theta * k](\rho\cos(\phi-\theta))\,d\theta \qquad (15)$$

where

$$k = \mathscr{F}^{-1}[|r|].$$

The tomographic reconstruction algorithm evaluates the convolutions (possibly in the Fourier domain) for each θ, and then approximates the integral as a sum of these results.

In the case of SAR, we have available $P_\theta(r)$, the Fourier transform of the projection, so let us consider using (14) rather than (15). Notice that (14) requires $P_\theta(r)$ for all r, but from (12), the SAR system determines $P_\theta(r)$ over only a relatively small interval centered at $r = 2\omega_0/c$. Therefore, a window function vanishing outside this interval is applied before inverting $P_\theta(r)$ via an inverse FFT. Furthermore, since the known segment of P_θ is offset from the origin, P_θ is translated to the origin and the FFT result compensated accordingly. This is based on rewriting (14) as

$$g(\rho\cos\phi, \rho\sin\phi) = \frac{1}{4\pi^2}\int_{-\pi/2}^{\pi/2}\left[\int_0^{X_2-X_1} P_\theta(r+X_1)|r+X_1|W(r) \cdot \exp[jr\rho\cos(\phi-\theta)]\,dr\right] \cdot \exp[jX_1\rho\cos(\phi-\theta)]\,d\theta \qquad (16)$$

where X_1 and X_2 are given by (12) and W is a window function which tapers to zero on the interval $[0, X_2 - X_1]$. Finally, note that (16) requires that the inverse transform be evaluated at $\rho\cos(\phi-\theta)$ for various ρ, ϕ, and θ. Therefore, some interpolation of the FFT results is required before summing over θ. In addition, P_θ is available in a SAR system for only a restricted set of look angles as illustrated in Fig. 6. Thus it is desirable to apply a second window before integration over θ.

We have simulated the modified convolution-backprojection algorithm and compared it with the direct Fourier domain algorithm using linear interpolation in each dimension. Initial results show that the backprojection algorithm produces images of somewhat better quality—we hope to make this the subject of a future paper.

IV. Additional Considerations in SAR

A. Resolution and Required Sampling Rate

A SAR system provides transform domain data in the small polar region of the X–Y plane shown in Fig. 6. A definition of resolution in the image (spatial) domain can be motivated by considering the polar region to be approximated by a rectangle of width ΔX and height ΔY (see Fig. 6). The transform of a point reflector at (x_0, y_0) with reflectivity γ is

$$G(X,Y) = \int_{-\infty}^{\infty}\int_{-\infty}^{\infty} \gamma\delta(x-x_0, y-y_0) \cdot \exp[-j(xX+yY)]\,dx\,dy$$
$$= \gamma\exp[-j(x_0X+y_0Y)]$$

so that the SAR system response to such a reflector is approximately

$$|\hat{g}(x,y)| = \frac{|\gamma|}{4\pi^2}\left|\int_{X_1}^{X_1+\Delta X}\int_{-\Delta Y/2}^{\Delta Y/2} \cdot \exp[-j(x_0X+y_0Y)][\exp j(xX+yY)]\,dY\,dX\right|$$
$$= \frac{|\gamma|}{4\pi^2}\left|\Delta X \operatorname{sinc}\left(\frac{1}{2}\Delta X(x-x_0)\right)\Delta Y \cdot \operatorname{sinc}\left(\frac{1}{2}\Delta Y(y-y_0)\right)\right|.$$

The first zero crossings in the response occur at $x - x_0 = 2\pi/\Delta X$ and $y - y_0 = 2\pi/\Delta Y$. Therefore, as a rough guide, resolution of two point reflectors having equal reflectivity requires that the reflectors be separated by more than $2\pi/\Delta X$ in the x dimension and $2\pi/\Delta Y$ in the y dimension. From (13) and Fig. 6 we have, for $\theta_M \ll 1$, that $\Delta X \approx 4\alpha T/c$ and $\Delta Y \approx 2[(2\omega_0/c)\sin\theta_M]$. Thus the system resolution is defined as[6]

$$\delta_x = \frac{2\pi}{\Delta X} \approx \frac{\pi c}{2\alpha T} \qquad (17)$$

$$\delta_y = \frac{2\pi}{\Delta Y} \approx \frac{\pi c}{2\omega_0\sin\theta_M}. \qquad (18)$$

It is common in the SAR literature to refer to an area of size δ_x by δ_y in the reconstructed image as a *resolution cell*. Notice that the range resolution depends on only the bandwidth of the

[6]Although (17) and (18) are useful in practice, it requires more than a single pair of numbers (δ_x, δ_y) to precisely characterize resolution, because an actual target is composed of many interfering reflectors of varying intensities. See [25], [29] for further discussion.

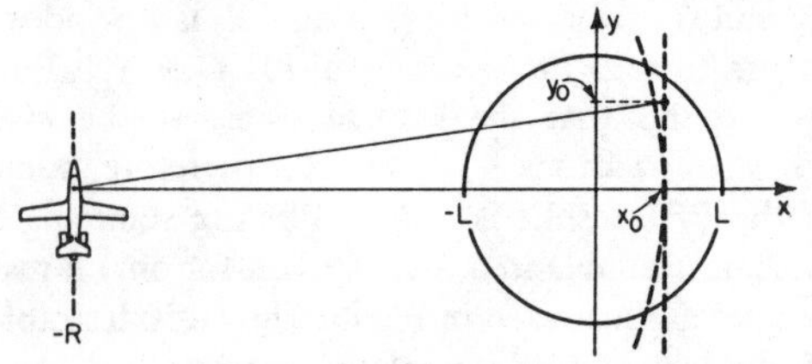

Fig. 9. Wavefront curvature over the target field.

transmitted signal, whereas the azimuth resolution depends on the center frequency and also on the range of look angles. Therefore, the resolving phenomena are entirely different in range and azimuth. Also, the necessary range of look angles to provide $\delta_y = \delta_x$ is actually quite small. For a typical SAR, ω_0 may be twenty times αT, giving $2\theta_M \approx 6°$.

Let us next consider an appropriate sampling rate for the transform in the polar region of Fig. 6. The usual sampling theorem will apply only if the inverse transform of the known segment of $G(X, Y)$ is spatially limited. Of course, this cannot be, since a signal cannot be both frequency limited and spatially limited. However, it is known that the true $g(x, y)$ is spatially limited by the radar antenna beam to a region of half-width L. Therefore, sampling rates in the X and Y directions should be *at least*

$$f_s = \frac{1}{2\pi}(2L) \text{ samples/rad.}$$

This corresponds to an image raster having

$$\left(\frac{L}{\pi}\cdot\Delta X = \frac{2L}{\delta_x}\right)\times\left(\frac{L}{\pi}\cdot\Delta Y = \frac{2L}{\delta_y}\right) = 4L^2/\delta_x\delta_y$$

samples or resolution cells in all. In practice, sampling rates somewhat higher than f_s are used.

B. Curvature of the Wavefront

As we have previously noted, SAR projections are not really taken over straight lines, but are taken over slight curves due to the outward propagation of waves from the transmitter. Two conditions must simultaneously be satisfied in order that this effect be neglected. First, the range error due to wavefront curvature over the target field must be less than a resolution cell, i.e.,

$$\frac{L^2}{2R} < \delta_x. \tag{19}$$

Condition (19) is obtained with the help of Fig. 9. The range error at a point (x_0, y_0) is

$$\left[(R + x_0)^2 + y_0^2\right]^{1/2} - (R + x_0)$$

$$= (R + x_0)\left\{\left[1 + \frac{y_0^2}{(R + x_0)^2}\right]^{1/2} - 1\right\}$$

$$\approx (R + x_0)\left\{1 + \frac{y_0^2}{2(R + x_0)^2} - 1\right\} \tag{20}$$

$$\approx \frac{y_0^2}{2R} \tag{21}$$

where the approximations hold since $R \gg x_0, y_0$. Evaluating (21) at the largest value of y_0 within the target patch gives (19).

A second condition arises since a SAR system coherently combines projections from many different angles. To preserve coherence, the range error due to wavefront curvature at a particular point must vary by no more than a small fraction of a wavelength through the full range of look angles. To analyze this effect, consider Fig. 10 which shows the wavefront curvature at the point (x_0, y_0) due to look angles 0 and θ. Evaluating the l_i gives

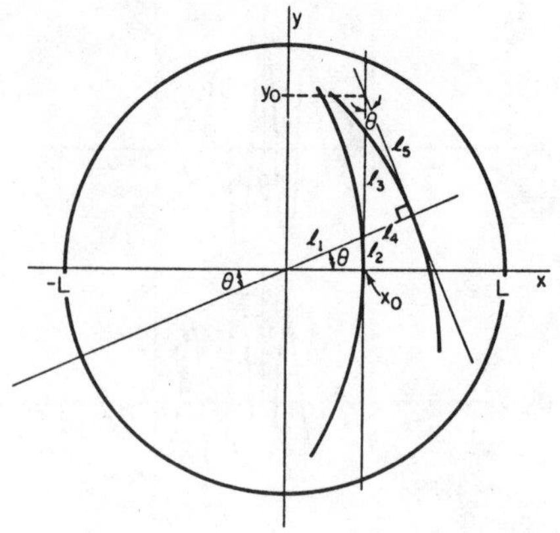

Fig. 10. Wavefront curvature for projection angles 0 and θ.

$$l_1 = \frac{x_0}{\cos\theta}$$

$$l_2 = l_1\sin\theta = x_0\tan\theta$$

$$l_3 = y_0 - l_2 = y_0 - x_0\tan\theta$$

$$l_4 = l_3\sin\theta = y_0\sin\theta - x_0\frac{\sin^2\theta}{\cos\theta}$$

$$l_5 = l_3\cos\theta = y_0\cos\theta - x_0\sin\theta.$$

The range error at point (x_0, y_0) for a projection at angle θ is

$$\left[(R + l_1 + l_4)^2 + l_5^2\right]^{1/2} - (R + l_1 + l_4) \approx \frac{l_5^2}{2(R + l_1 + l_4)} \tag{22}$$

similar to (20). For a typical SAR system $R \gg l_1 + l_4$, so, from (21) and (22), the difference in wavefront curvature from a projection at angle 0 and a projection at angle θ is approximately

$$D = \frac{1}{2R}\left(y_0^2 - l_5^2\right)$$

$$= \frac{1}{2R}\left(y_0^2\sin^2\theta + 2x_0y_0\sin\theta\cos\theta - x_0^2\sin^2\theta\right). \tag{23}$$

Given the constraints $\theta \leqslant \theta_M$ and $x_0^2 + y_0^2 \leqslant L^2$, numerical optimization could be applied to (23) to obtain the maximum deviation. Instead, however, let us examine two special cases which would appear to yield large deviations:

1) $x_0 = 0, \quad y_0 = L$

2) $x_0 = y_0 = \dfrac{L}{\sqrt{2}}.$

For case 1), the deviation (23) reduces to

$$D_1 = \frac{L^2\sin^2\theta}{2R} \leqslant D_1^{\max} \triangleq \frac{L^2\sin^2\theta_M}{2R}.$$

For case 2) we obtain

$$D_2 = \frac{L^2\sin\theta\cos\theta}{2R} = \frac{L^2\sin 2\theta}{4R} \leqslant D_2^{\max} \triangleq \frac{L^2\sin 2\theta_M}{4R}$$

where the inequality assumes $\theta_M \leqslant \pi/4$. For θ_M in this range, we have $D_2^{\max} \geqslant D_1^{\max}$. Hence, to preserve coherency we require that $D_2^{\max}$ be much smaller than a fraction of a wavelength, i.e.,

$$\frac{L^2 \sin 2\theta_M}{4R} \ll \frac{c}{8\omega_0} = \frac{\lambda}{8}. \tag{24}$$

Assuming $\delta_x = \delta_y$ in (17) and (18), condition (24) will be more severe than condition (19). If the target region to be mapped is too large to satisfy (24), the area may be mapped in smaller segments, using either antenna steering of a phased array or digital presumming methods [30]. Alternatively, the backprojection method may be applied even in the case of projections over curves in a manner described in [31].

C. Quadratic Phase Term

It was assumed earlier that the quadratic phase term in (10) can be removed. This can be partially accomplished by inverse transforming $C_\theta(t)$, multiplying by $\exp\{-j(4\alpha u^2/c^2)\}$, and then retransforming. However, there is some error in this procedure since $C_\theta(t)$ is not known for all t. If the step to remove the quadratic phase factor is omitted, the image suffers a consequent loss of resolution. However, if the inequality

$$\frac{4\alpha L^2}{c^2} \ll \frac{\pi}{2} \tag{25}$$

is satisfied, then the loss of resolution will be small. We previously saw that the number of resolution cells in the reconstructed image will be

$$N^2 = \frac{4L^2}{\delta_x \delta_y}$$

where δ_x and δ_y are given by (17) and (18). Assuming $\delta_y \approx \delta_x$ and substituting from (17) gives

$$N^2 = \frac{16L^2\alpha^2T^2}{\pi^2c^2}. \tag{26}$$

Defining the time–bandwidth product (TBW) of the transmitted chirp as

$$\text{TBW} = T \cdot \frac{2\alpha T}{2\pi} = \frac{\alpha T^2}{\pi}$$

and using (26), condition (25) becomes

$$N^2 \ll 2\,\text{TBW}. \tag{27}$$

Condition (27) implies that the quadratic phase term can be neglected as long as the TBW is sufficiently large compared with N^2.

These relations show that neglecting the quadratic phase term imposes a limit on the resolution using SAR processing with chirp waveforms. Compensating for this term, as suggested above, is necessary for improved resolution.

D. Effect of Doppler and Time-Varying Range

Thus far, little mention has been made of the fact that the radar unit is moving as measurements are made—the radar unit was simply assumed to be placed at discrete positions along the flight path as projection data were gathered. Due to the motion of the aircraft, there is a Doppler effect which must be considered.

From Fig. 5 there is a radial component of the aircraft velocity $v_r = v \sin\theta$, where v is the speed of the aircraft. Therefore, the received signal, $r_\theta(t)$ in (6), is more accurately given by

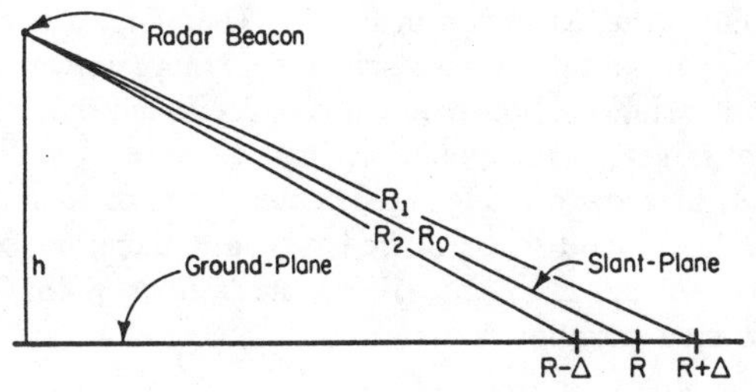

Fig. 11. Side view of radar flown at height h.

$$r_\theta(t) = A \cdot \text{Re}\left\{\int_{-L}^{L} p_\theta(u)\, s\left(at - \frac{2(R+u)}{c}\right) du\right\} \tag{28}$$

with

$$a = \left(1 + \frac{v_r}{c}\right)^2 \approx 1 + \frac{2v_r}{c}$$

where the square is due to two-way travel. Instead of (9), the reference used for demodulation should then be

$$\cos\left[\omega_0(at - \tau_0) + \alpha(at - \tau_0)^2\right]. \tag{29}$$

Equations (28) and (29) give the correction for a target point at the center of the ground patch. The Doppler shift will vary slightly from point to point within the target patch since the angle from the radar and, therefore, the radial component of the aircraft velocity will vary slightly from the center to either side of the patch. Thus the reference waveform given by (29) cannot completely compensate for the Doppler effect. Assuming $\theta = 0$ in Fig. 5, and considering an extreme case, the radial component of the aircraft velocity for a point p^* at the top of the target patch at a distance R from the radar is approximately

$$v_r^* = v \sin\left(\tan^{-1}\frac{L}{R}\right) \approx v\frac{L}{R}.$$

The difference between the round-trip propagation times to the center of the target patch and to p^* is

$$\frac{2R}{c} - \frac{2R}{c + v_R^*} = \frac{2R}{c}\left(1 - \frac{1}{1 + \frac{vL}{cR}}\right) \approx \frac{2vL}{c^2}. \tag{30}$$

To preserve the ability to measure fractional wavelengths (needed for high azimuth resolution), we require that c times (30) be less than one-quarter wavelength, i.e.,

$$\frac{2vL}{c} < \frac{\lambda_0}{4} \quad \text{or} \quad \frac{v}{c} < \frac{\lambda_0}{8L}$$

where $\lambda_0 = 2\pi c/\omega_0$. Thus for a given aircraft velocity, the variation in Doppler shift limits the target size. Notice that this condition is considerably different from the limitation (24) imposed by wavefront curvature.

Another factor that requires compensation is time-varying range. If the aircraft in Fig. 5 flies a straight path, the range to the center of the target patch will vary considerably. Thus R in (28) is actually a function of θ, and the demodulating waveform, (29), must be modified accordingly. Alternatively, a fixed high-frequency demodulator can be used, followed by a time-varying low-frequency demodulator that compensates for both time-varying range and Doppler. The second demodulator can be implemented digitally.

E. Modification of Results for Slant-Plane Geometry

In the previous analyses we have neglected the fact that the radar operates in the slant plane at some nonzero height h above the ground plane, as shown in Fig. 11. Therefore, it is necessary to modify our results by stretching the range dimension. The amount of compensation required is easily found by considering two point targets in the ground patch at distance R_1 and R_2 from the radar, as shown in Fig. 11. Let the locations in the ground plane be $R + \Delta$ and $R - \Delta$. The targets are, therefore, separated by a true distance of 2Δ. From the radar's point of view, however, the separation is

$$R_1 - R_2 = \sqrt{h^2 + (R+\Delta)^2} - \sqrt{h^2 + (R-\Delta)^2}$$
$$\approx \sqrt{h^2 + R^2 + 2R\Delta} - \sqrt{h^2 + R^2 - 2R\Delta}$$

for $R \gg \Delta$. This can be further approximated as

$$R_1 - R_2 = R_0\left[\sqrt{1 + \frac{2R\Delta}{R_0^2}} - \sqrt{1 - \frac{2R\Delta}{R_0^2}}\right]$$
$$\approx R_0\left[1 + \frac{R\Delta}{R_0^2} - \left(1 - \frac{R\Delta}{R_0^2}\right)\right]$$
$$= 2\Delta\frac{R}{R_0}.$$

Therefore, to obtain the true separation, 2Δ, it is necessary to stretch the range dimension by R_0/R. Taking this into account, the derivation in Section III can be easily modified, beginning with

$$r_\theta(t) = A \cdot \mathrm{Re}\left\{\int_{-L}^{L} p_\theta(u)\, s\left(t - \frac{2\left(R_0 + u\frac{R}{R_0}\right)}{c}\right) du\right\}$$

as a replacement for (6), where p_θ is the projection of the reflectivity density in the ground plane, R is the distance from the projection of the radar onto the ground plane to the center of the target patch, and R_0 is the true distance (in the slant plane) from the radar to the center of the target patch. Note that R/R_0 will vary as a function of range, and therefore as a function of θ for a straight-line flight path. Thus the modification required for each projection will, in general, be different.

V. Conclusion

A tomographic formulation of spotlight-mode SAR has been presented. We have shown that a chirped SAR operating in spotlight mode records a portion of the Fourier transform of a central projection of the ground patch at each look angle. This establishes that "polar format Doppler processing," used in SAR, is a form of tomographic reconstruction that may be conveniently described by the projection-slice theorem. The tomographic interpretation permits, we believe, a conceptually simpler understanding of SAR than an analysis based solely on Doppler concepts. An important benefit in making the connection between SAR and CAT is that algorithms and signal processing methods developed in one field may be applied to the other. For example, it was shown how the convolution-backprojection algorithm, that is widely used in commercial CAT scanners, can be modified for use in SAR.

Various issues concerning speckle, resolution, wavefront curvature, residual quadratic phase errors, and the effects of Doppler and time-varying range have been considered. In particular, it was shown that the terrain patch size is limited by the effects of wavefront curvature and Doppler.

Acknowledgment

The authors would like to thank the reviewers for making many helpful comments.

References

[1] H. J. Scudder, "Introduction to computer aided tomography," *Proc. IEEE*, vol. 66, pp. 628–637, June 1978.
[2] G. T. Herman, Ed., *Image Reconstruction from Projections*. New York: Springer, 1979.
[3] C. W. Sherwin, J. P. Ruina, and R. D. Rawcliffe, "Some early developments in synthetic aperture radar systems," *IRE Trans. Mil. Electron.*, vol. MIL-6, pp. 111–115, Apr. 1962.
[4] W. M. Brown and C. J. Palermo, "Theory of coherent systems," *IRE Trans. Mil. Electron.*, vol. MIL-6, pp. 187–196, Apr. 1962.
[5] W. M. Brown and L. J. Porcello, "An introduction to synthetic aperture radar," *IEEE Spectrum*, vol. 6, pp. 52–62, Sept. 1969.
[6] K. Tomiyasu, "Tutorial review of synthetic-aperture radar (SAR) with applications to imaging of the ocean surface," *Proc. IEEE*, vol. 66, pp. 563–583, May 1978.
[7] J. L. Walker, "Range-Doppler imaging of rotating objects," *IEEE Trans. Aerosp. Electron. Syst.*, vol. AES-16, pp. 23–52, Jan. 1980.
[8] W. M. Brown and R. J. Fredricks, "Range-Doppler imaging with motion through resolution cells," *IEEE Trans. Aerosp. Electron. Syst.*, vol. AES-5, pp. 98–102, Jan. 1969.
[9] "Advanced synthetic array radar techniques," First Interim Report, Radar and Optics Division, Environmental Research Institute of Michigan, unclassified excerpts, Mar. 1976, available from DTIC as Tech. Rep. AFAL-TR-75-87.
[10] R. M. Mersereau, "Recovering multidimensional signals from their projections," *Comput. Graph. Image Proc.*, vol. 1, pp. 179–195, Oct. 1973.
[11] R. M. Mersereau and A. V. Oppenheim, "Digital reconstruction of multidimensional signals from their projections," *Proc. IEEE*, vol. 62, pp. 1319–1338, Oct. 1974.
[12] M. I. Skolnik, *Introduction to Radar Systems*, 2nd ed. New York: McGraw-Hill, 1980, pp. 34–44.
[13] G. Wade, S. Elliott, I. Khogeer, G. Flesher, J. Eisler, D. Mensa, N. Ramesh, and G. Heidbreder, "Acoustic echo computer tomography," in *Acoustic Holography*, vol. 8, A. Metherell, ed. New York: Plenum, 1978.
[14] D. Mensa, G. Heidbreder, and G. Wade, "Aperture synthesis by object rotation in coherent imaging," *IEEE Trans. Nucl. Sci.*, vol. NS-27, pp. 989–998, Apr. 1980.
[15] D. Mensa and G. Heidbreder, "Bistatic synthetic-aperture radar imaging of rotating objects," *IEEE Trans. Aerosp. Electron. Syst.*, vol. AES-18, pp. 423–431, July 1982.
[16] C. Chen and H. C. Andrews, "Multifrequency imaging of radar turnable data," *IEEE Trans. Aerosp. Electron. Syst.*, vol. AES-16, pp. 15–22, Jan. 1980.
[17] G. T. Herman, *Image reconstruction from Projections*. New York: Academic Press, 1980.
[18] D. L. Mensa, S. Halevy, and G. Wade, "Coherent Doppler tomography for microwave imaging," *Proc. IEEE*, vol. 71, no. 2, pp. 254–261, Feb. 1983.
[19] W. M. Brown, "Walker model for radar sensing of rigid target fields," *IEEE Trans. Aerosp. Electron. Syst.*, vol. AES-16, pp. 104–107, Jan. 1980.
[20] H. H. Stark, J. W. Woods, I. Paul, and R. Hingorani, "Direct Fourier reconstruction in computer tomography," *IEEE Trans. Acoust., Speech, Signal Process.*, vol. ASSP-29, pp. 237–245, Apr. 1981.
[21] A. V. Oppenheim, G. V. Frisk, and D. R. Martinez, "An algorithm for the numerical evaluation of the Hankel transform," *Proc. IEEE*, vol. 66, pp. 264–265, Feb. 1978.
[22] A. J. Jerri, "Towards a discrete Hankel transform and its applications," *J. Appl. Anal.*, vol., 7, pp. 97–109, 1978.
[23] S. M. Candel, "Dual algorithms for fast calculation of the Fourier-Bessel transform," *IEEE Trans. Acoust., Speech, Signal Process.*, vol. ASSP-29, pp. 963–972, Oct. 1981.
[24] D. W. Dearholt and W. R. McSpadden, *Electromagnetic Wave Propagation*. New York: McGraw-Hill, 1973.
[25] A. W. Rihaczek, *Principles of High-Resolution Radar*. New York: McGraw-Hill, 1969, pp. 331–349.
[26] D. A. Schwartz, "Analysis and experimental investigation of three synthetic aperture radar formats," Coordinated Sci. Lab., Univ. of Illinois, Tech. Rep. T-94, Mar. 1980 (M.S. thesis).
[27] E. N. Leith, "Quasi-holographic techniques in the microwave region," *Proc. IEEE*, vol. 59, pp. 1305–1318, Sept. 1971.
[28] W. E. Kock, *Radar, Sonar, and Holography*. New York: Academic Press, 1973.
[29] R. O. Harger, *Synthetic Aperture Radar Systems: Theory and Design*. New York: Academic Press, 1970.
[30] J. C. Kirk, "A discussion of digital processing in synthetic aperture radar," *IEEE Trans. Aerosp. Electron. Syst.*, vol. AES-11, pp. 326–337, May 1975.
[31] B. K. P. Horn, "Density reconstruction using arbitrary ray-sampling schemes," *Proc. IEEE*, vol. 66, pp. 551–562, May 1978.

A Computational Study of Reconstruction Algorithms for Diffraction Tomography: Interpolation Versus Filtered Backpropagation

S. X. PAN AND AVINASH C. KAK

Abstract—From the standpoint of reporting a new contribution, this paper shows that by using bilinear interpolation followed by direct two-dimensional Fourier inversion, one can obtain reconstructions of quality which is comparable to that produced by the filtered-backpropagation algorithm proposed recently by Devaney. For an $N \times N$ image reconstructed from N diffracted projections, the former approach requires approximately $4N$ FFT's, whereas the backpropagation technique requires approximately N^2 FFT's.

We have also taken this opportunity to present the reader with a tutorial introduction to diffraction tomography, an area that is becoming increasingly important not only in medical imaging, but also in underwater and seismic mapping with microwaves and sound.

The main feature of the tutorial part is the statement of the Fourier diffraction projection theorem, which is an extension of the traditional Fourier slice theorem to the case of image formation with diffracting illumination.

I. Introduction

The imaging technique of computed tomography (CT) with X-rays has revolutionized biomedical imaging [13]. However, due to associated radiation hazards, X-ray based techniques are not suitable for such important applications as the mass screening of the female breast for the detection of cancerous tumors. As a result, in recent years much attention has been given to imaging with alternative forms of energy like ultrasound, low-level microwaves, and NMR (nuclear-magnetic resonance). Ultrasonic *B*-scan imaging [9] has already found widespread clinical applications; it, however, lacks the quantitative aspects of computed tomography.

There is a fundamental difference between tomographic imaging with X-rays on the one hand, and with ultrasound and microwaves on the other. X-rays, being nondiffracting, travel in straight lines, and therefore, the projection data measure the line integral of some object parameter along straight lines. This makes it possible to apply the Fourier slice theorem [15], [21] which says that the Fourier transform of a projection is equal to a slice of the two-dimensional Fourier transform of the object (Fig. 1).

On the other hand, when either microwaves or ultrasound are used for tomographic imaging, the energy often does not propagate along straight lines. When the object inhomogeneities are large compared to a wavelength, energy propagation is characterized by refraction and multipath effects. Moderate amounts of ray bending induced by refraction can be taken into account by combining algebraic reconstruction algorithms [2] with digital ray tracing and ray linking algorithms [1]. The multipath effects, caused by the arrival at the receiving transducer of more than one ray propagating through different parts of the refracting object, can usually be eliminated through homomorphic processing of the projection data [4].

When the object inhomogeneities become comparable in size

Manuscript received December 27, 1982; revised March 31, 1983. This work was supported by the Walter Reed Army Institute of Research.

S. X. Pan is with the Beijing Institute of Aeronautics and Astronautics, Beijing, China, on leave at the School of Electrical Engineering, Purdue University, West Lafayette, IN 47907.

A. C. Kak is with the School of Electrical Engineering, Purdue University, West Lafayette, IN 47907.

Reprinted from *IEEE Trans. Acoust., Speech, Signal Processing,* vol. ASSP-31, pp. 1262–1275, Oct. 1983.

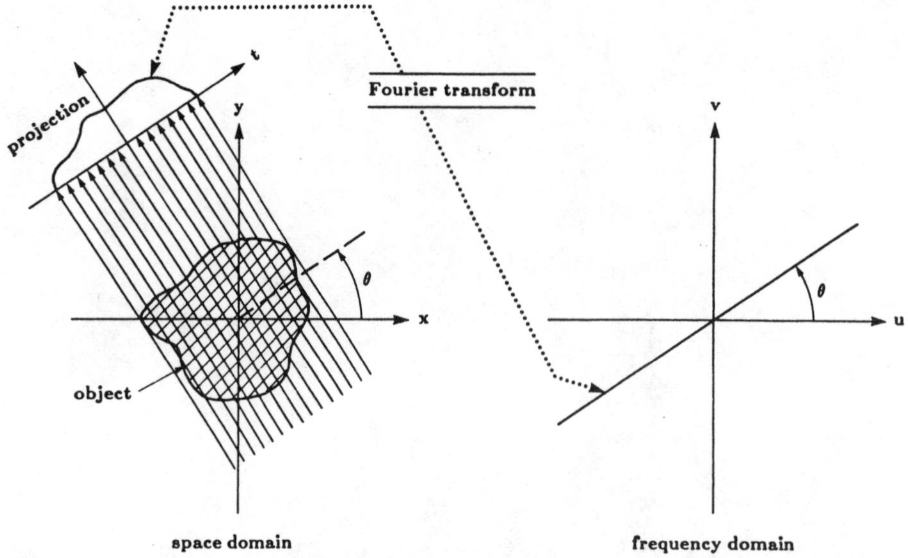

Fig. 1. The Fourier slice theorem.

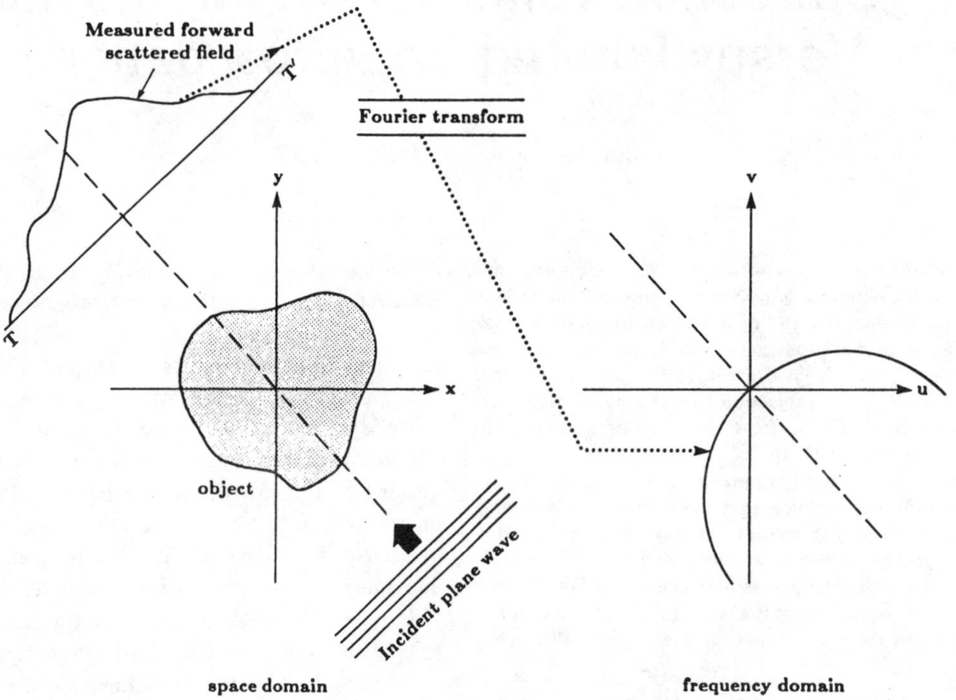

Fig. 2. The Fourier diffraction projection theorem.

to a wavelength, it is not even appropriate to talk about propagation along lines or rays, and energy transmission must be discussed in terms of wavefronts and fields scattered by the inhomogeneities. In spite of these difficulties, it has been shown [7], [16], [24] that within certain approximations a Fourier-slice-like theorem can be formulated. We will call this new theorem the Fourier diffraction projection theorem. It may simply be stated as follows.

When an object is illuminated with a plane wave as shown in Fig. 2, the Fourier transform of the forward scattered fields measured on a line perpendicular to the direction of propagation of the wave (line *TT* in Fig. 2) gives the values of the 2-D Fourier transform of the object along a circular arc as shown in the figure.

This theorem, which will be discussed in greater detail in Section II, *is valid when the inhomogeneities in the object are only weakly scattering.*

According to the Fourier diffraction projection theorem, by illuminating an object in many different directions and measuring the diffracted projection data, one can in principle fill up the Fourier space with the samples of the Fourier transform of the object over an ensemble of circular arcs and then reconstruct the object by Fourier inversion.

As is well known, the Fourier slice theorem for the X-ray case leads to a very accurate and computationally efficient filtered-backprojection algorithm for reconstruction [21]. It has recently been shown by Devaney [7] that the Fourier diffraction projection theorem leads to a conceptually similar

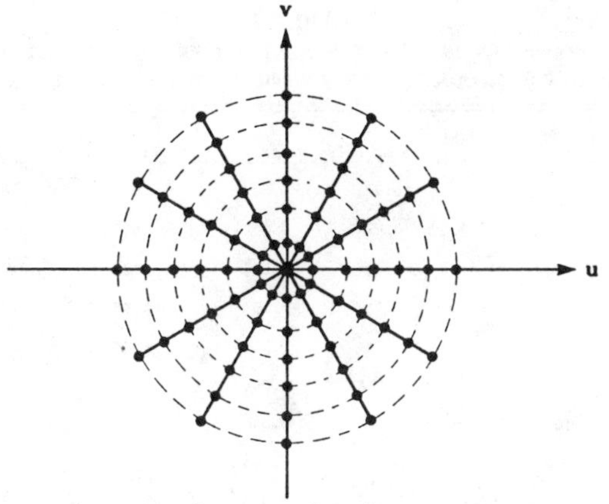

Fig. 3. With X-ray type projections, the Fourier transform of an object is known over a polar grid as shown here.

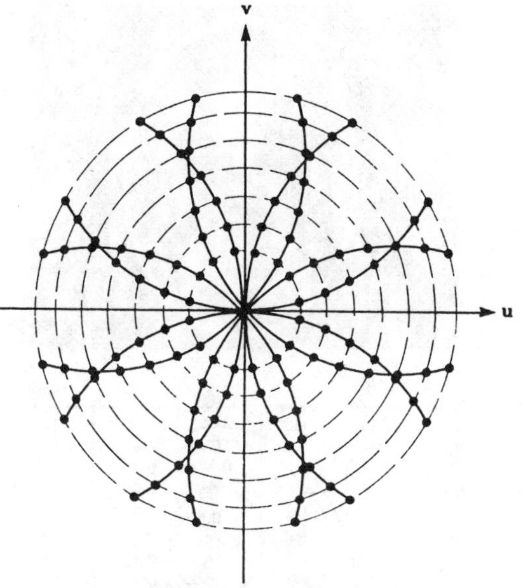

Fig. 4. With diffraction tomography under the assumption of weak scattering, the Fourier transform of an object is determined over a circular arc grid as shown here.

algorithm that he called the filtered-*backpropagation* algorithm. In spite of their conceptual similarities, for its full implementation, for an $N \times N$ reconstruction with N projections, the filtered-backprojection algorithm needs only N Fourier transforms, whereas the filtered-backpropagation algorithm requires about N^2 Fourier transforms.[1] Because of the computational burden associated with Devaney's approach, we have maintained an interest in direct Fourier reconstruction, in which one uses the measured projection data to estimate the values of the Fourier transform of the object over a rectangular grid by using appropriate interpolation, and then a 2-D inverse fast Fourier transform is performed to reconstruct the object.

Direct Fourier reconstruction techniques were tried in the early stages of the development of X-ray tomography. With X-ray type projections, the Fourier transform of the object is known over a polar grid as shown in Fig. 3, each spoke of the grid corresponding to one projection. In going from this polar grid to a rectangular grid required by the 2-D inverse FFT algorithms, any interpolation algorithm must contend with the fact that on a per-unit-area basis the density of samples generated by the polar grid becomes sparser the longer the distance from the frequency domain origin. This can degrade the quality of the estimation of high frequency components vis-à-vis the low frequency components. The resulting degradation in the image quality, and also all the computing required for interpolation, are the reasons why the direct Fourier methods were superceded by the filtered-backprojection algorithm.

Since in diffraction tomography, the Fourier transform of the object is only known over the sampling points of a circular arc grid as shown in Fig. 4, one again runs into interpolation errors in estimating the transform values over a rectangular grid. If the extensive demands on computer time are not a factor, one can use the filtered-backpropagation algorithm. However, in most practical applications one may have to resort to procedures that are computationally more efficient. It is for this reason that in this paper we have surveyed the different interpolation strategies available for direct Fourier reconstruction, and also compared the results obtained by direct Fourier reconstruction to those obtained by filtered backpropagation and modified filtered backpropagation. We will show that with bilinear interpolation, which is simple to implement, the direct methods are as accurate as the computationally intensive filtered-backpropagation algorithm. Direct reconstruction algorithms based on bilinear interpolation will also be shown to be superior to the modified filtered-backpropagation algorithm.

We should mention that in principle it is possible to completely avoid any frequency domain interpolation in the calculation of, say, N^2 points on a rectangular grid from N^2 points on a circular arc grid. This requires inverting a $N^2 \times N^2$ matrix, which in most cases of practical interest is computationally expensive and numerically unstable [5]. Interpolation could theoretically also be avoided by ensuring that at least one circular arc passes through every point on the rectangular grid [10]. This again is not practically possible because of the very large number of projections that are required for this purpose.

The paper is organized as follows. Section II presents a mathematical formulation of the Fourier diffraction projection theorem which constitutes the basis of the reconstruction algorithms for diffraction images. In Section III, we present the coordinate transformation equations for interpolation. In Sections IV and V we have shown reconstructions using the nearest-neighbor and the bilinear interpolation techniques. Section VI presents reconstruction results that are based on implementable approximations to the exact circular interpolation theorem discussed by Stark *et al.* [23]. In Section VII, we have discussed the filtered-backpropagation algorithm and have shown reconstructions made with it.

The computer simulation results in this paper will be shown

[1] By full implementation is meant that which would generate uniform accuracy over the entire image frame. Devaney has also shown that if one is only interested in generating locally accurate reconstructions, it is possible to match the computational effort required by the filtered-backprojection algorithm.

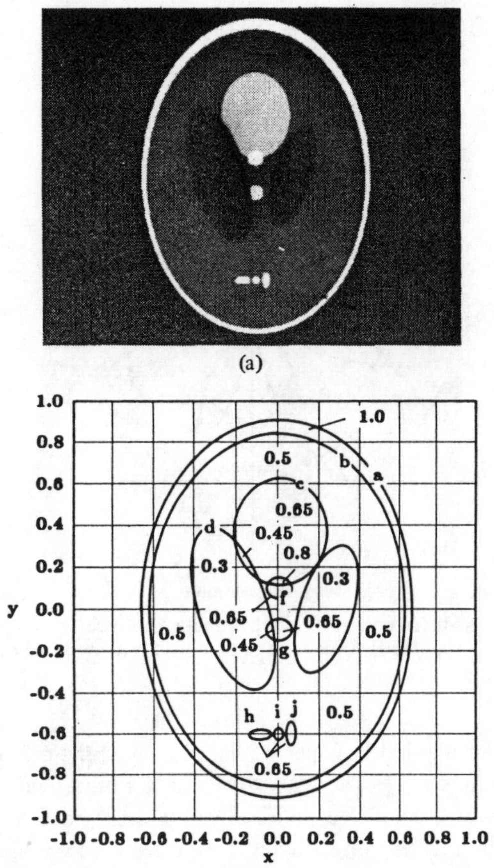

Fig. 5. (a) The test image used for reconstructions with different algorithms discussed in this paper. (b) Gray level distribution in (a).

for the image of Fig. 5(a). Visually this image looks identical to the well-known Shepp and Logan [22] "head phantom" for the case of X-ray computed tomography. However, we have changed the gray levels to those used by Devaney [8] for his circular test image. The phantom is a superposition of ellipses shown in Fig. 5(b). Each ellipse is assigned a gray level (the refractive index) as indicated in Table I. The composite gray levels are shown in Fig. 5(b).

A major advantage of using an image like Fig. 5(a) for computer simulation is that one can write analytical expressions for the transforms of the diffracted projections. The Fourier transform of an ellipse of semimajor and semiminor axes A and B, respectively, is given by

$$\frac{2\pi A J_1 \left[B\sqrt{(uA/B)^2 + v^2}\right]}{\sqrt{(uA/B)^2 + v^2}} \tag{1}$$

where u and v are spatial angular frequencies in the x- and y-directions, respectively; and J_1 is a Bessel function of the first kind and order 1. When the center of this ellipse is shifted to the point (x_1, y_1), and the angle of the major axis tilted by α, as shown in Fig. 6(b), its Fourier transform becomes

TABLE I

THE PARAMETERS OF THE COMPONENT ELLIPSES OF THE TEST IMAGE USED FOR RECONSTRUCTION WITH DIFFERENT ALGORITHMS

Ellipse	coordinates of the centers	A major axis	B minor axis	α rotation angle	ρ Gray level
a	(0, 0)	0.92	0.69	90°	1.0
b	(0, -0.0184)	0.874	0.6624	90°	-0.5
c	(0.22, 0)	0.31	0.11	72°	-0.2
d	(-0.22, 0)	0.41	0.16	108°	-0.2
e	(0, 0.35)	0.25	0.21	90°	0.1
f	(0, 0.1)	0.046	0.046	0	0.15
g	(0, -0.1)	0.046	0.046	0	0.15
h	(-0.08, -0.605)	0.046	0.023	0	0.15
i	(0, -0.605)	0.023	0.023	0	0.15
j	(0.06, -0.605)	0.046	0.023	90°	0.15

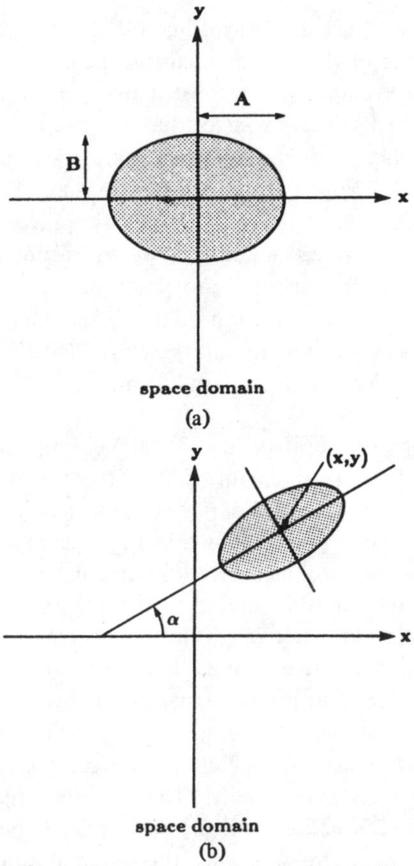

Fig. 6. (a) The test image of Fig. 5(a) is a superposition of ellipses. One such ellipse is shown here. (b) Rotated and shifted version of the ellipse in (a).

$$e^{-j(ux_1 + vy_1)} \cdot \frac{2\pi A J_1 \{B[((u\cos\alpha + v\sin\alpha)A/B)^2 + (-u\sin\alpha + v\cos\alpha)^2]^{1/2}\}}{[((u\cos\alpha + v\sin\alpha)A/B)^2 + (-u\sin\alpha + v\cos\alpha)^2]^{1/2}}. \tag{2}$$

Now consider the situation in which the ellipse is illuminated by a plane wave in the manner shown in Fig. 2(a). By the Fourier diffraction projection theorem discussed in the next section, the Fourier transform of the transmitted wave fields measured on a line like TT' shown in Fig. 2(a), will be given by the values of the above function on a circular arc as shown in Fig. 2(b). For the test object of Fig. 5(a), if we assume weak scattering and therefore no interactions among the ellipses, the Fourier transform of the total forward scattered field measured on the line TT', will be a sum of the values of functions like (2) over the circular arc. This procedure was used to generate the diffracted projection data for the test image.

II. The Fourier Diffraction Projection Theorem: Theoretical Background and Mathematical Statement[2]

In Fig. 7 a 2-D object is shown being illuminated by a plane wave propagating along a unit propagation vector $\vec{s}_0$. The object will be represented as a distribution function $o(x, y)$ in space which has a 2-D Fourier transform

$$O(\vec{w}) = \int_{-\infty}^{\infty}\int_{-\infty}^{\infty} o(\vec{r})\, e^{-j\vec{w}\cdot\vec{r}}\, d\vec{r} \tag{3}$$

where $\vec{w} = (u, v)$ and $\vec{r} = (x, y)$. For lossless media, the distribution function is related to the refractive index $n(\vec{r})$ by $o(\vec{r}) = n^2(\vec{r}) - 1$. The illuminating plane wave is assumed to be monochromatic at angular frequency ω. Also, the object is assumed to be immersed in a water medium. In the absence of the object, the plane wave propagates through the water medium obeying the following equation:

$$u_i(\vec{r}) = U_0 e^{-jk_0\vec{s}_0\cdot\vec{r}} \tag{4}$$

where U_0 is the complex amplitude of the illuminating wave at the origin of the space domain; k_0 is the wave number associated with the water medium and is equal to $2\pi/\lambda$ with λ being the wavelength.

When the object is immersed in the medium, scattering occurs, and the total field $u(\vec{r})$ at any position can be modeled as the superposition of the incident field $u_i(\vec{r})$ and the scattered field $u_s(\vec{r})$:

$$u(\vec{r}) = u_i(\vec{r}) + u_s(\vec{r}). \tag{5}$$

When the illuminating wave is incident at an angle ϕ as shown in Fig. 7, we will set up a rotated coordinate system (ξ, η) such that the η-axis is coincident with the unit propagation vector $\vec{s}_0$ as shown there, and measure the scattered field on a line $\eta = l$. The incident and the scattered fields will now be denoted by $u_{i,\phi}(\vec{r})$ and $u_{s,\phi}(\vec{r})$, respectively.

To obtain the relation between the scattered field and the scattering object, one must solve the inhomogeneous Helmholtz equation

$$(\nabla^2 + k_0^2)\, u(\vec{r}) = -k_0^2 o(\vec{r})\, u(\vec{r}). \tag{6}$$

It is often impossible to solve this equation exactly. However, under the assumption of weak scattering, i.e., if $|u_s| << |u_i|$, one may use the so-called Born approximation which consists of replacing u by u_i on the right-hand side of (6) [11], [12]. It can be shown that in this case the Fourier transform $U_s(\kappa)$ of the scattered fields measured on the line $\eta = l$

$$U_{s,\phi}(\kappa) = \int_{-\infty}^{\infty} u_{s,\phi}(\xi)\, e^{-j\kappa\xi}\, d\xi \tag{7}$$

[2]The reader is referred to [7] for a more detailed and a very elegant development of this material. The theorem is based on a result originally derived by Wolf [24].

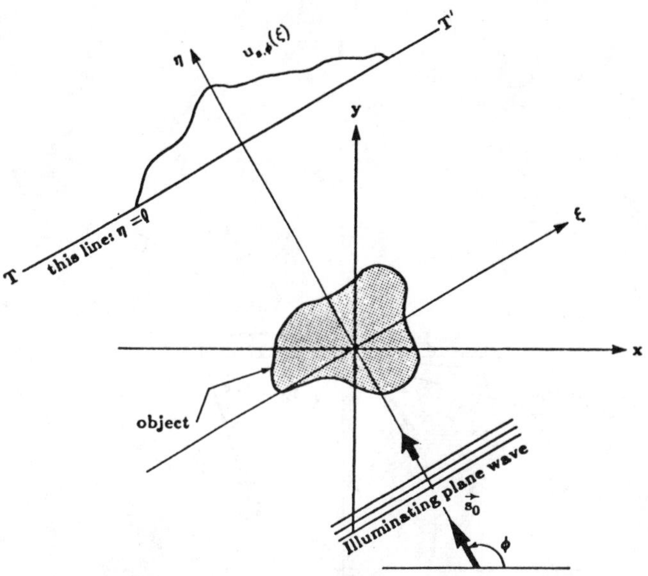

Fig. 7. An arbitrary object being illuminated by a plane wave propagating along the unit vector $\hat{s}_0$. This figure also illustrates the rotated coordinate system (ξ, η).

is related to the Fourier transform of the object by [6], [7], [16], [24]

$$U_{s,\phi}(\kappa) = \frac{k_0^2 U_0}{j2\gamma}\, e^{j\gamma l} Q_\phi(\kappa) \tag{8}$$

where

$$\gamma = \sqrt{k_0^2 - \kappa^2} \qquad |\kappa| \leqslant k_0 \tag{9}$$

and

$$Q_\phi(\kappa) = \int_{-\infty}^{\infty}\int_{-\infty}^{\infty} o(\vec{r})\, e^{-j[\kappa\xi + (\gamma - k_0)\eta]}\, d\vec{r} \tag{10}$$

is the Fourier transform of the object evaluated along a semicircle of radius k_0 and centered at $-k_0\vec{s}_0$ as shown in Fig. 8 (we will call such a semicircle the k_0-semicircle).

We will define a *unit* vector $\vec{s}$ as

$$\vec{s} = \frac{1}{k_0}(\kappa\vec{\xi} + \gamma\vec{\eta}) \tag{11}$$

which is also illustrated in Fig. 8. $\vec{\xi}$ and $\vec{\eta}$ are the unit vectors along the ξ- and η-axes, respectively. Since only the positive root is retained in (9), as κ varies from $-k_0$ to k_0, $\vec{s}$ will take all the directions such that $\vec{s}_0 \cdot \vec{s} \geqslant 0$. Since $\vec{s} \cdot \vec{r}$ is equal to $(\kappa\xi + \gamma\eta)/k_0$, and since $\vec{s}_0 \cdot \vec{r}$ equals η, (10) can be rewritten as

$$Q_\phi(\kappa) = \int_{-\infty}^{\infty}\int_{-\infty}^{\infty} o(\vec{r})\, e^{-jk_0(\vec{s} - \vec{s}_0)\cdot\vec{r}}\, d\vec{r}. \tag{12}$$

Since the locus of points defined by

$$\vec{w} = k_0(\vec{s} - \vec{s}_0) \tag{13}$$

is a semicircle having radius k_0 and centered at $-k_0\vec{s}_0$ in the Fourier domain, we conclude that $Q_\phi(\kappa)$ is the Fourier transform of the object along this semicircle.

The physics of wave propagation dictates that the highest angular spatial frequency in the measured scattered field on

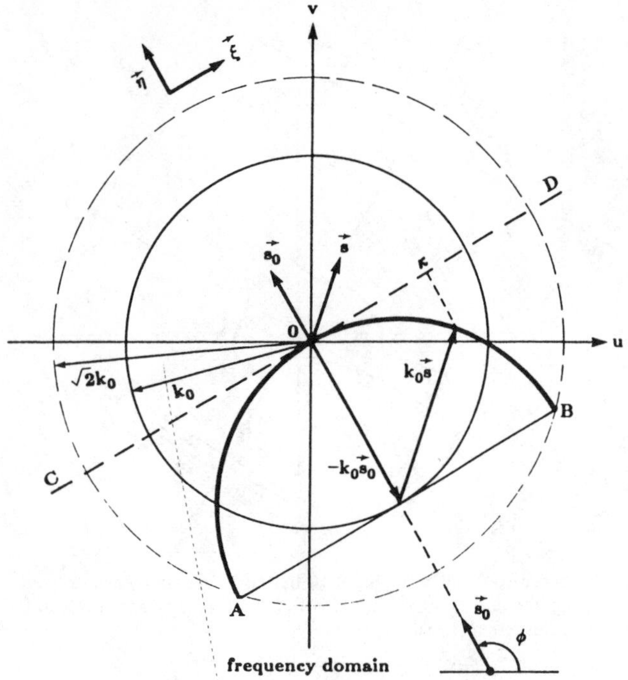

Fig. 8. The thick lined arc is the k_0-semicircle. The center of this semicircle is at $-k_0\vec{s}_0$. The ensemble of all such semicircles (for different $\vec{s}_0$) creates the $\sqrt{2}\ k_0$-disk.

the line $\eta = l$ be limited to k_0.[3] This fact is implicitly reflected in (9) where κ is not allowed to be greater than k_0. Therefore, the angular frequency in (7)-(13) spans the interval $(-k_0, k_0)$. This range of κ corresponds to one complete semicircle in the angular spatial frequency domain as shown in Fig. 8. The endpoints A and B of one such semicircle are at a distance of $\sqrt{2}\ k_0$ from the origin of the Fourier plane, and the ensemble of semicircles obtained by changing the illuminating angle ϕ will trace a disk having a radius $\sqrt{2}\ k_0$ and centered at the origin (we will call it the $\sqrt{2}\ k_0$-disk). Therefore, one can say that diffraction tomography determines the object up to the maximum spatial frequency of $\sqrt{2}\ k_0$. To this extent the reconstructed object is a low-pass version of the original. In practice, the loss of resolution caused by this band limit is negligible, it being more influenced by considerations such as the aperture sizes of the transmitting and receiving transducers, etc.

An alternative way to solve (6) under the condition of weak scattering is known as the Rytov approximation [12]. In the Rytov approximation, the total field is expressed as

$$u(\vec{r}) = U_0 e^{jk_0\vec{s}_0\cdot\vec{r} + \psi_s(\vec{r})}. \tag{14}$$

Thus, the scattered field is taken into account with an additional term in the complex phase. With the Rytov approximation, it can be shown that between the Fourier transform of $\psi_s(\vec{r})$ measured on the line $\eta = l$

$$\Psi_{s,\phi}(\kappa) = \int_{-\infty}^{\infty} \psi_{s,\phi}(\xi)\, e^{-j\kappa\xi}\, d\xi \tag{15}$$

[3]This statement is only true in the absence of evanescent waves, which make negligible contributions to scattered fields at distances greater than 10λ from the object.

and the Fourier transform of the object there exists the following relation [6], [16], [24]:[4]

$$\Psi_{s,\phi}(\kappa) = \frac{k_0^2 U_0}{j2\gamma}\, e^{j(\gamma - k_0)l}\, Q_\phi(\kappa). \tag{16}$$

As in the case of Born approximation, the Fourier transform of the object is determined within the $\sqrt{2}\ k_0$-disk in the frequency domain.

It can be shown that with noise-free measurements the Rytov approximation is numerically superior to the Born approximation [14]. However, whereas the Born approximation only requires the measurement of the complex amplitude of the scattered fields, the Rytov approximation requires the unwrapped phases of the scattered fields, and depending upon the noise level in the measurements, this might make the latter approach more difficult to implement.

To summarize, we have the following:

$$Q_\phi(\kappa) = \frac{j2\gamma}{k_0^2 U_0}\, e^{-j\gamma l}\, U_{s,\phi}(\kappa)$$

for the Born approximation (17)

$$Q_\phi(\kappa) = \frac{j2\gamma}{k_0^2 U_0}\, e^{-j(\gamma - k_0)l}\, \Psi_{s,\phi}(\kappa)$$

for the Rytov approximation. (18)

Equations (17) and (18) are the mathematical statements of the Fourier diffraction projection theorem.

In practice, only a finite number of projections are available, and each projection is measured over a finite set of sampling points only. This discretization of the ϕ and κ parameters is usually uniform and as a result the Fourier transform $O(u, v)$ of an object $o(x, y)$ is uniformly sampled in both ϕ and κ. Therefore, the problem of direct Fourier reconstruction in diffraction tomography may be stated as follows: to reconstruct the object $o(x, y)$ from a given set of sample points of its Fourier transform $O(u, v)$ distributed over a grid that is formed by the k_0-semicircles and the circles of constant κ as shown in Fig. 4.

III. Coordinate Transformation Equations for Interpolation

In order to discuss interpolation between the circular arc grids on which data are generated by diffraction tomography, and the rectangular grids suitable for image reconstruction, we must first select parameters for representing each grid and then write down a relation between the two sets of parameters.

As mentioned before, when the object is illuminated with a plane wave traveling along $\vec{s}_0$, the Fourier transform of the transmitted field lies on the arc AOB in Fig. 8. In most cases the transmitted data will be uniformly sampled in space, and a

[4]Here and in (5) it is seen that $\Psi_{s,\phi}(\kappa)$ and $U_{s,\phi}(\kappa)$ approach infinity as κ approaches k_0. This singularity is an unreasonable consequence resulting from the neglection of energy dissipation in the medium. It can be shown that when dissipation is taken into account, k_0 in (4) is replaced by a complex number $k = k_0 + j\epsilon$, with ϵ representing the attenuation coefficient of the medium, and the singularity disappears.

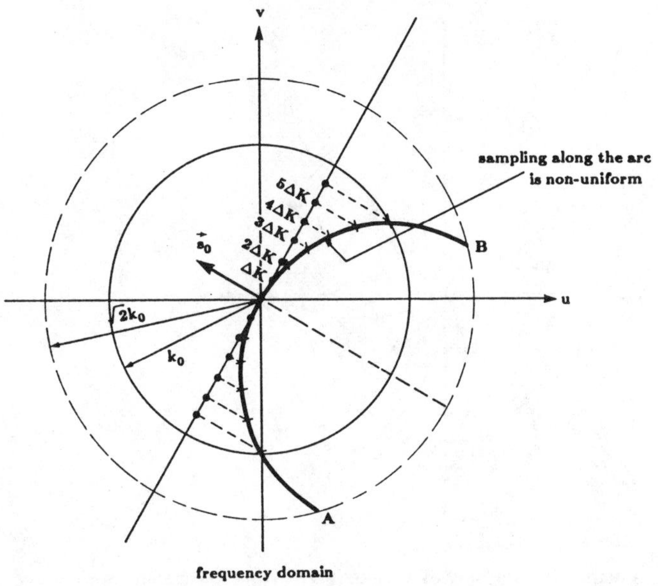

Fig. 9. This figure illustrates the fact that the Fourier transform of a diffracted projection generates frequency domain samples that are *not* equispaced along the corresponding semicircle.

discrete Fourier transform of these sampled data will generate uniformly spaced samples of $U_{s,\phi}(\kappa)$ [or $\psi_{s,\phi}(\kappa)$] in the κ domain. By (17) and (18), we will thus generate uniformly spaced samples of the $Q_\phi(\kappa)$ function in the κ domain. Since Q_ϕ is the Fourier transform of the object along the circular arc AOB in Fig. 8, and since κ is the projection of a point on the circular arc on the tangent line CD, the uniform samples of Q_ϕ in κ translate into nonuniform samples along the arc AOB as shown in Fig. 9.

Due to the fact that in an experimental measurement, it is the κ parameter that will be uniformly sampled, we will designate each point on the arc AOB by its (κ, ϕ) parameters, inspite of κ not being the radial distance of that point from the origin in the frequency domain. (What we are saying is that (κ, ϕ) are *not* the polar coordinates of a point on arc AOB in Fig. 8.) The rectangular coordinates in the frequency domain will be denoted by the symbols (u, v) introduced in (3).

Before we present relationships between (κ, ϕ) and (u, v), we would like to consider separately the points generated by the AO and OB portions of the arc AOB as ϕ is varied from 0 to 2π. We do this because as ϕ is varied from 0 to 2π, arc AOB generates a double coverage of the frequency domain, which is undesirable for discussing a one-to-one transformation between the (κ, ϕ) parameters and the (u, v) coordinates. (The reader might think that instead of breaking AOB into two halves, we might limit ϕ to the $0, \pi$ interval. Unfortunately, as the arc AOB is rotated through π, the frequency domain thus spanned has a hole in one region and a double coverage in another region.)

The arc grids generated by the portions OB and AO will be denoted by (κ_1, ϕ_1) and (κ_2, ϕ_2), respectively, as shown in Fig. 10. It is important to note that for the left arc grid in Fig. 10, κ varies from $-k_0$ to 0 and ϕ from 0 to 2π, and for the right arc grid κ goes from 0 to k_0 and ϕ from 0 to 2π. We will now present transformation equations between (κ_1, ϕ_1) and (u, v).

We first express (u, v) in polar coordinates (w, θ), as shown in Fig. 11:

$$w = \sqrt{u^2 + v^2} \tag{19}$$

$$\theta = \arc\tan \frac{v}{u}. \tag{20}$$

We now introduce a new angle β_1, which is the angular position of a point on the portion OB of the arc in Fig. 11. The parameter β_1 is the angular position on the arc of the point having parameter κ_1, as shown in the figure. The relationship between κ_1 and β_1 is given by

$$\kappa_1 = k_0 \sin \beta_1. \tag{21}$$

The following relationship exists between the polar coordinates (w, θ) on the one hand and the parameters β_1 and ϕ_1 on the other:

$$\beta_1 = 2 \arc\sin \frac{w}{2k_0} \tag{22}$$

$$\phi_1 - \frac{\beta_1}{2} = \theta + \frac{\pi}{2}. \tag{23}$$

From (19)-(23), we obtain the following transformation equations between (κ_1, ϕ_1) and (u, v):

$$\kappa_1 = \sin\left(2 \arc\sin \frac{\sqrt{u^2 + v^2}}{2k_0}\right) \tag{24}$$

$$\phi_1 = \arc\tan \frac{v}{u} + \arc\sin \frac{\sqrt{u^2 + v^2}}{2k_0} + \frac{\pi}{2}. \tag{25}$$

In order to generate the transformation equations between the (κ_2, ϕ_2) arc grid generated by the portion AO in Fig. 8 and the (u, v) coordinates, we can introduce an angle β_2 to denote the angular location of a point on AO. The relationship between κ_2 and β_2 is given by

$$\kappa_2 = -k_0 \sin \beta_2. \tag{26}$$

The relationships corresponding to (22) and (23) in this case are

$$\beta_2 = 2 \arc\sin \frac{w}{2k_0} \tag{27}$$

$$\phi_2 + \frac{\beta_2}{2} = \theta + \frac{3\pi}{2}. \tag{28}$$

Using (19) and (26)-(28), we obtain

$$\kappa_2 = -\sin\left(2 \arc\sin \frac{\sqrt{u^2 + v^2}}{2k_0}\right) \tag{29}$$

$$\phi_2 = \arc\tan \frac{v}{u} - \arc\sin \frac{\sqrt{u^2 + v^2}}{2k_0} + \frac{3\pi}{2}. \tag{30}$$

IV. Nearest-Neighbor Interpolation

Having presented the coordinate transformation equations, we will now compare different interpolation methods for direct Fourier reconstruction from diffracted projection data.

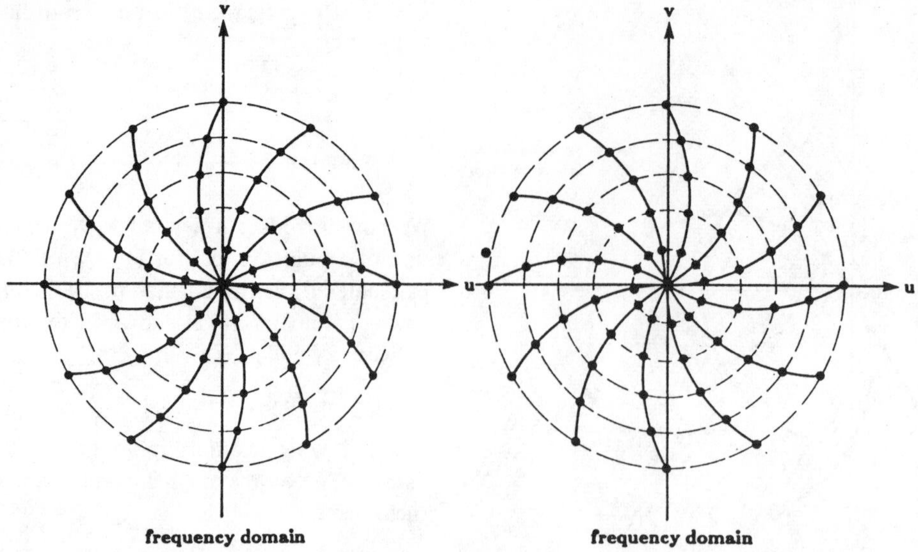

Fig. 10. The left arc grid is generated by intersections of equiangular rotations of the AO part of the k_0-semicircle in Fig. 8 and the constant κ circles. The set of points thus generated is represented by the (κ_1, ϕ_1) coordinate system. Similarly, the right arc grid is generated by the intersections of rotated versions of the OB part in Fig. 8 and the constant κ circle. the set of point on the right arc grid is represented by the (κ_2, ϕ_2) coordinate system.

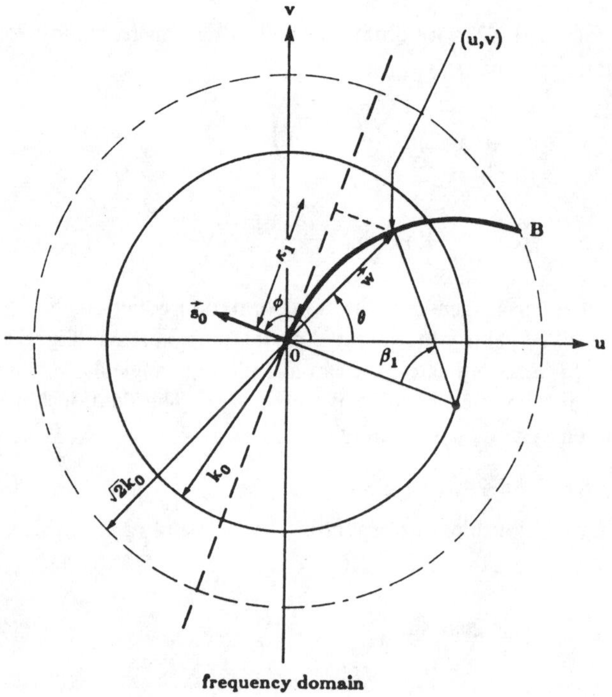

Fig. 11. The relationship between the (κ_1, ϕ_1) coordinate system and the (w, θ) polar coordinate system.

We will start with the nearest-neighbor (NN) interpolation technique. Although other authors have shown diffraction tomography reconstructions using this type of interpolation [3], [17], we have included the technique here so that the reader may compare its accuracy to those of other approaches.

The NN-interpolation can be implemented as follows: for each rectangular grid point (u, v) at which an interpolated value of the transform is desired, we first calculate the corresponding (κ_1, ϕ_1) and (κ_2, ϕ_2) parameters using either (24) and (25) or (29) and (30). We then find the two nearest neighbors of (κ_1, ϕ_1) and (κ_2, ϕ_2) on the two arc grids shown in Fig. 10(a) and (b), respectively. The nearer of these two grid points is retained and its value is assigned to the point (u, v).

The results obtained with both the nearest-neighbor interpolation discussed here and the bilinear interpolation discussed in the next section can be considerably improved if one first increases the sampling density in the (κ, ϕ)-plane by using the computationally efficient method of zero-extending the inverse two-dimensional FFT of the $Q(\kappa_i, \phi_j)$ matrix.[5] The technique consists of first taking a two-dimensional inverse FFT of the $N_\kappa \times N_\phi$ matrix consisting of the $Q(\kappa_i, \phi_j)$ values, zero-extending the resulting $N_h \times N_\phi$ array of numbers to, say, $mN_\kappa \times nM_\phi$, and then taking the FFT of this new array. The result is an mn-fold increase in the density of samples in the (κ, ϕ)-plane.

That digital signal processing techniques, such as the one discussed above, can be used to increase the sampling density in the (κ, ϕ)-plane raises the following question: what is the least number of experimental measurements that must be made in order to determine unaliased values of $Q(\kappa_i, \phi_j)$ over a minimal set of points in the (κ, ϕ)-plane? The minimal experimental sampling is obviously determined by the bandlimitedness of the fluctuations in the $Q(\kappa, \phi)$ function in the (κ, ϕ)-plane. We obviously require the 2D-transform of $Q(\kappa, \phi)$ with respect to the ϕ and κ variables to be a function of limited support. The transform of $Q(\kappa, \phi)$ with respect to κ yields the transmitted field measured on a line such as TT' in Fig. 7. To the extent this scattered field is *practically* zero outside a certain interval on this line, $Q(\kappa, \phi)$ may be considered to be bandlimited in the κ parameter. It is a little harder to bring out this finite support property for the transform of $Q(\kappa, \phi)$ with respect to ϕ. However, it may be shown to be a consequence of the object being space-limited. Suppose the object is confined within a circle of radius A, the highest sinusoidal fluctuation in

[5]For discussing interpolation it is more convenient to use the notation $Q(\kappa_i, \phi_j)$, as opposed to $Q_{\phi_i}(\kappa_i)$.

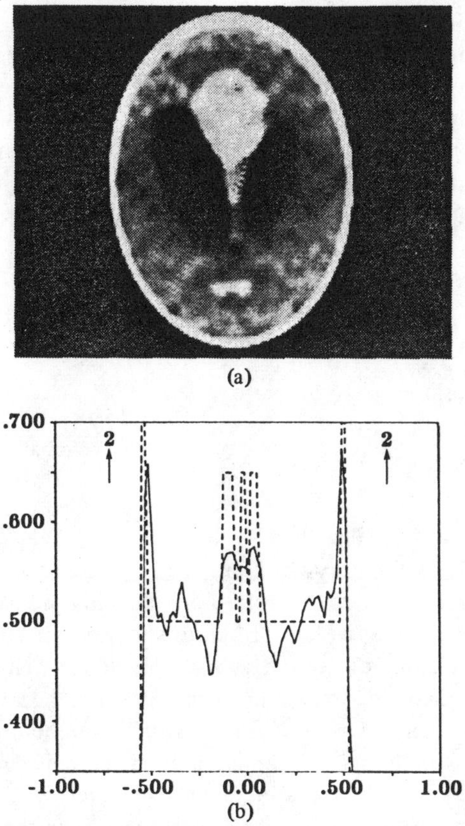

Fig. 12. (a) A 128 × 128 reconstruction of the test image by nearest-neighbor interpolation followed by direct Fourier inversion. The number of samples in each projection was 128, and the number of projections 64. Prior to nearest neighbor interpolation, the sampling density in the frequency domain was increased eightfold by the zero-padding technique. (b) A numerical comparison of true and reconstructed values on the line $y = -0.605$ through the test image.

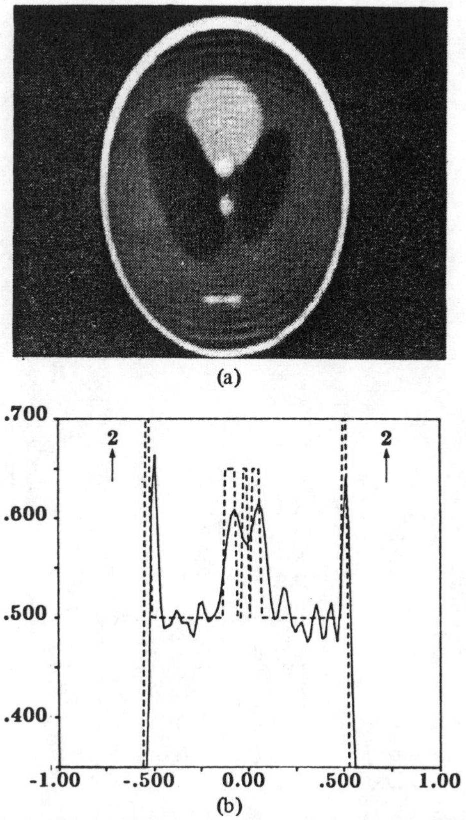

Fig. 13. (a) A 128 × 128 reconstruction obtained with bilinear interpolation from 64 projections and 90 samples per projection. Prior to bilinear interpolation, the frequency domain sampling density was increased eightfold by the zero-padding technique. (b) A numerical comparison of the true and reconstructed values on the line $y = -0.605$ through the test image.

$Q(\kappa, \phi)$ has a frequency less than A cycles/Hz.[6] Without any loss of generality, let us consider the fluctuations in Q in v-direction only. Then the fastest fluctuation in $Q(\kappa, \phi)$ may be written as $\cos(2\pi A v)$. Since $v = w \sin \phi$, as a function of ϕ this variation may be expressed as $\cos(2\pi A w \sin \phi)$. This function is periodic in ϕ with a period of 2π and therefore can be expanded as a Fourier series given by

$$\frac{1}{\pi} \int_{-\pi/2}^{\pi/2} \cos(\pi z \sin \phi) \cos 2\nu\phi \, d\phi = J_{2\nu}(\pi z) \tag{31}$$

where $z = 2Aw$; $J_{2\nu}(\pi z)$ is the Bessel function of first kind and order 2ν. For 2ν slightly greater than πz, $J_{2\nu}(\pi z)$ becomes very small. Therefore, we may truncate it for some ν implying that fluctuations in $Q(\kappa, \phi)$ with respect to ϕ may be considered to be of bandlimited nature.

Shown in Fig. 12(a) is a 128 × 128 reconstruction of the test image using the nearest-neighbor interpolation. Each projection was sampled over 128 points, and a total of 64 projections were used. The resulting 128 × 64 points in the (κ, ϕ)-plane were increased to 512 × 128 by using the space-domain zero-padding approach just discussed. This eight times denser set of points was then used for the nearest-neighbor interpolation. Fig. 12(b) shows a numerical comparison of the true and reconstructed values on the line $y = -0.605$ through the test image. For the reconstruction shown in Fig. 12(a), the CPU processing time on a VAX-11/780 was 2.2 min.

[6] Although cycles/Hz sounds somewhat odd, it does have the advantage of emphasizing that we are talking about the fluctuations in the frequency domain.

V. Bilinear Interpolation

Perhaps the most common technique used for interpolation in two dimensions is that of bilinear interpolation [20]. Given $N_\kappa \times N_\phi$ uniformly located samples, $Q(\kappa_i, \phi_j)$, the bilinearly interpolated values of this function may be calculated by using

$$Q(\kappa, \phi) = \sum_{i=1}^{i=N_\kappa} \sum_{j=1}^{j=N_\phi} Q(\kappa_i, \phi_j)\, h_1(\kappa - \kappa_i)\, h_2(\phi - \phi_j) \tag{32}$$

where

$$h_1(\kappa) = \begin{cases} 1 - \dfrac{|\kappa|}{\Delta\kappa} & |\kappa| \leq \Delta\kappa \\ 0 & \text{otherwise,} \end{cases} \tag{33}$$

$$h_2(\phi) = \begin{cases} 1 - \dfrac{|\phi|}{\Delta\phi} & |\phi| \leq \Delta\phi \\ 0 & \text{otherwise.} \end{cases} \tag{34}$$

$\Delta\phi$ and $\Delta\kappa$ are the sampling intervals for ϕ and κ, respectively. When expressed in the manner shown above, bilinear interpo-

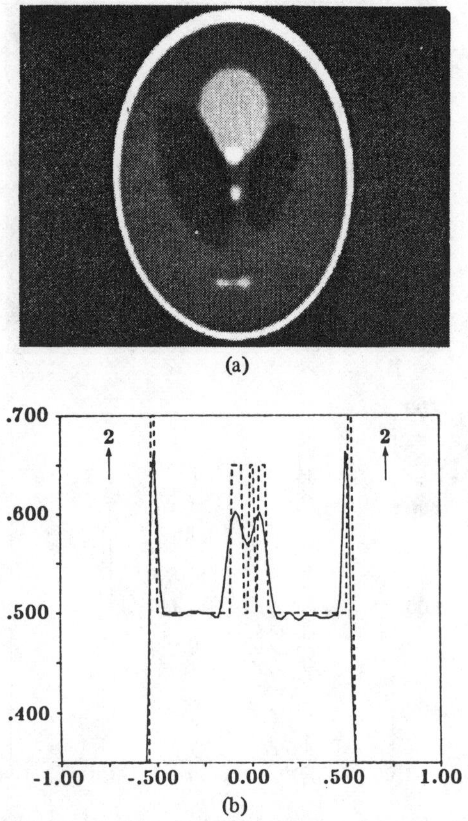

Fig. 14. (a) A 128 × 128 reconstruction obtained with bilinear interpolation from 64 projections and 128 samples per projection. Prior to applying bilinear interpolation, the sampling density in the (κ, ϕ) plane was increased eightfold by using the zero-padding technique. (b) A numerical comparison of the true and reconstructed values on the line $y = -0.605$ through the phantom.

lation may be interpreted as the output of a filter whose impulse response is $h_1 h_2$.

As in the preceding section, prior to bilinear interpolation, the sampling density in the (κ, ϕ)-plane was increased by zero-padding the inverse FFT of the $Q(\kappa_i, \phi_j)$ matrix, and then taking the FFT of the resulting enlarged array.

Computer simulation results are shown in Figs. 13 and 14. For Fig. 13, prior to bilinear interpolation, the space-domain zero-padding technique described in the preceding section was employed to increase the (κ, ϕ)-plane sampling density eightfold, from the original 90 × 64 sized array to a 360 × 128 array. When instead of using 90 sampling points per projection, we use 128, and then by zero-padding technique increase the size of the (κ, ϕ) array from 128 × 64 to 512 × 128 (again an eightfold increase in sampling density) prior to applying bilinear interpolation, the resulting reconstruction is shown in Fig. 14. Comparing Figs. 13(a) and 14(a), one sees the rings and other interference artifacts when only 90 sampling points are used for each projection. For the reconstructions shown in Figs. 13(a) and 14(a), the VAX-11/780 CPU processing time was 2.2 min.

VI. Interpolation Using the Circular Sampling Theorem

Since for a fixed κ, $Q(\kappa, \phi)$ is periodic in ϕ with period 2π and it may also be considered to be band limited in ϕ, we can use the circular sampling theorem for interpolation. To state the theorem, let $f(t)$ be a periodic signal with highest frequency K/T, where T is the period and K is an integer, then $f(t)$ can be recovered from its uniformly spaced samples with sampling interval no greater than $T/2K$ using the following interpolation equation [23]:

$$f(t) = \sum_{k=0}^{N-1} f\left(\frac{T}{N} k\right) \sigma \left(t - \frac{T}{N} k\right) \tag{35}$$

where

$$\sigma(t) = \frac{\sin \frac{N\pi}{T} t}{N \sin \frac{\pi}{T} t} \tag{36}$$

and $N \geq 2K + 1$. The theorem basically says that $f(t)$ can be recovered exactly everywhere provided we have at least $2K + 1$ samples in one period. Before proceeding to use this theorem in reconstructions, we would like to state that, in a sense, the circular sampling theorem is an example of the well-known z-transform sampling theorem, which states that the z-transform $X(z)$ of a sequence $x(k)$ of finite length N can be determined by its uniformly spaced samples around the unit circle [19]. When $X(z)$ is evaluated around the unit circle, i.e., $z = e^{j\omega}$, and the sequence is left-right symmetrical,[7] the theorem can be expressed by [19][8]

$$X(e^{j\omega}) = \sum_{k=0}^{N-1} X(k)\, \Phi \left(\omega - \frac{2\pi}{N} k\right) \tag{37}$$

where

$$X(k) = X(e^{j\omega})\big|_{\omega = (2\pi/N)k} \tag{38}$$

and

$$\Phi(\omega) = \frac{\sin (N\omega/2)}{N \sin (\omega/2)}. \tag{39}$$

Now a periodic signal $f(t)$ containing the highest frequency of K/T may be given a Fourier series representation with $2K + 1$ coefficients. These coefficients may be arranged as a left-right symmetrical sequence of length $2K + 1$. The z-transform of this sequence when evaluated on a unit circle defined by $z = \exp(-j2\pi t/T)$ will yield the original signal $f(t)$. According to (37), this z-transform can be determined exactly from the uniformly spaced samples around the unit circle, which is identical to implementing (35) and (36).

Using (35), we can write

$$Q(\kappa, \phi) = \sum_{n=1}^{N_\phi} Q\left(\kappa, \frac{2\pi}{N_\phi} n\right) \sigma \left(\phi - \frac{2\pi}{N_\phi} n\right) \tag{40}$$

where the interpolation function is

[7] A $2K + 1$ element long sequence $x(k)$ is left-right symmetrical if it is defined for indexes $k = -K, \cdots, -1, 0, 1, \cdots, K$.

[8] The equation for the z-transform sampling theorem given in [19] is for the case of a causal right sequence. Here we have adapted this equation for the case of a left-right symmetrical sequence.

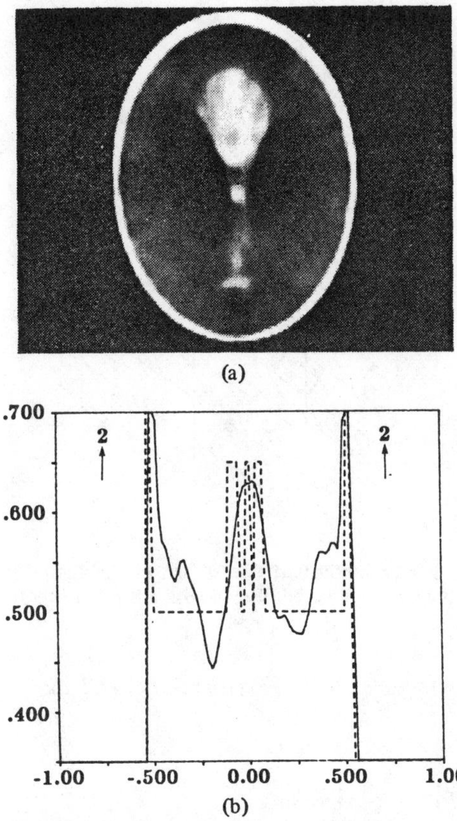

Fig. 15. (a) Reconstruction from 64 projections and 128 samples in each projection obtained by using the interpolation strategy derived from the circular sampling theorem. Fifteen neighbors were used for each interpolation. (b) A numerical comparison of the true and reconstructed values on the line $y = -0.605$.

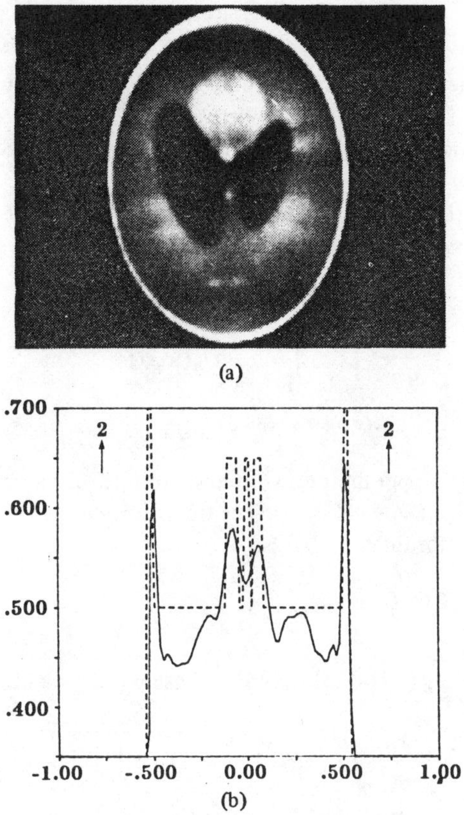

Fig. 16. (a) Same as in Fig. 15(a) except that now we have used 35 points for each interpolation. Note that although we are using a large number of points for each interpolation, the artifacts have not disappeared. (b) A numerical comparison of the true and reconstructed values on the line $y = -0.605$.

$$\sigma(\phi) = \frac{\sin\left(\dfrac{N_\phi}{2}\,\phi\right)}{N_\phi \sin\dfrac{\phi}{2}}. \tag{41}$$

Since $Q(\kappa, \phi)$ as a function of κ is also band limited, the following interpolation equation is a direct consequence of the sampling theorem:

$$Q(\kappa, \phi) = \sum_{p=1}^{N_\kappa} Q\left(\frac{2k_0}{N_\kappa}\,p, \phi\right) \operatorname{sinc}\left[\frac{\pi N_\kappa}{2k_0}\left(\kappa - \frac{2k_0}{N_\kappa}\,p\right)\right]. \tag{42}$$

Combining (41) and (42), we obtain

$$Q(\kappa, \phi) = \sum_{p=1}^{N_\kappa} \sum_{n=1}^{N_\phi} Q\left(\frac{2k_0}{N_\kappa}\,p, \frac{2\pi}{N_\phi}\,n\right) \sigma\left(\phi - \frac{2\pi}{N_\phi}\,n\right) \cdot \operatorname{sinc}\left[\frac{\pi N_\kappa}{2k_0}\left(\kappa - \frac{2k_0}{N_\kappa}\,p\right)\right]. \tag{43}$$

By (43), we can calculate the exact values for those (κ, ϕ) which correspond to points on the rectangular grid. In practice, however, we prefer to use a truncated version of (43) for computation savings:

$$Q(\kappa, \phi) = \sum_{p=i-K}^{i+K} \sum_{n=j-L}^{j+L} Q\left(\frac{2k_0}{N_\kappa}\,p, \frac{2\pi}{N_\phi}\,n\right)\left(\sigma - \frac{2\pi}{N_\phi}\,n\right) \cdot \operatorname{sinc}\left[\frac{\pi}{N_\kappa}\left(\kappa - \frac{2k_0}{N_\kappa}\,p\right)\right]. \tag{44}$$

That is, for each given (κ, ϕ), we find its nearest neighbor (i, j) on the arc grid; we then use the known values of the K and L neighbors of (i, j) that are around and centered at (i, j) on the arc grid in the κ- and the ϕ-directions, respectively, to interpolate the value for the given point. Thus, there are $K \times L$ points involved in the interpolation for calculating the values of each point on the rectangular grid. This leads to truncation errors.

In Fig. 15 we have shown the reconstruction obtained by using this procedure with $K = 5$ and $L = 3$. When K and L are increased to 7 and 5, respectively, the resulting reconstruction is presented in Fig. 16. The reader will notice that the artifacts have not disappeared (in Fig. 16) even though we are using 35 samples for the interpolation at each point. Both reconstructions shown here are based on 64 projections and 128 sampling points in each projection. The VAX-11/780 CPU processing time for the reconstruction in Fig. 15 was 2.0 min, and Fig. 16 was 4.0 min.

VII. Reconstruction with Filtered-Backpropagation Algorithm

For the sake of completeness we will first briefly illustrate the steps involved in the derivation of Devaney's filtered-propagation algorithm [7] from the equations in Sections II and III of this paper.

When the frequency domain is represented by the polar coordinates (w, θ) introduced in Section III, the inverse transform of (3) may be expressed as

$$o(x, y) = \frac{1}{4\pi^2} \int_0^{2\pi} \int_0^{\sqrt{2k_0}} O_p(w, \theta) \cdot e^{jw(x \cos \theta + y \sin \theta)}\, w\, dw\, d\theta \tag{45}$$

where the upper limit on w follows from the discussion in Section II. $O_p(w, \theta)$ is the object transform $O(u, v)$ represented in polar coordinate. That is,

$$O_p(w, \theta) = O(u, v) \Big|_{\substack{u = w \cos \theta \\ v = w \sin \theta.}} \tag{46}$$

Substituting (19)-(25) in (45) and using (10), we obtain

$$o(x, y) = \frac{k_0}{4\pi^2} \int_0^{2\pi} d\phi \int_0^{k_0} \frac{1}{\gamma} Q_\phi(\kappa) |\kappa| \cdot e^{j(\gamma - k_0)\eta}\, e^{j\kappa\xi}\, d\kappa. \tag{47}$$

Similarly, by using (26)-(30) and (10) in (45), we get

$$o(x, y) = \frac{k_0}{4\pi^2} \int_0^{2\pi} d\phi \int_{-k_0}^{0} \frac{1}{\gamma} Q_\phi(\kappa) |\kappa| \cdot e^{j(\gamma - k_0)\eta}\, e^{j\kappa\xi}\, d\kappa. \tag{48}$$

Combining (47) and (48) results in

$$o(x, y) = \frac{k_0}{8\pi^2} \int_0^{2\pi} d\phi \int_{-k_0}^{k_0} \frac{1}{\gamma} Q_\phi(\kappa) |\kappa| \cdot e^{j(\gamma - k_0)\eta}\, e^{j\kappa\xi}\, d\kappa. \tag{49}$$

In terms of the Fourier transform of the measured data, we will express (49) as

$$o(x, y) = \frac{1}{4\pi^2} \int_0^{2\pi} d\phi \int_{-k_0}^{k_0} \Gamma_\phi(\kappa) |\kappa| \cdot e^{j(\gamma - k_0)(\eta - l)}\, e^{j\kappa\xi}\, d\kappa \tag{50}$$

where from (17) and (18), we have

$$\Gamma_\phi(\kappa) = \begin{cases} \dfrac{j}{k_0 U_0} e^{-jk_0 l}\, U_{s,\phi}(\kappa) & \text{Born} \\[2ex] \dfrac{j}{k_0 U_0} \Psi_{s,\phi}(\kappa) & \text{Rytov.} \end{cases} \tag{51}$$

To bring out the similarities between (50) and the filtered-backprojection algorithm in X-ray tomography, following Devaney we write here separately the inner integration

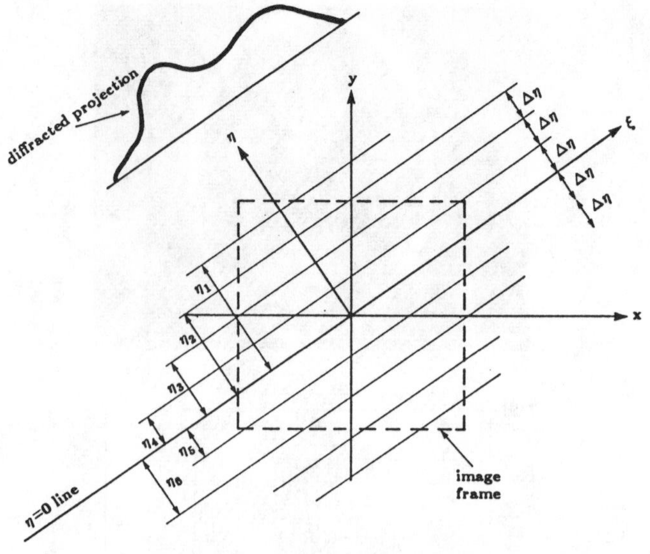

Fig. 17. For each η-constant line shown in this figure, the diffracted projection must be filtered with a transfer function.

$$\Pi_\phi(\xi, \eta) = \frac{1}{2\pi} \int_{-\infty}^{\infty} \Gamma_\phi(\kappa)\, H(\kappa)\, G_\eta(\kappa)\, e^{j\kappa\xi}\, d\kappa \tag{52}$$

where

$$\begin{aligned} H(\kappa) &= |\kappa| \qquad |\kappa| \leqslant k_0 \\ &= 0 \qquad |\kappa| \geqslant k_0 \end{aligned} \tag{53}$$

and

$$\begin{aligned} G_\eta(\kappa) &= \exp\,[j(\sqrt{k_0^2 - \kappa^2} - k_0)(\eta - l)] \qquad |\kappa| \leqslant k_0 \\ &= 0 \qquad |\kappa| > k_0. \end{aligned} \tag{54}$$

Without the extra filter function $G_\eta(\kappa)$, the rest of (52) would correspond to the filtering operation of the projection data in X-ray tomography. The filtering as called for by the transfer function $G_\eta(\kappa)$ is depth dependent due to the parameter η, which is equal to $x \cos \phi + y \sin \phi$.

In terms of the filtered projections $\Pi_\phi(\xi, \eta)$ in (52), the reconstruction equation (50) may be expressed as

$$o(x, y) = \frac{1}{2\pi} \int_0^{2\pi} d\phi\, \Pi_\phi(x \sin \phi - y \cos \phi, x \cos \phi + y \sin \phi). \tag{55}$$

The computational procedure for reconstructing an image on the basis of (52) and (55) may be presented in the form of the following steps.

Step 1: In accordance with (50), filter each projection with a separate filter for each depth in the image frame. For illustration, if we choose only nine depths as shown in Fig. 17, we would need to apply nine different filters to the diffracted projection shown there. In most cases for 128 X 128 reconstruction, the number of discrete depths chosen for filtering the projection will also be around 128. If they are much less than 128, spatial resolution will be lost.

Step 2: To each pixel (x, y) in the image frame, in accordance with (55) allocate a value of the filtered projection that corresponds to the nearest depth line.

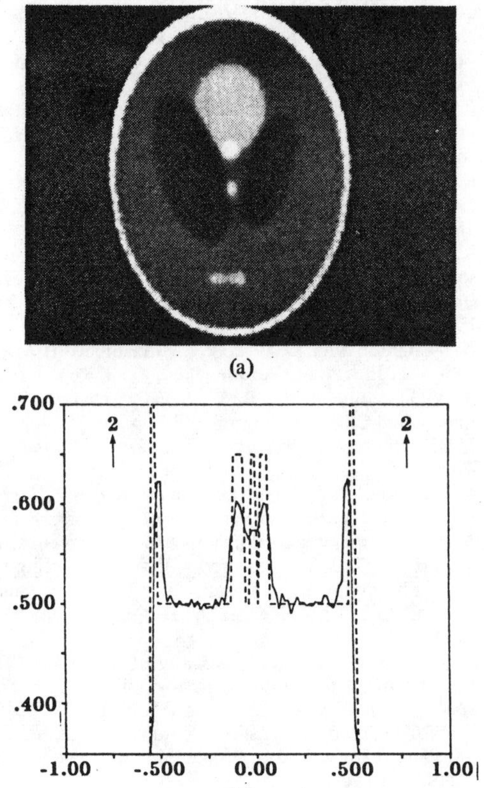

Fig. 18. (a) Reconstruction obtained by using the filtered-backpropagation algorithm on 64 projections and 128 samples in each projection. $N_\eta = 128$. (b) A numerical comparison of the true and the reconstructed values on the line $y = -0.605$.

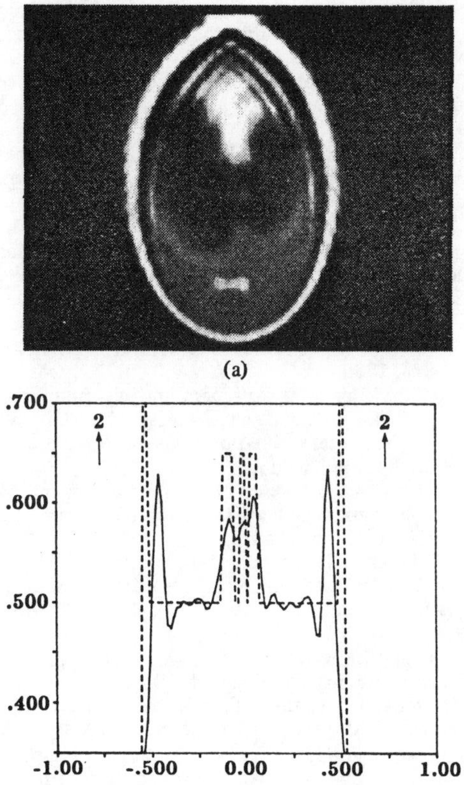

Fig. 19. (a) Reconstruction obtained by using the modified filtered-backpropagation algorithm on 64 projections and 128 samples in each projection. The filter function corresponds to what would yield local accuracy at the site of the three small ellipses. (b) A numerical comparison of the true and the reconstructed values on the line $y = -0.605$.

Step 3: Repeat the preceding two steps for all projections. As a new projection is taken up, add its contribution to the current sum at pixel (x, y).

The depth-dependent filtering in Step 1 makes this algorithm computationally very demanding. For example, if we choose N_η depth values, the processing of each projection will take $N_\eta + 1$ FFT's. If the total number of projections is N_ϕ, this translates into $(N_\eta + 1) N_\phi$ FFT's. For most $N \times N$ reconstructions, both N_η and N_ϕ will be approximately equal to N. Therefore, the filtered-backpropagation algorithm will require approximately N^2 FFT's compared to $4N$ FFT's for bilinear interpolation. (For precise comparisons, note that the FFT's for the case of bilinear interpolation are longer due to zero-padding.)

Fig. 18 shows the reconstruction obtained by using the filtered-backpropagation algorithm with $N_\eta = 128$ on the same original data set as was used for Fig. 14. The VAX-11/780 CPU processing time for the image in Fig. 18 was 30 min.

Devaney [8] has also proposed a modified filtered-backpropagation algorithm in which $G_\eta(\kappa)$ is simply replaced by a single $G_{\eta_0}(\kappa)$ where $\eta_0 = x_0 \cos\phi + y_0 \sin\phi$, (x_0, y_0) being the coordinates of the point where local accuracy in reconstruction is desired. (Elimination of depth dependent filtering reduces the number of FFT's to $2N_\phi$.) When this algorithm was implemented to yield accurate reconstruction in the vicinity of the three small ellipses in the test image, the overall reconstructed image is shown in Fig. 19. The VAX-11/780 processing time was 2.8 min.

VIII. Conclusions

Comparing Figs. 14 and 18, we see that the quality of the reconstruction obtained with bilinear interpolation is comparable to that achieved with the filtered-backpropagation algorithm. However, the amount of computing required by the two techniques is vastly different. Using identical projection data, the CPU processing time on a VAX 11/780 for the filtered-backpropagation algorithm was 30 min, whereas it was only 2.2 min for the reconstruction with bilinear interpolation. Therefore, our overall conclusion is that the simple technique of bilinear interpolation followed by direct Fourier inversion should be preferred for reconstructions from diffracted projection data, assuming, of course, the validity of Born and Rytov approximations.

In this paper we have discussed diffraction tomography with plane wave illumination. An attendant feature of such illumination is that it must be possible to measure the diffracted projections for a large number of different angles around the object. The reader is referred to [18] for a data collection strategy that is designed for *arbitrary* sources, and which calls for *only two* rotational positions of the object to fill up the Fourier space.

References

[1] A. H. Andersen and A. C. Kak, "Digital ray tracing in two-dimensional refractive fields," *J. Acoust. Soc. Amer.*, vol. 72, pp. 1593–1606, 1982.

[2] —, "Simultaneous algebraic reconstruction technique (SART): A superior implementation of the ART algorithm," *Ultrasonic Imaging*, to be published.
[3] W. H. Carter, "Computational reconstruction of scattering objects from holograms," *J. Opt. Soc. Amer.*, vol. 60, pp. 306-314, 1970.
[4] C. R. Crawford and A. C. Kak, "Multipath artifact corrections in ultrasonic transmission tomography," *Ultrasonic Imaging*, vol. 4, pp. 234-266, 1982.
[5] R. A. Crowther, D. J. DeRosier, and A. Klug, "The reconstruction of a three dimensional structure from projections and its applications to electron microscopy," *Proc. Roy. Soc. London*, ser. A317, pp. 319-340, 1970.
[6] A. J. Devaney, "A new approach to emission and transmission CT," in *Proc. 1980 Ultrasonics Symp.*, B. R. McAvoy, Ed., pp. 979-983, 1980.
[7] —, "A filtered backpropagation algorithm for diffraction images," *Ultrasonic Imaging*, vol. 4, pp. 336-350, 1982.
[8] —, "A computer simulation study of diffraction tomography," submitted for publication.
[9] M. Fatemi and A. C. Kak, "Ultrasonic *B*-scan imaging: Theory of image formation and a technique for restoration," *Ultrasonic Imaging*, vol. 2, pp. 1-48, 1980.
[10] A. F. Fercher, H. Bartelt, H. Becker, and E. Wiltschko, "Image formation by inversion of scattered data: Experiments and computational simulation," *Appl. Opt.*, vol. 18, pp. 2427-2439, 1979.
[11] A. Ishimaru, *Wave Propagation and Scattering in Random Media*, vol. 2. New York: Academic, 1978.
[12] K. Iwata and R. Nagata, "Calculation of refractive index distribution from interferograms using born and Rytov's approximation," *Japan J. Appl. Phys.*, vol. 14, Supp. 14-1, p. 383, 1975.
[13] A. C. Kak, "Computerized tomography with X-ray, emission, and ultrasound sources," *Proc. IEEE*, vol. 67, pp. 1245-1272, 1979.
[14] M. Kaveh, M. Soumekh, and R. K. Mueller, "A comparison of Born and Rytov approximation in acoustic tomography," *Acoust. Imaging*, vol. 10, 1981.
[15] R. M. Mersereau and A. V. Oppenheim, "Digital reconstruction of multidimensional signals from their projections," *Proc. IEEE*, vol. 62, pp. 210-229, 1974.
[16] R. K. Mueller, M. Kaveh, and G. Wade, "Reconstructive tomography and applications to ultrasonics," *Proc. IEEE*, vol. 67, pp. 567-587, 1979.
[17] R. K. Mueller, M. Kaveh and R. D. Iverson, "A new approach to acoustic tomography using diffraction techniques," in *Acoust. Holography*, vol. 8, A. F. Metherell, Ed. New York: Plenum, 1980, pp. 615-628.
[18] D. Nahamoo and A. C. Kak, "Ultrasonic diffraction imaging," School of Elec. Eng., Purdue Univ., Tech. Rep. TR-EE-82-20.
[19] A. V. Oppenheim and R. W. Schafer, *Digital Signal Processing.* Englewood Cliffs, NJ: Prentice-Hall, p. 96.
[20] P. M. Prenter, *Splines And Variational Methods.* New York: Wiley, ch. 5.
[21] A. Rosenfeld and A. C. Kak, *Digital Picture Processing*, 2nd ed., vol. 1. New York: Academic, 1982. (See ch. 8 for a survey of most of the major algorithms used today for reconstructing images from projection data measured with nondiffracting radiation.)
[22] L. A. Shepp and B. F. Logan, "The Fourier reconstruction of a head section," *IEEE Trans. Nucl. Sci.*, vol. NS-21, pp. 21-43, 1974.
[23] H. Stark, J. W. Woods, I. Paul, and R. Hingorani, "Direct Fourier reconstruction in computer tomography," *IEEE Trans. Acoust., Speech, Signal Processing*, vol. ASSP-29, pp. 237-244, 1981.
[24] E. Wolf, "Three-dimensional structure determination of semitransparent objects from holographic data," *Opt. Commun.*, vol. 1, pp. 153-156, 1969.

An Introduction to NMR Imaging: From the Bloch Equation to the Imaging Equation

WALDO S. HINSHAW AND ARNOLD H. LENT

Invited Paper

***Abstract*–The emerging technology of NMR imaging is introduced here as a problem in system identification. We show how selected families of signals may be input into the system ("system," in this case, is almost synonymous with "patient") in order that the system's responses to these inputs may be directly interpreted in terms of the system parameters. Once identified, a raster display of the system parameters provides an internal image of the patient.**

Inputs to the system are four-component functions of time. One component describes the strength of an RF signal, and the other three components govern the strength of three spatially varying, independently controlled magnetic fields (the *gradient* fields) in which the patient is immersed. In response to these inputs some of the protons in the patient, acting in concordance with the *Bloch equation*, give rise to local fluctuations in the magnetization which are detected with a tuned antenna and a sensitive receiver. The relationship between this output signal and the system parameters is summarized in the *imaging equation*.

I. Introduction

WE WILL TRY in this paper to introduce NMR (nuclear magnetic resonance) imaging to those who do not know NMR, but are somewhat familar with X-ray CT. Other articles in this issue should provide an adequate background. Here, we do not present all of the physical models and conceptual aids often used in introductory NMR papers, but present the mathematics in a reasonably accurate and consistent way. We do not discuss the complications resulting from departures from an ideal world, but do try to indicate where they occur. Finally, we stop short of discussing image reconstruction algorithms. This paper is intended as a resource for those who wish to develop their own reconstruction algorithms.

Even in a field as new as NMR imaging (the seminal paper appeared in 1973 [12]), subspecialties have already developed. In this paper we consider the imaging of protons (i.e., hydrogen nuclei) in stationary biological samples (people, mainly). We regretfully exclude from consideration the imaging of phosphorus, with its exciting potential for observing metabolism (but, for a review, see [11]), and dynamic imaging of the cardiovascular system (see, e.g., [18]). For introductory presentations of these (and other) clinical topics, as well as additional background reading in NMR, the recent conference proceedings [3] are recommended. Even more recent additions to the NMR tutorial literature are [21] and [22].

Before introducing the ideas of NMR imaging, we present a few representative NMR images. These images are intended to prove that the ideas, although sometimes subtle, are worth studying. Fig. 1 is a 1.5-cm-thick sagittal image of the torso and is included to demonstrate the ability of NMR to generate views of any orientation. Fig. 2 is a transverse slice through the head which demonstrates the high spatial resolution possible with the technique. Fig. 3 is a set of three images showing a coronal slice through the brain. The only difference between the images of this set is the timing of the applied magnetic fields. This figure demonstrates the ability of NMR to produce images based on more than one single property of the tissue.

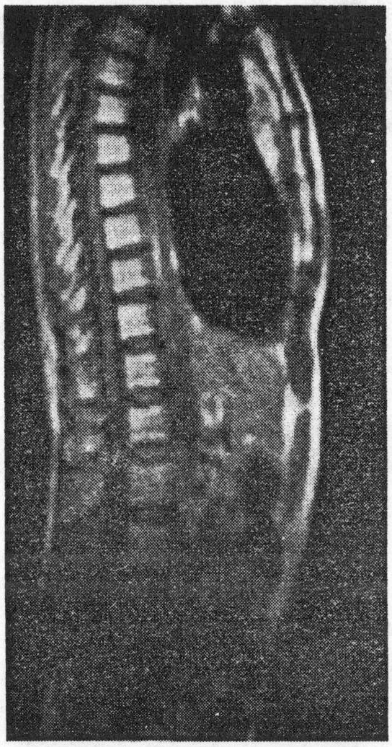

Fig. 1. This is a sagittal slice of the dorso-lumbar spine area. The image was obtained at 0.3 T using selective irradiation to define the slice, the saturation-recovery method to generate the NMR signal, and two-dimensional (2D) Fourier transformation to reconstruct the image. This image demonstrates the ability of NMR to provide views in any orientation.

All of these NMR images were produced on an experimental prototype NMR imaging system built by the Technicare Corporation. The system, shown in Fig. 4, was operated at a field strength of 0.3 T and a frequency of 12.8 MHz.

Even without knowing precisely how the images were produced, it is clear that valuable information has been obtained; add in the knowledge that, unlike X-ray CT, no ionizing radiation was used in forming these images, and the mind boggles.

Manuscript received October 5, 1982; revised December 15, 1982.
The authors are with the Technicare Corporation, Solon, OH 44139.

Reprinted from *Proc. IEEE,* vol. 71, pp. 338-350, Mar. 1983.

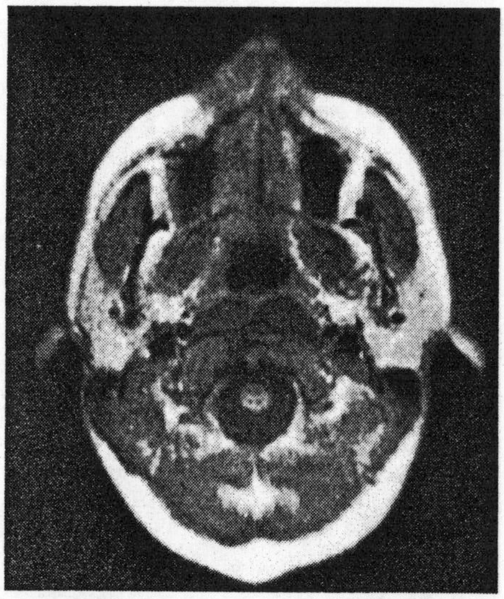

Fig. 2. This is a transverse slice of cranial anatomy at the level of the foramen magnum. It demonstrates the spatial resolution that can be obtained with NMR imaging. The methods used are the same as those for Fig. 1. The slice thickness is approximately 1 cm.

The remainder of Section I is devoted to introducing NMR imaging and comparing it with CT. Section II presents the Bloch equation, and Section III develops its consequences in an imaging system. Section IV discusses an idealized NMR receiver and its output signal $S(t)$. Finally, in Section V, the "imaging equations" of several different NMR imaging modalities are derived. At the end of Section I, for easy reference, is a partial list of symbols to aid the reader in sorting out notation.

A. CT and NMR: Physical Similarities

When one compares the NMR imaging system of Fig. 4 against a typical CT system, it is easy to be misled by their outward similarities. But CT relies on rotating mechanical gantries and the absorption of X-ray photons, while NMR is entirely electronic and is based on the interactions of small, rapidly varying, magnetic (the M of NMR) fields with loosely bound hydrogen nuclei (the N of NMR–R arrives in Section III) in the soft tissues of the body. Also, despite some brave attempts at three-dimensional reconstruction [20], CT is essentially a two-dimensional technology. NMR, on the other hand, seems to be intrinsically three-dimensional. One has a choice between collecting data from all of the three-dimensional objects and collecting data from only a single slice. In the first case, the data can be manipulated, using three-dimensional reconstruction techniques, to provide the image of a single slice. In the second case, special "slice-selection" techniques, which will be discussed later, are used to obtain data from only the selected slice. In either case, the slice can be in any orientation and at any level.

B. CT and NMR: Images

Let $\boldsymbol{x}$ stand for the vector of spatial coordinates. In CT, $\mu(\boldsymbol{x})$, the spatially varying X-ray attenuation coefficient, is the physical property being imaged. In NMR, *three* primary spatially varying physical properties (to be discussed), $M_0(\boldsymbol{x})$, $T_1(\boldsymbol{x})$, and $T_2(\boldsymbol{x})$, are necessary for a good description of an object. (In this sense, NMR resembles ultrasound imaging (see [6]) with its multiple imaging modalities.) However, these three properties are imaged differently. $M_0(\boldsymbol{x})$, which is

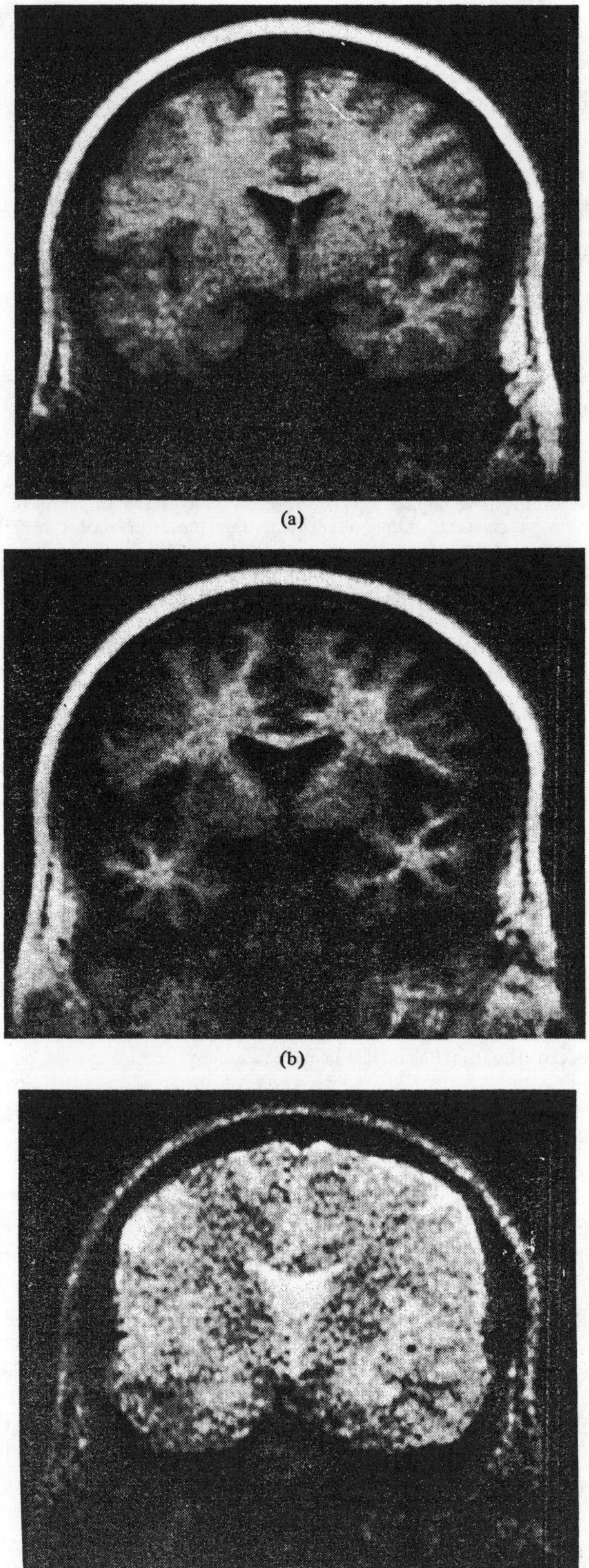

Fig. 3. This is a set of coronal head images showing the same slice with three different pulse sequences. Part (a) was obtained using the saturation-recovery method and part (b) using the inversion-recovery method. Note particularly the increased gray–white matter contrast in the inversion-recovery image. Part (c) was obtained using the same method as that used for (a) except that the data-acquisition interval occurred a significantly longer time after the first RF pulse. In this image, the material with long T_2 (the CSF) appears bright.

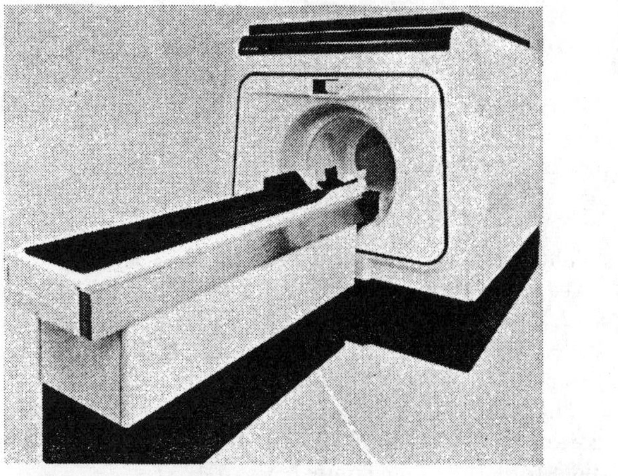

Fig. 4. Several NMR imaging systems, such as the one shown here, are now in use as clinical research tools. Note the deceptive resemblance to an X-ray CT scanner.

related to the distribution of "mobile" hydrogen nuclei (i.e., constituents of mobile molecules in liquid-like phases), has a role comparable to that of $\mu(\boldsymbol{x})$ in CT, in that it provides the overall image shape, while $T_1(\boldsymbol{x})$ and $T_2(\boldsymbol{x})$ (the "relaxation times") contribute significant local information. The distinction between gray matter and white matter in Fig. 3 is due almost entirely to T_1 differences, rather than to M_0 differences. It is possible (with considerable effort) to generate "pure" T_1 and T_2 images, but the usual NMR image is a composite combining M_0, T_1, and T_2; later, in Section V, we shall discuss some mechanisms for controlling their proportions.

C. CT and NMR: Input/Output Relationships

We are coming to some basic differences between CT and NMR.

1) In CT, as preceding articles have made clear, the inputs are X-ray pulses which are aimed at different parts of the sample. The information being collected is *spatially coded*. In NMR, a complicated signal $\boldsymbol{P}(t)$ (a four-component vector function of time) is transmitted into the NMR system, and a complex (as opposed to "real") signal $S(t)$ emerges. Information about the sample is *temporally coded*. Although the Fourier integral can be used to go back and forth between the time domain and the space domain, this easy transfer is conceptual, rather than real, and it is dangerous to consider NMR and CT "pretty much the same sort of thing."

2) In every CT system built, the input signals necessarily have a similar form; the X-ray pulses are uniform in strength, and vary in direction. (We do not mean to slight the major advances made between the first-generation CT systems and present-day systems.) With NMR, the input signals can be *tailored* by the observer so that their interaction with the sample is *controlled*, to produce strikingly diverse images. To draw an analogy, think of a microscope with some of its attachments (polarizers, phase contrast, dark-field illumination) and of the variety of images it can produce. It takes an intimate working knowledge of the properties of the multitude of possible input signals to make a competent zeugmatographer. (*Zeugmatography* is Lauterbur's [12] coinage for NMR imaging.)

D. Basic NMR Components

Having warned the reader that NMR and CT are different, we name, but do not yet explain, the basic components of an NMR imaging system. These are as follows:

1) A *magnet*, which provides a strong (about 0.5 T), uniform, steady, magnet field $\boldsymbol{H}_0$.

2) An *RF transmitter*, which delivers radiofrequency magnetic fields to the sample.

3) A *gradient system*, which produces time-varying magnetic fields of controlled spatial nonuniformity. The experimenter's controls for modulating the magnetic field experienced by the sample are lumped together in the vector $\boldsymbol{P}(t)$, which will be formally defined in Section III.

4) A *detection system*, which yields the output signal $S(t)$.

5) An *imager system*, including the computer, which reconstructs and displays the images.

Our objective in this paper is to establish the "imaging equation," which shows, for a few different input signals, how the output signal is related to the sample properties $M_0(\boldsymbol{x})$, $T_1(\boldsymbol{x})$, and $T_2(\boldsymbol{x})$. To achieve this objective, we must look at the physical model which underlies NMR.

Partial List of Symbols

$\omega_0 = -\gamma H_0$	Larmor frequency corresponding to the static field.
$\omega'(\boldsymbol{x}) = -\gamma(H_0 + \boldsymbol{G}\cdot\boldsymbol{x})$	Larmor frequency at $\boldsymbol{x}$ when gradient $\boldsymbol{G}$ is applied.
ω	The irradiation frequency of the RF field.
$\omega_1 = -\gamma H_1$	The frequency of RF-induced rotation.
$\omega_h = -\gamma h(\boldsymbol{x})$	"Additional" rotation frequency caused by gradients.
H_0	Strength of the static magnetic field.
$\boldsymbol{H}_0 = H_0\boldsymbol{k}$	The static magnetic field.
$\boldsymbol{H}(t, \boldsymbol{x})$	Total magnetic field experienced at $t, \boldsymbol{x}$.
$\boldsymbol{H}_1(t)$	Modulation function for the RF field.
$\boldsymbol{H}_1(t) = 2H_1(t)\cos\omega t i$	RF magnetic field.
$h(\boldsymbol{x}) = \boldsymbol{G}\cdot\boldsymbol{x}$	Gradient magnetic field.
$\boldsymbol{M}(t, \boldsymbol{x})$	Sample magnetization at $t, \boldsymbol{x}$.
M_x, M_y, M_z	Components of $\boldsymbol{M}$.
$M = M_x + iM_y$	(Complex) transverse magnetization.
$M_0\boldsymbol{k}$	The equilibrium nuclear magnetization.
$\boldsymbol{M}^0(\boldsymbol{x})$	Sample magnetization at $t = 0$.
M_x^0, M_y^0, M_z^0	Components of $\boldsymbol{M}^0$.
$M^0 = M_x^0 + iM_y^0$	Transverse magnetization at $t = 0$.

Arguments may be omitted (e.g., $\boldsymbol{M}$) for brevity or added (e.g., $\boldsymbol{M}(t, \boldsymbol{x})$) for clarity.

II. The Physical and Mathematical Model

The model of the spinning proton is as basic to NMR as the picture of absorbed and transmitted photons is to CT. In place of the Lambert–Beer law which describes photon absorption, NMR has the Bloch equation, which will be presented in this section and explored more fully in the next.

Physical Picture: Think of a top, spinning about its axis, which makes an angle α with the vertical. The rotation of the top's axis around the vertical is called *precession*. This spinning top is the simple, conceptual picture (the recent *Scientific*

American article on NMR imaging [16] has still more detail) of the proton and its spin, a subject treated rigorously only by quantum mechanics [1]. Under proper conditions, the proton (with its spin and associated magnetic field) precesses about a magnetic field (a *vector*, remember) as the top processes about the vertical.

In terms of this picture we can state some simple but important facts which, later, we derive from the Bloch equation.

Fact 1): The *rate* of precession (cycles per second) of a proton in a magnetic field is proportional to the *strength* of the field. (This is where the magnetic field $\boldsymbol{H}_0$ comes in.)

Fact 2): There is *no* signal emitted by the proton when it is sitting at equilibrium, with its spin lined up with the magnetic field. There *is* a signal when the proton has been knocked out of alignment so that it makes some angle α with the magnetic field. (This is where the RF transmitter comes in–*it* does the knocking.)

Fact 3): If the magnitude field can be made nonuniform in a *controlled manner*, then, by Fact 1), protons at different points in space will precess at different frequencies. (The gradient system takes care of this spatial encoding.)

A. Mathematical Model–The Bloch Equation

We begin with the Bloch equation [2], which gives an accurate but phenomenological description of the time dependence of the *nuclear magnetization* $\boldsymbol{M}(t)$ in the presence of an applied magnetic field $\boldsymbol{H}(t)$. The nuclear magnetization $\boldsymbol{M}(t)$ is the source of the "NMR signal" from which the image is ultimately constructed. We can think of the NMR sample as a black box with $\boldsymbol{H}(t)$ as the input signal, or stimulus, and $\boldsymbol{M}(t)$ as the output signal, or response. The black box is characterized by M_0, T_1, and T_2, and its behavior is governed by the Bloch equation.

$$d\boldsymbol{M}/dt = \gamma \boldsymbol{M} \times \boldsymbol{H} - (M_x \boldsymbol{i} + M_y \boldsymbol{j})/T_2 - (M_z - M_0)\boldsymbol{k}/T_1. \tag{2.1}$$

Initially, in order to illustrate some basic properties of the Bloch equation, we will consider the equation as applied to an ensemble of identical nuclei, each experiencing the same field $\boldsymbol{H}$. Then, fundamentals in hand, we will move on to heterogeneous samples and nonuniform fields.

B. Parameters of the Bloch Equation

The nuclear magnetization $\boldsymbol{M}(t)$, induced in the sample by the magnetic field $\boldsymbol{H}(t)$, is the local sum of the magnetic fields of the protons. The magnetization is a bulk property of the sample rather than a property of individual protons.

The *gyromagnetic ratio* γ is a physical property of the nucleus of the atom. Different chemical elements–in fact, different isotopes of the same element–exhibit significantly different gyromagnetic ratios. This makes it possible to observe protons and ignore phosphorus, for example. The gyromagnetic ratio of protons is 4.26×10^7 Hz $\cdot$ T^{-1} (or 2.68×10^8 rad $\cdot$ s^{-1} $\cdot$ T^{-1}).

$\boldsymbol{H}(t)$ is the magnetic field experienced by the nuclei being considered. We do not include in $\boldsymbol{H}(t)$ the contributions to the magnetic field resulting from the local interactions and collisions experienced by the nuclei. The effects of these "internal" fields are included in the parameters T_1 and T_2.

There is another phenomenon which affects $\boldsymbol{H}(t)$. The chemical electrons surrounding a given nucleus can shield the nucleus from the externally applied magnetic field by a very small amount. This shielding changes $\boldsymbol{H}(t)$, shifts the resonant frequency, and as a result, is called a "chemical shift." These shifts, which for protons cover a range of about 10 ppm, are the substance of NMR spectroscopy. Although the Bloch equations may be generalized to include chemical shifts, the effect is ignored in the following development. Thus for our purposes, $\boldsymbol{H}(t)$ is the field applied by the experimenter.

The coordinate system used in (2.1) is the *laboratory*, or *fixed*, reference frame. The $\boldsymbol{k}$ direction is taken to be parallel to $\boldsymbol{H}_0$, where $\boldsymbol{H}_0$ is the field of the large magnet. $\boldsymbol{H}_0$ is often called the *static field*. By convention, $\boldsymbol{k}$ is the "longitudinal direction" and $\boldsymbol{i}$ and $\boldsymbol{j}$ define the "transverse plane."

The *equilibrium magnetization* $M_0\boldsymbol{k}$ is the nuclear magnetization which exists if the sample is maintained at the static field for a time long compared to T_1.

The "relaxation times," T_1 and T_2, represent the effect of the "relaxation" processes. The constant T_1 is the "longitudinal" or "spin-lattice" relaxation time and governs the evolution of M_z toward its equilibrium value M_0. The physical process involved in this relaxation is the dissipation of energy from the collection of nuclei, the "spin system," to the atomic and molecular environment of the nuclei, the "lattice." The process can be thought of as the reorientation of the spins into alignment with $\boldsymbol{H}_0$.

The constant T_2 is the "transverse" or "spin-spin" relaxation time and governs the evolution of the magnitude of the transverse magnetization, $M_x\boldsymbol{i} + M_y\boldsymbol{j}$, toward its equilibrium value of zero. The physical processes which cause transverse relaxation include those which cause longitudinal relaxation, as well as the magnetic coupling between neighboring nuclei. The process can be thought of as the transverse orientation of the individual spins becoming randomized, or dephased, so that the sum of the transverse components of the fields of the nuclei in the collection goes to zero.

Although the Bloch equation accurately describes NMR imaging, it is, in fact, of limited validity. Perhaps the weakest assumption implicit in the equation is that the transverse relaxation is exponential. This assumption is fairly good for liquids, but is less valid for more "solid-like" samples. But solid-like samples have short T_2 values and, as a result, are not observed with the usual imaging methods.

C. The Bloch Equation as an Equation of Motion

The term $\boldsymbol{M} \times \boldsymbol{H}$ gives rise to the precessional motion mentioned earlier. To see how this comes about, we take $\boldsymbol{H} = H_0\boldsymbol{k}$, assuming for the moment that $\boldsymbol{H}$ is static, and, for convenience, temporarily ignore the T_1 and T_2 terms of (2.1). Initially, the magnetization has components M_x^0, M_y^0, M_z^0. Then

$$d\boldsymbol{M}/dt = \gamma \boldsymbol{M} \times H_0 \boldsymbol{k}. \tag{2.2}$$

In coordinate form

$$\begin{aligned} dM_x/dt &= \gamma H_0 M_y \\ dM_y/dt &= -\gamma H_0 M_x \\ dM_z/dt &= 0. \end{aligned} \tag{2.3}$$

These equations have the solution

$$\begin{aligned} M_x(t) &= M_x^0 \cos \omega_0 t - M_y^0 \sin \omega_0 t \\ M_y(t) &= M_x^0 \sin \omega_0 t + M_y^0 \cos \omega_0 t \\ M_z(t) &= M_z^0 \end{aligned}$$

where ω_0 is the *Larmor*, or *resonant*, frequency of the spin system given by

$$\omega_0 = -\gamma H_0. \tag{2.4}$$

These equations validate Facts 1) and 2).

The inclusion of the T_1 and T_2 terms complicates matters only a little. The solution becomes

$$M_x(t) = e^{-t/T_2}(M_x^0 \cos \omega_0 t - M_y^0 \sin \omega_0 t)$$
$$M_y(t) = e^{-t/T_2}(M_x^0 \sin \omega_0 t + M_y^0 \cos \omega_0 t)$$
$$M_z(t) = M_z^0 e^{-t/T_1} + M_0(1 - e^{-t/T_1}). \tag{2.5}$$

The longitudinal component decays from its initial value of M_z^0 toward its equilibrium value of M_0. The transverse component rotates at frequency ω_0 and decays toward zero.

There is a compact, convenient expression for the transverse component of the magnetization. Define

$$M = M_x + iM_y. \tag{2.6}$$

Then

$$M(t) = M^0 \exp(i\omega_0 t - t/T_2) \tag{2.7}$$

where $M^0 = M_x^0 + iM_y^0$.

D. The Bloch Equation for Heterogeneous Samples and Nonuniform Fields

We now move on to the case of physical interest, heterogeneous samples and nonuniform magnetic fields. First consider the heterogeneity of the sample. Imagine the Bloch equation applied to a small volume of the sample at a given point $\boldsymbol{x}$ in the sample. Within this small volume we can assume that $\boldsymbol{H}$ is uniform, but there is a distribution of values of T_1 and T_2. There are interactions between nuclei which complicate the picture, but we will approximate the true situation by assuming that there are a finite number of different "types" of nuclei in the volume, each with its own value of T_1 and T_2. With each type, we associate a different magnetization for the volume, which is written $\boldsymbol{M}_n(t, \boldsymbol{x})$, where $\boldsymbol{x}$ is the location of the small volume. Thus the total magnetization $\boldsymbol{M}(t, \boldsymbol{x})$ for the volume is

$$\boldsymbol{M}(t, \boldsymbol{x}) = \sum \boldsymbol{M}_n(t, \boldsymbol{x}). \tag{2.8}$$

Rather than carry this notion through the following discussions, it is easier to re-introduce it after the final result is obtained and, until then, consider only one particular value for T_1 and for T_2. Since the output signal is a linear function of the sample magnetization (this will be shown in Section IV), we can do the summation after the signal processing is completed.

As for nonuniform fields, all we do is regard every variable of the Bloch equation (except γ, of course) as a function of both $\boldsymbol{x}$ and t. γ is determined once and for all by the choice of the nuclide to be imaged.

III. Solutions of the Bloch Equation—The Basic NMR "Toolkit"

In this section we introduce the most important NMR "stimulus" signals, and, for each stimulus $\boldsymbol{H}(t, \boldsymbol{x})$, derive the resulting "response" magnetization $\boldsymbol{M}(t, \boldsymbol{x})$ of the sample. We continue to confine ourselves to a simplified and idealized world, with occasional asides to mention important deviations from ideality. The uniform field H_0 has been introduced in earlier sections, so we begin with the gradient field and continue with the RF field. As we shall see, both of these fields contribute to the field $\boldsymbol{H}(t, \boldsymbol{x})$ applied to the sample.

Both the gradient field and the RF field are under the control of the experimenter. The gradient field is governed by three input signals called $G_x(t)$, $G_y(t)$, and $G_z(t)$; the RF field is controlled by a signal called $H_1(t)$. We group these signals into a four-component vector $\boldsymbol{P}(t)$. Conventionally, $H_1(t)$ is taken as the first component of $\boldsymbol{P}$, followed by the three gradient controls. $\boldsymbol{P}(t)$ is a full description of what is usually called a "pulse sequence." Section V presents several types of pulse sequences used in NMR imaging.

A. The Gradient Field

The gradient system includes a set of three independently computer-controlled coils which generate a spatially varying and time-varying magnetic field within the sample. These coils are referred to as the x-gradient, the y-gradient, and the z-gradient coils, names which are slightly misleading. We consider only the x-gradient in detail; the other gradients behave similarly. Fig. 5 diagrams these three gradient fields and their relationship to $\boldsymbol{H}_0$ and the RF field.

The ideal x-gradient coil causes the z-component of the magnetic field to vary linearly with x as follows:

$$\boldsymbol{H}(t, \boldsymbol{x}) = H_0 \boldsymbol{k} + G_x(t) x \boldsymbol{k}.$$

The real x-gradient coil produces a field which has components in the x- and y-directions, but the uniform field in the z-direction is so strong that these components may be neglected by comparison. Thus we can call $G_x(t)$ the x-gradient and $G_x(t)x$ the x-gradient field even though the x-gradient coil produces other field components. When all three gradients coils are turned on, their superimposed fields yield

$$\boldsymbol{H}(t, \boldsymbol{x}) = H_0 \boldsymbol{k} + G_x(t) x \boldsymbol{k} + G_y(t) y \boldsymbol{k} + G_z(t) z \boldsymbol{k}.$$

The three gradients may be formally grouped into a gradient vector $\boldsymbol{G}(t)$ with components $G_x(t)$, $G_y(t)$, and $G_z(t)$, which summarizes the temporal variation of the gradients. Thus we can write

$$\boldsymbol{H}(t, \boldsymbol{x}) = (H_0 + \boldsymbol{G}(t) \cdot \boldsymbol{x}) \boldsymbol{k}.$$

It is this gradient field $\boldsymbol{G}(t) \cdot \boldsymbol{x}$ which induces the spatial coding of the resonance frequency that was alluded to in Fact 3). To see how this arises, take the special case of static gradients, $\boldsymbol{G}(t) = \boldsymbol{G}$. Then the Larmor frequency ω' of the infinitesimal sample at $\boldsymbol{x}$ is

$$\omega'(\boldsymbol{x}) = -\gamma(H_0 + \boldsymbol{G} \cdot \boldsymbol{x}). \tag{3.1}$$

For a given, fixed $\boldsymbol{G}$, two points in space which are displaced from each other by a vector which is orthogonal to $\boldsymbol{G}$ will have the same Larmor frequency, but, by transmitting a second stimulus signal with a different $\boldsymbol{G}$, the two points may be made distinguishable. By transmitting enough different stimulus signals, *all* points may be made (almost) distinguishable, and this is what NMR imaging systems do.

We now take a quick look at two useful, simple gradient inputs. As we shall see, it is the transverse magnetization components M_x and M_y which determine the output signal and, hence, are of interest. The gradient fields have no effect on M_z.

The solution of the Bloch equation in complex notation, (2.7), becomes

$$M(t, \boldsymbol{x}) = M^0(\boldsymbol{x}) \exp(i\omega'(\boldsymbol{x}) t - t/T_2(\boldsymbol{x})) \tag{3.2}$$

when we add the dependence upon $\boldsymbol{x}$ and change ω_0 to $\omega'(\boldsymbol{x})$.

1) The Readout Gradient: As we have just seen, if a static

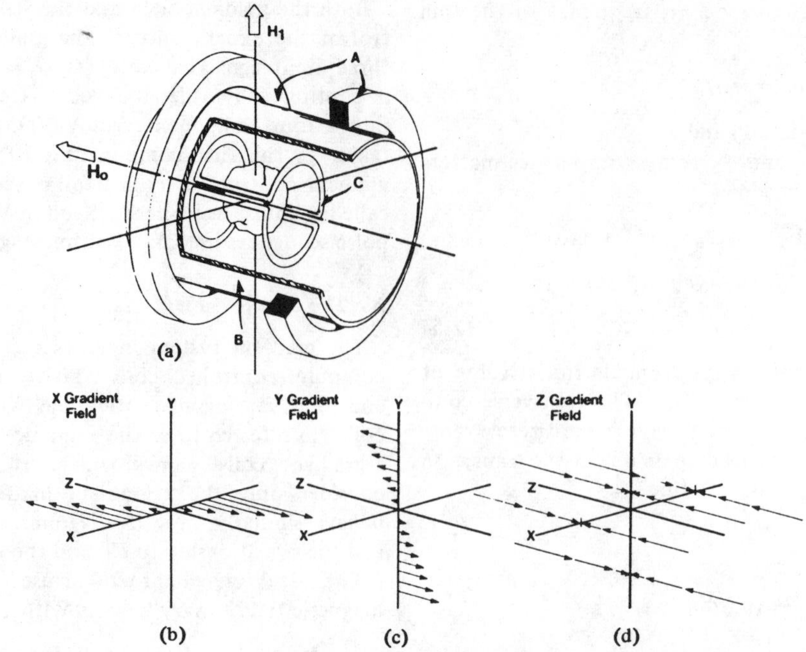

Fig. 5. This shows how the three major coil sets fit together, and indicates the geometry of their fields. The uniform static field H_0 is generated by the magnet A, drawn here as a large coil pair. The gradient field $G(x)$ is generated by a complex gradient coil set. In this drawing, they are shown as wound on the cylinder B. The RF field H_1 is generated by a "saddle coil" C. Part (b) is a representation of the x-gradient field. This field is parallel to H_0 and varies linearly with distance along the x axis but does not vary with distance along y or z. Parts (c) and (d) are similar representations of the y-gradient and z-gradient fields, respectively.

gradient $\boldsymbol{G}$ is applied while the signal is being observed, the frequency of the magnetization oscillations ω' becomes dependent upon $\boldsymbol{x}$. This spatial dependence is reflected in the behavior of the output signal. The use of this tool, the "readout" gradient, will be introduced in Section V.

2) The Gradient Pulse: Another tool is the "gradient pulse," which consists simply of turning on a gradient for a short interval τ which is chosen to be much shorter than the relaxation times, so that we may consider the gradient constant during the time of the pulse. We drop the T_2 term from (3.2), to find that, after the pulse, we have

$$M(t, \boldsymbol{x}) = M^0(\boldsymbol{x}) \exp [i(\omega_0 - \gamma(\boldsymbol{G} \cdot \boldsymbol{x}))\tau]. \qquad (3.2a)$$

B. The RF Field

In order to "activate" the nuclei so that they emit a signal, energy must be transmitted into the sample. That is what the RF transmitter does. The usual transmitter coil applies to the sample an RF magnetic field $\boldsymbol{H}_1(t)$, where

$$\boldsymbol{H}_1(t) = 2H_1(t) \cos \omega t\, \boldsymbol{i}.$$

Such a field is said to be *linearly polarized*, since it oscillates in a single direction. We have taken this direction as the definition of $\boldsymbol{i}$. ω is called the *irradiation frequency*; it is also the reference frequency of the RF transmitter and the detection system. A typical value for ω is 1.0×10^8 rad/s.

To discover how the RF field affects the magnetization, we would, in principle, just use the Bloch equation, setting $\boldsymbol{H} = H_0\boldsymbol{k} + \boldsymbol{H}_1(t)$. But if we do this, we find that the difficulty of the mathematics hides the phenomenon we are trying to examine. Fortunately, the physical insight of the early investigators [2] led them to a useful approximate solution. One writes

$$\boldsymbol{H}_1(t) = H_1(t)\,[\cos \omega t\, \boldsymbol{i} + \sin \omega t\, \boldsymbol{j}] + H_1(t)\,[\cos \omega t\, \boldsymbol{i} - \sin \omega t\, \boldsymbol{j}].$$

The two expressions in square brackets describe *circularly polarized* fields of opposite polarization. It can be shown that the nuclei will respond to one of the fields, and will be almost unaffected by the other. (It is the sign of γ that determines which of the circularly polarized fields affects the nuclei.) Thus the effective field is

$$\boldsymbol{H}_1(t) = H_1(t)\,[\cos \omega t\, \boldsymbol{i} + \sin \omega t\, \boldsymbol{j}].$$

We will look at the effect of this field on the nuclear magnetization using the simplifying assumption that the *modulation function* $H_1(t)$ turns the RF field on and off in a time short compared to both relaxation times. This is a fairly safe assumption for NMR imaging, where the shortest relaxation time observed is on the order of 40 ms and the longest RF pulse is on the order of 2 ms.

The Bloch equation, with the terms containing T_1 and T_2 omitted, is

$$d\boldsymbol{M}/dt = \gamma \boldsymbol{M} \times \boldsymbol{H}.$$

During the RF pulse, the magnetic field is

$$\boldsymbol{H}(t) = H_1(t) \cos \omega t\, \boldsymbol{i} + H_1(t) \sin \omega t\, \boldsymbol{j} + H\boldsymbol{k}.$$

Here we take $H = H_0 + h$, where h is the contribution from the gradients. Thus the Bloch equation becomes

$$dM_x/dt = \gamma M_y H - \gamma M_z H_1(t) \sin \omega t$$

$$dM_y/dt = \gamma M_z H_1(t) \cos \omega t - \gamma M_x H$$

$$dM_z/dt = \gamma M_x H_1(t) \sin \omega t - \gamma M_y H_1(t) \cos \omega t.$$

We solve this equation by a clever change of variables [2], commonly used in the study of uniformly rotating systems. This change of variables is able to transform a linear set of equations with time-varying coefficient to one with constant coefficients.

Let

$$M_x = u \cos \omega t - v \sin \omega t$$
$$M_y = u \sin \omega t + v \cos \omega t. \qquad (3.3)$$

Then the differential equation satisfied by u and v is

$$du/dt = (\gamma H + \omega) v$$
$$dv/dt = -(\gamma H + \omega) u + \gamma H_1 M_z$$
$$dM_z/dt = -\gamma H_1 v. \qquad (3.4)$$

If we introduce a complex function

$$c(t) = u(t) + iv(t)$$

then (3.3) expresses a simple relationship between $c(t)$ and the transverse magnetization of (2.6)

$$M(t) = \exp(i\omega t)\, c(t). \qquad (3.5)$$

1) The Rotating Frame: The u and v functions have an interesting interpretation. Consider an orthogonal set of unit vectors $\boldsymbol{i}'(t)$ and $\boldsymbol{j}'(t)$ such that

$$\boldsymbol{i}'(t) = \boldsymbol{i} \cos \omega t + \boldsymbol{j} \sin \omega t$$
$$\boldsymbol{j}'(t) = -\boldsymbol{i} \sin \omega t + \boldsymbol{j} \cos \omega t.$$

Using (3.3), one sees that

$$u(t)\,\boldsymbol{i}' + v(t)\,\boldsymbol{j}' = M_x \boldsymbol{i} + M_y \boldsymbol{j}.$$

The interpretation of this equation is that the functions $u(t)$ and $v(t)$ are the components of the vector $M_x(t)\,\boldsymbol{i} + M_y(t)\,\boldsymbol{j}$ with respect to the rotating axes $(\boldsymbol{i}'(t), \boldsymbol{j}'(t))$. A physicist, of course, would *start* with the rotating frame $(\boldsymbol{i}'(t), \boldsymbol{j}'(t))$ and *derive* the components of the magnetization in the rotating frame. We see now that the constant ω in the coordinate transformation (3.3) can be viewed as the rotation frequency of the rotating frame.

A convenient value to take for ω is ω_0, the Larmor frequency of the static field. Then, using the equality $\omega_0 = -\gamma H_0$, (3.4) becomes

$$du/dt = \gamma h v$$
$$dv/dt = -\gamma h u + \gamma H_1 M_z$$
$$dM_z/dt = -\gamma H_1 v \qquad (3.6)$$

where h is the field contribution from the gradients.

If we look at the special case where $h = 0$ and H_1 is constant, the solution of (3.6) becomes transparent. It is

$$u(t) = u(0)$$
$$v(t) = v(0) \cos \omega_1 t - M_z(0) \sin \omega_1 t$$
$$M_z(t) = v(0) \sin \omega_1 t + M_z(0) \cos \omega_1 t \qquad (3.7)$$

which says that the magnetization vector, as viewed in the rotating frame, rotates around the $\boldsymbol{i}'$ vector with angular frequency $\omega_1 = -\gamma H_1$. It is clear from this simple example, and true in general, that the Bloch equation is most simply solved and interpreted in the rotating frame.

This example also explains the R of NMR. The R stands for "resonance," and the matching of ω, the irradiation frequency, with ω_0, the natural frequency of precession of protons in the field $\boldsymbol{H}_0$, is the resonance alluded to. The physical result of the resonance is the rotation of the nuclear spins at a rate proportional to the strength of the RF field.

2) The "Short" RF Pulse: This simple example is a good model for the RF pulses applied in NMR imaging. It can be shown that, if the duration of the RF pulse τ is short compared to $1/(\gamma h)$, setting h to zero is a good approximation. The other assumption, that H_1 is static, serves to simplify the form of (3.7). Without this assumption, $\omega_1 t$ would be replaced by

$$-\gamma \int_0^\tau H_1(t)\, dt.$$

Thus for short pulses, the only property of $H_1(t)$ that is important is its integral over time—the pulse *shape* has no effect.

The RF pulse can be characterized by two parameters. The first is the tip angle α given by

$$\alpha = \gamma \int_0^\tau H_1(t)\, dt$$

where τ is the length of the pulse. Thus α is the angle through which the RF field rotates the magnetization. The pulses most used in NMR imaging are the 90°, or $\pi/2$ pulse, which rotates the magnetization from alignment with the main field $\boldsymbol{H}_0$ into the transverse plane, and the 180°, or π pulse, which inverts the magnetization, rotating it from alignment to antialignment with the main field. If we look at the magnetization just after a $\pi/2$ RF pulse at time τ, (3.7) becomes

$$u(\tau) = u(0)$$
$$v(\tau) = -M_z(0)$$
$$M_z(\tau) = v(0). \qquad (3.8)$$

Just after a π pulse, we have

$$u(\tau) = u(0)$$
$$v(\tau) = -v(0)$$
$$M_z(\tau) = -M_z(0). \qquad (3.9)$$

The second parameter characterizing the RF pulse is the orientation of the axis about which the magnetization is rotated. In this discussion, we considered only rotations about the $\boldsymbol{i}'$ axis in the rotating frame, but modern NMR transmitters can select any axis of rotation, and, in fact, many pulse sequences utilize more than one axis during the course of the sequence. It can be shown that selecting the axis in the rotating frame is the same as selecting the phase of the RF field in the lab frame.

3) Slice Selection: We would like to give a rough idea of the slice-selection, or tomographic, process. This process causes the magnetization within a selected slice to become activated and generate a signal, while the remainder of the sample remains quiet.

Two tomographic NMR methods have been proposed. One [8] involves applying time-dependent, or oscillating, gradients and using the steady-state-free-precession pulse sequence. It will not be discussed. The second is known as *selective irradia-*

tion or *selective excitation* and involves applying an RF pulse while a gradient is turned on and held steady. This method will be presented with the intention of showing only *how* it works—we will go out of our way to *avoid* the difficult and interesting mathematics associated with this technique.

As before, the RF pulse has the form

$$\boldsymbol{H}(t) = 2H_1(t) \cos \omega t \, \boldsymbol{i}$$

but this time we deal with a "long," or "shaped" pulse, and the variation of $H_1(t)$ with time is of special interest.

The gradient system contributes a field of the form $h(\boldsymbol{x})\boldsymbol{k}$, where

$$h(\boldsymbol{x})\boldsymbol{k} = (\boldsymbol{G} \cdot \boldsymbol{x})\boldsymbol{k}.$$

We shall see how, by choosing the vector $\boldsymbol{G}$, a slice of any desired orientation may be selected.

Combining the RF field, the gradient field, and the static field

$$\boldsymbol{H}(t, \boldsymbol{x}) = (H_0 + h(\boldsymbol{x}))\boldsymbol{k} + 2H_1(t) \cos \omega t \, \boldsymbol{i}.$$

The initial conditions are $M_x = 0$, $M_y = 0$, $M_z = M_0$. We simplify the analysis by assuming that $\omega = \omega_0$.

To analyze this experiment, we start with (3.6), which is the form the Bloch equation takes in the rotating frame. Such a system is difficult to analyze exactly. It has been the subject of a perturbation analysis by Hoult [9] and a simulation study by Locher [13]. What we shall do here is indicate what happens when a "weak" RF pulse is applied. This is the "small tip-angle approximation" and is reasonably valid as long as the RF field rotates the magnetization by less than about 30°.

A second simplifying approximation, the dropping of the relaxation terms, was made in deriving (3.6), our starting point. Thus any analysis based on this equation is suspect if it extends into times comparable to the relaxation times. Also, we again approximate the linearly polarized RF field with its "effective" circularly polarized field.

If we are considering times short enough and H_1 weak enough, then M_z deviates little from its starting value M_0. This simplifies (3.6) considerably, to the point of permitting an analytic solution. The simplified equations are

$$du/dt = -\omega_h v$$
$$dv/dt = \omega_h u + \gamma H_1(t) M_0 \tag{3.10}$$

where we have written ω_h for $-\gamma h(\boldsymbol{x})$.

We are now going to solve these differential equations using the proper initial conditions, $u = 0$ and $v = 0$. We use the now-familiar trick for converting (3.10) into a single differential equation. Reintroduce the complex function $c(t) = u(t) + iv(t)$. This function then satisfies the equation

$$dc/dt = i\omega_h c + i\gamma H_1(t) M_0$$

which has a simple integrating factor, $\exp(-i\omega_h t)$. Integrating

$$d(c \exp(-i\omega_h t))/dt = i\gamma M_0 \exp(-i\omega_h t) H_1(t).$$

The initial conditions specify the solution as

$$c(t) = i\gamma M_0 \exp(+i\omega_h t) \int_0^t \exp(-i\omega_h s) H_1(s)\, ds. \tag{3.11}$$

For ease of discussion, let us take $G_x = 0$, $G_y = 0$, and $G_z = G$, so that $(\boldsymbol{G} \cdot \boldsymbol{x}) = Gz$. When we reach the conclusion of the argument, it will be a simple matter to replace z with $(\boldsymbol{G} \cdot \boldsymbol{x})/G$ in order to demonstrate slice selection in any orientation. It is important to make explicit the z-dependence of c.

$$c(t, z) = i\gamma M_0 \exp(-i\gamma Gzt) \int_0^t \exp(i\gamma Gzs) H_1(s)\, ds.$$

Also, recall that $H_1(t)$ is supposed to describe a pulse, so it must turn off (i.e., equal zero) at some time τ. Then, for $t > \tau$, $|c(t, z)| = |c(\tau, z)|$. Accordingly, we will attempt to choose $H_1(t)$ so that $c(\tau, z)$ describes a "slice." For concreteness, we will choose the slice around $z = 0$. (Experimentally, it is easy to move the slice to any other z level).

Our problem is now to

1) make $|c(\tau, z)|$ "large" for $|z| \leqslant a$
2) make $|c(\tau, z)|$ "small" for $|z| > a$. (3.12)

The solvability of the problem becomes more plausible if we write

$$|c(\tau, z)| = \gamma M_0 \left| \int_{-\tau/2}^{\tau/2} \exp(i\gamma Gzs) H_1(s + \tau/2)\, ds \right| \tag{3.13}$$

because then the right-hand side of (3.13) looks very much like a Fourier transform, and this guides our choice of pulse shape. For example, if we choose a Gaussian pulse shape

$$H_1(s + \tau/2) = \exp(-(sa\gamma G)^2/8)$$

then, providing τ is long enough so that the integrand is negligible at the ends of the region of integration, a table of integrals shows that $|c(\tau, z)|$ is proportional to $\exp(-2z^2/a^2)$. Standard statistics formulas show that this function has 95 percent of its area in the region $|z| \leqslant a$, which makes it adequately "slice-like." For a still sharper pulse, we could apply the theory of prolate spheroidal wave functions developed by Slepian, Pollak, and Landau [19], but such refinements are not necessary here.

Returning briefly to the slice at an arbitrary orientation, we replace z with $(\boldsymbol{G} \cdot \boldsymbol{x})/G$, where $G^2 = G_x^2 + G_y^2 + G_z^2$. It is easy to verify that

$$|(\boldsymbol{G} \cdot \boldsymbol{x})| \leqslant Ga$$

does indeed define a slice of thickness $2a$, and it is orthogonal to $\boldsymbol{G}$.

IV. The NMR Detection System

In the previous section, we derived an expression for $\boldsymbol{M}(t, \boldsymbol{x})$, the nuclear magnetization. It is the function of the NMR detection system, the receiver, to detect $\boldsymbol{M}(t, \boldsymbol{x})$ and generate an output signal $S(t)$. In this section, we discuss, in terms of an idealized and simplified NMR receiver, how $S(t)$ is obtained. A block diagram of such a receiver is shown in Fig. 6.

A. The Receiver Coil

The receiver coil, which usually surrounds the sample, is an antenna which "picks up" the fluctuating nuclear magnetization of the sample and converts it to a fluctuating output voltage $V(t)$. (For this tutorial paper, we ignore the fine

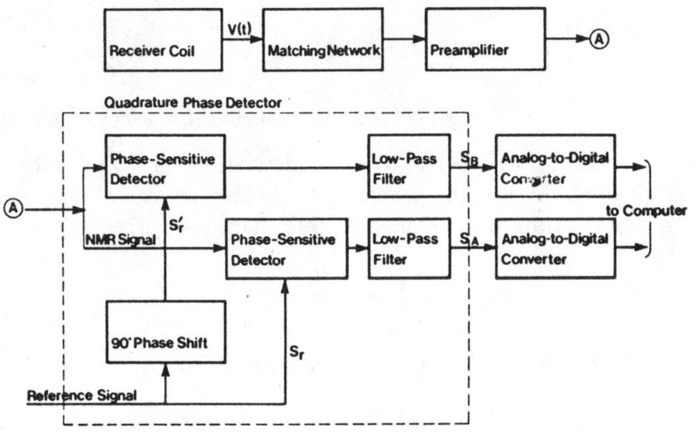

Fig. 6. Simplified block diagram of an NMR receiver system. The blocks are discussed in Section IV.

distinction between voltage and EMF.) Using some basic physics, it can be shown [10] that

$$V(t) = -\frac{d}{dt}\int \boldsymbol{M}(t, \boldsymbol{x}) \cdot \boldsymbol{B}_c(\boldsymbol{x})\, d\boldsymbol{x}. \tag{4.1}$$

The function $\boldsymbol{B}_c(\boldsymbol{x})$ describes the sensitivity of the receiver coil at different points in space. More specifically, $\boldsymbol{B}_c(\boldsymbol{x})$ is the ratio of the magnetic field produced by the receiver coil to the current in the coil. The function $\boldsymbol{B}_c(\boldsymbol{x})$ can be measured in a given coil, or calculated from the placement of the wires in the coil.

The primary objective of receiver coil design is to prescribe wire placements so that $\boldsymbol{B}_c(\boldsymbol{x})$ has the largest possible transverse component. The longitudinal component of $\boldsymbol{B}_c(\boldsymbol{x})$ contributes little to the output voltage, and can be ignored. This is a result of the fact that the time derivative of $M_z(t, \boldsymbol{x})$ is much less than that of the transverse component. $M_z(t, \boldsymbol{x})$ decays exponentially with time constant T_1, typically 0.1 to 1 s, while the transverse component is oscillatory with a period of, typically, 0.05 to 0.2 μs.

In order to proceed with the minimum of complicating detail, we will assume a "perfect" receiver coil and take $\boldsymbol{B}_c(\boldsymbol{x})$ to be

$$\boldsymbol{B}_c(\boldsymbol{x}) = a\boldsymbol{i} + b\boldsymbol{j} \tag{4.2}$$

where a and b are fixed, but unknown, constants. Equation (4.2) describes a coil which has uniform sensitivity over the sample, but whose direction of maximum sensitivity does not coincide with the direction of the applied RF field. Both of these assumptions are physically realistic: a carefully designed coil will not deviate from ideal uniformity by more than a few percent; however, in real coils, the direction of receiver sensitivity can be difficult to control and difficult to measure. But, as will become evident later, this indeterminacy in direction is absorbed into the "phase" of the output signal and causes little difficulty.

It is possible to use two receiver coils, one sensitive in the $\boldsymbol{i}$ direction and the other in the $\boldsymbol{j}$ direction, but a far more complicated receiver is required, and although this design provides a potential increase in signal strength, it does not change the relationship between the sample magnetization and the output signal of the receiver. Hence, we ignore this two-coil design.

With these simplifying assumptions, (4.1) becomes

$$V(t) = -\frac{d}{dt}\int (aM_x(t, \boldsymbol{x}) + bM_y(t, \boldsymbol{x}))\, d\boldsymbol{x}$$

$$= -\frac{d}{dt}\int \mathrm{Re}\,((a - ib)\, M(t, \boldsymbol{x}))\, d\boldsymbol{x}. \tag{4.3}$$

This can be simplified further by using the special form of $\boldsymbol{M}(t, \boldsymbol{x})$ appropriate to readout under steady gradients, (3.2). But, rewriting the (complex) transverse magnetization $M^0(\boldsymbol{x})$ in polar form

$$M^0(\boldsymbol{x}) = A(\boldsymbol{x}) \exp(i\theta(\boldsymbol{x}))$$

and, making the replacements

$$a = k\cos\phi' \quad \text{and} \quad b = -k\sin\phi'$$

$$V(t) = -\frac{d}{dt}\int A(\boldsymbol{x}) \exp(-t/T_2(\boldsymbol{x})) \cdot k\cos[\omega'(\boldsymbol{x})\, t + \theta(\boldsymbol{x}) + \phi']\, d\boldsymbol{x}. \tag{4.4}$$

We see that the receiver coil has modified $M_x(t, \boldsymbol{x})$ by a multiplicative factor k and by a *phase shift* ϕ'.

From here on, k is an arbitrary (fixed but unknown) constant, the *gain*, which depends upon the design of the receiver. Throughout the following development, we will retain the symbol k regardless of what factors it includes. Once a constant is arbitrary, it is *arbitrary*, and renaming it changes nothing. In practice, the absolute strength of the NMR signal, the numerical value of k, is difficult to obtain and is of little value.

We now simplify (4.4) still further by drawing on our knowledge of physical magnitudes. First, move the derivative inside the integral.

$$V(t) = k\omega_0 \int A(\boldsymbol{x}) \exp(-t/T_2(\boldsymbol{x}))\, \{(1 + (\omega_h(\boldsymbol{x})/\omega_0)) \cdot \sin[\;] + (1/(T_2(\boldsymbol{x})\,\omega_0)) \cos[\;]\}\, d\boldsymbol{x}.$$

Both the terms $\omega_h(\boldsymbol{x})/\omega_0$ and $1/(T_2(\boldsymbol{x})\,\omega_0)$ are on the order of 10^{-3}, so they are negligible compared to 1. Also, let $\phi' = \phi + \pi/2$. Then the working expression for $V(t)$ becomes

$$V(t) = k\int A(\boldsymbol{x}) \exp(-t/T_2(\boldsymbol{x})) \cdot \cos[(\omega_0 + \omega_h(\boldsymbol{x}))\, t + \theta(\boldsymbol{x}) + \phi]\, d\boldsymbol{x} \tag{4.5}$$

where we have absorbed all multiplicative constants into k.

Summarizing, the nuclear magnetization $\boldsymbol{M}(t, \boldsymbol{x})$ has induced an output voltage $V(t)$ in the receiver coil. We call $V(t)$ the *NMR signal.*

B. The Matching Network

The matching network couples the receiver coil to the preamplifier in order to maximize energy transfer into the amplifier. The only effect the matching network has on the NMR signal is to change the value of k and to introduce an unknown contribution to the phase of the signal. The phase shifts introduced here and in subsequent circuits (as well as simple time delays, which, in the RF stages, are equivalent to phase shifts) are arbitrary in exactly the same way that the gain k is arbitrary. We will incorporate this arbitrary, equipment-generated, phase into ϕ.

C. The Preamplifier

The preamplifier is a low-noise first-stage amplifier. The effect on the signal is simply a change in k and ϕ.

Discussion of noise—the noise from the sample, the noise introduced by the amplifiers, etc.—would be appropriate here. But we move on.

D. The Quadrature Phase Detector

The quadrature phase detector accepts the RF NMR signal, which consists of a distribution of frequencies centered around or near the irradiating frequency ω, and shifts the signal down in frequency by ω. Thus the distribution of frequencies is unchanged except that it is now centered about zero. By reducing the center frequency we reduce significantly the demands on the analog-to-digital converter and the computer.

First, let us look at the operation of a single phase-sensitive detector. This circuit accepts two inputs, the NMR signal and a reference signal, and multiplies them, so that the output is the product of the two inputs. The signal input is $V(t)$, given by (4.5). The reference signal S_r can be taken to be

$$S_r = a \cos(\omega t).$$

By choosing the frequency of the reference signal to be the same as that of the irradiating RF pulse, the receiver system is simplified considerably. Multiplying these two signals involves the product

$$\cos(\omega t) \cos(\omega_0 t + \beta)$$

which becomes

$$\tfrac{1}{2} \cos[(\omega + \omega_0) t + \beta] + \tfrac{1}{2} \cos[(\omega_0 - \omega) t + \beta]$$

where $\beta = \omega_h(\boldsymbol{x}) t + \theta(\boldsymbol{x}) + \phi$. But recall that the basic idea of resonance requires that ω and ω_0 are very close in value. Thus the output of the phase-sensitive detector consists of the sum of two components; one, a narrow range of frequencies centered at $2\omega_0$, and the other a narrow range centered at zero. (The output of nonideal detectors also includes a third component centered at ω_0.)

The low-pass filter following the phase-sensitive detector removes all components except those centered at zero. Then the signal $S_A(t)$ after the low-pass filter is

$$S_A(t) = k \int A(\boldsymbol{x}) \exp(-t/T_2(\boldsymbol{x})) \cos[(\omega_0 - \omega) t + \beta] \, d\boldsymbol{x}.$$

The 90° phase-shift circuit accepts the reference signal $S_r(t)$ and has as output a signal $S_r'(t)$, which, without a great loss of generality, we can take to be

$$S_r'(t) = a \sin(\omega t).$$

Following the same discussion for channel B, we obtain as the output of the low-pass filter

$$S_B(t) = k \int A(\boldsymbol{x}) \exp(-t/T_2(\boldsymbol{x})) \sin[(\omega - \omega_0) t - \beta] \, d\boldsymbol{x}.$$

If we view the two output signals provided by the receiver as one complex output signal $S(t) = S_A(t) - iS_B(t)$ we have

$$S(t) = k \int A(\boldsymbol{x}) \exp(-t/T_2(\boldsymbol{x})) \cdot \exp[+i(\omega_0 - \omega + \omega_h(\boldsymbol{x})) t + i(\theta(\boldsymbol{x}) + \phi)] \, d\boldsymbol{x}.$$

We can relate $S(t)$ to the magnetization by going back to (3.2), which gives the general form of the (complex) transverse magnetization under read gradients. This gives us

$$S(t) = K \int M(t, \boldsymbol{x}) \exp(-i\omega t) \, d\boldsymbol{x} \tag{4.6}$$

where $K = ke^{i\phi}$ is a complex "arbitrary constant," and

$$M(t, \boldsymbol{x}) = M_x(t, \boldsymbol{x}) + iM_y(t, \boldsymbol{x}).$$

E. The Analog-to-Digital Converters

Since the usual computer is digital and not analog, it is necessary to convert the complex (two-channel) signal to two strings of digital numbers $\{S_n\}$ given by

$$S_n = K \int M(t, \boldsymbol{x}) \exp(-i\omega n\Delta t) \, d\boldsymbol{x} \tag{4.7}$$

where the sampling interval is Δt. With this equation, we connect the sampled output of the receiver S_n with the nuclear magnetization of the sample $M(t, \boldsymbol{x})$.

This sampling of the output signal is the chief obstacle to easy transfer between the time domain and the spatial domain. The Fourier integral is no longer available, and the errors of its discrete approximations are well known.

V. The NMR "Imaging Equation"

This section introduces some simple "pulse sequences" and uses the equations introduced earlier to obtain the "imaging equation" for a given pulse sequence; i.e., the expression for the NMR signal in terms of the properties of the sample. The imaging equation is a representation of what the reconstruction algorithm has as an input. The few pulse sequences we will consider in this section are simplified, but show the principal ways of obtaining a set of signals which contain enough information to permit reconstruction of an image. The actual reconstruction will be left as an exercise for the reader. (But should the reader wish to check his results, there are many interesting papers in the literature. See, for example, Shepp [17], Louis [14], Grunbaum [7], Marr *et al.* [15], and a long report by Cho *et al.* [4].)

A. 3D Backprojection

The simplest method in terms of collecting data, but perhaps the most complicated in terms of reconstructing the image, is three-dimensional (3D) backprojection. In this method, a single $\pi/2$ RF pulse is applied. After each such pulse, the gradient is turned on and the signal is recorded. This "read sequence" is repeated after the spins have been allowed to recover. But for each new read sequence, the *direction* of the gradient is changed, although the gradient strength is kept the same. This sequence in shown in Fig. 7. After $\boldsymbol{G}$ has pointed throughout one hemisphere, enough data exist to reconstruct the 3D image [14]. The NMR signal as a function of $\boldsymbol{G}$ can be obtained by using the tools of Section III.

Experimental Deviations from Ideality: The pulse sequence

discussed here is considerably simpler than would be practical. In particular, the read gradient cannot be turned on or off instantaneously. Any of several tricks can be used to overcome this particular problem. For example, the spin-echo method (see, e.g., [5]) can be incorporated into the pulse sequence in order to move the signal away from the gradient transient. Alternatively, the gradient can remain on and constant, even during the RF pulse, but this requires that the RF pulse be strong. Finally, the data sampling times can be spaced to compensate for the changing gradient strength.

In order to keep the discussion to a reasonable length, the experimental problems will be ignored in the remainder of this section.

Derivation of the Imaging Equation for 3D Backprojection: If we take the magnetization just before the RF pulse to be $M_0(\boldsymbol{x})\,\boldsymbol{k}$, then, from (3.8), the magnetization just after the pulse in the rotating frame is $-M_0(\boldsymbol{x})\,\boldsymbol{j}'$. From (3.3), if we take $t = 0$ to be at the end of the RF pulse, we have that the magnetization in the lab frame at $t = 0$ is

$$M(0, \boldsymbol{x}) = -iM_0(\boldsymbol{x}).$$

Whether the read gradient is turned on at $t = 0$ or at some time later causes the magnetization $M(t, \boldsymbol{x})$ to differ only in the phase angle, which, we have already agreed, is arbitrary. We assume the gradients are turned on and stable at $t = 0$.

The evolution of the magnetization in the presence of the read gradient is given by (3.2), and the signal obtained from that magnetization is given by (4.6). If we combine these equations we obtain

$$S(t) = -iK \int M_0(\boldsymbol{x}) \exp\left[(i\omega'(\boldsymbol{x}) - i\omega - 1/T_2(\boldsymbol{x}))\, t\right] d\boldsymbol{x}. \tag{5.1}$$

Using (3.1)

$$\omega'(\boldsymbol{x}) = -\gamma \boldsymbol{G} \cdot \boldsymbol{x} + \omega_0$$

and incorporating $-i$ into K, we have

$$S(t) = K \int M_0(\boldsymbol{x}) \exp\left[-(i\gamma \boldsymbol{G} \cdot \boldsymbol{x} + 1/T_2(\boldsymbol{x}))\, t\right] d\boldsymbol{x}. \tag{5.2}$$

We have assumed a detector operating at resonance, so $\omega = \omega_0$.

This is the imaging equation for the 3D backprojection method. The equation gives the signal $S(t)$ for a given gradient $\boldsymbol{G}$. The constant K is independent of $\boldsymbol{G}$.

If T_2 is long compared to the time interval $(0, \tau)$ over which we collect data we can neglect the T_2 term. The imaging equation then takes the form

$$S(t) = K \int M_0(\boldsymbol{x}) \exp\left[-i\gamma t \boldsymbol{G} \cdot \boldsymbol{x}\right] d\boldsymbol{x}. \tag{5.3}$$

We recognize the right-hand side of (5.3) as the Fourier transform of $M_0(\boldsymbol{x})$, sampled at the point $\gamma t \boldsymbol{G}$. We are completely free to vary the vector $\boldsymbol{G}$ in any way that promises to give us sufficient data to approximate the inverse Fourier transform and, hence, $M_0(\boldsymbol{x})$.

One choice for a collection of pulse sequences is that of Shepp [17], who takes

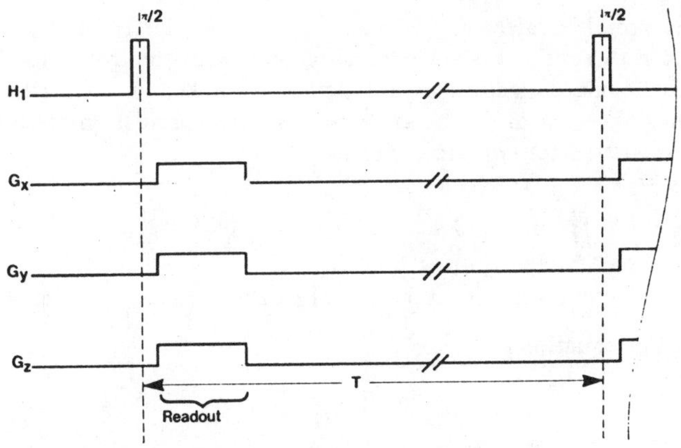

Fig. 7. Simplified timing sequence for the saturation-recovery method, using three-dimensional (3D) backprojection for image reconstruction. The sequence, consisting of an RF pulse followed by a gradient pulse, is repeated at intervals T. For each repetition, the direction of the gradient is changed. Data are collected during the time the gradient is applied.

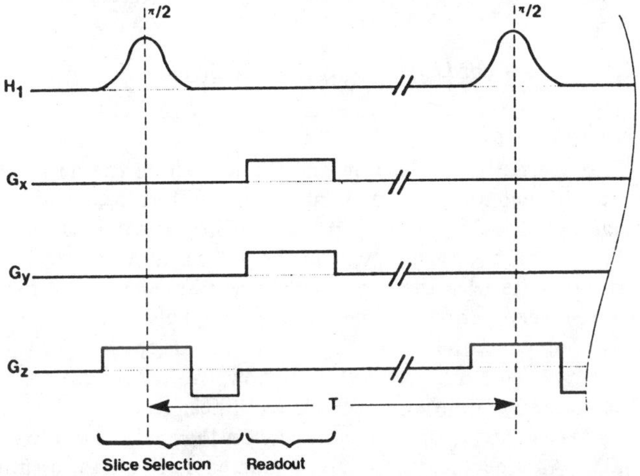

Fig. 8. Sequence for the saturation-recovery method using 2D backprojection for reconstruction. The sequence is the same as that in Fig. 7 with the short RF pulse replaced by the slice-selection RF and gradient pulses.

$$\begin{aligned}
G_{x(jk)}(t) &= G \cos(\theta_j) \cos(\phi_k), && 0 \leqslant t \leqslant \tau \\
G_{y(jk)}(t) &= G \cos(\theta_j) \sin(\phi_k), && 0 \leqslant t \leqslant \tau \\
G_{z(jk)}(t) &= G \sin(\theta_j), && 0 \leqslant t \leqslant \tau. \\
\theta_j &= (j - \tfrac{1}{2})\, \pi/n, && j = 1, \cdots, n \\
\phi_k &= k 2\pi/m, && k = 0, \cdots, m-1
\end{aligned}$$

and G is a fixed magnitude.

B. 2D Backprojection

If we modify the previous pulse sequence by using a *selective* $\pi/2$ RF pulse, we have a method for obtaining tomographic two-dimensional (2D) images. The pulse sequence is shown in Fig. 8. The data are collected as with 3D backprojection except that the direction of the gradient $\boldsymbol{G}$ stays in the plane of the selected slice.

As shown in Section III, the selective irradiation method, in a simple world, tips the magnetization within the slice into the x-y plane, leaving the magnetization in the remainder of the sample unaffected. For example, if the slice is normal to the z direction, we are concerned only with the distribution of the magnetization as a function of x and y. Thus the imaging equation is the same as in 3D backprojection except it is written for two dimensions:

$$S(t) = K \iint M_0(x,y) \cdot \exp\left[-(i\gamma(G_x x + G_y y)t\right] \,dx\,dy. \tag{5.4}$$

In this equation

$$M_0(x,y) = \int R(z)\, M_0(\boldsymbol{x})\, dz$$

where $R(z)$ is the response function resulting from the slice-selection pulse, and depicts the magnetization as a function of z. Essentially, $R(z)$ describes the shape of the slice carved out by the pulse.

Usually one samples data on a polar grid, using $\{P_k(t)\}$, $k = 0, \cdots, m-1$, with $H_1(t)$ and $G_z(t)$ independent of k, but taking

$$G_{x(k)}(t) = G\cos(\phi_k) \qquad G_{y(k)}(t) = G\sin(\phi_k), \qquad 0 \leqslant t \leqslant \tau.$$

C. 3D Fourier

This method of collecting and reconstructing the data uses an additional tool, the gradient pulse. An idealized pulse sequence $\boldsymbol{P}(t)$ is diagrammed in Fig. 9. There is a $\pi/2$ RF pulse, followed by a gradient pulse in the y-z plane, followed by a readout pulse with the x-gradient turned on. And then there is a recovery period before the next RF pulse. In the next pulse sequence, the y- and z-gradients are stepped slightly by the proper amount. We will show how this procedure can be used to sample Fourier space on a rectilinear grid.

As for the details, we draw heavily on the analysis leading to (5.2). Assume that the y- and z-gradients are on for a short time τ. Right after the pulse is turned off

$$M(\tau, \boldsymbol{x}) = -iM_0(\boldsymbol{x}) \exp\left[i(\omega_0 - \gamma(G_y y + G_z z)\,\tau)\right].$$

With the readout x-gradient turned on

$$\begin{aligned} M(t,\boldsymbol{x}) &= M(\tau,\boldsymbol{x}) \exp\left[i(\omega_0 - \gamma G_x x)(t-\tau) - t/T_2(\boldsymbol{x})\right]. \\ &= -iM_0(\boldsymbol{x}) \exp\left[i(\omega_0 - \gamma(G_y y + G_z z)\,\tau) \right. \\ &\quad \left. - \gamma G_x x(t-\tau) - t/T_2(\boldsymbol{x})\right]. \end{aligned}$$

Therefore,

$$S(t) = K \int M_0(\boldsymbol{x}) \exp\left[-t/T_2(\boldsymbol{x})\right) \cdot \exp\left(-i\gamma(G_y\tau y + G_z\tau z + G_x(t-\tau)\,x)\right)]\, d\boldsymbol{x}.$$

If we ignore the T_2 term, $S(t)$ is just the Fourier transform of $M_0(\boldsymbol{x})$ sampled at the point $(\gamma G_x(t-\tau), \gamma G_y \tau, \gamma G_z \tau)$. To reconstruct $M_0(\boldsymbol{x})$, we might use a double-indexed collection of pulse sequences of the form $\{\boldsymbol{P}_{jk}(t)\}$, where $-n \leqslant j \leqslant n$, $-n \leqslant k \leqslant n$. Each of these sequences would have the same $H_1(t)$ and $G_x(t)$, but

$$G_{y(jk)}(t) = j\Delta, \qquad 0 \leqslant t \leqslant \tau$$
$$G_{z(jk)}(t) = k\Delta, \qquad 0 \leqslant t \leqslant \tau.$$

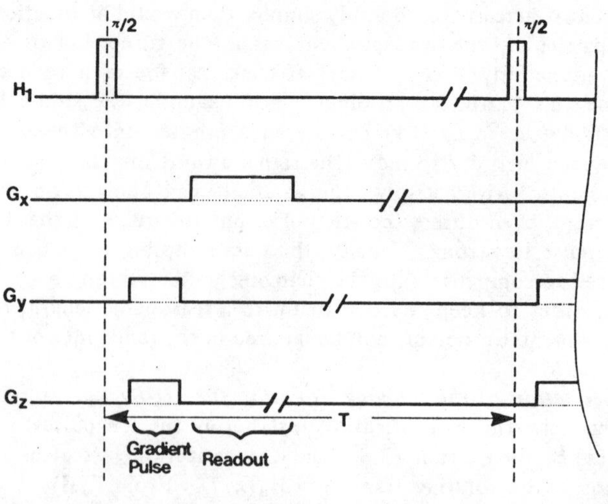

Fig. 9. Sequence for the saturation-recovery method using 3D Fourier transformation for image reconstruction. The direction and amplitude of the gradient pulse during the interval τ is different for each repetition. The data are collected during the time the x-gradient is applied.

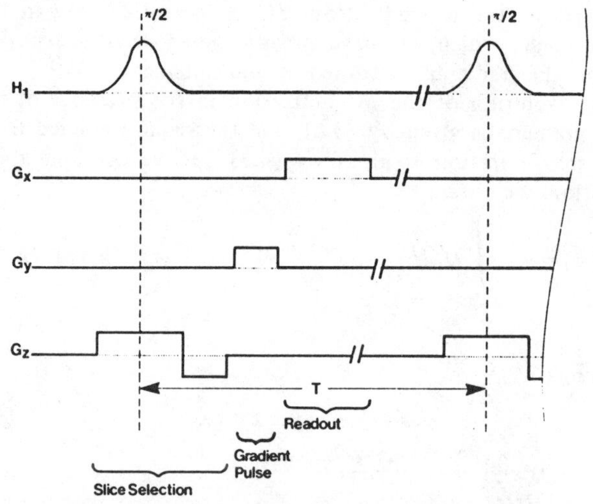

Fig. 10. Sequence for 2D imaging using Fourier transformation for image reconstruction. The amplitude of the y-gradient pulse is different for each repetition. The data are collected during the time the x-gradient is applied.

D. 2D Fourier

There are no surprises. The pulse sequence (Fig. 10) has a selective excitation pulse to delimit a z-slice, a y-gradient pulse, and an x-gradient readout pulse. A collection of pulse sequences $\{\boldsymbol{P}_j(t)\}$, $-n \leqslant j \leqslant n$, is used. For the jth sequence

$$G_{y(j)}(t) = j\Delta, \qquad 0 \leqslant t \leqslant \tau.$$

Again, $S_j(t)$ is the 2D Fourier transform of the slice, sampled at $(\gamma G_x(t-\tau), \gamma j\Delta\tau)$.

E. Saturation Recovery

Saturation recovery is not a method of collecting data and forming an image, as are the four methods discussed earlier. Saturation recovery and, in fact, inversion recovery are methods of controlling the contribution of T_1 to the images and can be used with all of the four imaging methods.

The $\pi/2$ RF pulse in the read sequence sets the longitudinal magnetization to zero, from which it recovers toward M_0 according to (2.5) with M_z^0 taken as zero.

In order to improve the data-collection efficiency, the second read sequence is usually applied without waiting for complete recovery of the magnetization to its equilibrium value. In fact, optimum efficiency occurs when the delay between repetitions T is roughly equal to T_1.

The saturation recovery method involves simply repeating the read sequence at regular intervals T. The earlier imaging equations are unchanged except for the replacement of $M_0(\boldsymbol{x})$ with the magnetization $M^0(\boldsymbol{x})$ occurring just before the read sequence

$$M^0(\boldsymbol{x}) = M_0(\boldsymbol{x})\,(1 - \exp[-T/T_1(\boldsymbol{x})]).$$

F. Inversion Recovery

This method is the same as saturation recovery except for the addition of a π, or inversion, RF pulse a time T_I before each read sequence. This has been found to increase the dependence of the image upon T_1.

At time T after the previous $\pi/2$ RF pulse, the sign of M_z is changed by the π RF pulse. Just after this pulse we have

$$M(T, \boldsymbol{x}) = -M_0(\boldsymbol{x})\,(1 - \exp[-T/T_1(\boldsymbol{x})]).$$

Thus just before the next read sequence, the longitudinal magnetization is not $M_0(\boldsymbol{x})$, but is given by (2.5) with M_z replaced by $M(T, \boldsymbol{x})$ and t replaced by T_I, giving

$$M^0(\boldsymbol{x}) = M_0(\boldsymbol{x})\,[\exp(-T_I/T_1)\,(-2 + \exp(-T/T_1)) + 1].$$

In practice, T is often made much larger than T_1 so that $M^0(\boldsymbol{x})$ becomes approximately

$$M^0(\boldsymbol{x}) = M_0(\boldsymbol{x})\,(1 - 2\exp(-T_I/T_1))$$

thus increasing the dependence of the image upon T_1 over that obtained in saturation recovery.

Acknowledgment

The authors are significantly indebted to many fellow workers. D. Kramer and H. Yeung provided the representative images and are responsible for their quality. J. Keller, R. Compton, E. Bomke, and others designed and built the imaging system. The referees, the editor, and other obliging readers of early versions of this paper contributed helpful comments.

References

[1] A. Abragam, *The Principles of Nuclear Magnetization.* Oxford, England: Oxford Univ. Press, 1961.

[2] F. Bloch, "Nuclear induction," *Phys. Rev.*, vol. 70, pp. 460–474, 1946.

[3] Bowman Gray School of Medicine, *NMR Imaging* (Proc. Int. Symp. on Nuclear Magnetic Resonance Imaging; Winston-Salem, NC, 1981).

[4] Z. H. Cho *et al.*, "Development of methods and algorithms for Fourier transform nuclear magnetic (NMR) imaging," Imaging System Science Lab., Dep. Elec. Sci., Korea Advanced Institute of Science, Seoul, Korea.

[5] C. T. Farrar and E. D. Becker, *Pulse and Fourier Transform NMR.* New York: Academic Press, 1971.

[6] J. F. Greenleaf, "Computerized tomography with ultrasound," this issue, pp. 330–337.

[7] F. A. Grunbaum, "Reconstruction with arbitrary directions: dimensions two and three," in *Mathematical Aspects of Computerized Tomography*, G. T. Herman and F. Natterer, Eds. Berlin, Germany: Springer, 1981.

[8] W. S. Hinshaw, P. A. Bottomley, and G. N. Holland, "Radiographic thin-section image of the human wrist by nuclear magnetic resonance," *Nature*, vol. 270, pp. 722–723, 1977.

[9] D. I. Hoult, "The solution of the Bloch equations in the presence of a varying B_1 field: an approach to selective pulse analysis," *J. Mag. Res.*, vol. 35, pp. 69–86, 1979.

[10] D. I. Hoult and R. E. Richards, "The signal-to-noise ratio of the nuclear magnetic resonance experiment," *J. Mag. Res.*, vol. 24, pp. 71–85, 1976.

[11] D. M. Kramer, "Imaging of elements other than hydrogen," in *Nuclear Magnetic Resonance Imaging in Medicine*, L. Kaufman, L. E. Crooks, and A. R. Margulis, Eds. Tokyo, Japan: Igaku-Shoin Med. Publ., 1981.

[12] P. C. Lauterbur, "Image formation by induced local interactions: Examples employing nuclear magnetic resonance," *Nature*, vol. 242, pp. 190–191, 1973.

[13] P. R. Locher, "Computer simulation of selective excitation in n.m.r. imaging," *Phil. Trans. Roy. Soc. London*, vol. 289, pp. 537–542, 1980.

[14] A. K. Louis, "Optimal sampling in nuclear magnetic resonance (NMR) tomography," *J. Comput. Assist. Tomog.*, vol. 6, pp. 334–340, 1982.

[15] R. B. Marr, C. Chen, and P. C. Lauterbur, "On two approaches to 3D reconstruction in NMR zeugmatography," in *Math. Aspects of Computerized Tomography*, G. T. Herman and F. Natterer, Eds. Berlin, Germany: Springer, 1981.

[16] I. L. Pykett, "NMR imaging in medicine," *Scient. Amer.*, May 1982.

[17] L. A. Shepp, "Computerized tomography and nuclear magnetic resonance," *J. Comput. Assist. Tomog.*, vol. 4, pp. 94–107, 1980.

[18] J. R. Singer, "Blood flow measurements by NMR," in *Nuclear Magnetic Resonance Imaging in Medicine*, L. Kaufman, L. E. Crooks, and A. R. Margulis, Eds. Tokyo, Japan: Igaku-Shoin Med. Publ., 1981.

[19] D. Slepian, H. O. Pollak, and H. T. Landau, "Prolate spheroidal wave functions, Fourier analysis, and the uncertainty principle, I and II," *Bell Syst. Tech. J.*, vol. 40, pp. 43–84, 1961.

[20] E. H. Wood, J. H. Kinsey, R. A. Robb, B. K. Gilbert, L. D. Harris, and E. L. Ritman, "Applications of high temporal resolution computerized tomography to physiology and medicine," in *Image Reconstruction from Projections: Implementations and Applications*, G. T. Herman, Ed. Berlin, Germany: Springer, 1979.

[21] P. Mansfield and P. G. Morris, *NMR Imaging in Biomedicine.* New York: Academic Press, 1982.

[22] Z. H. Cho, H. S. Kim, H. B. Song, and J. Cumming, "Fourier transform nuclear magnetic resonance tomographic imaging," *Proc. IEEE*, vol. 70, pp. 1152–1173, 1982.

Part VI
Image Processing

IN the past two decades, there have been considerable advances in the field of image processing. This is in part due to significant advances in hardware technology, which allow sophisticated image processing algorithms to be implemented in real time, and in part due to a large number of applications of image processing in such diverse areas as medicine, communications, consumer electronics, defense, law enforcement, robotics, geophysics, and agriculture. Image processing can be classified broadly into four areas, namely image restoration, enhancement, coding, and understanding. In this section, we have included six papers in the areas of image restoration and enhancement.

In image restoration, an image has been degraded in some manner and the objective is to reduce or eliminate the effect of the degradation. Typical degradations that occur in practice include image blurring, additive random noise, quantization noise, multiplicative noise, and geometric distortion. Two important problems in image restoration are determining the degradation characteristics in the degraded image and developing restoration algorithms. The paper by Anderson and Netravali considers the problem of reducing additive random noise in images. This is one of the first papers in which an image restoration algorithm is based on adapting to the local characteristics of the original undegraded image. Many papers published more recently exploit the notion that image restoration algorithms that adapt to the local image characteristics generally perform considerably better than algorithms that do not. The paper by Kuan, Sawchuck, Strand, and Chavel considers the problem of reducing signal-dependent noise such as multiplicative noise, film grain noise, and Poisson noise. The general approach is to adapt to the local image characteristics. The paper by Cannon considers the problem of determining the degradation characteristics from the degraded image when an image is blurred by uniform linear camera motion and an out-of-focus lens system. This paper also develops methods for deblurring, once the specific characteristics of blurring have been determined. The paper by Trussell and Civanlar develops a signal restoration approach which obtains a solution that satisfies all *a priori* constraints imposed by the real world system. The approach is applied to the problem of restoring a signal degraded by blurring and additive random noise.

Image enhancement is the processing of images to increase their usefulness. Methods and objectives vary with the application. When images are enhanced for human viewers, as in television, the objective may be to improve perceptual aspects: image quality, intelligibility, or visual appearance. In applications such as object identification by machine, an image may be preprocessed to aid machine performance. Because the objective of image enhancement is heavily dependent on the application context, and the criteria for enhancement are often subjective or too complex to be easily converted to useful objective measures, image enhancement algorithms tend to be simple, qualitative, and ad hoc. Image enhancement is closely related to image restoration. When an image is degraded, restoration and enhancement. In image restoration, an ideal There are, however, some important differences between restoration and enhancement. An image restoration, an ideal image has been degraded and the objective is to make the processed image resemble the original as much as possible. In image enhancement, the objective is to make the processed image better than the unprocessed image, in some sense. To illustrate this difference, note that the original, undegraded image cannot be further restored, but can be enhanced by increasing sharpness. The paper by Stockham proposes a model that accounts for the peripheral level of the human visual system. The model is used in processing images for various applications such as dynamic range reduction and contrast enhancement of images. The paper by Peli and Lim develops an algorithm that modifies the local contrast by adapting to the local characteristics of an image. The algorithm is applied in various situations including enhancing aerial photographs degraded by cloud cover.

In image coding, the objective is to represent an image with as few bits as possible, preserving a certain level of image quality and intelligibility acceptable for a given application. Image coding can be used in reducing the bandwidth of a communication channel when an image is transmitted, and in reducing the amount of required storage when an image needs to be retrieved at some future time. Image coding is related to image restoration and enhancement. If we can reduce the degradation such as quantization noise, that results due to an image coding algorithm, or enhance the visual appearance of the reconstructed image for example, we can reduce the number of bits required in representing an image at a given level of image quality and intelligibility.

In image understanding, the objective is to symbolically represent the contents of an image. Applications of image understanding include computer vision, robotics, and target identification. Image understanding differs from the other three areas discussed above in one major respect. In image restoration, enhancement, and coding, both the input and output are images; signal processing has been the backbone of many successful systems in these areas. In image understanding, the input is an image, but the output is typically some symbolic representation of the contents of the input image; successful development of a system in this area generally requires both signal processing and artificial intelligence concepts. In a typical image understanding system, signal processing is used to perform lower level processing, such as reduction of degradation and extraction of image features, such as edges; and artificial intelligence is used to perform higher level processing, such as symbol manipulation and knowledge-base management.

In the past two decades, literally hundreds of papers have been published on topics related to image processing. The six papers included in this section were chosen primarily for their signal processing point of view and represent only a small segment of the available literature on image processing. The readers who wish to have an overview of the overall image processing field are referred to [1]–[10].

Bibliography

[1] W. K. Pratt, *Digital Image Processing*. New York, NY: John Wiley and Sons, 1978.

[2] H. C. Andrews and B. R. Hunt, *Digital Image Restoration*. Englewood Cliffs, NJ: Prentice Hall, 1977.

[3] A. Rosenfeld and A. C. Kak, *Digital Picture Processing*. New York: NY: Academic Press, 1976.

[4] T. S. Huang, W. H. Schreiber, O. J. Tretiak, "Image processing," *Proc. IEEE.*, pp. 1586–1609, Nov. 1971.

[5] H. C. Andrews, "Monochrome digital image enhancement," *Appl. Opt.*, vol. 15, pp. 495–503, Feb. 1976.

[6] A. N. Netravali and J. O. Limb, "Picture coding: A review," *Proc. IEEE*, vol. 69, Mar. 1980.

[7] H. C. Barrow and J. Tenenbaum, "Computational vision," *Proc. IEEE*, vol. 69, pp. 572–595, May 1981.

[8] A. Rosenfeld, "Image pattern recognition," *Proc. IEEE*, vol. 69, pp. 596–605, May 1981.

[9] H. J. Scudder, "Introduction to computer aided tomography," *Proc. IEEE*, vol. 66, pp. 628–639, June 1978.

[10] W. H. Hinshaw and A. H. Lent, "An introduction to NMR imaging: From the Bloch equation to the imaging equation," *Proc. IEEE*, vol. 71, pp. 338–350, Mar. 1983 (reprinted in Part V).

Image Restoration Based on a Subjective Criterion

G. LEIGH ANDERSON, STUDENT MEMBER, IEEE, AND ARUN N. NETRAVALI

***Abstract*—The problem of removing random noise from gray tone images without significantly sacrificing the subjective resolution is considered. Based on a subjective visibility function, which gives the relationship between the visibility of a unit noise and a measure of local spatial detail (spatial masking), two procedures are developed to adapt continuously the finite impulse response of a two-dimensional, noncausal, linear digital filter. At sharp transitions in the image intensity, the filter operator is strongly peaked to preserve the resolution, whereas in flat areas it is flat to effectively average out the random noise. The first procedure (*S*-filter) is computationally more efficient, but does not perform as well as the second method (*SD*-filter) which requires solution of a new optimization problem at every picture element. Results of several simulations are presented to demonstrate the feasibility of our approach. Extensions are pointed out to incorporate different adaptation procedures and psychovisual criteria other than the type of spatial masking used here.**

Manuscript received April 12, 1976; revised July 21, 1976. This work was supported in part by the National Science Foundation under Grant ENG 74-17955 and in part by the AFOSR under Grant 75-2777.

G. L. Anderson was with Bell Laboratories, Holmdel, NJ. He is now with the Department of Mathematical Sciences, Rice University, Houston, TX 77001.

A. N. Netravali is with Bell Laboratories, Holmdel, NJ 07733.

I. Introduction

THE SUBJECT of digital image restoration has been receiving considerable attention since the advent of digital computers. Availability and use of digital computers allows the flexibility that is not afforded by most other techniques. Very simply, image restoration refers to recovery of an image which is subjected to some degrading phenomena. Thus if $s(x,y)$ is the "true" image intensity at spatial position (x,y) and if the observation $z(x,y)$ is related to $s(x,y)$ by

$$z(x,y) = F[s(x,y)] + n(x,y), \qquad a \le x, \quad y \le b \quad (1)$$

where $F[\cdot]$ is a mapping (not necessarily memoryless) and $n(x,y)$ is the corrupting measurement noise, then the image restoration problem is to recover $s(x,y)$ from the knowledge of $z(x,y)$ over the square $[a,b] \times [a,b]$.

The solution of the above problem obviously depends upon the type of models and *a priori* information assumed for the signal $s(x,y)$, noise $n(x,y)$, and the mapping $F[\cdot]$. We mention a few widely used techniques and refer the

Reprinted from *IEEE Trans. Syst., Man, Cybernetics,* vol. SMC-6, pp. 845–853, Dec. 1976.

reader to survey articles [1]–[4] for a more complete treatment of different approaches. Most of the techniques assume $F[\cdot]$ to be a spatial linear filter with response $h(\cdot,\cdot,\cdot,\cdot)$ and thus represent (1) as

$$z(x,y) = \iint h(x,y,\xi,\eta)s(\xi,\eta)\, d\xi\, d\eta + n(x,y) \qquad (2)$$

This equation can model several situations as pointed out in [1]–[4]. Assuming h to be space invariant, the technique of inverse filtering [5], [6] filters z by a filter whose Fourier transform is the inverse of the transform of h. This procedure obviously preserves the resolution of the image without regard to the removal of noise. Optimum filtering [7] on the other hand assumes statistical models for $s(x,y)$ and $n(x,y)$ and then obtains the restored image $\hat{s}(x,y)$ as the one which minimizes $E\{(s - \hat{s})^2\}$, where E denotes the expectation operator. Thus it gives no attention to the resolution of the restored image. Again depending on the assumptions made on the statistics of s, n, and h, different restoration schemes [8], [9] result.

In the case of restoration for human viewing, it is necessary to know the fidelity criterion of the human visual system so that the information most discernible to the human eye can be presented and the most discernible noise can be removed. While our understanding of the signal processing capabilities and limitations of the human visual system has increased [10], [11], its complexity has made optimum restoration under a complete visual criterion impractical, if not impossible. This makes interactive methods [12], in which iterative restoration is attempted such that every iteration depends upon a few parameters which in turn depend upon the human inspection of the image and the parameters of the previous iteration, very attractive. The concept of introducing a human being in the "loop," although cumbersome in some cases, may, if the iterative process converged, give an image resulting in a satisfactory tradeoff between the resolution and the noise.

Our approach in this paper is first to construct a measure of spatial detail. This we borrow from some of our earlier work on adaptive quantization of picture signals [13]. We then determine a "visibility function" which gives the relative visibility of a unit noise added to all points in the picture where the measure of spatial detail has a certain value. This visibility function, at least for a restricted class of images, represents a subjective tradeoff between the resolution and noise. Using this visibility function it is possible to construct a performance index which reflects a tradeoff between the noise removed and the blur introduced by the filter. Minimization of this performance index is done analogous to a method developed by Backus and Gilbert [14]. Although several different types of performance indexes are possible, we only consider two, one in which computation of the filter impulse response can be done off-line and stored, and another in which filter response must be computed at every picture element by solving a new optimization problem. The second method, although giving a better restoration, requires extensive computations.

The next section contains a brief description of the subjective tests and results relating noise visibility to our measure of spatial detail. Section III contains the formulation of the restoration problem and its solution. Discussion of our simulations on real pictures is contained in Section IV. Section V contains a critique of our approach and points out some limitations and extensions.

II. Masking and Visibility Functions

In this section we define our measure of local spatial detail in a picture. Although we have been guided by some work in psychophysics [15], [16] done under restrictive situations, our measure is certainly nonoptimum and somewhat arbitrary. However, such psychophysical experiments [17] and measures [13] have been used successfully for coding applications and do indeed produce remarkable restoration as we shall see later. It is well known that at sharp transitions in image intensity the contrast sensitivity of the human visual system decreases with the sharpness of the transition and increases somewhat exponentially (within limits) as a function of spatial distance from the transition. Using this, we define the "masking function" M_{ij} at coordinate i,j as a measure of spatial detail to be

$$M_{ij} = \sum_{p=i-k}^{i+k} \sum_{q=j-l}^{j+l} C^{\|(i,j)-(p,q)\|}[|m_{pq}{}^H| + |m_{pq}{}^V|] \qquad (3)$$

where $\|(i,j) - (p,q)\|$ denotes the Euclidean distance between positions (i,j) and (p,q); $m_{pq}{}^V$ and $m_{pq}{}^H$ are the vertical and horizontal slopes of the image intensity at (p,q), respectively; C is a constant controlling the rate of exponential decay of the effect of a transition of image intensity on its neighbors; and k,l are constants controlling the size of the relevant neighborhood around (i,j). Following the work of Harms and Aulhorn [15] we take C to be 0.35. It is clear that M_{ij} increases monotonically with the amount of spatial detail in a two-dimensional neighborhood surrounding a picture element (pixel). A pair of pictures which shows the original picture and its corresponding masking function is shown in Figs. 1 and 2. This was obtained by taking $k,l = 1$. Notice that, as expected, the masking function has high values not only at sharp transitions, but also at picture elements around it.

We now describe briefly the subjective tests that were performed to obtain the visibility function, i.e., the relationship between the visibility of noise and the masking function. Details of the testing procedure and conditions are given in [13]. The experiment was performed on a specially configured [18] PDP 11/45 computer with a high-speed drum capable of data transfers at 32 Mbits/s. The analog video signal with a bandwidth of 1 MHz was digitized by sampling at Nyquist rate with quantization of 8 bits/picture element. There were 256 picture elements in a line and 256 lines in a frame.

The test images were precomputed and stored on the drum. First, the masking function M_{ij} was computed for each (i,j)th pixel of the original image. Next, a certain

Fig. 1. Original Checker girl (256 × 256 array with signal having bandwidth of 1 MHz; each sample is linearly quantized with 8 bits).

Fig. 2. Masking function derived from noiseless image. 3 × 3 neighborhood of slopes is used.

masking value T and noise power V_n was chosen. Then to each pixel of the original image whose masking value M_{ij} is in the range $[T - \Delta T/2, T + \Delta T/2]$ white Gaussian noise of power V_n was added. Thus if T is large, noise is added only to pixels near edges; and if T is small, noise is added only to flat areas, etc. The test image so obtained was characterized by the values T and V_n. For each value of T at which we wish to measure the visibility $f(T)$, we generated three test images with noise powers V_{n1}, V_{n2}, V_{n3}.

During the subjective test, under program control, the experimenter randomly selected a test image, and the subject compared it with the original image to which he added just enough white noise everywhere in the picture until he reached a point of subjective equivalence between the two pictures. If V_{w1}, V_{w2}, and V_{w3} are the equivalent white noise powers selected by the subject corresponding

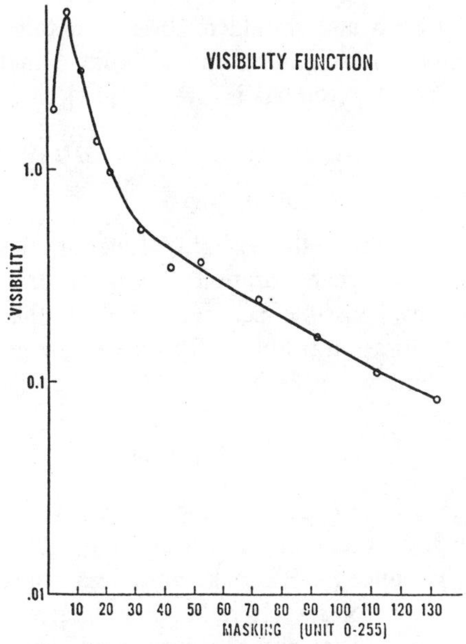

Fig. 3. Visibility function for checker girl with two-dimensional one-neighbor masking function.

to test images which have V_{n1}, V_{n2}, V_{n3} amounts of noise in all picture elements having the masking value in the range[1]

$$\left[T - \frac{\Delta T}{2}, T + \frac{\Delta T}{2}\right],$$

then assuming proportionality, the equivalent white noise (which is a function of T) for a unit noise is given by

$$f(T) = \frac{1}{\Delta T} \cdot \frac{V_{w1} + V_{w2} + V_{w3}}{V_{n1} + V_{n2} + V_{n3}}. \quad (4)$$

This, then, is the subjective visibility (in terms of the equivalent white noise) of a unit of noise added to all picture elements having masking equal to T or in a small neighborhood of T. The functional relationship between f and T was found for several values of T. Using the masking function of Fig. 2 the visibility function is plotted in Fig. 3. It is easy to see that in most regions of interest it can be approximated by a function of the form $ke^{-\lambda T}$ for suitable values of λ and k.

The visibility function decreases with respect to its argument because 1) at higher masking function values we are adding noise to a fewer picture elements, and 2) the perception of noise at picture points having higher masking value is decreased. Thus they combine both the statistical and perceptual effects together. However, this combination is indeed their drawback since different pictures have different statistics and since the visibility functions will vary from picture to picture. However, as we shall show later by using the visibility function of one picture on another, our results seem to be applicable to at least a class of pictures,

[1] This assumes that the visibility function is a constant in the interval

$$\left[T - \frac{\Delta T}{2}, T + \frac{\Delta T}{2}\right].$$

namely, the head and shoulder views. A detailed critique, the mechanics of obtaining the visibility functions, and their use in coding contexts is given in [13].

III. Formulation of Approach

A. The Restoration Problem

We wish to apply the visibility function developed in Section II to image restoration. To study its application in the clearest possible setting, we pose the following simplified restoration problem. Restore an image corrupted by independent white Gaussian additive noise. Explicitly, we observe

$$z(x,y) = s(x,y) + n(x,y) \tag{5}$$

where $s(x,y)$ is the original image and $n(x,y)$ is zero-mean white Gaussian noise independent of $s(x,y)$ and of unknown (constant) variance v_n. We make no assumptions about the statistical nature of $s(x,y)$; rather it is treated as a deterministic signal. From the observed signal $z(x,y)$ we will construct an estimate $\hat{s}(x,y)$ of the original image.

B. Design of the S-Type Filter

In order to simplify the exposition, we will develop only a one-dimensional version of the filter. Extension to two dimensions will be obvious. Thus we have

$$\boldsymbol{z} = \boldsymbol{s} + \boldsymbol{n}, \tag{6}$$

where $\boldsymbol{z}$, $\boldsymbol{s}$, and $\boldsymbol{n}$ are p-length column vectors. For the kth pixel of $\boldsymbol{s} = \text{col}\,(s_1,s_2,\cdots,s_p)$ we will obtain an estimate of $\hat{s}_k$ by computing a local average of the $(2q + 1)$ neighboring elements $(2q + 1 < p)$:

$$\hat{s}_k = \sum_{i=-q}^{q} a_i^{(k)} z_{k+i}. \tag{7}$$

This type of filter was chosen because the averaging weights $a_i^{(k)}$ may be changed for each $\hat{s}_k$ according to the visibility of noise at that pixel. For convenience of notation we drop the subscript/superscript k and consider contributions only from the elements of $\boldsymbol{z}$ for which the averaging coefficients are nonzero. Equation (7) can be written as

$$\hat{s} = \boldsymbol{a}^T\hat{\boldsymbol{z}} \tag{8}$$

where $\hat{s}$ is the estimate of s,

$$\hat{\boldsymbol{z}} = \text{col}\,(z_{k-q},\cdots,z_k,\cdots,z_{k+q}), \tag{9}$$

$$\boldsymbol{a} = \text{col}\,(a_{-q}^{(k)},\cdots,a_0^{(k)},\cdots,a_q^{(k)}), \tag{10}$$

and superscript T denotes the transpose. We require that the vector of filter weights satisfy the usual properties of an average, i.e.,

$$a_i \geq 0, \qquad i = -q,\cdots,q$$

and

$$\boldsymbol{a}^T\boldsymbol{u} = 1 \tag{11}$$

where

$$\boldsymbol{u} = \text{col}\,(1,\cdots,1). \tag{12}$$

The variance of noise in the noisy image is v_n, assumed to be constant over the entire image, and the amount of noise remaining after filtering is then $\text{var}\,\{\hat{s}\} = v_n\boldsymbol{a}^T\boldsymbol{a}$. Thus $v_a = \boldsymbol{a}^T\boldsymbol{a}$ gives the relative amount of noise passed by the filter. Using (11), we have

$$\frac{1}{2q+1} \leq v_a \leq 1. \tag{13}$$

The application of an averaging filter to an image inevitably results in some blurring. The tendancy of $\boldsymbol{a}$ to blur is quantified by measuring the "spread" of $\boldsymbol{a}$ defined by the quadratic form

$$w_a = \sum_{i=-q}^{q} a_i^2 i^2 = \boldsymbol{a}^T\boldsymbol{S}\boldsymbol{a} \tag{14}$$

where

$$\boldsymbol{S} = \begin{bmatrix} q^2 & & & & & & & & \\ & (q-1)^2 & & & & & & 0 & \\ & & \ddots & & & & & & \\ & & & 2^2 & & & & & \\ & & & & 1^2 & & & & \\ & & & & & 0 & & & \\ & & & & & & 1^2 & & \\ & & & & & & & 2^2 & \\ & 0 & & & & & & & \ddots \\ & & & & & & & & (q-1)^2 \\ & & & & & & & & & q^2 \end{bmatrix} \tag{15}$$

is the "spread" matrix. (It is interesting to note that $\boldsymbol{a}$ satisfies the requirements to be the probability density function of a discrete random variable. In a sense the spread w_a is similar (but not identical) to the variance of the distribution.)

Ideally, we would like the filter weights $\boldsymbol{a}$ to have good noise suppressing power (small v_a) but small tendency to blur (small w_a). These are somewhat opposing goals. We therefore propose to minimize the joint objective function

$$\begin{aligned} J(\boldsymbol{a}) &= \alpha v_a + (1-\alpha)w_a \\ &= \boldsymbol{a}^T[\alpha\boldsymbol{I} + (1-\alpha)\boldsymbol{S}]\boldsymbol{a} \end{aligned} \tag{16}$$

subject to $\boldsymbol{u}^T\boldsymbol{a} = 1$, where $\boldsymbol{I}$ is the identity matrix of appropriate size. $\alpha(0 \leq \alpha \leq 1)$ is a "tuning" parameter whose role is discussed below.

Using the Lagrange multiplier λ we have the solution

$$\boldsymbol{a} = \lambda[\alpha\boldsymbol{I} + (1-\alpha)\boldsymbol{S}]^{-1}\boldsymbol{u} \tag{17}$$

where the multiplier λ is adjusted to ensure $\boldsymbol{u}^T\boldsymbol{a} = 1$. (Note that $\boldsymbol{a}$ is very easy to compute since $\boldsymbol{S}$ is diagonal.)

This procedure, adapted from Backus and Gilbert [14], results in a one-parameter family of filters parameterized by α. The filters are bell-shaped, resembling a Gaussian probability density. As α varies from 0 to 1, the filter changes from a sharply peaked filter to a flat, equally-weighted averaging filter. We will call a member of this family an S-type filter. We observe that the filter performance measure $v_a = \boldsymbol{a}^T\boldsymbol{a}$ can be considered a function of α. Several filters for various values of α are shown in Fig. 4.

We remind the reader that we are restricting our exposition to one dimension. The real images, however, were processed

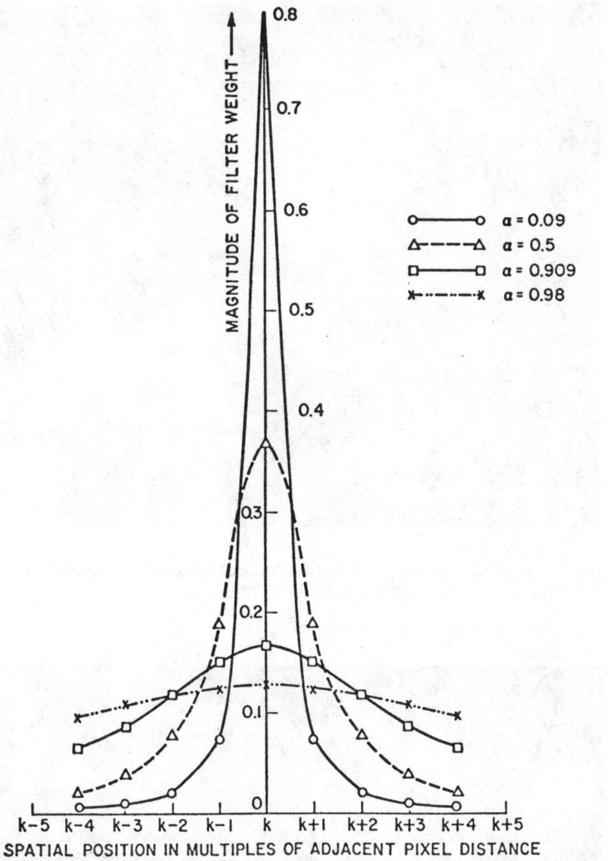

Fig. 4. Shapes of S-type filter. Values of α which tune filters are shown. At points of high detail in picture, α is small and filter is peaked, and in flat areas α is large giving a flat filter.

with two-dimensional filters. The two-dimensional S-type filter is obtained by rotating the one-dimensional filter about the center. It is computed by minimizing

$$J(a) = \alpha \sum_{i=-q}^{q} \sum_{j=-q}^{q} a_{ij}^2 + (1-\alpha) \sum_{i=-q}^{q} \sum_{j=-q}^{q} a_{ij}^2 (i^2 + j^2)$$

subject to the constraint

$$\sum_{i=-q}^{q} \sum_{j=-q}^{q} a_{ij} = 1.$$

C. Role of the Visibility Function

So far we have constructed a tunable filter to be applied to a neighborhood of a given pixel. We now use the visibility function to tune the filter via the parameter α. Assume that the visibility function f has been scaled so that $f(0) = 1$. At the (i,j)th pixel to be restored, there is an associated masking value M_{ij}. The relative visibility of noise at this pixel is given by the visibility function $f(M_{ij})$. We let Q be the number of elements in the two-dimensional filter, and choose a number ϕ, $(1/Q) \leq \phi \leq 1$, held constant for the whole image, which determines the amount of noise passed by the filter in a perfectly flat area of the image. That is, at a pixel where $M = 0, f(M) = 1$, we set α so that

$$v_a = \phi. \tag{18}$$

In a busy area of the image, the masking value M is greater than zero, and the relative noise visibility $f(M)$ is less than 1. Here we allow the filter to pass more noise until its subjective visibility is equal to that in the flat areas.

Specifically, we choose α such that

$$v_a \cdot f(M) = \phi. \tag{19}$$

This rule is applied at every pixel and results in uniform subjective noise visibility. Of course, as v_a is permitted to rise in the busy areas, the spread w_a goes down. Hence the filter is sharper and avoids blurring the edges.

We introduce another tuning factor γ to give some control over the way the filter responds to the visibility function. The rule which determine v_a at a particular pixel is

$$v_a \cdot f^{\gamma}(M) = \phi. \tag{20}$$

By setting $\gamma = 0$ we "turn off" the visibility function resulting in a constant filter of $v_a = \phi$. Setting $\gamma > 1$ results in an exaggerated response to edges, and $\gamma = 1$ gives the response desired in principle.

In summary, we choose two numbers ϕ and γ which remain constant over the entire image. ϕ governs the overall "amount" of filtering, and γ regulates the degree of adaptivity. At each pixel, (20) is solved for v_a. Then α is iteratively adjusted until a filter with the desired v_a is found.

D. The SD-Type Filter

We felt that it would be desirable if our procedure could recognize when the filter overlapped a prominent edge and automatically reduced the overlapping filter weights. This would tend to reduce the distorting influence of a nearby edge and hence preserve edge sharpness.

Again returning to the one-dimensional filters we define the "distortion" due to filter weights at the kth pixel to be

$$d_a = \sum_{i=-q}^{q} (z_{i+k} - z_k)^2 a_i^2 = a^T D a \tag{21}$$

where

$$D = \begin{bmatrix} (z_{k-q} - z_k)^2 & & \\ & \ddots & \\ & 0 & \\ & & \ddots \\ & & & (z_{k+q} - z_k)^2 \end{bmatrix}. \tag{22}$$

We then design the distortion-penalized filter (denoted as the SD-type) by minimizing the objective function

$$\begin{aligned} J(a) &= \alpha v_a + (1-\alpha)(w_a + \theta\, d_a) \\ &= a^T[\alpha I + (1-a)(S + \theta D)]a \end{aligned} \tag{23}$$

subject to $a^T u = 1$. θ is used as a tuning parameter.

Intuitively, diagonal elements of D represent a penalty applied to the corresponding elements of a. When a overlaps an edge, there is a large difference between the brightness at the center of the filter (z_k) and the brightness at a pixel across the edge (let us say, z_{k+i}). Then the penalty $(z_{k+i} - z_k)^2$ is large, forcing the coefficient a_i to be small.

The behavior of the SD-filter in one dimension is plotted in Fig. 5 as the filter approaches an ideal edge. In this figure, for the purposes of comparisons, θ and α were held constant. In the absence of large variations in brightness over the

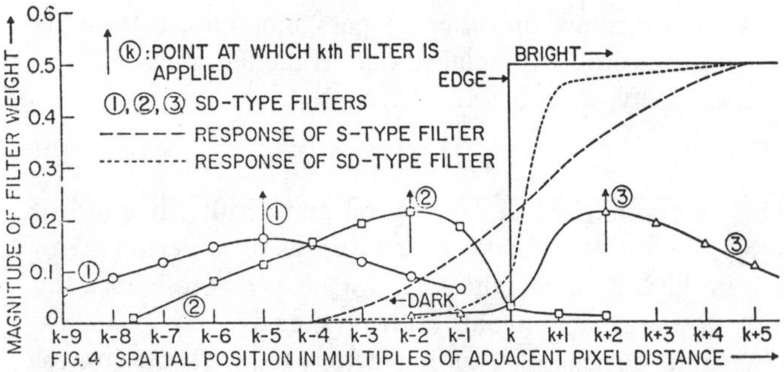

Fig. 5. Shapes of *SD*-type filter. $\alpha = 0.909$ and $\theta = 0.1$ are used. Edge has magnitude $0 \rightarrow 100$. Response of *SD*-type filter (- - -) preserves edge shapes much better than response of *S*-type filter (— — —).

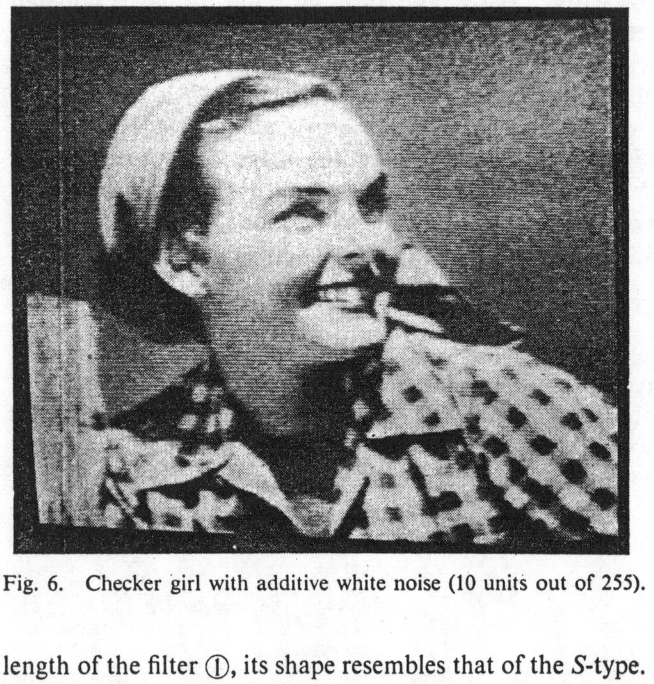

Fig. 6. Checker girl with additive white noise (10 units out of 255).

length of the filter ①, its shape resembles that of the *S*-type. Near an edge, the portion of the filter ② overlapping the edge is suppressed. As the edge is passed, the filter ③ again deemphasizes the dissimilar data values. Fig. 5 also displays the output of the *SD* filter (dots) and of the *S*-type filter (dashed) with equal α settings. Clearly the *SD*-type filter has better preserved the shape of the edge.

The parameter θ controls the sensitivity of the filter to brightness variations. The *S*-type filter is just the special case $\theta = 0$. If θ is chosen to be approximately one-half of the reciprocal of the standard deviation of the noise, then the penalties (elements of $\boldsymbol{D}$) are small for variations in brightness due to noise; however, a distinct edge results in large penalties due to the squaring of the difference.

IV. Simulations

A. Results of Tests

In order to test our procedure, we took the original image of Fig. 1, digitized it as described before, and added white noise with standard deviation of 10 units (out of 255) (see Fig. 6).

In Fig. 7 we tested the ideal performance of the *S*-filter. The noisy image of Fig. 6 was filtered using the masking

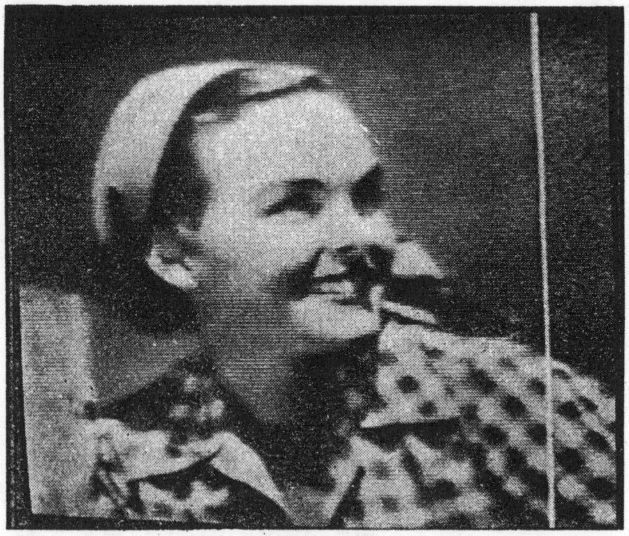

Fig. 7. Checker girl filtered with *S*-type filter using noiseless masking function, $\phi = 0.04$ and $\gamma = 0.5$.

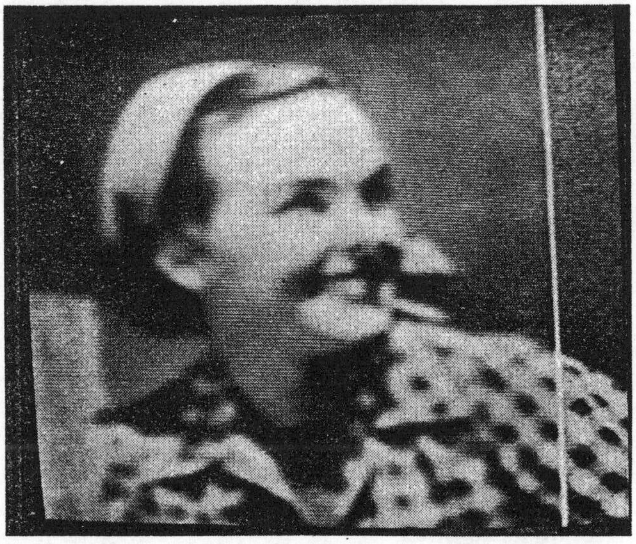

Fig. 8. Checker girl filtered with nonvarying filter with noise suppression same as Fig. 6.

data computed from the *noiseless* image (shown in Fig. 2). The filter parameters were $\phi = 0.04$, $\gamma = 0.5$. The filters in this and all other examples (except as noted) have the spread of 5×5 pixels.

Clearly Fig. 7 is a good restoration.[2] The *S*-type filter has greatly reduced the noise without degrading resolution. By way of comparison, let us filter Fig. 6 with a filter whose impulse response is held fixed over the entire image. That is, we do not vary the filter response in accordance to the masking function. Fig. 8 shows the result of a nonvarying filter chosen to provide noise suppression in the flat areas comparable to that in Fig. 7. Notice that the edges are

[2] All the remarks in this section regarding the subjective quality of pictures were obtained by showing the pictures to many observers skilled in television viewing. Some of the remarks may not be obvious due to imperfect photographic reproductions. A "better" set of photographs may be obtained by writing to the second author.

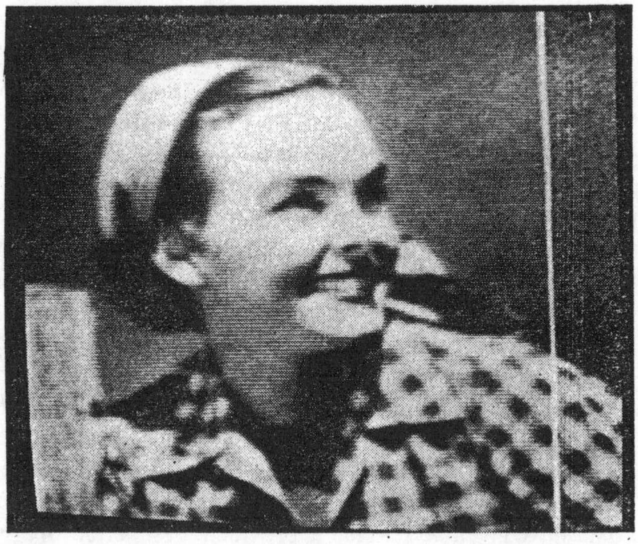

Fig. 9. Checker girl filtered with nonvarying filter with resolution same as Fig. 6.

Fig. 11. Masking function derived from noisy Checker girl of Fig. 6. 3 × 3 neighborhood of slopes is used.

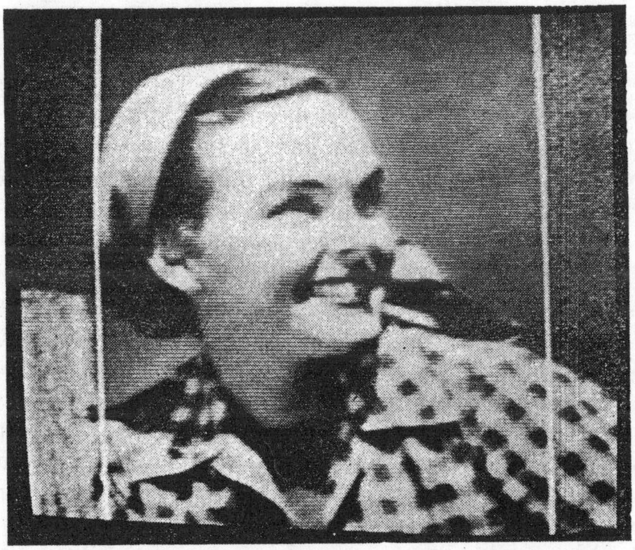

Fig. 10. Checker girl filtered with *SD*-type filter using noiseless masking function, $\phi = 0.04$, $\gamma = 0.5$, and $\theta = 0.5$.

Fig. 12. Checker girl filtered with *S*-type filter using noisy masking function, $\phi = 0.1$ and $\gamma = 0.5$.

severely blurred. In Fig. 9 we chose a nonvarying filter which gives edge resolution comparable to that in Fig. 7, but the noise removed is less than in Fig. 7. (Compare the background and neck areas of the two photographs.) It is clear from these examples that by varying the filter response in accordance with the masking function, we greatly improve the restoration. We also designed a minimum mean-square filter by assuming the image to be stationary with a given autocorrelation function (which was estimated from the image). The performance of this filter was similar to that shown in Fig. 8. Noise suppression was good in flat areas, but the edges were significantly blurred.

Applying the *SD*-type filter does not result in significant improvement when the *noiseless* masking data such as shown in Fig. 2 are available. Fig. 10 shows the image processed with parameters $\phi = 0.04$, $\gamma = 0.5$, $\theta = 0.5$ and the noiseless masking of Fig. 2.

In the next test we use the noisy image (Fig. 6) to compute the masking values; these are shown in Fig. 11. Fig. 12 shows the results of applying an *S*-type filter, with parameters $\phi = 0.1$, $\gamma = 0.5$, using the noisy masking data. Predictably, the restoration is inferior to that of Fig. 7. The noise in the masking data appears as spurious "edges," at which the filter reduces its degree of smoothing. Hence noise in the flat areas of the image is not effectively removed.

However, we regain much of the lost performance by applying the *SD*-type filter. Fig. 13 shows the *SD*-type results (parameters $\phi = 0.01$, $\gamma = 0.5$, $\theta = 0.07$) obtained using noisy masking data. We see that the background noise level is lower than in Fig. 12, and the resolution is better, especially around the teeth and eyes. Why the *SD*-type filter performs well when the masking data are unreliable is not quite clear. The effect of the ***D***-matrix penalty is similar to the effect of the masking function: the filter tends

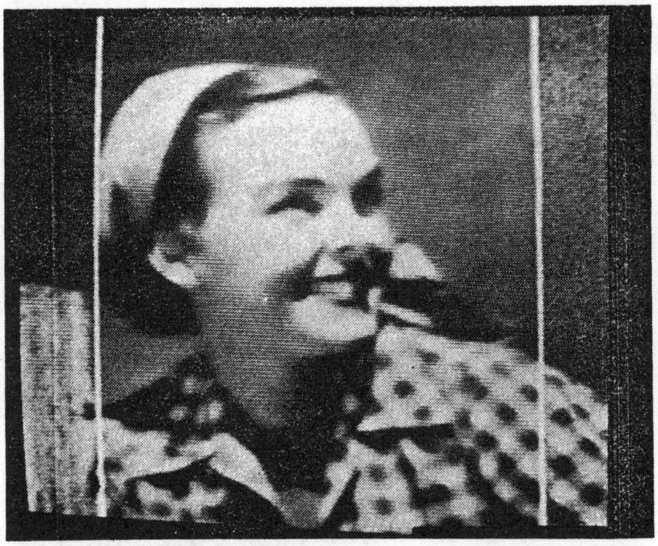

Fig. 13. Checker girl filtered with *SD*-type filter using noisy masking function, $\phi = 0.01$, $\gamma = 0.5$, and $\theta = 0.07$.

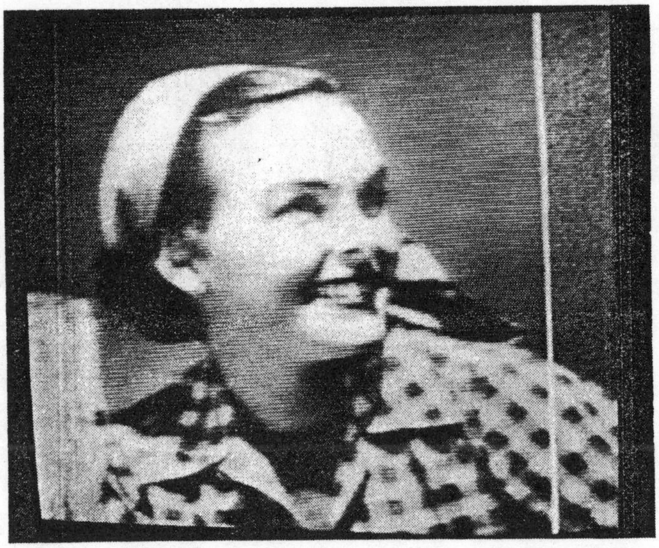

Fig. 14. Checker girl filtered with one dimensional *S*-type filter.

to be flat in flat areas and peaked near edges. However, the *D*-matrix effect is less disturbed by the presence of noise.

One further test was made to investigate the effectiveness of our technique when implemented as a one-dimensional filter. We wondered whether one could perform real-time filtering of a video signal along one scan line with good results. Fig. 14 shows the restoration obtained by the one-dimensional *S*-type filter using noiseless masking data. There is slight horizontal smearing, but the overall horizontal resolution is good. Vertical resolution, of course, is unaffected.

B. Masking Function and the Restoration Problem

The performance of the filter depends crucially on the nature of the masking function. The masking function used was originally formulated for use in the design of quantizers [13]. For the purpose of image restoration, the precise definition is not entirely satisfactory in two respects.

1) *Sensitivity to noise:* The value of the function is somewhat sensitive to uncorrelated additive noise. Although, to some extent all edge detection schemes are plagued by this problem, it may be possible to formulate a measure which would be more reliable in the presence of noise.

2) *Number of neighbors:* The above formulation employed a grid of 3 × 3 neighbors to compute M. A peculiar distortion is introduced when we apply a filter of 7 × 7 or larger governed by this masking function. Since the filter "sees" a wider area than the masking function, an artifact is introduced near every sharp edge. This artifact may be avoided by increasing the number of neighbors used in computing M. However, then the visibility function determined for the 3 × 3 masking function would have to be redetermined. Although we restricted ourselves to a 5 × 5 neighbor filter, it appears that spatially larger filters would have given better results provided they were properly tuned.

C. Visibility Function and the Restoration Problem

A natural question arises: How sensitive is the procedure to the visibility function? Perhaps any reasonable, monotonic decreasing function would give good results with proper tuning of ϕ,γ. We investigated this by filtering another image (also head and shoulder type) with the visibility function obtained from a different (and dissimilar) image. In each case attempted, the results were inferior to those obtained with the correct visibility function. However, the filter still performed better than a nonvarying filter on the same image. This suggests that the techniques may be applied to a wide variety of head and shoulder type images without knowing the exact visibility function for each image.

In each simulation the best restoration was obtained with the parameter γ set different from unity. This suggests that there are some psychovisual effects which are not accounted for in our approach. One such effect is the correlation of noise in the filtered image. The visibility function is valid, strictly speaking, only for white noise; the noise passed by the filter has nonzero spatial correlation. Another effect is that the masking values in the filtered image differ from that of the original image. The result is that the simple formula

$$v_a \cdot f(M) = \phi$$

is not strictly valid to control filtering. Introduction of the parameter γ is necessary to accommodate these unmeasured effects. Further research is needed on noise visibility in the context of image restoration.

D. Computational Effort

All the above-mentioned filters have been implemented on a PDP-11/45. The computational effort for the *S*-type filter is small. During an initialization phase we computed an ensemble of several (11) filters with variance $1/Q \leq \phi_j \leq 1, j = 1,\cdots,11$. This step took less than 10 s. During the filtering phase one of these filters is applied at each pixel; we selected the filter from the ensemble with variance most closely matching the desired variance. To filter one image required less than five minutes.[3]

The *SD*-type filter requires much more time since the filter must be redesigned at each pixel. The design effort greatly exceeds the time required to apply. To filter one frame requires nearly an hour of computer time.

V. Discussion

The results presented above suffice to show the merit of the approach. The main feature is a filter which is applied *locally* under the direction of a local fidelity criterion based on psychovisual principles. We discovered in the *S*-type filter a very tractable and convenient approach. Although the masking and visibility functions provided restoration in the right direction, they were certainly not the best suited for the restoration problem, and more subjective tests to determine the functions more suited to the noisy restoration problem are needed. Also needed is a study of the relation of visibility functions to picture content for pictures in a given class.

A. Extensions

1) *Improvements in the Psychovisual Model:* We noticed in our work (and indeed, the psychovisual literature supports this observation) that additive noise is less visible at higher luminance levels. In crude terms, the response to a perturbation in luminance at a point in the image decreases with increasing average luminance of the surround [19]. A natural extension then is to include a luminance term in the masking function and compute the visibility function corresponding to it. This would permit filters with sharper impulse response at high luminance levels.

Another extension of the fidelity criterion is to determine the subjective effect of filter blurring as a function of edge sharpness. At present we measure the resolution of the filter by the spread W_a defined in (14). The shortcoming of this measure is that it does not represent the subjective resolution of the filtered image. For example, in Fig. 8, where a nonvarying filter with constant W_a has been applied to all parts of the image, the filter resolution is acceptable in the background and in the flat skin texture, but it is unacceptable at the edges. What is needed is a measure of resolution which considers the degree of spatial detail at the pixel where the filter is applied. Formulating such a measure (perhaps as a function of W_a and masking value M_{ij}), we could vary the filter to achieve the best tradeoff of noise removal and subjective resolution.

2) *Restoration of Blurred Images:* In this study we considered for simplicity images degraded only by additive noise. That is, we took the mapping $F[\cdot]$ in (1) to be identity. It would be interesting to apply our restoration technique to cases when $F[\cdot]$ represents an arbitrary blurring filter. The inversion technique of Backus and Gilbert [14] is applicable in the case of general $F[\cdot]$. However, it is less clear how best to introduce the masking and visibility functions for blurred images. Interesting problems arise in connection with this extension.

[3] No effort was made to optimize the computer code.

B. Alternative Filter Designs

There are other filter techniques that can be applied locally under a visual fidelity criterion. An example is the Kalman filter approach developed by Nahi and Habibi [20]. A decision scheme is used to partition the image into subimages. Within each subimage the signal has approximately stationary autocovariance. Thus the image is represented as the switched output of several stationary sources. A Kalman filter is designed for each subimage. We can introduce the visibility function into this approach in the following way. In the Kalman filter equations we can modify the observation noise variance by the visibility function. That is, we can artificially set the variance of the observation noise equal to $f(M) \cdot V_n$, where V_n is the true noise variance. The degree of smoothing performed will then be governed by the subjective visibility of noise.

Acknowledgment

The authors would like to thank Barry Haskell, without whose help this study would not have been possible.

References

[1] T. S. Huang, W. F. Schreiber, and O. J. Tretiak, "Image processing," *Proc. IEEE*, vol. 59, pp. 1586–1608, Nov. 1971.
[2] M. M. Sondhi, "Image restoration: The removal of spatially invariant degradations," *Proc. IEEE*, vol. 60, pp. 842–853, July 1972.
[3] H. C. Andrews, "Digital image restoration: A survey," *Comput. J.*, pp. 36–45, May 1974.
[4] B. R. Hunt, "Digital image processing," *Proc. IEEE*, vol. 63, pp. 693–708, April 1975.
[5] J. Tsujiuchi, "Correction of optical images by compensation of aberrations and by spatial frequency filtering," in *Progress in Optics*, vol. 2. New York: Wiley, 1963, pp. 131–180.
[6] J. L. Harris, "Image evaluation and restoration," *J. of Opt. Soc. Amer.*, vol. 56, pp. 564–574, May 1966.
[7] C. W. Helstrom, "Image restoration by the method of least squares," *J. Opt. Soc. Amer.*, vol. 57, pp. 297–303, March 1967.
[8] D. Slepian, "Linear least-squares filtering of distorted images," *J. Opt. Soc. Amer.*, vol. 57, pp. 918–922, July 1967.
[9] B. R. Hunt, "The application of constrained least-squares estimation to image restoration by digital computer," *IEEE Trans. Comput.*, vol. C-22, pp. 805–812, Sept. 1973.
[10] Z. L. Budrikis, "Visual fidelity criterion and modeling," *Proc. IEEE*, vol. 60, pp. 771–779, July 1972.
[11] D. J. Sakrison, M. Halter, and H. Mostafavi, "Properties of the human visual system as related to the encoding of images," in *New Directions in Signal Processing in Communication and Control* (to be published).
[12] D. P. MacAdam, "Digital image restoration by constrained deconvolution," *J. Opt. Soc. Amer.*, vol. 60, pp. 1617–1627, Dec. 1970.
[13] A. N. Netravali and B. Prasada, "Adaptive quantization of picture signals based on spatial masking," (to appear in *Proc. IEEE*).
[14] G. Backus and F. Gilbert, "Uniqueness in the inversion of inaccurate gross earth data," *Phil. Trans. Roy. Soc. London*, vol. 266, pp. 123–192, March 1970.
[15] H. Harms and E. Aulhorn, "Studien Über den Grenzkontrast. I MiHeilung. Ein neueo Grenzphänomen," *Grafes Arch. Opthal.*, vol. 157, pp. 3–23, 1955.
[16] N. Weisstein, "Metacontrast," in *Visual Psychophysics*, D. Jameson and L. M. Hurvich, Eds. New York: Springer-Verlag, 1972, Ch. 10.
[17] J. O. Limb, "Source receiver encoding of television signals," *Proc. IEEE*, vol. 55, pp. 364–380, March 1967.
[18] J. D. Beyer and R. C. Brainard, "Picture processing computer system—Users manual," Bell Laboratories Memorandum, June 1974.
[19] P. Moon and D. E. Spencer, "The visual effect of non-uniform surrounds," *J. Opt. Soc. Amer.*, vol. 35, pp. 233–248, March 1945.
[20] N. E. Nahi and A. Habibi, "Decision-directed recursive image enhancement," *IEEE Trans. Circuits Syst.*, vol. CAS-22, pp. 286–293, March 1975.

Adaptive Noise Smoothing Filter for Images with Signal-Dependent Noise

DARWIN T. KUAN, MEMBER, IEEE, ALEXANDER A. SAWCHUK, SENIOR MEMBER, IEEE, TIMOTHY C. STRAND, MEMBER, IEEE, AND PIERRE CHAVEL

Abstract—In this paper, we consider the restoration of images with signal-dependent noise. The filter is noise smoothing and adapts to local changes in image statistics based on a nonstationary mean, nonstationary variance (NMNV) image model. For images degraded by a class of uncorrelated, signal-dependent noise without blur, the adaptive noise smoothing filter becomes a point processor and is similar to Lee's local statistics algorithm [16]. The filter is able to adapt itself to the nonstationary local image statistics in the presence of different types of signal-dependent noise. For multiplicative noise, the adaptive noise smoothing filter is a systematic derivation of Lee's algorithm with some extensions that allow different estimators for the local image variance. The advantage of the derivation is its easy extension to deal with various types of signal-dependent noise. Film-grain and Poisson signal-dependent restoration problems are also considered as examples. All the nonstationary image statistical parameters needed for the filter can be estimated from the noisy image and no *a priori* information about the original image is required.

Index Terms—Adaptive noise smoothing, image restoration, nonstationary image model, single-dependent noise.

I. Introduction

VARIOUS image restoration and enhancement methods have been proposed for removing degradations due to blurring and noise. The effectiveness of an image restoration algorithm depends on the validity of the image model, the criterion used to judge the quality of the restored image, and the statistical model for the noise process. Early techniques concentrated on nonrecursive algorithms implemented in the discrete frequency domain [1]. The necessary computations are carried out by fast Fourier transform (FFT) techniques. The use of the FFT makes it possible to do image restoration with reasonable computation time. However, these approaches cannot easily deal with images degraded by space-variant blur and assume a stationary image model. More recent work has centered on two-dimensional recursive filtering techniques extended from the one-dimensional Kalman filtering algorithm [2]-[4]. The main problems of two-dimensional recursive filtering are both the difficulty in establishing a suitable two-dimensional recursive model and the high dimensionality of the resulting state vector. Recent results on the reduced update Kalman filter by Woods *et al.* [5], [6] seem to overcome some of these difficulties and have potential for real applications. The advantages of two-dimensional recursive filters are that they require less computation time than nonrecursive algorithms and can handle space-variant blur easily.

The conventional stationary image model assumes that an image is a wide-sense stationary field. The statistical properties of an image are characterized globally rather than locally by its stationary correlation function. This assumption enables the use of FFT-based algorithms in the restoration procedure and consequently reduces the computation time dramatically. However, the restoration filter designed accordingly is insensitive to abrupt changes of the image intensity, and tends to smooth the edges where stationarity is not justified. To overcome these disadvantages and to have an improved restoration result, Hunt *et al.* [7], [8] proposed a nonstationary mean Gaussian image model. In this nonstationary image model, an image is modeled as stationary fluctuations about a nonstationary ensemble mean which has the gross structure that represents the context of the ensemble. Lebedev and Mirkin [9] suggested a composite image model that assumes an image is composed of many different stationary components. Each component has a distinct stationary correlation structure. Ingle and Woods [10] applied a reduced update Kalman filter to image restoration by using this composite image model. The result is a bank of Kalman filters running in parallel and an identification-estimation approach is used in which each point is assigned a stationary image model and filtered by the specific Kalman filter. Rajala *et al.* [11] derived a two-dimensional recursive filter based on a piecewise stationary image model. They first segment the image into disjoint regions according to the local spatial activity of the region and determine the covariance structure of different segments. In this framework, they also use a visibility adapted observation noise model. Then they use a different Kalman filter for each segment for nonstationary restoration. Compared with Ingle and Woods' approach, this method emphasizes nonstationarity within regions rather than nonstationarity at edges in the scene.

Another nonlinear approach to an improved restoration filter was proposed by Anderson and Netravali [12]. Instead of

Manuscript received October 13, 1982; revised June 10, 1983 and September 25, 1984. This work was supported in part by the Joint Services Electronics Program (JSEP) through the Air Force Office of Scientific Research under Contract F49620-81-C-0070, and by Battelle Pacific Northwest Laboratories, Richland, WA 99352.

D. T. Kuan is with Central Engineering Laboratories, FMC Corporation, Santa Clara, CA 95052.

A. A. Sawchuk is with the Image Processing Institute, Department of Electrical Engineering, University of Southern California, Los Angeles, CA 90089.

T. C. Strand is with IBM Corporation, San Jose, CA 95193.

P. Chavel is with Institut d'Optique, Universite de Paris-Sud, Orsay, France.

Reprinted from *IEEE Trans. Pattern Anal. Machine Intelligence,* vol. PAMI-7, pp. 165–177, Mar. 1985.

using a nonstationary image model, they used a subjective error criterion based on human visual system models and derived a nonrecursive filter that adapts itself to make a compromise between the loss of resolution and noise smoothing such that the same amount of subjective noise is suppressed throughout the image. Abramatic and Silverman [13] generalized this procedure and related it to the classical Wiener filter. In fact, nonlinear restoration is closely related to nonstationary restoration as we shall see in the following sections.

In contrast to the signal-independent, additive noise model assumed in most image restoration algorithms, many physical noise processes are inherently signal-dependent. Restoration algorithms based on a signal-independent noise model are not expected to be very effective in the signal-dependent noise environment. Naderi and Sawchuk [14] derived a nonstationary discrete Wiener filter for a signal-dependent film-grain noise model. In addition to the generality of the noise model, the filter is able to adapt itself to the local signal statistics given the conditional noise statistics. Lo and Sawchuk [15] developed a nonlinear MAP filter for images degraded by Poisson noise. The algorithm is nonrecursive and requires the solution of a set of nonlinear equations. The sectioning method is used to reduce computation and to do local processing. The MAP filter is optimal, but in some applications, it might be acceptable to trade off theoretical performance for ease of implementation. This becomes more appropriate when we consider the adaptive filtering techniques where the filter parameters have to be estimated from the noisy observations.

In this paper, we introduce a nonstationary mean, nonstationary variance (NMNV) image model. The nonstationary mean describes the gross structure of an image and the nonstationary variance characterizes edge and elementary texture information of the image. A local linear minimum mean square error (LLMMSE) filter for images degraded by blur and a class of signal-dependent, uncorrelated noise is derived based on the NMNV image model. If there is no blur degradation, the LLMMSE filter has a very simple structure and is a point processor. This property is due to the NMNV image model and the uncorrelated noise assumptions. Usually the nonstationary ensemble statistics are not available *a priori* and can only be estimated from the degraded image. With the substitution of the local statistics for the ensemble statistics in the LLMMSE filter, we have an adaptive noise smoothing filter that is able to change characteristics according to the local image statistics and to different types of signal-dependent noise. The methods used to calculate the local statistics from the degraded image are critical to the quality of the restored results. Various ways of estimating these ensemble statistics are discussed and their performance is compared. The adaptive noise smoothing filter has a form similar to Abramatic's filter, and the relationship between the nonstationary image modeling approach and the nonlinear subjective criterion approach becomes clear. Based on this connection, we show that the minimum mean square error criterion is not a bad criterion if it is used locally. The explicit structure of the adaptive noise smoothing filters for multiplicative noise, film-grain noise, and Poisson noise are derived for completeness. The comparison of the adaptive noise smoothing filter with Lee's local statistics algorithm [16] shows that the introduction of the NMNV image model is not only valuable for systematically deriving the optimal estimator structure for different signal-dependent noise models, but is also very useful for the extension of the adaptive noise smoothing filter to image restoration where the images are degraded by both blur and signal-dependent noise. This problem will be discussed in another paper [17].

In this paper we denote a two-dimensional image by $f(i,j) = f$, where (i, j) are spatial coordinates. The nonstationary statistical (ensemble) mean and variance of image f are denoted by $E(f(i, j))$ and $\sigma_f^2(i, j)$, respectively. The nonstationary local spatial mean and variance of f (described in detail later) are denoted by $\bar{f}(i, j)$ and $v_f(i,j)$. Various subscripts on σ^2 and v will refer to different two-dimensional functions.

II. Nonstationary Mean, Nonstationary Variance (NMNV) Image Model

In a conventional stationary image model, an image f is assumed to be a wide-sense stationary random field with constant mean vector and block Toeplitz covariance matrix. The joint probability density function is implicitly assumed to be multivariate Gaussian. All the statistical information of the image is carried by the covariance matrix. For a real world image as in Fig. 1(a), it is apparent from the picture that the image is not a stationary random field and its histogram [Fig. 1(b)] is not Gaussian. Thus the conventional stationary image model is an oversimplified model for its computational purposes. Hunt and Cannon [7] proposed a nonstationary mean and stationary covariance Gaussian image model. They assumed that an image f can be decomposed into a nonstationary statistical mean component $E(f)$, and a stationary residual component $f_0 = f - E(f)$. The nonstationary statistical mean describes the gross structure of an image and the residual component describes the detail variation of the image. For ease of computation and mathematical tractability, the covariance function C_f is assumed to be stationary and is defined by

$$C_f = E[(f - E(f))(f - E(f))^T]. \tag{1}$$

The joint probability density function (PDF) of f is assumed to be Gaussian, i.e.,

$$P(f) = ((2\pi)^{N^2} |C_f|)^{-1/2} \cdot \exp[-\tfrac{1}{2}(f - E(f))^T C_f^{-1}(f - E(f))] \tag{2}$$

where $|C_f|$ is the determinant of the covariance matrix C_f, and N is the size of the image. If we replace the nonstationary ensemble (statistical) mean $E(f)$ by a local (spatial average) mean estimate $\bar{f}$ that is calculated over a 3×3 uniform window, and subtract the local mean from the original image, we have the residual image f_0 that is shown for our example in Fig. 1(c). It is obvious from the picture that f_0 is still a correlated nonstationary process. However, the shape of the histogram is more Gaussian [Fig. 1(d)]. Because some of the structural information is now carried by the nonstationary mean, it is reasonable to assume that the covariance structure may be simplified compared with the conventional stationary image model. Trussell and Hunt [8] assumed that the covariance matrix of f_0 could be approximated by a constant diagonal

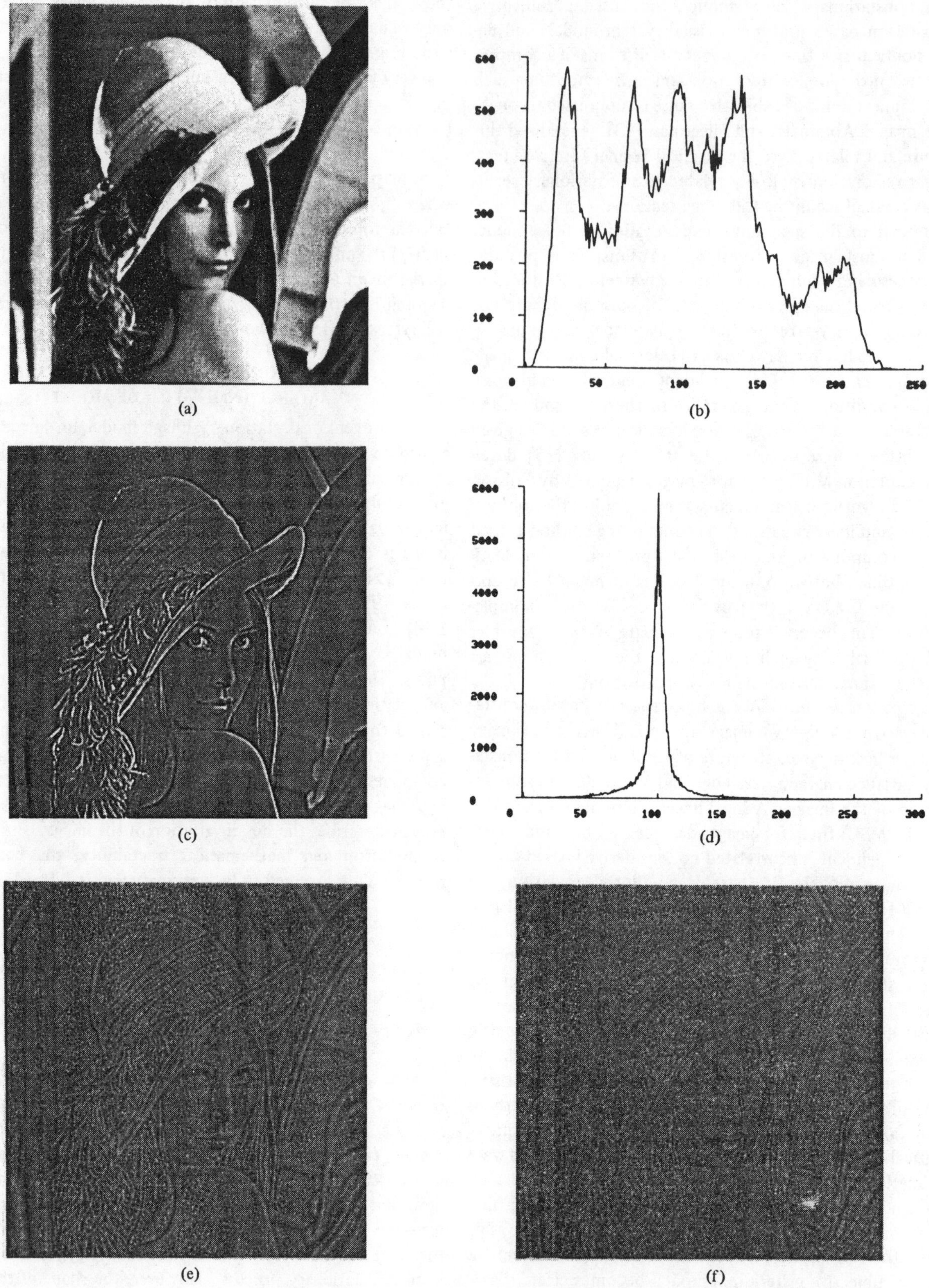

Fig. 1. Nonstationary mean and nonstationary variance image model. (a) Original image. (b) Histogram of (a). (c) Residual image of the original image and the local mean image. (d) Histogram of (c). (e) Normalized residual image with unit variance. (f) Normalized unit variance residual image with an edge detecting intelligent filter to calculate the local statistics.

matrix $\sigma_f^2 I$, where σ_f^2 is a scalar and I is the identity matrix. In this case, all the information is carried by the nonstationary mean. While letting the covariance matrix carry all the statistical information, as in a conventional stationary image model, is not satisfactory, neither is letting the nonstationary mean carry all the information. Therefore, it is heuristically reasonable to include more structure in the covariance matrix of the nonstationary mean Gaussian image model.

In order to consider the nonstationarity of the image and not to complicate the computation too much, we assume that $\boldsymbol{f}_0$ is a nonstationary white process. More specifically, $\boldsymbol{f}_0$ is statistically uncorrelated and is characterized by its nonstationary statistical variance $\sigma_f^2(i, j)$ at spatial position (i, j). This nonstationary mean and nonstationary variance (NMNV) image model has been used implicitly to some extent in the sectioned MAP method. There it was assumed that the nonstationary image can be divided into many sections. Each section has the covariance matrix $\sigma_f^{2(q)} I$, where $\sigma_f^{2(q)}$ is a scalar that varies from section to section. The section size is chosen according to the width of the point spread function and the extent of stationarity assumed. An overlap-save sectioning method is used to suppress the convolution wraparound effect at the section boundary. For the noise smoothing problem (no blur), the section size can be reduced to a single point and this becomes a NMNV image model. However, for the image restoration problem, the minimum section size cannot be reduced to a single point.

We can substitute the local (spatial average) variance $v_f(i,j)$ of f_0 at position (i, j) for the nonstationary variance $\sigma^2(i, j)$ and normalize f_0 in Fig. 1(c) by computing the normalized unit variance residual image

$$f_0'(i,j) = f_0(i,j)/[v_f(i,j)]^{1/2}. \tag{3}$$

This residual image f_0' is shown in Fig. 1(e). Note that in the uniform intensity region, there is no correlation and f_0' looks like white noise. This verifies that the NMNV image model is valid in those regions. However, in the neighborhood of an edge, visible correlations still exist in the normalized residual image. If we want to include these correlations into the image model, we need to specify a nonstationary mean and nonstationary covariance image model. While this is the direction to go, the possible performance improvement of using this model is hindered by the complexity of the restoration filter and the model identification procedure.

There is another factor that contributes to the correlation of the residual image in Fig. 1(e). Note that we used the local statistics to substitute for the ensemble statistics in the procedure. When there is a sharp edge in the image, the averaging window tends to blur the local mean estimate by using pixels on both sides of the edge. To avoid this we can use an intelligent filter that recognizes an edge and calculates the local mean only using those pixels on the one side of the edge. Fig. 1(f) is the normalized residual image where the local mean and local variance are calculated by such a filter. Note that the correlations around the edges are reduced, and the residual image is more like a white noise process.

From the discussion above, we know that the validity of the NMNV image model depends on the methods we use to estimate the local mean and variance. Therefore, the process of estimating local statistics can be thought of as a way to find an operation that transforms an image into a white noise process in order to facilitate the restoration procedure. This transformation is usually a nonlinear one, and the inherent correlation between adjacent pixels in an image is implicitly imbedded in the nonstationary mean.

III. Local Linear Minimum Mean Square Error Filter for a Class of Signal-Dependent Noise

Consider the observation equation

$$\boldsymbol{g} = H\boldsymbol{f} + \boldsymbol{n} \tag{4}$$

where $\boldsymbol{g}$ is the degraded observation, $\boldsymbol{f}$ is the original signal, $\boldsymbol{n}$ is a zero mean noise that can be signal-independent or signal-dependent, and H is the blurring matrix. The minimum mean square error (MMSE) estimate of $\boldsymbol{f}$ given observation $\boldsymbol{g}$ is the conditional mean estimate

$$\hat{\boldsymbol{f}} = E(\boldsymbol{f}|\boldsymbol{g}). \tag{5}$$

In general, the MMSE estimate is nonlinear and depends on the probability density functions of $\boldsymbol{f}$ and $\boldsymbol{n}$. The explicit form of the MMSE estimator is difficult to obtain for the general case. If we impose a linear constraint on the estimator structure, we have the linear minimum mean square error (LMMSE) estimator [18]

$$\hat{\boldsymbol{f}}_{\text{LMMSE}} = E(\boldsymbol{f}) + C_{fg}C_g^{-1}(\boldsymbol{g} - E(\boldsymbol{g})) \tag{6}$$

where C_{fg} is the cross-covariance matrix of $\boldsymbol{f}$ and $\boldsymbol{g}$. Also C_g and $E(\boldsymbol{g})$ are the covariance matrix and ensemble mean of $\boldsymbol{g}$, respectively. Unlike the MMSE filter, the LMMSE filter only requires the second order statistics of the signal and noise. Nonstationary ensemble mean and covariance statistics can be used in (6) if they are available. Because of the local nature of these nonstationary statistics, we refer to it as a local (or nonstationary) linear minimum mean square error (LLMMSE) filter. In this way, we distinguish it from the conventional stationary Wiener filter. Generally speaking, C_g is a nonstationary covariance matrix and matrix inversion has to be carried out without the help of FFT techniques. In a later section we replace the ensemble statistics with spatial averages obtained from the images themselves.

To see the structure of the LLMMSE filter more explicitly, we need to calculate the covariance matrices in (6). The cross-covariance matrix C_{fg} is given by

$$\begin{aligned} C_{fg} &= E[(\boldsymbol{f} - E(\boldsymbol{f}))(\boldsymbol{g} - E(\boldsymbol{g}))^T] \\ &= E\{(\boldsymbol{f} - E(\boldsymbol{f}))[H(\boldsymbol{f} - E(\boldsymbol{f})) + \boldsymbol{n}]^T\} \\ &= C_f H^T + E[(\boldsymbol{f} - E(\boldsymbol{f}))\boldsymbol{n}^T]. \end{aligned} \tag{7}$$

Similarly, the covariance matrix C_g can be calculated as

$$\begin{aligned} C_g &= E[(\boldsymbol{g} - E(\boldsymbol{g}))(\boldsymbol{g} - E(\boldsymbol{g}))^T] \\ &= E\{[H(\boldsymbol{f} - E(\boldsymbol{f})) + \boldsymbol{n}][H(\boldsymbol{f} - E(\boldsymbol{f})) + \boldsymbol{n}]^T \\ &= HC_f H^T + C_n + HE[(\boldsymbol{f} - E(\boldsymbol{f}))\boldsymbol{n}^T] \\ &\quad + E[\boldsymbol{n}(\boldsymbol{f} - E(\boldsymbol{f}))^T]H^T \end{aligned} \tag{8}$$

where C_n is the covariance matrix of $\boldsymbol{n}$.

The results of (7) and (8) simplify considerably if we can assume that the conditional mean of $\boldsymbol{n}$ given f is 0, i.e.,

$$E(\boldsymbol{n}|\boldsymbol{f}) = \boldsymbol{0}. \tag{9}$$

We now look at the term $E[(f - E(f))\,\boldsymbol{n}^T]$ in (7) and (8) and rewrite it as

$$E[(f - E(f))\,\boldsymbol{n}^T] = E_f[E_n[(f - E(f))\,\boldsymbol{n}^T|f]]. \tag{10}$$

Now the inner expectation on the right side of (10) can be written

$$E_n[(f - E(f))\,\boldsymbol{n}^T|f] = (f - E(f))\,E[\boldsymbol{n}^T|f] = [0] \tag{11}$$

from the condition of (9). Substituting this result back in (10) gives

$$E[(f - E(f))\,\boldsymbol{n}^T] = [0] \tag{12}$$

and using this result in (7) and (8) gives the simplified result

$$C_{fg} = C_f H^T$$
$$C_g = HC_f H^T + C_n. \tag{13}$$

Note that although (4) has the form of signal plus signal-dependent noise, the noise need not be strictly additive in the usual sense. Any noise degradation can always be expressed in terms of signal plus signal-dependent noise, although this may not be the most convenient way to represent it. If we examine

$$\begin{aligned} E(\boldsymbol{g}|f) &= E[(Hf + \boldsymbol{n})|f] \\ &= E[Hf|f] + E(\boldsymbol{n}|f) \\ &= HE(f|f) = Hf \end{aligned} \tag{14}$$

we see that the condition of (9) implies that the signal-dependent noise $\boldsymbol{n}$ has no bias. This condition is generally satisfied by many physical noise models such as multiplicative noise, film-grain noise, and Poisson noise. Instead of deriving a different filter for each individual case, we give a unified approach to design a noise smoothing filter for this class of signal-dependent observations.

Conditions for a Scalar (Point) Processor

If we consider the special case of no degradation by blurring, i.e., $H = I$, then the degraded image $g(i, j)$ can be expressed as the scalar equation

$$g(i,j) = f(i,j) + n(i,j) \tag{15}$$

where $n(i, j)$ is the (possibly) signal-dependent noise satisfying (9).

Another special case arises if the noise term $n(i, j)$ is uncorrelated, i.e.,

$$E[n(i,j)\,n(r,s)] = 0 \quad \text{for} \quad (i,j) \neq (r,s). \tag{16}$$

Assuming the condition of (15) and the NMNV assumption that C_f is diagonal, the cross-covariance matrix C_{fg} becomes diagonal. These assumptions together with the assumption of (16) imply that the covariance matrix C_g is diagonal. Under these conditions, the LLMMSE filter (6) becomes a scalar processor of the form

$$\hat{f}_{\text{LLMMSE}}(i,j) = E(f(i,j)) + \frac{\sigma_f^2(i,j)}{\sigma_f^2(i,j) + \sigma_n^2(i,j)} \cdot [g(i,j) - E(g(i,j))] \tag{17}$$

where $\sigma_n^2(i,j)$ is the nonstationary noise ensemble variance.

Interpretation of the Filter

If all the *a priori* image and noise statistics on the right side of (17) are known, then $\hat{f}_{\text{LLMMSE}}(i, j)$ is a function of measurements $g(i, j)$ only, and each estimated pixel can be produced in parallel, perhaps in real-time. Using the fact that the noise $\boldsymbol{n}$ is zero mean, we have

$$E(g(i,j)) = E(f(i,j)) \tag{18}$$

and we can rearrange (17) as

$$\hat{f}_{\text{LLMMSE}}(i,j) = (1 - w(i,j))\,E(f(i,j)) + w(i,j)\,g(i,j) \tag{19}$$

$$w(i,j) = \frac{\sigma_f^2(i,j)}{\sigma_f^2(i,j) + \sigma_n^2(i,j)}. \tag{20}$$

Thus, the LLMMSE estimate is a weighted sum of the ensemble mean $E(f(i, j))$ and the normalized observation $g(i, j)$, where the weight is determined by the ratio of the signal variance to the noise variance. For a low signal-to-noise (SNR) ratio, the LLMMSE filter puts more weight on the *a priori* mean $E(f(i, j))$ because the observation is too noisy to make an accurate estimate of the original image. Conversely, for high SNR, the LLMMSE estimate puts more weight on the noisy observation and the result is to preserve the edge sharpness.

It is interesting to compare these properties of the LLMMSE filter with the results obtained by Anderson and Netravali [12], in which they derived a nonlinear restoration filter based on a subjective visibility function to make a balance between noise smoothing and resolution. The nonlinear filter tends to average out the random noise in the flat areas and preserve edge sharpness so that the same amount of subjective noise is suppressed in the whole image. Thus, the filter response of their nonlinear approach is similar to that of the LLMMSE filter. The nonlinear approach uses a "masking function" to measure the spatial activity or the nonstationarity of the image and uses this measure to suppress the same amount of subjective noise over the image according to an exponentially decreased visibility function, whereas the LLMMSE filter directly uses a nonstationary image model and tries to minimize the local mean square error. The net effect of these two approaches are similar because one is varying the subjective noise variance according to the contextual information of the image, while the other imbeds the contextual information directly into a nonstationary image model. Therefore, the local image variance not only has its statistical meaning but also serves as a spatial "masking function," and the LLMMSE filter is similar to a subjective smoothing filter with a linearly decreasing visibility function. The two seemingly different approaches are now related and support the idea that the minimum mean square error criterion is a reasonable measure for image restoration if it is used locally with a nonstationary image model.

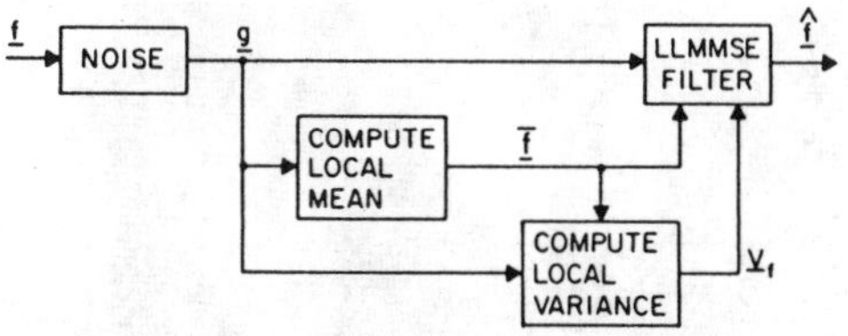

Fig. 2. Adaptive noise smoothing filter structure.

IV. Adaptive Noise Smoothing Filter

The LLMMSE filter shown in (17) requires the ensemble mean and variance of $f(i, j)$. Usually these statistics are not available *a priori* and can only be estimated from the degraded image. If we assume as before that the ensemble statistics can be replaced by local spatial statistics that are estimated from the degraded image, we have the adaptive noise smoothing filter

$$\hat{f}(i,j) = \bar{f}(i,j) + \frac{v_f(i,j)}{v_f(i,j) + \sigma_n^2(i,j)} (g(i,j) - \bar{g}(i,j)) \tag{21}$$

where $\bar{f}(i, j)$ and $\bar{g}(i, j)$ are the local spatial means of $f(i, j)$ and $g(i, j)$, respectively, and $v_f(i, j)$ is the local spatial variance of $f(i, j)$. Computation of these quantities is discussed in the following section. It is interesting to see that the adaptive noise smoothing filter is effectively a nonlinear filter even though it has the same linear form as the LLMMSE filter. The nonlinearity is introduced by the ratio of the local spatial variances in (21) and their estimation by nonlinear functions of the noisy observation $g(i, j)$. The performance of the adaptive noise smoothing filter depends heavily on the method used to calculate the local statistics. A block diagram of the adaptive noise smoothing filter is illustrated in Fig. 2.

Computation of the Local Statistics

The underlying assumption of the adaptive noise smoothing filter is that an image is locally ergodic such that the ensemble statistics can be replaced by the local spatial statistics. Therefore, the choice of method used for estimating local statistics is critical. One way to obtain the local mean and local variance is to calculate over a uniform moving average window of size $(2r + 1) \times (2s + 1)$. We then have

$$\bar{g}(i,j) = \frac{1}{(2r+1)(2s+1)} \sum_{p=i-r}^{i+r} \sum_{q=j-s}^{j+s} g(p,q) \tag{22}$$

and

$$v_g(i,j) = \frac{1}{(2r+1)(2s+1)} \sum_{p=i-r}^{i+r} \sum_{q=j-s}^{j+s} (g(p,q) - \bar{g}(i,j))^2 \tag{23}$$

where $g(i, j)$ and $v_g(i, j)$ are the local mean and local variance of $g(i, j)$, respectively. These statistics are commonly known as the sample mean and sample variance. They are widely used in statistical analysis and can be shown to be the maximum likelihood estimates of the unknown mean and variance of a Gaussian probability density function [19] by assuming that the samples in the summation are from the same ensemble. The local statistics of $f(i, j)$ can be calculated from the local statistics of $g(i, j)$ by assuming the relationship between their ensemble statistics also holds for their local statistics. Therefore, the functional forms of these transformations depend on the particular noise structure. The local statistics of $f(i, j)$ then feed into the filter to adapt the filter to the nonstationary content of the image and signal-dependent characteristics of the noise.

As an example of the computation of local statistics of $f(i, j)$, consider the signal-independent, additive statistically stationary noise model with no blurring. The local mean of $f(i, j)$ is equal to the local mean of $g(i, j)$ from (18). It is straightforward to show that the local variance of $f(i, j)$ is given by

$$v_f(i,j) = v_g(i,j) - \sigma_n^2(i,j) \tag{24}$$

where $\sigma_n^2(i, j)$ is the nonstationary noise variance. The function $\sigma_n^2(i, j)$ is assumed known from *a priori* measurements on the imaging system [1]. Substituting $v_f(i, j)$ into (21), we have the adaptive noise smoothing filter for the signal-independent, additive noise model.

The adaptive estimate is a balance between the local mean estimate $\bar{f}(i, j)$ and the noisy observation $g(i, j)$. The local variance $v_f(i, j)$ is an indication of our confidence in the local mean estimate. For the signal-independent, additive noise model, the adaptive noise smoothing filter with local statistics defined as in (22) and (23) is the same as Lee's local statistics algorithm [16], which is derived using a heuristic argument. Using the NMNV image model gives a better understanding of the problem and simplifies the extension of the filter to signal-dependent noise. For the multiplicative noise model and signal-dependent noise in general, we will show that Lee's approach is not optimal. Abramatic and Silverman's [13] nonlinear "signal equivalent" filter also has a similar structure to the adaptive noise smoothing filter. It can be treated as a special case of the adaptive noise smoothing filter where the local mean and local variance are estimated by a Wiener filter and a "masking function," respectively.

The local statistics calculated according to (22) and (23) assume that the samples within the averaging window are from the same ensemble. This is not true if there is a sharp edge within the window. The value of the sample variance near the edge will be larger than the ensemble variance because we use samples in two entirely different ensembles to calculate the local variance. The sample mean will tend to smear out compared with the ensemble mean. To avoid these effects, we could use an intelligent filter that can find edges and use the correct neighborhood of a pixel for calculating the local statistics. Various approaches for using edge detectors have been proposed for designing an anisotropic restoration filter and calculating the local statistics [12], [20], [24], and they do a reasonable job. Here, we introduce a simple function form for estimating the local variance by including the nonstationarity of the image in the function.

The inaccurate sample mean seems to have less effect on the filter output than the inaccurate sample variance because the NMNV image model only uses the nonstationary mean to describe the gross structure of the image, while the nonstationary

(a) (b) (c) (d)

Fig. 3. Adaptive noise smoothing filter for signal-dependent additive noise. (a) Original image. (b) Original degraded by additive noise with $\sigma_n^2 = 100$. (c) Adaptive noise smoothing estimate with local variance of (23). (The same as Lee's local statistics algorithm.) (d) Adaptive noise smoothing estimate with local variance of (25).

variance is used to characterize the edge information. In order to preserve the noise smoothing ability of the local mean, we still use the sample mean as our local mean estimate. The new local variance is defined as

$$v_g(i,j) = \frac{1}{(2r+1)(2s+1)} \sum_{p=i-r}^{i+r} \sum_{q=j-s}^{j+s} \cdot c(i-p, j-q)(g(p,q) - \bar{g}(p,q))^2 \tag{25}$$

where $c(i, j)$ is a weighting function. The weights $c(i, j)$ are chosen such that $\Sigma_i \Sigma_j c(i, j) = 1$, and $c(i, j)$ is a monotonically decreasing function (e.g., Gaussian function) to put more confidence on the center variance estimates. The sample variance estimate in (23) implicitly assumes that $g(i, j)$ is locally stationary such that $\bar{g}(i, j)$ can be used as the local mean for all $g(p, q)$ within the averaging window. In our new local variance estimate, the locally stationary assumption is removed and the local mean $\bar{g}(p, q)$ is allowed to vary for each $g(p, q)$ within the window. Therefore, this new local variance should have good performance near the edges.

Simulation Results

The original girl image is shown in Fig. 3(a). Fig. 3(b) is the original image degraded by a signal-independent, additive, white noise with variance 100. The degraded image is processed by the adaptive noise smoothing filter with the sample mean and sample variance calculated over a 5 X 5 window as the local statistics, and the smoothed image is shown in Fig. 3(c). In the uniform region, the estimate is close to the local mean estimate and the noise is smoothed by a large amount. Conversely, in the edge area, the restored image is close to the noisy observation and the edge sharpness is preserved. The squared error (SE) between the smoothed image and the original image is 36.7 per pixel. This value is very small compared with the SE of 94.1 per pixel for the local mean estimate that is essentially a spatially invariant low-pass filter. Fig. 3(d) shows the adaptive noise smoothing estimate of the original

Fig. 4. Adaptive noise smoothing filter for signal-dependent additive noise. (a) Original image. (b) Original degraded by additive noise with $\sigma_n^2 = 400$. (c) Adaptive noise smoothing estimate with local variance of (23). (The same as Lee's local statistics algorithm.) (d) Adaptive noise smoothing estimate with local variance of (25).

image using (25) to calculate the local variance. This estimate seems to have better visual quality than Fig. 3(c). It looks smoother both at the step edges and ramp edges. The SE of Fig. 3(d) is 34.6 per pixel. This is slightly smaller than for the sample variance case indicating the performance improvement of using the new local variance estimate. The same sequence of the simulation results for a low SNR case where the noise variance is equal to 400 is shown in Fig. 4.

V. Adaptive Noise Smoothing Filter for Various Signal-Dependent Noise Models

The general form of the adaptive noise smoothing filter for a class of uncorrelated, signal-dependent noise is shown in (21). It is useful to examine the explicit structure of this filter for some physical noise models that are frequently encountered in practical imaging systems. In this section, we derive the explicit structure of the adaptive noise smoothing filter for multiplicative noise, film-grain noise, and Poisson noise.

Multiplicative Noise

The degradation model for the multiplicative noise model can be written as

$$g'(i,j) = u(i,j)\, f(i,j) \tag{26}$$

where $u(i,j)$ is independent of $f(i,j)$. The multiplicative noise is assumed to have a stationary mean and variance given by

$$E(u(i,j)) = E(\boldsymbol{u}) \tag{27}$$

and

$$E[(u(i,j) - E(\boldsymbol{u}))^2] = \sigma_u^2, \tag{28}$$

respectively. We now define a normalized observation

$$g(i,j) = g'(i,j)/E(\boldsymbol{u}) \tag{29}$$

such that

$$E[g(i,j)] = E[f(i,j)]. \tag{30}$$

If we represent (29) in terms of signal plus signal-dependent

(a) (b) (c) (d)

Fig. 5. Adaptive noise smoothing filter for multiplicative noise. (a) Original image. (b) Original degraded by multiplicative noise with $\sigma_u^2 = 0.007$. (c) Adaptive noise smoothing estimate with local variance of (23). (Similar to Lee's local statistics algorithm.) (d) Adaptive noise smoothing estimate with local variance of (25).

additive noise, we have

$$g(i,j) = f(i,j) + \left[\frac{(u(i,j) - E(\boldsymbol{u}))}{E(\boldsymbol{u})}\right] \cdot f(i,j) \tag{31}$$

and we can identify

$$n(i,j) = (u(i,j)/E(\boldsymbol{u}) - 1)\, f(i,j) \tag{32}$$

from (15). It is straightforward to compute the variance

$$\sigma_n^2(i,j) = \frac{\sigma_u^2\,[(E(f(i,j)))^2 + \sigma_f^2(i,j)]}{[E(\boldsymbol{u})]^2} \tag{33}$$

and to verify that $n(i, j)$ satisfies (9). From these facts and (30) we can solve for the statistical variance

$$\sigma_f^2(i,j) = \frac{\sigma_g^2(i,j) - \sigma_u^2\,[E(g(i,j))^2]/[E(\boldsymbol{u})]^2}{1 + \sigma_u^2/[E(\boldsymbol{u})]^2} \tag{34}$$

in terms of statistics of $g(i,j)$.

Replacing ensemble statistics in (17), (30), and (34) by local spatial statistics gives the adaptive noise smoothing filter for the multiplicative noise model as

$$\hat{f}(i,j) = \bar{f}(i,j) + \frac{v_f(i,j)(g(i,j) - \bar{f}(i,j))}{v_f(i,j) + \sigma_u^2\,[(\bar{f}(i,j))^2 + v_f(i,j)]/[E(\boldsymbol{u})]^2} \tag{35}$$

where $g(i,j)$ is the normalized observation and $\sigma_u^2/[E(\boldsymbol{u})]^2$ is a parameter characterizing the multiplicative noise level. The term $(\bar{f}(i, j))^2 + v_f(i, j)$ in (35) shows the signal-dependent properties of multiplicative noise, and makes the adaptive noise smoothing filter change its characteristics adaptively to smooth the signal-dependent noise. Note that in the case of very low noise, $\sigma_u^2 \ll [E(\boldsymbol{u})]^2$, and the best estimate is the measurement $g(i,j)$ itself.

The result of (35) is a generalization of Lee's local statistics algorithm for multiplicative noise [20] in that an additional term $\sigma_u^2 v_f(i, j)/[E(\boldsymbol{u})]^2$ appears in the denominator. The term does not appear in Lee's derivation due to the linear approxi-

Fig. 6. Adaptive noise smoothing filter for multiplicative noise. (a) Original image. (b) Original degraded by multiplicative noise with $\sigma_u^2 = 0.04$. (c) Adaptive noise smoothing estimate with local variance of (23). (Similar to Lee's local statistics algorithm.) (d) Adaptive noise smoothing estimate with local variance of (25).

mation made there for the nonlinear multiplicative noise model. In the approach presented in this paper, we derive the optimal linear filter for the given classes of signal-dependent noise by first representing the noise model in signal plus additive noise form, and then obtaining the variance of the noise term. Thus, the linear constraint is applied to the estimator structure rather than to the nonlinear observation model. Of course, the significance of this additional term in actual application depends on the relative magnitude of σ_u^2, $E(\boldsymbol{u})$, $v_f(i, j)$ and $\bar{f}(i, j)$. Lee [21] has examined a multiplicative noise model of the form of (26) for synthetic aperture radar (SAR) imagery degraded by coherent speckle noise and has found the effects of the additional term to be small.

The simulation results are shown in Fig. 5. The original image is in Fig. 5(a). The image degraded by a multiplicative noise with unit mean and variance equal to 0.007 is shown in Fig. 5(b). Fig. 5(c) is the adaptive noise smoothing estimate of the original image by using the sample mean and sample variance as local statistics. Fig. 5(d) is the adaptive estimate using the new local variance estimate as in Fig. 3(d). These results are comparable with those of Fig. 3. Another set of simulation results for $\sigma_u^2 = 0.04$ are shown in Fig. 6. In these examples the effects of the additional term in the denominator of (35) are small.

Film-Grain Noise

Film-grain noise inherently exists in the process of photographic recording and reproduction. If we process the film in the linear region of the *D*-log *E* curve and ignore the blurring effect of the model [22], we have

$$g(i, j) = f(i, j) + \alpha f^{1/3}(i, j)\, u(i, j) \tag{36}$$

where $u(i, j)$ is a signal-independent noise and α is a proportionality constant. This model is very similar to the additive form of the multiplicative noise model except for the nonlinear effect of $f^{1/3}(i, j)$. Therefore, the derivation of the adaptive noise smoothing filter is similar to that for multiplicative noise.

(a) (b) (c) (d)

Fig. 7. Adaptive noise smoothing filter for Poisson noise. (a) Original image. (b) Original degraded by Poisson noise with $\lambda = 1$. (c) Adaptive noise smoothing estimate with local variance of (23). (d) Adaptive noise smoothing estimate with local variance of (25).

Poisson Noise

Photon noise is a fundamental limitation of images detected at low light levels [23]. The degradation model of Poisson noise is given by

$$g'(i,j) = \text{Poisson}_\lambda(f(i,j)) \tag{37}$$

where $\text{Poisson}_\lambda(\cdot)$ is a Poisson random number generator, and λ is a proportionality factor. The probabilistic description of a Poisson process is given by

$$P(g'(i,j)|f(i,j)) = \frac{(\lambda f(i,j)^{g(i,j)})\, e^{-\lambda f(i,j)}}{g(i,j)!}. \tag{38}$$

The conditional ensemble mean and variance of $g'(i,j)$ given $f(i,j)$ are

$$E[g'(i,j)|f(i,j)] = \lambda f(i,j) \tag{39}$$

$$\text{Var}\,[g'(i,j)|f(i,j)] = \lambda f(i,j). \tag{40}$$

We define the normalized observation as

$$g(i,j) = g'(i,j)/\lambda = \frac{\text{Poisson}_\lambda(f(i,j))}{\lambda}. \tag{41}$$

If we represent the normalized Poisson observation in terms of signal plus signal-dependent additive noise, we have

$$g(i,j) = f(i,j) + (g(i,j) - f(i,j)). \tag{42}$$

Therefore, the noise part has the form

$$n(i,j) = g(i,j) - f(i,j) = \frac{\text{Poisson}_\lambda(f(i,j))}{\lambda} - f(i,j) \tag{43}$$

and its variance can be shown to be

$$\sigma_n^2(i,j) = E(f(i,j))/\lambda. \tag{44}$$

From these equations, we solve for the ensemble variance of $f(i,j)$ and replace all ensemble statistics by local spatial statistics to obtain

$$v_f(i,j) = v_g(i,j) - (\bar{f}(i,j)/\lambda). \tag{45}$$

Thus, the adaptive noise smoothing filter for images with Poisson noise can be expressed as

$$\hat{f}(i,j) = \bar{f}(i,j) + \frac{v_f(i,j)(g(i,j) - \bar{f}(i,j))}{v_f(i,j) + (\bar{f}(i,j)/\lambda)} \tag{46}$$

(a) (b) (c) (d)

Fig. 8. Adaptive noise smoothing filter for Poisson noise. (a) Original image. (b) Original degraded by Poisson noise with $\lambda = 0.25$. (c) Adaptive noise smoothing estimate with local variance of (23). (d) Adaptive noise smoothing estimate with local variance of (25).

where $g(i, j)$ is the normalized observation and $\bar{f}(i, j)/\lambda$ is an indication of the Poisson noise level at point (i,j).

The simulation results are shown in Fig. 7. Fig. 7(a) is the original image. Fig. 7(b) is the original image degraded by a Poisson noise with $\lambda = 1$. Fig. 7(c) is the adaptive noise smoothing estimate of the original image obtained by using the sample mean and sample variance as local statistics. Fig. 7(d) is the result of using the new local variance of (25). The same sequence of simulation results for $\lambda = 0.25$ is shown in Fig. 8.

VI. Conclusions

The adaptive noise smoothing filter has a very simple structure and does not require any *a priori* information of the original image. All the parameters needed are estimated from the noisy image. The filter adopts a nonstationary image model and can be applied to different types of signal-dependent noise. The minimum mean square error criterion is used locally rather than globally and has some of the desirable properties of a subjective error criterion. The calculation of the local image statistics is critical to the quality of the restored image. The adaptive noise smoothing filter has the advantage of separating the estimation of local statistics from the image restoration filter. Thus, we can use various sophisticated methods to estimate the local image statistics while keeping the restoration filter structure fixed. Furthermore, the local statistics can be estimated by using a recursive filter to reduce the computation. The extension of adaptive noise smoothing filter to recursive image restoration is discussed in another paper [17].

References

[1] W. K. Pratt, *Digital Image Processing.* New York: Wiley-Interscience, 1978.

[2] N. E. Nahi and T. Assefi, "Bayesian recursive image estimation," *IEEE Trans. Comput.*, vol. C-12, pp. 734–738, July 1972.

[3] A. O. Aboutalib, M. S. Murphy, and L. M. Silverman, "Digital restoration of image degraded by general motion blurs," *IEEE Trans. Automat. Contr.*, vol. AC-22, pp. 294–302, June 1977.

[4] M. S. Murphy and L. M. Silverman, "Image model representation and line-by-line recursive restoration," *IEEE Trans. Automat. Contr.*, vol. AC-23, pp. 809–816, Oct. 1978.

[5] J. W. Woods and C. H. Radewan, "Kalman filtering in two dimensions," *IEEE Trans. Inform. Theory*, vol. IT-23, pp. 473–482, July 1977.

[6] J. W. Woods and V. K. Ingle, "Kalman filtering in two dimensions: Further results," *IEEE Trans. Acoust., Speech, Signal Processing*, vol. ASSP-29, pp. 188-196, Apr. 1981.
[7] B. R. Hunt and T. M. Cannon, "Nonstationary assumptions for Gaussian models of images," *IEEE Trans. Syst., Man, Cybern.*, vol. SMC-6, pp. 876-881, Dec. 1976.
[8] H. T. Trussell and B. R. Hunt, "Sectioned methods for image restoration," *IEEE Trans. Acoust., Speech, Signal Processing*, vol. ASSP-26, pp. 157-164, Apr. 1978.
[9] D. S. Lebedev and L. I. Mirkin, "Smoothing of two-dimensional images using the 'composite' model of a fragment," in *Iconics-Digital Holography-Image Processing.* USSR Inst. Inform. Transmission Problems, Academy of Sciences, pp. 57-62, 1975.
[10] V. K. Ingle and J. W. Woods, "Multiple model recursive estimation of images," in *Proc. IEEE Int. Conf. Acoust., Speech, Signal Processing*, Washington, DC, Apr. 1979, pp. 642-645.
[11] S. A. Rajala and R. J. P. de Figueiredo, "Adaptive nonlinear image restoration by a modified Kalman filtering approach," *IEEE Trans. Acoust., Speech, Signal Processing*, vol. ASSP-29, pp. 1033-1042, Oct. 1981.
[12] C. L. Anderson and A. N. Netravali, "Image restoration based on subjective criterion," *IEEE Trans. Syst., Man, Cybern.*, vol. SMC-6, pp. 845-853, Dec. 1976.
[13] J. F. Abramatic and L. M. Silverman, "Nonlinear restoration of noisy images," *IEEE Trans. Pattern Anal. Mach. Intell.*, vol. PAMI-4, pp. 141-149, Mar. 1982.
[14] F. Naderi and A. A. Sawchuk, "Estimation of images degraded by film-grain noise," *Appl. Optics.*, vol. 17, pp. 1228-1237, Apr. 1978.
[15] C. M. Lo and A. A. Sawchuk, "Nonlinear restoration of filtered images with Poisson noise," *Proc. SPIE Tech. Symp.–Appl. Digit. Image Processing-III*, San Diego, CA, Aug. 1979, vol. 207, pp. 84-95.
[16] J. S. Lee, "Digital image enhancement and noise filtering by use of local statistics," *IEEE Trans. Pattern Anal. Mach. Intell.*, vol. PAMI-2, pp. 165-168, Mar. 1980.
[17] D. T. Kuan, A. A. Sawchuk, T. C. Strand, and P. Chavel, "Nonstationary 2-D recursive image restoration," submitted for publication.
[18] A. P. Sage and J. L. Melsa, *Estimation Theory with Applications to Communications and Control.* New York: McGraw-Hill, 1971.
[19] J. S. Meditch, *Stochastic Optimal Linear Estimation and Control.* New York: McGraw-Hill, 1969.
[20] J. S. Lee, "Refined noise filtering using local statistics," *Comput. Graphics Image Processing*, vol. 15, pp. 380-389, 1981.
[21] —, "Speckle analysis and smoothing of synthetic aperture radar images," *Comput. Graphics Image Processing*, vol. 17, pp. 24-32, 1981.
[22] T. S. Huang, "Some notes on film-grain noise," Appendix 14, in *Restoration of Atmospherically Degraded Images*, NSF Summer Study Rep., Woods Hole, MA, 1966, pp. 105-109.
[23] J. W. Goodman and J. F. Belsher, "Fundamental limitations in linear invariant restoration of atmospherically degraded images," in *Proc. Soc. Photo-Opt. Instrum.*, 1976, pp. 141-154.
[24] M. Nagao and T. Matsuyama, "Edge preserving smoothing," *Comput. Graphics Image Processing*, vol. 9, pp. 394-407, 1979.

Blind Deconvolution of Spatially Invariant Image Blurs with Phase

MICHAEL CANNON, MEMBER, IEEE

Abstract—**This paper is concerned with the digital estimation of the frequency response of a two-dimensional spatially invariant linear system through which an image has been passed and blurred. For the cases of uniform linear camera motion and an out-of-focus lens system it is shown that the power cepstrum of the image contains sufficient information to identify the blur. Methods for deblurring are presented, including restoration of the density version of the image. The restoration procedure consumes only a modest amount of computation time. Results are demonstrated on images blurred in the camera.**

I. Introduction

TWO-DIMENSIONAL IMAGERY, be it digital, electronic, or photographic, has found use in many areas of modern endeavor. For example, many scientific space vehicles and other high-altitude craft rely heavily on two-dimensional data as a means of obtaining information. Objects of study may range from Nix Olympica on Mars, to cornfields in Iowa, to sections of a submarine hull in a Moscow steelyard. Aerial photography may also be used for geographic purposes, mensuration, conservation studies, and weather prediction. More down to earth uses of two-dimensional data include conventional medical X rays (radiographs), angiograms, stress photography, and flash radiography, to name but a few. There are also many signals that are not pictorial in nature, but are nevertheless two-dimensional, such as those obtained from many types of radar and sonar as well as some forms of geophysical seismic soundings.

All of the above signals, though perhaps not explicitly digital in nature, can be represented as such in a digital computer. This permits one to apply a variety of digital techniques to the data to achieve any of a wide range of goals such as image enhancement, data compression, scene analysis, or deblurring. It is the last of these operations which is explored in this paper.

II. Mathematical Model of Image Blurring

It will be assumed that the cause of the blurring can be modeled as a spatially invariant linear system. The blurred image $b(x,y)$ is then equal to the convolution of the object intensity function $i(x,y)$ and the point-spread function (PSF) $a(x,y)$ of the blurring system. All noise $n(x,y)$ is modeled as additive, although this may not always reflect reality (film-grain noise being a prime example). We therefore have

$$b(x,y) = \iint_{-\infty}^{\infty} i(x-\tau, y-\rho)\, a(\tau,\rho)\, d\tau d\rho + n(x,y) \qquad (1)$$

Manuscript received November 20, 1974; revised May 14, 1975 and September 3, 1975.

The author was with the Department of Electrical Engineering, University of Utah, Salt Lake City, UT. He is now with the Los Alamos Scientific Laboratory, Los Alamos, NM 87545.

Reprinted from *IEEE Trans. Acoust., Speech, Signal Processing,* vol. ASSP-24, pp. 58–63, Feb. 1976.

or

$$b(x,y) = i(x,y) \circledast a(x,y) + n(x,y). \tag{2}$$

In the blind deconvolution problem we are given $b(x,y)$ and must estimate $i(x,y)$ having incomplete knowledge of the blurring system, $a(x, y)$, which is hidden from view. In this paper $a(x,y)$ is restricted to be the PSF of either linear camera motion and/or an out-of-focus lens system. One then knows $a(x,y)$ exactly except for a small number of parameters which hopefully can be determined from the blurred image itself. The PSF resulting from uniform camera motion can be approximated by a rectangle, whose orientation and length are indicative of the direction and extent of the blur. In the case of a horizontal camera motion blur of length $2d$

$$a(x,y) = \begin{cases} 0; & y \neq 0, -\infty \leqslant x \leqslant \infty \\ 1/(2d); & y = 0, -d \leqslant x \leqslant d. \end{cases} \tag{3}$$

The PSF of a defocused lens system with a circular aperture can be approximated by a cylinder whose radius R depends on the focus defect extent.

In this case we have

$$a(x,y) = \begin{cases} 0; & \sqrt{x^2 + y^2} > R \\ 1/(\pi R^2); & \sqrt{x^2 + y^2} \leqslant R. \end{cases} \tag{4}$$

The effects of these blurs can be better understood by mapping the convolutional relationship of (2) into one of multiplication via the Fourier integral transform

$$B(u, v) = I(u, v) \cdot A(u, v) + N(u, v), \tag{5}$$

where $A(u, v)$ is the two-dimensional frequency response of the blurring system and B, I, and N are the transforms of the blurred image, original image, and noise, respectively. For a motion blur of length d whose direction is θ degrees off the horizontal, $A(u, v)$ has the form $\sin(\pi df)/(\pi df)$ where $f = u \cdot \cos(\theta) + v \sin(\theta)$. We see that higher frequencies are attenuated. The alternate lobes of $\sin(\pi df)/(\pi df)$ produce phase shifts of π radians. The frequency response for defocus blur is of the form $J_1(Rr)/(Rr)$, where R is the radius of the blur PSF and $r = \sqrt{u^2 + v^2}$. This is similar in shape to the motion blur response but is circularly symmetric.

III. Method of Blur Identification

One might be tempted to determine $a(x,y)$ by identifying the $A(u, v)$ component of $B(u, v)$ [1]. This might be done by searching for the zero crossings of the frequency response of the blurring system. If we neglect noise, the zeros of $A(u, v)$ are zeros of $B(u, v)$. In the case of motion blur these zeros occur along lines perpendicular to the direction of blur and which are spaced at intervals of $1/d$. In the case of defocus blur these zeros occur on concentric circles about the origin. The circles are nearly periodic in r.[1]

It is the author's experience that, except for images blurred in the computer, the technique of searching for zero crossings in $B(u, v)$ usually fails. Although the zero crossings are well-defined theoretically, they are obscured in practice by the extreme randomness of $I(u, v)$ and $N(u, v)$.

Another tempting approach to identify $a(x,y)$ is to compute the cepstrum [2] by $b(x,y)$. The cepstrum is defined as

$$C_b(p,q) = F\{\log|B(u, v)|\}, \tag{6}$$

where F denotes the Fourier integral transform. It follows from (5) and (6) that the convolutional effects of $a(x,y)$ are additive in the cepstral domain [3]. Neglecting the effects of noise, we have

$$C_b(p,q) = C_i(p,q) + C_a(p,q). \tag{7}$$

Rom [4] has pointed out that the periodic zeros in $A(u, v)$ (or nearly periodic in the case of focus blur) lead to large negative spikes in $C_a(p,q)$. For example, the zeros of motion blur fall along lines spaced $1/d$ apart. This periodic pattern results in a negative spike in $C_a(p,q)$, a distance d from the origin. The amount the spike has been rotated about the origin indicates the direction of the motion blur. The circular symmetry of the out of focus PSF results in rings of spikes in the cepstrum.

In the case in which both blurs are present in the same image, the two sets of spikes are simply added together. This follows from the fact that if

$$b(x,y) = i(x,y) \circledast a(x,y) \circledast g(x,y), \tag{8}$$

where $g(x,y)$ is the PSF of a second blur, then

$$C_b(p,q) = C_i(p,q) + C_a(p,q) + C_g(p,q). \tag{9}$$

It is again the author's experience that attempts to locate the spikes of $C_a(p,q)$ in $C_b(p,q)$ are often frustrated due to the effects of noise and the overlying structure of $C_i(p,q)$. It is with this in mind that we turn to an averaging scheme with the hope of mitigating the effects of both signal and noise on our attempts to identify the blur in both the spectral and cepstral domains.

Let us assume that the blurred signal under consideration (as well as the additive noise) are sample functions of a stationary random process. We can then express (5) in terms of power spectra

$$\Phi_b(u, v) = \Phi_i(u, v)\,|A(u, v)|^2 + \Phi_n(u, v). \tag{10}$$

It is further assumed that the spatial extent of the PSF of the blur is at least two orders of magnitude smaller than that of $i(x,y)$. We then proceed to estimate $\Phi_b(u, v)$ in the manner proposed by Welch [5]. The blurred image $b(x,y)$ is broken up into many smaller images $b_j(x,y)$, each of which is large enough to contain the PSF of the blur $a(x,y)$. If the b_j's are each 32 or 64 picture elements square, a 512 × 512 image can easily be broken up into one or two hundred (possibly overlapping) subsections. After each subsection is windowed to reduce edge effects, the following relationship holds approximately

$$b_j(x,y) = i_j(x,y) \circledast a(x,y) + n_j(x,y). \tag{11}$$

An average is then taken over the square of the magnitude of the Fourier transforms of the b_j's. This constitutes an estimate of $\Phi_b(u, v)$. It can be seen that such an estimate of the right-hand side of (10) will retain much of the flavor of

[1]The first five zeros of $J_1(r)/r$ occur at r = 3.83, 7.01, 10.2, 13.3, and 16.5, respectively.

$|A(u,v)|^2$. This is the case since $|A(u,v)|^2$ will be present in each one of the averaged subsections. The contribution from each i_j and n_j, on the other hand, will vary from section to section, thus resulting in a more subtle contribution to the average. This is especially true since the above assumption of stationarity is often not a very good one.

The search for the zeros of $|A(u,v)|$ is now much more successful than before, especially if one makes a rough guess as to the general form (say, an exponential) of $\Phi_i(u,v)$ and removes this bias from the right-hand side of (10). This can usually be done effectively in the low-frequency regions of $\Phi_b(u,v)$, where the effects of noise are small.

Since most blurred images have a reduced amount of high-frequency power, the search for the zeros of $|A(u,v)|$ must be carried out along the dc axes of $\Phi_b(u,v)$. Most other frequencies are usually dominated by noise. This masking of the zeros of $|A(u,v)|$ makes it difficult to determine the exact direction of a motion blur and nearly impossible to detect a combination of motion and focus blurs. This undesirable aspect of the power spectrum of $b(x,y)$ causes us to turn our attention to the power cepstrum.

The power cepstrum of $b(x,y)$ is defined as

$$P_b(p,q) = F\{\log \Phi_b(u,v)\}. \tag{12}$$

If we neglect the effects of noise, it follows from (10) and (12) that

$$P_b(p,q) = P_i(p,q) + 2C_a(p,q). \tag{13}$$

Since the right-hand side of (13) results from an averaging process, $P_i(p,q)$ retains little of the flavor of the original image, while $C_a(p,q)$ remains strongly characteristic of the spatially invariant blurring system.

The cepstral spikes which characterize the needed blur parameters are, as previously mentioned, negative ones. The ring of spikes resulting from focus blur is assumed to reflect radially periodic zeros in $J_i(r)/r$, even though that is not strictly true. A correction for this has not yet been made. In order for the cepstral spikes of motion and/or focus blur to be identifiable in spite of interference from noise and $P_i(p,q)$, the estimate of $\Phi_b(u,v)$ should be the result of an average based on about 100 image subsections. The resulting cepstral signatures of motion and focus blur are then sufficiently clear that automatic computer identification can be readily employed.

IV. Method of Image Restoration

Once the exact nature of the blur has been determined, many approaches can be taken to deconvolve it from the blurred image to produce a restoration [1], [2], [6], [13]. It is well known that restorations which are produced using the convolutional inverse of the blur are usually of very poor quality. This is the case because the inverse filter attempts to restore information lying in bands of frequency that may be dominated by noise. An attempt must therefore be made to limit the gain of the restoration filter, $H(u,v)$, in those areas of the image spectrum that have a low signal-to-noise ratio. The following constraint has been found [7] to be consistently useful:

$$\Phi_b(u,v)|H(u,v)|^2 = \Phi_r(u,v) = \Phi_i(u,v). \tag{14}$$

This constrains H so that the power spectrum, $\Phi_r(u,v)$, of the restored image is equal to that of the original, undegraded image. After substitution from (10), the power spectrum equalization filter becomes

$$|H(u,v)| = \sqrt{\frac{\Phi_i(u,v)}{\Phi_i(u,v)|A(u,v)|^2 + \Phi_n(u,v)}}. \tag{15}$$

The ergodic hypothesis is invoked to allow the estimation of the numerator of (15) from a scene believed to be statistically similar to $i(x,y)$. The choice of this scene does not seem to be critical [8]. The denominator of (15) is simply the power spectrum, $\Phi_b(u,v)$, of the blurred image itself. The above filter is particularly advantageous computationally in that $\Phi_b(u,v)$ was already estimated in the process of blur detection. It is also possible that $\Phi_i(u,v)$ could already be on hand, as it represents a whole ensemble of images. The above filter attempts to correct for the magnitude of $A(u,v)$. A correction for the phase of the blur is made by setting

$$\angle H(u,v) = -\angle A(u,v). \tag{16}$$

The phase of A is available as we have already identified the blur. Equations (15) and (16) are used to design a digital restoration filter [9].

V. Demonstration of Results

Figs. 1(a) and 2(a) were blurred through deliberate misuse of a real camera. The first was photographed out-of-focus, while the second suffers from camera motion. Figs. 1(b) and 2(b) show the respective power cepstra of the two images. The former shows the telltale ring of spikes resulting from focus blur. The latter shows clearly the twin peaks of motion blur. Figs. 1(c) and 2(c) are restorations of Figs. 1(a) and 2(a) based upon the relationships in (15) and (16). In each case the estimate of $\Phi_i(u,v)$ came from an image which was roughly similar to the blurred image in question.

The above restorations are marred by the presence of intense black areas in certain regions of the image. These black areas result from negative numbers in the processed image. Work by Lukosz [10] has shown that in order for the output of a linear system to be positive, given a positive input, the system must be roughly of a low-pass nature. This is clearly not the type of filter used to correct for motion or focus blur.

As a means of assuring a positive restoration, one may assume that it was a density version of the image that was blurred, not the intensity. Density refers to the concentration of silver in the developed photographic image, and is proportional to the logarithm of the original scene intensity which exposed the film. The last step of a density restoration is exponentiation. This converts the image to the intensity domain for viewing, and also ensures its positiveness. The results of restoring Figs. 1(a) and 2(a) in this manner are shown in Figs. 3(a) and (b). Fig. 4 demonstrates the density restoration of an image blurred by both camera motion and an out-of-focus lens. The processing of densities also allows for simultaneous contrast enhancement [11] as proposed by Stockham [12].

(a)

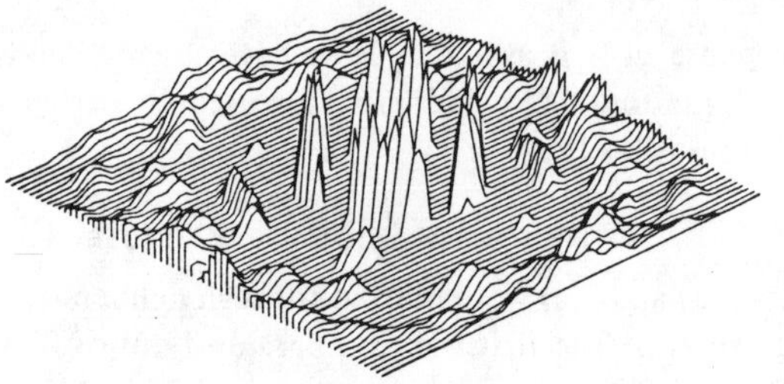

(b)

(c)

Fig. 1. (a) Original image, blurred in the camera by an out-of-focus lens. (b) Power cepstrum of the image in (a). In this picture the power cepstrum has been inverted and clipped at zero to disclose negative spikes that indicate the out-of-focus condition of the image. The radius of the ring of spikes reveals the severity of the blur. (c) Restoration of the image shown in (a). The image has been restored so that its power spectrum is now equal to that of the clear image before blurring occurred. Note the increased legibility of the fine text in the upper portion of the picture.

(a)

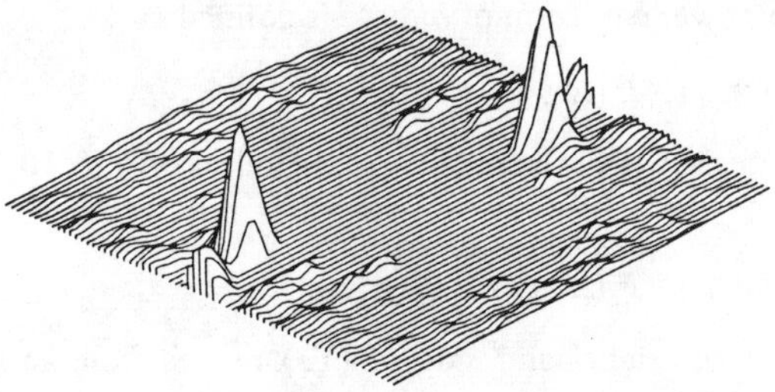

(b)

(c)

Fig. 2. (a) Original image, blurred by camera motion. The direction of motion was from left to right, at a slightly downward angle. (b) Inverted and clipped power cepstrum of the image in (a). The twin peaks indicate motion blur. Their separation and orientation about the origin indicate the severity and direction of the blur. (c) Restoration by power spectrum equalization of the image in (a). Note the increased clarity of the fine text. The ghost-like echoes are caused by having truncated the restoration kernel to fit on a 64×64 grid, instead of allowing it to extend to infinity, as it theoretically should.

(a)

(b)

Fig. 3. (a) and (b) These restorations were realized using the same restoration filters that created the restorations of Figs. 1(c) and 2(c). The difference is that the logarithm of the blurred image was taken before deblurring and the result exponentiated before display. This procedure has eliminated the dark "superblack" areas seen in the previous restorations.

VI. Conclusions

The blind deconvolution of image blurs with nonzero phase can be accomplished if the set of possible blurs is constrained to be a combination of linear camera motion and defocusing. Reliable cepstral parameters which determine the exact nature of the blur are attainable if the blur PSF is much smaller than the blurred image. The power spectrum equalization filter performs well in the presence of noise and produces meaningful restorations. The blur detection and image restoration

(a)

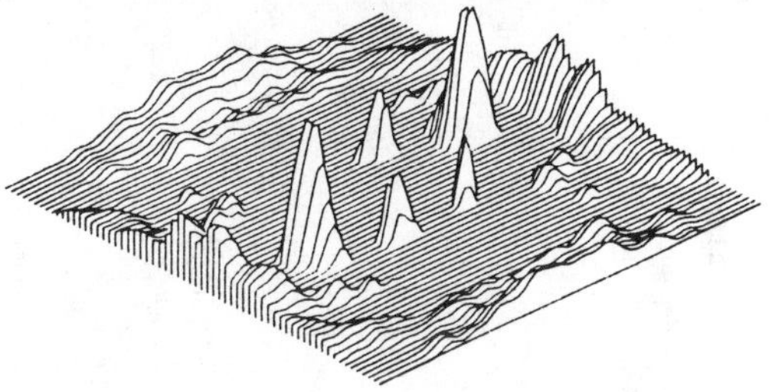

(b)

(c)

Fig. 4. (a) Original image, blurred in a moving, out-of-focus camera. (b) Power cepstrum (inverted and clipped) of the image in (a). Note the twin, telltale peaks of motion blur, as well as a smaller ring of spikes denoting out-of-focus blur. The position of the twin peaks denotes the direction and severity of the motion blur, while the radius of the ring of spikes indicates the severity of the focus blur. (c) Density restoration of the image in (a). Sufficient information to restore the fine text was lost due to the severity of the double blur, however, a marked improvement in the larger text can clearly be seen.

procedure requires only a modest amount of computing time.[2]

References

[1] D. Slepian, "Restoration of photographs blurred by image motion," *Bell Syst. Tech. J.*, vol. 46, pp. 2353-2362, Dec. 1967.

[2] M. M. Sondhi, "Image restoration: The removal of spatially invariant degradations," *Proc. IEEE* (*Special Issue on Digital Picture Processing*), vol. 60, pp. 842-853, July 1972.

[3] B. Bogart, M. Healy, and J. Tukey, "The quefrency analysis of time series for echoes," in *Proc. Symp. Time Series Analysis*, M. Rosenblatt, Ed. New York: Wiley, 1963, ch. 15.

[4] R. Rom, "On the cepstrum of two-dimensional functions," *IEEE Trans. Inform. Theory*, vol. IT-21, pp. 214-217, Mar. 1975.

[5] P. D. Welch, "The use of the fast Fourier transform for the estimation of power spectra," *IEEE Trans. Audio Electroacoust.* (*Special Issue on Fast Fourier Transform and its Application to Digital Filtering and Spectral Analysis*), vol. AU-15, pp. 70-73, June 1967.

[6] C. W. Helstrom, "Image restoration by the method of least squares," *J. Opt. Soc. Amer.*, vol. 57, pp. 297-303, Mar. 1967.

[7] T. G. Stockham, Jr., T. M. Cannon, and R. B. Ingebretsen, "Blind deconvolution through digital signal processing," *Proc. IEEE* (*Special Issue on Digital Signal Processing*), vol. 63, pp. 678-692, Apr. 1975.

[8] E. R. Cole, "The removal of unknown image blurs by homomorphic filtering," Dep. Comput. Sci., Univ. Utah, Salt Lake City, ARPA Tech. Rep. UTEC-CSc-74-029, 1974.

[9] H. D. Helms, "Nonrecursive digital filters: Design methods for achieving specifications on frequency response," *IEEE Trans. Audio Electroacoust.* (*Special Issue on Digital Filters: The Promise of LSI to Signal Processing*), vol. AU-16, pp. 336-342, Sept. 1968.

[10] W. Lukosz, "Uebertragung Nicht-negativer Signale durch Lineare Filter," *Opt. Acta*, vol. 9, pp. 335-364, 1962.

[11] T. M. Cannon, "Digital image deblurring by nonlinear homomorphic filtering," Dep. Comput. Sci., Univ. Utah, Salt Lake City, ARPA Tech. Rep. UTEC-CSc-74-091, 1974.

[12] T. G. Stockham, Jr., "Image processing in the context of a visual model," *Proc. IEEE* (*Special Issue on Digital Picture Processing*), vol. 60, pp. 828-842, July 1972.

[13] H. C. Andrews, "Outer product expansions and their uses in digital image processing," *Amer. Math. Mon.*, vol. 1, no. 82, pp. 1-13, Jan. 1975.

[2]The restorations shown in this paper were performed on a DEC PDP-10. The cepstral detection of blur parameters, filter generation, and restoration require about 20 min of computation (total) for a 512×512 image. The entire procedure can be carried out on a large scientific computer, such as a CDC 7600, in about 1 min.

The Feasible Solution in Signal Restoration

H. JOEL TRUSSELL, MEMBER, IEEE, AND MEHMET R. CIVANLAR, STUDENT MEMBER, IEEE

Abstract—**The feasible solution to the signal restoration problem is defined as the one which satisfies all constraints which can be imposed on the true solution. A very important set of constraints can be obtained by examining the statistics of the noise. These and other constraints can be described as closed convex sets. Thus, projection onto closed convex sets is the numerical method used to obtain a feasible solution. Examples of this method demonstrate its usefulness in one- and two-dimensional signal restoration. The limitations of the method are discussed.**

I. Introduction

The ultimate goal of signal restoration is to obtain the best estimate of a signal which has been degraded. A major problem is that there is no single definition of the "best" estimate. This paper proposes a solution which satisfies all *a priori* constraints imposed by the real world system, including knowledge of noise statistics.

Before defining any solution, it must be known how the data signal was formed. To proceed, we must have a mathematical model of the signal formation system. For this work, the standard linear model will be used. More realistic and more complex models may be used at the cost of additional computation time. The linear model may be written algebraically

$$g = Hf + n \tag{1}$$

where g is the recorded data of length N, f is the original signal to be estimated of length N, H is the impulse response or point spread function (psf) matrix ($N \times N$) usually representing convolution, and n is signal-independent noise of length N.

Once the model is specified, it remains to choose a restoration criterion. This criterion defines what is meant by the term "restoration." There are many criteria from which to choose. Maximum likelihood, minimum mean-square error, maximum *a posteriori* probability, and maximum entropy are a few criteria which have proven useful.

This paper will discuss a criterion which does not uniquely define a solution, but defines a class of solutions. If the class of solutions is sufficiently restrictive, any member of the class may be an acceptable estimate for the restoration problem. In other cases, the class may provide a restricted region over which some function may be optimized to produce a unique solution.

II. The Feasible Solution

In the mathematical programming environment, a solution is called feasible if it satisfies all the constraints placed on the system. The goal of mathematical programming is to find the optimal solution among all feasibles. This concept can be used in signal restoration. The problem can be divided into two parts: the appropriate constraints must be determined, and a definition of optimum must be quantified.

In this paper, we will be concerned with the first task. It will be shown that many times the restriction imposed by requiring a feasible solution is sufficient to produce a high-quality restoration. The problem of finding an optimal feasible solution will be left for future research.

In the most general case, we only know the model for the data formation. For this work, we have assumed the linear model of (1). A further assumption is the availability of the information about the parameters of the model; that is, the matrix H is given and the statistical properties of the noise are known. It would be beneficial if this limited knowledge could be used to maximal advantage. A method which uses the information about the noise statistics more effectively is the one which puts the appropriate constraints on the residual, defined by

$$r \triangleq g - H\hat{f} \tag{2}$$

where $\hat{f}$ is an estimate of the original signal in the model, (1).

A constraint which has been used previously is that the variance of the residual be approximately equal to the variance of the noise. This condition would hold for the true solution; hence, it is a reasonable constraint to place on the solution. This constraint is the only one used for constrained least squares restoration [1]. That method defines smoothness (minimum derivative) as the criterion to optimize. Other methods which use this constraint are maximum power [2] and maximum entropy [3].

There is usually more information about the noise which can be used. For example, if the noise is zero mean and has known autocorrelation, then the residual should be constrained to have approximately these same characteristics. The limits of these approximations can be determined by using confidence limits derived from sample statistics. If the probability density of the noise is known, e.g., Gaussian, it is possible to check the residual for outliers.

Checking the mean and outliers in the residual is straightforward. Checking the autocorrelation matrix is more difficult, and in addition, its modification is quite complex. However, since the power spectrum, which is directly related to the autocorrelation function, is more easily checked, the same information can be obtained by using it. The periodogram can be used to estimate the power spectrum. The statistics of the periodogram can be easily determined for many types of noise. A problem in using the periodogram is its inconsistency as an

Manuscript received August 2, 1983; revised October 14, 1983. This work was supported by the National Science Foundation under Grant ECS-8117499.

The authors are with the Department of Electrical and Computer Engineering, North Carolina State University, Raleigh, NC 27650.

Reprinted from *IEEE Trans. Acoust., Speech, Signal Processing,* vol. ASSP-32, pp. 201–212, Apr. 1984.

estimator for the power spectrum. Nevertheless, we will use the periodogram for reasons of mathematical tractability.

In addition to the noise parameters which are usually available, other characteristics of the signal may be known which provide additional constraints on the solution. It is clear that the more valid constraints that can be enforced upon the solution, the better the restoration will be. Often, these additional constraints will be more powerful than some of the noise constraints.

A. Variance of the Residual

As mentioned earlier, the variance or norm of the residual has been used in several restoration methods. It has been used as a convergence criterion in iterative restoration methods [4]. In order to use this statistic as a constraint, its variation caused by dealing with a finite number of samples must be considered. This has been done in [4].

To review briefly, if the noise is assumed to be Gaussian, the sample variance has a chi square distribution. The number of points in the signal is likely to be large, so that the Gaussian approximation to the chi square is valid. The confidence limit, usually 95 percent, is chosen by the user and is calculated by

$$\lim_v = \sigma^2 \; [\mp \lim_{0.95} + \sqrt{2(N-1)}]^2/2N \qquad (3)$$

where N is the number of points in the signal and $\lim_{0.95}$ is the 95 percent confidence limit for a standard normal distribution. Note that this gives upper and lower limits for the sample variance. Even if the noise is not Gaussian, N is usually large enough so that the central limit theorem can be applied. Thus, in most cases, calculating the limits for the sample variance is easily done.

The bounds on what is acceptable as an estimate of the variance is a subjective judgment. However, for any confidence interval, there is a finite probability that the sample variance lies outside the calculated limits. This probability can be taken into account in iterative methods by using successive relaxation of the limits. If the solution is not obtained in a reasonable amount of time using a particular set of limits, relax the limits and continue the iterations.

B. Mean of the Residual

It is usually assumed that the noise is a zero-mean process. If the noise is Gaussian, the sample mean is also Gaussian. As in the case of the sample variance, the large number of points in the signal will produce a statistic which is very close to having a Gaussian distribution. The confidence limits are easily found from the table of the standard normal distribution.

C. Outliers of the Residual

Not only should the mean of the residual be approximately zero, but the individual values of the residual should not deviate an unexpected distance from zero. Of course, if there is a large number of points, it is expected that more points in the residual will be farther away from zero. It may not be required that all points lie within some limit, but only that most of them do.

In this case, it is necessary to know the distribution of the noise. If it is assumed Gaussian, it is straightforward to calculate the confidence limits of the deviation from zero. The three sigma limits, which contain the individual elements with a probability larger than 99 percent, are frequently used for these limits.

D. Power Spectral Bounds

It is the goal of achieving a feasible solution to obtain a residual signal which has statistics similar to those of the noise. This means that the second- and higher order statistics should be matched. While this goal is valid, it is very difficult to test these higher order statistics for significant deviation. This is particularly true when the test must be done many times during an iterative restoration process. For this reason, some compromises must be made in which statistics to match.

If the noise is uncorrelated and Gaussian, the distribution of the periodogram is chi square with two degrees of freedom [5]. The confidence limits for the periodogram are easily obtained from tables. It is recognized that the periodogram is not a consistent estimator; however, the limits are easily calculated, and its modification is straightforward. Windowed estimators would be very difficult to modify.

E. Other Constraints for Feasibility

There are many other characteristics about a signal which may be useful in defining the feasible solution. The particular properties which can be used depend upon the specific signal in question. Some of the possibilities for constraints include band limits, region of support limits, regions where the value of the signal is known, and nonnegativity. Many of these types of constraints, such as smoothness or bounds on the signal's values, may be subjectively determined.

III. Obtaining a Feasible Solution

In the preceding section, the conditions for a feasible solution have been discussed. It remains to show how to produce a solution which satisfies all of these requirements. Individually, each of the constraints may be quite easy to satisfy. However, it is usually very difficult to satisfy all of the constraints simultaneously. One method which can be used here is projection onto closed convex sets.

A. Projection Onto Convex Sets

In a recent paper [6], Youla and Webb discussed applications of signal reconstruction by sequential projections onto convex sets. The method of projections onto convex sets (POCS) was first introduced by Bregman [7] and was extended by Gubin *et al.* [8]. It was noted in [6] that earlier work in alternating projections onto subspaces [9] is a special case of POCS. The primary application for this work has been in extrapolation of bandlimited signals [10], [11], [19] and tomographic image reconstruction [12], [19]. It will be shown that this theory can be applied to signal deconvolution by use of the feasible solution defined in the preceding section. The reason for this is that all of the constraints mentioned can be used to form convex sets.

The result of previous work on POCS which is of most interest here is stated as follows.

Let $C_0 = \cap_i C_i$ be nonempty where C_i is a closed convex set in a Hilbert space.

Let P_i be the projection operator which projects a vector onto the set C_i.

The iteration given by

$$f_{k+1} = P_1 P_2 \cdots P_M f_k \tag{4}$$

converges weakly to a point in C_0. Clearly, convergence is strong for finite-dimensional spaces.

This result states that if the constraint sets imposed for feasibility are closed and convex, then the method of sequential projection onto those sets will converge to a point in the intersection of the sets. Points in the intersection are the feasible solutions. Thus, a feasible solution can be produced. However, the solution is not necessarily unique, and is usually influenced by the order of projections and the initial estimate used to start the iterative process [13].

B. *Constraints as Convex Sets*

As a review, recall that a set is convex if for any two points $\boldsymbol{x}_1$ and $\boldsymbol{x}_2$ in the set, the point $\boldsymbol{x}_3 = \alpha \boldsymbol{x}_1 + (1 - \alpha)\boldsymbol{x}_2$ is also in the set for $0 \leqslant \alpha \leqslant 1$. It can be easily shown that all of the constraints mentioned in Section II can be used to form convex sets. These sets are defined below.

1) Variance of the Residual: The set based on this constraint is

$$C_v = \{f \mid \|g - Hf\|^2 \leqslant \delta_v\}. \tag{5}$$

As an example, we will show that this set is convex. Let $\boldsymbol{x}_1$ and $\boldsymbol{x}_2$ be two points in C_v and consider $\boldsymbol{x}_3 = \alpha \boldsymbol{x}_1 + (1 - \alpha)\boldsymbol{x}_2$. We need to show that $\boldsymbol{x}_3$ is in C_v. This is done as follows:

$$\| g - H\boldsymbol{x}_3 \| = \| g - H(\alpha \boldsymbol{x}_1 + (1 - \alpha)\boldsymbol{x}_2)\|$$

$$\| g - H\boldsymbol{x}_3 \| = \| \alpha(g - H\boldsymbol{x}_1) + (1 - \alpha)(g - H\boldsymbol{x}_2) \|$$

$$\| g - H\boldsymbol{x}_3 \| \leqslant \alpha \| g - H\boldsymbol{x}_1 \| + (1 - \alpha) \| g - H\boldsymbol{x}_2 \|$$

$$\| g - H\boldsymbol{x}_3 \| \leqslant \sqrt{\delta_v}.$$

It is noted that the constraint that is desired is that the variance of the residual be approximately equal to the variance of the noise. The set C_v defined by (5) requires only that the variance be less than of equal to the estimated variance. This less stringent condition is justified by noting that if a point f is not in the set, its projection is onto the boundary of the set. A point on the boundary satisfies the approximate equality constraint.

2) Mean of the Residual: The noise is assumed to be zero mean. Consider the set defined by

$$C_m = \left\{ f \;\middle|\; \left| \sum_i \left(g_i - \left[Hf \right]_i \right) \right| \leqslant \delta_m \right\} \tag{6}$$

where g_i and $[Hf]_i$ are the ith elements of the vectors $\boldsymbol{g}$ and $\boldsymbol{Hf}$, respectively. The bound δ_m represents the confidence limits on the sample mean. It is readily seen that the members of C_m satisfy the constraint that the mean of the residual be zero. It is also apparent that the set is convex.

3) Outliers of the Residual: The individual elements of the residual should have the same probability distribution as the noise. This distribution is usually known. Thus, it is possible to check for values in the residual which deviate an unlikely amount from the mean. For the most common case that the noise is Gaussian, the appropriate confidence limits are easily obtained from tables. The bound on the deviation from zero that is tolerated may depend on the number of points in the signal vectors. The convex set is defined by

$$C_0 = \{f \mid |g_i - [Hf]_i| \leqslant \delta_0\}. \tag{7}$$

It should be noted that, in fact, C_0 does not represent a single convex set, but N convex sets, one for each point in the signal. A single set could be defined by requiring that all the points in the residual lie within the δ_0 bounds. This projection could be made by again applying the sequential projection method. It was found that the single set was not necessary in practice and, thus, our algorithm was not implemented that way.

A second convex set based on outliers is one using the maximum deviation of the residual. If the distribution of the noise is Gaussian, then the distribution of the maximum deviation of the residual is

$$F(x) = \left[K \int_{-x}^{x} \exp\left[-\frac{1}{2} \frac{(\eta - \mu)^2}{\sigma^2} \right] d\eta \right]^N. \tag{8}$$

The projection onto this set is somewhat difficult, and so it was not used. The limits on the maximum can be used to set the bounds for the set C_0.

4) Power Spectral Bounds: The acceptable deviation of the power spectrum from its expected value is handled in the same way as the outliers in the time/space domain were. Since the periodogram has a chi square distribution, its confidence limits are obtained from tables. The convex set defined to enforce this constraint is

$$C_p = \left\{ f \;\middle|\; |G(k) - H(k)F(k)|^2 \leqslant \delta_p, \quad 1 \leqslant k \leqslant \frac{N}{2} - 1 \right\}. \tag{9}$$

The upper case letters represent the discrete Fourier transform of their lower case counterparts. Note that to use the DFT, we assume that the matrix H is circulant and the term $H(k)$ represents the kth frequency coefficient of the DFT of the first row of $\boldsymbol{H}$ [1].

5) Other Convex Sets: There are many other constraints which can reasonably be applied to the deconvolution problem. These constraints depend upon the characteristics of the specfic signal with which we are dealing. One of the most common constraints is nonnegativity. This is a very important property when the signal in question has many near zero values. The set defined by this property is obviously convex:

$$C_n = \{f \mid f_i \geqslant 0\}. \tag{10}$$

A list of useful convex sets is given in [6]. Many of these can be used in the deconvolution problem. Some of these sets are bandlimited functions, support-limited functions, functions whose Fourier transform is positive, and functions which have prescribed values on some region of support.

C. *The Projection Operators*

The projection of a point f onto a convex set C is the point f_p in the set C which is closest to the point f. For some of the sets defined in the preceding section, the projection operator is obvious; for others, a more detailed description is required.

1) Projection Onto C_v: The projection f_p of a vector f which is not in C_v is the solution of the following problem:

$$\underset{f_p}{\text{minimize}} \; \| f - f_p \|$$
$$\text{subject to } \| g - Hf \|^2 = \delta_v. \tag{11}$$

If f is in the set C_v, then f is its own projection, and no computation is necessary.

The projection operation derived from (11) is very similar to constrained least squares deconvolution [1]. This has been pointed out in previous work [13], [14]. The same modified Newton-Raphson method can be used to solve for f_p. The solution is usually obtained in a few iterations of a numerical scheme which does not require convolution. A derivation of this projection operator is given in the Appendix.

2) Projection Onto C_m: The project onto C_m is given by

$$f_p = \begin{cases} f + \left(\dfrac{\sum r_i - \delta_m}{\|h_c\|^2} \right) h_c & \sum r_i > \delta_m \\ f + \left(\dfrac{\sum r_i + \delta_m}{\|h_c\|^2} \right) h_c & \sum r_i < -\delta_m \\ f & \text{otherwise} \end{cases} \tag{12}$$

where $h_c^t = (\Sigma_i [H]_{i1}, \Sigma_i [H]_{i2}, \cdots, \Sigma_i [H]_{iN})$.

In the case of circular convolution, h_c is a constant vector.

3) Projection Onto C_0: The projection onto C_0 requires that each point of the residual which lies outside the specified limit be forced within that limit. This is done by considering one point at a time. Consider a point which violates the constraint imposed by C_0, i.e.,

$$| g_i - [Hf]_i | > \delta_0. \tag{13}$$

The formula for correcting the residual at this point is

$$f_p = \begin{cases} f + \dfrac{(r_i - \delta_0)}{\|h_i\|^2} h_i & \text{if } r_i > \delta_0 \\ f + \dfrac{(r_i + \delta_0)}{\|h_i\|^2} h_i & \text{if } r_i < -\delta_0 \\ f & \text{otherwise} \end{cases} \tag{14}$$

where h_i is the column vector containing the ith row of the matrix H. A derivation of (14) is given in the Appendix.

4) Projection Onto C_p: The projection onto C_p is derived in the Appendix. The operator is given by

$$F_p(k) = \frac{1}{H(k)} \left[G(k) - \sqrt{\delta_p} \, \frac{G(k) - H(k)F(k)}{| G(k) - H(k)F(k) |} \right] \tag{15}$$

if $| G(k) - H(k)F(k) | > \delta_p$ and $H(k) \neq 0$;

otherwise $F_p(k) = F(k)$.

This operation takes into account the fact that if the transfer function has a zero, there is nothing that can be done to alter the function to satisfy the constraint. If the transfer function has a zero at a particular frequency the residual is, in fact, the exact noise anyway. Thus, the goal of producing a residual with the same properties as the noise is satisfied.

5) Projection Onto C_n: The projection onto C_n is the simple clipper

$$f_{pi} = \begin{cases} f_i & \text{if } f_i \geq 0 \\ 0 & \text{otherwise.} \end{cases} \tag{16}$$

D. *Tolerances on the Projections*

In the mathematical sense, the method of successive projections onto convex sets converges when all of the constraints are satisfied, that is, when we find a vector in the intersection of all the convex sets. In reality, it is nearly impossible to find a true solution in this sense. This is because we are limited to a finite number of iterations. Thus, in order to obtain an estimate in a reasonable amount of time, it is necessary to accept a vector which is not in all of the sets, but is "close" to all of them. The determination of the acceptable limits to define the term "close" is left to the subjective judgment of the user. The guidelines which were used for this work are given below.

The convex sets, which are based on the properties of the noise, are usually defined in terms of confidence intervals. It has been found useful to have one set of limits to define the set and its corresponding projection operator, and a slightly more relaxed set of limits to define the set that is acceptable in the sense of convergence. For example, in defining C_v, the projection operator is defined using the 95 percent confidence limits, but in testing for convergence, we use 99 percent confidence limits. This same method can be used for all of the convex sets which are statistically based. Even in testing for nonnegativity, it was found that some tolerance was needed because of computational roundoff errors in the computer. For other sets, as those defined in [6], some tolerance is also required.

IV. Properties of the Feasible Solution

The feasible solution has been defined as the one that satisfies all of the constraints placed on it. Because the constraints are implemented as projections onto convex sets, this solution is one which lies in the intersection of all such sets. There are many factors which influence the particular solution that is obtained. Among the most important are parameters of the signal formation model equation (1), the impulse response H, and the noise n. The initial estimate f_0, used to start the iterative process, can be quite influential.

The size or volume of the intersection of all convex sets will determine the range of the feasible solution. Without being specific as to what is meant by size, it is intuitive that the larger the set C_0, the greater will be the range of possible solutions.

Mathematically, size could be defined as the largest distance between any two points in the set. There is a problem with this definition if the matrix H is singular. Components of f in the null space of H could be unbounded without affecting the residual at all. Because of this problem, our arguments will be restricted to the case where H is nonsingular. This approach is limited, but gives a heuristic understanding of the process.

In general, it is very difficult to determine the maximum distance for any two points in the intersection. The size of the intersection will be related to the size of its component sets. This relation may not be easily described mathematically, but it is intuitive that such a relation exists. The size of these sets can often be related to the model parameters.

A. Effect of the Impulse Response

It is intuitive that an increase in the spatial or temporal extent of the impulse response should increase the size of the set of feasible solutions, that is, it should be more difficult to obtain a good restoration. Clearly, an increase the size of the solution set means that more vectors are feasible, and so, the chances of obtaining a good one are diminished. Let us examine the quantitative effect of the temporal extent of the impulse response on the size of each of the convex sets discussed earlier.

The extent of the impulse response in space or time can be quantitatively measured by the matrix norm of the matrix $\boldsymbol{H}$. The norm which is used here is defined by

$$\|\boldsymbol{H}\| = \max \{\sqrt{\lambda_i}\} \tag{17}$$

where λ_i is the ith eigenvalue of $\boldsymbol{H}^t\boldsymbol{H}$. If $\boldsymbol{H}$ represents circular convolution, then these eigenvalues are the magnitudes squared of the DFT coefficients. It should be noted that this is the matrix norm induced by the Euclidean vector norm.

1) Extent of C_v: Let f_1 and f_2 be two vectors in C_v. We will show that the distance of each of these vectors from the true solution is bounded and, thus by the triangle inequality, their distance from each other is bounded. Consider one of the vectors, say f_1. Since f_1 is in C_v,

$$\| \boldsymbol{H}f + n - \boldsymbol{H}f_1 \| \leqslant \delta_v \tag{18}$$

where we have substituted g from the model equation (1). Since the variance of the noise is known, it is reasonable to assume that

$$\| n \| \leqslant \delta_v. \tag{19}$$

From this, it is true that

$$\| \boldsymbol{H}(f - f_1) \| \leqslant 2\delta_v. \tag{20}$$

Since we assumed that $\boldsymbol{H}$ is nonsingular, it can be stated that

$$\| f - f_1 \| \leqslant 2\delta_v \| \boldsymbol{H}^{-1} \|. \tag{21}$$

The norm of $\boldsymbol{H}^{-1}$ is $1/\min |H(k)|$ for the case of circular convolution. Thus, the extent of C_v depends on the maximum attenuation of the blurring function.

In the case that the matrix $\boldsymbol{H}$ is singular, we can restrict our interest to the range space of $\boldsymbol{H}$. For a starting vector f_0, in the range space, and projections defined by the residual, as was done above, the vector to which the iteration converges is in the range space also. For this case, the solutions produced by this method will vary as in the case of a nonsingular $\boldsymbol{H}$. The problem of recovering information in the null space can be addressed by incorporating further constraints on the solution, e.g., the nonnegativity constraint.

2) Extent of C_m: The restriction of this set is only that the sum of the elements of $\boldsymbol{H}f$ be equal to the sum of the elements of g. This set is unbounded. We can choose any two elements of f, measure their effect on the sum, and adjust their values so that they produce zero effect on the sum. Any multiple of this combination of two elements will have zero effect on the sum. Thus, vectors in C_m can be arbitrarily far apart.

3) Extent of C_0: To estimate the size of the set C_0, we will choose two vectors in this set f_1 and f_2 as in the case of C_v. These vectors both satisfy the condition (7). By using the same analysis as for C_v, it can be shown that

$$\|f_1 - f_2\| \leqslant 2\,\delta_0 \|\boldsymbol{H}^{-1}\|. \tag{22}$$

This gives the same qualitative results as in that earlier case.

4) Extent of C_p: Each of two vectors in C_p must satisfy the condition of (9). The complex components in the frequency domain satisfy

$$|F_1(k) - F_2(k)| \leqslant 2\delta_p / |H(k)|. \tag{23}$$

By applying Parseval's theorem, we can show that

$$\| f_1 - f_2 \| \leqslant 2\delta_p \| \boldsymbol{H}^{-1} \|. \tag{24}$$

Again, we see the same qualitative behavior as in the other cases.

5) Extent of Other Sets: The extents of other constraint sets for the feasible solution are usually infinite. Such is the case with those mentioned earlier. While the sets may be unbounded, they may have the effect of severely limiting the extent of the intersection. Nonnegativity is a good example of this type of behavior. In the final analysis, it is the intersection that must be reduced in extent as much as possible.

B. Effect of Noise

The noise in the model of (1) affects the size of the sets since it is the major factor in determining the confidence limits on which the sets are based. The bound used in the definition of all four convex sets based on the residual is a direct function of the noise. In the case of white Gaussian noise, the bound is proportional to the variance of the noise.

From the analysis in the previous section, it is seen that the size of the sets increases with the severity of the noise, that is, as the variance increases, so does the size of the set. This behavior is intuitive. As the noise increases in severity, the quality of the restored signal decreases. Heuristically, since the number of feasible solutions gets larger, it is harder to choose a good one.

C. Effect of the Initial Estimate

The set of feasible solutions can be very large or unbounded in many cases. Yet, even for the unbounded cases, the algorithms presented here produced reasonable solutions. The initial estimate plays a large role in making this method viable.

The method of successive projections onto convex sets projects each estimate to the closest point in the next convex set. This means that the final solution should be "close" in some sense to the initial estimate. In the very special case that all of the convex sets are linear varieties, the limit of the successive projections is the projection onto the intersection of all of the sets, that is, the solution is the closest point to the initial estimate in the intersection [6]. Unfortunately, the sets which have been defined to form the class of feasible solutions are not linear varieties.

Despite the lack of an exact definition of the term "close," work has been done which shows the role of the initial estimate in signal restoration [13]. If the estimate which starts the iteration is reasonable, then the solution to which the method converges is also reasonable. If additional information is available about the signal to be restored, many times that information can be used in selecting or creating the initial estimate. In the example used in [13], the positions of peaks in a simulated X-ray fluorescence spectrum were determined by a restoration using a relatively flat initial estimate; then the magnitudes of the peaks were refined by using an initial estimate which placed isolated peaks at the positions determined from the first restoration.

Several characteristics of signals can be considered by proper choice of the initial estimate. If a smooth restoration is desired, then a smooth initial estimate will help to limit the variation in the restoration. Peaks have been mentioned above. Hypotheses about the specific waveform, such as particular values within a specific region of support, can be tested. These properties can be combined to give a more generally useful and very adaptive tool.

D. Effect of Other Constraints

It is clear that the more valid constraints that can be imposed on the solution, the better that solution will be. To be effective, a constraint must be used. Requiring the solution to be nonnegative when the signal in question is not close to zero will gain nothing. However, if over much of the domain the signal is zero or near zero, then the addition of that constraint will result in much improved restorations.

For the signals used in most of the examples in this paper, nonnegativity was very important constraint. These signals are simulated X-ray fluorescence spectra and text images which have exactly the characteristics required to make the nonnegativity most effective. The set defined by nonnegativity is unbounded, but when combined with other constraints, drastically reduces the size of the intersection of all constraint sets.

Other constraints can be applied when further *a priori* knowledge about the signal is available. Another example, which can be applied if there is reason to believe that the initial estimate is a good one, is to limit the distance that the solution can move from that estimate. Such a constraint may be written mathematically:

$$C_i = \{f \mid \|f - f_0\| \leqslant \delta_i\}. \tag{25}$$

This type of constraint could be applied after a preliminary restoration had given an indication of the correct signal. The size of the set C_i is clearly determined by the user and his confidence in selecting an appropriate initial estimate.

V. Results

The signals which were used to demonstrate the effects of the feasible solution were signals that had characteristics which could be clearly defined. The X-ray fluorescence signal is a nearly perfect example for this method. Such signals have been used in several previous studies [2], [4], [16]. This signal is characterized by isolated peaks and large regions near zero. The text image also has large areas near zero and flat peaks. Both of these characteristics can be used to good advantage by constructing the appropriate constraint sets. The peaks can be estimated and included in a set such as C_i. It has already been mentioned that nonnegativity is a powerful constraint.

The feasible solution defined only by the statistical properties of the noise by itself is not a particularly good estimation method. This should not be too surprising. Noise can easily dominate a restoration; the inverse filter solution is a good example. The advantage that the feasible solution method gives is to use all of the information about the noise. The method is limited in its ability to restrict the solution space by the severity of the noise, as was seen in the sections which discussed the size of the sets. The power of the method is in limiting the solution space and letting other constraints be applied by the method of sequential projections onto convex sets.

While the methods were tested on many signals, the results presented here will concentrate on the original signal shown in Fig. 1. To test the method, several impulse response functions and noise levels were used. For the examples shown here, the impulse response was of Gaussian shape, with standard deviation 2.0. The noise was white Gaussian, with a variance of 0.001. The degraded signal obtained from the linear model, (1), using these parameters is shown in Fig. 2.

A. The Feasible Solution

The constraints which make up the feasible solution can be tested by applying each of the constraints individually. The projections onto the convex sets C_m and C_n are the only ones that cannot produce a reasonable restoration by themselves. The restorations based on the sets C_v, C_0, and C_p are shown in Fig. 3. While these restorations are not impressive, they compare favorably to other more common restoration techniques such as Wiener filtering and constrained least squares deconvolution [1], [15].

It should be noted that the Wiener filter requires additional information about the power spectrum of the orginal signal. This information may not be readily available. The constrained least squares method requires the same information as the feasible solution, but in addition, imposes a smoothness constraint. The constrained least squares restoration is shown in Fig. 4.

When the additional constraints are applied, the result should improve. For this example, the restorations in Fig. 3(a)-(c) were already contained in C_m. By including all of the projections in the algorithm, the restoration shown in Fig. 5 was obtained. This is an improvement in that the two central peaks are slightly better resolved and the negative ringing is reduced. The ringing in the regions where the true signal is zero is still quite pronounced and disturbing.

The addition of the nonnegativity constraint can help eliminate the ringing. The restoration based on each of sets C_v, C_0, and C_p together with C_n are shown in Fig. 6. From this estimate, it is seen how important the nonnegativity constraint is. The signal estimate obtained by requiring that the solution lie in the intersection of all of the above mentioned sets is shown in Fig. 7. The improvement is obvious.

It can be noted that without the nonnegativity constraint, the constraints C_p produced the best resolution of the central peaks. With nonnegativity, the constraint C_0 produced the best restoration. There is no *a priori* method of determining

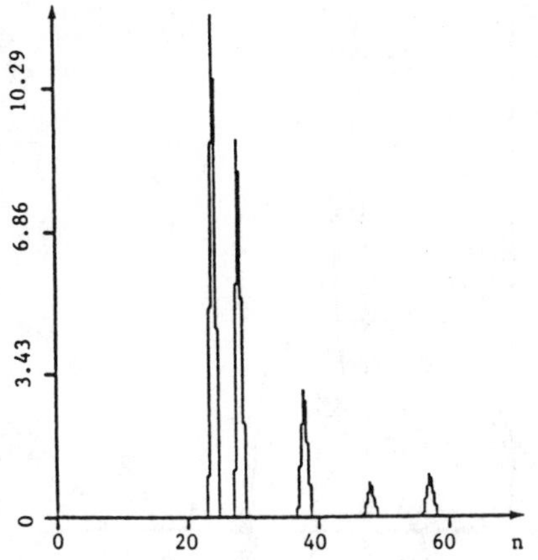

Fig. 1. Original signal.

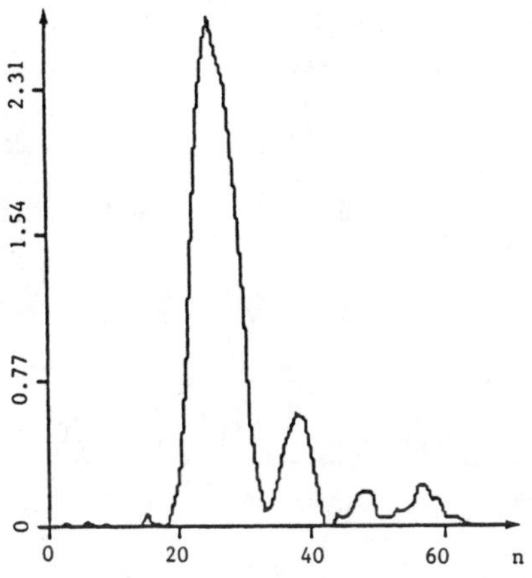

Fig. 2. Degraded signal.

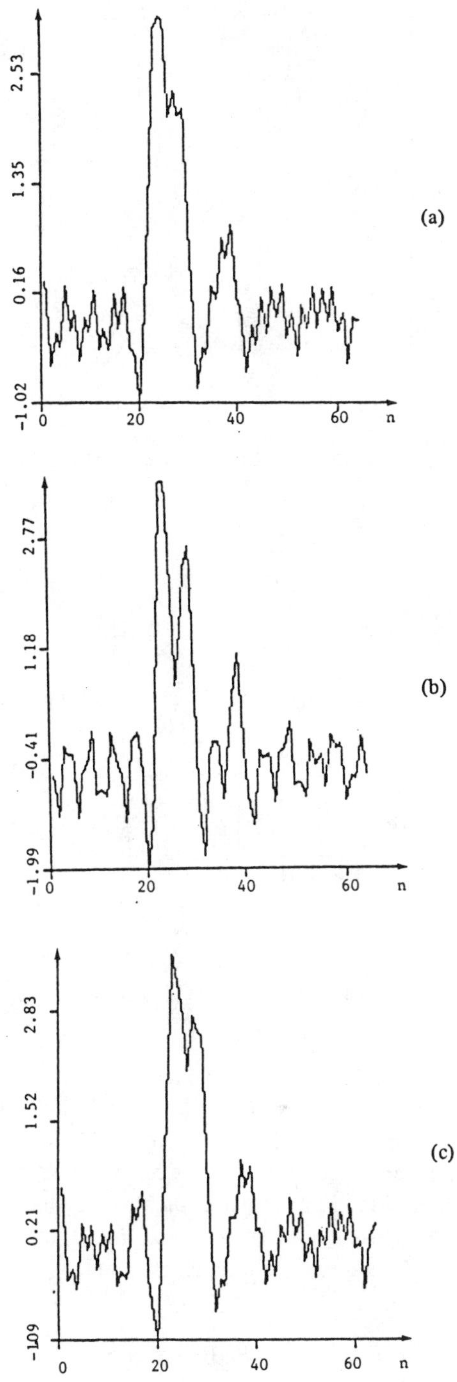

Fig. 3. (a) Restoration using only C_v. (b) Restoration using only C_p. (c) Restoration using only C_0.

which constraints will be dominant in producing high-quality restorations. An easy solution is to use all possible constraints.

B. Effect of Noise

As discussed earlier, as the variance of the noise decreases, the quality of the restoration increases, and vice versa. The restorations obtained by using the identical method as produced Fig. 7 from Fig. 2 was used, with signal degraded by the same impulse response, but with noise of variances 0.01 and 0.0001. The results of these restorations are shown in Figs. 8 and 9, respectively. The results are as expected.

C. Effect of the Initial Estimate

It was mentioned that the initial estimate could have a profound effect on the restoration obtained by the method of successive projections onto convex sets. The nature of the X-ray spectrum signal allows a testing ground for utilizing specialized knowledge about the signal. The positions and magnitudes of the peaks can be estimated and used as the initial estimate.

The restorations shown in Section V-A were all obtained using the degraded signal of Fig. 2 as the initial estimate. In this case, the noise was low enough so that almost any initial estimate would produce a good restoration. As was seen in the previous section, an increase in the noise level can seriously degrade the restoration. It is in this case that the initial estimate can play a very important role.

If the original signal is known to contain peaks, but the precise number of peaks is not known, a reasonable guess might be to estimate the locations of the peaks at the peaks of the degraded signal in Fig. 2. The magnitude could be estimated by the integated area under each peak. This would give an erroneous answer. The restoration obtained by using this guess is shown in Fig. 10. The splitting of the central peak is an indication that the initial estimate was in error.

The results of other restorations can be used to help in con-

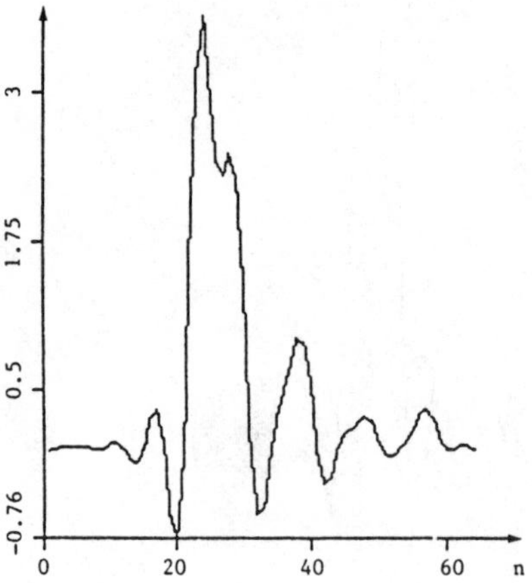

Fig. 4. Constrained least squares restoration.

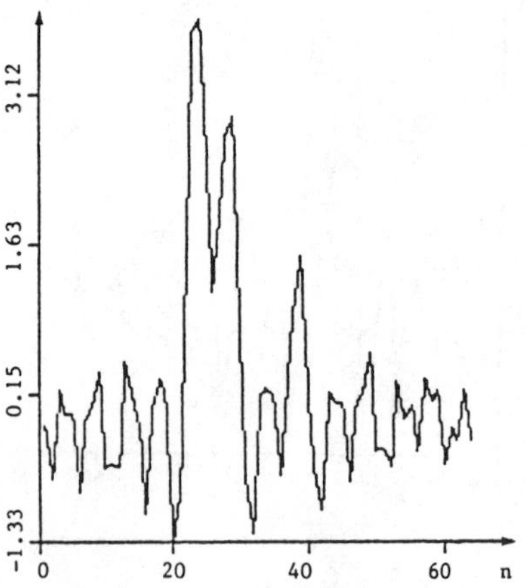

Fig. 5. Restoration using C_m, C_v, C_p, and C_0.

structing initial estimates for this method. The restoration in Fig. 8 showed that the central peak in the degraded signal was shifted slightly. Using this information, an initial estimate with a peak in the correct position can be constructed. The restoration in Fig. 11 was based on this starting vector. It shows the two peaks in the correct positions. If this restoration were used to form a two-peak initial estimate, the results would be almost exact.

D. Effect of Extent of Impulse Response

Just as in the case of the effects of noise, the effect of increasing and decreasing the severity of the impulse response is as predicted by earlier analysis. Fixing the noise level as in Fig. 2, two restorations were produced for degraded signals whose impulse responses were Gaussian with variances of 1.0 and 8.0. The restorations obtained are shown in Figs. 12 and 13, respectively. Again, the results are as predicted.

E. Effect of Outliers

Many of the sets which define the feasible solution are related to statistical parameters such as the noise. The limits of these sets are often chosen based on confidence intervals. Because the actual parameters are random variables, there is always a finite chance that the values which actually occurred to produce the recorded data are outside the confidence limits. This

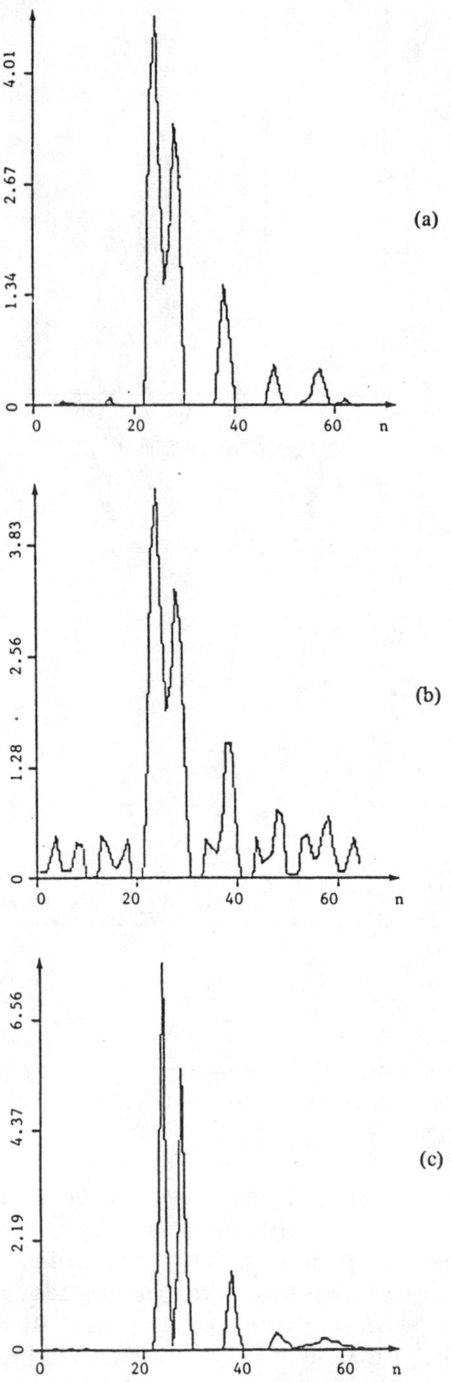

Fig. 6. (a) Restoration using C_v and C_n. (b) Restoration using C_p and C_n. (c) Restoration using C_0 and C_n.

means that the true solution to the restoration problem will not lie in the intersection of the convex sets. It is instructive to consider the results for this case.

There are two results which occur when the sets for the sequential projection method are incorrectly defined: the algorithm converges to a poor estimate of the original signal or the algorithm fails to converge at all. In the case that a poor estimate is found, there is little that can be done. Fortunately, the failure to converge is a good indicator that there is an error in the definition of the sets.

In practice, it has been encouraging that in the cases where

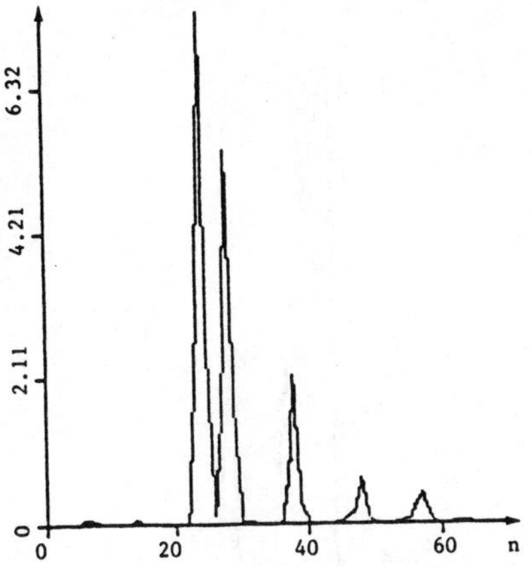

Fig. 7. Restoration using C_m, C_v, C_p, C_0, and C_n.

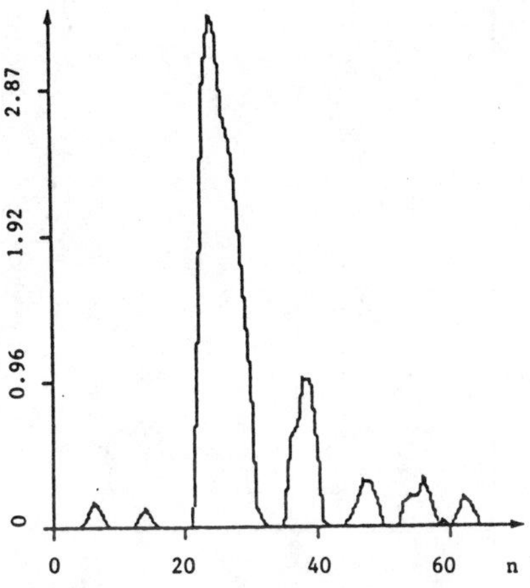

Fig. 8. Restoration when $\sigma_n^2 = 0.01$.

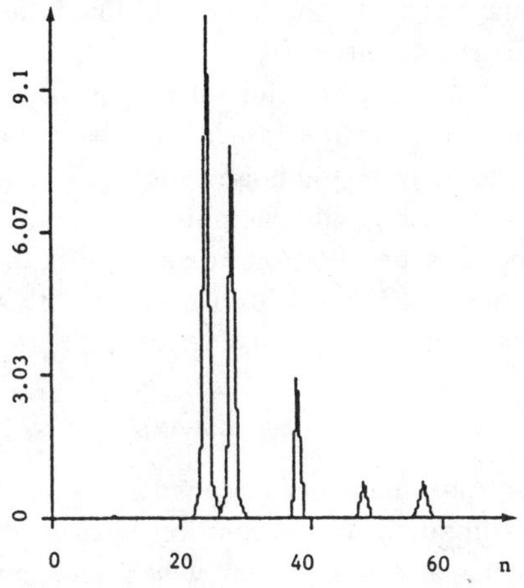

Fig. 9. Restoration when $\sigma_n^2 = 0.0001$.

outliers have caused the sets defined from confidence limits to be in error, the solution which has been obtained has never been very bad. This indicates that the method will still produce a solution somewhat close to the true solution.

When the algorithm fails to converge, the natural modification of the algorithm is to relax the constraining conditions. If this failure is due to outliers, this is the correct response. The relaxation technique can also be applied to constraints which are more subjective in nature. An example of this type of set is C_i, which restricts the distance of the solution from the initial estimate. This technique could be extended to any number of prototype signals which could restrict the range of the solution.

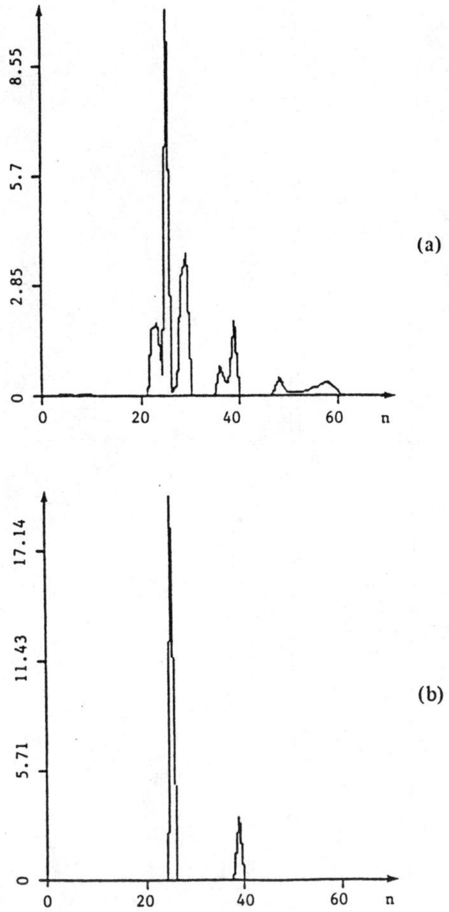

Fig. 10. (a) Restoration using the initial estimate of Fig. 10 (b). (b) Initial estimate derived from the degraded signal.

To show the effects of outliers on the restoration, the same model parameters were used as in the case that generated Fig. 2; however, the noise which was generated for this example contained one point in the periodogram which was outside the 99 percent confidence limits. The algorithm failed to converge. The estimate which was obtained at the last iteration when it was stopped is shown in Fig. 14. This estimate is not as good as those produced earlier, but it does recover the two central peaks.

F. Speed of Convergence

The computational expense of obtaining a feasible solution is a function of the number of constraints, the severity of the noise, and the extent of the impulse response. It makes sense that when the intersection of the convex sets is large, a solution will be easy to find. For most of the examples shown here, the number of projection sequences was less than 50. The only projection which is computationally expensive is the one onto C_v. This is caused by the convolution involved and the iterative method used to obtain the Lagrange multiplier [1].

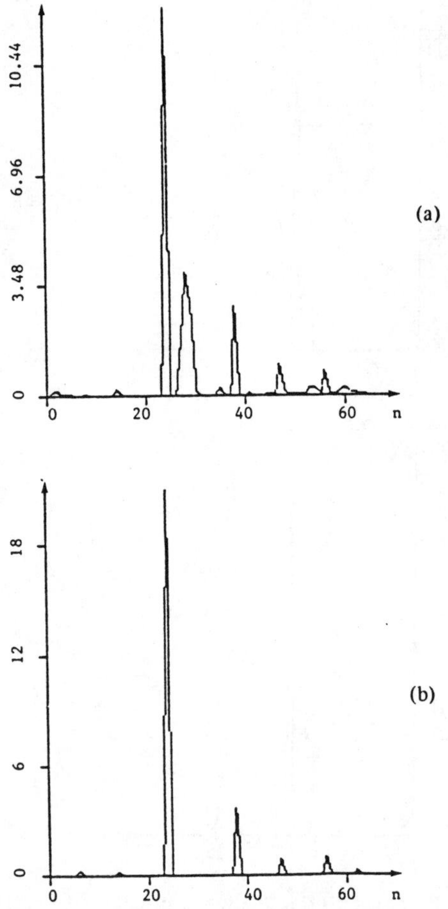

Fig. 11. (a) Restoration using the initial estimate of Fig. 11 (b). (b) Initial estimate derived from the restoration in Fig. 8.

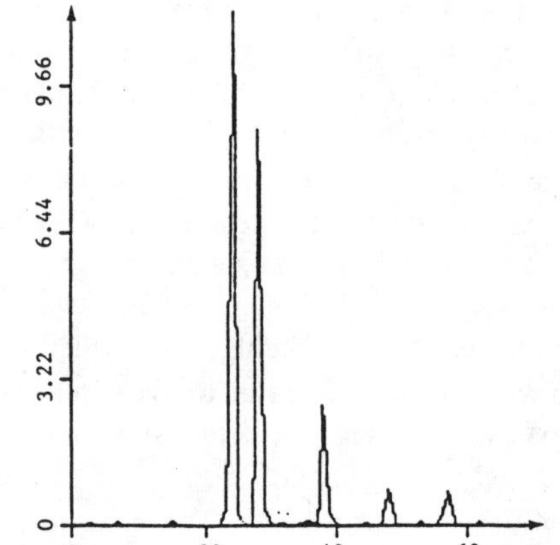

Fig. 12. Restoration when the variance of the Gaussian psf is 1.

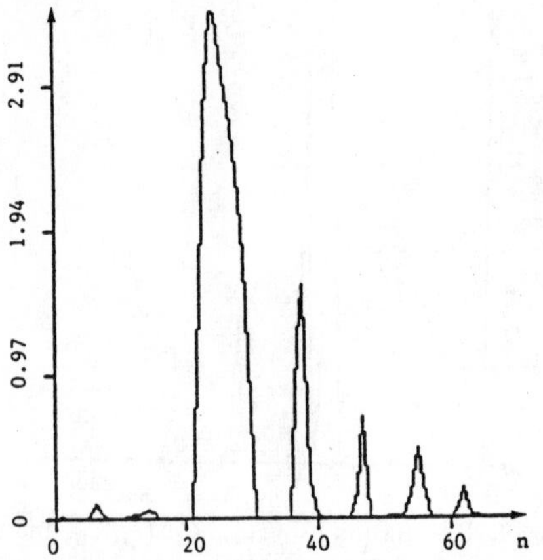

Fig. 13. Restoration when the variance of the Gaussian psf is 8.

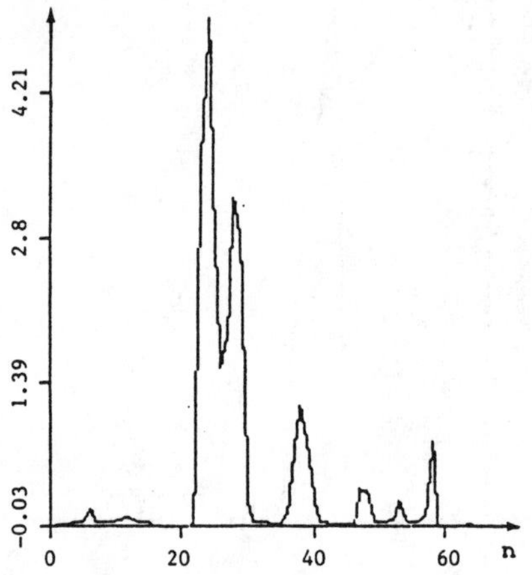

Fig. 14. Unconverged restoration result because of outliers.

G. Application to Image Restoration

There is no restriction on the dimensionality of the problem to which the feasible solution may be applied. The constraint set C_v has already been used in several image restoration applications [1], [17], [18]. The types of images which will benefit most from this approach will naturally be the ones which have the most enforceable constraints. Some images which would benefit from the positivity constraint as well as the noise constraints are starfields and text. Both of these types of images have large areas near zero.

An example of the restoration of a text image is shown in Fig. 15. The image on the top left has been blurred with a motion-type degradation and noise added. The top right image has been restored using only the noise constraints. The image on the bottom has been restored using the noise constraints and nonnegativity. The qualitative results are the same as were found in the one-dimensional example.

VI. Conclusions

The feasible solution has been defined as the one that satisfies all constraints imposed on the solution by *a priori* knowledge about the signal and the system which generated the data. With no knowledge about the signal, it was shown that it is possible to use knowledge about the sample statistics of the noise to produce a reasonable estimate of a degraded signal. If it is possible to add further information about the nature of the signal, this can improve the restoration.

The method used for obtaining the feasible solution was the sequential projection onto convex sets. It was shown that many useful constraints can be used to form the necessary sets. This algorithm was found to converge in a small number of iterations. The method is extendable in that any number of convex sets can be used to further refine the solution.

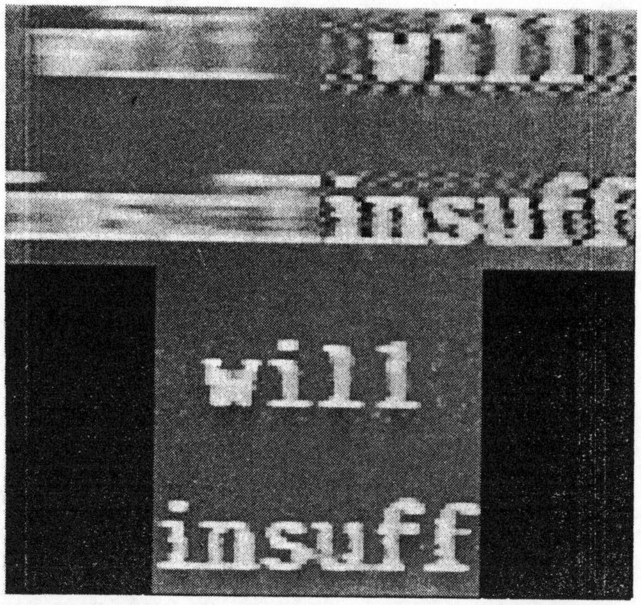

Fig. 15. Upper left–degraded image, upper right–restored image without nonnegativity constraint, lower–restored image with nonnegativity constraint.

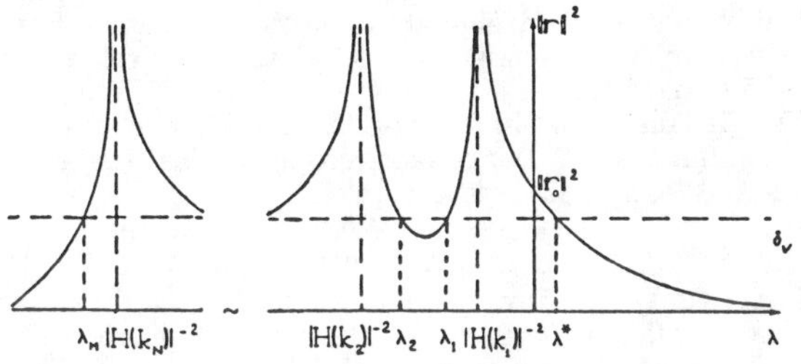

Fig. 16. Residual norm versus Lagrange multiplier.

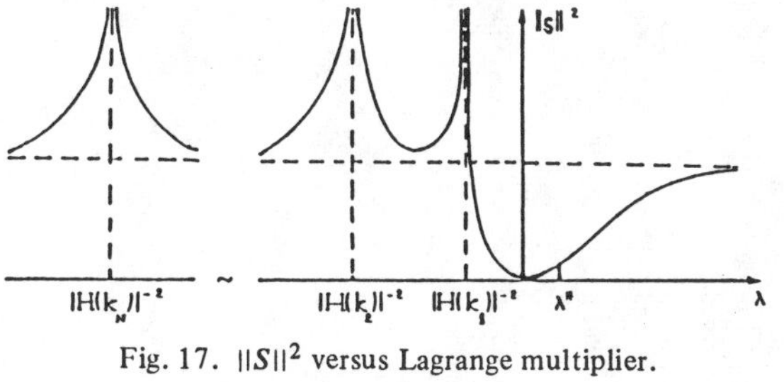

Fig. 17. $\|S\|^2$ versus Lagrange multiplier.

Appendix

Projection Onto C_v

For a given $f \notin C_v$, find the vector $f_p \in C_v$ which is closest to f. Mathematically, this can be formulated as follows.

Let

$$s = f_p - f;$$

$$\min s^t s \tag{A.1}$$

$$\text{subject to } [g - H(f+s)]^t [g - H(f+s)] = \delta_v.$$

The first-order necessary condition gives

$$(I + \lambda H^t H) s = \lambda H^t (g - Hf). \tag{A.2}$$

If $\lambda = 0$, then $s = 0$, which is possible if the initial point is already feasible.

If $\lambda \neq 0$, then f should be modified as

$$f_p = f + (H^t H + 1/\lambda I)^{-1} H^t (g - Hf). \tag{A.3}$$

The Lagrange multiplier λ is to be found from the constraint equation:

$$(g - Hf_p)^t (g - Hf_p) = r_0^t \left[I - \frac{1}{\lambda} H(H^t H + 1/\lambda)^{-2} \cdot H^t - H(H^t H + 1/\lambda)^{-1} H^t \right] r_0 = \delta_v \tag{A.4}$$

where $r_0 \triangleq g - Hf$.

At this point, assuming a circulant H, we can use DFT for the calculation of λ:

$$\|g - Hf_p\|^2 = \frac{1}{N} \|G(k) - Hf_p(k)\|^2 = \frac{1}{N} \sum_{k=0}^{N-1} |R_0(k)|^2 \frac{1}{(\lambda |H(k)|^2 + 1)^2} \tag{A.5}$$

$$\frac{\partial}{\partial \lambda} \|g - Hf_p\|^2 = \frac{1}{N} \sum_{k=0}^{N-1} - \frac{2 |R_0(k)|^2 |H(k)|^2}{(\lambda |H(k)|^2 + 1)^3}. \tag{A.6}$$

In Fig. 16, the projected residual norm versus Lagrange multiplier is shown. As seen from the figure, there are several critical points; however, only one of them corresponds to a positive λ. The optimality of λ^* can be seen from the cost function which is given in (A.7) and plotted in Fig. 17.

$$\|S\|^2 = \frac{1}{N} \sum_{\substack{k=0 \\ |H(k)| \neq 0}}^{N-1} \frac{|R_0(k)|^2 |H(k)|^2}{\left(|H(k)|^2 + \frac{1}{\lambda} \right)^2}. \tag{A.7}$$

An interesting feature of this projection is the monotonic dependence between the Lagrange multiplier and the distance between the initial and the projected signals: at the end of each projection, λ gives an idea about the modification done on the signal.

Projection Onto C_p

f_p is the solution of

$$\min \| f_p - f \|^2 \tag{A.8}$$

subject to $|G(k_0) - H(k_0) F_p(k_0)|^2 = \delta_p$.

The first-order necessary condition gives

$$f_p = f + 2\lambda \operatorname{Re} \left[\left(\frac{G(k_0)}{H(k_0)} - F(k_0) \right) \begin{bmatrix} W_N^0 \\ W_N^{k_0} \\ \vdots \\ W_N^{(N-1)k_0} \end{bmatrix} \right] \tag{A.9}$$

where $W_N = \exp(j(2\pi/N))$ or, in the DFT domain,

$$F_p(k_0) = F(k_0) + N\lambda \left[\frac{G(k_0)}{H(k_0)} - F_p(k_0) \right]$$

$$F_p(N - k_0) = F_p^*(k_0)$$

$$F_p(k) = F(k) \qquad \text{for } k \neq k_0. \tag{A.10}$$

The Lagrange multiplier can be solved directly, and the result is given in (15).

Projection Onto C_0

For this case, f_p is the solution of

$$\min \| f_p - f \|^2 \tag{A.11}$$

$$|g_i - [Hf_p]_i| \leq \delta_0.$$

The Kuhn-Tucker condition gives

$$f_p = f + \lambda h_i \tag{A.12}$$

where h_i is the ith row of H or the psf rotated $(i - 1)$ times in the case of convolution.

$$\lambda = \begin{cases} \frac{1}{\|h_i\|^2}(r_i - \delta_0) & \text{if } r_i > \delta_0 \\ 0 & \text{if } -\delta_0 \leq r_i \leq \delta_0 \\ \frac{1}{\|h_i\|^2}(r_i + \delta_0) & \text{if } r_i < -\delta_0 \end{cases} \tag{A.13}$$

solved from the constraint equation.

References

[1] B. R. Hunt, "The application of constrained least squares estimation to image reconstruction by digital computer," *IEEE Trans. Comput.*, vol. C-22, pp. 805–812, Sept. 1973.

[2] H. J. Trussell, "Maximum power restoration," *IEEE Trans. Acoust., Speech, Signal Processing*, vol. ASSP-29, pp. 1059–1061, Oct. 1981.

[3] H. C. Andrews and B. R. Hunt, *Digital Image Restoration.* Englewood Cliffs, NJ: Prentice-Hall, 1977, pp. 152–154.

[4] H. J. Trussell, "Convergence criteria for iterative restoration methods," *IEEE Trans. Acoust., Speech, Signal Processing*, vol. ASSP-31, pp. 129–136, Feb. 1983.

[5] G. M. Jenkins and D. G. Watts, *Spectral Analysis and Its Applications.* San Fransisco, CA: Holden-Day, 1968.

[6] D. C. Youla and H. Webb, "Image restoration by the method of convex projections: Part 1–Theory," *IEEE Trans. Med. Imaging*, vol. MI-1, pp. 81–94, Oct. 1982.

[7] L. M. Bregman, "The method of successive projection for finding a common point of convex sets," *Dokl. Akad. Nauk. SSSR*, vol. 162, no. 3, pp. 688–692, 1965.

[8] L. G. Gubin, B. T. Polyak, and E. V. Raik, "The method of projections for finding the common point of convex sets," *USSR Comput. Math. and Math. Phys.*, vol.7, no. 6, pp. 1–24, 1967.

[9] D. C. Youla, "Generalized image restoration by the method of alternating orthogonal projections," *IEEE Trans. Circuits Syst.*, vol. CAS-25, pp. 694–702, Sept. 1978.

[10] A. Papoulis, "A new algorithm in spectral analysis and band limited extrapolation," *IEEE Trans. Circuits Syst.*, vol. CAS-22, Sept. 1975.

[11] R. W. Gerchberg, "Super resoultion through error energy reduction," *Opt. Acta*, vol. 21, pp. 709–720, Sept. 1974.

[12] M. I. Sezan and H. Stark, "Image restoration by the method of convex projections: Part 2–Applications and numerical results," *IEEE Trans. Med. Imaging*, vol. MI-1, pp. 95–101, Oct. 1982.

[13] H. J. Trussell and M. R. Civanlar, "The initial estimate in constrained iterative restoration," presented at the 1983 IEEE Int. Conf. Acoust., Speech, Signal Processing, Boston, MA, Apr. 14–16, 1983.

[14] H. J. Trussell, "*A priori* knowledge in algebraic restoration methods," *Adv. Comput., Vision Image Processing*, vol. 1, 1984.

[15] A. Rosenfeld and A. C. Kak, *Digital Picture Processing, Vol. 1.* New York: Academic, 1982.

[16] R. W. Schafer, R. M. Mersereau, and M. A. Richards, "Constrained iterative restoration algorithms," *Proc. IEEE*, vol. 69, pp. 432–450, Apr. 1981.

[17] H. J. Trussell and B. R. Hunt, "Sectioned methods in image restoration," *IEEE Trans. Acoust., Speech, Signal Processing*, vol. ASSP-26, pp. 157–164, Apr. 1978.

[18] —, "Improved methods of maximum *a posteriori* image restoration," *IEEE Trans. Comput.*, vol. C-28, pp. 57–62, Jan. 1979.

[19] A. Lent and H. Tuy, "An iterative method for the extrapolation of band-limited functions," *J. Math. Anal. Appl.*, vol. 83, pp. 554–565, 1981.

Image Processing in the Context of a Visual Model

THOMAS G. STOCKHAM, JR., MEMBER, IEEE

Abstract—**A specific relationship between some of the current knowledge and thought concerning human vision and the problem of controlling subjective distortion in processed images are reviewed.**

I. Introduction

IMAGE QUALITY is becoming an increasing concern throughout the field of image processing. The growing awareness is due in part to the availability of sophisticated digital methods which tend to highlight the need for precision. Also there is a developing realization that the lack of standards for reading images into and writing images out of digital form can bias the apparent effectiveness of a process and can make uncertain the comparison of results obtained at different installations. Greater awareness and the desire to respond to it are partially frustrated, because subjective distortion measures which work well are difficult to find. Part of the difficulty stems from the fact that physical and subjective distortions are necessarily different.

The ideas presented here spring from our reevaluation of the relationship between the structure of images and 1) the problem of quantitative representation, 2) the effect of desired processing and/or unwanted distortion, and 3) the interaction of images with the human observer. They provide a framework in which we think about and perform our image processing tasks. By adding to our understanding of what is to be measured when dealing with images and by strengthening the bridge between the objective (physical) and the subjective (visual) aspects of many image processing issues, these ideas have clarified the meaning of image quality and thus have enhanced our ability to obtain it. We offer them with the hope that they may aid others as well.

In the course of the discussion it is noted that image processors which obey superposition multiplicatively instead of additively, bear an interesting resemblance both operationally and structurally to early portions of the human visual system. Based on this resemblance a visual model is hypothesized, and the results of an experiment which lends some support to and provides a calibration for the model are described. This tentative visual model is offered only for its special ability to predict approximate visual processing characteristics. (See footnote 11.)

In recent years there has been a large amount of quantitative work done by engineers and scientists from many fields

Manuscript received January 31, 1972; revised April 20, 1972. This research was supported in part by the University of Utah Computer Science Division monitored by Rome Air Development Center, Griffiss Air Force Base, N. Y. 13440, under Contract F30602-70-C-0300, ARPA order number 829.

The author is with the Computer Science Division, College of Engineering, University of Utah, Salt Lake City, Utah 84112.

Reprinted from *Proc. IEEE,* vol. 60, pp. 828–842, July 1972.

in support of a model for human vision. While many of these works are not referenced explicitly here, we have attempted to reference papers and texts which do a good job of collecting these references in a small number of places while providing a unifying interpretation [1]–[5].

II. Some Philosophy about Image Processing

The notion of processing an image involves the transformation of that image from one form into another. Generally speaking, two distinct kinds of processing are possible. One kind involves a form of transformation for which the results appear as a new image which is different from the original in some desirable way. The other involves a result which is not an image but may take the form of a decision, an abstraction, or a parameterization. The following discussion limits itself primarily to the first kind of processing.

The selection of a processing method for any particular situation is made easier when the available processes have some kind of mathematical structure upon which a characterization of performance can be based. For example, the bulwark for most of the design technology in the field of signal processing is the theory of linear systems. The fact that the ability to characterize and utilize these systems is as advanced as it is, stems directly from the fact that the defining properties of these systems guarantee that they can be analyzed. These analyses, based on the principle of superposition, lead directly to the concepts of scanning, sampling, filtering, waveshaping, modulation, stochastic measurement, etc.

Equally important, however, is the idea that the mathematical structure of the information being processed be compatible with the structure of the processes to which it is exposed. For example, it would be impossible to separate one radio transmission from another if it were not for the fact that the linear filters used are compatible with the additive structure of the composite received signal.

In the case of images the selection of processing methods has often been based upon tradition rather than upon a consideration of the ideas given above. In fields such as television and digital image processing where electrical technology is a dominating influence, the tradition has centered around the use of linear systems.

This situation is a very natural one since the heritage of electrical image processing stems from those branches of classical physics which employ linear mathematics as their foundation. Specifically, it is interesting to follow the development from electromagnetic field theory to electric measurements, circuit theory, electronics, signal theory, communications theory, and eventually to digital signal processing. The situation is similar when considering the role of optics in image processing, the laws of image formation and degradation being primarily those determined from linear diffraction theory.

The question that arises is whether this tradition of applying linear processing to images is in harmony with the ideas given above. The major point at issue cannot be whether the processors possess enough structure, because linear systems certainly do. The issue is then whether that structure is compatible with the structure of the images themselves. To clarify this issue the question of image structure must be elaborated upon.

III. The Structure of Images

As an energy, signal light must be positive and nonzero. This situation is expressed in (1)

$$\infty > I_{x,y} > 0 \tag{1}$$

where I represents energy, or intensity as it is commonly called, and x and y represent the spatial domain of the image. Furthermore, since images are commonly formed of light reflected from objects, the structure of images divides physically into two basic parts. One part is the amount of light available for illuminating the objects; the other is the ability of those objects to reflect light.

These basic parts are themselves spatial patterns, and like the image itself must be positive and nonzero as indicated in (2) and (3)[1]

$$\infty > i_{x,y} > 0 \tag{2}$$

$$1 > r_{x,y} > r_{\min} \approx 0.005. \tag{3}$$

These image parts, called the illumination component and the reflectance component, respectively, combine according to the law of reflection to form the image $I_{x,y}$. Since that law is a product law, (2) and (3) combine as in (4)

$$\infty > I_{x,y} = i_{x,y} \cdot r_{x,y} > 0 \tag{4}$$

which is in agreement with (1).

It follows from (4) that two basic kinds of information are conveyed by an image. The first is carried by $i_{x,y}$, and has to do primarily with the lighting of the scene. The second is carried by $r_{x,y}$, and concerns itself entirely with the nature of the objects in the scene. Although they are delivered in combination, these components are quite separate in terms of the nature of the message conveyed by each.

So far it has been assumed that the process of forming an image is carried out perfectly. Since ideal image forming methods do not exist and can only be approached, a practical image will only approximate that given in (4). Because most image forming methods involve linear mechanisms such as those which characterize optics, a practical image can be regarded as an additive superposition of ideal images. This fact is expressed in (5)

$$\infty > \tilde{I}_{x,y} = \int_{-\infty}^{\infty} I_{X,Y} h_{x,X;y,Y} \, dX dY > 0 \tag{5}$$

where $\tilde{I}_{x,y}$ represents a practical image and $h_{x,X;y,Y}$ represents the so-called point spread function of the linear image forming mechanism. In other words $h_{x,X;y,Y}$ is the practical image that an ideal image consisting of a unit intensity point of light located at $x=X$ and $y=Y$ would produce. Obviously h must be nonnegative.

If the point spread function is the same shape for all points of light in the ideal image, then the superposition integral (5) becomes a convolution integral (6)

$$\infty > \tilde{I}_{x,y} = \int_{-\infty}^{\infty} I_{X,Y} h_{x-X;y-Y} \, dX dY > 0 \tag{6}$$

[1] It is almost impossible to find a material that reflects less than about 1 percent of the incident light.

which is conventionally expressed using a compact notation as in (7)

$$\infty > \bar{I}_{x,y} = I_{x,y} * h_{x,y} > 0. \tag{7}$$

Combining (4) and (7) we obtain (8)

$$\infty > \bar{I}_{x,y} = (i_{x,y} \cdot r_{x,y}) * h_{x,y} > 0 \tag{8}$$

which under the assumption of a position invariant point spread function summarizes the essential structure of practical images as they are considered in most current efforts.

The expression (8) places in evidence the three essential components of a practical image. If $h_{x,y}$ is sufficiently small in its spatial extent, the practical image can be taken as an adequate approximation to the ideal. If $h_{x,y}$ fails in this respect, the practical image can be processed by any one of a variety of methods in an attempt to remedy the situation.[2]

Since the objective of the present discussion focuses primarily on the structure of an ideal image, it will be assumed in the following that the effect of $h_{x,y}$ can be neglected.[3] Primary concern here is thus redirected to (4).

We now return to the issue posed at the end of Section II as to whether or not the mathematical structure of linear processors is compatible with the structure of the images themselves. Since (4) indicates that the image components are multiplied to form the composite, and further since linear systems are compatible with signals possessing additive structure, it follows that there exists basic incompatibility. However, this incompatibility depends in a basic way upon some implicit assumptions which have been imposed upon the structure as described in (4).

An essential ingredient to the structure of images as expressed in (4) is the assumption that an image is an energy signal. This assumption really amounts to a choice of a representation for an image. The nature of that choice can be extremely important. To clarify this concept the question of representation must be elaborated upon.

IV. The Representation of Images

A key question in the transmission, storage, or processing of any information is that of representation. The reason that the choice of representation is important is that the problems of transmission, storage, and processing can be substantially effected by it.

If an ideal physical image is considered as a carrier of information, it follows that nature has already chosen a representation. It takes the form of light energy. Furthermore, if one takes nature literally when sensing an optical image, one will continue that representation by creating a signal proportional to the intensity of that light energy. Indeed this representation seems like a very natural one, and in fact as already indicated, it is commonly used in television and digital image processing.

Strangely enough representation by light intensity analogy is a relatively new practice in image technology. The process of photography, now over a century old, does not use it. It has only been with the advent of electrical imaging methods that it has received attention.

In order to clarify this point, imagine a black and white photographic transparency which portrays some optical image. In order to see the reproduction one must illuminate the transparency uniformly with some intensity i_0 and somehow view the transmitted pattern of light intensity $I_{x,y}$. The quantities of light which are transmitted are determined by the volume concentrations of amorphous silver suspended in a gelatinous emulsion. Thus it is these concentrations which represent the image in its stored form. Let these concentrations be expressed as $C_{x,y,z}$.

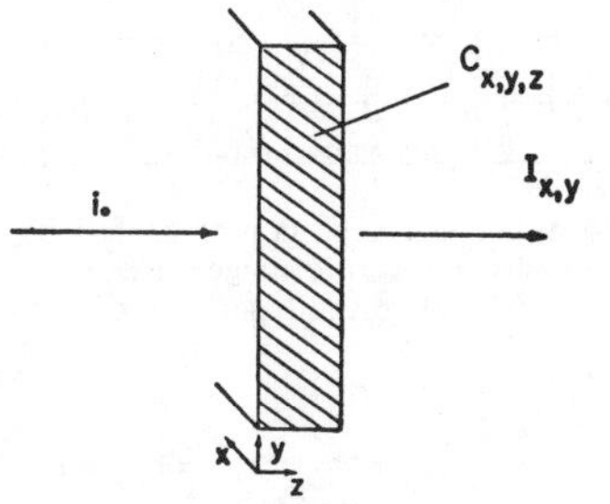

Fig. 1. An intensity image $I_{x,y}$ as reproduced by the transmission of light through a volume concentration of amorphous silver $C_{x,y,z}$.

Physically the situation is as depicted in Fig. 1. In order to derive the relationship between the reproduced image $I_{x,y}$ and $C_{x,y,z}$ we must consider the transmission of light through materials. The physics of the situation is given in (9)

$$\frac{di}{dz} = -kC_{x,y,z}i \tag{9}$$

where i is the intensity of the light at any point in the transmitting material and k is a constant representing the attenuating ability of a unit concentration of amorphous silver. Integration of (9) according to standard methods yields (10)

$$\int_{i_0}^{I_{x,y}} \frac{di}{i} = -k \int_0^{z_t} C_{x,y,z} dz \tag{10}$$

where z_t represents the thickness of the emulsion. Since the integral in the right-hand side of (10) represents the total quantity of silver per unit area of the transparency independent of how that silver is distributed in the z dimension, (10) can be rewritten as in (11)

$$\ln (I_{x,y}/i_0) = -kd_{x,y}. \tag{11}$$

A solution of (11) for $I_{x,y}$ yields (12)

$$I_{x,y} = i_0 e^{-kd_{x,y}}. \tag{12}$$

From (11) it can be seen that in the case of a photographic transparency, the physical representation of the image is actually $d_{x,y}$ which is proportional to the logarithm of the reproduced intensity image. In turn (12) reveals that the physical representation $d_{x,y}$ is exponentiated during its conversion to light intensity. Further, it follows that if $I_{x,y}$ is a faithful reproduction of the original intensity image from which the transparency was made, then the quantities of silver used to form the representation $d_{x,y}$ must have been

[2] For an excellent and recent summary, bibliography, and set of references representative of the many interesting efforts in this area, see Section II of a recent article by Huang *et al.* [1].

[3] There is still much to be learned both practically and theoretically about restoring practical images to the point where this is possible. Such restoration methods are very important; and since they attempt in part to compensate for distortions caused by linear mechanisms, linear processing is used extensively and often with great success.

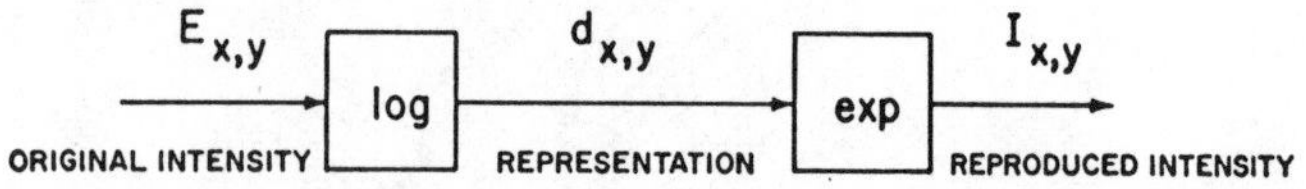

Fig. 2. In photography an image is represented by the total quantity $d_{x,y}$ of amorphous silver per unit image area. For faithful reproduction $d_{x,y}$ must be proportional to the logarithm of the image intensities.

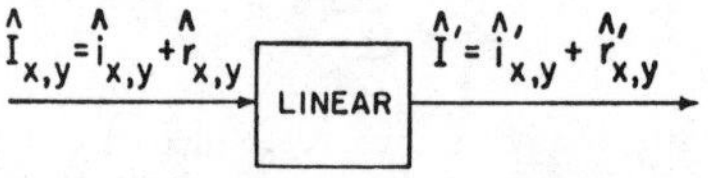

Fig. 3. A density image as processed by a linear system. Note that the basic structure of the image is preserved. The output is a processed illumination plus a processed reflectance regardless of what the process may be.

deposited in the emulsion by a process which was logarithmically sensitive to light energy.

This situation is summarized in Fig. 2 where the logarithmic and exponential transformations which mechanize the formation of a photographic image are placed in evidence. The variables i_0 and k which appear in (11) and (12) have been omitted for convenience since they are only scaling constants.[4]

The relationship of (12) is well known in photography but is usually presented in a somewhat altered form as in (13).

$$\log_{10} (i_0/I_{x,y}) = D_{x,y}. \tag{13}$$

Here the quantity $D_{x,y}$, called density, is proportional to $d_{x,y}$ but related directly to the common logarithm in a manner similar to that used in the definition of the decibel. Because $d_{x,y}$ and $D_{x,y}$ are both related to the popular notion of density it is reasonable to call any logarithmic representation of an image a density representation. As indicated above, all such representations are the same except for the choice of the two constant parameters.

Taking this into account (11) and (12) may be generalized to (14) and (15)

$$\hat{I}_{x,y} = \log (I_{x,y}) \tag{14}$$

$$I_{x,y} = \exp (\hat{I}_{x,y}) \tag{15}$$

where the hatted variables represent density and the unhatted variables represent intensity. All density representations are the same except for a scale factor and an additive constant.

V. Relationships Between Processing, Structure, and Representation

A study of the use of a density representation for images leads to a chain of interesting observations. These observations begin with the introduction of density representations into the previous discussion concerning the structure of ideal images. This introduction changes (1)–(4)[5]

$$\infty > \hat{I}_{x,y} = \log (I_{x,y}) > -\infty \tag{16}$$

$$\infty > \hat{i}_{x,y} = \log (i_{x,y}) > -\infty \tag{17}$$

$$0 > \hat{r}_{x,y} = \log (r_{x,y}) > \hat{r}_{\min} \tag{18}$$

and

$$\infty > \hat{I}_{x,y} = \hat{i}_{x,y} + \hat{r}_{x,y} > -\infty \tag{19}$$

where $\hat{i}_{x,y}$ and $\hat{r}_{x,y}$ represent illumination[6] and reflection densities, respectively.

It is obvious from these equations that a change from an energy representation to a density representation has introduced some interesting changes in the apparent structure of images. There is no longer a restriction upon the range of the representation. To see this fact compare (1) with (16). The manner in which the basic components of the scene are combined has been changed from multiplication to addition (compare (4) and (19)). Finally, the scene components themselves have been changed from an energy representation to a density representation.

In the case of the reflection component the transformation to a density representation is a very satisfactory one. This is so, because to a great extent the physical properties of an object which determine its ability to reflect light are the densities of the light blocking materials from which it is formed. The situation is similar to that of the photographic transparency as described in (9)–(12). Thus by using (19) the physical properties of an object are represented more directly than in (4).

The single most important effect of using a density representation is that it makes the structure of images compatible with the mathematical structure of linear processing systems. This fact is true, because linear systems obey additive superposition and from (19) we see that the basis for the structure of a density representation of an image is additive superposition.

To build upon this observation consider Fig. 3 in which a density image is being processed by a linear system. The input of the system is given as in (19). It follows from the property of superposition in linear systems that the output must be given in (20)

$$\infty > \hat{I}_{x,y}' = \hat{i}_{x,y}' + \hat{r}_{x,y}' > -\infty \tag{20}$$

where the primes indicate processed quantities. But (21) is in the same form as (19). What (20) says is that the basic structure of a density image is preserved by any linear processor. More specifically the illumination component of the processed image *is* the processed illumination component and the reflection component of the processed image *is* the processed reflection component.

For comparison consider the effect of a linear system upon an intensity image. The input is given in (4). It is clear that the notion of structure preservation cannot be maintained in this case. What is even more embarrassing is the fact that there is little guarantee that the output will be positive and nonzero which it must if it is to be regarded as an image at all.

Because an image carries information, and because information can be measured using concepts of probability, it is interesting to consider the probability density functions

[4] Actually i_0 is just a constant of proportionality on the image intensity and can be neglected if one considers normalized images only. Also k can be absorbed into the logarithmic and exponential transformations by adjusting the base being used.

[5] The minimum reflection density using the common logarithm would almost never exceed 2.0. See footnote 1.

[6] The concept of an illumination density may seem strange at the outset but proves to be an important mathematical concept even though it may be difficult to assign it any physical significance.

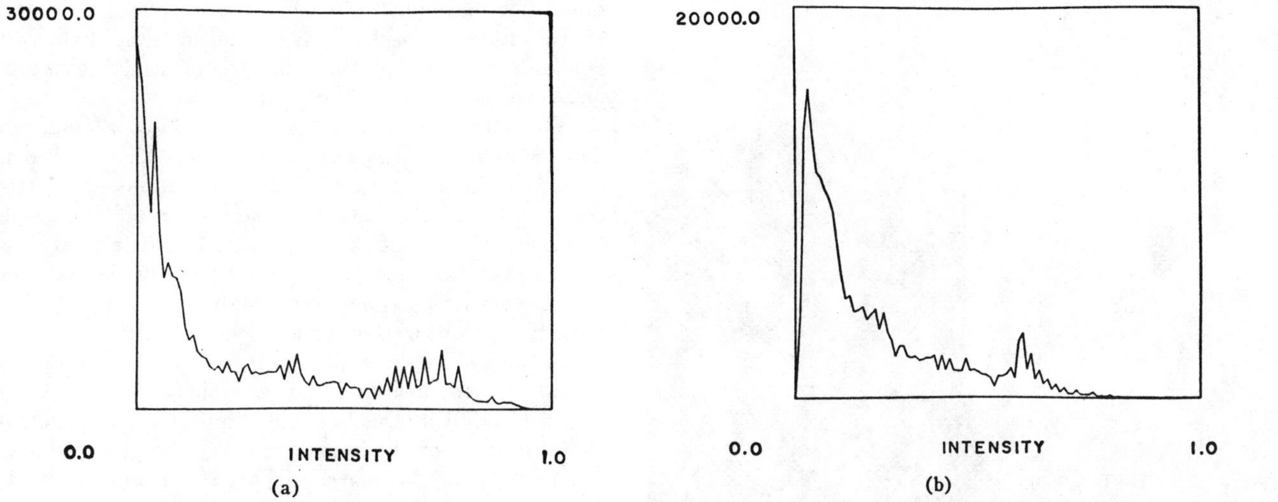

Fig. 4. Intensity histograms of 100 bins each obtained from high quality images carefully digitized to 340 by 340 samples using 12 bit/sample. (a) Three wide dynamic range scenes. (b) Two scenes of less dynamic range (approx. 30:1).

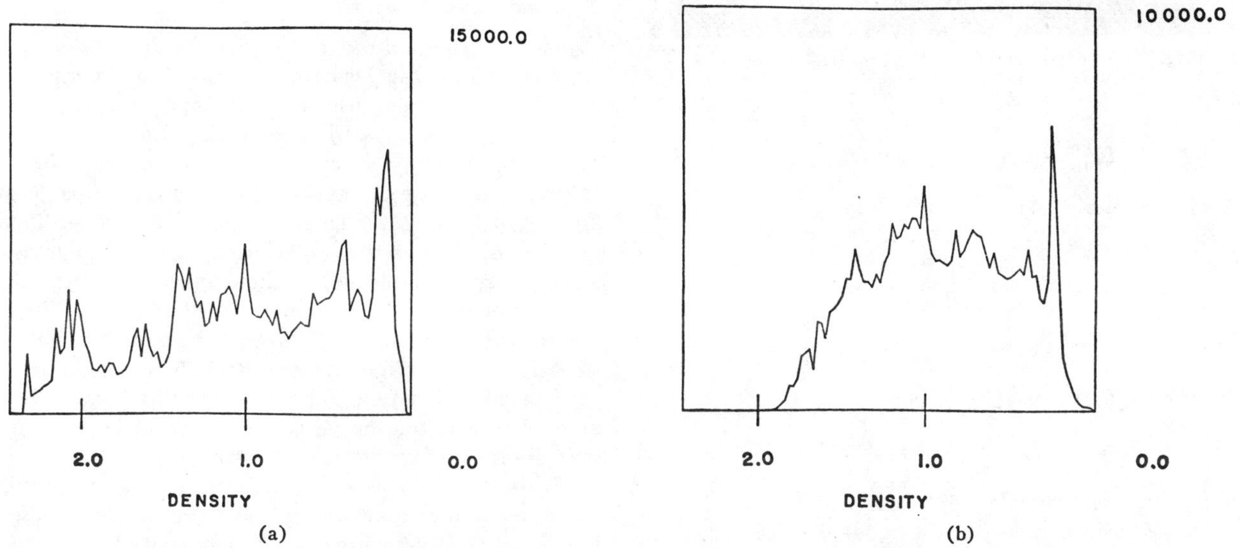

Fig. 5. Density histograms of 100 bins each obtained from the same images as in Fig. 4.

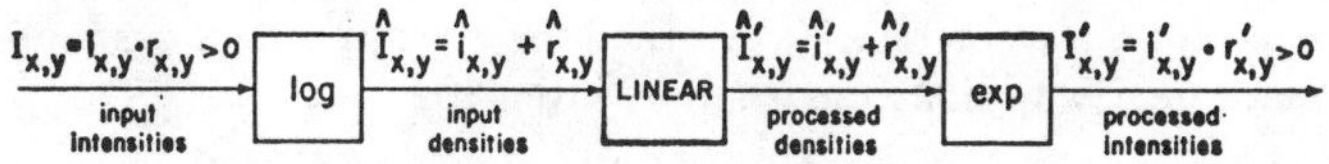

Fig. 6. An intensity image as processed by a multiplicative system. Again the basic structure of the image is preserved and the output is a processed illumination times a processed reflectance.

which are associated with both forms of representation. To this end Fig. 4 shows histograms for images which were represented by intensities and Fig. 5 shows histograms for the same images as represented by densities. These images were obtained using very careful methods from very high quality digital images.

It is instructive to compare the highly skewed distributions of Fig. 4 with the more nearly symmetric ones of Fig. 5. The fact that a density representation of an image tends to fill the representation space more uniformly than an intensity representation implies some important advantages for the former. For example, consider the problem of digitizing either representation by means of a quantizer using a binary code. The nearly symmetric distributions of Fig. 5 imply a more efficient use of the information carrying capacity of the binary code, a rectangular distribution being ideal in this respect. In addition, the symmetric distributions are more nearly aligned with the conventional assumptions associated with signals in many theoretical studies.

VI. Multiplicative Superposition in Image Processors

For some purposes it is important to be able to think of an image as represented by intensities. It is absolutely essential to do so when sensing an image to begin with or when reproducing an image for observation. In these cases it is possible to retain the match between the structure of images and the structure of processors by combining the concepts embodied in Figs. 2 and 3. This situation is depicted in Fig. 6. The input is given as in (4). It follows from (20) and (15) that

$$\infty > I_{x,y}' = \exp(\hat{I}_{x,y}') = \exp(\hat{i}_{x,y}' + \hat{r}_{x,y}') > 0 \quad (21)$$

which by the properties of the exponential function becomes

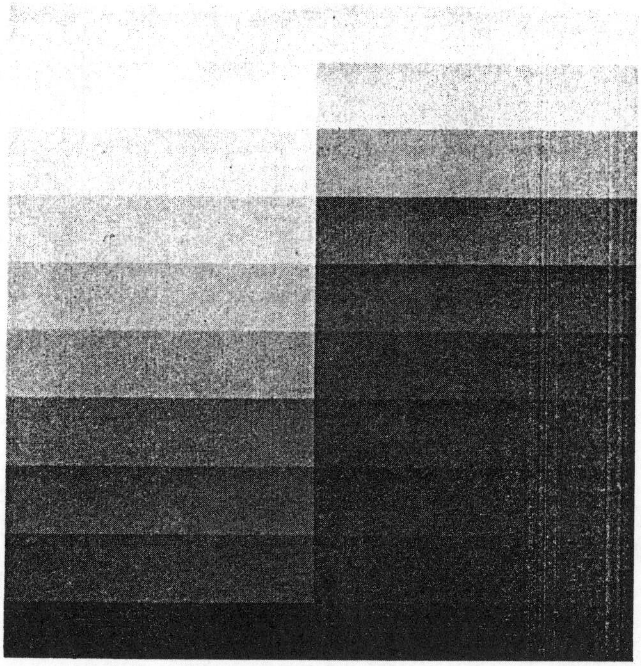

Fig. 7. Two grayscales.[8] (a) Linear intensity steps. (b) Linear density steps.

$$\infty > I_{x,y}' = \exp(\hat{\imath}_{x,y}') \cdot \exp(\hat{r}_{x,y}') > 0. \tag{22}$$

But in analogy with (21) we have

$$i_{x,y}' = \exp(\hat{\imath}_{x,y}') \tag{23a}$$

and

$$r_{x,y}' = \exp(\hat{r}_{x,y}'). \tag{23b}$$

So substituting (23) into (22) we get

$$\infty > I_{x,y}' = i_{x,y}' \cdot r_{x,y}' > 0 \tag{24}$$

which is in the same form as (4).

Again the basic structure of the image is preserved. However, this time the multiplicative superposition which characterizes the structure of an intensity image is compatible with the mathematical structure of the processor of Fig. 6. It follows that Fig. 6 depicts a class of systems which obey multiplicative superposition [2]. Besides demonstrating the preservation of structure for intensity images (24) also reveals the fact that a multiplicatively processed image is itself positive and nonzero and thus realizable. This later observation transcends the fact that the system used to process the input densities in Fig. 6 is linear, because the processed intensities are formed by exponentiating the processed densities regardless of how those densities were produced. The result of exponentiating a real density is always positive and nonzero. This property of density processing is called the realizable output guarantee.

VII. Multiplicative Superposition in Vision

Although a great deal of sophisticated and elaborate knowledge has been gained in the last several decades about the problem of communicating electrically between various sorts of automatic mechanisms, dissappointingly little has been done to match the ultimate source and receiver, namely the human being, to this body of knowledge and these systems. The basic obstacles have been a lack of understanding of the human mechanisms in terms describable by the available theory and the difficulty in studying the human mechanisms which are involved.

The philosophy that any communications system, whether man-made or natural, has structure and that that structure should be matched to the communications task at hand, seems to provide a stepping stone for understanding the operation of some of these systems. In this regard we would like to take the concept of a multiplicative image processor and explore its possible relationship to the known properties of early portions of the human visual system.

In many respects the multiplicative image processors previously described and their canonic form as represented in Fig. 6 bear an interesting resemblance to many operational characteristics of the human retina.[7] The presence of an approximately logarithmic sensitivity in vision has been known for some time [3]. Even more readily evident, and mechanized through the process of neural interaction, is the means for linear filtering [3], [4].

A. Logarithmic Sensitivity

The fact that light sensitive neurons fire at rates which are proportional to the logarithm of the light energy incident upon them has been measured for simple animal eyes [3, pp. 246–253]. Similar experiments with human beings are inconvenient to say the least, but there are some interesting experiments that serve as a partial substitute. The most convincing of these is the so called "just noticeable difference" experiment [5]. In this experiment an observer is asked to adjust a controllable light patch until it is just noticably brighter or darker than a reference light patch. The experimenter then steps his way through the gamut of light intensities from very bright to very dark. The step numbers are then plotted as a function of the intensity of the reference light. The resulting curve is very close to logarithmic over several orders of magnitude of intensity.

For a direct but less objective demonstration of this relationship consider the gray-scale steps[8] presented in Fig. 7. In Fig. 7(a) the scale consists of equally spaced intensity steps. In Fig. 7(b) the scale consists of exponentially spaced intensity steps which is the same as equally spaced density steps. The scale in Fig. 7(b) appears as a more nearly equally spaced scale than that of Fig. 7(a) so that the eye appears to respond more nearly to densities than to intensities.

B. Linear Filtering through Neural Interaction

The mechanism for linear spatial processing in vision is observed in the Hartline equations [4, pt. I, ch. 3], [3, ch. 11, pp. 284–310]. The effect of this processing can be observed by means of a number of simple optical illusions.

The simplest of these illusions is known as the illusion of simultaneous contrast[9] and can easily be observed in Fig. 8. In this image we observe two small squares surrounded by larger rectangles, one light, one dark. In fact the two small

[7] A recent, lucid, and elaborate discussion of these characteristics is presented by Cornsweet [3]. See especially chs. XI and XII.

[8] This and several other test images shown here should be presented using a calibrated display or calibrated photography. An uncertain but considerable distortion will have taken place during the printing of this paper. The reader must take this into account and estimate the possible degradation for himself.

[9] For a more complete discussion see [3, pp. 270–284].

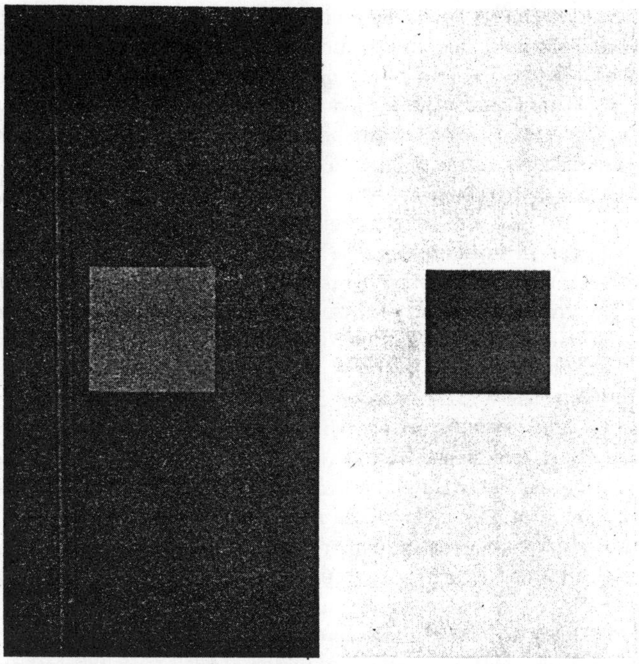

Fig. 8. The illusion of simultaneous contrast. The two small squares are of exactly the same intensity.

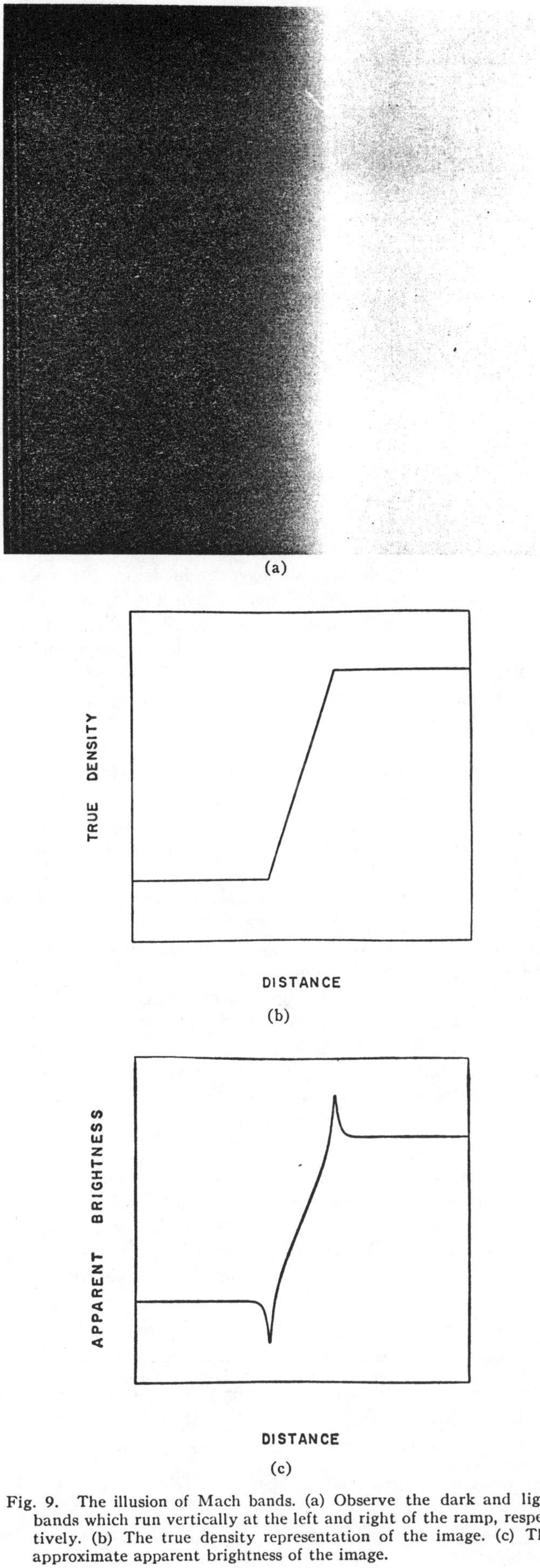

Fig. 9. The illusion of Mach bands. (a) Observe the dark and light bands which run vertically at the left and right of the ramp, respectively. (b) The true density representation of the image. (c) The approximate apparent brightness of the image.

squares are exactly the same shade of gray. They appear different, however, due to their surroundings. This illusion can be explained at least qualitatively by assuming that the image has been subject to linear spatial filtering in which low spatial frequencies have been attenuated relative to high spatial frequencies. Filters of this type cause the averages of different areas in one image to seek a common level. Since in Fig. 8 the area of the left has a darker average, it will be raised, making the left square brighter. Likewise, since the area on the right has a lighter average, it will be lowered, making the right square less bright.

Another illusion can be observed by returning attention to Fig. 7(b). Each rectangle in this gray scale is one uniform shade of gray. However, each rectangle appears to be darker near its lighter partner and lighter near its darker partner. Again the phenomenon can be explained at least qualitatively by the assumption of linear spatial filtering.[9]

The final illusion to be discussed here is presented in Fig. 9. It is known as the illusion of Mach bands [3, pp. 270–284], [4]. In this image[8] there are two large areas, one light and one dark but each of a uniform shade. These two areas are coupled by a linearly increasing density wedge (exponentially increasing intensity wedge) as indicated in Fig. 9(b). The observer will notice that immediately at the left and at the right of this wedge are a dark and light band as implied by Fig. 9(c). These bands, known as Mach bands, can also be explained at least qualitatively by linear processing.[10]

C. *Saturation Effects*

So far this discussion has implied that the linear spatial processing of densities can explain a number of visual phe-

[10] Quantitative studies of this illusion are common. Unfortunately, almost all of them employ a matching field or light which in turn perturbs the measurement considerably. Mach himself warned of this problem [4, pp. 50–54, 262, 305, 322] and suggested that there is no solution. The psychophysical experiment to be described later is offered as a possible counter example to this suggestion.

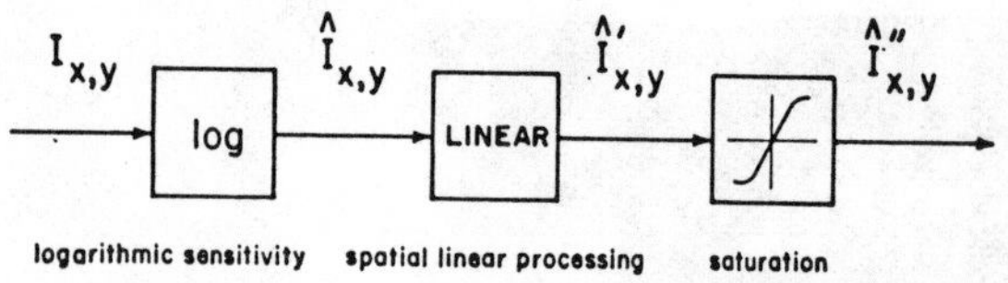

Fig. 10. A possible approximate model for the processing characteristics of early portions of the human visual system.

nomena. It is clear that these visual phenomena are only observable if there is a proper amount of light available for their presentation. It is common knowledge that below certain illumination levels one cannot see well if at all. The same is true if illumination levels become too great.

The physical limitations of any visual mechanism guarantee that saturation or threshold effects will occur if intensity levels are raised or lowered far enough. In this respect any consideration of the relationship between the processing of densities and properties of vision must eventually include the effects of saturation.

D. A Process Model for Early Portions of the Human Visual System

The preceding discussions suggest a model for the processing characteristics of early portions of the human visual system.[11] This model is shown in Fig. 10. The output $\hat{I}_{x,y}''$ is a saturated version of a linearly processed density representation. The linear processing is presumably of the form in which low spatial frequencies are attenuated relative to high spatial frequencies.

The most useful implications of this model do not come from its relationship to the optical illusions which we have already discussed as much as from the operational characteristics it embodies. The operational characteristics in question center around the ability of the human visual system to maintain its sensitivity to patterns of relatively low contrast in the context of a total image in which intensities are spread across a very large dynamic range,[12] and its ability to preserve an awareness of the true shades of an object in spite of huge differences in illumination. Moreover, these abilities are embodied without sacrificing the basic structure of images with respect to the separate physical components of illumination and reflectance!

If the illumination component of an image did not vary in space, (4) would become

$$I_{x,y} = i \cdot r_{x,y}. \tag{25}$$

In this case[13] the dynamic range of an image would be limited to about 100:1, because it would be determined by the reflection component[1] alone. Problems with saturation effects would be relieved if not avoided altogether. In addition the true shade of an object would be reproduced directly by $I_{x,y}$.

[11] This model is representative of approximate processing characteristics at early stages only. It is not intended as a biophysical or anatomical model for any specific visual mechanism or as an exact or complete processing representation. In image processing some such model must be assumed even if it is by default. The classical default assumption is that of fidelity reproduction namely that like an ideal camera the eye "sees" what it sees.

[12] The dynamic range of an image is the ratio of the greatest to the least intensity value therein contained. Ratios in excess of 1000:1 are often encountered by the eye or camera.

[13] This configuration, often sought at great expense in photographic and television studios, is called flat lighting.

Unfortunately, the illumination component of an image varies a great deal, often more than the reflectance component. For example a black piece of paper in bright sunlight will reflect more light than a white piece of paper in shadow. In the proper environment both situations could occur in the same image at the same time, but an observer would always call the white paper "white" and the black paper "black" in spite of the fact that the black paper would be represented by a higher intensity than the white paper. This visual phenomenon is called brightness constancy. Moreover, if there were low contrast markings on either sheet of paper they could be read in spite of their insignificance with respect to the total intensity scale.

With these facts in mind it is interesting to note that the system of Fig. 10 tends to produce an output in which the variations in illumination are indeed reduced. This is so, because the illumination component dominates the Fourier spectrum of a density image at low spatial frequencies while the reflectance component dominates at high spatial frequencies. As a result, the spatial linear filtering previously described reduces the illumination variations, because it attenuates low frequencies relative to high frequencies. At the same time the basic structure of images is preserved because the model operates linearly on a density representation.

The detailed consequences of this situation are described in more detail in [2, sec. V]. There the use of multiplicative processors for the purpose of simultaneous dynamic range reduction and detail contrast enhancement is discussed and demonstrated. An example of an image possessing some serious dynamic range problems is shown in Fig. 11 before and after such processing. Notice how the illumination is extremely variable from the outside to the inside of the building. In the unprocessed image, details within the room though present in the original are obscured by the limited dynamic range capabilities of the printing process you are now viewing. In the processed image these details are present in spite of this limitation.

E. Model and Process Compatibility

When the image of Fig. 11(b) is observed, the total processing system including the approximate visual model is that shown in Fig. 12 which combines Figs. 6 and 10. In Fig. 12(a) the two linear systems which characterize the processor and the visual system are labeled H and V, respectively. Fig. 12(b) shows the simplified exact equivalent system in which as much merging of subprocesses as is possible has been performed. The new composite linear system labeled $H \cdot V$ is merely the cascade of the two previous ones.

Fig. 12(b) demonstrates the compatibility of the visual model and the multiplicative image processor. It does so by placing in evidence the fact that within the validity of the model the experience of viewing a processed image is indistinguishable from that of viewing an unprocessed image except that it is possible to alter the linear processing performed through the manipulation of the linear system labeled H.

F. Model Testing and Calibration

The approximate visual model of Fig. 10 has been motivated in the above by studying certain illusions, noting certain aspects of neural structure and neural measurement, and by concentrating attention upon certain desirable and available performance characteristics. This motivation can be sup-

(a) (b)

Fig. 11. A large dynamic range scene. (a) Before processing. (b) After processing with a multiplicative processor adjusted to attenuate low and to amplify high frequency components of density. (Note: These and all other images in this paper are digital.)

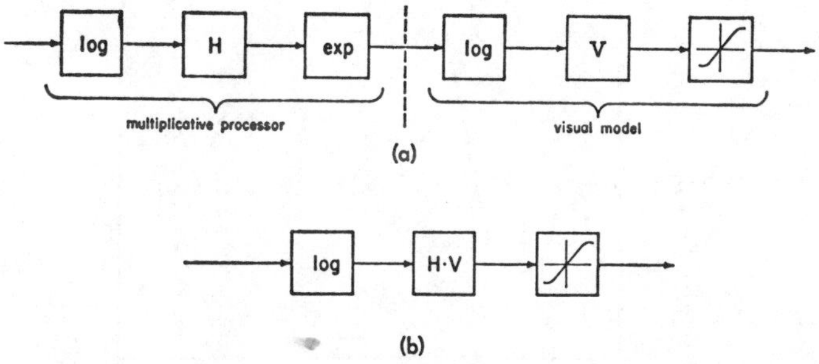

Fig. 12. Total processing system including visual model when viewing Fig. 11(b). (a) Unsimplified system. Processed intensities appear at the vertical dotted line. (b) Simplified system with processors merged.

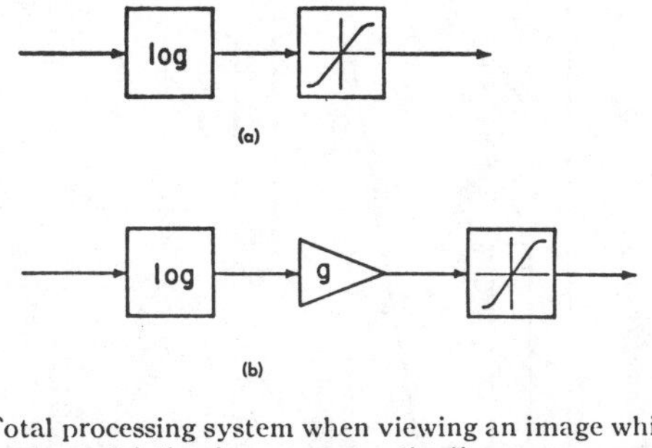

Fig. 13. Total processing system when viewing an image which has been subject to a multiplicative processor the linear component of which has been adjusted to be the inverse of the linear component of the visual model. (a) H is exactly the inverse of V. (b) H is the inverse of V except for a constant of proportionality g.

ported by a testing experiment which is suggested by the situation depicted in Fig. 12. If the system H were adjusted to become the inverse of the system V, the system of Fig. 12(b) could be further simplified as shown in Fig. 13. In this situation it should not be possible to observe the optical illusions described above and portrayed in Figs. 8 and 9.

An experiment designed to find an H which would simultaneously cancel the optical illusions described above can be carried out with significant success [6]. By comparing the pattern of Fig. 14 with Figs. 8 and 9 one can see that this pattern strongly induces the illusions in question.[8] If one processes this pattern by means of a multiplicative processor with the system H adjusted according to (26)

$$H = V^{-1} \tag{26}$$

one obtains a pattern which appears to have little remaining illusion phenomena.

Such a processed pattern[14] is shown in Fig. 15. The illusions have been significantly suppressed, and the apparent brightness of Fig. 15 follows the profile of true density of Fig. 14 remarkably well. The degree to which the illusions have been suppressed provides additional support for the model of Fig. 10. In addition an estimate of the system V results as a byproduct since (26) can be solved for V in terms of the actual H used in the experiment.

It should be noted that the above results support the logarithmic component of the model and its position in the system because the cancellation of the illusions depends upon the neutralization of the exponential component of the multiplicative processor. Without this neutralization Fig. 12(a) could not be reduced to Fig. 12(b).

Although one might find a system H that would cancel the illusions for a single fixed pattern, it has been shown that the experiment succeeds about equally well for all patterns such

[14] Here the comments of footnote 8 must be considered most seriously since the illusion cancelling experiment is a sensitive one and gray-scale distortions can upset it easily. The calibrated print sent to the publisher appears as described in the text. A limited number of such calibrated prints are available to readers with sufficient interest and requirements. As published here the pattern should be viewed approximately at arms length.

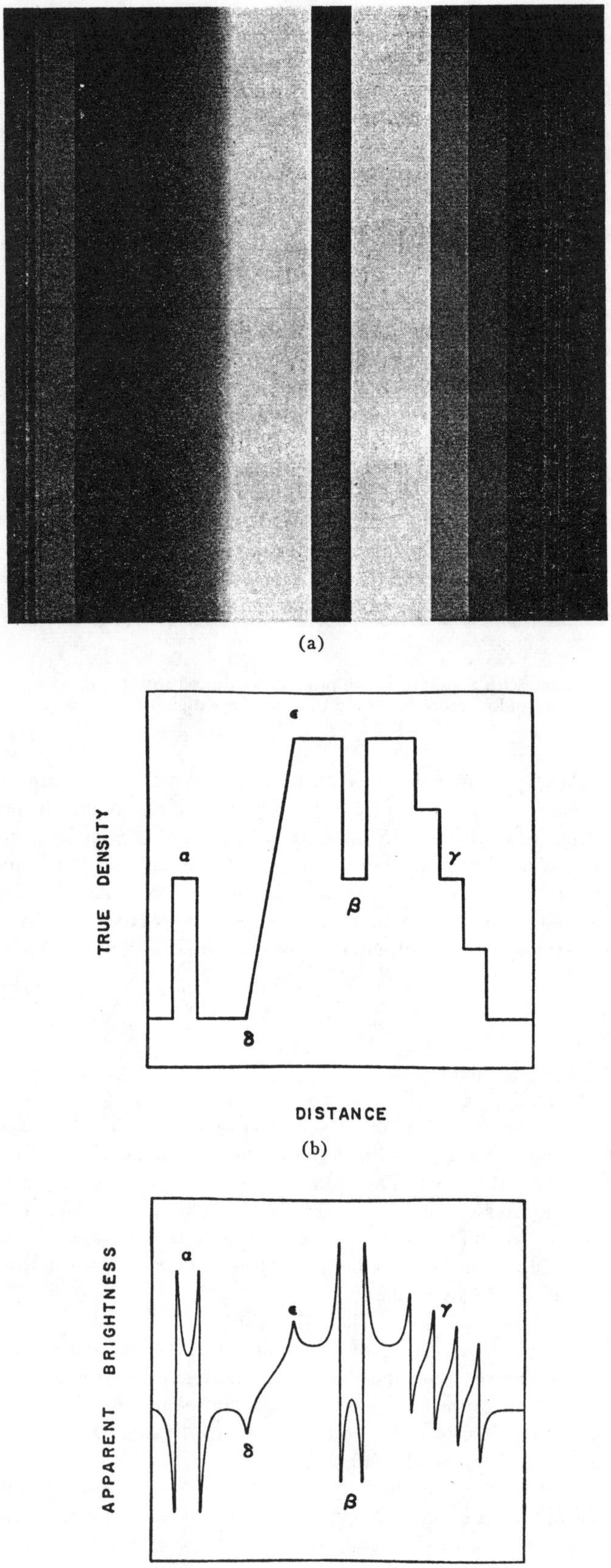

Fig. 14. Pattern for use in testing and calibrating the visual model. (a) Observe the illusions of simultaneous contrast α, β, γ, and Mach bands δ, ϵ. (b) The true density representation of the image. (c) The approximate apparent brightness of the image.

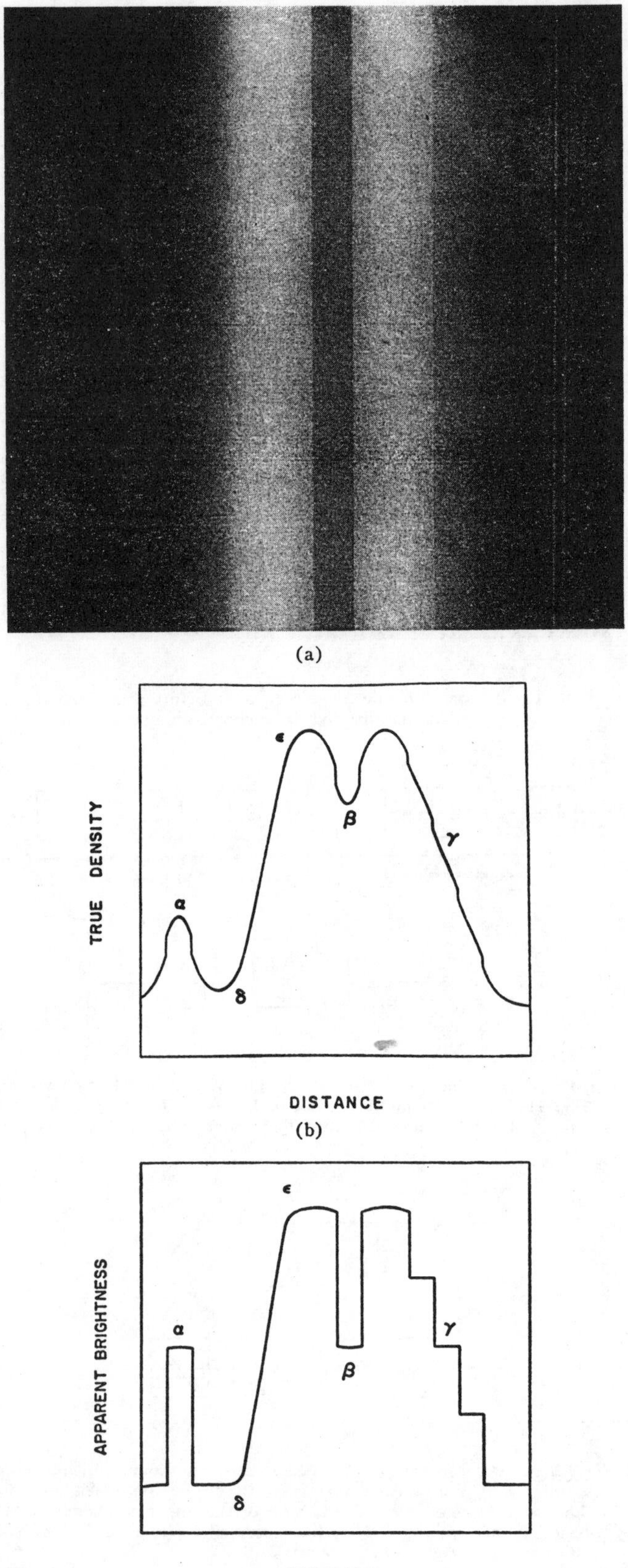

Fig. 15. The pattern of Fig. 14 processed for the suppression of optical illusions. Compare with Fig. 14. (a) Appraise the amounts of remaining simultaneous contrast α, β, γ, and Mach bands δ, ϵ. (b) The true density representation of the processed image. (c) The approximate apparent brightness of the processed image as observed from a calibrated print. Curve taken as a subjective consensus from five knowledgeable observers.

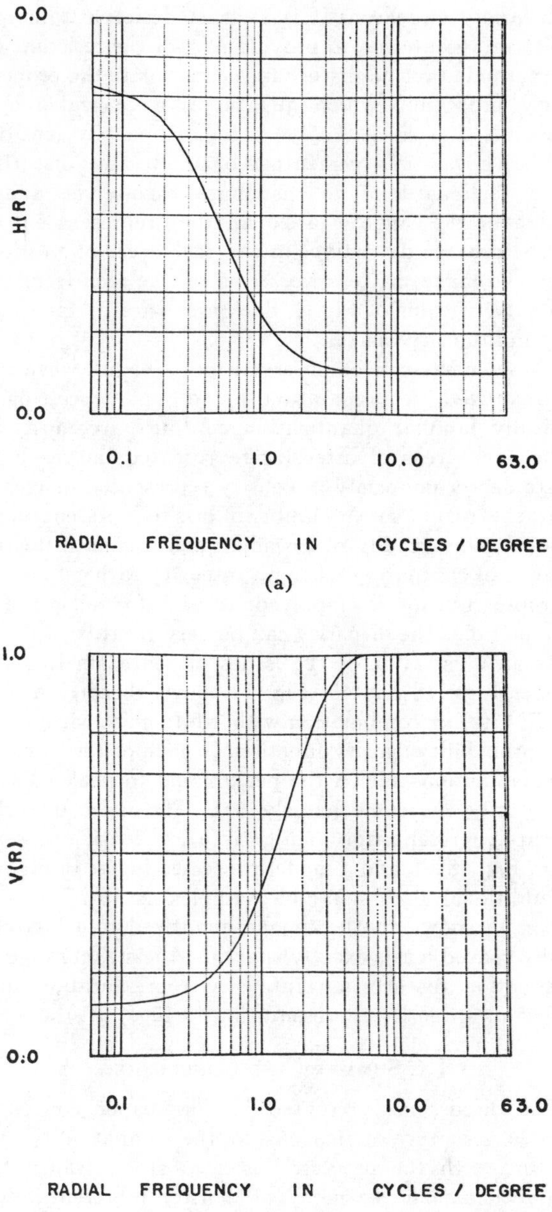

Fig. 16. Frequency response of one-dimensional systems used in test of eye model. (a) Response of system H for cancelling illusions. (b) Relative response of system V as estimated from H.

as Fig. 14 not just the one shown here. Alternately, it has been shown that the cancellation of Fig. 15 holds across a wide range of the constant of proportionality g in which the processed patterns have enough dynamic range to be clearly visible and not so much dynamic range so as to produce saturation effects.[15]

The actual linear system H used in the experiment described above was found by a cut-and-try procedure wherein an initial estimate was refined through successive rounds of processing, visual evaluation, and system redesign.

Since the test patterns varied only in one dimension, the development of a one-dimensional linear system for H was all that was required.[16] The one-dimensional frequency response of that system is shown along with its inverse in Fig. 16. It follows from two-dimensional Fourier analysis that under the assumption that the two-dimensional frequency response of the eye model has circular symmetry, the curve of Fig. 16(b) represents a radial cross section of that two-dimensional frequency response. Specifically

$$V(R) = V(X). \tag{27}$$

In addition the two-dimensional point spread function of the system V can be determined either from the Bessel transform of $V(R)$ or from the two-dimensional Fourier transform of the surface of revolution generated by $V(R)$.

It is interesting to compare the frequency response characteristics obtained here with those determined elsewhere. An excellent summary discussion and associated references are available [3, ch. 12, pp. 330–342]. In this respect there is a marked similarity between the approach taken here and the work of Davidson [3, ch. 12, pp. 330–342] in which problems with both logarithmic sensitivity and spatial interference between test patterns and matching fields are avoided.[17]

One might wonder what the world would look like if the eye did not create the illusions that we have been discussing. In this regard consider Fig. 17 which bears the same relation to Fig. 11(a) as Fig. 15(a) bears to Fig. 14(a).

VIII. Image Quality and the Visual Model

Image quality is a complicated concept and has been studied in a variety of ways and contexts. In most situations a final measure of quality can be defined only in the subjective sense. It can be measured only approximately and with difficulty by means of slow and expensive tests involving human observers. As the understanding of the human visual mechanism grows, objective measures become more feasible. So it is that with the aid of the visual model of Fig. 10 it is possible to define such a measure of image quality. By virtue of the discussions presented in Section VII one expects this measure to be related to some basic subjective considerations. An objective measure is defined by measuring the difference between a distorted image and its reference original, only after each has been transformed by the model. An example of such a definition based on a mean-square error measure is given in (28)

$$E^2 = \iint [V_{x,y} \circledast (\log I_{x,y} - \log R_{x,y})]^2 \, dxdy \tag{28}$$

[15] Since the cancellation of these illusions requires only that the apparent brightnesses of Fig. 15 take on a profile of a certain *relative* shape, the true value of g in (26) and in Fig. 13(b) cannot be determined. Thus V can only be estimated to within an unknown constant of proportionality.

[16] For the purpose of this experimental effort the linear system portion of the eye model was assumed to be position invariant. Since peripheral and central (foveal) vision possess quite different resolution properties, this assumption falls short of reality and leaves room for further refinements. For this reason and because the cancellation of illusions as shown in Fig. 15 might be improved we have not given an analytic expression for our present best estimate for $V(R)$ as part of (27). Tentatively we are using

$$V(R) = 742/(661 + R^2) - 2.463/(2.459 + R^2)$$

where R is the radial spatial frequency in cycles per degree. See Fig. 16(b). See also [7].

[17] One can still find fault with these methods, because the test patterns used do not fill the visual field and so there is still interaction between them and the surround which is uncontrolled. See also footnote 16.

Fig. 17. The scene of Fig. 11(a) processed for the suppression of optical illusions. Compare with Fig. 11(a).

where E is the objective measure, $V_{x,y}$ is the two-dimensional point spread function of the visual model, $I_{x,y}$ is the image being measured, and $R_{x,y}$ is the reference original. For examples of the use of such an objective measure see Sakrison and Algazi [7] and Davisson [8]. Since the model emphasizes certain aspects of an image and deemphasizes certain others in a manner approximately the same as early portions of the human visual system, distortions which are important to the observer will be considered heavily while those which are not will be treated with far less weight. This will be so even though the important distortions may be physically small and the unimportant ones physically large, which is frequently the case.

With the above ideas in mind it becomes clear that when an image is to be distorted as a result of the practical limitations which characterize all transmission, storage, and processing mechanisms it makes sense to allow such distortions to take place after the image has been transformed by the model. The image can then be transformed back again just before it is to be viewed. For example if an image bandwidth compression scheme is to be implemented it probably makes much better sense to invoke that scheme upon the model-transformed image than upon the physical intensity image. The motivations for this argument are not entirely subjective. Since the model transformation emphasizes the reflectance components and deemphasizes the illumination components of a scene, it renders that scene more resistant to disturbing influences on certain physical grounds as well, because it can be argued that the reflectance component is the more important one.

For some applications it may be inconvenient to transform an image by means of the complete visual model before exposing it to disturbing influences, because the processing power required to mechanize the linear portion of the model might be somewhat high in terms of the present technology. However, for a variety of reasons it is at least desirable to employ a density representation to provide part of the resistant effect. One reason is that no disturbance can violate the property of density processing which guarantees a realizable output. Another is that since the eye is logarithmically sensitive, it considers errors on a percentage basis. Because disturbances and distortions tend to distribute themselves uniformly throughout the range of a signal, they represent extremely large percentage distortions in the dark areas of an intensity image. To make matters worse, as can be seen from the intensity histograms of Fig. 4, dark areas are by far the most likely in intensity images.

These effects can be observed most readily when images are quantized in preparation for digital processing. The classically familiar quantization contours are most visible in the dark areas of intensity represented images but distribute nearly uniformly in density represented images. As a result, the use of a given number of bits to represent an image produces more readily observable quantization distortion in the form of contouring when an intensity rather than a density representation is employed. Indeed, for images of large dynamic range the disparity can be very great.[18]

As an illustration of the issues presented in this section consider Figs. 18 and 19. Fig. 18 shows the digital original of Fig. 11(a) in combination with white noise with a rectangular probability density function. In each of the three different combinations shown the peak signal to peak noise ratio was exactly the same namely 8:1. The noise disturbs an intensity representation in Fig. 18(a), a density representation in Fig. 18(b), and a model-processed image in Fig. 18(c). For additional discussion and examples see [6].

Fig. 19 shows another image quantized to 4 bit (i.e., 16 equally spaced levels exactly spanning the signal range). The quantization disturbs an intensity representation in Fig. 19(a), and a density representation in Fig. 19(b).

IX. Summary and Conclusions

The discussions presented in this paper concentrated upon the structure of images and the compatibility of that structure with the processes used to store, transmit, and modify them. The harmony of density representation and multiplicative processing with the physics of image formation was emphasized and special attention was drawn to the fact that early portions of the human visual system seem to enjoy that harmony. A visual model based upon these observations was introduced and a test yielding a calibration for the model was presented. Finally, an objective criterion for image quality based upon that model was offered and some examples of the use of the model for protecting images against disturbances were given.

During the past five years these concepts have been developed and employed in a continuing program of digital image processing research. Their constant use in guiding the

[18] The number of bits needed to represent an image cannot properly be determined without specifying at least the quality and character of the original, the kind of processing contemplated, the quality of the final display, the representation to be used, and the dynamic range involved. Similarly, the number of bits to be saved by using a density instead of an intensity representation given a fixed subjective distortion depends at least on the dynamic range in question. In the light of the quality obtainable with present technology the "rules of thumb" which have been popularly used in the past should be regarded with caution.

(a)

(b)

(c)

Fig. 18. Noisy disturbance in the context of three different representations. Peak signal to peak noise is 8:1 in all cases. (a) Disturbed intensities. (b) Disturbed densities. (c) Disturbed model-processed image. Compare with Fig. 11(a).

basic philosophy of the work has resulted in an ability to obtain high and consistent image quality and to enhance and simplify image processing techniques as they were proposed. Their ability to provide engineering insight and understanding complementary to existing ideas has been an invaluable aid in planning and in problem solving.

Continuing research is attempting to include within the model the aspects of color and time and to enlarge upon the model in the context of visual processes which take place at points farther along the visual pathway. It is hoped that enlargements and refinements of the model will continue to suggest useful image processing techniques and that digital

(a)

(b)

(c)

Fig. 19. Quantization distortion in the context of two different representations. In both cases 16 equally spaced levels exactly spanning the signal range were used. (a) Quantized intensities. (b) Quantized densities. (c) Original.

signal processing methods will continue to permit the investigation of those techniques which might be too complex to be explored without them.

Acknowledgment

I wish to thank the people who have helped me in the course of the image processing research which has led to the ideas presented here. I am grateful to A. V. Oppenheim for his theory of homomorphic filtering, which for me is the *sine qua non* of these views. Many thanks are also due to C. M. Ellison, D. M. Palyka, D. H. Johnson, P. Baudelaire, G. Randall, R. Cole, C. S. Lin, R. B. Warnock, R. W. Christensen, M. Milochik, Kathy Gerber, and to the many too numerous to name who have given encouragement, interest, and ideas. Special appreciation goes to my wife Martha who has given me unceasing support.

References

[1] T. S. Huang, W. F. Schrieber, and O. J. Tretiak, "Image processing," *Proc. IEEE*, vol. 59, pp. 1586–1609, Nov. 1971.
[2] A. V. Oppenheim, R. W. Schafer, and T. G. Stockham, Jr., "Nonlinear filtering of multiplied and convolved signals," *Proc. IEEE*, vol. 56, pp. 1264–1291, Aug. 1968.
[3] T. N. Cornsweet, *Visual Perception.* New York: Academic Press, 1970.
[4] F. Ratliff, *Mach Bands: Quantitative Studies on Neural Networks in the Retina.* San Francisco, Calif.: Holden-Day, 1965.
[5] L. M. Hurvich and D. Jameson, *The Perception of Brightness and Darkness.* Boston, Mass.: Allyn and Bacon, 1966, pp. 7–9.
[6] T. G. Stockham, Jr., "Intra-frame encoding for monochrome images by means of a psychophysical model based on nonlinear filtering of multiplied signals," in *Proc. 1969 Symp. Picture Bandwidth Compression*, T. S. Huang and O. J. Tretiak, Eds. New York: Gordon and Breach, 1972.
[7] D. J. Sakrison and V. R. Algazi, "Comparison of line-by-line and two-dimensional encoding of random images," *IEEE Trans. Inform. Theory*, Vol. IT-17, pp. 386–398, July 1971.
[8] L. Davisson, "Rate-distortion theory and applications," this issue, pp. 800–808.

Adaptive filtering for image enhancement

Tamar Peli
Jae S. Lim
Massachusetts Institute of Technology
Lincoln Laboratory
244 Wood Street
Lexington, Massachusetts 02173

Abstract. In this paper, we develop an image enhancement algorithm that modifies the local luminance mean of an image and controls the local contrast as a function of the local luminance mean of the image. The algorithm first separates an image into its lows (low-pass filtered form) and highs (high-pass filtered form) components. The lows component then controls the amplitude of the highs component to increase the local contrast. The lows component is then subjected to a nonlinearity to modify the local luminance mean of the image and is combined with the processed highs component. The performance of this algorithm when applied to enhance typical undegraded images, images with large shaded areas, and also images degraded by cloud cover will be illustrated by way of examples.

Keywords: digital image processing; adaptive filtering; cloud cover removal; contrast enhancement; image processing.

Optical Engineering 21(1), 108-112 (January/February 1982)

CONTENTS

1. INTRODUCTION

In a variety of practical problems, it is desirable to modify the local contrast and local luminance mean. For example, when an image with a large dynamic range is recorded on a medium with a smaller dynamic range, the details of the image in the very high and/or low luminance regions can not be well represented. One approach to solving such a problem is a simultaneous contrast enhancement and dynamic range reduction, which can be accomplished by modification of the local contrast and the local luminance mean. A one-dimensional example that illustrates this is shown in Fig. 1.

In this paper, we develop an algorithm which modifies the local contrast and the local luminance mean in a specific manner, and illustrate its performance by way of examples in several potential application problems of the algorithm. The algorithm will be seen to be simple both conceptually and computationally, and to have a wider variety of application problem areas than other approaches[1,2] to similar problems such as homomorphic filtering, unsharp masking, histogram equalization, etc. In Sec. 2 we develop the algorithm, and in Sec. 3 we discuss and illustrate the performance of the algorithm by way of examples in which the contrast of a typical image is increased, the details of shaded areas in an image are enhanced, and images degraded by cloud covers are improved.

Paper 1760 received Feb. 2, 1981; revised manuscript received May 26, 1981; accepted for publication June 2, 1981; received by Managing Editor June 22, 1981. This paper is a revision of a paper presented in the ICASSP conference, Atlanta, GA, March 1981.

2. ALGORITHM

A block diagram of an algorithm which modifies the local contrast and the local luminance mean in a specific manner is shown in Fig. 2. In the figure, $f(n_1,n_2)$ denotes the unprocessed digital image, and $f_L(n_1,n_2)$ which denotes the local luminance mean of $f(n_1,n_2)$ is obtained by low-pass filtering $f(n_1,n_2)$. The sequence $f_H(n_1,n_2)$ which denotes the local contrast is obtained by subtracting $f_L(n_1,n_2)$ from $f(n_1,n_2)$. The local contrast is modified by multiplying $f_H(n_1,n_2)$ with $k(f_L)$, a scalar which is a function of $f_L(n_1,n_2)$. The modified contrast is denoted by $f'_H(n_1,n_2)$. The specific functional form of $k(f_L)$ depends on the particular application under consideration, and $k(f_L)>1$ represents the local contrast increase while $k(f_L)<1$ represents local contrast decrease. The local luminance mean is modified by a point nonlinearity and the modified local luminance mean is denoted by $f'_L(n_1,n_2)$. The specific nonlinearity chosen depends on the particular application under consideration, and in most application problems the nonlinearity is chosen so that the overall dynamic range of the resulting image is approximately the same as the dynamic range of the recording medium. The modified local contrast and local luminance mean are then combined to obtain $g(n_1,n_2)$, the processed image.

As is clear from Fig. 2, the algorithm modifies the local contrast as a function of the local luminance mean, and also modifies the local luminance mean. This feature of the algorithm is useful in a variety of application problems, some of which are discussed in the next section.

3. EXAMPLES AND DISCUSSIONS

In this section, we discuss several potential application areas of the algorithm described in Sec. 2 and illustrate the performance of the algorithm by way of examples.

One potential application area of the above algorithm is the enhancement of undegraded images. For typical undegraded images,

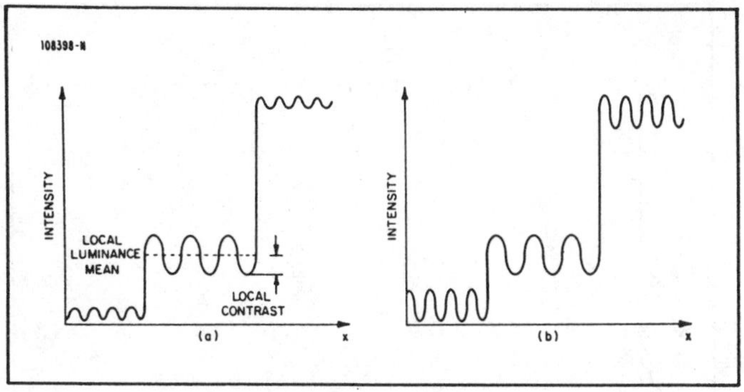

Fig. 1. (a) A degraded 1-D signal and (b) Desired signal.

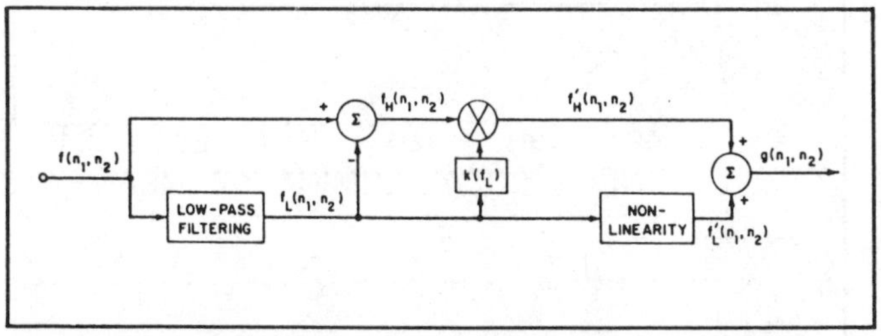

Fig. 2. Block diagram of an adaptive image enhancement algorithm.

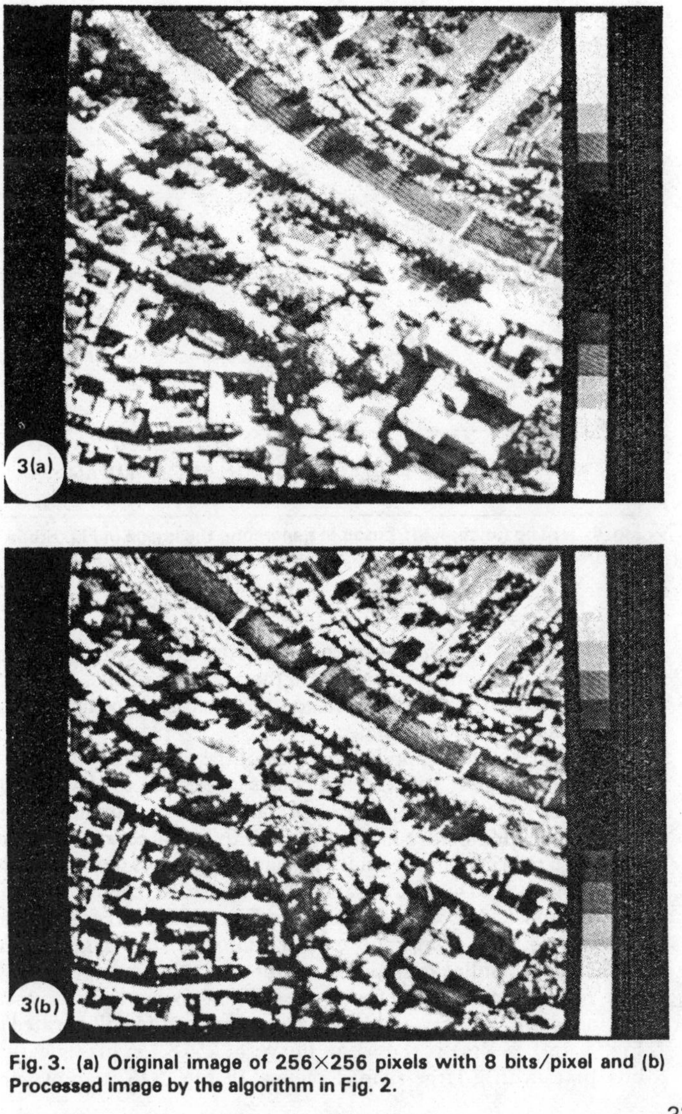

Fig. 3. (a) Original image of 256×256 pixels with 8 bits/pixel and (b) Processed image by the algorithm in Fig. 2.

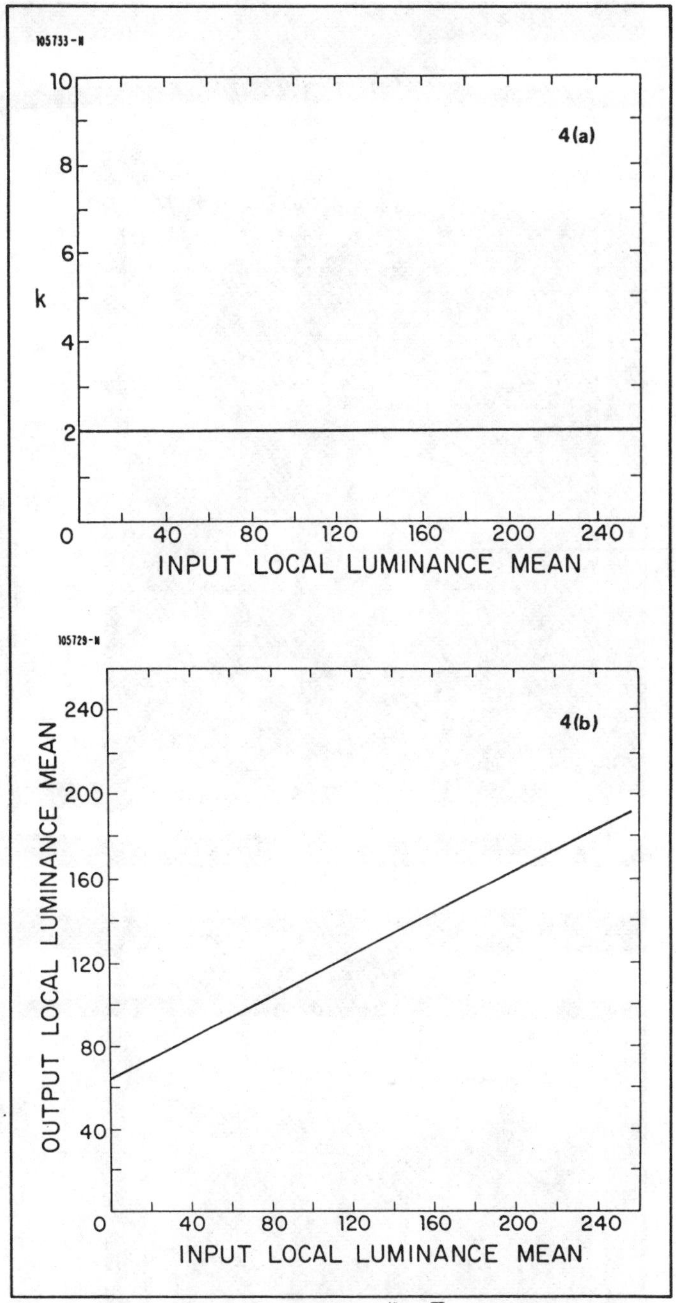

Fig. 4. (a) The function $k(f_L)$ used in generating the image in Fig. 3(b) and (b) The nonlinearity in the algorithm used in generating the image in Fig. 3(b).

it is often desirable[1,2] for visual purposes to increase the local contrast. One approach to achieve such an objective is to choose $k(f_L)$ in Fig. 2 to be a constant greater than 1, independent of local luminance mean f_L, and reduce the dynamic range of $f_L(n_1,n_2)$ through the nonlinearity to account for the dynamic range increase which results from the contrast increase. An example of such an application is shown in Fig. 3. In Fig. 3(a) is shown a typical image of 256×256 pixels with 8 bits/pixel. In Fig. 3(b) is shown the processed image by the algorithm described by Fig. 2. The function $k(f_L)$ and the nonlinearity used are shown in Fig. 4. The low-pass filtering operation used is a simple local averaging with mask size of $(2N_1+1)\times(2N_2+1)$ centered at (n_1,n_2):

$$f_L(n_1,n_2) = \frac{1}{(2N_1+1)(2N_2+1)} \sum_{k=n_1-N_1}^{n_1+N_1} \sum_{j=n_2-N_2}^{n_2+N_2} f(k,j) \;, \quad (1)$$

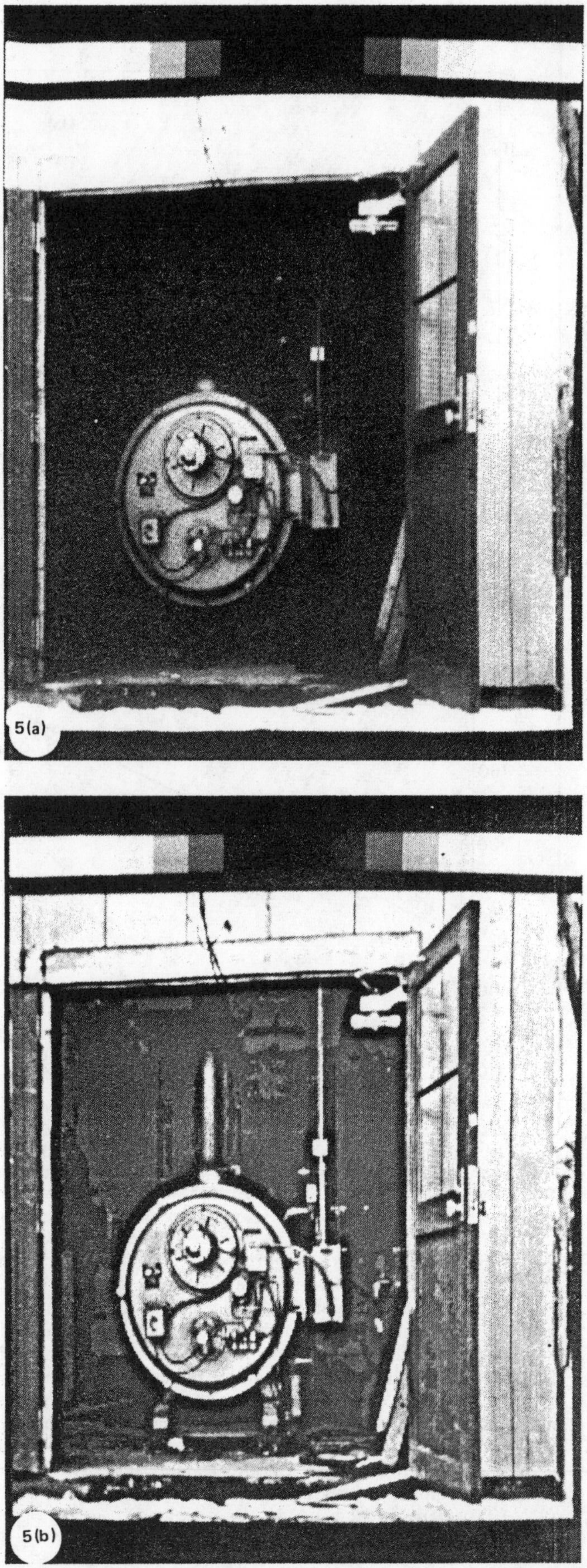

Fig. 5. (a) Original image of 256×256 with 6 bits/pixel with large shaded region and (b) Processed image by the algorithm in Fig. 2.

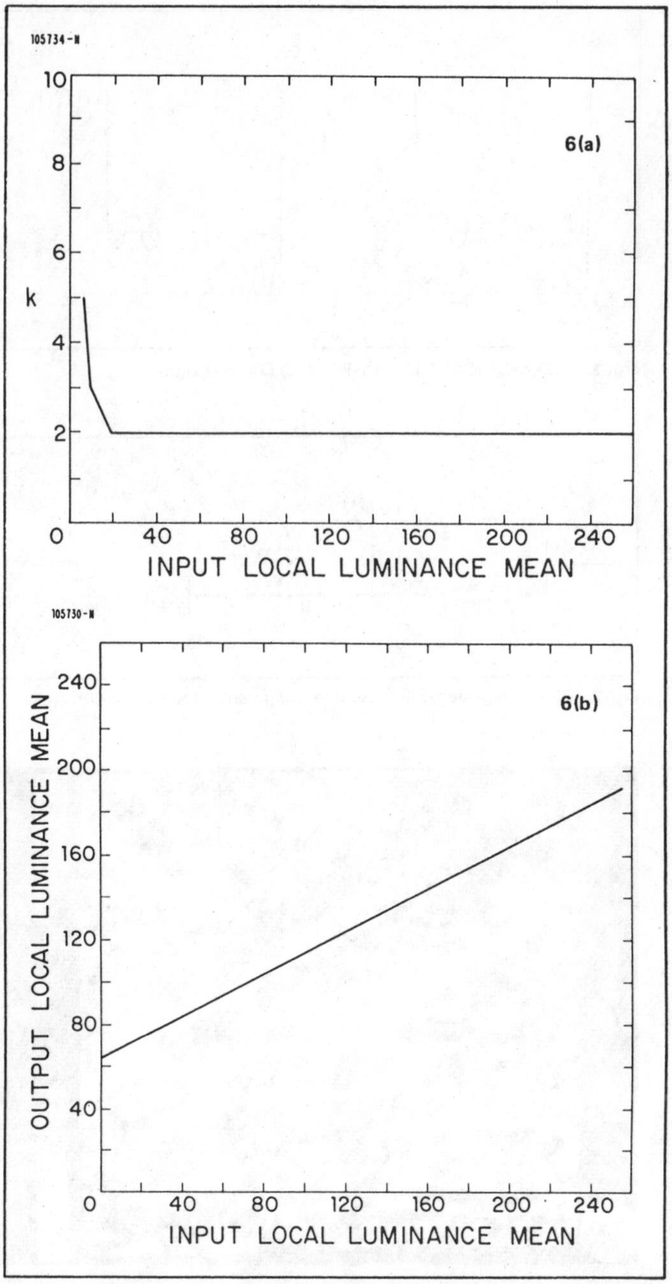

Fig. 6. (a) The function $k(f_L)$ used in generating the image in Fig. 5(b) and (b) The nonlinearity in the algorithm used in generating the image in Fig. 5(b).

with $N_1 = N_2 = 8$. For a point (n_1, n_2), at the edge of the image, the local average is computed only with the available points. It is clear from Fig. 3 that the processed image gives a sharper appearance than the original unprocessed image.

Another potential application area of the algorithm in Fig. 2 is the enhancement of an image which has large areas of shadows. In shaded regions of an image, the local contrast is typically suppressed and the local luminance mean is low relative to other regions. One approach to bring out the details in shaded regions is to increase the local contrast and local luminance mean in shaded regions. This can be accomplished by choosing $k(f_L)$ in Fig. 2 to be larger for small f_L and relatively small (still greater than one) for large f_L, and choosing the nonlinearity to account for the dynamic range increase that results from the contrast increase and to increase the local luminance mean. An example of such an application is illustrated in Fig. 5. In Fig. 5(a) is shown an image of 256×256 pixels with 6 bits/pixel with a large shaded region. In Fig. 5(b) is shown the

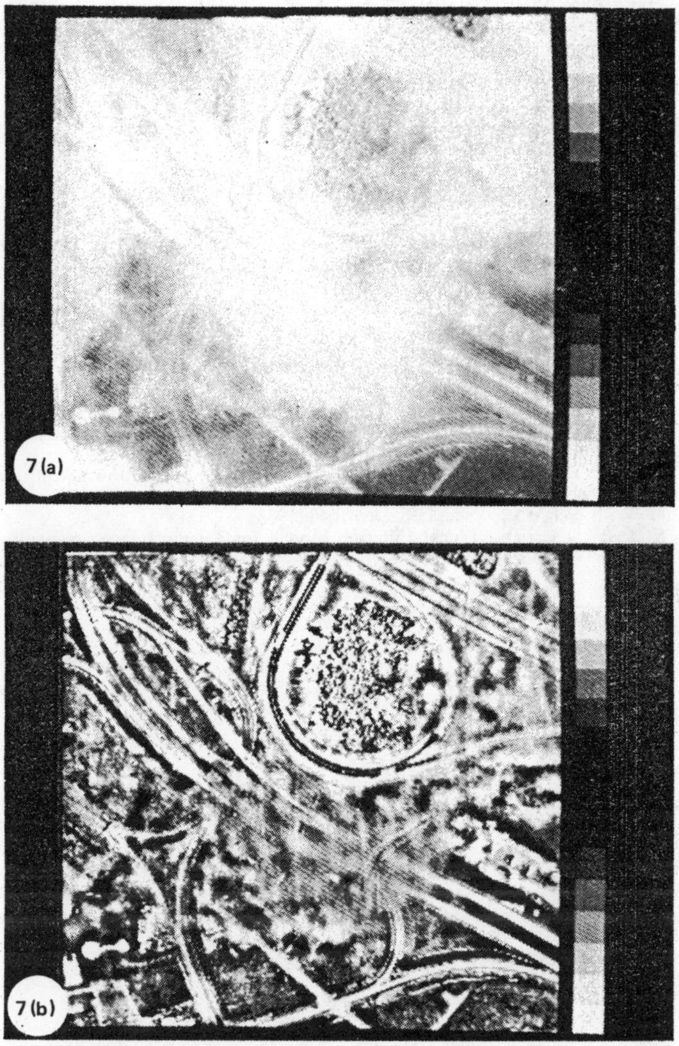

Fig. 7. (a) Image of 256×256 pixels with 8 bits/pixel degraded by cloud cover and (b) Processed image by the algorithm in Fig. 2.

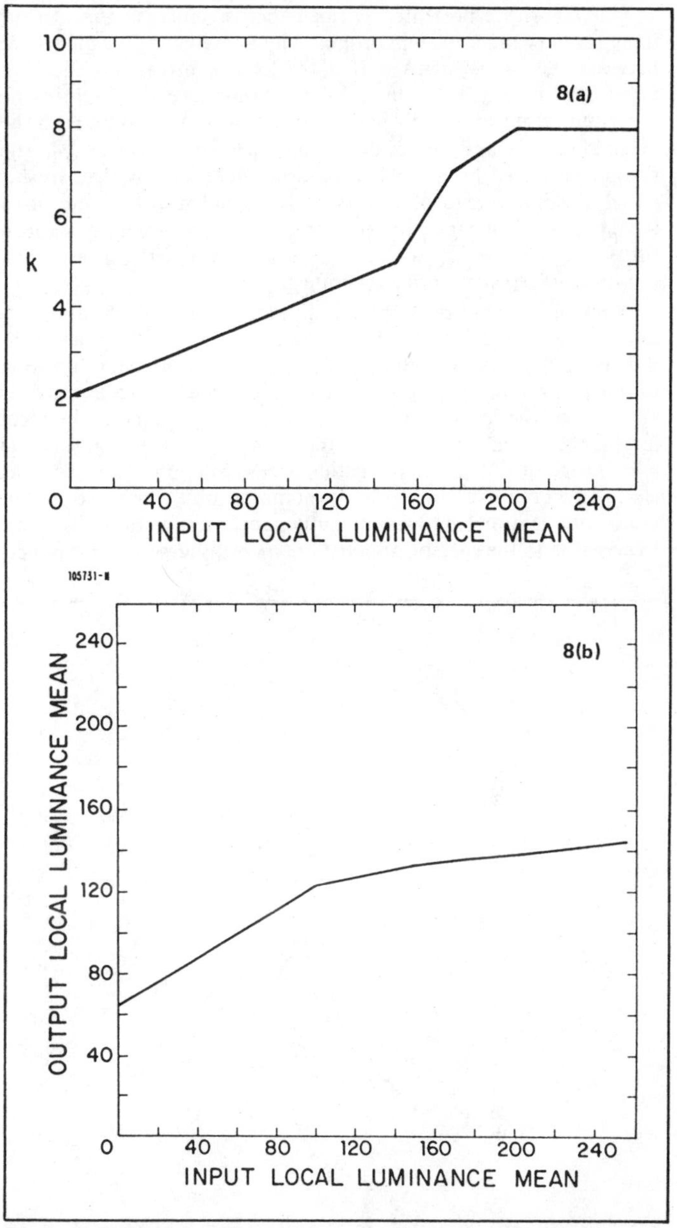

Fig. 8. (a) The function $k(f_L)$ used in generating the image in Fig. 7(b) and (b) The nonlinearity in the algorithm used in generating the image in Fig. 7(b).

processed image by the algorithm in Fig. 2. The function $k(f_L)$ and the nonlinearity used are shown in Fig. 6, and the low-pass filtering operation was performed using Eq. (1) with $N_1=N_2=5$. The function $k(f_L)$ was chosen to be greater than 1 for even large values of f_L, since some local contrast enhancement is generally desirable in typical undegraded images. It is clear from Fig. 5 that the details in the shaded region are much more visible in the processed image.

A third potential application of the algorithm in Fig. 2 is the enhancement of images degraded by cloud covers. According to one simplified model[3] of image degradation due to cloud cover, regions of an image covered by cloud have an increase in their local luminance mean and a decrease in the local contrast with the amount of change determined by the amount of the cloud cover. One approach to enhance images degraded by cloud covers, then, is to increase the local contrast and decrease the local luminance mean when the local luminance mean is high. This can be accomplished by choosing a larger $k(f_L)$ for a larger f_L and choosing the nonlinearity to account for the local luminance mean change and the contrast increase. An example of such an application is illustrated in Fig. 7. In Fig. 7(a) is shown an image of 256×256 pixels with 8 bits/pixel degraded by different amounts of cloud covers in different regions of the image. In Fig. 7(b) is shown the processed image by the algorithm in Fig. 2. The function $k(f_L)$ and the nonlinearity used are shown in Fig. 8, and the low-pass filtering operation was performed using Eq. (1) with $N_1=N_2=8$. It is clear from Fig. 6 that the details in the regions covered by cloud are much more visible in the processed image.

The specific choices of $k(f_L)$ and the nonlinearity depend on the application (shadows, clouds, etc.), and they have been chosen by trial and error. We have found, however, that the performance of the algorithm presented in this paper is not very sensitive to small variations in the specific choices of $k(f_L)$ and the nonlinearity, and thus it requires few trials and errors to achieve a desired performance. Specifically, when $k(f_L)$ and the nonlinearity are chosen so that they preserve the general characteristics but differ in small details from the ones used in this paper, the processed images are very similar to the results illustrated in this paper. In addition, other images with shaded regions were processed by using exactly the same functions of k and the nonlinearity in Fig. 6 and the results were similar to those presented. Significant differences in the processed images, of course, result if the general characteristics of $k(f_L)$ and the nonlinearity are changed.

In applications where $k(f_L)$ is a constant, as is the case for the example in Fig. 3, the system in Fig. 2 is similar to homomorphic filtering in that the gain of the highs signal is nonadaptively increased relative to the lows signal. Consequently, a result similar to

Fig. 3 has also been obtained by homomorphic filtering.

The algorithm presented in this paper is similar in form to the image enhancement system proposed by Schreiber.[4] One difference between the two systems is that the local contrast modification factor "k" is controlled by the local luminance mean f_L in the algorithm discussed in this paper, while k is controlled by both the original signal f and the result of a nonlinearity applied to f, in the system proposed by Schreiber. Another difference is that, in this paper, the modified local contrast f'_H is combined with the modified local luminance mean f'_L to generate the processed image g, while in the system proposed by Schreiber, f'_H is combined with the result of a nonlinearity applied to f to generate g.

As is clear from Fig. 2, the adaptive algorithm is both conceptually and computationally simple. In addition, unlike other techniques used for similar purposes such as homomorphic filtering, unsharp masking, histogram equalization, etc., the algorithm is adaptive to the local characteristics of an image, namely the local luminance mean. This feature of the algorithm is responsible for its wide range of potential application areas. In applications such as cloud cover removal, for example, homomorphic filtering or histogram equalization led to significantly poorer results than Fig. 7(b). In addition to the specific application areas discussed in this paper, the algorithm is potentially applicable to enhancement of images degraded by varying amounts of smoke cover, haze, fog, etc. in different regions of an image, and the local luminance mean equalization for image segmentation.[5]

4. ACKNOWLEDGMENTS

The authors acknowledge helpful discussions with Bill Schreiber, Charlie Therrien, Bob Lerner, J. Shapiro, F. Bucher, and Ted Bially. This work was sponsored by the Department of the Air Force. The U.S. Government assumes no responsibility for the information presented.

5. REFERENCES

1. W. K. Pratt, *Digital Image Processing,* John Wiley and Sons (1978).
2. H. C. Andrews, B. R. Hunt, *Digital Image Restoration,* Prentice-Hall, Englewood Cliffs, NJ (1977).
3. R. E. Danielson, D. R. Moor and H. C. Van De Hults, Journal of the Atmospheric Sciences, 26, 1078 (1969).
4. W. F. Schreiber, Proc. IEEE, 66 1640 (1978).
5. C. W. Therrien, "Linear Filtering Models for Terrain Image Segmentation," Lincoln Laboratory, M.I.T., TR-552, 1981.

Part VII
Geophysical Applications

THE field of geophysics presents a wide variety of areas of application of signal processing, since it encompasses the study of the physical aspects of the earth, its atmosphere, its oceans, and its associated gravitational and electromagnetic fields. As a result, any collection of papers must focus to some extent on a small subset of that broad spectrum of topics.

One of the more important areas of application of signal processing in geophysics relates to the analysis and interpretation of seismic signals. Such signals may be either natural in origin (e.g., earthquakes) or man-made (e.g., explosions). In either case, knowledge gained from analysis of these signals is primarily used to further our understanding of the structure and composition of the earth.

The papers collected in this section represent a wide variety of topics related to seismic signal processing. These range from detecting signals in noise using large aperture arrays, to solving the seismic inverse problem or modeling the response of an ideal earth, and finally, applying image reconstruction techniques to enhance our ability to interpret the subsurface structure of the earth using surface or borehole measurements. The scope of these papers is therefore large and diverse, and reflects to some extent the many sub-disciplines within signal processing which have been, or may be, brought to bear on the study of seismic signals.

The first paper, by Capon *et al.*, discusses the extension of maximum likelihood signal detection to the multidimensional case for the purpose of detecting a signal buried in noise using an array of receivers.

Whenever seismic data is available, it becomes of interest to attempt to invert the data in order to obtain an estimate of the structure of some portion of the earth's interior. The second paper, by Robinson, discusses a spectral estimation approach to geophysical inversion in exploration seismology. This paper introduces the Radon transform which is important to the next paper in the sequence. Finally, the application of the 2-D Fourier transform to a problem in seismic imaging (migration) is discussed.

In order to understand the complex interaction between an inhomogeneous medium like the earth and a seismic wavefield, modeling provides a useful tool for interpreting wave phenomena. The paper by Wenzel *et al.* considers an innovative approach to seismic modeling in the (Radon) transformed coordinates of intercept time and ray parameter. Many complex features of seismic wavefields separate nicely in this domain and may be more easily interpreted.

In an effort to extend our ability to correctly image the subsurface using both surface and borehole measurements, pioneering work is being done in applying the concepts of image reconstruction. These methods promise to have applications in diverse areas including civil engineering, mining, and hydrocarbon reservoir delineation. The last paper in this section, by Devaney, presents recent work in this very promising area of multidimensional signal analysis.

The reader interested in pursuing these topics further might wish to consult the following books and journals.

Bibliography

[1] K. Aki and P. G. Richards, *Quantitative Seismology*, 2 vols. San Francisco, CA: W. H. Freeman and Co., 1980.

[2] S. Haykin, J. H. Justice, N. L. Owsley, J. L. Yen, and A. C. Kak, *Array Signal Processing*. Englewood Cliffs, NJ: Prentice-Hall Signal Processing Series, 1985.

[3] E. R. Kanasewich, *Time Sequence Analysis in Geophysics*. Edmonton, Alberta, Canada: University of Alberta Press, 1981.

[4] E. A. Robinson and S. Treitel, *Geophysical Signal Analysis*. Englewood Cliffs, NJ: Prentice-Hall, Inc., 1980.

[5] W. M. Telford, L. P. Geldart, R. E. Sheriff, and D. A. Keys, *Applied Geophysics*. Cambridge, England: Cambridge University Press, 1976.

[6] *Geophysical Prospecting*. The Hague, Netherlands: European Association of Exploration Geophysicists.

[7] *Geophysics Journal of the Royal Astronomical Society*. Palo Alto, CA: Blackwell Scientific Publications Ltd.

[8] *Geophysics*. Tulsa, OK: Society of Exploration Geophysicists.

[9] K. H. Waters, *Reflection Seismology*, 2nd ed. New York, NY: John Wiley and Sons, 1981.

Multidimensional Maximum-Likelihood Processing of a Large Aperture Seismic Array

J. CAPON, MEMBER, IEEE, R. J. GREENFIELD, AND R. J. KOLKER

Abstract—The experimental Large Aperture Seismic Array (LASA) represents an attempt to improve the capability to monitor underground nuclear weapons tests and small earthquakes by making a large extrapolation in the existing art of building arrays of spaced and interconnected seismic transducers. The LASA is roughly equivalent to 21 separate subarrays, each consisting of 25 sensors, spread over an aperture of 200 km. The present work considers the problem of designing a linear filter which combines the outputs of the 25 sensors in a subarray so as to suppress the noise without distorting the signal, or event. This filter provides a minimum-variance unbiased estimate of the signal which is the same as the maximum-likelihood estimate of the signal if the noise is a multidimensional Gaussian process. An extensive discussion of the theory of multidimensional maximum-likelihood processing is given.

A computer program implementation of the maximum-likelihood filter is presented which employs the cross-correlation matrix of noise measured just prior to the arrival of the event. This time-domain synthesis procedure requires relatively large amounts of computer time to synthesize the filter, and is quite sensitive to the assumption of noise stationarity. The asymptotic theory of maximum-likelihood filtering is also given. An asymptotically optimum frequency-domain synthesis technique is given for two-sided multidimensional filters. This procedure is well suited to machine computation and has the advantage with respect to the time-domain procedure of requiring about 10 times less computation time.

A description of a computer program implementation of the frequency-domain synthesis method is given which employs the spectral matrix of the noise estimated just before the arrival of the event. The experimental results obtained by processing several events recorded at the LASA are presented, as well as a comparison of the performance of the frequency-domain method relative to the time-domain synthesis technique. It is found that the signal-to-noise improvement given by the frequency-domain procedure is within 2 dB of the gain obtained with the time-domain procedure, and that the frequency-domain method is relatively insensitive to the assumption of noise stationarity.

Manuscript received July 7, 1966; revised November 16, 1966.
The authors are with Lincoln Laboratory, Massachusetts Institute of Technology, Lexington, Mass. (Operated with support from the U. S. Advanced Research Projects Agency.)

I. Introduction

IT HAS BEEN recognized that a considerable reduction of seismic noise is possible by employing multidimensional filtering for seismic arrays. The multidimensional Wiener filtering approach to array processing has been employed previously [1]. In this approach, the signal and noise are represented as stationary multidimensional random processes with known cross-correlation functions and a linear filter is designed which produces an output that is a minimum mean-squared-error version of the

Reprinted from *Proc. IEEE,* vol. 55, pp. 192–211, Feb. 1967.

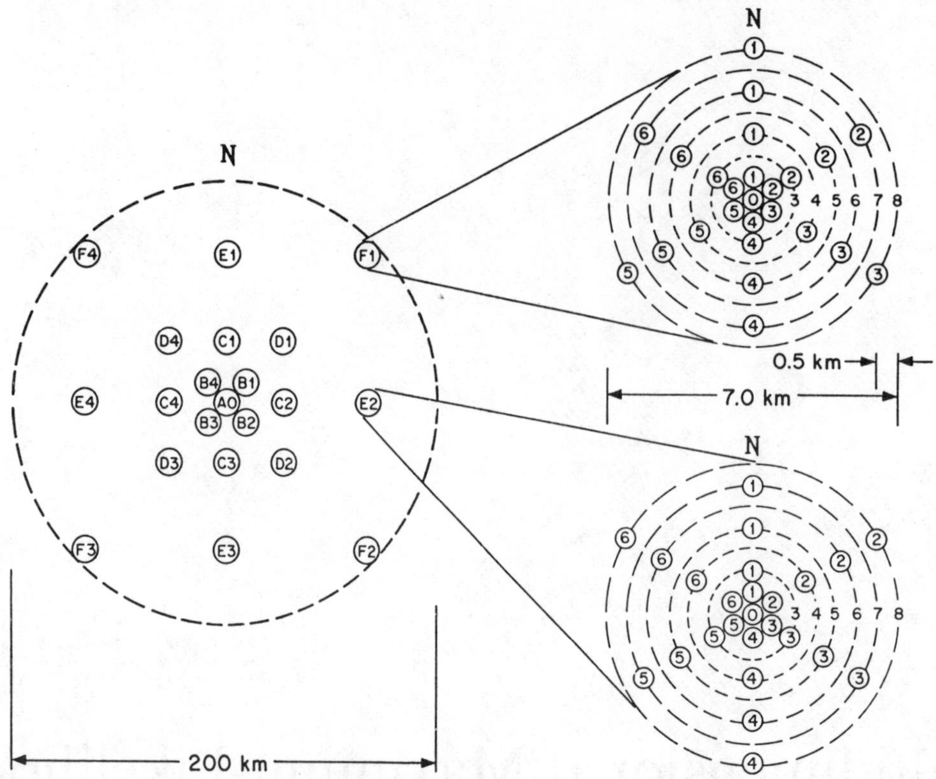

Fig. 1. General arrangement of the Large Aperture Seismic Array.

signal. A second approach [2] is to employ the noise cross-correlation functions to design a linear processor for the seismometer outputs which provides a minimum mean-squared-error prediction of the noise at some one seismometer a short interval, say 0.3 second, ahead. This prediction is subtracted from the actual seismometer output, thus greatly reducing the noise level but distorting the signal waveform after the prediction interval.

We will consider an arbitrary two-dimensional horizontal array of identical single-component seismometers. It will be assumed that the seismic noise is a zero-mean, time-stationary, multidimensional random process. There is considerable evidence that this is a reasonable representation for the noise. Since the form of the signal at a seismometer cannot be forecast before its arrival, it cannot be treated as a known waveform or even as a known function of some unknown parameters. In addition, the finite duration of the seismogram of an event and the variations in its character as different phases arrive suggest that the representation of the signal as a stationary random process is not realistic. In the present work it will be assumed that a single signal is present and that it is a plane wave propagating in a homogeneous, linear, and nondispersive medium. The signal waveform is taken to be the same in each of the seismometers except for a time delay due to the finite propagation velocity. This model differs from previous contributions in the assignment of the signal waveform as a completely unknown time function which is to be estimated. Hence, no a priori assumptions regarding the shape of the signal are made. This model for the signal is due to Kelly and Levin [3].

The primary concern in the present work will be the consideration of the multidimensional filtering problem for the experimental Large Aperture Seismic Array [4] (LASA) located in eastern Montana. This array consists of 21 subarrays of 25 seismometers each arranged as indicated in Fig. 1. In particular, we will be interested in the maximum-likelihood or minimum-variance unbiased estimator approach to seismic array processing. This multidimensional filtering method forms a single output waveform which serves as an estimator of the unknown signal coming from a fixed direction.

The basic assumption in our analysis of multidimensional filters is that the output, $X_k(t)$, of the kth seismometer, may be written as

$$X_k(t) = S(t) + N_k(t) \tag{1}$$

where $S(t)$ is the signal waveform which is assumed to be the same in each seismometer and $N_k(t)$ is the noise present in seismometer k, $k = 1, \cdots, K$. In writing (1) it is assumed that the azimuth and horizontal velocity of the event, or signal, have already been determined with sufficient accuracy to allow the signal waveforms from each seismometer to be shifted to bring them into time coincidence. In most applications, the outputs of the seismometers are given in sampled form in which case (1) becomes

$$X_{km} = S_m + N_{km} \qquad \begin{array}{l} k = 1, \cdots, K \\ m = 0, \pm 1, \pm 2, \cdots. \end{array} \tag{2}$$

Only the sampled-data multidimensional filtering problem for seismic arrays will be considered.

A time-domain technique can be used to design the maximum-likelihood filter. In this method, the cross-correlation matrix of the noise is measured in what is called a fitting interval, usually three minutes long, immediately preceding the event, and using this measured matrix the filter is synthesized by employing a recursive synthesis procedure. The filter designed in this manner is optimum in the sense that it provides a maximum-likelihood estimate of the signal, provided the noise is a multidimensional Gaussian process. It is also optimum in the sense that the output noise power is minimized subject to the constraint that the signal be undistorted by the filter.

However, in spite of these advantages, the time-domain synthesis method is subject to several severe criticisms. The most important criticism is that the method requires a large amount of computer time, typically about 30 minutes of 7094 computing time to synthesize a filter for a 25-channel subarray of seismometers. In addition, this method tends to develop a "supergain" inside the fitting interval which is not maintained outside the fitting interval where the filtering action is most important. Perhaps another way of saying this is that the technique is too sensitive to the assumption that the noise is stationary. Still another disadvantage is that the technique is at times sensitive to the assumption that the signal is identical across the array of sensors. This drawback manifests itself in the form of a precursor, when two-sided or infinite-lag filters are used, which precedes the event. This can be troublesome if first motion is to be preserved for use as a discrimination criterion between natural seismic events and nuclear explosions.

A frequency-domain synthesis procedure for two-sided, or infinite-lag, multidimensional maximum-likelihood filters will be presented which does not have the disadvantages of the time-domain synthesis method just mentioned. The theory for the method will be presented subsequently. In brief, the frequency-domain method is optimum in the same sense as the time-domain method when the length, or memory, of the filter is large. The greatest advantage of the frequency-domain method is that it requires about one order of magnitude less computing time than the time-domain technique in the synthesis of a filter for a 25-channel subarray of seismometers. In addition, the frequency-domain method is less sensitive to the assumptions of noise stationarity and identical signals across the array than the time-domain technique. However, the output noise power reduction for the frequency-domain method is typically about two out of a total of 20 dB worse than that of the time-domain technique. This disadvantage, however, would seem to be offset by the advantages cited previously, making the frequency-domain technique the more practical method for performing multidimensional filtering of sampled data obtained from an array consisting of a large number of sensors.

II. Derivation of the Maximum-Likelihood and Minimum-Variance Unbiased Estimators

The derivation of the maximum-likelihood estimator of the signal requires the assumption that the noise components have a multidimensional Gaussian distribution. We assume, for simplicity, that the noise components have zero mean, and that the covariance matrix is

$$\rho_{jk}(m, n) = \rho_{kj}(n, m) = E\{N_{jm}N_{kn}\}, \qquad \begin{matrix} 1 \leq j, k \leq K \\ -v \leq m, n \leq v \end{matrix} \tag{3}$$

where E denotes expectation, and it assumed that the estimator is to use $2v+1$ samples extending in time from $-v$ to v. Thus, the likelihood function can be written

$$L = (2\pi)^{-K/2(2v+1)}|\rho|^{-1/2} \cdot \exp\left\{-\frac{1}{2}\sum_{j,k=1}^{K}\sum_{m,n=-v}^{v}\rho_{jk}^{-1}(m, n)(X_{jm}-S_m)(X_{kn}-S_n)\right. \tag{4}$$

where $|\rho|$ denotes the determinant of the matrix ρ which is a matrix of $K\times K$ submatrices, the mnth submatrix has the elements $\rho_{jk}(m, n)$, $j, k=1, \cdots, K$, $m, n=-v, \cdots, v$, with a corresponding notation for the inverse matrix ρ^{-1} whose elements are $\rho_{j,k}^{-1}(m, n)$. We assume throughout that the matrix ρ is positive definite. Differentiating the logarithm of the likelihood function with respect to S_n and equating the result to zero, we obtain

$$\sum_{j,k=1}^{K}\sum_{m=-v}^{v}\rho_{jk}^{-1}(m, n)(\hat{S}_{vm} - X_{jm}) = 0, \quad n = -v, \cdots, v \tag{5}$$

where $\hat{S}_{vm}$ is the maximum-likelihood estimator for S_m.

We can rewrite (5) as follows

$$\sum_{m=-v}^{v}\hat{S}_{vm}\sum_{j,k=1}^{K}\rho_{jk}^{-1}(m, n) = \sum_{j,k=1}^{K}\sum_{m=-v}^{v}X_{jm}\rho_{jk}^{-1}(m, n), \quad n = -v, \cdots, v. \tag{6}$$

Let us define the $(2v+1)\times(2v+1)$ matrix $\mathscr{A}(m, n)$ in terms of its inverse $\mathscr{A}^{-1}(m, n)$ whose elements are defined as

$$\alpha^{-1}(m, n) = \sum_{j,k=1}^{K}\rho_{jk}^{-1}(m, n), \qquad m, n = -v, \cdots, v. \tag{7}$$

We can solve for $\hat{S}_{vn}$ as

$$\hat{S}_{vn} = \sum_{m=-v}^{v}\sum_{j=1}^{K}\theta_j(m|n)X_{jm} \tag{8}$$

where

$$\theta_j(m|n) = \sum_{k=1}^{K}\sum_{m'=-v}^{v}\rho_{jk}^{-1}(m, m')\alpha(m', n) \qquad \begin{matrix} j=1, \cdots, K \\ m = -v, \cdots, v. \end{matrix} \tag{9}$$

It should be noted that a different set of weights is obtained from (9) for different values of v and also for different values of n, which can vary from $-v$ to v. The fact that $\theta_j(m|n)$ depends on v has not been brought out by the notation. It is easily seen from the quadratic nature of the logarithm of the likelihood function that the solution given in (9) is unique.

We now consider the minimum-variance unbiased estimator of S_n, denoted by $\hat{S}'_{vn}$, expressible as

$$\hat{S}'_{vn} = \sum_{m=-v}^{v}\sum_{j=1}^{K}\theta'_j(m|n)X_{jm} \tag{10}$$

with the constraints

$$\sum_{j=1}^{K} 0'_j(m|n) = \delta_{mn}, \qquad m = -\nu, \cdots, \nu \tag{11}$$

where

$$\begin{aligned}\delta_{mn} &= 1, \quad m = n \\ &= 0, \quad \text{otherwise.}\end{aligned} \tag{12}$$

The variance of $\hat{S}'_{\nu n}$ is, assuming for simplicity $S_n = 0$,

$$E(\hat{S}'_{\nu n})^2 = \sum_{m,m'=-\nu}^{\nu} \sum_{j,k=1}^{K} 0'_k(m'|n) 0'_j(m|n) \rho_{jk}(m, m'). \tag{13}$$

Using the calculus of variations, we obtain that the minimum-variance unbiased estimator has weights which satisfy the system of equations

$$\sum_{j=1}^{K} \sum_{m'=-\nu}^{\nu} 0'_j(m'|n)\rho_{jk}(m, m') + \lambda_{\nu mn} = 0, \qquad \begin{aligned} j &= 1, \cdots, K \\ m &= -\nu, \cdots, \nu\end{aligned} \tag{14}$$

where the $\lambda_{\nu mn}$ are $2\nu+1$ Lagrangian multipliers chosen to satisfy the constraints given in (11).

If we define

$$0'_{K+1}(m|n) = \lambda_{\nu mn} \tag{15}$$

$$\begin{aligned}\rho_{k,K+1}(m, m') = \rho_{K+1,k}(m, m') = \delta_{mm'}(1 - \delta_{k,K+1}), \\ k = 1, \cdots, K+1 \\ m, m' = -\nu, \cdots, \nu\end{aligned} \tag{16}$$

then (11) and (14) may be written as a single set of equations as follows

$$\sum_{k=1}^{K+1} \sum_{m'=-\nu}^{\nu} 0'_k(m'|n)\rho_{jk}(m, m') = \delta_{j,K+1}\delta_{mn} \qquad \begin{aligned} j &= 1, \cdots, K+1 \\ m &= -\nu, \cdots, \nu.\end{aligned} \tag{17}$$

This set of equations is equivalent to the set given previously (cf. Levin [5], pp. 21–24). The reason for writing the equations in the above form is that we now have a Toeplitz matrix of submatrices, and a recursive procedure [6] can be used in the solution of filter weighting coefficients, which is more efficient, i.e., requires about $(2\nu+1)$ less computations, than the procedure which would invert directly the matrix given in (17).

We can write the system of (14) as

$$0'_j(m|n) = -\sum_{k=1}^{K} \sum_{m'=-\nu}^{\nu} \rho_{jk}^{-1}(m, m')\lambda_{\nu m'n} \qquad \begin{aligned} j &= 1, \cdots, K \\ m &= -\nu, \cdots, \nu.\end{aligned} \tag{18}$$

Using (11) in (18) we get

$$-\sum_{j,k=1}^{K} \sum_{m'=-\nu}^{\nu} \rho_{jk}^{-1}(m, m')\lambda_{\nu mn} = \delta_{mn}, \quad m = -\nu, \cdots, \nu \tag{19}$$

and using (7)

$$\begin{aligned}\lambda_{\nu mn} &= -\sum_{m'=-\nu}^{\nu} \alpha(m, m')\delta_{mn} \\ &= -\alpha(m, n), \qquad m = -\nu, \cdots, \nu.\end{aligned} \tag{20}$$

Therefore, we see from (18) and (20) that

$$0'_j(m|n) = \sum_{k=1}^{K} \sum_{m'=-\nu}^{\nu} \rho_{jk}^{-1}(m, m')\alpha(m', n), \qquad \begin{aligned} j &= 1, \cdots, K \\ m &= -\nu, \cdots, \nu\end{aligned} \tag{21}$$

which is the same as (9). Hence the maximum-likelihood and minimum-variance unbiased estimators are identical. It is easily seen from the quadratic nature of the expression in (13) that the solution given in (21) is unique.

Using (11), (13), and (14) we obtain

$$\begin{aligned}E\{\hat{S}_{\nu n} - S_n\}^2 &= \sum_{j=1}^{K} \sum_{m=-\nu}^{\nu} 0_j(m|n)(-\lambda_{\nu mn}) \\ &= -\sum_{m=-\nu}^{\nu} \lambda_{\nu mn}\delta_{mn} = -\lambda_{\nu nn} \\ &= \alpha(n, n), \qquad n = -\nu, \cdots, \nu.\end{aligned} \tag{22}$$

In addition, we have

$$\begin{aligned}E\{(\hat{S}_{\nu m} - S_m)(\hat{S}_{\nu n} - S_n)\} \\ = \sum_{j=1}^{K} \sum_{m'=-\nu}^{\nu} 0_j(m'|m)[\alpha(m', n)] \\ = \alpha(m, n) = -\lambda_{\nu mn} \qquad m, n = -\nu, \cdots, \nu.\end{aligned} \tag{23}$$

It should be noted that

$$E\{\hat{S}_{\nu m} - S_m\}^2 \le E\{\hat{S}_{\nu' m} - S_m\}^2, \qquad \nu' \le \nu. \tag{24}$$

This follows from the fact that the minimum value of a quadratic form, subject to certain constraints, in a $K(2\nu'+1)$-dimensional space cannot be increased if the dimension of the space is increased to $K(2\nu+1) \ge K(2\nu'+1)$.

III. Description of Computer Programs for Time-Domain Synthesis Method

A FORTRAN II computer program implements the synthesis of the maximum-likelihood filters in the time domain according to the algorithm given in Section II. The outputs of this program are the above filters and CALCOMP plots of the processed traces. The program works with data duplicated from LASA tapes onto a FORTRAN compatible tape. The basic sampling rate of the data is 20 samples per second. The parameters used in the program are as follows.

NRSKP is the number of data samples to be skipped, starting from the beginning of data tape, before estimating the cross-correlation matrix of the noise inside the fitting interval which usually immediately precedes the event.

NS is the number of seismometers to be processed and must be less than or equal to 25; note that NS $= K$.

NFP is the number of filter points to be used and must be less than or equal to 41; note that NFP $= 2\nu + 1$.

NIM: Every NIMth point is used as data in the fitting interval; the others are skipped. Data samples are skipped to save on computation time.

NMFACT is a normalizing factor and is numerically the same as the number of data samples used in the cross-correlation matrix estimation procedure.

AZ, HVEL are the azimuth and horizontal velocity, respectively, of the event being processed.
RS(I), THETA(I) are the polar coordinates of the Ith seismometer.

The output of the program consists of CALCOMP plots of the center seismometer, denoted $A\phi$, of the subarray being processed, WDS, DS, and FS. FS denotes filter and sum and is the designation for the maximum-likelihood filter which employs NFP filter points, WDS denotes weighted delay and sum and designates a maximum-likelihood filter with NFP=1, and DS denotes the sum of the delayed data. If NMFACT=1800 (corresponding to three minutes of noise, 20 samples per second and every other data sample skipped, i.e., NIM=2), NFP=21, NS=25, and six minutes of data are processed, the 7094 computer running time is about 45 minutes, consisting of 20 minutes to measure the cross-correlation matrix, 10 minutes to synthesize the filter, and 15 minutes to obtain all the processed traces.

A detailed description of the flow of the program is as follows.

1) Compute the delays for the NS channels, using the azimuth and horizontal velocity of the event and the channel coordinates, assuming that the event propagates across the subarray as a plane wave.

2) Apply the computed delays to the data and let the delayed data sample in the jth channel at time point m be denoted by X_{jm} to be consistent with previous notation. Let $m=1$ designate the beginning of the fitting interval.

3) Compute

$$\bar{X}_j = \frac{1}{L}\sum_{m=1}^{L} X_{j,m\cdot \mathrm{NIM}}, \qquad j = 1, \cdots, \mathrm{NS}$$

where $L=$NMFACT.

4) Let $X'_{jm}=X_{jm}-\bar{X}_j$ and compute the estimate of the cross-correlation coefficient

$$\hat{\rho}_{jk}(m) = \frac{1}{L}\sum_{\substack{0\le i\le L\\ 0\le i+m\le L}} X'_{j,i+m}X'_{k,i}, \qquad \begin{matrix} j,k = 1,\cdots,K \\ m = 0, 1, \cdots, 2v. \end{matrix}$$

5) Border the matrices $\{\hat{\rho}_{jk}(m)\}$, j, $k=1, \cdots$, NS with zeros and ones as indicated in (17), $m=0, \cdots, 2v$, to obtain NFP matrices of order (NS+1)×(NS+1).

6). Compute the filter coefficients for FS and WDS from (17) by essentially inverting the bordered cross-correlation matrix by the recursive procedure given in Robinson and Wiggins [6]. Lagrangian multipliers are also obtained in this computation which are numerically the same as the output power of FS and WDS in the fitting interval, denoted by P_{OUT} and P_{WDS}, respectively.

7) Compute signal-to-noise ratio gain of FS and WDS as

$$G_{\mathrm{FS}} = 10 \log_{10} \frac{P_{\mathrm{IN}}}{P_{\mathrm{OUT}}}$$

$$G_{\mathrm{WDS}} = 10 \log \frac{P_{\mathrm{IN}}}{P_{\mathrm{WDS}}}$$

where

$$P_{\mathrm{IN}} = \frac{1}{\mathrm{NS}}\sum_{j=1}^{\mathrm{NS}} \hat{\rho}_{jj}(0).$$

8) Compute the output power of DS

$$P_{\mathrm{DS}} = (\mathrm{NS})^{-2}\sum_{j,k=1}^{\mathrm{NS}} \hat{\rho}_{jk}(0)$$

and signal-to-noise ratio gain of DS

$$G_{\mathrm{DS}} = 10 \log_{10}\frac{P_{\mathrm{IN}}}{P_{\mathrm{DS}}}.$$

9) Apply the filter coefficients to obtain the FS trace

$$\hat{S}_{\mathrm{NFP},n} = \sum_{m=-((\mathrm{NFP}-1)/2)}^{(\mathrm{NFP}-1)/2}\ \sum_{j=1}^{\mathrm{NS}} \theta_{jm} X'_{j,m(\mathrm{NIM})+n}, \qquad n = 1, 2, 3, \cdots.$$

10) The DS and WDS traces are obtained in a manner similar to the computation in (9), with the appropriate weights replacing the θ_{jm}'s. For DS the weights are $\theta_j=1/\mathrm{NS}$, $j=1, \cdots$, NS, and for WDS the θ_j's are as computed in step 6). The experimental results obtained with this program will be presented subsequently.

We will close this section by showing that the estimated cross-correlation matrix is nonnegative-definite, since

$$\begin{aligned}
\sum_{j,k=1}^{K}\ \sum_{m,n=-v}^{v} \hat{\rho}_{jk}(m,n)\alpha_{jm}\alpha^*_{kn}
&= \frac{1}{L}\sum_{j,k=1}^{K}\ \sum_{m,n=-v}^{v}\ \sum_{\substack{0\le i\le L\\ 0\le i+m-n\le L}} N_{j,i+m-n}N_{k,i}\alpha_{jm}\alpha^*_{kn}\\
&= \frac{1}{L}\sum_{j,k=1}^{K}\ \sum_{m,n=-v}^{v}\ \sum_{\substack{0\le i+m\le L\\ 0\le i+n\le L}} N_{j,i+m}N_{k,i+n}\alpha_{jm}\alpha^*_{kn}\\
&= \frac{1}{L}\sum_{i=0}^{L}\ \sum_{j,k=1}^{K}\ \sum_{m,n=-v}^{v} e(i+m)e(i+n)N_{j,i+m}N_{k,i+n}\alpha_{jm}\alpha^*_{kn}\\
&= \frac{1}{L}\sum_{i=0}^{L}\left|\sum_{j=1}^{K}\ \sum_{m=-v}^{v} e(i+m)\alpha_{jm}N_{j,i+m}\right|^2 \ge 0
\end{aligned}$$

where α^* denotes the complex conjugate of α and

$$\begin{aligned} e(i+m) &= 1, \quad 0 \le i+m \le L\\ &= 0, \quad \text{otherwise.}\end{aligned}$$

IV. Theoretical Justification for the Asymptotic Approach

We now wish to show that the use of filters based on only a part of either the past, future, or past and future, can be used to approximate the performance of filters based on, respectively, the full past, the full future, or the full past and future. In order to do this it will be convenient to introduce new random variables, as indicated by Rozanov [7],

$$\begin{aligned} Y_{k^*m} &= X_{k^*m} \\ Y_{km} &= X_{k^*m} - X_{km} \qquad \begin{matrix} k\ne k^*, k=1,\cdots,K\\ m = 0, \pm 1, \pm 2, \cdots. \end{matrix}\end{aligned} \tag{25}$$

Let us denote the set of random variables Y_{km}, $k \neq k^*$, $m = -\nu, \cdots, \nu$, by Y_ν. It should be noted that the random variables Y_{km}, $k \neq k^*$, do not depend on S_m, i.e.,

$$Y_{km} = N_{k^*m} - N_{km}, \qquad k \neq k^* \tag{26}$$

and that

$$Y_{k^*m} = S_m + N_{k^*m}. \tag{27}$$

It therefore follows that the minimum-variance unbiased estimator of S_n may be written as

$$\hat{S}_{\nu n} = X_{k^*n} - \hat{N}_{\nu k^*n} \tag{28}$$

where $\hat{N}_{\nu k^*n}$ is a linear combination of the Y_{km}'s, $k \neq k^*$, $m = -\nu, \cdots, \nu$, i.e., $\hat{N}_{\nu k^*n}$ depends only on Y_ν. When written in this form, it follows easily that $\hat{N}_{\nu k^*n}$ is given by

$$\hat{N}_{\nu k^*n} = E\{N_{k^*n} | Y_\nu\} \tag{29}$$

i.e., $\hat{N}_{\nu k^*n}$ is the conditional expectation of N_{k^*n} with respect to Y_ν (cf. Doob [8], pp. 150–155). However, $\hat{N}_{\nu k^*n}$ is a martingale since (cf. Doob [8], pp. 91–94)

$$\begin{aligned} E\{\hat{N}_{\nu+1, k^*n} | Y_\nu\} &= E\{E[N_{k^*n} | Y_{\nu+1}] | Y_\nu\} \\ &= E\{N_{k^*n} | Y_\nu\} = \hat{N}_{\nu k^*n}. \end{aligned} \tag{30}$$

Thus, it follows from a martingale convergence theorem (cf. Doob [8], p. 167, Theorem 7.4, and also pp. 560–562) that

$$\hat{N}_{k^*n} = \underset{\nu \to \infty}{\text{l.i.m.}}\ \hat{N}_{\nu k^*n} \tag{31}$$

for fixed n. If we wish to consider sequences of physically realizable filters for which $n = \nu$, we obtain in a similar manner

$$\hat{N}_{k^*\infty} = \underset{\nu \to \infty}{\text{l.i.m.}}\ \hat{N}_{\nu k^*\nu} \tag{32}$$

and for filters using future values, $n = -\nu$, and

$$\hat{N}_{k^*, -\infty} = \underset{\nu \to \infty}{\text{l.i.m.}}\ \hat{N}_{\nu k^*, -\nu}. \tag{33}$$

The minimum-variance unbiased estimators for the above three cases are, respectively,

$$\hat{S}_n = \underset{\nu \to \infty}{\text{l.i.m.}}\ S_{\nu n} = X_{k^*n} - \hat{N}_{k^*n} \tag{34}$$

$$\hat{S}_\infty = \underset{\nu \to \infty}{\text{l.i.m.}}\ S_{\nu\nu} = X_{k^*\infty} - \hat{N}_{k^*\infty} \tag{35}$$

$$\hat{S}_{-\infty} = \underset{\nu \to \infty}{\text{l.i.m.}}\ S_{\nu, -\nu} = X_{k^*, -\infty} - \hat{N}_{k^*, -\infty}. \tag{36}$$

Thus, we have shown that filters based on a large part of either the past, future, or past and future, can approximate the performance of filters based on, respectively, the full past, the full future, or the full past and future. It is easily seen that all results remain valid when k^* is any integer between 1 and K.

We have already shown that the maximum-likelihood estimator for S_n is identical with the minimum-variance unbiased estimator for S_n; cf. (9) and (21). However, at this point it is extremely simple to show that this is true. The joint probability density of X_{k^*n}, Y_ν is

$$p_{S_n}(X_{k^*n}, Y_\nu) = p_{S_n}(X_{k^*n} | Y_\nu) p(Y_\nu) \tag{37}$$

since Y_ν is independent of S_n. In addition, we have for the Gaussian multivariate case

$$p_{S_n}(X_{k^*n} | Y_\nu) = (2\pi)^{-1/2} [\sigma^2(X_{k^*n} | Y_\nu)]^{-1/2} \cdot \exp\left\{-\frac{1}{2}\left[\frac{X_{k^*n} - E(X_{k^*n} | Y_\nu)}{\sigma(X_{k^*n} | Y_\nu)}\right]^2\right\} \tag{38}$$

where $\sigma(X_{k^*n} | Y_\nu)$ is independent of S_n and

$$E\{X_{k^*n} | Y_\nu\} = E\{\hat{N}_{\nu k^*n} | Y_\nu\} + \hat{S}_{\nu n}. \tag{39}$$

The maximum-likelihood estimator for S_n is obtained by differentiating, with respect to S_n, the probability density function in (37), or equivalently that in (38), from which we get, when using (39),

$$\hat{S}_{\nu n} = X_{k^*n} - E\{\hat{N}_{\nu k^*n} | Y_\nu\} \tag{40}$$

which agrees with (28) and (29). Thus, all asymptotic results derived for the minimum-variance unbiased estimator remain true, of course, for the maximum-likelihood estimator.

V. Synthesis of Physically Realizable Filter by Means of a Spectral Matrix Factorization Method

We now consider the synthesis of the filter for the physically realizable case, $n = \nu$, $\nu \to \infty$. Hereafter, it will be assumed that the noise is wide-sense stationary, so that

$$\rho_{jk}(m, n) = \rho_{jk}(m - n) \tag{41}$$

$$= \int_{-\pi}^{\pi} f_{jk}(x) \varepsilon^{-i(m-n)x} \frac{dx}{2\pi} \tag{42}$$

and

$$f_{jk}(x) = \sum_{m=-\infty}^{\infty} \rho_{jk}(m) \varepsilon^{imx} \qquad j, k = 1, \cdots, K \tag{43}$$

is the sampled cross-power spectral density function, $x = \omega T$, T is the sampling interval, and ω is the frequency in radians per second. In this case we have

$$A_k(x) = \sum_{j=0}^{\infty} \theta_k(j | \infty) \varepsilon^{-ijx} \tag{44}$$

and

$$\int_{-\pi}^{\pi} \varepsilon^{ilx} \left\{\sum_{k=1}^{K} f_{jk}(x) A_k(x) + \Lambda(x)\right\} dx = 0 \qquad \begin{matrix} l = 0, 1, 2, \cdots \\ j = 1, \cdots, K. \end{matrix} \tag{45}$$

Let us define $z = \varepsilon^{-ix}$ and

$$\psi_j(z) = \sum_{k=1}^{K} f_{jk}(z) A_k(z) + \Lambda(z), \qquad j = 1, \cdots, K. \tag{46}$$

We have from (44)

$$A_k(z) = \sum_{j=0}^{\infty} \theta_k(j | \infty) z^j, \qquad k = 1, \cdots, K. \tag{47}$$

Since $A_k(z)$ is a power series in ascending powers of z,

$A_k(z)$ must be analytic in the unit circle of the z-plane. In order to have $\psi_k(z)$ satisfy (45) we must have

$$\psi_k(z) = \sum_{j=1}^{\infty} b_j z^{-j}, \qquad k = 1, \cdots, K \tag{48}$$

so that $\psi_k(z)$ is analytic outside and on the unit circle of the z-plane. If we subtract the first equation in (46) from all the others and substitute

$$A_1(z) = 1 - \sum_{k=2}^{K} A_k(z) \tag{49}$$

we obtain a new system of equations

$$\sum_{k=2}^{K} [f_{11}(z) + f_{jk}(z) - f_{j1}(z) - f_{1k}(z)]A_k(z)$$
$$= f_{11}(z) - f_{j1}(z) + \psi_1(z) - \psi_j(z), \qquad j = 2, \cdots, K. \tag{50}$$

The functions $\psi_1(z) - \psi_j(z)$, $j=2, \cdots, K$ are analytic outside and on the unit circle.

Thus, we have a system of $(K-1)$ equations in $(K-1)$ unknowns from which we can solve for $A_2(z), \cdots, A_K(z)$. We then obtain $A_1(z)$ from (49). The system of equations in (50) may be written in matrix notation as

$$A(z)f(z) = h(z) + \psi(z). \tag{51}$$

The matrix $f(x)$ is easily seen to be a spectral matrix. It will be assumed that $f(z)$ has a spectral matrix factorization

$$f(z) = P(z)P'(z^{-1}) \tag{52}$$

where the matrices $P(z)$, $[P(z)]^{-1}$ have matrix Laurent expansions on $|z|=1$ with no negative powers of z, and $P'(z)$ denotes transpose. We have

$$A(z)P(z)P'(z^{-1}) = h(z) + \psi(z) \tag{53}$$

and

$$A(z)P(z) = h(z)[P'(z^{-1})]^{-1} + \psi(z)[P'(z^{-1})]^{-1}. \tag{54}$$

The matrix on the left-hand side of (54) has a matrix Laurent expansion with no negative powers of z which is assumed to converge in some annulus containing the unit circle. The second term on the right has only negative powers of z. Equating coefficients we obtain (cf. Whittle [9], pp. 67, 100)

$$A(z) = \{h(z)[P'(z^{-1})]^{-1}\}_+[P(z)]^{-1} \tag{55}$$

where the operation $\{\ \}_+$ indicates that only the nonnegative powers of z in the Laurent expansion of the matrix within the braces are to be retained. Equation (55) represents the complete solution to the synthesis problem for multidimensional physically realizable filters.

It is now necessary to show how the factorization in (52) can be obtained. A procedure for obtaining the spectral matrix factorization for rational spectral matrices has been given by Whittle [9], pp. 101–103, and is similar to that used in the one-dimensional case. As an example, let us consider that $K=2$ and

$$f_{11}(z) - f_{21}(z) = \frac{1}{(1-\alpha z)(1-\alpha z^{-1})}, \qquad |\alpha| < 1 \tag{56}$$

$$f_{11}(z) + f_{22}(z) - f_{12}(z) - f_{21}(z)$$
$$= \frac{(1-\beta z)(1-\beta z^{-1})}{(1-\alpha z)(1-\alpha z^{-1})}, \qquad |\beta| < 1. \tag{57}$$

We have

$$P(z) = \frac{1-\beta z}{1-\alpha z} \tag{58}$$

and

$$A_2(z) = \left[\frac{1}{(1-\alpha z)(1-\beta z^{-1})}\right]_+ \frac{1-\alpha z}{1-\beta z}. \tag{59}$$

Since $|\alpha|<1$, the Laurent expansion of $1/(1-\alpha z)$ may be made in positive powers of z and will converge in the circle $|z|<(1/|\alpha|)$, which encloses the unit circle. The Laurent expansion of $1/(z-\beta)$ must be made in negative powers of z, i.e., $z^{-1}+\beta z^{-2}+\cdots$, and will converge in the annulus $|\beta|<|z|<\infty$, which also encloses the unit circle. Thus, in order to perform the required operation on the term in the brackets in (59), we perform a partial fraction expansion and neglect the term $1/(z-\beta)$, to obtain

$$A_2(z) = \frac{1}{1-\alpha\beta}\,\frac{1}{1-\beta z} \tag{60}$$

$$A_1(z) = \frac{\beta}{1-\alpha\beta}\,\frac{z(\alpha\beta-1)-\alpha}{1-\beta z}. \tag{61}$$

Another method for achieving the spectral matrix factorization which is valid for general spectral matrices is due to Wiener and Masani [10]. Unfortunately, the spectral matrix factorization can be expressed only as an infinite sum of matrices, the rate of convergence of which is difficult to determine.

It is readily appreciated that the techniques for synthesizing physically realizable filters by means of a spectral matrix factorization method for rational matrices are impractical for machine computation due to the requirement of having to determine roots of polynomials. In addition, the first step would require approximating all measured spectral matrices with rational spectral matrices. This step could also be quite complicated and could entail a serious loss with respect to the amount by which the noise power can be reduced.

VI. Frequency-Domain Synthesis Procedure for Two-Sided Filters

A frequency-domain synthesis procedure will now be presented for maximum-likelihood filters which use a large part of the past and future. Such filters which use past as well as future values will be termed two-sided, or infinite-lag, filters. We begin by letting $\nu=\infty$, $n=0$, so that (13) and (14) may be written as

$$E\{\hat{S}_0^2\} = \lim_{v \to \infty} E\{\hat{S}_{v0}^2\} = \int_{-\pi}^{\pi} \left[\sum_{j,k=1}^{K} f_{jk}(x) A_j^*(x) A_k(x) \right] \frac{dx}{2\pi} \tag{62}$$

and

$$\int_{-\pi}^{\pi} \sum_{k=1}^{K} f_{jk}(x) A_k(x) \varepsilon^{-ilx} \frac{dx}{2\pi} + \lambda_l = 0, \quad \begin{matrix} l = 0, \pm 1, \pm 2, \cdots, \\ j = 1, \cdots, K \end{matrix} \tag{63}$$

where

$$A_k(x) = \sum_{m=-\infty}^{\infty} \theta_k(m|0) \varepsilon^{imx}, \quad k = 1, \cdots, K,$$

$$\rho_{jk}(l) = \int_{-\pi}^{\pi} f_{jk}(x) \varepsilon^{-ilx} \frac{dx}{2\pi}, \quad \begin{matrix} j, k = 1, \cdots, K \\ l = 0, \pm 1, \pm 2, \cdots \end{matrix} \tag{64}$$

and

$$f_{jk}(x) = \sum_{l=-\infty}^{\infty} \rho_{jk}(l) \varepsilon^{ilx}, \quad j, k = 1, \cdots, K \tag{65}$$

is the sampled cross-power spectral density function, $x = \omega T$, T is the sampling interval. We have from (64) that

$$\theta_k(m|0) = \int_{-\pi}^{\pi} A_k(x) \varepsilon^{-ijx} \frac{dx}{2\pi}, \quad \begin{matrix} k = 1, \cdots, K \\ m = 0, \pm 1, \pm 2, \cdots. \end{matrix} \tag{66}$$

The constraint (11) becomes

$$\sum_{k=1}^{K} A_k(x) = 1. \tag{67}$$

If we let

$$\Lambda(x) = \sum_{k=-\infty}^{\infty} \lambda_k \varepsilon^{ikx}, \tag{68}$$

then (63) may be written as

$$\int_{-\pi}^{\pi} \varepsilon^{-ilx} \left[\sum_{k=1}^{K} f_{jk}(x) A_k(x) + \Lambda(x) \right] dx = 0, \quad \begin{matrix} l = 0, \pm 1, \pm 2, \cdots \\ j = 1, \cdots, K. \end{matrix} \tag{69}$$

According to (69), the Fourier coefficients of the quantity in the brackets in (69) must all be zero. It follows that the quantity itself must be zero so that

$$\sum_{k=1}^{K} f_{jk}(x) A_k(x) + \Lambda(x) = 0, \quad \begin{matrix} -\pi \le x \le \pi \\ j = 1, \cdots, K. \end{matrix} \tag{70}$$

Thus

$$A_k(x) = \frac{\sum_{j=1}^{K} q_{kj}(x)}{\sum_{j,k=1}^{K} q_{kj}(x)}, \quad k = 1, \cdots, K \tag{71}$$

$$-\Lambda(x) = \left[\sum_{j,k=1}^{K} q_{kj}(x) \right]^{-1} \tag{72}$$

where $\{q_{jk}(x)\}$ is the inverse of the spectral matrix $\{f_{jk}(x)\}$, $j, k = 1, \cdots, K$. We note that

$$q_{jk}(x) = q_{kj}^*(x) \tag{73}$$

since

$$f_{jk}(x) = f_{kj}^*(x) \tag{74}$$

and

$$q_{jk}(x) = q_{jk}^*(-x) \tag{75}$$

since

$$f_{jk}(x) = f_{jk}^*(-x). \tag{76}$$

Therefore

$$A_k(x) = A_k^*(-x), \quad k = 1, \cdots, K. \tag{77}$$

We obtain from (62) and (71)

$$E\{(\hat{S}_0 - S_0)^2\} = \int_{-\pi}^{\pi} \frac{dx}{2\pi} \left[\sum_{j,k=1}^{K} q_{kj}(x) \right]^{-1}. \tag{78}$$

The filter weighting coefficients are obtained from (66) as

$$\theta_k(m|0) = \int_{-\pi}^{\pi} \frac{dx}{2\pi} [\mathrm{Re}\, A_k(x) \cos mx + \mathrm{Im}\, A_k(x) \sin mx], \quad \begin{matrix} k = 1, \cdots, K \\ m = 0, \pm 1, \pm 2, \cdots. \end{matrix} \tag{79}$$

It is easily seen that the constraint conditions are satisfied since

$$\sum_{k=1}^{K} \theta_k(m|0) = \int_{-\pi}^{\pi} \cos mx \frac{dx}{2\pi} = \delta_{m0}. \tag{80}$$

The filter weights given by (79) would have to be obtained by inverting the spectral matrix at all points on the frequency axis between $-\pi$ and π. This is clearly impractical, so that an approximation would have to be used in practice. Such an approximation is most easily obtained by approximating the integral in (79) by a finite sum

$$\theta_k(m|0) = \frac{1}{2v} \sum_{l=-v+1}^{v} \left[\mathrm{Re}\, A_k\left(l\frac{\pi}{v}\right) \cos ml \frac{\pi}{v} + \mathrm{Im}\, A_k\left(l\frac{\pi}{v}\right) \sin ml \frac{\pi}{v} \right], \quad \begin{matrix} k = 1, \cdots, K \\ m = -v, \cdots, v. \end{matrix} \tag{81}$$

The symmetry properties of $A_k(x)$ expressed in (77) enable us to simplify the above equation

$$\theta_k(m|0) = \frac{1}{2v} \Bigg[A_k(0) + (-1)^m A_k(\pi) + 2 \sum_{l=1}^{v-1} \mathrm{Re}\, A_k\left(l\frac{\pi}{v}\right) \cos ml \frac{\pi}{v} + \mathrm{Im}\, A_k\left(l\frac{\pi}{v}\right) \sin ml \frac{\pi}{v} \Bigg] \quad \begin{matrix} k = 1, \cdots, K \\ m = -v, \cdots, v. \end{matrix} \tag{82}$$

The constraint equations are still satisfied since

$$\sum_{k=1}^{K} \theta_k(m|0) = \frac{1}{2\nu}\left[(-1)^m + 2\left(\frac{1}{2} + \sum_{l=1}^{\nu-1} \cos ml\frac{\pi}{\nu}\right)\right]$$

$$= \frac{1}{2\nu}\left[(-1)^m + \frac{\sin\left(\nu - \frac{1}{2}\right)m\frac{\pi}{\nu}}{\sin\frac{1}{2}m\frac{\pi}{\nu}}\right]$$

$$= \frac{1}{2\nu}[+1 + 2\nu - 1] = 1, \quad m = 0$$

$$= \frac{1}{2\nu}\left[(-1)^m - \frac{\sin\left(\frac{1}{2}m\frac{\pi}{\nu} - m\pi\right)}{\sin\frac{1}{2}m\frac{\pi}{\nu}}\right]$$

$$= \frac{1}{2\nu}[(-1)^m - (-1)^m] = 0, \quad m \neq 0 \qquad (83)$$

where we have used the identity

$$\frac{1}{2} + \sum_{k=1}^{n} \cos ku = \frac{1}{2}\,\frac{\sin(n + \frac{1}{2})u}{\sin\frac{1}{2}u}.$$

It is easy to see that as $\nu \to \infty$, the estimator based on the approximate filter weights given by (82) converges in the mean to $\hat{S}_0$.

It is difficult in general to determine how well a physically realizable filter performs relative to a two-sided filter. The two-sided filter must always be better than the physically realizable filter since

$$E\{\hat{S}_0^2\} \leq E\{\hat{S}_\infty^2\}.$$

This follows from the fact that the optimum weighting functions in the two-sided case are subject to fewer constraints than those in the physically realizable case.

We now wish to give an example in which we will compare the performance of the physically realizable filter with that of the two-sided filter. For simplicity we consider that $K=2$ and

$$f_{11}(z) = \frac{1}{(1 - \alpha z)(1 - \alpha z^{-1})}, \qquad 0 < \alpha < 1$$

$$f_{22}(z) = \frac{1}{(1 - \beta z)(1 - \beta z^{-1})}, \qquad 0 < \beta < 1$$

$$f_{22}(z) = f_{21}(z) = 0$$

so that

$$f_{11}(z) - f_{21}(z) = \frac{1}{(1 - \alpha z)(1 - \alpha z^{-1})}$$

$$f_{11}(z) + f_{22}(z) - f_{12}(z) - f_{21}(z) = \frac{(1 - \alpha z)(1 - \alpha z^{-1}) + (1 - \beta z)(1 - \beta z^{-1})}{(1 - \alpha z)(1 - \alpha z^{-1})(1 - \beta z)(1 - \beta z^{-1})}.$$

In addition, let us define

$$(1 - \theta z)(1 - \theta z^{-1}) \equiv (1 - \alpha z)(1 - \alpha z^{-1}) + (1 - \beta z)(1 - \beta z^{-1})$$

so that

$$\theta = \alpha + \beta$$

and

$$\theta^2 = 1 + \alpha^2 + \beta^2.$$

Thus, α and β must satisfy the equation

$$2\alpha\beta = 1$$

and

$$\theta = \alpha + \frac{1}{2\alpha}.$$

Hence α and β must lie in the open interval $(\frac{1}{2}, 1)$, θ is in the open interval $(\sqrt{2}, \frac{3}{2})$, and we have

$$f_{11}(z) + f_{22}(z) - f_{12}(z) - f_{21}(z) = \frac{\theta^2(1 - \theta^{-1})(1 - \theta^{-1}z^{-1})}{(1 - \alpha z)(1 - \alpha z^{-1})(1 - \beta z)(1 - \beta z^{-1})}.$$

Therefore, we can write

$$P(z) = \frac{\theta(1 - \theta^{-1}z)}{(1 - \alpha z)(1 - \beta z)}$$

and

$$A_2(z)\left[\frac{\theta(1 - \theta^{-1}z)}{(1 - \alpha z)(1 - \beta z)}\right] = \left[\frac{(1 - \alpha z^{-1})(1 - \beta z^{-1})}{\theta(1 - \alpha z)(1 - \alpha z^{-1})(1 - \theta^{-1}z^{-1})}\right]_+$$

$$= \frac{\alpha}{1 - \alpha z},$$

$$A_2(z) = \frac{1}{2}\,\frac{z - 2\alpha}{z - \theta},$$

$$A_1(z) = \frac{1}{2}\,\frac{z - \alpha^{-1}}{z - \theta}.$$

We have from (13)

$$E\{\hat{S}_\infty^2\} = \int_{-\pi}^{\pi} \sum_{j,k=1}^{K} A_j^*(x)A_k(x)f_{jk}(x)\,\frac{dx}{2\pi}$$

$$= \frac{1}{2\pi i}\int_{\text{unit circle}} \sum_{j,k=1}^{K} A_j(z^{-1})A_k(z)f_{jk}(z)\,\frac{dz}{z}$$

$$= \sum_{\substack{\text{inside unit}\\ \text{circle}}} \text{Residues}\left\{\sum_{j,k=1}^{K} \frac{1}{z}\,A_j(z^{-1})A_k(z)\,f_{jk}(z)\right\}$$

$$= \operatorname*{Res}_{z=\theta^{-1}}\left\{\frac{z - \alpha^{-1}}{4\alpha(z - \theta)(\theta z - 1)(1 - \alpha z)} + \frac{\alpha}{2}\,\frac{z - 2\alpha}{(z - \theta)(\theta z - 1)(1 - \beta z)}\right.$$

$$= \frac{1}{\theta^2 - 1}\left[\alpha^2 + \frac{1}{4\alpha^2}\right].$$

For the two-sided filter, we obtain

$$q_{11}(z) = (1 - \alpha z)(1 - \alpha z^{-1})$$
$$q_{22}(z) = (1 - \beta z)(1 - \beta z^{-1})$$
$$q_{12}(z) = q_{21}(z) = 0$$
$$\left[\sum_{j,k=1}^{2} q_{jk}(z)\right]^{-1} = \frac{1}{(1 - \theta z)(1 - \theta z^{-1})}.$$

Using (78) we have

$$\begin{aligned} E\{\hat{S}_0^2\} &= \int_{-\pi}^{\pi} \left[\sum_{j,k=1}^{K} q_{jk}(x)\right]^{-1} \frac{dx}{2\pi} \\ &= \frac{1}{2\pi i} \int_{\text{unit circle}} \sum_{j,k=1}^{K} q_{jk}(z)\Bigg]^{-1} \frac{dz}{z} \\ &= \sum_{\substack{\text{inside unit}\\ \text{circle}}} \text{Res}\left\{\frac{1}{z}\left[\sum_{j,k=1}^{K} q_{jk}(z)\right]^{-1}\right\} \\ &= \operatorname*{Res}_{z=\theta^{-1}} \frac{1}{(1 - \theta z)(z - \theta)} \\ &= \frac{1}{\theta^2 - 1}. \end{aligned}$$

Thus

$$\frac{E\{\hat{S}_\infty^2\}}{E\{\hat{S}_0^2\}} = \alpha^2 + \frac{1}{4\alpha^2}.$$

The minimum value of the above ratio is unity and occurs when $\alpha = 1/\sqrt{2}$. The maximum value, in the permissible range for α, is $\frac{5}{4}$ and occurs when $\alpha \to \frac{1}{2}$ or $\alpha \to 1$. Thus, there can be a loss in noise variance reduction of between zero and approximately 1 dB by using the physically realizable filter rather than the two-sided filter, when $K=2$. It has been observed from experimental results obtained using computer runs that losses of about 6 dB occur when $K=25$.

The minimum-variance unbiased estimator, or, equivalently, the maximum-likelihood estimate of S_n, denoted by $\hat{S}_{vn}$, can be written as a moving average

$$\hat{S}_{vn} = \sum_{m=-v}^{v} \sum_{k=1}^{K} \theta_{km} X_{k,m+n}$$

where the filter weights satisfy the constraint

$$\sum_{k=1}^{K} \theta_{km} = \delta_{m0}, \qquad m = -v, \cdots, v$$

and it is assumed that a two-sided filter is to be used.

It has been shown [cf. (82)] that for two-sided filters, asymptotically optimum filter weights are given by

$$\begin{aligned} \theta_{km} = \frac{1}{2v}\Bigg[A_k(0) + (-1)^m A_k(\pi) + 2 \sum_{n=1}^{v-1} \text{Re}\, A_k\left(n\frac{\pi}{v}\right) \cos mn\frac{\pi}{v} \\ + \text{Im}\, A_k\left(n\frac{\pi}{v}\right) \sin mn\frac{\pi}{v}\Bigg], \qquad k = 1, \cdots, K \\ m = -v, \cdots, v. \end{aligned}$$

If v is large, the spectral density of $(\hat{S}_{vn} - S_n)$ is given approximately by [cf. (23), (68), (72)]

$$p(x) = \left[\sum_{j,k=1}^{K} q_{jk}(x)\right]^{-1}. \tag{84}$$

In essence, the frequency-domain synthesis procedure operates as follows. The spectral matrices $\{f_{jk}(x)\}$ are inverted at $v+1$ points in frequency, namely $n(\pi|v)$, $n=0, 1, \cdots, v$ to yield the matrices $\{q_{jk}(n(\pi/v))\}$. The frequency functions $A_k(n(\pi/v))$ are then obtained according to (71). In order to obtain the filter functions, $A_k'(x)$, for all $-\pi \leq x \leq \pi$, a bank of filter functions of the type

$$\left(\sin\frac{N}{2}x\right) \Big/ \left(\sin\frac{1}{2}x\right)$$

are used as follows:

$$A_k'(x) = \frac{1}{2v} \sum_{n=-v+1}^{v} A_k\left(n\frac{\pi}{v}\right) \frac{\sin\left(v + \frac{1}{2}\right)\left(x - \frac{n\pi}{v}\right)}{\sin\frac{1}{2}\left(x - \frac{n\pi}{v}\right)},$$
$$k = 1, \cdots, K.$$

It is easily seen that the Fourier transform of these frequency filter functions leads to the filter coefficients given in (82). Thus, the first step in the frequency-domain synthesis procedure must be to estimate the spectral matrices of the noise $\{f_{jk}(n(\pi/v))\}$, $n=0, \cdots, v$ inside a fitting interval known to contain only noise. The design of a spectral matrix estimation procedure will be discussed in the next section.

VII. The Spectral Matrix Estimation Procedure

There is a large amount of literature available on power spectral density estimation techniques. A book by Grenander and Rosenblatt [11] treats many of the theoretical problems encountered in spectral estimation. Two books which are motivated more by the practical considerations in spectral estimation are due to Blackman and Tukey [12] and Blackman [13]. These contributions deal mainly with the estimation of the power spectrum of a single random process. A discussion of the problems involved in the estimation of the spectral matrix of a multidimensional random process has been given by Goodman [14] and Rosenblatt [15].

It is possible to divide the spectral estimation procedures into two broad categories, namely direct and indirect methods. In the direct method the data are transformed immediately into the frequency domain and then the spectrum is measured, using the transformed data. The indirect method first estimates the correlation function of the data and transforms this to obtain an estimate of the spectrum. This latter method has been discussed extensively by Blackman and Tukey [12]. Some of the criteria which have been used to judge the merits of a spectral estimation procedure are the bias and variance of the estimate, which will be discussed in detail subsequently, and the amount of computer time required to obtain the estimate. In the one-dimensional case it is largely a matter of taste as to which method to use, as both the direct and indirect methods yield estimates with roughly the same bias and variance, and the computation time is comparable for both methods. However, in the multidimensional case it has been recognized that much less computation time is required by the direct method than the in-indirect method, and both methods can be made to yield estimates with approximately the same bias and variance; cf. for example Jones [16]. In the present application, data from the LASA, in which there are 25 sensors in each of the 21 subarrays, are to be filtered. Thus, the spectral matrix of a 25-dimensional random process is to be estimated, since a single subarray is processed at a time, and in order to keep the computation time within reasonable bounds a direct method of spectral matrix estimation is necessary.

The present method of spectral matrix estimation may be termed a direct segment method. The number of data points in each channel which is to be used in the estimation, namely L, is divided into M segments of $N=2\nu+1$ data points. It should be noted that $2\nu+1$ is equal to the number of filter points used. The data in each segment and each channel are transformed into the frequency domain, and these transforms are used to obtain an estimate of the cross spectra in the segment. The stability of the estimate is then increased by averaging over the M segments. We now describe the method in some detail.

The transform of the noise data in the nth segment, jth channel, and at the lth frequency, is

$$S_{jn}(l) = (N)^{-1/2} \sum_{m=1}^{N} w_m N_{j,m+(n-1)N} \varepsilon^{iml(2\pi/N-1)}, \qquad j = 1, \cdots, K \quad l = 0, \cdots, \frac{N-1}{2} \quad n = 1, \cdots, M \tag{85}$$

where w_m, $m=1, \cdots, N$ are the coefficients of the weighting function. Let

$$\lambda(l) = l\frac{2\pi}{N-1}, \qquad l = 0, \cdots, \frac{N-1}{2} \tag{86}$$

so that

$$S_{jn}(\lambda) = (N)^{-1/2} \sum_{m=1}^{N} w_m N_{j,m+(n+1)N} \varepsilon^{im\lambda}. \tag{87}$$

As an estimate for $f_{jk}(\lambda)$ we take

$$\hat{f}_{jk}(\lambda) = \frac{1}{M} \sum_{n=1}^{M} S_{jn}(\lambda) S_{kn}^*(\lambda), \qquad j, k = 1, \cdots, K. \tag{88}$$

The mean value of this estimate is

$$\begin{aligned} E\{\hat{f}_{jk}(\lambda)\} &= E\frac{1}{M}\sum_{n=1}^{M} S_{jn}(\lambda)S_{kn}^*(\lambda) \\ &= E\frac{1}{MN}\sum_{n=1}^{M}\sum_{m,m'=1}^{N} w_m w_{m'} \\ &\quad N_{j,m+(n+1)N} N_{k,m'+(n-1)N} \varepsilon^{i(m-m')\lambda} \\ &= \frac{1}{N}\sum_{m,m'=1}^{N} w_m w_{m'} \rho_{jk}(m-m') \varepsilon^{i(m-m')\lambda} \\ &= \int_{-\pi}^{\pi} f_{jk}(x) |W_N(x-\lambda)|^2 \frac{dx}{2\pi} \end{aligned} \tag{89}$$

where the frequency window $W_N(x)$ is defined as

$$W_N(x) = (N)^{-1/2} \sum_{m=1}^{N} w_m \varepsilon^{-imx} \tag{90}$$

and thus

$$w_m = N^{1/2} \int_{-\pi}^{\pi} W_N(x) \cdot \varepsilon^{imx} \frac{dx}{2\pi}. \tag{91}$$

As usual, we require that $W_N(x)$ be a reasonably good window, in the sense that $|W_N(x)|^2$ approaches a delta function in such a manner that

$$\int_{-\pi}^{\pi} |W_N(x)|^2 \frac{dx}{2\pi} = 1. \tag{92}$$

In this case $\hat{f}_{jk}(\lambda)$ is a reasonably good estimate for $f_{jk}(\lambda)$.

Since $f_{jk}(\lambda)$ is, in general, complex, it is convenient to define

$$f_{jk}(\lambda) = c_{jk}(\lambda) + iq_{jk}(\lambda) \qquad j, k = 1, \cdots, K \tag{93}$$

where $c_{jk}(\lambda)$, $q_{jk}(\lambda)$ are real-valued functions and are known as the cospectrum and quadrature spectrum, respectively. Similarly,

$$\hat{f}_{jk}(\lambda) = \hat{c}_{jk}(\lambda) + i\hat{q}_{jk}(\lambda). \tag{94}$$

It thus follows that $\hat{c}_{jk}(\lambda)$, $\hat{q}_{jk}(\lambda)$, are reasonable estimates for $c_{jk}(\lambda)$, $q_{jk}(\lambda)$, respectively.

The mean-square value of $\hat{c}_{jk}(\lambda)$ is

$$\begin{aligned} E\{\hat{c}_{jk}(\lambda)\}^2 = E\frac{1}{M^2N} \sum_{n,n'=1}^{M} \sum_{m,m',m'',m'''=1}^{N} w_m w_{m'} w_{m''} w_{m'''} \\ \cdot N_{j,m+(n-1)N} N_{k,m'+(n-1)N} N_{j,m''+(n'-1)N} N_{k,m'''+(n'-1)N} \\ \cdot \cos(m-m')\lambda \cos(m''-m''')\lambda. \end{aligned} \tag{95}$$

In order to proceed with the analysis, we must at this point introduce the assumption that $\{N_{km}\}$ is a multidimensional Gaussian process, so that

$$\begin{aligned} E\{N_{j,m+(n-1)N} N_{k,m'+(n-1)N} N_{j,m''+(n'-1)N} N_{k,m'''+(n'-1)N}\} \\ = \rho_{jk}(m-m')\rho_{jk}(m''-m''') + \rho_{jj}[m-m''+(n-n')N] \\ \cdot \rho_{kk}(m'-m'''+(n-n')N) + \rho_{jk}[m-m'''+(n-n')N] \\ \cdot \rho_{kj}[m'-m''+(n-n')N]. \end{aligned} \tag{96}$$

Using the above result in (93), we obtain, after some manipulations,

$$\begin{aligned} \text{VAR}\{\hat{c}_{jk}(\lambda)\} \; \frac{1}{2}\text{Re}\int_{-\pi}^{\pi}\int_{-\pi}^{\pi} \{f_{jj}(x)f_{kk}(x') + f_{jk}(x)f_{jk}(x')\} \\ \cdot \left(\frac{\sin\frac{M}{2}N(x-x')}{M\sin\frac{N}{2}(x-x')}\right)^2 [|W_N(x-\lambda)W_N(x'-\lambda)|^2 \\ + W_N(x-\lambda)W_N^*(x+\lambda)W_N(x'+\lambda)W_N^*(x'-\lambda)] \frac{dx}{2\pi}\frac{dx'}{2\pi}. \end{aligned} \tag{97}$$

At this point it is necessary to place some regularity conditions on $W_N(x)$ which guarantee that $|W_N(x)|^2$ will behave like a delta function at $x=0$, as $N\to\infty$; i.e., in addition to (92) we require

$$\int_{|x|<\varepsilon} |W_N(x)|^2 \frac{dx}{2\pi} \to 1 \tag{98}$$

for any ε, as $N\to\infty$, and

$$|W_N(x)| \to 0 \tag{99}$$

uniformly when $|x|\geq\varepsilon$, as $N\to\infty$. Under these conditions it may be shown, in a manner similar to that given by Grenander and Rosenblatt [11], pp. 137–145, that (97) becomes, as $M, N\to\infty$,

$$\begin{aligned}\text{VAR}\,\{\hat{c}_{jk}(\lambda)\} &\\ &\cong \frac{1}{2M}\{f_{jj}(\lambda)f_{kk}(\lambda) + c_{jk}^2(\lambda) - q_{jk}^2(\lambda)\}, \quad \lambda\neq 0,\pi\\ &\cong \frac{1}{M}\{f_{jj}(\lambda)f_{kk}(\lambda) + c_{jk}^2(\lambda) - q_{jk}^2(\lambda)\}, \quad \lambda = 0,\pi.\end{aligned} \tag{100}$$

Similarly,

$$\begin{aligned}\text{VAR}\,\{\hat{q}_{jk}(\lambda)\} &\\ &\cong \frac{1}{2M}\{f_{jj}(\lambda)f_{kk}(\lambda) + q_{jk}^2(\lambda) - c_{jk}^2(\lambda)\}, \quad \lambda\neq 0,\pi\\ &= 0, \qquad \lambda = 0,\pi\end{aligned} \tag{101}$$

since $q_{jk}(0)=\hat{q}_{jk}(0)=0$, and $q_{jk}(\pi)=\hat{q}_{jk}(\pi)=0$.

In the important case of uniform weighting, $w_m=1$, $m=1,\cdots,N$ and

$$|W_N(x)|^2 = \frac{1}{N}\left|\sum_{m=1}^{N}\varepsilon^{-imx}\right|^2 = \frac{1}{N}\left|\frac{\sin\frac{N}{2}x}{\sin\frac{1}{2}x}\right|^2. \tag{102}$$

It is easily seen that this weighting function satisfies the conditions in (92), (98), and (99). Therefore it follows from (100), (101) that $\hat{c}_{jk}(\lambda)$ and $\hat{q}_{jk}(\lambda)$ are consistent estimates for $c_{jk}(\lambda)$, $q_{jk}(\lambda)$, respectively, and that the variances of these estimators go to zero as $M\to\infty$.

The bias of the estimate of the cospectrum is defined as

$$b_{Njk}(\lambda) = E\{\hat{c}_{jk}(\lambda)\} - c_{jk}(\lambda) \tag{103}$$

and provides an indication of the error between the mean value of the estimate and the true value, at each frequency. We may evaluate the bias for the uniform weighting function by using (89) and (94)

$$\begin{aligned}&b_{Njk}(\lambda)\\
&= \frac{1}{N}\sum_{m,m'=1}^{N}\rho_{jk}(m-m')\cos(m-m')\lambda - \sum_{m=-\infty}^{\infty}\rho_{jk}(m)\cos m\lambda\\
&= \sum_{m=-N+1}^{N-1}\left(1-\frac{|m|}{N}\right)\rho_{jk}(m)\cos m\lambda - \sum_{m=-\infty}^{\infty}\rho_{jk}(m)\cos m\lambda\\
&\cong -\sum_{m=-N+1}^{N-1}\frac{|m|}{N}\rho_{jk}(m)\cos m\lambda\\
&= -2\sum_{m=0}^{N-1}\frac{|m|}{N}\rho_{jk}(m)\cos m\lambda\\
&= \frac{2}{N}\sum_{m=0}^{N-1}\cos m\lambda\int_{-\pi}^{\pi}c'_{jk}(x)\sin mx\frac{dx}{2\pi}\\
&= \frac{2}{N}\sum_{m=0}^{N-1}\int_{-\pi}^{\pi}\frac{\sin m(x-\lambda)+\sin m(x+\lambda)}{2}c'_{jk}(x)\frac{dx}{2\pi}\\
&= \frac{2}{N}\int_{-\pi}^{\pi}\frac{1}{2}\,\text{Im}\left\{\frac{1-\varepsilon^{iN(x-\lambda)}}{1-\varepsilon^{i(x-\lambda)}}+\frac{1-\varepsilon^{iN(x+\lambda)}}{1-\varepsilon^{i(x+\lambda)}}\right\}c'_{jk}(x)\frac{dx}{2\pi}\\
&= \frac{2}{N}\int_{-\pi}^{\pi}\left\{\frac{\sin(x-\lambda)-\sin N(x-\lambda)+\sin(N-1)(x-\lambda)}{4[1-\cos(x-\lambda)]}\right.\\
&\qquad\left.+\frac{\sin(x+\lambda)-\sin N(x+\lambda)+\sin(N-1)(x+\lambda)}{4[1-\cos(x+\lambda)]}\right\}c'_{jk}(x)\frac{dx}{2\pi}\\
&\cong \frac{1}{N}P\int_{-\pi}^{\pi}\frac{\sin(x-\lambda)}{1-\cos(x-\lambda)}c'_{jk}(x)\frac{dx}{2\pi}\end{aligned} \tag{104}$$

where P denotes that the principal part of the last integral is to be taken. Similarly,

$$\begin{aligned}b'_{Njk}(\lambda) &= E\{\hat{q}_{jk}(\lambda)\} - q_{jk}(\lambda)\\ &\cong \frac{1}{N}P\int_{-\pi}^{\pi}\frac{\sin(x-\lambda)}{1-\cos(x-\lambda)}q'_{jk}(x)\frac{dx}{2\pi}.\end{aligned} \tag{105}$$

Thus, the estimators of the cospectrum and quadrature spectrum are asymptotically unbiased and their biases decrease to zero at the rate of N^{-1}.

If the regularity conditions for $W_N(x)$ given in (92), (98), and (99) are satisfied, and if we assume that $\{N_{jm}\}$ is a multidimensional Gaussian process in order to avoid some complicated regularity conditions which make the following result valid for non-Gaussian processes, then a straightforward extension of a limit theorem of Rosenblatt [15] shows that the $\hat{c}_{jk}(\lambda)$, $\hat{q}_{jk}(\lambda)$, $j, k=1,\cdots,K$ are jointly asymptotically normally distributed with variances and covariances given by

$$\begin{aligned}&\lim_{M,N\to\infty}(2M)\,\text{COV}\,\{c_{\alpha\beta}(\lambda), c_{\gamma\delta}(\mu)\}\\
&\quad= \delta_{\lambda\mu}\,\text{Re}\,[f^*_{\alpha\gamma}(\lambda)f_{\beta\delta}(\lambda)+f^*_{\alpha\delta}(\lambda)f_{\beta\gamma}(\lambda)], \qquad \lambda\neq 0,\pi\\
&\quad= 2\delta_{\lambda\mu}[f_{\alpha\gamma}(\lambda)f_{\beta\delta}(\lambda)+f_{\alpha\delta}(\lambda)f_{\beta\gamma}(\lambda)], \qquad \lambda=0,\pi\\
&\lim_{M,N\to\infty}(2M)\,\text{COV}\,\{\hat{c}_{\alpha\beta}(\lambda), \hat{q}_{\gamma\delta}(\mu)\}\\
&\quad= \delta_{\gamma\mu}\,\text{Im}\,[f^*_{\alpha\gamma}(\lambda)f_{\beta\delta}(\lambda)-f^*_{\alpha\delta}(\lambda)f_{\beta\gamma}(\lambda)], \qquad \lambda\neq 0,\pi\\
&\quad= 0, \qquad \lambda=0,\pi\\
&\lim_{M,N\to\infty}(2M)\,\text{COV}\,\{\hat{q}_{\alpha\beta}(\lambda), \hat{q}_{\gamma\delta}(\mu)\}\\
&\quad= \delta_{\lambda\mu}\,\text{Re}\,[f_{\alpha\gamma}(\lambda)f^*_{\beta\delta}(\lambda)-f_{\alpha\delta}(\lambda)f^*_{\beta\gamma}(\lambda)], \qquad \lambda\neq 0,\pi\\
&\quad= 0, \qquad \lambda=0,\pi.\end{aligned}$$

It is possible to use these formulas to establish asymptotic confidence intervals for the spectral estimates. A different approximation for the asymptotic distribution of the spectral estimates has been given by Goodman [14] in terms of the complex Wishart distribution.

We obtain from (87) and (88) that

$$\sum_{j,k=1}^{K} a_j a_k^* \hat{f}_{jk}(x) = \frac{1}{L} \sum_{n=1}^{M} \left| \sum_{j=1}^{K} \sum_{m=1}^{N} w_m a_j N_{j,m+(n-1)N} e^{imx} \right|^2 \geq 0 \qquad -\pi \geq x \geq \pi \tag{106}$$

so that the spectral matrix is nonnegative-definite at all frequencies. This is a highly desirable property, since inverses of spectral matrices must be computed in the frequency-domain synthesis procedure. If the spectral matrix is singular at a particular frequency, $M \geq K$, then, physically speaking, this means that infinite suppression of the noise at that frequency is possible. As a practical matter, it is not possible to achieve infinite suppression, so that the spectral matrix will not be singular. This conclusion has been borne out by computer experimentation, as no singular spectral matrices have ever been encountered.

In general, the estimated spectral matrix will be nonnegative-definite and hermitian. Let us denote such a matrix by $C = A + iB$, where A, B are real matrices. It is easily seen that A is symmetric, that B is skew symmetric, and that neither A nor B need be nonnegative-definite if C is nonnegative-definite. Let the inverse of C be denoted by $D + iE$, where D, E are real matrices. It is easily seen that

$$\left[\begin{array}{c|c} D & E \\ \hline E & D \end{array}\right] = \left[\begin{array}{c|c} A & B \\ \hline -B & A \end{array}\right]^{-1}. \tag{107}$$

Thus, the inverse of the matrix C may be obtained by inverting the real matrix

$$\left[\begin{array}{c|c} A & B \\ \hline -B & A \end{array}\right].$$

The advantage of performing the matrix inversion in this manner is that this matrix is invertible if and only if the matrix C is invertible; cf. Goodman [17]. It is for this reason that this method of inverting the estimated spectral matrices is used in the computer program implementation of the frequency-domain synthesis procedure.

The amount of computer time required to estimate the cross-correlation matrix for N lags is given approximately by

$$T_c = LK^2N(\mu + \alpha) \text{ seconds} \tag{108}$$

where μ, α are the multiplication and addition times, respectively, in seconds. It is seen from Section VII that the amount of computer time required to estimate the spectral matrix at $(N+1)/2$ frequencies is approximately

$$T_s = LK(K + N)(\mu + \alpha) \text{ seconds} \tag{109}$$

and

$$\frac{T_c}{T_s} = \frac{KN}{K + N}. \tag{110}$$

If K and N are large, the ratio T_c/T_s can be very large. If $K = 25$, $N = 21$, then $T_c/T_s = 11.4$. Thus, we see that the amount of computer time required to estimate the spectral matrix is inherently much smaller than that required to estimate the cross-correlation matrix when K and N are large. An even greater saving is possible by using the Cooley-Tukey [18] algorithm to transform the data into the frequency domain. In this case, the amount of computer time becomes

$$T_s' = LK(K + F_N)(\mu + \alpha) \text{ seconds} \tag{111}$$

and

$$\frac{T_c}{T_s'} = \frac{KN}{K + F_N} \tag{112}$$

where F_N is the sum of the factors of $N-1$. If $N-1 = 20 = 5 \cdot 2 \cdot 2$, then $F_N = 2+2+5 = 9$ and $T_c/T_s = 15.5$.

If an indirect spectral matrix estimation procedure is used in which the estimates of the cross-correlation function are transformed into the frequency domain, the amount of computer time becomes

$$\begin{aligned} T_I &= (LK^2N + K^2N^2)(\mu + \alpha) \\ &\cong LK^2N(\mu + \alpha) \text{ seconds} \\ &= T_c, \qquad L \gg N. \end{aligned} \tag{113}$$

It has been assumed that the Cooley-Tukey algorithm is not employed, since the saving would be incurred in the second term above which is already negligible compared to the first term. Thus, T_I/T_s is quite large and T_I/T_s' is even larger. This disadvantage makes the indirect method undesirable for spectral matrix estimation.

The amount of computer time required to synthesize the maximum-likelihood filter in the time domain is 2.5 $N^2K^3(\mu+\alpha)$ seconds [6], and in the frequency domain is $2NK^3(\mu+\alpha)$, where it is assumed that $8K^3$ operations are required to invert a matrix with complex elements, according to the procedure given previously. Thus, the total computing time required in the time domain is

$$T_T = (LK^2N + 2.5N^2K^3)(\mu + \alpha) \text{ seconds} \tag{114}$$

and in the frequency domain is

$$T_F = [LK(K + N) + 2NK^3](\mu + \alpha) \text{ seconds,} \tag{115}$$

assuming the Cooley-Tukey algorithm is not used. Thus

$$\frac{T_T}{T_F} = \frac{LKN + 2.5N^2K^2}{L(K + N) + 2NK^2} \tag{116}$$

which, if L is 1800 (corresponding to three minutes of noise, 20 data points per second and every other data point skipped), $K = 25$, $N = 21$, becomes $T_T/T_F = 15$. In practice this large saving is not achieved, of course, due to indexing operations and tape reading which are common to both methods, but savings of approximately a factor of ten have been realized.

The input noise power at the lth frequency, inside the fitting interval, is obtained from the spectral estimation program as follows, when the uniform weighting function is used.

$$P_{IN}(l) = \frac{1}{(N-1)K} \sum_{j=1}^{NS} \hat{f}_{jj}(l) \tag{117}$$

and the total noise power is

$$P_{IN} = 2 \sum_{l=1}^{(N-3)/2} P_{IN}(l) + P_{IN}(0) + P_{IN}\left(\frac{N-1}{2}\right). \tag{118}$$

The justification for this definition of $P_{IN}(l)$ is apparent and the justification for defining P_{IN} in (118) is that P_{IN} may be written, after some manipulation, as

$$P_{IN} = \frac{1}{LK} \sum_{j=1}^{K} \sum_{m=1}^{L} N_{jm}^2 \tag{119}$$

which is numerically equal to the average input noise power.

Let us consider the power measurement which is made by summing the squares of the processed noise samples at the midpoints of the segments used in the spectral matrix measurement, i.e.,

$$P_{OUT} = \frac{1}{M} \sum_{n=1}^{M} \left(\sum_{j=1}^{NS} \sum_{m=-((N-1)/2)}^{(N-1)/2} 0_{jm} N_{j,m+(n-1)N} \right)^2. \tag{120}$$

We may show, after some manipulation, that

$$P_{OUT} = \frac{N}{N-1} \frac{1}{N-1} \left\{ \sum_{l=-((N-3)/2)}^{(N-1)/2} \left[\sum_{j,k=1}^{K} \hat{q}_{jk}(l) \right]^{-1} \right\} \tag{121}$$

where $\{\hat{q}_{jk}(l)\}$ is the inverse of the estimated spectral matrix $\{\hat{f}_{jk}(l)\}$. The frequency-domain computer program uses the inverse spectral matrices to compute P_{OUT} as indicated in (121). It should be noted that this estimate is based on M out of a possible L data samples in the fitting interval and must be regarded as an approximation for the true output power in the fitting interval, which is equal to the sum of the squares of the L data points in the filtered output trace. In a similar manner, the output power at the lth frequency is computed as

$$P_{OUT}(l) = \frac{N}{N-1} \frac{1}{N-1} \left[\sum_{j,k=1}^{NS} \hat{q}_{jk}(l) \right]^{-1} \tag{122}$$

which can be shown to be equal to the following

$$= \frac{1}{M(N-1)^2} \sum_{n=1}^{K} \left| \sum_{k=1}^{K} \sum_{m=1}^{N} A_k(l) N_{k,m+(n-1)N} \cdot \varepsilon^{-iml(2\pi/N-1)} \right|^2. \tag{123}$$

This latter quantity is what is obtained by transforming the noise data in the segments used in the spectral matrix estimation, applying the filter $A_k(l)$, and then summing the squares of the magnitudes of results in each segment. This, then, is the justification for the definition of $P_{OUT}(l)$. In addition, we have

$$P_{OUT} = \sum_{l=-((N-3)/2)}^{(N-1)/2} P_{OUT}(l) \tag{124}$$

which is a desirable property for power estimates. It is interesting to compare (121) and (122) with (84), as this provides further justification for the computations.

VIII. Description of Computer Program for Frequency-Domain Synthesis Procedure

A FORTRAN IV computer program for the 7094 implements the synthesis of the maximum-likelihood filters in the frequency domain according to the algorithm given previously. The outputs of this program are the above filters and SC-4020 plots of the processed traces. The program works with data duplicated from LASA tapes on a FORTRAN compatible tape. The parameters used in the program are as follows.

NSKP is the number of data samples skipped initially.

WARMUP is the number of data samples used to determine average dc levels for each seismometer. The interval over which the average dc is estimated immediately precedes the fitting interval, in order to eliminate a tape rewind.

NSAMB is the number of sample blocks $= M$.

NIM: Every NIMth point is used as data, in the fitting interval; the others are skipped. Data samples are skipped to save on computation time.

NS is the number of seismometers to be processed and must be less than or equal to 25.

NFP is the number of filter points and is an odd integer which must be less than or equal to 33.

XC(I), YC(I) are the rectangular coordinates of the Ith seismometer.

AZ = azimuth of event.

HVEL = horizontal velocity of event.

The output consists of SC-4020 hard-copy plots of the center seismometer, WDS, DS, FS, and measured variances of FS, WDS, and average power of all seismometers taken over blocks of 100 consecutive data points. FS denotes filter and sum and is the designation for the maximum-likelihood filter which employs NFP filter points, WDS denotes weighted delay and sum and designates a one-point maximum-likelihood filter (i.e., NFP = 1), and DS denotes the sum of the delayed data.

For WARMUP = 600, NSAMB = 90, NIM = 2, NFP = 21, NS = 25, six minutes of data processed, the 7094 computer running time is about 10 minutes, consisting of 3.5 minutes to measure the spectral matrix, 0.5 minute to synthesize the filter, and six minutes to obtain all the processed traces.

A detailed description of the flow of the program is as follows. Note that the uniform weighting function was used and that the Cooley-Tukey algorithm was not employed. However, a later version of the program will use this algorithm, making the program running time even smaller. The bordering method was used to invert the spectral matrices.

1) Compute the delays for the NS channels, using the azimuth and horizontal velocity of the event and the channel coordinates, assuming that the event propagates across the subarray as a plane wave.

2) Apply the computed delays to the data, and let the delayed data sample in the jth channel at time point m be

denoted by X_{jm}, to be consistent with previous notation. Let $m=1$ designate the beginning of the warmup period.

3) Compute

$$\overline{X}_j = \frac{1}{L'}\sum_{m=1}^{L'} X_{jm}, \qquad j = 1, \cdots, \text{NS},$$

where $L'=\text{WARMUP}$, which usually consists of 600 data points immediately preceding the fitting interval. No data samples are skipped in this computation.

4) Let $X'_{jm}=X_{jm}-\overline{X}_j$ and compute

$$S_{jn}(l) = (\text{NFP})^{-1/2}\sum_{m=1}^{\text{NFP}} X'_{j,(m)(\text{NIM})+L'+(n-1)(\text{NFP})(\text{NIM})} \cdot e^{iml(2\pi/\text{NFP}-1)}$$

$$l = 0, \cdots, \frac{\text{NFP}-1}{2}, \qquad j = 1, \cdots. \text{NS},$$

$$n = 1, \cdots, M, \qquad M = \text{NSAMB}.$$

5) Compute

$$\hat{f}_{jk}(l) = \frac{1}{M}\sum_{n=1}^{M} S_{jn}(l)S^*_{kn}(l) \qquad l = 0, \cdots, \frac{\text{NFP}-1}{2},$$

$$j = 1, \cdots, \text{NS}, \ k \geq j.$$

Set $\hat{f}_{jk}(l)=\hat{f}^*_{kj}(l)$, $k<j$, $j=1, \cdots, \text{NS}$.

6) Compute the inverse of $\{\hat{f}_{jk}(l)\}$, which we denote

$$\{\hat{q}_{jk}(l)\}, \qquad l = 0, \cdots, \frac{\text{NFP}-1}{2}.$$

7) Compute

$$A_k(l) = \frac{\sum_{j=1}^{\text{NS}} \hat{q}_{kj}(l)}{\sum_{j,k=1}^{\text{NS}} \hat{q}_{kj}(l)}, \qquad l = 0, \cdots, \frac{\text{NFP}-1}{2}, \quad k = 1, \cdots, \text{NS}.$$

Note that, for each l, the denominator need be computed only once for all k.

8) Compute

$$\theta_{jm} = \frac{1}{\text{NFP}-1}\Bigg[A_k(0) + (-1)^m A_k\,\frac{\text{NFP}-1}{2} + 2\sum_{l=1}^{(\text{NFP}-3)/2} \text{Re}\,\{A_k(l)\}\cos\left(ml\frac{2\pi}{\text{NFP}-1}\right) + \text{Im}\,\{A_k(l)\}\sin\left(ml\frac{2\pi}{\text{NFP}-1}\right)\Bigg],$$

$$j = 1, \cdots, \text{NS}$$

$$m = 0, \pm 1, \cdots, \pm\left(\frac{\text{NFP}-1}{2}\right).$$

9) Compute the zero-lag correlation coefficients

$$\hat{\rho}_{jk}(0) = (\text{NFP}-1)^{-1}\sum_{l=-((\text{NFP}-3)/2)}^{(\text{NFP}-1)/2} \hat{f}_{jk}(l), \qquad j = 1, \cdots, \text{NS}, \quad k \geq j.$$

Set $\hat{\rho}_{jk}(0)=\hat{\rho}_{kj}(0)$, $k<j$, $j=1, \cdots, \text{NS}$. Note that $\rho_{jk}(0)$ is numerically equal to the estimate obtained by correlating data directly, namely

$$\hat{\rho}_{jk}(0) = \frac{1}{L}\sum_{m=1}^{L} X'_{j,m(\text{NIM})+L}X'_{k,m(\text{NIM})+L}.$$

10) The filter coefficients for WDS, θ_k, $k=1, \cdots, \text{NS}$, are obtained by inverting the zero-lag correlation matrix bordered by zeroes and ones, just as in time-domain method. A Lagrangian multiplier is also obtained in this computation which is numerically the same as the output power of WDS in fitting interval, P_{WDS}.

11) Compute

$$P_{\text{IN}}(l) = \frac{1}{\text{NFP}-1}\,\frac{1}{\text{NS}}\sum_{j=1}^{\text{NS}} \hat{f}_{jj}(l), \qquad l = 0, \cdots, \frac{\text{NFP}-1}{2}$$

and

$$P_{\text{IN}} = P_{\text{IN}}(0) + P_{\text{IN}}\left(\frac{\text{NFP}-1}{2}\right) + 2\sum_{l=1}^{(\text{NFP}-3)/2} P_{\text{IN}}(l).$$

12) Compute

$$P_{\text{OUT}}(l) = \frac{\text{NFP}}{(\text{NFP}-1)^2}\left[\sum_{j,k=1}^{\text{NS}} \hat{q}_{jk}(l)\right]^{-1}, \ l = 0, \cdots, \frac{\text{NFP}-1}{2}$$

and

$$P_{\text{OUT}} = P_{\text{OUT}}(0) + P_{\text{OUT}}\left(\frac{\text{NFP}-1}{2}\right) + 2\sum_{l=1}^{(\text{NFP}-3)/2} P_{\text{OUT}}(l).$$

13) Compute signal-to-noise ratio gain of FS at lth frequency

$$G_{\text{FS}}(l) = 10\log_{10}\frac{P_{\text{IN}}(l)}{P_{\text{OUT}}(l)}$$

and total gain

$$G_{\text{FS}} = 10\log_{10}\frac{P_{\text{IN}}}{P_{\text{OUT}}}.$$

14) Compute signal-to-noise ratio gain of WDS

$$G_{\text{WDS}} = 10\log_{10}\frac{P_{\text{IN}}}{P_{\text{WDS}}}.$$

15) Compute the output power of DS

$$P_{\text{DS}} = (\text{NS})^{-2}\sum_{j,k=1}^{\text{NS}} \hat{\rho}_{jk}(0)$$

and signal-to-noise ratio gain for DS

$$G_{\text{DS}} = 10\log_{10}\frac{P_{\text{IN}}}{P_{\text{DS}}}.$$

16) Apply the filter coefficients to obtain the FS trace

$$\hat{S}_{\text{NFP},n} = \sum_{m=-((\text{NFP}-1)/2)}^{(\text{NFP}-1)/2}\ \sum_{j=1}^{\text{NS}} \theta_{jm}X'_{j,m(\text{NIM})+n}, \qquad n = 1, 2, 3, \cdots.$$

17) The DS and WDS traces are obtained in a manner similar to the computation in (16), with the appropriate weights replacing the θ_{jm}'s. For DS the weights are

$\theta_j = 1/\text{NS}, j = 1, \cdots, \text{NS}$, and for WDS the θ_j's are as computed in step 10).

IX. Comparison of Time-Domain and Frequency-Domain Synthesis Procedures

We will now present some of the experimental results obtained using the time-domain and frequency-domain synthesis computer programs to synthesize two-sided filters. The original data were obtained from short-period vertical seismometers at LASA which were sampled every 1/20 second. The data were quantized by using 14 bits, one bit for sign, and single-precision accuracy was used in all of the computations. Every other sample was used in the fitting interval, i.e., NIM = 2, so that the sampling rate was 10 Hz corresponding to an aliasing frequency of 5 Hz. Although the filters were designed as if the sampling interval were 1/10 second, the filters were applied to the data with the sampling interval of 1/10 second, but in steps of 1/20 second. The choice of NIM = 2 was made because it was found that there was very little noise energy above 5 Hz. A larger value for NIM tends to lead to serious aliasing due to frequency foldover. A value of NFP = 21, leading to a filter length of two seconds, was chosen as a compromise among several factors. The filter length should be made long enough to provide effective filtering. However, the signals in the sensors within the various subarrays do not tend to be identical, after the time-delay correction, for more than about two seconds. In addition, the computer time requirements become excessive if NFP is too large. Thus, the optimum compromise among all these factors was felt to be NFP = 21. The length of the fitting interval was set at three minutes as a compromise between stability of the estimates of noise characteristics and computation time.

In Figs. 2, 4, 6, 8, 10 are shown the results of the frequency-domain processing. The top trace is the center seismometer of the 25 sensor subarray, the next trace is WDS, the next is DS, and the bottom trace represents FS, all defined previously. The corresponding results for the time-domain synthesis are shown in Figs. 3, 5, 7, 9, 11. It should be noted that only the FS trace changes between these two sets of figures. Similar results for the frequency-domain synthesis and for a different event are given in Figs. 12–16. The manner in which the signal-to-noise ratio gain drops outside the fitting interval for the time and frequency-domain methods is given in Fig. 17, for various NFP and fitting interval lengths. These results indicate that the time-domain method tends to be more sensitive to the assumption of noise stationarity than the frequency-domain method. The fitting interval estimates for P_{OUT} obtained in the frequency-domain program were compared with measured values and were found to be in agreement within ± 1 dB. A plot of the frequency window $|W_N(x)|^2$, and $-10 \log_{10} |W_N(x)/W_N(0)|^2$, $N = 21$, used in the spectral estimation program, are shown in Figs. 18 and 19, respectively.

Unfortunately, the 0.5 Hz resolution obtained with this frequency window is not adequate to provide accurate estimates of the input noise spectra, due to a very large and narrow spectral peak located at about 0.2 Hz. This narrow peak is difficult to resolve with only 0.5 Hz resolution, and also causes windowing to occur at other frequencies. A comparison of this coarse resolution estimate with fine resolution estimates, using 0.1 Hz resolution, shows that the estimates can be in error by as much as ± 5 dB due to the lack of resolution. However, in spite of this, the resolution appears to be adequate to obtain effective filtering. The spectral estimates of the processed trace, FS, are much more accurate, due to the greater uniformity of the spectrum of the output trace. Comparison of the coarse resolution with fine resolution estimates in this case indicate errors of only about ± 1 dB. This is quite good agreement, considering that the spectral estimate is obtained from computed inverses of the input spectral matrices and not by operating on the output trace. However, the serious errors introduced into the input spectral estimates cause large errors in the curves for signal-to-noise ratio gain vs. frequency, as indicated in Fig. 20. This figure also shows that the signal-to-noise ratio gain vs. frequency for the frequency-domain filter is approximately the same outside as inside the fitting interval, for periods of at least five minutes after the end of the fitting interval.

The results in Figs. 12 through 16 for the Central Kazakh event of February 1, 1966, are rather interesting since the epicenter for this event is located about 900 miles to the southwest of a presumed nuclear test site. The raw data traces do not show clearly the PcP and pP phases, as can be seen from the top trace in these figures. However, the FS traces do show these phases quite clearly. The arrival of pP at about seven seconds after the initial P phase establishes the focal depth of the event, according to seismological travel time tables, as approximately 33 km. This in turn establishes the event as being definitely an earthquake. This conclusion was also reached by the U. S. Coast and Geodetic Survey on the basis of recordings from a worldwide network of stations.

The experimental results show that the frequency-domain filter tends to produce a much smaller precursor, in general, than the time-domain filter. This can be seen by comparing the results on the Rat Island event of November 11, 1965. In addition, the signal-to-noise ratio gain of the frequency-domain filter tends to remain about the same outside the fitting interval as it is inside the fitting interval, for periods of up to 14 minutes after the fitting interval. This is not true of the time-domain filter, which for NFP = 21 and three minute fitting intervals tends to drop about 4 dB just outside the fitting interval. However, the gain of the time-domain filter is still about 2 dB better than that of the frequency-domain filter outside the fitting interval, for NFP = 21 and three minute fitting interval. This loss is quite acceptable when it is recognized that the 7094 computer running time to measure the noise characteristics and to synthesize the filter is only about four minutes for the frequency-domain method, as opposed to 30 minutes for the time-domain technique.

It is also interesting to consider the case of maximum-likelihood processing of traces which have been bandpass

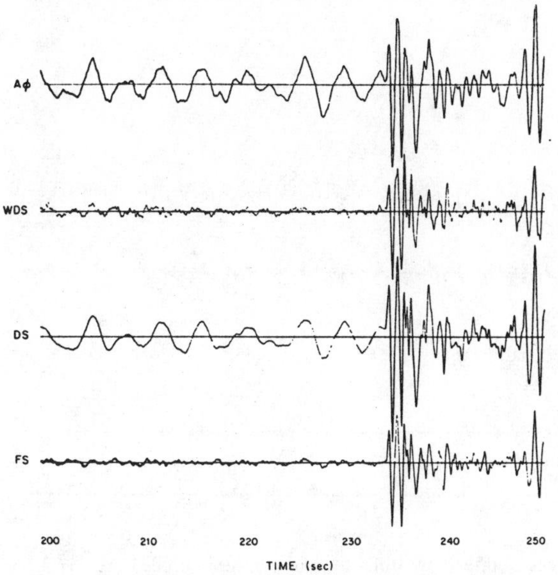

Fig. 2. Processed traces obtained by frequency-domain synthesis procedure, 11/11/65 Rat Island event, subarray B1.

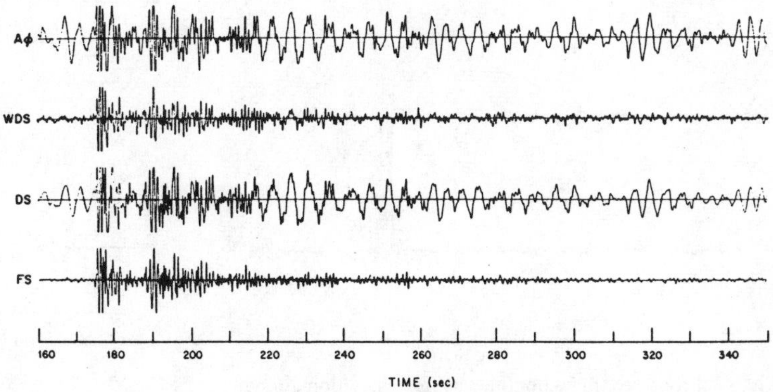

Fig. 3. Processed traces obtained by time-domain synthesis procedure, 11/11/65 Rat Island event, subarray B1.

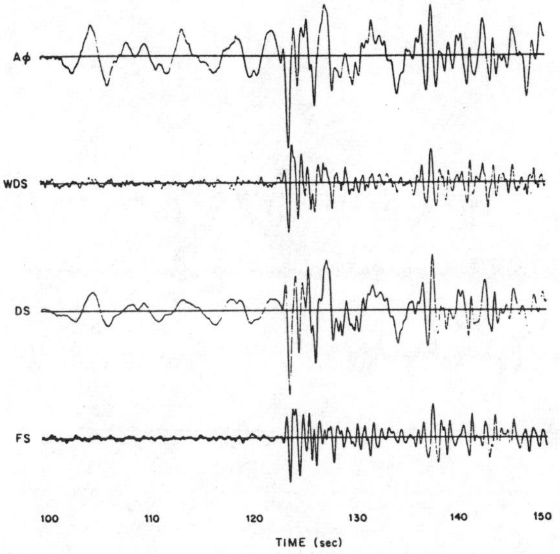

Fig. 4. Processed traces obtained by frequency-domain synthesis procedure, 11/11/65 Rat Island event, subarray Aφ.

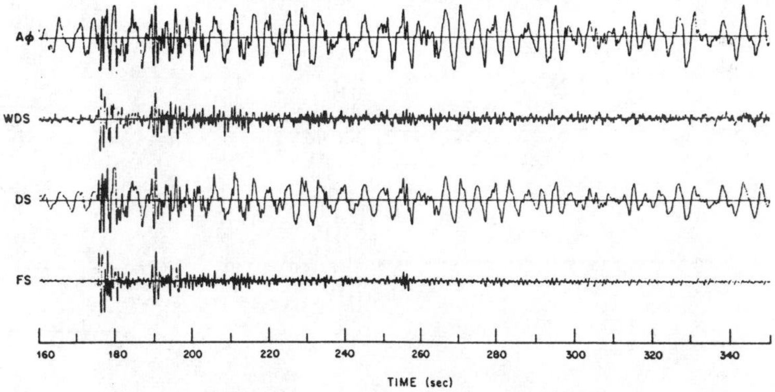

Fig. 5. Processed traces obtained by time-domain synthesis procedure, 11/11/65 Rat Island event, subarray Aφ.

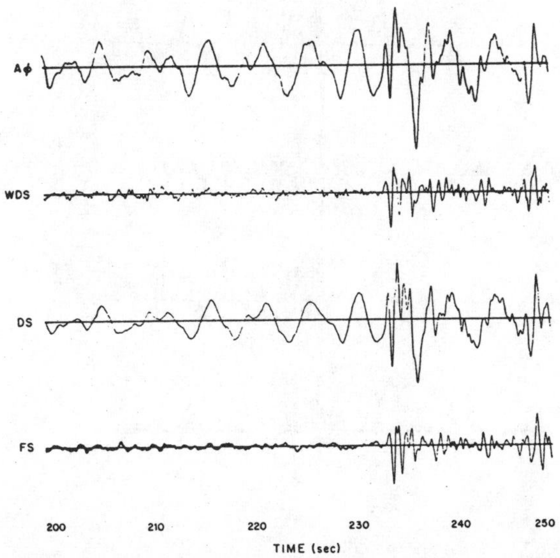

Fig. 6. Processed traces obtained by frequency-domain synthesis procedure, 11/11/65 Rat Island event, subarray B3.

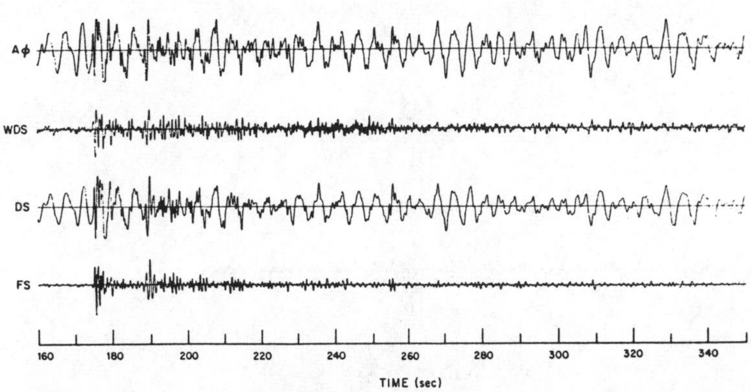

Fig. 7. Processed traces obtained by time-domain synthesis procedure, 11/11/65 Rat Island event, subarray B3.

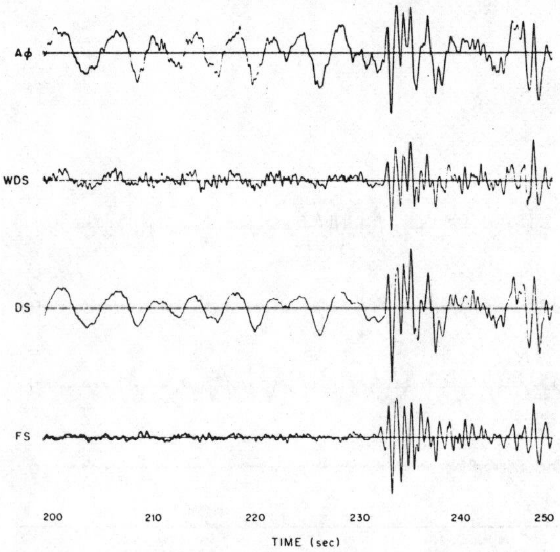

Fig. 8. Processed traces obtained by frequency-domain synthesis procedure, 11/11/65 Rat Island event, subarray B4.

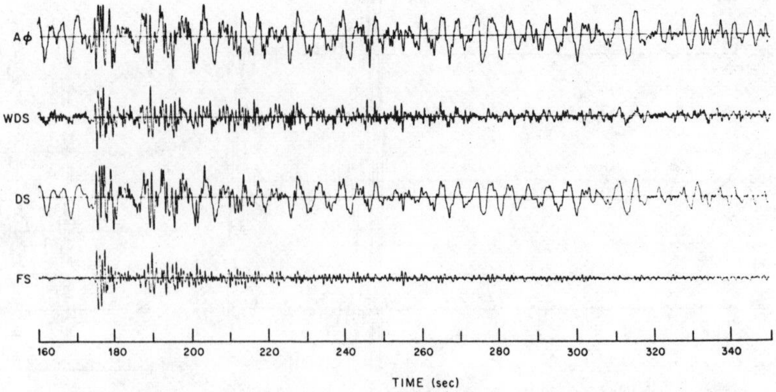

Fig. 9. Processed traces obtained by time-domain synthesis procedure, 11/11/65 Rat Island event, subarray B4.

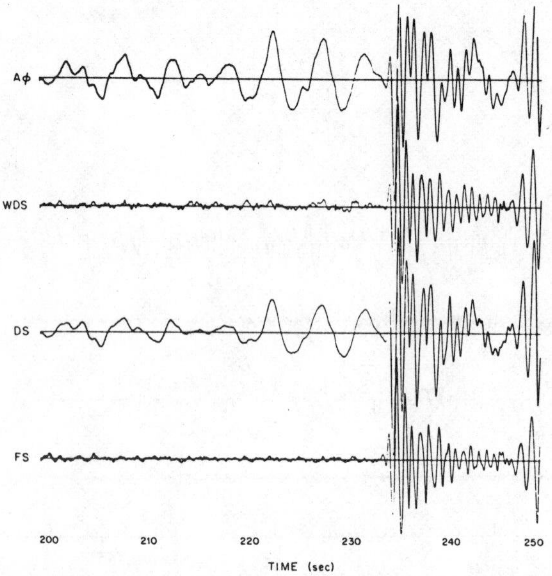

Fig. 10. Processed traces obtained by frequency-domain synthesis procedure, 11/11/65 Rat Island event, subarray B2.

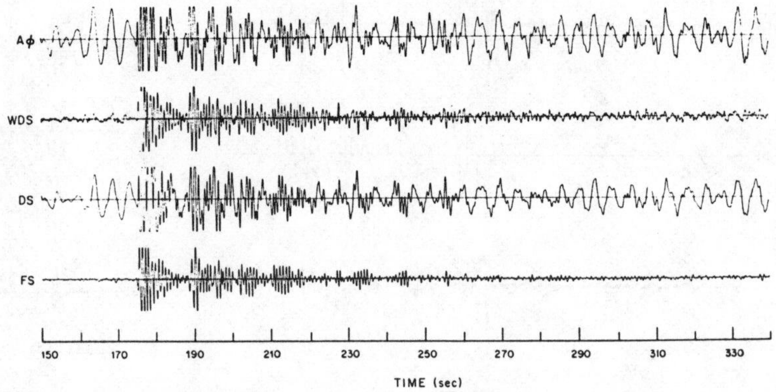

Fig. 11. Processed traces obtained by time-domain synthesis procedure, 11/11/65 Rat Island event, subarray B2.

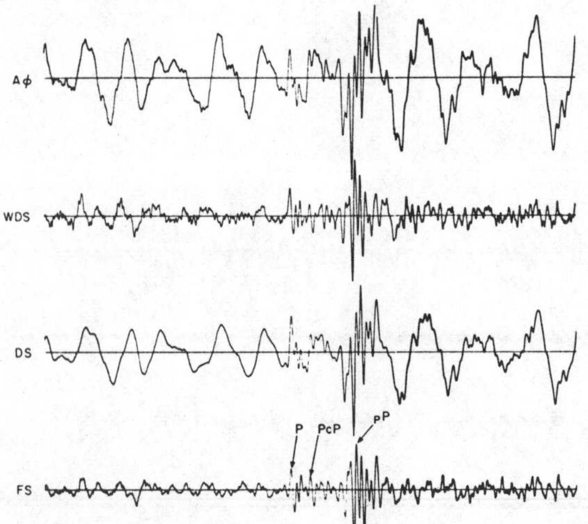

Fig. 12. Processed traces obtained by frequency-domain synthesis procedure, 2/1/66 Central Kazakh event, subarray B1.

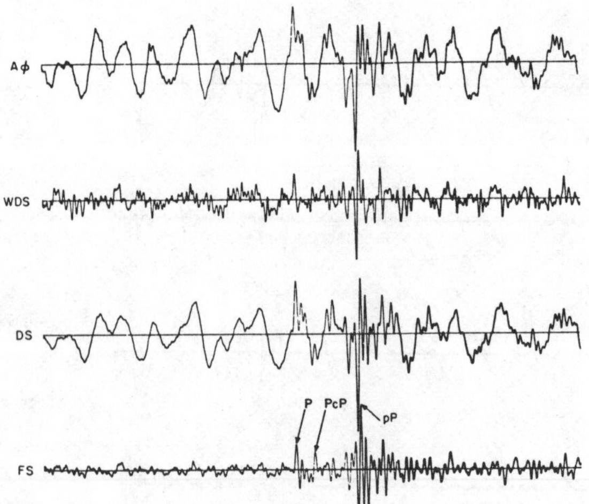

Fig. 13. Processed traces obtained by frequency-domain synthesis procedure, 2/1/66 Central Kazakh event, subarray Aϕ.

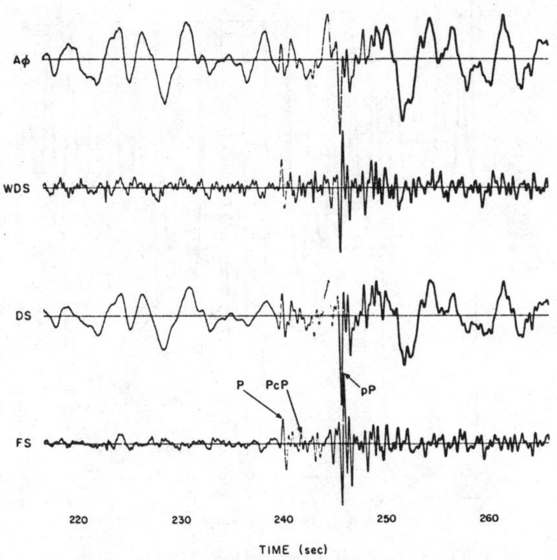

Fig. 14. Processed traces obtained by frequency-domain synthesis procedure, 2/1/66 Central Kazakh event, subarray B3.

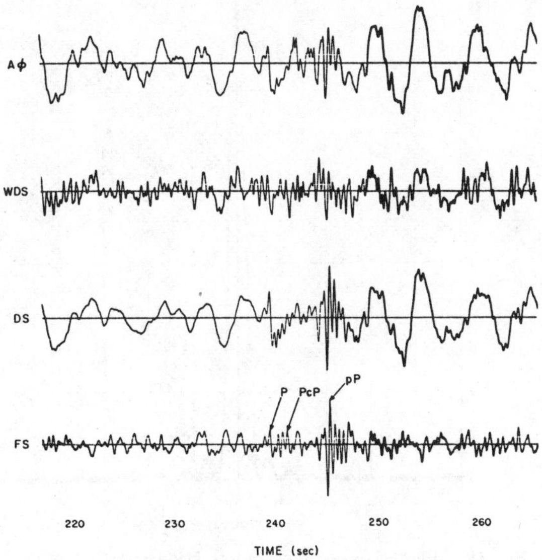

Fig. 15. Processed traces obtained by frequency-domain synthesis procedure, 2/1/66 Central Kazakh event, subarray B4.

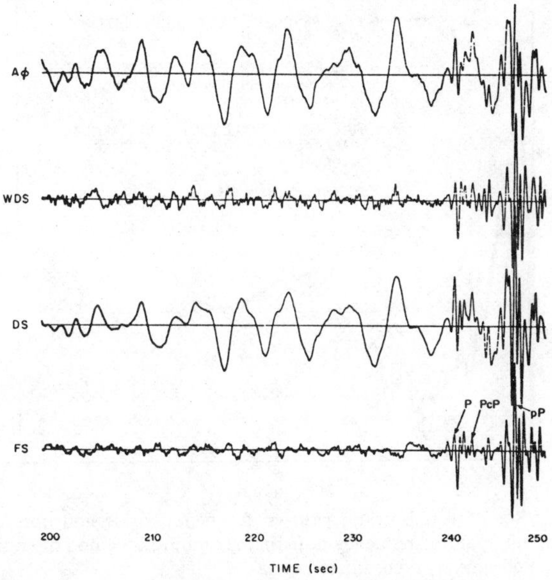

Fig. 16. Processed traces obtained by frequency-domain synthesis procedure, 2/1/66 Central Kazakh event, subarray B2.

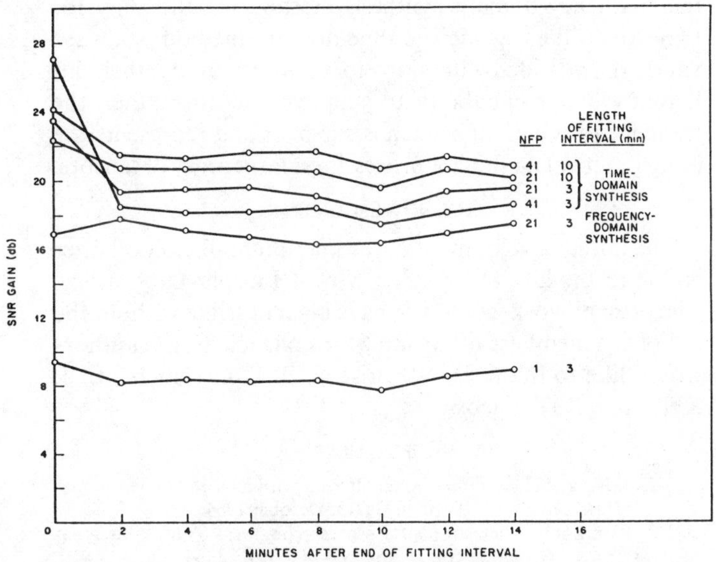

Fig. 17. Signal-to-noise ratio gain of maximum-likelihood filter outside the fitting interval for various NFP and fitting interval lengths, noise from February 4, 1966, subarray Aφ.

prefiltered, say 0.6 to 2.0 Hz, so as to allow only a small amount of signal distortion while effectively suppressing the noise outside of the signal frequency band. In this case, within each subarray, both the time- and frequency-domain methods provide an average signal-to-noise ratio gain of about 12 dB, outside the fitting interval, while DS achieves about 6 dB. An extensive discussion of such results has been given recently [19].

X. Conclusions

A detailed discussion of the theory for the maximum-likelihood multidimensional filter has been presented, as well as a description of a computer program implementation for the filter in the time domain. This design procedure is subject to several severe criticisms, of which the most serious is that the computer time requirements are exorbitant. However, the time-domain program does have the advantage of being capable of designing realizable as well as two-sided filters.

It has been shown that, basically, the reason for the equivalence of the maximum-likelihood and minimum-variance unbiased estimators is that they are based on a conditional expectation. The martingale property of conditional expectation then assures that the asymptotic properties of these estimators are well defined. Synthesis procedures for physically realizable filters, in the asymptotic case, have also been given. These procedures were based on a spectral matrix factorization technique and are not well adapted to machine computation.

An asymptotically optimum frequency-domain synthesis

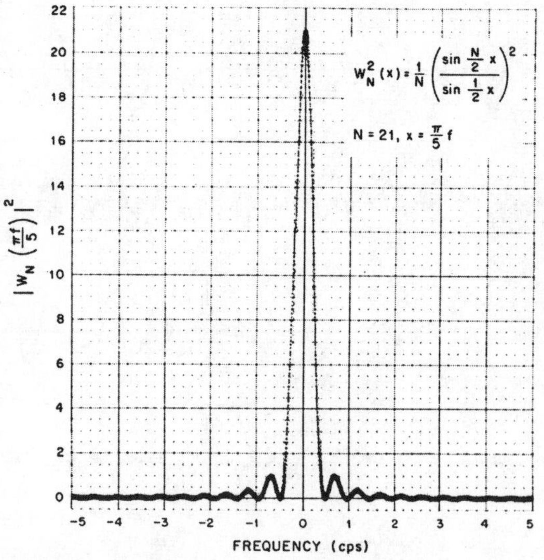

Fig. 18. Plot of frequency window $[W_N(\pi f/5)]^2$, $N=21$.

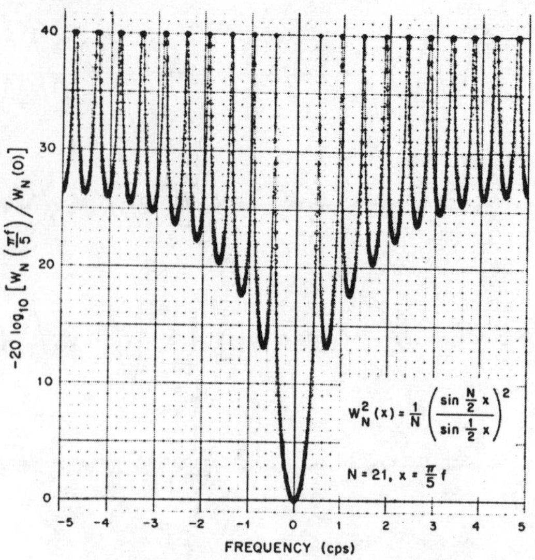

Fig. 19. Plot of frequency window $-10 \log_{10} [W_N(\pi f/5)/W_N(0)]^2$, $N=21$.

procedure has been presented, as well as a computer program implementation, which requires much less computing time than the time-domain synthesis technique. However, the frequency-domain synthesis method is valid only for two-sided filters, while the time-domain method is always valid. If two-sided filters are to be used, due to their inherently better capability to suppress the noise, then the frequency-domain approach is superior and represents the more practical method of processing a large array of sensors.

Acknowledgment

The initial development of the maximum-likelihood filter is due to the late Dr. M. J. Levin of Lincoln Laboratory. The present work would not have been possible without the aid of the members of Group 64. In particular, the authors would like to thank H. W. Briscoe, P. E. Green, Jr., E. J. Kelly, and R. T. Lacoss.

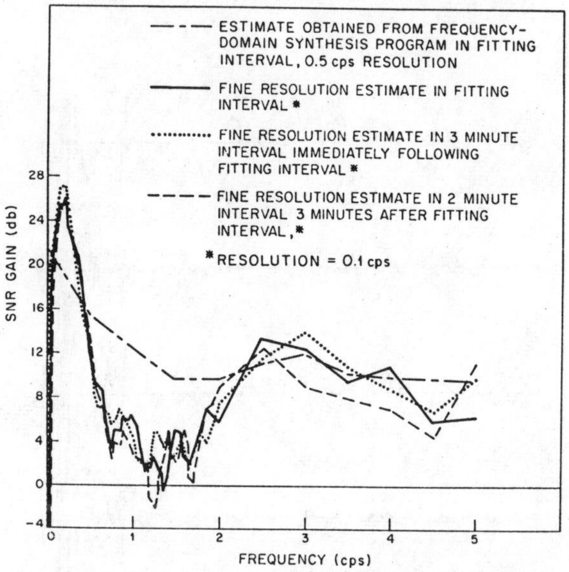

Fig. 20. Signal-to-noise ratio gain vs. frequency inside and outside of fitting interval for frequency-domain maximum-likelihood filter, noise from February 4, 1966, subarray Aϕ.

References

[1] J. P. Burg, "Three-dimensional filtering with an array of seismometers," *Geophys.*, vol. 29, pp. 693–713, October 1964.

[2] J. Claerbout, "Detection of *P*-waves from weak sources at great distances," *Geophys.*, vol. 29, pp. 197–211, April 1964.

[3] E. J. Kelly, Jr., and M. J. Levin, "Signal parameter estimation for seismometer arrays," Mass. Inst. Tech. Lincoln Lab., Lexington, Mass., Tech. Rept. 339, January 8, 1964.

[4] P. E. Green, Jr., R. A. Frosch, and C. F. Romney, "Principles of an experimental Large Aperture Seismic Array (LASA)," *Proc. IEEE*, vol. 53, pp. 1821–1833, December 1965.

[5] M. J. Levin, "Maximum-likelihood array processing," Mass. Inst. Tech. Lincoln Lab., Lexington, Mass., semiannual Tech. Summary Rept. on seismic discrimination, December 31, 1964.

[6] E. A. Robinson and R. A. Wiggins, "Recursive solution to the multichannel filtering problem," *J. Geophys. Res.*, vol. 70, pp. 1885–1891, April 15, 1965.

[7] Y. A. Rozanov, "A problem of optimal control of a complex of instruments" (in Russian), *Tr. Inst. Math. imeni V. A. Steklova, Akad. Nauk SSSR*, vol. LXXI, pp. 88–101, 1964. English translation available from Lincoln Laboratory Library, Lexington, Mass.

[8] J. L. Doob, *Stochastic Processes*. New York: Wiley, 1953.

[9] P. Whittle, *Prediction and Regulation by Linear Least-Squares Methods*. Princeton, N. J.: Van Nostrand, 1963.

[10] N. Wiener and P. Masani, "Prediction theory of multivariate stochastic processes, Part I. The regularity condition," *Acta Math.*, vol. 93, pp. 111–150, November 1957; "Part II. The linear predictor," *Acta Math.*, vol. 99, pp. 93–137, April 1958.

[11] U. Grenander and M. Rosenblatt, *Statistical Analysis of Stationary Time Series*. New York: Wiley, 1957.

[12] R. B. Blackman and J. W. Tukey, *The Measurement of Power Spectra from the Point of View of Communications Engineering*. New York: Dover, 1959.

[13] R. B. Blackman, *Data Smoothing and Prediction*. Reading, Mass.: Addison-Wesley, 1965.

[14] N. R. Goodman, "On the joint estimation of the spectrum, cospectrum and quadrature spectrum of a two-dimensional stationary Gaussian process," Engrg. Statistics Lab., New York University, New York, N. Y., Scientific Paper 10, 1957.

[15] M. Rosenblatt, "Statistical analysis of stochastic processes with stationary residuals," in *Probability and Statistics (The Harald Cramér Volume)*, U. Grenander, Ed. New York: Wiley, 1959.

[16] R. H. Jones, "A reappraisal of the periodogram in spectral analysis," *Technometrics*, vol. 7, pp. 531–542, November 1965.

[17] N. R. Goodman, "Statistical analysis based on a certain multivariate complex Gaussian distribution," *Annals Math. Stat.*, vol. 34, pp. 152–177, March 1963.

[18] J. W. Cooley and J. W. Tukey, "An algorithm for the machine calculation of complex Fourier series," *Math. of Comp.*, vol. 19, pp. 297–301, April 1965.

[19] J. Capon, R. J. Greenfield, and R. T. Lacoss, "Off-line signal processing results for the large aperture seismic array," Mass. Inst. Tech. Lincoln Lab., Lexington, Mass., Tech. Note 1966-37, July 11, 1966.

Spectral Approach to Geophysical Inversion by Lorentz, Fourier, and Radon Transforms

ENDERS A. ROBINSON

Invited Paper

Abstract–Geophysical inversion seeks to determine the structure of the interior of the earth from data obtained at the surface. In reflection seismology, the problem is to find inverse methods that give structure, composition, and source parameters by processing the received seismograms. The pioneering work of Jack Cohen and Norman Bleistein on general inverse methods has caused a revolution in the direction of research on long-standing unsolved geophysical problems. This paper does not deal with such general methods, but instead gives a survey of some production-type data processing methods in everyday use in geophysical exploration. The unifying theme is the spectral approach which provides methods for the approximate solution of some simplified inverse problems of practical importance.

This paper is divided into two parts, one dealing with one-dimensional (1-D) inversion, the other with two-dimensional (2-D) inversion. The 1-D case treated is that of a horizontally layered earth (Goupillaud model) with seismic raypaths only in the vertical direction. This model exhibits a lattice structure which corresponds to the lattice methods of spectral estimation. It is shown that the lattice structure is mathematically equivalent to the structure of the Lorentz transformation of the special theory of relativity. The solution of this 1-D inverse problem is the discrete counterpart of the Gelfand–Levitan inversion method in physics. A practical computational scheme to carry out the inversion process is the method of dynamic deconvolution. It is based on a generalization of the Levinson recursion, and involves the interacting recursions of two polynomials P and Q. This paper treats only much simplified 2-D models. One 2-D method gives the forward and inverse solution for a horizontally layered earth (Goupillaud model) with slanting seismic raypaths. This method involves the Radon transform which is often called "slant stacking" by geophysicists. The other 2-D methods given in this paper are concerned with the process of wavefield reconstruction and imaging known as "migration" in the geophysical industry. A major breakthrough occurred in 1978 when Stolt introduced spectral migration which makes the use of the fast Fourier transform. Another method, the slant-stack migration of Hubral, is based on the Radon transform.

Manuscript received February 17, 1982; revised May 6, 1982.

The author was with the Departments of Geological Sciences and Theoretical and Applied Mechanics, Cornell University, Ithaca, NY 14853. He is now at 100 Autumn Lane, Lincoln, MA 01773.

I. Introduction

THE DIRECT problem in geophysics may be thought of as the determination of how seismic waves propagate on the basis of a known makeup of the subsurface of the earth. The inverse problem is to determine the subsurface makeup on the basis of wave motion observed at the surface of the earth. The inverse problem is not unique to geophysics. A large part of our physical contact with our surroundings depends upon an intuitive solution of inverse problems. In many other real-world problems, we must also infer the size, shape, and texture of remote objects from the way they trans-

Reprinted from *Proc. IEEE,* vol. 70, pp. 1039–1054, Sept. 1982.

mit, reflect, and scatter traveling waves. For example, in X-ray computerized tomography [1], it is necessary to combine X-ray scans taken at different angles to form a cross-sectional image which represents the internal details of the scanned structure. In nondestructive testing [2], a reconstruction of three-dimensional (3-D) refractive-index field is made from holographic measurements taken at different angles. In electron microscopy [3], 3-D biological structures are deduced from 2-D electron micrographs taken at different tilt angles. Optimization techniques are being developed for digital image reconstruction from various types of projections [4].

In 1877, Lord Rayleigh [5] treated a physical inverse problem; he was one of the first scientists to do so. Rayleigh considered the problem of finding the density distribution of a string from knowledge of the vibrations. Kac [6] aptly described this inverse problem in the title of his well-known lecture "Can one hear the shape of a drum?" Because seismic waves are sound waves in the earth, we can describe our inverse problem (i.e., seismic exploration) as "Can we hear the shape of an oil field?"

With the introduction of the Schrödinger equation to describe the microphysical world, the scope of inverse problems in physics became enormously enlarged. Over the years, many general inversion procedures were formulated. One of the most elegant approaches was given by Gelfand and Levitan [7] on the solution of a differential equation by its spectral function. This solution leads to an integral equation, the Gelfand–Levitan equation. We will treat the discrete form of the Gelfand–Levitan integral equation in this paper, as well as give a computational scheme, known in geophysics as dynamic deconvolution, for its solution. The word "dynamic" is used to differentiate this deconvolution process from the usual time-invariant deconvolution methods.

The spectral methods used in geophysics are closely connected with the inverse problem [8]. Basically, the spectral approach provides methods for the approximate solution of many simplified inverse problems. We will be concerned in this paper with such specialized solutions and not with the general case. In this sense, this paper deals with the production-type data processing methods in everyday use in geohysical exploration, which serve not perfectly but well.

Since the energy crisis, a greatly increased research effort has been devoted to problems in the geophysical exploration for oil and natural gas. General inverse scattering methods from the mainstream of physics and applied mathematics have been introduced to the geophysical exploration industry. The pioneering work of Jack Cohen and Norman Bleistein has caused a revolution in the direction of research on the long-standing unsolved problems in the seismic exploration for oil. We do not treat these more powerful methods in this paper, but refer the reader to the recent work of Silvia [9], Silvia and Weglein [10], Cohen and Bleistein [11], [12], Mendel and Habibi-Ashrafi [13], Coen [14], Carroll and Santosa [15], Driessel and Symes [16], Stolt [17], Bube and Burridge [18], Zemanian and Subramaniam [19], Gazdag [20], Bamberger, Chavent, and Lailly [21], and Treitel, Gutowski, Hubral, and Wagner [22]. Larner [23] gives a balanced understanding of the entire field of exploration geophysics, from the most difficult physical and mathematical methods to the practical field implementation.

Because we are at a turning point in the methods used in seismic exploration, it is well to summarize where we stand today. Some of the methods presently in use will be replaced by the newer, more powerful methods now being developed, and others may survive, albeit in some modified form.

We divide this paper into two parts, 1-D and 2-D. The 1-D problems are concerned with a laterally homogeneous earth, so the only variation is in the depth dimension z. The basic method which we treat here is the Gelfand–Levitan solution of the inverse problem and the related dynamical-deconvolution scheme. These techniques are closely related to lattice approaches to spectral estimation [24], which use the Levinson recursion [25] and the Burg method [26]. Briefly, dynamic deconvolution is based on the interacting recursions of P and Q polynomials. These polynomials are related to the polynomial A appearing in the Levinson recursion by means of the equation $A_k = P_k - c_0 Q_k$, where $|c_0| = 1$.

We do not give any general solution of the 2-D problem, but instead treat special simplified cases which have proved useful. The 3-D problem is not treated at all because the simplified 2-D methods which we present can readily be extended to the 3-D case. Of course, in such an extension, many facets which depend upon dimension must be respected, but the basic ideas are the same. Today the exploration industry is using 3-D data (lateral dimensions x and y, and depth dimension z pointing downward) more and more, and eventually all exploration will be done with 3-D models. It is here that the general inverse approaches introduced by the pathfinding work of Cohen and Bleistein will reach their greatest effectiveness.

The specialized 2-D methods which we present make use of spectral analysis to solve a much simplified problem, known in the industry as migration. The two spectral techniques used are the 2-D Fourier transform and its related Radon transformation. The Fourier migration method is due to Stolt [27], and the Radon migration method is due to Hubral [28]. The Stolt method, which takes advantage of the efficiency of the fast Fourier transform, was a major breakthrough in seismic migration, and it has led to many other developments.

In brief, the limelight in geophysical exploration is on inverse methods. However, when one looks closer, one sees that underlying these methods are the associated spectral methods. Basically, the spectral approach can be used in conjunction with inversion theory to obtain solutions which give us both physical insight and practical exploration tools. The spectral approach, as exemplified by the 1-D lattice (i.e., layered-earth) methods and the 2-D Fourier and Radon migration methods, has provided practical computational schemes for seismic data processing. These spectral methods work and can be used routinely to solve exploration problems successfully in a large number of cases, even though the complete mathematical inverse solution is yet beyond our reach.

II. One-Dimensional Inversion

A. The Goupillaud Layered-Earth Model

The determination of the properties of the earth from waves that have been reflected from the earth is the clasic problem of reflection seismology. As a first step in mathematical analysis, the problem is usually simplified by assuming that the earth's crust is made up of a sequence of sedimentary layers. The well-known *Goupillaud model* [29] (Fig. 1) approximates the heterogeneous earth with a sequence of horizontal layers, each of which is homogeneous, isotropic, and nonabsorptive. This stratified model is subjected to vertically traveling plane compressional waves, and thus it is a normal incidence model. It is assumed that the two-way travel time in each layer is the same and is equal to one time unit. In other words, the one-way

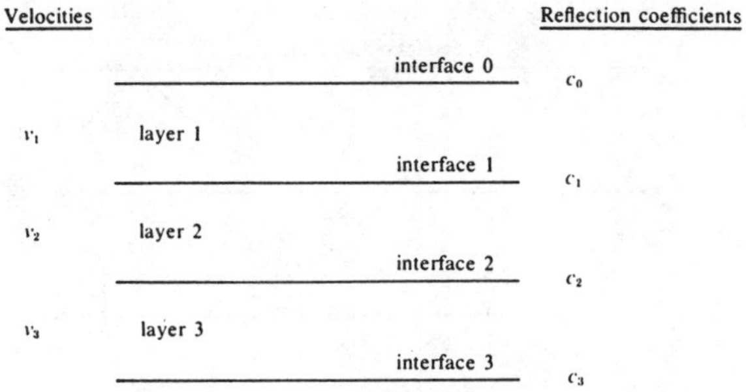

Fig. 1. The Goupillaud model.

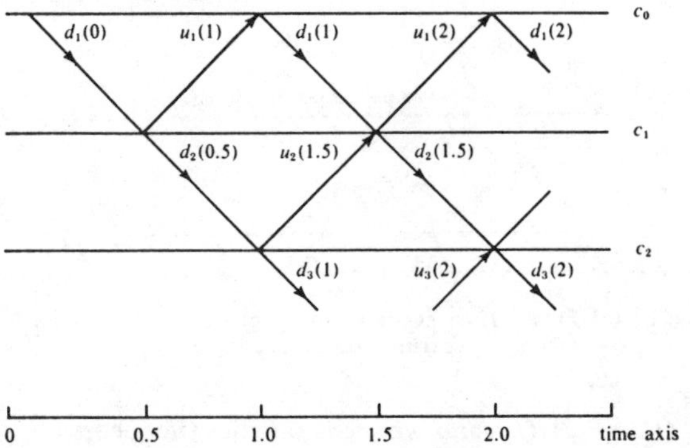

Fig. 2. Space-time lattice diagram.

travel time in each layer is taken to be one-half of the discrete unit of time. The upper half-space (the air) is called half-space 0, the first layer underneath is called layer 1, the next layer underneath is called layer 2, and so on. Interface 0 is the interface at the bottom of half-space 0, interface 1 is the interface at the bottom of layer 1, and so on. Let c_k be the reflection coefficient for downgoing waves striking interface k. The reflection coefficient for upgoing waves striking interface k is thus equal to $-c_k$. We will assume that the amplitudes of our waves are measured in units such that squared amplitude is proportional to energy. Then the transmission coefficient through interface k is equal to $(1 - c_k^2)^{1/2}$ for either upgoing or downgoing waves. All waves are digitized with unit time spacing. Although the waves exist throughout the layers, we will only be concerned with them as measured at the tops of the layers; see Fig. 2. This is done in order to keep the bookkeeping as simple as possible. If the number of the layer is odd, then time is measured at integer values: $n = 0, 1, 2, \cdots$. However, if the layer is even, then time is measured at integer-plus-one-half values: $n + 0.5 = 0.5, 1.5, 2.5, \cdots$. This is because it takes only 0.5 time unit for a wave to traverse a layer. For example, if a downgoing impulse is introduced at time 0 at the top of layer 1, then it arrives at the top of layer 2 at time 0.5, at the top of layer 3 at time 1, at the top of layer 4 at time 1.5, and so on. Because z is used for the depth dimension in geophysics, we will use the Laplace generating function instead of the z-transform. The variable s in the Laplace generating function corresponds to the variable z^{-1} in the z-transform. The downgoing wave at the top of layer k is denoted by $d_k(n)$ if k is odd, and by $d_k(n + 0.5)$ if k is even, where n is an integer. The respective generating functions are

$$D_k(s) = \sum_n d_k(n)s^n \qquad (k \text{ odd})$$

$$D_k(s) = \sum_n d_k(n + 0.5)s^{n+0.5} \qquad (k \text{ even}).$$

The corresponding z transforms are obtained by letting $s = z^{-1}$. Similarly, the upgoing wave $u_k(n)$ for k odd, and $u_k(n + 0.5)$ for k even, is also measured at the top of the layer. Its generating function is $U_k(s)$.

We have thus described the Goupillaud-layered media. Besides its interest in exploration geophysics, the Goupillaud model is of interest in spectral estimation, for it is mathematically the same as a *lattice network*. These networks are useful in building models of many processes which occur in engineering practice, such as the acoustic tube model for digital speech processing, as developed by Gray and Markel [30], and others. As shown by Makhoul [31] and Durrani, Murukutla, and Sharman [32], the lattice model provides methods for adaptive spectral estimation. We now want to discuss some of the properties of this model that have been useful in exploration geophysics.

B. The Lorentz Transformation

When Maxwell derived the electromagnetic wave equation, it soon became known that it is not invariant under the Galilean transformation. However, it is invariant under the Lorentz transformation, and this observation was a key factor in Einstein's development of the special theory of relativity [33]. The *Lorentz transformation* can be written as

$$D_2 = \frac{1}{(1 - c_1^2)^{1/2}} [D_1 - c_1 U_1]$$

$$U_2 = \frac{1}{(1 - c_1^2)^{1/2}} [-cD_1 + U_1]$$

where D_1 and U_1 are, respectively, the time and space coordinates of an event in frame 1, where D_2 and U_2 are, respectively, the time and space coordinates of the event in frame 2, and c_1 (where $|c_1| < 1$) is the velocity (in natural units, such that the velocity of light is unity) between the two frames. The Lorentz transformation is a consequence of the invariance of the interval between events. By direct substitution, it can be shown that the coordinates of two events must satisfy the equation

$$D_2^2 - U_2^2 = D_1^2 - U_1^2$$

on transition from one frame of reference to the other.

We now want to find the relationship between the waves in the Goupillaud model. Instead of the conventional treatment, we will try to put this relationship in a more general setting. We know that the waves in each layer obey their respective wave equation. Let $D_1(s)$ and $U_1(s)$ be, respectively, the generating functions of the downgoing wave and the upgoing wave at the top of layer 1, and let $D_2(s)$ and $U_2(s)$ be the corresponding functions for layer 2. We then say that *wave motion* must be related by the *Lorentz transformation*

$$D_2(s) = \frac{1}{(1 - c_1^2)^{1/2}} [s^{1/2}D_1(s) - c_1 s^{-1/2}U_1(s)]$$

$$U_2(s) = \frac{1}{(1 - c_1^2)^{1/2}} [-c_1 s^{1/2}D_1(s) + s^{-1/2}U_1(s)].$$

The constant c_1 (where $|c_1| < 1$) is the reflection coefficient of the interface between the two layers. This Lorentz transformation is a consequence of the invariance of the net downgoing energy in the layers. By direct substitution, it can be shown that

$$D_2\overline{D}_2 - U_2\overline{U}_2 = D_1\overline{D}_1 - U_1\overline{U}_1$$

(where the bar indicates that s is to be replaced by s^{-1}; i.e., $\overline{D(s)} = D(s^{-1})$). This equation says that the net downgoing energy in each layer is the same. Because there is no absorption, this energy relation is a physical fact implied by the model.

C. Polynomial Recursions

The Lorentz transformation between two adjacent layers can be written in matrix form as

$$\begin{bmatrix} D_{k+1} \\ U_{k+1} \end{bmatrix} = \frac{s^{-1/2}}{t_k} \begin{bmatrix} s & -c_k \\ -c_k s & 1 \end{bmatrix} \begin{bmatrix} D_k \\ U_k \end{bmatrix}$$

where we have used the symbol t_k to denote the transmission coefficient $(1 - c_k^2)^{1/2}$. Robinson [34] defines the polynomials $P_k(s)$ and $Q_k(s)$, and the reverse polynomials (with superscript R for reverse) given by

$$P_k^R(s) = s^k P_k(s^{-1})$$
$$Q_k^R(s) = s^k Q_k(s^{-1}).$$

These polynomials are defined by the equation

$$\begin{bmatrix} P_k^R & Q_k^R \\ Q_k & P_k \end{bmatrix} = \begin{bmatrix} s & -c_k \\ -c_k s & 1 \end{bmatrix} \begin{bmatrix} s & -c_{k-1} \\ -c_{k-1} s & 1 \end{bmatrix} \cdots \begin{bmatrix} s & -c_1 \\ -c_1 s & 1 \end{bmatrix}.$$

By inspection, we can find the first and last coefficients of these polynomials. We have

$$P_k(s) = 1 + \cdots + c_1 c_k s^{k-1}$$
$$Q_k(s) = -c_1 s + \cdots - c_k s^k$$
$$P_k^R(s) = c_1 c_k s + \cdots + s^k$$
$$Q_k^R(s) = -c_k + \cdots - c_1 s^{k-1}.$$

The polynomials for adjacent layers are related by

$$\begin{bmatrix} P_k^R & Q_k^R \\ Q_k & P_k \end{bmatrix} = \begin{bmatrix} s & -c_k \\ -c_k s & 1 \end{bmatrix} \begin{bmatrix} P_{k-1}^R & Q_{k-1}^R \\ Q_{k-1} & P_{k-1} \end{bmatrix}.$$

This equation gives the *Robinson recursion* [34]

$$P_k = P_{k-1} - c_k s\, Q_{k-1}^R$$
$$Q_k = Q_{k-1} - c_k s\, P_{k-1}^R$$

and its inverse recursion

$$P_{k-1} = \frac{1}{1 - c_k^2}(P_k + c_k Q_k^R)$$

$$Q_{k-1} = \frac{1}{1 - c_k^2}(Q_k + c_k P_k^R).$$

Let us now subtract the two recursion equations to obtain

$$(P_k - Q_k) = (P_{k-1} - Q_{k-1}) - c_k s(Q_{k-1}^R - P_{k-1}^R).$$

Let c_0 be the reflection coefficient of interface 0, where $|c_0| = 1$. To be definite, let $c_0 = 1$. If we define the polynomial A_k as $A_k = P_k - c_0 Q_k = P_k - Q_k$, we obtain the *Levinson recursion* [25]

$$A_k = A_{k-1} + c_k s\, A_{k-1}^R.$$

The *polynomial of the second kind* is defined as $B_k = P_k + c_0 Q_k = P_k + Q_k$, and we see it satisfies the recursion

$$B_k = B_{k-1} - c_k s\, B_{k-1}^R.$$

The inverse of the Levinson recursion is

$$A_{k-1} = \frac{1}{1 - c_k^2}(A_k - c_k A_k^R).$$

We will now make use of the two recursions given in this section for the analysis of layered-earth (lattice) models; the A recursion for the free-surface case and the P, Q recursion for the non-free-surface case.

D. Free-Surface and Non-Free-Surface Reflection Seismograms

We now want to consider an idealized seismic experiment. The source is a downgoing unit impulse introduced at the top of layer 1 at time instant 0. This pulse proceeds downward where it undergoes multiple reflections and refractions within the layered system. Some of the energy is returned to the top of layer 1, where it is recorded in the form of a seismic trace, which we denote by the sequence $r_1, r_2, r_3, \cdots$, where the subscript indicates the discrete time index; see Fig. 3.

There are two types of boundary conditions commonly imposed on the top interface (interface 0 with reflection coefficient c_0). One is the free-surface condition, which says that interface 0 (the air-earth interface) is a perfect reflector; that is, the free-surface condition is that $|c_0| = 1$. The free-surface condition approximately holds in the case of a marine seismogram taken in a very smooth, calm sea, so the surface of the water (interface 0) is virtually a perfect reflector.

The other condition is the non-free-surface case. For notational convenience, we will choose the non-free surface as interface 1; that is, the air-earth interface is taken as interface 1, which can have an arbitrary (nonperfect) reflection coefficient c_1, where $|c_1| < 1$. Since interface 1 represents the surface of the ground, it follows that interface 0 is not present, that is, $c_0 = 0$. Thus the non-free-surface condition is that $c_0 = 0$ and c_1 is arbitrary.

In summary, an ideal marine seismogram is generated by the

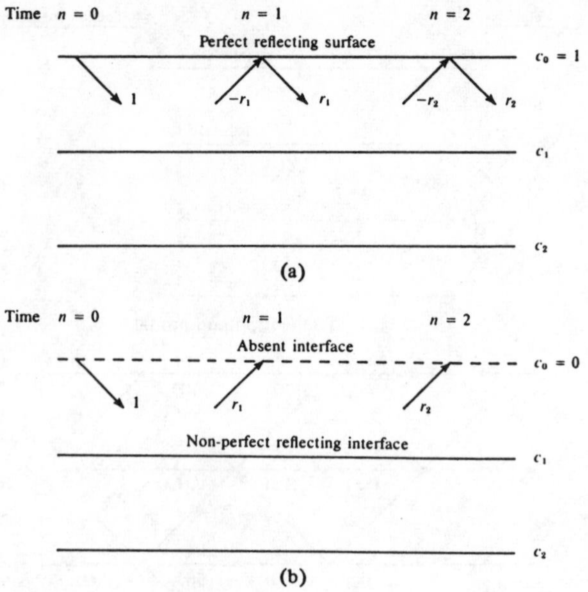

Fig. 3. (a) Free-surface siesmogram $r_1, r_2, r_3, \cdots$. (b) Non-free-surface seismogram $r_1, r_2, r_3, \cdots$.

Goupillaud model with the free-surface condition $|c_0| = 1$. To be definite, we will use $c_0 = 1$. Thus the reflection coefficient for upgoing waves is $-c_0$, which is -1, so an upcoming pulse $-r_n$ is reflected into the downgoing pulse r_n. A typical land seismogram is generated by the Goupillaud model with the non-free-surface condition, which for convenience of notation, we take as $c_0 = 0$ and $|c_1| < 1$ (so interface 1 is the surface of the earth). The well-known acoustic tube model [35] for human speech may be described as a Goupillaud-type model with the free-surface condition at the lips.

Kunetz [36] gave the solution for the inversion of a free-surface reflection seismogram. This inversion method yields the reflection coefficient series $c_1, c_2, c_3, \cdots$, from which the impedance function of the earth as a function of depth may be readily computed. Robinson [34] reformulated the Kunetz solution in terms of the Levinson [25] recursion and gave a computer program to do both the forward process (generation of the synthetic seismogram) and the inverse process (inversion of the seismogram to obtain the reflection coefficients). The Kunetz method is covered in the next section.

The celebrated inversion method of Gelfand and Levitan [7] represents the solution of the inversion problem for non-free-surface reflection seismograms. This method has been in the mainstream of physics for many years, and has been further developed and extended by many physicists and mathematicians. The discrete form of the Gelfand–Levitan equation is derived by Aki and Richards [38] for the case of a finite inhomogeneous medium, that is, an inhomogeneous medium bounded by homogeneous media at both ends. In Section II-F, we treat the discrete Gelfand–Levitan equation and give a derivation which holds for an unbounded inhomogeneous medium. We then discuss dynamic deconvolution, which is a means of solving the Gelfand–Levitan equation. Dynamic deconvolution makes use of the interactive recursion of the P and Q polynomials [34]. This recursion for the non-free-surface case represents the counterpart of the Levinson recursion for the free-surface case.

E. The Kunetz Inversion of Free-Surface Reflection Seismograms

Let the source in the Goupillaud model with the free-surface condition $c_0 = 1$ be a unit spike at time 0; see Fig. 3(a). The source gives rise to an upgoing wave in the first layer as a result of reflections and refractions from the interfaces below. We denote this upgoing wave by $-r_1, -r_2, -r_3, \cdots$; that is,

$$u_1(n) = -r_n \qquad (\text{for } n = 1, 2, 3, \cdots)$$

represents the wave motion striking the free surface from below. The free surface is a perfect reflector (with upgoing reflection coefficient $-c_0 = -1$). The upgoing wave is reflected back to produce the downgoing wave

$$d_1(n) = r_n \qquad (\text{for } n = 1, 2, 3, \cdots).$$

The entire downgoing wave at the top of layer 1 is made up of this reflected portion together with the initial source pulse

$$d_1(0) = 1.$$

We will call $r_1, r_2, r_3, \cdots$, the reflection seismogram. Thus

$$R(s) = r_1 s + r_2 s^2 + r_3 s^3 + \cdots$$

is the generating function of the reflection seismogram. Then it follows that

$$U_1(s) = -R(s)$$

$$D_1(s) = 1 + R(s).$$

We will now use the invariance property of the Lorentz transformation. The net downgoing energy in layer 1 is

$$D_1\overline{D}_1 - U_1\overline{U}_1 = (1 + R)(1 + \overline{R}) - R\overline{R} = 1 + R + \overline{R}.$$

Here we use the convention that a bar over a function indicates that each s is replaced by s^{-1}. If we go very deep, we can assume that we reach a depth where no waves are reflected upward, so we can write

$$U_\infty = 0.$$

Thus at this infinite depth, we have

$$D_\infty\overline{D}_\infty - U_\infty\overline{U}_\infty = D_\infty\overline{D}_\infty.$$

We now come to an important point, which makes the layered earth model (i.e., a lattice network) useful for spectral analysis. Because $\Phi(\omega)$ defined as

$$\Phi(\omega) = D_\infty(e^{-i\omega})D_\infty(e^{i\omega})$$

is a *bona fide* spectral density function (i.e., nonnegative function of ω), we can use the invariance of the net downgoing energy from layer to layer to establish that

$$\Phi(\omega) = 1 + R(e^{-i\omega}) + R(e^{i\omega})$$

is the same spectral density function. Therefore, the seismogram, completed by the initial pulse and by symmetry

$$\cdots, r_{-3}, r_{-2}, r_{-1}, 1, r_1, r_2, r_3, \cdots$$

is a *bona fide* autocorrelation function. That *the seismogram is the right-half side of an autocorrelation function* is the celebrated result of *Kunetz* [36].

The problem of finding the reflection coefficients $c_1, c_2, c_3, \cdots$ from the free-surface reflection seismograms $r_1, r_2, r_3, \cdots$ represents the inverse problem. The earth's acoustic impedance function is readily computed from the reflection coefficient series.

The Lorentz transformation from layer 0 to layer $k + 1$ may be written as

$$\begin{bmatrix} D_{k+1} \\ U_{k+1} \end{bmatrix} = \frac{s^{-k/2}}{\sigma_k} \begin{bmatrix} P_k^R & Q_k^R \\ Q_k & P_k \end{bmatrix} \begin{bmatrix} 1 + R \\ -R \end{bmatrix}$$

where $\sigma_k = t_1 t_2 \cdots t_k$ is the one-way transmission coefficient through the k interfaces. Using this matrix equation, we solve for D_{k+1}. Then, by replacing s by s^{-1}, we form $\overline{D}_{k+1}$. We also solve for U_{k+1}. We thus find that

$$\overline{D}_{k+1} - U_{k+1} = \frac{s^{-0.5k}}{\sigma_k} A_k(1 + R + \overline{R})$$

where A_n is defined as $A_n = P_n - Q_n$. Since $\Phi = 1 + R + \overline{R}$, we have

$$A_k\Phi = \sigma_k s^{0.5k}(\overline{D}_{k+1} - U_{k+1}).$$

Let us appeal to the physics of the situation; see Fig. 4. The function D_{k+1} is the generating function of the downgoing wave at the top of layer $k + 1$. This downgoing wave is made up of the direct pulse $d_{k+1}(0.5k)$ together with the following pulses: $d_{k+1}(0.5k + 1)$, $d_{k+1}(0.5k + 2), \cdots$. Because the

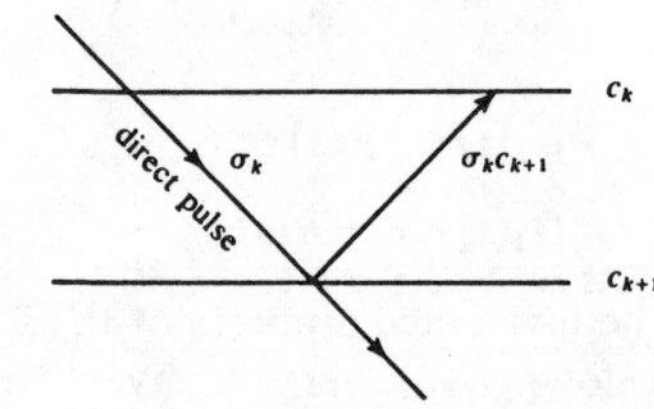

Fig. 4. Reflection of the direct pulse.

direct pulse is the result of only the transmissions through the first k interfaces, we see that the direct pulse is the product of these k transmission coefficients, that is,

$$d_{k+1}(0.5k) = t_1 t_2 \cdots t_k = \sigma_k.$$

Because the time instant of this direct pulse is $0.5k$, we see that

$$D_{k+1}(s) = \sigma_k s^{0.5k} + (\text{terms in higher powers of } s).$$

The function U_{k+1} is the generating function of the upgoing wave at the top of layer $k+1$. The first pulse in the upgoing wave is the reflection of the direct downgoing pulse at interface $k+1$. Thus the magnitude of this first upgoing pulse is $\sigma_k c_{k+1}$; that is, the magnitude is equal to the magnitude of the direct downgoing pulse times the reflection coefficient c_{k+1}. Because one time unit elapses for the round trip in layer $k+1$, the first pulse of the upgoing wave in layer $k+1$ occurs at one time unit later than the first pulse of the downgoing wave in layer $k+1$. That is, the first upgoing pulse at the top of layer $k+1$ occurs at time $0.5k+1$. Thus

$$U_{k+1}(s) = \sigma_k c_{k+1} s^{0.5k+1} + (\text{terms in higher powers of } s).$$

Using the above expressions, we obtain

$$A_k \Phi = \sigma_k s^{0.5k} \,[(\text{terms in lower powers of } s) + \sigma_k s^{-0.5k} - \sigma_k c_{k+1} s^{0.5k+1} + (\text{terms in higher powers of } s)]$$

or

$$A_k \Phi = \sigma_k^2 \,[(\text{terms in negative powers}) + 1 - c_{k+1} s^{k+1} + (\text{terms in higher powers})]$$

where $\sigma_k^2 = (t_1 t_2 \cdots t_k)^2$ is the two-way transmission coefficient through the k interfaces. Now comes the critical observation. The powers of s from 1 to k are missing on the right-hand side of the above equation. We will exploit this fact. We equate coefficients on each side of this equation for the powers of s from 0 to $k+1$. We thus obtain the equations (one for each power from 0 to $k+1$) given by

$$a_{k0} r_0 + a_{k1} r_1 + \cdots + a_{kk} r_k = \sigma_k^2$$

$$a_{k0} r_1 + a_{k1} r_0 + \cdots + a_{kk} r_{k-1} = 0$$

$$\cdots$$

$$a_{k0} r_k + a_{k1} r_{k-1} + \cdots + a_{kk} r_0 = 0$$

$$a_{k0} r_{k+1} + a_{k1} r_k + \cdots + a_{kk} r_1 = \sigma_k^2 c_{k+1}.$$

Here $a_{k0}, a_{k1}, \cdots, a_{kk}$ are the coefficients of the polynomial A_k. We note that $a_{k0} = 1$ and $r_0 = 1$. Now let us look at these equations. As we have seen, the Kunetz result says that r_k is an autocorrelation function. Thus these equations are *normal equations*, and hence the Levinson [25] recursion can be used to solve them. The result is the inversion program [34] that finds the reflection coefficients from the free-surface (marine) reflection seismogram.

F. The Gelfand–Levitan Inversion of Non-Free-Surface Reflection Seismograms

Let us now turn to the non-free-surface reflection seismogram, that is, the seismogram produced by the Goupillaud model with the non-free-surface condition $c_0 = 0$ and c_1 arbitrary; see Fig. 3(b). Let the source be a unit spike. The resulting seismogram is taken to be the upgoing wave in the first layer, that is, by

$$u_1(n) = r_n \qquad (\text{for } n = 1, 2, 3, \cdots).$$

Because interface 0 is absent (i.e., $c_0 = 0$) the upgoing wave is not reflected back into the medium. Thus the downgoing wave at the top of layer 1 is simply the initial source pulse

$$d_1(0) = 1, \quad d_1(n) = 0 \qquad (\text{for } n = 1, 2, 3, \cdots).$$

Let $R(s) = r_1 s + r_2 s^2 + \cdots$ be the generating function of the reflection seismogram. Thus $U_1(s) = R(s)$ and $D_1(s) = 1$. The Lorentz transformation is

$$\begin{bmatrix} D_{k+1} \\ U_{k+1} \end{bmatrix} = \sigma_k^{-1} s^{-0.5k} \begin{bmatrix} P_k^R & Q_k^R \\ Q_k & P_k \end{bmatrix} \begin{bmatrix} 1 \\ R \end{bmatrix}.$$

Thus we obtain

$$D_{k+1} = \sigma_k^{-1} s^{-0.5k} (P_k^R + Q_k^R R)$$

$$U_{k+1} = \sigma_k^{-1} s^{-0.5k} (Q_k + P_k R).$$

We add these two equations to obtain

$$D_{k+1} + U_{k+1} = \sigma_k^{-1} s^{-0.5k} (G_k^R + R G_k)$$

where G_k is defined as

$$\begin{aligned} G_k(s) &= P_k(s) + Q_k^R(s) \\ &= g_{k0} + g_{k1} s + \cdots + g_{k,k-1} s^{k-1} \\ &= (1 - c_k) + g_{k1} s + \cdots + (c_1 c_k - c_1) s^{k-1}. \end{aligned}$$

Because $g_{k0} = 1 - c_k$, we can find c_k as soon as we can determine g_{k0}. Thus the reflection coefficient series, and hence the impedance function of the earth, can be found directly from the sequence $g_{10}, g_{20}, g_{30}, \cdots$. Let us now show how this sequence is determined.

The direct pulse $d_{k+1}(k/2)$ is the product of the transmission coefficients $\sigma_k = t_1 t_2 \cdots t_k$ and it arrives at time $0.5k$. The first term of U_{k+1} arrives at time $0.5k+1$. Thus $D_{k+1} + U_{k+1}$ has the form

$$D_{k+1} + U_{k+1} = \sigma_k s^{0.5k} + (\text{terms in higher powers of } s).$$

Thus we have

$$\sigma_k s^{0.5k} + (\text{higher power terms}) = \sigma_k^{-1} s^{-0.5k} (G_k^R + R G_k)$$

which gives

$$G_k^R + R G_k = \sigma_k^2 s^k + (\text{higher power terms}).$$

This equation shows that the coefficients of $G_k^R + R G_k$ for the powers $1, 2, \cdots, k-1$ are zero and the coefficient for the power k is equal to σ_k^2; that is,

$$g_{k,k-1} + g_{k0} r_1 = 0$$

$$g_{k,k-2} + g_{k0} r_2 + g_{k1} r_1 = 0$$

$$\cdots$$

$$g_{k1} + g_{k0} r_{k-1} + \cdots + g_{k,k-2} r_1 = 0$$

$$g_{k0} + g_{k0} r_k + \cdots + g_{k,k-2} r_2 + g_{k,k-1} r_1 = \sigma_k^2.$$

This set of equations is the discrete version of the *Gelfand–Levitan equation.*

Given the free-surface reflection seismogram $r_1, r_2, r_3, \cdots$, the Gelfand–Levitan inversion method involves solving the above set of equations for each of $k = 1, 2, \cdots N$. For $k = 1$ the set is

$$g_{10} + g_{10} r_1 = \sigma_1^2.$$

For $k = 2$ the set is

$$g_{21} + g_{20} r_1 = 0$$

$$g_{20} + g_{20} r_2 + g_{21} r_1 = \sigma_2^2.$$

After we solve for $g_{10}, g_{20}, \cdots$, then we can find the reflection coefficients by means of

$$c_1 = 1 - g_{10}$$

$$c_2 = 1 - g_{20}$$

and so on.

If we define $a_{k1} = g_{k,k-1}, a_{k2} = g_{k,k-2}, \cdots, a_{k,k-1} = g_{k1}$, $a_{kk} = g_{k0} - 1$, then the Gelfand–Levitan equations are

$$\begin{bmatrix} a_{k1} \\ a_{k2} \\ \cdots \\ a_{kk} \end{bmatrix} + \begin{bmatrix} 0 & 0 & 0 & r_1 \\ 0 & 0 & r_1 & r_2 \\ & & & \\ r_1 & r_2 & r_{k-1} & r_k \end{bmatrix} \begin{bmatrix} a_{k1} \\ a_{k2} \\ \\ a_{kk} \end{bmatrix} + \begin{bmatrix} r_1 \\ r_2 \\ \\ r_k \end{bmatrix} = 0$$

which we recognize as this discrete counterpart of the *Gelfand–Levitan integral equation* [7], [9], [38]

$$a(\tau, t) + \int_{-t}^{\tau} a(\tau, \beta)\, r(t + \beta)\, d\beta + r(t + \tau) = 0.$$

G. Seismic Inversion by Dynamic Deconvolution

The Gelfand–Levitan method of inversion, together with its many related methods, has received wide recognition. However, instead of the Gelfand–Levitan approach given in the preceding section, the seismic industry approaches the problem from a different point of view. This alternative inversion computational scheme is the method of *dynamic deconvolution* (dy-decon) [39], [40], [8]. Dy-decon inversion is based upon the physical structure of the reflection seismogram. The key fact is that the reflection seismogram is generated from the reflection coefficient by means of the Einstein addition formula [33]. The recognition of this fact makes the inversion of a reflection seismogram very simple from a computational point of view.

Let us now give the dy-decon computation scheme for the inversion of a non-free-surface reflection seismogram. We use the same conventions as in the preceding section. The field-recorded reflection seismogram $r_1, r_2, r_3, \cdots$ is represented by its generating function $R(s)$, which we now will denote by $R_1(s)$ because the field-recorded seismogram occurs in layer 1. We have

$$R_1(s) = r_1 s + r_2 s^2 + r_3 s^3 + \cdots$$

$$= c_1 s + (\text{terms in higher powers of } s).$$

We know $R_1(s)$ must have this form, because $c_1 s$ represents the first bounce from interface 1. No multiple reflections can appear at the time of the first bounce. Now suppose that layer 2 expands to fill the whole upper half-space, so there is no interface 1. Now the top interface is interface 2. Let the resulting seismogram in this expanded layer 2 be represented by its generating function

$$R_2(s) = c_2 s + (\text{terms in higher powers of } s).$$

Here $c_2 s$ represents the first bounce. Next, expand layer 3 to fill up the whole upper half-space. The resulting reflection seismogram has generating function

$$R_3(s) = c_3 s + (\text{terms in higher powers of } s)$$

where $c_3 s$ represents the first bounce. Thus conceptually we have a suite of reflection seismograms ($k = 1, 2, 3, \cdots$) with generating functions

$$R_k(s) = c_k s + (\text{terms in higher powers of } s)$$

where $c_k s$ represents the first bounce from interface k. We can, therefore, make the following important conclusion. Given the reflection seismogram for layer k, we can immediately find the reflection coefficient c_k for layer k, because c_k is simply the first coefficient appearing in the seismogram. This conclusion represents the solution of one-half of the inversion problem. The other half of the problem involves determining the suite of reflection seismograms. The given information is the top seismogram $R_1(s)$; this seismogram is the one physically recorded in the field by a seismic crew.

If the current in a river has velocity c_1, and if our motor boat in still water has a velocity R_1, then our motor boat headed upstream will have a velocity R_2 given by

$$R_2 = R_1 - c_1.$$

This equation is the *Newton addition formula.* However, if one attempts to apply Newtonian mechanical laws to ultra-high-speed charged particles, then an insurmountable contradiction is encountered. That is, the simple addition of velocities, as used in the boat example, does not apply in electrodynamics. Instead, one should use the Einstein addition formula for combining velocities. The Einstein formula guaranteed that the resulting velocity will never exceed the velocity of light. Thus the above Newton addition formula should be replaced by the *Einstein addition formula,* which is

$$R_2 = \frac{R_1 - c_1}{1 - R_1 c_1}.$$

Here we assume that all velocities are measured in natural units (i.e., in units such that the velocity of light is unity). Suppose $c_1 = -0.5$ and $R_1 = 0.8$. Then, according to the Newton formula, $R_2 = 1.3$, so R_2 is greater than unity; that is, R_2 is greater than the velocity of light. According to the Einstein formula

$$R_2 = \frac{0.8 + 0.5}{1 + 0.8(0.5)} = 0.929$$

which (necessarily) is less than the velocity of light. The Einstein formula is the correct one to use for ultra-high velocities.

Now let us make the following important observation, which led to the dynamic deconvolution process. A reflection coefficient in magnitude can never exceed unity. A reflection coefficient greater than unity in magnitude is just as impossible from the physics as a velocity greater than the velocity of light. Thus in combining the reflection coefficients of a system of layers, it follows that the Einstein addition formula must be used. As a result, the resulting reflectivity will never exceed unity in magnitude, as required in any physical system.

Now we want to head our boat upstream; that is, we want

to dynamically deconvolve down into the earth. We could use the Newton addition formula and write

$$sR_2 = R_1 - c_1 s.$$

That is, the Newton formula says that we merely subtract the first bounce from R_1 in order to obtain R_2. (The s represents a time shift.) In fact, this formula is approximately true if the reflection coefficients are very small in magnitude, just as the Newton formula for velocities is true if the velocities are very small with respect to the velocity of light. Otherwise, we must use the Einstein addition formula

$$R_2 = \frac{R_1 - c_1 s}{s - R_1 c_1}.$$

Let us now give the inversion algorithm of dynamic deconvolution. We know R_1 (i.e., the field-recorded reflection seismogram). In step 1, we find c_1 as the first bounce of R_1, and then we compute R_2 by the above Einstein addition formula. This computation is easily done by making use of subroutine POLYDV in [34]. So ends step 1. In step 2, we find c_2 as the first bounce of R_2, and then we use the Einstein addition formula to find R_3

$$R_3 = \frac{R_2 - c_2 s}{s - R_2 c_2}.$$

So ends step 2. In step 3, we find c_3 as the first bounce of R_3, and then we find R_4 as

$$R_4 = \frac{R_3 - c_3 s}{s - R_3 c_3}.$$

So ends step 3. Thus given R_1, we can perform the entire deconvolution, and so obtain the sequence of reflection coefficients $c_1, c_2, c_3, \cdots$ and the suite of reflection seismograms $R_2, R_3, R_4, \cdots$. From the sequence of reflection coefficients, we can compute the impedance function of the earth.

Dynamic deconvolution gives the same impedance function as that found by the Gelfand–Levitan discrete inversion method. However, dynamic deconvolution is carried out in terms of physically meaningful quantities (i.e., reflection coefficients and reflection seismograms), so interactive computation can be used to reduce the effects of noise. Also, there are many other features of dynamic deconvolution which make it attractive, such as the option [39] of the determination of the reflection coefficients in the reverse order, i.e., in the order $c_n, c_{n-1}, \cdots, c_3, c_2, c_1$. This reverse order can be useful because in many cases, the most harmful noise appears at the beginning of the reflection seismogram. In such cases, it is better to work backwards in time on the reflection seismogram, leaving its noisy beginning to the end of the computations.

Let us now establish the Einstein addition formula by means of the Lorentz transformation. This proof can be found in most physics textbooks. We have

$$\begin{bmatrix} D_2 \\ U_2 \end{bmatrix} = \frac{s^{-1/2}}{t_1} \begin{bmatrix} s & -c_1 \\ -c_1 s & 1 \end{bmatrix} \begin{bmatrix} 1 \\ R_1 \end{bmatrix}$$

which gives

$$t_1 s^{1/2} D_2 = s - c_1 R_1$$
$$t_1 s^{1/2} U_2 = -c_1 s + R_1.$$

The reflection seismogram R_2 is the result obtained by deconvolving the upgoing wave U_2 by the downgoing wave D_2; that is,

$$R_2 = \frac{U_2}{D_2} = \frac{R_1 - c_1 s}{s - R_1 c_1}.$$

This equation is the Einstein addition formula.

H. Polynomial Recursion for Dynamic Deconvolution Inversion

Further insight on dynamic deconvolution can be obtained by examining the role played by the polynomials P_k and Q_k. Using the Lorentz transformation for the first k layers, we have

$$\begin{bmatrix} D_{k+1} \\ U_{k+1} \end{bmatrix} = \frac{s^{-k/2}}{\sigma_k} \begin{bmatrix} P_k^R & Q_k^R \\ Q_k & P_k \end{bmatrix} \begin{bmatrix} 1 \\ R_1 \end{bmatrix}.$$

As usual, σ_k is the one-way transmission factor

$$\sigma_k = t_k t_{k-1} \cdots t_1.$$

This equation gives

$$D_{k+1} = \sigma_k^{-1} s^{-k/2} (P_k^R + Q_k^R R_1)$$
$$U_{k+1} = \sigma_k^{-1} s^{-k/2} (Q_k + P_k R_1).$$

The direct downgoing pulse arrives at layer $k + 1$ at time $k/2$ and with a transmission loss of $\sigma_k = t_1 t_2 \cdots t_k$. Thus D_{k+1} has the form

$$D_{k+1} = \sigma_k s^{k/2} + (\text{terms in higher powers of } s).$$

This direct downgoing pulse is reflected at interface $k + 1$ at time $(k + 1)/2$ and produces the initial pulse of the upgoing wave. Because the incident pulse has amplitude σ_k and the reflection coefficient is c_{k+1}, it follows that the initial upgoing pulse has amplitude $\sigma_k c_{k+1}$. This initial upgoing pulse arrives at the top of layer $k + 1$ at time $(k/2) + 1$. Thus U_{k+1} has the form

$$U_{k+1} = \sigma_k c_{k+1} s^{(k/2)+1} + (\text{terms in higher powers of } s).$$

We now have two equations for U_{k+1} Equating them, we have

$$\sigma_k c_{k+1} s^{(k/2)+1} + (\text{higher power terms}) = \sigma_k^{-1} s^{-k/2} (Q_k + P_k R_1).$$

Let us now just pick out the $s^{(k/2)+1}$ term on the right-hand side. The right-hand side is

$$\sigma_k^{-1} s^{-k/2} [(q_{k1} s + \cdots + q_{kk} s^k) + (p_{k0} + \cdots + p_{k,k-1} s^{k-1})(r_1 s + \cdots + r_{k+1} s^{k+1} + \cdots)]$$

so the $s^{(k/2)+1}$ term is displayed as

$$\sigma_k^{-1} s^{-k/2} [(p_{k0} r_{k+1} + \cdots + p_{k,k-1} r_2) s^{k+1} + (\text{higher power terms})].$$

Equating the $s^{(k/2)+1}$ terms on the left and right, we have

$$\sigma_k c_{k+1} s^{(k/2)+1} = \sigma_k^{-1} s^{-k/2} (p_{k0} r_{k+1} + \cdots + p_{k,k-1} r_2) s^{k+1}$$

which gives

$$c_{k+1} = \frac{p_{k0} r_{k+1} + p_{k1} r_k + \cdots + p_{k,k-1} r_2}{\sigma_k^2}.$$

We thus have the following scheme based on the Robinson [34] recursion for P and Q in order to invert the non-free-surface reflection seismogram $r_1, r_2, r_3, \cdots$. As the initial step, we set $c_1 = r_1, \sigma_1^2 = 1 - c_1^2, P_1(s) = 1, Q_1(s) = -c_1 s$. Then we perform the following DO loop from $k = 1$ to the total number of layers that we wish to consider

Compute c_{k+1} by the above formula

$$\text{Compute } \sigma_{k+1}^2 = (1 - c_{k+1}^2)\sigma_k^2$$

$$\text{Compute } P_{k+1}(s) = P_k(s) - c_{k+1} s\, Q_k^R(s)$$

$$\text{Compute } Q_{k+1}(s) = Q_k(s) - c_{k+1} s\, P_k^R(s).$$

We thus obtain the reflection coefficient series $c_1, c_2, c_3, \cdots$ from which the required impedance function can be calculated.

Finally, let us write down the expression for the reflection seismogram R_{k+1} at layer $k + 1$. It is

$$R_{k+1} = \frac{U_{k+1}}{D_{k+1}} = \frac{Q_k + P_k R_1}{P_k^R + Q_k^R R_1}$$

$$= \frac{\sigma_k c_{k+1} s^{(k/2)+1} + (\text{higher power terms})}{\sigma_k s^{k/2} + (\text{higher power terms})}$$

$$= c_{k+1} s + (\text{higher power terms}).$$

We have thus confirmed the statement made in the preceding section that the first bounce of R_{k+1} is $c_{k+1}\, s$.

In closing our discussion of 1-D inversion, let us mention the *fixed-entropy estimate* [41] of the seismic acoustic log. This scheme is an autocorrecting iterative method, based on a forward model, to compute the acoustic impedance of the earth from the reflection seismogram. This method was offered commercially by Digicon Inc. during 1968–1969 under the trade name S.A.L. (for seismic acoustic log), and it has proved successful as a practical inversion scheme in seismic data processing centers.

III. Two-Dimensional Inversion

A. Plane-Wave Decomposition and Migration

We will now consider two spatial dimensions: lateral coordinate x and depth coordinate z, where the z axis points downward. The other dimension is that of time t. The 2-D methods discussed here involve extremely simple models, so these methods are not truly 2-D in any general sense. We discuss two methods. One treats the forward and inverse problem for a horizontally layered media. This method involves the Radon transform [42], which is often called "slant stacking" by geophysicists. The Radon transform produces a plane-wave decomposition of a wavefield, and thus we can operate on each plane wave separately in order to produce our desired results. The other method we present is a process for wavefield reconstruction and imaging. This process is known as "migration" in geophysics. Although various migration methods have been in constant use by the seismic industry for about forty years, a major breakthrough occurred in 1978, when Stolt [27] disclosed his spectral method of migration. This pioneering work of Stolt is based upon the fast Fourier transform, and it is often called frequency-domain migration or *f-k* migration (frequency-wavenumber migration). Several universities have seismic research projects which devote large efforts to migration and related problems.

B. The Radon Transform

Let $u(x, t)$ represent the wave motion observed on the surface of the ground ($z = 0$). For fixed horizontal position x, the 1-D function $u(x, t)$ of time t represents a seismic trace.

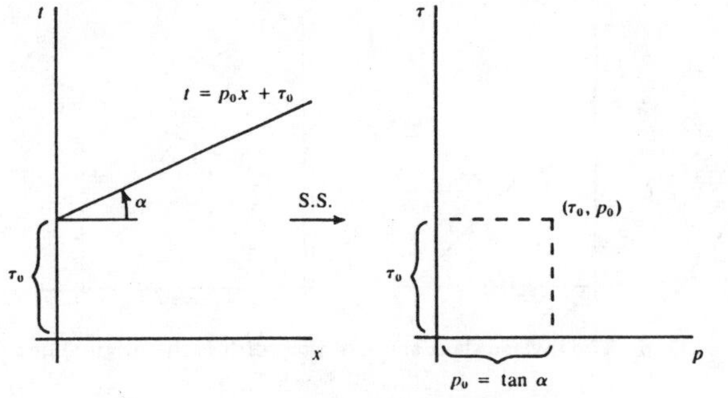

Fig. 5. The slant stack of a line is a point.

The entire suite of seismic traces for all positions x represents the seismic section.

The *Radon transform* [42] of $u(x, t)$ is defined as

$$U_R(p, \tau) = \int_{-\infty}^{\infty} u(x, px + \tau)\, dx.$$

We see that the Radon transform is a function of two variables; τ is called the *intercept* and p is called the *slope*. The reason is that $t = \tau + px$ represents a line in the (x, t) plane (where x is the horizontal axis and t is the vertical axis). This line has slope $p = \tan \alpha$ (i.e., the line makes an angle α with the x axis) and t-intercept τ (i.e., the line cuts the t axis at τ). Geophysicists call the Radon transformation a *slant stack*, as this concept has been in use in exploration seismology since the work of the MIT Geophysical Analysis Group [43] in the 1950's. We may regard the terms *Radon transform* and *slant stack* as meaning the same thing.

The reason for the name "slant stack" is the following. We can perform the Radon transform by sweeping over the wavefield $u(x, t)$ with lines, each given by its slope p and intercept τ. We then add (integrate) all the values on each line and associate the sum (integral) with its slope p and intercept τ. That is, we "stack" all the values of the wavefield on each "slant" line; that is, we "slant stack."

Thus the Radon transform involves summing all the amplitudes $u(x, t)$ along a given line of slope p_0 and intercept τ_0, and plotting that sum at the corresponding point (p_0, τ_0) on the p, τ plane. In brief, the Radon transform has this characteristic; it takes a line of intercept τ_0 and slope p_0 in the x, t space and transforms it into a point (p_0, τ_0) in the new space. That is, the Radon transform converts a line into a point; see Fig. 5. No information is lost because a line is completely described by its intercept and slope.

Now let us look at the *inverse Radon Transform*. This is an inverse slant stack. To be mathematically correct, we must first take the time derivative of the Hilbert transform of $U_R(p, \tau)$. We then slant stack along lines of slope $\tan \beta$ and intercept t; that is, lines $\tau = t + (\tan \beta)p$. Let us find the inverse slant stack of the point (p_0, τ_0). The only slant lines that will contribute are the ones that go through this point. Such a contributing line satisfies

$$\tau_0 = t + (\tan \beta) p_0$$

so its intercept is $t = \tau_0 - p_0 \tan \beta$. For the inverse slant stack, we let x be the negative slope, that is,

$$x = -\tan \beta.$$

Thus the point (p_0, τ_0) for each β gets transformed into the point $(x, t) = (-\tan \beta, \tau_0 - p_0 \tan \beta)$ in the original (x, t) plane.

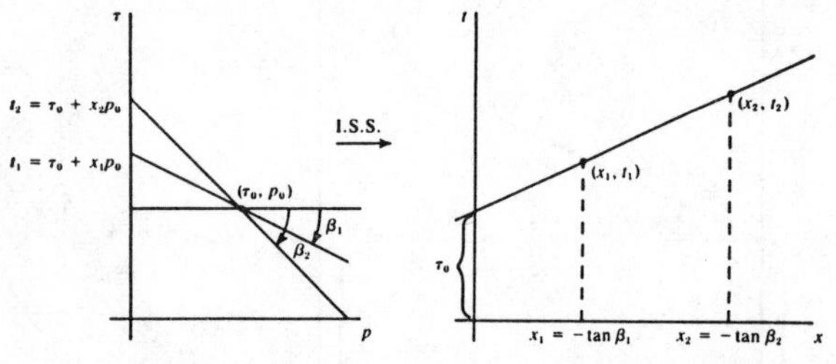

Fig. 6. The inverse slant stack of the point is the original line.

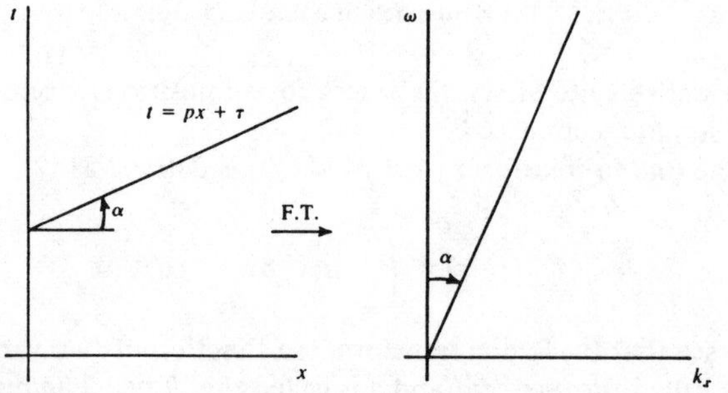

Fig. 7. The Fourier transform of a line is a line through the origin with reciprocal slope. The intercept of the original line is contained in the phase spectrum of the transform line.

As β sweeps out all values, the locus of all these points is the line $t = \tau_0 + p_0 x$, that is, the line of slope p_0 and intercept τ_0 in the (x, t) plane. This is the original line, so we see that the inverse slant stack works; see Fig. 6.

The above heuristic explanation of the inverse Radon transform must be made more precise mathematically. An equation for our heuristic description is

$$u(x, t) = \int_{-\infty}^{\infty} \frac{d}{dt} HU_R(p, t - px)\, dp$$

where H denotes the Hilbert transform.

C. The Fourier Transform

The Radon transform takes a line into a point, and the inverse Radon transform takes the point back into the original line. The 2-D Fourier transform takes a line into another line, and the inverse Fourier transform takes it back into the original line.

Let the function be a delta function $\delta(t - px - \tau)$ along the line $t = px + \tau$ in the (x, t) plane. This line has slope $p = \tan \alpha$ and intercept τ. The *Fourier transform* is

$$\int_{-\infty}^{\infty} \int_{-\infty}^{\infty} \delta(t - px - \tau) e^{-i(\omega t - k_x x)}\, dx\, dt$$

$$= \int_{-\infty}^{\infty} e^{-i[\omega(px+\tau) - k_x x]}\, dx$$

$$= 2\pi e^{-i\omega\tau}\, \delta(p\omega - k_x).$$

Thus the Fourier transform has magnitude given by a delta function along the line $\omega = k_x/p$, and has phase given by $-\omega\tau$. The line makes an angle α with the ω axis and goes through the origin (intercept = 0). The slope of the given line is the reciprocal of the slope of the tranformed line, and the intercept of the given line is locked up in the phase spectrum; see Fig. 7.

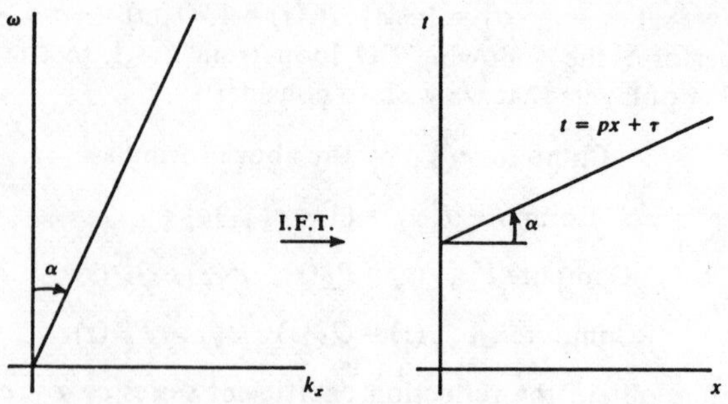

Fig. 8. The inverse Fourier transform gives back the original line.

By taking the *inverse Fourier transform*, we regain the original line; that is,

$$\frac{1}{2\pi} \int_{-\infty}^{\infty} \int_{-\infty}^{\infty} e^{-i\omega\tau}\, \delta(p\omega - k_x) e^{i(\omega t - k_x x)}\, dk_x\, d\omega$$

$$= \delta(t - px - \tau)$$

see Fig. 8.

Because of this interplay between lines, the Radon transform is closely related to the Fourier transform. Let us evaluate the Fourier transform $U(k_x, \omega)$ of the seismic wavefield $u(x, t)$ along the line $\omega = k_x/p$. We have $k_x = p\omega$ so $U(k_x, \omega)$ becomes

$$U(p\omega, \omega) = \int_{-\infty}^{\infty} \int_{-\infty}^{\infty} u(x, t) e^{-i(\omega t - p\omega x)}\, dx\, dt.$$

Because $\tau = t - px$, we have

$$U(p\omega, \omega) = \int_{-\infty}^{\infty} e^{-i\omega\tau} \left[\int_{-\infty}^{\infty} u(x, \tau + px)\, dx \right] d\tau.$$

We recognize the expression within the brackets as the Radon transform $U_R(p, \tau)$. Thus

$$U(p\omega, \omega) = \int_{-\infty}^{\infty} e^{-i\omega\tau}\, U_R(p, \tau)\, d\tau$$

which says that *the Fourier transform of the Radon transform with respect to the intercept variable τ is equal to the 2-D Fourier transform evaluated on the line $k_x = p\omega$.*

D. The Ray Parameter p and the Intercept Parameter τ

Let us now consider a horizontally stratified layered medium in 2-D space (x, z). We assume that the depth axis z points downward, with $z = 0$ indicating the surface. Each interface is represented by a flat (i.e., nontilted) line; line z = positive constant. Let us assume there are N earth layers in total, with interfaces at $z_0 = 0, z_1, z_2, \cdots, z_N$. Layer k lies between interface z_{k-1} (on top) and z_k (on bottom). The thicknesses of the layers are

$$h_1 = z_1 - z_0, h_2 = z_2 - z_1, \cdots, h_N = z_N - z_{N-1}.$$

We let $v_1, v_2, \cdots, v_N$ denote the compressional velocities in the respective layers.

Let an impulsive source be initiated at the surface. Waves will travel down into the earth where they will suffer reflections and refractions at the interfaces. At each surface point $(x, z = 0)$, the resulting wave motion can be recorded in the form of a seismic trace. The seismic trace will include primary reflections (i.e., impulses that have traveled paths directly

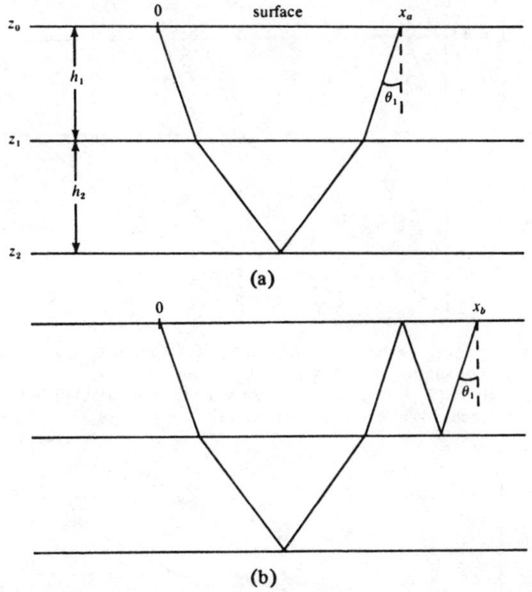

Fig. 9. Primary (a) and peg-leg multiple (b) each with the same value θ_1 of the emergence angle but with different offsets x_a and x_b.

down and back up) and multiple reflections (i.e., impulses that have bounced back and forth within various layers and combinations of layers before returning to the surface). Because the interfaces are flat, there is a symmetry; for each downgoing leg there is a corresponding upgoing leg. Thus we can consider merely the upgoing legs of the ray paths. We thus work with one-way times. The complete times, or two-way times, would then be twice the one-way times that we obtain.

The primary reflection from layer k would have one leg from each layer from interface k up to the surface. A multiple reflection includes extra bounces, so it would have more than one leg in some or all of the layers. For example, a peg-leg multiple has one long leg (corresponding to the primary) and a short leg (corresponding to an extra bounce); see Fig. 9.

Using geometric seismics, we can say that the waves travel along rays. At a given receiver point ($x_1, z = 0$) on the surface of the earth, the waves will come in from all directions. Let θ_1 represent the angle of emergence of a ray. (All angles are measured from the normal; that is, all angles are measured from the z-axis to the ray.) For example, at a given receiver point (x_1, 0), the primary from interface k would come in at one angle, the primary from another interface would come in at a different angle, and each possible multiple would come in at its own angle. At a different receiver point (x_2, 0), all the angles would be different. To make any sense at all from the received signals, we must do some sorting.

One way to sort is by angle. Let us pick one given angle θ_1, and then move along the horizontal x-axis, and pick out each ray that comes in at that given angle. We throw away all the other rays. We now appeal to Snell's law, which says that p is a constant where

$$p = \frac{\sin\theta_1}{v_1} = \frac{\sin\theta_2}{v_2} = \cdots = \frac{\sin\theta_N}{v_N}.$$

All the ray paths that we have picked out have the same value of the parameter p, which is called the *ray parameter* or *Snell's parameter.* A wave which travels along a path with a fixed value of p is called a *Snell wave.* Thus we have sorted out all the Snell waves for one given value of p, namely, $p = \sin\theta_1/v_1$. What good does this do us?

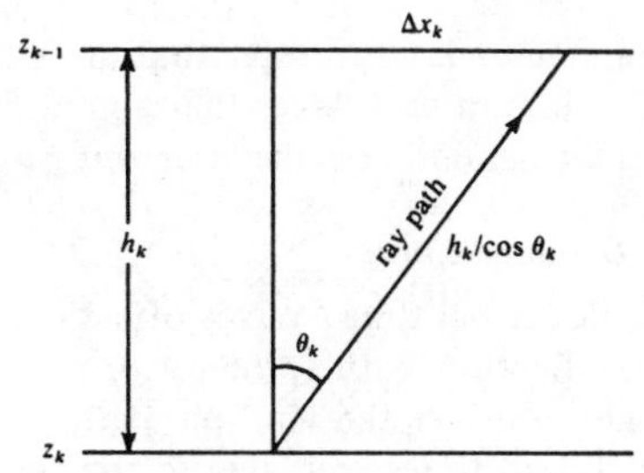

Fig. 10. Upgoing ray path in layer k.

The answer is that we can find an associated parameter τ which has nice properties. Each ray path that we have saved is characterized by the same value of p. We are going to construct our model (i.e., we are going to pick the interface positions $z_1, z_2, \cdots, z_N$) to fit this value of p. In other words, we are going to pick a model (characterized by the value p) to fit our data (all the received signals with a fixed value p).

Let us look at the ray path of one leg in layer k; see Fig. 10. The thickness of the layer is h_k, and the angle that the ray makes with the vertical is θ_k. Thus we have a right triangle with angle θ_k and adjacent, or vertical, side h_k. The travel distance of the wave is the slant distance (i.e., the hypotenuse). The offset distance Δx_k is the projection of the travel distance on the x-axis (i.e., Δx_k is the opposite, or horizontal, side).

The offset distance Δx_k is

$$\Delta x_k = h_k \tan\theta_k.$$

The slant distance is $h_k/\cos\theta_k$, so the travel time is

$$\Delta t_k = \frac{h_k}{v_k \cos\theta_k}.$$

Let us now consider the parameter τ. Its increment $\Delta\tau_k$ for layer k is defined as

$$\Delta\tau_k = \Delta t_k - p\,\Delta x_k.$$

Let us compute $\Delta\tau_k$. It is

$$\Delta\tau_k = \frac{h_k}{v_k\cos\theta_k} - p h_k \frac{\sin\theta_k}{\cos\theta_k} = \frac{h_k\cos\theta_k}{v_k}.$$

The angle θ_k is fixed because $\sin\theta_k = v_k p$. We will now choose the interfaces of our model so that the thicknesses of the layers are

$$z_k - z_{k-1} = h_k = \frac{v_k}{\cos\theta_k}.$$

In other words, we fit the model to the data; the data consist of all the ray paths with a fixed value of p. Thus for the chosen model, we have

$$\Delta\tau_k = 1$$

for each layer k. The primary from interface k goes through k layers; thus the τ value for this primary is

$$\tau = \sum_{j=1}^{k} \Delta\tau_j = k.$$

Suppose a multiple path goes through layer 1 four times, layer 2 three times, and layer 3 once. Then the τ value for this multiple reflection is

$$\tau = 4\Delta\tau_1 + 3\Delta\tau_2 + \Delta\tau_3 = (4 + 3 + 1) = 8.$$

Thus the τ parameter is an integer that gives the sum of the total number of legs in each layer that a multiple path makes. The name of the τ parameter is the *intercept parameter.*

E. The Time-Distance Curve

Let us now plot travel time t versus offset distance x for any given type of reflection path (primary or multiple). For example, we might consider the multiple path (2, 1) which goes through layer 1 two times and layer 2 one time; see Fig. 11. We consider this type of path (a path with 2, 1 bounces in layers 1, 2, respectively) for all values of the ray parameter p; that is, for all possible emergence angles θ_1. For each value of offset x there will be a travel time t. A plot of travel time t versus offset x is called the *time-distance curve* for the multiple path (2,1). Let us designate this curve as $t = f(x; 2, 1)$. This curve is shown in Fig. 12. Also shown are the time-distance curve $t = f(x; 1)$ for the primary from interface 1, the time-distance curve $t = f(x; 1, 1)$ for the primary from interface 2, and the time-distance curve $t = f(x; 1, 2)$ for the multiple with 1, 2 bounces in layers 1, 2, respectively. Except for the case of the first primary $t = f(x; 1)$, there is no explicit formula for such a curve. However, we can find a parametric representation in terms of the ray parameter p. Since $\sin\theta_k = pv_k$, it follows that $\cos\theta_k = (1 - p^2v_k^2)^{1/2}$ and $\tan\theta_k = pv_k/(1 - p^2v_k^2)^{1/2}$. Thus the parametric representation of the above time-distance curve $t = f(x; n_1, n_2, \cdots, n_k)$ is

$$x = \sum_k \Delta x_k = \sum_k \frac{h_k p v_k}{(1 - p^2 v_k^2)^{1/2}}$$

$$t = \sum_k \Delta t_k = \sum_k \frac{h_k}{v_k(1 - p^2 v_k^2)^{1/2}}$$

where the index k runs over the value $k = 1$, n_1 times; $k = 2$, n_2 times; and so on, up to $k = k$, n_k times.

Because the ray emerges at angle θ_1 with the vertical, it follows that the wavefront makes an angle θ_1 with the horizontal, as the wavefront is at right angles to the ray. In a small increment of time dt, the wave travels a distance $v_1 dt$ along the ray path. In the same amount of time, the wavefront sweeps out a distance dx along the x-axis, where $dx \sin\theta_1 = v_1\, dt$, as shown in Fig. 13. Thus the slope of the time-distance curve is

$$\frac{dt}{dx} = \frac{\sin\theta_1}{v_1}.$$

That is, *the slope of the time-distance curve at any point x is equal to the ray parameter p of the ray path that emerges at that point*; i.e.,

$$\frac{dt}{dx} = p.$$

In summary, each type of multiple path has a time-distance curve whose slope at any value of x is equal to the ray parameter characterizing the ray emerging at that point. For example, for the time-distance curve $t = f(x; 2, 1)$, we have

$$p = \frac{dt}{dx} = \frac{df(x; 2, 1)}{dx}$$

where the ray emerging at x has angle $\theta_1 = \sin^{-1}(pv_1)$.

Let us now find out what the parameter τ is. We know that the slope of the curve $f(x; 2, 1)$ is p. That is, the tangent to this curve is a line with slope p. Let τ be the t-intercept of this

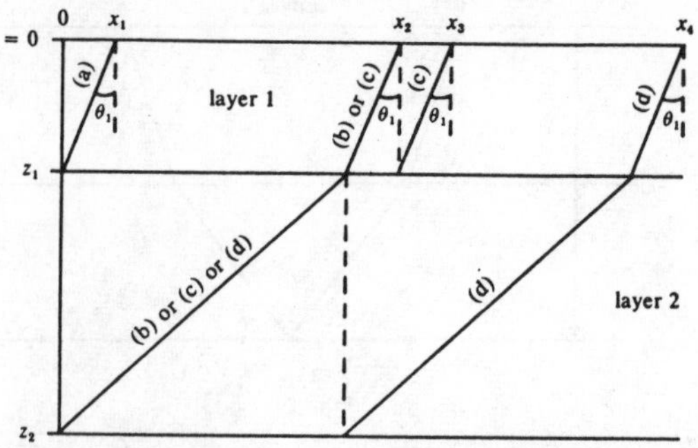

Fig. 11. Upgoing paths only are shown: (*a*) first primary (1 bounce in layer 1); (*b*) second primary (1, 1 bounces in layers 1, 2, respectively); (*c*) multiple (2, 1 bounces in layers 1, 2, respectively); (*d*) multiple (1, 2 bounces in layers 1, 2, respectively).

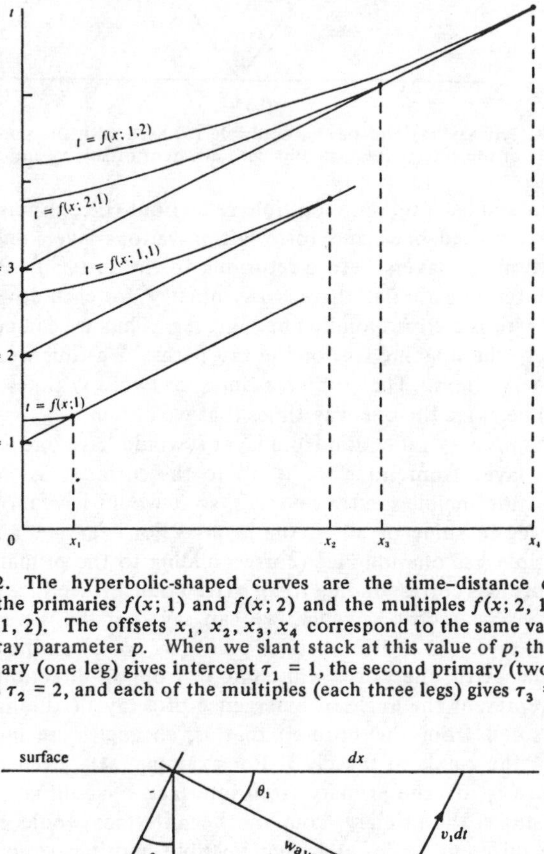

Fig. 12. The hyperbolic-shaped curves are the time-distance curves for the primaries $f(x; 1)$ and $f(x; 2)$ and the multiples $f(x; 2, 1)$ and $f(x; 1, 2)$. The offsets x_1, x_2, x_3, x_4 correspond to the same value of the ray parameter p. When we slant stack at this value of p, the first primary (one leg) gives intercept $\tau_1 = 1$, the second primary (two legs) gives $\tau_2 = 2$, and each of the multiples (each three legs) gives $\tau_3 = 3$.

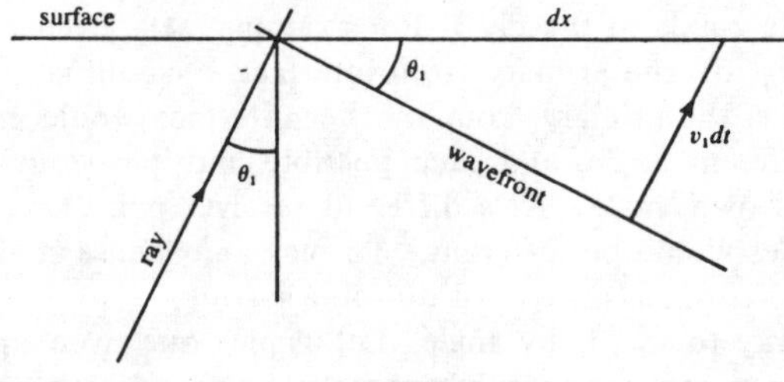

Fig. 13. Geometric demonstration that the slope of the time-distance curve is p.

tangent line, so the equation of the tangent line is

$$t = px + \tau.$$

In the preceding section, we defined $\Delta\tau_k$ as

$$\Delta\tau_k = \Delta t_k - p\Delta x_k$$

so

$$\sum_k \Delta\tau_k = \sum_k \Delta t_k - p\sum_k \Delta x_k$$

which is

$$\tau = t - px.$$

Thus we see that the parameter τ given in the preceding section is the same as the τ given in this section. This explains why we have called τ the *intercept parameter*. The parameter τ is the intercept of the tangent line with the t-axis.

What happens if we slant stack a time–distance curve? A time–distance curve may be thought of as the envelope of its tangent lines. Slant stacking converts each tangent line into a point (p, τ), where p is the slope and τ is the intercept of the line. As an example, let us consider the case of the first primary reflection. The ray-path travel time is t, so the ray-path distance is $v_1 t$, which is the hypotenuse of a right triangle with sides h_1 and x. Thus by the Pythagorean theorem, we have

$$v_1^2 t^2 = x^2 + h_1^2$$

so

$$t = \frac{1}{v_1}(x^2 + h_1^2)^{1/2}.$$

which shows that the first-primary time-distance curve is a hyperbola. This is the one time–distance curve we can write explicitly (instead of implicitly in parametric form). The curve in parametric form is

$$x = \frac{h_1 p v_1}{(1 - p^2 v_1^2)^{1/2}}$$

$$t = \frac{h_1}{v_1(1 - p^2 v_1^2)^{1/2}}.$$

The intercept parameter τ is

$$\tau = t - px = \frac{h_1 \cos \theta_1}{v_1}$$

where

$$\cos \theta_1 = (1 - p^2 v_1^2)^{1/2}.$$

Thus the above equation for τ together with the Snell equation are

$$\frac{\tau}{h_1} = \frac{\cos \theta_1}{v_1}$$

$$p = \frac{\sin \theta_1}{v_1}$$

which give

$$\frac{\tau^2}{h_1^2} + p^2 = \frac{1}{v_1^2}.$$

Thus the (p, τ) curve is an *ellipse*. This ellipse is the Radon transform (i.e., slant stack) of the time–distance hyperbola

$$t^2 - \frac{x^2}{v_1^2} = \frac{h_1^2}{v_1^2}.$$

F. Dynamic Deconvolution (Inversion) and Dynamic Reconvolution (Construction)

A seismic section $u(x, t)$ gives the amplitude of the received waves as a function of receiver position x and time t. A seismic section represents the received data on the surface of the earth. Each type of path (primary and multiple) appear as an event on the seismic section. These events lie along the time–distance curves of the respective paths.

Given the seismic data $u(x, t)$, we now want to pick out all the paths that have the same value of the ray parameter p. That is, we want to pick out all the ray paths that emerge at some given angle. As we have said, if we fix the receiver position x, then all the different types of paths come in at different angles. On the other hand, if we fix the angle, then all the different types of ray paths with that emergent angle come in at different receiver positions x.

We recall that we choose the layers in our model for a particular value of p. Consider all the ray paths with the value p. Each ray path will emerge at a different x value. At that x value, the slope of the corresponding time–distance curve will be p. Thus the tangent line at this point will have slope p and intercept τ. The value of τ will depend upon the type of ray path. More specifically, τ is an integer equal to the sum of the number of legs through each layer. For example, for the multiple path with four passes through layer 1, three passes through layer 2, and one pass through layer 3, the value of τ is $4 + 3 + 1 = 8$.

Let us now slant stack the data with slope p. Then the stacked data as a function of τ will be as follows. Every one of the paths has ray parameter with the given value p:

$\tau_1 = 1$: (1) First primary

$\tau_2 = 2$: (1) Second primary, and
(2) multiple with two passes through layer 1

$\tau_3 = 3$: (1) Third primary,
(2) multiple with three passes through layer 1,
(3) multiple with two passes through layer 1 and one pass through layer 2, and
(4) multiple with one pass through layer 1 and two passes through layer 2.

Thus for fixed p, we obtain a trace as a function of τ. This slant-stack trace satisfies the requirement that the timing of all primaries and multiples are integer valued. Thus this slant-stack trace is the same as a normal incidence trace generated by a Goupillaud model where the reflection coefficients are the ones specified by the ray parameter p. In particular, the multiple arrivals of the slant-stack trace have the same time structure as the 1-D seismic trace generated by the Goupillaud model.

It follows, therefore, we can take this slant stack and invert it by the 1-D methods given in the 1-D part of this paper. In particular, we can use Gelfand–Levitan inversion as implemented by the dynamic deconvolution method.

Conversely, we can construct a forward scheme which generates a seismic section $u(x, t)$ from a given layered model. For each ray parameter p, we find the non-normal-incidence reflection coefficients for each of the layers. We then use the forward Gelfand–Levitan scheme (i.e., dynamic reconvolution) and generate the 1-D seismic trace. This seismic trace will be a function of τ for the given p. Repeating this process for many p values, we obtain $U_R(p, \tau)$. We then apply the inverse Radon transform to obtain the synthetic seismic section $u(x, t)$.

G. Migration

So far we have considered flat layers. Now we want to consider a sloping interface in a constant velocity medium. Sup-

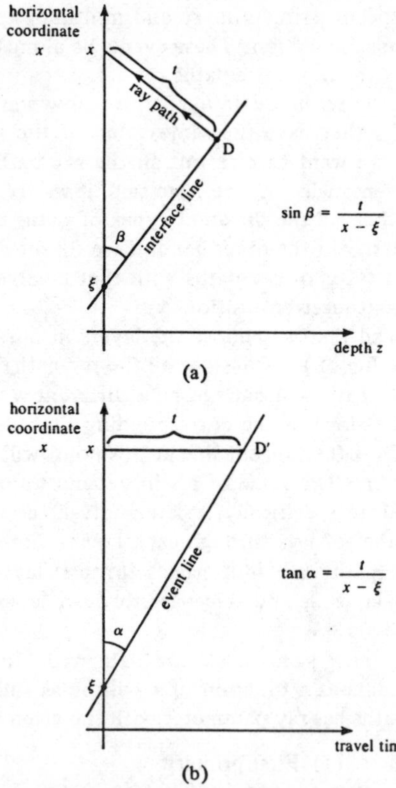

Fig. 14. Wave propagation (a) to (b). Migration (b) to (a). (a) Earth cross section. With velocity $v = 1$, a wave takes time t to travel up from depth point D to receiver x. (b) Seismic section. The depth point appears at D' on the seismic section. The interface with dip $\tan\beta$ and intercept ξ appears as the event with dip $\tan\alpha$ and same intercept. We see that $\sin\beta = \tan\alpha$.

pose the first interface is a line which makes the angle β with the horizontal. We consider a source and receiver at the same surface point x. By Fermat's theorem of least time, the ray path must be a *least time path.* Thus the ray path will be a straight line from the source to the reflecting point on the interface. The same path is used by the reflected wave back to the receiver, and this common ray path will be at right angles to the interface. Thus the ray path will make an angle β with the vertical.

Let the medium have constant velocity v, which for simplicity we take to be $v = 1$. Let us refer to Fig. 14. The interface cuts the surface of the ground at point ξ. We call ξ the intercept. Let x be the coordinate of the receiver. Thus we have a right triangle with hypotenuse $x - \xi$ and angle β. The side opposite β is the ray path $vt = t$. Here t is the one-way travel time along the ray path. Thus

$$\sin\beta = \frac{t}{x - \xi}.$$

On the seismic section, we will plot t versus x. Then the seismic trace at receiver position x will have an event at t. As we vary x, these events will fall on a straight line which intercepts the x axis at ξ; that is, the intercept of the interface line and the intercept of the event line are the same. Let the event line make an angle α with the x-axis. Again we have a right triangle. The side opposite α has length t and the (non-hypotenuse) side adjaccent to α has length $x - \xi$. Thus we have

$$\tan\alpha = \frac{t}{x - \xi}.$$

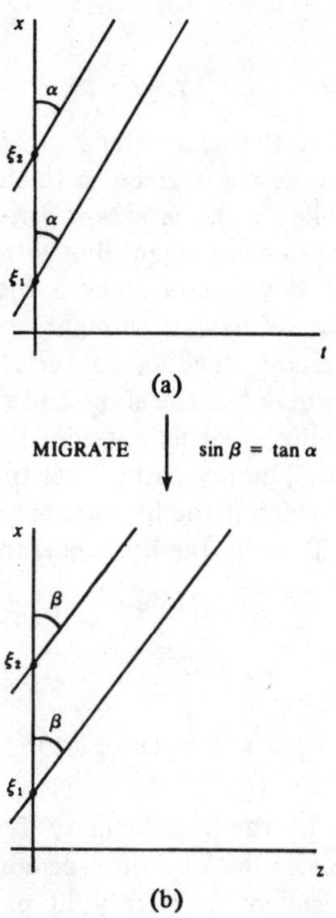

Fig. 15. Mechanical migration: (a) events on a seismic section with dips $\tan\alpha$; (b) each event is rotated about its intercept from $\tan\alpha$ to $\tan\beta$ to produce interfaces of dip $\tan\beta$.

Comparing the above two equations, we have

$$\sin\beta = \tan\alpha.$$

This is the basic equation of migration. *Migration* is the conversion of the event line in the (x, t) plane to the interface line in the (x, z) plane. Each of these lines have the same x-intercept ξ. Thus migration involves rotating the event line around this intercept ξ so as to change the event angle α to the interface angle β, where

$$\beta = \sin^{-1}\tan\alpha.$$

Prior to 1960, migration was performed by mechanical devices; see Fig. 15.

H. Spectral Migration

Each of the events on the seismic section (i.e., the observed data) is represented by a curve in the (x, t) space. Each curve may be considered as the envelope of its tangent lines. Let the equation for a given tangent line be

$$x = \frac{1}{\tan\alpha}\, t + \xi.$$

Let us now Fourier transform the data; see Fig. 16. The above line will be transformed to line

$$k_x = (\tan\alpha)\,\omega$$

and with phase spectrum $k_x\xi$. The process of migration transforms the above line into the line

$$k_x = (\tan\beta)\,k_z$$

and with same phase spectrum. We now inverse Fourier transform. The result gives the corresponding line

$$x = \frac{1}{\tan\beta}\, z + \xi$$

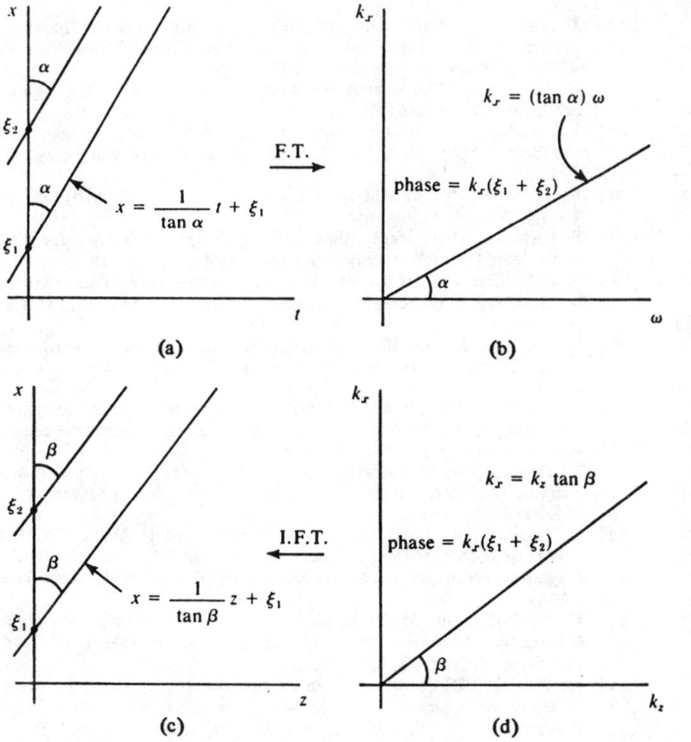

Fig. 16. Spectral migration: (a) events on seismic section with dips tan α; (b) Fourier transform converts all events into same line, with intercepts preserved in the phase; (c) migration changes dip from tan α to tan β; (d) inverse Fourier transform produces interfaces of dip tan β.

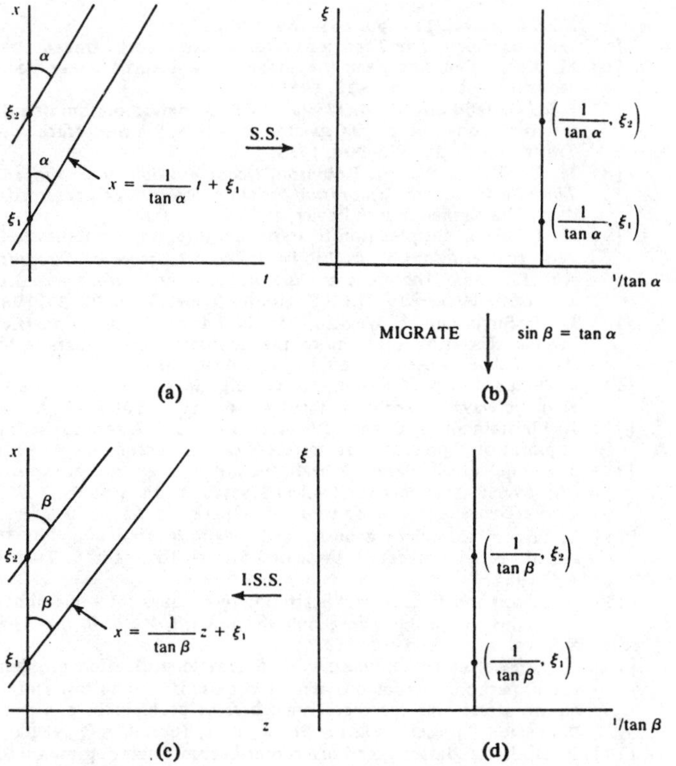

Fig. 17. Slant-stack migration: (a) events on seismic section with dips tan α; (b) slant stacking converts all events into same vertical line; (c) migration moves vertical line to left, from 1/tan α to 1/tan β; (d) inverse slant stacking produces interfaces of dip tan β.

in the (x, z) space. The interfaces appear as the envelopes of these lines. We have thus described the process of *spectral migration*, or *Stolt migration* [27].

I. Slant-Stack Migration

Each of the events on the seismic section (i.e., the observed data) is represented by a curve in (x, t) space. Each curve may be considered as the envelope of its tangent lines. Let the equation for a given tangent line be

$$\tan\alpha = \frac{t}{x - \xi}$$

which is

$$x = \frac{1}{\tan\alpha} t + \xi.$$

Let us now slant stack the data; see Fig. 17. The above line will be transformed into the point $(1/\tan\alpha, \xi)$. All the tangent lines with slope $\tan\alpha$ will fall along the transform line $(1/\tan\alpha) =$ constant in the transform plane. We now move this line left to the position $1/\tan\beta$. We thus shift all such transform lines. This shifting is the process of migration; i.e., the changing of dips. We then inverse slant stack. The result gives the corresponding lines

$$x = \frac{1}{\tan\beta} z + \xi$$

in the (x, z) space. The interfaces appear as the envelopes of these lines. We have thus described the process of *slant-stack migration* (often called *S-two migration*) which is due to Hubral [28].

IV. Conclusion

The problem of the geophysicist is to determine the structure of the interior of the earth from data obtained at the surface of the ground. Ultimately the problem is to find a method that will give structure, composition, and source parameters by processing the whole seismogram. Such a problem is an inverse problem. The forward problem is one which a model of the earth structure and seismic source is used to give properties of seismic motion. When the solution of the forward problem is known, an iteration based on trial and error represents one method of inversion. The parameters of the model are readjusted according to some criterion until some satisfying agreement between the data and the computed wave motion is discovered. This iterative type of approach to the inverse problem has proven to be successful in many applications. Today far more sophisticated inversion methods are being developed for seismic data. In this paper, we have described the present state of the art in which spectral methods in the form of lattice methods, the Fourier transform, and the Radon transform play the key role. An understanding of these spectral methods will provide a firm basis for the appreciation of the exciting new inversion methods which the future holds.

References

[1] H. J. Schudder, "Introduction to computer aided tomography," *Proc. IEEE*, vol. 66, pp. 628–637, 1978.
[2] D. W. Sweeney and C. H. West, "Reconstruction of three-dimensional refractive index fields from multidimensional interferometric data," *Appl. Opt.*, vol. 11, pp. 2649–2664, 1973.
[3] P. F. Gilbert, "The reconstruction of a three-dimensional structure from projections and its applications to electron microscopy," *Proc. R. Soc. London*, ser. B, vol. 182, pp. 89–102, 1972.
[4] T. S. Durrani and C. E. Goutis, "Optimization techniques for digital image reconstruction from their projections," *Proc. Inst.*

Elec. Eng., vol. 127, pp. 161–169, 1980.

[5] Lord Rayleigh, *The Theory of Sound.* New York: Dover, 1950.

[6] M. Kac, "Can one hear the shape of a drum?" *Amer. Math. Monthly*, vol. 73, pp. 1–23, 1966.

[7] I. M. Gelfand and B. M. Levitan, "On the determination of a differential equation by its spectral function," *Amer. Math. Soc. Trans.*, vol. 1, pp. 253–304, 1955.

[8] M. T. Silvia and E. A. Robinson, *Deconvolution of Geophysical Time Series in the Exploration for Oil and Natural Gas.* Amsterdam, The Netherlands: Elsevier, 1979.

[9] M. T. Silvia, "Application of statistical filtering techniques to inverse scattering problems," in *Proc. 2nd Int. Symp. on Computer Aided Seismic Analysis and Discrimination at Southeastern Massachusetts University* (IEEE Computer Society), pp. 20–27, 1981.

[10] M. T. Silvia and A. Weglein, "Method for obtaining a nearfield inverse scattering solution to the acoustic wave equation," *J. Acoust. Soc. Amer.*, vol. 69, pp. 478–489, 1981.

[11] J. Cohen and N. Bleistein, "A velocity inversion procedure for acoustic waves," *Geophysics*, vol. 44, pp. 1077–1087, 1979.

[12] N. Bleistein and J. Cohen, "Velocity inversion. A tool for seismic exploration," presented at the *SIAM Conf.*, Tucson, AZ, 1981.

[13] J. Mendel and F. Habibi-Ashrafi, "A survey of approaches to solving inverse problems for lossless layered media systems," *IEEE Trans. Geosci. Remote Sensing*, vol. GE-18, pp. 320–330, 1980.

[14] S. Coen, "Complete acoustic and elastic layered modeling for seismic reflection data," presented at the SIAM Conf., Tucson, AZ, 1981.

[15] R. Carroll and F. Santosa, "Scattering techniques for a one-dimensional inverse problem in geophysics," *Math. Meth. in the Appl. Sci.*, vol. 3, pp. 145–171, 1981.

[16] K. Driessel and W. Symes, "Coefficient identification problems for hyperbolic partial differential equations. Some fast and accurate algorithms for the seismic inverse problem in one space dimension," presented at the SIAM Conf., Tucson, AZ, 1981.

[17] R. H. Stolt, "Imaging and inversion of seismic data," presented at the SIAM Conf., Tucson, AZ, 1981.

[18] K. P. Bube and R. Burridge, "Solution of the one-dimensional inverse problem of reflection seismology by downward continuation of surface data," presented at the SIAM Conf., Tucson, AZ, 1981.

[19] A. H. Zemanian and P. Subramaniam, "The application of the theory of infinite networks to the geophysical exploration of layered strata," presented at the SIAM Conf., Tucson, AZ, 1981.

[20] J. Gazdag, "Migration of seismic data by phase shift plus interpolation," presented at the SIAM Conf., Tucson, AZ, 1981.

[21] A. Bamberger, G. Chavent, and P. Lailly, "About the stability of the inverse problem in 1-D wave equations. Application to the Interpretation of Seismic Profiles," *Appl. Math. Optim.*, vol. 5, pp. 1–47, 1979.

[22] S. Treitel, P. Gutowski, P. Hubral, and D. Wagner, "Plane wave decomposition of seismograms," presented at the SEG Meet., Los Angeles, CA, 1981.

[23] K. L. Larner, "Computational problems resulting from the discreteness of data sets obtained in seismic exploration," presented at the SIAM Conf., Tuscon, AZ, 1981.

[24] J. Makhoul, "Stable and efficient lattice methods for linear prediction," *IEEE Trans. Acoust., Speech, Signal Processing*, vol. ASSP-25, pp. 423–428, Oct. 1977.

[25] N. Levinson, "The Wiener RMS error criterion in filter design and prediction," *J. Math. Phys.*, 1947.

[26] J. P. Burg, "Maximum entropy spectral analysis," presented at the Soc. Exploration Geophysicists Meet., Oklahoma City, OK, 1967.

[27] R. H. Stolt, "Migration by Fourier transform," *Geophysics*, vol. 43, pp. 23–48, 1978.

[28] P. Hubral, "Slant stack migration," in *Festischrift Theodor Krey.* Hannover, Germany: Prakla-Seismos, 1980, pp. 72–78.

[29] P. Goupillaud, "An approach to inverse filtering of near surface layer effects from seismic records," *Geophysics*, vol. 26, pp. 754–760, 1961.

[30] A. Gray and J. Markel, "Digital lattice and ladder filter synthesis," *IEEE Trans. Audio Electroacoust.*, vol. AU-21, pp. 491–500, 1973.

[31] J. Makhoul, "A class of all-zero lattice digital filters: Properties and applications," *IEEE Trans. Acoust., Speech, Signal Processing*, vol. ASSP-26, pp. 304–314, Aug. 1978.

[32] T. Durrani, N. Murukutla, and K. Sharman, "Constrained algorithms for multi-input adaptive lattices in array processing," in *IEEE ICASSP Proc.*, 1981.

[33] H. Lorentz, A. Einstein, H. Minkowski, and H. Weyl, *The Principles of Relativity: A Collection of Original Memoirs.* London, England: Methuen, 1923 (reprinted by Dover Publ., New York, 1958).

[34] E. A. Robinson, *Multichannel Time Series Analysis with Digital Computer Programs.* San Francisco, CA: Holden-Day, 1967. (Revised edition, 1978.)

[35] H. Wakita, "Direct estimation of the vocal tract shape by inverse filtering of acoustic speech waveforms," *IEEE Trans. Audio Electroacoust.*, vol. AU-21, pp. 417–427, 1973.

[36] G. Kunetz, "Généralisation des opérateurs d'antirésonance à un nombre quelconque de réflecteurs," *Geophys. Prospecting*, vol. 12, pp. 283–289, 1964.

[37] G. Kunetz and I. D'Erceville, "Sur certaines propriétés d'une onde plane de compréssion dans un milieu stratifié," *Ann. Geophysique*, vol. 18, pp. 351–359, 1962.

[38] K. Aki and P. G. Richards, *Quantitative Seismology*, vol. 2. San Francisco, CA: W. H. Freeman, 1980.

[39] E. A. Robinson, "Dynamic predictive deconvolution," *Geophys. Prospecting*, vol. 23, pp. 779–797, 1975.

[40] V. Bardan, "Comments on dynamic predictive deconvolution," *Geophys. Prospecting*, vol. 25, pp. 569–572, 1977.

[41] E. A. Robinson, "Iterative identification of non-invertible auto-regressive moving-average systems with seismic applications," *Geoexploration*, vol. 16, pp. 1–19, 1978.

[42] J. Radon, "Uber die Bestimmung von Funktionen durch ihre Integralwerte langs gewisser Manningfaltigkeiten," *Berichte der Sächsischen Akademie der Wissenschaften*, vol. 69, pp. 262–277, 1917.

[43] "The MIT Geophysical Analysis Group Reports," *Geophysics*, vol. 32, 1967.

Seismic Modeling in the Domain of Intercept Time and Ray Parameter

FRIEDEMANN WENZEL, PAUL L. STOFFA, MEMBER, IEEE, AND PETER BUHL

Abstract—In recent years, seismic modeling has been based primarily on two methods: generalized ray theory and the reflectivity method. Variations within these methods are characterized by the domains of computation and the approximations applied. One common feature of both approaches is the straightforward translation of a mathematical solution of the wave equation into a computational algorithm, which develops the seismic response for a given earth structure in the observation plane of seismic experiments, source-receiver offset and travel time (X-T). We present an alternative approach in which we develop acoustic and seismic models in the domain of vertical delay, or intercept, time, and horizontal ray parameter (τ-p). This is accomplished by a modified version of the reflectivity method followed by a Fourier transform over frequency. In these models, the τ-p seismic response for compressional and shear waves is readily interpreted for all primary and multiple arrivals. The τ-p response can also be used to compute broadband synthetic X-T seismograms by an appropriate integration through the τ-p plane. A fast method of computing seismograms is suggested, based on a weighted integration designed to automatically evaluate the points of stationary phase.

I. Introduction

In 1971, Fuchs and Muller presented a method of computing the seismic response as a function of source-receiver offset (X) and travel time (T) for a stack of horizontal homogeneous layers (Fig. 1) called the "reflectivity" method. In this approach, the classical Thomson [14] and Haskell [8] matrix method is used, but instead of equispaced wavenumbers and frequency, the medium response is determined for equispaced angles of incidence and frequency. In this approach the final X-T seismic response is found by an integration over all angles of incidence instead of an inverse spatial Fourier transform. As an extension of this method, we evaluate the seismic response for equispaced ray parameters and frequency. An inverse Fourier transform over frequency results in the intercept or vertical delay time and ray parameter response $R(\tau, p)$, described by Chapman [3]. A final integration is then required over all ray parameters to determine the seismic X-T data.

The τ-p plane appears particularly attractive for seismic inversion and modeling; see Bessonova [1], Chapman [3], Fryer [6], Diebold and Stoffa [4], Stoffa *et al.* [12], and Phinney *et al.* [11]. In this plane, the loci of seismic arrivals are easily determined. Amplitudes can be described using plane wave reflection coefficients, and multiple arrivals are readily recognized by their intercept time periodicity. Since densely sampled X-T data can be transformed directly to τ-p without aliasing or the artificial suppression of high frequencies, Stoffa *et al.* [12], we investigate some practical aspects of developing and comparing seismic models directly in this domain. We also find that many of the seismic properties of a model can be clearly observed in τ-p, and illustrate these with several simple examples (Fig. 2). In this investigation, we do not suppress numerical problems by band limiting the derived seismograms. Instead, we find that using the maximum bandwidth for a specified temporal sampling interval, and computing the seismic response for all arrivals, necessitates a discussion of numerical noise and sampling requirements. By analogy with the transformation of observational X-T seismic data to τ-p, Stoffa *et al.* [12], we find that, when computing X-T seismograms from τ-p data, a considerable savings in computational effort can be obtained by using a weighted integration designed to automatically evaluate the points of stationary phase.

After a discussion of numerical difficulties, we investigate τ-p and X-T seismograms for the models of Fig. 2 and Table I. We begin with an acoustic three-layer model, model A, then move to the elastic case, model C, and then consider the case of velocity gradients, model D. For both the elastic case and gradient case, we compare the direct and weighted integration methods of computation.

II. Discrete τ-p and X-T Seismograms

We assume a stratified medium overlaid by a liquid half space, Fig. 1. The potential of a unit amplitude point source excitation at the origin may be expressed in the time domain as

$$\Phi_s(T, R) = \frac{1}{R}\,\delta\left(T - \frac{R}{\alpha_1}\right) \tag{1}$$

where $R = (X^2 + Y^2 + Z^2)^{1/2}$ and δ is Dirac's delta function. This is a spherical wave traveling with velocity α_1, representing a solution of the wave equation in a homogeneous medium. Its Fourier transform can be written as the Sommerfield integral

$$\frac{1}{R}\,e^{-j(\omega/\alpha_1)R} = \int_0^\infty J_o(k \cdot r)\,e^{-j\nu_1|Z|}\,\frac{k}{j\nu_1}\,dk \tag{2}$$

Manuscript received April 1, 1981; revised December 15, 1981. This work was supported in part by the Office of Naval Research under Contract N00014-80-C-0098 and by Lamont-Doherty Geological Observatory Contribution 3195.

F. Wenzel was with the Lamont-Doherty Geological Observatory, Columbia University, Palisades, NY 10964. He is now with the Geophysikalisches Institut, Universität Karlsruhe, 7500 Karlsruhe-West, Herzstrabe 16, Germany.

P. L. Stoffa was with the Lamont-Doherty Geological Observatory, Columbia University, Palisades, NY 10964. He is now with the Gulf Center for Marine Crustal Studies, Pearl River, NY 10965.

P. Buhl is with the Lamont-Doherty Geological Observatory, Columbia University, Palisades, NY 10964.

Reprinted from *IEEE Trans. Acoust., Speech, Signal Processing,* vol. ASSP-30, pp. 406–422, June 1982.

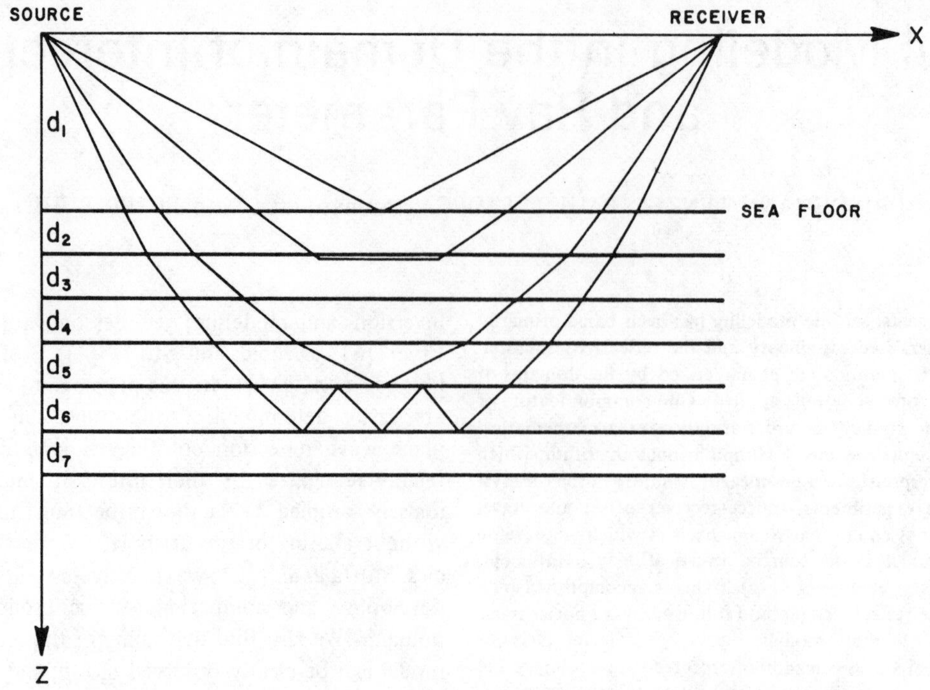

Fig. 1. Schematic diagram for an elastic layered half space. In this paper, we assume that the first layer of thickness d_1 is a fluid.

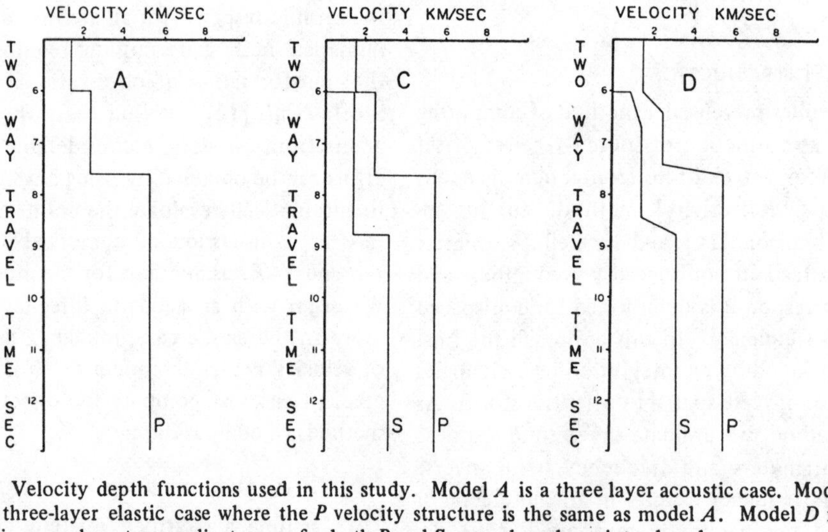

Fig. 2. Velocity depth functions used in this study. Model *A* is a three layer acoustic case. Model *C* is a three-layer elastic case where the *P* velocity structure is the same as model *A*. Model *D* is an elastic case where two gradient zones for both *P* and *S* waves have been introduced.

TABLE I

MODEL	LAYER	A	B	T	RHO
A	1	1.500	0.000	6.000	1.000
	2	2.500	0.000	7.600	1.000
	3	5.500	0.000		1.000
C	1	1.500	0.000	6.000	1.000
	2	2.500	1.443	7.600	1.500
	3	5.500	3.175		3.000
D	1	1.500	0.000	6.000	1.000
	2	1.520	0.878	6.008	1.010
	51	2.500	1.443	6.400	1.500
	52	2.500	1.443	7.400	1.500
	102	5.500	3.175	7.600	3.000
	103	5.500	3.175		3.000

where

$$\nu_1 = \begin{cases} \left(\dfrac{\omega^2}{\alpha_1^2} - k^2\right)^{1/2} & \dfrac{\omega^2}{\alpha_1^2} \geq k^2 \\ -j\left(k^2 - \dfrac{\omega^2}{\alpha_1^2}\right)^{1/2} & \dfrac{\omega^2}{\alpha_1^2} \leq k^2 \end{cases}$$

and $r = (X^2 + Y^2)^{1/2}$. J_o is the zeroth-order Bessel function, ω is the angular frequency, and k is the horizontal wavenumber. If we consider the integral representation for $J_o(kr)$,

$$J_o(kr) = \frac{1}{2\pi}\int_0^{2\pi} e^{-jk(X\cos\rho + Y\sin\rho)}\, d\rho, \tag{3}$$

it is clear that (2) is a decomposition of the spherical wave into plane waves.

This plane wave decomposition can be used to describe the reflection response of a layered medium observed at $Z = 0$:

$$\Phi(\omega, k) = \int_0^\infty R(\omega, k) J_o(kX) \frac{k}{j\nu_1} dk, \tag{4}$$

where $R(\omega, k)$ is the plane wave reflection response for the complete stack of layers beneath the liquid, including all internal multiples and compressional and shear wave conversions. $R(\omega, k)$ also includes the time delay required for a plane wave to travel from the source to the first interface at depth d and back to the receiver at $Z = 0$:

$$e^{-j\nu_1 2d_1}.$$

In addition, we have replaced r by X because, for the case of cylindrical symmetry, there is no azimuthal preference.

The potential Φ can be transformed to pressure by the relation

$$P(\omega, X) = \rho_1 \omega^2 \Phi(\omega, X), \tag{5}$$

where ρ_1 is the density of the half space. Changing variables from horizontal wavenumber K to horizontal ray parameter p where $p = K/\omega$, we find

$$\Phi(\omega, X) = \int_0^\infty (-j\omega)\tilde{R}(\omega,p) J_o(\omega pX) \frac{p}{\eta_1} dp \tag{6}$$

where

$$\eta_1 = \begin{cases} \left(\dfrac{1}{\alpha_1^2} - p^2\right)^{1/2} & p \leqslant \dfrac{1}{\alpha_1} \\ -j\left(p^2 - \dfrac{1}{\alpha_1^2}\right)^{1/2} & p \geqslant \dfrac{1}{\alpha_1}. \end{cases}$$

Using the standard high frequency approximation for the Bessel function and considering only outward traveling waves,

$$J_o(\omega pX) \simeq \frac{1}{\sqrt{2\pi|\omega|pX}} e^{-j\omega pX} e^{j(\pi/4)\frac{\omega}{|\omega|}} \tag{7}$$

we can rewrite (6) in the following form:

$$\Phi(\omega, X) = \frac{1}{\pi\sqrt{2X}} \int_0^\infty \frac{p^{1/2}}{\eta_1} \left[(-j\omega)\left(\frac{\pi}{|\omega|}\right)^{1/2} e^{+j(\pi\omega/4|\omega|)}\right] \cdot [\tilde{R}(\omega,p) e^{-j\omega pX}] \, dp. \tag{8}$$

To determine the time response, we take the inverse Fourier transform and find

$$\Phi(T, X) = \frac{1}{\pi\sqrt{2X}} \int_0^\infty \frac{p^{1/2}}{\eta_1} \left(-\frac{d}{dT} \frac{\theta(-T)}{(-T)^{1/2}}\right) * R(T - pX, p) \, dp \tag{9}$$

where $\theta(T)$ is the Heaviside or unit step function, R is the Fourier transform of R, and $*$ denotes convolution. The term $\theta(-T)/(-T)^{1/2}$ results from approximating the Bessel function. The convolutional term in brackets removes the leading noncasual part of the pulse that would appear if we performed the integration just over $p^{1/2}/\eta_1 \cdot R(T - pX, p)$. (For very small offsets, however, this approximation equation (7) for the Bessel function is incorrect, and the resulting seismograms (9) are also in error.) To derive the plane wave reflection response from seismograms, the above procedure may be reversed, and we find

$$\tilde{R}(\omega, p) = \frac{\eta_1}{\pi\sqrt{2p}} \int_0^\infty X^{1/2} j\omega \left[\frac{\pi}{|\omega|}\right]^{1/2} \cdot e^{-j(\pi\omega/4|\omega|)} \Phi(\omega, X) e^{j\omega pX} dX \tag{10}$$

or

$$R(\tau, p) = \frac{\eta_1}{\pi\sqrt{2p}} \int_0^\infty X^{1/2} \left(\frac{d}{d\tau} \frac{\theta(\tau)}{\tau^{1/2}}\right) * \Phi(\tau + pX, X) \, dX. \tag{11}$$

Equations (9) and (11) demonstrate how seismograms are derived from the plane wave reflection response, and how the plane wave reflection response can be derived from seismograms transformed from the observational plane to τ-p. (For detailed discussion, see Chapman [3] and Phinney *et al.* [11].)

Synthetic X-T seismograms are calculated using a discrete form of (9). Numerical difficulties will be encountered since discretization implies limited bandwidth. Previous studies minimized these numerical difficulties by considering only a limited suite of arrivals, frequencies, ray parameters (or angles of incidence) and offsets; see Fuchs and Muller [7]. In this study, we investigate the computation of all P and SV primary and multiple arrivals for a dense spatial sampling, and the maximum possible bandwidth for a fixed temporal sampling. This objective leads immediately to numerical problems in the calculation of the discrete plane wave response, and in the final discrete X-T seismograms.

A. The Reflection Response, $R(\tau, p)$

To calculate the plane wave elastic reflection response $R(\tau, p)$ for a layered half space, we compute its Fourier transform $\tilde{R}(\omega, p)$ using the method of Thomson [14] and Haskell [8]. By computing the subdeterminant of the individual layer matrices (see, for example, Dunkin, [5]) and an additional scaling, we obtain high numerical stability even in the kHz range. For details of the procedure, see Buhl and Wenzel [2].

Since $\tilde{R}(\omega, p)$ is not a band-limited function, its discrete Fourier transform $R(i, p)$ (where i is the time sample index) will have oscillations at the Nyquist frequency. This can be demonstrated for a simple two-layer case where the postcritical response is

$$\tilde{R}(\omega, p) = r(p) e^{j\Phi(p)\frac{\omega}{|\omega|}}. \tag{12}$$

The inverse Fourier transform, $R(\tau, p)$ is

$$R(\tau, p) = r(p) \left[\cos \Phi(p)\delta(\tau) - \frac{1}{\pi\tau} \sin \Phi(p)\right], \tag{13}$$

and the discrete Fourier transform for a sampled $R(\omega, p)$ is

$$R(0,p) = \frac{1}{N}\,[R_0 + (N-2)\,r(p)\cos\Phi(p) + R_{N/2}], \qquad i = 0$$

$$R(i,p) = \frac{1}{N}\,[R_0 - 2r(p)\cos\Phi(p) + R_{N/2}], \qquad \text{for even } i\text{'s}$$

$$R(i,p) = \frac{1}{N}\left[R_0 - 2r(p)\sin\Phi(p)\,\mathrm{ctg}\left(\pi\frac{i}{N}\right) - R_{N/2}\right], \qquad \text{for odd } i\text{'s where} \quad (14)$$

the length of the time series is N, and R_0 and $R_{N/2}$ are the dc and Nyquist terms, respectively. If $R_0 = R_{N/2} = r(p)\cos\Phi(p)$, we have

$$R(0,p) = r(p)\cos\Phi(p), \qquad \text{for } i = 0$$

$$R(i,p) = 0, \qquad \text{for even } i\text{'s}$$

$$R(i,p) = \frac{-1}{\pi}\,r(p)\sin(p)\,\Phi\,\frac{2\pi}{N}\,\mathrm{ctg}\left(\pi\frac{i}{N}\right), \qquad \text{for odd } i\text{'s where} \quad (15)$$

$0 < i < N/2$. Since $(2\pi/N)\,\mathrm{ctg}\,(\pi i/N)$ approaches $2/i$ for large values of N, we observe the $1/\tau$ decay for postcritical reflections.

A similar phenomenon occurs for precritical reflections when the event arrival time does not correspond to an integer number of samples. For example, in a two-layer case where the thickness of the layer is d_1, the precritical reflection response is

$$R(\omega,p) = r(p)\,e^{-j\omega 2d_1\eta_1}. \qquad (16)$$

For a fixed sampling rate ΔT, the linear phase shift associated with the time delay $2d_1\eta_1$ will not, in general, be an integer multiple of π at the Nyquist frequency. In this case, the discrete Fourier transform for a sampled $R(\omega, p)$ is

$$R(i,p) = r(p)\,\frac{\sin\left(\pi\left(i - \frac{2d_1\eta_1}{\Delta T}\right)\right)\cos\left(\pi\left(i - \frac{2d_1\eta_1}{\Delta T}\right)\right)}{N\sin\left(\frac{\pi}{N}\left(i - \frac{2d_1\eta_1}{\Delta T}\right)\right)}. \qquad (17)$$

To minimize these artificial oscillations, we apply a frequency domain sinc filter, whose first zero crossing is at the Nyquist frequency:

$$F_o = 1$$

$$F_k = \frac{\sin\left(\frac{2\pi}{N}k\right)}{\left(\frac{2\pi}{N}k\right)}, \qquad k = 1, 2, \cdots, \frac{N}{2} \qquad (18)$$

where k is the discrete frequency sample index. This corresponds to smoothing over two samples in the time domain and is preferable to a low-pass filter since the pulse shape is preserved; see Fig. 3.

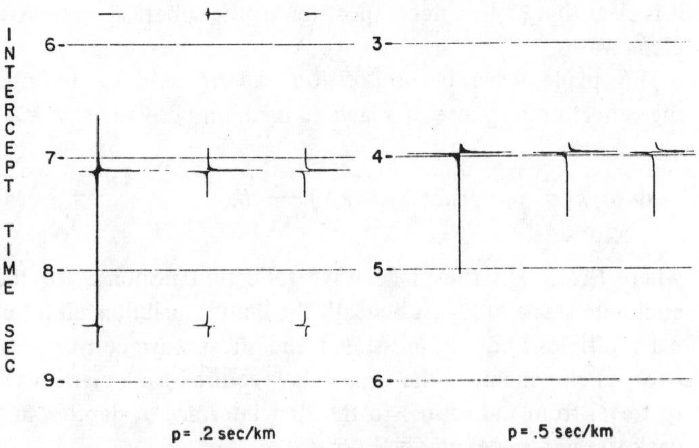

Fig. 3. Broad-band 0–125 Hz pulse shapes for two reflections of model *A* at ray parameters of 0.2 s/km (left) and 0.5 s/km (right). In all cases, the reflection response was computed in the frequency domain. For both ray parameters, the seismograms on the left were computed using the discrete Fourier transform, and consequently, oscillations appear at the Nyquist frequency, while the seismograms on the right were first multiplied by a frequency domain sinc filter, which corresponds to averaging over two time samples. This produces a superior result compared to the seismograms in the center, which were filtered with a zero phase low-pass filter from 0–80 Hz, with a 20 Hz cosine taper.

Another problem occurs when a free surface is included and the reflection response is computed in the frequency domain. A free surface modifies the half-space reflection response by requiring perfect feedback of the medium response:

$$\sum_{n=0}^{N_w} (-1)^n \tilde{R}^{n+1}(\omega,p)\,e^{-j\omega 2d_1(n+1)\eta_1(p)} \qquad (19)$$

where d_1 is the thickness of the first layer and N_w is the number of water column multiples to be included. Performing this computation in the frequency domain will result in cyclic time-domain convolution, generating artificial arrivals which should be outside the time window of interest. We avoid this problem by computing the response for a half space overlying a stack of horizontal layers $\tilde{R}(\omega, p)$, performing a Fourier transform to time, and then including only the feedback required for the times of interest.

Figs. 9(a), 10(a), and 12(a) are displays of τ-p seismograms for the models of Fig. 2. In all cases, the seismograms were computed in the frequency domain for each ray parameter using the modified reflectivity method. The frequency response was then modified using a frequency domain sinc filter (18). In all cases, the temporal sampling interval is 4 ms, and except for the sinc smoothing, the τ-p seismograms represent the 0–125 Hz plane wave reflection response.

B. X-T Seismograms by Direct Integration

To calculate X-T seismograms from the τ-p response $R(i, p)$ requires a discrete form of (9). Before the integration over ray parameter, we apply the operator $(-d/dT)(\theta(-T)/(-T))^{1/2}$ in the frequency domain. That is, the seismogram $\Phi(T, X)$ is

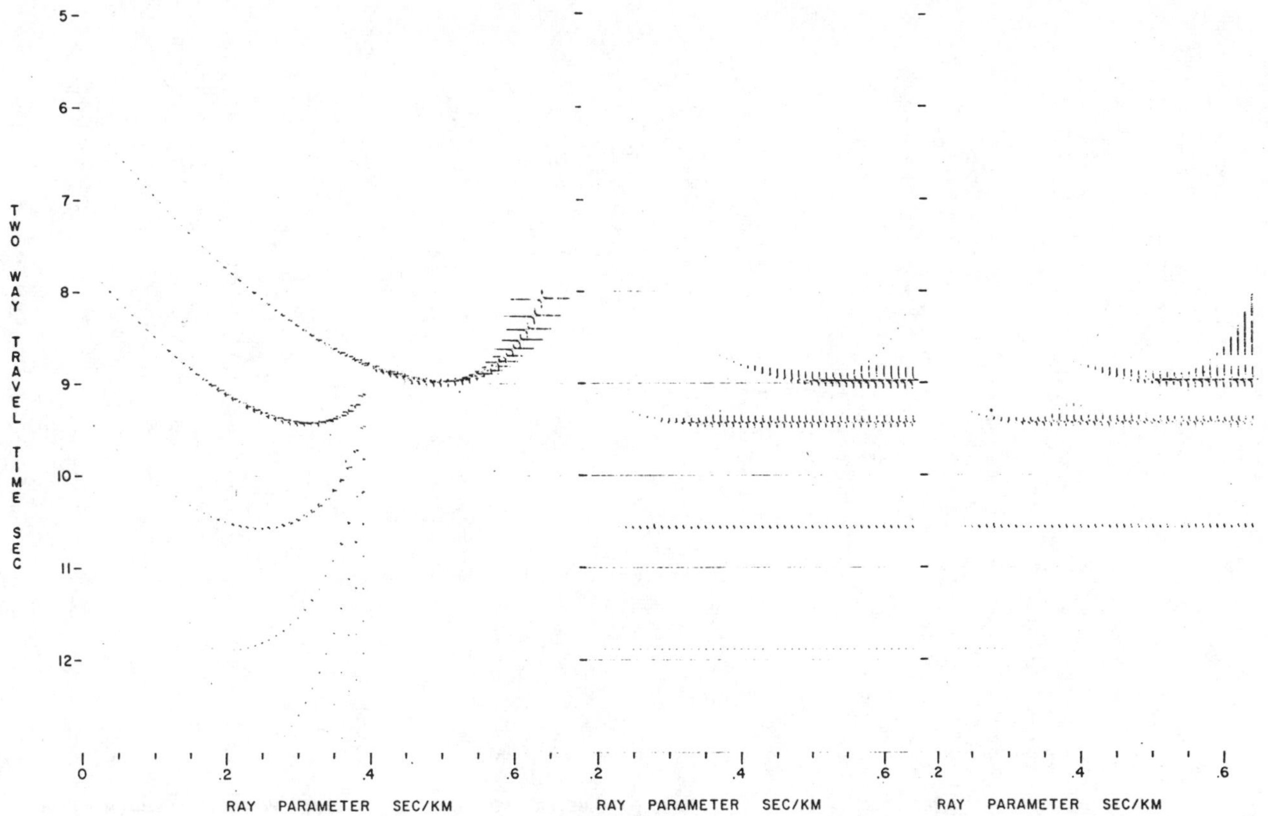

Fig. 4. The integrands for acoustic model A are plotted on the left for a range of 10 km. In the center are the partial integrals for a ray parameter sampling interval Δp of 3.125×10^{-4} s/km plot displayed every 0.01 s/km, and on the right for a Δp of 1.25×10^{-3} s/km. The decreased ray parameter sampling introduces significant numerical noise. All three plots indicate that only the arrivals corresponding to the geometric ray paths should contribute significantly to the seismogram.

$$\Phi(T, X) = \frac{1}{\sqrt{2\pi X}} \int_0^\infty R'(T - pX, p) \frac{p^{1/2}}{\eta_1} dp \tag{20}$$

where

$$R'(\tau, p) = \frac{1}{2\pi} \int_{-\infty}^{\infty} |\omega|^{1/2} e^{+j(\pi\omega/4|\omega|)} \tilde{R}(\omega, p) e^{+j\omega\tau} d\omega. \tag{21}$$

If we want to include a source function $S(t)$, we would multiply $\tilde{R}(\omega, p)$ with the Fourier transform of the source $\tilde{S}(\omega, p)$.

In discrete form, the seismogram $\Phi(i, X)$ is found by approximating the integral of (20) by the sum

$$\Phi(i, X) = \frac{1}{\sqrt{2\pi X}} \sum_{j=0}^{NP} R'(i - j\Delta pX, j\Delta p) \frac{(j\Delta p)^{1/2}}{\eta_1} \Delta p. \tag{22}$$

To see how a seismogram at a specific distance X is generated and the influence of successive ray parameters, we plot the integrands of (22) and the partial integrals for successive upper limits of the integration NP.

The integrands for the acoustic example, model A (Fig. 4) and an elastic example, model C (Fig. 5), show that only the stationary points as determined by the geometric ray paths should contribute substantially to the final seismogram. All the other contributions of the integrand should cancel after integration. For example, the left panel of Fig. 4 is the integrand for a seismogram at 10 km. The center and right panels show the partial integrals computed for a ray parameter sampling interval Δp of 3.125×10^{-4} and 1.25×10^{-3} s/km. For both sampling intervals, the partial sums are displayed every 0.01 s/km. Increasing the ray parameter sampling interval introduces significant numerical noise. For the coarse ray parameter sampling interval, the refraction arrival in Fig. 5 is almost completely lost in numerical noise. Although the effect of spherical spreading (particularly the 45° phase shift) facilitates cancellation, Figs. 4 and 5 indicate that, for broad-band data, a dense ray parameter sampling is required, even though only the stationary points should contribute significantly to the final seismogram.

The requirement for fine ray parameter sampling results in significantly increased computation time. A fine ray parameter sampling increases the time required for the integration, but the major increase in time comes from defining the plane wave reflection response $R(i, j\Delta p)$. This computational effort can be significantly reduced by first locating the stationary points, and then integrating only over the appropriate region. Unfortunately, for multilayered models which include all possible P and S conversions and all multiple arrivals, considerable effort must be made to define the locations of all stationary points. Thus, we suggest below a method of automatically defining the points of stationary phase, which permits seismograms to

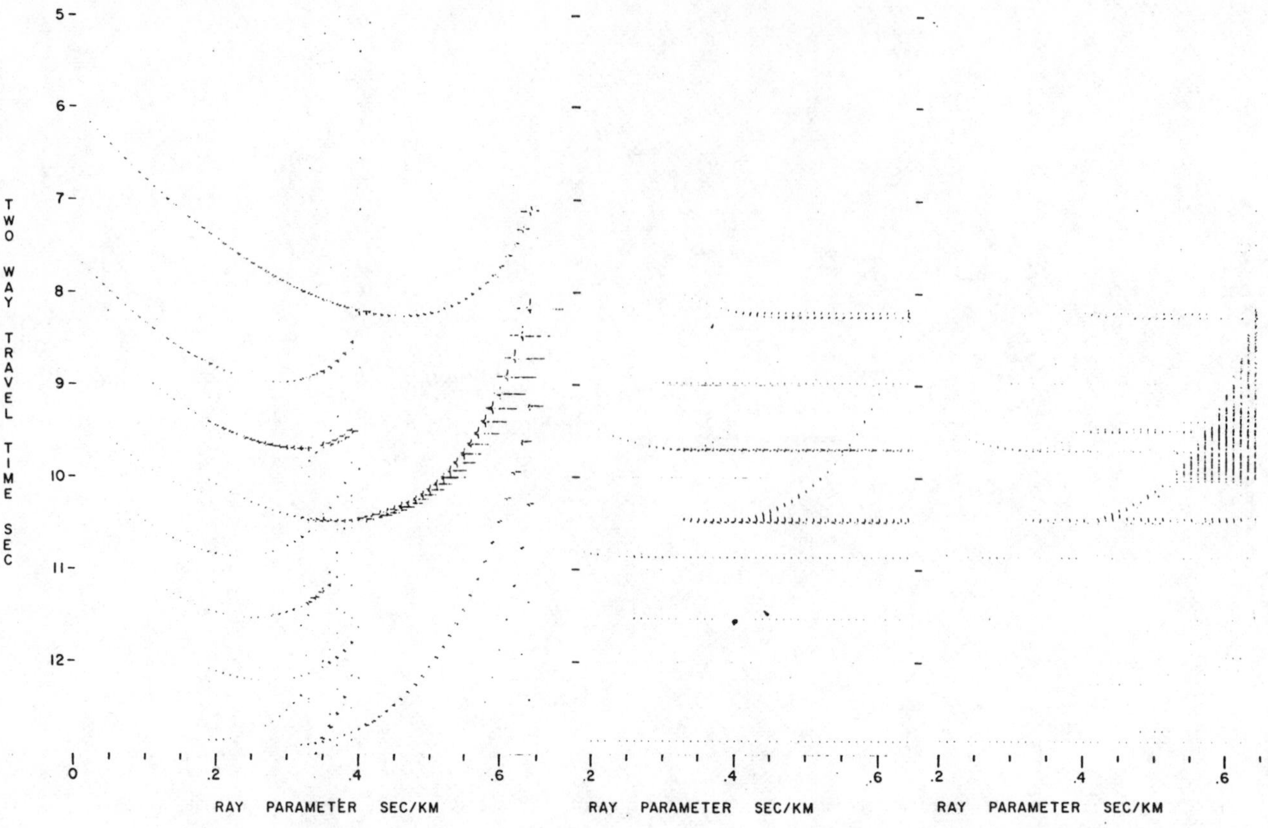

Fig. 5. Integrands for elastic model C are plotted on the left for a range of 8.5 km. In the center are the partial integrals for a ray parameter sampling interval Δp of 3.125×10^{-4} s/km, and on the right for a Δp of 1.25×10^{-3} s/km. In both cases, the partial sums are displayed every 0.01 s/km. In this example, the first arrival is a refraction from just below the sea floor, followed by the sea floor reflection. The decreased ray parameter sampling results in incomplete cancellation, introducing significant numerical noise which overwhelms the weak refraction arrival.

be calculated using a relatively coarse ray parameter sampling.

Figs. 4 and 5 also show that a simple truncation of the integration introduces artificial, but predictable, arrivals that may have significant amplitudes. To minimize the effect of this rectangular ray parameter window, we continue the integration beyond the maximum ray parameter of interest. Before including these additional ray parameter traces in the integration, we apply cosine weights as a function of ray parameter to window the data. This results in a considerable reduction of these arrivals.

We have also noticed slightly different results when the integration is performed in the frequency domain, as in the original reflectivity method. Fig. 6 is a comparison of several integration methods for the seismogram of model C at 8 km. In method 1, the seismograms were integrated over ray parameter (1a) and angle (1b) in the frequency domain. In method 2, the seismograms were integrated in the time domain as a function of ray parameter (2a) and angle (2b). In all cases, 2048 ray parameters or angles were used in the integration with a sampling interval of 3.125×10^{-4} s/km or, equivalently, 0.042°. The sampling interval in time was 4 ms. Since the time delay of each ray parameter trace required in the integration will not necessarily correspond to an integer number of samples, the ray parameter traces were delayed to the nearest sample. In method 3, the seismograms were integrated in the time domain, but the data were linearly interpolated in time before being included in the integration. This results in a considerable improvement, and the resulting seismogram is now in good agreement with the seismograms calculated in the frequency domain. The differences between integrating in the time or frequency domain (methods 1a and 3a) arise solely from the different interpolation methods. A noninteger sample time shift corresponds in the frequency domain to a linear phase shift, which is not an exact multiple of π at the Nyquist frequency. In the time domain, this would be equivalent to interpolating with the time function of (13) where $r(p)$ is defined as unity and $2d_1\eta_1$ is the time delay. Because the differences are minor, we prefer to interpolate linearly and then integrate in the time domain, since less computational effort is required. Also, we did not find any significant difference between calculating the integration over ray parameter or angle of incidence. We prefer to integrate over ray parameters only because once the reflectivity is evaluated in terms of τ and p, we can compare it directly with seismic data decomposed into plane waves (see Stoffa *et al.* [12]).

At small ranges (<1 km), an additional problem develops from the use of the standard high frequency approximation to the Bessel function, (7). At small source-receiver offsets and ray parameters, or for low frequencies, this approximation subjects the spectrum of the reflected pulses to an improper phase shift. In Fig. 7 we plot $J_0(\omega pX)$, the standard high frequency approximation and the absolute value of their differ-

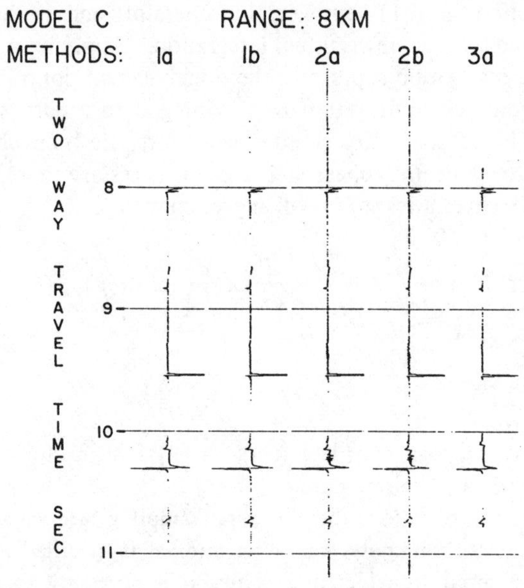

Fig. 6. A comparison of integration methods for the seismogram of model C at a range of 8 km. The seismograms for method 1 were computed by integrating over ray parameter, method 1a; and angle, method 1b, in the frequency domain. In method 2 the seismograms were integrated over ray parameter, method 2a, and angle, method 2b, in the time domain. The data were not linearly interpolated for noninteger time shifts. In method 3, the data were integrated over ray parameter in the time domain, but the data were linearly interpolated for the required time shift before being included in the integration. The result is in good agreement with the frequency domain, method 1a and 1b.

ence (dashed line) multiplied by 10. For values of the argument below π there is significant error in this approximation. For a fixed X and p, this will result in an incorrect phase shift over a range of frequencies that distorts the reflected pulse shape. For example, at a range of 1 km and a ray parameter of 0.01 s/km, almost all of the frequencies below 50 Hz will have a phase shift error. This error can only be avoided by computing the Bessel function exactly.

B. Weighted Integration

Figs. 4 and 5 clearly demonstrate that the principal τ-p contributions to an X-T seismogram should be, from the stationary phase points, associated with each source-receiver offset. That is, the integrand of (18) should have significant contributions only from those time-delayed ray parameter traces which appear along horizontal τ-p trajectories. Knowledge of the velocity-depth function makes it possible to find the stationary phase points from purely geometrical considerations. Unfortunately, this requires ray tracing for each arrival selected. An alternative approach is to define a quantitative measure of stationarity that represents an acceptable tradeoff in computational effort and accuracy, and automatically includes all contributions acceptable to this measure in the computation of the seismogram. Based on the ray parameter and temporal sampling interval, this can be accomplished by a weighted rather than direct integration.

By analogy with the transformation of seismic array data to vertical delay time and ray parameter (Stoffa *et al.* [12]), we propose that the weights used in the integration be based on

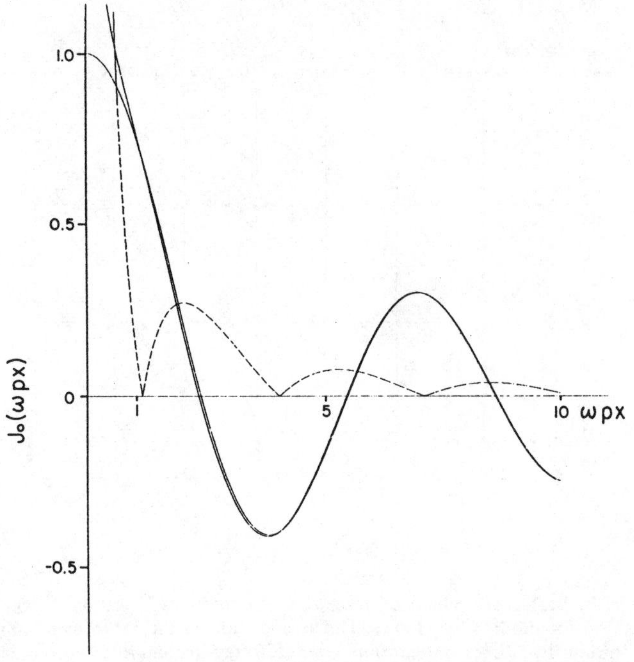

Fig. 7. Zero-order Bessel function, high frequency approximation [(3)], and the absolute error (dashed line) of the latter exaggerated by a factor of 10. The approximation goes to infinity for small arguments. In our calculation, we use only that part of the approximation that represents an outward traveling wave. Errors in this approximation for a specified X and p result in an incorrect phase shift.

the local ratio of coherent to total energy. Neidell and Taner [10] show that semblance, as defined by Taner and Koehler [13], is this ratio. In our application, we compute semblance on a unit sample basis for $NS + 1$ delayed ray parameter traces centered about the ray parameter $j\Delta p$:

$$S(i, X, j\Delta p) = \frac{\left[\sum_{k=j-NS/2}^{j+NS/2} R'(i - k\Delta p) \frac{(k\Delta p)^{1/2}}{\eta_1}\right]^2}{(NS+1) \sum_{k=j-NS/2}^{j+NS/2} \left[R'(i - k\Delta pX, k\Delta p) \frac{(k\Delta p)^{1/2}}{\eta_1}\right]^2}. \tag{23}$$

Semblance is a normalized quantity ranging from 0 to 1 and measures the coherent energy in the vicinity of the ray parameter $j\Delta p$ for each offset X. For a given model, the usefulness of this measure will be dependent upon the ray parameter and temporal sampling intervals, offset, and on the width of the ray parameter window considered. We have used semblance directly as weights in the integration or have derived weights of unity for values of semblance above a specified minimum. That is,

$$\begin{aligned} W(i, X, j\Delta p) &= S(i, X, j\Delta p), \quad &\text{for } S \geqslant S_{\min} \\ &= 0, \quad &\text{for } S \leqslant S_{\min} \end{aligned} \tag{24}$$

or

$$\begin{aligned} W(i, X, j\Delta p) &= 1, \quad &\text{for } S \geqslant S_{\min} \\ &= 0, \quad &\text{for } S < S_{\min}. \end{aligned} \tag{25}$$

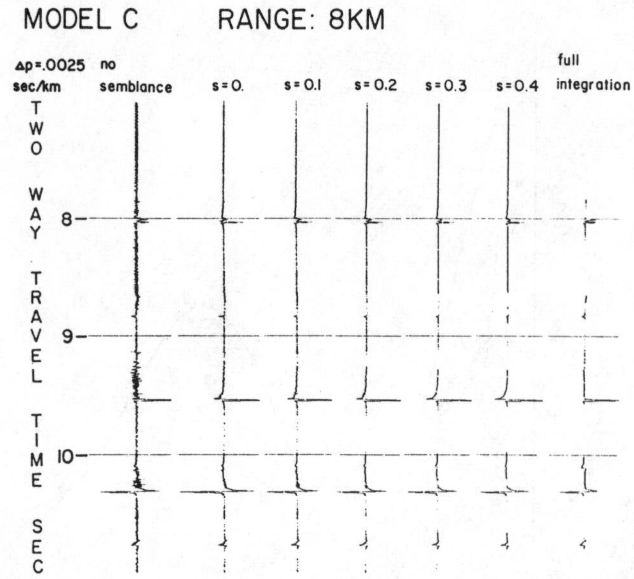

Fig. 8. Stationary phase seismograms computed by weighted integration for model C at 8 km. The seismogram on the far left was computed by direct integration over 256 ray parameters with a ray parameter sampling interval of 0.0025 s/km. On the far right, the seismogram was computed by a direct integration of 2048 ray parameters with a ray parameter sampling interval of 3.125×10^{-4} s/km. This fine ray parameter sampling interval is required to approximate the analytic solution. The seismograms in the center were computed by a weighted integration. In this example, the semblance values above the indicated minimum values were used as weights, and zero was used as the weight for values of semblance below the minimum. Semblance was computed over subray parameter intervals of 0.02 s/km as a measure of coherent to total energy in the interval. A limit of 0.1 indicates that only arrivals with a 10 percent coherent to total energy ratio over each subinterval are included in the seismogram. As the minimum acceptable value is raised, the seismograms begin to approximate the seismogram on the right for the reflection arrivals. The nonstationary refraction arrival just below 8 s of travel time is, however, not properly recovered.

Using these weights in the integration we derive the approximate seismogram:

$$\Phi'(i, X) = \frac{1}{\sqrt{2\pi X}} \sum_{j=0}^{NP} W(i, X, j\Delta p) \cdot R(i - j\Delta p X, j\Delta p) \frac{(j\Delta p)^{1/2}}{\eta_1} \Delta p. \tag{26}$$

In order to investigate the effect of the weighted integration on the resulting seismograms and to justify that the frequency correct according to (21) is now unnecessary, we evaluate (26) in the continuous case for an onset from an interface whose travel time is $T(X)$, using the method of stationary phase. This yields

$$\phi'(T, X) = \frac{1}{\pi\sqrt{X}} \frac{p}{\eta_1} \left(\frac{d^2T}{dX^2}(X)\right) \left\{\mathrm{Re}\,(R(p))\,\delta(T - T(X)) - \frac{\mathrm{Im}(R(p))}{\pi} \frac{1}{(T - T(X))}\right\} * \frac{H(T - T(X))}{(T - T(X))^{1/2}} \tag{27}$$

where $p = dT(X)/dX$ for a given X. The convolutional operator $H(T - T(X))/(T - T(X))^{1/2}$ is introduced by including the incoherent pulses around the stationary point. The frequency correction in (21) removes this operator, and is therefore required in the unweighted integration. Applying a weight in the integration suppresses these inchoherent contributions, and confines contribution to the integral to a narrow range around the stationary point. Therefore, the convolutional operator does not appear and it is not necessary to remove it. The weighted integration will approximate

$$\phi'(T, X) = \frac{1}{\pi\sqrt{X}} \frac{p}{\eta_1} \left(\frac{d^2T}{dX^2}(X)\right) \left\{\mathrm{Re}\,(R(p))\,\delta(T - T(X)) - \frac{\mathrm{Im}(R(p))}{\pi} \frac{1}{(T - T(X))}\right\} \tag{28}$$

which we recognize as Helmberger's [9] first motion solution for a reflection seismogram.

In our application, the temporal sampling interval is fixed and the ray parameter sampling interval is specified to minimize computational effort. Thus, it is necessary to specify only the size of the ray parameter subinterval NS or, equivalently, the minimum acceptable semblance value. Either parameter can be varied as a function of source-receiver offset or reference ray parameter.

Fig. 8 is an example in which semblance was used directly as weights in the integration. In this example, the ray parameter sampling interval is 0.0025 s/km, and the seismogram on the far left is the result of direct integration. To the right of this seismogram, the semblance values above the indicated minimums were used as weights in the integration. In all cases, eight ray parameter traces were used to define the ray parameter subinterval. On the right is the seismogram computed by direct integration using 2048 ray parameters, with a sampling interval of 3.125×10^{-4} s/km. Significant improvement over the direct integration with a comparable ray parameter sampling interval (left) is obtained by simply using the semblance values as weights ($S_{\min} \geqslant 0$). As the minimum acceptable value of semblance is raised (or the degree of required stationarity is increased), the integration noise is significantly reduced. However, the amplitude of some events are changed. For example, in all cases the amplitude of the refraction event just below 8 s of travel time is diminished.

The reasonable agreement of our preliminary results for broad-band seismograms indicates that this approach may prove useful in many applications. Computation time can be minimized and approximate synthetic seismograms can be generated for all P and SV primary and multiple arrivals, without specifying the desired ray paths. These data can then be used to iteratively refine a trial earth model before performing a more exact analysis. It may also be possible to improve the results presented here by including some minor knowledge about the model in the definition of the ray parameter subinterval or semblance threshold.

Figs. 10(d) and 12(d) are examples of seismograms computed using this method for the elastic models C and D. In both cases, only 256 ray parameter traces were used in the computation. In Figs. 10(d) and 12(d), the results of the weighted integration are in good agreement with the full integration, Figs. 10(c) and 12(c) (where 2048 ray parameter traces were

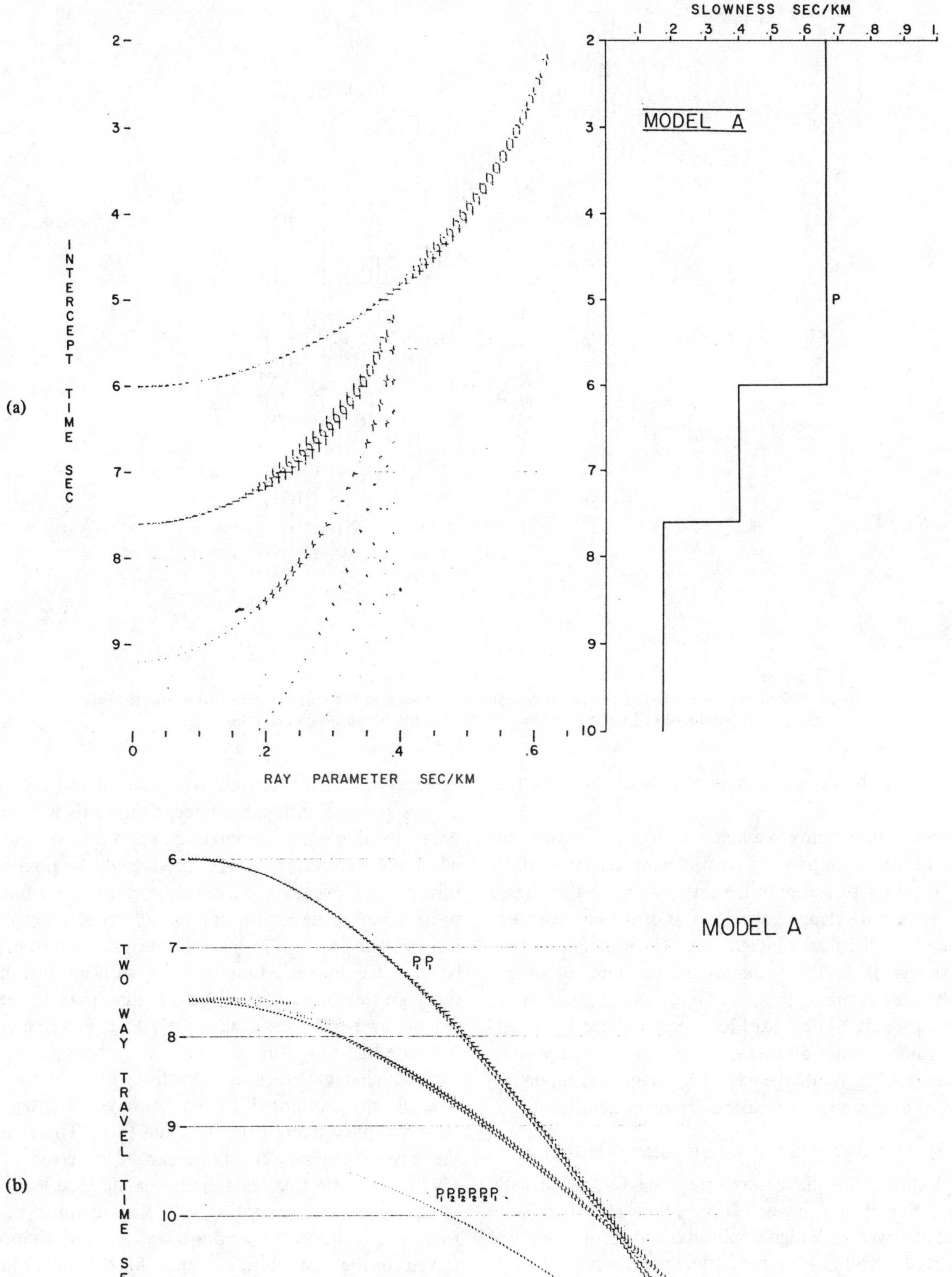

Fig. 9. (a) The τ-p response of the acoustic three-layer model A without water column multiples. (b) Seismograms for model A. P_1P_1 is the sea floor reflection, $P_1P_2P_2P_2P_1$ is the first multiple of layer 2.

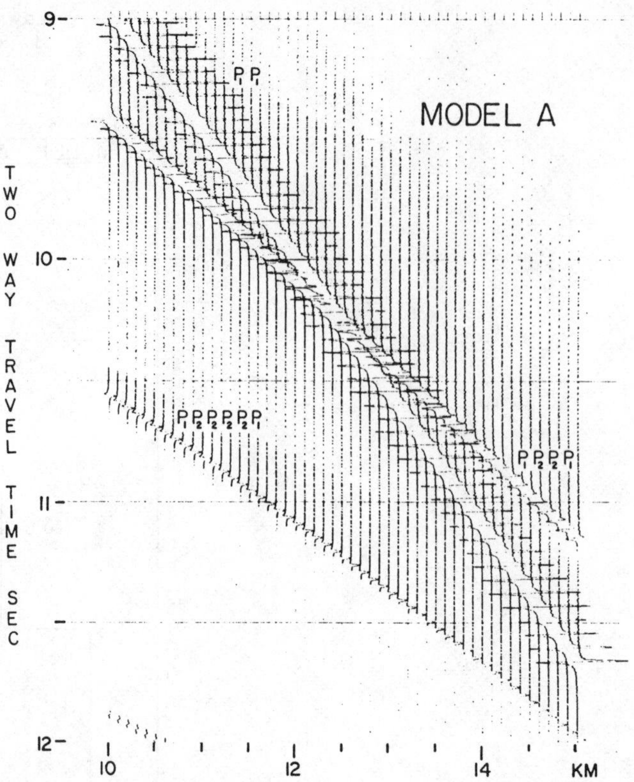

(c)

Fig. 9. (*Continued.*) (c) Detail of the seismograms of model *A* in the range 10 km–15 km and the time window 9–12 s, showing the crossover of the two primary reflections.

used) except for the decreases in amplitude of the refraction arrivals.

The weighted integration method results in a considerable savings (factor of 8) compared to the direct integration method. The time required to compute the integration weights, based on semblance for overlapping windows, is of the same order as that required for the direct integration. The principle computational savings is in the reduction of the number of ray parameter traces required to approximate the exact solution. In a Floating Point System AP120-B, the microcoded reflectivity algorithm requires 60 μs for every layer, ray parameter, and frequency. For multilayered or high frequency modeling, this factor of 8 computational savings becomes significant.

III. τ-p and X-T Response for Simple Models

Fig. 9(a) shows the plane wave response for an acoustic model consisting of three layers. The sea floor reflection is the first arrival observed with high amplitudes and increasing phase shift, observed at and beyond the critical point at $p = 0.4$ s/km. The primary response of the second layer begins at 7.6 s for $p = 0$ s/km and reaches its critical value at 0.18 s/km. This arrival is limited by the sea floor response with its critical point at 0.4 km/s, which corresponds to a grazing ray in the second layer. This pulse, at 5.2 s and 0.4 s/km, has the expected π phase shift. At vertical incidence, the internal multiples from within this layer have intercept times of 6, 7.6, and 9.2 s, but converge to 4.8 s at the critical point.

From τ-p seismograms we are able to immediately derive some properties of the X-T seismograms. For example, the tangent $-d\tau/dp$ to any of the τ-p trajectories will be the distance at which the arrivals are observed and the intersection of this tangent with the intercept time axis will be its arrival time. By this purely geometrical approach, we can anticipate when the significant seismic arrivals will be observed. Using this approach, we expect a strong postcritical sea floor response to be observed beyond 8 km and at arrival times greater than 8 s. We also expect a high amplitude reflection from the second layer at 6.5 km at a time of 8.5 s, with internal multiples at times greater than 10 s. We also predict that the sea floor and second layer will cross over at $T = 10.2$ s, and that both arrivals will have high amplitudes.

These characteristics are exactly what is observed in the seismograms computed by an unweighted integration over 2048 ray parameters [Fig. 9(b) and (c)]. The phase shifts of the beyond-critical reflections can be observed in the range 10–15 km. In this region, the sea floor reflection exhibits phase shifts of from $\pi/4$ to $\pi/2$, the second layer reflection, from $\pi/2$ to almost π, and a phase shift of almost 2π is observed for the first multiple within this layer at 15 km.

In this acoustic model, the refractions are very weak and appear in Fig. 9(b) more as "noise" that precedes the reflections. Once the refracted and reflected arrivals are separated, however, we do observe a small step function for the refracted pulse. This is not surprising, since at a given postcritical distance, every ray is completely backscattered. The rays between the critical point and the ray parameter of the reflection contribute to the refraction and, since all the energy goes back to the surface, the reflected pulse will be strong. This result is in agreement with the first motion approximation of the wave equation; see Helmberger [9].

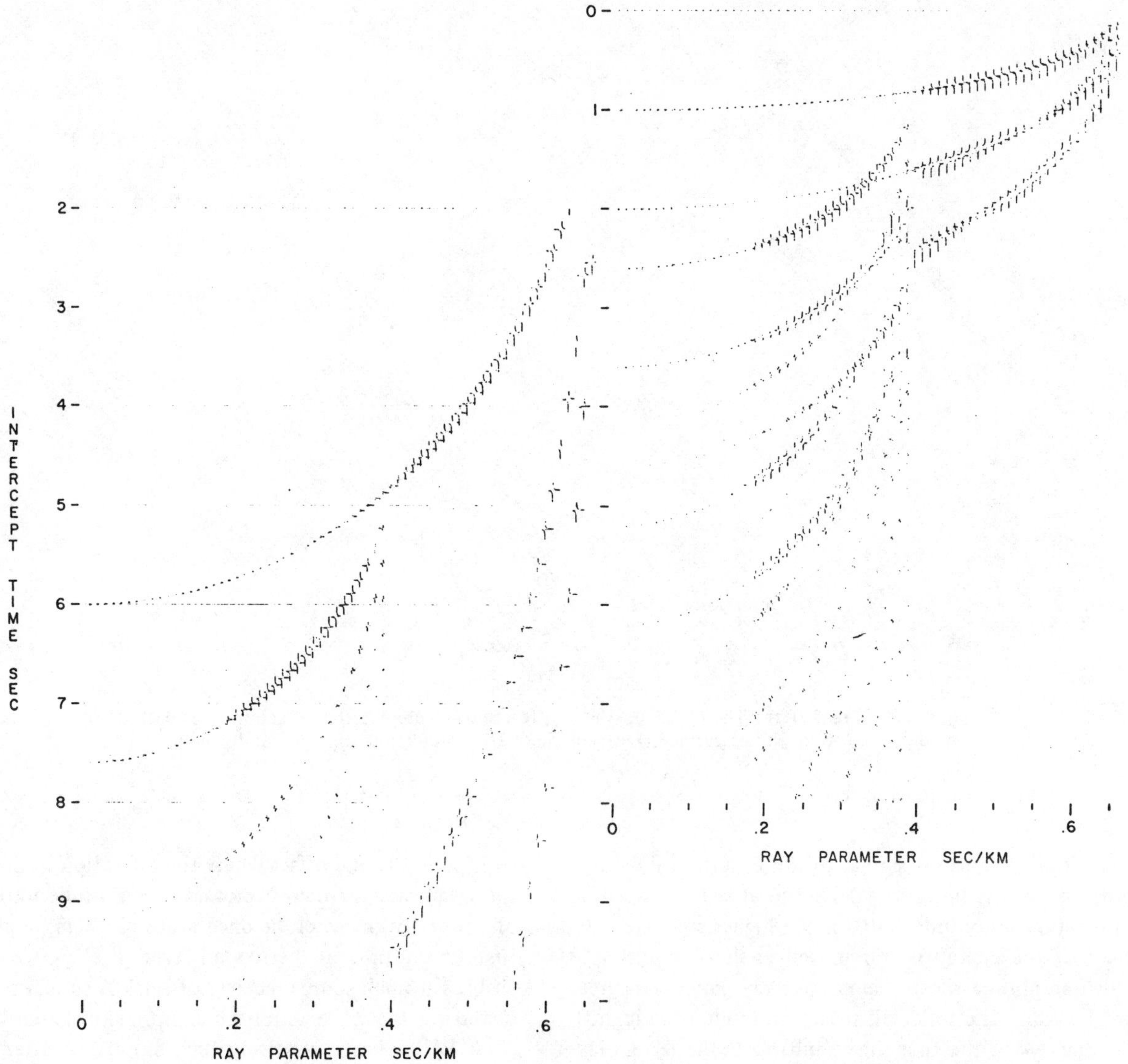

(d)

Fig. 9. (*Continued.*) (d) Left: τ-p response of model A with two water column multiples; the vertical two-way time of the first layer is 6 s. Right: τ-p response of model A with water column multiples when first layer is decreased to 1 s of travel time.

Fig. 9(d) is an example for model A that includes water column multiples (left), while on the right is a shallow-water example for the same sub-sea floor structure. That is, the acoustic parameters are the same, but the water depth has been reduced to 0.75 km. For the water column multiples, we see the same phase relation for subsequent multiples as we did for the multiples of the second layer. However, their amplitudes do not decrease as rapidly with the order of the multiple, because of total reflection at the sea surface. In the shallow water case, this is particularly clear, since beyond $p = 0.4$ s/km, the multiple reflections have about constant amplitude, but are reversed in sign. Even this simple example indicates that in contrast to the deep water case, the multiples observed in shallow water areas introduce significant interpretational and analysis problems.

Fig. 9(e) shows the effect of dipping the structure of model A relative to the sea surface. On the left is an up dip of 10°; in the middle is the no dip case; and on the right is a down dip of 10°. Horizontal ray parameters for the dipping case p_{dip} are related to the horizontal ray parameter p in the no dip case by

$$p_{dip} = p \cos\theta + \eta_1 \cdot \sin\theta$$

where $\theta > 0$ indicates an up dip and $\theta < 0$ a down dip, and η_1 is defined in (6). This results in the sea floor ellipse being rotated clockwise for $\theta > 0$ and counterclockwise for $\theta < 0$. Consequently, the critical point for the reflection is shifted to 0.3 s/km for $\theta = +10°$ and to 0.49 s/km for $\theta = -10°$. In the X-T seismograms, we then expect the transition from a sub-to-postcritical pulse at smaller X and T values in the up dip case, and at higher ones in the down dip case.

In model C, Figs. 10(a), (b), and (c), we switch from the acoustic to the elastic case with the first layer remaining fluid. For small ray parameters, only the compressional response is significant. At 9.2 s, the change in polarity of the first compressional (P wave) multiple within the second layer is observed.

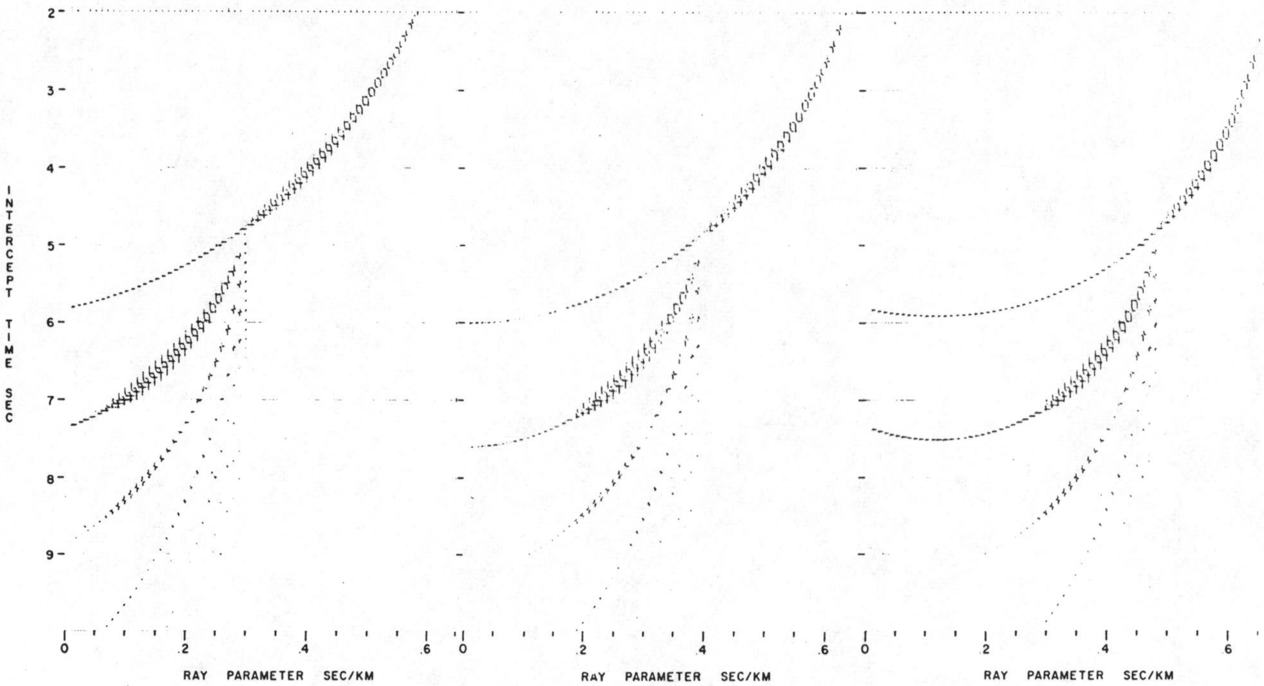

(e)

Fig. 9. (*Continued.*) (e) The effect on the τ-p response of dipping the reflecting layer package of model A relative to the sea surface. Left: up dip of 10°. Middle: no dip. Right: down dip of 10°.

The sea floor response is similar to the acoustic model up to the critical ray parameter, $p = 0.4$ s/km; at critical, we have a high reflection amplitude, but for the higher ray parameters, the bulk of the energy is transmitted as shear, resulting in a very high amplitude shear response for ray parameters greater than 0.4 s/km. The reflected shear amplitude is so high that we can clearly see the first shear multiple in the second layer. The amplitude of the shear wave reflected from the base of the second layer increases from zero to a high value at the ray parameter $p = 0.32$ s/km, which corresponds to the critical shear velocity in the lower half space. The amplitude then decreases and then becomes strong after the sea floor critical ray parameter $p = 0.4$ s/km is reached.

All waves traveling in the second medium as *PP*, *PS*, or their multiples are restricted to $p < 0.4$ s/km. The *PP* arrival, for instance, with $\tau = 7.6$ s at $p = 0$ s/km, has high amplitude at its critical ray parameter $p = 0.18$ s/km, but rapidly decreases and then increases again as it approaches the sea floor. The next phase is a converted mode that has traveled either up or down as P, either down or up as S. The high amplitudes of this converted phase occur after the shear critical point of the lower half space 0.32 s/km. The first P multiple in the second layer is only visible in the near vertical range.

From the τ-p seismograms, the amplitudes of the X-T seismograms in Figs. 10(b) and (c) can be explained. Strong sea floor reflections occur before the refraction appears. Strong shear reflection from the base of the second layer $P_1S_2S_2P_1$ occur once the critical shear ray parameter 0.32 s/km is reached and continues over a large distance with increasing ray parameter. The P response of the second layer is high in amplitude from $X = 0$ to the offset where the refraction of the lower half space appears. It then decreases in amplitude until a distance of about 13 km where it once again has high amplitude. The first *PP* multiple of the second layer ($P_1P_2P_2P_2P_2P_1$) is only visible for small source-receiver offsets. A similar feature is observed for the wave with both compressional and shear phases. ($P_1P_2S_2P_1$) has a relatively high amplitude after the first P refraction, and again after the S refraction from the lower half space. Fig. 10(c) shows that the sea floor reflection suffers almost no phase shift after the refractions appear. The P reflection from the half space increases in amplitude at 10 km, and weak but visible S refractions are observed after this point. Compared to the acoustic model A, the influence of shear velocities results in an impulsive refraction even in the case of first-order discontinuities, [Fig. 10(b) and (c)]. The partition of the reflected and transmitted energy at an interface explains this behavior. In the acoustic case, all the energy is reflected once the critical angle is reached, while at a liquid-solid interface, only at the critical point is a significant part of the energy refracted. Once the critical point is exceeded, the bulk of the energy is transmitted as a shear wave (Fig. 11).

Fig. 10(d) is the result obtained using the weighted integration method discussed earlier. The major reflection arrivals P_1P_1, $P_1P_2P_2P_1$, and $P_1P_2S_2P_1$ are all in good agreement with that of Fig. 10(c). The most obvious difference in the two methods is the decreased amplitude of the refraction event $P_1R_2P_1$ in the weighted integration method. This event is still observed, particularly as it emerges from the P_1P_1 reflection at about 8 km, but its amplitude decreases more rapidly with increasing offset. This example is a severe test for both

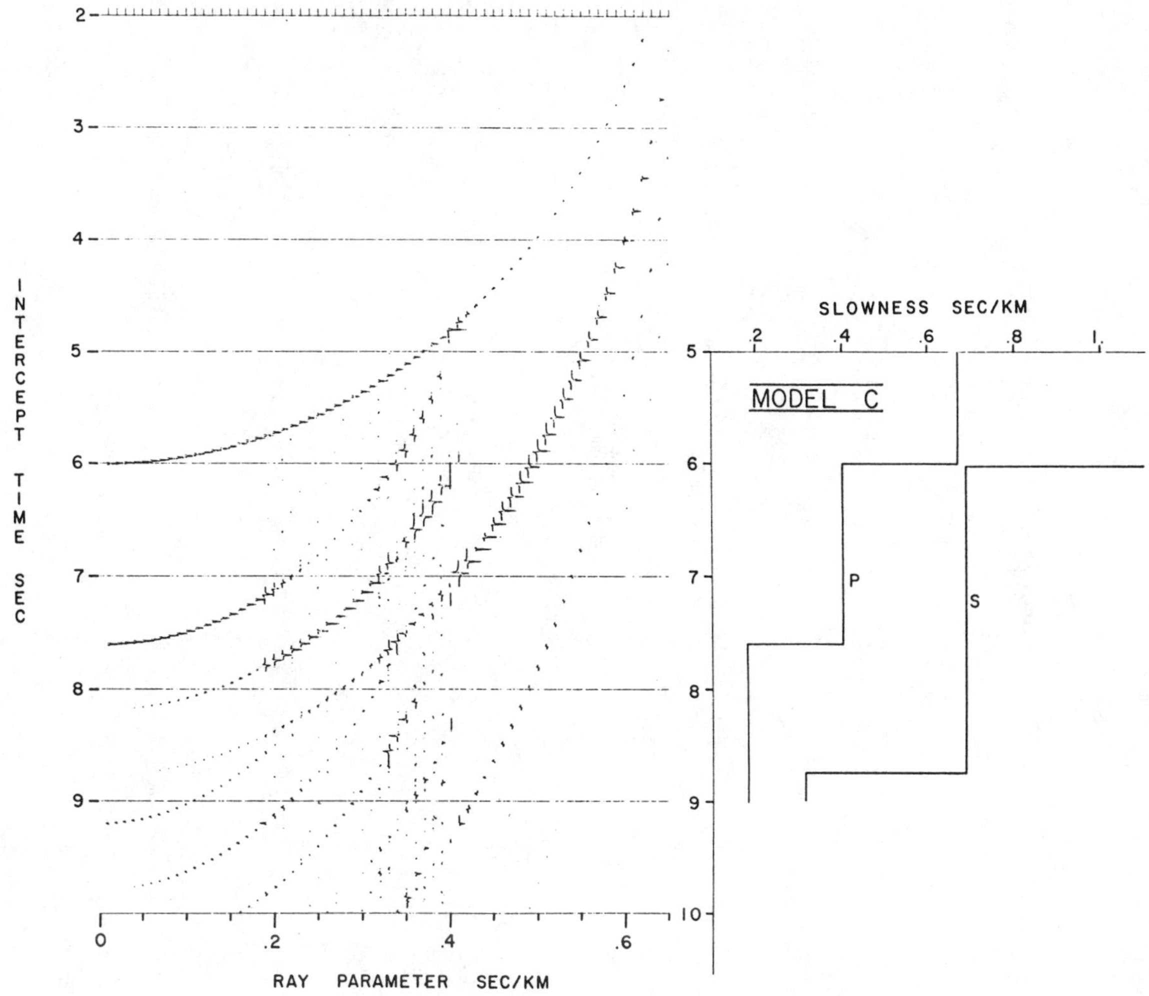

(a)

Fig. 10. (a) The τ-p response of the elastic model C.

methods, since band limiting these data or convolving with a typical seismic waveform would average (in time) the minor pulse shape differences, making the minor differences observed difficult to detect.

If we change the transitions between the layers of model C from sharp boundaries to linear vertical gradients in vertical travel time, we get the results presented in Fig. 12(a). In addition to the reflection from the lower level of the gradient zone, we also see a strong amplitude turning ray whose pulse before $p = 0.5$ s/km has a $\pi/2$ phase shift and beyond this ray parameter exhibits dispersion. Multiple P reflections from the lower half space are also visible. In this model, the shear waves play an important role once the ray parameter corresponding to the shear velocity in the half space 0.32 s/km is exceeded and two multiples are readily seen. Within the primary shear wave arrival, we can distinguish the reflection from the upper and lower levels of the gradient zone for ray parameters around $p = 0.4$ s/km. These arrivals also have pulse dispersion beyond 0.5 s/km. The converted mixed mode waves traveling in the second layer have weak amplitudes. For ray parameters beyond 0.2 s/km, these arrivals are seen between the primary P and primary S response of the half space.

The corresponding X-T seismograms appear in Fig. 12(b), (c), and (d). Sea floor arrivals consist of reflections from the top of the first gradient zone as well as turning rays and their internal multiples for distances less than 15 km. At larger offsets, these phases merge to form a dispersive pulse. A similar suite of arrivals is observed for the S waves. The first multiple of the P reflection from the lower half space ($P_1P_2P_2P_2P_2P_1$) is clearly visible. We also observe P refractions with velocities of 2.5 and 5.5 km/s. In Fig. 12(c), we see that the mixed mode ($P_1P_2S_2P_1$) is weak as expected and also strongly dispersed.

Fig. 12(d) is the result using the weighted integration method. As in Fig. 10, all the major arrivals are observed and are in good agreement with the direct integration method. Again, the major difference is in the decrease in amplitude of the refraction event $P_1R_2P_1$, particularly for the larger offsets.

IV. Summary

We have shown that broad-band seismic waveform data can be computed as a function of frequency and ray parameter. An inverse Fourier transform over frequency gives the τ-p response, which can be used immediately to predict the major features of X-T seismic data that will result after the required integration over ray parameter. We have not limited our investigation to low frequencies or to a limited suite of seismic arrivals. This approach has served to illustrate many of the computational problems inherent in the creation of synthetic

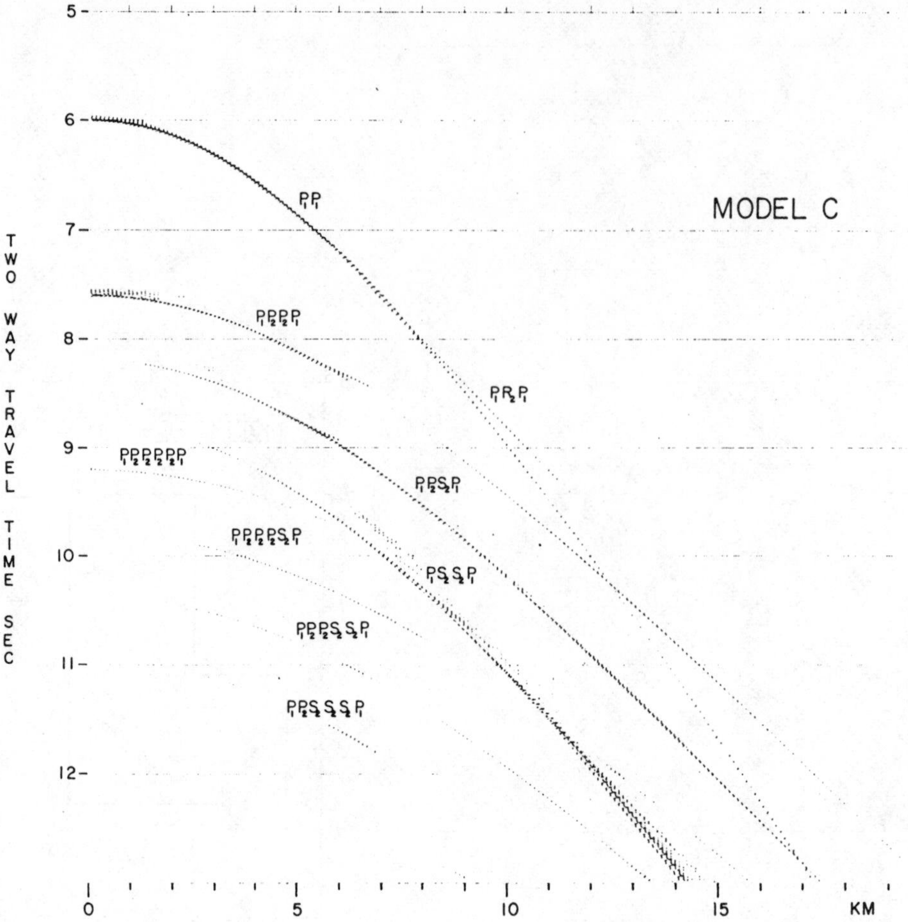

(b)

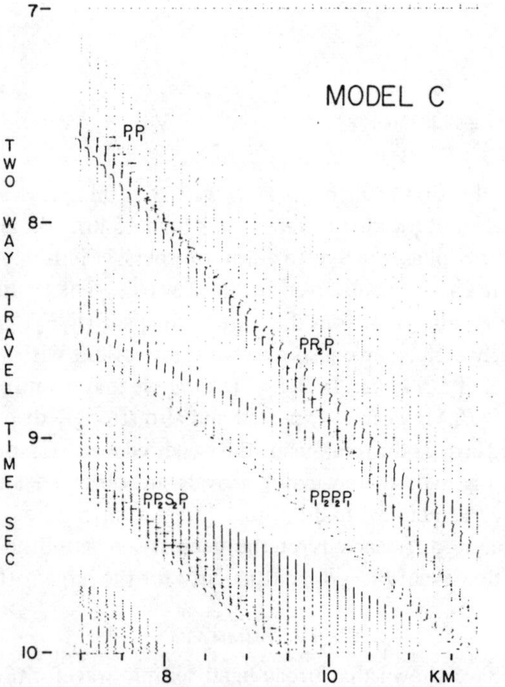

(c)

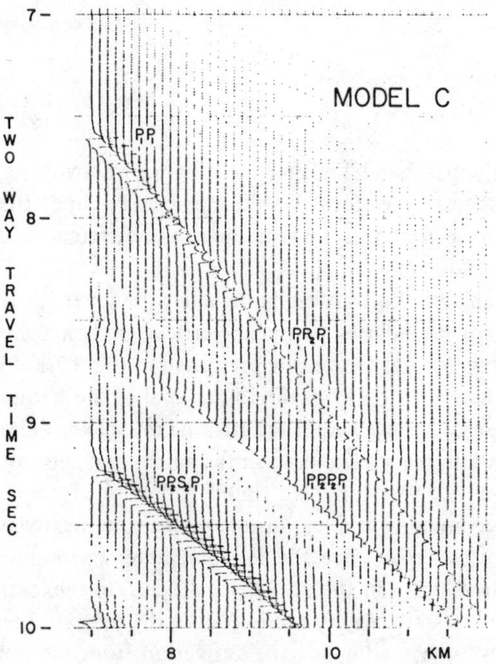

(d)

Fig. 10. (*Continued.*) (b) Seismograms for model *C* using the direct integration method. P_1P_1 is the sea floor reflection; $P_1R_2P_1$ is the sea floor refraction; $P_1P_2P_2P_1$ is the *P* reflection from the lower half space; $P_1P_2P_2P_2P_2P_1$ is a *P* wave with one multiple reflection in layer 2; $P_1P_2S_2P_1$ is a wave traveling in layer 2 down as *P* and up as *S* (or vice versa); $P_1S_2S_2P_1$ is the shear response from the lower half space; $P_1P_2P_2P_2S_2P_1$ is a multiple reflected wave in layer 2 traveling one way as *S* and three ways as *P*. (c) Detail of the seismograms of model *C* for the direct integration method; 2048 ray parameters were used in the computation. In addition to the sea floor refraction, the refraction from the lower half space can be seen from 8.5 s at 7 km to 9.3 s at 12 km; a multiple of it occurs at 9.1 s and 7 km to 9.9 s at 12 km. The *S* refraction from the lower half space emerges from the $P_1P_2P_2P_1$ onsets at 9.6 s and 10.5 km. (d) Detail of seismograms for model *C* computed by using semblance to determine the points of stationary phase and performing a weighted integration over ray parameter. In contrast to Fig. 10(c), 256 ray parameters were used in the computation. The results are in good agreement with the full integration except that the refractions are decreased in amplitude.

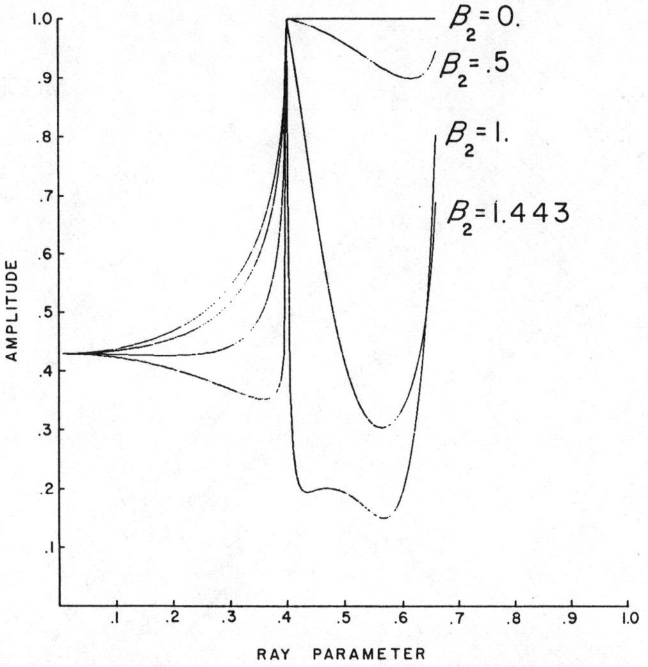

Fig. 11. Modulus of the reflection coefficients for an interface where the compressional velocities are 1.5 km/s and 2.5 km/s above and below, and with a constant density of 1 g/cm^3. The shear velocity of the upper layer is 0, but for the lower layer is 0, 0.5, 1.0, 1.443 km/s. The sharpest peak at the critical ray parameter of 0.4 s/km is for β_2 = 1.443 km/s as in model C; this gives a high amplitude and impulsive refraction in the seismograms.

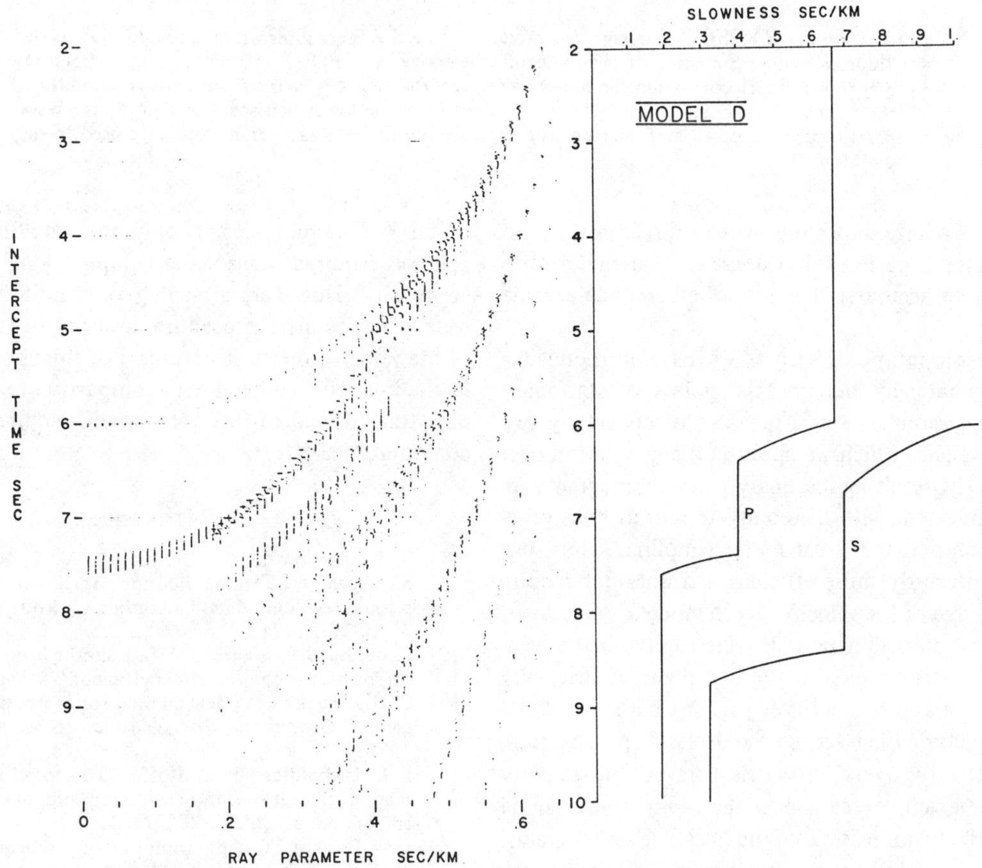

(a)

Fig. 12. (a) The τ-p response of the elastic model D with gradients as transitions. The gradients are linear in terms of velocity and vertical two-way travel time.

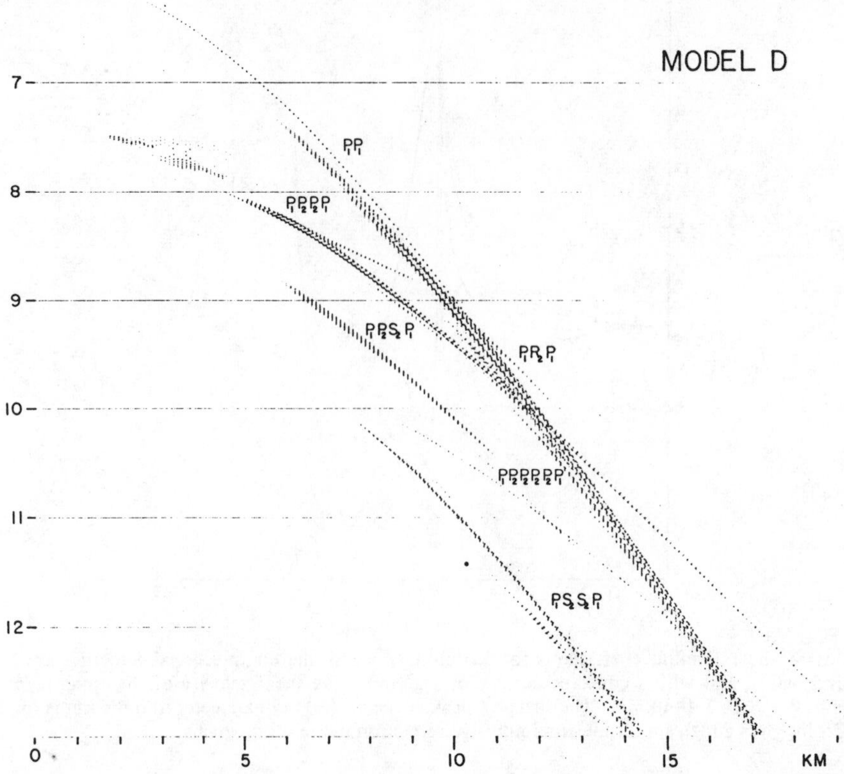

(b)

Fig. 12. (*Continued.*) (b) Seismograms for model D using the direct integration method. P_1P_1 is the sea floor reflection consisting of a bundle of converging rays; $P_1R_2P_1$ is the sea floor refraction; $P_1P_2P_2P_1$ is a P reflection from the lower half space; $P_1P_2P_2P_2P_2P_1$ is a P wave with one multiple reflection in layer 2; $P_1S_2S_2P_1$ is a shear reflection from the lower half space; $P_1P_2S_2P_1$ is a wave traveling in layer 2 down as P and up as S (or vice versa); 2048 ray parameters were used in the computation.

waveform data. We have shown that when a direct integration over ray parameter is performed, a dense ray parameter sampling is required to accurately compute the complete seismic response.

To minimize computational effort, we have introduced a method of automatically defining the points of stationary phase. By using semblance as a measure of coherency, we create a set of weights which we apply in the integration over ray parameter. The resulting stationary phase seismograms are in reasonable agreement with the complete seismograms generated with a much denser ray parameter sampling. Thus, our approach is significantly more efficient and does not require any prior knowledge of the velocity-depth model.

Finally, we have also illustrated the interpretational advantage of modeling seismic data in the τ-p plane as originally suggested by Chapman [3] and Fryer [6]. We have considered three examples, comparing acoustic and elastic propagation, and constant velocity layers and vertical travel time-velocity gradients. These examples show how the τ-p response can be used to predict the major features of the final X-T seismograms. This suggests that observational seismic waveform data transformed to the domain of intercept time and ray parameter can be quickly compared to models in this domain, without the additional computational effort required to generate synthetic X-T seismograms. If only one suite of synthetic seismograms is required, there is no computational advantage in this approach. However, modeling of seismic waveform data is usually an iterative procedure designed to refine the estimate of the velocity-depth structure. For this application, modeling in the τ-p plane coupled with comparison to seismic waveform data transformed to this domain will require significantly less computational effort.

References

[1] E. N. Bessonova, V. M. Fishman, V. Z. Ryaboyi, and G. A. Sitnikova, "The tau method for inversion of travel times–I. Deep seismic sounding data," *Geophys. J. Roy. Astronom. Soc.*, vol. 36, p. 377, 1974.

[2] P. Buhl and F. Wenzel, "A fast and high precision calculation of the plane wave reflection coefficient," submitted to *Geophysics*.

[3] C. H. Chapman, "A new method for computing synthetic seismograms," *Geophys. J. Roy. Astronom. Soc.*, vol. 54, pp. 481–518, 1978.

[4] J. B. Diebold and P. L. Stoffa, "The travel time equation, tau-p mapping and inversion of common midpoint data," *Geophysics*, vol. 46, no. 3, pp. 238–254, 1981.

[5] J. W. Dunkin, "Computation of model solutions in layered elastic media at high frequencies," *Bull. Seismol. Soc. Amer.*, vol. 55, no. 2, pp. 335–358, 1965.

[6] G. J. Fryer, "A slowness approach to the reflectivity method of seismogram synthesis," *Geophys. J. Roy. Astronom. Soc.*, vol. 63, pp. 747–758, 1980.

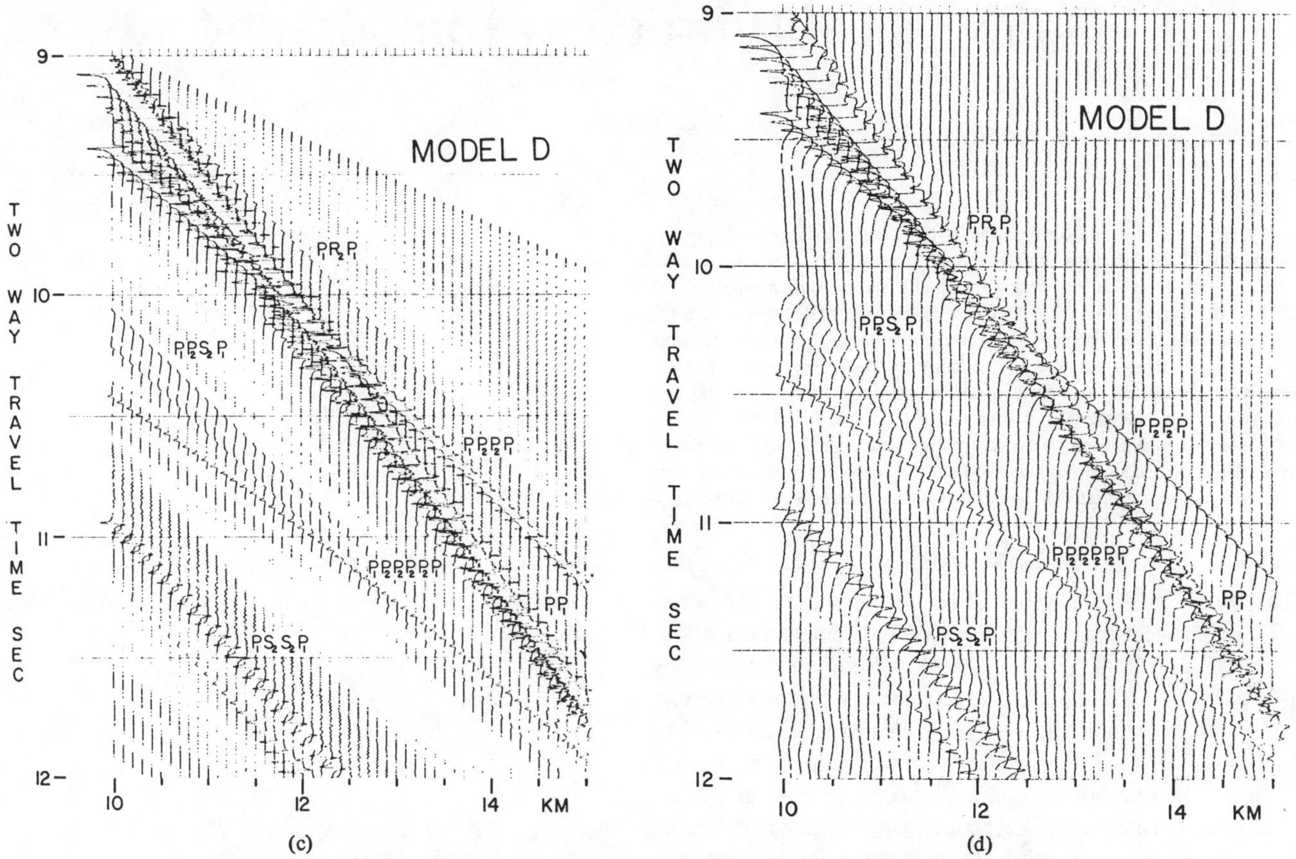

Fig. 12. (*Continued.*) (c) Detail of the seismograms for model D with the direct integration method. In addition to the labeled onsets, the sea floor refraction can be seen at 9 s and 10.2 km to 9.5 s at 15 km. (d) Detail of the seismograms for model D computed using semblance to determine the points of stationary phase and performing a weighted integration over ray parameter. As in Fig. 10(d), only 256 ray parameters were used. The results are in good agreement with the full integration.

[7] K. Fuchs and G. Muller, "Computation of synthetic seismograms with the reflectivity method and comparison with observations," *Geophys. J. Roy. Astronom. Soc.*, vol. 23, pp. 417-433, 1971.

[8] N. A. Haskell, "The dispersion of surface waves in multilayered media," *Bull. Seismol. Soc. Amer.*, vol. 43, pp. 17-34, 1953.

[9] D. V. Helmberger, "The crust mantle transition in the Bering Sea," *Bull. Seismol. Soc. Amer.*, vol. 58, pp. 179-214, 1968.

[10] N. S. Neidell and M. T. Taner, "Semblance and other coherency measures for multichannel data," *Geophysics*, vol. 36, no. 3, pp. 482-497, 1971.

[11] R. A. Phinney, K. R. Chowdhury, and L. N. Frazer, "Transformation and analysis of record sections," *J. Geophys. Res.*, vol. 86, no. B1, pp. 359-377, 1981.

[12] P. L. Stoffa, P. Buhl, J. B. Diebold, and F. Wenzel, "Direct mapping of seismic data to the domain of intercept time and ray parameter: A plane wave decomposition," *Geophysics*, vol. 46, no. 3, pp. 255-267, 1981.

[13] M. T. Taner and F. Koehler, "Velocity spectra-digital derivation and applications of velocity functions," *Geophysics*, vol. 34, no. 6, pp. 859-881, 1969.

[14] W. T. Thomson, "Transmission of elastic waves through a stratified medium," *J. Appl. Phys.*, vol. 21, pp. 89-93, 1950.

Geophysical Diffraction Tomography

A. J. DEVANEY

Abstract—Diffraction tomography is the generalization of X-ray tomography to applications such as seismic exploration where diffraction effects must be taken into account. In this paper, the foundations of diffraction tomography for offset vertical seismic profiling and well-to-well tomography are presented for weakly inhomogeneous formations for which the Born or Rytov approximations can be employed. Reconstruction algorithms are derived for approximately determining the acoustic or electromagnetic velocity profile of such formations from borehole measurements of acoustic or electromagnetic fields generated by sources located on the surface or in an adjacent borehole. Computer simulations are presented for the case of offset vertical seismic profiling.

I. Introduction

There exists a number of remote sensing procedures in geophysical exploration that closely resemble the methods of computed tomography employed in X-ray medical imaging applications. Two important examples are well-to-well tomography [1], [2] and offset vertical seismic profiling [3]. In well-to-well tomography sonic or electromagnetic sources are deployed in one borehole and associated receiver arrays are deployed in an adjacent borehole. The wavefields generated by the sources are transmitted through the geological formation to the receiver array and the received signals are used to generate an "image" of the intervening formation. In offset vertical seismic profiling (offset VSP), the transmitter array is deployed along a straight line extending outward from the borehole on the Earth's surface (a second configuration has the receivers on the surface and the sources in the borehole). Again, as in well-to-well tomography, the wavefields detected across the receiver array are employed to generate an "image" of the intervening formation. The two geometries are illustrated in Fig. 1.

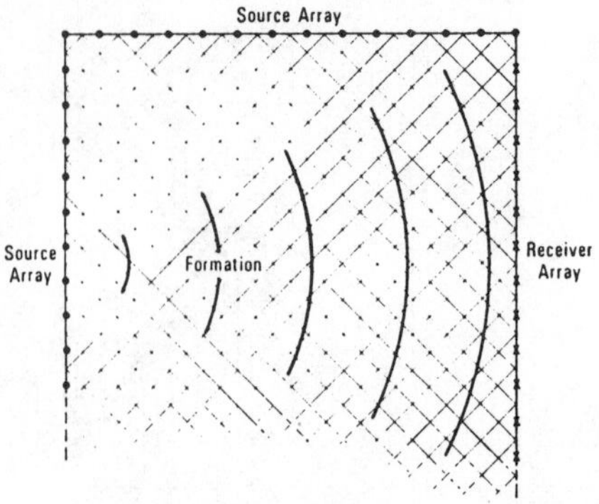

Fig. 1. Geometries for well-to-well tomography and offset VSP.

Although well-to-well tomography and offset VSP are similar in some respects to X-ray medical tomography, the two types of applications differ in important ways. For example, a standard geometry for X-ray medical imaging tomography employs an X-ray source arranged opposite a linear array of detectors as illustrated in Fig. 2. The source and receiver array are free to rotate in a plane about the object so that multiple X-ray exposures of the object are obtained—one exposure for each angular position of the source and receiver array. In practice, as many as 2000 exposures may be obtained over a full 180° range of view angles. Clearly, in geophysical applications such as offset VSP, it is not possible to rotate the sources and receivers about the object. Rather, the sources and locations of the receivers are fixed in space, being determined by the specific application in hand, and do not completely surround the formation being imaged. We can then expect the reconstruction algorithms for the two types of applications to differ. Moreover, due to the limited range of view angles achievable in geophysical applications, we can also expect the reconstructions obtainable in these applications to be generally inferior to those obtained in X-ray tomography.

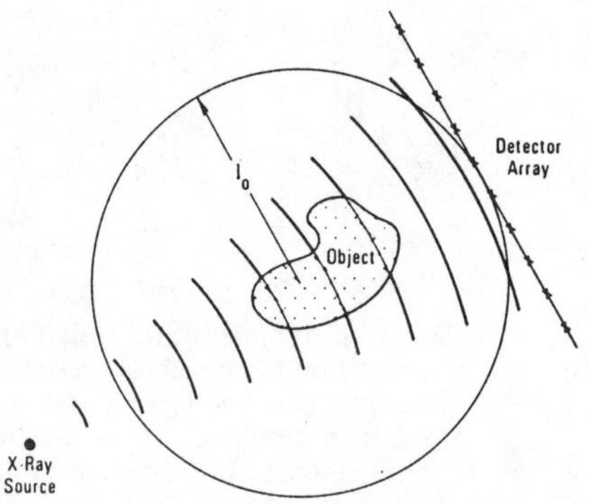

Fig. 2. Geometry for parallel beam X-ray transmission tomography.

A second and much more important difference between geophysical and X-ray tomography has to do with the types of wavefields used in the two applications. In geophysical tomography the relative size of the mean wavelength of the wavefields to typical structural detail in the geological formations will be much larger than that occurring in X-ray tomography. Because of this, wave phenomenon, such as diffraction and scattering, will play a dominant role in the interaction process between the formation and the wavefields. The simple straight line ray model adequate for X-ray tomographic imaging will thus not be appropriate for geophysical tomography. A consequence of this is that the standard reconstruction (imaging) algorithms of X-ray tomography will not be suitable for geophysical tomography and new algorithms, which account for the wave features of the interaction process, must be employed.

A reconstruction algorithm has been developed recently [4], [5] for ultrasound medical tomography where, as in geophysi-

Manuscript received March 22, 1983; revised September 16, 1983.
The author is with Schlumberger-Doll Research, Ridgefield, CT 06877.

Reprinted from *IEEE Trans. Geosci. Remote Sensing,* vol. GE-22, pp. 3–13, Jan. 1984.

cal tomography, the wave features of the interaction process between the ultrasonic wavefield and the object being imaged play a dominant role. This algorithm is based on an exact inversion of the wave equation within either the Born or Rytov approximations and thus provides a rigorously correct inversion algorithm for ultrasound tomography of weakly inhomogeneous bodies for which these approximations are valid. The algorithm was shown in [4] to be a natural generalization of the filtered backprojection algorithm of X-ray tomography to cases such as ultrasound tomography where diffraction of the wavefield must be taken into account. It differs from the filtered backprojection algorithm in that the back*projection* operation has to be replaced by a back*propagation* operation whereby the *phase* of the measured field data is made to propagate (migrate) back through the object space. For this reason the algorithm has been named the filtered backpropagation algorithm.

In this paper we investigate well-to-well tomography and offset VSP within the Born and Rytov approximations. Following the nomenclature employed in ultrasound tomography [4] we have coined the term "geophysical diffraction tomography" for the methods proposed here. For the sake of simplicity we shall restrict our attention to the two-dimensional case where the formation properties are constant in one direction (say the z-direction) and electromagnetic or sonic *line* sources, aligned parallel to the z-direction, are employed. Because the formation properties and sources are taken to be constant along the z-direction the induced fields will be independent of z which will simplify considerably the analysis presented in the paper.

The principle result in the paper is the derivation of filtered backpropagation algorithms for well-to-well tomography and offset VSP. As in [4] the algorithms are based on a rigorous inversion of the wave equation within either the Born or Rytov approximations and thus will be rigorously correct only for weakly inhomogeneous formations. However, as in the case of ultrasound tomography, the algorithms are natural generalizations of conventional tomographic reconstruction procedures and thus can be expected to perform at least as well as these methods.

The algorithms differ from the filtered backpropagation algorithm of medical ultrasound tomography because of the differing geometries of the two types of applications. As in medical X-ray tomography, medical ultrasound tomography employs a source arranged opposite a receiver array as illustrated in Fig. 2. The source and reciever array are free to rotate about the object so that a full 360° range of view angles can be obtained. In geophysical applications the receiver array is fixed in space (usually along the borehole axis) and the sources do not completely surround the formation being imaged. These changes in geometry have a pronounced effect on the form of the filtered backpropagation algorithm and also on the resulting image quality of the reconstructions derived from the algorithm.

The organization of the paper is as follows. Section II presents the theoretical foundations of geophysical diffraction tomography within the Born and Rytov approximations. The filtered backpropagation algorithms are presented in Section III and tested in computer simulations of offset VSP within the Rytov approximation in Section IV. The final section summarizes the results presented in the paper and discusses future research directions for geophysical diffraction tomography.

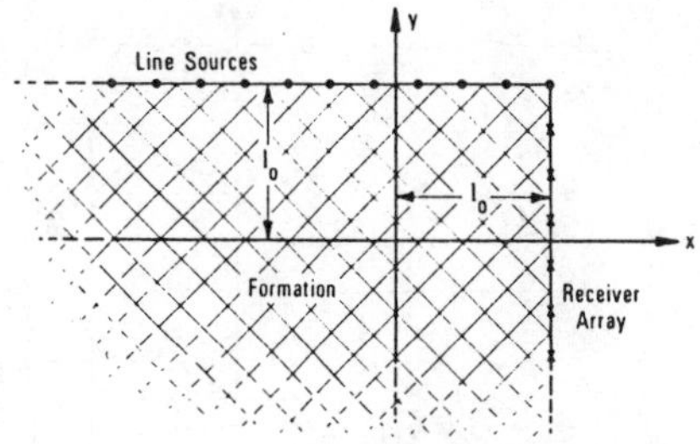

Fig. 3. Offset VSP for a two-dimensional formation using cylindrical wave sources.

II. Problem Formulation

As mentioned in the introduction we shall restrict our attention in this paper to two-dimensional cases where the formation properties and the electromagnetic or sonic sources are constant in one direction (say the z-direction). Such model formations can provide good approximations over limited volumes to real formations when we take the x-y plane to be perpendicular to the Earth's surface as for the case illustrated in Fig. 3. In this figure the x-y plane is the plane formed by the axes of a linear array of electromagnetic or sonic line sources located along the Earth's surface and aligned parallel to the z-axis and a linear array of point receivers located in a nearby borehole. Because the formation and sources are taken to be constant along the z-direction the induced fields will be independent of z which will simplify considerably the analysis presented later.

Besides restricting our attention to the two-dimensional case, we will also employ a scalar wave model for describing the propagation and scattering of the wavefields. Our results will thus be strictly applicable only for the case of isotropic electromagnetic media and for the constant density acoustic case. However, the results can be extended to more general types of media without too much difficulty.

Within the scalar wave model that we shall employ, the formation is characterized by a (possibly complex) velocity profile $C(\boldsymbol{r}, \omega)$ which varies as a function of position $\boldsymbol{r} = (x, y)$ within the formation and as a function of the frequency ω of the Fourier components of the transmitted wavefield. The inverse problem consists of determining $C(\boldsymbol{r}, \omega)$ from specification (measurements) of the waveforms received across the receiver array. Since we have taken the transmitter array to be an array of line sources which generate cylindrical waves the inverse problem can be thought of as being that of determining $C(\boldsymbol{r}, \omega)$ from the "cylindrical wave response" of the formation. Alternatively, by collectively pulsing the sources comprising the transmitter array or by employing the well-known method of slant stacking [6], it is possible to determine the "plane wave response" of the formation. The inverse problem can thus also be posed as being that of determining the formation's velocity profile from the plane wave response—this latter information being directly measured using collective excitation of the transmitter array or by post processing the cylindrical wave response using a slant stacking algorithm [6].

In this paper we shall pose the inverse problem in terms of the plane wave response rather than the cylindrical wave response. In place of the actual physical geometry such as is illustrated in Fig. 3 for offset VSP we thus shall consider the hypothetical (but physically equivalent) case illustrated in Fig. 4 of a section

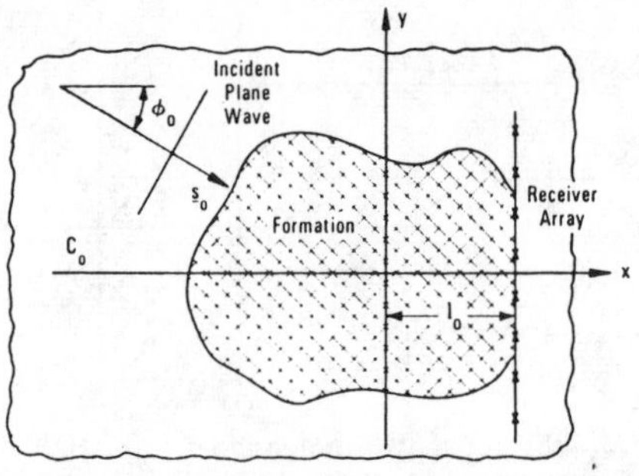

Fig. 4. Canonical problem for treating offset VSP and well-to-well tomography.

of formation insonified by plane waves having unit propagation vectors s_0. We imagine the formation to be surrounded by a homogeneous isotropic medium having a constant (possibly frequency dependent), real valued, velocity $C_0(\omega)$. The embedding medium is introduced for the sake of convenience in later calculations and in no way limits the generality of the model. The field is assumed to be known (measured) along the line $x = l_0$ for some set S_0 of unit propagation vectors s_0 lying in the x-y plane. In the offset VSP case, the plane wave response is to be synthesized from cylindrical wave sources lying along the line $y = l_0$ so that the set S_0 contains only unit vectors s_0 whose y components are negative (corresponding to plane waves propagating *into* the formation). In the case of well-to-well tomography the cylindrical waves are generated by line sources lying along a line parallel to the receiver axis (say the line $x = -l_0$) so that the set S_0 contains only unit vectors s_0 whose x components are positive (corresponding to plane waves propagating from the transmitter array to the receiver array). The inverse problem consists of estimating $C(\boldsymbol{r}, \omega)$ from specification (measurement) of the fields along the line $x = l_0$ generated in the set of "experiments" employing insonifying plane waves whose unit propagation vectors span the set S_0.

Within the scalar wave model employed here the Fourier amplitude

$$U(\boldsymbol{r}, \omega; \boldsymbol{s}_0) = \int_{-\infty}^{\infty} dt\, \hat{U}(\boldsymbol{r}, t; \boldsymbol{s}_0)\, e^{i\omega t} \tag{1}$$

of the time dependent wavefield $\hat{U}(\boldsymbol{r}, t; \boldsymbol{s}_0)$ generated by a plane wave incident to a section of formation as illustrated in Fig. 4 satisfies the reduced wave equation

$$(\nabla^2 + k^2)\, U(\boldsymbol{r}, \omega; \boldsymbol{s}_0) = k^2 O(\boldsymbol{r}, \omega)\, U(\boldsymbol{r}, \omega; \boldsymbol{s}_0). \tag{2}$$

In this equation, $\boldsymbol{r} = (x, y)$ is the position vector, $k = 2\pi/\lambda_0 = \omega/C_0$ is the wavenumber in the medium surrounding the section of formation (embedding medium), and $O(\boldsymbol{r}, \omega)$ is the "object profile" of the formation and is related to the velocity profile according to the equation

$$O(\boldsymbol{r}, \omega) = 1 - \frac{C_0^2(\omega)}{C^2(\boldsymbol{r}, \omega)}. \tag{3}$$

The argument $\boldsymbol{s}_0$ is included in the wavefields $\hat{U}(\boldsymbol{r}, t; \boldsymbol{s}_0)$ and $U(\boldsymbol{r}, \omega; \boldsymbol{s}_0)$ to denote their dependence on the direction of propagation of the incident plane wave. From this point on we shall work primarily in the frequency domain and refer to the Fourier amplitude $U(\boldsymbol{r}, \omega; \boldsymbol{s}_0)$ as simply the "wavefield". We can, of course, return to the time dependent wavefield $\hat{U}(\boldsymbol{r}, t; \boldsymbol{s}_0)$ via an inverse Fourier transformation.

The wavefield $\hat{U}(\boldsymbol{r}, t; \boldsymbol{s}_0)$ represents in the electromagnetic case the z-component of the electric field vector induced in the formation by a TE polarized incident plane wave. This incident field is synthesized from the cylindrical electromagnetic waves generated by line sources carrying current along the z-direction. In the acoustic case $\hat{U}(\boldsymbol{r}, t; \boldsymbol{s}_0)$ represents the total pressure field induced in the formation by an incident plane pressure wave.

The partial differential equation (2) is readily converted to an integral equation by incorporating Sommerfeld's radiation condition [7] (causality condition in the time domain). We obtain [4]

$$U(\boldsymbol{r}, \omega; \boldsymbol{s}_0) = U_0(\omega)\, e^{ik\boldsymbol{s}_0 \cdot \boldsymbol{r}} - i\frac{k^2}{4}\int d^2r'\, O(\boldsymbol{r}', \omega) \cdot U(\boldsymbol{r}', \omega; \boldsymbol{s}_0)\, H_0(k\,|\boldsymbol{r} - \boldsymbol{r}'|) \tag{4}$$

where $H_0(kR)$ is the zero order Hankel function of the first kind and where $U_0(\omega) \exp(ik\boldsymbol{s}_0 \cdot \boldsymbol{r})$ is the Fourier amplitude of the incident plane wave $\hat{U}_0(\boldsymbol{r}, t; \boldsymbol{s}_0)$; i.e.,

$$\hat{U}_0(\boldsymbol{r}, t; \boldsymbol{s}_0) = \frac{1}{2\pi}\int_{-\infty}^{\infty} d\omega\, U_0(\omega)\, e^{ik\boldsymbol{s}_0 \cdot \boldsymbol{r}}\, e^{-i\omega t}. \tag{5}$$

Equation (4) is a Fredholm integral equation relating the object profile $O(\boldsymbol{r}', \omega)$ to the field $U(\boldsymbol{r}, \omega; \boldsymbol{s}_0)$. In the so-called *forward* or *direct* scattering problem the object profile is taken to be known and the total field is to be calculated. Although computationally difficult, this problem is, at least in principle, quite straight-forward mathematically; reducing, as it does, to the problem of solving a Fredholm integral equation of the second kind.

In the inverse problem the field $U(\boldsymbol{r}, \omega; \boldsymbol{s}_0)$ is known over some set of space points that generally lie outside the support volume of the object profile (object volume), and for some set of unit propagation vectors. For example, in the geophysical problems under consideration here, $U(\boldsymbol{r}, \omega; \boldsymbol{s}_0)$ is assumed to be specified (measured) over the line $x = l_0$ for unit propagation vectors $\boldsymbol{s}_0$ contained in the set S_0. A set of simultaneous equations relating the measured field data to the object profile can be obtained from (4) and the inverse problem then becomes that of solving this set of coupled equations for the object profile in terms of the measured data.

Unfortunately, in the coupled set of equations governing the inverse problem, both the object profile *and* the field within the object volume are unknown. Because of this, the inverse problem is considerably more complicated than the direct problem and, except in certain simple cases, does not admit a computationally feasible solution. A way out of this difficulty is to employ approximations for the field within the object volume and thereby remove this quantity as an unknown from the problem. The simplest such approximation is the so-called first Born approximation [8] which sets the field within the object volume equal to the incident field. Within this approximation then the geophysical inverse problem consists of solving the set of equations.

$$U(\mathbf{r}_0, \omega; \mathbf{s}_0) = U_0(\omega)\, e^{ik\mathbf{s}_0 \cdot \mathbf{r}_0} - i\frac{k^2}{4} U_0(\omega) \cdot \int d^2r' O(\mathbf{r}', \omega)\, e^{ik\mathbf{s}_0 \cdot \mathbf{r}'} H_0(k|\mathbf{r}_0 - \mathbf{r}'|) \tag{6}$$

where $\mathbf{r}_0 = (x = l_0, y)$ is the coordinate of a receiver location and $\mathbf{s}_0$ is contained in the set S_0.

The first Born approximation is a "weak scattering approximation" and requires that the total scattered field (second term on the right-hand side of (4)) be small in comparison with the incident field [8], [9]. This, in turn, requires that both the magnitude of the object profile and the total extent of the object volume be small [8], [9]. This second condition can be relaxed somewhat by employing the Rytov approximation [9] rather than the first Born approximation. While the Born approximation results in a linear mapping between the object profile and the field $U(\mathbf{r}, \omega; \mathbf{s}_0)$ the Rytov approximation yields a linear mapping between the object profile and the complex *phase* of the field. In particular, defining the complex phase $\Psi(\mathbf{r}, \omega; \mathbf{s}_0)$ via the equation

$$U(\mathbf{r}, \omega; \mathbf{s}_0) \equiv U_0(\omega) \exp[\Psi(\mathbf{r}, \omega; \mathbf{s}_0)] \tag{7}$$

the Rytov approximation to $\Psi(\mathbf{r}, \omega; \mathbf{s}_0)$ is given by [4] [1]

$$\Psi(\mathbf{r}, \omega, \mathbf{s}_0) = ik\mathbf{s}_0 \cdot \mathbf{r} - i\frac{k^2}{4} e^{-ik\mathbf{s}_0 \cdot \mathbf{r}} \cdot \int d^2r' O(\mathbf{r}', \omega)\, e^{ik\mathbf{s}_0 \cdot \mathbf{r}'} H_0(k|\mathbf{r} - \mathbf{r}'|). \tag{8}$$

The inverse problem within the Rytov approximation then consists of solving the set of coupled equations (8) for $O(\mathbf{r}', \omega)$ in terms of the complex phase specified over some set of measurement points (the line $x = l_0$ in the geophysical problem) and for the set of unit propagation vectors employed in generating the field data (e.g., the set S_0).

On comparing (6) and (8) it is seen that the mathematical structure of the inverse problem within the first Born and Rytov approximations is virtually identical. It is important to keep in mind, however, that the assumptions underlying these two approximations are quite different so that one of the two formulations may be preferable over the other in a given application. For example, in crystal and protein structure determination using X-rays, the first Born approximation is universally employed [10] while in ultrasound medical tomography the Rytov approximation is more appropriate [4], [11]. In the geophysics problems under consideration here the Rytov approximation is probably to be preferred over the Born approximation although at present no hard data is available to confirm this conjecture. In any case, the inverse method developed in the following section is quite general in that it applies within both formulations so that the question of which approximation is most appropriate need not concern us here.

[1]The expression for the phase $\Psi(\mathbf{r}, \omega; \mathbf{s}_0)$ given in [4] incorrectly included the factor $U_0(\omega)$ multiplying the second term in (8). This error resulted in the factor of $U_0(\omega)$ incorrectly appearing in all equations involving the Rytov approximation. The correct expressions are given in [5] and are obtained from those given in [4] by simply setting $U_0(\omega)$ equal to unity in all equations involving the Rytov approximation.

Equations (6) and (8) can be cast into the unified form

$$D(y, \omega, \mathbf{s}_0) = \tfrac{1}{2} \int d^2r' O(\mathbf{r}', \omega)\, e^{ik\mathbf{s}_0 \cdot \mathbf{r}'} H_0(k|\mathbf{r}_0 - \mathbf{r}'|) \tag{9}$$

where $D(y, \omega; \mathbf{s}_0)$ is related to the data according to the equations

$$D(y, \omega; \mathbf{s}_0) = i\frac{2}{k^2} \begin{cases} \dfrac{U(\mathbf{r}_0, \omega; \mathbf{s}_0)}{U_0(\omega)} - e^{ik\mathbf{s}_0 \cdot \mathbf{r}_0} & \text{Born approximation} \quad (10a) \\ [\Psi(\mathbf{r}_0, \omega; \mathbf{s}_0) - ik\mathbf{s}_0 \cdot \mathbf{r}_0]\, e^{ik\mathbf{s}_0 \cdot \mathbf{r}_0} & \text{Rytov approximation} \quad (10b) \end{cases}$$

and where $\mathbf{r}_0 = (x = l_0, y)$ is the line over which the data is taken (the receiver axis). The factor of $\frac{1}{2}$ is included in (9) for later notational convenience. We shall occasionally refer to $D(y, \omega; \mathbf{s}_0)$ as simply "the data."

Equation (9) is best treated in Fourier transform space where we shall now show that it relates the one-dimensional Fourier transform of $D(y, \omega; \mathbf{s}_0)$ to the two-dimensional Fourier transform of the object profile evaluated over certain semicircular arcs. The Fourier transform of $D(y, \omega; \mathbf{s}_0)$ is readily calculated by making use of the plane wave expansion of the Hankel function [12]

$$H_0(k|\mathbf{r}_0 - \mathbf{r}'|) = \frac{1}{\pi} \int_{-\infty}^{\infty} \frac{d\kappa}{\gamma} e^{i[\kappa(y-y') + \gamma(l_0 - x')]} \tag{11}$$

where

$$\gamma = \begin{cases} \sqrt{k^2 - \kappa^2} & \text{if } |\kappa| \leq k \\ i\sqrt{\kappa^2 - k^2} & \text{if } |\kappa| > k. \end{cases} \tag{12}$$

On substituting (11) into (9) and Fourier transforming the resulting expression we obtain

$$\tilde{D}(\kappa, \omega; \mathbf{s}_0) \equiv \int_{-\infty}^{\infty} dy\, D(y, \omega; \mathbf{s}_0)\, e^{-i\kappa y} = \frac{e^{i\gamma l_0}}{\gamma} \tilde{O}(\gamma - ks_{0x}, \kappa - ks_{0y}, \omega) \tag{13}$$

where

$$\tilde{O}(\mathbf{K}, \omega) \equiv \int d^2r\, O(\mathbf{r}, \omega)\, e^{-i\mathbf{K} \cdot \mathbf{r}} \tag{14}$$

is the two-dimensional Fourier transform of the object profile and where (13) holds for all κ values lying in the interval $-k \leq \kappa \leq k$.[2]

Equation (13) states that the one-dimensional Fourier transform $\tilde{D}(\kappa, \omega; \mathbf{s}_0)$ is directly proportional, with proportionality factor $[\exp(i\gamma l_0)/\gamma]$, to the two-dimensional Fourier transform $\tilde{O}(\mathbf{K}, \omega)$ over the locus of $\mathbf{K}$ values given by

$$\mathbf{K} \equiv (\gamma - ks_{0x})\hat{\mathbf{x}} + (\kappa - ks_{0y})\hat{\mathbf{y}} \tag{15}$$

[2]The κ values are constrained to lie in the interval $-k \leq \kappa \leq k$ due to the inherent low pass filtering effect of wave propagation in homogeneous media [13]. This result can be traced to the fact that γ in (12) is pure imaginary for $|\kappa| > k$.

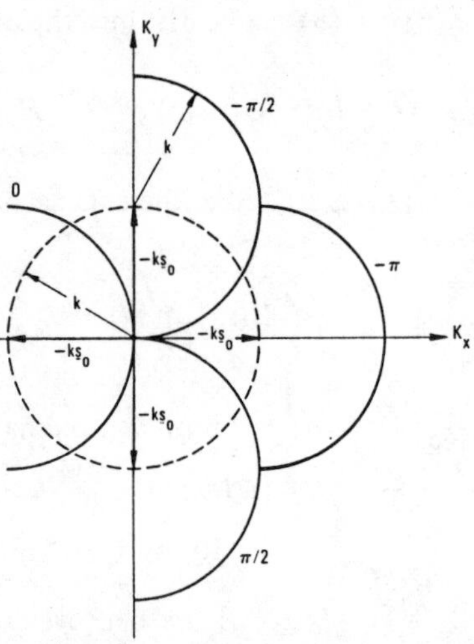

Fig. 5. Locus of spatial frequencies over which plane wave response determines object profile transform. The four curves correspond to angles $\phi_0 = -\pi$, $-\pi/2$, 0, and $\pi/2$ rad. The centers of all semicircles lie on dashed circle.

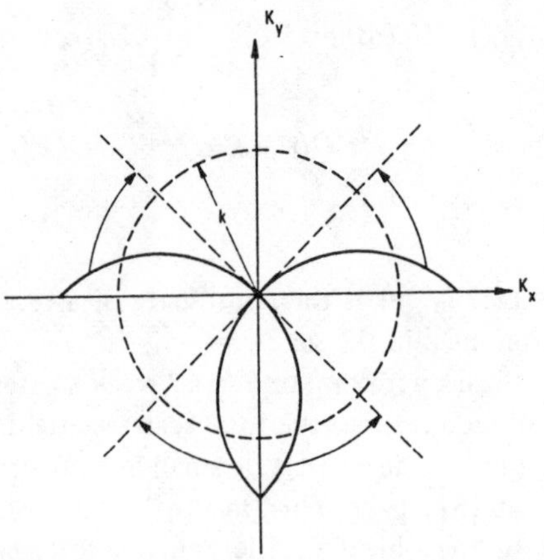

Fig. 6. Representative loci of spatial frequencies over which data determines object profile transform in ultrasound tomography. The two curves correspond to propagation vectors $\boldsymbol{s}_0$ forming angles of $-45°$ and $-135°$ with the x axis. Dashed lines represent the limiting case of zero wavelength which corresponds to conventional tomography.

where $\hat{\boldsymbol{x}}, \hat{\boldsymbol{y}}$ are unit vectors along the x, y coordinate axes, respectively. In order to examine the locus of points defined by (15) it is convenient to introduce the unit vector

$$\boldsymbol{s} \equiv \frac{1}{k}\left[\gamma\hat{\boldsymbol{x}} + \kappa\hat{\boldsymbol{y}}\right]. \tag{16}$$

In terms of $\boldsymbol{s}$, (15) becomes

$$\boldsymbol{K} = k(\boldsymbol{s} - \boldsymbol{s}_0) \tag{17}$$

and (13) can then be expressed in terms of $\boldsymbol{s}$ as follows

$$\tilde{D}(\kappa, \omega; \boldsymbol{s}_0) = \frac{e^{i\gamma l_0}}{\gamma}\tilde{O}[k(\boldsymbol{s} - \boldsymbol{s}_0), \omega]. \tag{18}$$

Equation (17) is seen, for fixed $\boldsymbol{s}_0$, to define a semicircle in $\boldsymbol{K}$ space of radius k, centered at $\boldsymbol{K} = -k\boldsymbol{s}_0$ and over which $K_x \geqslant -ks_{0x}$; this last condition being a consequence of $\gamma \equiv \sqrt{k^2 - \kappa^2} \geqslant 0$. Shown in Fig. 5 are four such semicircles corresponding to propagation vectors $\boldsymbol{s}_0$ forming angles ϕ_0 of $-\pi, -\pi/2, 0$, and $\pi/2$ rad with the positive x axis. The semicircles resulting from angles of $-\pi, -\pi/2$, and 0 rad are generated in offset VSP experiments while those resulting from angles of $-\pi/2$, and $\pi/2$ rad are obtained in the case of well-to-well tomography.

We conclude from the analysis presented above that offset VSP and well-to-well tomography within the Born or Rytov approximations consist ultimately of determining a function $O(\boldsymbol{r}, \omega)$ from specification of its Fourier transform $\tilde{O}(\boldsymbol{K}, \omega)$ over a set of semicircular arcs. One solution to this problem is to employ an interpolation procedure to estimate the transform over a rectangular sampling grid in Fourier space from the available sample values and then perform a two-dimensional discrete Fourier transformation to recover $O(\boldsymbol{r}, \omega)$. Unfortunately, this method suffers the same limitations as do the analogous Fourier space interpolation methods employed in conventional tomography [14]. An alternative and preferable procedure is to employ a reconstruction algorithm that operates directly on the available sample values as is achieved in conventional tomography by the filtered backprojection algorithm [15]. As discussed in the introduction, the appropriate generalization of the filtered backprojection algorithm is the filtered backpropagation algorithm which will be derived in the following section.

Before closing this section, it is worthwhile to digress briefly and consider the case of medical ultrasound tomography. As discussed in the introduction, medical ultrasound tomography differs from geophysical tomography in that the receiver axis is not fixed but, rather, rotates with the direction of propagation of the incident plane wave, remaining at all times perpendicular to $\boldsymbol{s}_0$ (see Fig. 2). Nevertheless, the data is still related to the object profile through the set of equations (9) where, however, the y axis is not fixed in space but rather is taken to be a straight line perpendicular to $\boldsymbol{s}_0$ and located at a distance l_0 from the center of the object. For any given orientation of the receiver array the one-dimensional Fourier transform of the data $\tilde{D}(\kappa, \omega; \boldsymbol{s}_0)$ is related to the two-dimensional Fourier transform of the object profile according to (18) where, however, $\boldsymbol{s} \cdot \boldsymbol{s}_0 \equiv \gamma = \sqrt{k^2 - \kappa^2} \geqslant 0$.

As in the geophysical case, the loci of points over which the data specifies the object profile's two-dimensional Fourier transform consists of a semicircle centered at $\boldsymbol{K} = -k\boldsymbol{s}_0$ and having a radius equal to the wavenumber k. Unlike the geophysical case, however, each semicircle is bisected by the vector $-k\boldsymbol{s}_0$—this later result being a consequence of the fact that the receiver array is always perpendicular to the $\boldsymbol{s}_0$ vector so that $\boldsymbol{s} \cdot \boldsymbol{s}_0 \geqslant 0$. Two such semicircles are shown in Fig. 6 together with the degenerate forms of these semicircles obtained in the zero wavelength limit ($k \to \infty$).

In the zero wavelength limit the circular arcs degenerate to straight lines as illustrated in Fig. 6 and $\gamma \equiv \sqrt{k^2 - \kappa^2} \to k$. In this limit then the transform of the data $\tilde{D}(\kappa, \omega; \boldsymbol{s}_0)$ is proportional, with *constant proportionality factor* $[\exp(ikl_0)/k]$, to the two-dimensional Fourier transform of the object profile $\tilde{O}(\boldsymbol{K}, \omega)$ evaluated along the straight line *slice* passing through the origin and making the angle $\phi \equiv \phi_0 + \pi/2$ with the positive x axis (see Fig. 6). This result is the well-known *projection-slice theorem* of X-ray tomography [15] and shows that medical ultrasound tomography within the Rytov approximation reduces to conventional X-ray tomography in the short-wavelength limit [4].

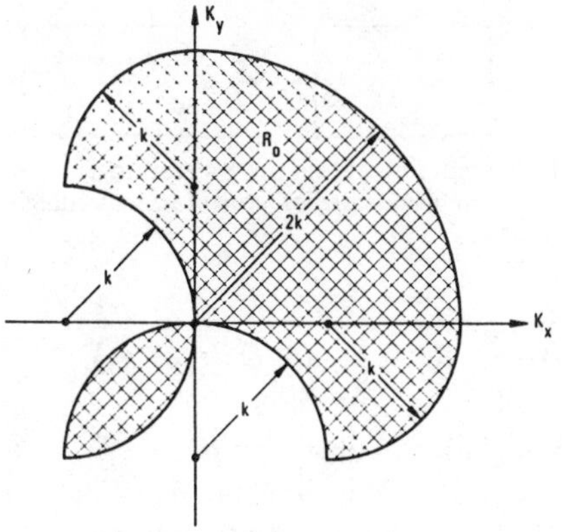

(a)

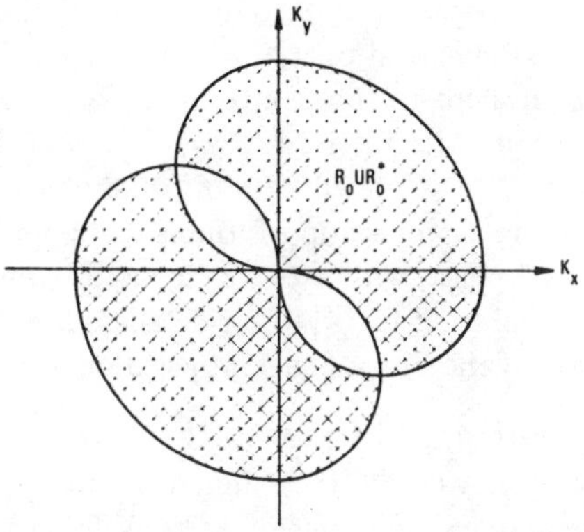

(b)

Fig. 7. Regions in Fourier space over which object profile transform is determined from a full set of offset VSP experiments. The region R_0 is obtained in the case of complex profiles while $R_0 \cup R_0^*$ is obtained in the case of real profiles.

III. Reconstruction Algorithms

We showed in the preceding section that the inverse geophysical problem reduces mathematically to estimating the object profile $O(\boldsymbol{r}, \omega)$ from specification of its Fourier transform over an ensemble of semicircular arcs such as shown in Fig. 5. In this section we shall derive reconstruction (imaging) algorithms for estimating $O(\boldsymbol{r}, \omega)$ from this information. The derived algorithms are completely analogous to the filtered backpropagation algorithm derived recently for medical ultrasound tomography [4] and are derived using a method entirely analogous to that employed in [4].

Let us consider first offset VSP where the set S_0 consists of unit propagation vectors making angles ϕ_0 with the positive x axis which vary from 0 to $-\pi$ rad. Now, as the angle ϕ_0 varies continuously from 0 to $-\pi$ rad, the semicircular arcs defined by (17) with $\gamma \geqslant 0$ cover the region R_0 in $\boldsymbol{K}$ space illustrated in Fig. 7(a). Thus within the Born or Rytov models employed here it is possible to determine Fourier components of $\tilde{O}(\boldsymbol{K}, \omega)$ lying within this region. We note, however, that for *real* object profiles

$$\tilde{O}^*(\boldsymbol{K}, \omega) = \tilde{O}(-\boldsymbol{K}, \omega) \tag{19}$$

so that for such profiles $\tilde{O}(\boldsymbol{K}, \omega)$ can be determined over the region $R_0 \cup R_0^*$ formed by the union of R_0 and its reflection about the origin (see Fig. 7(b)). In general, however, the set S_0 will include unit propagation vectors whose angles ϕ_0 span only a segment of the interval $[0, -\pi]$. For such cases then, the regions R_0 and $R_0 \cup R_0^*$ over which $\tilde{O}(\boldsymbol{K}, \omega)$ is known will lie within the regions illustrated in Fig. 7.

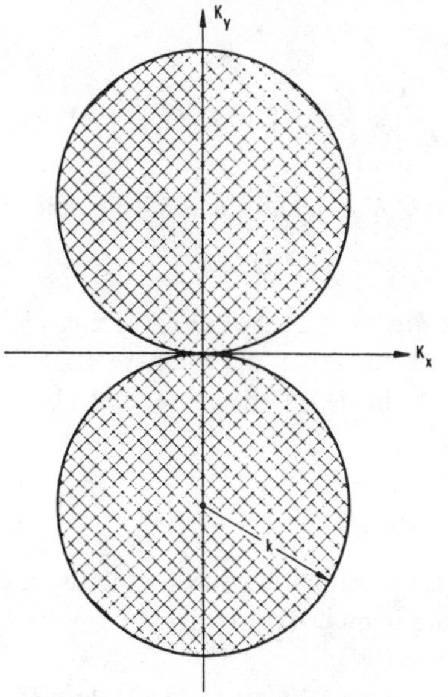

Fig. 8. Region in Fourier space over which object profile transform is determined from a full set of well-to-well tomographic experiments.

Now, consider well-to-well tomography. In this case the angle ϕ_0 can vary between $-\pi/2$ and $\pi/2$ rad so that the semicircular arcs map out the region in $\boldsymbol{K}$ space illustrated in Fig. 8. Note that in this case the region R_0 is symmetrical with respect to the origin so that $R_0 \cup R_0^* \equiv R_0$ and no advantage is gained for real valued object profiles. Again, as in the case of offset VSP, if ϕ_0 does not range over the entire interval $[-\pi/2, \pi/2]$ the region R_0 over which $\tilde{O}(\boldsymbol{K}, \omega)$ is known will lie within the region illustrated in Fig. 8.

We define the filter

$$\tilde{H}_0(\boldsymbol{K}) \equiv \begin{Bmatrix} 1 & \text{if } \boldsymbol{K} \in D_0 \\ 0 & \text{otherwise} \end{Bmatrix} \tag{20}$$

where D_0 is the region in $\boldsymbol{K}$ space over which $\tilde{O}(\boldsymbol{K}, \omega)$ is specified by the data. Thus for example, if the object profile is complex, the region $D_0 \equiv R_0$ as defined above. However, if in offset VSP the object profile is known to be real, the region $D_0 = R_0 \cup R_0^*$. Now, consider the quantity $\hat{O}_0(\boldsymbol{r}, \omega)$ defined by

$$\hat{O}_0(\boldsymbol{r}, \omega) \equiv \frac{1}{(2\pi)^2} \int d^2K \, \tilde{H}_0(\boldsymbol{K}) \, \tilde{O}(\boldsymbol{K}, \omega) \, e^{i\boldsymbol{K} \cdot \boldsymbol{r}}. \tag{21}$$

Clearly, the two-dimensional Fourier transform of $\hat{O}_0(\boldsymbol{r}, \omega)$ is equal to $\tilde{O}(\boldsymbol{K}, \omega)$ over the region D_0 and is zero elsewhere. $\hat{O}_0$ is then an object profile that *will generate identical data to that generated by* $O(\boldsymbol{r}, \omega)$ for all $\boldsymbol{s}_0$ values lying in the set S_0. In other words, $\hat{O}_0(\boldsymbol{r}, \omega)$ is one possible solution to the inverse problem.

Although (21) is a solution to the inverse problem it is not the only solution. Indeed, since the data specifies the object profile transform only over the region D_0, we are free, in general, to specify the transform arbitrarily outside this region. Thus we can always add to $\hat{O}_0(\boldsymbol{r}, \omega)$ any function of the general form

$$\Delta(\boldsymbol{r}, \omega) = \frac{1}{(2\pi)^2} \int d^2K [1 - \tilde{H}_0(\boldsymbol{K})]\, \tilde{\Delta}(\boldsymbol{K}, \omega)\, e^{i\boldsymbol{K}\cdot\boldsymbol{r}} \tag{22}$$

with arbitrary $\tilde{\Delta}(\boldsymbol{K}, \omega)$, and the resulting object profile

$$\hat{O}(\boldsymbol{r}, \omega) \equiv \hat{O}_0(\boldsymbol{r}, \omega) + \Delta(\boldsymbol{r}, \omega) \tag{23}$$

will *generate identical data as* $O(\boldsymbol{r}, \omega)$ and $\hat{O}_0(\boldsymbol{r}, \omega)$. In this connection, we note that the estimate $\hat{O}_0(\boldsymbol{r}, \omega)$ is the minimum energy estimate in the sense that $\hat{O}_0(\boldsymbol{r}, \omega)$ minimizes the functional

$$E \equiv \int d^2r |\hat{O}(\boldsymbol{r}, \omega)|^2 \tag{24}$$

among all possible estimates consistent with the data. This conclusion follows immediately by employing Parseval's theorem [16] to evaluate (24).

We can evaluate the estimate $\hat{O}_0(\boldsymbol{r}, \omega)$ defined in (21) directly in terms of the data by following a procedure analogous to that employed in [4]. Although the procedure is relatively straightforward, the calculations are somewhat long and have, thus, been relegated to the Appendix. We find that $\hat{O}_0(\boldsymbol{r}, \omega)$ can be expressed as a superposition of *partial reconstructions*

$$\hat{O}_0(\boldsymbol{r}, \omega; \boldsymbol{s}_0) \equiv \frac{k}{2\pi} e^{-ik\boldsymbol{s}_0\cdot\boldsymbol{r}} \int_{-\infty}^{\infty} d\kappa\, \tilde{F}(\kappa, \phi_0) \cdot \Theta(\kappa, \phi_0)\, \tilde{D}(\kappa, \omega; \boldsymbol{s}_0)\, e^{i[\gamma(x - l_0) + \kappa y]} \tag{25}$$

generated from data obtained from a single plane wave response i.e., at a fixed $\boldsymbol{s}_0$. Here, as in Section II $\gamma = \sqrt{k^2 - \kappa^2}$. The filter $\tilde{F}(\kappa, \phi_0) \equiv 0$ for $|\kappa| \geq k$ and

$$\tilde{F}(\kappa, \phi_0) \equiv \begin{cases} 0 & \text{if} \quad \kappa \geq -k \sin\phi_0 \\ & \text{with} \quad -\pi \leq \phi_0 \leq -\dfrac{\pi}{2} \\ |\kappa \cos\phi_0 - \gamma \sin\phi_0| & \text{otherwise} \end{cases} \tag{26}$$

for $|\kappa| < k$. $\Theta(\kappa, \phi_0)$ is unity for cases of complex profiles and

$$\Theta(\kappa, \phi_0) = \begin{cases} 0 & \text{if} \quad \kappa \leq k \sin\phi_0 \quad \text{with} \quad -\pi/2 \leq \phi_0 \leq 0 \\ 1 & \text{otherwise} \end{cases} \tag{27}$$

for offset VSP with real profiles. Each partial reconstruction $\hat{O}_0(\boldsymbol{r}, \omega; \boldsymbol{s}_0)$ is obtained by linearly filtering the data by a cascade of the stationary (convolutional) filter $\tilde{F}(\kappa, \phi_0)\, \Theta(\kappa, \phi_0)$ with the space varying filter $\exp\{i[\gamma(x - l_0) - k\boldsymbol{s}_0 \cdot \boldsymbol{r}]\}$ as illustrated in the block diagram in Fig. 9. The final reconstruction of the object profile is then obtained from the partial reconstructions according to the equation

$$\hat{O}_0(\boldsymbol{r}, \omega) = \frac{1}{2\pi} \int_{S_0} d\phi_0\, \hat{O}_0(\boldsymbol{r}, \omega; \boldsymbol{s}_0) \tag{28a}$$

for complex object profiles and

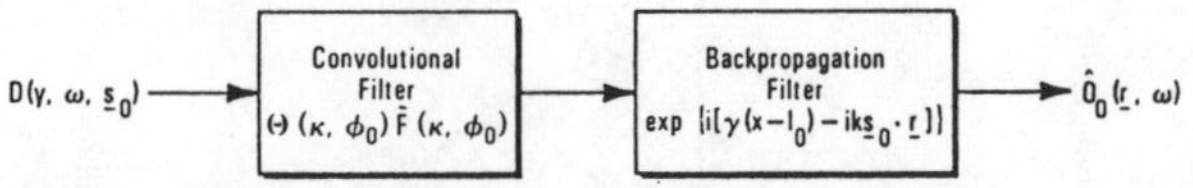

Fig. 9. Block diagram illustrating the generation of a partial reconstruction by means of a cascade of convolutional filtering followed by backpropagation.

$$\hat{O}_0(\boldsymbol{r}, \omega) = \frac{1}{\pi} \text{ Real} \int_{S_0} d\phi_0\, \hat{O}_0(\boldsymbol{r}, \omega; \boldsymbol{s}_0) \tag{28b}$$

for offset VSP with real object profiles.

The reconstruction algorithms embodied in (25) and (28) are the geophysical equivalents of the filtered backpropagation algorithm derived in [4] for ultrasound tomography. The term *filtered backpropagation* is a consequence of the interpretation of the mathematical operation embodied in (25) as being physically equivalent to first convolutionally *filtering* the data $D(y, \omega; \boldsymbol{s}_0)$ with the filter $\tilde{F}(\kappa, \phi_0)\, \Theta(\kappa, \phi_0)$, followed by *backpropagating* this filtered data into the object space (e.g., formation) using a propagation formula that approximates the propagation of phase within the Rytov approximation [17]. The reader is referred to [4] for a detailed discussion of this interpretation of the filtered backpropagation algorithm.

IV. Computer Simulations of Offset VSP

In this section we test the filtered backpropagation algorithm in computer simulations of offset VSP with real valued object profiles. In these simulations, our primary goal is to evaluate the performance of the algorithm on "perfect data", i.e., on the noise-free plane wave response computed according to the Rytov approximation. We shall also restrict our attention to simple object profiles composed of linear superpositions of circular disks having constant grey levels. Although such objects are not representative of real geological formations, they have the advantage that their plane wave response within the Rytov approximation can be calculated analytically which simplifies considerably the complexity of the computer simulations. We refer the reader to [5] for details of the calculation of the plane wave response of such objects.

In the simulations we approximated the integral over angles ϕ_0 in (28b) with a summation with $\Delta\phi_0$ set equal to 0.0314 rad (0.18°). The filtering operations required to generate each partial reconstruction $\hat{O}_0(\boldsymbol{r}, \omega; \boldsymbol{s}_0)$ were implemented using a 128-point FFT algorithm on a DEC VAX 11/780 digital computer. The images were generated on a RAMTEK image display system and photographed with a 35-mm camera mounted in a Dunn camera assembly.

We consider first a single circular disk whose radius R_0 is much smaller than the wavelength λ_0 of the insonifying wavefield. The image of such an object approximates the point spread function (impulse response) of the filtered backpropagation algorithm. Shown in Fig. 10 are partial reconstructions, as a function of view angle range, of a uniform disk whose radius is one-fifth of the wavelength. Shown in Fig. 11 are the spectrums (magnitude of the two-dimensional Fourier transforms) for each of the cases shown in Fig. 10. These spectrums were generated by Fourier transforming the images shown in Fig. 10 and displaying the magnitude of the respective transforms on the RAMTEK.

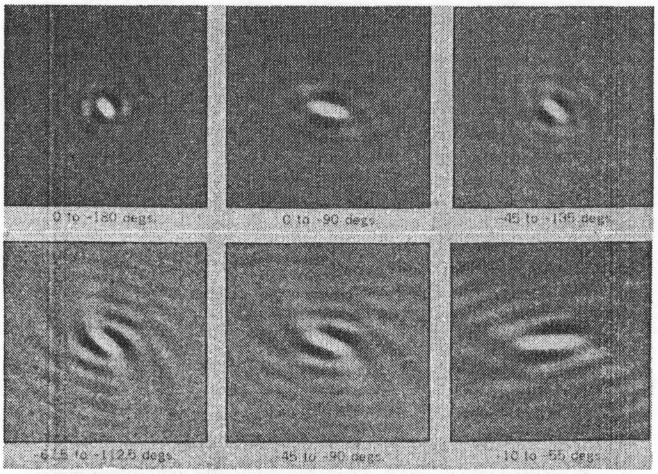

Fig. 10. Point spread functions for offset VSP.

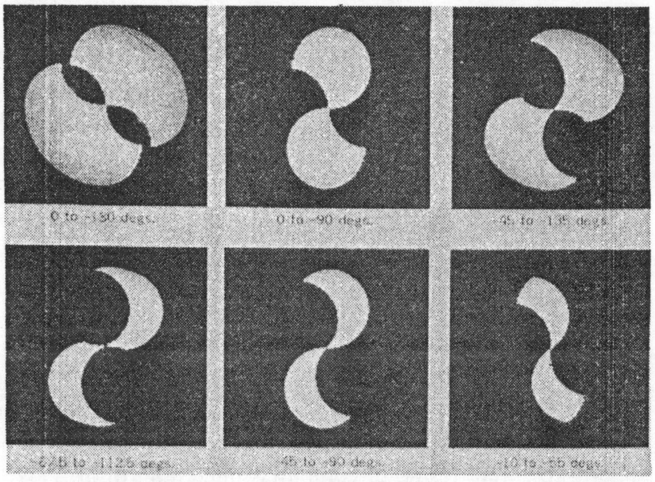

Fig. 11. Spectrums of the point spread functions shown in Fig. 10.

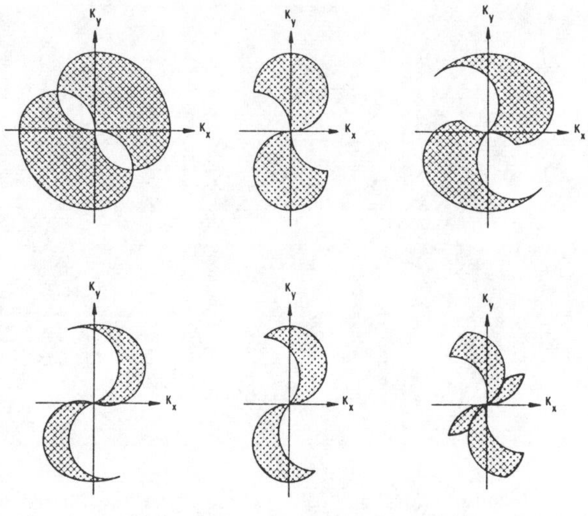

Fig. 12. Theoretical spectrums corresponding to those illustrated in Fig. 11.

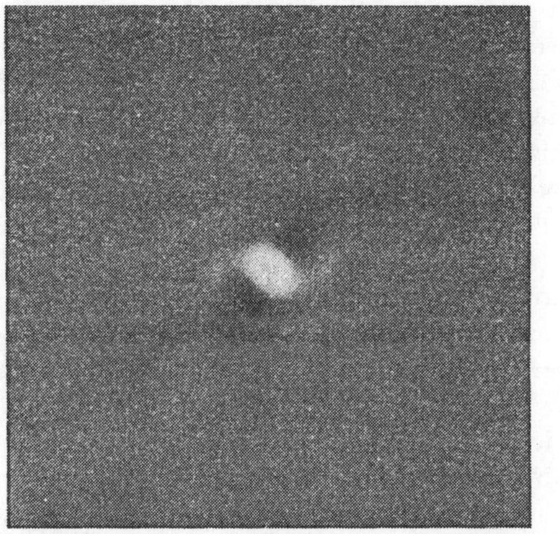

Fig. 13. Point spread function for full ϕ_0 coverage after low pass filtering with a three-term Blackman-Harris filter.

The spectrums shown in Fig. 11 should, ideally, be equal to the filter $\tilde{H}_0(\boldsymbol{K})$ defined in (20). They will, of course, differ from $\tilde{H}_0(\boldsymbol{K})$ due to the approximation of the integral over angles ϕ_0 by a summation and the use of a discrete Fourier transform to implement the filtering operations. The case shown in Fig. 11(a) corresponds to a full range of view angles varying from 0° to -180° and is seen to compare favorably with the theoretical spectrum shown in Fig. 7(b).

Shown in Fig. 12 are theoretical spectrums for all of the point spread functions shown in Fig. 10. Except for the case corresponding to a ϕ_0 range of -10° to -55° these spectrums are seen to be quite close to those displayed in Fig. 11. The reason for the disagreement between the theoretical and observed spectrums for the case of ϕ_0 ranging from -10° to -55° can be traced back to the use of an incorrect Θ filter for this case. In the simulations we employ the Θ filter defined in (27). This filter was derived for the full range ϕ_0 coverage case and *happens* to be correct in most situations. However, in certain cases, such as occurs for the ϕ_0 range of -10° to -55°, this filter is not optimal and results in incomplete coverage in Fourier space. The correct filter for any given case can be derived following the general lines employed for the full coverage case but will not be presented here since the full coverage Θ filter is optimal or near optimal in the vast majority of cases.

Referring to Fig. 10 we see that the point spread functions do not fall off smoothly from their peak values, but, rather have pronounced oscillations which gradually fall off in amplitude with increasing distance from the peak values. This oscillatory behavior is, essentially, a Gibbs phenomena [15] and is caused by the sharp cutoffs in Fourier space experienced by their associated spectrums as can be seen in Figs. 11 and 12. This oscillatory behavior in the point spread functions introduces "image artifacts" when reconstructing complicated structures. It can be corrected by tapering the spectrums with a smooth filter that falls gradually to zero at or in the immediate vicinity of the boundary in $\boldsymbol{K}$ space where the spectrums go to zero.

Shown in Fig. 13 is the point spread function for full ϕ_0 coverage resulting from tapering its spectrum with a circularly symmetric low pass filter whose radial dependence is that of a three-term Blackman-Harris filter [18] with cutoff frequency equal to $3k$. As is seen from this figure, the low pass filter has the desired effect of removing the oscillations from the point spread function. Unfortunately, the resulting increase in cosmetic quality of the point spread function is accompanied by a decrease in resolution so that the value of such filters when

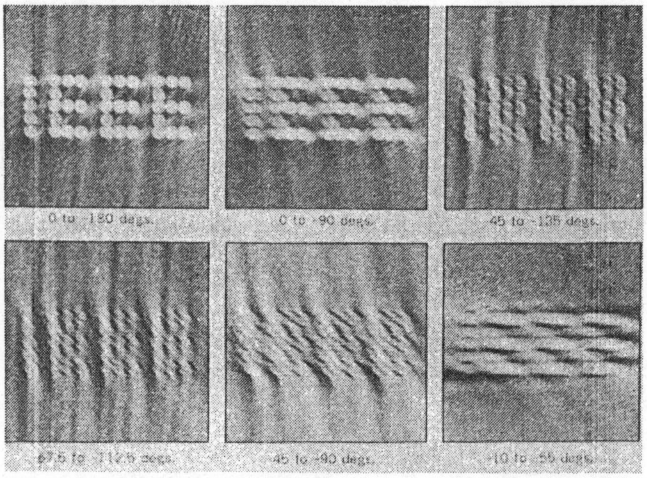

Fig. 14. Reconstructions of "IEEE" obtained via the filtered backpropagation algorithm. Each disk has a radius of four units and the wavelength is 5 units.

imaging complicated structures is questionable. We shall return to this question shortly.

Referring again to Fig. 10 we see that as the view angle range is increased the point spread function becomes smaller in overall extent which, of course, means that the resolution of the algorithm increases with increasing view angle range. It is important to note that in most cases shown and, indeed, as a general rule, the resolution of the algorithm will be best along the y (vertical) direction. This conclusion, which is obvious also from the spectrums shown in Figs. 11 and 12, means that good image quality can be expected in geophysical applications since geological formations tend to be layered in the y direction so that most of the high-frequency content of such structures lies in the immediate vicinity of the K_y axis.

As an example of a complicated structure we chose an object consisting of 38 circular disks, each having a radius of four units, forming the letters "IEEE". This structure is ideally suited for the simulations since it has high-frequency content along both the K_x and K_y axes. The differing resolution of the filtered backpropagation algorithm with respect to these two axes then becomes readily apparent in the simulations.

In Fig. 14 we show the sequence of reconstructions of the object for the set of view angle ranges employed in Figs. 10 and 11. The "donut-like" appearance of the circular disks in these reconstructions is caused by the oscillations in the point spread functions and can be removed, with an attendant loss in resolution, by appropriate low pass filtering as discussed above. Shown in Fig. 15 is a full view angle range reconstruction corresponding to the smoothed point spread function shown in Fig. 13. Note that although the donut-like appearance of the disks has been removed, the image looks "blurry" which, of course, is due to an overall loss in resolution caused by the low pass filtering.

The differing resolution of the algorithm along the x and y directions is readily apparent from Fig. 14. This is especially true for the $-10°$ to $-55°$ view angle range case. Since in actual applications it will be possible to determine the plane wave response only over a limited range of view angles, the character of the algorithm to yield high vertical resolution in these cases will become important in practice.

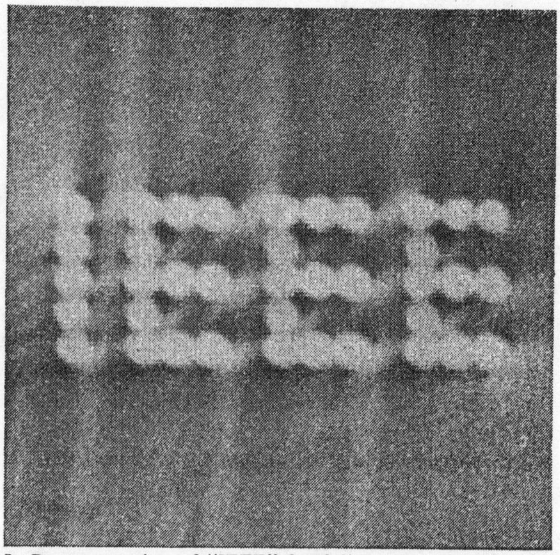

Fig. 15. Reconstruction of "IEEE" for full ϕ_0 coverage after low pass filtering with a three-term Blackman–Harris filter.

V. Summary and Future Directions

We have, in this paper, shown how the theory and algorithms of medical diffraction tomography [4] can be generalized to geophysical applications such as offset VSP and well-to-well tomography. This generalized theory, which we have called *geophysical diffraction tomography*, is based on the Born or Rytov approximations and is thus applicable to weakly inhomogeneous formations. The reconstruction algorithms were shown to be generalizations of the filtered back*projection* algorithm of X-ray tomography where the process of backprojection is replaced by back*propagation* and where the tomographic filter becomes view angle dependent. Finally, the reconstruction algorithms, called filtered backpropagation algorithms, were tested in computer simulations of offset VSP within the Rytov approximation.

The goal of this paper was to provide an introduction to the use of diffraction tomography in geophysical applications. We have avoided discussing a number of important technical details and generalizations of the material presented here that are useful, or possibly even necessary, in practical applications. For example, we have tacitly assumed that a slant stack of the cylindrical wave response could be performed. This operation is sometimes not feasible or desirable in practice and, consequently, reconstruction algorithms that operate directly on the cylindrical wave response—so-called "fan beam" algorithms—may need to be employed in place of the plane wave algorithms discussed here. Similarly, we have not discussed the "phase unwrapping" operation required to obtain the phase of the measured field from its complex amplitude nor have we discussed the effects of noise and finite transmitting and receiver array lengths on image resolution.

Most of the extensions required to reduce the material presented here to practice are rather straightforward. Some, however, are not. For example, the generalization of the filtered backpropagation algorithms to the fan beam case is quite involved and has only been recently completed for ultrasound tomography [19]. Particularly difficult is determining the effect on image quality of a breakdown of the underlying assumptions of the theory; namely, the assumptions of two-dimensional formations and wavefields and the weak scattering assumption implicit in the Born and Rytov approximations.

One of the most important areas for future work will be in testing the filtered backpropagation algorithms on "exact data"; i.e., on experimental data or in computer simulations where the scattered field is accurately computed using, say, a finite element code. Other areas for future work include extending the theory and algorithms beyond the weak scattering approximation and to the full elastic wave case.

Appendix

Derivation of the Reconstruction Algorithms

We consider first well-to-well tomography and offset VSP in cases where the object profile is complex. For these cases the region $D_0 \equiv R_0$ and (21) can be expressed in the form

$$\hat{O}_0(\boldsymbol{r}, \omega) = \frac{1}{(2\pi)^2} \int_{R_0} d^2K\, \tilde{O}(\boldsymbol{K}, \omega)\, e^{i\boldsymbol{K}\cdot\boldsymbol{r}} \tag{A1}$$

where R_0 is the region illustrated in Fig. 7(a) for offset VSP employing a full set of incident plane waves and is the region shown in Fig. 8 for well-to-well tomography employing a full set of incident plane waves. More generally, the region R_0 will lie within the regions illustrated in these figures for cases where a full set of incident plane waves are not employed.

We can evaluate $\hat{O}_0(\boldsymbol{r}, \omega)$ as given in (A1) directly in terms of the data by following a procedure analogous to that employed in [4]. In particular, we make the change of integration variables from $\boldsymbol{K}$ to κ, ϕ_0 where

$$K_x = \gamma - ks_{0x} = \gamma - k\cos\phi_0 \tag{A2a}$$

$$K_y = \kappa - ks_{0y} = \kappa - k\sin\phi_0 \tag{A2b}$$

with $\gamma = \sqrt{k^2 - \kappa^2}$. In terms of κ, ϕ_0 we then have that

$$\boldsymbol{K} \equiv K_x\hat{\boldsymbol{x}} + K_y\hat{\boldsymbol{y}} = (\gamma - k\cos\phi_0)\hat{\boldsymbol{x}} + (\kappa - k\sin\phi_0)\hat{\boldsymbol{y}} = k(\boldsymbol{s} - \boldsymbol{s}_0). \tag{A3}$$

The Jacobian of the transformation is readily found to be

$$\frac{\partial(K_y, K_y)}{\partial(\kappa, \phi_0)} = \frac{k}{\gamma}|\kappa\cos\phi_0 - \gamma\sin\phi_0| \tag{A4}$$

so that

$$d^2K = d\phi_0\, d\kappa\, \frac{k}{\gamma}|\kappa\cos\phi_0 - \gamma\sin\phi_0|. \tag{A5}$$

The upper κ limit of integration in the integral for fixed ϕ_0 is determined by the points of tangency between the semicircular arcs $\boldsymbol{K} = k(\boldsymbol{s} - \boldsymbol{s}_0)$ and the boundary of the region of integration R_0. Referring to Figs. 5, 7, and 8 it is not difficult to deduce that the limits of integration for κ are $-k \leqslant \kappa \leqslant k$ for $-\pi/2 \leqslant \phi_0 \leqslant \pi/2$, and $-k \leqslant \kappa \leqslant -k\sin\phi_0$ for $-\pi \leqslant \phi_0 \leqslant -\pi/2$. Thus defining the filter function

$$\hat{\Theta}(\kappa, \phi_0) \equiv \begin{cases} 0 & \text{if } \kappa \geqslant -k\sin\phi_0 \text{ with } -\pi \leqslant \phi_0 \leqslant -\frac{\pi}{2} \\ 1 & \text{otherwise} \end{cases} \tag{A6}$$

we find the (A1) can be expressed in the form

$$\hat{O}_0(\boldsymbol{r}, \omega) = \frac{1}{(2\pi)^2} \int_{S_0} d\phi_0 \int_{-k}^{k} d\kappa\, \frac{k}{\gamma}|\kappa\cos\phi_0 - \gamma\sin\phi_0| \cdot \hat{\Theta}(\kappa, \phi_0)\, \tilde{O}[k(\boldsymbol{s} - \boldsymbol{s}_0)]\, e^{ik(\boldsymbol{s}-\boldsymbol{s}_0)\cdot\boldsymbol{r}} \tag{A7}$$

where S_0 is a sub-interval of $-\pi \leqslant \phi_0 \leqslant 0$ for the offset VSP and is a sub-interval of $-\pi/2 \leqslant \phi_0 \leqslant \pi/2$ for well-to-well tomography. Note that $\hat{\Theta}(\kappa, \phi_0) \equiv 1$ for well-to-well tomography. Finally, on making use of (16) and (18) we can rewrite (A7) in terms of $\tilde{D}(\kappa, \omega; \boldsymbol{s}_0)$ as follows

$$\hat{O}_0(\boldsymbol{r}, \omega) = \frac{k}{(2\pi)^2} \int_{S_0} d\phi_0\, e^{-ik\boldsymbol{s}_0\cdot\boldsymbol{r}} \int_{-k}^{k} d\kappa \cdot |\kappa\cos\phi_0 - \gamma\sin\phi_0|\, \hat{\Theta}(\kappa, \phi_0)\, \tilde{D}(\kappa, \omega; \boldsymbol{s}_0) \cdot e^{i[\gamma(x - l_0) + \kappa y]}. \tag{A8}$$

Consider now, offset VSP in cases where the object profile is real. For such cases $D_0 = R_0 \cup R_0^*$ and (21) becomes

$$\hat{O}_0(\boldsymbol{r}, \omega) = \frac{1}{(2\pi)^2} \int_{R_0 \cup R_0^*} d^2K\, \tilde{O}(\boldsymbol{K}, \omega)\, e^{i\boldsymbol{K}\cdot\boldsymbol{r}} \tag{A9}$$

where $R_0 \cup R_0^*$ is the region illustrated in Fig. 7(b) if a full set of incident plane waves are employed and is contained within this region otherwise. Defining the function

$$\Theta(\boldsymbol{K}) \equiv \begin{cases} 0 & \text{if } K_x < 0, K_y < 0 \\ 1 & \text{otherwise} \end{cases} \tag{A10}$$

we can write (A9) in the form

$$\hat{O}_0(\boldsymbol{r}, \omega) = \frac{1}{(2\pi)^2} \int_{R_0} d^2K\, \Theta(\boldsymbol{K})\, \tilde{O}(\boldsymbol{K}, \omega)\, e^{i\boldsymbol{K}\cdot\boldsymbol{r}} + \frac{1}{(2\pi)^2} \int_{R_0^*} d^2K\, \Theta(-\boldsymbol{K})\, \tilde{O}(\boldsymbol{K}, \omega)\, e^{i\boldsymbol{K}\cdot\boldsymbol{r}}. \tag{A11}$$

Making the change of integration variables $\boldsymbol{K} \to -\boldsymbol{K}$ in the second integral and using (19) we obtain

$$\hat{O}_0(\boldsymbol{r}, \omega) = 2\,\text{Real}\left\{\frac{1}{(2\pi)^2} \int_{R_0} d^2K\, \Theta(\boldsymbol{K})\, \tilde{O}(\boldsymbol{K}, \omega)\, e^{i\boldsymbol{K}\cdot\boldsymbol{r}}\right\}. \tag{A12}$$

With the exception of the function $\Theta(\boldsymbol{K})$ occuring in the integral in (A12), this integral is identical to the right-hand side of (A1). Consequently, all the steps leading to (A8) apply to this integral so that we can rewrite (A12) in terms of the data as follows

$$\hat{O}_0(\boldsymbol{r}, \omega) = 2\,\text{Real}\left\{\frac{k}{(2\pi)^2} \int_{S_0} d\phi_0\, e^{-ik\boldsymbol{s}_0\cdot\boldsymbol{r}} \cdot \int_{-k}^{k} d\kappa\, |\kappa\cos\phi_0 - \gamma\sin\phi_0|\, \hat{\Theta}(\kappa, \phi_0) \cdot \Theta(\kappa, \phi_0)\, \tilde{D}(\kappa, \omega; \boldsymbol{s}_0)\, e^{i[\gamma(x - l_0) + \kappa y]}\right\} \tag{A13}$$

where $\hat{\Theta}(\kappa, \phi_0)$ is defined in (A6), and where $\Theta(\kappa, \phi_0)$ is readily determined from (A2) and (A10). We find that

$$\Theta(\kappa, \phi_0) = \begin{cases} 0 & \text{if } \kappa \leqslant k \sin \phi_0 \text{ with } -\pi/2 \leqslant \phi_0 \leqslant 0 \\ 1 & \text{otherwise.} \end{cases} \quad \text{(A14)}$$

Equations (A8) and (A13) can be unified by introducing *partial reconstructions* $\hat{O}_0(\boldsymbol{r}, \omega, \boldsymbol{s}_0)$ defined by

$$\hat{O}_0(\boldsymbol{r}, \omega; \boldsymbol{s}_0) \equiv \frac{k}{2\pi} e^{-ik\boldsymbol{s}_0 \cdot \boldsymbol{r}} \int_{-\infty}^{\infty} d\kappa\, \tilde{F}(\kappa, \phi_0)\, \Theta(\kappa, \phi_0) \cdot \tilde{D}(\kappa, \omega; \boldsymbol{s}_0)\, e^{i[\gamma(x-l_0)+\kappa y]} \quad \text{(A15)}$$

where $\tilde{F}(\kappa, \phi_0) = |\kappa \cos \phi_0 - \gamma \sin \phi_0|\, \hat{\Theta}(\kappa, \phi_0)$ if $|\kappa| \leqslant k$ and is zero otherwise and where $\Theta(\kappa, \phi_0)$ is unity for complex profiles and is defined in (A14) for offset VSP with real profiles. In terms of the partial reconstructions we then have

$$\hat{O}_0(\boldsymbol{r}, \omega) = \frac{1}{2\pi} \int_{S_0} d\phi_0\, \hat{O}_0(\boldsymbol{r}, \omega; \boldsymbol{s}_0) \quad \text{(A16a)}$$

for complex object profiles and

$$\hat{O}_0(\boldsymbol{r}, \omega) = \frac{1}{\pi} \text{Real} \int_{S_0} d\phi_0\, \hat{O}_0(\boldsymbol{r}, \omega; \boldsymbol{s}_0) \quad \text{(A16b)}$$

for offset VSP with real object profiles.

References

[1] K. A. Dines and R. J. Lytle, "Computerized geophysical tomography," *Proc. IEEE*, vol. 67, pp. 471-480, 1979.

[2] E. J. Witterholt, J. L. Kretzsehmar, and K. L. Jay, "The application of crosshole electromagnetic wave measurements to mapping of a steam flood," in *Proc. Petroleum Society of CIM*, 1982.

[3] E. I. Gal'perin, *Vertical Seismic Profiling*. Tulsa, OK: Society of Exploration Geophysicists, 1974.

[4] A. J. Devaney, "A filtered backpropagation algorithm for diffraction tomography," *Ultrasonic Imaging*, vol. 4, pp. 336-350, 1982.

[5] —, "A computer simulation study of diffraction tomography," *IEEE Trans. Biomed. Eng.*, vol. BME-30, pp. 377-386, 1983.

[6] R. A. Phinney, K. Kay Chowdhury, and L. Neil Frazer, "Transformation and analysis of record sections," *J. Geophys. Res.*, vol. 86, pp. 359-377, 1981.

[7] A. Sommerfeld, *Partial Differential Equations in Physics*. New York: Academic Press, 1967, p. 189.

[8] L. Schiff, *Quantum Mechanics*, 3rd ed. New York: McGraw-Hill, 1968, ch. 8.

[9] A. J. Devaney, "Inverse scattering within the Rytov approximation," *Opt. Lett.*, vol. 6, pp. 374-376, 1981.

[10] B. K. Vainshtein, *Diffraction of X-rays by Chain Molecules*. New York: Wiley, 1974.

[11] R. K. Mueller, M. Kaveh, and G. Wade, "Reconstructive tomography and applications to ultrasonics," *Proc. IEEE*, vol. 67, pp. 567-587, 1979.

[12] P. M. Morse and H. Feshbach, *Methods of Theoretical Physics*. New York: McGraw-Hill, 1953, p. 823.

[13] J. W. Goodman, *Introduction to Fourier Optics*. New York: McGraw-Hill, Sec. 3-7, 1968.

[14] H. Stark and I. Paul, "An investigation of computerized tomography by direct Fourier inversion and optimum interpolation," *IEEE Trans. Biomed. Eng.*, vol. BME-28, pp. 496-505, 1981.

[15] A. C. Kak, "Computerized tomography with X-ray emission and ultrasound sources," *Proc. IEEE*, vol. 67, pp. 1245-1272, 1979.

[16] A. Papoulis, *Systems and Transforms with Applications in Optics*. New York: McGraw-Hill, 1968, p. 92.

[17] A. J. Devaney, H. J. Liff, and S. Apsell, "Spectral representations for free space propagation of complex phase perturbations of optical fields," *Opt. Commun.*, vol. 15, pp. 1-5, 1975.

[18] F. J. Harris, "On the use of windows for harmonic analysis with the discrete Fourier transform," *Proc. IEEE*, vol. 66, pp. 51-83, 1978.

[19] A. J. Devaney and G. Beylkin, "A filtered backpropagation algorithm for fan beam diffraction tomography," in *Proc. 13th Int. Symp. on Acoustical Imaging*, 1983.

Part VIII
Sensor-Array Processing

THE problem of processing signals received by an array of sensors can be treated in several ways. In its most general form it is a multidimensional problem concerned with filtering a signal that is a function of both space and time to remove unwanted interference and extract the desired information. In some cases it can be phrased as a multidimensional spectrum estimation problem, where it is desirable to estimate all or part of the wavenumber-frequency spectrum of the sensed wave field. (The reader interested in this aspect of sensor-array processing is encouraged to see Part II of this book.)

In its more pragmatic forms sensor-array processing strives to estimate the direction of propagation (bearing estimation) of waves that pass by the sensor array or to estimate the difference in arrival time of a propagating signal at two spatially separated sensors. A classical approach to the bearing estimation problem involves delaying the sensor signals relative to each other and then summing the delayed signals. This will cause signals propagating across the array at a particular velocity to be added in phase while noise and other signals will be attenuated. This approach is called beamforming, and it is analogous to constructing a tapped delay line bandpass filter for 1-D signals. Modern bearing estimation techniques try to analyze the correlation matrix estimated from sensor cross-correlation coefficients to obtain a power density estimate as a function of bearing.

The six papers reprinted in this section were chosen to represent a variety of approaches to various sensor-array processing problems. The papers by Halpeny and Childers and by Dudgeon serve as discussions of the multidimensional aspects of the problem. The first paper discusses the problem of estimating the number and the velocity vectors of propagating plane waves in the presence of interference. The results are related to other array processing techniques by examining the equivalent space-time filtering of the wavenumber-frequency spectrum. The second paper discusses the relationship between the frequency spectra of the sensor signals, the frequency spectra of the beam signals obtained by delay-and-sum beamforming, and the wavenumber-frequency spectrum of the array of sensor signals for the case of a linear array of equally spaced sensors. The analogy between beamforming and bandpass filtering is discussed.

The next two papers discuss some practical aspects of sensor-array processing. The Cox paper focuses on the performance of optimal array processors for narrow-band signals, examining their resolution and gain characteristics. It contains a detailed analysis of the performance of two optimal array processors when their parameters have been determined under incorrect assumptions about the propagating wave field. For comparison the results obtained with a conventional beamformer are also discussed.

The paper by Mucci examines efficient beamforming algorithms in the context of modern signal processing hardware and techniques. It discusses delay-and-sum beamforming, partial sum beamforming, interpolation beamforming, and shifted sideband beamforming, as well as frequency-domain implementations.

The paper by Carter introduces the problem of time-delay estimation. The objective is to estimate the difference in arrival times of a propagating wave at two spatially separated sensors. Accurate estimation of the relative time delay helps to deduce the direction of propagation. For arrays of many sensors, estimation of relative time delay can be used to measure the curvature of the wavefront from a nearfield source. This measurement can be used in turn to estimate the range from the receiver to the source.

The final paper in this section, by Johnson and DeGraaf, discusses improvements in resolution for the bearing estimation problem by using eigenvector/eigenvalue analysis of the spatial correlation matrix. This research area in particular is currently very active because it holds the promise of increased performance by using matrix analysis algorithms that are amenable to implementation by high-speed special purpose digital multiprocessors. (See, for example, Sessions 15 and 46 of the *Proceedings of the 1985 IEEE International Conference on Acoustics, Speech, and Signal Processing*.)

Space limitations have precluded the inclusion of papers on other aspects of sensor-array processing and related topics. In particular, we have included nothing about computed imaging problems such as ultrasonic imaging, radar imaging, and computer tomographic imaging. (The reader interested in these applications should consult Part V of this book and its associated references.) Geophysical applications of array processing techniques are covered to some extent in Part VII of this book. The following short bibliography should be useful for literature searches into aspects of sensor-array processing not discussed in the papers reprinted here.

Bibliography

[1] G. C. Carter, Ed., Special Issue on Time Delay Estimation, *IEEE Trans. Acoust., Speech, Signal Processing*, vol. ASSP-29, June 1981.

[2] D. G. Childers, Ed., *Modern Spectrum Estimation*. New York, NY: IEEE PRESS, 1978.

[3] D. E. Dudgeon and R. M. Mersereau, *Multidimensional Digital Signal Processing*, chs. 6 and 7. Englewood Cliffs, NJ: Prentice-Hall, 1984.

[4] S. Haykin and J. A. Cadzow, Eds., Special Issue on Spectral Estimation, *Proc. IEEE*, vol. 70, September 1982.

[5] G. T. Herman, Ed., Special Issue on Computerized Tomography,

Proc. IEEE, vol. 71, Mar. 1983.
[6] R. A. Monzingo and T. W. Miller, *Introduction to Adaptive Arrays*. New York, NY: John Wiley & Sons, 1980.
[7] A. V. Oppenheim, Ed., *Applications of Digital Signal Processing*. Englewood Cliffs, NJ: Prentice-Hall, 1978.
[8] K. Y. Wang, Ed., Special Issue on Acoustic Imaging, *Proc. IEEE*, vol. 67, Apr. 1979.
[9] M. Wax, T.-J. Shan, and T. Kailath, "Spatio-temporal spectral analysis by eigenstructure methods," *IEEE Trans. Acoust., Speech, Signal Processing*, vol. ASSP-32, pp. 817–827, Aug. 1984.

Composite Wavefront Decomposition Via Multidimensional Digital Filtering of Array Data

OWEN S. HALPENY, MEMBER, IEEE, AND DONALD G. CHILDERS, SENIOR MEMBER, IEEE

***Abstract*—A solution is obtained to the problem of estimating the number, vector velocity, and waveshape of overlapping planewaves in the presence of interfering planewaves and channel noise, where previous solutions have assumed one or more of these quantities as known. A general optimum solution is not found; instead, a heuristic solution is presented along with a complete working implementation program for large scale computers. For the case where the number of waves and the vector velocities are known, the solution is optimum.**

The detection of waves and the estimation of their bearing, velocity, and waveshape is accomplished via digital filtering of the frequency-wavenumber power spectrum, which is computed via an efficient estimator, of the array sensed data. A new approach to the multiwave estimation problem is to reduce it to a succession of single wave problems using especially developed frequency-wavenumber filters.

Special attention is given throughout the study to computationally efficient approaches.

The results of the paper are placed in perspective by showing how the historically important approaches to the processing of array data such as delay and sum, weighted delay and sum, array prewhitening, beam forming, inverse filtering, least mean-square estimation, and maximum likelihood estimation are related via the spatio-temporal filtering of the frequency-wavenumber spectrum.

The spectral estimation, digital filtering, and the multiwave maximum likelihood estimator developments are demonstrated by the processing of a set of simulated planewaves of various bearings, velocities, and frequencies, as well as by processing electroencephalographic (brain wave) data monitored via an array of scalp electrodes.

I. INTRODUCTION

DIGITAL SIGNAL processing of spatio-temporal data is a topic of current interest in many fields from both a theoretical and practical viewpoint. Typical research includes the radar and sonar problem to locate and classify targets that emit or reflect energy. Radio astronomy investigates electromagnetic radiation from different regions of the universe. Seismologists are frequently interested in locating earthquake sites and differentiating between natural and man-made seismological events. There is even an interest amongst electrophysiologists to delineate propagating wavefronts in the electroencephalogram (EEG).

Accordingly, this study is directed toward techniques which have definite practical merit, but which nontheless have applicability to many areas. The signal model is a sum of planewaves which overlap both spatially and temporally. The noise consists of both ambient channel noise and interfering planewaves. It is generally assumed that the noise statistics are available, but the signals are completely unknown *a priori*.

In one dimension, the planewave may be defined as $s(t - x/|\nu|)$ where t is time, x is distance as measured from the spatial origin along the direction of propagation, $|\nu|$ is the speed of the wave, and $s(t)$ is the functional form of the wave at the spatial origin. Typical approximations to planewaves are far-field electromagnetic radiation, explosion shock waves, and tidal waves caused by earthquakes.

In the Cartesian coordinate system the planewave becomes $s(t - \boldsymbol{\alpha} \cdot \boldsymbol{r})$ where $\boldsymbol{r} = (r_x, r_y, r_z)$ is the position vector at which $s(\cdot)$ is evaluated and $\boldsymbol{\alpha}$ is the inverse velocity vector, i.e., $\boldsymbol{\alpha} = \boldsymbol{\nu}/|\boldsymbol{\nu}|^2$. A monochromatic planewave in complex exponential form is given by

$$\exp\left[2\Pi i f_0(t - \boldsymbol{\alpha}_0 \cdot \boldsymbol{r})\right] = \exp\left[2\Pi i(f_0 t - \boldsymbol{k}_0 \cdot \boldsymbol{r})\right] \quad (1)$$

where the vector wavenumber is $\boldsymbol{k}_0 = f_0 \boldsymbol{\alpha}_0$. Any transformable spatio-temporal function may be expressed as a sum of monochromatic planewaves,

$$s(t,\boldsymbol{r}) = \iint S(f,\boldsymbol{k}) \exp\left[2\Pi i(ft - \boldsymbol{k} \cdot \boldsymbol{r})\right] df\, d\boldsymbol{k} \quad (2)$$

where $s(t,\boldsymbol{r})$ and $S(f,\boldsymbol{k})$ form a multidimensional Fourier transform pair.

To determine the waveshape of a single wave of known vector velocity in a noiseless environment, one simply observes the wave at any point in space and applies the appropriate delay (or advance) to produce $s(t)$. If a single planewave is spatially sampled using a discrete array of sensors and each sensor adds white noise, an estimate of the wave is obtained by aligning the signals on each of the channels and summing the results. This is the simplest form of array processing. If, in addition to the channel noise, interfering planewaves are present, then more complicated processing must be employed. This is a common problem in radar [1]. In this case, the estimation problem is usually solved by inserting amplitude weighting in each sensor channel prior to the delay and sum [2].

To obtain the optimum estimates for various signal parameters or waveshapes, decision and estimation theory are

Manuscript received December 10, 1973; revised May 13, 1974. This work was supported in part by the Office of Naval Research under Contract N00014-68-A-0173-0014, Task NR 042-278, and in part by the Alfred P. Sloan Foundation under Grant 74-2-9.

O. S. Halpeny is with the Defense Communications Agency, Washington, D.C.

D. G. Childers is with the Department of Electrical Engineering, University of Florida, Gainesville, Fla. 32611.

Reprinted from *IEEE Trans. Circuit Syst.*, vol. CAS-22, pp. 552–562, June 1975.

applied. For most cases of practical interest, the array processor requires a different linear filter applied to each sensor channel before the delay and sum operation. The filters generally depend upon the array geometry and noise statistics. The resulting estimates are referred to as maximum-likelihood estimates. The problem was first solved by Kelly and Levin [3]. A closed form solution is not obtained to the general problem of estimating the waveshape and vector velocity but, rather, an integral equation is solved by iteration. The parameter being varied is the vector velocity. If the vector velocity is known, a closed form solution is possible. Capon successfully implements this solution, applies the results to seismological data [4] and discusses the wave detection problem [5]. For one wave and $\boldsymbol{v}$ known the optimum estimate is known [6], [7].

A constrained least mean-squares filter for an array pointed in the look-direction is known [2] which reduces to that in [3] when the filter frequency characteristic in the look-direction is all-pass and linear-phase when the noise is Gaussian and the angle of arrival is known. The algorithm in [2] does not assume that the input correlation matrix is known *a priori* but must be learned adaptively.

The multiwave maximum-likelihood estimation problem, foı the special case where the number of signal and noise planewaves is known, is solved in [8]. Again, a closed form solution is not obtained. Here it is necessary to vary all of the vector velocities simultaneously in order to solve an integral equation by iteration. When compared to the one-wave problem, this is referred to as a multidimensional search. If the vector velocities are known, closed form solutions are possible. An implementation is given in [9] for two planewaves of known vector velocities. The multiwave detection problem, i.e., determining the number of waves, has not received much attention in the literature.

The general problem considered in this paper is the determination of the number, vector velocity, and waveshape of overlapping planewaves in the presence of additive noise where previous authors have assumed one or more of these quantities as known. A general optimum solution is not found. Instead, a heuristic solution is presented along with a complete working implementation scheme for large-scale computers. For the case where the number of waves and the vector velocities are known, the solution is optimum.

The detection of waves and the estimation of the vector velocities are accomplished by filtering the frequency-wavenumber power spectral density function [4], [10]. Generalized nonrecursive frequency domain digital filters are developed which can select or reject planewaves on the basis of their bearing, or velocity, or frequency composition, or any combination thereof. A formal solution to this part of the problem requires maximum-likelihood detection and an exhaustive search algorithm.

To reduce the multiwave problem to a succession of single wave estimation problems, frequency-wavenumber filters are employed (and discussed) which pass only a single planewave. This avoids the difficult multidimensional search required by the solution in [8]. Thus a series of single wave maximum-likelihood estimates is obtained and discussed along with the important properties of the maximum-likelihood estimators.

The thrust of this research was provided by [8] which suggested that implementation is more important, at this time, than further theoretical development. But, as often happens, such an attack resulted in the new approach to the multiwave estimation problem mentioned above and to be described subsequently.

Examples of processed array data are presented and analyzed which demonstrate the effectiveness of spatio-temporal spectral estimation and filtering.

II. Multidimensional Power Spectral Density Estimation

The literature discussing one-dimensional power spectrum density (PSD) estimation is extensive for both continuous and discrete data [11]–[19]. Frequency-wavenumber power spectrum estimation can be approached in a similar manner.

A. Frequency-Wavenumber PSD Estimation

The multidimensional PSD function can be defined as [10]

$$P(f,\boldsymbol{k}) \equiv E\{R(\tau,\boldsymbol{r})\}$$

$$\equiv \iint_{-\infty}^{+\infty} R(\tau,\boldsymbol{r}) \exp\left[-i2\pi(f\tau - \boldsymbol{k}\cdot\boldsymbol{r})\right] d\tau\, d\boldsymbol{r} \quad (3)$$

where the multidimensional autocorrelation function is given by

$$R(\tau,\boldsymbol{r}) = E[s(t,\boldsymbol{r}_0)s(t+\tau, \boldsymbol{r}_0+\boldsymbol{r})] \quad (4)$$

and $\boldsymbol{k}$ is the vector "spatial frequency" or wavenumber, i.e., $\boldsymbol{k} = f\boldsymbol{\alpha} = f\boldsymbol{v}/|\boldsymbol{v}|^2$, and $\boldsymbol{\alpha}$ is referred to as the inverse phase velocity which is a projection of the actual (group) velocity onto a plane, e.g., in two-dimensional space the $x-y$ plane. $\boldsymbol{\alpha}$ is oriented in the direction of propagation of the wave and represents the vector of delays incurred per unit distance as measured along each coordinate axis. It is sometimes referred to as the slowness vector. The vector product $\boldsymbol{\alpha}\cdot\boldsymbol{r}$ is known as the moveout in seismology.

The standard Fourier transform pairs exist for $P(f,\boldsymbol{k})$ and $P(f,\boldsymbol{r})$.

For computational purposes, as will be discussed later the multidimensional spectrum can be estimated directly by calculating the n-dimensional Fourier transform of the truncated $s(t,\boldsymbol{r})$ using proper windowing.

Consider a monochromatic planewave of inverse velocity $\boldsymbol{\alpha}_0$ and frequency f_0 expressed as

$$s(t,\boldsymbol{r}) = A\cos(2\pi(f_0t - \boldsymbol{\alpha}_0\cdot\boldsymbol{r}) + \theta)$$

$$= A\cos(2\pi f_0(t - \boldsymbol{k}_0\cdot\boldsymbol{r}) + \theta) \quad (5)$$

where $\boldsymbol{k}_0 = f_0\boldsymbol{\alpha}_0$ and θ is a random variable uniformly

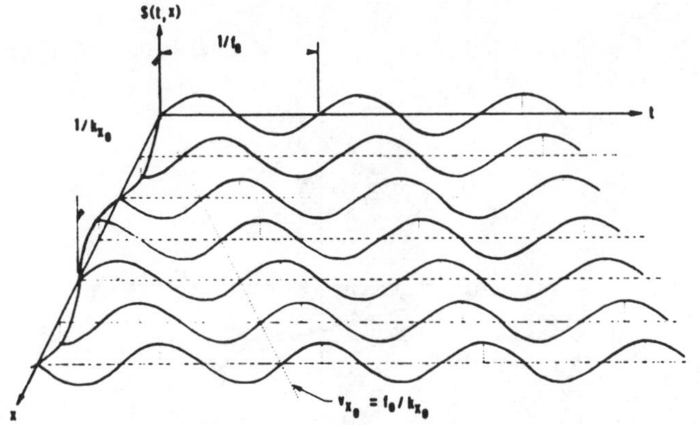

Fig. 1. Two-dimensional monochromatic spatio-temporal function with frequency f_0, wavenumber k_{x0}, and velocity v_{x0}.

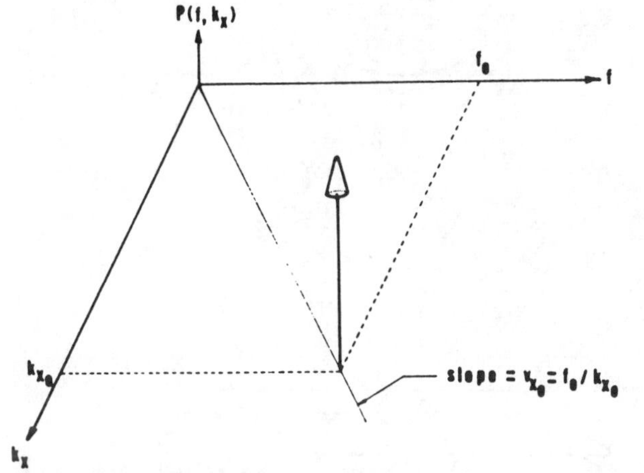

Fig. 2. Frequency-wavenumber spectrum of two-dimensional monochromatic spatio-temporal function with frequency f_0, wavenumber k_{x0}, and velocity v_{x0}.

distributed on $[0,2\pi]$. For one spatial dimension, $s(t,x)$ is illustrated in Fig. 1.

From (3), the frequency-wavenumber PSD function is the two-dimensional transform of the autocorrelation function

$$P(f,k) = \iint_{-\infty}^{+\infty} A^2/2 \cos (2\pi(f_0\tau - k_0 \cdot p)) \cdot \exp[-i2\pi(f\tau - k \cdot p)]\, d\tau\, dp$$

$$= \frac{A^2}{4}[\delta(f - f_0, k - k_0) + \delta(f + f_0, k + k_0)]. \quad (6)$$

Fig. 2 shows the frequency-wavenumber spectrum for $s(t,x)$.

The frequency-wavenumber spectrum concisely summarizes the defining parameters of the planewave. Note that the magnitude velocity of the wave is given by the slope of the line from the origin to the delta function, i.e., $|v_0| = 1/|\alpha_0| = f_0/|k_0|$.

If $s(t - x/|v|)$ is wideband instead of monochromatic, the energy distribution in the frequency-wavenumber domain is contained in the plane perpendicular to the $k_x - f$ plane defined by the line with slope f_0/k_0, since the Fourier transform of $s(t - x/|v|)$ is $S(f)\, \delta[k_x - (f/v)] = S(f)\, \delta(k_x - k)$. It is also of interest to note that the familiar one-dimensional PSD is obtained when $\lim_{v\to\infty} k_0 = 0$. In practical terms, this condition implies that the velocity is so large that the wave appears at every point in space simultaneously and that the power spectrum analysis reduces to an analysis of a function of one variable, time.

B. Wave Parameter Estimation

As noted in [20], beam-forming and one-dimensional spectral analysis can be combined as the time domain method to accomplish a study of the velocity and frequency structure of the wave without any direct suggestion that estimates of frequency-wavenumber spectra are being obtained. But it is difficult to achieve selective velocity filtering by this method.

The beam-forming frequency domain method [20] is termed the conventional method by Capon [4] and has been used by the latter author with success.

The procedure presented in this study is based upon estimating the frequency-wavenumber spectrum directly from a multidimensional discrete Fourier transform (DFT).

This method of estimating the frequency-wavenumber spectrum will be referred to as the direct fast Fourier transform (FFT) method. This procedure is similar to that advanced by Smart and Flinn [21]. The difference is mainly due to the method of calculation. The reason for choosing this estimate is to utilize the computational efficiency of the multidimensional FFT [22].

The Fortran programs written to implement the estimate include provisions for multidimensional windows. Whether or not special windows are included, the computations required by direct frequency-wavenumber PSD estimation are significantly less than those required by other estimates.

Multidimensional window design is discussed in [23], [24], [34] which is a problem common to image processing, optical systems, and two-dimensional antenna theory.

C. Example

The performance of the Fortran subroutines is illustrated as follows: seven truncated, monochromatic planewaves are generated, and summed and the frequency-wavenumber spectrum of the resulting spatio-temporal function is estimated and plotted. These input functions are listed in Table I.

The temporal sampling rate is 50 Hz, which yields a folding frequency of 25 Hz and a time interval of 0.02 s. The record length is 32 points or 640 ms. Spatial sampling took the form of a 4×4 square array with a spatial interval of 2 cm. The "spatial folding frequency" was then 0.25 cm^{-1} with wavenumber interval, 0.125 cm^{-1}. The spatial dimensions of the array are 6 cm $\times$ 6 cm. These parameters are selected since they are used later to process EEG data.

The raw spectral estimates are computed and the result listed. An actual output is reproduced in Fig. 3. This figure

TABLE I
LISTING OF OVERLAPPING PLANEWAVES

Number	f_0 (Hz)	v_0 (cm/s)	$[<v_0]$ (degrees)
1	3.1	30.0	120.0
2	7.8	42.0	90.0
3	7.8	125.0	90.0
4	12.5	75.0	45.0
5	12.5	75.0	−45.0
6	17.2	100.0	0.0
7	21.9	100.0	0.0

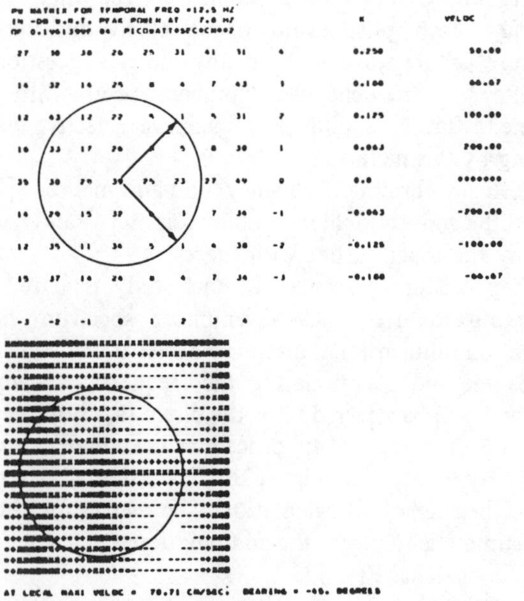

Fig. 3. Frequency-wavenumber spectrum for waves listed in Table I for $f = 12.5$ Hz.

corresponds to a segment of the frequency-wavenumber space for which the temporal frequency is constant. Thus Fig. 3 gives power as a function of k_x and k_y for 12.5 Hz. The power level is recorded at each point in the space in negative decibels with respect to recorded peak power density. The minus signs are suppressed for simplicity. The peak power density is calculated for an input given in microvolts or microamperes. The power density scale is limited at −90 dB. Power levels below this point are recorded as −90 dB. The coding of the power density regions into gray tones is produced by overprinting specially selected characters. Linear interpolation is used to determine the level between given data points.

In Fig. 3 the circle corresponds to a constant velocity of 75 cm/s and the two peaks account for the fourth and fifth entries in Table I, i.e., two planewaves of equal velocity propagating in the ±45° directions. The superimposed arrows correspond to the true velocity, while the estimated velocity for one of the waves is given below the gray plot. For the sake of brevity, we have omitted the figures which correspond to the other waves in Table I, but these same waves will be used later to illustrate angle and velocity filtering.

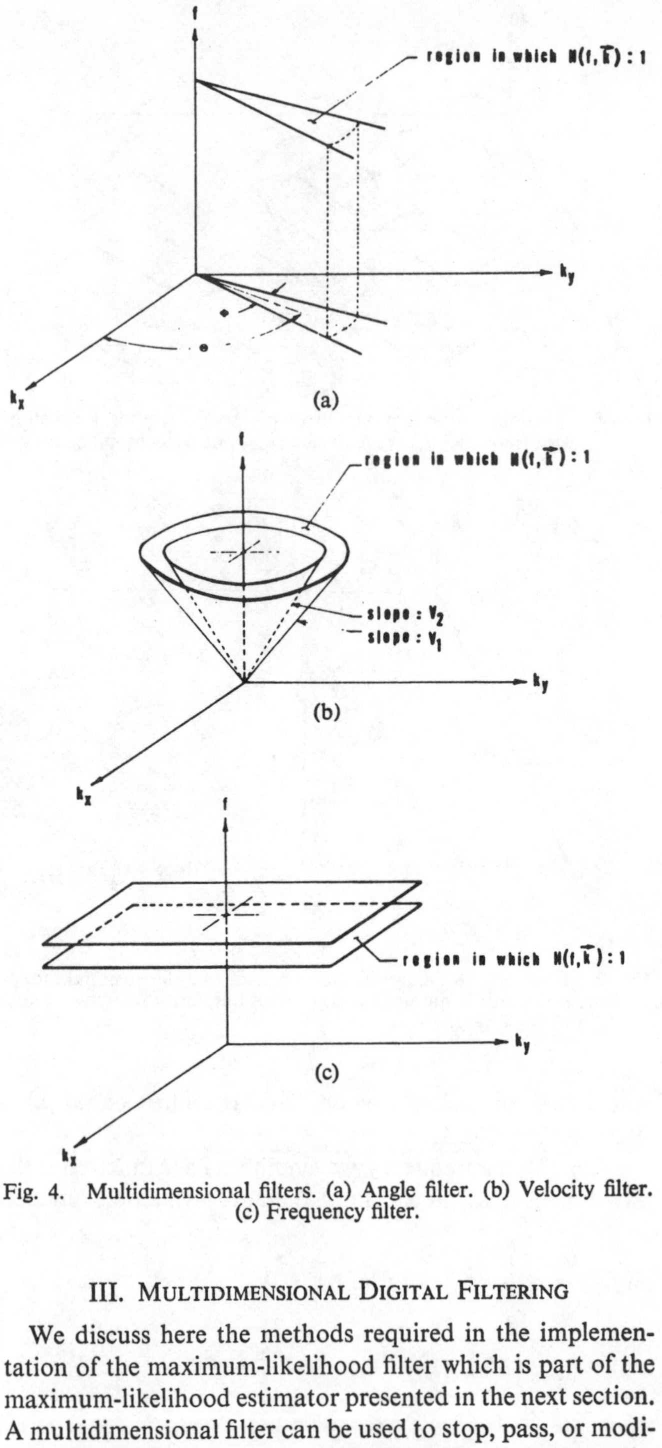

Fig. 4. Multidimensional filters. (a) Angle filter. (b) Velocity filter. (c) Frequency filter.

III. MULTIDIMENSIONAL DIGITAL FILTERING

We discuss here the methods required in the implementation of the maximum-likelihood filter which is part of the maximum-likelihood estimator presented in the next section. A multidimensional filter can be used to stop, pass, or modify a traveling planewave and is used later to reduce the multiwave maximum-likelihood estimation problem to that of a single wave estimation problem. Ideal filters are shown in Fig. 4. Practically, these filters must be smoothed in the transition regions. Of the various methods available for reducing the sidelobes of digital nonrecursive filters, the window and sampling methods are quite useful [25], [26]. We have designed filters via both procedures.

In many cases, sensor data are available in a spatially discrete form. Here generalized multichannel filtering may be applied, i.e., each channel may be filtered separately.

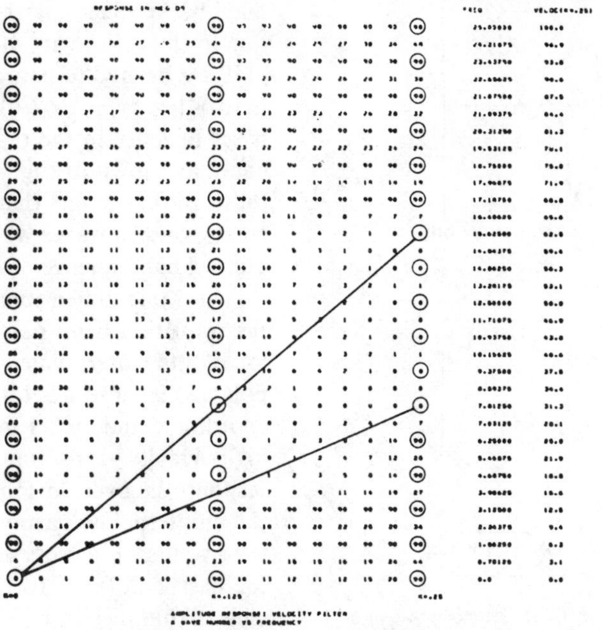

(a)

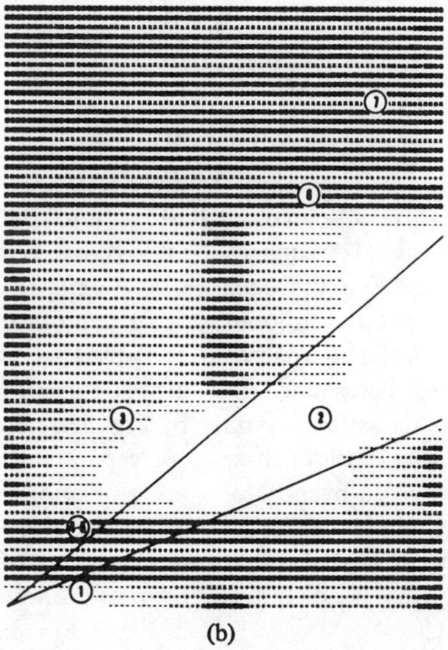

(b)

Fig. 5. Velocity filter. High velocity cutoff, 61.5 cm/s; low velocity cutoff, 31.25 cm/s. (a) Numerical array output. (b) Gray tone contour output.

Delay and sum, weighted delay and sum, and even maximum-likelihood filtering can be put into this form.

Alternatively, the data may be considered as spatiotemporal and a multidimensional filter used. Such a technique is commonly employed in seismology to effect velocity filtering of planewaves [27]–[29].

The filter is usually developed in the multidimensional domain, transformed to one-dimensional filters, and implemented using discrete convolution. In this section, an alternative solution is presented which consists of digital filtering directly in the multidimensional domain. This method offers substantial savings in computational effort if the multidimensional FFT is employed.

As stated earlier, the purpose of the filter, which precedes the single wave maximum-likelihood estimator, is to pass a single planewave and stop all others. Thus, for example, if the signal is a single wideband planewave with specified vector velocity and the noise is the sum of all other planewaves, then the filter is [29], [30]

$$H(f,k) = \begin{cases} 1, & k = \boldsymbol{\nu} f/|\boldsymbol{\nu}|^2 \\ 0, & \text{elsewhere} \end{cases} \tag{7}$$

for all f.

The response pattern is a ray emanating from the origin in the (f,k_x,k_y) space with a locus of points $H(f,k) = 1$ and corresponds to points for which $\boldsymbol{\nu} = f\boldsymbol{k}/|\boldsymbol{k}|^2$. For all other

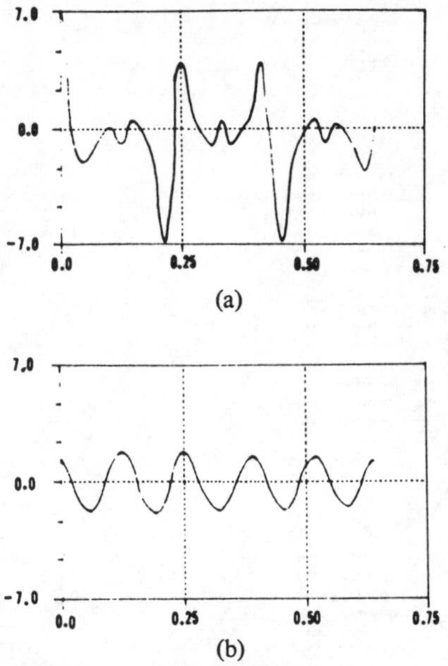

Fig. 6. Velocity filtering for one channel. (a) Composite waveform of waves in Table I. (b) Post-velocity filter estimate of wave 2.

points in the frequency-wavenumber space, $H(f,k) = 0$. The resultant filter is a vector velocity bandpass filter.

A. Implementation

The digital implementation chosen for this study involves approximating (7) by using multidimensional frequency domain specifications by the window method. Our approach is to approximate the idealized response based on continuous data of unlimited extent in all dimensions. As an alternative, the problem could be reformulated to determine the optimum linear processor which operates on sampled data of fixed extent in all dimensions. The relative merit of these two approaches is an issue aside from the main purpose of maximum-likelihood estimation and is not investigated here. The frequency domain method was chosen because it is consistent with techniques previously discussed and is efficient from a computational viewpoint.

B. Velocity Filtering

This aspect of multidimensional filtering has not received much attention, since in many fields such as radar, radio astronomy, and some sonar situations the velocity of propagation is essentially constant and simple filtering operations such as delay and sum or weighted delay and sum are adequate [2]. One area where the velocity of propagation is important and not constant for all waves is seismology [27]–[30]. The velocity filter introduced here differs from others in that the filtering is done directly in the frequency-wavenumber domain.

The purpose of a velocity filter is to pass or stop all planewaves which have a magnitude velocity lying within a specified band. The ideal velocity discrimination is made independent of the bearing and frequency composition of planewaves. The ease with which the filtering operation may be specified in the frequency-wavenumber space is due to the colinearity of $\boldsymbol{k}$ and $\boldsymbol{\nu}$.

If it is desired to pass all planewaves with velocity $\boldsymbol{\nu}$, such that $|\nu_1| \leq |\nu| \leq |\nu_2|$, the complex frequency-wavenumber is set to unity in the region between the two right circular cones as shown in Fig. 4. All other points in the space are set to zero. The fact that all points on the surface of a cone correspond to waves whose velocity is given by the slope is a direct consequence of the definition of wavenumber.

As for one-dimensional filtering, it is advisable to smooth the transition from passband to stopband so that the sidelobes of the interpolated response are reduced. For the interpolated response, a 2:1 interpolation is done in the frequency coordinate and an 8:1 interpolation is given for k. With the high-velocity transition set to 61.5 cm/s, the interpolated response is given in Fig. 5. The circled numbers represent the uninterpolated array. The triangular region enclosed by the two lines is a cross section of the two cones which define the passband (refer to Fig. 4). If perfect filtering were possible all numbers inside this region would be zero and those outside would be -90 dB. Fig. 5 also provides the gray tone contour output for the same filter specification. The complete performance of the digital filter can be obtained by generating this output for filtering specifications of interest.

For the particular 4×4 sensor array being considered, the velocity resolution is about 15 cm/s from approximately 10 cm/s to 200 cm/s. Beyond 200 cm/s discrimination is impossible. With an 8×8 sensor array the resolution is near 5 cm/s from 5 cm/s to 400 cm/s.

To demonstrate the velocity filter, the seven planewaves of Table I are summed and then velocity filtered to eliminate all but wave number 2. The filtering specifications depicted by Fig. 5 are used. The numbers superimposed on Fig. 5(b) correspond to the numbered waves. Waves 4 and 5 are in the same position because they have the same velocity. Waves 2 and 4 obviously have the same center frequency.

The waves of Fig. 6(a) and (b) are the composite waveform of the waves in Table I and postfilter estimate of wave 2. Fig. 6(b) is nearly identical to the actual waveform, thus the effectiveness of velocity filter is demonstrated.

The prefilter and postfilter estimates of the other waves are not displayed. But, as expected, waves 6 and 7 are heavily attenuated. Wave 3 is only attenuated by approximately 3 dB.

C. Angle Filtering

Angle filtering, illustrated in Fig. 4, is to be accomplished independent of the velocity magnitude and bandwidth of the planewaves. The relative ease with which this may be carried out stems from the fact that the vector wavenumber $\boldsymbol{k}$ and the vector velocity have the same direction.

As for one-dimensional filtering, it is usually advisable to smooth the transition from stopband to passband. Two-dimensional smoothing need only be applied to each $k_x - k_y$ plane since the angle filter response is independent of frequency.

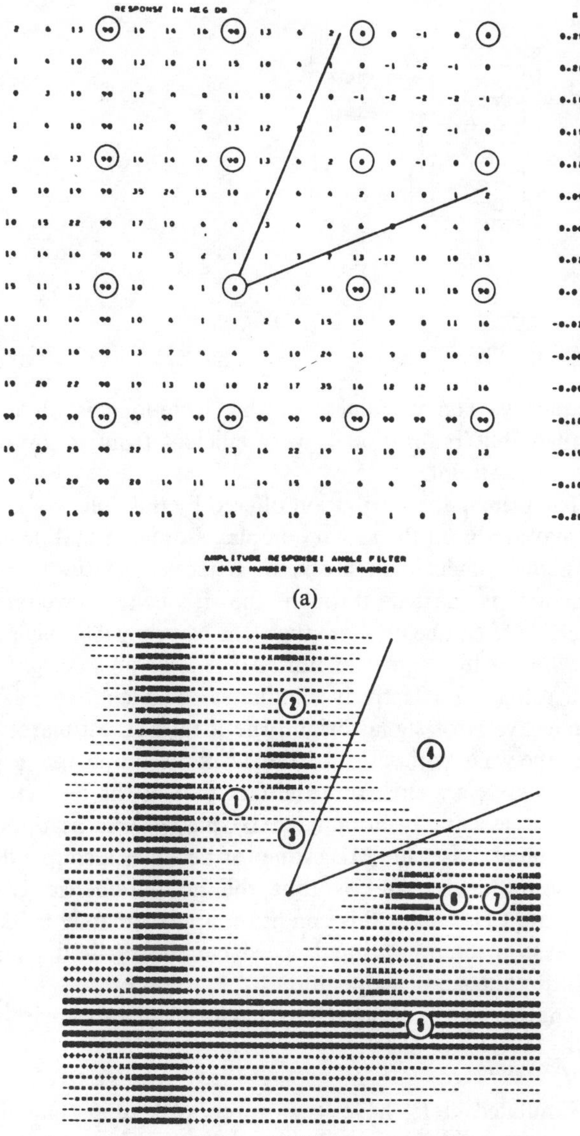

(a) (b)

Fig. 7. Angle filter. Bearing, 45°; beamwidth, 45°. (a) Numerical array output. (b) Gray tone contour output.

An angle filter with the bearing and beamwidth set at 45° is shown in Fig. 7. The expanded (interpolated) response array in negative decibels with respect to a specified point in the passband is listed in Fig. 7(a). To avoid printing excess minus signs, the sign of each number is complemented before printing. Thus the -2's found on the column corresponding to $k_x = 0.21875$ cm^{-1} and those found on the row corresponding to $k_y = 0.18750$ cm^{-1} indicate a 2-dB rise above the specified response (0 dB). The region in the first quadrant, enclosed by the two lines superimposed on the printer output, is the beam pattern. If perfect response were possible, all numbers in the region would be zero and those outside would be 90. The lowest level printed is -90 dB, as before. The sixteen circled numbers of Fig. 7(a) are the numbers generated initially by the angle filter program and returned for actual filtering operation. The remaining numbers are all part of the interpolated response. Fig. 7(b) presents the same data in gray tone format.

The filter response and resolution in bearing are improved by increasing the spatial array size. For rectangular arrays the resolution is a function of bearing since circular symmetry is not present.

In the interests of brevity, the effects of the angle filter are not demonstrated but they can be as good as those shown for velocity filtering.

The standard frequency filter has not been discussed because of its familiarity and vast literature.

If an angle filter is cascaded with a velocity filter, the result is obviously a vector velocity filter. These two filtering specifications are kept distinct in order to maintain a greater degree of flexibility.

The study of design methods for two-dimensional filters has received considerable attention [31]–[34]. A comparison of computational aspects of spatial filtering appears in [35].

IV. Multiwave Maximum-Likelihood Estimation

This study is motivated by a need to obtain estimates of spatio-temporal signals which are completely unknown *a priori.* The investigation is limited to planewaves because any stationary spatio-temporal function can be decomposed into a sum (possibly infinite) of Fourier components which are complex planewaves. The maximum-likelihood (ML) estimation is chosen because it is the optimum solution in the decision theoretic sense [36], [37].

A. Single-Wave ML Estimation

The ML estimate of single planewaves with *known* vector velocity can be approached by either the time domain method [4] or the frequency domain method [3]. The former assumes the data are limited and sampled both spatially and temporally. The latter assumes the sensor time functions are continuous. Both methods assume the noise is additive and multidimensional Gaussian and is temporally stationary but not necessarily spatially stationary. The samples may be correlated in time and space in accordance with a known covariance matrix. The computer implementation of the solution for the time domain method is quite sensitive to the assumption of noise stationarity [4], and it is computationally inefficient with respect to the frequency domain method [3].

The general ML estimator for a single planewave by the frequency domain method is an array prewhitener, a beam former, and an inverse filter in cascade [3], [38], [39]. The beam former is a delay and sum processor which is a least mean-square error (LMSE) estimator. For the special case when the off-diagonal terms of the crosspower spectral density matrix are zero and the noise is common to all sensors, then the LMSE and ML estimates are the same.

The general ML estimator is reduced to a form suitable for numerical calculations in [4]. We use this form here with α negated to conform to the definition in [3].

If the noise is spatially stationary and the wave being estimated is monochromatic the ML estimate is the frequency-wavenumber high resolution estimate [10].

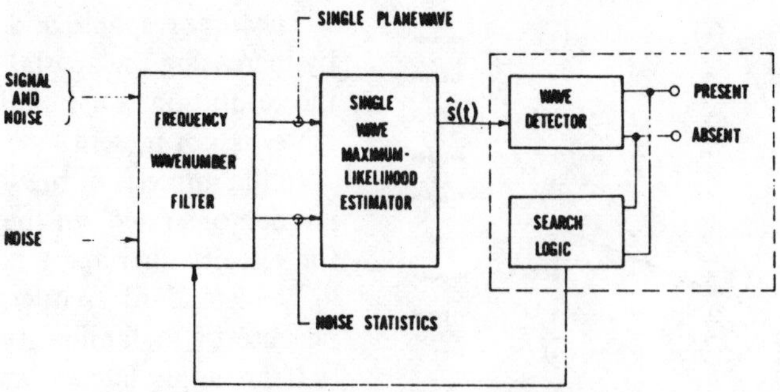

Fig. 8. General multiwave ML estimator.

If the noise is entirely interfering planewaves all with different vector velocities, then the ideal ML estimator output noise is zero. This property is variously referred to as superdirectivity [38]–[40], infinite sidelobe suppression, singular detection, etc. [3]–[5], [8], [27]. This is the most significant property of the ML estimator.

It can be shown that, if the estimator is designed under the multidimensional Gaussian assumption but, in fact, the noise is not Gaussian, the resultant estimator still provides an unbiased and minimum variance estimate [4], [41].

The above discussion applies to the problem of estimating the waveshape $s(t)$. In general, the wave must be detected and its velocity estimated. ML detection [42] is possible. Instead the frequency-wavenumber power spectrum is used here to detect planewaves and to estimate their vector velocity.

B. *Multiwave ML Estimation*

The multiwave ML estimator is given in Fig. 8. It detects the number of planewaves J, their vector velocities $\boldsymbol{v}$, and their functional form $s(t)$. In simplest terms, it is a cascade of a frequency-wavenumber filter which reduces the multiwave problem to a single wave problem followed by a single-wave ML estimator. The frequency-wavenumber filters are those developed previously, and the ML estimator for a given vector velocity is that described above [4]. To determine the number of waves and their vector velocities, one must scan the frequency-wavenumber space and repetitively determine if a wave is present or not. Since the waveshape is unknown *a priori*, an ML detector is used. The search logic could be a computer algorithm which is designed to find the extrema of a function of several variables. Director [43] has given a survey of the techniques which have proven to be of practical value. If the noise level is high, one may alternatively utilize stochastic approximation methods [44]–[46].

The elements enclosed by the broken lines in Fig. 8, i.e., those elements that are, in effect, responsible for the detection and determination of the vector velocities, will require very sophisticated and elaborate designs which are not attempted in this study. Instead, the use of the frequency-wavenumber PSD function is advanced. If the SNR is sufficiently high, the local maxima of the PSD can be attributed to the presence of signal planewaves. The vector velocities are readily determined from the wavenumber and frequency. The usefulness of this technique is obviously limited, but realization is very efficient from a computational viewpoint.

The principal contribution offered by this solution is that we have reduced the rather complex J-dimensional problem to a one-dimensional problem. After the spatio-temporal function is passed through the frequency-wavenumber filter, there can be no more than one wave present regardless of whether it is signal or noise. If it is of noise origin, the ML estimator effectively eliminates the planewave. If the planewave is of signal origin, the ideal ML estimator will pass the wave without distortion. Another important aspect of this system stems from the fact that, if the background noise is essentially dominated by planewaves as opposed to channel noise, the ML estimator becomes a simple delay and sum device and is thus the nonparametric LMSE estimate [3]. Under this condition approximately a 1000:1 increase in computational speed can be realized by substituting the simpler operation.

Another recent approach to this problem appears in [47].

C. *Examples*

Simulated data have been processed quite effectively using the single-wave ML estimator for cases where the interference is another planewave and channel noise is present. The standard of comparison is the LMSE estimate.

For the case when the amplitude of the interfering planewave is varied and the sum of desired signal plus channel noise is fixed, many examples have been run for signal-to-interference ratios (SIR) from 2 to 0.1, with fewer examples outside this range. When SIR = 2 the LMSE and ML estimates are the same, as the SIR is reduced the ML estimate becomes superior to the LMSE estimate and is essentially perfect until SIR = 0.1. For SIR's below this figure both estimates deteriorate rapidly.

Another case is to have the interfering wave present and of the same amplitude as the desired wave but to vary the channel noise. For SNR's which exceed the range 0.5–5.0, the LMSE and ML estimates are nearly identical, but improve with increasing SNR.

As an example, five of the seven planewaves of Table I (waves 1, 3, 5, 6, and 7) are summed to form the noise. Channel noise is not added. The remaining two planewaves are taken to be the signals to be estimated. The frequency-wavenumber filters presented in Section III are used to limit the number of planewaves. A typical result is summarized

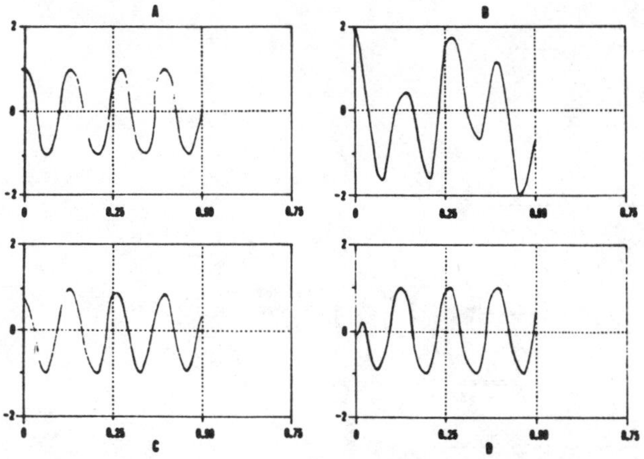

Fig. 9. Multiwave estimation of wave 2. (a) Wave 2. (b) Prefilter LMSE estimate. (c) Postfilter LMSE estimate. (d) ML estimate.

in Fig. 9. From this example it appears that the ML estimate is better than the prefilter LMSE estimate. The postfilter LMSE estimate appears superior, however, to the ML estimate. Since the background noise consists entirely of planewaves, the postfilter estimate and the ML estimate are identical in the ideal case. In the actual realization, however, the added complexity of the general ML estimator introduces errors which can be avoided by using the simpler delay and sum operation.

V. Computer Processing of Spatio-Temporal EEG Data

The techniques presented in the preceding sections are well suited for the analysis of data in fields such as sonar and seismology. In this section we apply the techniques to the analysis of EEG's, or recorded brain waves.

A particular EEG, the visual evoked response (VER), is by definition the bioelectrical brain activity resulting from the presentation of a visual stimulus. Recordings can be made from the surface of the scalp. Various experiments vary the stimulus as well as the recording location using subjects with normal vision and visual dysfunctions in an effort to form correlations between the recorded response and visual and/or cerebral defects as well as to study the basic visual processing system [48]–[50].

The data are collected using standard techniques [50], [51]. An electrode array is shown in Fig. 10 along with the basic elements of the data collection system. The stimulus is presented to the subject repetitively. The signals are amplified (approximately 50 dB), digitized, averaged, and recorded. Our investigations are the first to apply true array processing to the EEG. Many possible areas of interest are being investigated. For illustration, the effect of low-pass velocity filtering will be demonstrated.

The first step is to determine where in the frequency-wavenumber space the energy is concentrated. A typical data set for binocular white light stimulation is shown in Fig. 11 for subject JGM. The frequency-wavenumber spectrum for this data is shown in Fig. 12(a), and the high-energy region is concentrated in the center of each frame. This indicates a very high-velocity wave or, in the limiting case, the major components of the time function are identical at each sensor. The effect can be observed in either the spatio-temporal domain or the frequency-wavenumber domain.

To determine if there are slow planewaves present, the high-velocity component must be removed. The data are low-pass velocity filtered at 200 cm/s. The results are displayed in Fig. 12(b). This process has obviously revealed components which were completely masked before filtering. This has in effect removed a high-velocity data trend. To obtain an estimate of a wave of particular interest, a delay and sum or ML operation is performed.

One phenomenon observed in the data in Fig. 12 is that, after velocity filtering, two waves propagating in different directions ($\pm 90°$) can be observed. A similar result has been found in a study of penicillin induced focal epileptic discharge data recorded from rat neocortex [51].

The two waves propagating in different directions might be the result of sequential excitation of neuronal populations, one population successively triggering another across the array at 90° and at $-90°$. Thus the epileptic focus would consist of activity arising simultaneously from all neuronal populations within the vicinity of the electrode array as well as activity arising from sequential triggering of neuronal populations at two directions across the array.

Since similar results have been observed for the human VER data, but with different directions of propagation for different frequency bands a similar explanation may be advanced. However, here one may even further conjecture that the successive triggering of neuronal populations across the array is a result of the communication of information to and from associative and primary visual areas, respectively, since this is the region monitored by the scalp array.

An alternative explanation for the observed results is that one or more rotating dipoles or alternating dipole-like layers of neuronal populations exist within the vicinity of the monitoring array. The dipoles then change their orientation during the period of observation giving rise to electrical activity which has the appearance of propagating wavefronts when monitored by the array. Numerous such models have been advanced in the literature [52].

For many years researchers have been conjecturing about the origin of the EEG and about certain characteristic phenomena, e.g., apparent traveling waves of various frequencies propagating across the scalp. Are these waves electrical signatures of neurochemical reactions? Are they indicators of information transfer between neuronal populations? Can they be used as a reliable diagnostic aid? The answers to these and related questions are not available. But we believe new analysis procedures such as the one described here will help provide the answers; e.g., we are now in the process of tracking selected waves propagating across the array of scalp electrodes to determine their trajectories and hopefully gain further insight into their origin and significance.

Future experiments specifically designed to test these conjectures and others should prove interesting and assist us in interpreting brain functioning via the remotely sensed EEG.

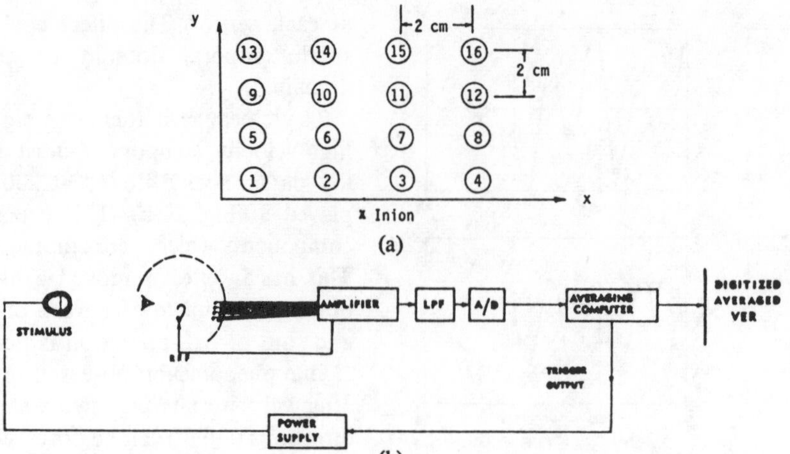

Fig. 10. Electrode array and basic elements of averaging system.

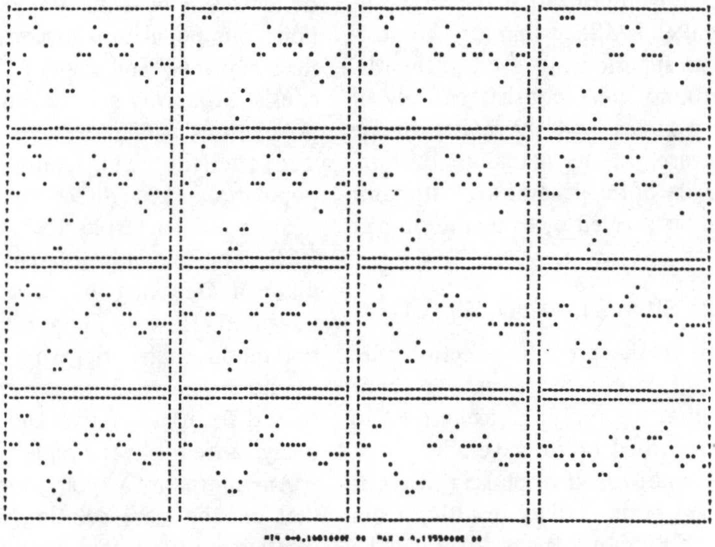

Fig. 11. 16-channel VER data for binocular white light stimulation (JGM). The data arrangement corresponds to that shown for the electrode array in Fig. 10.

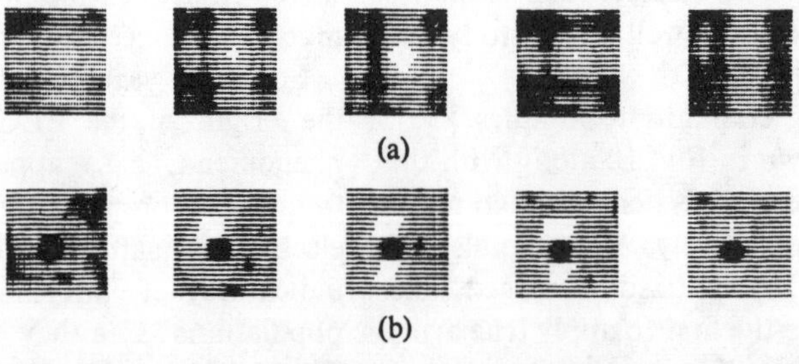

Fig. 12. Frequency-wavenumber spectrum of VER data from 1.6 Hz to 8.0 Hz in steps of 1.6 Hz from left to right. (a) Before low-pass velocity filtering. (b) After low-pass velocity filtering at 200 cm/s.

References

[1] J. L. Allen, "Array antennas: new applications for an old technique," *IEEE Spectrum*, vol. 1, pp. 115–130, Nov. 1964.

[2] O. L. Frost, "An algorithm for linearly constrained adaptive array processing," *Proc. IEEE*, vol. 60, pp. 926–935, Aug. 1972.

[3] E. J. Kelly and M. J. Levin, "Signal parameter estimation for seismic arrays," M.I.T. Lincoln Lab., Lexington, Mass., Tech. Rep. 339, Jan. 1964.

[4] J. Capon, R. J. Greenfield, and R. J. Kolker, "Multidimensional maximum-likelihood processing of a large seismic array," *Proc. IEEE*, vol. 55, pp. 192–211, Feb. 1967.

[5] J. Capon, "Applications of detection and estimation theory to large array seismology," *Proc. IEEE*, vol. 58, pp. 760–770, May 1970.

[6] D. Middleton and H. L. Groginsky, "Detection of random acoustic signals," *J. Acoust. Soc. Amer.*, vol. 38, pp. 737–739, Nov. 1965.

[7] D. Middleton, "Multidimensional detection and estimation of signals in random media," *Proc. IEEE*, vol. 58, pp. 696–706, May 1970.
[8] F. C. Schweppe, "Sensor-array data processing for multiple-signal sources," *IEEE Trans. Inform. Theory*, vol. IT-14, pp. 294–305, Mar. 1968.
[9] H. Kobayashi and P. D. Welch, "The detection and estimation of two simultaneous seismic events," in *Symp. Computer Processing in Communications*, J. Fox, Ed., Polytechnic Institute of Brooklyn, Brooklyn, N.Y., Apr. 1969, pp. 757–777.
[10] J. Capon, "High-resolution frequency-wavenumber spectrum analysis," *Proc. IEEE*, vol. 57, pp. 1408–1418, Aug. 1969.
[11] J. B. Thomas, *Statistical Communication Theory*. New York: Wiley, 1969.
[12] J. S. Bendat and A. G. Piersol, *Measurement and Analysis of Random Data*. New York: Wiley, 1966.
[13] R. B. Blackman and J. W. Tukey, *The Measurement of Power Spectra*. New York: Dover, 1958.
[14] P. I. Richards, "Computing reliable power spectra," *IEEE Spectrum*, vol. 4, pp. 83–90, Jan. 1967.
[15] P. Welch, "The use of the fast Fourier transform for the estimation of power spectra: a method based on time averaging over short, modified periodograms," *IEEE Trans. Audio Electroacoust.*, vol. AU-15, pp. 70–73, June 1967.
[16] C. Bingham, M. D. Godfrey, and J. W. Tukey, "Modern techniques of power spectrum estimation," *IEEE Trans. Audio Electroacoust.*, vol. AU-15, pp. 56–66, June 1967.
[17] B. Gold and C. M. Rader, *Digital Processing of Signals*. New York: McGraw-Hill, 1969.
[18] J. W. Cooley, P. A. W. Lewis, and P. D. Welch, "The fast Fourier transform algorithm and its applications," IBM Watson Res. Ctr., Yorktown, N.Y., Res. Paper RC-1743, Feb. 1967.
[19] G. D. Bergland, "A guided tour of the fast Fourier transform," *IEEE Spectrum*, vol. 6, pp. 41–52, July 1969.
[20] R. T. Lacoss, E. J. Kelly, and M. N. Toksoz, "Estimation of seismic noise structure using arrays," *Geophysics*, vol. 34, pp. 21–38, Feb. 1969.
[21] E. Smart and E. A. Flinn, "Fast frequency-wavenumber analysis and Fisher signal detection in real-time infrasonic array data processing," *Geophys. J. Roy. Astron. Soc.*, vol. 26, pp. 279–284, 1971.
[22] J. W. Cooley and J. W. Tukey, "An algorithm for the machine calculation of complex Fourier series," *Math. Comput.*, vol. 19, pp. 297–301, Apr. 1965.
[23] A. Papoulis, *Systems and Transforms With Applications in Optics*. New York: McGraw-Hill, 1968.
[24] J. W. Goodman, *Introduction to Fourier Optics*. New York: McGraw-Hill, 1968.
[25] H. D. Helms, "Nonrecursive digital filters: design methods for achieving specifications on frequency response," *IEEE Trans. Audio Electroacoust.*, vol. AU-16, pp. 336–342, Sept. 1968.
[26] L. R. Rabiner, B. Gold, and C. A. McGonegal, "An approach to the approximation problem for nonrecursive digital filters," *IEEE Trans. Audio Electroacoust.*, vol. AU-18, pp. 83–106, June 1970.
[27] E. A. Robinson, *Statistical Communication and Detection with Special Reference to Digital Data Processing of Radar and Seismic Signals*. New York: Hafner, 1967.
[28] M. Backus, J. Burg, D. Baldwin, and E. Bryan, "Wide-band extraction of mantle *P* waves from ambient noise," *Geophysics*, vol. 29, pp. 672–692, Oct. 1964.
[29] R. L. Sengbush and M. R. Foster, "Design and application of optimal velocity filters in seismic exploration," *IEEE Trans. Comput.*, vol. C-21, pp. 648–654, July 1972.
[30] J. P. Burg, "Three-dimensional filtering with an array of seismometers," *Geophysics*, vol. 29, pp. 693–713, Oct. 1964.
[31] J. L. Shanks, S. Treitel, and J. H. Justice, "Stability and synthesis of two-dimensional recursive filters," *IEEE Trans. Audio Electroacoust.*, vol. AU-20, pp. 115–128, June 1972.
[32] T. S. Huang, "Stability of two-dimensional recursive filters," *IEEE Trans. Audio Electroacoust.*, vol. AU-20, pp. 158–163, June 1972.
[33] J. V. Hu and L. R. Rabiner, "Design techniques for two-dimensional digital filters," *IEEE Trans. Audio Electroacoust.*, vol. AU-20, pp. 249–257, Oct. 1972.
[34] T. S. Huang, "Two-dimensional windows," *IEEE Trans. Audio Electroacoust.*, vol. AU-20, pp. 88–89, Mar. 1972.
[35] E. L. Hall, "A comparison of computations for spatial frequency filtering," *Proc. IEEE*, vol. 60, pp. 887–891, July 1972.
[36] C. W. Helstrom, *Statistical Theory of Signal Detection*. New York: Pergamon, 1968.
[37] H. L. Van Trees, *Detection, Estimation, and Modulation Theory*, pt. 1. New York: Wiley, 1968.
[38] W. Vanderkulk, "Optimum processing for acoustic arrays," *J. Brit. Inst. Radio Eng.*, pp. 285–292, Oct. 1963.
[39] D. G. Childers and I. S. Reed, "On the theory of continuous array processing," *IEEE Trans. Aerosp. Navig. Electron.*, vol. ANE-12, pp. 103–109, June 1965.
[40] F. Bryn, "Optimum signal processing of three-dimensional arrays operating on gaussian signals and noise," *J. Acoust. Soc. Amer.*, vol. 34, pp. 289–297, Mar. 1962.
[41] E. J. Kelly, "A comparison of seismic array processing schemes," M.I.T. Lincoln Lab., Lexington, Mass., Tech. Note 1965-21, June 1965.
[42] E. J. Kelly, I. S. Reed, and W. L. Root, "The detection of radar echoes in noise," *J. Soc. Ind. Appl. Math.*, vol. 8, pp. 309–341, June 1965.
[43] S. W. Director, "Survey of circuit-oriented optimization techniques," *IEEE Trans. Circuit Theory*, vol. CT-18, pp. 3–10, Jan. 1971.
[44] R. T. Lacoss, "Adaptive combining of wideband array data for optimal reception," *IEEE Trans. Geosci. Electron.*, vol. GE-6, pp. 78–86, May 1968.
[45] L. J. Griffiths, "A simple adaptive algorithm for real-time processing in antenna arrays," *Proc. IEEE*, vol. 57, pp. 1696–1704, Oct. 1969.
[46] H. Kobayashi, "Iterative synthesis methods for a seismic array processor," *IEEE Trans. Geosci. Electron.*, vol. GE-8, pp. 169–178, July 1970.
[47] E. J. Mercado, "Maximum-likelihood filtering of reflection seismograph data," *Proc. Comput. Image Processing and Recognition*, vol. 2, pp. 8-2-1–8-2-12, Aug. 1972.
[48] J. R. Bourne, D. G. Childers, and N. W. Perry, Jr., "Topological characteristics of the visual evoked response in man," *Electroencephalogr. Clin. Neurophysiol.*, vol. 30, pp. 423–436, 1971.
[49] D. G. Childers, N. W. Perry, Jr., O. S. Halpeny, and J. R. Bourne, "Spatio-temporal measures of cortical functioning in normal and abnormal vision," *Comput. Biomed. Res.*, vol. 5, pp. 114–130, 1972.
[50] N. W. Perry, Jr., and D. G. Childers, *The Human Visual Evoked Response, Method and Theory*. Springfield, Ill.: Charles C. Thomas, 1969.
[51] L. J. Pinson and D. G. Childers, "Frequency-wavenumber spectrum analysis of EEG multielectrode array data," *IEEE Trans. Biomed. Eng.*, vol. BME-21, pp. 192–206, May 1974.
[52] D. G. Childers, W. Mesa, and O. S. Halpeny, "A neuronal model for and simulation of spatio-temporal evoked EEG," *IEEE Trans. Syst., Man, Cybern.*, vol. SMC-3, pp. 336–348, July 1973.

Fundamentals of Digital Array Processing

DAN E. DUDGEON, MEMBER, IEEE

***Abstract*—With the advent of high-speed digital electronics, it has become feasible to use digital computers and special purpose digital processors to perform the computational tasks associated with signal reception using an antenna or directional array. The purpose of this paper is mainly tutorial, to describe mathematically and intuitively the fundamental relationships necessary to understand digital array processing. It is hoped that those readers with a background in antenna theory or array processing will see some of the advantages digital processing can offer, while those with a background in digital signal processing will recognize the array processing area as a potential application for multidimensional signal processing theory.**

I. INTRODUCTION

MUCH of the theoretical work being done today in the area of multidimensional signal processing is motivated by the need to process signals carried by propagating wave phenomena. For radar to be successful, it was necessary to develop directional transmitting and receiving antennas so that azimuth as well as range and range rate information could be ascertained from the radar return. Similarly, this problem is also encountered in active sonar and ultrasonics applications. In applications where the source signal is not precisely controlled (such as exploratory seismology) or where the received signal is externally generated (such as passive sonar, bioelectrical measurements, or earthquake seismology), it is desired to elicit characteristics of the received signal (its signature) as well as its direction and speed of propagation.

In recent times, it has become more and more feasible to perform the signal processing operations associated with array processing using digital computers or special purpose digital processing hardware. Correspondingly, digital signal processing theory has grown to encompass these various applications. The following references are representative of recent articles of digital processing in the fields of radar [1], seismology [2], sonar [3], ultrasound [4], and bioelectrical measurement [5].

The point of this paper is to examine the fundamental array processing techniques, in particular the concept of beamforming to determine the speed and direction of propagation of an incoming wave, from the point of view of a multidimensional signal processing problem. We shall see the close relationship between conventional sampled-data systems and the sensor array as a receiver sampling the waveform in space. Accordingly, Section II reviews some essential points about sampled-data systems and digital signal processing techniques. In Section III, a linear array of sensors is used as a basis for discussing the weighted delay-and-sum beamformer with attention given to how to choose the appropriate weights. In Section IV, the relationship between the computation of beam spectra and the computation of a two-dimensional (2-D) discrete Fourier transform is examined. Section V looks at extending the results of Section III to higher dimensions. As an example of results from digital signal processing which can be applied to digital beamforming, the problem of designing the sensor weights for a multidimensional beamformer is discussed in Section VI. In the case of a Cartesian array of sensors, an ingenious mapping due to McClellan [6] can be used to design and implement beams with nearly spherically symmetric main lobes in a computationally efficient manner.

Manuscript received July 26, 1976; revised November 12, 1976.
The author is with the Computer Systems Division, Bolt Beranek and Newman, Inc., 50 Moulton Street, Cambridge, MA 02138.

II. IMPORTANT CONCEPTS IN DIGITAL PROCESSING

In this section, several important concepts from digital signal processing theory will be reviewed. These concepts will be presented in terms of a one-dimensional (1-D) signal for ease of understanding, but they are easily generalized to multidimensional signals. The reader is directed to [7] as a text on digital signal processing and to [8] as a review of 2-D filtering concepts.

The fundamental assumption of digital processing is that input signals are bandlimited to frequencies below one-half the sampling rate. If a continuous-time signal is sampled at a rate too slow (undersampling) for the frequency content of the signal, the Nyquist sampling theorem tells us that frequencies above one-half the sampling rate in the continuous-time signal will act like frequencies below one-half the sampling rate. This phenomenon is known as aliasing, and it is explained in detail in [7] as well as in a variety of texts and papers on sampled-data systems. Although we have been speaking of a 1-D time signal, the same statements apply to signals which are a function of distance or other continuous independent variables.

Reprinted from *Proc. IEEE,* vol. 65, pp. 898–904, June 1977.

We can represent a 1-D digital signal by $s(n)$ where n is an integer. By doing so we are effectively normalizing the sampling rate to be unity. The Fourier transform of such a digital signal is defined by

$$S(\omega) = \sum_{n} s(n)e^{-j\omega n}. \tag{1}$$

The reader will quickly recognize that the Fourier transform is continuous and periodic in the radian frequency variable ω with period 2π.

If only samples of $S(\omega)$ are desired, (1) can be evaluated at the points $\omega = (2\pi k)/N$, where k is an integer variable taking on values from 0 to $N - 1$ and N is an integer constant

$$S\left(\frac{2\pi k}{N}\right) = \sum_{n} s(n) \exp\left(-j\frac{2\pi nk}{N}\right). \tag{2}$$

If the signal $s(n)$ is zero for n outside the range from 0 to $N - 1$, then the sum over n in (2) extends only from 0 to $N - 1$. In this case (2) defines a discrete Fourier transform (DFT) which is invertible; that is, the samples $s(n)$ may be recovered from the values $S[(2\pi k)/N]$ by

$$s(n) = \frac{1}{N}\sum_{k=0}^{N-1} S\left(\frac{2\pi k}{N}\right) \exp\left(j\frac{2\pi nk}{N}\right) \quad \text{for } n = 0, N - 1.$$

The DFT of a signal may be computed by an efficient algorithm (an FFT), the details of which are contained in texts [7], [9], [10], and papers [11], [12]. The advantage of the FFT algorithm is that the computation of $S[(2\pi k)/N]$ is proportional to $N \log_2 N$ rather than N^2 as in a direct evaluation of the DFT [see (2)].

One type of digital filter which is important to the understanding of digital array processing is the finite impulse-response (FIR) filter. Again we shall briefly review the 1-D case which is covered in detail in [7], [9], [13], and [14]. The name FIR refers to the fact that the impulse response of the filter is nonzero only over a finite domain of the independent variable. For example, if a filter has an impulse response $h(n)$ such that

$$h(n) = 0 \text{ for } n < 0 \quad \text{and} \quad n \geqslant N$$

where n and N are integer-valued, then the filter is said to be an FIR filter.

A special class of FIR filters are those which are said to be "linear phase." The impulse response of such a filter (assumed to be purely real) possesses even (or odd) symmetry about the midpoint of its nonzero region. An example of such an impulse response is

$$h(n) = 0 \quad \text{for } n < 0 \text{ and } n \geqslant N$$
$$h(n) = h(N - n - 1). \tag{3}$$

Because of this symmetry, the phase response of such a filter is exactly linear and produces no phase distortion. A further specialization can be made by requiring $h(n)$ to be even about $h(0)$. In that case, the frequency response of the filter is purely real.

In addition to the perfect phase characteristics of linear phase FIR filters, they have the additional advantage of being easily and quickly designed by a computer program [15]. This program approximates a given ideal frequency response optimally in a weighted mini-max sense; that is, the weighted maximum deviation from the ideal is minimized.

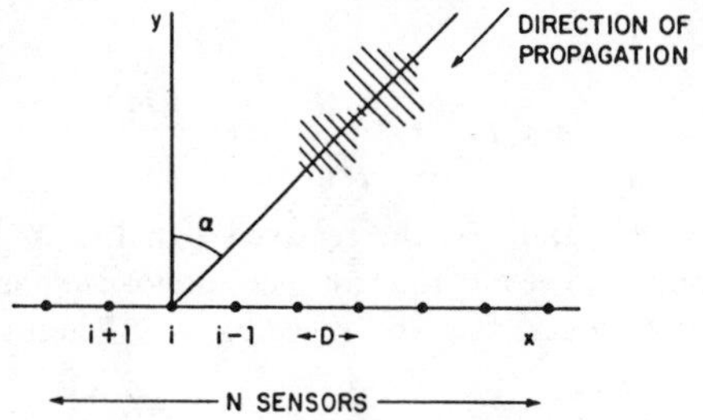

Fig. 1. A plane wave impinges upon a linear array of N sensors at an angle α.

III. Beamforming

The reasons for studying the formation of beams from an array of sensors are straightforward. In several of the applications mentioned in Section I, particularly passive sonar, the objective is to use the signals received by the sensors in a phased manner so as to preferencially detect signals coming from a particular direction (i.e., signals coming in on a particular beam). In addition, by averaging over many sensors, the signal-to-noise ratio (SNR) is increased to aid in the measurement of other signal parameters [19].

An appropriate analogy is that beamforming is related to multidimensional spectral analysis in the same way that bandpass filtering and 1-D spectral analysis are related. Both beamforming and spectral analysis can be used to segregate received energy by frequency, direction and speed of propagation. (An excellent discussion of the latter approach is contained in [30].) Our discussion will concentrate on the beamforming approach, since it is the author's opinion that that formulation more accurately reflects the type of processing done in a real-time digital array processing system.

In many physical situations, the signal one is interested in receiving and analyzing can be modeled as a propagating plane wave. In such a signal model, the value along a line (or plane) perpendicular to the direction of propagation is constant. If we assume that the plane wave is propagating with a speed c and in a direction at an angle α to the y-axis (Fig. 1), then the signal value r at a particular place (x, y) and time t may be written

$$r(x, y, t) = s\left(t - \left(\frac{x \sin\alpha + y \cos\alpha}{c}\right)\right).$$

Note that this is really a function of one independent variable. Consequently, the function $s(\cdot)$ along with the direction and speed of propagation completely specifies the model signal.

In order to focus on the structure of the beamforming computation, we shall assume that the signal $s(\cdot)$ is deterministic, not stochastic, and that the SNR is high enough that we may ignore the contribution of the noise. Later we shall indicate the way in which knowledge of the signal and noise statistics can be used in beamformer design.

A simple way to try to measure $s(\cdot)$ and the direction of propagation is by using a delay-and-sum beamformer [19]. In Fig. 1, a plane wave impinges upon an array of N sensors uniformly separated by a distance D. If we let α denote the angle of incidence of the wave, then we would expect the signal received at the $i + 1$st sensor to be a delayed version of the signal received at the ith sensor (in the absence of noise and other waves). The amount of delay is $D \sin \alpha/c$. If we want to look for a signal coming from an angle α, we can form the sum

$$g(a, t) = \frac{1}{N} \sum_{i=0}^{N-1} r_i \left(t - \frac{iDa}{c}\right)$$

where $a = \sin \alpha$ and r_i is the received signal from the ith sensor. Suppose, however, that an incoming wave has a different angle of incidence $\alpha_0 \neq \alpha$ and speed $c_0 \neq c$. Then

$$r_i(t) = s\left(t + \frac{iDa_0}{c_0}\right).$$

Substituting for $r_i(t)$

$$g(a, t) = \frac{1}{N} \sum_{i=0}^{N-1} s\left(t - iD\left(\frac{a}{c} - \frac{a_0}{c_0}\right)\right).$$

If we let $s(t)$ be represented by its continuous Fourier transform $S(\omega)$

$$s(t) = \frac{1}{2\pi} \int_{-\infty}^{\infty} S(\omega)\, e^{j\omega t}\, d\omega$$

and let $k = \omega a/c$, then

$$g(a, t) = \frac{1}{2\pi} \int_{-\infty}^{\infty} S(\omega)\, W(k - k_0)\, e^{j\omega t}\, d\omega \tag{4}$$

where

$$W(\nu) = \frac{1}{N} \sum_{i=0}^{N-1} e^{-ji\nu D}.$$

We shall call $W(\nu)$ the array pattern. Note that (4) has the form of an output signal being equal to the inverse Fourier transform of the product of the Fourier transform of the input signal and the frequency response of a filter. Thus for particular values of a, a_0, D, c, and c_0, the pattern W tells us how the frequencies in the input signal are weighted to form the output signal.

For the case at hand, it is easily shown that

$$W(\nu) = \frac{\sin(N\nu D/2)}{N \sin(\nu D/2)} \exp\left[-j\frac{(N-1)\nu D}{2}\right].$$

The magnitude of $W(\nu)$ is plotted in Fig. 2 for $N = 7$. Notice that it is periodic in ν with a period of $\nu = (2\pi)/D$. Since

$$\nu = \omega\left(\frac{a}{c} - \frac{a_0}{c_0}\right) \tag{5}$$

the width of the central lobe of the array pattern decreases with increasing temporal frequency (ω) and with increasing sensor spacing D. However, if ω or D become too large, the array pattern will exhibit other large lobes in addition to the main lobe because of the periodicity of $W(\nu)$. Historically, these extraneous lobes have been called "grating" lobes because of the analogy with optical diffraction gratings.

We shall review the phenomena of grating lobes from the point of view of spatial sampling of a propagating wave. First, we shall assume that the wave $s(t)$ is of a single frequency ω_0 (monochromatic). This is no real restriction since a more general waveform can be decomposed into an integral of weighted sinusoids. If such a wave is traveling at a speed c_0, then it will have a wavelength $\lambda_0 = 2\pi c_0/\omega_0$. If the wave were incident at an angle α (Fig. 1), and we were to measure its value along the line connecting the sensors at one instant of time, we

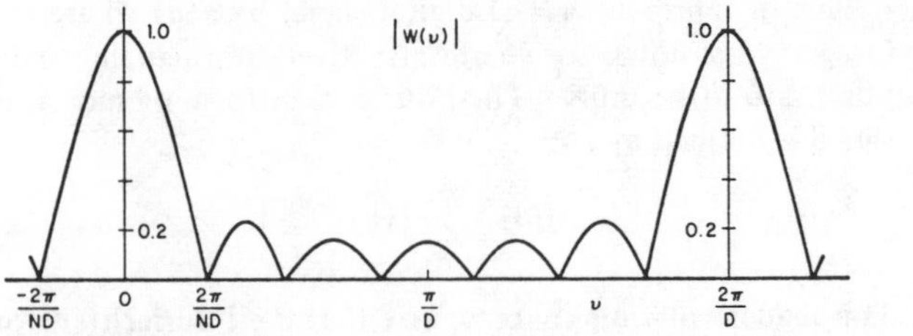

Fig. 2. The array pattern for a delay-and-sum beamformer with $N = 7$ sensors.

would observe a sinusoidal variation as a function of position along the line of sensors. The spatial period of this variation can be shown to be $\lambda_0/\sin \alpha$. If the spacing D happened to equal $\lambda_0/\sin \alpha$, the sensor measurements would all be identical and one might mistakenly conclude that the wave arrived perpendicular to the array rather than at the correct angle α. This is precisely the problem of aliasing described in the previous section except that here we are sampling a waveform that is a function of position (by choosing the sensor spacing D) rather than sampling a waveform that is a function of time (by choosing the sampling period T). Consequently, if it is expected that a signal will have a component with a wavelength as short as λ_{min}, then the sensor spacing should be at most $\lambda_{min}/2$ to avoid the effects of spatial aliasing (e.g., grating lobes).

In designing a beamformer, the objective (in the most straightforward case) is to have $W(\nu)$ be as close to an impulse as possible using only a finite number of sensors. Traditionally, measures of closeness are the width of the central lobe (the smaller the better) and the height of the side lobes (also the smaller the better). One way to decrease the central lobe width and the sidelobe height is to increase N, the number of sensors. Obviously, this expedient can be used only so much before economic constraints or a breakdown in the signal model come into play. Another way to alter the performance of the beamformer is to weight the sensor signals individually before summing them. If $w(i)$ is the sensor weighting for the ith sensor, then the beamformer output can be represented by

$$g(a, t) = \frac{1}{N} \sum_{i=0}^{N-1} w(i)\, s\left(t - iD\left(\frac{a}{c} - \frac{a_0}{c_0}\right)\right).$$

Following the previous derivation we can again write

$$g(a, t) = \frac{1}{2\pi} \int_{-\infty}^{\infty} S(\omega)\, W(k - k_0)\, e^{j\omega t}\, d\omega$$

where now

$$W(\nu) = \frac{1}{N} \sum_{i=0}^{N-1} w(i)\, e^{-ji\nu D}$$

is a generalization of the previous definition of W which includes the sensor weights $w(i)$. The problem of determining the sensor weights so that the array pattern has some desired characteristic is the same as designing a good data window for spectral estimation [16], or designing a prototype low-pass filter for use in a digital filter bank [29]. Either problem may be stated in terms of an FIR filter design problem and the FIR mini-max design technique [13], as well as others, brought to bear on it.

Array patterns can also be designed to take advantage of knowledge about the expected distribution of noise which a array is likely to see. The sensor weights $w(i)$ can be adjusted to maximize the SNR. This is analogous to designing a Wiener filter given the spectral estimates of the signal spectrum and noise spectrum. Intuitively, the filter will have a frequency response which passes parts of the spectrum where the SNR is high and rejects the parts where the SNR is poor [17]. Furthermore, it is possible to adapt the sensor weights as the received signal varies, thus attempting to maintain a high SNR under nonstationary conditions.

IV. Digital Beamforming and 2-D DFT's

Thus far we have treated time as a continuous independent variable and we have treated the spatial independent variable as being discrete, since measurements can only be obtained at the sensor positions. Now, however, we shall constrain the time variable to be discrete as well by insisting that sensor measurements be made at intervals of T seconds. In doing so, we must remain aware that there will be aliasing problems if received signals possess any frequency components above $1/2T$ Hz as seen by the sensor. We shall further assume that all of the sensors are sampled simultaneously. Thus we can denote the output of the N sensors as $r_i(nT), i = 0, \cdots, N-1$. As before we can form beams by weighting, delaying, and summing the sensor measurements.

$$g(a, nT) = \frac{1}{N} \sum_{i=0}^{N-1} w(i)\, r_i(nT - id_n T)$$

where $a = \sin\alpha$ and $d_n T = (D \sin\alpha)/c$. Because the time variable has been discretized, d_n must be an integer. This puts constraints on the values of $\sin\alpha$ which are allowed, namely

$$\sin\alpha = \frac{d_n Tc}{D} \qquad d_n \text{ an integer.}$$

Consequently, only a finite number of beams may be formed.

It should be mentioned here that it is possible to interpolate other beams between these beams. This is equivalent to interpolating the sensor outputs $r_i(nT)$ to a higher sampling rate [18] so that T is reduced and the inter-sensor delay $d_n T$ can have a higher number of possible values. Making T small, however, increases the number of samples to be processed and correspondingly, the amount of computation per unit time. For any practical beamforming processor, a lower limit for T is dictated by computation speed and the amount of available data storage.

For notational convenience, in the following derivations we shall assume $T = 1$. This may be viewed as taking our unit of time measurement to be one sample period. Consequently, frequencies will be measured in rad/sample period rather than rad/s.

Quite often, one is more interested in the time evolution of the short-period spectrum of a beam signal than in the beam signal itself. This is equivalent to passing the beam signal through a bank of bandpass filters and examining their output. Early on it was recognized that the FFT could be used to make such computations efficiently [26], [27]. Recent work in the speech area has further demonstrated an efficient way of realizing a digital filter bank using the FFT [29].

We shall proceed to derive the formula for the time evolution of a beam spectrum showing its relationship to the beam signal, the sensor spectra, and finally to the multidimensional spectrum of the sensor signals. Recalling that $T = 1$, we see that the beam signal may be written as before

$$g\left(\frac{d_n c}{D}, n\right) = \frac{1}{N} \sum_{i=0}^{N-1} w(i)\, r_i(N - id_n).$$

The short-period spectrum of such a beam signal may be written as

$$G\left(\frac{2\pi k}{M}, \frac{d_n c}{D}, n\right) = \sum_{m=n}^{M+n-1} v(m-n)\, g\left(\frac{d_n c}{D}, m\right) \cdot \exp\left[-j\left(\frac{2\pi km}{M}\right)\right]$$

where $v(\cdot)$ is a spectral window as discussed in [16], [29]. Because of the limits of the summation, the FFT algorithm cannot be applied directly, but the formula is easily rewritten as

$$G\left(\frac{2\pi k}{M}, \frac{d_n c}{D}, n\right) = \exp\left[-j\frac{2\pi kn}{M}\right] \sum_{m=0}^{M-1} v(m)\, g\left(\frac{d_n c}{D}, m+n\right) \cdot \exp\left[-j\frac{2\pi km}{M}\right].$$

As indicated in [29], the FFT may be used to calculate the above sum. The exponential term external to the summation may even be incorporated into the FFT by making use of the circular shift property [7].

Using similar reasoning we may write the short-period sensor spectrum for the ith sensor as

$$R_i\left(\frac{2\pi k}{M}, n\right) = \exp\left[-j\frac{2\pi kn}{M}\right] \sum_{m=0}^{M-1} v(m)\, r_i(m+n) \cdot \exp\left[-j\frac{2\pi km}{M}\right].$$

Consequently, we may write the beam spectrum in terms of the sensor spectra as

$$G\left(\frac{2\pi k}{M}, \frac{d_n c}{D}, n\right) = \frac{1}{N} \sum_{i=0}^{N-1} w(i)\, R_i\left(\frac{2\pi k}{M}, n - id_n\right) \cdot \exp\left[-j\frac{2\pi kid_n}{M}\right]. \tag{6}$$

The form of the above equation also suggests an FFT, but the form of the exponential term is a bit troublesome since kid_n/M is not necessarily an integer multiple of $1/N$.

To gain more insight into the structure of the computation of the beam spectrum, we shall turn to a formulation using the 2-D short-period spectrum of the sensor signals. The usefulness of thinking in terms of the multidimensional spectrum when approaching array processing problems has been previously recognized [28], [30]. We define the 2-D short-period spectrum of the sensor signals as

$$R\left(\frac{2\pi k}{M}, \frac{2\pi l}{N}, n\right) = \frac{1}{N} \sum_{i=0}^{N-1} w(i) \sum_{m=n}^{n+M-1} v(m-n)\, r_i(m) \cdot \exp\left[-j2\pi\left(\frac{li}{N} + \frac{km}{M}\right)\right].$$

As before this can be written as

$$R\left(\frac{2\pi k}{M}, \frac{2\pi l}{N}, n\right) = \exp\left[-j\frac{2\pi kn}{M}\right] \sum_{i=0}^{N-1} \sum_{m=0}^{M-1} \frac{1}{N} w(i)v(m) \cdot r_i(m+n) \exp\left[-j2\pi\left(\frac{li}{N} + \frac{km}{m}\right)\right].$$

In this form, R may be evaluated using a 2-D FFT in the same manner that G was calculated by a 1-D FFT. [As an aside, note that the separable window function $w(i)v(m)/N$ could be generalized to a 2-D window $w(i, m)$.] Using the relationship for R_i, the above equation may be written more concisely

$$R\left(\frac{2\pi k}{M}, \frac{2\pi l}{N}, n\right) = \frac{1}{N} \sum_{i=0}^{N-1} w(i)R_i\left(\frac{2\pi k}{M}, n\right) \exp\left[-j\frac{2\pi li}{N}\right]. \quad (7)$$

By comparing (6) and (7), the relationship between the short-period spectrum of the beam signal and the 2-D short-period spectrum of the sensor signals becomes evident. First, R must be evaluated for $l = kd_nN/M$. This value for l may not be an integer, so we are faced with the problem of interpolating between FFT points as we were in (6). The problem can be circumvented by adjusting the FFT lengths so that M divides N evenly.

Second, we must make the approximation

$$R_i\left(\frac{2\pi k}{M}, n\right) \simeq R_i\left(\frac{2\pi k}{M}, n - id_n\right).$$

By returning to the defining relation for R_i, the reader will readily see that this approximation requires that the short-period spectrum of $r_i(n)$ be the same over the M points from n to $n + M - 1$ as it is over the M points from $n - id_n$ to $n - id_n + M - 1$. (Note that there is no rotating phase factor in the approximation, since both short-period spectra are referenced to the same time origin at $n = 0$.) Intuitively, one would expect the approximation to be valid for well-behaved signals if $id_n \ll M$. The approximation will be exact if $r_i(n)$ is periodic with a period of M samples.

With these two points in mind, we can write

$$G\left(\frac{2\pi k}{M}, \frac{d_n c}{D}, n\right) \simeq R\left(\frac{2\pi k}{M}, \frac{2\pi k d_n}{M}, n\right)$$

relating the short-period beam spectrum and the 2-D short-period spectrum.

V. Multidimensional Beamforming

In the previous section, the fundamental techniques for processing signals received by a linear array of sensors was described. The advantages of spacing the sensor locations a uniform distance from one another were discussed, namely that a uniform spacing allows one to take full advantage of signal processing techniques (filter design algorithms, FFT's, etc.) developed for sampled-data systems where the sampling period is uniform.

In this section, we shall outline the similarities between multidimensional array processing and the techniques of multidimensional signal processing. The primary emphasis will be to point out how existing techniques could be applied to the design of beamformers and how problems in multidimensional array processing can be approached by reformulating them in ways easily understood by 2-D signal processing researchers.

Based on the discussion of the last section, it is easy to imagine a 2-D or 3-D array of sensors located in space with the intention of detecting and recording propagating wave disturbances. As before we shall assume that the signals we are interested in may be adequately modeled by a plane wave propagating with a speed c in a direction $\boldsymbol{a} = a_x i_x + a_y i_y + a_z i_z$ whose amplitude varies as a function of time $s(t)$ if we record the wave from a single stationary sensor [23]. (The vectors i_x, i_y, and i_z are of unit length in the direction of the x, y, and z axes.) Thus the received signal for the ith sensor located at position $p_i = x_i i_x + y_i i_y + z_i i_z$ will be

$$r_i(p_i, t) = s\left(t - \frac{a \cdot p_i}{c}\right).$$

(Note that a is a unit length vector so that $a_x^2 + a_y^2 + a_z^2 = 1$.)

The quantity (a/c) is often called the "slowness" vector since it points in the direction of propagation with a magnitude of one over the speed [30].

If we want to look for signals coming from a particular direction a_0 at a speed c_0, we can add up the weighted and delayed sensor signals

$$g(a, t) = \sum_i w(i)\, r_i\left(p_i, t + \frac{a_0 \cdot p_i}{c_0}\right) = \sum_i w(i)\, s\left(t - p_i \cdot \left(\frac{a}{c} - \frac{a_0}{c_0}\right)\right).$$

Making the substitution

$$s(t) = \frac{1}{2\pi}\int_{-\infty}^{\infty} S(\omega)\, e^{j\omega t}\, d\omega$$

and using the definition of the wave number vector [20], [30] $k = \omega a/c$ we see that

$$g(a, t) = \frac{1}{2\pi}\int_{-\infty}^{\infty} S(\omega)\, W(k - k_0)\, e^{j\omega t}\, d\omega \quad (8)$$

where the multidimensional array pattern is now given by

$$W(\boldsymbol{v}) = \sum_i w(i) \exp(-jp_i \cdot \boldsymbol{v}) = \sum_i w(i) \exp\left[-j(v_x x_i + v_y y_i + v_z z_i)\right].$$

As pointed out by Kelley [20], a wide-band signal processed by a beamforming operation will be altered since the argument of array pattern function depends on frequency. Equation (8) demonstrates this. It has the same interpretation as its counterpart (4), namely that the signal spectrum S is weighted by the array pattern W, which is a function of the difference in wave-number vectors $k - k_0$, and implicitly the sensor positions p_i. The multidimensional beamforming operation is therefore performing the task of a multidimensional bandpass filter in frequency-wave number space, rejecting signals whose direction and speed of propagation are sufficiently different from the bandpass center. Contrast this to the processing method of Halpeny and Childers [30], where a spectrum analysis approach is used to the same end.

Let us now assume that the sensors are located on a Cartesian grid. For simplicity, we shall further assume that the intersensor spacing is D in all dimensions. In general, the spacing can be different for each dimension. The sensors will be indexed on n_x, n_y, n_z, and their positions will be

$$p(n_x, n_y, n_z) = (n_x D i_x + n_y D i_y + n_z D i_z).$$

The array pattern becomes

$$W(\nu) = \sum_{n_x} \sum_{n_y} \sum_{n_z} w(n_x, n_y, n_z) \cdot \exp\,[-jD(n_x \nu_x + n_y \nu_y + n_z \nu_z)]. \quad (9)$$

VI. Designing Multidimensional Digital Array Patterns

In this section we shall borrow multidimensional FIR filter design techniques from the discipline of digital signal processing and apply them to the problem of designing an array pattern when the sensors are positioned at points on a Cartesian grid.

As before, the array pattern should have a narrow central lobe and small sidelobes. Ideally, it would be an impulse. A variety of design techniques for 2-D FIR filters (which can be extended to higher dimensions) are reviewed in [8], but we shall restrict ourselves to two techniques which should suffice in most cases.

The easiest way to design a multidimensional array pattern is to consider separable solutions of the form

$$W(\nu) = W_x(\nu_x)\, W_y(\nu_y)\, W_z(\nu_z)$$

then

$$w(n_x, n_y, n_z) = w_x(n_x)\, w_y(n_y)\, w_z(n_z).$$

Consequently the problem has been broken in a number of 1-D design problems which may be solved as indicated earlier.

A separable design technique is suited to the symmetry of the Cartesian grid (resulting in central lobes which are roughly rectangular) but not necessarily suited to the desired array pattern. In particular, for certain applications, it may be desirable to have an array pattern which exhibits circular (or spherical) symmetry. In the past this has been accomplished by using sensors arranged in circular (more accurately, polygonal) arrays [19]-[22]. To design the sensor weights in general, an optimization must be applied to (9) to force W to approximate some ideal array pattern. However, an ingenious technique due to McClellan [6] was developed to design 2-D linear phase FIR filters from 1-D linear phase FIR filters with nearly circular symmetry. McClellan's technique can be easily extended to higher dimensions as shown in the example below, and thus applied to the beamformer design problem when the sensors are equally spaced on a Cartesian grid.

Let us assume that we have designed a 1-D zero-phase array pattern (by zero-phase, recall that we mean that the sensor weights have symmetry)

$$w(i) = w(-i) \text{ for } i = 0, \cdots, \frac{N-1}{2}.$$

We shall further assume that N is odd so that the array pattern can be classified as a type 1 zero-phase FIR filter [14]. Then

$$W(\nu) = \sum_{i=0}^{(N-1)/2} a(i) \cos \nu i$$

where $a(0) = w(0)$, and $a(i) = 2w(i)$ for $i = 1, 2, \cdots, (N-1)/2$. Following McClellan's derivation, W can be written as

$$W(\nu) = \sum_{i=0}^{(N-1)/2} a'(i)\, [\cos \nu]^i \quad (10)$$

by using the appropriate trigonometric identities. Now, for

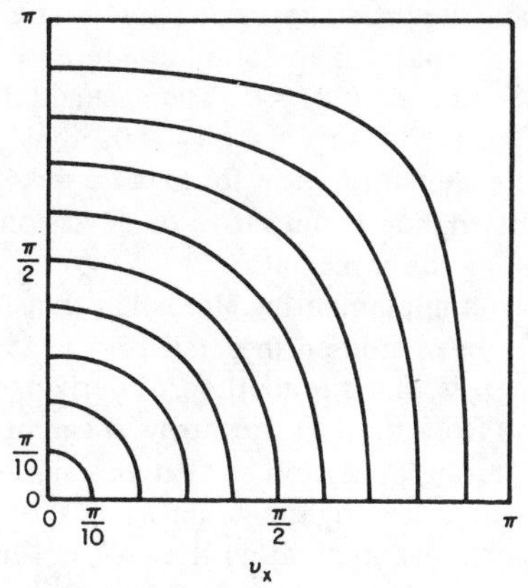

Fig. 3. Contours for McClellan's transformation with $\nu_z = 0$ (after [6], [24]).

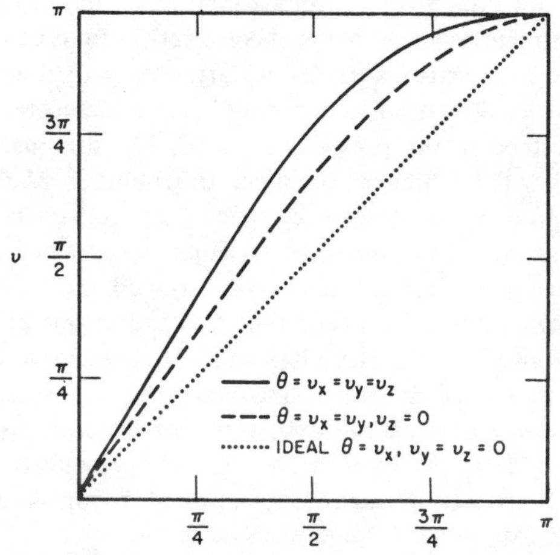

Fig. 4. Deviation from spherical symmetry of a 3-D McClellan transformation. The dotted line represents the ideal and the actual mapping along the axes. The dashed line represents the mapping along the path $\nu_x = \nu_y$ in the $\nu_z = 0$ plane. The solid line represents the mapping along the path $\nu_x = \nu_y = \nu_z$.

the 3-*D* case, we make the substitution

$$\begin{aligned}\cos \nu = {} & 1/4 \cos \nu_x + 1/4 \cos \nu_y + 1/4 \cos \nu_z \\ & + 1/4 \cos \nu_x \cos \nu_y + 1/4 \cos \nu_x \cos \nu_z + 1/4 \cos \nu_y \cos \nu_z \\ & + 1/4 \cos \nu_x \cos \nu_y \cos \nu_z - \frac{3}{4}. \end{aligned} \quad (11)$$

Equation (11) is such that if $\nu_z = 0$, the transformation of variables is identical to McClellan's 2-D circular transformation. Furthermore, the expression is symmetric in ν_x, ν_y, and ν_z. Substituting (11) into (10) will yield the 3-D array pattern

$$W(\nu_x, \nu_y, \nu_z) = \sum_{n_x=0}^{(N-1)/2} \sum_{n_y=0}^{(N-1)/2} \sum_{n_z=0}^{(N-1)/2} a''(n_x, n_y, n_z) \cdot \cos n_x \nu_x \cos n_y \nu_y \cos n_z \nu_z \quad (12)$$

where the a'' coefficients are derivable from a', the transformation (11), and trigonometric identities. Finally the sensor weights $w(n_x, n_y, n_z)$ can be derived from the a'' coefficients by comparing (12) and (9). A plot of constant values of ν in the (ν_x, ν_y)-plane is shown in Fig. 3. Fig. 4 shows the frequency variable ν plotted against a parameter Θ for three paths in (ν_x, ν_y, ν_z) - space. The dotted line represents the transformation for $\Theta = \nu_x, \nu_y = \nu_z = 0$. The dashed line represents the transformation for $\Theta = \nu_x = \nu_y, \nu_z = 0$. Finally, the solid line represents the transformation for $\Theta = \nu_x = \nu_y = \nu_z$. Deviation from the dotted line is indicative of deviation from spherical symmetry along the three paths.

Designing and implementing McClellan transformation filters have recently been studied in detail [24], [25]. The remarkable fact emerges that the method of deriving the 3-D weights $w(n_x, n_y, n_z)$ from the 1-D weights $w(i)$ can be combined with the actual filtering operation so that the amount of computation needed to calculate the beam signals is significantly reduced (proportional to N rather than N^3 in this case). We see, therefore, that one advantage of locating sensors on a Cartesian grid is the availability of design and implementation techniques for array patterns which exhibit good circular symmetry and reduce computation.

As before, in certain applications one may be more interested in short-period beam spectra rather than beam signals. If the sensor locations are on a Cartesian grid, then the FFT algorithm may be used to compute beam spectra similar to the way described in the previous section. The dimensionality of the FFT will be higher to reflect the number of degrees of freedom in direction-frequency space a signal may have.

The use of multidimensional filter design techniques (in particular, McClellan's transformation) and multidimensional filter banks using FFT's represent two important examples of results from the field of digital signal processing which can be applied to digital beamforming and array processing. The reader should bear in mind that the preceeding discussion is more an illustrative than comprehensive presentation of digital signal processing techniques applied to array processing problems. Much work remains to be done.

Acknowledgment

The author wishes to express his thanks to H. Briscoe and R. Estrada of BBN for several educational discussions of practical beamforming systems.

References

[1] P. E. Blankenship and E. M. Hofstetter, "Digital pulse compression via fast convolution," *IEEE Trans. Acoust. Speech, Signal Processing*, vol. ASSP-23, pp. 189-201, Apr. 1975.

[2] L. C. Wood and S. Treitel, "Seismic signal processing," *Proc. IEEE*, vol. 63, pp. 649-661, Apr. 1975.

[3] F. J. Harris, "A maximum entropy filter," Naval Undersea Center, Rep. TP 441, Jan. 1975.

[4] K. R. Erikson, F. J. Fry, and J. P. Jones, "Ultrasound in Medicine–A Review," *IEEE Trans. Sonics Ultrason.*, vol. SU-21, pp. 144-170, July 1974.

[5] L. J. Pinson and D. G. Childers, "Frequency-wavenumber spectrum analysis of EEG multielectrode array data," *IEEE Trans. Biomed. Eng.*, vol. BME-21, pp. 192-206, May 1974.

[6] J. H. McClellan, "The design of two-dimensional digital filters by transformations," in *Proc. 7th Annu. Princeton Conf. Information Sciences and Systems*, 1973, pp. 247-251.

[7] A. V. Oppenheim and R. W. Schafer, *Digital Signal Processing*. Englewood Cliffs, NJ: Prentice-Hall, 1975.

[8] R. M. Mersereau and D. E. Dudgeon, "Two-dimensional digital filtering," *Proc. IEEE*, vol. 63, pp. 610-623, Apr. 1975.

[9] L. R. Rabiner and B. Gold, *Theory and Application of Digital Signal Processing*. Englewood Cliffs, NJ: Prentice-Hall, 1975.

[10] B. Gold and C. M. Rader, *Digital Processing of Signals*. New York: McGraw-Hill, 1969.

[11] W. T. Cochran *et al.*, "What is the fast Fourier transform?," *IEEE Trans. Audio Electroacoust.*, vol. AU-15, pp. 45-55, June 1967.

[12] G. D. Bergland, "A guided tour of the fast Fourier Transform," *IEEE Spectrum*, vol. 6, July 1969.

[13] J. H. McClellan and T. W. Parks, "A unified approach to the design of optimum FIR linear phase digital filters," *IEEE Trans. Circuit Theory*, vol. CT-20, pp. 697-701, Nov. 1973.

[14] L. R. Rabiner, J. H. McClellan, and T. W. Parks, "FIR digital filter design techniques using weighted chebyshev approximation," *Proc. IEEE*, vol. 63, pp. 595-610, Apr. 1975.

[15] J. H. McClellan, T. W. Parks, and L. R. Rabiner, "A computer program for designing optimum FIR linear phase digital filters," *IEEE Trans. Audio Electroacoust.*, vol. AU-21, pp. 506-526, Dec. 1973.

[16] G. M. Jenkins and D. G. Watts, *Spectral Analysis and its Applications*. San Francisco, CA: Holden-Day, 1968, Ch. 7.

[17] J. P. Burg, "Three-dimensional filtering with an array of seismometers," *Geophysics*, vol. 29, no. 5, pp. 693-713, 1964.

[18] R. W. Schafer and L. R. Rabiner, "A digital signal processing approach to interpolation," *Proc. IEEE*, vol. 61, pp. 692-702, June 1973.

[19] J. Capon, R. J. Greenfield, and R. T. Lacoss, "Design of seismic arrays for efficient on-line beamforming," Lincoln Lab. Tech. Note 1967-26, June 27, 1967.

[20] E. J. Kelly, "Response of seismic arrays to wide-band signals," Lincoln Lab. Tech. Note 1967-30, June 29, 1967.

[21] J. L. Allen, "The theory of array antennas," Lincoln Lab. Tech. Rep. no. 323, July 25, 1963.

[22] D. K. Cheng, "Optimization techniques for antenna arrays," *Proc. IEEE*, vol. 59, pp. 1664-1674, Dec. 1971.

[23] E. J. Kelley, Jr., "The representation of seismic waves in frequency-wave number space," Lincoln Lab. Tech. Note 1964-15, Mar. 6, 1964.

[24] R. M. Mersereau, W. F. G. Mecklenbrauker, and T. F. Quatieri, Jr., "McClellan Transformations for two-dimensional digital filtering: I. Design," *IEEE Trans. Circuits and Systs.*, vol. CAS-23, pp. 405-414, July 1976.

[25] W. F. G. Mecklenbrauker and R. M. Mersereau, "McClellan transformations for two-dimensional digital filtering: II. Implementation," *IEEE Trans. Circuit and Systs.*, vol. CAS-23, pp. 414-422, July 1976.

[26] J. R. Williams, "Fast beam forming algorithm," *Acoust. Soc. Amer.*, vol. 44, no. 5, pp. 4154-55, 1968.

[27] P. Rudnick, "Digital beam forming in the frequency domain," *J. Acoust. Soc. Amer.*, vol. 46, no. 5, (Part I), pp. 1089-1090, 1969.

[28] S. Haykin and J. Kesler, "Relation between the radiation pattern of an array and the two-dimensional discrete fourier transform," *IEEE Trans. Antennas Propagat.*, vol. AP-23, no. 3, pp. 419-420, May 1975.

[29] M. R. Portnoff, "Implementation of the digital phase vocoder using the fast Fourier transform," *IEEE Trans. Acoust. Speech, Signal Processing*, vol. ASSP-24, pp. 243-248, June 1976.

[30] O. S. Halpeny and D. G. Childers, "Composite wavefront decomposition via multidimensional digital filtering of array data," *IEEE Trans. Circuits Systs.*, vol. CAS-22, pp. 552-563, June 1975.

Reprinted from: The Journal of the Acoustical Society of America

Resolving power and sensitivity to mismatch of optimum array processors

Henry Cox*

University of California, San Diego, Marine Physical Laboratory of the Scripps Institution of Oceanography, San Diego, California 92152

(Received 9 January 1973; revised 2 February 1973)

Mismatch in a beamformer occurs when the knowledge of the signal directional properties is imprecise. The effects of mismatch on a conventional beamformer and two optimum beamformers are compared. One optimum beamformer is based on inversion of the noise cross-spectral matrix while the other is based on inversion of the signal-plus-noise cross-spectral matrix. When there is mismatch, the inclusion of the signal in the matrix inversion process can lead to dramatic reductions in the output signal-to-noise ratio when the output signal-to-noise ratio of a perfectly matched beamformer would be greater than unity. However, the corresponding effect on the total beamformer output is less dramatic since an increase in the noise response partially offsets the decrease in signal response. The question of suppressing mismatched signals is closely related to the question of resolving closely spaced sources. Exact conditions are presented for resolution of closely spaced sources by an optimum beamformer. These results are applied to a line array and compared with the resolution capability of a conventional beamformer. It is found, for example, that an output signal-to-noise ratio of about 47 dB is required to achieve a resolving power with an optimum processor which is ten times that given by the classical Rayleigh limit. Conditions are also presented for the resolution of two sources of unequal strength.

Subject Classification: 15.3.

INTRODUCTION

Optimum array processors have received a great deal of attention in the last decade in a variety of application areas.[1–12] A fairly general array processing configuration is shown in Fig. 1. It consists of a beamformer which filters the output of each sensor and then sums the filtered outputs, followed by a single-channel postsummation processor. It is well known that the optimum processors under a variety of detection and estimation criteria can be structured to use the same beamformer and differ only in the postsummation processor. The optimum set of filters on the individual sensors depends on the inverse of the noise cross-spectral density matrix and the directional properties of the signal.

In many cases of practical importance it is not possible to obtain a signal-free estimate of the noise cross-spectral matrix. The measurable quantity is the cross-spectral matrix of the sensor outputs which in general contains signal plus noise. Indeed when there are multiple signal-like waves arriving from different directions even the definitions of signal and noise become somewhat arbitrary. While the use of the signal-plus-noise spectral matrix can also be optimum when the signal directional characteristics are known exactly, a problem of signal suppression arises when the knowledge of the signal directional characteristics is imperfect.

This paper is devoted to the analysis of two fundamental questions which arise in the use of the signal-plus-noise matrix in optimum beamforming. The first is the effect of "mismatch" which arises when there is imperfect knowledge about the signal directional characteristics. The second is the question of resolution, that of determining conditions under which the processor will indicate that the array is responding to waves from two spatially separated sources rather than to a wave from a single source.

In spite of the importance of these questions, they have largely been ignored in the literature and few quantitative results have been reported. Capon[13] coins the phrase "high resolution frequency-wavenumber estimation" for a procedure based on inversion of the signal-plus-noise matrix. He concludes that resolution greater than that of a conventional beamformer is attainable when the output signal-to-noise ratio is sufficiently high but provides no quantitative indication of how much signal-to-noise ratio is sufficient for resolution. Seligson[14] in commentating on Capon's paper concludes that the "high resolution estimator" may display less angular resolution than the conventional beamformer under conditions of mismatch. Actually, he examines the relative sharpness in the peaks of the bearing response patterns of the two

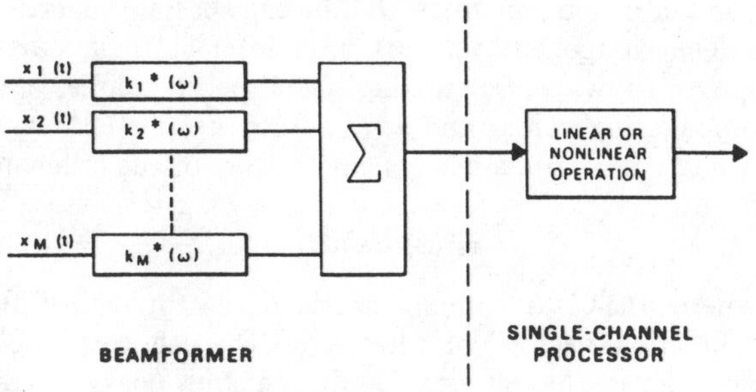

FIG. 1. Array processor configuration.

processors for the case of a single source. McDonough,[15] using a statistical analysis, concludes that Capon's processor may exhibit anomalous behavior when there are small random errors in the signal model if a certain measure of sensitivity is high. Cox[16] presents a general sensitivity analysis for beamformers of the type shown in Fig. 1, including some initial results on the question of mismatch.

In this paper the sensitivity to mismatch of the conventional beamformer, the beamformer based on the inverse of the noise cross-spectral matrix, and the beamformer based on inversion of the signal-plus-noise matrix are compared. The analysis of mismatched performance provides the basis for examining the more difficult problem of determining exact conditions under which two spatially separated sources will be resolved by an optimum beamformer based on inversion of the signal-plus-noise cross-spectral matrix. Resolving power of this optimum beamformer is compared with that of a conventional beamformer.

The approach in the paper emphasizes geometric aspects of these problems. The geometric approach leads to significant simplification of many of the basic results so that they are more easily understood. It also permits us to obtain new results with relative ease, in cases in which other approaches are discouragingly complicated.

I. BACKGROUND

Consider an array consisting of M omnidirectional sensors in an arbitrary known spatial configuration. The output of the ith sensor is a time series denoted by $x_i(t)$. The cross-correlation matrix of the M sensor outputs is defined to be the matrix with the following entry in its ith row and jth column:

$$r_{ij}(\tau)=\overline{x_i(t)x_j(t-\tau)}, \tag{1}$$

where the overbar denotes ensemble average. The cross-spectral matrix $\mathbf{R}(\omega)$ is the matrix which has as its i,jth entry the Fourier transform of the corresponding entry of the cross-correlation matrix. That is,

$$R_{ij}(\omega)=\int r_{ij}(\tau)\exp(-i\omega\tau)d\tau. \tag{2}$$

$\mathbf{R}(\omega)$ may be thought of as being composed of a number of components with differing spatial properties.

The components of particular interest to us correspond to waves emanating from point sources and propagating across the array. Such a coherent wave would contribute a component to $\mathbf{R}(\omega)$ of the following form:

$$\sigma_d^2(\omega)\mathbf{d}(\omega)\mathbf{d}^*(\omega),$$

where $\mathbf{d}(\omega)$ is a column matrix or vector, called the "direction vector" of the wave. The jth component of $\mathbf{d}(\omega)$ is the relative amplitude and phase of the component at the jth sensor. The notational convention used in this paper is that column matrices or vectors are represented by boldface lower-case letters. The asterisk denotes conjugate transposition. Boldface upper-case letters are used to denote Hermitian matrices. Thus $\mathbf{d}^*(\omega)$ is a row vector and $\mathbf{d}(\omega)\mathbf{d}^*(\omega)$ is a rank-one Hermitian matrix. The inner product $\mathbf{d}^*(\omega)\mathbf{d}(\omega)$ is normalized to be equal to M so that $\sigma_d^2(\omega)$ is the average input power spectrum of the wave, the average being made across the M sensors. For a plane wave received at all the sensors with equal intensity, $\mathbf{d}(\omega)$ would have the following form:

$$\mathbf{d}(\omega)=\begin{bmatrix}\exp\{i(\mathbf{p}_1\cdot\boldsymbol{\mu}_d)\omega/c\}\\ \exp\{i(\mathbf{p}_2\cdot\boldsymbol{\mu}_d)\omega/c\}\\ \vdots\\ \exp\{i(\mathbf{p}_M\cdot\boldsymbol{\mu}_d)\omega/c\}\end{bmatrix},$$

where $\mathbf{p}_j$ is the three-dimensional vector of position coordinates of the jth sensor, $\boldsymbol{\mu}_d$ is a unit vector in the direction from which the wave is propagating, and c is the velocity of propagation.

In most of the subsequent analysis it is not necessary to assume that the waves of interest are planar. The general results are equally applicable, for example, to nearfield focusing of arrays for imaging.

In the special case of plane waves, a fairly mild form of array symmetry leads to certain simplifications. An array will be called pairwise symmetric if for each sensor located away from the origin at position coordinates $\{x_j,y_j,z_j\}$ there is also a sensor located at $\{-x_j,-y_j,-z_j\}$. This type of symmetry is also known as inversion symmetry. Pairwise symmetric arrays have the property that if $\mathbf{d}(\omega)$ and $\mathbf{b}(\omega)$ are direction vectors corresponding to plane waves, then the inner product $\mathbf{d}^*(\omega)\mathbf{b}(\omega)$ is real.

The power spectrum of the output of the beamformer shown in Fig. 1 is given by the following Hermitian form:

$$z(\omega)=\mathbf{k}^*(\omega)\mathbf{R}(\omega)\mathbf{k}(\omega), \tag{3}$$

where $\mathbf{k}^*(\omega)$ is the row vector $\{k_1^*(\omega),\cdots,k_M^*(\omega)\}$ of the filter transfer function for each sensor.

Suppose that $\mathbf{R}(\omega)$ consists of signal-plus-noise so that we may express $\mathbf{R}(\omega)$ as the following sum:

$$\mathbf{R}(\omega)=\sigma_0^2(\omega)\mathbf{Q}(\omega)+\sigma_1^2\mathbf{d}(\omega)\mathbf{d}^*(\omega), \tag{4}$$

where $\sigma_0^2(\omega)\mathbf{Q}(\omega)$ is the noise component and $\mathbf{Q}(\omega)$ is normalized to have its trace equal to the number of sensors M. Then $\sigma_0^2(\omega)$ is the input noise spectral level averaged across the sensors. Recalling that $\mathbf{d}(\omega)$ is also normalized so that $\mathbf{d}^*(\omega)\mathbf{d}(\omega)=M$, it follows that $\sigma_1^2(\omega)/\sigma_0^2(\omega)$ is the input signal-to-noise spectral ratio. Substituting from Eq. 4 into Eq. 3 yields

$$z(\omega)=\sigma_0^2(\omega)\mathbf{k}^*(\omega)\mathbf{Q}(\omega)\mathbf{k}(\omega)+\sigma_1^2(\omega)\,|\mathbf{k}^*(\omega)\mathbf{d}(\omega)|^2. \tag{5}$$

Henceforth, we shall confine our attention to a single frequency and the dependence of various quantities on ω will not be shown explicitly. Many of the results expressed in general vector–matrix form may be reinterpreted and applied directly to related problems.[10] The ratio of the two terms in Eq. 5,

$$(S/N)_0 = \sigma_1^2 |\mathbf{k}^*\mathbf{d}|^2 / \sigma_0^2 \mathbf{k}^*\mathbf{Q}\mathbf{k}, \tag{6}$$

is the output signal-to-noise spectral ratio and the quantity

$$G = |\mathbf{k}^*\mathbf{d}|^2 / (\mathbf{k}^*\mathbf{Q}\mathbf{k}) \tag{7}$$

is the array gain or improvement in signal-to-noise ratio due to beamforming. We shall frequently write $G(\mathbf{k})$ and $z(\mathbf{k})$ to denote the array gain and output spectrum for a particular choice of the filter vector $\mathbf{k}$. From Eqs. 6 and 7 it is evident that $(S/N)_0$ and G do not change if $\mathbf{k}$ is multiplied by a scalar.

In many cases of practical importance the available information about the signal direction vector $\mathbf{d}$ is imprecise. This leads to the definition of the steering vector $\mathbf{m}$ as the vector which represents the assumed signal characteristics. The filter vector $\mathbf{k}$ is always a function of the steering vector $\mathbf{m}$. When $\mathbf{m}=\mathbf{d}$ the processor is said to be perfectly matched to the signal directional characteristics. Mismatch occurs when $\mathbf{m}\neq\mathbf{d}$. The steering vector $\mathbf{m}$ is also normalized so that $\mathbf{m}^*\mathbf{m}=M$. Some possible causes of mismatch are distortion in the wavefront during propagation, amplitude, phase and position errors in the sensors, sampling and quantization. In addition to these sources of mismatch, $\mathbf{m}$ will usually be scanned over some set of steering vectors which correspond to a class of interesting signals. For example, this set might correspond to plane waves from some angular region of interest or to spherical waves emanating from a spatial region near the array. Scanning away from the signal direction increases mismatch. This usually has the effect of producing a peak in the scanned array response at the steering vector $\mathbf{m}$ which corresponds most closely to the actual signal direction vector $\mathbf{d}$.

Three particular choices of the filter vector $\mathbf{k}$ will be examined in this paper. These are

$$\mathbf{k}_1 = \mathbf{m}/M, \tag{8}$$

$$\mathbf{k}_2 = \mathbf{Q}^{-1}\mathbf{m}/(\mathbf{m}^*\mathbf{Q}^{-1}\mathbf{m}), \tag{9}$$

and

$$\mathbf{k}_3 = \mathbf{R}^{-1}\mathbf{m}/(\mathbf{m}^*\mathbf{R}^{-1}\mathbf{m}). \tag{10}$$

The denominators of Eqs. 8, 9, and 10 provide an additional normalization so that $\mathbf{k}^*\mathbf{m}=1$. Thus, the three processors all have a unit response to a unit signal from the assumed signal direction so they provide unbiased estimates of signals from that direction.

The first choice, $\mathbf{k}_1=\mathbf{m}/M$, is the conventional beamformer, which simply matches to the assumed signal directional characteristics.

The second choice, $\mathbf{k}_2=\mathbf{Q}^{-1}\mathbf{m}/(\mathbf{m}^*\mathbf{Q}^{-1}\mathbf{m})$, when perfectly matched and followed by an appropriate single-channel processor is optimum for a variety of detection and estimation problems. It reduces to the conventional processor when the noise is uncorrelated from sensor to sensor so that $\mathbf{Q}=\mathbf{I}$, the identity matrix. When perfectly matched, the gain of this processor is seen from Eq. 7 to be

$$G(\mathbf{k}_2) = \mathbf{d}^*\mathbf{Q}^{-1}\mathbf{d}, \quad \text{for } \mathbf{m}=\mathbf{d}. \tag{11}$$

The output signal-to-noise ratio is also maximized by the choice $\mathbf{k}=\mathbf{k}_2$. This maximum signal-to-noise is an important parameter in our analysis. Substituting from Eq. 9 into 6 yields

$$(S/N)_{\max} = \mathbf{d}^*\mathbf{Q}^{-1}\mathbf{d}\sigma_1^2/\sigma_0^2. \tag{12}$$

The third choice, $\mathbf{k}_3=\mathbf{R}^{-1}\mathbf{m}/(\mathbf{m}^*\mathbf{R}^{-1}\mathbf{m})$, produces the same output spectrum as $\mathbf{k}_2$ does when both are perfectly matched. Hence, when perfectly matched, $\mathbf{k}_2$ and $\mathbf{k}_3$ produce the same gain and the same output signal-to-noise ratio. However, when there is mismatch these processors differ in interesting and important ways which we shall examine in some detail. The output spectrum for $\mathbf{k}_3$ may be found by substituting from Eq. 10 into Eq. 3 as follows:

$$z(\mathbf{k}_3) = \left(\frac{\mathbf{m}^*\mathbf{R}^{-1}}{\mathbf{m}^*\mathbf{R}^{-1}\mathbf{m}}\right)\mathbf{R}\left(\frac{\mathbf{R}^{-1}\mathbf{m}}{\mathbf{m}^*\mathbf{R}^{-1}\mathbf{m}}\right) = (\mathbf{m}^*\mathbf{R}^{-1}\mathbf{m})^{-1}. \tag{13}$$

With the advent of high-speed digital circuitry and FFT algorithms, it may be desirable to make direct computations on the sensor outputs rather than building a configuration resembling Figure 1. A procedure suggested by Eq. 13 is to obtain an estimate of $\mathbf{R}$ denoted by $\hat{\mathbf{R}}$ using a finite-length data sample and to compute

$$(\mathbf{m}^*\hat{\mathbf{R}}^{-1}\mathbf{m})^{-1}$$

for the set of steering vectors $\mathbf{m}$ of interest. This procedure has been called a "high resolution" estimator by Capon,[13] who suggests its use in seismic applications. Some results concerning bias and variance of this estimator have been given by Capon and Goodman.[17] In the limit as $\hat{\mathbf{R}}\rightarrow\mathbf{R}$ this procedure is equivalent to computing the output spectrum of a beamformer with $\mathbf{k}=\mathbf{k}_3$. Thus the analysis of this paper will provide asymptotic properties of this estimation procedure.

II. MATHEMATICAL PRELIMINARIES

This section presents some mathematical concepts and results which will be used repeatedly. The approach will be to emphasize geometric aspects of various relationships in an attempt to provide simple interpretations of what initially appear to be quite complicated expressions.

A. Sines and Cosines

A useful concept is that of a generalized angle between two vectors. Let $\mathbf{a}$ and $\mathbf{b}$ be M-component complex column vectors and let $\mathbf{C}$ be a positive definite Hermitian matrix. Then we may define an inner product between **a** and **b** by $\mathbf{a}^*\mathbf{C}\mathbf{b}$ and let $H(\mathbf{C})$ denote the space of M-dimensional complex vectors with the inner product so defined. In this space it is natural to define the cosine-squared of the generalized angle between **a** and **b** as follows:

$$\cos^2(\mathbf{a},\mathbf{b};\mathbf{C}) = |\mathbf{a}^*\mathbf{C}\mathbf{b}|^2/\{(\mathbf{a}^*\mathbf{C}\mathbf{a})(\mathbf{b}^*\mathbf{C}\mathbf{b})\}. \quad (14)$$

By the Schwarz inequality,

$$0 \leq \cos^2(\mathbf{a},\mathbf{b};\mathbf{C}) \leq 1. \quad (15)$$

The length of a vector **b** in $H(\mathbf{C})$ is $(\mathbf{b}^*\mathbf{C}\mathbf{b})^{\frac{1}{2}}$. When $\cos^2(\mathbf{a},\mathbf{b};\mathbf{C})=0$, the vectors **a** and **b** are orthogonal in $H(\mathbf{C})$. When $\cos^2(\mathbf{a},\mathbf{b};\mathbf{C})=1$, **a**, and **b** are in perfect alignment in the sense that one is a scalar multiple of the other. It is natural to define $\sin^2(\mathbf{a},\mathbf{b};\mathbf{C})$ through the identity

$$\sin^2(\cdot) = 1 - \cos^2(\cdot). \quad (16)$$

Among the cases of interest in this paper are $\mathbf{C}=\mathbf{I}$, the identity matrix, and $\mathbf{C}=\mathbf{Q}^{-1}$, the inverse of the normalized noise cross-spectral matrix. When $\mathbf{C}=\mathbf{I}$ we will write $\cos^2(\mathbf{a},\mathbf{b})$ instead of $\cos^2(\mathbf{a},\mathbf{b};\mathbf{I})$ since no confusion will result.

In order to follow the developments of this paper it is sufficient to know the definitions of $\cos^2(\mathbf{a},\mathbf{b};\mathbf{C})$ and $\sin^2(\mathbf{a},\mathbf{b};\mathbf{C})$ given above. However, since the approach has broad applicability a few words of further explanation may be useful.

In order to better understand the effect of the metric $\mathbf{Q}^{-1}$ it is worthwhile to compare $\cos^2(\mathbf{a},\mathbf{b})$ and $\cos^2(\mathbf{a},\mathbf{b};\mathbf{Q}^{-1})$. To do this, consider the eigenvalues and corresponding orthonormal eigenvectors of **Q**, which we denote by $\{\lambda_1,\cdots,\lambda_M\}$ and $\{\mathbf{e}_1,\cdots\mathbf{e}_M\}$, respectively. Since **Q** is normalized to have its trace equal to M, $\sum \lambda_i = M$. Representing **a** and **b** in terms of their projections on the eigenvectors of **Q** as follows:

$$\mathbf{a} = \sum_{i=1}^{M} h_i \mathbf{e}_i \quad \mathbf{b} = \sum_{i=1}^{M} g_i \mathbf{e}_i,$$

where

$$h_i = \mathbf{e}_i^*\mathbf{a} \quad g_i = \mathbf{e}_i^*\mathbf{b};$$

then

$$\cos^2(\mathbf{a},\mathbf{b}) = |\sum h_i^* g_i|^2/(\sum|h_i|^2)(\sum|g_i|^2) \quad (17)$$

and

$$\cos^2(\mathbf{a},\mathbf{b};\mathbf{Q}^{-1}) = \frac{|\sum h_i^* g_i/\lambda_i|^2}{(\sum |h_i|^2/\lambda_i)(\sum |g_i|^2/\lambda_i)}. \quad (18)$$

From Eqs. 17 and 18 it is apparent that the effect of the metric $\mathbf{Q}^{-1}$ is equivalent to scaling h_i and g_i by $(1/\lambda_i)^{\frac{1}{2}}$. This scaling emphasizes components of **a** and **b** corresponding to small eigenvalues of **Q**, and deemphasizes components corresponding to large eigenvalues of **Q**. Since the small eigenvalues of **Q** correspond to components with less noise, it is natural that the metric $\mathbf{Q}^{-1}$ arises in optimization considerations.

The space $H(\mathbf{Q}^{-1})$ arises naturally when optimum processors are considered. It is closely related to the reproducing kernel Hilbert space approach which has been pioneered by Parzen[18] and emphasized by Kailath[19] in treating continuous parameter detection and estimation problems. Hence, by a suitable reinterpretation, many of the results of this paper can be extended to a more general abstract case.

B. Some Useful Relationships

In this section some relationships are presented which apply when **R** is of the following form:

$$\mathbf{R} = \sigma_0^2\mathbf{Q} + \sigma_1^2\mathbf{d}\mathbf{d}^*. \quad (19)$$

When **R** is given by Eq. 19, its inverse may be obtained using the following matrix identity:

$$\mathbf{R}^{-1} = (1/\sigma_0^2)\{\mathbf{Q}^{-1} - \mathbf{Q}^{-1}\mathbf{d}\mathbf{d}^*\mathbf{Q}^{-1}(\sigma_1^2/\sigma_0^2) \times (1+\mathbf{d}^*\mathbf{Q}^{-1}\mathbf{d}\sigma_1^2/\sigma_0^2)^{-1}\}. \quad (20)$$

Using the definition of $(S/N)_{max}$ given in Eq. 12, the following relationships follow directly from Eqs. 20, 14, and 16 as shown in Appendix A:

$$\mathbf{m}^*\mathbf{R}^{-1}\mathbf{m} = \frac{\mathbf{m}^*\mathbf{Q}^{-1}\mathbf{m}}{\sigma_0^2} \times \left\{\frac{1+(S/N)_{max}\sin^2(\mathbf{m},\mathbf{d};\mathbf{Q}^{-1})}{1+(S/N)_{max}}\right\}, \quad (21)$$

$$|\mathbf{m}^*\mathbf{R}^{-1}\mathbf{d}|^2 = \frac{(\mathbf{m}^*\mathbf{Q}^{-1}\mathbf{m})(\mathbf{d}^*\mathbf{Q}^{-1}\mathbf{d})\cos^2(\mathbf{m},\mathbf{d};\mathbf{Q}^{-1})}{\sigma_0^4[1+(S/N)_{max}]^2}, \quad (22)$$

$$\mathbf{m}^*\mathbf{R}^{-1}\mathbf{Q}\mathbf{R}^{-1}\mathbf{m} = \frac{\mathbf{m}^*\mathbf{Q}^{-1}\mathbf{m}}{\sigma_0^4} \times \left\{\frac{1+[2(S/N)_{max}+(S/N)_{max}^2]\sin^2(\mathbf{m},\mathbf{d};\mathbf{Q}^{-1})}{[1+(S/N)_{max}]^2}\right\}. \quad (23)$$

III. MISMATCH

In this section expressions are obtained for the array gain and output spectrum of beamformers with filter vectors given by Eqs. 8, 9, and 10 under the assumption that **R** is given by Eq. 19.

A. Conventional Beamforming ($\mathbf{k}_1 = \mathbf{m}/M$)

Substituting from Eq. 8 into Eq. 7 yields the following expression for gain of a conventional beamformer:

$$G(\mathbf{k}_1) = |\mathbf{m}^*\mathbf{d}|^2/(\mathbf{m}^*\mathbf{Q}\mathbf{m}) = M^2\cos^2(\mathbf{m},\mathbf{d})/(\mathbf{m}^*\mathbf{Q}\mathbf{m}). \quad (24)$$

For spatially uncorrelated noise ($\mathbf{Q}=\mathbf{I}$), Eq. 24 reduces to

$$G(\mathbf{k}_1) = M\cos^2(\mathbf{m},\mathbf{d}). \quad (25)$$

Similarly, substituting from Eq. 8 into Eq. 5 results in the following expression for the output spectrum of a conventional beamformer:

$$z(\mathbf{k}_1) = \sigma_1^2\cos^2(\mathbf{m},\mathbf{d}) + \sigma_0^2(\mathbf{m}^*\mathbf{Q}\mathbf{m})/M^2. \quad (26)$$

The first term in Eq. 26 is due to the signal and the second term is due to the noise. From Eq. 26 it is evident that the quantity $\cos^2(\mathbf{m},\mathbf{d})$ is simply the power response of a conventional beamformer steered in "direction" **m** to a unit signal from "direction" **d**. Plotting $\cos^2(\mathbf{m},\mathbf{d})$ when the signal direction vector **d** is varied over the set of plane waves for a fixed steering direction vector **m** results in the familiar beam pattern.

Notice that the signal response of a conventional beamformer depends on the cosine squared of the generalized angle between **m** and **d** in $H(\mathbf{I})$, and is not influenced by the noise matrix **Q**. Moreover, mismatch does not affect the noise component directly but only through the factor $(\mathbf{m}^*\mathbf{Q}\mathbf{m})$ appearing in Eq. 26 instead of the factor $(\mathbf{d}^*\mathbf{Q}\mathbf{d})$. Either of these two quantities may be larger, depending on whether the direction of **m** is noisier or quieter than the direction of **d**. Because of its dependence on $\cos^2(\mathbf{m},\mathbf{d})$, the signal response is relatively insensitive to small mismatch between **m** and **d**.

B. Noise-Alone Matrix Inverse ($\mathbf{k}_2=\mathbf{Q}^{-1}\mathbf{m}/\mathbf{m}^*\mathbf{Q}^{-1}\mathbf{m}$)

For the optimum beamformer based on inversion of the noise-alone cross-spectral matrix, substituting from Eq. 9 into Eq. 7 results in the following expression for array gain:

$$G(\mathbf{k}_2)=|\mathbf{m}^*\mathbf{Q}^{-1}\mathbf{d}|^2/(\mathbf{m}^*\mathbf{Q}^{-1}\mathbf{m})$$
$$=\mathbf{d}^*\mathbf{Q}^{-1}\mathbf{d}\cos^2(\mathbf{m},\mathbf{d};\mathbf{Q}^{-1}). \quad (27)$$

Again, the array gain depends on the cosine squared of a generalized angle between **m** and **d**, but this time it is the angle in the space $H(\mathbf{Q}^{-1})$, where the metric is the inverse of the noise cross-spectral matrix. Notice that Eq. 27 reduces to Eq. 25 for the case of spatially uncorrelated noise $\mathbf{Q}=\mathbf{I}$. When **m** and **d** are perfectly matched, Eq. 27 reduces to Eq. 11.

The output spectrum of this beamformer is given by the following expression obtained by substituting from Eq. 9 into Eq. 5:

$$z(\mathbf{k}_2)=\sigma_1^2\{(\mathbf{d}^*\mathbf{Q}^{-1}\mathbf{d})/(\mathbf{m}^*\mathbf{Q}^{-1}\mathbf{m})\}\cos^2(\mathbf{m},\mathbf{d};\mathbf{Q}^{-1})$$
$$+\sigma_0^2/(\mathbf{m}^*\mathbf{Q}^{-1}\mathbf{m}). \quad (28)$$

The first term in Eq. 28 is the signal response and the second term is the noise response. Two distinct effects of the metric $\mathbf{Q}^{-1}$ on the signal response are evident. First is the effect of $\mathbf{Q}^{-1}$ on the angular separation between **m** and **d** manifested through the quantity $\cos^2(\mathbf{m},\mathbf{d};\mathbf{Q}^{-1})$. The angular separation and hence sensitivity to mismatch may be either larger or smaller in $H(\mathbf{Q}^{-1})$ than in $H(\mathbf{I})$, depending on how the metric $\mathbf{Q}^{-1}$ alters the vector space in the vicinity of **d**. For example, if the eigenvalues of **Q** had considerable spread and if **d** nearly corresponded to an eigenvector associated with a large eigenvalue of **Q**, then the optimum processor would usually be more sensitive to mismatch than the conventional processor, since differences in small projections of **m** and **d** on eigenvectors associated with small eigenvalues of **Q** would be emphasized by the metric $\mathbf{Q}^{-1}$. The second effect of the metric $\mathbf{Q}^{-1}$ on the signal response is manifested in the ratio $\{(\mathbf{d}^*\mathbf{Q}^{-1}\mathbf{d})/(\mathbf{m}^*\mathbf{Q}^{-1}\mathbf{m})\}$, which is the ratio of the length squared of **d** to the length squared of **m** in $H(\mathbf{Q}^{-1})$. Either of these two "lengths" may be the larger, depending on which corresponds to the quieter direction. The ratio may also be interpreted as the ratio of the maximum possible gain for the true signal direction to the maximum possible gain for the assumed signal direction.

C. Signal-plus-Noise Matrix Inverse ($\mathbf{k}_3=\mathbf{R}^{-1}\mathbf{m}/\mathbf{m}^*\mathbf{R}^{-1}\mathbf{m}$)

The array gain of the beamformer based on inversion of the signal-plus-noise cross-spectral matrix is given by the following equation obtained by substituting from Eq. 10 into Eq. 7:

$$G(\mathbf{k}_3)=|\mathbf{m}^*\mathbf{R}^{-1}\mathbf{d}|^2/(\mathbf{m}^*\mathbf{R}^{-1}\mathbf{Q}\mathbf{R}^{-1}\mathbf{m}). \quad (29)$$

Using Eqs. 22 and 23, this may be rewritten as

$$G(\mathbf{k}_3)=\frac{\mathbf{d}^*\mathbf{Q}^{-1}\mathbf{d}\cos^2(\mathbf{m},\mathbf{d};\mathbf{Q}^{-1})}{1+[2(S/N)_{\max}+(S/N)_{\max}^2]\sin^2(\mathbf{m},\mathbf{d};\mathbf{Q}^{-1})}. \quad (30)$$

The numerator of Eq. 30 may be recognized as $G(\mathbf{k}_2)$, the array gain of the beamformer based on inversion of the noise-alone cross-spectral matrix given in Eq. 27. Thus, the quantity in the denominator of Eq. 30,

$$G(\mathbf{k}_2)/G(\mathbf{k}_3)=1+[2(S/N)_{\max}+(S/N)_{\max}^2]$$
$$\times\sin^2(\mathbf{m},\mathbf{d};\mathbf{Q}^{-1}), \quad (31)$$

gives the effect on array gain and output signal-to-noise ratio of including the signal in the matrix inversion process. Equation 31 is an important result with many practical implications.

Figure 2 presents plots of the gain ratio $G(\mathbf{k}_2)/G(\mathbf{k}_3)$ versus $\sin^2(\mathbf{m},\mathbf{d};\mathbf{Q}^{-1})$ for various values of $(S/N)_{\max}$. Because the gain ratio depends on $(S/N)_{\max}^2\sin^2(\mathbf{m},\mathbf{d};\mathbf{Q}^{-1})$, mismatch can cause a dramatic signal suppression when $(S/N)_{\max}$ is greater than unity. For example, when $(S/N)_{\max}=10$, and $\sin^2(\mathbf{m},\mathbf{d};\mathbf{Q}^{-1})=\cos^2(\mathbf{m},\mathbf{d};\mathbf{Q}^{-1})=0.5$, the gain ratio $G(\mathbf{k}_2)/G(\mathbf{k}_3)=61$. Then, the output signal-to-noise ratio of the $\mathbf{k}_2$-beamformer is $(0.5)(S/N)_{\max}=5.0$, while the output signal-to-noise ratio of the $\mathbf{k}_3$-beamformer is $(5.0)/(61)=0.082$. The reduction is nearly 18 dB. If $(S/N)_{\max}$ had been 2 instead of 10, the output signal-to-noise ratios of the $\mathbf{k}_2$ and $\mathbf{k}_3$ beamformers would have been 1 and 0.2, respectively. It is particularly significant that mismatch in the $\mathbf{k}_3$-beamformer can cause strong signals to be suppressed to such an extent that they have smaller output signal-to-noise ratios than weak signals with comparable mismatch. Also shown in Fig. 2 is the contour $G(\mathbf{k}_2)/G(\mathbf{k}_3)=(S/N)_{\max}$. This locus is of interest because when $G(\mathbf{k}_2)/G(\mathbf{k}_3)\geq(S/N)_{\max}$, the output signal-to-noise ratio of the $\mathbf{k}_3$-processor will be less than unity. For

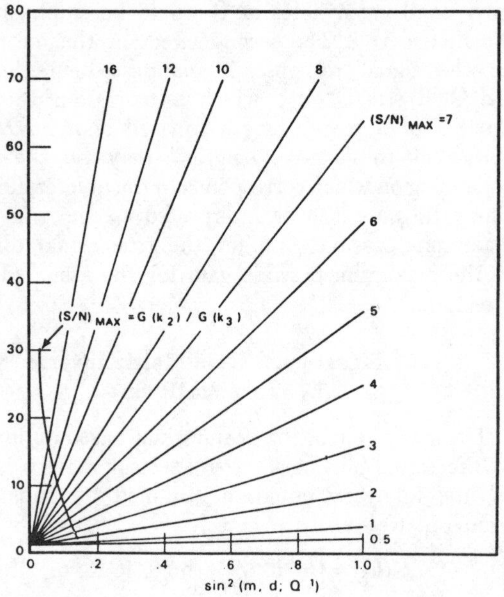

FIG. 2. Ratio of array gains $G(\mathbf{k}_3)/G(\mathbf{k}_2)$ vs amount of mismatch for various maximum possible output signal-to-noise ratios.

example, when $(S/N)_{max}=6$, the output signal-to-noise ratio of the $\mathbf{k}_3$-beamformer will be less than unity unless $\sin^2(\mathbf{m},\mathbf{d};\mathbf{Q}^{-1})$ is less than about 0.1.

From Eq. 31 it is evident that signal suppression cannot be large when $(S/N)_{max}$ is less than unity. However, values of $(S/N)_{max}$ larger than unity, when signal suppression can be a major problem, can correspond to small values of the input signal-to-noise ratio (σ_1^2/σ_0^2), especially when M is large.

One practical implication of Eq. 31 is that a $\mathbf{k}_3$-processor requires more closely spaced beams than a $\mathbf{k}_2$-, or $\mathbf{k}_1$-processor in order to avoid serious signal suppression effects being introduced on signals arriving from directions between the beams. Recalling that $\cos^2(\mathbf{m},\mathbf{d})$ is simply the beampattern of the conventional beamformer, suitable beam spacings can be obtained from Fig. 2 for anticipated values of $(S/N)_{max}$ for the case $\mathbf{Q}=\mathbf{I}$. To put the situation in perspective, we observe that the 3-dB down points on the main lobe response of a conventional beamformer correspond to $\sin^2(\mathbf{m},\mathbf{d})=0.5$.

Signal-to-noise ratio is not the only measure of performance. The behavior of the output is also of interest. Since the output spectrum of this beamformer is given by Eq. 13, the following expression for $z(\mathbf{k}_3)$ may be obtained by taking the reciprocal of Eq. 21:

$$z(\mathbf{k}_3)=\frac{\{\sigma_0^2/(\mathbf{m}^*\mathbf{Q}^{-1}\mathbf{m})\}\{1+(S/N)_{max}\}}{\{1+(S/N)_{max}\sin^2(\mathbf{m},\mathbf{d};\mathbf{Q}^{-1})\}}. \quad (32)$$

The effect of mismatch on the output spectrum may be seen more directly by examining the ratio of the output spectrums for the mismatched and perfectly matched $\mathbf{k}_3$-processor:

$$\frac{z(\mathbf{k}_3;\mathbf{m}\neq\mathbf{d})}{z(\mathbf{k}_3;\mathbf{m}=\mathbf{d})}=\{(\mathbf{d}^*\mathbf{Q}^{-1}\mathbf{d})/(\mathbf{m}^*\mathbf{Q}^{-1}\mathbf{m})\}\times\{1+(S/N)_{max}\sin^2(\mathbf{m},\mathbf{d};\mathbf{Q}^{-1})\}^{-1}. \quad (33)$$

In Eq. 33 we again see two distinct effects of mismatch. First is the effect of the ratio $\{(\mathbf{d}^*\mathbf{Q}^{-1}\mathbf{d})/\mathbf{m}^*\mathbf{Q}^{-1}\mathbf{m})\}$ which was discussed earlier, following Eq. 28. Second is the direct effect of mismatch embodied in the quantity $\{1+(S/N)_{max}\sin^2(\mathbf{m},\mathbf{d};\mathbf{Q}^{-1})\}^{-1}$. Notice that the effect of mismatch on the output spectrum depends on $(S/N)_{max}\sin^2(\mathbf{m},\mathbf{d};\mathbf{Q}^{-1})$ and not $(S/N)_{max}^2\times\sin^2(\mathbf{m},\mathbf{d};\mathbf{Q}^{-1})$ as did the effect on the output signal-to-noise ratio. From Eq. 32 it is apparent that increasing the signal strength and thereby increasing $(S/N)_{max}$ will always cause an increase in $z(\mathbf{k}_3)$ unless $\sin^2(\mathbf{m},\mathbf{d};\mathbf{Q}^{-1})$ is equal to 1. While this increase will be small when $(S/N)_{max}\sin^2(\mathbf{m},\mathbf{d};\mathbf{Q}^{-1})\gg 1$, the effect is unlike the effect on the output signal-to-noise ratio, where weak signals could lead to higher output signal-to-noise ratios than strong signals.

This apparent discrepancy can be clarified by obtaining an expression for $z(\mathbf{k}_3)$ which shows the effects of mismatch on the signal and noise responses individually. Such an alternative expression for $z(\mathbf{k}_3)$ may be obtained by rewriting Eq. 13 as follows:

$$z(\mathbf{k}_3)=\mathbf{m}^*\mathbf{R}^{-1}[\sigma_0^2\mathbf{Q}+\sigma_1^2\mathbf{d}\mathbf{d}^*]\mathbf{R}^{-1}\mathbf{m}/(\mathbf{m}^*\mathbf{R}^{-1}\mathbf{m})^2$$

or

$$z(\mathbf{k}_3)=\sigma_0^2\{\mathbf{m}^*\mathbf{R}^{-1}\mathbf{Q}\mathbf{R}^{-1}\mathbf{m}/(\mathbf{m}^*\mathbf{R}^{-1}\mathbf{m})^2\}+\sigma_1^2|\mathbf{m}^*\mathbf{R}^{-1}\mathbf{d}|^2/(\mathbf{m}^*\mathbf{R}^{-1}\mathbf{m})^2. \quad (34)$$

Substituting for $\mathbf{m}^*\mathbf{R}^{-1}\mathbf{m}$, $\mathbf{m}^*\mathbf{R}^{-1}\mathbf{Q}\mathbf{R}^{-1}\mathbf{m}$, and $|\mathbf{m}^*\mathbf{R}^{-1}\mathbf{d}|^2$ in Eq. 34 from Eqs. 21, 22, and 23 yields

$$z(\mathbf{k}_3)=\sigma_1^2\left(\frac{\mathbf{d}^*\mathbf{Q}^{-1}\mathbf{d}}{\mathbf{m}^*\mathbf{Q}^{-1}\mathbf{m}}\right)\times\frac{\cos^2(\mathbf{m},\mathbf{d};\mathbf{Q}^{-1})}{\{1+(S/N)_{max}\sin^2(\mathbf{m},\mathbf{d};\mathbf{Q}^{-1})\}^2}+\frac{\sigma_0^2}{\mathbf{m}^*\mathbf{Q}^{-1}\mathbf{m}}\times\left\{\frac{1+[2(S/N)_{max}+(S/N)_{max}^2]\sin^2(\mathbf{m},\mathbf{d};\mathbf{Q}^{-1})}{[1+(S/N)_{max}\sin^2(\mathbf{m},\mathbf{d};\mathbf{Q}^{-1})]^2}\right\}. \quad (35)$$

While Eq. 35 is considerably more complicated than Eq. 32, it does present explicit expressions for the signal response (first term in Eq. 35) and the noise response (second term in Eq. 35). By comparing these terms with the corresponding terms of Eq. 28, the effects of including the signal in the matrix inversion process may be perceived directly. Most interesting is the factor in large braces which multiples the average input noise power spectral level σ_0^2 in Eq. 35. The quantity in the numerator of this factor differs from the quantity in the denominator in that the numerator contains the term $(S/N)_{max}^2 \sin^2(\mathbf{m},\mathbf{d};\mathbf{Q}^{-1})$ in place of the term $(S/N)_{max}^2 \sin^4(\mathbf{m},\mathbf{d};\mathbf{Q}^{-1})$ in the denominator. Thus, the factor is always equal to or greater than unity, with it being equal to unity only when $\sin^2(\mathbf{m},\mathbf{d};\mathbf{Q}^{-1})$ is equal to zero or 1. Hence, the noise response of the $\mathbf{k}_3$-beamformer is a function of the amount of mismatch. The nature of the noise response can be seen more easily by considering the case $\mathbf{Q}=\mathbf{I}$ so that the quantity $\mathbf{m}^*\mathbf{Q}^{-1}\mathbf{m}$ is constant. Then, as $\sin^2(\mathbf{m},\mathbf{d})$ is varied from zero to 1, the noise response increases from the value σ_0^2/M at $\sin^2(\mathbf{m},\mathbf{d})=0$ until it attains a maximum of $(\sigma_0^2/M)[2+(S/N)_{max}]^2[4+4(S/N)_{max}]^{-1}$ at $\sin^2(\mathbf{m},\mathbf{d}) =[2+(S/N)_{max}]^{-1}$ and then decreases until it again reaches the value (σ_0^2/M) at $\sin^2(\mathbf{m},\mathbf{d})=1$. One explanation of this unusual behavior of the noise response is that the $\mathbf{k}_3$-beamformer treats the mismatched signal as an unwanted interference and performs a compromise between suppressing it and rejecting the real noise. The stronger the mismatched signal, the more importance the processor puts on suppressing it. In suppressing the mismatched signal it accepts a lesser rejection of the noise. Near the point $\sin^2(\mathbf{m},\mathbf{d})=0$, the constraint $\mathbf{k}^*\mathbf{m}=1$ inhibits the suppression of the signal. As $\sin^2(\mathbf{m},\mathbf{d})$ is increased, the effect of the constraint decreases so that greater suppression of the signal is possible with a corresponding increased penalty in noise response. Eventually, the mismatch reaches the point where the signal suppression is sufficient that a further penalty in noise response is not justified. The processor then reverses the trend and places greater emphasis on rejecting the noise.

The increase in the noise response partially offsets the effect of the signal suppression on the output $z(\mathbf{k}_3)$. However, these two effects work together in decreasing the output signal-to-noise ratio.

Care should be taken in applying these asymptotic results to systems which adapt their parameters based on real-time measurements. The derivation of Expression 32 for $z(\mathbf{k}_3)$ involves a cancellation of field and filter parameters as can be seen in Eq. 13. Hence Eq. 32 does not allow for any change in the field which is not compensated for by a corresponding change in the filter vector $\mathbf{k}_3$. The same sort of cancellation is implicit in the use of the estimator $(\mathbf{m}^*\hat{\mathbf{R}}^{-1}\mathbf{m})^{-1}$. However, Eqs. 30 and 35 do not involve this type of cancellation. For example, suppose that the filter $\mathbf{k}_3$ was based on a cross-spectral matrix $\mathbf{R}_{T_1}$ measured during a particular time interval T_1 when the input signal and noise levels were $\tilde{\sigma}_1^2$ and $\tilde{\sigma}_0^2$ so that

$$\mathbf{R}_{T_1}=\tilde{\sigma}_0^2\mathbf{Q}+\tilde{\sigma}_1^2\mathbf{d}\mathbf{d}^*.$$

If the filter vector $\mathbf{k}_3$ which was determined from $\mathbf{R}_{T_1}$ were subsequently applied at a later time T_2 when the signal and noise levels were different so that

$$\mathbf{R}_{T_2}=\sigma_0^2\mathbf{Q}+\sigma_1^2\mathbf{d}\mathbf{d}^*,$$

then Eqs. 30 and 35 properly describe the array gain and output provided that $(S/N)_{max}$ in these equations is defined in terms of the signal-to-noise ratio $(\tilde{\sigma}_1^2/\tilde{\sigma}_0^2)$ used in determining $\mathbf{k}_3$.

Another view of the effect on the output spectrum of including the signal in the matrix inversion can be obtained by examining the ratio $z(\mathbf{k}_2)/z(\mathbf{k}_3)$. Dividing Eq. 28 by Eq. 32 and rearranging yields the following:

$$\frac{z(\mathbf{k}_2)}{z(\mathbf{k}_3)} =1+\frac{(S/N)_{max}^2 \sin^2(\mathbf{m},\mathbf{d};\mathbf{Q}^{-1})\cos^2(\mathbf{m},\mathbf{d};\mathbf{Q}^{-1})}{1+(S/N)_{max}}. \quad (36)$$

The interpretation of Eq. 36 may be simplified through the use of the trigonometric identity

$$\sin^2 a \cos^2 a=(1/4)\sin^2(2a). \quad (37)$$

Thus Eq. 36 may be written as

$$\frac{z(\mathbf{k}_2)}{z(\mathbf{k}_3)}=1+\frac{(S/N)_{max}^2 \sin^2[2(\mathbf{m},\mathbf{d};\mathbf{Q}^{-1})]}{4[1+(S/N)_{max}]}, \quad (38)$$

where Eq. 37 serves as the definition of $\sin^2[2(\mathbf{m},\mathbf{d};\mathbf{Q}^{-1})]$. From Eqs. 36 and 38 it is apparent that the level of the output spectrum of the $\mathbf{k}_2$-beamformer will be equal to or greater than that of the $\mathbf{k}_3$-beamformer. The two are equal only when there is perfect match $(\mathbf{m}=\mathbf{d})$ or when $\mathbf{m}$ and $\mathbf{d}$ are orthogonal in $H(\mathbf{Q}^{-1})$, $(|\mathbf{m}^*\mathbf{Q}^{-1}\mathbf{d}|^2=0)$. The relative behavior of the two beamformers as $\mathbf{m}$ is scanned in the vicinity of $\mathbf{d}$ may be determined from Eq. 38. As $\mathbf{m}$ is scanned through the set Ω of interest it will pass through some position of best match to $\mathbf{d}$ in the sense that $\cos^2(\mathbf{m},\mathbf{d};\mathbf{Q}^{-1})$ will achieve a local maximum. The maximum value may be less than unity since $\mathbf{d}$ may not correspond exactly to any of the steering vectors in the set through which $\mathbf{m}$ is scanned. From Eq. 38 we see that if the angle between $\mathbf{m}$ and $\mathbf{d}$ in $H(\mathbf{Q}^{-1})$ at the point of best match is sufficiently small, the ratio $z(\mathbf{k}_2)/z(\mathbf{k}_3)$ will initially increase as the angle is increased by scanning $\mathbf{m}$ away from the point of best match. This initial increase in $z(\mathbf{k}_2)/z(\mathbf{k}_3)$ means that the output of the $\mathbf{k}_3$-processor is decreasing more rapidly than that of the $\mathbf{k}_2$-processor and hence the peak on the output of the $\mathbf{k}_3$-processor will be sharper than that of the $\mathbf{k}_2$-processor. The condition that the angle between $\mathbf{m}$ and $\mathbf{d}$ be sufficiently small at the position of best match may be expressed as

$$\max_{\mathbf{m}\in\Omega} \cos^2(\mathbf{m},\mathbf{d};\mathbf{Q}^{-1})>\tfrac{1}{2}. \quad (39)$$

Whenever Expression 39 is satisfied, the peak in the scanned output of the $\mathbf{k}_3$-processor will be sharper than that of the $\mathbf{k}_2$-processor. Conversely, if

$$\max_{\mathbf{m}\in\Omega} \cos^2(\mathbf{m},\mathbf{d};\mathbf{Q}^{-1})<\tfrac{1}{2}, \quad (40)$$

$z(\mathbf{k}_2)/z(\mathbf{k}_3)$ will initially decrease and the $\mathbf{k}_2$-processor will have the sharper peak in its scanned response.

The preceding argument concerning the ratio $z(\mathbf{k}_2)/z(\mathbf{k}_3)$ and the relative sharpness of the peaks in the scanned outputs generalizes a result of Seligson,[14] who compared $z(\mathbf{k}_1)/z(\mathbf{k}_3)$ for the case of spatially uncorrelated noise so that $z(\mathbf{k}_2)$ and $z(\mathbf{k}_1)$ coincided. He derived an expression equivalent to Eq. 36 in that case and presented the argument concerning the relative sharpness of the peaks.

A simple expression for $z(\mathbf{k}_3)$ which will be useful in later discussions may be obtained by substituting from Eq. 19 into Eq. 3:

$$z(\mathbf{k}_3)=\{\mathbf{m}^*[\sigma_0^2\mathbf{Q}+\sigma_1^2\mathbf{dd}^*]^{-1}\mathbf{m}\}^{-1}. \tag{41}$$

IV. RESOLUTION

The classical concept of resolution consists of recognizing that an observed effect is due to two separate sources rather than a single source. High resolution is the ability to separate the effects of two closely spaced sources. In this section we will consider primarily the resolving power of the $\mathbf{k}_3$-beamformer. Some results will be presented for the conventional $\mathbf{k}_1$-beamformer to provide a basis for comparison.

A. Qualitative Discussion

Before determining exact conditions for resolution it is worthwhile to use the results already obtained concerning mismatch in considering the simpler problem of determining the effect of one source on the output of a $\mathbf{k}_3$-beamformer which is perfectly matched to a second source. Let

$$\mathbf{R}=\sigma_0^2\mathbf{Q}+\sigma_1^2\mathbf{dd}^*+\sigma_2^2\mathbf{bb}^*, \tag{42}$$

where $\mathbf{d}$ and $\mathbf{b}$ are the direction vectors for the first and second sources, respectively. When the beamformer is perfectly matched to $\mathbf{b}$, it follows from Eq. 32 that its output spectrum is given by the following equation:

$$z_b=[\mathbf{b}^*\mathbf{R}^{-1}\mathbf{b}]^{-1}=\sigma_2^2+\{\mathbf{b}^*[\sigma_0^2\mathbf{Q}+\sigma_1^2\mathbf{dd}^*]^{-1}\mathbf{b}\}^{-1}. \tag{43}$$

The subscript b on z_b is a reminder that the processor is perfectly matched to $\mathbf{b}$, that is, $\mathbf{k}_3=\mathbf{R}^{-1}\mathbf{b}/(\mathbf{b}^*\mathbf{R}^{-1}\mathbf{b})$.

The first term in Eq. 43 is due to the $\mathbf{b}$-component which passes through the beamformer without distortion. The second term is the combined result of the noise and the $\mathbf{d}$-component. If $\mathbf{m}$ in Eq. 41 is replaced with $\mathbf{b}$, the second term in Eq. 43 becomes identical to Eq. 41. Thus the results developed earlier for a single source with a mismatched $\mathbf{k}_3$-beamformer may be applied directly to the case of two sources with the beamformer perfectly matched to one of them.

For example, defining

$$(S_1/N)_{\max}=\mathbf{d}^*\mathbf{Q}^{-1}\mathbf{d}\sigma_1^2/\sigma_0^2 \tag{44}$$

so that $(S_1/N)_{\max}$ is the maximum output signal-to-noise ratio for the $\mathbf{d}$-component in the absence of the $\mathbf{b}$-component, the expression given in Eq. 32 may be substituted for the second term in Eq. 43, yielding

$$z_b=\sigma_2^2+\{\sigma_0^2/(\mathbf{b}^*\mathbf{Q}^{-1}\mathbf{b})\}\{1+(S_1/N)_{\max}\}/\{1+(S_1/N)_{\max}\sin^2(\mathbf{b},\mathbf{d};\mathbf{Q}^{-1})\}. \tag{45}$$

When the first term in Eq. 45 is much larger than the second, the output signal-to-background ratio for the $\mathbf{b}$-component is high when the processor is perfectly matched to $\mathbf{b}$.

An alternate form of Eq. 45 may be obtained by adding $\sigma_0^2/(\mathbf{b}^*\mathbf{Q}^{-1}\mathbf{b})$ to the first term in Eq. 45 and compensating by subtracting it from the second term. Then Eq. 45 becomes

$$z_b=\{\sigma_2^2+\sigma_0^2/(\mathbf{b}^*\mathbf{Q}^{-1}\mathbf{b})\}+\frac{\sigma_0^2}{\mathbf{b}^*\mathbf{Q}^{-1}\mathbf{b}}\left\{\frac{(S_1/N)_{\max}\cos^2(\mathbf{d},\mathbf{b};\mathbf{Q}^{-1})}{1+(S_1/N)_{\max}\sin^2(\mathbf{d},\mathbf{b};\mathbf{Q}^{-1})}\right\}. \tag{46}$$

The term within the first pair of braces in Eq. 46 may be recognized as $\{\mathbf{b}^*[\sigma_0^2\mathbf{Q}+\sigma_2^2\mathbf{bb}^*]^{-1}\mathbf{b}\}^{-1}$, which is what the output would be in the absence of the $\mathbf{d}$-component. When this term is much larger than the second term in Eq. 46, the presence of the $\mathbf{d}$-component has little effect on the beamformer output. Defining

$$(S_2/N)_{\max}=\mathbf{b}^*\mathbf{Q}^{-1}\mathbf{b}\sigma_2^2/\sigma_0^2,$$

the condition that the first term in Eq. 46 be much larger than the second term may be written as

$$1+(S_2/N)_{\max}\gg\frac{(S_1/N)_{\max}\cos^2(\mathbf{d},\mathbf{b};\mathbf{Q}^{-1})}{1+(S_1/N)_{\max}\sin^2(\mathbf{d},\mathbf{b};\mathbf{Q}^{-1})}. \tag{47}$$

When $(S_1/N)_{\max}\sin^2(\mathbf{b},\mathbf{d};\mathbf{Q}^{-1})>1$ this may be simplified to the following:

$$(S_2/N)_{\max}\gg\frac{\cos^2(\mathbf{d},\mathbf{b};\mathbf{Q}^{-1})}{\sin^2(\mathbf{d},\mathbf{b};\mathbf{Q}^{-1})}=\cot^2(\mathbf{d},\mathbf{b};\mathbf{Q}^{-1}). \tag{48}$$

For the case $\mathbf{Q}=\mathbf{I}$, Capon[13] argues that, when the first source has little effect on the output of the $\mathbf{k}_3$-processor which is perfectly matched to the second source, the two sources will be resolved. He suggests a relationship equivalent to Eq. 48 and an analogous one for $(S_1/N)_{\max}$ as criteria for resolution. A weakness in this argument is that it attempts to infer the behavior of the scanned processor output from only the magnitude of the two terms at the single point $\mathbf{m}=\mathbf{b}$. This, together with the imprecise nature of the "much greater than" type of condition makes desirable a more quantitative treatment of the resolution question.

In spite of the above remarks concerning their use as a basis of a resolution criterion, Eqs. 45 and 46 do provide useful information concerning the effect of one signal on the output of a $\mathbf{k}_3$-processor which is matched to the other. Another equivalent expression for z_b may be obtained by using Eq. 35 to replace the second term in Eq. 43 similarly to the way Eq. 32 was used in obtaining Eq. 45. Rather than pursue this straightforward substitution, we simply note that it provides explicit expressions for the contributions of the two signals and noise to the output.

B. Exact Conditions for Resolution

In developing exact conditions under which two components will be resolved by a $\mathbf{k}_3$-beamformer, we will consider the case of uncorrelated noise so that $\mathbf{Q}=\mathbf{I}$. This assumption will make the analysis slightly simpler and will avoid the necessity of stating restrictions on $\mathbf{Q}$ in order to avoid anomalous situations which are possible for a general $\mathbf{Q}$ matrix. For example, a general noise matrix $\mathbf{Q}$ could contain a third component between the two we are trying to resolve. The assumption of uncorrelated noise also avoids confusion between the super-gain phenomenon[16] which arises in the isotropic noise case and the resolution phenomenon which is our prime concern. We shall initially consider the case when the two components to be resolved are of equal strength. Resolution of signals of unequal strength is considered in a later section. Thus $\mathbf{R}$ is assumed to have the following form:

$$\mathbf{R}=\sigma_0^2\mathbf{I}+\sigma_1^2\mathbf{dd}^*+\sigma_1^2\mathbf{bb}^*. \tag{49}$$

From the symmetry of Eq. 49 it is evident that the output will be the same when $\mathbf{m}$ is perfectly matched to either $\mathbf{d}$ or $\mathbf{b}$.

In order to study the question of resolution, we shall compare the output spectrum when $\mathbf{m}$ is perfectly matched to one of the two signals with the output spectrum when $\mathbf{m}$ is steered to a value $\mathbf{m}_0$ which corresponds to a position midway between the two sources. Rather than work with the outputs directly, it is easier to work with the ratio

$$z_d/z_m=(\mathbf{m}^*\mathbf{R}^{-1}\mathbf{m})/(\mathbf{d}^*\mathbf{R}^{-1}\mathbf{d}), \tag{50}$$

where $\mathbf{R}$ is given by Eq. 49. An expression for this ratio may be obtained from the reciprocal of Eq. 33 by using $(\sigma_0^2\mathbf{I}+\sigma_1^2\mathbf{bb}^*)$ in place of $\sigma_0^2\mathbf{Q}$. This yields

$$z_d/z_m=\{\mathbf{m}^*[\sigma_0^2\mathbf{I}+\sigma_1^2\mathbf{bb}^*]^{-1}\mathbf{m}\}\times\{\mathbf{d}^*[\sigma_0^2\mathbf{I}+\sigma_1^2\mathbf{bb}^*]^{-1}\mathbf{d}\}^{-1}\{1+\sigma_1^2\mathbf{d}^*[\sigma_0^2\mathbf{I}+\sigma_1^2\mathbf{bb}^*]^{-1}\mathbf{d}\times\sin^2(\mathbf{m},\mathbf{d};[\sigma_0^2\mathbf{I}+\sigma_1^2\mathbf{bb}^*]^{-1})\}. \tag{51}$$

After considerable algebraic manipulation (shown in Appendix B), Eq. 51 may be reduced to the following simplified form:

$$z_d/z_m=\{1+(M\sigma_1^2/\sigma_0^2)[1-\cos^2(\mathbf{m},\mathbf{b})-\cos^2(\mathbf{m},\mathbf{d})-\alpha\cos^2(\mathbf{d},\mathbf{b})+(2\alpha/M^3)\,\mathrm{Re}(\mathbf{m}^*\mathbf{dd}^*\mathbf{bb}^*\mathbf{m})]\}\times\{1-\alpha\cos^2(\mathbf{d},\mathbf{b})\}^{-1}, \tag{52}$$

where $\mathrm{Re}(\cdot)$ denotes the real part and α is defined as follows:

$$\alpha=\frac{(M\sigma_1^2/\sigma_0^2)}{(1+M\sigma_1^2/\sigma_0^2)}. \tag{53}$$

Notice that $M\sigma_1^2/\sigma_0^2$ is the maximum output signal-to-noise ratio for either source in the absence of the other. In this case $(\mathbf{Q}=\mathbf{I})$, $M\sigma_1^2/\sigma_0^2$ is the output signal-to-noise ratio for either source of a perfectly matched conventional beamformer in the absence of the other source. The parameter α is nearly unity when $M\sigma_1^2/\sigma_0^2\gg1$.

An important simplification occurs when $\mathbf{m}$, $\mathbf{b}$, and $\mathbf{d}$ are direction vectors corresponding to plane waves and the array is composed of symmetric pairs of sensors. In this case $(\mathbf{m}^*\mathbf{d})$, $(\mathbf{d}^*\mathbf{b})$, and $(\mathbf{b}^*\mathbf{m})$ are all real. Then, since $(\mathbf{m}^*\mathbf{m})$, $(\mathbf{b}^*\mathbf{b})$, and $(\mathbf{d}^*\mathbf{d})$ are all normalized to be equal to M, we can define

$$\cos(\mathbf{m},\mathbf{b})=(\mathbf{m}^*\mathbf{b})/M,\quad \text{for } (\mathbf{m}^*\mathbf{b}) \text{ real}, \tag{54}$$

consistent with the definition of $\cos^2(\mathbf{m},\mathbf{b})$. Then

$$(2\alpha/M^3)\,\mathrm{Re}(\mathbf{m}^*\mathbf{dd}^*\mathbf{bb}^*\mathbf{m})=2\alpha\cos(\mathbf{m},\mathbf{d})\cos(\mathbf{d},\mathbf{b})\cos(\mathbf{b},\mathbf{m}) \tag{55}$$

and Eq. 52 becomes

$$z_d/z_m=\{1+(M\sigma_1^2/\sigma_0^2)[1-\cos^2(\mathbf{m},\mathbf{b})-\cos^2(\mathbf{m},\mathbf{d})-\alpha\cos^2(\mathbf{d},\mathbf{b})+2\alpha\cos(\mathbf{m},\mathbf{d})\cos(\mathbf{d},\mathbf{b})\cos(\mathbf{m},\mathbf{b})]\}\times\{1-\alpha\cos^2(\mathbf{d},\mathbf{b})\}^{-1}. \tag{56}$$

In order to study the question of resolution, we examine the ratio (z_d/z_m) when $\mathbf{m}$ is equal to $\mathbf{m}_0$ which is the midpoint between $\mathbf{b}$ and $\mathbf{d}$ in the sense that

$$\cos^2(\mathbf{m}_0,\mathbf{b})=\cos^2(\mathbf{m}_0,\mathbf{d}). \tag{57}$$

Substituting from Eq. 57 into Eq. 56 gives the following surprisingly simple expression for (z_d/z_{m_0}) for plane wave signals and pairwise symmetric arrays:

$$z_d/z_{m_0}=\frac{1+(M\sigma_1^2/\sigma_0^2)\{1-\alpha\cos^2(\mathbf{d},\mathbf{b})-2\cos^2(\mathbf{m}_0,\mathbf{b})[1-\alpha\cos(\mathbf{d},\mathbf{b})]\}}{1-\alpha\cos^2(\mathbf{d},\mathbf{b})}. \tag{58}$$

Equation 58 provides the basis for determining the ability of pairwise symmetric arrays to resolve plane wave components. If $(z_d/z_{m_0})<1$, the response at the midpoint is greater than at either of the signal directions and the effects of the two sources have merged into a single peak in the scanned output. When

$$(z_d/z_{m_0})>1, \tag{59}$$

the response at the "midpoint" is less than at either of the "on-target" directions and we shall say that signals are resolved. Moreover, Eq. 58 enables us to determine the depth of "valley" between the peaks associated with the resolved signals.

In practice, it may be desirable to set a slightly higher threshold on the minimum (z_d/z_{m_0}) for which signals will be called resolved, so that the "valley" is easily seen. However, unlike Expression 59, the setting of such a higher threshold is rather arbitrary. For example, $(z_d/z_{m_0})=\pi^2/8\approx 1.23$ corresponds to the classical Rayleigh limit for a line array, but the value of (z_d/z_{m_0}) at the Rayleigh limit is different for different geometries.[20]

Recalling that $\cos^2(\cdot,\cdot)$ may be interpreted as the beam pattern of the array with a conventional beamformer, it is seen that Eq. 58 provides the exact resolution information of the optimum $\mathbf{k}_3$-processor explicitly in terms of conventional beam patterns and the output signal-to-noise ratio $(M\sigma_1^2/\sigma_0^2)$ of a perfectly matched conventional beamformer with only one source present.

In the absence of the pairwise symmetry and plane wave assumptions, Eq. 57 may still be used in Eq. 52. Then

$$z_d/z_{m_0}=\{1+(M\sigma_1^2/\sigma_0^2)[1-\alpha\cos^2(\mathbf{d},\mathbf{b})-2\cos^2(\mathbf{m}_0,\mathbf{b}) +(2\alpha/M^3)\operatorname{Re}(\mathbf{m}_0{}^*\mathbf{dd}^*\mathbf{bb}^*\mathbf{m}_0)]\} \times\{1-\alpha\cos^2(\mathbf{d},\mathbf{b})\}^{-1}$$

may be used in Expression 59 as the condition for resolution. Noting that

$$\operatorname{Re}(\mathbf{m}_0{}^*\mathbf{dd}^*\mathbf{bb}^*\mathbf{m}_0)\leq|\mathbf{m}_0{}^*\mathbf{d}|\,|\mathbf{d}^*\mathbf{b}|\,|\mathbf{b}^*\mathbf{m}_0| =|\mathbf{m}_0\mathbf{d}|^2|\mathbf{d}^*\mathbf{b}| \quad (60)$$

or

$$(2\alpha/M^3)\operatorname{Re}(\mathbf{m}_0{}^*\mathbf{dd}^*\mathbf{bb}^*\mathbf{m}_0) \leq 2\alpha\cos^2(\mathbf{m}_0,\mathbf{b})[\cos^2(\mathbf{d},\mathbf{b})]^{\frac{1}{2}}, \quad (61)$$

we obtain the following bound on z_d/z_{m_0}:

$$z_d/z_{m_0}\leq\frac{1+(M\sigma_1^2/\sigma_0^2)(1-\alpha\cos^2(\mathbf{d},\mathbf{b})-2\cos^2(\mathbf{m}_0,\mathbf{b})\{1-\alpha[\cos^2(\mathbf{d},\mathbf{b})]^{\frac{1}{2}}\})}{1-\alpha\cos^2(\mathbf{d},\mathbf{b})}, \quad (62)$$

which closely resembles Eq. 58. The condition that the right side of Eq. 62 be greater than unity is clearly necessary for resolution.

C. Conventional Beamformer

For purposes of comparison it is useful to consider the corresponding resolution question for a conventional beamformer. When $\mathbf{R}$ is given by Eq. 49, the output spectrum of the conventional beamformer is

$$z(\mathbf{k}_1)=(\sigma_0^2/M)+\sigma_1^2\cos^2(\mathbf{m},\mathbf{b})+\sigma_1^2\cos^2(\mathbf{m},\mathbf{d}). \quad (63)$$

The ratio of "on-target" to "midpoint" response is

$$\frac{z_d(\mathbf{k}_1)}{z_{m_0}(\mathbf{k}_1)}=\frac{\mathbf{d}^*\mathbf{R}\mathbf{d}}{\mathbf{m}_0{}^*\mathbf{R}\mathbf{m}_0}=\frac{1+\cos^2(\mathbf{d},\mathbf{b})+(M\sigma_1^2/\sigma_0^2)^{-1}}{2\cos^2(\mathbf{m}_0,\mathbf{b})+(M\sigma_1^2/\sigma_0^2)^{-1}}. \quad (64)$$

The signal-to-noise ratio parameter $(M\sigma_1^2/\sigma_0^2)$ cannot alter the result whether or not $z_d(\mathbf{k}_1)/z_{m_0}(\mathbf{k}_1)$ is larger than unity. For $(M\sigma_1^2/\sigma_0^2)>1$, the ratio $z_d(\mathbf{k}_1)/z_{m_0}(\mathbf{k}_1)$ is quite insensitive to changes in $(M\sigma_1^2/\sigma_0^2)$.

From Eq. 64 a necessary and sufficient condition for resolution by a conventional beamformer is

$$[1+\cos^2(\mathbf{d},\mathbf{b})]/[2\cos^2(\mathbf{m}_0,\mathbf{b})]>1. \quad (65)$$

Condition 65 will be satisfied whenever

$$\cos^2(\mathbf{m}_0,\mathbf{b})<\tfrac{1}{2}, \quad (66)$$

so that Expression 66 is a sufficient condition for resolution by a conventional beamformer. Condition 66 simply states that resolution is guaranteed if the spacing of the sources is such that the response at "midpoint" steering from each source is more than 3 dB below the response for "on-target" steering.

D. Resolving Power of a Line Array

In this section the results of the preceding analysis are applied to an array of sensors uniformly distributed on a straight line.

The geometry of the line array example is shown in Fig. 3. Two targets are located in the farfield of the line array at small angles of $\theta/2$ and $-\theta/2$ from broadside, respectively. The length L of the array is at least several wavelengths and the sensors are spaced at intervals of less than $\lambda/2$, one-half the wavelength.

Under these conditions, the following approximate expressions relate the mathematical quantities $\cos^2(\mathbf{m}_0,\mathbf{b})$ and $\cos(\mathbf{d},\mathbf{b})$ to the physical quantities L, θ, and λ:

$$\cos(\mathbf{d},\mathbf{b})=\frac{\sin(\pi L\theta/\lambda)}{(\pi L\theta/\lambda)} \quad (67)$$

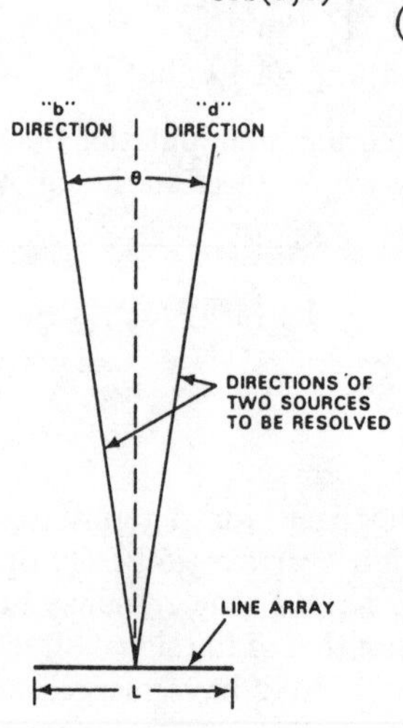

FIG. 3. Geometry of line array example.

and

$$\cos^2(\mathbf{m}_0,\mathbf{b}) = \frac{\sin^2[\pi L(\theta/2)/\lambda]}{[\pi L(\theta/2)/\lambda]^2} \tag{68}$$

For this geometry, the classical Rayleigh resolution criterion for conventional beamforming is

$$\theta \geq \lambda/L. \tag{69}$$

The equality in Expression 69 holds when the separation of the two sources is such that one source is located at the position of the first null of the beampattern of a conventional beamformer steered at the second source. That is, using $\theta = \lambda/L$ in Eq. 67 gives $\cos(\mathbf{d},\mathbf{b}) = 0$. Using $\theta = \lambda/L$ in Eq. 68 gives $\cos^2(\mathbf{m}_0,\mathbf{b}) = (2/\pi)^2 = 0.41$, which satisfies Expression 66. Using Eq. 64, we see that, when $M\sigma_1^2/\sigma_0^2 \gg 1$, the separation $\theta = \lambda/L$ yields a ratio of "on-target" to "midpoint" response of $z_d(\mathbf{k}_1)/z_{m_0}(\mathbf{k}_1) = (\pi^2/8) \approx 1.23$. Thus the Rayleigh limit is more than the bare minimum criterion of $z_d(\mathbf{k}_1)/z_{m_0}(\mathbf{k}_1) > 1$. The quantity $(1/\theta)$ at the resolution limit is known as the resolving power of the system.[20] These classical results for conventional beamformers serve as a benchmark for the resolving power of the optimum $\mathbf{k}_3$-processor.

The ratio of "on-target" to "midpoint" response obtained by substituting from Eqs. 67 and 68 into 58 is plotted in Fig. 4, for various values of the signal-to-noise ratio parameter $M\sigma_1^2/\sigma_0^2$. Also shown is the same ratio for a conventional beamformer obtained by substituting from Eqs. 67 and 68 into Eq. 64 for large $M\sigma_1^2/\sigma_0^2$.

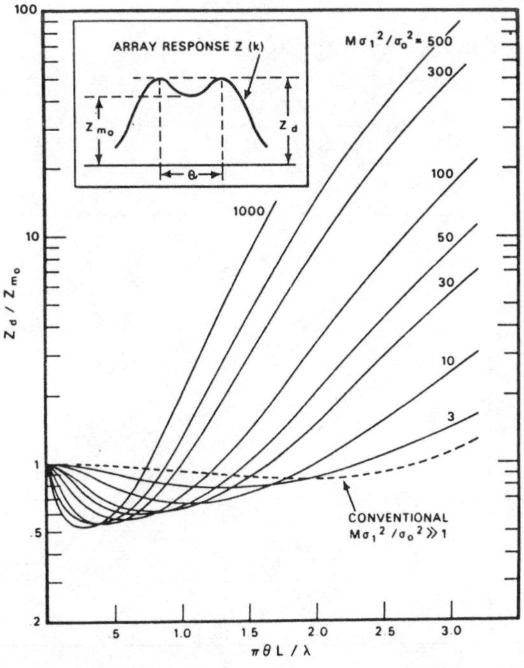

FIG. 4. Ratio of "on target" to "mid-point" response ratio of optimum ($\mathbf{k}_3$) beamformer on a line array for various values of the output signal-to-noise parameter $M\sigma_1^2/\sigma_0^2$ compared with a conventional ($\mathbf{k}_1$) beamformer.

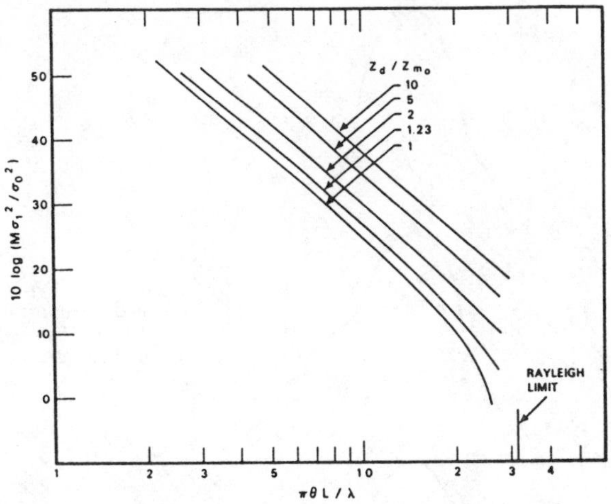

FIG. 5. Required value of the output signal-to-noise ratio parameter $M\sigma_1^2/\sigma_0^2$ versus normalized angular separation for resolution by a line array for various levels of the "on target" to "mid-point" response ratio.

The use of some value larger than unity for a minimum level of z_d/z_{m_0} for practical resolution is facilitated by the presentation of Fig. 4. A direct comparison with the classical Rayleigh limit may be made by using $z_d/z_{m_0} = 1.23$. For example, when $M\sigma_1^2/\sigma_0^2 = 500$ or 27 dB, the $\mathbf{k}_3$-processor achieves the ratio $z_d/z_{m_0} = 1.23$ at about $\theta = \lambda/(\pi L)$ compared with $\theta = \lambda/L$ for the conventional processor, and the $\mathbf{k}_3$-processor can be said to have slightly more than three times the resolving power of the conventional beamformer.

The effect of signal-to-noise ratio on resolution by the $\mathbf{k}_3$-processor may be seen more directly in Fig. 5, which presents the required value of $M\sigma_1^2/\sigma_0^2$ versus angular separation for various levels of z_d/z_{m_0}. From Fig. 5, it is apparent that increasing signal-to-noise ratio by about 13 dB improves the resolving power of the $\mathbf{k}_3$-processor by a factor of 2. For a fixed separation angle θ, Fig. 5 shows that the ratio z_d/z_{m_0} increases rapidly as the signal-to-noise ratio is increased. Thus the $\mathbf{k}_3$-processor will usually have clearly defined peaks in its response when the signal-to-noise ratio is slightly higher than that required for resolution.

E. Resolution of Signals of Unequal Strength

The case in which the two signals to be resolved are of unequal strength is somewhat more complicated than when they are of equal strength. Indeed, the very definition of resolution is subject to question. Three hypothetical scanned beamformer responses to two spatially separated sources plus noise are shown in Fig. 6. For the situation shown in Fig. 6(a) only a single peak is visible and it is obvious that the two sources are not resolved. For the situation shown in Fig. 6(c) there is a definite valley between two peaks of different

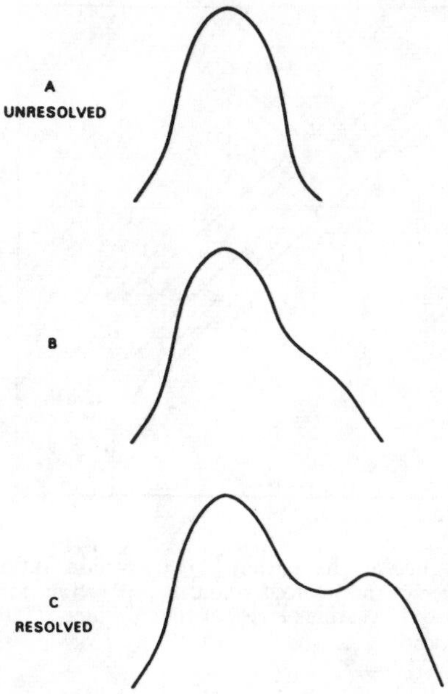

FIG. 6. Typical scanned beamformer responses to two unequal sources.

heights, and it is natural to say that the two sources are resolved. An intermediate situation is illustrated in Fig. 6(b), in which the presence of the weaker source alters the shape of the response due to the stronger source but no separate peak exists. The definition which we shall use for resolution will require a definite valley between the two peaks as in Fig. 6(c). Thus we shall say that in the situation shown in Fig. 6(b) the sources are not resolved.

Before proceeding, it is worthwhile to consider what the results already developed imply about the resolution of signals of unequal strength. Suppose that there are two sources of equal strength at some fixed separation. If the strength of the second source is increased, then from Eq. 32 the output $z(\mathbf{k}_3)$ will increase for all steering directions except those which are orthogonal in the appropriate space to the direction vector for that source. The amount of increase of the output depends on the sine squared of the generalized angle between the steering vector and the direction vector of the second source. The dependence is monotonic, so that the smaller the sine squared, the greater the increase in the output $z(\mathbf{k}_3)$. Thus, if the sine squared of that angle increases monotonically as the steering vector is scanned from the second source to the first source, then increasing the strength of the second source will cause an increase in the output for all steering directions between the two sources with the amount of increase being less for steering directions which are farther from the second source. In this situation, which is normal for closely spaced (within the Rayleigh limit) sources, a necessary condition for resolution is that the two sources were originally resolvable when the strength of the second source was equal to that of the weaker first source. Thus a necessary condition for resolution by a $\mathbf{k}_3$-beamformer of two closely unequal sources is that the two sources would be resolvable if both had the strength of the weaker source.

To be precise, let

$$\mathbf{R}=\sigma_0^2\mathbf{I}+\sigma_1^2\mathbf{dd}^*+\sigma_2^2\mathbf{bb}^*,\quad \sigma_2^2\geq\sigma_1^2, \tag{70}$$

and let

$$\mathbf{W}=\sigma_0^2\mathbf{I}+\sigma_1^2\mathbf{dd}^*. \tag{71}$$

If $\sin^2(\mathbf{m},\mathbf{b};\mathbf{W}^{-1})$ increases monotonically as $\mathbf{m}$ is scanned from $\mathbf{b}$ to $\mathbf{d}$, then a necessary condition for resolution is that the sources be resolvable when

$$\mathbf{R}=\sigma_0^2\mathbf{I}+\sigma_1^2\mathbf{dd}^*+\sigma_1^2\mathbf{bb}^*.$$

While the above condition is necessary for resolution, it is not sufficient, since increasing the strength of the second source could result in the situation depicted in Fig. 6(b). In order to develop a sufficient condition for resolution we shall proceed in a manner similar to that used for the case of signals of equal strength and consider the ratio z_d/z_m given by Eq. 50. Since $\mathbf{d}$ is the direction vector for the weaker source a value of z_d/z_m greater than unity at some point between $\mathbf{d}$ and $\mathbf{b}$ in the scanned response is necessary and sufficient for resolution. The following generalization of Eq. 52 for the situation in which $\mathbf{R}$ is given by Eq. 70 is derived in Appendix B:

$$\begin{aligned} z_d/z_m=\{1+(M\sigma_1^2/\sigma_0^2)[1-(\alpha_2/\alpha_1)\cos^2(\mathbf{m},\mathbf{b}) \\ -\cos^2(\mathbf{m},\mathbf{d})-\alpha_2\cos^2(\mathbf{d},\mathbf{b}) \\ +(2\alpha_2/M^3)\,\mathrm{Re}(\mathbf{m}^*\mathbf{dd}^*\mathbf{bb}^*\mathbf{m})]\} \\ \times\{1-\alpha_2\cos^2(\mathbf{d},\mathbf{b})\}^{-1}, \end{aligned} \tag{72}$$

where in analogy with Eq. 53

$$\alpha_i=\frac{(M\sigma_i^2/\sigma_0^2)}{(1+M\sigma_i^2/\sigma_0^2)},\quad i=1,2. \tag{73}$$

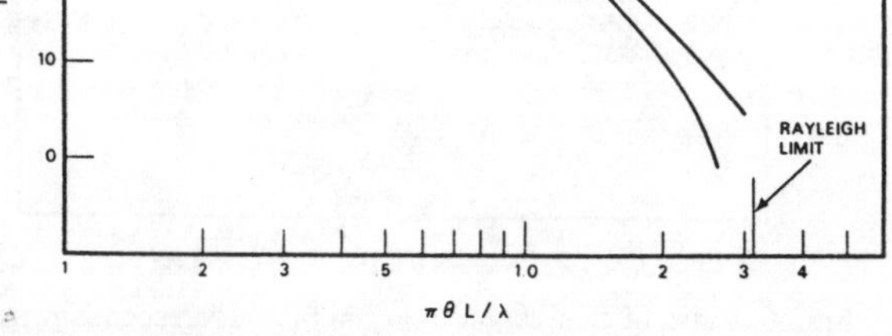

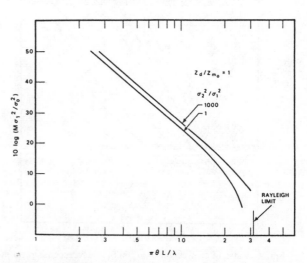

FIG. 7. Comparison of sufficient condition for resolution of unequal sources with a line array when $\sigma_2^2/\sigma_1^2=1000$ with the necessary condition.

For the case of plane waves and pairwise symmetric arrays, Eq. 72 reduces to the following generalization of Eq. 56:

$$z_d/z_m=\{1+(M\sigma_1^2/\sigma_0^2)[1-(\alpha_2/\alpha_1)\cos^2(\mathbf{m},\mathbf{b}) -\cos^2(\mathbf{m},\mathbf{d})-\alpha_2\cos^2(\mathbf{d},\mathbf{b}) +2\alpha_2\cos(\mathbf{m},\mathbf{d})\cos(\mathbf{d},\mathbf{b})\cos(\mathbf{m},\mathbf{b})]\} \times\{1-\alpha_2\cos^2(\mathbf{d},\mathbf{b})\}^{-1}. \quad (74)$$

Unlike the situation for signals of equal strength, we do not expect that when the two signals are resolved the largest value of z_d/z_m will occur at the midpoint $\mathbf{m}=\mathbf{m}_0$. Thus the condition

$$z_d/z_{m_0}>1 \quad (75)$$

used previously is not a necessary condition for resolution when the signals are of unequal strength, but it clearly is a sufficient condition for resolution. Substituting from Eq. 57 into Eq. 74 yields the following generalization of Eq. 58:

$$z_d/z_{m_0}=\frac{1+(M\sigma_1^2/\sigma_0^2)\{1-\alpha_2\cos^2(\mathbf{d},\mathbf{b})-\cos^2(\mathbf{m}_0,\mathbf{b})[1+(\alpha_2/\alpha_1)-2\alpha_2\cos(\mathbf{d},\mathbf{b})]\}}{1-\alpha_2\cos^2(\mathbf{d},\mathbf{b})}. \quad (76)$$

Substituting from Eq. 76 into Expression 75 yields the desired sufficient condition for resolution of unequal sources.

Having obtained a necessary condition and a sufficient condition for resolution, it is of interest to examine whether these conditions provide tight bounds on the signal-to-noise ratio required for resolution or these conditions leave a large middle ground where the resolution situation is in question. To examine this issue, we return to the line array geometry of the previous section but this time with $\mathbf{R}$ given by Eq. 70. Figure 7 presents the sufficient condition obtained from Eq. 76 when $\sigma_2^2/\sigma_1^2=1000$, so that the strong signal is three orders of magnitude larger than the weak signal.

Also presented for comparison is the necessary condition $\sigma_2^2/\sigma_1^2=1$ which was presented earlier in Fig. 5. Even in this extreme case in which one source is 1000 times stronger than the other, the two curves differ by only about 2 dB over a large range of $M\sigma_1^2/\sigma_0^2$. The necessary condition and the sufficient condition are seen to provide tight bounds on the signal-to-noise ratio required for resolution. Thus, we conclude that two closely spaced sources of unequal strength will be resolved by a $\mathbf{k}_3$-beamformer if the strength of the weaker source is slightly larger than that required for resolution of two sources of equal strength in a similar geometric configuration.

V. CONCLUSION

The inclusion of the signal in the matrix inversion process leads to signal suppression when the maximum possible output signal-to-noise ratio is greater than unity. The effect on output signal-to-noise ratio is particularly dramatic in that strong signals may produce lower output signal-to-noise ratios than weak signals. The effect on the total beamformer output is less dramatic since the decrease in signal response is partially offset by an increase in noise response. Thus, the importance of the signal suppression effect can depend on how the beamformer outputs are to be used. A coherent processor operating on the beamformer output could be seriously affected by the reduced signal-to-noise ratio, while the effects on a direct comparison of beam outputs would be less serious.

With an understanding of this phenomenon, techniques presumably can be developed to overcome the difficulties. Since the effect is not present in the $\mathbf{k}_2$-processor, there seems to be potential payoff in techniques which strive to obtain signal-free estimates of the noise cross-spectral matrix.

The same effect which leads to anomalous suppression of mismatched signals leads to the possibility of resolving closely spaced sources. High output signal-to-noise ratios are required in order to achieve resolving power significantly better than the Rayleigh limit. Optimum processing can lead to much better definition of peaks in the scanned output with much deeper valleys between adjacent peaks.

Closely spaced unequal sources can be resolved by a $\mathbf{k}_3$-beamformer if the strength of the weaker source is slightly larger than that required for resolution of two equistrength sources in the same geometric configuration.

The results of this paper are applicable in a variety of situations in which arrays of sensors are used to determine directional properties of propagating waves. These include seismology, radio astronomy, radar, sonar, and measurement of directional properties of ocean waves. In addition, since spectral analysis is essentially the same problem as beamforming with a line array, the results of this paper can easily be reinterpreted and applied to the problem of computing spectrums from correlation functions.

APPENDIX A: DERIVATIONS OF EQS. 21, 22, AND 23

Let

$$\mathbf{R}=\sigma_0^2\mathbf{Q}+\sigma_1^2\mathbf{d}\mathbf{d}^* \quad (A1)$$

so that

$$\mathbf{R}^{-1}=(1/\sigma_0^2)\{\mathbf{Q}^{-1}-\mathbf{Q}^{-1}\mathbf{d}\mathbf{d}^*\mathbf{Q}^{-1}(\sigma_1^2/\sigma_0^2) \times(1+\mathbf{d}^*\mathbf{Q}^{-1}\mathbf{d}\sigma_1^2/\sigma_0^2)^{-1}\}. \quad (A2)$$

Define

$$(S/N)_{max} = \mathbf{d}^*\mathbf{Q}^{-1}\mathbf{d}\sigma_1^2/\sigma_0^2. \tag{A3}$$

To derive Eq. 21, we proceed as follows: From Eqs. A2 and A3,

$$\mathbf{m}^*\mathbf{R}^{-1}\mathbf{m} = (1/\sigma_0^2)\{\mathbf{m}^*\mathbf{Q}^{-1}\mathbf{m} - |\mathbf{m}^*\mathbf{Q}^{-1}\mathbf{d}|^2(\sigma_1^2/\sigma_0^2) \times [1+(S/N)_{max}]^{-1}\}. \tag{A4}$$

Factoring and using Eq. A3 yields

$$\mathbf{m}^*\mathbf{R}^{-1}\mathbf{m} = \frac{\mathbf{m}^*\mathbf{Q}^{-1}\mathbf{m}}{\sigma_0^2} \times \left\{1 - \frac{|\mathbf{m}^*\mathbf{Q}^{-1}\mathbf{d}|^2(S/N)_{max}}{(\mathbf{m}^*\mathbf{Q}^{-1}\mathbf{m})(\mathbf{d}^*\mathbf{Q}^{-1}\mathbf{d})[1+(S/N)_{max}]}\right\}. \tag{A5}$$

Using the definition

$$\cos^2(\mathbf{m},\mathbf{d}; \mathbf{Q}^{-1}) = |\mathbf{m}^*\mathbf{Q}^{-1}\mathbf{d}|^2/(\mathbf{m}^*\mathbf{Q}^{-1}\mathbf{m})(\mathbf{d}^*\mathbf{Q}^{-1}\mathbf{d}), \tag{A6}$$

Eq. A5 becomes

$$\mathbf{m}^*\mathbf{R}^{-1}\mathbf{m} = \frac{\mathbf{m}^*\mathbf{Q}^{-1}\mathbf{m}}{\sigma_0^2} \times \left\{1 - \frac{(S/N)_{max}\cos^2(\mathbf{m},\mathbf{d}; \mathbf{Q}^{-1})}{1+(S/N)_{max}}\right\}. \tag{A7}$$

Using the relationship

$$\sin^2(\mathbf{m},\mathbf{d}; \mathbf{Q}^{-1}) = 1 - \cos^2(\mathbf{m},\mathbf{d}; \mathbf{Q}^{-1}), \tag{A8}$$

$$\mathbf{m}^*\mathbf{R}^{-1}\mathbf{m} = \frac{\mathbf{m}^*\mathbf{Q}^{-1}\mathbf{m}}{\sigma_0^2} \times \left\{\frac{1+(S/N)_{max}\sin^2(\mathbf{m},\mathbf{d}; \mathbf{Q}^{-1})}{1+(S/N)_{max}}\right\}. \tag{A9}$$

Equation A7 is the same as Eq. 21, which was to be proven.

To derive Eq. 22, we proceed as follows: From Eqs. A2 and A3,

$$|\mathbf{m}^*\mathbf{R}^{-1}\mathbf{d}|^2 = \frac{1}{\sigma_0^4}\left|\mathbf{m}^*\mathbf{Q}^{-1}\mathbf{d}\left(1 - \frac{(S/N)_{max}}{1+(S/N)_{max}}\right)\right|^2. \tag{A10}$$

Simplifying,

$$|\mathbf{m}^*\mathbf{R}^{-1}\mathbf{d}|^2 = (1/\sigma_0^4)|\mathbf{m}^*\mathbf{Q}^{-1}\mathbf{d}|^2\{1+(S/N)_{max}\}^{-2}. \tag{A11}$$

Using Eq. A6 yields the following result, which is identical with Eq. 22:

$$|\mathbf{m}^*\mathbf{R}^{-1}\mathbf{d}|^2 = \frac{(\mathbf{m}^*\mathbf{Q}^{-1}\mathbf{m})(\mathbf{d}^*\mathbf{Q}^{-1}\mathbf{d})\cos^2(\mathbf{m},\mathbf{d}; \mathbf{Q}^{-1})}{\sigma_0^4[1+(S/N)_{max}]^2}. \tag{A12}$$

To derive Eq. 23, we proceed as follows: From Eq. A1,

$$\mathbf{Q} = (1/\sigma_0^2)\mathbf{R} - (\sigma_1^2/\sigma_0^2)\mathbf{d}\mathbf{d}^*. \tag{A13}$$

Hence,

$$\mathbf{m}^*\mathbf{R}^{-1}\mathbf{Q}\mathbf{R}^{-1}\mathbf{m} = (1/\sigma_0^2)\mathbf{m}^*\mathbf{R}^{-1}\mathbf{m} - (\sigma_1^2/\sigma_0^2)|\mathbf{m}^*\mathbf{R}^{-1}\mathbf{d}|^2. \tag{A14}$$

Substituting from Eqs. A9 and A12 into Eq. A14 and using Eq. A3 yields

$$\mathbf{m}^*\mathbf{R}^{-1}\mathbf{Q}\mathbf{R}^{-1}\mathbf{m} = \frac{\mathbf{m}^*\mathbf{Q}^{-1}\mathbf{m}}{\sigma_0^4}\left\{\frac{1+(S/N)_{max}\sin^2(\mathbf{m},\mathbf{d}; \mathbf{Q}^{-1})}{1+(S/N)_{max}} - \frac{(S/N)_{max}\cos^2(\mathbf{m},\mathbf{d}; \mathbf{Q}^{-1})}{\{1+(S/N)_{max}\}^2}\right\}. \tag{A15}$$

Multiplying and dividing the first term in braces in Eq. A15 by $\{1+(S/N)_{max}\}$, and using Eq. A8 yields the following result, which is identical to Eq. 23:

$$\mathbf{m}^*\mathbf{R}^{-1}\mathbf{Q}\mathbf{R}^{-1}\mathbf{m} = \frac{\mathbf{m}^*\mathbf{Q}^{-1}\mathbf{m}}{\sigma_0^4} \times \left\{\frac{1+[2(S/N)_{max}+(S/N)_{max}^2]\sin^2(\mathbf{m},\mathbf{d}; \mathbf{Q}^{-1})}{\{1+(S/N)_{max}\}^2}\right\}. \tag{A16}$$

APPENDIX B: DERIVATION OF EQS. 52 AND 72

Let

$$\mathbf{R} = \sigma_0^2\mathbf{I} + \sigma_1^2\mathbf{d}\mathbf{d}^* + \sigma_2^2\mathbf{b}\mathbf{b}^* \tag{B1}$$

and

$$z_d/z_m = (\mathbf{m}^*\mathbf{R}^{-1}\mathbf{m})/(\mathbf{d}^*\mathbf{R}^{-1}\mathbf{d}). \tag{B2}$$

Define

$$\mathbf{V} = \sigma_0^2\mathbf{I} + \sigma_2^2\mathbf{b}\mathbf{b}^*. \tag{B3}$$

Then, letting $\mathbf{V}$ play the role of $\sigma_0^2\mathbf{Q}$ in Eq. A9, Eq. B2 becomes

$$z_d/z_m = \{(\mathbf{m}^*\mathbf{V}^{-1}\mathbf{m})/(\mathbf{d}^*\mathbf{V}^{-1}\mathbf{d})\} \times \{1+\sigma_1^2\mathbf{d}^*\mathbf{V}^{-1}\mathbf{d}\sin^2(\mathbf{m},\mathbf{d}; \mathbf{V}^{-1})\}. \tag{B4}$$

By definition,

$$\sin^2(\mathbf{m},\mathbf{d}; \mathbf{V}^{-1}) = \frac{(\mathbf{m}^*\mathbf{V}^{-1}\mathbf{m})(\mathbf{d}^*\mathbf{V}^{-1}\mathbf{d}) - |\mathbf{m}^*\mathbf{V}^{-1}\mathbf{d}|^2}{(\mathbf{m}^*\mathbf{V}^{-1}\mathbf{m})(\mathbf{d}^*\mathbf{V}^{-1}\mathbf{d})}, \tag{B5}$$

so that Eq. B4 may be rewritten as follows:

$$z_d/z_m = \{\mathbf{m}^*\mathbf{V}^{-1}\mathbf{m} + \sigma_1^2[(\mathbf{m}^*\mathbf{V}^{-1}\mathbf{m})(\mathbf{d}^*\mathbf{V}^{-1}\mathbf{d}) - |\mathbf{m}^*\mathbf{V}^{-1}\mathbf{d}|^2]\}/(\mathbf{d}^*\mathbf{V}^{-1}\mathbf{d}). \tag{B6}$$

Let

$$\alpha_2 = \frac{(M\sigma_2^2/\sigma_0^2)}{(1+M\sigma_2^2/\sigma_0^2)} \tag{B7}$$

then

$$\mathbf{V}^{-1} = (1/\sigma_0^2)\{\mathbf{I} - \mathbf{b}\mathbf{b}^*\alpha_2/M\}. \tag{B8}$$

Each of the quantities in Eq. B6 will now be examined individually. Recall that **d**, **b**, and **m** are normalized, so that

$$\mathbf{m}^*\mathbf{m} = \mathbf{b}^*\mathbf{b} = \mathbf{d}^*\mathbf{d} = M. \tag{B9}$$

Using Eq. B8,

$$\mathbf{m}^*\mathbf{V}^{-1}\mathbf{m} = (1/\sigma_0^2)[M - |\mathbf{m}^*\mathbf{b}|^2\alpha_2/M] \tag{B10}$$

or

$$\mathbf{m}^*\mathbf{V}^{-1}\mathbf{m} = (M/\sigma_0^2)[1-\alpha_2 \cos^2(\mathbf{m},\mathbf{b})]. \tag{B11}$$

Similarly,

$$\mathbf{d}^*\mathbf{V}^{-1}\mathbf{d} = (M/\sigma_0^2)[1-\alpha_2 \cos^2(\mathbf{d},\mathbf{b})]. \tag{B12}$$

Again using Eq. B8,

$$|\mathbf{m}^*\mathbf{V}^{-1}\mathbf{d}|^2 = (1/\sigma_0^4)|\mathbf{m}^*\mathbf{d} - (\alpha_2/M)\mathbf{m}^*\mathbf{b}\mathbf{b}^*\mathbf{d}|^2$$

or

$$|\mathbf{m}^*\mathbf{V}^{-1}\mathbf{d}|^2 = (1/\sigma_0^4)\{|\mathbf{m}^*\mathbf{d}|^2 + (\alpha_2/M)^2|\mathbf{m}^*\mathbf{b}|^2|\mathbf{b}^*\mathbf{d}|^2 - (2\alpha_2/M)\,\mathrm{Re}[\mathbf{m}^*\mathbf{b}\mathbf{b}^*\mathbf{d}\mathbf{d}^*\mathbf{m}]\}, \tag{B13}$$

where Re denotes the real part. Equivalently,

$$|\mathbf{m}^*\mathbf{V}^{-1}\mathbf{d}|^2 = (M/\sigma_0^2)^2\{\cos^2(\mathbf{m},\mathbf{d}) + \alpha_2^2 \cos^2(\mathbf{m},\mathbf{b}) \cos^2(\mathbf{b},\mathbf{d}) - (2\alpha_2/M^3)\,\mathrm{Re}[\mathbf{m}^*\mathbf{b}\mathbf{b}^*\mathbf{d}\mathbf{d}^*\mathbf{m}]\}. \tag{B14}$$

Substituting from Eqs. B11, B12, and B14 into Eq. B6 and simplifying using Eq. B7 results in the following equation which is identical with Eq. 72:

$$z_d/z_m = \{1 + M\sigma_1^2/\sigma_0^2[1-(\alpha_2/\alpha_1)\cos^2(\mathbf{m},\mathbf{b}) - \cos^2(\mathbf{m},\mathbf{d}) - \alpha_2\cos^2(\mathbf{d},\mathbf{b}) + (2\alpha_2/M^3)\,\mathrm{Re}(\mathbf{m}^*\mathbf{b}\mathbf{b}^*\mathbf{d}\mathbf{d}^*\mathbf{m})]\} \times \{1-\alpha_2\cos^2(\mathbf{d},\mathbf{b})\}^{-1}, \tag{B15}$$

where

$$\alpha_1 = \frac{(M\sigma_1^2/\sigma_0^2)}{(1+M\sigma_1^2/\sigma_0^2)}. \tag{B16}$$

When $\sigma_1^2 = \sigma_2^2$, $\alpha_1 = \alpha_2 = \alpha$ and Eq. B15 reduces to Eq. 52.

*This work was done while the author was a visiting Research Associate under the sponsorship of the U. S. Navy Professional Development Program and was supported in part by the Office of Naval Research.

[1]F. Bryn, "Optimum Signal Processing of Three-Dimensional Arrays Operating on Gaussian Signals and Noise," J. Acoust. Soc. Am. **34,** 289–297 (1962).

[2]W. Vanderkulk, "Optimum Processing of Acoustic Arrays," J. Br. Inst. Radio Eng. **26,** 285–292 (1963).

[3]J. P. Burg, "Three-Dimensional Filtering with an Array of Seismometers," Geophysics **29,** 693–713 (1964).

[4]H. Mermoz, "Filtrage Adapté et Utilization Optimale d'une Antenne," Proc. NATO Advanced Study Institute on Signal Processing with Emphasis on Underwater Acoustics, Grenoble (1964).

[5]D. Middleton and H. Groginski, "Detection of Random Acoustic Signals by Receiver with Distributed Elements," J. Acoust. Soc. Am. **38,** 727 (1965).

[6]P. E. Green, Jr., E. J. Kelly, Jr., and M. J. Levin, "A Comparison of Seismic Array Processing Methods," Geophys. J. R. Astron. Soc. **11,** 67–84 (1966).

[7]Y. T. Lo, S. W. Lee, and Q. H. Lee, "Optimization of Directivity and Signal-to-Noise Ratio of an Arbitrary Antenna Array," Proc. IEEE **54,** 1033–1045 (1966).

[8]D. J. Edelblute, J. M. Fisk, and G. L. Kinnison, "Criteria for Optimum-Signal-Detection Theory for Arrays," J. Acoust. Soc. Am. **41,** 199 (1967).

[9]J. Capon, R. J. Greenfield, and R. J. Kolker, "Multidimensional Maximum-Likelihood Processign of Large Aperture Seismic Array," Proc. IEEE **55** (2) (1967).

[10]H. Cox, "Interrelated Problems in Estimation and Detection I & II," Proc. NATO Advanced Study Institute on Signal Processing with Emphasis on Underwater Acoustics, Enschede, the Netherlands (Aug. 1968).

[11]A. H. Nuttall and D. W. Hyde, "A Unified Approach to Optimum and Suboptimum Processing for Arrays," U. S. Navy Underwater Sound Lab. Rep. No. 992 (Apr. 1969).

[12]J. B. Lewis and P. M. Schultheiss, "Optimum and Conventional Detection Using a Linear Array,"J. Acoust. Soc. Am. **49,** 1083–1091 (1971).

[13]J. Capon, "High-resolution Frequency-Wavenumber Spectrum Analysis," Proc. IEEE **57,** 1408–1418 (1969).

[14]C. D. Seligson, "Comments on High Resolution Frequency-Wave Number Spectrum Analysis," Proc. IEEE **58** (6), 947–949 (1970).

[15]R. N. McDonough, "Degraded Performance of Nonlinear Array Processors in the Presence of Data Modeling Errors," J. Acoust. Soc. Am. **51,** 1186–1193 (1972).

[16]H. Cox, "Sensitivity Considerations in Adaptive Beamforming," Proc. NATO Advanced Study Institute on Signal Processing, Loughborough, U.K. (Aug. 1972).

[17]J. Capon and N. R. Goodman, "Probability Distributions of Estimators of Frequency-Wavenumber Spectrum," Proc. IEEE **58** (10), 1785–1786 (1970); correction, Proc. IEEE **59** (1), 112 (1971).

[18]E. Parzen, "Extraction and Detection Problems in Reproducing Kernel Hilbert Spaces," J. SIAM Control **1** (1), 35–62 (1962).

[19]T. Kailath, "RKHS Approach to Detection and Estimation Problems—Part I: Deterministic Signals in Gaussian Noise," IEEE Trans. Inf. Theory **IT-17** (5), 530–549 (1971).

[20]M. Born and E. Wolf, *Principles of Optics* (Pergamon, London, 1970), 4th ed.

A Comparison of Efficient Beamforming Algorithms

RONALD A. MUCCI, MEMBER, IEEE

Abstract—**Time domain and frequency domain concepts which aid in the design and efficient implementation of a digital beamformer have been described at various times in the literature. The numerous beamformer structures that result are discussed with an emphasis on hardware requirements and spectral areas of application. Time domain procedures which include delay-sum, partial-sum, interpolation and shifted-sideband beamforming, and frequency domain techniques which include the application of discrete Fourier transforms and phase shift beamforming are considered. Hardware considerations are primarily in the areas of analog-to-digital conversion, data storage, and computational throughput requirements.**

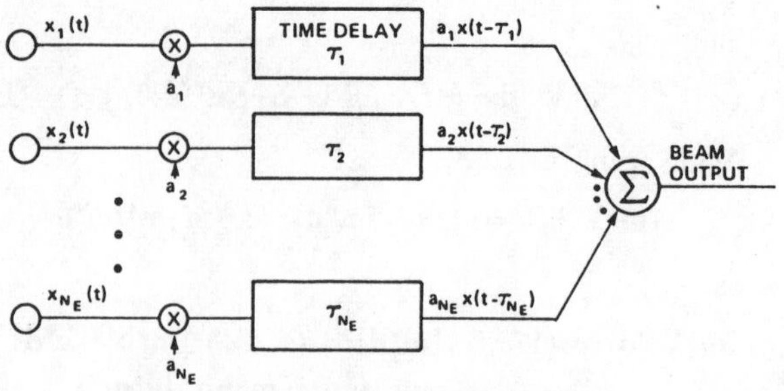

Fig. 1. Time domain analog beamformer.

I. Introduction

In a conventional beamformer the measurement of a propagating, coherent wavefront, relative to ambient background noise and spatially localized interferences is enhanced by coherent processing of data collected with an array of sensors, i.e., by time delaying and summing weighted sensor data as shown in Fig. 1. Sophisticated beamforming applications involving large numbers of sensors frequently are implemented digitally, due to the availability of relatively low cost, small size, and high speed digital circuitry, such as analog to digital (A/D) converters, random access memory, and microprocessors.

Time domain and frequency domain techniques which aid in the design and implementation of a digital beamformer and in the minimization of hardware requirements are discussed in this paper. Emphasis is placed on nonadaptive, conventional procedures such as delay-sum, partial-sum, interpolation with first order sampling, interpolation with complex sampling, shifted sideband, and phase shift beamforming. Adaptive procedures for noise cancellation [1], [2] and procedures based on principal component [3] and maximum entropy [4]-[6] concepts are not discussed.

The techniques presented have played an important role in reducing the cost of beamforming in large sonar systems for underwater applications [7], [8]. However, due to the versatility of a digital, microprocessor based configuration, these concepts are well suited to seismic applications, aeroacoustic applications, and ultrasonic imaging in medical applications [9], [10]. Due to recent advances in very large scale integration (VLSI) and very high speed integrated circuit (VHSIC) technologies, these procedures are also being considered for radar applications.

In the next section a concise description, summarizing each technique, is presented in a manner that addresses important issues relating to digital beamformer implementation and spectral areas of application. A summary and conclusions are presented in Section III. An overview is presented in Table I, summarizing the beamforming procedures examined and the associated hardware considerations and spectral areas of application.

Manuscript received May 5, 1982; revised December 22, 1982 and October 27, 1983.

The author is with Bolt Beranek and Newman, Inc., Cambridge, MA 02238.

II. Discussion of Beamforming Techniques

In a conventional time domain beamformer, the beam output $b(t)$ is given by

$$b(t) = \sum_{n=1}^{N_E} a_n x_n(t - \tau_n) \tag{1}$$

where N_E denotes the number of sensors, the x_n denote the sensor outputs, the a_n denote the constant gains applied to the sensor outputs, and the τ_n denote the time delays required to point or steer the beam to the specified direction.

The τ_n steering delays compensate for differences in propagation times from the signal source to the individual array elements or sensors. For distant sources, frequently the propagating signal is modeled as a plane wave and the τ_n are proportional to the projection of the sensor position vector r_n, relative to a reference point, onto a unit vector K_θ pointing in the direction of wave propagation θ, i.e.,

$$n = r_n \cdot K_\theta / C \tag{2}$$

where C denotes the speed of wave propagation in the medium and $\cdot$ denotes dot or inner product of two vectors. Generally, the reference point is selected so that the minimum value of τ_n is zero.

The spatial shading or weighting coefficients, the a_n, are generally applied to the individual sensors to control the spatial response or beam pattern achieved with the given array. Numerous procedures for synthesis of the weighting coefficients have been presented in the literature [11], [12] which provide the designer with control over mainlobe width and sidelobe level. However, a reduction in the level of the sidelobes of the

Reprinted from *IEEE Trans. Acoust., Speech, Signal Processing,* vol. ASSP-32, pp. 548–558, June 1984.

beam pattern generally produces an increase in the width of the mainlobe.

In a discrete-time (i.e., digital) implementation of the beamforming process, the sensor data from N_E sensors are sampled in time, weighted, delayed, and summed to form the discrete-time beam output $b(kT_0)$ given by

$$b(kT_0) = \sum_{n=1}^{N_E} a_n x_n (kT_0 - M_n T_i) \tag{3}$$

where F_i $(= 1/T_i)$ and F_0 $(= 1/T_0)$ denote the input and output sampling rates, respectively, the a_n and x_n are as previously defined, and M_n is an integer such that

$$\tau_n - T_i/2 < M_n T_i \leqslant \tau_n + T_i/2, \tag{4}$$

i.e., the exact steering delays, the τ_n, are approximated by an integer number of increments of the input sampling interval T_i. This process is depicted in Fig. 2.

In order to permit reconstruction of the desired waveform, the beam output $b(kT_0)$ must be computed at the Nyquist rate [13] or greater; for low-pass signals, this minimum rate required to avoid objectionable frequency aliasing is twice the highest frequency of the signal spectrum. However, an input sampling rate F_i, significantly greater than that required for waveform reconstruction, is needed to achieve acceptable approximations to the exact steering delays; frequently, the input sampling rate is five to ten times that required for waveform reconstruction.

Delay-Sum Beamformer

In a *delay-sum* beamformer implementation, the digitized sensor data, sampled at the rate of F_i samples/s, are buffered or stored prior to beamforming. When the last data sample required for $b(kT_0)$ arrives, the beam output can be computed. The obvious disadvantages of this direct approach to implementation are as follows.

1) The sensor data must be sampled at a rate much greater than the Nyquist rate to approximate the time delays required for beam steering.

2) Large amounts of memory or storage are generally needed to achieve long delays associated with large arrays and high input sampling rates.

3) The cable bandwidth required for transmission of high data rates may be difficult to achieve if a large physical distance separates the A/D converters and the beamformer.

The costs associated with achieving the required A/D conversion throughput of $N_E F_i$ samples/s and the required storage of $N_E \max_\theta (\tau_n) F_i$ words, where $\max_\theta (\tau_n)$ denotes the maximum delay required for beam steering, generally represent an appreciable percentage of the overall cost of implementing a delay-sum beamformer. This is especially true if ten or more bits of accuracy are required for input data representation. Since the requirements are proportional to F_i, a reduction in the input sampling rate can produce significant hardware savings. It is important to note, however, that the computational operations necessary to perform the shading and summing generally can be performed at the output sampling rate F_0, where F_0 is consistent with the Nyquist rate,

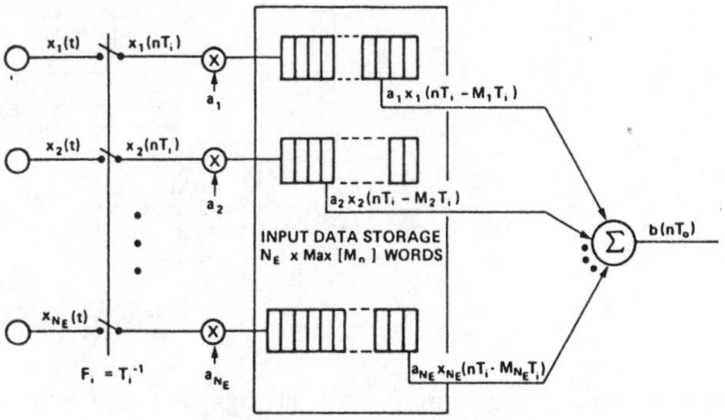

Fig. 2. Discrete-time, delay-sum beamformer.

i.e., only the value of $b(kT_0)$ required at the time intervals kT_0 are computed.

Partial-Sum Beamformer

The significant amount of memory needed in a conventional delay-sum beamformer to store the sensor data for the purpose of beam steering is reduced considerably in a *partial-sum* or *sum-delay* beamformer which utilizes the sensor data immediately after sampling, as shown in Fig. 3.

The partial-sum technique can be explained as follows. Let $b(kT_0)$ denote the beam output at time kT_0 and let $M_{n1} T_i$ denote the maximum delay, i.e., $M_{n1} T_i \geqslant M_n T_i$ for $n = 1, \cdots, N_E$. At time $kT_0 - M_{n1} T_i$, the partial sum $P(kT_0)$ is initialized by shading and summing all sensors which are to be delayed $M_{n1} T_i$ s. If $M_{n2} T_i$ is the maximum of the remaining delays, at time $kT_0 - M_{n2} T_i$ the partial sum $P(kT_0)$ is incremented by the shaded sum of all sensors which are to be delayed $M_{n2} T_i$ s. At time kT_0 the continuation of this process will produce a partial sum $P(kT_0)$ which is equal to $b(kT_0)$, i.e., all sensor data will have been correctly delayed, shaded, and summed in the partial sum storage element.

In order to produce a beam output at times kT_0, it is necessary to initialize a partial sum every T_0 s. Conceptually, this is similar to a pipelining process. However, since the initialization and accumulation process is repeated in exactly the same way for every output sample of a given beam, the control of the data flow in the pipeline is relatively simple.

The number of partial sums S_j which must be maintained for a given beam pointing in direction θ_j is given by

$$S_j = \max_{\theta_j} (\tau_n)/T_0 \tag{5}$$

where $\max_{\theta_j} (\tau_n)$ denotes the maximum delay corresponding to the jth beam. Hence, the total number of partial sums S_T required to form N_B beams is given by

$$S_T = \sum_{j=1}^{N_B} S_j \leqslant N_B \max_\theta (\tau_n)/T_0 \tag{6}$$

where $\max_\theta (\tau_n)$ denotes the maximum delay required for all beams. The amount of memory generally required to store the sensor data for a conventional delay-sum beamformer is equal to $N_E \max_\theta (\tau_n)/T_i$. Hence, if $N_E > N_B$ and T_0 is significantly

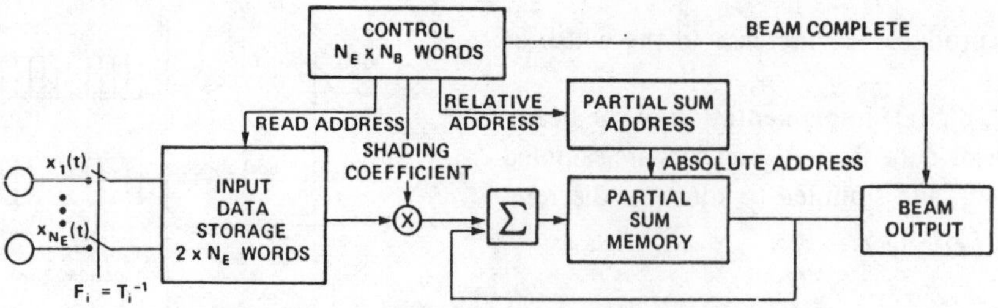

Fig. 3. Discrete-time, partial-sum beamformer.

greater than T_i, an appreciable savings in memory can be achieved with the partial-sum technique.

The microprogrammable digital beamformer developed by Raytheon Company [14], incorporates the partial-sum concept and is configured as shown in Fig. 3. An incremental counter is used in conjunction with programmable microcode to address the partial-sum memory in an ingenious manner. It is worth noting that as a result of its programmability, this off-the-shelf beamformer module can be used in numerous applications involving various array geometries and it can be easily modified to accommodate on-line changes in the number of beams, number of sensors, shading, beam steering, etc.

The partial-sum concept also has been used in conjunction with charge coupled devices (CCD's) for beamforming purposes as shown in Fig. 4. The CCD's are operated as a series of storage elements which can be clocked in a delay line manner. A charge which is proportional to the sensor voltage is injected into an element of this delay line by properly illuminating the photosensitive exposed surface of the storage element. The injected charge combines (adds) with an existing charge which is the accumulation of previously injected charges that are being clocked or transferred along the delay line. By the time the charge (partial sum) reaches the end of the delay line, a charge accumulates that is proportional to the beam output. Spatial shading can be accomplished by controlling the amount of charge injected with a translucent optical mask that covers the photosensitive surfaces of the delay line elements.

Interpolation Beamforming–Low-Pass Applications

Although the partial sum concept reduces the input data storage requirements, generally a high input sampling rate is still required for beam steering. However, if temporal interpolation of the sensor data is performed in conjunction with the beamforming operation, as shown in Fig. 5, the sensor data need only be sampled at the Nyquist rate. Using this approach, commonly referred to as *interpolation beamforming* [15], an appreciable savings in A/D converter circuitry and connecting cable bandwidth for data transmission can be realized.

Spatial interpolation, for the purpose of achieving the desired beam steering resolution, also can be performed by properly combining the outputs of a set of preformed beams. Due to the linearity of both spatial and temporal interpolation processes a close examination will reveal a host of similarities. However, the present discussion focuses on the concept of temporal interpolation of sensor data.

The temporal interpolation process, simply stated, pro-

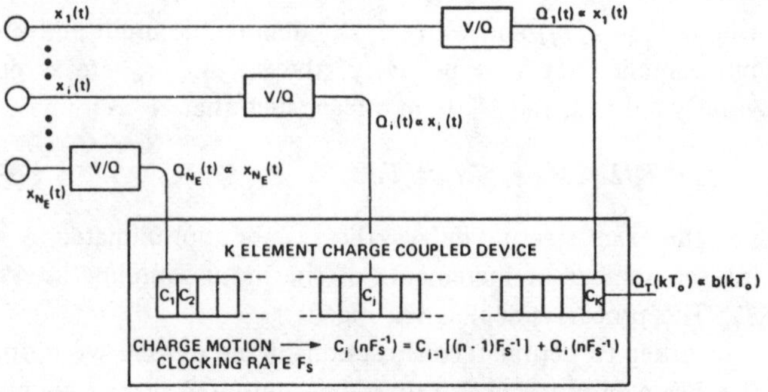

Fig. 4. Charge coupled device beamformer implementation: $Q_i(T)$ denotes change proportional to signal amplitude $x_i(t)$.

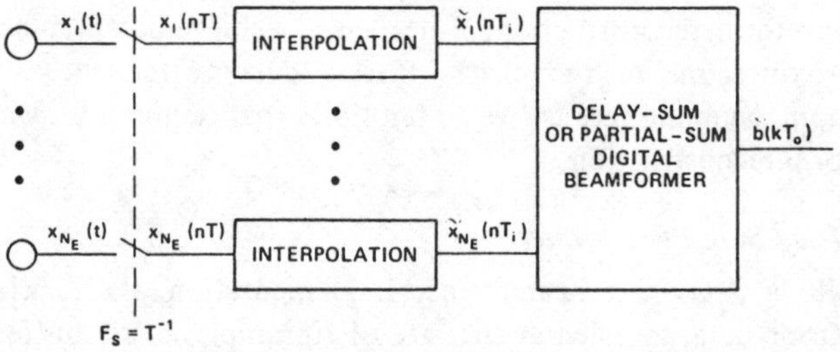

Fig. 5. Interpolation beamforming for low-pass applications.

duces estimates of the sensor data, at a higher sampling rate F_i, from which the time delays can be realized with sufficient accuracy for beam steering. Although the interpolation can be accomplished in various ways, it can be performed with computational efficiency by zero padding the original sequence, $x_n(kT_0)$, and passing the resulting sequence $v_n(kT_i)$, through a finite impulse response (FIR) digital low-pass filter as shown in Fig. 6 [16], [17]. That is, first K-1 zeros are inserted between samples of the original sequence obtained at the Nyquist rate in order to realize an effective sampling rate which is K times greater. Next, the low-pass filtering is performed with an FIR digital filter which is characterized by a nonrecursive input–output relationship such that

$$y_k = \sum_{n=1}^{N_c} h_n x_{k-n} \tag{7}$$

where x_k and y_k denote the kth samples of the filter input and output sequences, respectively, and the h_n denote the N_c filter coefficients.

Conceptually, the FIR filter is implemented as shown in Fig. 7. The desired filter bandpass and stopband character-

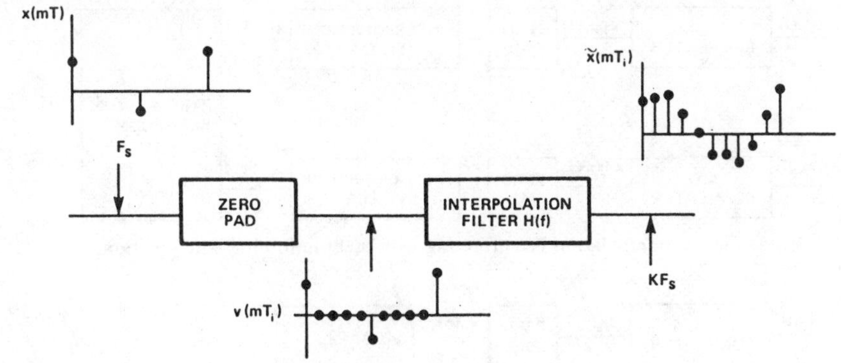

Fig. 6. Two-step interpolation process.

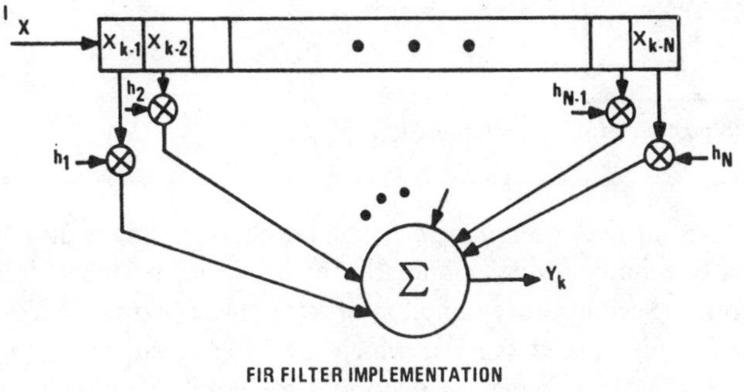

$$Y_k = \sum_{n=1}^{N} h_n X_{k-n}$$

Fig. 7. Finite impulse response, nonrecursive digital filter.

istics can be obtained through the proper design or synthesis of the filter coefficient set.

This two-step process incorporating a nonrecursive filter structure offers the following advantages.

1) An interpolation filter with a linear phase response can be achieved with a symmetric set of coefficients; generally approximately $5K$ (i.e, five times the interpolation factor K) coefficients are sufficient to achieve the filter response required.

2) A computationally efficient filter implementation can be realized by not performing the calculations involving the padded zeros.

3) Various algorithms have been presented in the literature for synthesizing FIR filter coefficients for interpolation [18]-[21].

The optimum placement of the interpolation filter also is an important consideration. Since both the interpolation and beamforming operations are linear, the placement of the two processes can be interchanged as shown in Figs. 8 and 9.

The appropriate placement of the interpolation filters depends upon the beamforming scenario [15]. For example, if N_B beams are formed simultaneously, each requiring the data of N_E sensors, then approximately $2N_E N_c F_0$ arithmetic operations (additions and multiplications) per second are required for prebeamforming interpolation assuming computations involving padded zeros are not performed. Approximately $2N_B N_c F_0$ calculations per second are required for postbeamforming interpolation assuming the upsampled input to the interpolation filter consists primarily of nonzeros. Hence, if N_E and N_B differ significantly, one approach will be computationally more efficient than the other since the same N_c filter coefficients are used for prebeamforming or postbeamforming interpolation.

Whether the partial-sum or delay-sum concept is used is also a consideration in the placement of the interpolation filters. For example, if postbeamforming interpolation is performed in conjunction with delay-sum beamforming, the input data storage requirements are proportional to the lower input sampling rate consistent with the Nyquist frequency. However, if postbeamforming interpolation is used in conjunction with partial-sum beamforming, then the partial-sum storage requirements are proportional to the higher sampling rate required for beam steering, since the reduction in the sampling rate occurs at the output of the interpolation filter.

Filter dynamic range and arithmetic quantization errors are also a consideration in filter placement, since each of the postbeamforming interpolation filters must accommodate peak amplitudes corresponding to the coherent addition of the weighted signals of the N_E sensors whereas the prebeamforming interpolation filters need only accommodate peaks corresponding to data of a single channel.

Although additional circuitry is required to perform the interpolation, the cost can be minimized by proper design and configuration, and generally is offset by significant savings in A/D converter circuitry, data storage, and data transmission bandwidth.

Interpolation Beamforming–Bandpass Applications

The techniques discussed previously are concerned primarily with the implementation of beamformers for low-pass frequency applications. Additional savings in A/D converters, connecting cable bandwidths, data storage, and computational throughput can be realized in bandpass applications [22]. Such savings result from the sampling of the bandpass sensor data at a rate consistent with the signal bandwidth rather than at a rate consistent with the highest frequency; these savings can be appreciable if the bandwidth is significantly less than the center frequency of the band.

In order to avoid objectionable frequency aliasing, complex sampling procedures must be used to sample the bandpass sensor data. Three sampling procedures that are frequently used, analytic signal sampling, second-order sampling, and

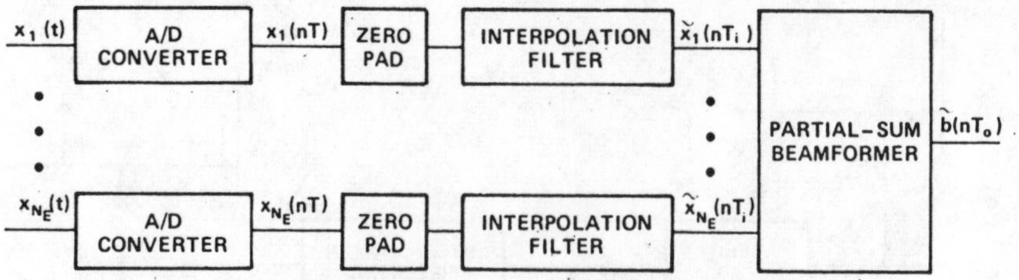

Fig. 8. Interpolation beamforming with prebeamforming interpolation.

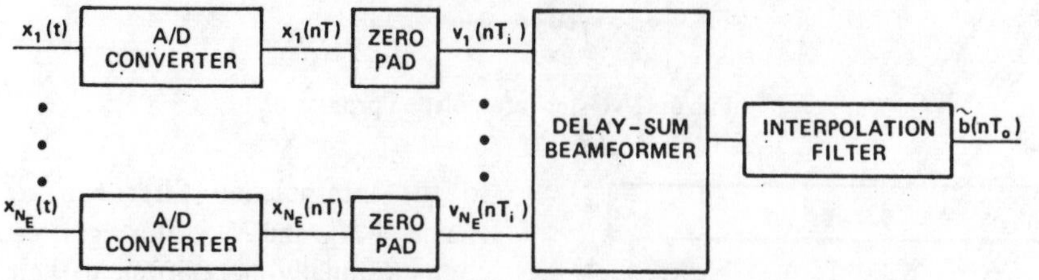

Fig. 9. Interpolation beamforming with postbeamforming interpolation.

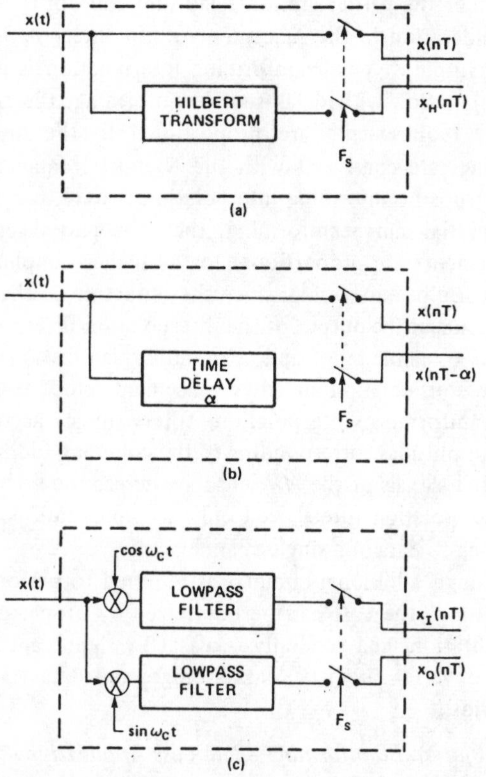

Fig. 10. Complex sampling procedures for bandpass applications.

quadrature sampling [23], are shown in Fig. 10. Each is briefly reviewed here. For more detail see [22]-[26].

Analytic Signal Sampling: The analytic signal is formed by removing the negative frequency components of the signal. This can be done with the aid of the Hilbert transform. The Fourier transform of the sampled analytic signal consists of replicas of the positive frequency spectra of the original bandpass signal repeated every F_s Hz where F_s denotes the sampling rate in complex samples/s. If F_s is greater than the width of the signal passband, the positive spectra do not overlap and the signal content is preserved.

Second-Order Sampling: Second-order sampling is an alternate technique which eliminates the need for the Hilbert transform. Second-order sampling of $x(t)$ yields two sequences of uniformly spaced samples which are interleaved, i.e., $x(mT)$ and $x(mT + \alpha)$, where α denotes a temporal offset and T denotes the sampling interval. The original waveform can be reconstructed from the two staggered sequences if $\alpha \neq T/2$, and the sampling rate is greater than the signal bandwidth. However, an ideal bandpass filter is required with relatively complex frequency transfer characteristics for the interpolation process [22], [23].

Quadrature Sampling [25], [26]: This procedure characterizes a bandpass signal $x(t)$ with uniformly spaced samples of its quadrature components, i.e., $x(t)$ can be expressed as

$$x(t) = x_I(t) \cos 2\pi f_0 t - x_Q(t) \sin 2\pi f_0 t \tag{8}$$

where $x_I(t)$ and $x_Q(t)$ denote the in-phase and quadrature components of $x(t)$ and f_0 denotes the center of the passband. If $x(t)$ occupies a frequency band of width W Hz, then $x_I(t)$ and $x_Q(t)$ occupy the frequency band from $-W/2$ to $W/2$ and each can be sampled at a rate of W samples/s without aliasing. Thus, it is evident that the signal can be recovered from the sequences $x_I(mT)$ and $x_Q(mT)$ where $T \leq W^{-1}$. However, exact reconstruction of $x(t)$ requires ideal low-pass filters.

Generally, an interpolation process is used in conjunction with the sampling procedure to obtain estimates of the data at the times required for beam steering since the sensor data are sampled at a relatively low rate. Descriptions of the frequency characteristics of the bandpass interpolation filters, which are functions of the sampling procedure, are contained in [22], [23].

As discussed for low-pass applications, the additional circuitry required for interpolation can be minimized by performing the filtering either prior to or subsequent to beamforming as shown in Figs. 11 and 12. Also, either the partial-sum or delay-sum concept can be employed.

It is important to note that the beamformer output is com-

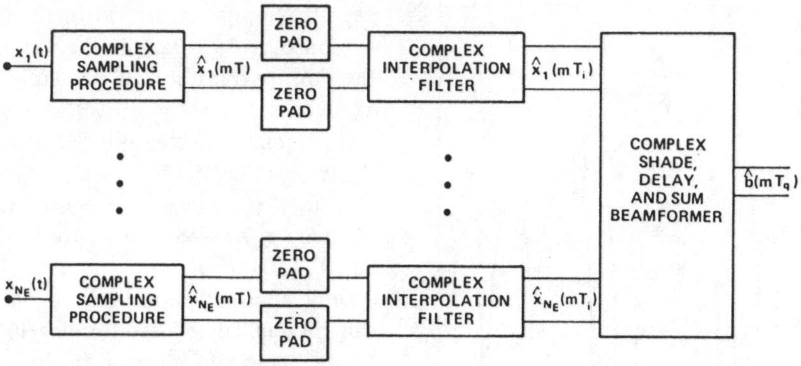

Fig. 11. Interpolation beamforming with complex sampling and prebeamforming interpolation.

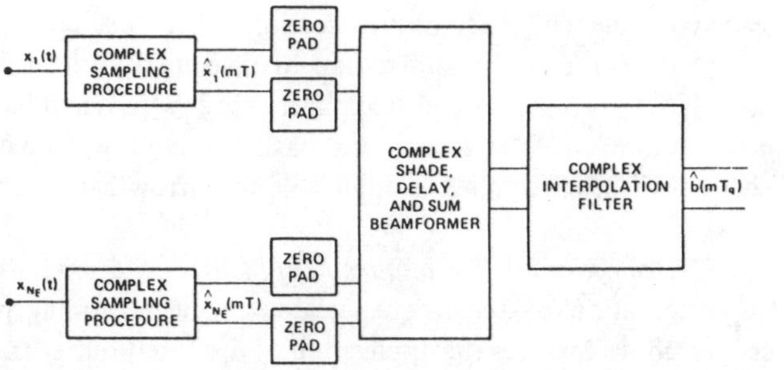

Fig. 12. Interpolation beamforming with complex sampling and postbeamforming interpolation.

plex, and hence only needs to be computed at a rate consistent with the signal bandwidth rather than with the highest frequency of the original bandpass signal. Hence, a significant reduction in the computational throughput required to shade and sum the sensor data can also be realized when the signal bandwidth is much less than the center frequency of the band.

Shifted Sideband Beamformer

Another technique, which is intended primarily for bandpass applications and is commonly referred to as *shifted sideband beamforming,* forms beams using frequency translated, single sideband, complex sensor data [27]. Since the sensor data is translated down in frequency, the signals are less rapidly varying functions of time than the original signals, and hence they can be aligned more coarsely in time in the beam steering operation. This results in a lowering of the input sampling rate required for beam steering and an associated hardware savings in the areas of A/D conversion, data transmission bandwidth, and input data storage.

The sampled beam output, $\hat{b}(mT_q)$, is merely the frequency translation of the conventional beam output, i.e.,

$$\hat{b}(mT_q) = e^{-i2\pi f_q t} b(t)|_{t=mT_q}$$

$$= \sum_{n=1}^{N_E} a_n \hat{x}_n (nT_q - M_n T_i) e^{-i2\pi f_q \tau_n} \tag{9}$$

where the $\hat{x}_n$ denote the sensor data translated down in frequency by an amount f_q, and T_i and T_q denote the input and output sampling rates, respectively. The phase shift by $2\pi f_q \tau_n$ compensates in the phase modulation process associated with frequency translation for the differences in signal arrival times at each sensor. For beam steering purposes, the complex data, i.e., the $\hat{x}_n$, frequently are sampled at a rate of ten or more times the highest signal frequency after translation. However, if the band has been translated down in frequency by a significant amount, then this sampling rate is proportionately less than the sampling rate needed in a conventional digital approach. Therefore, the maximum reduction in sampling rate and associated hardware savings occur when the sensor data are basebanded and sampled, for example, at five times the signal bandwidth. However, in certain applications it might be more important to operate at an intermediate frequency rather than at baseband to avoid contamination or distortion of the data by the signal conditioning electronics that occur near dc.

It is important to note that the accuracy of the beam-dependent phase shifts, the $2\pi f_q \tau_n$, affects performance. However, frequently a total of 16 bits is sufficient for representation of the real and imaginary parts of exp $\{-i2\pi f_q \tau_n\}$, and since the phase shifts are achieved independently of the time delays, this accuracy is not affected by the input sampling rate.

There is a grating lobe phenomenon associated with the shifted sideband beamformer which occurs for certain array geometries, e.g., planar and line arrays of uniformly spaced sensors. These grating lobes occur when, due to the coarse time delay quantization, small groups of two or more adjacent sensors receive the same steering delay. The net effect is an equivalent sparse array of subgroups of two or more adjacent sensors which spatially undersamples the wave. Fig. 13 is a beam pattern demonstrating this effect for a line array of twenty-two equispaced sensors. The sensor data have been translated from 11 kHz to 1 kHz and sampled at a rate of $15K$ samples/s. As a result of the quantization associated with a sampling rate of $15K$ samples/s for a beam pointing direction 45 degrees from broadside, pairs of sensors receive the same time delays. This produces the grating lobes shown in Fig. 13 corresponding to the beam pattern or spatial response for a frequency of 11 kHz. The grating lobes roll off as a result of the spatial directivity achieved by combining the outputs of the individual sensors in forming the subgroups. For a more detailed discussion of this and a comparison with phase shift beamforming see [28].

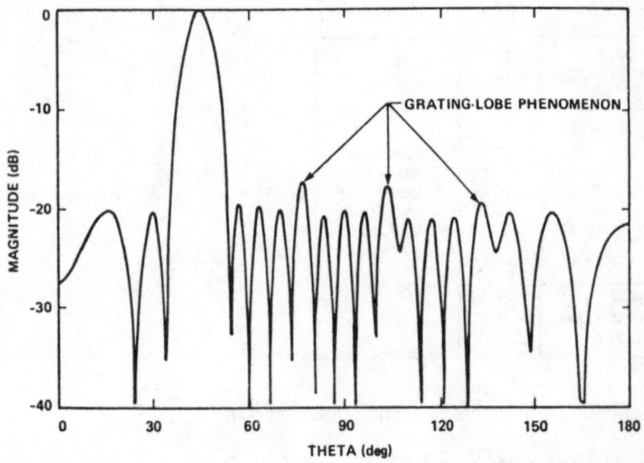

Fig. 13. Beam pattern for shifted sideband beamforming with a line array consisting of 22 equispaced omnidirectional sensors: The sensor data are sampled at 15 kHz. The beam is steered 45 from broadside, the center of the band is 10 kHz, and the single frequency pattern is shown for a frequency of 11 kHz.

The shifted sideband beamformer can be implemented digitally as shown in Fig. 14. Both frequency translation of the sensor data and beam dependent phase correction for each input channel are required. However, the cost of this additional complexity can be offset by hardware savings associated with a lower input sampling rate. There is also a savings in the computational throughput required for beamforming since the resulting beam output is translated down in frequency and can be computed at a relatively low rate. These savings can be appreciable when the signal bandwidth is significantly less than the center of the original band.

Although a shifted sideband beamformer can be implemented without interpolation, a further reduction in the input sampling rate can be realized if interpolation is used in conjunction with the shifted sideband concept. With interpolation, the complex sensor data only need to be sampled at a rate of approximately twice the highest signal frequency after translation, or at a rate consistent with the bandwidth if complex sampling procedures are used. Estimates of the data at the times required for beam steering are obtained with the interpolation process. As with the other beamformer implementations, the interpolation can be performed prior to or subsequent to beamforming, and in conjunction with the partial-sum or delay-sum concept in order to minimize hardware cost.

Frequency Domain Beamforming

Frequency domain beamforming concepts are the result of the application of Fourier transform techniques to the beamforming process. Due to the linearity of the beamforming process and certain properties of the Fourier transform [29], the Fourier transform of the beam output can be obtained from the Fourier transform of the sensor outputs as follows:

$$B(f,\theta_m) = F\{b(t,\theta_m)\}$$

$$= F\left\{\sum_{n=1}^{N_E} a_n x_n[t-\tau_n(\theta_m)]\right\}$$

$$= \sum_{n=1}^{N_E} a_n X_n(f)\, e^{-i2\pi f \tau_n(\theta_m)}, \qquad (10)$$

i.e., the Fourier transform of the beam output steered to direction θ_m, $B(f, \theta_m)$, is the weighted linear combination of the Fourier transform of the received waveforms. In (10), the a_n denote the shading coefficients, the $\tau_n(\theta_m)$ denote the delays required to steer the beam in the direction θ_m, the phase shifts, the $2\pi F\tau_n(\theta_m)$, are the frequency domain counterpart to the time delays required for beam steering, and the X_n denote the Fourier transforms of the received waveforms, i.e., $X_n(f) = F\{x_n(t)\}$.

Only approximations to the Fourier transform operations can be applied to beamforming in practice since calculation of the transform requires knowledge of the waveform for all time. However, two beamforming procedures which are commonly employed and based on the above concept will be discussed in the remainder of this section. The first approach results from the direct application of discrete Fourier transform (DFT) techniques and is applicable to low-pass and band-pass conditions. The second approach, commonly known as phase shift beamforming, is applicable to narrow-band signals only.

DFT Beamforming Techniques [29], [30]: A discrete-time implementation of the frequency domain beamforming procedure above involves the application of discrete Fourier transform (DFT) techniques. If the K-point DFT of the sample sensor data is given by

$$X_n(j) = \sum_{k=1}^{K} x_n(kT_i)\, e^{-i2\pi jk/K} \qquad (11)$$

where $X_n(j)$ is an estimate of $X_n(2\pi jF_i/K)$ and F_i denotes the input sampling rate, then the frequency domain representation of the beam output, $B(j, \theta_m)$, is given by

$$\hat{B}(j,\theta_m) = \sum_{n=1}^{N_E} a_n X_n(j)\, e^{-i2\pi j \tau_n(\theta_m) F_i/K}. \qquad (12)$$

This procedure can be implemented as shown in Fig. 15.

Note that generally the inverse Fourier transform is not required. In fact, often the frequency domain beam output representation is highly conducive to postbeamforming processing objectives, such as signal detection, localization, and classification [7].

Clearly, one advantage of this approach is that the input sampling frequency F_i does not have an impact on beam steering resolution, i.e., F_i must only satisfy the Nyquist criterion. Beam steering is accomplished by the proper selection of $\tau_n(\theta_m)$ in the phase shifting operation subsequent to computing the DFT of the sampled sensor data. Therefore, the sensor data can be sampled at a low rate which is comparable to the sampling rate required for interpolation beamforming. The other obvious advantage is that the DFT can be computed efficiently using a fast Fourier transform (FFT) algorithm [31], [32].

One of the major disadvantages of this approach is associated with the physical constraint on K, the size of the DFT. Since no actual time delays are inserted in this beamforming procedure, and the data transformed are of finite duration K/F_i, the DFT of each sensor output will act on a different temporal region of the signal portion of the received data as a result of the propagation delay across the array. Hence, the transform size must be consistent with the array size and signal

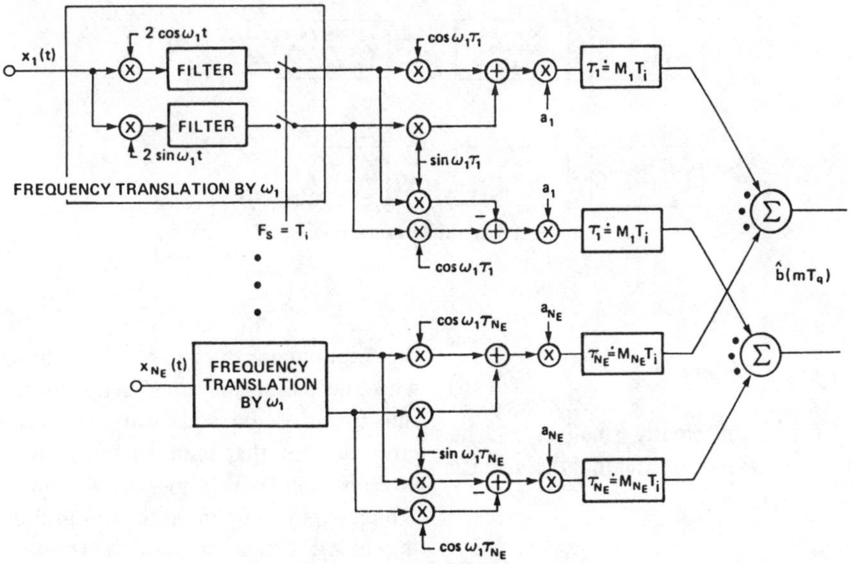

Fig. 14. Shifted sideband beamformer.

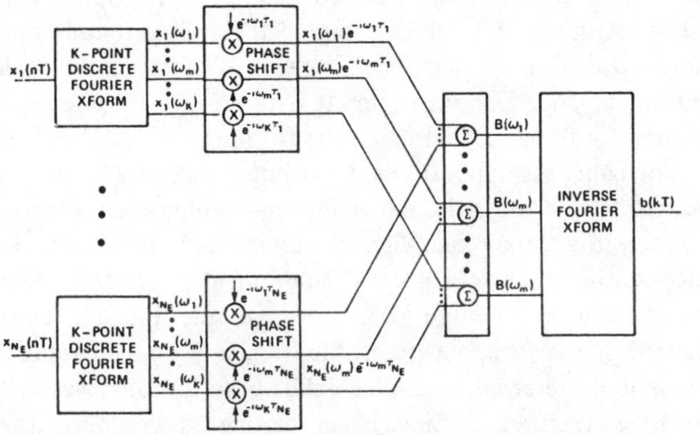

Fig. 15. Discrete-time, single beam, full band frequency domain beamformer.

bandwidth, e.g., the maximum propagation delay across the array should be less than ten percent of the duration of data transformed, K/F_i, where F_i is a function of signal bandwidth. Since the DFT of each sensor is required, this can have a significant impact on input storage requirements even though the input may be sampled at a rate consistent with the Nyquist criterion. Note that as a result of this temporal misalignment of sensor data, $B(j, \theta_m)$ is only an approximation of the DFT of the sampled sequence of $b(t, \theta_m)$ except in the limit as K becomes very large.

There is also a spatial aspect of the frequency domain approach that can be exploited. For certain array geometries (e.g., see [32]) the steering delays have a linear dependence on n, i.e.,

$$\tau_n(\theta_m) = ng(\theta_m) \tag{13}$$

and the expression for the beam output $B(j, \theta_m)$ is in the form of a discrete Fourier transform [see (11)], i.e.,

$$B(j, \theta_m) = \sum_{n=1}^{N_E} a_n X_n(j)\, e^{-i2\pi jng(\theta_m) F_i/K}$$

$$= \sum_{n=1}^{N_E} a_n X_n(j)\, e^{-i2\pi jnm/N_E} \tag{14}$$

where

$$g(\theta_m) F_i N_E/K = m \quad \text{for } m = 1, \cdots, N_E. \tag{15}$$

For a line array of uniformly spaced sensors $g(\theta_m)$ is proportional to the cosine of the beam pointing direction θ_m, measured from the array axis passing through the sensors, i.e.,

$$g(\theta_m) = \frac{d}{C} \cos\theta_m \tag{16}$$

and

$$\theta_m = \cos^{-1}\left(\frac{mKC}{F_i N_E d}\right) \quad \text{for } m = 1, \cdots, N_E \tag{17}$$

where d denotes the intersensor spacing.

The major advantage of the spatial aspect of frequency domain beamforming is that a computationally efficient FFT algorithm can be used to compute the multidimensional DFT over space and time. However, for the computational saving to be significant, the number of sensors, i.e., N_E, must be large, e.g., 100 or more, and highly composite (factorable). Also, the application produces beams with a fixed set of nonuniformly spaced pointing directions which satisfy (17). Hence, if not all N_E beams are required, or the resulting θ_m require spatial interpolation to achieve the desired set of beams, there may be little or no computational savings.

Phase Shift Beamformer [33], [34]: The phase shift beamformer is the second important technique based on frequency domain concepts to be considered. Since this is a frequency domain approach, the steering delays are realized with phase shifts and consequently the beam steering resolution is not a function of the input sampling rate for a digital implementation. However, this approach is intended primarily for narrowband applications.

The phase shift beamformer results from the approximation of each linear phase shift $2\pi f\tau_n(\theta_m)$ by a constant phase shift $2\pi f_0 \tau_n(\theta_m)$ which is frequently matched to the center frequency of the narrow band of interest, i.e.,

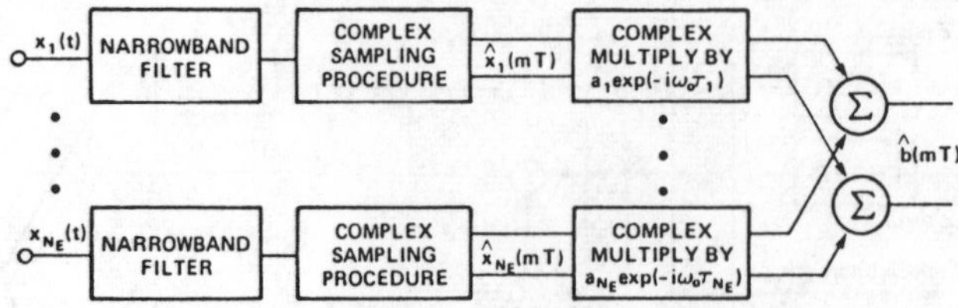

Fig. 16. Phase shift beamformer.

$$B(f, \theta_m) = \sum_{n=1}^{N_E} a_n X_n(f) e^{-i2\pi f_0 \tau_n(\theta_m)} \tag{18}$$

for $f_0 - \Delta f \leqslant f \leqslant f_0 + \Delta f$ and sufficiently small Δf. The time domain equivalent is obtained by taking the inverse Fourier transform of $B(f, \theta_m)$, i.e.,

$$b(t, \theta_m) = F^{-1}\{B(f, \theta_m)\}$$

$$= \sum_{n=1}^{N_E} a_n x_n(t) e^{-i2\pi f_0 \tau_n(\theta_m)}. \tag{19}$$

Hence, the phase shift beamformer can be implemented digitally by merely sampling the narrow-band sensor data at a rate consistent with the bandwidth of interest and summing the sampled data, shaded by the beam dependent weighting $a_n e^{-i2\pi f_0 \tau_n(\theta_m)}$, as shown in Fig. 16.

Note that the phase shift beamforming concept is related to the DFT approach, described previously, in the following way. Each bin of the DFT of the sensor data, which represents a narrow frequency band, is phase shifted by a constant phase $2\pi f_j \tau_n(\theta_m)$ which depends on the beam pointing direction and is matched to the center frequency f_j of the jth bin. If a beam is formed, using only a single frequency bin from each sensor, then the phase shift beamformer structure results. The phase shift beamformer is also equivalent to the shifted sideband beamformer when the propagation delay across the entire array is less than one half of the sampling interval [28], i.e.,

$$\max\,[\tau_n(\theta_m)] < T_i/2. \tag{20}$$

Although the phase shift beamformer requires a relatively small amount of analog or digital circuitry to implement, good results are achieved for only very narrow frequency bands since degradation in the beam pattern occurs at frequencies other than band center. For example, errors in the beam pointing direction are shown in [28] to occur for frequencies other than band center such as line arrays and planar arrays. For other array geometries such as conformal arrays, there is a loss in array gain due to the phase errors at frequencies other than band center.

III. Summary and Conclusions

Various important concepts, which aid in the efficient implementation of discrete-time beamformers, were presented and the associated hardware considerations, primarily in the areas of A/D conversion, input data storage, and beamformer computational throughput, were discussed. The delay-sum technique was shown to require a large amount of data storage and a high input sampling rate for most applications. Although the input data storage requirement is reduced significantly with the partial-sum approach, this approach requires partial-sum memory and addressing, and also a high input sampling rate equal to that required for delay-sum beamforming. The interpolation beamformer was shown to eliminate the need for a high input sampling rate, and in most applications, the savings in A/D converters, data transmission, and input data storage generally offset the additional complexity required for interpolation. The application of frequency domain concepts was shown to eliminate the need for a high input sampling rate also. Although the DFT's required for a discrete-time implementation can be performed efficiently with an FFT algorithm, a large amount of input data storage is generally required for frequency domain beamforming.

For bandpass applications, both interpolation with complex sampling and shifted sideband beamforming were shown to reduce the input sampling requirements significantly and, hence, the input data storage requirements, data transmission bandwidth, and computational throughput. Finally, for very narrow bandpass applications, the frequency domain approach, commonly referred to as phase shift beamforming, was shown to be very efficient. Since beam steering is accomplished with constant phase shifts, the input sampling rate and input data storage requirements are minimized. However, for other than very narrow bandpass applications, performance of the phase shift beamformer degrades rapidly and it becomes advisable to consider interpolation with complex sampling or shifted sideband techniques. These results are summarized in Table I.

It is worth noting that frequently an optimum method of implementation is not easily determined. Generally, system specifications such as the number of sensors, the number of beams formed simultaneously, beam steering resolution, the frequency band(s) of interest, postbeamforming processing, dynamic range, fault tolerance, etc., contribute to decisions affecting the implementation. For example, in certain bandpass applications the bandwidth is not significantly less than the center frequency, and interpolation beamforming with first-order sampling may be the most cost effective approach. In certain low-pass applications, the savings in A/D circuitry and associated hardware may not offset the additional complexity required for interpolation, and partial sum beamforming may be the most cost effective approach. The intent of this paper was to provide an overview and understanding of important concepts which in themselves should not affect the overall system objectives or performance, but rather, will aid the system designer in the design of a cost effective beamformer architecture and subsequent implementation.

TABLE I
BEAMFORMING TECHNIQUES AND ASSOCIATED HARDWARE CONSIDERATIONS AND SPECTRAL AREA OF APPLICATION: H—HIGH; M—MODERATE; L—LOW

	Spectral Characterization			Hardware Considerations		
Beamforming Technique	Lowpass	Bandpass	Narrowband	Analog to Digital Converter[1]	Data Storage[2]	Additional Computational Complexity[3]
Delay-sum	*	*		H	H	—
Partial-sum	*	*		H	L-M	L
Interpolation	*			L	L-M	L-M
Interpolation with Complex Sampling		*		L	L	M
Shifted-sideband		*	*	L-M	L-M	L-M
Discrete Fourier Transform	*	*		L	H	M
Phase-shift			*	L	L	L

[1] A low sampling rate is consistent with the Nyquist sampling criterion. A high sampling rate indicates five times the Nyquist criterion for low-pass applications and many more times this rate for bandpass applications.
[2] Low data storage requirements are consistent with the number of sensors and a sampling rate consistent with the Nyquist sampling criterion. High data storage requirements in excess of five times the low rate are common for delay-sum and full band Fourier transform techniques.
[3] Amount of additional circuitry required relative to delay-sum approach.

ACKNOWLEDGMENT

The author wishes to acknowledge the important recommendations provided by the reviewers and incorporated into this paper.

REFERENCES

[1] B. Widrow, J. R. Glover, Jr., J. M. McCool, J. Kaunitz, C. S. Williams, R. H. Hearn, J. R. Zeidler, E. Dong, Jr., and R. C. Goodlin, "Adaptive noise cancelling: Principles and applications," *Proc. IEEE*, vol. 63, pp. 1692-1716, Dec. 1975.
[2] W. R. Gabriel, "Adaptive arrays–An introduction," *Proc. IEEE*, vol. 64, pp. 329-272, Feb. 1976.
[3] R. R. Kneipfer, E. S. Eby, and H. S. Newman, "An eigenvector interpretation of an array's bearing-response pattern," Tech. Rep. 1098, Navy Underwater Sound Laboratory, May 1970.
[4] D. P. Skinner, S. M. Hedicka, and A. D. Matthews, "Maximum entropy array processing," *J. Acoust. Soc. Amer.*, vol. 66, no. 2, pp. 488-493, Aug. 1979.
[5] S. Holm and J. M. Hovem, "Estimation of scalar ocean wave spectra by the maximum entropy method," *IEEE J. Ocean. Eng.*, vol. OE-4, pp. 76-83, July 1979.
[6] T. E. Barnard, "Two maximum entropy beamforming algorithms for equally spaced line arrays," *IEEE Trans. Acoust., Speech, Signal Processing*, vol. ASSP-30, pp. 175-189, Apr. 1982.
[7] W. C. Knight, R. G. Pridham, and S. M. Kay, "Digital signal processing for sonar," *Proc. IEEE*, vol. 69, pp. 1451-1507, Nov. 1981.
[8] T. E. Curtis and R. J. Ward, "Digital beam forming for sonar systems," *Proc. IEE*, vol. 127, pp. 257-265, Aug. 1980.
[9] D. White and E. A. Lyons, *Ultrasound in Medicine.* New York: Plenum, 1978.
[10] A. Macovski, "Ultrasonic imaging using acoustic arrays," *Proc. IEEE*, vol. 67, pp. 484-495, Apr. 1979.
[11] S. A. Schelkunoff, "A mathematical theory of linear arrays," *Bell Syst. Tech. J.*, vol. XXII, no. 1, pp. 80-107, 1943.
[12] F. J. Harris, "On the use of windows for harmonic analysis with the discrete Fourier transform," *Proc. IEEE*, vol. 66, pp. 51-83, Jan. 1978.
[13] C. E. Shannon, "Communication in the presence of noise," *Proc. IRE*, vol. 37, pp. 10-21, 1942.
[14] W. Martin, "The microprogrammable beamformer," Tech. Rep. Raytheon Co., 1974.
[15] R. G. Pridham and R. A. Mucci, "A novel approach to digital beamforming," *J. Acoust. Soc. Amer.*, vol. 63, pp. 425-434, Feb. 1978.
[16] R. W. Schafer and L. R. Rabiner, "A digital signal processing approach to interpolation," *Proc. IEEE*, vol. 61, pp. 692-720, June 1973.
[17] R. E. Crochiere and L. R. Rabiner, "Optimum FIR digital filter implementation for decimation, interpolation and narrowband filtering," *IEEE Trans. Acoust., Speech, Signal Processing*, vol. ASSP-23, pp. 444-456, Oct. 1975.
[18] R. A. Gabel and R. R. Kurth, "Synthesis of efficient digital beamformers," *J. Acoust. Soc. Amer.*, vol. 70, no. S1, p. 517, 1981.
[19] H. S. Hersey, D. W. Tufts, and J. T. Lewis, "Interactive minimax design of linear-phase nonrecursive digital filters subject to upper and lower function constraints," *IEEE Trans. Audio Electroacoust.*, vol. AU-20, pp. 171-173, June 1972.
[20] R. A. Gabel, "On asymmetric FIR interpolators with minimum LP error," presented at IEEE Int. Conf. Acoust., Speech, Signal Processing, Denver, CO, 1980.
[21] J. H. McClellan, T. W. Parks, and L. R. Rabiner, "A computer program for designing optimum FIR linear phase digital filters," *IEEE Trans. Audio Electroacoust.*, vol. AU-21, pp. 506-516, Dec. 1973.
[22] R. G. Pridham and R. A. Mucci, "Digital interpolation beamforming for lowpass and bandpass signals," *Proc. IEEE*, vol. 67, pp. 904-919, June 1979.
[23] D. A. Linden, "A discussion of sampling theorems," *Proc. IRE*, vol. 47, pp. 1219-1226, July 1959.
[24] C. L. Byrne, "Reconstruction from partial information with applications to tomography," *SIAM J. Appl. Math*, vol. 42, pp. 933-940, Aug. 1982.
[25] O. D. Grace and S. P. Pitt, "Sampling and interpolation of bandlimited signals by quadrature methods," *J. Acoust. Soc. Amer.*, vol. 48, no. 6, pp. 1311-1318, 1970.
[26] J. L. Brown, Jr., "On quadrature sampling of bandpass signals," *IEEE Trans. Aerospace Electron. Syst.*, vol. AES-15, pp. 366-371, May 1979.
[27] R. G. Pridham and R. A. Mucci, "Shifted sideband beamforming," *IEEE Trans. Acoust., Speech, Signal Processing*, vol. ASSP-27, pp. 713-722, Dec. 1979.

[28] R. A. Mucci and R. G. Pridham, "Impact of beam steering errors on shifted sideband and phase shift beamforming techniques," *J. Acoust. Soc. Amer.*, vol. 69, pp. 1360-1368, May 1981.

[29] P. Rudnick, "Digital beamforming in the frequency domain," *J. Acoust. Soc. Amer.*, vol. 46, pp. 1089-1090, 1969.

[30] G. L. DeMuth, "Frequency domain beamforming techniques," in *Proc. IEEE Int. Conf. Acoust., Speech, Signal Processing*, 1977, pp. 713-715.

[31] J. W. Cooley, P. A. W. Lewis, and P. D. Welch, "Applications of the fast Fourier transform to computation of Fourier integrals, Fourier series, and convolution integrals," *IEEE Trans. Audio Electroacoust.*, vol. AU-15, pp. 79-84, June 1967.

[32] J. R. Williams, "Fast beamforming algorithm," *J. Acoust. Soc. Amer.*, vol. 44, pp. 1454-1455, 1968.

[33] W. J. Hughes and W. Thompson, Jr., "Tilted directional response patterns formed by amplitude weighting and a single 90 degree phase shift," *J. Acoust. Soc. Amer.*, vol. 59, pp. 1040-1045, May 1976.

[34] S. P. Pitt, W. T. Adams, and J. K. Vaughan, "Design and implementation of a digital phase shift beamformer," *J. Acoust. Soc. Amer.*, vol. 64, pp. 808-814, Sept. 1978.

Time Delay Estimation for Passive Sonar Signal Processing

G. CLIFFORD CARTER, SENIOR MEMBER, IEEE

Abstract–An overview of applied research in *passive sonar* signal processing estimation techniques for naval systems is presented. The naval problem that motivates *time delay estimation* is the source state estimation problem. A discussion of this problem in terms of estimating the position and velocity of a moving acoustic source is presented. Optimum bearing and range estimators are presented for the planar problem and related to the optimum time delay vector estimator. Suboptimum realizations are considered together with the effects of source motion and receiver positional uncertainty.

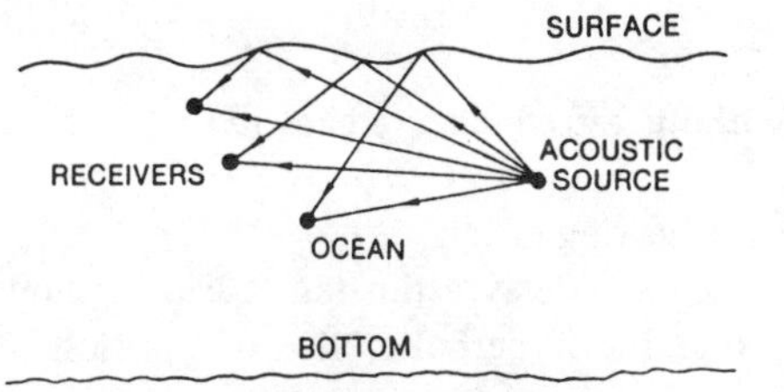

Fig. 1. Model of direct and surface-reflected sound ray paths received at three sensors.

I. Introduction

THE purpose of this paper is to provide a tutorial review of *time delay estimation* in the *passive sonar* signal processing field.

In the passive sonar problem of interest here, signals received at two or more receiving sensors or hydrophones are used to estimate the position and velocity of a detected acoustic source. Passive systems, unlike radar or active sonar systems, cannot control the amount of transmitted energy to be reflected off the source; however, the covertness of passive systems can be advantageous, both in military and biomedical applications. A discussion of radar and active sonar can be found in Altes [3]. In practice, the number of receiving sensors, the observation time and the ratio of the background noise to the source signal strength after propagation loss, when balanced against total system cost, dictate the feasibility of passive systems.

In the ocean, sound usually arrives at each individual omnidirectional receiver through more than one path. For example, sound may arrive through the direct and surface-reflected paths, as shown in Fig. 1. A more complete discussion of recent research in underwater acoustics is available in the texts edited by Oppenheim [44] and Bjørnø [7]. To make the problem mathematically tractable from a signal processing point of view, it is convenient to decouple the problem into multipath and planar problems. For multipath signals, a simplistic model of the received signal is that each receiver sees a signal plus an attenuated and delayed signal corrupted by additive uncorrelated noise. For the planar problem (i.e., when all the receivers and the source are in the same plane), the time delay to be estimated is the travel time of the acoustic wavefront between pairs of receivers, so that the source position and velocity can be estimated. The type of signal processing to be used and the bounds on performance for such processing are important subjects of this paper.

Manuscript received September 26, 1980; revised November 20, 1980. This work summarizes research completed at the University of Connecticut, portions of which were presented at the 1976 North Atlantic Treaty Organization (NATO) Advanced Study Institute (ASI) on Underwater Acoustics and Signal Processing (UASP) in Portovenere, Italy. It also summarizes work completed since that time by the author and others working in this field. Portions of this work were also presented at IEEE EASCON '79, Washington, DC, at the 13th Annual Asilomar Conference on Circuits, Systems, and Computers, Monterey, CA, and at the 1980 NATO ASI on UASP, Kollekolle, Denmark.

The author is with the Naval Underwater Systems Center, New London, CT 06320.

II. The Mathematical Model

For the purpose of this paper, only the planar problem will be considered. In particular, it will be assumed that acoustic energy arrives at each receiver through only one propagation path in the same plane with all the receivers and source.

Pictorially, we are interested in the delay of the signal from one receiving sensor to the next, as shown in Fig. 2. For zero relative time delay, the source bearing is broadside to the sensor pair; and for maximum time delay, the source bearing is endfire. A certain amount of caution must be exercised in applying the results directly to "real world" problems, since the ocean medium is more complex than this simple model. More sophisticated propagation modeling and associated signal processing is considered by Owsley [45]. However, considerable insight can be gained in dealing with the passive bearing estimation problem in a simplified decoupled fashion.

A simple mathematical model for the received signals in the planar case is

$$r_1(t) = s(t) + n_1(t) \tag{1a}$$

$$r_2(t) = s(t - D) + n_2(t) \tag{1b}$$

where the source signal $s(t)$ and the noises $n_1(t)$ and $n_2(t)$ are Gaussian, stationary, and mutually uncorrelated. After the signal has been detected, an important passive sonar problem is to estimate the time delay D between two sensors separated by length L. This time delay estimate then is used to estimate the bearing angle shown in Fig. 2. The bearing estimate is given by the approximate rule

$$\hat{B} \cong \cos^{-1} (C\hat{D}/L) \tag{2}$$

Reprinted from *IEEE Trans. Acoust., Speech, Signal Processing,* vol. ASSP-29, pp. 463–470, June 1981.

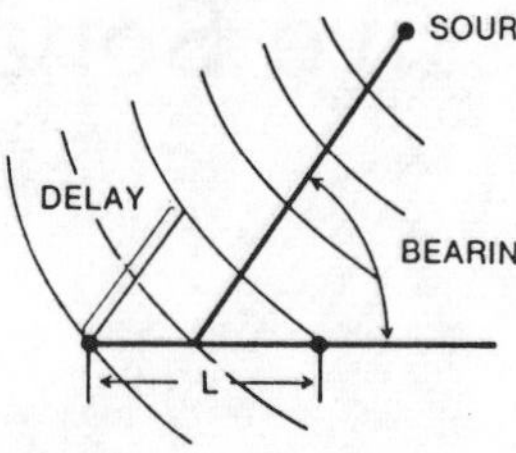

Fig. 2. Planar model of two receivers separated by a distance L.

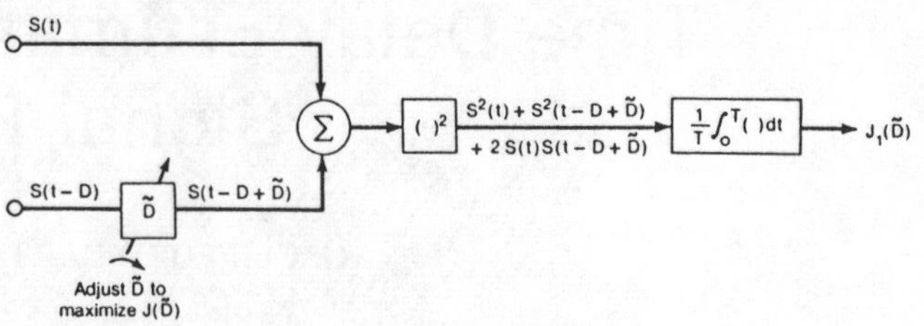

Fig. 3. Conceptual delay, sum, square and integrate configuration.

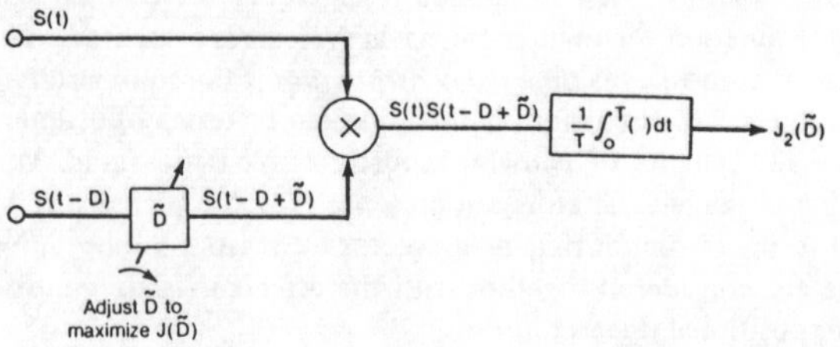

Fig. 4. Conceptual cross correlator configuration.

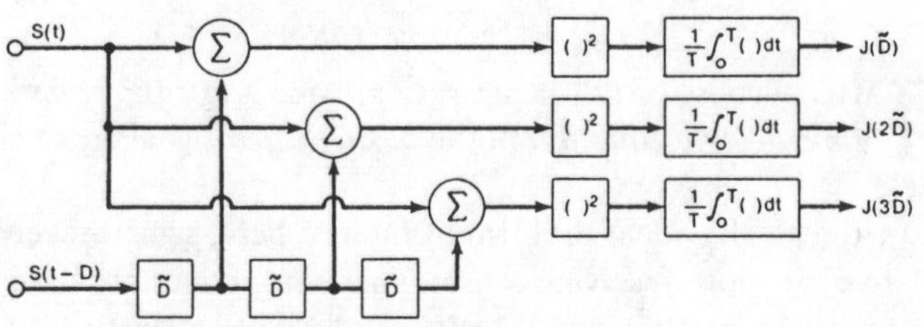

Fig. 5. Time-domain beam former instrumentation of one conceptual configuration for three hypothesized delays.

where

1) C (without a frequency argument) is the speed of sound in water;

2) $\hat{B}$ is the bearing estimate;

3) $\hat{D}$ is the time delay estimate. It can be shown that B is the angle that the hyperbolic "line of position" makes with the axis of the receivers; hence the approximation equation (2) is increasingly accurate as the range to the acoustic source increases.

III. System Configurations for Time Delay Estimation

The critical part of the passive bearing estimation problem is the accurate estimation of time delay. Two conceptual configurations come to mind based on either 1) an intuitive approach and a familiarity with detection theory, or 2) a rigorous application of the *maximum likelihood* (ML) approach for white signals in white noise. In both conceptual configurations we attempt to "advance" the delayed received signal by a hypothesized amount in order to align it with the other received signal. Then we either sum, square, and average, as shown in Fig. 3, or we multiply and average as shown in Fig. 4. In both cases the hypothesized delays are adjusted in order to maximize the configuration output. As shown in the figures, both configuration outputs consist of "signal-cross-signal" terms. Further, both configurations are ML estimators for time delay under the assumptions that the signal and noises are white and mutually uncorrelated. A further discussion of the cross correlator configuration for time delay estimation is given by Bendat and Piersol [6]. When the signal and noise spectral characteristics are nonwhite, the received waveforms must be prefiltered with particular equiphase filters (i.e., the prefilters must have the same phase characteristics) to accentuate the frequency bands with good *signal-to-noise ratio* (SNR). It is of interest to note that a signal detector can be realized by comparing either configuration output to a threshold. Moreover, *in terms of detection* (but not estimation) in the presence of noise, the system in Fig. 3 outperforms the system in Fig. 4; however, the system in Fig. 3 requires prior knowledge of power levels in order to set the proper threshold. The system in Fig. 4 has a zero mean output in the signal absent, noise present case.

The conceptual systems in Figs. 3 and 4 can be achieved in a number of different ways. The system in Fig. 3 can be instrumented as shown (for three fixed hypothesized delays) in Fig. 5. Fig. 5 also is called a time delay *beam former*. It is presumed that $s(t)$ is a sampled version of the original broadband time signal and that the sampling rate is large in comparison with the required Nyquist rate. In particular, the sampling rate is much greater than twice the bandwidth. (A further discussion of beam formers can be found in Pridham and Mucci [47] and D'Assumpcao [16].) An example of applying the configuration of Fig. 5 to a pulse signal is shown in Fig. 6. Note that in the example, Fig. 6, when the hypothesized time delay equals the true time delay, the system output power achieves a maximum. The other system configuration, shown in Fig. 4, including prefilters can be instrumented by a *generalized cross-correlation* (GCC) function. This function can be estimated using the Fortran digital computer program in *Programs for Digital Signal Processing*, edited by the DSP committee [17]. Briefly, it consists of taking the inverse *fast Fourier transform* (FFT) of the product of a weighting function and the estimated complex cross-power spectrum.

IV. Attaining the Cramer-Rao Lower Bound By a GCC Function

The paper on generalized cross correlators by Knapp and Carter [33] puts in perspective several suggested methods of filtering and cross correlation for time delay estimation including the often useful *smoothed coherence transform* (SCOT) or "*rehocence*" method that prewhitens and then cross correlates. (See, for example, the paper by Kostić [32] and references contained therein.) Recent extensions to that work, including simulation experiments, have been reported by Hassab and Boucher [27], and detection performance of the SCOT by Kuhn [35]. In related work Chan, Hattin, and Plant [13], using the results of Hamon and Hannan [25], give the minimum variance of time delay error in terms of the number of time samples, the number of frequency points to average,

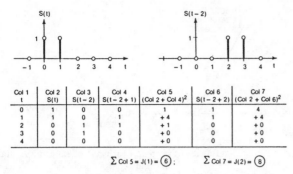

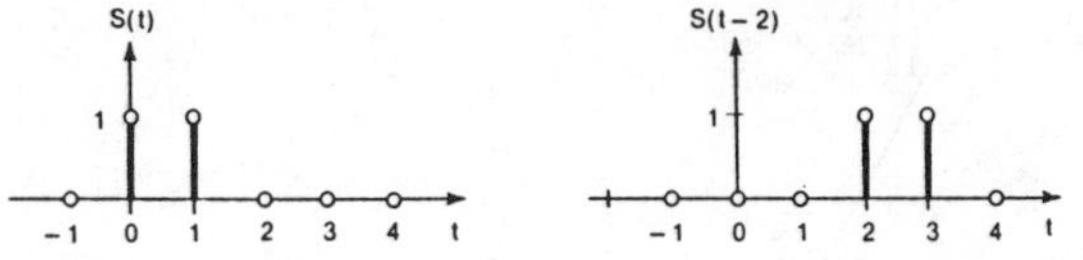

Col 1 t	Col 2 S(t)	Col 3 S(t - 2)	Col 4 S(t - 2 + 1)	Col 5 (Col 2 + Col 4)^2	Col 6 S(t - 2 + 2)	Col 7 (Col 2 + Col 6)^2
0	1	0	0	1	1	4
1	1	0	1	+ 4	1	+ 4
2	0	1	1	+ 1	0	+ 0
3	0	1	0	+ 0	0	+ 0
4	0	0	0	+ 0	0	+ 0

Σ Col 5 = J(1) = (6); Σ Col 7 = J(2) = (8)

Fig. 6. Example of pulsed arrival for one instrumentation.

and a dimensionless SNR-type term. That work is consistent with the results of Carter [8], that the *Cramér-Rao lower bound* (CRLB) for the variance of time delay errors for the physical problem modeled by (1) is given by

$$\sigma^2(\hat{D} - D) \geqslant \frac{2\pi}{\left\{2T \int_0^{\omega_H} \frac{C(\omega)}{[1 - C(\omega)]} \omega^2 d\omega\right\}} \tag{3a}$$

where

1) ω_H is the highest source frequency after propagation to the receiver,

2) T is the observation time, and

3) $C(\omega)$ is the *magnitude-squared coherence* (MSC). Equation (3a) can be expressed in terms of SNR by observing that the MSC is defined by

$$C(\omega) = \frac{|G_{r_1 r_2}(\omega)|^2}{[G_{r_1 r_1}(\omega) G_{r_2 r_2}(\omega)]} \leqslant 1, \quad \forall \omega. \tag{3b}$$

The MSC in (3b) for the model in (1) gives

$$C(\omega) = \frac{G_{ss}^2(\omega)}{\{[G_{ss}(\omega) + G_{n_1 n_1}(\omega)]\ [G_{ss}(\omega) + G_{n_2 n_2}(\omega)]\}} \tag{3c}$$

where $G(\omega)$ is the auto- or cross-power spectrum at frequency ω of the subscripted random process(es). Equation (3a) agrees with estimated variances for selected cases of a digital simulation by Scarbrough, Ahmed, and Carter [50]. Utilizing (3c), (3a) is plotted as a function of center frequency and SNR in decibels (10 $\log_{10}$ of power) by Quazi [49] for the case of SNR independent of frequency in the signal band.

Several clarifying facts should be pointed out about (3). First, it is dimensionally correct. Second, in the underwater acoustics problem it is an implicit function of range; this is because signal power at the receiver is a function of propagation loss (conceptually owing to effects like spreading loss), which itself is a (or can be an extremely complicated) function of range. Third, the CRLB given by (3a) presumes that the signal and noise power spectral densities are known and that prior to cross correlation the received waveforms $r_1(t)$ and $r_2(t)$ in (1) are filtered by filters with transfer functions $H_1(f)$ and $H_2(f)$, respectively; the filters both have the same phase and the characteristic that the product of $H_1(f)$ with the complex conjugation of $H_2(f)$ forms the real weighting function given by

$$W(f) = H_1(f) H_2^*(f) = C(f)/\{|G_{r_1 r_2}(f)|[1 - C(f)]\}. \tag{4}$$

Equation (4) agrees with the equation between (17) and (18) of Hannan and Thomson [26]; following proper manipulation, (4) agrees with (17) of Hahn [23]. See Carter [8] for the specific filter characteristics for the case of more than two filters. (In the interest of completeness, we note that there is a typographical error in the filter transfer function given by (5) in Hahn [23].) In order to estimate the GCC function the estimated (complex) cross-power spectrum between the two received signals is multiplied by a weighting function $W(f)$, and then the inverse discrete Fourier transform (DFT) is computed via digital signal processing techniques such as the FFT. (As noted by Knapp and Carter [33], since the variance of the phase error is inversely proportional to MSC over one minus MSC, the GCC function with the ML weighting of (4) attaches most weight to the estimated phase when the variance of the estimated phase error is lowest. The work by Cleveland and Parzen [15] is one of the early references to using phase to determine time delay.) Another reference to using phase is [13]. A fourth item that must be considered in practice when using (3) is that often $W(f)$ is a function of unknown spectral quantities that must be estimated; in this case, the process of time delay estimation involves estimating the complex cross-power spectrum and the *coherence* function. The derivation of (3) presumes the spectral quantities needed for $W(f)$ are known. Finally (3) presumes Gaussian received signals. When the source radiates a sinusoidal signal, a tighter *Barankin bound* (poorer performance) occurs at low SNR. A more complete discussion is given in this issue by Chow and Schultheiss [14].

V. Standard Deviation of Bearing Error

The trigonometric relationship between time delay and bearing yields the result (see, e.g., Tacconi [60]) that the standard deviation of the bearing error in radians is given by

$$\sigma(\hat{B} - B) = \sigma(\hat{D} - D)C/(L \sin B) \tag{5}$$

where

1) B is the bearing in radians;

2) L is the distance between the two sensors (or subarrays); and

3) C (without a frequency argument) is the speed of sound in water.

This result does not take into account uncertainty in sensor position which increases the variance by an additive amount and will be discussed later. (Note that (5) is dimensionally correct.) Substituting (3) into (5) gives the two-hydrophone result of MacDonald and Schultheiss [37]. Equations (3) and (5) show the effects of array length, source bearing, observation time, and coherence on bearing accuracy. It is noteworthy that when the pairwise sensor coherence is low we can still theoretically attain good performance if the source bandwidth is large or if we cluster a number of sensors together and beam form to increase the observed (or output) SNR. Extensions to (5) for endfire approximations are given by Hinich [29].

It should be noted that these results [(3) and (5)] pertain to a *stationary processes large observation time* (the so-called SPLOT condition; see, e.g., Van Trees [62, Parts I and III]) of one member function of a nominally broad-band Gaussian random process observed in a nondispersive medium with sensors at known positions and with known SNR. Thus, in practical applications, these results will serve as a bound on performance and not necessarily an absolute indicator of performance. Work with non-Gaussian signals is discussed by Hilliard and Pinkos [28]. When the noises are correlated, the results are more complicated; this problem has been treated by Howell [30] and Bangs [4]. For dispersive medium effects, see Kirlin [31]. Correlator configurations that both detect and estimate bearing for narrow-band signals are considered by Miller [40]. The problem of processes with random relative phase is treated by Wax [63].

The problem of estimating time delay between two received signals in noise is closely related to the work by Hahn [20] and Hahn and Tretter [22]. In that work, the problem is to estimate the $M - 1$ components of the relative time delay parameter vector $(D_2 - D_1, D_3 - D_1, \cdots, D_M - D_1)$ from the M received waveforms,

$$r_i(t) = s(t - D_i) + n_i(t), \quad i = 1, M. \tag{6}$$

Other work related to the reduction in pairwise time delay error due to additional sensors has been reported by Schultheiss [54] and Bjørnø [7]. For passive localization in a plane, three unique sensors (or subarray beam former outputs) are required. In general, we define the (relative) time delay as

$$D_{ij} = D_i - D_j, \quad i,j = 1, 2, 3. \tag{7}$$

By knowing (or estimating) the relative delays $D_{21} = D_2 - D_1$ and $D_{32} = D_3 - D_2$ and the sensor positions, we can estimate the position of the acoustic source. In particular, we find the position in terms of polar coordinates, i.e., range R and bearing B from the central sensor according to the exact relationship (see, e.g., Griffiths, Stocklin, and Van Schooneveld [19], Carter [12], and references contained therein)

$$R = \frac{\left\{L_1\left[1 - \left(\frac{CD_{21}}{L_1}\right)^2\right] + L_2\left[1 - \left(\frac{CD_{32}}{L_2}\right)^2\right]\right\}}{\left[2\left(\frac{CD_{32}}{L_2} - \frac{CD_{21}}{L_1}\right)\right]} \tag{8a}$$

and an expression for bearing is

$$B = \cos^{-1}\left(\frac{[L_2^2 - 2R_2 CD_{32} - (CD_{32})^2]}{[2R_2 L_2]}\right) \tag{8b}$$

where L_1 and L_2 are the distances between sensor pairs depicted in Fig. 7. For large R the middle term in the arcosine expression, (8), becomes dominant, consistent with (2).

VI. Variance of Range Error

The variance of the range error has been shown by Hahn [21] to be dominated by the errors in estimating the difference in time delays. In particular, the variance of the range errors is

$$\sigma^2(\hat{R} - R) = C^2 \sigma^2 [\hat{D}_{32} - \hat{D}_{21} - (D_{32} - D_{21})] (R/L_e)^4 \tag{9}$$

where

1) R is the range;

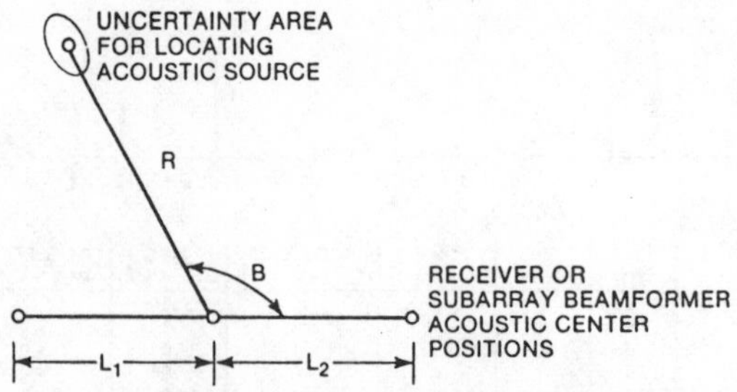

Fig. 7. Pictorial definition of range R, bearing B and subarray separations L_1 and L_2.

2) L_e is the effective *half* array length (i.e., the sine of the bearing times the interarray separation);

3) C is the speed of sound in water; and

4) σ^2 ( ) is the variance of the error in time delay difference measurements.

When comparing (9) with other work, it is noteworthy that by rewriting that equation in terms of the total array length, it apparently becomes 16 times larger. Using Carter [9] we readily can show that for the *optimum* range and bearing sonar signal processor the variance of the range error is given by

$$\sigma^2(\hat{R} - R) = 12\sigma^2(\hat{B} - B)R^4/L_e^2 \tag{10}$$

where the bearing errors are in radians. Using (5) to relate time delay errors to bearing errors and observing that for an equispaced three-sensor array we have twice the array length and 1.5 times as many sensors as used to obtain (5), we postulate that the variance of bearing errors can be approximated by

$$\sigma^2(\hat{B} - B) = C^2 \sigma^2(\hat{D}_{31} - D_{31}) (2/3) (1/[2L_e])^2 \tag{11}$$

where σ^2 ( ) is the variance of the error when estimating the time delay from sensor 3 to 1. We note that (11) depends on the effective array length but not on the range to the source R. Substituting (11) into (10) yields

$$\sigma^2(\hat{R} - R) = 2C^2 \sigma^2(\hat{D}_{31} - D_{31}) (R/L_e)^4. \tag{12}$$

The variance of the error in estimating the difference of time delays (used in the denominator of (8), the ranging equation) is approximately twice as great as the variance of the error in estimating D_{31}. (With regard to the approximation see Hahn [23] for an evaluation of the covariance of the errors in estimating D_{21} and the errors in estimating D_{31}.) Therefore, we conclude that (12) and (9) agree. Comparing (12) with (11) note that the range error variance and bearing error variance both depend linearly on time delay error variance; however, unlike bearing errors *ranging errors are highly dependent on the ratio of the true source range to the length of the effective baseline*. Indeed, in passive ranging, it is the fourth power of the range relative to the effective array length that is important to the range error variance. This mathematical result helps develop physical insight into the difficulties of estimating the position of an acoustic source. Specifically, we see that inherent in the physics of the passive ranging problem is the need to have long baselines, extremely accurate time delay estimates, or small (true) range from the receiving array to the acoustic source, or all three. In addition, we point out here that owing to propagation loss the SNR or coherence at the receivers is a function of range from the source to the receivers. Hence, for certain underwater acoustic applications, (12) is a stronger function of range than might otherwise be explicitly apparent.

An empirical method for estimating the variance necessary in (12) was found by Schneider [52] of ENSCO, Inc., who observed from (7) that

$$D_{21} + D_{32} - D_{31} = 0. \tag{13}$$

That is, in the absence of estimation noise independent of source and receiver geometries, the time it takes an acoustic wave to travel from sensor 3 to 2 plus the travel time from sensor 2 to 1 should equal the travel time from sensor 3 to 1. Hence, a useful internal consistency check for the estimated time delay parameters is $\hat{D}_{21} + \hat{D}_{32} - \hat{D}_{31}$. The mean-square value of this three-term sum represents a practical guide to performance bounds without consideration for source and receiver motion. Schneider [52] and Beam and Carter [5] have found this guide useful for actual sea data. An extension to a *generalized broadband Doppler* or *relative time compression* (RTC) internal consistency check for moving sources has also been done by Schneider [51].

VII. Performance Bounds Found by Hypothetical Arrays

In the work of MacDonald and Schultheiss [37] the optimum processor for bearing estimation was derived assuming plane-waves, Gaussian signals and uncorrelated Gaussian noises of equal strength. The performance of such a system was found together with a bound obtained by considering a *hypothetical* system in which half the elements are placed at each end of the available aperture. (In practice, though, we postulate that we would require half-wavelength spacing at the design frequency for each subarray; see, e.g., Fig. 8.) An extension (in concert with Pryor [48]) to unequal SNR was reported by Carter [10]. These results showed that at low unequal SNR the available hydrophones should be divided into two equal groups. Thus we see that passive sonar systems designed to be optimum bearing estimators for equal SNR are also optimum for unequal SNR in the important case when the SNR is low.

When the acoustic source is sufficiently close to the center of the receiving array relative to the length of the array, then, unlike Fig. 8, the received wavefronts appear curved as shown in Fig. 9. A bound on the best ranging performance can be calculated by distributing a quarter of the receiving elements at each end and the remaining half of the elements at the middle of the array aperture (see Carter [9]). We emphasize that the bound is good even though the array configuration is theoretical in nature. These results hold for known sensor positions, and extensions are underway to the moving sensor. In particular, Schultheiss and Ianniello [56] consider an ensemble of statically deformed arrays and correctly infer that an equispaced line array might in some situations be preferable to three clustered subarrays. Work on a moving array is in progress. When the source uncertainty area is to be minimized and sensor positions are known, a bound on performance is calculated by a sensor distribution that places one third of the available sensing hydrophones at each end and the middle of the available aperture [9]. Of course, if sensor positions are not known exactly, performance is also expected to be worse than the performance bound found when sensor position is known.

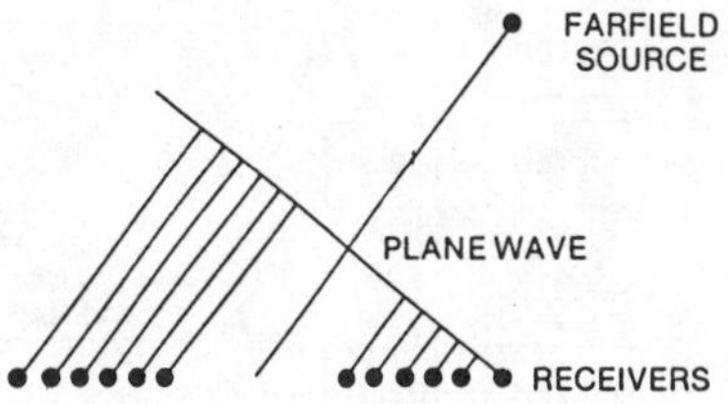

Fig. 8. Plane wave model for bearing estimation.

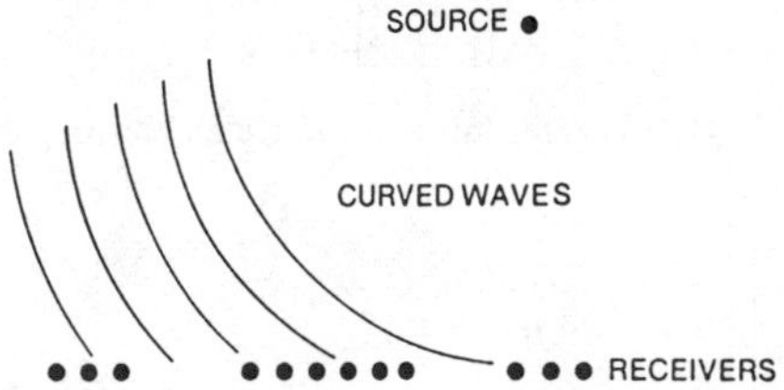

Fig. 9. Curved wave model for range estimation.

VIII. Configuration for Optimum Position Estimation

Configurations for the optimum processor for range and bearing estimation are discussed by Hahn [23] and Carter [9]. One configuration presented by Carter [9] is shown in Fig. 10. We refer to this configuration as a *focused beam former* because it constrains the hypothesized time delays to focus the acoustic energy from a hypothesized position. It effectively presumes that the receiving sensor positions are known. Unlike the focused beam former, the ML estimate for the time delay vector maximizes the output power J of this network and does not constrain the power to be focused at one hypothesized range R and bearing B. A second realization is to form all possible GCC pairs and combine the delay estimates from each pair according to the rules of Hahn [23]. The input signals are filtered with particular signal-enhancing filters, all with the same phase (e.g., symmetric digital finite impulse response (FIR) filters, all with the same bulk or group time delay); and the delayed outputs are summed, squared, and averaged. The particular filter transfer functions needed to yield the best performance are given by Carter [8]. The ML range and bearing estimates are achieved by adjusting hypothesized range and bearing parameters that, in turn, give rise (with knowledge of the receiving hydrophone positions) to the appropriate time delays to be inserted after filtering. When the total system output power J is maximized, the selected hypothesized range and bearing estimates are the ML source position estimates. In the limit of large observation time these ML estimators are the minimum variance source-position estimators.

For systems with the available elements grouped into three subarrays, a suboptimum realization is shown in Fig. 11; it can be achieved at lower cost and is postulated to have performance similar to that of the optimum system in Fig. 10. In the Fig. 11 suboptimum system we first beam form each of the three subarrays. Subsequently, we perform a GCC on the forward and middle and the middle and aft beam former outputs and read off the correlator abscissa showing the peak. Many commercially available correlators perform similar functions. We also can form a GCC on the forward and after beam former output and form the internal consistency check. For purposes of analysis, we model the GCC as prefilters, a variable delay,

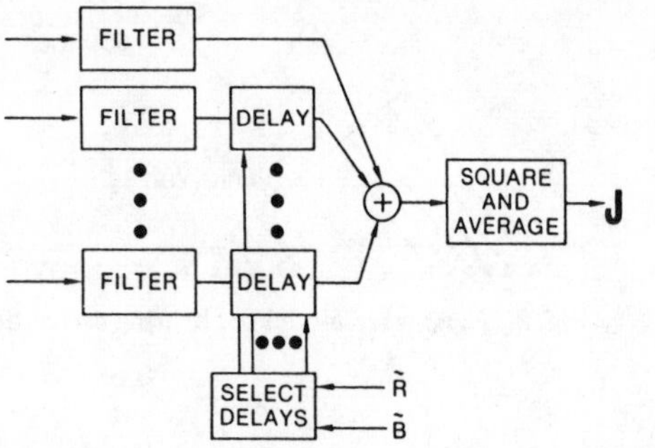

Fig. 10. Focused beam former for range and bearing estimation.

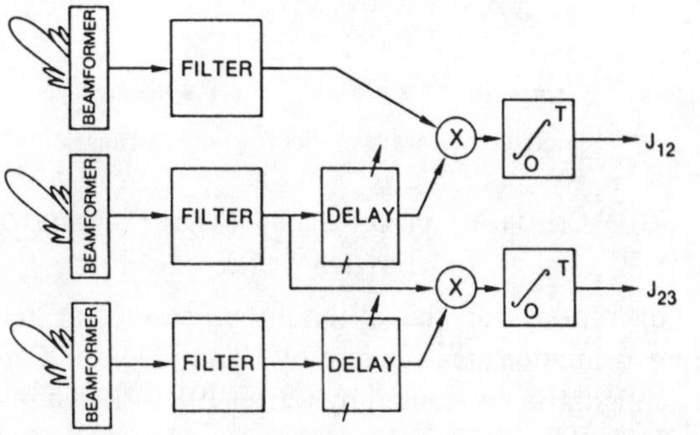

Fig. 11. Suboptimum realization of focused beam former.

and T-second integrator. Then the delay estimates found by the Fig. 11 realization are inserted in the ranging equation (8) to obtain a range estimate.

IX. Other Limitations to Performance

In practice, both the optimum and suboptimum realizations are degraded by realistic factors. A most significant factor can be that signal processing implementations derived from treating the planar problem may not suit even the simple multipath model shown earlier in Fig. 1. These and other significant environmental factors, such as the sound speed profile, the bottom depth, slope, and layering together with bottom loss, often need to be considered. Limitations due to source motion have been studied and reported by Gerlach [18]. These types of studies provide information on the fundamental bounds on performance.

Two other potentially significant effects on passive sonar localization are uncertainty in sensor position and failure to compensate for broad-band Doppler or RTC. The first cause of error to be discussed is depicted in Fig. 12 and has been studied and reported by Carter [11]. As mentioned earlier, additional work has been conducted and is continuing by Ianniello and Schultheiss. In bearing estimation two receivers of known position are needed. Consider the two left receivers indicated by squares for the actual positions. If the receiver were perturbed by an amount P (small in comparison with the sensor separation L) from the position of the square in Fig. 12 to the solid dot position, then the bearing estimate would be in error by approximately an amount of P/L radians. Of course, if P were a random perturbation, the variance of the bearing error could be reduced by averaging a series of independent bearing measurements over a very large observation time. The total bearing error then consists of the sum of two terms, one having

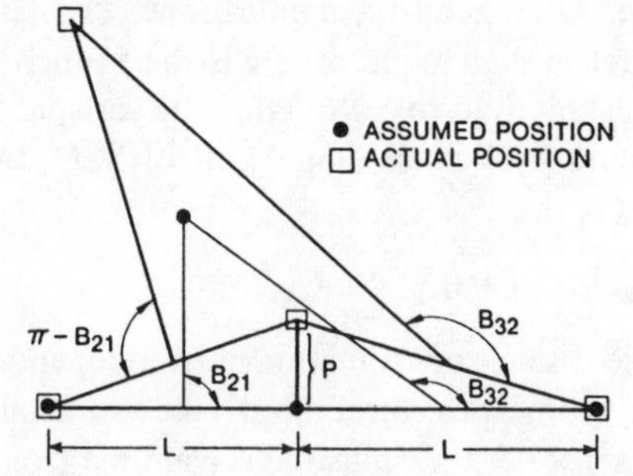

Fig. 12. Effect of receiver positional uncertainty on localization.

to do with SNR and the other having to do with array perturbation. Fundamentally, the relative ranging errors due to unknown random sensor perturbation P about some known mean P are approximately given by

$$\sigma(\hat{R} - R)/R \cong (\sigma(\hat{P} - P)/L)\,(2R/L_e) \tag{14}$$

where

1) $\sigma(\hat{P} - P)$ is the standard deviation of the middle sensor perturbation from a line connecting the two end sensors;
2) R is the range to the source;
3) L_e is the effective half array length; and
4) L is half the total array length.

Thus, the ratio of the perturbations relative to the array length and the range relative to the effective array length are the dominant contributors to this type of relative ranging error. We note that array perturbations can be viewed as time delay errors with a simple trigonometric mapping for small perturbations. Not unexpectedly then, the form of (14) is consistent with (9).

X. Velocity Estimation

Another important type of degradation is caused by failure to compensate for the "deterministic" part of source motion. A simple but useful model of received signals is discussed by Abraham and Carter [1]. A more complete discussion of the model and the ML estimator for the fundamental motion parameter, tantamount to time delay rate or bearing rate, is given by Knapp and Carter [34]. Theoretically though, a suboptimum realization is depicted in Fig. 13. Work in this area has been done at the Naval Ocean Systems Center (NOSC) by Stradling [58], McCarthy [38], Mohnkern [41], and Trueblood [61]. Other models include one discussed by Adams, Kuhn, and Whyland [2], Schultheiss [53], Schultheiss and Weinstein [55], and Stein [57]. They treat the moving source problem by considering a truncated Taylor series expansion. Such models physically include time delay, time delay rate, and time delay acceleration. Related work includes an extended MSC estimator, similar to an algorithm by Trueblood [61], that was reported by Patzewitsch, Srinath, and Black [46]. Other related work on tracking of moving time delays for velocity estimation has been reported by Meyr [39] and Lindsey and Meyr [36]. Work in this area, including sensing systems for determining train speed, is being continued by Meyr and his students. The work by Moura [43] and Moura and Baggeroer [42] also is concerned with passive velocity estimation. For many practical

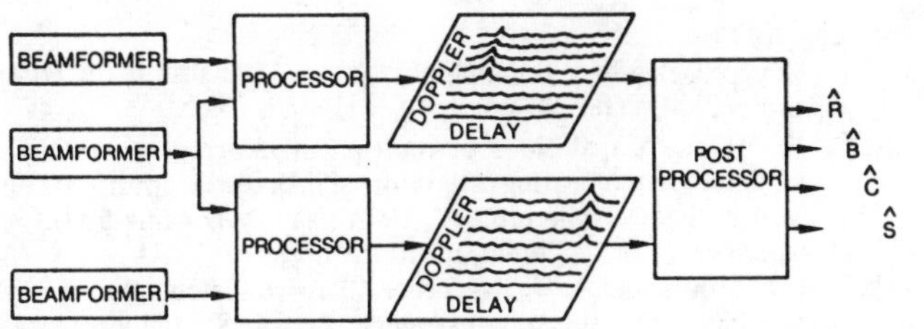

Fig. 13. Theoretical procedure for range, bearing, course and speed estimation.

geometries when the source range is long compared with the receiver baseline and the SNR is large so that T can be kept small, the estimation of RTC or generalized Doppler is not necessary in order to estimate time delay. For other geometries, RTC or time delay rate estimation is necessary. Even when it is not necessary, RTC estimation can provide rapid early indications of source velocity.

It is analytically useful in this estimation procedure to decouple velocity into two orthogonal components: bearing rate or *cross line-of-sight speed* (CLSS) and range rate or *radial line-of-sight speed* (RLSS). In particular, the CLSS is defined as the projection of the velocity vector on the perpendicular to the bearing line, and the RLSS is the projection on the bearing line. The problem is to determine the accuracy with which source velocity can be determined. Theoretical considerations indicate that CLSS is much easier to estimate than RLSS. The RTC can be estimated using Knapp and Carter [34]; see also Stein [57]. Bearing can be estimated using MacDonald and Schultheiss [37], and range can be estimated by means of the triangulation methods of Hall and Hayford [24] and Hilliard and Pinkos [28].

From these results Carter [12] has shown the variance in CLSS errors is dominated by range relative to the baseline projection and the variance of the RTC estimation error.

Further, the standard deviation of the CLSS errors is predominantly a function only of the range relative to the effective baseline steered at the source and the standard deviation of the RTC errors. In most practical ways, it is not a function of the course or speed of the source or errors in time delay. However, when the coherence or SNR is low and the observation time T is short, errors in both time delay and RTC or CLSS will be large. Theriault and Berkman [59] have derived an approximate expression for the standard deviation of the errors in the radial component of velocity.

RLSS errors, like CLSS errors, are not predominantly a function of the course or speed of the source. However, unlike CLSS errors, RLSS errors depend on the square of the range relative to the effective baseline.

It was corroborated by computer simulation that the CLSS is much easier to estimate than the RLSS. Also, doubling the range-to-effective-baseline ratio doubled the CLSS errors and quadrupled the RLSS errors. Further, the errors in CLSS and RLSS did not appear to depend on either time delay errors explicitly or source velocity.

XI. Summary

Time delay estimation is an important part of passive sonar signal processing. An overview of applied research in passive sonar signal processing estimation techniques has been presented. One problem that motivates time delay estimation is the estimation of the position and velocity of a moving acoustic source. A discussion of this problem in terms of estimating time delay has been presented. In order to develop an understanding of the signal processing required, an approach of decoupling the problem into multipath and planar components was followed. Optimum estimators for acoustic source position were presented for the planar problem and related to the optimum time delay vector estimator with a focusing constraint. In particular, the focused beam former with appropriate prefilters is, in some sense, the best range and bearing position estimator.

References

[1] P. B. Abraham and G. C. Carter, private communications, 1978; see also G. C. Carter and P. B. Abraham, "Estimation of source motion from time delay and time compression measurements," *J. Acoust. Soc. Amer.*, vol. 67, no. 3, pp. 830–832, 1980.

[2] W. B. Adams, J. P. Kuhn, and W. P. Whyland, "Correlator compensation requirements for passive time delay estimation with moving source or receivers," *IEEE Trans. Acoust., Speech, Signal Processing*, vol. ASSP-28, no. 2, pp. 158–168, 1980.

[3] R. A. Altes, "Target position estimation in radar and sonar, and generalized ambiguity analysis for maximum likelihood parameter estimation," *Proc. IEEE*, vol. 67, no. 6, pp. 920–930, 1979.

[4] W. J. Bangs, II, "Array processing with generalized beamformers," Ph.D. dissertation, Yale Univ., New Haven, CT, 1971; see also the work by Bangs and Schultheiss appearing in Griffiths, Stocklin, and Van Schooneveld, 1973.

[5] J. P. Beam and G. C. Carter, private communication on passive localization sea test results, 1980.

[6] J. S. Bendat and A. G. Piersol, *Engineering Applications of Correlation and Spectral Analysis.* New York: Wiley, 1980.

[7] L. Bjørnø, Ed., in *Proc. 1980 NATO Advanced Study Institute on Underwater Acoustics and Signal Processing.* Boston, MA: Reidel, 1981.

[8] G. C. Carter, "Time delay estimation," Ph.D. dissertation, Univ. Connecticut, Storrs, 1976 (also available as NUSC TR 5335 through NTIS under AD A025408).

[9] —, "Variance bounds for passively locating an acoustic source with a symmetric line array," *J. Acoust. Soc. Amer.*, vol. 62, no. 4, pp. 922–926, 1977.

[10] —, "Optimum element placement for passive bearing estimation in unequal signal-to-noise ratio environments," *IEEE Trans. Acoust., Speech, Signal Processing*, vol. ASSP-26, no. 4, pp. 365–366, 1978.

[11] —, "Passive ranging errors due to receiving hydrophone position uncertainty," *J. Acoust. Soc. Amer.*, vol. 65, no. 2, pp. 528–530, 1979.

[12] —, "Sonar signal processing for source state estimation," in *EASCON '79 Rec.*, IEEE Pub. 79CH 1476-1-AES, 1979.

[13] Y. T. Chan, R. V. Hattin, and J. B. Plant, "The least squares estimation of time delay and its use in signal detection," *IEEE Trans. Acoust., Speech, Signal Processing*, vol. ASSP-26, no. 3, pp. 217–222, 1978; see also Y. T. Chan, J. M. Riley, and J. B. Plant, "A parameter estimation approach to time delay estimation and signal detection," *IEEE Trans. Acoust., Speech, Signal Processing*, vol. ASSP-28, no. 1, pp. 8–16, 1980.

[14] S. K. Chow and P. M. Schultheiss, "Delay estimation using narrowband processes," this issue, pp. 478–484.

[15] W. S. Cleveland and E. Parzen, "The estimation of coherence, frequency response and envelope delay," *Technometrics*, vol. 17, no. 2, pp. 167–172, 1975.

[16] H. A. D'Assumpcao, "Some new signal processors for arrays of sensors," *IEEE Trans. Inform. Theory*, vol. IT-26, no. 4, pp. 441–453, 1980.

[17] Digital Signal Processing Committee, Ed., *Programs for Digital Signal Processing.* New York: IEEE Press (dist. Wiley), 1979.

[18] A. A. Gerlach, "Acoustic transfer function of the ocean for a motional source," *IEEE Trans. Acoust., Speech, Signal Processing*, vol. ASSP-26, no. 6, pp. 493–501, 1978.

[19] J. W. R. Griffiths, P. L. Stocklin, and C. Van Schooneveld, Eds., *Signal Processing*. New York: Academic, 1973 (including P. Heimdal and F. Bryn, "Passive ranging techniques"; E. B. Lunde, "Wavefront stability in the ocean"; W. J. Bangs and P. M. Schultheiss, "Space-time processing for optimal parameter estimation").

[20] W. R. Hahn, "Optimum estimation of a delay vector caused by a random field propagating across an array of noisy sensors," Ph.D. dissertation, Univ. Maryland, College Park, 1972 (available as NOL TR 72-120 through NTIS AD 751555).

[21] —, private communication on passive ranging, 1973.

[22] W. R. Hahn and S. A. Tretter, "Optimum processing for delay-vector estimation in passive signal arrays," *IEEE Trans. Inform. Theory*, vol. IT-19, no. 5, pp. 608-614, 1973.

[23] W. R. Hahn, "Optimum signal processing for passive sonar range and bearing estimation," *J. Acoust. Soc. Amer.*, vol. 58, no. 1, pp. 201-207, 1975.

[24] J. B. Hall and R. I. Hayford, private communications on certain processing techniques, 1977.

[25] B. V. Hamon and E. J. Hannan, "Spectral estimation of time delay for dispersive and non-dispersive systems," *Appl. Statist.*, vol. 23, no. 2, pp. 134-142, 1974.

[26] E. J. Hannan and P. J. Thomson, "Estimating group delay," *Biometrika*, vol. 60, no. 2, pp. 241-253, 1973; see also E. J. Hannan and P. J. Thomson, "Delay estimation," this issue, pp. 485-490.

[27] J. C. Hassab and R. E. Boucher, "A quantitative study of optimum and suboptimum filters in the generalized correlator," in *IEEE 1979 Int. Conf. Acoust., Speech, Signal Processing Rec.*, 79CH1379-7 ASSP, pp. 124-127, see also J. C. Hassab and R. E. Boucher, "Performance of the generalized cross correlator in the presence of a strong spectral peak in the signal," this issue, pp. 549-555.

[28] E. J. Hilliard, Jr. and R. F. Pinkos, "An analysis of triangulation ranging using beta density angular errors," *J. Acoust. Soc. Amer.*, vol. 65, pp. 1218-1228, May 1979.

[29] M. J. Hinich, "Estimating bearing when the source is endfire to an array," *J. Acoust. Soc. Amer.*, vol. 65, no. 3, pp. 845-846, 1979.

[30] L. R. Howell, "Passive sonar bearing estimation in the presence of highly anisotropic noise fields," Ph.D. dissertation, Catholic Univ. America, Washington, DC, 1978.

[31] R. L. Kirlin, "Augmenting the maximum likelihood delay estimator to give maximum likelihood direction," *IEEE Trans. Acoust., Speech, Signal Processing*, vol. ASSP-26, no. 1, pp. 107-108, 1978.

[32] L. Kostić, "Local steam transit time estimation in a boiling water reactor," this issue, pp. 555-560.

[33] C. H. Knapp and G. C. Carter, "The generalized correlation method for estimation of time delay," *IEEE Trans. Acoust., Speech, Signal Processing*, vol. ASSP-24, no. 4., pp. 320-327, 1976.

[34] —, "Estimation of time delay in the presence of source or receiver motion," *J. Acoust. Soc. Amer.*, vol. 61, no. 6, pp. 1545-1549, 1977.

[35] J. P. Kuhn, "Detection performance of the smooth coherence transform," in *1978 IEEE Int. Conf. Acoust., Speech, Signal Processing Rec.*, 78CH1285-6ASSP, pp. 678-683.

[36] W. C. Lindsey and H. Meyr, "Complete statistical description of the phase-error process generated by correlative tracking systems," *IEEE Trans. Inform. Theory*, vol. IT-23, no. 2, pp. 194-202, 1977.

[37] V. H. MacDonald and P. M. Schultheiss, "Optimum passive bearing estimation in a spatially incoherent noise environment," *J. Acoust. Soc. Amer.*, vol. 46, pp. 37-43, 1969.

[38] S. J. McCarthy, "Likelihood filtering of passive ambiguity surfaces," Naval Ocean System Center, San Diego, CA, Tech. Rep. 408, 1979.

[39] H. Meyr, "Delay lock tracking of stochastic signals," *IEEE Trans. Commun.*, vol. COM-24, no. 3, pp. 331-339, 1976.

[40] L. E. Miller, "Capabilities of multiplicative array processing as signal detector and bearing estimator," Ph.D. dissertation, Catholic Univ. America, Washington, DC, 1974 (NTIS AD A004-587).

[41] G. Mohnkern, private communication, 1975.

[42] J. M. F. Moura and A. G. Baggeroer, "Passive systems theory with narrow-band and linear constraints: Part I–Spatial diversity," *IEEE J. Ocean Eng.*, vol. OE-3, pp. 5-13, Jan. 1978.

[43] J. M. F. Moura, "Passive systems theory with narrow-band and linear constraints: Part II–Temporal diversity," *IEEE J. Ocean Eng.*, vol. OE-4, pp. 19-30, Jan. 1979.

[44] A. V. Oppenheim, Ed., *Applications of Digital Signal Processing*. Englewood Cliffs, NJ: Prentice-Hall, 1978 (especially ch. 6 on sonar signal processing by A. B. Baggeroer).

[45] N. L. Owsley, "Source location with an adaptively focused array," in *Proc. IEEE Conf. Decision and Control*; see also N. L. Owsley and G. R. Swope, "Time delay estimation in a sensor array," this issue, pp. 519-523, and paper by Owsley in Bjørnø [7].

[46] J. T. Patzewitsch, M. D. Srinath, and C. I. Black, "Nearfield performance of passive correlation processing sonars," *J. Acoust. Soc. Amer.*, vol. 64, no. 5, pp. 1412-1423, 1978; see also J. T. Patzewitsch, M. D. Srinath, and C. I. Black, "Near field performance of passive coherence processing sonars," *IEEE Trans. Acoust., Speech, Signal Processing*. vol. ASSP-27, Part I of II, no. 2, pp. 573-582, 1979.

[47] R. G. Pridham and R. A. Mucci, "Digital interpolation beamforming for low-pass and bandpass signals," *Proc. IEEE*, vol. 67, no. 6, pp. 904-919, 1977.

[48] C. N. Pryor, private communications, 1978.

[49] A. H. Quazi, "An overview on time delay estimate in active and passive systems for target localizations," this issue, pp. 527-533.

[50] K. Scarbrough, N. Ahmed, and G. C. Carter, "On the simulation of a class of time delay estimation algorithms," this issue, pp. 534-540.

[51] S. M. Schneider, private communication on coherent interarray processing, 1978.

[52] —, private communication, 1979.

[53] P. M. Schultheiss, private communication on estimation of Doppler shift, 1977.

[54] —, "Locating a passive source with array measurements: A summary of results," in *1979 IEEE Int. Conf. Acoust., Speech, Signal Processing Rec.*, 79CH1379-7ASSP, pp. 967-970.

[55] P. M. Schultheiss and E. Weinstein, "Source tracking using passive array data," this issue, pp. 600-607.

[56] P. M. Schultheiss and J. P. Ianniello, "Optimum range and bearing estimation with randomly perturbed arrays," *J. Acoust. Soc. Amer.*, vol. 68, no. 1, pp. 167-173, 1980.

[57] S. Stein, "Algorithms for ambiguity function processing," this issue, pp. 588-599.

[58] C. Stradling, private communication, 1975.

[59] K. B. Theriault and E. F. Berkman, private communications, 1978.

[60] G. Tacconi, Ed., *Aspects of Signal I*. Boston, MA: Reidel, 1977 (including G. C. Carter, "The role of coherence in time delay estimation," pp. 252-256).

[61] R. Trueblood, private communication, 1975.

[62] H. L. Van Trees, *Detection, Estimation and Modulation Theory*. New York: Wiley, 1968; see also Part III, 1971.

[63] M. Wax, "The estimation of time delay between two signals with random relative phase shift," this issue, pp. 497-501; see also related manuscript submitted in 1980 to *IEEE Trans. Inform. Theory*.

Improving the Resolution of Bearing in Passive Sonar Arrays by Eigenvalue Analysis

DON H. JOHNSON, MEMBER, IEEE, AND STUART R. DEGRAAF

Abstract—A method of improving the bearing-resolving capabilities of a passive array is discussed. This method is an adaptive beamforming method, having many similarities to the minimum energy approach. The evaluation of energy in each steered beam is preceded by an eigenvalue-eigenvector analysis of the empirical correlation matrix. Modification of the computations according to the eigenvalue structure results in improved resolution of the bearing of acoustic sources. The increase in resolution is related to the time-bandwidth product of the computation of the correlation matrix. However, this increased resolution is obtained at the expense of array gain.

I. INTRODUCTION

An array of acoustic sensors is placed in a known spatial pattern to record the acoustic environment. Measurement of the acoustic field by an array offers two basic improvements over the signal processing capabilities of a single sensor. The first is the determination of the bearing of the acoustic source(s). Bearing cannot be obtained with a single sensor, whereas an array offers some bearing-resolving capability. This capability is usually measured by the just detectable separation of two equistrength sources for a given signal-to-noise ratio at one sensor in the array [6]. The determination of the bearing of a remote source of acoustic energy remains one of the fundamental problems of passive sonar systems. The waveforms recorded by each sensor can be acquired with as much fidelity as desired. In contrast, the size and geometry of the array are usually limited by physical considerations; these limitations restrict the spatial resolution of source bearing. The second improvement is the increase of signal-to-noise ratio. If the noise field is uncorrelated at each sensor location with respect to all other locations, the signal-to-noise ratio in the array output is increased by a factor equal to the number of sensors comprising the array. This factor decreases when the noise field is correlated at the sensor locations. The measure of improvement in the signal-to-noise ratio is array gain: the ratio of the signal-to-noise ratio at the array output to that obtained with a single sensor.

Adaptive beamforming methods (ABF) are known to have superior bearing-resolution capabilities when compared to conventional beamformers (in a theoretical sense) [6]. Specifically, the minimum energy method has been analyzed extensively in this regard. However, there is no theoretical basis indicating that this method has the best possible bearing-resolution properties. On the other hand, the array gain provided by this method is optimum: no other beamforming technique can yield a larger increase in signal-to-noise ratio. This paper is concerned with a new ABF scheme which is similar in many respects to the minimum energy method. It can demonstrate greatly increased bearing-resolution properties, but at the expense of array gain. The scheme is based on an eigenvector-eigenvalue decomposition of the empirical correlation matrix, which is then truncated so as to retain only those terms which best contribute to increased bearing resolution. This approach is similar to those described by Schmidt [13] in his MUSIC system, by Owsley [12] in his modal decomposition approach, and by Bienvenu [2], [3]. Analytic results are presented here which contrast the bearing-resolving properties of these various eigenvector methods and the minimum energy method.

Manuscript received August 14, 1981; revised November 17, 1981. This work was supported in part by an American Society for Engineering Education Fellowship and by the Office of Naval Research under Contract N00014-81-K-0565. Portions of this paper were presented at the Acoustics, Speech, and Signal Processing Spectral Estimation Workshop, McMaster University, Hamilton, Ont., Canada, August 1981.

The authors are with the Department of Electrical Engineering, Rice University, Houston, TX 77001.

II. PRELIMINARIES

Let $x_m(t)$ denote the outputs taken from an array of sensors having a known geometry. *Beamforming* consists of computing the quantity

$$y(t) = \sum_{m=0}^{M-1} a_m x_m(t - \tau_m) \tag{1}$$

where a_m is the amplitude weighting (shading) applied to the mth sensor output, τ_m is the delay applied to the mth sensor output, and M is the number of sensors in the array. The parameters $\{a_m\}$ and $\{\tau_m\}$ of a beam are chosen according to some desired criterion (e.g., steering the beam in a particular direction, minimizing sidelobe height, etc.).

Beamforming can also be viewed as a type of multidimensional spectral analysis [1], [9]. Evaluating the Fourier transform of the beam $y(t)$, we have

$$Y(f) = \sum_{m=0}^{M-1} a_m e^{-j2\pi f \tau_m} X_m(f). \tag{2}$$

Therefore, at each temporal frequency f, the Fourier transform of the beam can be written as the dot product of two vectors

$$Y(f) = \boldsymbol{A}'\boldsymbol{X} \tag{3}$$

where $\boldsymbol{A}$ denotes the steering vector consisting of the elements

$$A_m = a_m e^{+j2\pi f \tau_m}$$

Reprinted from *IEEE Trans. Acoust., Speech, Signal Processing,* vol. ASSP-30, pp. 638–647, Aug. 1982.

and $\boldsymbol{X}$ denotes the vector comprised of the Fourier transforms of the sensor outputs. Here, A' denotes the conjugate transpose of $\boldsymbol{A}$. Assume that we have a linear array of equally spaced sensors; in this instance, the delays τ_m will be of the form $\tau_m = mT$. Equation (2) becomes

$$Y(f, T) = \sum_{m=0}^{M-1} a_m e^{-j2\pi m f T} X_m(f). \tag{4}$$

Consequently, $Y(f, T)$ is the Fourier transform of the sequence $a_m X_m(f)$. For a linear array, computing a transform along the index m is identical to computing the transform in space across the array. $Y(f, T)$ is, therefore, the spatial transform of $a_m X_m(f)$ evaluated at the spatial frequency $k = fT$. The result of applying a particular shading a_m is to convolve the true (infinite aperture) spatial spectrum with the Fourier transform of a_m, $m = 0, \cdots, M-1$, thereby smearing the true spectrum and limiting the resolution that can be obtained.

III. High-Resolution Techniques

One can obtain a set of weights (or, equivalently, a steering vector) to achieve better resolution by adapting them to the particular noise field and signal field impinging on the array. In these adaptive beamforming schemes, the steering vector is the solution to an optimization problem [4], [11]: find the steering vector which minimizes the energy in the beam subject to the constraint $\boldsymbol{A}'\boldsymbol{Z} = 1$ where $\boldsymbol{Z}$ is the constraint vector. The energy contained in a beam can be expressed by the quadratic form $\boldsymbol{A}'\mathrm{R}\boldsymbol{A}$ where R denotes the empirical spatial correlation matrix of the Fourier transforms of the sensor outputs. The correlation matrix R is usually estimated from the sensor outputs by a variation of the Bartlett procedure. The output of each sensor is sectioned and windowed. The Fourier transform of each section is evaluated and the vector $X_i(f_0)$ of Fourier transform values at the frequency f_0 across the array for the ith section is formed. Assuming that K sections are available, R is computed according to

$$\mathrm{R} = \frac{1}{K} \sum_{i=0}^{K-1} X_i(f_0) X_i'(f_0). \tag{5}$$

Usually K is taken to be the number of statistically independent terms used in the empirical computation of the correlation matrix R; K is frequently referred to as the time-bandwidth product.

The solution to this optimization problem is

$$\boldsymbol{A} = \frac{\mathrm{R}^{-1}\boldsymbol{Z}}{\boldsymbol{Z}'\mathrm{R}^{-1}\boldsymbol{Z}}. \tag{6}$$

The resulting value of the energy in the beam is

$$\boldsymbol{A}'\mathrm{R}\boldsymbol{A} = (\boldsymbol{Z}'\mathrm{R}^{-1}\boldsymbol{Z})^{-1}. \tag{7}$$

In the so-called high-resolution [5] or *minimum energy* scheme, the constraint vector $\boldsymbol{Z}$ is a plane-wave direction-of look vector $\boldsymbol{W}$, each element of which is given by $W_m = e^{j2\pi mk}$ where k corresponds to a specific spatial frequency. As $k = fT$ and f is known, each value of k corresponds to a specific per-channel delay. Defining θ to the bearing of the source relative to array-broadside, $T = (d/f\lambda)\sin\theta$. Consequently, spatial frequency k is related to bearing θ as $k = (d/\lambda)\sin\theta$. The constraint $\boldsymbol{A}'\boldsymbol{W} = 1$ fixes the gain of the steering vector in the direction-of-look $\boldsymbol{W}$ to be unity. Forcing the energy to be minimum thereby reduces the contributions from plane waves arriving from other directions and from the noise field. The energy in the beam corresponding to the spatial frequency k is expressed by

$$S_{ME}(k) = (\boldsymbol{W}'\mathrm{R}^{-1}\boldsymbol{W})^{-1}. \tag{8}$$

The bearing of the target(s) is determined by finding the spatial frequency(s) at which the quantity in (8) achieves maxima. The maximum likelihood spectral estimate is closely related to the minimum energy estimate, but differs from it in an important way. The maximum likelihood estimate uses the noise-only correlation matrix in its evaluation of the beam energy:

$$S_{ML}(k) = (\boldsymbol{W}'\mathrm{R}_n^{-1}\boldsymbol{W})^{-1}. \tag{9}$$

Here $\mathrm{R}_n = E[NN']$, the theoretical correlation matrix of the noise component of $\boldsymbol{X}$. Note specifically that the matrix R_n is assumed to be a known quantity and is not computed from empirical data. The maximum likelihood method yields the optimum array gain [6], [7]. The maximum likelihood and minimum energy methods yield the same array gain when the beam is steered toward the acoustic source [6].

IV. Improving Resolution

From the viewpoint presented here, there is great flexibility in choosing the constraint vector $\boldsymbol{Z}$. One wonders if there is a particular choice for the constraint vector which can maximize the spatial resolution of source bearing. For example, choose a constraint vector of the form

$$\boldsymbol{Z} = \mathrm{C}\boldsymbol{W} \tag{10}$$

where C is a matrix to be described. The energy in the beam when steered toward the source would then be

$$S(k) = (\boldsymbol{W}'\mathrm{C}'\mathrm{R}^{-1}\mathrm{C}\boldsymbol{W})^{-1}. \tag{11}$$

Suppose the vector $\boldsymbol{W}$ corresponded to an actual plane-wave source. If C were a matrix having the property that this choice of $\boldsymbol{W}$ lay in the null space of the matrix ($\mathrm{C}\boldsymbol{W} = \boldsymbol{0}$), the energy in the beam when steered in this direction would be infinite. If the direction vector corresponding to the plane-wave source were the *only* direction vector that lay in the null space of C, one would therefore obtain a marked indication of the bearing of the source.

While it is theoretically possible to have a perfect indication of source bearing by this approach, the difficulty lies in finding the matrix C. This matrix must have the property that direction vectors lying in its null space correspond only to plane waves eminating from actual sources. Assuming that the bearing of the sources is not known, construction of the matrix C would seem impossible. However, one can construct a matrix having a null space consisting of vectors which closely resemble direction vectors of source plane waves. The key idea of this procedure is to carefully analyze the eigenvectors and eigenvalues of the correlation matrix.

A. The Eigenvector Method

The eigenvectors of the matrix R are defined by the property

$$\mathrm{R} V_i = \lambda_i V_i \qquad i = 1, \cdots, M \tag{12}$$

where λ_i is the eigenvalue associated with the eigenvector V_i. As correlation matrices are conjugate-symmetric (Hermitian), the eigenvectors form an orthonormal set. Assume that the sensor outputs contain one signal and noise uncorrelated with the signal ($X = S + N$). The result of computing R according to (5) for sufficiently large K is

$$\mathrm{R} = \sigma_n^2 \mathrm{Q} + \sigma_s^2 SS'. \tag{13}$$

Q is the noise correlation matrix normalized to have a trace equal to the dimension M of the matrix R. S is the direction vector of a plane wave source and has a squared norm equal to M. The cross terms involving signal and noise are assumed to be negligible. If the noise correlation matrix equals the identity matrix (i.e., only sensor noise is assumed to be present), the eigenvector corresponding to the largest eigenvalue (hereby referred to as the "largest eigenvector") is equal to the vector S with eigenvalue equal to $\sigma_n^2 + M\sigma_s^2$. The remaining $M - 1$ eigenvectors of R consist of those vectors orthogonal to S and each has eigenvalue σ_n^2. If p *incoherent* linearly independent signals S_i, $i = 1, \cdots, p$, are present so that the correlation matrix is of the form

$$\mathrm{R} = \sigma_n^2 \mathrm{I} + \sum_{i=1}^{p} \sigma_{s,i}^2 S_i S_i', \tag{14}$$

the p largest eigenvectors correspond to the signal terms and the $M - p$ smallest are orthogonal to all of the signal direction vectors. Note that the largest eigenvectors are not necessarily equal to the signal direction vectors in this case; these eigenvectors comprise an orthonormal basis for the vector space containing the signal vectors. Consequently, one cannot always inspect the largest eigenvectors and determine the signal vectors directly.

Define C_{EV} to be the sum of the outer products of the $M - p$ smallest eigenvectors:[1]

$$\mathrm{C}_{EV} = \sum_{i=1}^{M-p} V_i V_i'. \tag{15}$$

As the p largest eigenvectors are orthogonal to each of the $M - p$ smallest, $\mathrm{C}_{EV} V_i = \mathbf{0}$, $i = M - p + 1, \cdots, M$. As the p largest eigenvectors span the space containing the signal vectors, each signal lies in the null space of C_{EV} and $\mathrm{C}_{EV} S_i = \mathbf{0}$. In this manner, perfect resolution of the bearing of multiple sources can be obtained from an eigenvector analysis of the correlation matrix R.

Note that in computing the beam energy (11), the matrix C_{EV} need never be computed. The correlation matrix R can be expressed in terms of its eigenvectors as

$$\mathrm{R} = \sum_{i=1}^{M} \lambda_i V_i V_i' \tag{16}$$

[1]By convention, the smallest eigenvector is denoted by the subscript 1, the next smallest by 2, etc.

and the inverse of R as

$$\mathrm{R}^{-1} = \sum_{i=1}^{M} \frac{1}{\lambda_i} V_i V_i'. \tag{17}$$

Because of the orthogonality property of the eigenvectors of a correlation matrix, the quantity $\mathrm{C}_{EV}' \mathrm{R}^{-1} \mathrm{C}_{EV}$ appearing in (11) becomes

$$\mathrm{C}_{EV}' \mathrm{R}^{-1} \mathrm{C}_{EV} = \sum_{i=1}^{M-p} \frac{1}{\lambda_i} V_i V_i'. \tag{18}$$

It is this matrix which is computed in the evaluation of the quadratic form of (11).

$$S_{EV}(k) = (W' \mathrm{C}_{EV}' \mathrm{R}^{-1} \mathrm{C}_{EV} W)^{-1} = \left[\sum_{i=1}^{M-p} \frac{1}{\lambda_i} |W' V_i|^2 \right]^{-1}. \tag{19}$$

The following sequence of computations constitute the *eigenvector method.*

1) Compute the correlation matrix R.

2) Decompose the matrix R into its eigenvectors and eigenvalues.

3) Determine the number p of sources present in the acoustic field.

4) Compute the energy in the beams corresponding to all possible bearings (19).

5) The major peaks in this spectrum correspond to acoustic sources.

The methods of Owsley [12], Bienvenu [2], [3], and Schmidt [13] are similar, but differ somewhat from the eigenvector method just described. The expression for R^{-1} is truncated as in (5); however, the small eigenvalues are set to the same value (taken here to be unity). Instead of evaluating the quadratic form of (11), the spectral estimate of the MUSIC method can be expressed by

$$S_{\mathrm{MUSIC}}(k) = (W' \mathrm{C}_{EV}' \mathrm{C}_{EV} W)^{-1} = \left[\sum_{i=1}^{M-p} |W' V_i|^2 \right]^{-1}. \tag{20}$$

This estimate can also be evaluated in the same manner as (11) if the matrix C is redefined to be

$$\mathrm{C}_{\mathrm{MUSIC}} = \sum_{i=1}^{M-p} \sqrt{\lambda_i}\, V_i V_i'.$$

A steering vector can be defined (6) when the latter approach is used to express the MUSIC spectral estimate. Under the conditions just described, this spectral estimate also has the capability of resolving source bearing perfectly.

The presumption of the preceding analysis has been that the noise correlation matrix Q is equal to the identity matrix. It is this key assumption which leads to perfect bearing resolution. Imperfection in bearing resolution occurs when this presumption is false. The noise field may contain more than just sensor noise: for example, isotropic noise may also be present. In addition, a finite amount of averaging is used to compute the correlation matrix R. Even if the noise field were spatially white, an empirical noise correlation matrix would not be an identity matrix. Either of these situations can reduce the resolution of the eigenvector and MUSIC methods.

The performances of the eigenvector method and of the MUSIC method under these more realistic conditions are analyzed mathematically in the Appendix. From this analysis, an approximate lower limit on the energy in the beam when steered on-target using the eigenvector method is found to be

$$S(k) = \frac{\sigma_s^2}{\frac{\sigma_n^2}{\sigma_s^2}\frac{\gamma^2}{M} + \frac{M}{K}}. \tag{21}$$

A similar result is obtained for the MUSIC method. The quantity γ is a measure of the spread of the eigenvalues α_i of Q. For the eigenvector method, this quantity is given by

$$\gamma_{EV}^2 = \frac{1}{M}\sum_{i=1}^{M}\left(\alpha_i - \frac{1}{M}\sum_m \alpha_m^2\right)^2 \tag{22}$$

and for the MUSIC method by

$$\gamma_{\text{MUSIC}}^2 = \frac{1}{M}\sum_{i=1}^{M}\alpha_i\left(\alpha_i - \frac{1}{M}\sum_m \alpha_m^2\right)^2. \tag{23}$$

γ_{MUSIC}^2 tends to be smaller than γ_{EV}^2. In either case, the quantity γ is an implicit function of the time-bandwidth product K. Generally speaking, γ will decrease with increasing K, tending toward the spread of the eigenvalues of the spatial correlation matrix of the underlying noise process. If the theoretical noise correlation matrix equals the identity matrix (i.e., spatially white noise), the spread of its eigenvalues is zero. Otherwise, the spread is nonzero.

The first term in the denominator of (21) determines how large the energy peak will be if an infinite time-bandwidth product were available. The second term describes how the size of the energy peak depends on K. The larger of these will dominate the expression in (21). When steered off-target, the energy in the beam produced by the eigenvector method coincides with that produced with the minimum energy approach. The ratio ν of the on-target to off-target beam energy can be used to assess the size of the peak in the beam energy as the direction vector $\boldsymbol{W}$ is scanned through all possible bearings. When only sensor noise is present, this quantity is given in the minimum energy method by [10]

$$\nu_{ME} = 1 + \frac{M\sigma_s^2}{\sigma_n^2}. \tag{24}$$

In the eigenvector method and the MUSIC method, this quantity is given by

$$\nu = \frac{1}{\frac{\sigma_n^2}{\sigma_s^2}\frac{\gamma^2}{M^2} + \frac{1}{K}} \cdot \frac{\sigma_s^2}{\sigma_n^2}. \tag{25}$$

From simulation studies, this quantity can be somewhat larger (a few decibels) in the MUSIC method than in the eigenvector method. Comparing (24) and (25), one concludes that when the eigenvector method is used, the array appears to consist of a number of elements $\tilde{M}$ given by

$$\tilde{M} = \frac{1}{\frac{\sigma_n^2}{\sigma_s^2}\frac{\gamma^2}{M^2} + \frac{1}{K}}. \tag{26}$$

In most instances, this quantity is larger than the actual number of sensors M.

B. Simulation Results

Computer simulations were used to evaluate the eigenvector and MUSIC methods and to compare them with the minimum energy method. In these studies, sequences of the data vector $\boldsymbol{X}_i = \boldsymbol{S} + \boldsymbol{N}_i$ were produced. The parameters of the signal vector $\boldsymbol{S}$ were defined as described in Section II. The noise vector $\boldsymbol{N}_i$ consisted of identically distributed complex Gaussian noise components. The covariance matrix of this random vector could be specified by the user. Each noise vector was generated to be statistically independent of all other noise vectors. The empirical correlation matrix R was then computed as in (5), and its eigenvectors and eigenvalues were computed according to a *QL* algorithm [8].

With one exception, each step of the procedure outlined above for the eigenvector method was followed. As the number p of sources present in the acoustic field was known by the user, it was supplied by him. In a physical situation, this parameter would not be so readily known. However, the purpose of the simulations was to determine the validity of the theory and to test how well the methods could perform. The effects of inaccurate choices for p are described in a later section.

A comparison of the beam energies produced by the minimum energy method and by the eigenvector method is shown in Fig. 1(a) when one source is present in the acoustic field. A similar comparison is found in Fig. 1(b) for the MUSIC method. Note that the height of the main peak relative to the background noise level varies with K, the time-bandwidth product, in both methods, whereas it does not in the minimum energy method. In this example, sensor noise (i.e., spatially white) is present. For the array length used, the second term in the denominator of (21) was larger than the first. Simulated and theoretical values of ν_{EV} are compared in Table I. The theoretical prediction of the value of this quantity is close to that obtained from the simulations.

The spectra obtained from the eigenvector method and the MUSIC method differ only slightly in these examples. The latter method tends to produce much flatter off-target spectra. The small eigenvalues of the correlation matrix R tend to represent the noise field [see the discussion following (13)], and the equalization of these values will "whiten" the background noise, thereby resulting in a flat spectrum. This effect is further illustrated by considering a case where isotropic noise is present in the acoustic field (Fig. 2).

C. Resolution of Multiple Targets

Cox [6] derives analytic expressions for the limits to which the minimum energy method applied to a linear array can resolve two equistrength incoherent acoustic sources. There, "resolution" is defined as the minimum bearing separation at broadside at which two targets can be distinguished by evaluating beam energy. Here, a slight dip in the beam energy is re-

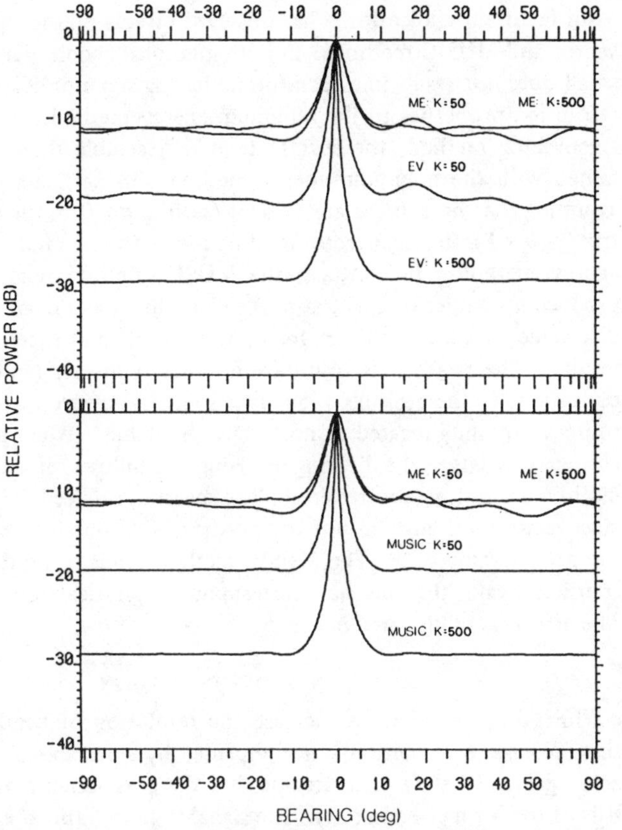

Fig. 1. Energy in beams formed by the minimum energy method (*ME*) and by the eigenvector method (*EV*) are plotted against bearing for a linear array. Energy is expressed in decibels relative to the peak value. Bearing is expressed in degrees with zero corresponding to broadside. The ten sensors are equally spaced and separated by half a wavelength. The source was assumed to be narrow-band, with all of its source energy concentrated in one temporal-frequency analysis bin. In each subfigure, sensor noise and one source located at 0° are present in the sound field; the sensor signal-to-noise ratio is 0 dB. The results obtained when two time-bandwidth products (*K*) are used are shown in each subfigure for each method.

TABLE I
EMPIRICAL AND THEORETICAL PEAK-TO-BACKGROUND RATIOS

K	$\widetilde{M}$	$\hat{\nu}_{ME}$(dB)	$\hat{\nu}_{EV}$(dB)	ν_{EV}(dB)
20	18	10	13	13
50	46	11	18	17
100	90	12	22	20
200	183	11	27	23
500	461	11	29	27
1000	940	11	32	30
2000	1784	11	34	33
5000	4597	11	39	37

The results of computer simulations and theoretical predictions of the value of ν are shown. In the simulations, an equally spaced linear array containing ten sensors was assumed to be present in an acoustic field containing sensor noise and a plane-wave source. The sensor signal-to-noise ratio was 0 dB. Values of γ were computed in separate simulations from noise-only correlation matrices having the same time-bandwidth product. The source was assumed to be narrow-band, with all of its source energy concentrated in one temporal-frequency analysis bin. The sensor spacing is one-half wavelength. Plot of beam energy versus bearing were obtained and empirical values of ν estimated. The quantities $\hat{\nu}_{ME}$ and $\hat{\nu}_{EV}$ correspond to these empirical values.

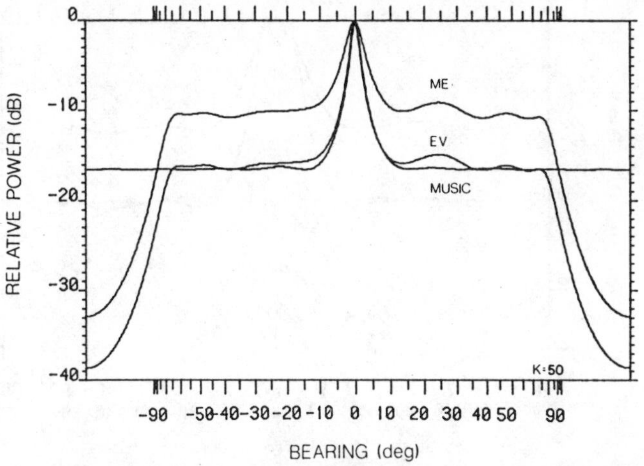

Fig. 2. Beam energy is plotted against bearing when one source (located at 0°) and isotropic noise are present in the sound field. The array configuration is similar to that described in Fig. 1, the only exception being that the sensor spacing is three-eighths of a wavelength. The sensor signal-to-noise ratio is 0 dB. The results of applying the minimum energy (*ME*), eigenvector (*EV*), and MUSIC methods are shown. The time-bandwidth product here is 50; the theoretical values of ν corresponding to this situation are $\nu_{EV} = 15.7$ dB and $\nu_{\text{MUSIC}} = 16.4$ dB.

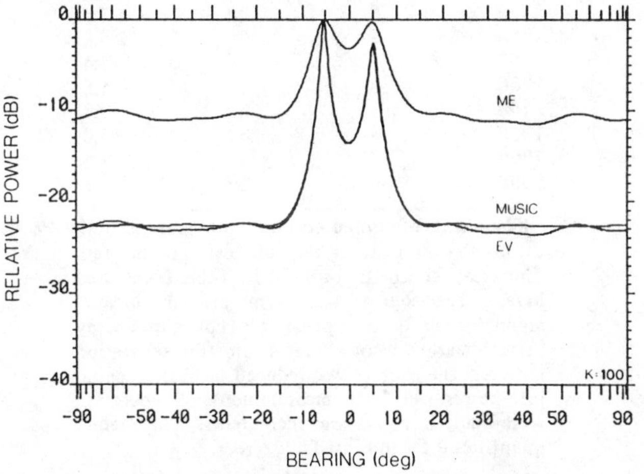

Fig. 3. Beam energy is plotted against bearing when two sources are present in the acoustic field. The conventions defined in Fig. 1 apply to this plot. Here, the sensor signal-to-noise ratio of each source is 0 dB and the sources are located at -5° and +5°.

quired when the array is steered between the sources. The critical factors determining the resolution of beams formed by the minimum energy approach are aperture [defined as the spatial extent of the array relative to a wavelength - (D/λ)], the number of elements in the array (M), and the sensor signal-to-noise ratio (σ_s^2/σ_n^2). Cox's results are summarized in [6, Fig. 5]. An approximation to those results is

$$\frac{M\sigma_s^2}{\sigma_n^2} \propto \left(\theta \frac{D}{\lambda}\right)^4 \tag{27}$$

where θ is bearing separation of the two targets.

When the eigenvector method is applied in situations such as these, targets are more easily resolved, and furthermore, the resolution capabilities of the array are increased. Figs. 3 and 4

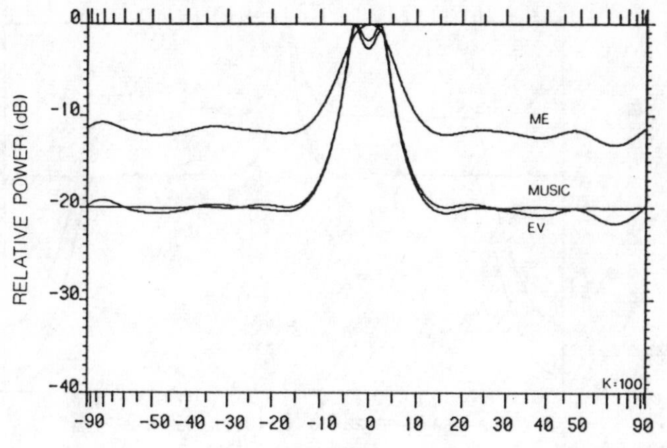

Fig. 4. Beam energy is plotted against bearing when two sources are present in the acoustic field. The conventions defined in Fig. 1 apply to this plot. Here, the sensor signal-to-noise ratio of each source is 0 dB and the sources are located at −3° and +3°.

TABLE II
EMPIRICAL AND THEORETICAL RESOLUTION LIMITS

K	θ_{ME}	$\hat{\theta}_{EV}$	θ_{EV}
20	8	7.5	7.5
50	8	6	5.8
100	8	6	5.4
200	8	4.5	4.6
500	8	3.5	3.7
1000	8	3.5	3.0
2000	8	3	2.6
5000	8	2.5	2.1

The results of computer simulations and theoretical predictions of resolution limits are summarized. The array configuration used in Table I was used here. The sources were symmetrically located about broadside (0 degrees). Plots of beam energy versus bearing were obtained, and the separation between the sources was reduced until they could just be resolved. The measurements of separation were made in half-degree increments. The angular quantities θ are indicated in degrees.

display typical examples of these cases. A theoretical prediction of the degree to which resolution is increased can be obtained from (27). If one substitutes $\tilde{M}$ evaluated by (26) for M, the value of θ thus obtained is the resolution limit of the eigenvector method and the MUSIC method. A comparison of the resolution obtained from some of the simulations with that predicted by the theory is shown in Table II. The degree of agreement between theory and simulation results implied by Table II is valid for all of the simulations.

D. Effect of an Improper Choice for p

The theoretical and simulation results presented thus far are valid only when the parameter p equals the actual number of acoustic sources. In practice, this quantity may not be known, and one questions the sensitivity of the eigenvector and MUSIC methods to an incorrect choice of p. This sensitivity was studied through the simulations; no analytic results were obtained on this issue.

An incorrect choice for p has different effects on the eigenvector and MUSIC methods. In both methods, choosing p too small does *not* result in a beamformer having superior bearing-resolution properties to the minimum energy method. In the eigenvector method, the spectra tend to resemble those obtained with the minimum energy method. In particular, by assuming that no sources are present (setting $p = 0$ or the matrix $C_{EV} = I$), the eigenvector method is exactly the minimum energy method. In contrast, the MUSIC method tends to produce a number of spectral peaks equal to p. For example, if a value of zero is chosen for p, a uniformly flat spectrum results. The peaks that result from nonzero choices tend to correspond to the bearings of acoustic sources; however, which sources are thus located is not easily predicted. When p is chosen too large, the bearing-resolving capabilities of either method are not greatly reduced. The spectra produced by the eigenvector method tend not to vary from that obtained with the proper value of p. The MUSIC method tends to produce spurious peaks that do not correspond to physical sources. The effects are illustrated in Fig. 5.

V. RESOLUTION AND ARRAY GAIN

While these approaches increase the resolution of bearing, this increase in resolution is accompanied by a decrease in the array gain. To show this, assume that the correlation matrix R is of the form given in (13). The signal-to-noise ratio at each sensor is therefore σ_s^2/σ_n^2. The signal-to-noise ratio in the beam output is the quantity

$$\frac{\sigma_s^2 A'SS'A}{\sigma_n^2 A'QA}. \tag{28}$$

The array gain G is the ratio of these signal-to-noise ratios:

$$G = \frac{A'SS'A}{A'QA}. \tag{29}$$

Recalling that the steering vector in this case is given by

$$A = \frac{R^{-1}CW}{W'C'R^{-1}CW} \tag{30}$$

$$G = \frac{S'C'R^{-1}SS'R^{-1}CS}{S'C'R^{-1}QR^{-1}CS}. \tag{31}$$

and setting the direction vector to correspond to the source ($W = S$), the array gain becomes

As the matrix Q is given by

$$Q = \frac{1}{\sigma_n^2}(R - \sigma_s^2 SS'), \tag{32}$$

the denominator of (31) becomes

$$S'C'R^{-1}QR^{-1}CS = \frac{1}{\sigma_n^2}(S'C'R^{-1}CS - \sigma_s^2 S'C'R^{-1}S \cdot S'R^{-1}CS) \tag{33}$$

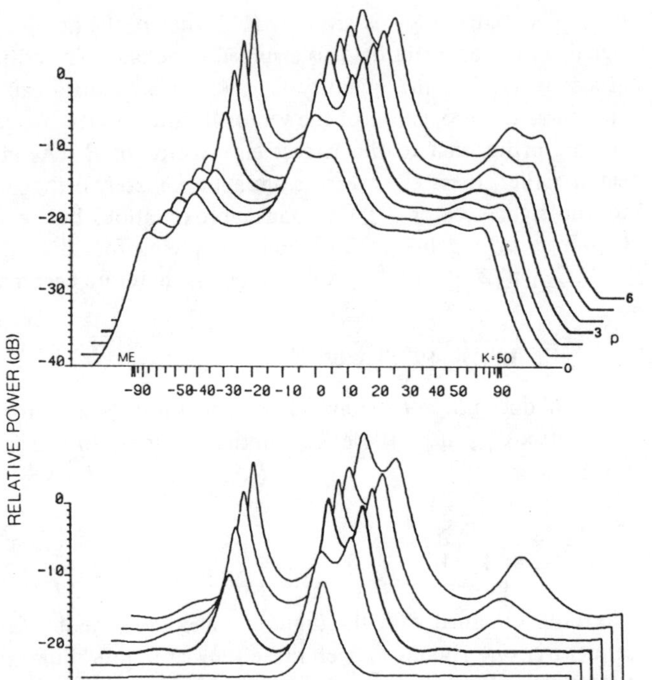

Fig. 5. Beam energy is plotted against bearing with the number p of terms truncated from the eigenvector expansion of R^{-1} as a parameter. The parametric beam energy functions in each panel is plotted with the same vertical scaling. A linear array of ten equally spaced sensors (spacing equal to three-eighths of a wavelength) is present in an acoustic field. Three incoherent sources are present in the acoustic field: two have unity amplitude and are located at bearings +5° and -5°, while the third has an amplitude of one-half and a bearing of -40°. Isotropic noise is also present in the acoustic field; the sensor signal-to-noise ratio (relative to the larger signals) is 0 dB. The time-bandwidth product in both panels is 50. The upper panel displays the result of applying the eigenvector method, and the bottom panel illustrates the result of applying the MUSIC method for the same set of data. Note that the proper value of p for these data is $p = 3$.

so that we obtain

$$G = \frac{\sigma_n^2}{\dfrac{\boldsymbol{S}'\mathrm{C}'\mathrm{R}^{-1}\mathrm{C}\boldsymbol{S}}{\boldsymbol{S}'\mathrm{C}'\mathrm{R}^{-1}\boldsymbol{S}\cdot\boldsymbol{S}'\mathrm{R}^{-1}\mathrm{C}\boldsymbol{S}} - \sigma_s^2}. \tag{34}$$

For the eigenvector method,

$$\mathrm{C}'_{EV}\mathrm{R}^{-1} = \mathrm{R}^{-1}\mathrm{C}_{EV} = \mathrm{C}'_{EV}\mathrm{R}^{-1}\mathrm{C}_{EV} \tag{35}$$

and the array gain becomes

$$G_{EV} = \frac{\sigma_n^2\boldsymbol{S}'\mathrm{C}'_{EV}\mathrm{R}^{-1}\mathrm{C}_{EV}\boldsymbol{S}}{1 - \sigma_s^2\boldsymbol{S}'\mathrm{C}'_{EV}\mathrm{R}^{-1}\mathrm{C}_{EV}\boldsymbol{S}}. \tag{36}$$

In the MUSIC method, the ratio appearing in the denominator of (34) can be bounded using the Schwarz inequality:

$$\frac{\boldsymbol{S}'\mathrm{C}'_{\mathrm{MUSIC}}\mathrm{R}^{-1}\mathrm{C}_{\mathrm{MUSIC}}\boldsymbol{S}}{\boldsymbol{S}'\mathrm{C}'_{\mathrm{MUSIC}}\mathrm{R}^{-1}\boldsymbol{S}\cdot\boldsymbol{S}'\mathrm{R}^{-1}\mathrm{C}_{\mathrm{MUSIC}}\boldsymbol{S}} \geqslant \frac{1}{\boldsymbol{S}'\mathrm{C}'_{EV}\mathrm{R}^{-1}\mathrm{C}_{EV}\boldsymbol{S}}. \tag{37}$$

Equality occurs only when the $M - 1$ smallest eigenvalues of R are identical (i.e., Q = I). This bound can be used in (34) to obtain an upper bound on G_{MUSIC} if the bound is not smaller than σ_s^2. As the expression thus obtained equals G_{EV} (36), this condition is satisfied. Consequently,

$$G_{\mathrm{MUSIC}} \leqslant G_{EV}. \tag{38}$$

Considering (36), G_{EV} is a monotonically increasing function of the quadratic form $\boldsymbol{S}'\mathrm{C}'_{EV}\mathrm{R}^{-1}\mathrm{C}_{EV}\boldsymbol{S}$. Therefore, whenever one decreases this quadratic form to improve the indication of bearing (19), the array gain decreases in the eigenvector method. Because of the relationship given in (38), the array gain obtained with the MUSIC method also decreases. In the limit, perfect indication of bearing (a zero-valued quadratic form) corresponds to zero array gain with either method.

VI. Conclusions

The eigenvector method can enhance the bearing-resolving capabilities of an array. Here, the eigenvectors and eigenvalues of the correlation matrix must be found and the weighted sum of Fourier transforms of the eigenvectors computed. In the minimum energy method, the inverse of the correlation matrix must be found and the quadratic form of (8) computed. Roughly speaking, the computational complexities involved in the use of the eigenvector method are not excessive when compared to those required in the minimum energy method.

The degree to which the resolving power is increased is related to the quantity $\tilde{M}$. Because of (27), this increase is proportional to $(\tilde{M}/M)^{1/4}$. Consequently, to increase the bearing resolution by a factor of two requires the virtual number of sensors $\tilde{M}$ to be 16 times the actual number. Referring to Table I, such large virtual array lengths can be obtained only when large time-bandwidth products are possible. Under these circumstances, enhanced bearing resolution is possible. For a given time-bandwidth product, the smaller the number of elements in the array, the greater the increase in bearing resolution.

The eigenvector method and the MUSIC method produce quite similar results. They differ in at least two respects, however. The first is that a nonzero value of p, the number of assumed sources in the acoustic field, must be chosen in the MUSIC method. If the value of p is not close to the actual value, the spectra thus obtained can differ from that obtained with the proper value: spurious peaks appear and/or peaks can be missed. In contrast, the eigenvector method is less sensitive to the choice of p. Second, the shape of the spectrum of the background "noise" is drastically altered in the MUSIC method. For example, the variations due to low-level sources or to the physical noise spectrum are lost (see Fig. 2, for example). This portion of the spectrum can also vary as p is changed; this effect is much less pronounced in the eigenvector method.

The increase in bearing resolution is obtained at the expense of array gain. Consequently, if more than bearing information is required, other techniques should probably be used to obtain them. One can conceive of the eigenvector method being used to acquire source bearing, and this information being used to steer a beam with the minimum energy method so as to analyze the waveform produced by the source. Note that this two-step procedure need only be sequential in a conceptual manner.

Because of the close relationship between the two methods, obtaining the steering vector for the minimum energy beamformer means including more terms in the eigenvector decomposition of R.

The decrease of array gain with increasing resolution raises many theoretical issues. The minimum energy method is known to yield the optimal value of array gain. Consequently, any method which has greater resolution capabilities cannot also increase array gain. Can array gain be maintained while increasing resolution or is increased resoltuion always obtained at the expense of array gain? The present method has the latter property. A theoretical understanding of the limits to which array gain and resolution can be traded against each other would be of interest.

The main issue not addressed in this study is the determination of the number of sources, p. From the analysis presented in Section IV, the number of sources corresponds to the number of dominant eigenvalues in the matrix R. Determining p in this way can be difficult when small time-bandwidth products are available and isotropic noise is present. Reasonably accurate methods of determining p from the eigenvalues of R are not known at this time.

APPENDIX

Let the vector X of sensor outputs be of the form $X = \sigma_s S + \sigma_n N$ where S denotes the source direction vector as before and N denotes additive noise. The correlation matrix R is computed empirically according to (5) to yield

$$\mathrm{R} = \sigma_n^2 Q + \sigma_n \sigma_s S\tilde{N}' + \sigma_n \sigma_s \tilde{N} S' + \sigma_s^2 SS' \tag{A1}$$

where $\mathrm{Q} = (1/K) \sum_{i=1}^{K} N_i N_i'$, a statistical estimate of the noise correlation matrix, and $\tilde{N} = (1/K) \sum_{i=1}^{K} N_i$, an estimate of the average noise component. The vectors N_i are assumed to be statistically independent random vectors; each component of N_i has zero mean and unity variance. Consequently, the components of the vector $\tilde{N}$ have zero mean and energy $1/K$. Define the matrix P to be the noise-related terms in (A1):

$$\mathrm{P} = \mathrm{Q} + \frac{\sigma_s}{\sigma_n} S\tilde{N}' + \frac{\sigma_s}{\sigma_n} \tilde{N} S'. \tag{A2}$$

Consequently, the expression for R in (A1) can be written more simply as

$$\mathrm{R} = \sigma_n^2 \mathrm{P} + \sigma_s^2 SS'. \tag{A3}$$

The estimate of the energy in the beam pointed in the W direction is given by (11). Following Cox [6], this expression can be written as

$$S(k) = \frac{\sigma_n^2}{W'\mathrm{C}'\mathrm{P}^{-1}\mathrm{C}W} \left[\frac{1 + \left(\frac{S}{N}\right)_{\max}}{1 + \left(\frac{S}{N}\right)_{\max} \sin^2 (\mathrm{C}W, S; \mathrm{P}^{-1})} \right] \tag{A4}$$

where $(S/N)_{\max} = S'\mathrm{P}^{-1} S (\sigma_s^2/\sigma_n^2)$ is the signal-to-noise ratio of the beam output obtained with the optimally chosen steering vector and $\sin^2 (\mathrm{C}W, S; \mathrm{P}^{-1})$ is the sine squared of the angle between the vectors $\mathrm{C}W$ and S with respect to the matrix P^{-1}. The matrix C_{EV} is given by (15) when p, the number of signals in the acoustic field, is assumed to be one. The critical aspect of (A4) is the quantity $\mathrm{C}_{EV} W$. This quantity can be viewed as the projection of the vector W onto the set of eigenvectors orthogonal to the largest eigenvector of R. As indicated earlier, these eigenvectors are approximately orthogonal to the signal vector. To a good approximation, the vector $\mathrm{C}_{EV} W$ is orthogonal to S, thereby implying for all W that $\sin^2 (\mathrm{C}_{EV} W, S; \mathrm{P}^{-1}) = 1$. Expression (A4) therefore becomes

$$S(k) = \frac{\sigma_n^2}{W'\mathrm{C}_{EV}'\mathrm{P}^{-1}\mathrm{C}_{EV} W}. \tag{A5}$$

When W does not correspond to the signal direction vector S, the matrix C_{EV} has little effect on the vector W. In this case, one obtains

$$S(k) = \frac{\sigma_n^2}{W'\mathrm{P}^{-1} W},$$

the result obtained with the minimum energy estimate. Consequently, one should expect the eigenvector procedure and the minimum energy procedure to yield the same numerical values when steered off-target. As the beam is steered toward the source, the two estimates will begin to differ as the matrix C_{EV} begins to affect the vector W.

In the succeeding analysis, the inverse of the matrix P is assumed to be approximately equal to the inverse of Q in the computation of the quadratic form appearing in (A5). Inspecting (A2), this approximation will be less accurate as the signal-to-noise ratio (σ_s/σ_n) increases and as the amount of averaging (K) decreases. The energy estimate can be written approximately as

$$S(k) \approx \frac{\sigma_n^2}{W'\mathrm{C}_{EV}'\mathrm{Q}^{-1}\mathrm{C}_{EV} W}. \tag{A6}$$

The quadratic form in (A6) can be interpreted as the squared length of the vector $\mathrm{C}_{EV} W$ with respect to the norm induced by the matrix Q^{-1}. We therefore seek an expression for this quantity when $W = S$.

Assume that V_M, the largest eigenvector of R, is given by $V_M = S + \epsilon$ where is a vector orthogonal to S. Consider the vector diagram shown in Fig. 6. The vector a is defined to be orthogonal to the eigenvector $S + \epsilon$. What is sought in (A6) is the square of the length L of the vector S projected onto a. As shown below, the length of ϵ is small compared to S; therefore, the quantity L will be approximately equal to the length of ϵ. To a good approximation, the energy in the beam steered on-target is given by

$$S(k) = \frac{\sigma_n^2}{\|\epsilon\|_{\mathrm{Q}^{-1}}^2} \tag{A7}$$

where $\|\epsilon\|_{\mathrm{Q}^{-1}}^2$ denotes the squared norm of ϵ with respect to Q^{-1}:

$$\|\epsilon\|_{\mathrm{Q}^{-1}}^2 = \epsilon' \mathrm{Q}^{-1} \epsilon.$$

The definition of an eigenvector implies that this vector must satisfy

$$\mathrm{R}(S + \epsilon) = \lambda_M (S + \epsilon). \tag{A8}$$

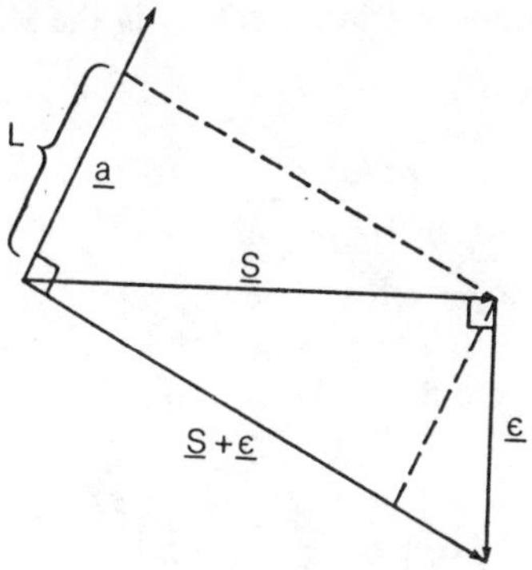

Fig. 6. Relationship of the signal vector and the largest eigenvector.

As the vectors S and ϵ are assumed to be orthogonal, the eigenvalue λ_M can be found through the relationship

$$S'R(S+\epsilon) = \lambda_M S'(S+\epsilon) = M\lambda_M.$$

Using (A3) for R, we have

$$\lambda_M = \frac{\sigma_n^2 S'QS}{M} + \sigma_s\sigma_n(S'\tilde{N} + \tilde{N}'S + \tilde{N}'\epsilon) + M\sigma_s^2. \tag{A9}$$

An expression for the vector ϵ is obtained by evaluating the quantity $R(S+\epsilon) - \lambda_M S$. After some manipulation and assuming that the length of ϵ is small compared to the length of S, we have

$$\epsilon = (\lambda_M I - \sigma_n^2 Q)^{-1} \left(\sigma_n^2 QS - \sigma_n^2 \frac{S'QS}{M} S + M\sigma_s\sigma_n\tilde{N} - \sigma_s\sigma_n S'\tilde{N}S\right). \tag{A10}$$

This expression for the vector ϵ consists of a matrix $(\lambda_M I - \sigma_n^2 Q)^{-1}$ times the sum of two terms. The first term, denoted by B_1, is comprised only of the signal-related terms and the matrix Q:

$$B_1 = \sigma_n^2 QS - \sigma_n^2 \frac{S'QS}{M} S. \tag{A11}$$

The second term contains the terms depending on the average noise vector $\tilde{N}$:

$$B_2 = M\sigma_s\sigma_n\tilde{N} - \sigma_s\sigma_n(S'\tilde{N})S. \tag{A12}$$

If one assumes the noise vector $\tilde{N}$ to be zero (implying infinite statistical averaging), the vector ϵ is given by the quantity $\epsilon = (\lambda_M I - \sigma_n^2 Q)^{-1} B_1$. Note that if $Q = I$, the quantity $B_1 = 0$ which implies $\epsilon = 0$. Therefore, the signal vector S corresponds to the largest eigenvector of R; this result is consistent with the analysis described while leading to (5). The term B_2 expresses the effect on the eigenvector of the statistical averaging process. Note that if noise field can be described as containing only sensor noise ($Q = I$), the expression for ϵ is dominated by B_2. The vector B_2 can be interpreted as the noise vector which results when all of its components in the direction of the signal vector S are eliminated (A12). The squared magnitude of this vector therefore depends on the "angle" between the vectors $\tilde{N}$ and S. This angle will be a random quantity when the computation of R is completed. This vector will be largest when $\tilde{N}$ and S are assumed to be orthogonal. In this case, $S'\tilde{N} = 0$, and the expression for B_2 becomes

$$B_2 = M\sigma_s\sigma_n\tilde{N}. \tag{A13}$$

Define $\tilde{B}_1$ ($\tilde{B}_2$) to be the product of $(\lambda_M I - \sigma_n^2 Q)^{-1}$ and B_1 (B_2). The vector ϵ is therefore written as

$$\epsilon = \tilde{B}_1 + \tilde{B}_2. \tag{A14}$$

An expression for the norm of ϵ can now be obtained. Let U_i denote an eigenvector of the matrix Q and α_i the associated eigenvalue. As these eigenvectors are orthonormal, any vector can be expressed as a linear combination of them. As U_i is also an eigenvector of $(\lambda_M I - \sigma_n^2 Q)^{-1}$, $\tilde{B}_1$ and $\tilde{B}_2$ can be written as

$$\tilde{B}_1 = \sigma_n^2 \sum_i \frac{\left(\alpha_i - \frac{1}{M}\sum_m \alpha_m |S'U_m|^2\right) U_i'S}{\lambda_M - \alpha_i\sigma_n^2} U_i \tag{A15a}$$

$$\tilde{B}_2 = M\sigma_s\sigma_n \sum_i \frac{U_i'\tilde{N}}{\lambda_M - \alpha_i\sigma_n^2} U_i. \tag{A15b}$$

The quantity $\lambda_M - \alpha_i\sigma_n^2$ can be simplified. Using (A9) and assuming that both ϵ and $\tilde{N}$ are small compared to S, we have

$$\lambda_M - \alpha_i\sigma_n^2 = M\sigma_s^2 + \left(\frac{S'QS}{M} - \alpha_i\right)\sigma_n^2.$$

The quantities within the parentheses are comparable, and further assuming that $M\sigma_s^2 > \sigma_n^2$, we have

$$\lambda_M - \alpha_i\sigma_n^2 \approx M\sigma_s^2. \tag{A16}$$

The norm of ϵ depends upon the angle between these two vectors. Based on statistical arguments, a zero-mean vector obtained from averaging ($\tilde{B}_2$ in this case) will be nearly orthogonal to any fixed vector. Consequently,

$$\|\epsilon\|_{Q^{-1}}^2 = \|\tilde{B}_1\|_{Q^{-1}}^2 + \|\tilde{B}_2\|_{Q^{-1}}^2. \tag{A17}$$

Now

$$\|\tilde{B}_1\|_{Q^{-1}}^2 = \sigma_n^4 \sum_i \frac{\left(\alpha_i - \frac{1}{M}\sum_m \alpha_m |S'U_m|^2\right)^2}{\alpha_i(\lambda_M - \alpha_i\sigma_n^2)^2} |S'U_i|^2.$$

To evaluate this expression, the relationship between the signal direction vector S and the eigenvectors of Q must be specified. If S were proportional to an eigenvector of Q, the quantity $\|\tilde{B}_1\|_{Q^{-1}}^2$ would be zero. As the norm of $\tilde{B}_1$ will appear in the denominator of the expression for beam energy (A7), one can obtain a lower bound on the energy in the beam by assuming the largest possible value for its length. To approximate the maximal length of $\tilde{B}_1$, assume that S does not prefer *any* of the eigenvector directions of Q. A reasonable mathematical description of this situation is that $|S'U_i|^2 = \alpha_i$. In this instance, we have, using (A16), that

$$\|\tilde{B}_1\|_{Q^{-1}}^2 = \frac{\sigma_n^4}{\sigma_s^4} \frac{1}{M^2} \sum_i \left(\alpha_i - \frac{1}{M}\sum_m \alpha_m^2\right)^2. \tag{A18}$$

The quantity in the summation depends only on the eigenvalues of the matrix Q. Define the quantity γ_{EV}^2 to be

$$\gamma_{EV}^2 = \frac{1}{M}\sum_i \left(\alpha_i - \frac{1}{M}\sum_m \alpha_m^2\right)^2. \tag{A19}$$

Then (A18) becomes

$$\|\tilde{B}_1\|^2_{Q^{-1}} = \frac{\sigma_n^4}{\sigma_s^4} \cdot \frac{\gamma_{EV}^2}{M}. \tag{A20}$$

As the term $\tilde{B}_2$ is a random quantity, its squared norm is defined to be

$$\|\tilde{B}_2\|^2_{Q^{-1}} = E\left[\frac{\sigma_n^2}{\sigma_s^2} \sum_i \frac{|\tilde{N}'U_i|^2}{\alpha_i}\right]. \tag{A21}$$

where $E[\cdot]$ denotes expected value. To a good approximation, U_i is an eigenvector of the correlation matrix associated with the random vector $\tilde{N}$. Consequently,

$$E[|\tilde{N}'U_i|^2] = \frac{\alpha_i}{K}$$

so that

$$\|\tilde{B}_2\|^2_{Q^{-1}} = \frac{\sigma_n^2}{\sigma_s^2} \cdot \frac{M}{K}. \tag{A22}$$

Substituting equations (A18), (A22), and (A17) into (A7), we have finally

$$S_{EV}(k) = \frac{\sigma_s^2}{\frac{\sigma_n^2}{\sigma_s^2}\frac{\gamma_{EV}^2}{M} + \frac{M}{K}} \tag{A23}$$

as an expression for the energy in the beam when steered toward the source.

The analysis for the MUSIC method differs only in detail from that just described. In this method, the spectral estimate is given by

$$S_{\text{MUSIC}}(k) = (W'C_{EV}'C_{EV}W)^{-1}. \tag{A24}$$

Off-target, $C_{EV}W$ approximately equals W, implying that $S_{\text{MUSIC}}(k) = M^{-1}$. When steered on-target, the expression for the MUSIC estimate differs little from that given in (A7). The significant difference is that the norm of ϵ is computed with respect to the identity matrix instead of Q^{-1}. The quantity of interest is therefore

$$\|\epsilon\|^2 = \|\tilde{B}_1\|^2 + \|\tilde{B}_2\|^2.$$

The norm of $\tilde{B}_2$ with respect to I equals that computed with respect to Q^{-1}; the expression for the norm of $\tilde{B}_1$ is

$$\|\tilde{B}_1\|^2 = \frac{\sigma_n^4}{\sigma_s^4}\frac{1}{M^2}\sum_i \alpha_i\left(\alpha_i - \frac{1}{M}\sum_m \alpha_m^2\right)^2.$$

Therefore, the on-target beam energy in the MUSIC method is given by

$$S_{\text{MUSIC}}(k) = \frac{\sigma_s^2/\sigma_n^2}{\frac{\sigma_n^2}{\sigma_s^2}\frac{\gamma_{\text{MUSIC}}^2}{M} + \frac{M}{K}} \tag{A25}$$

where γ_{MUSIC}^2 is defined to be

$$\gamma_{\text{MUSIC}}^2 = \frac{1}{M}\sum^M \alpha_i\left(\alpha_i - \frac{1}{M}\sum \alpha_m^2\right)^2.$$

ACKNOWLEDGMENT

The authors are indebted to D. J. Edelblute of the Naval Ocean Systems Center for many fruitful discussions concerning this research.

REFERENCES

[1] A. B. Baggeroer, "Sonar signal processing," in *Applications of Digital Signal Processing*, A. V. Oppenheim, Ed. Englewood Cliffs, NJ: Prentice-Hall, 1978, pp. 331–437.

[2] G. Bienvenu and L. Kopp, "Adaptivity to background noise spatial coherence for high resolution passive methods," in *Proc. IEEE ICASSP*, Denver, CO, 1980, pp. 307–310.

[3] —, "Source power estimation method associated with high resolution bearing estimation," in *Proc. IEEE ICASSP*, Atlanta, GA, 1981, pp. 153–156.

[4] J. A. Cadzow and T. P. Bronez, "An algebraic approach to superresolution adaptive array processing," in *Proc. IEEE ICASSP*, Atlanta, GA, 1981, pp. 302–305.

[5] J. Capon, "High-resolution frequency-wavenumber spectrum analysis," *Proc. IEEE*, vol. 57, pp. 1408–1418, Aug. 1969.

[6] H. Cox, "Resolving power and sensitivity to mismatch of optimum array processors," *J. Acoust. Soc. Amer.*, vol. 54, no. 3, pp. 771–785, 1973.

[7] D. J. Edelbute, J. M. Fisk, and G. L. Kinnison, "Criteria for optimum-signal-detection theory for arrays," *J. Acoust. Soc. Amer.*, vol. 41, no. 1, pp. 199–205, 1967.

[8] B. S. Garbow, J. M. Boyle, J. J. Dongarra, and C. B. Moler, "Matrix eigensystem routines–EISPACK guide extension," in *Lecture Notes in Computer Science*, vol. 51, G. Goos and J. Hartmanis, Eds. New York: Springer-Verlag, 1977.

[9] M. J. Hinich, "Frequency-wavenumber array processing," *J. Acoust. Soc. Amer.*, vol. 69, no. 3, pp. 732–737, 1981.

[10] D. H. Johnson, "On improving the resolution of bearing in passive sonar arrays," Naval Ocean Syst. Cen., San Diego, CA, Tech. Note 914, 1980.

[11] R. T. Lacoss, "Data adaptive spectral analysis methods," *Geophysics*, vol. 36, no. 4, pp. 661–675, 1971.

[12] N. L. Owsley, "Modal decomposition of data adaptive spectral estimates," presented at the Yale Univ. Workshop on Appl. of Adaptive Syst. Theory, New Haven, CT, 1981.

[13] R. Schmidt, "Multiple emitter location and signal parameter estimation," in *Proc. RADC Spectral Estimation Workshop*, Rome, NY, 1979, pp. 243–258.

Part IX
Image Analysis

IMAGE ANALYSIS: PROBLEMS, PROGRESS AND PROSPECTS*

AZRIEL ROSENFELD
Center for Automation Research, University of Maryland, College Park, MD 20742, U.S.A.

(*Received* 10 *February* 1983; *received for publication* 6 *May* 1983)

Abstract—Over the past 25 years, many *ad hoc* techniques for analyzing images have been developed and the subject has gradually begun to develop a scientific basis. This paper outlines the basic steps in a general image analysis process. It summarizes the state of the art with respect to each step, points out limitations of present methods and indicates potential directions for future work.

Image analysis Scene analysis Pictorial pattern recognition Computer vision

1. INTRODUCTION

Over the past 25 years, much of the work on applications of pattern recognition, and a significant fraction of the work in artificial intelligence, has dealt with the analysis and interpretation of images. This subject has been variously known as pictorial pattern recognition, image analysis, scene analysis, image understanding and computer vision. Its applications include document processing (character recognition, etc.), microscopy, radiology, industrial automation (inspection, robot vision), remote sensing, navigation and reconaissance, to name only the major areas.

Many *ad hoc* techniques for analyzing images have been developed, so that a large assortment of tools is now available for solving practical problems in this field. More important, during the past few years the field has begun to develop a scientific basis. This paper outlines the basic steps in a general image analysis process. It summarizes the state-of-the-art with respect to each step, points out limitations of present methods and indicates potential directions for future work.

2. AN IMAGE ANALYSIS PARADIGM

The goal of image analysis is the construction of scene descriptions on the basis of information extracted from images or image sequences. With reference to Fig. 1, the following are some of the major steps in the image analysis process. We consider here only images obtained by optical sensors, though some of the discussion is also applicable to other types of sensors.

Many types of scenes are essentially two-dimensional (documents are an obvious example), but two-dimensional treatment is often quite adequate in applications such as remote sensing (flat terrain seen from very high altitudes), radiology (where the image is a "shadow" of the object) or microscopy (where the image is a cross-section of the object). In such situations, the image analysis process is basically two-dimensional. We extract "features", such as edges, from the image or segment the image into regions, thus obtaining a map-like representation, which Marr[1] called the "primal sketch", consisting of image features labelled with their property values. Grouping processes may then be used to obtain improved maps from the initial one. The maps may be represented by abstract relational structures in which, for example, nodes represent regions, labelled with various property values (color, texture, shape, etc.), and arcs represent relationships among regions. Finally, these structures are matched against stored models, which are generalized relational structures representing classes of maps that correspond to general types of images. Successful matches yield identifications for the image parts, and a structural description of the image, in terms of known entities.

In other situations, notably in robot vision applications, the scenes to be described are fundamentally three-dimensional, involving substantial surface relief and object occlusion. Successful analysis of images of such scenes requires a more elaborate approach in which the three-dimensional nature of the underlying scenes is taken into account. Here, the key step in the analysis is to infer the surface orientation at each image point. Clues to surface orientation can be derived directly from shading (i.e. gray level variation) in the

* The support of the Defense Advanced Research Projects Agency and the U.S. Army Night Vision Laboratory under Contract DAAG-53-76C-0138 (DARPA Order 3206) is gratefully acknowledged, as is the help of Janet Salzman in preparing this paper. A different version of this paper was presented at the Sixth International Conference on Pattern Recognition in Munich, FRG, in October 1982.

Reprinted with permission from *Pattern Recognition,* vol. 17, pp. 3–12, 1984.

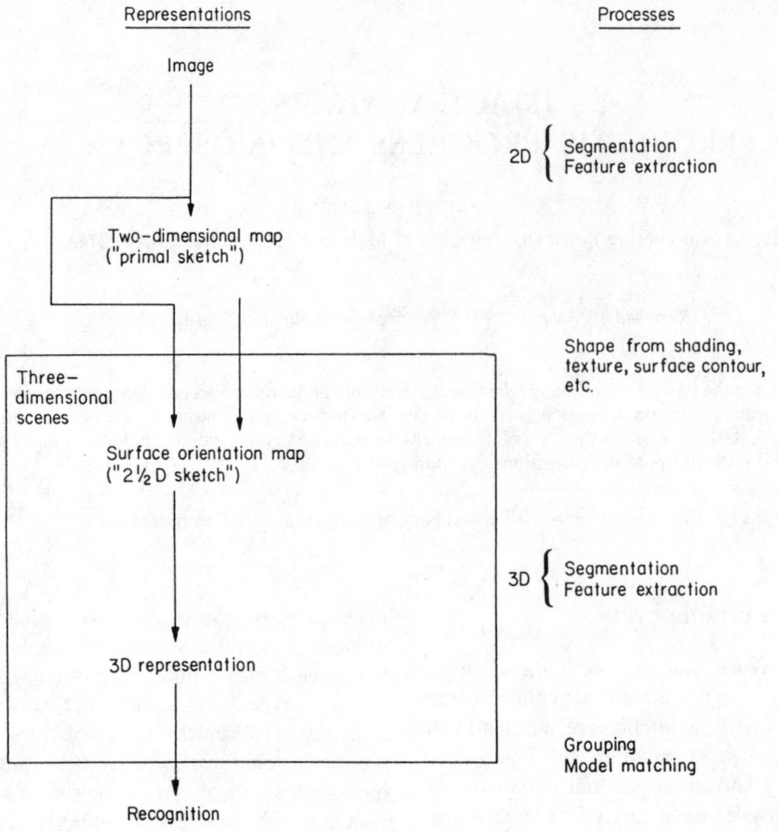

Fig. 1. Simplified diagram of an image analysis system.

image. Alternatively, two-dimensional segmentation and feature extraction techniques can first be applied to the image to extract such features as surface contours and texture primitives, and surface orientation clues can then be derived from contour shapes or from textural variations. Using the surface orientation map, which Marr called the "$2\frac{1}{2}$D sketch", feature extraction and segmentation techniques can once again be applied to yield a segmentation into (visible parts of) bodies or objects, and these can in turn be represented by a relational structure. Finally, the structure can be matched against models to yield an interpretation of the scene in terms of known objects. Note that the matching process is more difficult in the three-dimensional case, since the image only shows one side of each object and objects may partially occlude one another.

The image analysis paradigm described in the last two paragraphs, and illustrated in Fig. 1, is highly simplified in several respects. The following are some of the directions in which it needs to be extended or generalized.

(a) Ideally, the value (gray level or spectral signature) at each point of an image represents the light received from the scene along a given direction, but these values will not be perfectly accurate because of degradations arising in the process of imaging (for example blur and noise introduced by the environment or the sensor) or digitization. Image restoration techniques should be used to correct the image values before performing the steps outlined in Fig. 1. (Feature extraction may be useful as an aid in estimating the degradations in order to perform effective restoration.)

(b) We have assumed in Fig. 1 that only a single image of the scene is available as input. If two images taken from different viewpoints are available, stereomapping techniques can be used to construct the surface orientation map by matching either image values or extracted features on the two images and measuring their parallaxes. If images taken at different times are available, comparing them yields information about the motion of the sensor or of objects in the scene. In this case, the processes of segmentation, model matching, etc., should be performed on the image sequence rather than on the individual images.

(c) Figure 1 shows a "one-way" process in which we start with the image and successively construct a 2D map, a $2\frac{1}{2}$D map, etc. More realistically, the arrows in Fig. 1 should point both ways. Knowledge about the expected results of a

process (segmentation, etc.) should be used to criticize the actual results and modify the process so as to improve them.
(d) The model matching process may be hierarchical, with objects composed of sub-objects, etc. Hierarchical models are extensively used in syntactic approaches to pattern recognition.

A discussion of image restoration techniques is beyond the scope of this article, but stereomapping, time-varying imagery analysis, syntactic methods and the use of feedback in image analysis will all be briefly discussed in later sections.

This paper reviews the basic stages in the image analysis process from a technique-oriented, rather than application-oriented, standpoint. Methods currently used at each stage are reviewed, their shortcomings are discussed, and approaches that show promise of yielding improved performance are described. The specific areas covered are feature extraction, texture analysis, surface orientation estimation, image matching, range estimation, segmentation, object representation and model matching.

3. FEATURE EXTRACTION

The extraction of features such as edges and curves from an image is useful for a variety of purposes. Edges and similar locally defined features play important roles in texture analysis (see Section 4). The interpretation of image edges as arising from various types of discontinuities in the scene (occluding edges are discontinuities in range; convex or concave edges are discontinuities in surface slope; shadow edges are discontinuities in illumination) plays an important role in the inference of 3D surface structure from an image (Section 5). Edges are useful in image matching for obtaining sharp matches that are insensitive to grayscale distortions (but quite sensitive to geometric distortion) (see Section 6). Edges can be used in conjunction with various segmentation techniques to improve the quality of a segmentation (Section 8). It should also be mentioned that linear features (curves) are often of importance in their own right, e.g. roads or drainage patterns on low-resolution remote sensor imagery.

The classical approach to edge detection makes use of digital (finite-difference) versions of standard isotropic derivative operators, such as the gradient or Laplacian. A closely related approach is to linearly convolve the image with a set of masks representing ideal step edges in various directions. Lines and curves can be similarly detected by linear convolutions. However, linear operators are not specific to features of a given type. They also respond in other situations involving local intensity changes. An alternative approach[2] is to use "gated" (nonlinear) operators that respond only when specific relationships hold among the local intensities, e.g. all intensities along the line higher than all the flanking intensities on both sides, and similarly for edges. In all of these approaches, the output is a quantitative edge or curve *value*. The final detection decision can be made, if desired, by thresholding this value. Similar methods can be used to detect edges defined by discontinuities in color, rather than in intensity.

Several important improvements to the edge detection process have been made over the past decade. New classes of operators have been defined based on fitting polynomial surfaces to the local image intensities and using the derivatives of these polynomials (which can in turn be expressed in terms of the local intensities) as edge value estimates. This method, which was first proposed by Prewitt,[3] allows edges to be located (at maxima of the surface gradient for example) to subpixel accuracy. Another important idea, first proposed by Hueckel,[4,5] is to find a best-fitting step edge (or edge-line) to the local intensities.

Classical edge detectors were based on small image neighborhoods, typically 3 × 3. A more powerful approach[2,6] is to use a set of first- or second-difference operators based on neighborhoods having a range of sizes (e.g. increasing by factors of 2) and combine their outputs, so that discontinuities can be detected at many different scales. Here the edges are localized at maxima of first differences or at zero-crossings of second differences. Operators based on large neighborhoods can also be used to detect texture edges, at which the statistics of varous local image properties change abruptly. Cooperation between operators in different positions can be used to enhance the feature values at points lying on smooth edges or curves. This was one of the first applications of "relaxation" methods[7,8] to image analysis at the pixel level. Cooperation among operators of different sizes is also important in characterizing edge types.

The standard approaches to edge detection are implicitly based on a very simple model in which the image is regarded as ideally composed of essentially "constant" regions separated by step edges. Recent work by Haralick[9] is based on the more general assumption of a piecewise linear, rather than piecewise constant, image, which allows simple shading effects to be taken into account. Even this approach, however, does not directly address the problem of edge detection in three-dimensional scenes. Research is needed on the development of algorithms designed to detect intensity edges resulting from specific types of scene discontinuities, including shadow edges, slope edges and range edges. Detection of texture edges (see below) should be based on models for surface texture, rather than for image texture.

4. TEXTURE ANALYSIS

Textural properties of image regions are often used for classification (e.g. of terrain types or materials) or for segmentation of the image into differently textured regions. Changes in texture "coarseness" also provide important cues about surface slant and relative range. The direction in which coarseness is changing most rapidly corresponds, for a uniformly textured surface, to the slant direction, while an abrupt change in coarseness indicates the possibility of an occluding edge.

Classically, texture properties have been derived from the autocorrelation or Fourier power spectrum, e.g. the coarser the texture (in a given direction), the slower its autocorrelation falls off in that direction from the origin (zero displacement) and the faster its power spectrum falls off in that direction from zero frequency. A related approach[10,11] characterizes textures by their second-order intensity statistics, i.e. by the frequencies with which given pairs of gray levels occur at given relative displacements. It has long been realized, however, that first order statistics of various local property values (e.g. responses to operators sensitive to local features such as edges, lines, line ends, etc.) are at least equally effective in texture discrimination.

More recent work suggests that local processes of linking between local features, giving rise to "texture elements" or "primitives", also play a significant role in the perception of texture differences. Texture discrimination based on second-order statistics of local features (e.g. occurrences of edge elements in given relative positions and orientations) has begun to be investigated.[12] Texture analysis based on explicit extraction of primitives has also been explored.[13] Here, statistics derived from properties of the primitives, or of pairs of adjacent primitives, are used as textural properties.

All of this work has dealt with texture as an image property and has been primarily concerned with uniformly textured regions, such as might arise from non-perspective views of uniformly textured surfaces. Research is needed on the development of texture analysis methods that take surface geometry into account and that perform cooperative estimation of surface slant and surface texture characteristics, leading to better estimates of both. Similarly, methods of texture-based segmentation or texture edge detection should consider both surface geometry differences and texture differences, and in stereomatching of textured regions, one should use surface slant estimates to correct for the effects of perspective on the quality of the match.

5. SURFACE ORIENTATION ESTIMATION

If only a single image of a scene is available, clues about the orientations and relative ranges of the visible surfaces in the scene can still be derived from a number of sources. One source of such information is the shapes of edges in the image, representing occluding contours at the boundaries of surfaces or contours that lie on the surfaces. The early work of Waltz[14] and his predecessors, as well as the more recent work of Kanade,[15] developed methods of inferring the nature of edges in the scene (e.g. convex, concave or occluding) from the shapes of the junctions at which the edges meet, as seen in the image. A variety of constraints on a scene can be derived from global properties of contours in the image. For example, if an edge is continuous or straight or two edges are parallel or two features coincide in the image, we assume that the same is true in the scene, and if a shape in the image could be the result of perspective distortion of a simpler shape in the scene, we assume that this is actually the case.[15,16] Other work[17,18] has dealt with the three-dimensional interpretation of occluding and surface contours. For example, given a curve in the image, we might assume that it arises from a space curve of the least possible curvature. Constraints of these types, and others still to be formulated, often yield unambiguous three-dimensional interpretations of the surfaces that appear in the images. Most of this work has as yet been applied only to idealized line drawings, but some parts of it have been successfully applied to noisy real-world images.

The inference of information about the surfaces in a scene from the shapes of edges in an image is known as "shape from shape" (i.e. 3D shape from 2D shape). A closely related problem is that of inferring surface shape from textural variations in the image. Gibson pointed out over 30 years ago that changes in texture coarseness arise from changes in range. Thus it should be possible to infer changes in range from changes in coarseness. It has recently been demonstrated[19] that the 3D orientation of a surface can be inferred from the anisotropy in its texture—note that here again, as in the case of shape from contour, we are assuming that if the anisotropy could have arisen from perspective distortion, then it actually did. To obtain good results, one should use edge-based or primitive-based, rather than pixel-based, texture descriptors. The richer the descriptors, the more likely it is that the inferences will be reliable.

In the absence of discriminable features, surface orientation in the scene can also be inferred from intensity variations ("shading") in the image. The pioneering work on the inference of "shape from shading" was done by Horn.[20,21] Given the position of the (small, distant) light source and the surface reflectance function, surface shape is still not unambiguously determined, but it is strongly constrained and can be estimated based on additional information, such as the shapes of surface contours, the restriction that the surface be a surface of revolution or the requirement that the surface curvature be as uniform as possible. Surface shape becomes unambiguous if we are given several images taken from the same position, but with light sources in different positions ("photometric stereo").[22] Much of this work has assumed diffuse reflectance and needs to be extended to reflectance functions that have strong specular components. In such cases the shapes of highlights may provide additional information.

6. IMAGE MATCHING

Image matching and registration are used for a number of different purposes. By registering two images of a scene obtained from different sensors, one can obtain the multisensor (e.g. multispectral) characteristics of each scene point, which can then be used to classify the points. By comparing images obtained from different locations, one can compute the stereoscopic parallaxes of scene points that are visible on both images and thus determine their 3D positions. By comparing images taken at different times, one can detect changes that have taken place in the scene, e.g. due to motion of the sensor or motions of objects in the scene. In all of these tasks, registration is carried out by finding pairs of subimages that match one another closely. Subimage matching is also used to detect the occurrence of specific patterns ("templates" or "control points") in an image, for the purposes of location (e.g. navigation) or object detection.

Classically, image matching has used match measures derived from cross-correlation computation, or sometimes mismatch measures based on sums of absolute differences. Both of these approaches involve point-by-point intensity comparison of the images being matched. Such processes are unsatisfactory for several reasons: they often yield unsharp matches, making it difficult to decide when a match has been detected; they are sensitive to distortions in both grayscale and geometry; they are computationally expensive. Match sharpness and grayscale insensitivity can be greatly increased by applying derivative operators, possibly followed by thresholding, to the images before matching—for example, taking first derivatives (e.g. gradient magnitudes) of both images or the second derivative (e.g. Laplacian magnitude) of one image. Geometric insensitivity can be improved by matching smaller pieces or local features (which are less affected by geometric distortion) and then searching for combinations of such matches in approximately the correct relative positions[(23)] or using relaxation methods to reinforce such combinations.[(24–26)] This hierarchical approach also serves to reduce the computational cost of the matching process. An alternative idea[(27)] is to segment the image into parts, represent the parts and their relationships by a graph structure and match these graph structures (see Section 10). Here too, relaxation methods are useful.

Another approach to pattern matching makes use of geometric transformations that map instances of a given pattern into peaks in a transform space. This "Hough transform" approach was originally developed to handle simple classes of shapes such as straight lines or circles, but it was recently extended[(28)] to arbitrary shapes in both two and three dimensions.

The matching techniques described above are all two-dimensional. They do not take into account the fact that the subimages or patterns being matched may differ not only in position and orientation, but also in size and perspective if the images are related by a three-dimensional transformation. Research is needed on matching techniques that take estimated surface orientation explicitly into account, thus making possible correction for the distorting effects of perspective, as well as for the associated intensity differences. The difficulty of matching images of objects that differ by three-dimensional rotations will be further discussed in Section 10.

7. RANGE ESTIMATION

If a high-resolution range sensor is available, the shapes of the visible surfaces in a scene can be obtained directly by constructing a range map. In this section we assume that range information is not directly available. In its absence, range can be inferred from stereopairs, by measuring stereo parallax, or relative range can be inferred from image sequences obtained from a moving sensor, by analyzing the motions of corresponding pixels from frame to frame ("optical flow").

Stereomapping is based on identifying correspnding points in the two images using image matching techniques. As indicated in Section 6, matching performance is improved if we match features such as edges, rather than intensity values. The MIT approach to stereo[(29)] is based on applying a set of edge operators, having a range of sizes, to the images, matching the edges produced by the coarsest operator, to yield a rough correspondence between the images, and then refining this correspondence by using successively finer edges. Edge-based approaches may still yield ambiguous results in heavily textured regions where edges are closely spaced. The ambiguity can be reduced by using intensity matching as a check or by classifying the edges into types (e.g. discontinuities in illumination, range or orientation) and requiring that corresponding edges be of the same type. In general, matching should be based on feature descriptions, rather than on raw feature response values. Work is needed on the development of matching methods based on other feature types and particularly on features derived from surface orientation maps—e.g. matching of surface patches. Matching yields a set of range values at the positions of features. Grimson[(29)] has developed methods of fitting smooth range surfaces to these values. For wide-angle stereo, where there is significant perspective distortion, derivation of a camera model and rectification of the images prior to matching are very desirable.

When a static scene is viewed by a moving sensor, yielding a succession of images, the relative displacements of pixels from one image to the next are known as "optical flow". If these displacements could be computed accurately, it would be possible in principle to infer the motion of the sensor relative to the scene and the relative distances of the scene points from the sensor (but note that there is an inherent speed/range ambiguity). Ideally, the displacements can be estimated by comparing the space and time rates of change of the image intensity, but in practice these estimates are quite noisy. Horn[30] has developed an iterative method of estimating a smooth displacement field, but it yields inaccurate results at object boundaries. It should be possible to obtain improved results by combining the rate of change approach with edge detection and matching. For larger displacements, a matching approach can be used to determine corresponding points in successive frames. Ullman[31] has shown that the motion of a rigid object is completely determined if we know the correspondence between a few points on the object as they appear in two or three successive images. Extensions of this work to jointed objects have also been investigated.

8. SEGMENTATION

Descriptions of an image generally refer to significant parts (regions; global features such as contours or curves) of which the image is composed. Thus image description requires segmenting it into such parts. A much more challenging task is to segment the image into parts corresponding to the surfaces or bodies of which the underlying scene is composed. This is often very hard to do, since variations in image intensity may not be good indicators of physical variations in the scene and, conversely. physical variations do not always give rise to intensity variations.

The most commonly used approach to image segmentation involves classification of the individual image points (pixels) into subpopulations. The parts obtained in this way are just the subsets of pixels belonging to each class. The classification can be done on the basis of intensity alone ("thresholding"), of color or spectral signature, or of local properties derived from the neighborhood of the given pixel. The last approach is used in feature detection (e.g. classify a pixel as on an edge if the value of some locally computed derivative operator is high in its neighborhood) and it can also be used to segment an image into differently textured regions. Pixels can be classified using a set of properties simultaneously or the properties can be used one at a time to recursively refine the segmentation.[32]

Pixels are usually classified independently, which allows fast implementation on parallel hardware. Better results can be obtained by classifying sequentially, so that regions composed of pixels belonging to a given class can be "grown" in accordance with given constraints, but such approaches are inherently slow and would not be appropriate for use in real-time systems. Another possibility is to use a relaxation approach in which pixels are classified fuzzily and the class memberships are then adjusted to favor local consistency. This approach[33, 34] requires a short sequence of iterations each of which can be implemented in parallel. In addition to local consistency, other sources of convergent evidence can be used to improve the quality of segmentation. For example, the classification criteria can be adjusted so as to maximize the edge strengths around the resulting region borders.

An alternative approach to segmentation is region-based rather than pixel-based. An example is the split-and-merge approach advocated by Pavlidis,[35] where the goal is to partition an image into homogeneous connected regions by starting with an initial partition and modifying it by splitting regions if they are not sufficiently homogeneous and merging pairs of adjacent regions if their union is still homogeneous. In this approach, "homogeneous" might mean approximately constant in intensity or, more generally, it might mean a good fit to a polynomial of some degree greater than 0, as in the facet model. Still more generally, the merging and splitting can be controlled by a "semantic" model which estimates probable interpretations of the regions and performs merges or splits so as to increase the likelihood and consistency of the resulting image interpretation.[36, 37] Note, however, that these methods still make no explicit use of surface orientation estimation. They should be based on object semantics rather than region semantics. Grouping locally detected features (edges or lines) into global contours or curves can be done on the basis of global shape (Hough transforms; see Section 6), but if this is not known in advance one can use methods analogous to split-and-merge—e.g. break a curve at branch points or sharp turns; link curves if they continue one another smoothly. Here again, it would be desirable to modify these criteria to take surrace curvature into account.

Image models should play an important role in image segmentation, but the models used in practice are usually much too simple. In segmenting an image by pixel classification, it is always assumed that the subpopulations are homogeneous, i.e. have essentially constant feature values (intensity, color, etc.). For scenes containing curved surfaces, this assumption is very unrealistic. Even if variations in illumination are ignored, changes in surface orientation will give rise to changes in feature values on the image of the surface. Similar remarks apply to region-based segmentation schemes. Haralick's facet model allows certain types of variations in feature values (e.g. linear), but the role of surface orientation needs to be made more explicit. By making local orientation estimation an integral part of the segmentation process and using these estimates to correct the feature values, it should be possible to cooperatively compute orientation estimates that optimize the clustering of feature values into subpopulations representing homogeneous surfaces (and not merely homogeneous regions). Spatial consistency constraints, as well as other types of convergent evidence, can also be incorporated into this process.

Another major drawback of segmentation based on pixel classification, particularly when it is implemented in parallel, is the difficulty of incorporating geometric knowledge about the desired regions into the segmentation process. The standard approach is to segment, measure geometric properties of the resulting regions and then attempt to improve the values of these properties by adjusting parameters of the segmentation process. However, it would be much preferable to make use of geometric constraints in the segmentation process itself. In region-based segmentation, since the units being manipulated are (pieces of) regions rather than pixels, somewhat greater control over region geometry can be achieved, by biasing the choices of splits and merges to favor the desired geometry. Another possibility is to perform segmentation using a multi-resolution ("pyramid") image representation, in which region geometry is coarsely represented by local patterns of "pixels" at the low-resolution levels. Here, segmentation is based on a cooperative process of pixel linking, which can be designed so that the linking is facilitated if it will give rise to the desired types of low-resolution local patterns. This approach too should be combined with surface orientation estimation, perhaps carried out at multiple resolutions.

9. OBJECT REPRESENTATION

The algorithms used to measure properties of image or scene parts depend on the data structures used to represent the parts. In this section we review the basic types of representations, both two- and three-dimensional, that are commonly used in image analysis systems.

Digital images are 2D arrays in which each pixel's value gives the intensity (in one or more spectral bands) of the radiation received by the sensor from the scene in a given direction. Other viewer-centered representations of the scene are also conveniently represented in array form, with the value of a "pixel" representing illumination, reflectivity, range or components of surface slant at the scene point located along a given direction. Various types of image transforms, as well as symbolic "overlay" images defining the locations of features (contours, curves, etc.) or regions, are other examples of 2D arrays that are often used in image processing.

Features and regions in any image can also be represented in other ways which are usually more compact than the overlay array representation and which also may make it easier to extract various types of information about their shapes. The following representations are all two-dimensional and are appropriate only if 3D shape information is not known. One classical approach is to represent regions by border codes, defining the sequence of moves from neighbor to neighbor that must be made in order to circumnavigate the border. Curves can also be represented by such move sequences ("chain codes").[38] Another standard way of representing regions is as unions of maximal "blocks" contained in them—for example, maximal "runs" of region points on each row of the image or maximal upright squares contained in the region; the set of run lengths on each row, or the set of centers and radii of the squares (known as the "medial axis"),[39] completely determines the region. The square centers tend to lie on a set of arcs or curves that constitute the "skeleton" of the region. If we specify each such arc by a chain code and also specify a radius function along the arc, we have a representation of the region as a union of "generalized ribbons" (see the end of this section).

There has been recent interest in the use of hierarchically structured representations that incorporate both coarse and fine information about a region or feature. One often-used hierarchical maximum-block representation is based on recursive subdivision into quadrants, where the blocks can be represented by the nodes of a degree-4 tree (a "quadtree").[40] Hierarchical border or curve representations can be defined based on recursive polygonal approximation, with the segments represented by the nodes of a "strip tree",[41] or based on quadrant subdivision.[42]

At a higher level of abstraction, a segmented image is often represented by a graph in which the nodes correspond to regions (or parts or surfaces if 3D information is available) or features, labelled with property names or values, and the arcs are labelled with relation values or names. A problem with this type of representation is that it does not preserve the details of region geometry and so can only provide simplified information about geometrical properties and relations, many of which have no simple characterizations. An ideal representation should provide information at multiple resolution, so that both gross geometry and important local features are easily available, together with the topological and locational constraints on the features' positions, where these constraints may have varying degrees of fuzziness. It should also be easy to modify the representation to reflect the effects of 3D geometrical transformations, so that representations of objects viewed from different positions can be easily compared.

Representations of surfaces and objects, i.e. "$2\frac{1}{2}$-dimensional" and "3-dimensional" scene representations, are also an important area of study. The visible surfaces in a scene can be represented by an array of slope vectors. The histogram of these vectors is known as the "gradient space" map. The range to each point in the scene is another important type of viewer-centered array representation.

In order to identify the objects in a scene, it is desirable to relate the viewer-centered representations of the visible surfaces to object-centered representations that describe the objects on a three-dimensional level. A variety of object representations can be defined, generalizing the representations of two-dimensional regions described above. An object can be represented by a series of slices and a 2D representation can be used for each slice. Alternatively, an object can be represented as a union of maximal blocks—e.g. by an "octree" (based on recursive subdivision of space into octants) or by a 3D "medial axis". If this axis is approximated by a set of space curves, each represented by a 3D chain code, and we also specify a radius function along each curve, we have a representation of the object as a union of "generalized cylinders" or "generalized cones".(43)

10. MODEL MATCHING

The image analysis processes described up to now give rise to a decomposition of the image into regions or of the scene into objects. A "literal" description of the image or scene can thus be given in the form of a relational structure in which the nodes correspond to features, regions or objects, labelled by lists of their property values (shape, texture, color, etc.). However, this type of "semantics-free" description is usually not what is wanted. Rather, one wants a description in terms of a known configuration of known objects. This requires "recognizing" the objects by comparing their descriptions to stored "models", which are generalized descriptions defining object classes.

Even in two dimensions, such models are often very difficult to formulate, since the constraints on the allowable property values and relationships are hard to define. In three dimensions, the problem is rendered even more difficult by the fact that only one side of an object can be visible in an image. The image description is two-dimensional, while the stored object models are presumably three-dimensional, object-centered representations.

The most extensive work on recognition of three-dimensional objects from images is embodied in the ACRONYM system.(44) This system incorporates methods of predicting the two-dimensional appearance (shape, shading, etc.) of a given object in an image taken from a given point of view. Conversely, it provides means of defining constraints on the three-dimensional properties of the object that could give rise to a given image and for manipulating sets of such constraints. These capabilities are incorporated in a prediction/verification process which uses the image to make predictions about the object and verifies that the image could in fact have arisen from an object that satisfies the resulting set of constraints. Thus far, ACRONYM has been implemented only in restricted domains, but it is based on very general principles and should be widely extendable.

It is often appropriate to model regions or objects hierarchically, i.e. as composed of parts arranged in particular ways, where the parts themselves are arrangements of subparts, and so on. There is an analogy between this type of hierarchical representation and the use of grammars to define languages. Here, a sentence is composed of clauses which are in turn composed of phrases, etc. Based on this observation, the process of recognizing an object as belonging to a given hierarchically defined class of objects is analogous to the process of recognizing a well-formed sentence as belonging to a given language, by *parsing* it with respect to a grammar for that language. This "syntactic" approach to object (or pattern) recognition has been extensively studied by Fu and his students.(45) It has been used successfully for recognition of two-dimensional shapes, patterns and textures, but it is less appropriate for three-dimensional object recognition, since it is not obvious how to incorporate in it mechanisms for relating 2D images to 3D objects.

Many difficult problems are associated with the model matching task. It is not trivial to define models for given classes of patterns or objects. (In the case of syntactic models, the problem of inferring them from sets of examples is known as *grammatical inference*. The pioneering work on the inference of relational structure models from examples was done by Winston.(46) Given a large set of models, it is not obvious how to determine the right one(s) with which to compare a given object. This is known as the *indexing* problem. Even if the correct model is known, comparing it with the descriptions of a given object may involve combinatorial search. (Here, however, relaxation or constraint satisfaction methods can often be used to reduce the search space.) The best approach is to use the model(s) to control the image analysis process and to design this process in such a way that most of the possible models are eliminated at early stages of the analysis. Unfortunately, there exists as yet no general theory of how to design image analysis processes based on given sets of models. The control structures used have been designed largely on heuristic grounds.

11. CONCLUDING REMARKS

We have seen that image analysis involves, in general, many different processes that incorporate many different types of information about the class of images being analyzed. There is no general theory of control in image analysis. In other words, there are no general principles that specify how these processes should interact in carrying out a given task. In particular, when a number of methods exist for performing a given task, e.g. feature detection, or inference of surface orientation, it would usually be desirable to implement several of the methods in order to obtain a consensus. However, there is no general theory of how to combine evidence from multiple sources.

Most of the successful applications of image analysis have involved relatively simple domains and have been primarily two-dimensional. For example, in robot vision, systems that recognize parts on a belt (well illuminated, non-overlapping, in specific 3D orientations) are not hard to build, but systems that recognize parts in a bin (shadowed, overlapping, arbitrarily oriented) are still a research issue. Techniques exist that will in principle handle such complex situations, but they need to be refined and extensively tested before they can be used in practice.

The discussion of the image analysis process in this paper has been quite general-purpose, without emphasis on particular domains of application. It is also possible to build "specialist" or "expert" systems tailored to a specific domain, which make use of methods especially designed for that domain. From a practical standpoint, successful applications of image analysis are likely to be of this specialized nature. It is the general approach, however, that makes image analysis at least potentially a science and that will continue to provide a theoretical background for the design of application-oriented systems.

REFERENCES

1. D. Marr, Visual information processing: the structure and creation of visual representations, *Proc. 6th Int. Joint Conf. on Artificial Intelligence*, pp. 1108–1126 (1979).
2. A. Rosenfeld and M. H. Thurston, Edge and curve detection for visual scene analysis, *IEEE Trans. Comput.* **C-20,** 562–568 (1971).
3. J. M. S. Prewitt, Object enhancement and extraction, *Picture Processing and Psychopictorics*, B. S. Lipkin and A. Rosenfeld, eds., pp. 75–149. Academic Press, New York (1970).
4. M. H. Hueckel, An operator which locates edges in digital pictures, *J. Ass. comput. Mach.* **18,** 113–125 (1971).
5. M. H. Hueckel, A local visual operator which recognizes edges and lines, *J. Ass. comput. Mach.* **20,** 634–647 (1973).
6. D. Marr and E. Hildreth, Theory of edge detection, *Proc. R. Soc.* **B207,** 187–217 (1980).
7. S. W. Zucker, R. A. Hummel and A. Rosenfeld, An application of relaxation labeling to line and curve enhancement, *IEEE Trans. Comput.* **C-26,** 394–403, 922–929 (1977).
8. B. J. Schachter, A. Lev, S. W. Zucker and A. Rosenfeld, An application of relaxation methods to edge reinforcement, *IEEE Trans. Syst. Man Cybernet.* **SMC-7,** 813–816 (1977).
9. R. M. Haralick and L. Watson, A facet model for image data, *Comput. Graphics Image Process.* **15,** 113–129 (1981).
10. B. Julesz, Visual pattern discrimination, *IRE Trans. Inf. Theory* **IT-8,** 84–92 (1962).
11. R. M. Haralick, K. Shanmugam and I. Dinstein, Textural features for image classification, *IEEE Trans. Syst. Man Cybernet.* **SMC-3,** 610–621 (1973).
12. L. S. Davis, S. A. Johns and J. K. Aggarwal, Texture analysis using generalized cooccurrence matrices, *IEEE Trans. Pattern Anal. Mach. Intell.* **PAMI-1,** 251–259 (1979).
13. J. T. Maleson, C. M. Brown and J. A. Feldman, Understanding natural texture, *Proc. DARPA Image Understanding Workshop*, pp. 19–27 (1977).
14. D. Waltz, Understanding line drawings of scenes with shadows, *The Psychology of Computer Vision*, P. H. Winston, ed., pp. 19–91. McGraw-Hill, New York (1975).
15. T. Kanade, Recovery of the three-dimensional shape of an object from a single view, *Artif. Intell.* **17,** 409–460 (1981).
16. T. O. Binford, Inferring surfaces from images, *Artif. Intell.* **17,** 205–244 (1981).
17. H. R. Barrow and J. M. Tenenbaum, Interpreting line drawings as three-dimensional surfaces, *Artif. Intell.* **17,** 75–116 (1981).
18. K. A. Stevens, The visual interpretation of surface contours, *Artif. Intell.* **17,** 47–73 (1981).
19. A. P. Witkin, Recovering surface shape and orientation from texture, *Artif. Intell.* **17,** 17–45 (1981).
20. B. Horn, Obtaining shape from shading information, *The Psychology of Computer Vision*, P. H. Winston, ed., pp. 115–155. McGraw-Hill, New York (1975).
21. K. Ikeuchi and B. K. P. Horn, Numerical shape from shading and occluding boundaries, *Artif. Intell.* **17,** 141–184 (1981).
22. R. J. Woodham, Analyzing images of curved surfaces, *Artif. Intell.* **17,** 117–140 (1981).
23. M. A. Fischler and R. A. Elschlager, The representation and matching of pictorial structures, *IEEE Trans. Comput.* **C-22,** 67–92 (1973).
24. L. S. Davis and A. Rosenfeld, An application of relaxation labelling to spring-loaded template matching, *Proc. 3rd Int. Joint Conf. on Pattern Recognition*, pp. 591–597 (1976).
25. S. Ranade and A. Rosenfeld, Point pattern matching by relaxation, *Pattern Recognition* **12,** 268–275 (1980).
26. O. D. Faugeras and K. E. Price, Semantic description of aerial images using stochastic labelling, *Proc. 5th Int. Conf. on Pattern Recognition*, pp. 352–357 (1980).
27. K. Price and R. Reddy, Matching segments of images, *IEEE Trans. Pattern Anal. Mach. Intell.* **PAMI-1,** 110–116 (1979).
28. D. H. Ballard, Generalizing the Hough transform to detect arbitrary shapes, *Pattern Recognition* **13,** 111–122 (1981).
29. W. E. L. Grimson, *From Images to Surfaces: A Computational Study of the Human Early Visual System.* MIT Press, Cambridge, MA (1981).
30. B. K. P. Horn and B. C. Schunk, Determining optical flow, *Artif. Intell.* **17,** 185–203 (1981).
31. S. Ullman, *The Interpretation of Visual Motion.* MIT Press, Cambridge, MA (1979).
32. R. Ohlander, K. Price and D. R. Reddy, Picture segmentation using a recursive region splitting method, *Comput. Graphics Image Process.* **8,** 313–333 (1978).
33. J. O. Eklundh, H. Yamamoto and A. Rosenfeld, A relaxation method for multispectral pixel classification, *IEEE Trans. Pattern Anal. Mach. Intell.* **PAMI-2,** 72–75 (1980).
34. A. Rosenfeld and R. C. Smith, Thresholding using relaxation, *IEEE Trans. Pattern Anal. Mach. Intell.* **PAMI-3,** 598–606 (1981).
35. T. Pavlidis, *Structural Pattern Recognition.* Springer, New York (1977).
36. J. A. Feldman and Y. Yakimovsky, Decision theory and artificial intelligence: I. A semantics-based region analyzer, *Artif. Intell.* **5,** 349–371 (1974).
37. J. M. Tenenbaum and H. R. Barrow, Experiments in interpretation-guided segmentation, *Artif. Intell.* **8,** 241–274 (1977).
38. H. Freeman, Computer processing of line-drawing images, *Comput. Surv.* **6,** 57–97 (1974).
39. H. Blum, A transformation for extracting new descriptors of shape, *Models for the Perception of Speech and Visual Form*, W. Wathen-Dunn, ed., pp. 362–380. MIT Press, Cambridge, MA (1967).
40. H. Samet and A. Rosenfeld, Quadtree representation of binary images, *Proc. 5th Int. Conf. on Pattern Recognition*, pp. 815–818 (1980).
41. D. Ballard, Strip trees: a hierarchical representation for curves, *Commun. Ass. comput. Mach.* **24,** 319–321 (1981).
42. M. Shneier, Two hierarchical linear feature representations: edge pyramids and edge quadtrees, *Comput. Graphics Image Process.* **17,** 211–224 (1981).
43. R. Nevatia and T. O. Binford, Description and recognition of curved objects, *Artif. Intell.* **8,** 77–98 (1977).
44. R. A. Brooks, Symbolic reasoning among 3-D models and 2-D images, *Artif. Intell.* **17,** 285–348 (1981).
45. K. S. Fu, *Syntactic Pattern Recognition and Applications.* Prentice-Hall, Englewood Cliffs, NJ (1982).
46. P. H. Winston, Learning structural descriptions from examples, *The Psychology of Computer Vision*, P. H. Winston, ed., pp. 157–209. McGraw-Hill, New York (1975).

Author Index

Subject Index

T

W